The Dictionary of
CELL AND MOLECULAR BIOLOGY

This edition is dedicated to the memory of Dr Geoffrey Moores, who died in 2006. He was a good colleague and a committed teacher of Cell Biology who will long be remembered by his pupils.

The Dictionary of
CELL AND MOLECULAR BIOLOGY

Fourth Edition

J. M. Lackie

Plumbland Consulting Ltd

Authors Contributing to Earlier Editions

S. E. Blackshaw*
C. T. Brett
A. S. G. Curtis
J. A. T. Dow**
J. G. Edwards
A. J. Lawrence
G. R. Moores

* 2nd Edition only
** Co-Editor, Editions 1–3

AMSTERDAM · BOSTON · HEIDELBERG · LONDON
NEW YORK · OXFORD · PARIS · SAN DIEGO
SAN FRANCISCO · SINGAPORE · SYDNEY · TOKYO

OCM137221396

Academic Press is an imprint of Elsevier

Academic Press is an imprint of Elsevier
30 Corporate Drive, Suite 400, Burlington, MA 01803, USA
525 B Street, Suite 1900, San Diego, California 92101-4495, USA
84 Theobald's Road, London WC1X 8RR, UK

Library of Congress Cataloging-in-Publication Data
A catalogue record for this book is available from Library of Congress

British Library Cataloguing-in-Publication Data
A catalogue record for this book is available from the British Library.

ISBN: 978-0-12-373986-5

For information on all Academic Press publications
visit our Web site at www.books.elsevier.com

Printed and bound by MPG Ltd, Bodmin, Cornwall, Gt Britain
07   08   09   10   11   9   8   7   6   5   4   3   2   1

Table of Contents

Note about entries

The main entry word (headword) is in bold **Helvetica** and is followed by synonyms that are set in *Times italic*. Within the definition words in **bold** are cross-references to other entries that might contribute further information regarding the entry being consulted – although this does not mean that other words within the definition may not also have their own entries. Note that because synonyms are set in italic, any italic portion of the synonym will appear as roman (the standard convention). Generally, entry words with a Greek-letter or numeric prefix have been alphabetised ignoring the prefix. Where I think the abbreviation is more likely to be looked up than the full name I have associated the definition with the abbreviation – and cross-referenced accordingly – but the choice may sometimes seem idiosyncratic. Other conventions used are that names of genes are in italic and their products in roman (thus the gene is *src* but the kinase is src), although the botanical convention is slightly different (it would be SRC for the protein). Proprietary names of drugs are usually relegated to being synonyms and the generic name used for the entry. Diseases do not usually get an entry unless there is some understanding of the underlying molecular defect.

John Lackie

Tables

Preface

Modern biology continues to evolve at an astonishing rate, and the boundaries between the old subdisciplines are becoming so blurred as to be almost irrelevant – we are all 'bioscientists' now. But there is a welcome trend towards integration across the continuum, and 'translational research', the process of moving from a molecule directed against a cellular target to a therapeutic drug being administered to patients, has become a fashionable term. This dictionary tries to serve the needs of bioscientists or clinicians who are unfamiliar with the terminology from adjacent subspecialities, and is therefore extensive in its coverage rather than specialist. It is, however, a dictionary and not an encyclopaedia; it does not have large chunks of text to cut and paste into an essay.

As before, the choice of new headwords has been determined to a significant extent by the logging of abortive searches on the web-based version; people continue to misspell things in imaginative ways (for example, more than 50 recognizable but incorrect versions of 'mitochondrion' have been typed in the search box!). The task of defining the new headwords has been greatly facilitated by the resources available on the web: NCBI's Entrez PubMed has probably been the major source of information, supplemented by the OMIM (Online Mendelian Inheritance in Man) database and various other 'professional' databases. The definitions have been distilled from multiple sources – essentially, the dictionary summarizes the searching that any web-connected scientist could do – and the aim has been to provide succinct definitions with sufficient cross-referencing to enable the context to be inferred.

Searches for three-letter abbreviations (TLAs) were also common, and as a general principle I have defined those that came up in more than two Entrez PubMed abstracts, reasoning that this meant they were in sufficiently general use to justify being defined. It is noticeable that some are very popular – thus there are eight possible interpretations of CPA, DMP, CSP, etc. This is a problem when authors forget that others may not recognize which version they intend, and fail to specify in full on first use.

The most difficult task, in some ways, was the revision of the older entries; some named proteins have apparently disappeared from the literature, but proving disappearance is almost impossible. Where appropriate, I have drawn attention to such obsolescence but retained the entries, since a few rare souls do read the older literature and may find the definitions useful.

Thus, having tried to define things people failed to find and choosing not to delete older entries, the dictionary has grown yet again: this edition has more than 10 000 headwords, some with multiple definitions. Once the printed version has been published the new edition will appear on the web, and again abortive searches will be logged to inform the inevitable fifth edition. But feedback is always welcome and, since it would be extraordinary if some entries were not less than perfect, do not hesitate to let me know. My apologies if you fail to find what you are looking for or consider my attempt at a definition to be inaccurate or incomplete! I do hope you find it useful.

John Lackie
2007
john@lackie.nildram.co.uk

Preface to the Third Edition

Although the title has changed, we as Editors have talked of this as the Third Edition of *The Dictionary of Cell Biology* – and the change in title simply reflects the changes that have happened to Cell Biology in the last decade. In the Preface to the First Edition, we commented that the boundaries of 'cell biology' were difficult to define and this has certainly not changed – if anything the territory has grown! But, the molecular entries have continued to increase in number as more detail of cell structure and the complexity of signaling pathways is known. The new title reflects the content more accurately – though we have also added a number of entries that are neither cellular nor molecular.

Now that the Human Genome Project is nearing completion and we have the complete genome sequence of yeast and the nematode *Caenorhabditis elegans*, the big problem becomes that of assigning function to genes. Inevitably this will draw heavily on cell biology and many molecular biologists will find themselves entering new territory: and cell biologists they meet will need to talk the language of molecular biology. Cell biology encompasses an extremely wide range of experimental systems and has a very diverse vocabulary: this Dictionary should help to provide some guidance.

In preparing this volume, we have been much influenced by the usage of the Internet version of the Second Edition of *The Dictionary of Cell Biology*. Putting the Dictionary on the Net was an experiment – and one that has proved fascinating. For the first time, perhaps, it has been possible to monitor how people used the 'book' and what they searched for. Approximately one-third of a million visits have been made to the site and it has been cross-referenced from various other web pages. We have maintained a log of abortive searches and also gave an e-mail address for feedback. Though we had put a tear-out sheet for comment in the First and Second paper editions, we had almost no response, however, by e-mail it was different!

Many abortive searches were a result of inability to spell or perhaps to type accurately – we can do little about that! But many searches were for things that we felt really should have been in – sometimes we had omitted to write an entry, sometimes the search was for something new, sometimes the search was for things that are part of the wider vocabulary of those trained in the older disciplines. Thus there were quite a lot of searches for fixatives well known to the histologist, for Latin names of species where the common name is well known, for syndromes that have emerged or been diagnosed (we suspect). We have tried to put in some entries to help on these aspects, tried to put in new things from the literature but, in the case of diseases, we have tended to put entries only where there is a known cell or molecular basis for the disease. This edition has more than 7000 entries and we have been more comprehensive with cross-referencing of synonyms and from the text making this Dictionary almost double the size of the First Edition! The entries from the First and Second editions have been scrutinized and modified where necessary, particularly when there has been comment from readers, and many of the new entries are for words sought by on-line users. A variety of sources have been used and a brief list is appended. Again, Dr Ann Lackie helped with the final preparation and cross-checking.

We have always tried to provide short, clear definitions that are helpful to people with the widest range of backgrounds, and the huge volume of feedback we have received from the Internet edition suggests that these efforts are well received. Not only career cell biologists, but school teachers, high school students and journalists have all sent glowing testimonials. Our aim is to continue to develop this resource, both on-line and, given the speed at which the discipline is evolving, to produce another in due course. Meanwhile we hope that people will find this edition useful and find it much easier, and quicker to search adjacent entries. We also hope that people will not hesitate to send new entries, suggest amendments and help in the evolution of this unusually interactive resource.

The on-line Dictionary is to be found on *http://www.mblab.gla.ac.uk/dictionary*

John Lackie
Julian Dow
1999

Preface to the Second Edition

In the preface to the first edition we commented that 'the subject was far from static', and we have not been disappointed. The last few years have seen rapid progress and an increasing reliance upon the powerful tools of molecular biology. In this second edition we have extended the coverage, particularly in molecular biology and neurobiology, though the sense of barely keeping up with the emergence of new names for proteins is ever more pressing. 'Cytokine' has overtaken 'Interleukin' as the numbers of cytokines have almost doubled and chemokines have arrived. The table of CD numbers has grown impressively and the G-proteins have proliferated; far more genes have been named and new proteins are legion. A measure of the rate of change is that we have had to include more than 1000 new entries; by the next revision we will have to start deleting entries – and anybody who scans the old literature (pre-1990!) will find names that have disappeared without trace.

As before, the choice of entries partly reflects our own interests, but we have tried to be broadminded with our inclusions. This time Dr Susanna Blackshaw has written entries rather than acting as an external adviser, in order to increase the neurobiological content, and both editors have become much more involved with molecular biology. In the first edition we included a tear-off page for users to send us their neologisms and revisions but, disappointingly, had little response. We continue to hope that interested users of this book will provide suggestions for future improvements.

Several people have helped with the revision of the manuscript itself and in particular we wish to thank Mrs Lynn Sanders and Mrs Joanne Noble for help with typing. Many colleagues who have not actually been directly involved have nevertheless been plagued with questions, and we are grateful for their patience. We hope that this new edition will be as useful as the first and that it will date no faster. On past experience, however, our rapidly growing subject will doubtless keep this project alive for years to come. We have taken the unusual step of making a searchable version of the Dictionary available on the Internet. Although undoubtedly less convenient than the paper version, this will carry incremental changes and updates – including those provided electronically by readers – until the third edition is published. Details are provided on the revision form at the back of the book: we hope you find this service useful.

John Lackie
Julian Dow
July 1994

Preface to the First Edition

The stimulus to write this dictionary came originally from our teaching of a two-year Cell Biology Honours course to undergraduates in the University of Glasgow. All too often students did not seen to know the meanings of terms we felt were commonplace in cell biology, or were unable, for example, to find out what compounds in general use were supposed to do. But before long it become obvious that although we all considered ourselves to be cell biologists, individually we were similarly ignorant in areas only slightly removed from our own – though collectively the knowledge was there. It was also clear that many of the things we considered relevant were not easy to find, and that an extensive reference library was needed. In that we have found the exercise of preparing the Dictionary informative ourselves, we feel that it may serve a useful purpose.

An obvious problem was to decide upon the boundaries of the subject. We have not solved this problem: modern biology is a continuum and any attempt to subdivide it is bound to fail. 'Cell Biology' implies different things to zoologists, to biochemists, and indeed to each of the other species of biologists. There is no sensible way to set limits, nor would we wish to see our subject crammed into a well-defined niche. Inevitably, therefore the contents are somewhat idiosyncratic, reflecting our current teaching, reading, prejudices, and fancies.

It may be of some interest to explain how we set about preparing the Dictionary. The list of entry words was compiled largely from the index pages of several textbooks, and by scanning the subject indexes of cell-biological journals. To this were added entries for words we cross-referenced. The task of writing the basic entries was then divided amongst us roughly according to interests and expertise. We all wrote subsets of entries which were then compiled and alphabetized before being edited by one of us. Marked copies were then sent out to a panel of colleagues who scrutinized entries in their own fields. All entries were looked at by one or more of this panel, and then the annotated entries were re-edited, corrections made on disc, and the files copy-edited for consistency of style. A very substantial amount of the handling of the compiled text and the preparation of the final discs was done by Dr A M Lackie who also acted as copy-editor.

Glasgow is a major centre for Life Sciences, and we are fortunate in having many colleagues to whom we could turn for help. We are very grateful to them for the work which they put in and for the speed with which they checked the entries that we sent. Although we have tried hard to avoid errors and ambiguities, and to include everything that will be useful, we apologise at this stage for the mistakes and omissions, and emphasise that blame lies with the authors and not with our panel (though they have saved us from many embarrassments).

Since there is no doubt Cell Biology is developing rapidly as a field, it is inevitable that usages will change, that new terms will become commonplace, that new proteins will be christened on gels, and that the dictionary will soon have omissions. Were the subject static this dictionary would not be worth compiling – and we cannot anticipate new words.

Because the text is on disc, it will be relatively easy to update: please let us have your comments, suggestions for entries (preferably with a definition), and (perhaps) your neologisms. A sheet is included at the back of the dictionary for this purpose.

John Lackie

Acknowledgements

I am grateful to **Julian Dow**, my former co-editor, for supplying the list of abortive searches and for organising the web-hosting of the Third Edition at *http://onto/dictionary*.

Although all the new entries and amendments of old entries for this edition were written by me, there remain some entries written wholly or in part by contributors to earlier editions, my former colleagues in Glasgow: **S.E. Blackshaw, C.T. Brett, A.S.G. Curtis, J.A.T. Dow, J.G. Edwards, A.J. Lawrence and G.R. Moores**. I remain extremely grateful to them – but by now they can be absolved of responsibility for any mistakes, which will be mine alone.

John Lackie

1, 2, 3, . . .

5q-syndrome A type of **myelodysplasia** associated with deletion of the long arm of chromosome 5 and therefore with multiple gene dysfunctions. Genes involved include the tumour suppressor *IRF*1, and those for IL5, CDC25C, IL3, CSF2 and POP2.

9E3 *pCEF-4* A **cytokine** produced by chicken cells infected with **Rous sarcoma virus**. Has significant amino acid identity with both human IL8 and human GROalpha.

14-3-3 proteins Family of adapter proteins able to interact with a range of signalling molecules including c-Raf, Bcr, PI-3-kinase, polyoma middle T-antigen. Bind to phosphorylated serine residues in cdc25C and block its further activity, may bind Bad (death inducer) thereby blocking heteromeric interaction with Bcl-XL, and in plants bind and inhibit activity of phosphorylated nitrate reductase. Basic mode of action may be to block specific protein–protein interactions. There are seven highly conserved human 14-3-3 proteins that are involved in cellular proliferation, checkpoint control and apoptosis. More than 200 14-3-3 target proteins have been identified, including proteins involved in mitogenic and cell survival signalling, cell cycle control and apoptotic cell death.

464.1 *744* Human **macrophage inflammatory protein 1 β** (MIP1β).

744 See **464.1**.

A

A *Alpha* Entry prefix is generally given as 'alpha'; alternatively, look for main portion of word.

A4 protein See **amyloidogenic glycoprotein**.

A9 cells Established line of heteroploid mouse fibroblasts that are deficient in **HGPRT**.

A20 A stress-response gene in endothelial cells that encodes a dual function enzyme with de-ubiquitinating activity towards Lys63-linked poly-ubiquitin chains and Lys48 E3 ligase activity. The enzyme is an inhibitor of TNF signalling and acts by triggering degradation of RIP (**receptor-interacting protein kinase**). Deficiency in A20 and TNF or TNF-R leads to spontaneous inflammation in mice.

A260 Spectrophotometric absorbance at 260 nm. The ratio of absorbance at 260 nm to that at 280 nm is often used as a quick assessment of the purity of nucleic acid samples, since nucleic acids absorb strongly at 260 nm and proteins at 280 nm.

A431 A line of human epidermoid carcinoma cells.

A23187 A monocarboxylic acid extracted from *Streptomyces chartreusensis* that acts as a mobile-carrier calcium **ionophore**. See Table I3.

AA-tRNA See **aminoacyl tRNA**.

Aarskog–Scott syndrome *Faciogenital dysplasia (FGDY)* An X-linked developmental disorder characterized by disproportionately short stature and by facial, skeletal and urogenital anomalies. Molecular genetic analyses mapped FGDY to chromosome Xp11.21. See **FGD1** and **frabin**.

Ab Common abbreviation for antibody. See **immunoglobulins**.

AB toxin Multisubunit toxin in which there are two major components, an active (A) portion and a portion that is involved in binding (B) to the target cell. The A portion can be effective in the absence of the B subunit(s) if introduced directly into the cytoplasm. In the well-known examples, the A subunit has **ADP-ribosylating** activity. See **cholera toxin, diphtheria toxin, pertussis toxin**.

abaecins Proline-rich basic antibacterial peptides (4 kDa) found in the **haemolymph** of the honeybee. See **apidaecins**.

A-band That portion of the **sarcomere** in which the thick myosin filaments are located. It is anisotropic in polarized light.

abaxial Located on the side away from the axis or facing away from the axis of stem or root. Typically, the lower surface of leaves. *cf.* **adaxial**.

ABC 1. Antigen-binding cell or antigen-binding capacity. 2. Avidin–biotin peroxidase complex. Used in visualizing antigen. Primary (antigen-specific) antibody is bound first, a second biotinylated anti-immunoglobulin antibody is then bound to the first antibody, then ABC complex that has excess biotin-binding capacity is bound to the biotin on the second antibody and, finally, the peroxidase used to catalyse a colorimetric reaction generating brown staining. The method gives substantial signal enhancement. **3.** See **ABC proteins, ABC-excinuclease**.

ABC proteins Membrane proteins involved in active transport or regulation of ion channel function and having an ATP-binding cassette. Most examples are prokaryotic, but important eukaryotic examples are the P-glycoprotein (multidrug resistance transporter), the cystic fibrosis transmembrane conductance regulator (**CFTR**) and **SUR**.

ABC-excinuclease Enzyme complex, product of *uvrA*, *uvrB* and *uvrC* genes from *E. coli* that mediates incision and excision steps of DNA excision repair. Enzyme has the ability to recognize distortion in DNA structure caused by, for example, ultraviolet irradiation.

ABCD1 An ATPase-binding cassette protein in the same family of transporter proteins such as **CFTR** and **MDR** proteins. See **adrenoleucodystrophy**.

Abelson leukaemia virus A replication-defective virus originating from the Moloney murine leukaemia virus by acquisition of *c-abl*. The virus induces B-cell lymphoid leukaemias within a few weeks. The *v-abl* product has tyrosine kinase activity.

abenzyme See **catalytic antibody**.

aberration Departure from normal. In microscopy, two common forms of optical aberration cause problems: **spherical aberration,** in which there is distortion of the image of the magnified object, and **chromatic aberration,** which leads to coloured fringes – a consequence of the unequal refraction of light of different wavelength.

abetalipoproteinaemia Autosomal-recessive defect in which there is total absence of apoprotein B (a component of LDL, VLDL and chylomicrons). Characteristic feature is presence of **acanthocytes**; later in life neurological disorders and retinitis pigmentosa develop, and death is usually a consequence of cardiomyopathy.

abiogenesis Spontaneous generation of life from non-living material.

abiotic Non-living.

abl An oncogene identified in Abelson murine leukaemia virus that encodes a non-receptor tyrosine kinase, **abl**. c-Abl contains both a G-actin-binding site and an independent F-actin site. See also **ABLV** and Table O1.

ABLV The Abelson murine leukaemia virus, a mammalian **retrovirus**. Its transforming gene, *abl*, encodes a protein with **tyrosine kinase** activity closely related to **src**.

ABM paper Aminobenzyloxy methylcellulose paper: paper to which single-stranded nucleic acid can be covalently coupled.

ABO blood group system Probably the best known of the blood group systems, involves a single gene locus that codes for a fucosyl transferase. If the H-gene is expressed then fucose is added to the terminal galactose of the precursor oligosaccharide on the red cell surface and the A- or

B-gene products, also glycosyl-transferases, can then add *N*-acetyl galactosamine or galactose to produce the A or B-antigens, respectively. Antibodies to the ABO-antigens occur naturally and make this an important set of antigens for blood transfusion. Transfusion of mismatched blood with surface red cell antigens that elicit a response leads to a transfusion reaction. The natural antibodies are usually IgM. See **Rhesus**, **Kell**, **Duffy** and **MN blood group antigens**.

abortive infection Viral infection of a cell in which the virus fails to replicate fully or produces defective progeny. Since part of the viral replicative cycle occurs, its effect on the host can still be cytopathogenic.

abortive transformation Temporary transformation of a cell by a virus that fails to integrate into the host DNA

ABP See **actin-binding proteins**. A useful web resource, the *Encyclopedia of Actin-Binding Proteins* has been put together by S. MacIver of Edinburgh University.

Abp1p An actin-binding protein originally from yeast and associated with the membrane. Has an **ADF domain** and an **SH3** domain. Mammalian homologues are thought to have a role in endocytosis. Abp1p binds actin through a region homologous to ADF/**cofilin** and to **dynamin** through the SH3 domain.

ABP-50 Actin-binding protein (50 kDa) from *Dictyostelium* that cross-links actin filaments into tight bundles. Identical to **elongation factor** EF-1a. Calcium insensitive; localized near cell periphery and in protrusions from moving cells.

ABP-67 Early name for **fimbrin**. In yeast encoded by *SAC6* gene, mutations which lead to disruption of the actin **cytoskeleton**.

ABP-120 Dictyostelium *gelation factor* Actin-binding protein (857 amino acids; 92 kDa) from *Dictyostelium*. A small rod-shaped molecule (35–40 nm long), dimeric, capable of cross-linking filaments. Has strong sequence similarities with **ABP-280**.

ABP-280 Actin-binding protein (2647 amino acids; 280 kDa) originally isolated from *Dictyostelium*, but very similar to filamin from other sources. A long rod-shaped phosphoprotein (80 nm long) present in the periphery of the cytoplasm, dimeric, with the two monomers associated end-to-end making a very long cross-linker of microfilaments. Associates with the third cytoplasmic loop of dopamine D(2) and D(3) receptors, but not with D(4) receptors. Has actin-binding domain similar to that in **ABP-120**, **spectrin**, filamin, **alpha actinin** and **fimbrin**, also has binding site for platelet **von Willebrand factor**.

abrin Toxic **lectin** from seeds of *Abrus precatorius* (jequirity bean) that has a binding site for galactose and related residues in carbohydrate but, because it is monovalent, is not an **agglutinin** for **erythrocytes**. Abrin-α A-chain (ABRαA) has *N*-glycosylase activity towards eukaryotic 28S rRNA.

abscess A cavity within a tissue occupied by pus (chiefly composed of degenerating inflammatory cells), generally caused by bacteria that resist killing by phagocytes.

abscisic acid A growth-inhibiting plant hormone found in vascular plants. Originally believed to be important in abscission (leaf fall), now known to be involved in a number of growth and developmental processes in plants, including, in some circumstances, growth promotion. Primary plant hormone that mediates responses to stress. Downstream signalling through **cyclic ADP-ribose** as a second messenger.

absolute lethal concentration, LC_{100} Lowest concentration of a substance that kills 100% of test organisms or species under defined conditions. This value is dependent on the number of organisms used in its assessment.

absorption coefficient 1. Any of four different coefficients that indicate the ability of a substance to absorb electromagnetic radiation. Absorbance is defined as the logarithm of the ratio of incident and transmitted intensity, and thus it is necessary to know the base of the logarithm used. Scattering and reflectance are generally ignored when dealing with solutions. **2.** Ratio of the amount of a substance absorbed (uptake) to the administered quantity (intake).

absorption spectrum Spectrum of wavelengths of electromagnetic radiation (usually visible and UV light) absorbed by a substance. Absorption is determined by existence of atoms that can be excited from their ground state to an excited state by absorption of energy carried by a photon at that particular wavelength.

ABTS *2,2'-Azino-bis(3-ethylbenzthiazoline-6-sulfonic acid)* Compound that will produce a water-soluble, green-coloured product upon reaction with horseradish peroxidase; used in enzyme-linked immunoassays. It is light sensitive and must be kept in the dark both as a stock solution and as a working solution.

abzyme A catalytic antibody, one that has enzymic activity. Catalytic antibodies have two distinct advantages, they can be selected to catalyse a reaction that is not catalysed by endogenous enzymes, and, to minimize immunogenicity, humanization is possible.

acamprosate *Calcium acetylaminopropane sulfonate* A synthetic compound with a chemical structure similar to that of the endogenous amino acid homotaurine, which is a structural analogue of the amino acid neurotransmitter γ-aminobutyric acid and the amino acid neuromodulator taurine. A drug thought to be beneficial in maintaining abstinence in alcohol-dependent patients although the mechanism of action is unclear.

Acanthamoeba Soil amoebae 20–30 μm in diameter that can be grown under **axenic** conditions and have been extensively used in biochemical studies of cell motility. They have been isolated from cultures of monkey kidney cells, and are pathogenic when injected into mice or monkeys.

acanthocyte Cell with projecting spikes; most commonly applied to erythrocytes, where the condition may be caused naturally by **abetalipoproteinaemia** or experimentally by manipulating the lipid composition of the plasma membrane.

acanthocytosis Condition in which red cells of the blood show spiky deformation; symptomatic of **abetalipoproteinaemia**. See **acanthocyte**.

acanthosis nigricans A rare disease characterized by pigmentation and warty growths on the skin. Often associated with cancer of the stomach or uterus.

acanthosome 1. Spinous membranous **organelle** found in skin **fibroblasts** from **nude mice** as a result of chronic ultraviolet irradiation. **2.** Sometimes used as a synonym for **coated vesicle** (should be avoided).

Acanthus Genus of spiny-leaved Mediterranean plants.

acapnia Medical condition in which there is a low concentration of carbon dioxide in the blood.

ACAT *Acyl coenzyme A: cholesterol acyltransferase; EC 2.3.1.26; sterol o-acyltransferase* An enzyme, located in the ER, that catalyses cholesterol ester formation from cholesterol and fatty acyl CoA substrates. Sterol esterification by ACAT or homologous enzymes is highly conserved in evolution. The human cDNA for ACAT has been cloned, and an ACAT gene family identified.

ACC *1-Aminocyclopropane-1-carboxylic acid* Immediate precursor of the plant hormone ethylene in most vascular plants. Synthesized from *S*-adenosyl methionine by **ACC synthase**.

ACC synthase *ACC methylthioadenosine lyase; EC 4.1.1.14* Enzyme (65 kDa) that catalyses conversion of *S*-adenosylmethionine to **ACC**, the first step in production of the plant hormone ethylene.

accelerin Activated blood factor V which acts on prothrombin to generate thrombin during blood coagulation.

accessory cells Cells that interact, usually by physical contact, with T-lymphocytes and that are necessary for induction of an immune response. Include antigen-presenting cells, antigen-processing cells, etc. They are usually MHC class II positive (see **histocompatibility antigens**). Monocytes, macrophages, dendritic cells, **Langerhans cells**, B-lymphocytes may all act as accessory cells.

accessory pigments In **photosynthesis**, pigments that collect light at different wavelengths and transfer the energy to the primary system.

ACE See **angiotensin**.

ACE inhibitor See **angiotensin-converting enzyme inhibitors**.

A-cells *α cells* Cells of the endocrine pancreas (Islets of Langerhans) that form approximately 20% of the population; their opaque spherical granules may contain **glucagon**. See **B-cells, D-cells.**

acellular Not made of cells; commonest use is in reference to **acellular slime moulds** such as *Physarum* that are multinucleate syncytia.

acellular slime moulds Protozoa of the Order **Eumycetozoida** (also termed true slime moulds). Have a multinucleate plasmodial phase in the life cycle.

acentric Descriptive of pieces of **chromosome** that lack a **centromere**.

Acetabularia Giant single-celled **alga** of the Order Dasycladaceae. The plant is 3–5 cm long when mature, and consists of **rhizoids** at the base of a stalk, at the other end of which is a cap that has a shape characteristic of each species. The giant cell has a single nucleus, located at the tip of one of its rhizoids, which can easily be removed by cutting off that rhizoid. Nuclei can also be transplanted from one cell to another.

Acetobacter Genus of aerobic bacilli that will use ethanol as a substrate to produce acetic acid – thus will convert wine to vinegar.

acetocarmine A solution of carmine, a basic dye prepared from the insect *Coccus cacti*, in 45% glacial acetic acid; used for the staining of plant chromosomes by the squash method.

acetonitrile *CH₃CN; cyanomethane; ethyl nitrile; ethanenitrile; methanecarbonitrile; methyl cyanide* A colourless, toxic, inflammable liquid used extensively in industrial organic synthesis and wherever a polar solvent having a high dielectric constant is required. Can break down in the body to produce cyanide.

acetosyringone *1-(4-Hydroxy-3,5-dimethoxyphenyl)-ethanone* A phenolic inducer of the virulence genes (***vir* genes**) of *Agrobacterium tumefaciens*, produced by wounding of plant tissue. Acetosyringone can act as a chemical attractant *in vitro* and thus may act as chemotactic agent in nature.

acetylation Addition, either chemically or enzymically, of acetyl groups.

acetylcholine *ACh* Acetyl ester of choline. Perhaps the best-characterized **neurotransmitter**, particularly at neuromuscular junctions. ACh can be either excitatory or inhibitory and its receptors are classified as **nicotinic** or **muscarinic**, according to their pharmacology. In **chemical synapses** ACh is rapidly broken down by **acetylcholine esterases**, thereby ensuring the transience of the signal.

acetylcholine esterase An enzyme, found in the **synaptic clefts** of cholinergic synapses, that cleaves the **neurotransmitter** acetylcholine into its constituents, acetate and choline, thus limiting the size and duration of the **postsynaptic potential**. Many nerve gases and insecticides are potent acetylcholine esterase inhibitors and thus prolong the timecourse of postsynaptic potentials.

acetylcholine receptor See **nicotinic acetylcholine receptor, muscarinic acetylcholine receptor**.

acetyl CoA Acetylated form of coenzyme A that is a carrier for acyl groups, particularly in the **tricarboxylic acid cycle**.

A-chain Shorter of the two polypeptide chains of insulin (21 residues compared to 30 in the B-chain). Many other heterodimeric proteins have their smaller chain designated the A-chain, so the term cannot be used without qualification. Other A-chains: **abrin, activin**, C1q (see **complement**), **diphtheria toxin, inhibin, laminin**, mistletoe lectin, **PDGF, relaxin, ricin, tPA**.

A-channel Type of potassium-selective ion channel that is activated by depolarization but only after a preceding hyperpolarization, i.e. it is inactivated at rest. Important for repetitive firing of cells at low frequencies – see **Shaker** mutant.

achene A type of simple, dry, one-seeded fruit, produced by many flowering plants. Do not split (dehisce) when mature.

achondroplasia *Chondrodystrophia fetalis* Failure of endochondral ossification responsible for a form of dwarfism; caused by an **autosomal-dominant mutation** in the gene for FGFR3 on chromosome 4. (In 98% of cases, the mutation is a G380R substitution, resulting from a G-to-A point mutation at nucleotide 1138.) Relatively high incidence (1:20 000 live births), mostly (90%) new mutations.

Achyla Genus of aquatic fungi with a branched coenocytic mycelium.

acid-citrate-dextrose *ACD* Citric acid/sodium citrate-buffered glucose solution used as an anticoagulant for blood (citrate complexes calcium).

acid hydrolases *EC 3* Hydrolytic enzymes that have a low pH optimum. The name usually refers to the **phosphatases**, **glycosidases**, **nucleases** and **lipases** found in the **lysosome**. They are secreted during **phagocytosis**, but are considered to operate as intracellular digestive enzymes.

acidic FGF See **fibroblast growth factor**.

acidophilia Having an affinity for acidic dyes, particularly eosin; may be applied either to tissues or bacteria.

acidophilic **1.** Easily stained with acid dyes. **2.** Flourishing in an acidic environment.

acidophils One class of cells found in the pars distalis of the **adenohypophysis**.

acidosome Non-lysosomal vesicle in which receptor–ligand complexes dissociate because of the acid pH.

acid phosphatase *EC 3.1.3.2* Enzyme with acidic pH optimum that catalyses cleavage of inorganic phosphate from a variety of substrates. Found particularly in **lysosomes** and **secretory vesicles**. Can be localized histochemically using various forms of the **Gomori procedure**.

acid protease Rather imprecise term for a proteolytic enzyme with an acid pH optimum, characteristically found in lysosomes. See **peptidase**.

acid-secreting cells Large specialized cells of the epithelial lining of the stomach (parietal or oxyntic cells) that secrete 0.1N HCl by means of H$^+$ **antiport** ATPases on the luminal cell surface.

acinar cells Epithelial **secretory cells** arranged as a ball of cells around the lumen of a gland (as in the pancreas).

Acinetobacter A Gram-negative bacterium commonly found in soil and water, but also on the skin of healthy people. *Acinetobacter*, particularly *Acinetobacter baumannii*, is an important opportunistic, nosocomial (hospital acquired) pathogen and is very resistant to antimicrobials; relatively few antibiotics are effective.

acinus Small sac or cavity surrounded by **secretory cells**.

acoelomate Animal without a **coelom**. The acoelomate phyla include sponges, coelenterates, and lower worms such as nematodes and platyhelminths.

aconitase Enzyme of the **tricarboxylic acid cycle** that catalyses isomerization of citrate/isocitrate. Isoforms are found both in mitochondrial matrix and cytoplasm.

acou- Combining form meaning 'related to hearing', as in acoustic and the acoustico-lateralis system.

ACP *Acyl carrier protein* Small acidic proteins associated with fatty acid synthesis in many pro- and eukaryotic organisms. They are functional only when modified by attachment of the prosthetic group, 4′-phosphopantetheine (4′-PP), which is then transferred from CoA to the hydroxyl group of a specific serine residue. In eukaryotes they are part of the multi-enzyme complex, **fatty acid synthase;** in prokaryotes they exist as separate proteins of around 8.8 kDa.

acquired immune deficiency syndrome See **AIDS**.

acquired immunity Classically, the reaction of an organism to a new antigenic challenge and the retention of a memory of this, as opposed to innate immunity. In modern terms, the **clonal selection** and expansion of a population of immune cells in response to a specific antigenic stimulus and the persistence of this clone.

acrasiales See **Acrasidae**.

Acrasidae Order of **Protozoa** also known as the cellular slime moulds. They normally exist as free-living phagocytic soil amoebae (vegetative cells), but when bacterial prey become scarce they aggregate to form a pseudoplasmodium (*cf.* true **plasmodium** of Eumycetozoida) that is capable of directed motion. The grex, or slug, migrates until stimulated by environmental conditions to form a fruiting body or sorocarp. The slug cells differentiate into elongated stalk cells and spores, where the cells are surrounded by a cellulose capsule. The spores are released from the sporangium at the tip of the stalk and, in favourable conditions, an **amoeba** emerges from the capsule, feeds, divides, and so establishes a new population. They can be cultured in the laboratory and are widely used in studies of cell–cell adhesion, cellular **differentiation**, **chemotaxis** and **pattern formation**. The commonest species studied are *Dictyostelium discoideum*, *D. minutum* and *Polysphondylium violaceum*.

acrasin Name originally given to the **chemotactic** factor produced by cellular slime moulds (**Acrasidae**); now known to be **cyclic AMP** (cAMP) for *Dictyostelium discoideum*.

acridine orange A fluorescent vital dye that intercalates into nucleic acids. The nuclei of stained cells fluoresce green; cytoplasmic RNA fluoresces orange. Acridine orange also stains acid mucopolysaccharides, and is widely used as a pH-sensitive dye in studies of acid secretion. May be carcinogenic.

acridines Heterocyclic compounds with a pyridine nucleus. Usually fluorescent and reactive with double-stranded DNA as intercalating agents at very low concentrations; hence dsDNA can be detected on gels by fluorescence after acridine staining. Mutagenic (causing frame-shift mutations), cytostatic (and hence antimicrobial). They also affect RNA synthesis and have been used for cell marking.

acritarchs Small organic structures found as fossils and presumed to be remnants of simple unicellular organisms (e.g. bacteria, dinoflagellates, marine algae). Found in sedimentary rocks from as early as the Precambrian era.

acrocentric See **metacentric**.

acrolein *2-Propenal; acraldehyde; allyl aldehyde; acryl aldehyde* Inflammable liquid with a sharp, disagreeable odour that will readily polymerize to form a plastic solid. Used in a qualitative test for glycerol. Acrolein is principally used as a chemical intermediate in the production of acrylic acid and its esters, but also directly as an aquatic herbicide and algicide in irrigation canals. Can be produced by fires.

acromegaly Enlargement of the extremities of the body as a result of the overproduction of growth hormone (**somatotrophin**), e.g. by a pituitary tumour.

acropetal Transport or differentiation occurring from the base towards the apex.

acrophase The time at which the peak of a rhythm occurs. Originally referred to the phase angle of the peak of a cosine wave fitted to the raw data of a rhythm.

acroplaxome A marginal ring containing 10-nm-thick filaments of keratin-5 and F-actin that anchors the developing **acrosome** to the nuclear envelope. The ring is closely associated with the leading edge of the acrosome and with the nuclear envelope during the elongation of the spermatid head.

acrosin *EC 3.4.21.10* **Serine peptidase** stored in the **acrosome** of a sperm as an inactive precursor; involved in penetration by the sperm of the outer layers of the egg.

acrosomal process A long process actively protruded from the acrosomal region of the spermatozoon following contact with the egg and that assists penetration of the gelatinous capsule. See **acrosome**, **acrosin**.

acrosome Vesicle at the extreme anterior end of the spermatozoan, derived from the **lysosome**.

ACT *Artemisinin-based combination therapy* See **artemisinin**, **lumefantrine**.

ACT1, ACT2 Actin genes from yeast; ACT1 is the essential (conventional) actin, 89% homologous in sequence with mouse cytoplasmic actin; *ACT2* encodes a 44-kDa protein 47% identical to yeast actin that is required for vegetative growth. Divergence from conventional actin by ACT2 is in regions associated with actin polymerization, DNAase I and myosin binding.

Act-2 Human **macrophage inflammatory protein 1** β.

ActA Major surface protein (90 kDa) of *Listeria monocytogenes* that acts as the nucleating site for actin polymerization at one pole of the bacterial cell; assembly of the bundle of microfilaments pushes the bacterium through the cell – though the appearance is like a comet with a tail. ActA spans both the bacterial membrane and the peptidoglycan cell wall.. Interacts with several mammalian proteins, including the phosphoprotein **VASP**, actin and the **Arp2/3 complex**. A functionally similar protein, **IcsA**, is found in *Shigella*.

ACTH See **adrenocorticotrophin**.

actin A protein of 42 kDa, very abundant in eukaryotic cells (8–14% total cell protein) and one of the major components of the **actomyosin** motor and the cortical microfilament meshwork. First isolated from **striated muscle** and often referred to as one of the muscle proteins. G-actin is the globular monomeric form of actin, 6.7 × 4.0 nm: it polymerizes to form filamentous F-actin. See also **actin-RPV**.

actin-binding proteins A diverse group of proteins that bind to **actin** and that may stabilize F-actin filaments, nucleate filament formation, cross-link filaments, lead to bundle formation, etc. See Table A1. A useful web resource, the *Encyclopedia of Actin-Binding Proteins*, has been put together by S. MacIver of Edinburgh University.

actinfilin An actin-binding protein (75 kDa) expressed predominantly in brain. Actinfilin has the same overall structure as mayven, **kelch** and Enc-1, other actin-binding proteins. Self-associates through an amino-terminal POZ domain.

actin-fragmin kinase *AFK* Kinase found in *Physarum polycephalum* that specifically phosphorylates actin in the EGTA-resistant 1:1 actin–fragmin complex and may thereby regulate cytoskeletal dynamics.

Actinia equina Common beadlet anemone; a **coelenterate**. See **equinatoxins**.

actinic keratosis Thickened area of skin as a result of excessive exposure to sunlight – particularly common in those with very fair skin. See **keratoses**.

actin meshwork **Microfilaments** inserted proximally into the plasma membrane and cross-linked by **actin-binding proteins** to form a mechanically resistive network that may support protrusions such as **pseudopods** (sometimes referred to as the **cortical meshwork**).

actinogelin An **actin-binding protein** (115 kDa) from Ehrlich ascites cells that gelates and bundles **microfilaments**.

Actinomycetales Order of Gram-positive bacteria, widespread in soil, compost and aquatic habitats. Most are saprophytic, but there are a few pathogens; some produce important antibiotics. Important genera include *Actinomyces, Corynebacterium, Frankia, Mycobacterium, Streptomyces.*

actinomycin C A mixture of **antibiotics**, actinomycins C1, C2 and **actinomycin D** elaborated by a species of *Streptomyces*.

actinomycin D Antibiotic from *Streptomyces* spp. that, by binding to DNA, blocks the movement of **RNA polymerases** and prevents RNA synthesis in both pro- and eukaryotes.

actinomycosis A chronic granulomatous infection caused by various filamentous bacteria of the genus *Actinomyces*. Actinomycosis also occurs in cattle and pigs.

Actinophrys sol Species of **Heliozoa** often used in studies on microtubule stability: the **axopodia** are supported by a bundle of cross-linked microtubules arranged in a complex double-spiral pattern when viewed in cross-section.

Actinopterygii Teleost fishes, a subclass of the Gnathostomata. Includes cod and herring. The fins of the Actinopterygii are webs of skin supported by bony or horny spines.

actin-RPV *Alpha-centractin; centrosome-associated actin homologue; Arp1* See **centractin**.

Actinosphaerium Genus that is a member of the Class **Heliozoa**, Order Actinophryida, Family Actinosphaeridae: multinucleate cells, 80–200 μm in size; remarkable for long radial protruding **axopodia** that contain complex double-spiral arrangements of many microtubules. It catches prey by protrusion and retraction of the axopodia. Similar to *Actinophrys*.

actinotrichia *Sing. actinotrichium* Aligned **collagen** fibres (*ca.* 2 μm diameter) that provide a guidance cue for **mesenchymal** cells in the developing fin of teleost fish.

action potential An electrical pulse that passes along the membranes of excitable cells, such as **neurons, muscle cells,** fertilized eggs and certain plant cells. The precise shape of action potentials varies, but action potentials always involve a large **depolarization** of the cell membrane, from its normal **resting potential** of −50 to −90 mV. In a neuron, action potentials can reach +30 mV and last

TABLE A1. Actin-binding proteins

Protein	M_r(kDa)	Source
(i) *Monomer sequestering* (bind G-actin) or depolymerizing		
19-kDa protein	19	Pig brain
Actobindin	9.7	*Acanthamoeba castellani*
Actolinkin	20	Echinoderm eggs
Actophorin	13–19	*Toxoplasma gondii*
ADF	19	Various
Coactosin	17	*Dictyostelium*
Cofilin	19	Various. Depolymerizes
Depactin	18	Starfish oocytes
Destrin	19	Various
DNAase I	31	Pancreas
Drebin	140	Neuronal cells
Profilin	12–15	Various
Twinfilin	37	Yeast
Vit D-binding protein	57	Plasma
(ii) *End-blocking and nucleating*		
ActA	90	*Listeria*/promotes polymerization
β–Actinin	37 and 35	Kidney and striated muscle/end-blocker
Acumentin	65	Mammalian leucocytes/end-blocker
Adseverin	74	Adrenal medulla/calcium sensitive
Aginactin	70	*Dictyostelium*/barbed-end cap
Capping protein	31 and 28	*Acanthamoeba*
Fragmin/severin	40–45	*Physarum, Dictyostelium*, sea urchin eggs
Gelsolin	90	Mammalian cells; same as brevin and ADF/calcium sensitive
Hisactophilin	13.5	*Dictyostelium*/promotes polymerization
IcsA		*Shigella flexneri*/promotes assembly
Ponticulin	17	*Dictyostelium*
Radixin	82	Adherens junction and cleavage furrow
Tensin/insertin	30	Focal adhesions. Bind barbed end
Villin	95	Amphibian eggs, avian and mammalian epithelium

Protein	Subunits (kDa)	Source
(iii) *Cross-linking*		
Isotropic gelation		
Actin-binding protein	2×270	Macrophages, platelets, *Xenopus* eggs
ABP-120	120	*Dictyostelium*
ABP-280	280	*Dictyostelium*
Cortactin	80/85	Various. Has SH3 domain
Filamin	2×250	Smooth muscle
Spectrin	$2\times240, 2\times220$	Erythrocytes
Fodrin	$2\times260, 2\times240$	Brain
Transgelin	21	Highly conserved
TW 260/240	$2\times260, 2\times240$	Intestinal epithelium
Anisotropic bundling		
ABP-50	50	*Dictyostelium*
ABP-67	67	Yeast
α−Actinin	2×95	Various
Actinogelin	2×115	Ehrlich ascites tumour cells
Coronin	55	*Dictyostelium*
Dematin	52	Erythrocytes
Fascin	53–57	Pig brain, echinoderm gametes
Fimbrin	68	Intestinal epithelium
Plastin/p65	65	Various mammalian cells
Villin	95	Intestinal epithelium
(iv) *Miscellaneous*		
Caldesmon	88	Non-muscle cells
Calpactins	35 and 36	Various
Connectin	70	Binds both laminin and actin
Gelactins	23–38	Four types; from *Acanthamoeba*
Hsp90	90	Stressed cells
MAP2	300	Brain, microtubule associated
MARCKS	80	Various, substrate for PKC
Nebulin	600–900	N-line of sarcomere
Nuclear actin BP	2×34	Various
Scruin	102	*Limulus* acrosomal process
Tau	50–68	Microtubule associated

Specific entries should be seen for further details.

1 ms. In muscles, action potentials can be much slower, lasting up to 1 s.

action spectrum The relationship between the frequency (wavelength) of a form of radiation and its effectiveness in inducing a specific chemical or biological effect.

activated macrophage A macrophage (mononuclear phagocyte) that has been stimulated by lymphokines and that has greatly enhanced cytotoxic and bactericidal potential.

activation (of eggs) Normally brought about by contact between spermatozoon and egg membrane. Activation is the first stage in development, and occurs independently of nuclear fusion. The first observable change is usually the cortical reaction, which may involve elevation of the fertilization membrane; the net result is a block to further fusion and thus to polyspermy. In addition to the morphological changes, there are rapid changes in metabolic rate and an increase in protein synthesis from maternal mRNA.

activation energy The energy required to bring a system from the ground state to the level at which a reaction will proceed.

activation-induced deaminase AID; EC 3.5.4.5 Enzyme (24 kDa) that acts on single-stranded DNA during replication and deaminates cytosine to uracil (the mismatch is then processed by base-excision or mismatch repair systems); a mechanism for generating diversity in B-cells for antibody production but strictly controlled in most cells. See **ADAR**.

activation loop A region on protein kinases of the **AGC kinase** class containing a tyrosine residue (termed the activation loop site), phosphorylation of which is critical for their activity. In some cases autophosphorylation may occur. Analogous 'activation regions' have been described in, for example, **histone acetyltransferase**.

activation tagging A method used extensively in plant molecular biology; a **T-DNA** tagging vector containing four transcriptional enhancers derived from the cauliflower mosaic virus is used to randomly insert this T-DNA into the plant genome through *Agrobacterium* infection and leads to the overexpression of genes near the inserted T-DNA. Activation-tagging vectors that confer resistance to the antibiotic kanamycin or the herbicide glufosinate have been developed, and the method has allowed the generation of a wide range of gain of function mutants.

active immunity Immunity resulting from the normal response to antigen. Only really used to contrast with **passive immunity** in which antibodies or sensitized lymphocytes are transferred from the reactive animal to the passive recipient.

active site The region of a protein that binds to substrate molecule(s) and facilitates a specific chemical conversion. Produced by juxtaposition of amino acid residues as a consequence of the protein's **tertiary structure**.

active transport Often defined as transport up a gradient of **electrochemical potential**. More precisely defined as unidirectional or vectorial transport produced within a membrane-bound protein complex by coupling an energy-yielding process to a transport process. In primary active transport systems the transport step is normally coupled to **ATP** hydrolysis within a single protein 'complex'. In secondary active transport the movement of one species is coupled to the movement of another species down an electrochemical gradient established by primary active transport.

active zone Site of transmitter release on presynaptic terminal at chemical synapses. At the neuromuscular junction active zones are located directly across the synaptic cleft from clusters of **acetylcholine receptors**. Evidence from **conotoxin**-binding studies suggests that presynaptic **calcium channels** are exclusively localized at active zones.

activin Dimeric growth factors of the TGF family with effects on a range of cell types in addition to its original role (FSH releasing) in gonadal sites. Composed of two of the β subunits of **inhibin** (which is an αβ-heterodimer); since there are two isoforms, A and B, there are three forms of activin, AA, BB and AB. Receptor has serine/threonine kinase activity in its cytoplasmic domain.

activin-response factor ARF Induced in early *Xenopus* blastomeres by **activin**, Vg-1 and **TGFβ**, binds to activin-response element in the *mix*-2 homeobox gene. The ARF complex contains *XMAD2*, a *Xenopus* homologue of the *Drosophila* **Mad** gene product, and **FAST**-1, a winged-helix transcription factor.

actobindin Protein (9.7 kDa) from *Acanthamoeba castellani*. Potent inhibitor of actin polymerization under certain conditions, possibly by binding to actin oligomers thus rendering them non-nucleating. Eighty-eight amino acids with two **beta-thymosin repeats**. Homologous proteins in *Drosophila* and *C. elegans* have three beta-thymosin repeats.

actolinkin Monomeric protein (20 kDa) from echinoderm eggs. Seems to link actin filaments to inner surface of plasma membrane by their barbed ends.

actomere Site of actin filament nucleation in sperm of some echinoderms in which the **acrosomal process** is protruded by rapid assembly of a parallel microfilament bundle.

actomyosin Generally: a motor system that is thought to be based on **actin** and **myosin**. The essence of the motor system is that myosin makes transient contact with the actin filaments and undergoes a conformational change before releasing contact. The hydrolysis of ATP is coupled to movement through the requirement for ATP to restore the configuration of myosin prior to repeating the cycle. More specifically: a viscous solution formed when actin and myosin solutions are mixed at high salt concentrations. The viscosity diminishes if ATP is supplied and rises as the ATP is hydrolysed. Extruded threads of actomyosin will contract in response to ATP.

actophorin An actin-depolymerizing factor (ADF) (13–15 kDa) from protozoa (*Toxoplasma gondii, Acanthamoeba castellanii*) that will sever actin filaments (longer filaments being more rapidly severed) and sequester G actin. A member of the ADF/cofilin family. Binds ADP-G-actin with higher affinity than ATP-actin, and binding is very sensitive to the divalent cation present on the actin. Has high sequence homology with vertebrate **cofilin** and **destrin**, echinoderm **depactin** and some plant ADFs, but lacks the nuclear localization sequence found in the vertebrate ADFs.

acumentin Protein (65 kDa) originally thought to cap pointed end of microfilaments isolated from vertebrate **macrophages**.

acute 1. Sharp or pointed. 2. Of diseases: coming rapidly to a crisis, not persistent (chronic).

acute inflammation Response of vertebrate body to insult or infection; characterized by redness (rubor), heat (calor), swelling (tumour), pain (dolour) and sometimes loss of function. Changes occur in local blood flow, and **leucocytes** (particularly **neutrophils**) adhere to the walls of postcapillary venules (margination) and then move through the **endothelium** (diapedesis) towards the damaged tissue. Although usually acute inflammation is short term, there are situations in which acute-type inflammation persists.

acute lymphoblastic leukaemia *ALL* See **leukaemia**.

acute myeloblastic leukaemia *AML* See **leukaemia**.

acute phase protein Proteins found in the serum of organisms showing **acute inflammation**. In particular, **C-reactive protein** and **serum amyloid** A protein.

acute phase reaction Response to acute inflammation involving the increased synthesis of various plasma proteins (**acute phase proteins**).

acutely transforming virus Retrovirus that rapidly **transforms** cells, by virtue of possessing one or more **oncogenes**. Archetype: **Rous sarcoma virus**.

ACV 1. Acyclovir. 2. Alpha-aminoadipylcysteinyl-valine, precursor for isopenicillin synthesis.

ACV synthase *Alpha-aminoadipylcysteinyl-valine synthase* Enzyme responsible for an early step in cephalosporin synthesis. ACV is acted upon by **IPNS** to produce isopenicillin N.

acyclovir *ACV; 2-(hydroxyethoxy)methyl guanine; zovirax* Antiviral agent that is an analogue of **guanosine** and inhibits **DNA replication** of viruses. Particularly successful against herpes simplex infections.

acylation Introduction of an acyl (RCO-) group into a molecule: for example, the formation of an ester between glycerol and fatty acid to form mono-, di- or triacylglycerol or the formation of an aminoacyl tRNA during protein synthesis. Acyl–enzyme intermediates are transiently formed during covalent catalysis.

acyltransferase Enzymes of the class EC 2.3.1 that catalyse the transfer acyl groups from a carrier such as acetyl CoA to a reactant.

adalimumab *Humira* Monoclonal antibody directed against **TNF**α for the treatment of inflammatory disease.

ADAM family *(a-Disintegrin and metalloprotease) family* Family of membrane-anchored peptidases that regulate cell behaviour by proteolytically modifying the cell surface and ECM. In some cases proteolytically release membrane-bound growth factors. Have both cell adhesion and protease activities. Members of the family include *C. elegans* MIG-17, **alpha-secretase**, **TACE**. A similar and related family is the ADAM metallopeptidases with thrombospondin type 1 motif, ADAMTS.

adaptation A change in sensory or excitable cells upon repeated stimulation that reduces their sensitivity to continued stimulation. Those cells that show rapid adaptation are known as phasic; those that adapt slowly are known as tonic. Can also be used in a more general sense for any system that changes responsiveness with time – for example, by downregulation of receptors (**tachyphylaxis**) or through internal modulation of the signalling system, as in **bacterial chemotaxis**.

adaptins Proteins of 100–110 kDa found as part of the **adaptor** complex.

adaptor A protein complex associated with coated vesicles. Adaptors have been shown to promote the *in vitro* assembly of **clathrin** cages and to bind to the cytoplasmic domains of specific membrane proteins. It has been proposed that adaptors link selected membrane proteins to clathrin, causing them to be packaged into a **coated vesicle**. Two adaptors have been identified (HA-1 and HA-2); they are hetero-tetramers composed of two proteins of 100–110 kDa, termed adaptins, and two smaller polypeptides of around 50 and 20 kDa. The HA-1 subunits are γ-adaptin, β'-adaptin with polypeptides of 50 and 17 kDa. HA-1 is associated with coated pits formed from the Golgi complex. The HA-2 subunits are α-adaptin, β-adaptin with polypeptides of 47 and 20 kDa. HA-2 is associated with coated pits formed from the plasma membrane. (β'- and β-adaptin are closely related.)

ADAR *Adenosine deaminase acting on RNA; EC 3.5.4.-* Small family of adenosine deaminases (ADAR1-3; editases) that edit adenosine residues to inosine in double-stranded RNA (dsRNA). Although this editing recodes and alters the functions of several mammalian genes, its most common targets are non-coding repeat sequences, indicating the involvement of this editing system in currently unknown functions. ADAR2 is widely expressed in brain and other tissues, and knockout mice die young. Mice can be rescued by exonically edited AMPA receptor, suggesting that mRNA for this receptor is a major substrate for the enzyme.

adaxial The surface of a plant organ, such as a leaf or petal, that during early development faced towards the axis. In the case of leaves the upper surface is usually adaxial. *cf.* **abaxial**.

ADDA *3-Amino-9-methoxy-10-phenyl-2,6,8 trimethyl deca, 4,6 dienoic acid* An unusual hydrophobic amino acid found in **microcystins** and **nodularins** and essential for their toxicity.

Addison's disease Chronic insufficiency of the adrenal cortex as a result of tuberculosis or specific **autoimmune** destruction of the **ACTH**-secreting cells. Characterized by extreme weakness, wasting, low blood pressure and pigmentation of the skin. Not to be confused with Addison's anaemia or pernicious anaemia.

additive effect An effect that is simply the sum of the effects of separate exposures to two (or more) agents under the same conditions – there is no synergistic effect. Proving synergy requires demonstrating that the agents together produce an effect that is greater than the maximum either can produce alone.

addressins Cell surface molecules, particularly on endothelial cells, believed to be involved in controlling the location of migrating cells, especially in lymphocyte homing. Probably act as cell-adhesion molecules or their receptors. All have lectin, EGF and complement-regulatory domains extracellularly in addition to the membrane-spanning and cytoplasmic domains. Examples are ELAM-1, GMP-140 (PADGEM), gp90mel; also known as **selectins**.

adducin Calmodulin-binding protein associated with the membrane skeleton of erythrocytes. A substrate for **protein kinase C**, it binds to **spectrin–actin** complexes (but only weakly to either alone) and promotes the assembly of spectrin onto spectrin–actin complexes unless micromolar calcium is present. There are two similar subunits, α-adducin (103 kDa) and β-adducin (97 kDa) which together form higher order structures. Is distinguishable from band 4.1.

adductor muscle Large muscle of bivalve molluscs that is responsible for holding the two halves of the shell closed. Its unusual feature is its ability to maintain high tension with low-energy expenditure by using a 'catch' mechanism and the high content of **paramyosin**.

adenine *6-aminopurine* One of the bases found in **nucleic acids** and **nucleotides**. In DNA, it pairs with **thymine**.

adenitis Inflammation of a gland.

adeno- Prefix indicating association with, or similarity to, glandular tissue.

adenocarcinoma Malignant neoplasia of a glandular epithelium or **carcinoma** showing gland-like organization of cells.

adenohypophysis Anterior lobe of the **pituitary** gland, responsible for secreting a number of hormones and containing a comparable number of cell types.

adenoma Benign tumour of glandular epithelium.

adenomatous polyposis coli Inherited disorder in which there is a defect in the *APC* gene, a tumour suppressor. See **polyposis coli**.

adenomyoma See **endometrioma**.

adenomyosis Presence of endometrial tissue within the **myometrium** of the uterus. Nodules of ectopic tissue, not usually capsulated, may be diffuse or focal, and the glandular tissue is generally of basal type and inactive.

adenopathy Disease or disorder of glandular tissue. Usually refers to lymphatic gland enlargement.

adenosine *9-β-D-ribofuranosyladenine* The **nucleoside** formed by linking adenine to **ribose**.

adenosine deaminase *ADA; EC 3.5.4.4* An enzyme which deaminates adenosine and 2′-deoxyadenosine to inosine or 2′-deoxyinosine respectively. A rare genetic defect in this enzyme is responsible for 20–30% of cases of **severe combined immunodeficiency** (SCID), which became the first candidate disease for gene replacement therapy. ADA deficiency causes an increase of dATP, which inhibits *S*-adenosylhomocysteine hydrolase, causing an increase in *S*-adenosylhomocysteine; both are particularly toxic to lymphocytes.

adenosine deaminase deficiency See **adenosine deaminase**.

adenosine diphosphate See **ADP**.

adenosine monophosphate See **AMP, cAMP**.

adenosine receptors Four adenosine receptors have been identified: A1, A2A, A2B and A3. All are seven-membrane-spanning G-protein-coupled receptors. There is considerable difference in properties of receptors from different species. A2A receptors regulate voltage-sensitive **calcium channels**.

adenosine triphosphate See **ATP**.

adenosquamous Description of a benign tumour of epithelial origin (**adenoma**) in which cells have flattened morphology, as opposed to being cuboidal or columnar.

adenoviral vector Vector used for gene transfer, usually replication defective due to a deletion in the E1 region (early genes). Many vectors also have deletions in E3 or E4. To produce infectious particles, plasmids containing the defective adenovirus genome and the gene to be expressed are introduced into cells constitutively expressing the E1A genes, such as human 293 cells. Have been used clinically for gene therapy (for **ornithine transcarbamylase deficiency**), but with untoward side-effects in some cases and a fatality in one.

Adenoviridae Large group of viruses first isolated from cultures of adenoids. The **capsid** is an icosahedron of 240 hexons and 12 pentons, and is in the form of a base and a fibre with a terminal swelling; the genome consists of a single, linear molecule of double-stranded DNA. They cause various respiratory infections in humans. Some of the avian, bovine, human and simian adenoviruses cause tumours in newborn rodents, generally hamsters. They can be classified into highly, weakly and non-oncogenic viruses according to their ability to induce tumours *in vivo*, though all of these groups will transform cultured cells. The viruses are named after their host species and subdivided into many serological types – e.g. human adenovirus type 3. See **adenoviral vector**.

adenylate cyclase Enzyme that produces cAMP (**cyclic AMP**) from **ATP** and acts as a signal-transducer coupling hormone binding to changes in cytoplasmic cAMP levels. The name strictly refers to the catalytic moiety, but it is usually applied to the complex system that includes the hormone receptor and the GTP-binding modulator protein (see **GTP-binding proteins**).

ADF 1. Actin-depolymerizing factor (19 kDa); regulates actin polymerization in developing chick skeletal muscle. Has the ability to depolymerize microfilaments and bind G-actin, but does not cap filaments. Destrin is apparently identical, **cofilin** is similar in function but the product of a different gene. Has a nuclear localization domain, and interaction with actin is regulated by phosphoinositides. See **ADF domain**. 2. Adult T-cell leukaemia-derived factor. Inducer of interleukin 2 receptor-a (IL-2Rα). Homologue of **thioredoxin**. Autocrine growth factor produced by HTLV-1 or **EBV**-transformed cells.

ADF domain *ADF-homology domain* An actin-binding module found in an extensive family of proteins with three phylogenetically distinct classes: ADF/**cofilins**, **twinfilins** and **drebin**/ABP-1s.

ADH Antidiuretic hormone. See **vasopressin**.

adherens junction Specialized cell–cell junction into which are inserted **microfilaments** (in which case it is also known as the **zonula adherens**) or **intermediate filaments** (macula adherens or spot **desmosomes**).

adhesins General term for molecules involved in adhesion, but its use is restricted in microbiology where it refers to bacterial surface components. Generally seem to behave as lectins, binding to surface carbohydrates.

adhesion plaque Another term for a **focal adhesion**, a discrete area of close contact between a cell and a noncellular substratum, with cytoplasmic insertion of **microfilaments** and considerable electron density adjacent to the contact area. On the cytoplasmic face are local concentrations of various proteins such as **vinculin** and **talin**.

adhesion site 1. In Gram-negative bacteria, a region where the outer membrane and the plasmalemma appear to fuse. May be important in export of proteins or viral entry. **2.** Used rather generally to refer to any region of a cell specialized for adhesion.

Adiantum capillus-veneris A species of fern, the source of a number of drugs.

adipocere White or yellowish waxy substance formed postmortem by the conversion of body fats to higher fatty acids.

adipocyte Mesenchymal cell in fat tissue that has large lipid-filled vesicles. There may be distinct types in white and brown fat. 313-L1 cells are often used as a model system; they can be induced to differentiate with **dexamethasone/insulin/IBMX** treatment.

adipofibroblasts Adipocytes from subcutaneous fat that have lost their fat globules and developed a fibroblastic appearance when grown in culture. Unlike skin fibroblasts, they will take up fat from serum taken from obese donors and probably retain a distinct differentiated state.

adiponectin *AdipoQ; apM1; GBP28; adipocyte complement related protein 30; Acrp30* A protein hormone (247 aa) secreted by **adipocytes** following activation of the nuclear receptor PPAR-gamma. The C-terminal globular domain has significant homology to the TNF (tumour-necrosis factor) family. Adiponectin circulates as a homotrimer with normal serum levels of approximately 5–10 μg/ml, but levels are significantly decreased in both obesity and Type II diabetes.

adipose tissue Fibrous connective tissue with large numbers of fat-storing cells, (**adipocytes**).

adipsin A **serine peptidase** with complement factor D activity. Synthesized by **adipocytes**. Altered levels are characteristic of some genetic and acquired obesity syndromes.

adjuvant Additional components added to a system to affect action of its main component, typically to increase **immune response** to an **antigen**. See **Freund's adjuvant**.

ADME Commonly used abbreviation for *Absorption, Distribution, Metabolism* and *Excretion*: the **pharmacokinetics** of a drug.

A-DNA Right-handed double-helical **DNA** with approximately 11 residues per turn. Planes of basepairs in the helix are tilted 20° away from perpendicular to the axis of the helix. Formed from **B-DNA** by dehydration.

adnexa Appendages; usually in reference to ovaries and Fallopian tubes, but can, more generally, also be applied to accessory or adjoining organs.

adocia-sulphate-2 *AS-2* Inhibitor of **kinesin**, isolated from sponge (*Haliclona* spp.). Binds to motor domain of kinesin, mimicking tubulin.

adoptive immunity Immunity acquired as a result of the transfer of lymphocytes from another animal.

ADP Adenosine diphosphate. Unless otherwise specified, the nucleotide 5′ ADP, **adenosine** bearing a diphosphate (pyrophosphate) group in ribose-O-phosphate ester linkage

at position 5′ of the ribose moiety. Adenosine 2′5′and 3′5′ diphosphates also exist, former as part of **NADP** and the latter in **coenzyme A**.

ADP-ribosylation A form of **post-translational** modification of protein structure involving the transfer to protein of the ADP-ribosyl moiety of **NAD**. Believed to play a part in normal cellular regulation as well as in the mode of action of several bacterial toxins.

ADP-ribosylation factor *ARF* Ubiquitous **GTP-binding protein**, approximately 20 kDa, N-myristoylated, stimulates **cholera toxin** ADP-ribosylation. Mediates binding of non-clathrin-coated vesicles and AP1 (adaptor protein 1) of **clathrin-coated vesicles** to Golgi membranes. At least six isoforms have been identified.

adrenal Endocrine gland adjacent to the kidney. Distinct regions (cortex and medulla) produce different ranges of hormones.

adrenaline *Epinephrine* A hormone secreted (with **noradrenaline**) by the medulla of the **adrenal** gland and by **neurons** of the **sympathetic nervous system** in response to stress. The effects are those of the classic 'fight or flight' response, including increased heart function, elevation in blood sugar levels, cutaneous vasoconstriction making the skin pale, and raising of hairs on the neck.

adrenergic neuron A neuron is adrenergic if it secretes **adrenaline** at its terminals. Many neurons of the **sympathetic nervous system** are adrenergic.

adrenergic receptors Receptors for **noradrenaline** and **adrenaline**. All are seven membrane-spanning G-protein-coupled receptors linked variously either to adenylate cyclase or phosphoinositide second messenger pathways. Three subgroups are usually recognized, the β-adrenergic receptors linked to Gs, the α1 linked to Gi and the α2 linked to Gq. The β-adrenergic receptor gene is unusual in having no introns.

adrenocorticotrophin *ACTH* A peptide hormone produced by the pituitary gland in response to stress (mediated by corticotrophin-releasing factor, a 41-residue peptide, from the hypothalamus). Stimulates the release of adrenal cortical hormones, mostly **glucocorticoids**. Derived from a larger precursor, **pro-opiomelanocortin**, by the action of an endopeptidase that also releases β-**lipotropin**. See also Table H2.

adrenoleucodystrophy *X-linked Schilder's disease* Demyelinating disease with childhood onset due to a mutation in the *ABCD1* gene, with the result that there is an apparent defect in peroxisomal beta-oxidation and the accumulation of saturated very long-chain fatty acids in all tissues of the body.

adrenomedullin *ADM* A vasodilator peptide (52 aa) related to **calcitonin gene-related peptide** (CGRP). In addition to vasodilatory effects it has been reported that adrenomedullin protects a variety of cells against oxidative stress, induced by stressors such as hypoxia, ischaemia/reperfusion and hydrogen peroxide, through the phosphatidylinositol 3-kinase (PI3K)-dependent pathway. See **proadrenomedullin-20**.

adriamycin An antibiotic that **intercalates** into RNA and DNA. Related to **daunomycin**.

adseverin Actin-regulating protein (74 kDa) isolated from adrenal medulla. Has severing, nucleating and capping activities similar to those of **gelsolin**, but does not cross-react immunologically. Has phospholipid-binding domain and its properties are regulated by phosphatidyl inositides and by calcium. May be identical to **scinderin**.

adsorption coefficient A constant, under defined conditions, that relates the binding of a molecule to a matrix as a function of the weight of matrix, for example, in a column.

adult respiratory distress syndrome *Acute respiratory distress syndrome; ARDS* See **septic shock**.

adventitia Outer coat of the wall of **vein** or **artery**, composed of loose **connective tissue** that is vascularized. Generally used to mean the outer covering of an organ.

Aedes Genus of mosquitoes, several of which transmit diseases of man. *A. aegypti* is the vector of the **yellow fever virus**.

Aequorea victoria Hydrozoan jellyfish (a **coelenterate**) from which **green fluorescent protein** (GFP) was isolated. **Aequorin** can be isolated from *A. victoria* and *A. forskaolea*.

aequorin Protein (30 kDa) extracted from jellyfish (*Aequorea victoria*) that emits light in proportion to the concentration of calcium ions. Used to measure calcium concentrations, but has to be microinjected into cells. Contains **EF-hand** motif. See also **bioluminescence**.

aerenchyma Form of **parenchyma** with large air spaces that gives buoyancy to aquatic plants.

aerobes Organisms that rely on oxygen.

aerobic respiration Controlled process by which carbohydrate is oxidized to carbon dioxide and water, using atmospheric oxygen, to yield energy.

aerolysin Channel-forming bacterial **exotoxin** produced by *Aeromonas hydrophila* as a 50-kDa protoxin that later has a 43-residue C-terminal peptide cleaved off to generate the active toxin. Binds to specific receptor on target cells (probably glycophorin on human erythrocytes, but may be other proteins in other cells) and polymerizes to form a heptameric complex that inserts into the plasma membrane and has a pore of approximately 1.5 nm diameter with some properties similar to **porin** channels.

Aeromonas Genus of Gram-negative bacteria some species of which are pathogenic. *Aeromonas salmonicida* causes furunculosis in fish.

aerotaxis A **taxis** in response to oxygen (air).

aesculin A hydroxycoumarin extracted from the horse chestnut (*Aesculus hippocastanum*), used in homeopathic medicine to thin the blood, acting as an anticoagulant and possibly hypodiuretic. Has toxic effects at higher doses. Aesculin is also used as an additive in agar media for the isolation of glycopeptide-resistant enterococci.

aetiology *Etiology (US)* The study of the causation of disease.

AF2 1. Activation function domain 2 of steroid receptors. 2. Antiflammin-2, a synthetic peptide inhibitor of PLA2.

afadin *AF-6 protein* Protein containing a single **PDZ domain** that forms a peripheral component of cell membranes at specialized sites of cell–cell junctions. The carboxyl termini of the **Eph-related receptor tyrosine kinases** EphA7, EphB2, EphB3, EphB5 and EphB6 interact with the PDZ domain of the ras-binding protein AF6. Bind **nectins**.

AFAP-110 *Actin filament-associated protein-110* An actin-binding protein (110 kDa) with a **PH domain**; has an alpha-helical N-terminal region capable of self-association through a leucine zipper interaction. AFAP-110 is a **Src** substrate and phosphorylation regulates self-association; a major function of AFAP-110 may be to relay signals from PKCalpha via activation of c-Src leading to the formation of podosomes.

afferent Leading towards; afferent nerves lead towards the central nervous system, afferent lymphatics towards the lymph node. The opposite of efferent.

affinity An expression of the strength of interaction between two entities, e.g. between receptor and ligand or between enzyme and substrate. The affinity is usually characterized by the equilibrium constant **association constant** or **dissociation constant** for the binding, this being the concentration at which half the receptors are occupied.

affinity chromatography **Chromatography** in which the immobile phase (bed material) has a specific biological affinity for the substance to be separated or isolated, such as the affinity of an antibody for its antigen or an enzyme for a substrate analogue.

affinity labelling Labelling of the active site of an enzyme or the binding site of a receptor by means of a reactive substance that forms a covalent linkage once having bound. Linkage is often triggered by a change in conditions, for example in photoaffinity labelling as a result of illumination by light of an appropriate wavelength.

affinity maturation Following exposure to a novel antigen, antibody-forming cells produce antibody of progressively higher affinity and memory lymphocytes capable of producing high-affinity antibody upon re-exposure to the antigen. Affinity maturation occurs by the early selective differentiation of high-affinity variants into antibody-forming cells that persist in the bone marrow.

aflatoxins A group of highly toxic substances produced by the fungus *Aspergillus flavus*, and other species of *Aspergillus*, in stored grain or mouldy peanuts. They cause enlargement and death of liver cells if ingested, and may be carcinogenic.

AFP *Alpha-fetoprotein* See **alpha-fetoprotein**.

Agalenopsis aperta American funnel web spider. See **agatoxins**.

agammaglobulinaemia Sex-linked genetic defect that leads to the complete absence of immunoglobulins (IgG, IgM and IgA) in the plasma as a result of the failure of pre-B-cells to differentiate. Failure to produce a humoral antibody response leads to high incidence of opportunistic infections, but cell-mediated immunity is unimpaired.

agar A **polysaccharide** complex extracted from seaweed (Rhodophyceae) and used as an inert support for the growth of cells, particularly bacteria and some cancer cell lines (e.g. sloppy agar).

agarose A galactan polymer purified from **agar** that forms a rigid gel with high free water content. Primarily used as an electrophoretic support for separation of macromolecules. Stabilized derivatives are used as 'macroporous' supports in **affinity chromatography**. See **Sepharose**.

agatoxins Toxins from the American funnel web spider, *Agalenopsis aperta*. The μ-toxins are 36–38 residue peptides that act on insect but not vertebrate voltage-sensitive sodium channels. The ω-agatoxins are more diverse (5–9 kDa) and act on various calcium channels, mostly in neuronal cells, blocking release of neurotransmitters. Affect vertebrate and invertebrate channels.

AGC kinases A large subclass of protein kinases that includes protein kinases A, G and C, also protein kinase B (PKB)/akt, p70 and p90 ribosomal S6 kinases and phosphoinositide-dependent kinase-1 (PDK-1). All have an activation loop with a tyrosine phosphorylation site that is important for activity.

AGE *Advanced glycation endproducts* Modified proteins (glycation adducts) that arise as a result of the reaction of reducing sugars with proteins; analogous reactions occur during cooking (Maillard reaction products). They form at an accelerated rate in diabetes and contribute to the development of vascular disease. Receptor is **RAGE**.

agglutination The formation of adhesions by particles or cells to build up multicomponent aggregates, otherwise termed agglutinates or flocs. Distinguished from **aggregation** by the fact that agglutination phenomena are usually very rapid. Usually caused by addition of extrinsic agents such as **antibodies**, **lectins** or other bi- or polyvalent reagents. See **aggregation**.

agglutinins Agents causing **agglutination**, e.g. antibodies, **lectins**, **polylysine**.

agglutinogen The antigen (in the case of antibody) or ligand (in the case of lectin) with which an **agglutinin** reacts.

aggrecan The major structural **proteoglycan** of cartilage. It is a very large and complex macromolecule, comprising a core protein of 210 kDa to which are linked around 100 chondroitin sulphate chains and several keratan sulphate chains as well as O- and N-linked oligosaccharide chains. It binds to a link protein (around 40 kDa) and to hyaluronic acid, forming large aggregates – hence its name.

aggregation The process of forming adhesions between particles such as cells. Aggregation is usually distinguished from **agglutination** by the slow nature of the process; not every encounter between the cells is effective in forming an adhesion.

aggresome *Sequestosome* A perinuclear inclusion body containing aggregated, misfolded, ubiquitinated protein; it has been suggested that aggresome formation is a general cellular response when the capacity of the **proteosome** is exceeded. May share a common biogenesis with **Lewy bodies**. Cells that overexpress wild-type **parkin** form fewer aggresomes, although parkin is found within aggresomes induced by stress in neuroblastoma cells in culture. Disruption of microtubules blocks the formation of aggresomes.

aginactin Agonist-regulated actin filament barbed-end capping protein (70 kDa) that inhibits microfilament polymerization, isolated from *Dictyostelium*. Interestingly it is regulated by cAMP, a chemotactic factor for *Dictyostelium*.

Turns out to be a complex of Hsc70 and **Capping protein** (Cap32/34).

agitoxins Scorpion toxins (peptides) that inhibit potassium channels. Closely related to **kaliotoxin**.

aglycone *Aglycon; aglucon* The portion of a glycoside that remains when the sugar moiety is removed.

agmatinase *EC 3.5.3.11; agmatine ureohydrolase* Enzyme that degrades **agmatine** to **putrescine**.

agmatine *1-Amino-4-guanidobutane* A metabolite of arginine via arginine decarboxylase is metabolized to **putrescine** by **agmatinase**. Suppresses polyamine biosynthesis and polyamine uptake by cells by inducing **antizyme**. Binds to imidazoline receptors and alpha-2-adrenoreceptors but with different affinities, and is thought to be the endogenous ligand for imidazoline (I1) receptors. On the basis of its distribution in the brain, is proposed to be a neurotransmitter involved in behavioural and visceral control. An increasing range of biological activities are being described.

Agnatha A superclass of anguilliform (eel-shaped) chordates without jaws or pelvic fins. Lampreys and hagfishes.

agnoprotein A small, highly basic protein (71 aa) encoded by neurotropic JC virus, SV40 and BK virus. It apparently plays a critical role in regulation of viral gene expression, inhibits DNA repair after DNA damage and interferes with DNA damage-induced cell cycle regulation. Agnoprotein is expressed during lytic infection of glial cells by JCV in progressive multifocal leucoencephalopathy (PML) and also in some JCV-associated human neural tumours, particularly medulloblastoma.

agonist **1.** In neurobiology, of a **neuron** or **muscle**; one that aids the action of another. If the two effects oppose each other, then they are known as antagonistic. **2.** In pharmacology, a compound that acts on the same receptor and with a similar effect to the natural ligand. **3.** In ethology, 'agonistic behaviour' means aggressive behaviour towards a conspecific animal.

agorins Major structural proteins of the membrane matrix, constituting approximately 15% of total plasma membrane proteins of P815 mastocytoma cells. They form large detergent-insoluble structures when the membranes are extracted with Triton X-100 and EGTA. Agorin I, 20 kDa; Agorin II, 40 kDa.

agouti Central American rodent that has given its name to a grey-flecked coat colouration in mice caused by alternate light and dark bands on individual hairs. The gene codes for a 131-residue-secreted protein that regulates phaeomelanin synthesis in melanocytes, but associated with the locus are genes important in embryonic development. Mice with dominant mutation at the *Ay* locus develop diabetes and obesity. The agouti gene product binds to **melanocortin** receptor 1 but does not antagonize α-**MSH** and has similar antiproliferative effects on melanoma cells in culture. The dark agouti (DA) rat has a high susceptibility to developing arthritis.

agouti-related peptide *AGRP* A neuropeptide, produced by neurons in the arcuate nucleus of the brain, which increases appetite and decreases metabolism. Secretion of AGRP and **neuropeptide Y** is stimulated when the amount of stored fat and **leptin** decreases. Agouti-related protein is an endogenous antagonist at the melanocortin types 3 and 4 receptors.

AGP *Arabinoglycan-protein* A class of extracellular **proteoglycan** found in many higher-plant tissues and secreted by many suspension-cultured plant cells. Contains 90–98% **arabinogalactan** and 2–10% protein. Related to arabinogalactan II of the cell wall.

agranular vesicles Synaptic vesicles that do not have a granular appearance in EM; 40–50 nm in diameter, with membrane only 4–5 nm thick. Characteristic of peripheral cholinergic **synapses** (see also **neurotransmitter**). Some are located very close to presynaptic membrane.

agranulocytosis Severe acute deficiency of **granulocytes** in blood.

agretope Portion of antigen that interacts with an **MHC** molecule.

agrin Secreted protein (200 kDa) isolated from the synapse-rich electric organ of *Torpedo californica* that induces the formation of synaptic specializations on **myotubes** in culture. Present in muscle cells before innervation, and concentrated at the **neuromuscular junction** once AChR-clustering occurs. The release of agrin from motor axon terminals is thought to trigger the formation of the post-synaptic apparatus at developing and regenerating neuromuscular junctions. Has several EGF repeats and a protease inhibitor-like domain.

Agrobacterium tumefaciens A Gram-negative, rod-shaped flagellated bacterium responsible for **crown gall** tumour in plants. Following infection the T1 **plasmid** from the bacterium becomes integrated into the host plant's DNA, and the presence of the bacterium is no longer necessary for the continued growth of the tumour. See **T-DNA**.

agropine One of the three **opines**.

AGS cells A line of human gastric **adenocarcinoma** cells.

AH receptor *AHR* Cytoplasmic receptor for aryl hydrocarbons. See **AHR**.

AHNAK Protein (700 kDa), apparently identical to **desmoyokin**, that is widely expressed in smooth and striated muscle cells and in epithelial cells of most lining epithelium, though absent in epithelium with more specialized secretory or absorptive functions. In all adult tissues, the main localization of AHNAK is at the plasma membrane.

AHR *Aryl hydrocarbon receptor* Cytosolic protein (~800 aa) encoded by the *Ahr* gene that binds a range of aryl hydrocarbons and dioxin and is then translocated to the nucleus where it forms a dimer with **ARNT** and binds to the **xenobiotic response element**. Has a basic helix–loop–helix motif. In development AHR plays an important role in the closure of the ductus venosus, and the nuclear form of the activated AHR/aryl hydrocarbon nuclear translocator complex is responsible for alterations in immune, endocrine, reproductive, developmental, cardiovascular and central nervous system functions.

AIDS *Acquired immune deficiency syndrome* Disease caused by infection with HIV (human immunodeficiency virus, also called LAV or HTLV-3 in the early literature), resulting in a deficiency of **T-helper cells** and thus **immunosuppression**; as a result, opportunistic infections are likely to occur and there is predisposition to certain types of tumour, particularly **Kaposi's sarcoma**.

AIF See **apoptosis-inducing factor**.

AIP1 1. Actin-interacting protein-1. Actin-binding protein from yeast, subsequently found elsewhere. Interacts with ADF/**cofilin** to produce bundles in *Dictyostelium* in response to osmotic stress. Aip1 and cofilin cooperate to disassemble actin filaments *in vitro* and are thought to promote rapid turnover of actin networks *in vivo*. **2.** Alix/AIP1 (ALG-2-interacting protein X/apoptosis-linked-gene-2-interacting protein 1), an adaptor protein that controls the production of endosomes and trafficking through the endosomal system.

air-lift fermenter A fermenter in which circulation of the culture medium and aeration is achieved by injection of air into some lower part of the fermenter. Usually not suitable for animal cell production. Related to gas-lift systems where an inert gas is used to achieve circulation in anaerobic conditions.

AKAP family *A-kinase-anchoring protein family* Family of giant scaffolding proteins that can assemble and compartmentalize multiple signalling and structural molecules. See **pericentrin** and **AKAP79**.

AKAP79 *A-kinase-anchoring protein* Scaffold protein from mammalian cells to which **Protein Kinase A** (PKA), **calcineurin** and **Protein Kinase C** (PKC) all bind. PKC apparently binds at a different site so that PKC and calcineurin can, for example, simultaneously be bound. Through anchorage to AKAP79, kinases and phosphatases are co-localized at postsynaptic densities in neurons. AKAP79 resembles yeast STE5.

A kinase *PKA; EC 2.7.1.37* cAMP-regulated protein kinase, sometimes abbreviated PKA to contrast with **protein kinase C**.

Akt *PKB* Product of the normal gene homologue of *v-akt*, the transforming oncogene of AKT8 virus. A serine/threonine kinase (58 kDa) with SH2 and PH domains, activated by PI3kinase downstream of insulin and other growth factor receptors. AKT will phosphorylate **GSK3** and is involved in stimulation of Ras and control of cell survival. Three members of the Akt/PKB family have been identified, Akt/PKBα, AKT2/PKBβ and AKT3/PKBγ. Only AKT2 has been shown to be involved in human malignancy.

AKV A replication-competent murine leukaemia virus occurring endogenously in some mouse strains.

alae Flat, wing-like processes or projections, especially of bone. *Adj.* alar, alary.

ALADIN See **triple A syndrome**.

alamethicin A polyene pore-forming **ionophore** that forms relatively non-specific anion or cation transporting pores in plasma membranes or artificial lipid membranes. The pores formed by alamethicin are potential gradient sensitive.

alanine Normally refers to L-α-alanine, the aliphatic **amino acid** found in proteins. See Table A2. The isomer β-alanine is a component of the vitamin **pantothenic acid** and thus also of **coenzyme A.**

alanine aminotransferase *ALT; EC 2.6.1.2* A cytoplasmic hepatocellular enzyme, the increase of which in blood is highly indicative of liver damage and is often taken as a possible sign of infection with non-A non-B hepatitis. Catalyses L-alanine and 2-oxoglutarate conversion into pyruvate and L-glutamate.

TABLE A2. Amino acids

Name	Abbreviation	Single letter	Side chain	pK_a^a	M_r (Da)	Hydropathy index^b (Kyte and Doolittle)	Codons
Alanine	ala	A	–CH$_3$		89.1	1.8	GC(X)
Arginine	arg	R	–CH$_2$ CH$_2$ CH$_2$ NH (CN$^+$H$_2$) NH$_2$	12	174.2	–4.5	CG(X) AGA AGG
Asparagine	asn	N	–CH$_2$ CONH$_2$		132.2	–3.5	AAU AAC
Aspartic acid	asp	D	–CH$_2$ COO$^-$	4.4	133.1	–3.5	GAU GAC
Cysteine	cys	C	–CH$_2$ SH	8.5	121.2	2.5	UGU UGC
Glutamic acid	glu	E	–CH$_2$ CH$_2$ COO$^-$	4.4	147.2	–0.4	GG(X)
Glutamine	gln	Q	–CH$_2$ CH$_2$ CONH$_2$		146.2	–3.5	CAA CAG
Glycine	gly	G	–H		75.1	–3.5	GG(X)
Histidine	his	H		6.5	155.2	–3.2	CAU CAC
Isoleucine	ile	I	–CH (CH$_3$) CH$_2$CH$_3$		131.2	4.5	AUU AUC AUA
Leucine	leu	L	–CH$_2$ CH (CH$_3$)$_2$		131.2	3.8	CU(X) UUA UUG
Lysine	lys	K	–CH$_2$ CH$_2$ CH$_2$ CH$_2$ NH$_3^+$	10	146.2	–3.9	AAA AAG
Methionine	met	M	–CH$_2$ CH$_2$ SCH$_3$		149.2	1.9	AUG
Phenylalanine	phe	F			165.2	2.8	UUU UUC
Proline	pro	P			115.1	–1.6	CC(X)
Serine	ser	S	–CH$_2$ OH		105.1	–0.8	UC(X)
Threonine	thr	T	–CH (OH) CH$_3$		119.1	–0.7	AC(X)
Tryptophan	trp	W			204.2	–0.9	UGG (UGA mitochondria)
Tyrosine	tyr	Y		10	181.2	–1.3	UAU UAC
Valine	val	V	–CH (CH$_3$)$_2$		117.2	4.2	GU(X)

L-amino acids specified by the biological code for proteins.

^{a} The value for side chain ionization when the amino acid residue is present in a polypeptide.

^{b} A measure of the tendency for the residue to be buried within the interior of a folded protein.

alarmone A small signal molecule in bacteria that induces an alteration of metabolism as a response to stress. Many metabolic responses may be altered by a single alarmone.

ALA synthase *EC 2.3.1.37; 5-aminolevulinate synthase* Enzyme responsible for the synthesis of 5-aminolevulinic acid. Defects in the enzyme cause microcytic anaemia because activity is essential for haem formation. The erythrocyte isoform is distinct from that in other tissues.

A layer S-layer in *Aeromonas* sp.

Albers–Schoenberg disease See **osteopetrosis**.

albinism Condition in which no **melanin** (or other pigments) is present.

albino Organism deficient in **melanin** biosynthesis. Hair and skin are unpigmented and the retinal pigmented epithelium is transparent, making the eyes appear red.

Albino3 An *Arabidopsis* nuclear gene essential for chloroplast differentiation, encodes a chloroplast protein that is the translocase responsible for the Sec-independent insertion of light-harvesting chlorophyll-binding protein into the chloroplast membrane. Has similarities to **Oxa1** and **YidC**, proteins present in bacterial membranes and yeast mitochondria.

alboaggregin Alboaggregin A is a C-type lectin (50 kDa) from the white-lipped tree viper (*Trimeresurus albolabris*); causes platelet aggregation through acting as a platelet GPIb agonist. Alboaggregin B also has platelet-aggregating activity but is calcium independent. Similar in action to **botrocetin**.

albolabrin See **disintegrin**.

albumen *ovalbumin* Major protein of the white of bird's eggs. See **albumin**.

albumin The term normally refers to serum albumins, the major protein components of the serum of vertebrates. They have a single polypeptide chain, with multidomain structure containing multiple binding sites for many lipophilic metabolites, notably fatty acids and bile pigments. In the embryo their functions are fulfilled by **alpha-fetoproteins**. The viability of analbuminaemic mutants (those deficient in albumin) raises serious questions about the biological role of albumin. See **albumen**.

Alcaligenes Widespread genus of Gram-negative aerobic bacilli found in the digestive tract of many vertebrates and on skin. Occasionally cause opportunistic infections.

Alcian blue *Alcian blue 8GX* Water-soluble copper phthalocyanin stain used to demonstrate acid **mucopolysaccharides**. By varying the ionic strength some differentiation of various types is possible. Similar dyes, Alcian green and Alcian yellow, have comparable properties.

aldolase *EC 4.1.2.13* A glycolytic enzyme which, like phosphofructokinase, enolase and hexokinase, binds actin filaments, probably in order to increase efficiency by locally concentrating enzyme and substrate. The glucose transporter **GLUT4** is connected to actin via aldolase.

aldose reductase *EC 1.1.1.21* Enzyme that mediates conversion of glucose to sorbitol and the rate-limiting enzyme in the polyol pathway. Altered activity of aldose reductase is thought to play a part in the alterations in the vasculature seen as a complication of diabetes.

aldosterone A steroid hormone produced by the **adrenal** cortex that controls salt and water balance in the kidney.

aldosterone secretion inhibiting factor A natriuretic factor isolated from chromaffin cells that is an agonist at NPR receptors that inhibit aldosterone production. Closely related in structure to brain natriuretic peptide (BNP). Atrial natriuretic factor (ANF) also inhibits **renin** and **aldosterone** secretion.

alemtuzumab A humanized monoclonal antibody used in the treatment of B-cell chronic lymphocytic leukaemia that has proved refractory to cytopathic drugs such as **fludarabine**.

alendronic acid Bisphosphonate drug used for treatment of **osteoporosis**.

aleurone grain *Aleurone body* Membrane-bounded **storage granule** within plant cells that usually contains protein. May be an **aleuroplast** or just a specialized **vacuole**.

aleuroplast A semiautonomous organelle (**plastid**) within a plant cell that stores protein.

Aleutian disease A chronic, fatal disease of mink, caused by a parvovirus; recognized originally in mink homozygous for the aleutian gene that controls fur colour, but also affects raccoons, skunks and ferrets.

Alexander disease Rare and usually fatal neurodegenerative disorder caused by a *de novo* mutation in the glial fibrillary acidic protein (**GFAP**) gene. There is an abundance of protein aggregates (Rosenthal fibres) in astrocytes that contain the protein chaperones alpha B-crystallin and HSP27 as well as glial fibrillary acidic protein (GFAP).

Alexandrium spp. Genus of **dinoflagellates** that produce toxins associated with shellfish poisoning.

alexithymia Psychiatric term for the state of having no words for emotions. A clinical construct, not constituting an illness in its own right.

algae A non-taxonomic term used to group several phyla of the lower plants, including the **Rhodophyta** (red algae), **Chlorophyta** (green algae), **Phaeophyta** (brown algae) and **Chrysophyta** (diatoms). Many algae are unicellular or consist of simple undifferentiated colonies, but red and brown algae are complex multicellular organisms, familiar to most people as seaweeds. Blue-green algae are a totally separate group of **prokaryotes**, more correctly known as Cyanophyta or **Cyanobacteria**.

alginate Salts of alginic acids, occurring in the cell walls of some algae. Commercially important in food processing, swabs, some filters, fire-retardants, etc. Calcium alginates

form gels. Alginic acid is a linear polymer of mannuronic and glucuronic acids.

algorithm A process or set of rules by which a calculation or process can be carried out, usually referring to calculations that will be done by a computer.

aliphatic Carbon compound in which the carbon chain is open (non-cyclic).

aliphatic amino acids The naturally occurring amino acids with aliphatic side chains are glycine, alanine, valine, leucine and isoleucine.

aliphatic esterase *Aliesterase* An enzyme that hydrolyses ester linkages, particularly in aliphatic esters.

aliquot Small portion. It is common practice to subdivide a precious solution of reagent into aliquots that are used when needed, without handling the total sample.

aliskiren Potent and selective orally available inhibitor of human **renin** designed by a combination of molecular modelling and crystallographic structure analysis.

alkaline phosphatase *EC 3.1.3.1* Enzyme catalysing cleavage of inorganic phosphate non-specifically from a wide variety of phosphate esters and having a high (>8) pH optimum. Found in bacteria, fungi and animals but not in higher plants.

alkaloid A nitrogenous base. Usually refers to biologically active (toxic) molecules, produced as allelochemicals by plants to deter grazing. Examples: **ouabain**, **digitalis**.

alkaptonuria Congenital absence of homogentisic acid oxidase, an enzyme that breaks down tyrosine and phenylalanine. Accumulation of **homogentisic acid** in homozygotes causes brown pigmentation of skin and eyes and damage to joints; urine blackens on standing.

alkylating agent A reagent that places an alkyl group, e.g. propyl in place of a nucleophilic group, in a molecule. See **alkylating drug**.

alkylating drug Cytotoxic drug which acts by damaging DNA in various ways: addition of alkyl groups to guanine leads to fragmentation during frustrated repair; nucleotides may be cross-bridged, blocking strand separation; mispairing leading to mutation may occur during replication. There are six groups of **alkylating agents**: nitrogen mustards, ethylenimes, alkylsulfonates, triazenes, piperazines and nitrosureas. Common examples are **cyclophosphamide**, chlorambucil, busulphan and mustine.

ALL Acute lymphoblastic **leukaemia**.

all-or-nothing Of an action potential, meaning that action potentials once triggered are of a stereotyped size and shape, irrespective of the size of stimulus that triggered them.

allantoin *5-Ureidohydantoin* A derivative of uric acid; occurs in allantoic fluid and in urine of certain mammals, and is also excreted by certain insects and gastropods. Bizarrely, is used in various cosmetics and in various oral hygiene preparations and eye drops.

allantois Outgrowth from the ventral side of the hindgut in embryos of reptiles, birds and mammals. Serves the embryo as a store for nitrogenous waste and in chick embryos fuses with the chorion to form the **chorioallantoic membrane** (CAM).

allatostatins Peptide hormones produced by the **corpora allata** of insects that reversibly inhibit the production of **juvenile hormone**. Similar peptides are found in other phyla. Allatostatin-4, smallest of the family, is DRLYSFGL-amide. Allatostatins may also be produced in other insect tissues, particularly midgut.

alleles Different forms or variants of a **gene** found at the same place or **locus**, on a **chromosome**. Assumed to arise by **mutation**.

allelic exclusion The process whereby one or more loci on one of the **chromosome** sets in a **diploid** cell is inactivated (or destroyed) so that the locus (or loci) is not expressed in that cell or a clone founded by it. For example, in mammals one of the X chromosome pairs of females is inactivated early in development (see **Lyon hypothesis**) so that individual cells express only one allelic form of the product of that locus. Since the choice of chromosome to be inactivated is random, different cells express one or other of the X chromosome products, resulting in mosaicism. The process is also known to occur in **immunoglobulin** genes so that a clone expresses only one of the two possible allelic forms of immunoglobulin.

allelic imbalance A situation where one **allele** of a heterozygous gene pair is lost (loss of heterozygosity) or amplified; the molecular basis of aneuploidy and a frequent finding in tumours. The mechanisms leading to allelic imbalance are uncertain, but it is thought to result in dysregulation of oncogenes or tumour-suppressor genes near the sites of imbalance.

allelochemical Substances effecting allelopathic reactions. See **allelopathy**.

allelomorph One of several alternative forms of a gene: commonly shortened to **allele**.

allelopathic Harmful interaction between individuals of a species – for example, in plants by the production of a chemical inhibitor such as a terpenoid or phenolic. A clash between strains or variants that differ in their alleles.

allelopathy The deleterious interaction between two organisms or cell types that are **allogeneic** to each other (the term is often applied loosely to interactions between **xenogeneic** organisms). Allelopathy is seen between different species of plant, between various individual sponges and between sponges and gorgonians.

allelotype Occurrence of an **allele** in a population or an individual with a particular allele. The allelotype of a tumour, the expression of particular microsatellite markers or isoenzymes, can indicate whether it is of polyclonal or monoclonal origin and the extent to which there is development of **aneuploidy**.

Allen video-enhanced contrast *AVEC-microscopy* A method for enhancing microscopic images, pioneered by R. D. Allen. The digitized image has the background (an out-of-focus image of the same microscopic field with comparable unevenness of illumination, etc.) subtracted and the contrast expanded to utilize the potential contrast range. Interestingly, it is possible to produce images of objects that are below the theoretical limit of resolution – **microtubules**, for example.

allergenic Capable of provoking an allergic response (see **allergy**). Commonly used to describe substances (allergens) that cause immediate-type hypersensitivity reactions, such as pollens or insect venoms.

allergens See **allergenic**.

allergic encephalitis See **experimental allergic encephalomyelitis**.

allergic rhinitis A response to inhaled allergens causing swelling of the mucous membranes of the nose and upper respiratory tract. Hayfever is a common seasonal form, often a response to grass or tree pollen.

allergy In an animal, a **hypersensitivity** response to some **antigen** that has previously elicited an immune response in the individual, producing a large and immediate **immune response**. Allergies, for example to bee venom, are occasionally fatal in humans.

allicin *2-Propene-1-sulfinothioc acid* S-*2-propenyl ester* An antibacterial compound, with a strong odour, produced when raw garlic (*Allium sativum*) is either crushed or somehow injured. The enzyme, alliinase, converts alliin in raw garlic to allicin. It is rapidly degraded and its medical efficacy is doubtful.

Allium Genus that includes onions (*A. cepa*), leeks (*A. porrum*) and garlic (*A. sativum*).

alloantibody *Alloserum* Antibody raised in one member of a species that recognizes genetic determinants in other individuals of the same species. Common in multiparous women and multiply-transfused individuals who tend to have alloantibodies to **MHC** or **blood group antigens**.

alloantigen Individuals of a species differ in alleles (are **allogeneic**) and the antigenic differences will cause an immune response to allografts. The antigens concerned are often of the **histocompatibility complex** and are referred to as alloantigens.

allochthonous Anything found at a site remote from that of its origin.

allogeneic Two or more individuals (or strains) are stated to be allogeneic to one another when the genes at one or more loci are not identical in sequence in each organism. Allogenicity is usually specified with reference to the locus or loci involved.

allograft Grafts between two or more individuals allogeneic at one or more loci (usually with reference to **histocompatibility** loci). As opposed to **autograft** and **xenograft**.

allolactose *β-D-galactopyranosyl-(1→6)-D-glucopyranose* An isomer of lactose and the natural intracellular inducer of the **lac operon**.

allometric 1. Differing in growth rate. **2.** Relating to **allometry**.

allometric growth Pattern of growth such that the mass or size of any organ or part of a body can be expressed in relation to the total mass or size of the entire organism according to the allometric equation: $Y = axb$ where $Y =$ mass of the organ, $x =$ mass of the organism, $b =$ growth coefficient of the organ and $a =$ a constant.

allometric scaling Adjustment of data either to allow for a change in proportion between an organ or organs and other body parts during the growth of an organism, or to allow for differences and make comparisons between species having dissimilar characteristics, for example in size and shape.

allometry **1.** Study of the relationship between the growth rates of different parts of an organism. **2.** Change in the proportion of part of an organism as it grows.

allomone Compound produced by one organism that affects, detrimentally, the behaviour of a member of another species. If the benefit is to the recipient, the substance is referred to as a **kairomone**; if both organisms benefit then it is a **synomone**.

allopatric speciation The accumulation of genetic differences in a geographically isolated subpopulation leading to the evolution of a new species.

allophycocyanin **Phycobilin** found in some Rhodophyceae and Cyanobacteria.

allopolyploidy **Polyploid** condition in which the contributing genomes are dissimilar. When the genomes are doubled, fertility is restored and the organism is an amphidiploid. Common in plants but not in animals.

allopurinol *4-Hydroxypyrazolo-(3,4-d) pyrimidine* A **xanthine oxidase** inhibitor used in the treatment of gout.

allosomes *Accessory chromosomes, heterochromosomes, sex chromosomes* One or more chromosomes that can be distinguished from **autosomes** by their morphology and behaviour.

allospecific **1.** In taxonomic terms allospecificity implies having the status of a distinct species, genetically distinct and isolated from other similar species, whereas there can be genetic exchange (successful interbreeding) between subspecies and the gene-pool is not 'closed'. **2.** In immunological usage it is shorthand for allele specific; antigenically distinct and thus capable of being recognized by the immune system. Thus individuals, unless genetically identical, are allospecific and capable of rejecting grafts that differ in antigenic features (allotypic determinants).

allostasis Adaptive processes that actively maintain stability through change; this concept has become popular in analysing the way in which psychological stress is handled and the longer-term cost that this may bring (allostatic load), although clearly the concept is generalizable.

allosteric Of a binding site in a protein, usually an enzyme. The catalytic function of an enzyme may be modified by interaction with small molecules, not only at the **active site**, but also at a spatially distinct (allosteric) site of different specificity. Of a protein, a protein possessing such a site. An allosteric effector is a molecule bound at such a site that increases or decreases the activity of the enzyme.

allosteric activator A compound that activates an enzyme through an **allosteric** interaction.

allotetraploidy Example of **allopolyploidy** in which the hybrid diploid genome (formed from two chromosome sets) doubles in chromosome number.

allotope *Allotypic determinant* The structural region of an **antigen** that distinguishes it from another **allotype** of that antigen.

allotype Products of one or more **alleles** that can be detected as inherited variants of a particular molecule. Usually the usage is restricted to those **immunoglobulins** that can be separately detected antigenically. See also **idiotype**. In humans, light-chain allotypes are known as Km (Inv) allotypes and heavy chain allotypes as Gm allotypes.

alloxan Used to produce **diabetes mellitus** in experimental animals. Destroys **B-cells** in the pancreas by a mechanism involving **superoxide** production.

allozyme Variant of an enzyme coded by a different allele. See **isoenzyme**.

alopecia Baldness – can take various forms; alopecia areata in which hair loss is patchy, alopecia universalis in which loss is complete.

alpha-2-macroglobulin *α-2-macroglobulin* Large (725-kDa) plasma antipeptidase with very broad spectrum of inhibitory activity against all classes of proteases. Apparently works by trapping the peptidase within a cage that closes when the peptidase-sensitive bait sequence is cleaved. The peptidase is still active against small substrates that can diffuse into the cage, and the conformational change that closes the trap alters the properties of the α-2-macroglobulin molecule so that it is rapidly removed from circulation. **plasminogen activator** is one of the few peptidases against which α-2-macroglobulin is ineffective.

alpha-actinin *α-actinin* A protein of 100 kDa normally found as a dimer and that may link actin filaments end-to-end with opposite polarity. Originally described in the **Z-disc**, now known to occur in **stress-fibres** and at **focal adhesions**. Non-muscle isoform contains **EF-hand** motif. An actinin-like actin-binding domain has been found in the N-terminal region of many different actin-binding proteins (e.g. beta-chain of spectrin (or fodrin), dystrophin, ABP-120, filamin and fimbrin.). See **calponin homology domain**.

alpha-amylase *α-amylase; EC 3.2.1.1* An endo-amylase enzyme that rapidly breaks down starch to dextrins.

alpha blockers *α-adrenoceptor blockers* Class of vasodilatory drugs that block the effect of noradrenaline, which is a vasoconstrictor, on peripheral blood vessels. Examples: doxazosin, phentolamine, prazosin, tamsulosin.

alpha-cell *α-cell* See **A-cells** of endocrine pancreas.

alpha complementation complementation of assembly-incompetent mutants of *E. coli* β-galactosidase by a small (26 aa) amino-terminal fragment of the lacZ product (the so-called alpha-polypeptide) allows assembly of a functional tetramer that will convert **X-gal** to a blue product. By putting a polycloning site within the alpha-polypeptide gene fragment, successful insertion of a sequence that is being cloned prevents complementation and restores the inactivity of β-gal. Colonies with inserts are therefore white and can be selected. This strategy requires *E. coli* host strains, such as DH5a, with mutations that are subject to alpha complementation, but has the advantage that the vector is small, allowing correspondingly large inserts.

alpha factor *α-factor* Oligopeptide (WHWLQLKPGQP-MY) mating pheromone of *S. cerevisiae*; exposure to this pheromone arrests yeast in G1 of the cell cycle and induces the **shmoo** phenotype. Binds to receptor coded by *STE2*.

alpha-fetoprotein *α-fetoprotein* Proteins from the serum of vertebrate embryos, which probably fulfil the function of **albumin** in the mature organism. Found in both glycosylated and non-glycosylated forms. Presence in the fluid of the **amniotic sac** is diagnostic of spina bifida in the human foetus.

alpha-glucosidase I *EC 3.2.1.20; α-glucosidase I* Enzyme that catalyses the splitting of α-D-glucosyl residues from the non-reducing end of substrates to release α-glucose.

alpha-helix *α-helix* A particular helical folding of the polypeptide backbone in protein molecules (both fibrous and globular), in which the carbonyl oxygens are all hydrogen-bonded to amide nitrogen atoms three residues along the chain. The translation of amino acid residues along the long axis is 0.15 nm and the rotation per residue 100°, so that there are 3.6 residues per turn.

alpha-neurotoxins *α-neurotoxins* Postsynaptic neurotoxins, many varieties of which are found in snake venoms. Two subclasses, short (four disulphides, 60–62 residues) and long (five disulphides, 66–74 residues). Examples include alpha-**bungarotoxin**, alpha-**cobratoxin**, **erabutotoxins**.

alpha-sarcin *α-sarcin* Antitumour factor (ribotoxin; 17 kDa) from *Aspergillus giganteus* that is generally cytotoxic and has high specificity as an RNAase, binding to the 28S rRNA. One of a family that includes **mitogillin**, **restrictocin** and AspFl.

alpha-secretase *ADAM10* A **disintegrin** metalloprotease that cleaves **Amyloid Precursor Protein (APP)** within the amyloid beta domain to produce the soluble non-amyloidogenic product sAPP alpha, and precludes amyloid beta peptide production.

alpha-synuclein *α-synuclein* Protein that accumulates in the brain in Parkinson's disease, particularly in **Lewy bodies**. Mutations in the gene are associated with some familial forms of **Parkinsonism**.

Alphavirus Genus of the Togaviridae family of RNA viruses. Sindbis and Semliki Forest viruses are the best-known examples.

Alport's syndrome Commonest of the hereditary nephropathies. Associated with nerve deafness and variable ocular disorders. The X-linked phenotype is the result of mutation in the gene for the alpha-5 chain of **basement membrane** collagen.

Alsever's solution A solution used for preserving red blood cells. 2.05% glucose, 0.42% sodium chloride, 0.8% tri-sodium citrate, adjusted to pH 6.1 with citric acid.

alsin The gene product mutated in three juvenile-onset neurodegenerative disorders, including **amyotrophic lateral sclerosis** 2 (ALS2). Sequence motif searches within alsin predict the presence of Vps9, DH and PH domains, implying that alsin may function as a guanine nucleotide exchange factor (**GEF**) for Rab5 and a member of the Rho GTPase family.

altered-self hypothesis The hypothesis that the **T-cell** receptor in MHC-mediated phenomena recognizes a **syngeneic** MHC class I or class II molecule after modification by a virus or certain chemicals. See **MHC restriction**.

alternative oxidase pathway Pathway of mitochondrial electron transport in higher plants, particularly in fruits and seeds, that does not involve **cytochrome oxidase** and thus is resistant to cyanide.

alternative pathway See **complement** activation.

alternative splicing Eukaryotic genes are composed of **exon**s and **intron**s, the latter being removed by **RNA** splicing before transcribed **mRNA** leaves the nucleus. Commonly, a single gene can encode several different mRNA transcripts, caused by cell- or tissue-specific combination of different exons. This is known as alternative splicing.

altricial Descriptive of birds whose young are hatched in a very immature condition and unable to leave the nest.

Alu 1. Type II **restriction endonuclease**, isolated from *Arthrobacter luteus*. The recognition sequence is 5′-AG/CT-3′. **2.** *Alu* sequences are highly repetitive sequences found in large numbers (100–500 000) in the human genome and that are cleaved more than once within each sequence by the Alu endonuclease. The *Alu* sequences look like DNA copies of mRNA because they have a 3′**poly-A tail** and flanking repeats.

alveolar cell Cell of the air sac of the lung.

alveolar macrophage Macrophage found in lung and that can be obtained by lung lavage; responsible for clearance of inhaled particles and lung surfactant. Metabolism slightly different from peritoneal macrophages (more oxidative metabolism), often have **multivesicular bodies** that may represent residual undigested lung surfactant.

Alzheimer's disease A presenile dementia characterized cellularly by the appearance of unusual helical protein filaments in nerve cells (neurofibrillary tangles) and by degeneration in cortical regions of brain, especially frontal and temporal lobes. See also **senile plaques**.

amacrine cell A class of **neuron** of the middle layer of the **retina**, with processes parallel to the plane of the retina. They are thought to be involved in image processing.

Amanita phalloides Poisonous fungus, the Death Cap; contains **amanitin** and **phalloidin**.

amanitin *α-, β-, γ- amanitin* Group of cyclic peptide toxins. The most toxic components of *Amanita phalloides* (Death Cap toadstool). Specific inhibitors of **RNA polymerase** II in eukaryotes, thus inhibiting protein synthesis by blocking the production of **mRNA**.

amantadine Compound used as an antiviral agent (especially against influenza virus). Produces some symptomatic relief in **Parkinsonism**.

amassin Olfactomedin family member, from the sea urchin *Strongylocentrotus purpuratus*. Mediates a rapid cell-adhesion event resulting in a large aggregation of **coelomocytes**.

amastigote Stage in the life cycle of trypanosomatid protozoa; resembles the typical adult form of members of the genus *Leishmania*, in which the oval or round cell has a nucleus, kinetoplast and basal body but lacks a flagellum.

amber codon One of the three **termination codons**. Its sequence is UAG. See also **ochre codon**, **opal codon**.

amber suppressor A mutation in a tRNA allele, altering it so that an amino acid is inserted at the **amber codon** and termination does not occur.

Ambystoma mexicanum Mexican axolotl (amphibian). A salamander that shows **neoteny**. The adult may retain the larval form, but can reproduce. The neotenous, aquatic axolotl will metamorphose into the terrestrial form if injected with thyroid or pituitary gland extract.

AMD Age-related macular degeneration, a complex multi-factorial disease that affects the central region of the retina. See **hemicentin**.

amelia Congenital abnormality in which one or more limbs are completely absent. Tetra-amelia is a rare autosomal-recessive human genetic disorder characterized by complete absence of all four limbs, and other anomalies; a *WNT3* mutation in tetra-amelia indicates that WNT3 is required at the earliest stages of human limb formation.

ameloblasts Columnar epithelial cells that secrete the enamel layer of teeth in mammals. Their apical surfaces are tapering (Tomes processes) and are embedded within the enamel matrix.

amelogenins Extracellular matrix proteins (20 and 25 kDa) of developing dental enamel; regulate form and size of hydroxyapatite crystallites during mineralization. Hydrophobic and proline-rich, produced by **ameloblasts**.

Ames test One of a number of procedures used to test substances for likely ability to cause cancer that combines the use of animal tissue (usually liver-derived) to generate active metabolites of the substance with a test for muta-genicity in bacteria.

amethopterin See **aminopterin**.

amidation site A C-terminus consensus sequence, required for C-terminus amidation of peptides. Consensus is glycine, followed by two basic amino acids (arg or lys).

amiloride Drug that blocks sodium/proton **antiport**; used clinically as a potassium-sparing diuretic.

aminergic Generally a description of receptors that respond to amines. Term usually applied to neurons that release nor-adrenaline, dopamine or serotonin. *cf.* **adrenergic, cholinergic**.

amino acid permease A widely distributed group of large integral membrane proteins required for the entry of amino acids into cells.

amino acid receptors Ligand-gated **ion channel**s with specific receptors for amino acid **neurotransmitter**s. An extended protein superfamily that also includes subunits of the **nicotinic acetylcholine receptor**.

amino acid transmitters Amino acids released as neu-rotransmitter substances from nerve terminals and acting on postsynaptic receptors, e.g. γ-aminobutyric acid (**GABA**) and glycine, which are fast inhibitory transmitters in the mammalian central nervous system. Glutamate and aspar-tate mediate fast excitatory transmission. **Strychnine** (for glycine) and bicuculline (for GABA) are blocking agents for amino acid action.

amino acids Organic acids carrying amino groups. The L-forms of about 20 common amino acids are the compo-nents from which proteins are made. See Tables A2 and C5 for the **codon** assignment.

amino sugar Monosaccharide in which an OH^- group is replaced with an amino group; often acetylated. Com-mon examples are D-galactosamine, D-glucosamine, neu-raminic acid, muramic acid. Amino sugars are important constituents of bacterial cell walls, some antibiotics, blood group substances, milk oligosaccharides and chitin.

amino-transferases *Transaminase* A family of enzymes (EC 2.6.1.x) that transfer an amino group from an amino acid to an α-keto acid, as in the transfer from glutamic acid to oxaloacetic acid, to form aspartic acid and α-ketoglutarate. One reactant is often glutamic acid and the reactions employ **pyridoxal phosphate** as coenzyme.

aminoacyl tRNA Complex of an **amino acid** with its tRNA, formed by the action of aminoacyl tRNA synthetase. Requires ATP, which forms the linkage between the two molecules.

aminoacyl tRNA synthetases Enzymes that attach an amino acid to its specific tRNA. An intermediate step is the formation of an activated amino acid complex with AMP; the AMP is released following attachment to the tRNA.

aminoacylation Addition of an aminoacyl group (formed by the removal of hydroxyl group from α-carbonyl group of an α-amino acid) to a substrate, the best known of which is tRNA (forming an aminoacyl tRNA).

aminocaproic acid ε-*aminocaproic acid; EACA* Alter-native name for 6-aminohexanoic acid, a lysine side chain mimic that inhibits the activity of plasmin, enhances the activation of **plasminogen** by urinary-type and tissue-type **plasminogen activators** but inhibits plasminogen activa-tion induced by streptokinase. Used clinically as an antifib-rinolytic, antihaemorrhagic agent to reduce bleeding in, for example, haemophiliacs.

aminoglycoside antibiotics Group of antibiotics active against many aerobic Gram-negative and some Gram-positive bacteria. Composed of two or more amino sugars attached by a glycosidic linkage to a hexose nucleus; polycationic and highly polar compounds. Inhibit bacterial protein synthesis by binding to a site on the 30S ribo-somal subunit thereby altering codon–anticodon recogni-tion. Common examples are **streptomycin**, **gentamicin**, amikacin, **kanamycin**, **tobramycin**, netilmicin, neomycin, framycetin. See also Table A3.

aminopeptidase Enzymes that remove the N-terminal amino acid from a protein or peptide.

aminophylline An inhibitor of cAMP **phosphodi-esterase**.

aminopterin A **folic acid** analogue and inhibitor of **dihy-drofolate reductase**. A potent cytotoxic agent used in the treatment of acute **leukaemia**.

amiodarone Antiarrhythmic drug used for treatment of tachycardia and atrial fibrillation.

amitosis An unusual form of nuclear division, in which the nucleus simply constricts, rather like a cell, without chromosome condensation or spindle formation. Partition-ing of daughter chromosomes is haphazard. Observed in some Protozoa.

amitriptyline Tricyclic antidepressant drug used in the treatment of moderate to severe depression. Has sedative effects.

AML Acute myeloblastic **leukaemia**.

ammodytoxins *atx A, B and C; ammodytin L* Group II secretory phospholipases A2 (122 residues) found in the venom of *Vipera ammodytes*. Act specifically at peripheral nerve endings in the neuromuscular junction. Binding site may be a subunit of potassium channels.

amnesic shellfish poisoning See **domoic acid**.

TABLE A3. Mode of action of various antibiotics

Antibiotic	Source organism	Mode of action
Penicillins (β-lactams)	*Penicillium notatum*	Inhibit enzyme that cross-links the peptidoglycan wall of Gram-positive bacteria. Resistant bacteria produce penicillinases (β-lactamase) that cleave the four-membered β-lactam ring
Amoxicillin	Semisynthetic	Broad spectrum
Ampicillin	Semisynthetic	Inhibitors of cell wall synthesis
Benzylpenicillin (benzathine benzylpenicillin)		Often referred to as penicillin G
Methicillin	Synthetic	Beta-lactamase resistant
Phenoxymethyl penicillin	Semisynthetic	Orally active
Cephalosporins		Inhibitors of cell wall synthesis. Similar mode of action to penicillins
Cephaloridine	*Cephalosporium acremonium* (fungi)	
Cephalothin	*Cephalosporium* species	
Aminoglycosides		Bind to the 30S subunit of the 70S (bacterial) ribosome, though at a number of different sites. They prevent the transition from an initiating complex to a chain-elongating ribosome
Gentamicin	*Micromonospora purpurea*	
Hygromycin B	Streptomyces hygroscopicus	
Kanamycin	*Streptomyces kanamyceticus*	
Neomycin	*Streptomyces fradiae*	
Streptomycin	*Streptomyces griseus*	
Tobramycin	*Streptomyces* sp.	
Aurovertins		
Aurovertins, A, B, D	*Calcarisporium arbuscula*	Inhibit the proton-pumping **F-type ATP synthase**
Macrolides		
Bafilomycin	*Streptomyces griseus*	Specific inhibitor of the V-type ATPase
Cycloheximide	*Streptomyces noursei/Streptomyces griseus*	Inhibits eukaryotic, but not prokaryotic, protein synthesis by preventing the peptidyl transferase reaction
Erythromycin	*Streptomyces erythraeus*	Inhibits bacterial protein synthesis by binding to 50S subunit preventing the translocation step
Lincomycin	*Streptomyces lincolnensis*	Clindamycin is a derivative
Midecamycin	*Streptomyces mycarofaciens*	Inhibits staphylococcal enterotoxin B-induced mRNA expression of IL-4 and IL-5
Oleandomycin	*Streptomyces antibioticus*	As erythromycin
Rapamycin	*Streptomyces hygroscopicus*	Immunosuppressive by inhibiting TOR
Oligosaccharides		
Avilamycin	*Streptomyces* spp.	Used in animal feed
Everninomicins	*Micromonospora carbonacea* var. *africana*	Bind to 50S bacterial ribosome subunit and block protein synthesis
Peptides		
Bacitracin	*Bacillus subtilis*	Inhibitor of cell wall synthesis
Bialaphos	*Streptomyces hygroscopicus*	Used as a herbicide
Blasticidin	*Streptomyces griseochromogenes*	Inhibits pro- and eukaryotic protein synthesis

TABLE A3. (Continued)

Antibiotic	Source organism	Mode of action
Bleomycin	*Streptomyces verticillus*	A glycopeptide: blocks cell division in G2
Daptomycin	*Streptomyces lividans*	Lipopeptide; inhibits lipoteichoic acid synthesis
Gramicidin	*Bacillus brevis*	Ionophore. Forms 'pore' in cell membrane, causing loss of K^+
Ionomycin	Streptomyces conglobatus	A diacidic polyether; Ca^{2+} ionophore
Katanosin B, plusbascin A3	Strains of *Cytophaga* and *Pseudomonas*, respectively	Cyclic depsipeptides; block transglycosylation in bacterial cell wall peptidoglycan synthesis
Lantibiotics (ancovenin, mersacidin, nisin, subtilin)	Various	Lanthionine-containing peptides; various modes of action
Mannopeptimycins	*Streptomyces hygroscopicus* LL-AC98	Cyclic glycopeptides; inhibit cell wall biosynthesis through lipid II binding
Netropsin	*Streptomyces*	Binds selectively in minor groove of B-DNA
Polymyxin B	*Bacillus polymyxa*	Acts on cell membrane causing leakage of small molecules
Ramoplanin	Actinomycetes of the genus *Actinoplanes*	Sequesters lipid intermediates for peptidoglycan biosynthesis
Teicoplanin	*Actinoplanes teichomyceticus*	Glycopeptide; interferes with formation of links in bacterial cell wall
Valinomycin	*Streptomyces fulvissimus*	Ionophore. Causes leakage of K^+
Vancomycin	*Amycolatopsis orientalis*	Glycopeptide
Polyenes		Form complex with cholesterol in the plasma membrane
Amphotericin B	*Streptomyces nodosus*	
Filipin	*Streptomyces filipinensis*	
Nystatin	*Streptomyces noursei*	
Quinolones		Inhibit DNA gyrase
Ciprofloxacin, ofloxacin, levofloxacin	Synthetic	Fluoroquinolones related to nalidixic acid
Nalidixic acid	Synthetic	
Tetracyclines		
Chlortetracycline	*Streptomyces aureofaciens*	Inhibits bacterial protein synthesis by preventing aminoacyl tRNA binding to the A-site of the 30S ribosomal subunit
Tetracycline	*Streptomyces aureofaciens* (mutant)	
Others		
Actinomycin D	*Streptomyces parvullus*	Inhibits RNA synthesis by binding to DNA, blocking movement of RNA polymerase
Brefeldin A	*Penicillium brefeldianum*	Macrocyclic lactone; blocks secretion in a pre-Golgi compartment
Chloramphenicol	*Streptomyces venezuelae*	Inhibits prokaryote, but not eukaryote, protein synthesis by preventing the peptidyl transferase reaction
Clavulanic acid	*Streptomyces clavuligerusa*	Inhibits beta-lactamase, so used in conjunction with beta-lactams
Daunomycin	*Streptomyces coeruleorubidus.*	Anthracycline; inhibits topoisomerase II
Daunorubicin, doxorubicin	*Streptomyces peucetius*	Anthracycline; cytotoxic by intercalating into DNA, used in cancer chemotherapy
Deoxynojirimycin, nojirimycin	*Bacillus* spp.; *Streptomyces* spp.	Inhibits α-glucosidases
Fumagillin	*Aspergillus fumigatus*	Inhibits endothelial cell proliferation; binding to methionine aminopeptidase 2

TABLE A3. (Continued)

Antibiotic	Source organism	Mode of action
Fusidic acid	*Fusidium coccineum*	Blocks bacterial protein synthesis
Geldanamycin	*Streptomyces hygroscopicus*	Benzoquinone ansamycin; binds to HSP90
Griseofulvin	*Penicillium griseofulvum*	Polyketide; blocks microtubule assembly
Leptomycin B	*Streptomyces* sp.	Unsaturated, branched-chain fatty acid with antifungal activity; binds to exportin-1
Mitomycin C	*Streptomyces caespitosus*	Inhibits DNA synthesis by cross-linking the strands
Monensin		Polyether; sodium ionophore; nigericin is similar
Mupirocin	*Pseudomonas fluorescens*	Polyketide derivative; targets isoleucyl-tRNA synthase
Novobiocin	*Streptomyces niveus*	Aromatic ether; inhibits DNA gyrase
Pactamycin	*Streptomyces pactum*	Inhibits translation in pro- and eukaryotes
Puromycin	*Streptomyces albo-niger*	Inhibits protein synthesis. Analogue of $3'$ end of aminoacyl tRNA
Rifamycin/rifampicin	*Streptomyces mediterranei*/semisynthetic derivative	Inhibits bacterial RNA synthesis. Binds to RNA polymerase and prevents initiation of transcription. Effective against acid-fast as well as Gram-positive bacteria
Saccharomicins	*Saccharothrix espanaensis*	Heptadecaglycosides; antibacterial and antifungal
Sparsomycin	*Streptomyces sparsogenes*	Inhibits peptidyl transferase in pro- and eukaryotes
Spectinomycin	*Streptomyces spectabilis*	Aminocyclitol; binds to the 30S subunit of the bacterial ribosome
Streptogramins	Various Streptomycetes	Bind bacterial ribosome
Streptovaricins	various Actinomycetes	Block initiation of transcription in bacteria
Sulfonamides	Synthetic	Competitive inhibitors of *p*-aminobenzoic acid in the folic acid metabolism cycle
Tautomycin	*Streptomyces spiroverticillatus*	Inhibits type 1 and type 2a protein phosphatases
Teicoplanin	*Actinoplanes teichomyceticus*	Glycopeptide; interferes with formation of links in bacterial cell wall
Trichostatin A	*Streptomyces platensis*	Reversible histone deacetylase (HDAC) inhibitor
Tunicamycin	*Streptomyces lysosuperificus*	Nucleoside; inhibits N-glycosylation

amniocentesis Sampling of the fluid in the **amniotic sac**. In humans this is carried out, between the twelfth and sixteenth weeks of pregnancy, by inserting a needle through the abdominal wall into the uterus. By **karyotyping** the cells and determining the proteins present, it is possible to determine the sex of the foetus and whether it is suffering from certain congenital diseases, such as **Down's syndrome** or **spina bifida**.

amniocyte Cell type found floating freely in the amnion sac; following **amniocentesis**, amniocytes can be subcultured and used for prenatal genetic diagnosis.

amnion Terrestrial vertebrates have embryos that develop in fluid-filled sacs formed by the outgrowth of the extraembryonic **ectoderm** and **mesoderm** as projecting folds. These folds fuse to form two epithelia separated by the mesoderm and **coelom**. The inner layer is the amnion and encloses the amniotic sac in which the embryo is suspended. The outer layer is the **chorion**.

amniote egg Egg type produced by **Amniota** that allows development in a terrestrial environment because the egg is self-contained and provides a stable fluid environment, the embryo being surrounded and protected by the amnion, filled with amniotic fluid.

amniotic sac Sac, enclosing the embryo of amniote vertebrates that provides a fluid environment to prevent dehydration during development of land-based animals. See **amnion**.

amoeba Genus of protozoa, but also an imprecise name given to several types of free-living unicellular phagocytic organism. Giant forms (e.g. *Amoeba proteus*) may be up to 2 mm long and crawl over surfaces by protruding **pseudopods**

(**amoeboid movement**). Amoebae exhibit great plasticity of form and conspicuous **cytoplasmic streaming**.

amoebiasis Dysentery caused by *Entamoeba histolytica*.

amoebocytes Phagocytic cells found circulating in the body cavity of coelomates (particularly annelids and molluscs) or crawling through the interstitial tissues of sponges. A fairly non-committal classification.

amoeboid movement Crawling movement of a cell brought about by the protrusion of **pseudopods** at the front of the cell (one or more may be seen in monopodial or polypodial amoebae, respectively). The pseudopods form distal anchorages with the surface.

amoxicillin *Amoxycillin* Broad-spectrum **beta-lactam** antibiotic. Proprietary names include Amoxil and Augmentin.

AMP *Adenosine monophosphate* Unless otherwise specified, 5′AMP, the nucleotide bearing a phosphate in ribose-*O*-phosphate ester linkage at position 5 of 3′ derivatives also exist. See also **cyclic AMP** moiety. Both 2′ and 3′ derivatives also exist. See also **cyclic AMP** (adenosine 3′5′-cyclic monophosphate).

AMP-PNP *5′-aadenylyl imidodiphosphonate* Non-hydrolysable analogue of ATP used in isolation of some motor proteins.

AMPA *α-amino-3-hydroxy-5-methyl-4-isoxazoleproprionate* Synthetic agonist for metabotropic **glutamate receptors**.

AMPA receptor Glutamate-operated **ion channel**. See **excitatory amino acid** receptor channels.

amphetamine Drug of abuse that acts by increasing extraneuronal **dopamine** in midbrain. Thought to displace dopamine in **synaptic vesicles**, leading to increased synaptic levels.

amphibolic Description of a pathway that functions not only in **catabolism**, but also to provide precursors for **anabolic** pathways.

amphimixis Sexual reproduction resulting in an individual having two parents. Invariably the case in most animals, with the exception of a few hermaphrodite organisms, but not uncommon in plants, where a single individual may produce both male and female gametes (be monoecious) and be self-fertile.

Amphioxus Obsolete generic name for **Branchiostoma**. Name is descriptive – 'sharp at both ends'.

amphipathic Of a molecule, having both **hydrophobic** and **hydrophilic** regions. Can apply equally to small molecules, such as phospholipids, and macromolecules such as proteins.

amphiphilic Having affinity for two different environments – for example, a molecule with hydrophilic (polar) and lipophilic (non-polar) regions. Detergents are classic examples. Antonym of **amphipathic**.

amphiphysin Protein of the nerve terminal that associates with synaptic vesicles, probably through AP2 and **synaptotagmin**.

amphiploid *Amphidiploid* A polyploid organism in which chromosomes come from two different species. Not uncommon in plant breeding.

amphiregulin A heparin-binding **growth factor** containing an **EGF-like domain**. See **HB-EGF**.

amphitrichous Having a flagellum at both ends of the cell.

amphitrophic Of organisms that can grow either photosynthetically or chemotrophically.

ampholyte Substance with **amphoteric** properties. Most commonly encountered as descriptive of the substances used in setting up electrofocusing columns or gels.

amphoteric Having both acidic and basic characteristics. This is true of proteins since they have both acidic and basic side groups (the charges of which balance at the **isoelectric point**).

amphotericin B *Fungizone* Polyene antibiotic from *Streptomyces* spp. Used as a fungicide, it is cytolytic by causing the formation of pores (5–10 molecules of amphotericin in association with cholesterol) that allow passage of small molecules through the plasma membrane and thus to cytolysis. Only acts on membranes containing sterols (preferentially ergosterol, hence selectivity for fungi). See also **filipin**.

amphoterin *HMG-1; DEF* Heparin-binding protein (30 kDa) that enhances nerve growth cone migration and neurite outgrowth in the developing brain. Ligand of the receptor for advanced glycation end products (**RAGE**). High levels of amphoterin are released to serum during septic shock. Identical with the non-histone chromosomal protein HMG-1 and has also been isolated as DEF (differentiation-enhancing factor).

amphotropic virus A virus that does not produce disease in its natural host, but will replicate in tissue culture cells of the host species and cells from other species. Amphotropic murine leukaemia virus has been extensively used as a vector in experimental gene transfer.

ampicillin A semisynthetic **penicillin** derivative with a broad range of activity against bacteria causing bronchitis, pneumonia, gonorrhoea, some forms of meningitis, enteritis, biliary and urinary tract infections. Ampicillin resistance is often used as a marker for **plasmid** transfer in genetic engineering (e.g. **pBR322** is ampicillin resistant).

AMPK *AMP-activated protein kinase* An enzyme that plays a key role in regulating energy homeostasis. AMPK-mediated phosphorylation switches cells from ATP consumption towards ATP production. AMPK is itself regulated by physiological stimuli which lead to its activation by AMPK kinases (AMPKK). Mammalian AMPK is a trimeric enzyme composed of a catalytic α subunit and the non-catalytic β and γ subunits.

amplicon The DNA product of a **polymerase chain reaction**.

ampulla A small membranous vesicle. *Adj.* ampullary.

amydricaine hydrochloride A local anaesthetic used particularly in ophthalmic practice.

amygdala Almond-shaped body in the lateral ventricle of the brain.

amylase *EC 3.2.1.1* An enzyme that cleaves polysaccharides; there are two isoforms of human amylase: pancreatic and salivary amylase (ptyalin). Alpha-amylase randomly cleaves the α(1-4)glycosidic linkages of amylose to yield dextrin, maltose or glucose. Beta-amylase also catalyses the hydrolysis of α-1,4 glycosidic bonds but only from the non-reducing end to yield maltose molecules.

amylin *Islet amyloid polypeptide (IAPP)* Natural hormone (37 residues) produced by pancreatic **beta-cells** that moderates the glucose-lowering effects of insulin. Co-secreted with insulin, controls nutrient intake as well as nutrient influx to the blood by an inhibition of food intake, gastric emptying and glucagon secretion. One of the **calcitonin family peptides**.

amyloid **Glycoprotein** deposited extracellularly in tissues in **amyloidosis**. The glycoprotein may either derive from **light chain of immunoglobulin** (AIO (amyloid of immune origin): 5- to 18-kDa glycoprotein; product of a single clone of **plasma cells**, the N-terminal part of lambda or kappa **L chain**) or, in what used to be referred to as AUO, amyloid of unknown origin, from serum amyloid A (SAA), one of the acute phase proteins that increases manyfold in inflammation. The polypeptides are organized as a **beta-pleated sheet** making the material rather inert and insoluble. Minor protein components are also found. Should be distinguished from β-amyloid deposited in the brain and that is derived from **amyloid precursor protein** (see **amyloidogenic glycoprotein**.

amyloid precursor protein *APP* Individuals with **Alzheimer's disease** are characterized by extensive accumulation of amyloid in the brain, referred to as **senile plaques**. These consist of a core of amyloid fibrils surrounded by dystrophic neurites. The principal component of the amyloid fibrils is B/A4, a peptide derived from the larger APP. The specific role of amyloid protein is unclear but it is thought that amyloid deposits may cause neurons to degenerate. Amyloid deposits also occur in brains of older **Down's syndrome** patients.

amyloidogenic glycoprotein *A4 protein* An integral membrane **glycoprotein** of the brain and related to the *Drosophila vnd*-gene product. A precursor of β-amyloid that accumulates in **Alzheimer's disease** and **Down's syndrome**. See **amyloid precursor protein**.

amyloidosis Deposition of **amyloid**. A common complication of several diseases (leprosy, tuberculosis); often associated with perturbation of the immune system, although there may be immunosuppression or enhancement.

amylopectin Component of **starch** in which the glucose chain is α-1,4 linked (α-1,6 at branch points).

amyloplast A plant **plastid** involved in the synthesis and storage of starch. Found in many cell types, but particularly storage tissues. Characteristically has starch grains in the plastid **stroma**.

amylose A linear polysaccharide formed from α-D-glucopyranosyl units in α-1,4 linkage. Found both in starch (starch amylose) and glycogen (glycogen amylose).

amyotrophic lateral sclerosis *Lou Gehrig's disease; motor neuron disease* Progressive degenerative disease of motor neurons in the brain stem and spinal cord that leads to weakening of the voluntary muscles. Some (15–20%) cases of familial amyotrophic lateral sclerosis (which is itself only 10% of all cases) are associated with mutations in the superoxide dismutase-1 gene. Susceptibility has also been associated with mutations in the genes encoding the heavy neurofilament subunit, **peripherin**, **dynactin** and **angiogenin**. Various other genes have also been implicated.

amytal *Amylobarbitone; amobarbital* A barbiturate that inhibits respiration.

Anabaena A genus of **Cyanobacteria** that forms filamentous colonies with specialized cells (**heterocysts**), capable of nitrogen fixation. Ecologically important in wet tropical soils and forms symbiotic associations with the fern *Azolla*.

anabolic Of a process, route or reaction. **Metabolic** pathways are classically divided into anabolic and **catabolic** types. The former are synthetic processes, frequently requiring expenditure of phosphorylating ability of **ATP** and reductive steps; the latter, degradative processes, often oxidative, with attendant regeneration of ATP.

anabolic steroids Synthetic forms of male sex hormones (androgens) that promote tissue growth, especially of muscle. Have been misused by bodybuilders and athletes, e.g. Stanozolol.

anabolism Synthesis; opposite of **catabolism**.

anacharis Common name for a aquarium plant, mostly *Elodea* spp.

anadromous Having the habit of migrating from more dense to less dense water to breed, usually from oceanic to coastal waters or from salt water to fresh water.

anaemia *Anemia (US)* Reduced level of **haemoglobin** in blood for any of a variety of reasons including abnormalities of mature red cells (**sickle cell anaemia, spherocytosis**), iron deficiency, haemolysis of erythrocytes, reduced **erythropoiesis**, haemorrhage (to name the most common).

anaerobic The absence of air (specifically of free oxygen). Used to describe a biological habitat or an organism that has very low tolerance for oxygen.

anaerobic respiration Metabolic processes in which organic compounds are broken down to release energy in the absence of oxygen. Requires inorganic oxidizing agents or accumulation of reduced coenzymes.

anagenesis Progressive evolution of species through alterations in gene frequency in an entire population so that eventually the new population would be recognized as distinct from the ancestral species, as opposed to cladogenesis in which two species emerge.

analeptic Having restorative or strengthening properties. An example of an analeptic drug is doxapram hydrochloride, a respiratory stimulant used in situations of ventilatory failure when artificial ventilation is impossible. Usually act on the central nervous system.

analgesia A state of insensitivity to pain, even though the subject is fully conscious.

analogous Of genes or gene products performing a similar role in different organisms. *cf.* **homologous**.

analyte Substance or compound for which an analysis is being carried out.

anamnestic response Archaic term now replaced by such terms as **secondary immune response, immunological memory**.

anandamide *Arachidonyl ethanolamide* Endogenous agonist for **cannabinoid receptors**.

anaphase The stage of **mitosis** or **meiosis** beginning with the separation of sister **chromatids** (or homologous **chromosomes**) followed by their movement towards the poles of the **spindle**.

anaphase-promoting complex *Cyclosome* An unusually complicated **ubiquitin** ligase, composed of 13 core subunits (total 1.5 MDa) and either of two loosely associated co-activators, Cdc20 and Cdh1, that is responsible for initiation of sister chromatid separation and the inactivation of cyclin-dependent kinases. The largest subunit, Apc1, serves as a scaffold that associates independently with two separable subcomplexes, one that contains Apc2 (**cullin**), Apc11 (RING) and Doc1/Apc10, and another that contains the three tetratricopeptide repeat (**TPR motif**)-containing subunits (Cdc27, Cdc16 and Cdc23). In S-phase, the APC is inactivated by binding of cyclin A.

anaphylatoxin Originally used as an antigen that reacted with an **IgE** antibody, thus precipitating reactions of **anaphylaxis**. Now restricted to defining a property of **complement** fragments C3a and C5a, both of which bind to the surfaces of **mast cells** and **basophils** and cause the release of inflammatory mediators.

anaphylaxis As opposed to **prophylaxis**. A system or treatment that leads to damaging effects on the organism. Now reserved for those inflammatory reactions resulting from combination of a soluble antigen with **IgE** bound to a **mast cell** that leads to degranulation of the mast cell and release of **histamine** and histamine-like substances, causing localized or global immune responses. See **hypersensitivity**.

anaplasia Lack of differentiation, characteristic of some tumour cells.

anaplasmosis Bovine disease caused by protozoa of the genus *Anaplasma* that are found in red blood cells. Transmitted by ticks, biting flies and mosquitoes.

anaplerotic Anaplerotic reactions replenish **TCA cycle** intermediates and allow respiration to continue; for example, carboxylation of **phosphoenolpyruvate** in plants.

Anas platyrhynchos Mallard – from which domestic duck is derived by breeding (traditional genetic engineering).

anastomosis Joining of two or more cell processes or multicellular tubules to form a branching system. Anastomosis of blood vessels allows alternative routes for blood flow.

anastrozole A non-steroidal **aromatase** inhibitor that inhibits the production of estrogen, used in the treatment of advanced breast cancer. It significantly lowers serum oestradiol concentrations, and has no detectable effect on the formation of adrenal corticosteroids or aldosterone.

anatoxins A group of low-molecular weight neurotoxic alkaloids first described in the freshwater cyanobacteria *Anabaena flos-aquae*, but subsequently found in other species. Anatoxin-a and homoanatoxin-a are secondary amines that bind and irreversibly activate nicotinic acetylcholine receptors; anatoxin-a(s) is the only natural organophosphate known and inactivates acetylcholine esterase in a similar fashion to synthetic organophosphate pesticides such as parathion and malathion. All cause muscle exhaustion by overstimulation and death through respiratory failure.

ANCA *Antineutrophil cytoplasmic antibodies* ANCA positivity is seen in patients with a variety of inflammatory disorders, including IBD (inflammatory bowel disease), **Wegener's granulomatosis** and hepatobiliary disorders. Two forms are recognized: peripheral ANCA (p-ANCA), where the antigen seems to reside at the periphery of the nucleus; and cytoplasmic ANCA (c-ANCA), where the antigen is distributed throughout the cytoplasm of the neutrophil.

anchorage Attachment, not necessarily adhesive in character; because the mechanism is not assumed the term ought to be more widely used.

anchorage dependence The necessity for attachment (and spreading) in order that a cell will grow and divide in culture. Loss of anchorage dependence seems to be associated with greater independence from external growth control and is probably one of the best correlates of **tumorigenic** events *in vivo*. Anchorage independence is usually detected by **cloning** cells in soft agarose; only anchorage-independent cells will grow and divide (as they will in suspension).

anchored PCR *Anchored polymerase chain reaction* Variety of **polymerase chain reaction** in which only enough information is known to make a single primer. A known sequence is thus added to the end of the DNA, perhaps by enzymic addition of a polynucleotide stretch or by ligation of a known piece of DNA. The PCR can then be performed with the gene-specific primer and the anchor primer.

ancovenin An inhibitor of angiotensin I converting enzyme isolated from the culture broth of a *Streptomyces* species; a 16-residue **lantibiotic** containing unusual amino acids such as threo-beta-methyllanthionine, meso-lanthionine and dehydroalanine. *Streptoverticillium cinnamoneum* produces a lantibiotic with similar properties, lanthiopeptin.

AND-34 A member of a novel family of proteins (NSP1, NSP2, and NSP3) that have an amino-terminal SH2 domain but bind by a carboxy-terminal GEF (Cdc25)-like domain to the carboxy-terminus of the focal adhesion adapter protein **p130cas**. Overexpression of AND-34 in epithelial breast cancer cells leads to activation of Rac and Cdc42 by a PI3K-dependent mechanism.

Andersen's syndrome An inherited human ion channel disorder (**channelopathy**) that has been linked to muscle abnormalities and developmental defects. The mutations affect a potassium channel called Kir2.1. Mutation is in gene *KCNJ2* on Chr17.

Androctonus mauretanicus mauretanicus Moroccan scorpion. See **kaliotoxin**.

androgen General term for any male sex hormone in vertebrates.

androgenesis 1. Development from a male cell. 2. Development of an egg after entry of male germ cell but without the participation of the nucleus of the egg.

androstenedione *4-androstene-3,17-dione* Precursor of testosterone and estrone produced in the testis or ovary from 17a-hydroxyprogesterone or from dehydroepiandrosterone. Also secreted into the circulation by the adrenal glands.

anemone toxins Polypeptide toxins (around 5 kDa) from sea anemones (anthozoan **coelenterates**), most of which act on voltage-gated sodium channels. Some, however, block voltage-regulated potassium channels.

anergy Generally, a lack of energy. In immunology, failure of lymphocytes that have been primed to respond to second exposure to the antigen. Consequence is a depression or lack of normal immunological function. *Adj.* anergic.

aneugenic Agents that induce changes in chromosome numbers, aneuploidy or polyploidy, rather than mutation. A range of chemicals and treatments (e.g. X-irradiation) have been shown to be aneugenic although there are no universally accepted standard test methods as yet.

aneuploid Having a chromosome complement that is not an exact multiple of the haploid number. Chromosomes may be present in multiple copies (e.g. trisomy) or one of a homologous pair may be missing in a diploid cell.

aneurysm Balloon-like swelling in the wall of an **artery**.

ANF See **atrial natriuretic peptide**.

Angelman syndrome Syndrome in which there is severe mental retardation and ataxic movement associated with absence of maternal 15q11q13 and the absence of the β3 subunit of **GABA receptor**-A.

angina pectoris A condition caused by narrowing of the coronary arteries and reduced blood supply to the heart. Characterized by the sudden onset of pain or crushing sensation in the chest which may radiate to the throat and arms, often provoked by exercise. Often referred to simply as angina.

angioedema The development of large welts below the surface of the skin, especially around the eyes and lips, usually associated with allergic responses. Postcapillary venule inflammation results in fluid leakage and oedema in the layers of the skin below the dermis, whereas urticaria is localized superficial to the dermis.

angiogenesis The process of **vascularization** of a tissue involving the development of new capillary blood vessels.

angiogenin Polypeptide (14 kDa) that induces the proliferation of endothelial cells; one of the components of **tumour angiogenesis factor**. It has sequence homology with pancreatic ribonuclease and has ribonucleolytic activity, although the biological relevance of this is unclear. It has also been suggested that angiogenin binds to an actin-like molecule present on the surface of endothelial cells.

angioma A knot of distended blood vessels atypically and irregularly arranged. Most are not tumours but haematomas.

angiomyolipoma Rare, slow-growing benign lesions of kidney, often asymptomatic, composed of varying amounts of blood vessels, smooth muscle and fat. A small proportion of cases are associated with **tuberous sclerosis**.

angioneurotic oedema An immunologically mediated disease in which there is dramatic, occasionally life-threatening, swelling of the eyelids, lips, mucous membranes of the mouth and respiratory tract.

angiopathy Any disease of blood vessels. Microangiopathy affects small blood vessels and is, for example, a complication of diabetes, particularly in the vasculature of the retina. Macroangiopathy affects large blood vessels lead-ing to coronary artery disease, cerebrovascular disease and peripheral vascular disease.

angioplasty *PTCA* Surgical distension of an occluded blood vessel. Percutaneous transluminal coronary angioplasty (PTCA) is commonly used as a method for restoring patency to occluded coronary arteries (the cause of **angina pectoris**); a catheter is passed from a vessel in the arm through to the coronary vessels and a balloon at the end of the catheter is then inflated to dilate the vessel.

angiopoietin Angiopoetin-1 (498 residues) is the ligand for **Tie2**; angiopoietin-2 (496 residues) is a natural antagonist. Angiopoietin-1, but not Ang-2, is chemotactic for endothelial cells: neither have effects on proliferation.

angiosperm A large class of flowering plants that bear seeds in a closed fruit.

angiostatin Potent angiogenesis inhibitor, a proteolytic fragment of plasminogen containing the first three or four kringle domains (K1–4). Mode of action unclear.

angiotensin A peptide **hormone**. Angiotensinogen (renin substrate) is a 60-kDa polypeptide released from the liver and cleaved in the circulation by **renin** to form the biologically inactive decapeptide angiotensin I. This is in turn cleaved to form active angiotensin II by angiotensin converting enzyme (ACE). Angiotensin II causes contraction of vascular smooth muscle, and thus raises blood pressure and stimulates **aldosterone** release from the adrenal glands. Angiotensin is finally broken down by angiotensinases.

angiotensin-converting enzyme *ACE* See **angiotensin**.

angiotensin II receptor antagonists Group of drugs that work by blocking binding of **angiotensin** II to its receptor and thus have effects similar to those of **angiotensin-converting enzyme inhibitors**; used to treat hypertension. Examples are candesartan, irbesartan, losartan and valsartan.

angiotensin-converting enzyme inhibitors *ACE inhibitors* Drugs that inhibit the enzymatic conversion of inactive **angiotensin**-I to the active form (angiotensin II); used in the treatment of hypertension and heart failure. Captopril and Enalapril are common examples.

angiotensinase See **angiotensin**.

angiotensinogen See **angiotensin**.

ångström unit Small unit of measurement (10^{-10} m) named after Swedish physicist and astronomer. Much used as a unit in early electron microscopy though, since it is not in the approved SI system, should probably be avoided (but nanometres, which are 10 times larger, are sometimes less convenient).

anguilliform Eel-like in shape.

anidulafungin An **echinocandin** antifungal agent with broad-spectrum activity against *Candida* and *Aspergillus*.

anillin Actin-binding protein first identified in *Drosophila* embryo extracts as ABP8. The cDNA encodes a 1201-amino acid protein. During anaphase–telophase it becomes highly enriched in the cleavage furrow along with myosin II. Seems to organize and/or stabilize the cleavage furrow and other cell cycle-regulated, contractile domains of the actin cytoskeleton.

animal pole In most animal **oocytes** the nucleus is not centrally placed and its position can be used to define two poles. That nearest to the nucleus is the animal pole and the other is the **vegetal pole**, with the animal–vegetal axis between the poles passing through the nucleus. During **meiosis** of the **oocyte** the **polar bodies** are expelled at the animal pole. In many eggs there is also a graded distribution of substances along this axis, with pigment granules often concentrated in the animal half and yolk, where present, largely in the vegetal half.

animalized cells The 8- to 16-cell early blastula of sea urchins has **animal** and **vegetal poles**; by manipulating the environmental conditions it is possible to shift more cells from vegetal to animal in their characteristics.

anion exchanger Family of integral membrane proteins that perform the exchange of chloride and bicarbonate across the plasma membrane. Best known is **band III** of the red blood cell.

anionic detergents Detergents in which the hydrophilic function is fulfilled by an anionic grouping. **Fatty acids** are the best-known natural products in this class, but it is doubtful if they have a specific detergent function in any biological system. The important synthetic species are aliphatic sulphate esters, e.g. sodium dodecyl sulphate (**SDS** or SLS).

aniridia Rare, congenital absence or partial absence of the iris; caused by an autosomal-dominant mutation in PAX6 (oculorhombin), or an identifiable chromosome deletion of the short arm of chromosome 11, including band p13, or, in sporadic cases, by mutations in *WT* (Wilm's tumour gene) and *AN2* (aniridia 2 gene) or only in *AN2*.

anisogamy Mode of sexual reproduction in which the two gametes are of different sizes.

anisotropic Not the same in all directions.

ANK repeat Amino acid motif found in diverse proteins, including **ankyrin** (hence the name), the *Notch* product, transcriptional regulators, cell cycle regulatory proteins and a toxin produced by the black widow spider. The motif is about 33 amino acids long and is generally found as a tandem array of 2–7 repeats, though ankyrins contain 24 repeats. Their role is not established, but they may be involved in protein–protein binding.

ankylosing spondylitis **Polyarthritis** involving the spine, which may become more or less rigid. Interestingly, the disease seems to be associated with HLA-B27; those with this **histocompatibility antigen** are 300 times more likely to get the disease, while 90% of sufferers have HLA-B27.

ankylosis Fusion of bones across a joint. Complication of **chronic inflammation**. See **ankylosing spondylitis**.

ankylostomiasis *Hookworm disease* Infection of the small intestine by parasitic nematode (*Ankylostoma duodenale* or *A. americanum*); can cause iron deficiency for those on an inadequate diet.

ankyrin Globular protein (200 kDa) that links **spectrin** and an integral membrane protein (**band III**) in the erythrocyte plasma membrane. Isoforms exist in other cell types.

ankyrin repeat See **ANK repeat**.

anlage Region of the embryo from which a specific organ develops.

annealing 1. Toughening upon slow cooling. **2.** Used in the context of DNA renaturation after temperature dissociation of the two strands. Rate of annealing is a function of complementarity. **3.** Fusion of microtubules or microfilaments end-to-end.

Annelida Phylum of segmented (metameric) coelomate worms. Common earthworm (*Lumbricus terrestris*) is a familiar example of the phylum.

annexins *ANXA* Group of calcium-binding proteins that interact with acidic membrane phospholipids in membranes. There are 12 mammalian annexin genes, now classified as *ANXA 1–13* with *A*12 not assigned. Non-vertebrate annexins are classified as B (invertebrates), C (fungi and some unicellular organisms), D (plants) and E (protists). All the mammalian annexins have four conserved repeats of a 61-amino acid domain that folds into five a-helices, except ANXA 6, which has eight. Also known by several other names (e.g. lipocortins, endonexins), reflecting the history of their discovery in different contexts. See Table A4, **lipocortin**, **endonexin** I and II, **calpactin**, **p70**, and **calelectrin**.

annulate lamellae Organelles described in the oocytes of several animal species. Associated with the nuclear envelope; may be associated with tubulin synthesis from mRNA accumulated in these organelles.

annulus Ring-like structure. *Adj.* annulate.

anoikis **Apoptosis** in normal epithelial and endothelial cells. Process is important in the regulation of cell number in skin.

anomers The α- and β-forms of hexoses. Interconversion (**mutarotation**) is anomerization and is promoted by mutarotases (aldose epimerases).

Anopheles Genus of mosquitoes (order **Diptera**) which carry the **Plasmodium** parasites that causes malaria.

anosmia Condition of being unable to smell. Can be transiently induced by osmic acid.

ANOVA *Analysis of variance* Statistical technique used for analysis of the source of variability in a set of data. Allows variation due to specific cause and to random variation to be estimated.

anoxia Total lack of oxygen, *cf.* **hypoxia**.

ANP See **atrial natriuretic peptide**.

ANP receptor Family of three receptors for **atrial natriuretic peptide**. ANP-A and ANP-B have intracellular **guanylate cyclase** and **protein kinase**-like domains. ANP-C, shares the extracellular ligand-binding and transmembrane domains, but lacks the functional intracellular domains, and is not thought to be involved in **signal transduction**.

antagonist Compound that inhibits the effect of a hormone or drug; the opposite of **agonist**.

antennal complex Light-harvesting complexes (LHC) of protein and pigment in or on photosynthetic membranes of bacteria are organized into arrays, called antennae. They transfer photon energy to reaction centres.

TABLE A4. Vertebrate annexins

Name	Synonyms	Expression
Annexin 1	Lipocortin 1, Calpactin 2, p35, Chromobindin 9	Ubiquitous
Annexin 2	Lipocortin 2, Calpactin 1, Protein I, p36, Chromobindin 8	Ubiquitous
Annexin 3	Lipocortin 3, PAP-III	Neutrophils
Annexin 4	Lipocortin 4, Endonexin I, Protein II, Chromobindin 4	Ubiquitous
Annexin 5	Lipocortin 5, Endonexin 2, VAC-a, Anchorin CII, PAP-I	Ubiquitous
Annexin 6	Lipocortin 6, Protein III, Chromobindin 20, p68, p70	Ubiquitous
Annexin 7	Synexin	Ubiquitous
Annexin 8	Vascular anticoagulant-b, VAC-b	Placenta and skin
Annexin 9		Tongue
Annexin 10		Stomach
Annexin 11		Ubiquitous
Annexin 13		Small intestine

There are 12 mammalian annexin genes, now classified as ANXA 1–13 with A12 not assigned.

antennapedia antp **Homeotic gene** of *Drosophila*, controlling thoracic/head fate determination. In addition to the **homeobox**, there is also a 6-amino acid antennapedia-specific consensus shared by a range of homeotic genes in human, mouse, chicken, *Xenopus*, newt, zebrafish and *Caenorhabditis*.

anterograde transport Movement of material from the cell body of a **neuron** into axons and dendrites (retrograde axoplasmic transport also occurs).

anthelminthic Drug used against parasitic worms.

anthocyanidin Molecule (an aglycone) that when complexed with a sugar moiety forms an **anthocyanin**.

anthocyanins Red plant pigments not directly involved in **photosynthesis**. Can mask the green of **chlorophyll** and give the plant a red-purple colour.

Anthopleura Genus of sea anemone (an anthozoan **coelenterate**). See **anthopleurins**.

anthopleurins Peptide toxins (anthopleurin A, B and C, 49, 50, 47 residues) from sea anemone, *Anthopleura*. Affect sodium channel of nerve and muscle, increase the duration of the action potential.

anthozoa *Actinozoa* Class of Cnidaria in which alternation of generations does not occur and the medusoid phase is entirely suppressed. Sea anemones, corals, sea pens.

anthracosis 'Coal-miner's lung', caused by inhalation of coal dust.

anthrax Highly contagious disease of man and domestic animals caused by *Bacillus anthracis*. Onset is rapid and disease often fatal. A variety of **anthrax toxins** are known.

anthrax toxins **1.** (Anthrax oedema factor) Multisubunit toxin produced by *Bacillus anthracis*. Active subunit is a calmodulin-dependent adenylyl cyclase. **2.** Anthrax lethal toxin (LeTx). Toxin responsible for the shock-like effects of infection with *B. anthracis*. Two subunits, protective antigen (PA, 83 kDa) and lethal factor (LF, 90 kDa). PA binds to the target cell, is proteolytically cleaved to a 63-kDa subunit that then oligomerizes to form a ring that allows LF access to the cytoplasm. LF acts on macrophages to induce a massive oxidative burst and also the release of IL-1b and TNFα followed by lysis of the cell.

antibiotic Substance produced by one microorganism that selectively inhibits the growth of another. Many wholly synthetic antibiotics have been produced; see also Table A3.

antibiotic resistance gene Gene that encodes an enzyme that degrades or excretes an antibiotic, thereby conferring resistance. Frequently found in cloning vectors like plasmids, and sometimes in natural populations of bacteria. Example: bacterial ampicillin resistance is conferred by expression of the **beta-lactamase** gene.

antibody General term for an **immunoglobulin**.

antibody-dependent cell-mediated cytotoxicity *ADCC* Killing of target cells by lymphocytes or other leucocytes that carry antibody specific for the target cell, attached to their **Fc receptors**. The cell involved in the killing may be a passive carrier of the antibody.

antibody-induced lysis See **complement** lysis, also **natural killer cells**. The term is imprecise and should not be used because there is confusion as to which mechanism is involved, i.e. natural killing or complement lysis.

antibody-producing cell A **lymphocyte** of the B series synthesizing and releasing **immunoglobulin**. Equivalent to plasmacyte and **plasma cell**.

anticlinal Perpendicular to the nearest surface. Used particularly in botanical anatomy to describe the orientation of cell walls with respect to the surface of the plant or organ. If a cell divides anticlinally, the daughter cells will be separated by an anticlinal wall and will both lie in the same plane with respect to the surface. *cf.* **periclinal**.

anticoagulant Substance that inhibits the clotting of blood. The most commonly used are EDTA and citrate (both of which work by chelating calcium) and **heparin** (that interferes with **thrombin**, probably by potentiating **antithrombins**). Other compounds such as **warfarin** and dicoumarol act as anticoagulants *in vivo* by interfering with clotting factors.

anticodon Nucleotide triplet on **transfer RNA** that is complementary to the **codon** of the **messenger RNA**.

anticonvulsants Drugs used to prevent or reduce the severity and frequency of seizures that occur when electrical activity in the brain that controls motor systems becomes chaotic and paroxysmal. Commonest form of seizure is epilepsy but not all seizures cause convulsions and not all convulsions are due to epileptic seizures. Commonly used drugs are carbamazepine, **phenytoin**, **valproate** and **diazepam**.

antidepressants Drugs that relieve the symptoms of moderate to severe depression. Main classes are **tricyclic antidepressants**, **monoamine-oxidase inhibitors**, **SSRIs** and **lithium salts**.

antidiuretic Inhibiting the formation of urine; an antidiuretic drug. See **vasopressin**.

antidiuretic hormone *ADH* See **vasopressin**.

antidromic Running in the opposite direction: most common usage is in neurophysiology for the passage of an action potential in the opposite direction to that in which it would normally travel, i.e. from presynaptic region towards the cell body.

antiemetic drugs Drugs that stop vomiting and, to a lesser extent, nausea. They are used for motion sickness, for the side-effects of chemotherapy and for some gastrointestinal disorders. Examples include **hyoscine**, **antihistamines**, **phenothiazines**, metoclopramide and ondantseron.

antigen A substance inducing and reacting in an **immune response**. Normally antigens have molecular weights greater than about 1 kDa. The **antigenic determinant** group is termed an **epitope**, and the association of this with a carrier molecule (that may be part of the same molecule) makes it active as an antigen. Thus dinitrophenol-modified human serum albumin is antigenic to humans, dinitrophenol being the hapten. Usually antigens are foreign to the animal in which they produce immune reactions.

antigen presentation See **antigen-presenting cell**.

antigen-presenting cell A cell that carries on its surface antigen bound to MHC class I or class II molecules and presents the antigen in this 'context' to T-cells. Includes macrophages, endothelium, **dendritic cells** and **Langerhans cells** of the skin. See also **MHC restriction, histocompatibility antigens**.

antigen processing Modification of an antigen by **accessory cells**. This usually involves endocytosis of the antigen and either minimal cleavage or unfolding. The processed antigen is then presented in modified form by the accessory cell.

antigen shift Abrupt change in surface antigens expressed by a species or variety of organisms. Usually seen in microorganisms, where the change may allow escape from immune recognition. Antigenic drift is a more gradual change. See **antigenic variation**.

antigen–antibody complex *Immune complex* The product of the reaction of **antigen** and **immunoglobulin**. If the antigen is polyvalent the complex may be insoluble; see also **glomerulonephritis, Arthus reaction**, type III

hypersensitivity. Immune complexes activate **complement** through the classical pathway.

antigenic determinant *Epitope* That part of an antigenic molecule against which a particular immune response is directed – for instance, a tetra- to pentapeptide sequence in a protein, a tri- to penta-glycoside sequence in a polysaccharide. See also **hapten**. In the animal, most **antigens** will present several or even many antigenic determinants simultaneously.

antigenic variation The phenomenon of changes in surface **antigens** in parasitic populations of *Trypanosoma* and *Plasmodium* (and some other parasitic protozoa) in order to escape immunological defence mechanisms. At least 100 different surface proteins have been found to appear and disappear during antigenic variation in a clone of trypanosomes. Each antigen is encoded in a separate gene. Antigenic variation is also known to occur in free-living Protozoa and certain bacteria.

antihistamine A substance or drug which inhibits the actions of **histamine** by blocking its site of action. See **histamine receptors**.

anti-idiotype antibody An antibody directed against the antigen-specific part of the sequence of an antibody or T-cell receptor. In principle, an anti-idiotype antibody should inhibit a specific immune response.

anti-inflammatory drugs Drugs that inhibit the inflammatory response. There are two major classes, the **non-steroidal anti-inflammatory drugs** (NSAIDs) and the **glucocorticoids**.

antilymphocyte serum Immunoglobulins raised **xenogeneically** against lymphocyte populations. Referring particularly to **antisera** recognizing one or more **antigenic determinants** on **T-cell** populations. Of use in experimental **immunosuppression**.

antimere A part on a bilaterally or radially symmetrical organism corresponding to a similar structure on the other side. *Adj.* antimeric.

antimetabolite Drugs used in treatment of cancer which are incorporated into new nuclear material and prevent normal cell division. Common examples are **methotrexate**, cytoarabinose and **fluorouracil**.

antimicrobial peptides *AMPs* Peptides found in the **haemolymph** of insects. In *Drosophila* there are eight classes of AMPs that can be grouped into three families based on their main biological targets, Gram-positive bacteria (**defensin**), Gram-negative bacteria (**cecropins, drosocin, attacins, diptericin, MPAC**) or fungi (**drosomycin, metchnikowin**). *Drosophila* AMPs are synthesized by the fat body in response to infection and secreted into the haemolymph.

antimitotic drugs Drugs that block mitosis; the term is often used of those which cause metaphase arrest such as **colchicine** and the **vinca alkaloids**. Many antitumour drugs are antimitotic, blocking proliferation rather than being cytotoxic.

antimorph A **dominant negative** mutant expressing some agent that antagonizes a normal gene product.

anti-Müllerian hormone *AMH, Müllerian-inhibiting substance (MIS)* A dimeric glycoprotein hormone belonging to the TGFβ superfamily. Synthesized as a large precursor with a short signal sequence followed by the prepro hormone that forms homodimers. Prior to secretion, the mature hormone undergoes glycosylation and dimerization to produce a 140-kDa dimer of identical disulphide-linked subunits. High levels of AMH are produced by Sertoli cells during foetal and postnatal testicular development. In the human female, AMH is produced by ovarian granulosa cells from 36 weeks of gestation to the menopause.

antimuscarinic drugs *Anticholinergic drugs* Class of drugs that block the action of acetylcholine at the **muscarinic** subclass of acetylcholine receptors. Effect is generally to relax smooth muscle of gut (antispasmodic) or airways (bronchodilatory). Example: **tolterodine**.

antimycin Inhibitor of **QH2-cytochrome C reductase**.

antineoplaston A naturally occurring cytodifferentiating agent that has been tested for antitumour activity and used to induce differentiation of astrocytes in rat models of neurodegenerative disease.

anti-oncogene See **tumour suppressor** gene.

antioxidant Any substance that inhibits oxidation, usually because it is preferentially oxidized itself. Common examples are vitamin E (α-**tocopherol**) and vitamin C. Important for trapping free radicals generated during the **metabolic burst** and possibly for inhibiting ageing.

antipain [(S)-*1-Carboxy-2-Phenyl]-carbamoyl-Arg-Val-arginal* Peptidase inhibitor that inhibits **papain**, **trypsin** and, to a lesser extent, **plasmin**. More specific for papain and trypsin than **leupeptin**.

antiparallel Having the opposite **polarity** (e.g. the two strands of a DNA molecule).

antiplasmin α-2-*Antiplasmin* Plasma protein (65 kDa) that inhibits **plasmin** (and Factors XIa, XIIa, **plasma kallikrein**, **thrombin** and **trypsin**) and therefore acts to regulate **fibrinolysis**.

antiplectic Pattern of **metachronal** coordination of the beating of **cilia**, in which the waves pass in the opposite direction to that of the active stroke.

antipodal cells Three cells of the **embryo sac** in angiosperms, found at the end of the embryo, away from the point of entry of the pollen tube.

antiport A membrane protein that transports two different ions or molecules, in opposite directions, across a lipid bilayer. Energy may be required, as in the **sodium pump**, or may not, as in Na$^+$/H$^+$ antiport.

antiproteases *Antiproteinases, antipeptidases* Substances that inhibit proteolytic enzymes (peptidases).

antipsychotic drugs (neuroleptic drugs) Group of drugs sometimes referred to as major tranquillizers. Used short term to calm or sedate disturbed patients, to control acute symptoms of mania and relieve severe positive symptoms of schizophrenia. Most act by reducing levels of the neurotransmitter **dopamine** in the central nervous system (e.g. **chlorpromazine**, **haloperidol**, flupentixol), but the atypical antipsychotics (e.g. **risperidone**, clozapine) act by interfering with **serotonin**-based neurotransmission.

antipyretic Counteracting fever; a remedy for fever.

antiretroviral 1. Acting to counteract or control a **retrovirus**. **2.** An antiretroviral agent or drug.

Antirrhinum majus A flowering plant (the common snapdragon), widely used as a model system for plant molecular genetics.

antisense In general, the complementary strand of a coding sequence of DNA or of mRNA. Antisense RNA hybridizes with and inactivates mRNA.

antisepsis Processes, procedures or chemical treatments that kill or inhibit microorganisms; in contrast with **asepsis,** where microorganisms are excluded.

antiserum Serum containing **immunoglobulins** against specified **antigens**.

antispasmodic drugs Drugs that relax smooth muscle of the gut wall and relieve symptoms of indigestion, irritable bowel syndrome and diverticular disease. Most are **antimuscarinic drugs**.

antitermination During transcription, failure of an **RNA polymerase** to recognize a termination signal: can be of significance in regulation of gene expression.

antithymocyte globulin *ATG* A polyclonal IgG fraction that selectively targets and destroys T-lymphocytes; used as an immunosuppressive agent.

antithrombins Plasma **glycoproteins** of the α-2-globulin class that inhibit the proteolytic activity of **thrombin** and serve to regulate the process of blood clotting. See also Table F1.

antitoxin An **antibody** reacting with a toxin, e.g. anti-**cholera toxin** antibody.

antitrypsin α-*1-Antitrypsin; alpha-1-peptidase inhibitor; serpin A1* A major protein (54 kDa) of blood **plasma** (3 mg/ml in human), part of the α-globulin fraction and able to inhibit a wide spectrum of **serine peptidases**.

antitussive drug A cough suppressant. Most work by suppressing the cough reflex, rather than treating the cause of the cough.

antiviral drugs Drugs that inhibit viral infection. Two main categories have been developed so far; those which inhibit or interfere with the replication of viral nucleic acid (nucleoside analogues such as **acyclovir** and **AZT**), and those that interfere with virus-specific enzymes such as proteases (e.g. **Saquinavir**, Ritonavir, Indinavir) or neuraminidases (e.g. **Relenza**, **Tamiflu**) which are important for processing of viral proteins to produce infective particles.

antizyme Repressor of **ornithine decarboxylase** (ODC). Antizyme (29 kDa) is a polyamine-inducible protein involved in feedback regulation of cellular polyamine levels. The N-terminus of antizyme is not required for the interaction with ODC but is necessary to induce its degradation. Antizyme can be induced by IL-1. The elaborate regulation of ODC activity in mammals still lacks a defined developmental role but an antizyme-like gene in *Drosophila*, gutfeeling (*guf*), is required for proper development of the embryonic peripheral nervous system.

Antp See **antennapedia**.

antral Relating to an **antrum**.

antral mucosa *Pyloric mucosa* Type of mucosa found in the gastric **antrum** which has coiled and branching antral glands that are lined by mucus cells interspersed with endocrine cells (chiefly G and D types) and a few parietal (**oxyntic**) cells.

antrum **1.** A cavity or chamber, especially in bone. **2.** The lower third of the stomach which lies between the body of the stomach and the pyloric canal. **3.** In the ovary, the fluid-filled space within the follicle.

Antrycide Proprietary name for an antitrypanocidal drug used in veterinary practice.

anucleate Having no nucleus.

anucleolate Literally, having no nucleoli. An anucleolate mutant of *Xenopus* (viable when **heterozygous**) is used in nuclear transplantation experiments because nuclei are of identifiable origin.

Anura Class of amphibians; the frogs and toads.

anuresis Failure to secrete urine.

anxiogenic Causing anxiety; something that enhances fear/anxiety reactions.

anxiolytic Drug that reduces anxiety, for example **benzodiazepines** and **barbiturates**.

AP-1 **1.** A **transcription factor**, formed from a heterodimer of the products of the **proto-oncogenes** *fos* and *jun*. Binds the palindromic DNA sequence TGACTCA. See also Table O1. **2.** Adaptor protein found in the trans-Golgi network that links membrane proteins to clathrin (see **AP-2**).

AP-2 **1.** Cis-acting transcription activator. **2.** One of the multimeric adaptor proteins (APs; ca. 270 kDa) found in **clathrin**-associated complexes. AP-2 is found at the plasma membrane and may bind preferentially to the cytoplasmic tail of the EGF receptor. Also associates with the EGF-R tyrosine kinase substrate eps15. See **adaptins**.

AP3, AP4, AP5 *Amino-3-phosphonopropanoate, 2-amino-4-phoshonobutanoate, 2-amino-5-phosphonopentanoic acid* Selective antagonists for **NMDA receptors**.

APA *Amino pimelic acid* Low affinity, rapidly dissociating competitive antagonist of **NMDA receptors**.

APAF-1 *Apoptosis protease activating factor 1* Protein which binds to cytochrome *c* that has been released from mitochondria and links with **caspase**-9, which then activates caspase-3, initiating a cascade of events that ends in apoptotic death of the cell.

apamin A small (2027-Da) basic peptide present in the venom of the honey bee (*Apis mellifera*). Blocks calcium-activated potassium channels and has an inhibitory action in the central nervous system.

APC **1.** **Antigen-presenting cell**. **2.** Adenomatous **polyposis coli**.

apelin A bioactive peptide that is the endogenous ligand for the G-protein-coupled receptor **APJ**. The peptide is produced from a pre-preprotein consisting of 77 amino acid residues and exists in multiple molecular forms (ranging from 12 to 36 aa). Apelin partially suppresses cytokine production from mouse spleen and is involved in the regulation of blood pressure and blood flow.

aperient Drug that has a laxative or purgative effect.

APH-1 APH-1 (anterior pharynx defective) along with **nicastrin** and **PEN-2** (presenilin enhancer) are essential components of the **presenilin**-dependent **gamma-secretase** complex. APH-1 apparently stabilizes the complex. Aph-1 is present at the cell surface, presumably in active gamma-secretase complexes and interacts with the **Notch** receptor.

aphasia Loss or defect in language function due to brain lesion. May be receptive (understanding speech or written words) or expressive (defective in writing or speech).

aphidicolin Reversible inhibitor of eukaryotic **DNA polymerases**, a tetracyclic diterpenoid from *Cephalosporium*.

apical bud Bud at the tip of a growing plant; usually dominant, repressing the development of lower buds by the production of **auxin**.

apical dominance Growth-inhibiting effect exerted by actively growing **apical bud** of higher-plant shoots, preventing the growth of buds further down the shoot. Thought to be mediated by the basipetal movement of **auxin** from the apical bud.

apical ectodermal ridge Ridge of tissue at the developing limb-bud of the vertebrate embryo, a transient structure critical to maintaining limb outgrowth.

apical meristem *Eumeristem* The **meristem** at the tips of stems and roots. Composed of undifferentiated cells, many of which divide to add to the plant body but the central mass (the quiescent centre) remains inert and only becomes active if the meristem is damaged.

apical plasma membrane The term used for the cell membrane on the apical (inner or upper) surface of transporting epithelial cells. This region of the cell membrane is separated, in vertebrates, from the basolateral membrane by a ring of **tight junctions** that prevents free mixing of membrane proteins from these two domains.

Apicomplexa Large group of parasitic protists, many of which were formerly classed as Sporozoa. Characterized by an apical complex of microtubules within the cell in the sporozoite and **merozoite** stages but lack flagella. Includes *Plasmodium*, *Cryptosporidium*, *Babesia*, *Toxoplasma*.

apidaecins Proline-rich basic antibacterial peptides (2 kDa) found in the immune **haemolymph** of the honeybee.

apigenin *Naringenin chalcone; 5,7,4′-trihydroxyflavone* A bioflavone, considered to have a beneficial effect on human health as an antioxidant, radical scavenger and anti-inflammatory. Present in leafy plants and vegetables (e.g. parsley, celery) and said to have chemopreventive activity against UV radiation.

Apis mellifera Common honeybee, a Hymenopteran.

APJ An orphan G-protein-coupled receptor APJ that is a co-receptor for human and simian immunodeficiency virus (HIV and SIV) strains. **Apelin** is an endogenous ligand for the APJ receptor; apelin peptides inhibit the entry of some HIV-1 and HIV-2 into the NP-2/CD4 cells expressing APJ. The inhibitory efficiency has been found to be in the order of apelin-36 > apelin-17 > apelin-13 > apelin-12.

aplasia Defective development of an organ or tissue so that it is totally or partially absent from the body.

aplastic anaemia Anaemia due to loss of most or all of the **haematopoietic** bone marrow. Usually all haematopoietic cells are equally diminished in number.

aplyronine *Aplyronine A* A cytotoxic macrolide from the sea hare *Aplysia kurodai* that has actin-depolymerizing activity and antitumour activity.

Aplysia *Sea hare* Opisthobranch mollusc with reduced shell; favourite source of ganglia for neurophysiological study.

apoA, etc. *apoA, apoB, apoC, apoE* Plasma **apolipoproteins**. ApoE is the specific ligand for uptake of lipoprotein by the LDL receptor and different alleles of the ApoE gene are associated with variations in plasma cholesterol levels.

apocrine Form of secretion in which the apical portion of the cell is shed, as in the secretion of fat by cells of the mammary gland. The fat droplet is surrounded by apical plasma membrane and this has been used experimentally as a source of plasma membrane.

apocynin 4′-Hydroxy-3′-methoxyacetophenone A selective inhibitor, isolated from dogbane (*Apocynum cannabinum*), of NADPH-oxidase in activated polymorphonuclear (PMN) leucocytes, prevents the generation of reactive oxygen species.

apoenzyme An enzyme without its **co-factor**. See **apoprotein**.

apolipoprotein The protein component of serum lipoproteins. Small proteins containing multiple copies of the **kringle** domain. Late onset **Alzheimer's disease** is unusually frequent in persons with a particular variant form of apolipoprotein E termed ApoE4, although there seems to be no causative linkage.

apomixis Type of degenerate sexual reproduction in some fungi: meiosis and gamete formation do not occur even though an **ascus** containing identical diploid spores is formed.

apomorphic Derived or advanced characteristics that arose relatively late in members of a group and therefore differ among them. These are useful in assessing genealogical links among taxa; the more recent the common ancestor, the more apomorphic traits are shared. Apomorphic characters for the primates, like humans, include a large brain size and binocular vision. Plesiomorphic characteristics are ancient homologies (e.g. pentadactyl limbs), symplesiomorphic characteristics are recently derived. Synaptomorphic characters are shared between taxa.

apomorphine An alkaloid of the **morphine** series. Not a narcotic, but an expectorant and emetic. It is a relatively non-selective **dopamine receptor** agonist, having possible slightly higher affinity for D2-like dopamine receptors.

apopain *Caspase-3; Yama protein; CASP3; SCA-1* Peptidase responsible for the cleavage of poly(ADP-ribose) polymerase and necessary for apoptosis. Has two subunits of relative molecular mass (M_r) 17 and 12 kDa that are derived from a common proenzyme identified as CPP32.

apoplast Since the **protoplasts** of cells in a plant are connected through **plasmodesmata**, plants may be described as having two major compartments: the apoplast, which is external to the plasma membrane and includes cell walls, xylem vessels, etc. through which water and solutes passes freely; and the **symplast**, the total cytoplasmic compartment.

apoprotein When a protein can exist as a complex between polypeptide and a second moiety of non-polypeptide nature, the term apoprotein is sometimes used to refer to the molecule divested of the latter. For example, **ferritin** lacking its ferric hydroxide core may be referred to as apoferritin.

apoptin A small proline rich protein derived from the chicken anaemia virus, induces cell death selectively in cancer cells.

apoptosis The most common form of physiological (as opposed to pathological) cell death. Apoptosis is an active process requiring metabolic activity by the dying cell; often characterized by shrinkage of the cell, cleavage of the DNA into fragments, which gives a so-called 'laddering pattern' on gels, and by condensation and margination of chromatin. Often called **programmed cell death**, though this is not strictly accurate. Cells that die by apoptosis do not usually elicit the inflammatory responses that are associated with necrosis, though the reasons are not clear. See also: **apoptosis-inducing factor**, **apoptosome**, **ced mutant**, **bcl-2**.

apoptosis-inducing factor *AIF* A flavoprotein, 57 kDa, which shares homology with the bacterial **oxidoreductases**; it is normally confined to mitochondria but translocates to the nucleus when apoptosis is induced. Interacts with DNA by virtue of positive charges clustered on the AIF surface and can cause condensation of single and double-stranded DNA. Has a role in mitochondrial respiration.

apoptosome Complex of approximately 700 kDa (occasionally 1.4 MDa) assembled by rapid oligomerization of **Apaf-1** followed by a slower process of procaspase-9 recruitment and cleavage to form the p35/34 forms. **XIAP** binds to caspase-9 and inhibits further proteolytic activity.

App(NH)p *(β, γ-Imido) ATP* Non-hydrolysable analogue of ATP.

applagin See **disintegrin**.

apple domain A consensus sequence, composed of 90 amino acids including six cysteines, that forms a characteristic, vaguely apple-shaped, pattern via disulphide bridges. Shared by **plasma kallikrein** and coagulation Factor XI, both **serine peptidases**.

appressorium A flattened outgrowth which attaches a fungal parasite to its plant host.

aprataxin Product of the *APTX* gene that is mutated in patients with the neurological disorder ataxia with oculomotor apraxia type 1 (AOA1). Interacts with several nucleolar proteins, including **nucleolin**, **nucleophosmin** and upstream binding factor-1 (UBF-1).

aprotinin *Trasylol* Basic polypeptide, originally isolated from bovine lung although now produced as a recombinant protein, that inhibits several **serine peptidases** (including **trypsin**, **chymotrypsin**, **kallikrein** and **pepsin**).

aptamer *Aptamere* Double-stranded DNA or single-stranded RNA molecule that binds to a specific molecular target.

aptamere See **aptamer**.

apterous Without wings.

APUD cells Acronym for amine-precursor uptake and decarboxylation cells: **paracrine** cells, of which **argentaffin cells** are an example. Usage neither helpful nor memorable.

apurinic sites Sites in DNA from which purines have been lost by cleavage of the deoxy-ribose N-glycosidic linkage.

APV 1. Avian polyoma virus. **2.** An antiviral protease inhibitor, amprenavir. **3.** The anterior periventricular nucleus. **4.** Avian pneumonvirus, an emerging disease in turkeys. **5.** NMDA antagonist, (DL-2-amino-5-phosphonopentanoic acid).

apyrase *EC 3.6.1.5* Enzyme that catalyses breakdown of ATP to AMP; usually extracted from plants, but aortic and placental forms have also been described.

aquaporin *AQP; CHIP28* Integral membrane protein (28 kDa) with six transmembrane domains that greatly increases water permeability. Found especially in kidney, red blood cells. AQP1 forms a homotetramer of four independent channels. *Arabidopsis* has at least 20 distinct AQP genes. Members of the **major intrinsic protein** family.

Aquifex aeolicus Thermophilic, hydrogen-oxidizing, microaerophilic, obligate chemolithoautotrophic bacterium, able to grow at a remarkable 96°C. Genome, approximately one-third that of *E. coli*, has been sequenced.

ara operon Operon involved in **arabinose** metabolism, especially the *araBAD* operon of *E. coli*. AraA encodes arabinose isomerase, which converts arabinose to ribulose; *araB* encodes ribulokinase, which phosphorylates ribulose; *araD* encodes ribulose-5-phosphate epimerase, which converts ribulose-5-phosphate to xylulose-5-phosphate, which is then metabolized via the **pentose phosphate pathway**. Regulation of the operon is by the *araC* gene product that binds to the araC site and represses its own transcription from the PC promoter. In the presence of arabinose, AraC bound at this site helps to activate expression of the PBAD promoter.

Arabidopsis thaliana The common wall cress. A small plant, much used as a model system for plant molecular biology, because of its small genome (7×10^7 bp) and short generation time (5–8 weeks). Genome now fully sequenced.

arabinogalactans Plant cell-wall polysaccharides containing predominantly arabinose and galactose. Two main types are recognized: arabinogalactan 1, found in the pectin portion of angiosperms and containing α(1-4)-linked galactan and α-arabinose side chains; arabinogalactan II, a highly branched polymer containing β(1-3)- and β(1-6)-linked galactose and peripheral α-arabinose residues. Arabinogalactan II is found in large amounts on some gymnosperms, especially larches, and is related to **AGP**.

arabinoglycan-protein See **AGP**.

arabinose A **pentose** monosaccharide that occurs in both D- and L- configurations. D-arabinose is the 2-epimer of **D-ribose**, i.e. differs from D-ribose by having the opposite configuration at carbon 2. D-arabinose occurs *inter alia* in the polysaccharide arabinogalactan, a neutral **pectin** of the cell wall of plants, and in the metabolites **cytosine arabinoside** and adenine arabinoside.

arabinoxylan Polysaccharide with a backbone of **xylose** (β-1,4 linked) with side chains of **arabinose** (α-1,3 linked): constituent of **hemicellulose** of angiosperm cell wall.

araC In bacteria, the arabinose **ara operon** regulatory protein. One of a large group of bacterial **transcription factors** with the **helix–turn–helix** motif. See **ara operon**.

arachidonic acid *5,8,11,14 eicosatetraenoic acid* An essential dietary component for mammals. The free acid is the precursor for biosynthesis of the signalling molecules **prostaglandins, thromboxanes,** hydroxyeicosatetraenoic acid derivatives including **leukotrienes,** and is thus of great biological significance. Within cells the acid is found in the esterified form as a major acyl component of membrane **phospholipids** (especially **phosphatidyl inositol**), and its release from phospholipids is thought to be the limiting step in the formation of its active metabolites.

arboviruses Diverse group of single-stranded RNA viruses that have an envelope surrounding the **capsid**. Arthropod borne, hence the name, and multiply in both invertebrate and vertebrate host, causing, e.g. yellow fever and encephalitis. The group is very heterogeneous and three major families are recognized: **Togaviridae, Bunyaviridae,** and **Arenaviridae.**

arbuscular 1. Having the characteristics of a dwarf tree or shrub of tree-like habit. **2.** Characteristic of the much-branched haustorium formed within the host cells by some endophytic fungi in **vesicular-arbuscular mycorrhiza**.

Arcella A small amoeba of the Phylum Sarcodina that has a chitinous test (shell), dome-like on the top and concave on the bottom. Around 50–200 μm wide.

Archaea Alternative name suggested for the **Archaebacteria** to emphasize the difference of this subkingdom from the **Eubacteria,** and now generally adopted. Archaea are one of two major subdivisions of the prokaryotes. There are three main orders: extreme **halophiles, methanobacteria** and sulphur-dependent extreme **thermophiles**. Archaebacteria differ from **Eubacteria** in ribosomal structure, the possession (in some cases) of **introns** and a number of other features, including membrane composition.

Archaebacteria See **Archaea**.

archaeocyte An amoeboid cell type of sponges (Porifera).

archegonium Female sex organ of liverworts, mosses, ferns and most gymnosperms.

archenteron *Archentron* Cavity in the **gastrula** that opens to the exterior at the **blastopore**.

arcuate Bent like a bow.

arcuate nucleus Region of the hypothalamus containing two major types of neurons. One type secretes **neuropeptide Y** and **agouti-related peptide,** which act to increase appetite and decrease metabolism. The other type of neurons, called POMC/CART neurons, produce alpha-melanocyte-stimulating hormone (alpha-**MSH**), which inhibits eating.

ARD-1 Bifunctional protein (64 kDa) that has 18-kDa GTP-binding ADP-ribosylation factor (**ARF**) domain and 46-kDa GTPase-activating (GAP) domain.

ARE 1. Activin-response element to which **ARF** binds. **2. AU-rich element**.

area opaca In developing tetrapods, a whitish peripheral zone of **blastoderm** in contact with the yolk. The area pellucida is a central clear zone of blastoderm that does not have direct contact with the yolk; the area vasculosa is the region of extraembryonic blastoderm in which the blood vessels develop.

area pellucida See **area opaca**.

area vasculosa See **area opaca**.

arecoline An alkaloid isolated from *Areca catechu* (betel palm) and one of the major pharmacologically active components; a **muscarinic** acetylcholine receptor agonist. Effects on the central nervous system are similar to those of nicotine. Has also been used medicinally as an anthelminthic.

Arenaviridae Family of ssRNA viruses including **Lassa virus**, lymphocytic choriomeningitis virus and the **Tacaribe** group of viruses; not all require arthropods for transmission, despite their inclusion in the **arbovirus** group.

areolar connective tissue Loose **connective tissue** of the sort found around many organs in vertebrates. Does not have marked anisotropy, nor particularly pronounced content of any particular matrix protein.

ARF 1. **ADP-ribosylation factor**. 2. **Auxin response factor**.

arg *Abelson-related gene; ABL2; v-abl; Abelson murine leukaemia viral oncogene homologue 2* **Oncogene**, related to *abl*, that encodes a **tyrosine kinase**.

argentaffin cells So called because they will form cytoplasmic deposits of metallic silver from silver salts. Their characteristic histochemical behaviour arises from 5-HT (**serotonin**), which they secrete. Found chiefly in the epithelium of the gastrointestinal tract (though possibly of neural crest origin) their function is rather obscure, although there is a widely distributed family of such **paracrine** (local endocrine) cells (**APUD cells**).

argentation chromatography Modified form of standard thin layer chromatography in which the solid phase includes silver salts. Used for lipid analysis.

arginine *Arg; R; 174 Da* An essential amino acid; a major component of proteins and contains the guanido group that has a pKa of greater than 12, so that it carries a permanent positive charge at physiological pH. See Table A2.

argireline *Acetyl hexapeptide-3* Compound claimed, on fairly skimpy evidence, to relax facial tension (wrinkles) by inhibition of neurotransmitter release.

argonaute *AGO1, AGO2* Argonaute 1 and 2 are proteins concentrated in **GW bodies** and key components of RNA-induced silencing complexes (**RISC**), binding the single-stranded siRNA and miRNA. RNA is required for the integrity of GW bodies and RNase eliminates argonaute 2 localization. In *Drosophila*, unlike in humans, both AGO1 and AGO2 have **slicer activity**. Prokaryotic Ago proteins have unknown function but are similar to eukaryotic proteins. RNA interference and related RNA-silencing phenomena use short antisense guide RNA molecules to repress the expression of target genes. Contain amino-terminal **PAZ** (for PIWI/Argonaute/Zwille) domains and carboxy-terminal **PIWI** domains.

argyrophil cells Neuroendocrine cells that take up silver ions from a staining solution but require the addition of a reducing agent to precipitate metallic silver (unlike **argentaffin cells** which do not). **Carcinoids** of the foregut tend to be argyrophilic whereas those of the lower intestine tend to be argentaffinic.

ARIA *Acetylcholine receptor inducing activity* Polypeptide (**HRG1** residues 177–244) purified from chick brain that stimulates transcription of **acetylcholine receptor** subunits in nuclei that underlie the developing synapses in chick **myotubes**. ARIA, NDF, **heregulin** and GGF are encoded by alternatively spliced transcripts of the same gene. ARIA activates **erbB** receptor tyrosine kinases.

ARID *AT-rich interaction domain* ARID is an ancient DNA-binding domain conserved throughout the evolution of higher eukaryotes. The consensus sequence spans about 100 amino acid residues, and structural studies identify the major groove contact site as a modified helix–turn–helix motif. Characteristic of a family that includes 15 distinct human proteins with important roles in development, tissue-specific gene expression and proliferation control.

Ark1 1. Ark1p; A yeast actin-regulating serine/threonine kinase; see **Prk1p**. 2. betaark1: The most abundant cardiac G-protein-coupled receptor kinase (GRK2) also known as beta AR kinase 1 (beta ARK1), involved in desensitizing and downregulating beta adrenoreceptors (betaARs).

armadillo *β-Catenin; beta-catenin; arm* *Drosophila* gene encoding β-**catenin**, a component of **adherens junctions**. Links junctional complex to the cytoskeleton.

armadillo repeat Protein motif comprising 42 amino acids, originally described in the *Drosophila* **armadillo** protein. Usually found in multiple repeats that form a super-helix of helices with a positively charged groove. Mediates interactions with proteins such as **cadherins**, Tcf-family transcription factors and the tumour-suppressor gene, *APC* (adenomatous **polyposis coli**).

ARNO *ARF nucleotide-binding site opener* Guanine nucleotide exchange factor (**GEF**) for **ARFs** that will stimulate nucleotide exchange on both ARF1 and ARF6. Closely related to **cytohesin** and **GRP-1**. Exchange-factor activity resides in the **sec7**-like domain.

ARNT *Aryl hydrocarbon receptor nuclear translocator* A promiscuous bHLH-PAS (Per-ARNT-Sim) protein (~790 aa) that forms heterodimeric transcriptional regulator complexes with several other bHLH-PAS subunits to control a variety of biological pathways. In association with **hypoxia-inducible transcription factors** is important in cellular adaptation to low oxygen environments; it is also a dimeric partner for the Ah receptor (**AHR**) and this complex is essential in regulating the adaptive metabolic response to polycyclic aromatic hydrocarbons. See **ArnT**, which is different.

ArnT *4-Amino-4-deoxy-l-arabinose transferase* An inner membrane protein that catalyses covalent addition of 4-amino-4-deoxy-l-arabinose (l-Ara4N) groups to lipid A in the outer membranes of bacteria such as *Salmonella typhimurium* and *E. coli*, the final step in the polymyxin-resistance pathway. Highly alpha-helical with an apparent 62-kDa MW on SDS-PAGE. See **ARNT**.

aromatase *EC 1.14.14.1* Microsomal enzyme complex that converts testosterone to oestradiol. One of the cytochrome P450 enzymes, of which various isoforms exist (CYP2D6, CYP2E1, CYP3A4).

ARP-1 *Apolipoprotein regulatory protein-1* Nuclear receptor that binds to response element with two core motifs, 5′-RG(G/T)TCA, as do various receptors such as **COUP-TFI** and **PPAR**.

Arp2/3 complex A stable assembly of two actin-related proteins (Arp2 and Arp3) with five other subunits; a central player in the cellular control of actin assembly. Arp2/3 caps the pointed end of actin filaments and nucleates actin polymerization with low efficiency. The GTPase **Cdc42** acts in concert with **WASP** family proteins to activate the Arp2/3 complex. See also **ActA**, IcsA.

arrestins Family of inhibitory proteins that bind to tyrosine-phosphorylated receptors, thereby blocking their interaction with G-proteins and effectively terminating the signalling. Arrestin (S antigen; 48 kDa, from retinal rods) competes with **transducin** for light-activated **rhodopsin**, thus inhibiting the response to light (adaptation). Immune responses to arrestin lead to autoimmune uveitis. Similarly, β-arrestin binds to phosphorylated β-**adrenergic receptors**, inhibiting their ability to activate the **G-protein** Gs.

Arrhenius plot A plot of the logarithm of reaction rate against the reciprocal of absolute temperature. For a single-stage reaction this gives a straight line from which the activation energy and the frequency factor can be determined. Often applied to data from complex biological systems when the form observed is frequently a series of linear portions with sudden changes of slope. Great caution must be observed in interpreting such slopes in terms of activation energies for single processes.

arrhythmia Lack of normal ordered rhythm, particularly in the case of the heart where arrhythmia can be a prelude to cardiac arrest.

arrowheads Fanciful description given to the pattern of **myosin** molecules attached to a filament of **F-actin**. Easier to see if tannic acid is added to the fixative. The arrowheads indicate the polarity of the filament; the barbed (attachment) end is the site of major subunit addition.

ARS *Autonomously replicating sequence* A DNA sequence originally isolated from *S. cerevisiae* that, when linked to a non-replicating sequence, can confer on the latter the ability to be replicated in a yeast cell. Transformations effected with the use of ARS occur at relatively high frequency but are unstable. Homologous recombination of the DNA of interest with the host cell chromosomes is not required for expression when ARS routes are used.

arsenite Arsenite, the trivalent form of arsenic, is a thiol-reactive oxidative stressor that is toxic, co-carcinogenic and known to inhibit protein synthesis. It activates multiple stress signalling pathways and has effects on gene control, such as the interruption of cell cycle control by initiating G(2)/M arrest. At low doses may have useful antitumour activity. It is used as a herbicide and pesticide.

Artemether A rapid-acting antimalarial drug of the same class as **artemisinin**. Often used in combination with **lumefantrine**.

Artemia salina Brine shrimp, a crustacean of the Order Anostraca.

artemisinin Antimalarial drug extracted from the Chinese herb *Artemisia annua* (qinghaosu or sweet wormwood), used to treat uncomplicated falciparum malaria; a sesquiterpene lactone. More potent derivatives such as artemether and artesuna have been developed. Artemisinin is often used in combination therapy with, for example, **lumefantrine**.

arteriole Finest branch of an artery upstream of the capillary bed.

arteriosclerosis Imprecise term for various disorders of arteries, particularly hardening due to fibrosis or calcium deposition; often used as a synonym for **atherosclerosis**.

artery Blood vessel carrying blood away from the heart; walls have smooth muscle and are innervated by the **sympathetic nervous system**.

arthritis General term for inflammation of one or more joints. Many diseases may cause arthritis, although in most cases the cause of the inflammation is not understood. This is particularly true of rheumatoid arthritis, though knowledge of other forms is not much better.

Arthrobacter Genus of obligate aerobic bacteria of irregular shape, found extensively in soil.

arthropathy Any disease affecting a joint – care should be taken not to confuse arthrosclerosis (stiffness of joints) with atherosclerosis.

arthropod The largest phylum of the animal kingdom, containing several million species. Arthropods are characterized by a rigid external skeleton, paired and jointed legs and a haemocoel. The Phylum Arthropoda includes the major classes Insecta, Crustacea, Myriapodia and Arachnida.

Arthus reaction A localized **inflammation** due to injection of **antigen** into an animal that has a high level of circulating **antibody** against that antigen. A haemorrhagic reaction with oedema occurs due to the destruction of small blood vessels by thrombi. It may occur, as in 'Farmer's Lung', as a reaction to natural exposure to antigen.

articulins Membrane-associated protein complex of *Euglena*; two isoforms of 80 and 86 kDa, completely unlike spectrin, though functionally analogous. Have a core domain of 12-residue repeats, rich in valine and proline. May attach directly to membrane proteins.

artificial selection Conventional genetic engineering, the selection of progeny by a plant or animal breeder, rather than by environmental factors (**natural selection**). Quite remarkable phenotypic alterations can be brought about, a classic example being the variety of breeds of dogs.

Artiodactyla Order of herbivorous, even-toed mammals that includes antelopes, pig, cow, giraffe and hippopotamus.

aryl hydrocarbon receptor See **AHR**.

aryl sulphatase *EC 3.1.6.1* Aryl sulphatases A, B and C comprise a group of enzymes originally assayed by their ability to hydrolyse *O*-sulphate esters of aromatic substrates. Aryl sulphatase A, substrate cerebroside 3-sulphate, is deficient in metachromatic leucodystrophy. Aryl sulphatase B, substrate acetylhexosamine 4-sulphate in glycosaminoglycans, is deficient in **Maroteaux–Lamy syndrome**. Aryl sulphatase C hydrolyses estrogen sulphates. All three are deficient in multiple sulphatase deficiency.

arylation Addition of an aryl group to a substrate, often catalysed by palladium. Not an enzymatic reaction, although the removal of aryl groups (for example by **aryl sulphatase**) can be carried out in biological systems.

AS-2 See **adocia-sulphate-2**.

asbestosis Fibrosis of the lung as a result of the chronic inhalation of asbestos fibres. The needle-like asbestos fibres are phagocytosed by **alveolar macrophages** but burst the phagosome (**phagocytic vesicle**) and kill the macrophage, and the cycle is repeated. **mesothelioma**, a rare tumour of the mesothelial lining of the pleura, is associated with intense chronic exposure to asbestos dust, particularly that of crocidolite asbestos.

Ascaris Genus of nematodes (Aschelminthes). *Ascaris suum* is the common roundworm of pigs; *Ascaris lumbricoides* causes ascariasis in man; although the worms are restricted to the gut they divert a substantial proportion of food intake, and heavy infestation can cause growth retardation in children or, in extreme cases, intestinal blockage. One of the WHO's six major diseases. Used in developmental studies.

Aschelminthes Cluster of invertebrate phyla of which the best known are nematodes and acanthocephala. All have a pseudocoelom and an unsegmented elongated body with a terminal anus and a non-muscular gut.

Aschheim–Zondek test Obsolete pregnancy testing method that involved injecting a specimen of urine into mice: a positive sample will cause swelling of the ovaries.

Aschoff bodies Small **granulomas** composed of **macrophages**, **lymphocytes** and multinucleate cells grouped around eosinophilic **hyaline** material derived from collagen. Characteristic of the **myocarditis** of rheumatic fever.

Ascidiacea Class of simple or compound **tunicates** that have a motile larva but sedentary adult form that filter-feeds. Sea squirts are the commonly known examples.

ascites Accumulation of fluid in the peritoneal cavity causing swelling; causes include infections, portal hypertension and various tumours.

ascites tumour Tumour that grows in the peritoneal cavity as a suspension of cells. Obviously such cells have lost **anchorage dependence** and they can easily be isolated and passaged. **Hybridomas** are sometimes grown as ascites tumours, and the ascites fluid can then be used as the crude 'antiserum'.

ascocarp Fruiting body of the ascomyce fungi.

Ascomycotina Ascomycete fungi that produce spores, usually eight, in a structure known as an ascus. Includes yeasts and *Neurospora*.

ascorbic acid *Vitamin C* A requisite in the diet of man and guinea pigs. May act as a reducing agent in enzymic reactions, particularly those catalysed by **hydroxylases**. See also Table V1.

ascospore Diploid spore formed by ascomycete fungi, contained within an **ascus**.

ascus Elongated spore case containing four or eight haploid sexual ascospores of ascomycete fungi (which include most yeasts).

asepsis State in which harmful microorganisms are absent. Aseptic technique aims to avoid contamination of sterile systems.

asexual Reproducing without a sexual process and thus without formation of gametes or reassortment of genetic characters.

ASF/SF-2 *SRp30a* Mammalian SR-type splicing factors (see **SR-proteins**), ASF/SF 2 and SC-35, that play crucial roles in pre-mRNA splicing and have been shown to shift splice site choice *in vitro*.

ASGP Membrane-associated mucin present on rat mammary carcinoma cells. ASGP-1 and ASGP-2 are generated from a single precursor; ASGP-2 acts as a membrane anchor for ASGP-1 Though ASGP is thought to have similar functions, has no sequence homology with episialin.

asialoglycoprotein The carbohydrate moiety of many vertebrate glycoproteins bears terminal residues of **sialic acid**. If such residues are removed, e.g. by treatment with a **neuraminidase**, the resulting proteins are known as asialoglycoproteins. In the case of certain plasma proteins, the asialo-derivatives are specifically bound by a receptor on the surface of liver parenchymal cells (the scavenger receptor).

ASIP 1. Agouti-signalling protein (ASIP) is the human homologue of mouse **agouti**, inhibits melanogenesis and the response of human melanocytes to alpha-melanotropin. **2.** Atypical protein kinase C isotype-specific interacting protein (ASIP) specifically interacts with the atypical protein kinase C isozymes PKCγ and PKCζ. Overexpression of ASIP inhibits insulin-induced glucose uptake by specifically interfering with signals transmitted through PKCλ. ASIP is the mammalian homologue of *C. elegans* polarity protein PAR-3. ASIP and PAR-3 share three **PDZ domains** and can both bind to aPKCs.

A-site Site on the **ribosome** to which aminoacyl tRNA attaches during the process of peptide synthesis. See also **P-site**.

ASK Arabidopsis *SHAGGY-related protein kinase* Gene family, *Arabidopsis* SHAGGY-related protein kinases, that have homology to mammalian **GSK**3 and *Drosophila* SHAGGY. There are at least 10 *ASK* genes in the haploid *Arabidopsis* genome.

ASK1 *Apoptosis signal-regulating kinase* A mitogen-activated protein kinase kinase kinase (MAPKKK) that is regulated under conditions of cellular stress. ASK1 phosphorylates c-Jun N-terminal kinase (**JNK**) and elicits an apoptotic response.

Askenazy cells *Hurthle cells; oxyphil cells; oncocytes* Abnormal thyroid epithelial cells found in autoimmune **thyroiditis**. The cubical cells line small **acini** and have **eosinophilic** granular cytoplasm and often bizarre nuclear morphology.

asparaginase *EC 3.5.1.1* Enzyme that hydrolyses L-asparagine to L-aspartate and ammonia that is used as an antitumour agent especially against lymphosarcoma and lymphatic leukaemia.

asparagine *β-Asparagine; Asn; N; 132 Da* The β-amide of aspartic acid; the L-form is one of the 20 amino acids directly coded in proteins. Coded independently of aspartic acid. See Table A2.

aspartame Proprietary name for Asp-Phe methyl ester, an artificial sweetener.

aspartate *Aspartic acid; Asp; D; 133 Da* L-aspartate is one of the 20 amino acids directly coded in proteins; the free amino acid is a neurotransmitter. See Table A2.

aspartate transaminase See **SGOT**.

aspartic peptidase *EC 3.4.23 class; aspartic protease, aspartic proteinase, aspartyl proteinase* Relatively small family of endopeptidases, the best-known member being

pepsin A. Have a pH optimum below 5. The catalytic centre is formed by two Asp residues that activate a water molecule, and this mediates the nucleophilic attack on the peptide bond. They have significant roles in human diseases (e.g. **renin** in hypertension, **cathepsin** D in metastasis of breast cancer, **beta-secretase** in Alzheimer's disease).

aspartokinase EC 2.7.2.4 Enzyme that phosphorylates L-aspartate to produce aspartyl phosphate.

aspartyl protease See **aspartic peptidase**.

aspergillins Family of toxins (17 kDa) produced by *Aspergillus*. All are ribonucleases and disrupt protein biosynthesis. Includes **alpha-sarcin**, **mitogillin**, **restrictocin** and Asp fl.

aspergillosis Lung disease caused by fungi of the genus *Aspergillus*.

Aspergillus A genus of common ascomycete fungi found in soil. Industrially important in production of organic acids and a popular fungus for genetic study (esp. *A. niger*).

aspirin *Acetyl salicylate* An analgesic, antipyretic and anti-inflammatory drug. It is a potent **cyclo-oxygenase inhibitor** and blocks the formation of **prostaglandins** from arachidonic acid. See **COX-1**.

association constant K_a; K_{ass} Reciprocal of **dissociation constant**. A measure of the extent of a reversible association between two molecular species at equilibrium.

astacin A zinc endopeptidase from crayfish (*Astacus*), the prototype for the astacin family of metalloendopeptidases. Family includes **BMP**-1, Meprin A, **stromelysin** 1 and **thermolysin**.

astaxanthin *3,3'-Dihydroxy-4,4'-diketo-b-carotene* A naturally occurring red carotenoid pigment with antioxidant properties. Most crustaceans are tinted red by accumulated astaxanthin, and the pink flesh of a healthy salmon is due to accumulated astaxanthin. This is added to feed in fish farms to substitute for the astaxanthin in the diet of a wild salmon.

aster Star-shaped cluster of microtubules radiating from the polar **microtubule-organizing centre** at the start of mitosis.

asthma Inflammatory disease of the airways involving marked eosinophil infiltration and remodelling of the airways. Attacks can be triggered by allergic responses, physical exertion, inhaled chemicals or stress, and involve wheezing, breathlessness and coughing.

astrin A microtubule-associated protein (134 kDa) that localizes with mitotic spindles in M-phase and is essential for progression through mitosis, although astrin's function is unclear. Present in most tissues but highly expressed in the testis (see **ODF**).

astroblast An embryonic **astrocyte**.

astrocyte A **glial cell** found in vertebrate brain, named for its characteristic star-like shape. Astrocytes lend both mechanical and metabolic support for neurons, regulating the environment in which they function. See **oligodendrocytes**.

astrocytoma A neuroectodermal tumour (**glioma**) arising from **astrocytes**. Probably the commonest glioma, it has a tendency to **anaplasia**.

astroglia See **astrocytes**.

astrogliosis Hypertrophy of the **astroglia**, usually in response to injury.

Astropectinidae Family of echinoderms that includes many starfish species with long spines.

astrotactin Neuronal surface glycoprotein (100–105 kDa; three **EGF-like** repeat domains, two **fibronectin** III repeats) that functions in murine cerebellar granule cell migration *in vitro*, acting as the ligand for neuron–glial cell binding. Message has been detected in neuronal precursors in the cerebellum, hippocampus, cerebrum and olfactory bulb in the **brain**. See **weaver** mutant.

astroviruses *Astroviridae* Spherical viruses with five- or six-pointed star-shaped surface pattern. May be associated with enteritis in various vertebrates. Genome is a single molecule of linear, positive-sense, single-stranded RNA.

AT hook An AT-rich DNA-binding domain that occurs three times in mammalian high-mobility-group chromosomal proteins and in DNA-binding proteins from plants.

ataxia Imbalance of muscle control. See **ataxia telangiectasia**.

ataxia telangiectasia *Louis-Bar syndrome* A hereditary **autosomal** recessive disease in humans characterized by a high frequency of spontaneous chromosomal aberrations, neurological deterioration and susceptibility to various cancers. In part an immune deficiency disease and in part one of DNA repair; it is believed to be due to hypersensitivity to background ionizing radiation as a result of mutations in the **ATM** gene.

ATCase EC 2.1.3.2; *aspartate transcarbamylase* Enzyme that catalyses the first step in pyrimidine biosynthesis, condensation of aspartate and carbamyl phosphate. Positively allosterically regulated by ATP and negatively by CTP; classic example of an allosterically regulated enzyme. Bacterial ATCases exist in three forms: class A (*ca.* 450–500 kDa), class B (*ca.* 300 kDa) and class C (*ca.* 100 kDa).

ATCC The American Type Culture Collection, repository of many eukaryotic cell lines (which may be purchased). Comparable collections of microorganisms, protozoa, etc. are kept.

atelocollagen A highly purified pepsin-treated type I collagen from calf dermis which is of low immunogenicity because telopeptides are absent. It is used clinically for a wide range of purposes, including wound healing, vessel prosthesis and as a bone cartilage substitute, and more recently has been used as a complex with DNA or siRNA because these complexes are efficiently transduced into cells.

atenolol Beta adrenoreceptor antagonist or blocker with greater affinity for β1-receptors. It has more cardiac effects and less effect on airway tone.

ATF *Activating transcription factor* Commonly: **1.** Member of the **CREB**/ATF family of **bZip** transcription factors. A number of ATF variants (ATF-1, ATF-2, etc.) have been described and are, in general, involved in mediating transcription in response to intracellular signalling. Occasionally: **2.** Artificial transcription factor. **3.** Amino-terminal fragment.

ATG 1. See **antithymocyte globulin**. 2. See **ATG genes**.

ATG genes Large group of genes (16 in *S. cerevisiae*) required for autophagy-related processes that transport proteins/organelles destined for proteolytic degradation to the vacuole. Most Atg proteins (coded by the *ATG* genes) are co-localized at the pre-autophagosomal structure (PAS).

athanogene Apparently a neologism for a gene that has antiapoptotic function. Has been applied particularly to **BAG-1**.

atheroma Degeneration of the walls of the arteries because of the deposition of fatty plaques in the **intima** of the vessel wall, and scarring and obstruction of the lumen.

atherosclerosis Condition caused by the deposition of lipid in the wall of arteries in atheromatous plaques. Migration of smooth muscle cells from media to intima, smooth muscle cell proliferation, the formation of **foam cells** and extensive deposition of extracellular matrix all contribute to the formation of the lesions that may ultimately occlude the vessel or, following loss of the endothelium, trigger the formation of thrombi. Should be distinguished from **arteriosclerosis**, which is a more general term usually applied to arterial hardening through other causes. Atherosclerosis is a major medical problem in most of the developed world.

ATM Protein product of the gene mutated in **ataxia telangiectasia** (AT), a member of the phosphatidylinositol-3-kinase family. ATM constitutively binds to the SH3 domain of the tyrosine kinase c-**Abl** in normal but not AT cells and ATM seems to activate DNA-damage-induced activation of c-Abl (which is deficient in AT cells).

atomic force microscopy *AFM* A form of **scanning probe microscopy**, in which a microscopic probe is mechanically tracked over a surface of interest in a series of *x–y* scans and the force encountered at each coordinate measured with piezoelectric sensors. This provides information about the chemical nature of a surface at the atomic level.

atopy Allergic (**hypersensitive**) response at a site remote from the stimulus (e.g. food-induced dermatitis).

atorvastatin A **statin** used to reduce blood cholesterol levels.

ATP *Adenosine 5′ triphosphate* Often referred to as 'the energy currency of the cell', a compound synthesized in cells from ADP by energy-yielding processes. Enzymic transfer of the terminal phosphate or pyrophosphate from ATP to a wide variety of substrates provides a means of transferring chemical free energy from metabolic to catabolic processes.

ATP-binding site *'A' motif* A consensus domain found in a number of ATP or GTP-binding proteins, for example **ATP synthase**, **myosin heavy chain**, **helicases**, **thymidine kinase**, **G-protein** α-subunits, GTP-binding **elongation factors**, **Ras** family. Consensus is: (A or G)-XXXXGK-(S or T); this is thought to form a flexible loop (the P-loop) between α-helical and **beta-pleated sheet** domains.

ATP synthase A proton-translocating **ATPase**, found in the inner membrane of **mitochondria**, **chloroplasts** and the plasmalemma of bacteria. It can be known as the F_1/F_0 or CF_1/CF_0 ATPase or as the class of F-type ATPases. In all these cases, the enzyme is driven in reverse by the large **proton motive force** generated by the **electron transport chain**, and thus synthesizes, rather than uses, **ATP**. See also **chemiosmosis**, **V-type ATPase**, **P-type ATPase**.

ATPase Any enzyme capable of releasing the terminal (γ) phosphate from ATP, yielding **ADP** and inorganic phosphate. The description could mislead, because in most cases the enzymic activity is not a straightforward hydrolysis but part of a coupled system for achieving an energy-requiring process, such as ion-pumping or the generation of motility.

atrial natriuretic factor Obsolete name for **atrial natriuretic peptide**.

atrial natriuretic peptide *ANP* A polypeptide hormone found mainly in the **atrium** of the heart of many species of vertebrates. It is released in response to atrial stretching and thus to elevated blood pressure. ANP acts to reduce blood pressure through stimulating the rapid excretion of sodium and water in the kidneys (reducing blood volume), by relaxing vascular smooth muscle (causing vasodilation) and through actions on the brain and adrenal glands.

atriopeptin General term originally used for the polypeptide hormones **atrial natriuretic peptide** (ANF, ANP) and **brain natriuretic peptide** (BNP).

atrium *Pl.* atria A cavity in the body, especially either of the two upper chambers of the heart in higher vertebrates.

atrophic Undergoing **atrophy** or shrinkage in size and usually function; relating to or characterized by atrophy.

atrophy Wasting away of tissue with loss of mass and/or function.

atropine An alkaloid, isolated from deadly nightshade (*Atropa belladonna*), that inhibits **muscarinic acetylcholine receptors**. Applied to the eye it causes dilation of the pupil, which is said to enhance the beauty of a woman – hence belladonna as the specific name of the plant from which the ancients extracted the drug.

attachment constriction See **centromere**.

attachment plaques Specialized structures at the ends of a chromosome by which it is attached to the nuclear envelope at **leptotene** stage of mitosis.

attacins Antibacterial proteins (20–22 kDa) produced by insect haemocytes following bacterial challenge. May be basic or acidic and are fairly highly conserved between species. Induction of these **antimicrobial peptides** in *Drosophila* involves NF-kappaB elements.

attenuation Viruses that have been **passaged** extensively may become attenuated (non-virulent) and can be used as a vaccine.

A-type particles Retrovirus-like particles found in cells. Non-infectious. The mouse genome contains around 1000 copies of homologous sequences.

AU-rich element *ARE* Cytoplasmic mRNA stability is mediated by proteins that bind to AU-rich elements (AREs) in the 3′ untranslated region of transcripts. This has been shown for mRNA encoding oncoproteins, cytokines and transcription factors.

AUG The **codon** in **messenger RNA** that specifies initiation of a polypeptide chain, or within a chain, incorporation of a **methionine** residue.

Augmentin Proprietary name for an antibiotic preparation consisting of a mixture of **amoxicillin** and **clavulanic acid**.

Aujesky's disease An encephalomyelitis affecting domestic animals and pets; caused by a thermostable **herpes** virus.

Aurelia aurita Common jellyfish – transparent disc with four blue/purple horseshoe-shaped gonads clearly visible. Phylum Cnidaria; Class Scyphozoa.

aurintricarboxylic acid *ATA; Aluminon* A general inhibitor of nucleases, shown to inhibit (*in vitro*): **DNAse** 1, **RNAse** A, **S1 nuclease**, exonuclease III, various **restriction endonucleases**, DNA **topoisomerase** II and protein–nucleic acid interactions. Stimulates the tyrosine phosphorylation of **MAP kinases**, inhibits both major **calpain** isoforms. Potent and selective inhibitor of **SARS** coronavirus replication. Apoptosis inhibitor.

auristatins Synthetic members of the **dolastatin** class of tubulin polymerization inhibitors.

Aurora-A A centrosomal serine–threonine kinase that regulates mitosis. Overexpression of Aurora-A has been found in a wide range of tumours and has been implicated in oncogenic transformation.

aurosome Gold-containing secondary lysosome found in patients treated with gold complexes.

aurovertins A family of related antibiotics from the fungus *Calcarisporium arbuscula* that inhibits oxidative phosphorylation in mitochondria and in many bacterial species. Aurovertins B and D have identical biological properties and are more potent than aurovertin A. Inhibit the proton-pumping **F-type ATP synthase** by binding to β-subunits in its F1 catalytic sector.

Australia antigen *HbsAg* A viral envelope antigen of **hepatitis B** virus. Appearance of the antigen in serum is associated with a phase of high infectivity. First identified in the serum of an Australian aborigine.

autacoids Local hormones such as **histamine, serotonin, angiotensin, eicosanoids**.

autapomorphy A derived trait unique to any given taxon.

autapses Synapses formed by a neuron with itself. The functional significance is unclear. *Adj.* autaptic.

autoantibody Antibody that reacts with an antigen that is a normal component of the body. Obviously this can lead to some problems, and autoimmunity has been proposed as a causative factor in a number of diseases such as rheumatoid arthritis. See also **systemic lupus erythematosus, Hashimotos thyroiditis, myasthenia gravis**.

autocatalytic A compound that catalyses its own chemical transformation. More commonly, a reaction that is catalysed by one of its products or an enzyme-catalysed reaction in which one of the products functions as an enzyme activator.

autochthonous Found in the place where it was originally formed, indigenous.

autoclave 1. An apparatus for sterilization by steam at high pressure. 2. To sterilize by using steam at high pressure. A high temperature, e.g. 121°C for 15 min, is necessary to ensure killing of bacterial spores.

autocrine Secretion of a substance, such as a **growth factor**, that stimulates the secretory cell itself. One route to independence of **growth control** is by autocrine growth factor production.

autofluorescence Property of a compound or material that will fluoresce in its own right – without the addition of an exogenous fluorophore. A common problem in fluorescence microscopy and in assays where the read-out is fluorescence.

autogamy Self-fertilization, common in plants and also in some ciliate protozoa, where gametic nuclei from a single micronucleus subsequently fuse to form the zygote nucleus.

autogenous Generated without external influence or input.

autograft Graft taken from one part of the body and placed in another site on the same individual.

autoimmune Adjective describing a situation in which the immune system responds to normal components of the body. Several diseases are thought to have an autoimmune component, but it is often not clear whether this is causative.

autoinducer-2 A 5-carbon sugar that spontaneously cyclizes from 4,5-dihydroxy-2,3-pentanedione (DPD), a product of the LuxS enzyme in the catabolism of *S*-ribosylhomocysteine. It appears to be a universal signal molecule mediating interspecies communication among bacteria. See **quorum sensing**.

autologous Derived from an organism's own tissues or DNA. *cf.* **heterologous, homologous**.

autolysis Spontaneous **lysis** (rupture) of cells or organelles produced by the release of internal hydrolytic enzymes. Normally associated with the release of **lysosomal enzymes**.

autonomic nervous system **Neurons** that are not under conscious control, comprising two antagonistic components, the **sympathetic** and **parasympathetic nervous systems**. Together, they control the heart, viscera, smooth muscle, etc.

autophagy Removal of cytoplasmic components, particularly membrane-bounded organelles, by digesting them within **secondary lysosomes** (autophagic vacuoles). Particularly common in embryonic development and senescence.

autophosphorylation Addition of a phosphate to a protein kinase (possibly affecting its activity) by virtue of its own enzymic activity.

autoradiography Technique in which a specimen containing radioactive atoms is overlaid with a photographic emulsion, which, after an appropriate lapse of time, is developed, revealing the localization of radioactivity as a pattern of silver grains. Resolution is determined by the path length of the radiation, and so the low-energy β-emitting isotope, tritium, is usually used.

autoregulation Regulation of a gene encoding a **transcription factor** by its own gene product: a feedback process.

autosomal dominant Gene located on an **autosome** that has a dominant effect – even though two copies of the gene exist, one of them is normal. Often attributed to a **gain of function mutation**.

autosomal recessive Mutation carried on an **autosome** that is deleterious only in **homozygotes**.

autosomes Chromosomes other than the sex chromosomes.

autotaxin *ATX* An enzyme originally associated with melanoma cells, known to stimulate motility in tumour cells, that turns out to be **lysophospholipase D** and works by producing **LPA**.

autotroph *Lithotroph* Organisms that synthesize all their organic molecules from inorganic materials (carbon dioxide, salts, etc.). May be photo-autotrophs or chemo-autotrophs, depending upon the source of the energy. Also known as lithotrophic organisms.

auxesis Growth by increase in cell size rather than by increasing cell numbers.

auxilin A novel **adaptin** found associated with the coated vesicles isolated from brain cells.

auxin response factors *ARFs* Transcription factors that mediate responses to the phytohormone, **auxin**. ARF2 promotes transitions between multiple stages of Arabidopsis development, ARF1 acts in a partially redundant manner with ARF2.

auxins A group of **plant growth substances** (often called phytohormones or plant hormones), the most common example being **indole acetic acid** (IAA), responsible for raising the pH around cells, making the cell wall less rigid and allowing elongation.

auxotroph Mutant that differs from the wild-type in requiring a nutritional supplement for growth. A deficiency mutant.

auxotyping Method for strain-typing *Neisseria* by checking their requirements for specific nutrients in defined media.

available nutrients Nutrients in the extracellular environment that are accessible to uptake, not bound irreversibly to some other component or in insoluble complexes.

Avena sativa Cultivated oat.

avermectin B1 *Abamectin* Metabolite of *Streptomyces avermitilis* used as an acaricide, insecticide and anthelminthic.

avian erythroblastosis virus See **avian leukaemia virus**.

avian flu *Avian influenza, bird flu* A highly contagious strain of influenza that affects poultry and can be transmitted to humans; having switched species it could become extremely virulent (a similar phenomenon occurred with the 1918 strain of flu, which caused more deaths than the preceding 4 years of war). The **H5N1** strain is currently causing considerable concern.

avian leukaemia virus Group of C-type RNA tumour viruses (**Oncovirinae**) that cause various leukaemias and other tumours in birds. The acute leukaemia viruses, which are replication-defective and require helper viruses, include avian erythroblastosis (AEV), myeloblastosis (AMV) and myelocytomatosis viruses. AEV carries two transforming genes, *v-erbA* and *v-erbB*; the cellular homologue of the latter is the structural gene for the **epidermal growth factor** receptor. AMV carries *v-myb* and causes a myeloid leukaemia; avian myelocytomatosis virus carries *v-myc*. The avian lymphatic leukaemia viruses (ALV) are also **Retroviridae,** but are replication-competent and induce neoplasia only after several months; they often occur in conjunction with replication-defective leukaemia viruses.

avian myeloblastosis virus *AMV* Retrovirus of the Subfamily Oncornaviridae. Causes myelocytomatosis, osteopetrosis, lymphoid leucosis and nephroblastoma. May be a mixture of viruses.

avidin **Biotin**-binding protein (68 kDa) from egg white. Binding is so strong as to be effectively irreversible – a diet of raw egg white leads to biotin deficiency.

avidity Strength of binding, usually of a small molecule with multiple binding sites by a larger one; particularly the binding of a complex antigen by an antibody. (**Affinity** refers to simple receptor–ligand systems.)

avilamycin Oligosaccharide antibiotic that has been used in animal feed for many years. See **everninomicins**.

avirulent Organism or virus that does not cause infection or disease.

axenic A situation in which only one species is present. Thus an axenic culture is uncontaminated by organisms of other species, an axenic organism does not have commensal organisms in the gut, etc. Some organisms have obligate symbionts and cannot be grown axenically.

axial filaments The central filaments, of which there may be several hundred, of the periplasmic flagella of spirochaetes that rotate within the periplasmic space and cause the whole bacterium to flex like a corkscrew and thus to move. The central filaments are composed of at least three proteins (FlaB1, FlaB2 and FlaB3) that have significant homology with flagellin; the sheath is composed of FlaA protein (~43 kDa).

axil Member of the **axin** family. Interacts with **GSK**3 and β-**catenin**. By enhancing phosphorylation and thus the subsequent degradation of β-catenin, inhibits axis formation in *Xenopus* embryos.

axin A negative regulator of the **Wnt** signalling pathway. Binds to APC (adenomatous **polyposis coli** protein) and to β-**catenin**. Interacts directly with glycogen synthase kinase-3 (**GSK**3) and promotes GSK3 phosphorylation of β-catenin, which is then degraded. Axin is encoded by the fused locus in mice that is required for normal vertebrate axis formation. Dvl (dishevelled protein), axin and GSK can form a ternary complex bridged by axin, and **Frat1** can be recruited into this complex probably by Dvl. Wnt-1 appears to promote the disintegration of the Frat1–Dvl–GSK–Axin complex, resulting in the dissociation of GSK from axin.

axokinin Axonemal protein (56 kDa) that, when phosphorylated by a cAMP-dependent protein kinase, reactivates the **axoneme**.

axolemma **Plasma membrane** of an axon.

axon Long process, usually single, of a **neuron**, that carries efferent (outgoing) **action potentials** from the cell body towards target cells. See **dendrite**.

axon hillock Tapering region between the cell body of a **neuron** and its axon. This region is responsible for summing the graded inputs from the **dendrites** and producing **action potentials** if the threshold is exceeded.

axonal guidance General term for mechanisms that ensure correct projections by nerve cells in developing and regenerating nervous systems. Implies accurate navigation by **growth cones**, the highly motile tips of growing neuronal processes. See **growth cone collapse**.

axoneme The central **microtubule** complex of eukaryotic cilia and flagella with the characteristic '9 + 2' arrangement of tubules when seen in cross-section.

axonin Chick homologue of TAG-1. See **tax-1**.

axonogenesis The growth and differentiation of axonal processes by developing neurons. See **axon**.

axoplasm The **cytoplasm** of a **neuron**.

axopod *Pl.* axopodia Thin processes (a few μm in diameter but up to 500 μm long), supported by complex arrays of microtubules that radiate from the bodies of **heliozoa**.

axostyles Ribbon-like bundles of **microtubules** found in certain parasitic protozoa that may generate bending waves by **dynein**-mediated sliding of microtubules.

azacytidine *5-Azacytidine; β-ribofuranosyl 5-azacytidine* The ribonucleoside of **5-azacytosine**.

azacytosine *5-Azacytosine* An analogue of the **pyrimidine** base cytosine, in which carbon 5 is replaced by a nitrogen. In DNA, unlike cytosine, it cannot be methylated.

azaserine An analogue of **glutamine** that competitively inhibits various pathways in which glutamine is metabolized, hence an antibiotic and antitumour agent.

azide Usually the sodium salt NaN_3, an inhibitor of electron transport that blocks electron flow from cytochrome oxidase to oxygen. Frequently used to prevent growth of microorganisms in, e.g. refrigerated antisera or chromatography columns.

azidothymidine *AZT* See **AZT**.

azoospermia Absence of spermatozoa in the semen.

azothioprine *Azidothioprine* Immunosuppressive drug used to prevent graft rejection and to treat a variety of connective tissue disorders.

Azotobacter Genus of free-living rod-shaped bacilli capable of fixing atmospheric nitrogen.

AZT *Azidothymidine; zidovudine* An antiviral drug derived from thymidine, used in treatment of AIDS. Blocks the enzyme that stimulates growth and multiplication of the human immunodeficiency virus (HIV).

azurin 1. Blue copper-containing protein from *Pseudomonas aeruginosa*. **2.** Histochemical dye.

azurophil granules **Primary lysosomal** granules found in **neutrophil granulocytes**; contain a wide range of hydrolytic enzymes. Sometimes referred to as primary granules to distinguish them from the secondary or **specific granules**.

B

β *Beta* Entry prefix is given as 'beta'; alternatively look for the main portion of the word.

B7 Superfamily of co-stimulatory molecules that bind to CD28 or **CTLA-4** and regulate T-cell responses. They are part of the **immunoglobulin superfamily** and have an extracellular Ig variable-like (IgV) and constant-like (IgC) domains. B7-1 is CD80, B7-2 is CD86. Other members are ICOS-L, B7-H1, B7-DC. Several B7 homologues are expressed on cells other than professional antigen-presenting cells.

B12 Vitamin B12. See Table V1.

B220 The mouse CD45R antigen, predominantly expressed on B-cells. See Table C2.

Babes–Ernst granules Metachromatic intracellular deposits of polyphosphate found in *Corynebacterium diphtheriae* when the bacteria are grown on suboptimal media. Stain reddish with methylene blue or toluidine blue.

Babesia Genus of protozoa that are found as parasites within red blood cells of mammals and are transmitted by ticks.

babesiosis *Piroplasmosis* Disease caused by infection with protozoa of the genus *Babesia*.

BAC library *Bacterial artificial chromosome library* Library constructed in a vector with an **origin of replication** that allows its propagation in bacteria as an extra chromosome. Advantageous in constructing **genomic libraries** with relatively large DNA fragments (100–300 kb). See also **bacterial artificial chromosome**.

Bac7 Proline- and arginine-rich antimicrobial peptide (7 kDa) isolated from bovine neutrophils. Bac-5 is similar. The upstream region of proBac-5 and proBac-7 both have sequence homology with similar regions of other neutrophil antimicrobial peptides (CAP18 from rabbit neutrophils and bovine indolicidin). The pro-region also has similarity to porcine **cathelin**. Member of the **protegrin** family of peptides.

Bacille Calmette-Guerin An attenuated **mycobacterium** derived from *Mycobacterium tuberculosis*. The bacterium is used in tuberculosis vaccination. Extracts of the bacterium have remarkable powers in stimulation of lymphocytes and leucocytes and are used in **adjuvants**.

Bacillus Cylindrical (rod-shaped) bacterium. Bacilli are usually 0.5–1.0 μm long, 0.3–1 μm wide.

Bacillus cereus A Gram-positive, facultatively aerobic spore-forming bacterium. *B. cereus* food poisoning is caused by two distinct metabolites; the diarrhoeal type of illness by a large molecular weight protein, the vomiting (emetic) type of illness by a low-molecular weight, heat-stable peptide.

Bacillus megaterium A Gram-positive, spore producing, rod-shaped bacterium found in the soil. It is one of the largest Eubacteria and is extensively used in biotechnology due to its size and cloning abilities. Enzymes produced by *B. megaterium* are used in production of synthetic penicillin, modification of corticosteroids and include several amino acid dehydrogenases.

Bacillus thuringiensis Soil-living bacterium that produces a **delta-endotoxin** that is deadly to insects. Many strains exist, each with great specificity as to target Orders of insects. In general, the mode of action involves solubilization at high pH within the target insect's gut, followed by proteolytic cleavage; the activated peptides form pores in the gut cell apical plasma membranes, causing lysis of the cells. The toxin has been genetically engineered into various plant species (GM plants) to confer insect resistance, although many human consumers have perceived this as being unacceptable.

bacitracin Branched cyclic peptides produced by strains of *Bacillus licheniformis*. Interfere with murein **peptidoglycan** synthesis in Gram-positive bacteria.

baclofen γ-*Amino-β−*(p-*chlorophenyl*) *butyric acid* Skeletal muscle relaxant, a derivative of **GABA**, used to relieve muscle spasm in trauma, multiple sclerosis and cerebral palsy. Selectively binds GABA-B receptor and inhibits release of other neurotransmitters in the CNS.

bactenecin Highly cationic polypeptides found in lysosomal granules of bovine neutrophil granulocytes. They are thought to be involved in bacterial killing and occur in a third class of granules, the large granules, not found in the neutrophils of most species.

bacteraemia *Bacteremia (USA)* The presence of living bacteria in the circulating blood; usually implies the presence of small numbers of bacteria that are transiently present without causing clinical effects, in contrast to **septicaemia**.

bacteria One of the two major classes of prokaryotic organisms (the other being the **Cyanobacteria**). Bacteria are small (linear dimensions of around 1 μm), non-compartmentalized, with circular DNA, and ribosomes of 70S. Protein synthesis differs from that of eukaryotes, and many anti bacterial antibiotics interfere with protein synthesis, but do not affect the infected host. Recently bacteria have been subdivided into **Eubacteria** and **Archaebacteria**, although some would consider the Archaebacteria to be a third kingdom, distinct from both Eubacteria and Eukaryotes. The Eubacteria can be further subdivided on the basis of their staining using **Gram stain**. Since the difference between Gram positive and Gram negative depends upon a fundamental difference in cell wall structure, it is therefore more soundly based than classification on gross morphology alone (into cocci, bacilli, etc.).

bacterial artificial chromosome *BAC* Method of construction of a **genomic library**, in which the vector contains sites necessary for the DNA to be handled and replicated as a bacterial chromosome. Like **YACs**, this allows clones to contain very large pieces of DNA (around 200 kb), so aiding rapid, low resolution **physical mapping**.

bacterial cell wall Bacterial cells walls are of two major types; those that take up Gram stain (Gram-positive bacteria) and those that do not (Gram-negative bacteria).

Gram-positive bacteria have a wall approximately 50 nm thick containing **teichoic acid** and **peptidoglycan**, made up of repeating *N*-acetyl-glucosamine and *N*-acetyl-muramic acid. The wall of Gram-negative bacteria is separated from the cell membrane by the periplasmic space, is much thinner, contains a different peptidoglycan, and has an outer lipid bilayer (containing lipid A) resembling a membrane. Some bacteria, notably **mycoplasmas**, do not produce a cell wall. **Archaea** generally have rigid walls but the peptidoglycan composition differs.

bacterial chemotaxis The response of bacteria to gradients of attractants or repellents. In a gradient of attractant the probability of deviating from a smooth forward path is reduced if the bacterium is moving up-gradient. Since the opposite is true if moving down-gradient, the effect is to bias displacement towards the source of attractant. Strictly, should perhaps be considered a **klinokinesis** with adaptation.

bacterial flagella Thin filaments composed of **flagellin** subunits that are rotated by the basal motor assembly and act as propellers. If rotating anticlockwise (as viewed from the flagellar tip), the bacterium moves in a straight path; if rotating clockwise, the bacterium 'tumbles'. The direction of rotation is controlled through the bacterial chemotactic receptor system (see **bacterial chemotaxis**).

bacteriochlorophyll Varieties of **chlorophyll** (bacteriochlorophylls a, b, c, d, e and g) found in **photosynthetic bacteria** and differing from plant chlorophyll in the substituents around the tetrapyrrole nucleus of the molecule, and in the absorption spectra.

bacteriocide A substance that kills bacteria.

bacteriocins **Exotoxins**, often **plasmid** coded, produced by bacteria and which kill other bacteria (not eukaryotic cells). **Colicins** are produced by about 40% of *E. coli* strains: colicin E2 is a DNAase, colicin E3 an RNAase. See **lantibiotics**.

bacteriophaeophytin-b One of the components of the bacterial photosynthetic **reaction centre**. (See also **ubiquinone**).

bacteriophages *Phages* Viruses that infect bacteria. The bacteriophages that attack *E. coli* are termed coliphages; examples of these are lambda phage and the T-even phages, T2, T4 and T6. Basically, phages consist of a protein coat or **capsid** enclosing the genetic material, DNA or RNA, that is injected into the bacterium upon infection. In the case of virulent phages, all synthesis of host DNA, RNA and proteins ceases and the phage genome is used to direct the synthesis of phage nucleic acids and proteins using the host's transcriptional and translational apparatus. These phage components then self-assemble to form new phage particles. The synthesis of a phage lysozyme leads to rupture of the bacterial cell wall, releasing, typically, 100–200 phage progeny. The temperate phages, such as **lambda**, may also show this lytic cycle when they infect a cell, but more frequently they induce **lysogeny**. The study of bacteriophages has been important for our understanding of gene structure and regulation. Lambda has been extensively used as a **vector** in recombinant DNA studies.

bacteriorhodopsin A light-driven proton-pumping protein (248 residues, 26 kDa), similar to **rhodopsin**, found in 'purple patches' in the cytoplasmic membrane of the bacterium *Halobacterium halobium*. It is composed of seven transmembrane helices, and contains the light-absorbing chromophore, **retinal**. Light absorption maxima: 568 nm (light-adapted); 558 nm (dark-adapted). Each photon results in the movement of two protons from cytoplasmic to extracellular sides of the membrane. The resulting proton gradient is used (amongst other things) to drive synthesis of ATP by **chemiosmosis**.

bacteriostatic Adjective applied to substances that inhibit the growth of bacteria without necessarily killing them.

bacteroid Small, often irregularly rod-shaped bacterium, e.g. those found in root nodules of nitrogen-fixing plants.

Baculovirus Viruses specialized as pathogens of lepidopteran larvae. Widely used as eukaryotic **expression vectors** for proteins requiring post-translational modifications such as **glycosylation**, proteolytic **cleavage** and fatty acylation.

Bad A pro-apoptotic member of the **Bcl-2** family that is tightly regulated by survival factors. Several major signalling pathways influence cell death through their direct effects on the phosphorylation state of Bad.

bafilomycin Microbial toxin, a macrolide antibiotic from *Streptomyces griseus* that is a specific inhibitor of the **V-type ATPase**; also blocks lysosomal cholesterol transport in macrophages.

bag cell neurons Cluster of electrically coupled neurons in the abdominal ganglion of *Aplysia* that are homogeneous, easily dissected out and release peptides that stimulate egg-laying.

BAG-1 Bcl-2 *associated athanogene-1* A multifunctional protein that exists in three major isoforms, p50, p46 and p36. A fourth isoform of 29 kDa also exists, but its function remains mostly unknown. It is a co-chaperone of heat-shock proteins (Hsc70/Hsp70) that is expressed in most cells, attenuates glucocorticoid receptor (GR) nuclear translocation, activates ERK//MAP kinases, and potentiates anti-apoptotic functions of **Bcl-2,** to which it binds. Frequently deregulated in a variety of malignancies.

bagassosis Respiratory disease, similar to **farmer's lung**, caused by inhalation of dust from mouldy sugar cane. A Type III **hypersensitivity** reaction to mould spores.

Bak Member of the pro-apoptotic **bcl-2** family, a mitochondrial membrane protein. Loss of the interaction between Bak and the anti-apoptotic Bcl2-family member Mcl1 allows interaction with **p53**, oligomerization of Bak and release of cytochrome c from mitochondria.

BALB/c Inbred strain of white (albino) mice. Used as a source for one of the various 3T3 cell lines.

Balbiani ring The largest **puffs** seen on the **polytene chromosomes** of Diptera are called Balbiani rings after the nineteenth century microscopist who first described polytene chromosomes.

balloon cell Non-specific description of any cell with abundant clear cytoplasm. May arise through a variety of causes; includes some **carcinoid** cells, hepatocytes following some forms of toxic insult, neurons or other cells in **storage diseases**, and cells in some **melanomas**.

Balo's concentric sclerosis A rare variant of **multiple sclerosis**, described by Balo in 1928, characterized by alternating rings of demyelination and spared myelin.

BamH I *BamHI* Common **restriction enzyme** (from *Bacillus amyloliquefaciens* H) that cuts the sequence G|GATCC. See Table R1.

BAN *British Approved Name* Formal name for a medicinal substance. Several BANs were recently (2003) changed to match the recommended International Non-Proprietary Names (rINNs). In general, BANs have been used for the main entry and the old name is given as a synonym.

band III A 90-kDa protein of the human erythrocyte membrane, identified as the major anion transport/exchange protein. Analogous proteins exist in other erythrocytes. A dimeric transmembrane glycoprotein, with binding sites for many cytoplasmic proteins, including **ankyrin**, on its cytoplasmic domain.

band 4.1 domain *Band 4.1/JEF domain* The band 4.1 domain was first identified in the red blood cell protein band 4.1, and subsequently in **ezrin**, **radixin** and **moesin** (ERM proteins) and in other proteins, including the tumour suppressor **merlin**/schwannomin, **talin**, unconventional myosins VIIa and X, and **protein tyrosine phosphatases**. A structurally related domain has also been demonstrated in the N-terminal region of two groups of tyrosine kinases: the **focal adhesion kinases** (FAK) and the **Janus kinases** JAK. Additional proteins containing the 4.1/JEF (JAK, ERM, FAK) domain include plant kinesin-like calmodulin-binding proteins (KCBP). Additional properties common to band 4.1/JEF domains of several proteins are binding of phosphoinositides and regulation by GTPases of the **Rho** family.

band cells Immature **neutrophils** released from the bone-marrow reserve in response to acute demand.

banding patterns Chromosomes stained with certain dyes, commonly quinacrine (Q banding) or Giemsa (G banding) , show a pattern of transverse bands of light and heavy staining that is characteristic for the individual chromosome. The basis of the differential staining, which is the same in most tissues, is not understood: each band represents 5–10% of the length, about 10^7 base pairs, although this is not true for **polytene chromosomes** in *Drosophila*, which show more than 4000 bands.

bandshift assay *Gel shift assay* An assay for proteins, such as **transcription factors**, that bind specific DNA sequences. A labelled oligonucleotide corresponding to the recognition sequence is incubated with an appropriate nuclear protein extract, and run on a non-denaturing acrylamide gel. Oligonucleotides that have been bound by proteins are retarded relative to those that are unbound.

BAPTA *1,2-Bis(o-aminophenoxy)ethane tetraacetate* Calcium chelator with low affinity for magnesium. Absorption maximum shifts when calcium is bound so it can be used as an indicator of intracellular calcium concentration (though it will chelate calcium and therefore alter the situation. See **MAPTAM**.

barbital *Barbitone; veronal; Diethyl-malonyl-urea; 5,5-diethylbarbituric acid* Once widely used, in the form of the sodium salt, as a sedative and hypnotic.

barbiturates Class of drugs that depress activity of the central nervous system, largely superseded by **benzodiazepines**.

BARD1 Originally identified as a **BRCA1**-interacting protein, but has also been described in tumour suppressive functions independent of BRCA1. Forms a heterodimeric complex with BRCA1 that is important for the cellular response to DNA damage.

bariatrics The medical treatment of obesity.

barnase *EC 3.1.27.-* A 110-residue extracellular ribonuclease produced by *Bacillus amyloliquefaciens*. It is inhibited intracellularly by barstar, a 90-residue polypeptide.

baroreceptor *Baroceptor* In an organism, a receptor that is sensitive to pressure.

Barr body Small, dark-staining inactivated X chromosome seen in female (XX) cells. According to the **Lyon hypothesis**, random inactivation occurs.

barrier-to-autointegration factor *BAF* A conserved 10-kDa chromatin protein essential in proliferating cells. BAF dimers bind dsDNA, histone H3, histone H1.1, **lamin** A and transcription regulators, plus **emerin** and other **LEM domain** nuclear proteins. Binding to emerin and lamin A is inhibited by phosphorylation at serine-4 of BAF, and this is important in nuclear envelope disintegration at the start of mitosis.

BARS *Brefeldin A-ADP-riboslyated substrate* Two cytosolic proteins of 38 and 50 kDa that become ADP-ribosylated following treatment of cells by **brefeldin A**.

barstar See **barnase**.

basal body Structure found at the base of eukaryotic **cilia** and **flagella** consisting of a continuation of the nine outer sets of axonemal **microtubules** but with the addition of a C-tubule to form a triplet (like the **centriole**). May be self-replicating, and serves as a nucleating centre for axonemal assembly. Anchored in the cytoplasm by **rootlet system** Synonymous with **kinetosome**.

basal cell carcinoma *BCC; rodent ulcer* Common **carcinoma** derived from the basal cells of the epidermis. Often a consequence of exposure to sunlight, and much more common in those with fair skin; rarely metastatic.

basal cells General term for relatively undifferentiated cells in an epithelial sheet that give rise to more specialized cells (act as **stem cells**). In the stratified **squamous epithelium** of mammalian skin, the basal cells of the epidermis (stratum basale) give rise by an unequal division to another basal cell and to cells that progress through the spinous, granular and horny layers, becoming progressively more keratinized, the outermost being shed as **squames**. In olfactory mucosa the basal cells give rise to olfactory and sustentacular cells. In the epithelium of epididymis their function is unclear, but they probably serve as stem cells.

basal disc **1.** Portion of the stalk of a cellular slime mould fruiting body that is attached to the substratum. **2.** General name for the conical-shaped structure that anchors the stalk of a fungal fruiting body, a hydroid, or any other sessile organism, to the substratum.

basal ganglia Three large subcortical nuclei of the vertebrate brain: the putamen, the caudate nucleus and the globus pallidus. They participate in the control of movement, along with the **cerebellum**, the corticospinal system and other descending motor systems. Lesions of the basal ganglia occur in a variety of motor disorders, including **Parkinsonism** and **Huntington's chorea**.

basal lamina See **basement membrane**.

base analogues Purine and pyrimidine bases that can replace normal bases used in DNA synthesis and hence can be included in DNA, e.g. 5-bromouracil (replacing thymine) or 2-aminopurine (replacing adenine). May be used for inducing mutations, including point mutations.

base-pairing The specific hydrogen-bonding between **purines** and **pyrimidines** in double-stranded nucleic acids. In DNA the pairs are **adenine** and **thymine**, and **guanine** and **cytosine**, while in RNA they are adenine and **uracil**, and guanine and cytosine. Base-pairing leads to the formation of a DNA double helix from two complementary single strands.

Basedow's disease Thyrotoxicosis. See **Graves' disease**.

basement membrane Extracellular matrix characteristically found under epithelial cells. There are two distinct layers: the basal lamina, immediately adjacent to the cells, is a product of the epithelial cells themselves and contains collagen type IV; the **reticular lamina** is produced by fibroblasts of the underlying **connective tissue** and contains fibrillar collagen.

baseplate A hypothetical cell adhesion molecule possibly involved in sponge cell adhesion, existence unproven.

basic leucine zipper *bZIP* Family of proteins having a basic region and a **leucine zipper**. The basic region is the DNA-binding domain and the leucine zipper is involved in protein-protein interactions to form homo- or heterodimers. Includes **AP-1**, **ATF** and **CREB** transcription factors.

basidiocarp *Basidioma* The fruiting body of basidiomycete fungi. (**Basidiomycotina**).

Basidiomycetes See **Basidiomycotina**.

Basidiomycotina *Basidiomycetes* Subdivision or class of Eumycota (true fungi) in which the sexual spores (basidiospores) are formed on a basidium. Includes the Tiliomycetes comprising the rusts (Uredinales) and smuts (Ustilaginales), the Hymenomycetes and the Gasteromycetes.

basidiospore Spores of basidiomycete fungi. These spores are usually uninucleate and haploid.

basidium Club-shaped organ involved in sexual reproduction in basidiomycete fungi (mushrooms, toadstools etc.). Bears four haploid basidiospores at its tip.

basilar membrane A thin layer of tissue covered with mesothelial cells that separates the cochlea from the scala tympani in the ear.

basket cells Cerebellar neurons with many small dendritic branches that enclose the cell bodies of adjacent **Purkinje cells** in a basket-like array.

basolateral plasma membrane The plasma membrane of epithelial cells that is adjacent to the **basal lamina** or to the adjoining cells of the sheet. Differs both in protein and phospholipid composition from the **apical plasma membrane,** from which it is isolated by **tight junctions**.

basophil Mammalian **granulocyte** with large heterochromatic basophilic granules that contain **histamine** bound to a protein and a heparin-like mucopolysaccharide matrix. They are not phagocytic. Very similar to **mast cells,** though it is not clear whether they have common lineage.

basophilia **1.** Having an affinity for basic dyes. **2.** Condition in which there is an excess of **basophils** in the blood.

Bateman function Equation used in toxicology that expresses the build-up and decay in concentration of a substance (usually in plasma) based on first-order uptake and elimination in a one-compartment model.

Batesian mimicry A form of defensive colouration in which an animal is protected from predators by its resemblance to another which is dangerous or unpalatable.

batrachotoxin *BTX* **Neurotoxin** from the Columbian poison frog *Phyllobates*. A steroidal alkaloid that affects sodium channels; batrachotoxin R is more effective than related batrachotoxin A.

batroxostatin See **disintegrin**.

Batten's disease *Juvenile neuronal ceroid lipofuscinosis (JNCL); Spielmeyer–Vogt–Sjogren disease* Severe autosomal recessive disorder which causes blindness, deafness, loss of muscle control and early death. A storage disease in which lipopigment accumulates in neurons and tissues because of mutations in the *CLN3* gene (chromosome 16).

battenin See *CLN3*.

bax Protein related to **bcl-2**; Homodimers seem to promote apotosis in cultured cells, but heterodimers with Bcl-2 or Bcl-X$_L$ block cell death. Mice deficient in bax have selective hyperplasias. Bax seems to act as a tumour suppressor and is induced by **p53**, though is not solely responsible for p53-mediated apoptosis. Gene potentially encodes three proteins: bax-α, 21 kDa; bax-β, 24 kDa and bax-γ, 5 kDa.

Bayer's patches *Bayer's junctions* Sites of adhesion between the outer and cytoplasmic membranes of Gram-negative bacteria.

Bayesian statistics Statistical theory, based on Bayes' decision rules, that outlines a framework for producing decisions based on relative payoffs of different outcomes. Used in genetic counselling.

b-c1 complex A part of the **mitochondrial electron transport chain** that accepts electrons from **ubiquinone**, and passes them on to **cytochrome c**. The b/c1 complex consists of 2 cytochromes.

BCA **1.** BCA-1 is a CXC chemokine (CXCL13) that regulates B cell migration in lymphoid tissues and binds to the chemokine receptor CXCR5. Also known as B-lymphocyte chemoattractant (BLC). **2.** Benzethonium chloride assay (BCA), used for protein estimation. **3.** Blackcurrant anthocyanin (BCA). **4.** Beta-cyanoalanine (BCA).

BCECF *2',7'-Bis(2-carboxyethyl)-5(6)-carboxyfluorescein* Fluorescent dye used to monitor intracellular pH.

B-cell receptor A membrane-bound form of immunoglobulin that binds antigen. Each B-cell (B-lymphocyte) expresses one immunoglobulin and can recognize one antigen; the population of B-cells provides diversity. Following binding of antigen the B-cell receptor complex, which includes Ig-alpha (CD79a) and Ig-beta (CD79b) chains noncovalently associated with it, signals through a kinase cascade (**lyn**, **btk**, etc.) and stimulates proliferation and the production of **plasma cells** and memory B-cells.

B cells **1.** Cells within discrete endocrine islands (**Islets of Langerhans**) embedded in the major exocrine tissue of vertebrate pancreas. The B or β-cells (originally distinguished by differential staining from A, C and D), are responsible for synthesis and secretion into the blood of the hormone insulin. **2.** Casual term for **B-lymphocytes**.

BCG See **Bacille Calmette-Guerin**.

B chromosome Small **acentric** chromosome; part of the normal genome of some races and species of plants.

BCIP *5-Bromo-4-chloro-3-indolyl phosphate* A substrate for **alkaline phosphatase** that forms a precipitate in the presence of nitroblue tetrazolium (NBT). Used in various colorimetric assays.

bcl *Bcl-1, -2, -3* are oncogenes rearranged by translocations to the immunoglobulin genes in human B-cell malignancies (B-cell lymphomas, hence 'bcl'). *Bcl-2* encodes a plasma membrane protein that inhibits programmed cell death (**apoptosis**) and is homologous with the worm gene *ced-9*; see *ced* **mutant**. The protein product of *bcl-3* contains seven 'ankyrin-repeats' very similar to those found in **IkB,** and interferes with binding of 50-kDa subunit of NFkB to DNA.

Bcl-2 homology domain 3 *BH3* Proteins in the BH3 domain-only family (Bim, Bmf, Bik, **Bad**, Bid, Puma, Noxa and Hrk) are pro-apoptotic through the activation of Bax-like relatives.

bcr *Breakpoint cluster region* Region on chromosome 22 involved in the **Philadelphia chromosome** translocation.

B-DNA The structural form of **DNA** originally described by Crick and Watson. It is the form normally found in hydrated DNA, and is strictly an average, approximate structure for a family of B forms. In B-DNA, the double helix is a right-handed helix with about 10 residues per turn and has a major and a minor groove. The planes of the base pairs are perpendicular to the helix axis.

BDNF **Brain-derived neurotrophic factor.**

beaded filaments *Beaded-chain filaments* Intermediate filaments found in the lens fibre cells of the eye: composed of **filensin** and **phakinin**.

Becker muscular dystrophy Benign X-linked muscular dystrophy with later onset and lower severity than **Duchenne muscular dystrophy**.

Beckwith–Wiedemann syndrome Rare developmental disorder with a complex pattern of inheritance suggesting a defect in maternal **genomic imprinting**. Characteristics are all growth abnormalities – enlarged tongue, gigantism, enlarged adrenal glands, enlarged visceral organs, advanced ageing and predisposition to childhood tumours. Possibly due to a defect in the **cyclin-dependent kinase inhibitor, p57** KIP2, though in some cases there are two copies of the IGF-2 gene (see **insulin-like growth factor**).

Becquerel (Bq) The Systeme Internationale (SI, MKS) unit of radioactivity, named after the discoverer of radioactivity, and equal to 1 disintegration per second. Supersedes the Curie (Ci). 1 Ci = 37 GBq.

Bee1p *Bee1p/Las17p* A member of the Wiskott–Aldrich syndrome protein (**WASP**) family.

Beggiatoales An order of chemosynthetic sulphur-oxidizing gliding bacteria that occur mostly as filaments. Sulphur granules occur intracellularly.

beige mouse A mouse strain typified by beige hair and **lymphadenopathy**, reticulum cell neoplasms and giant lysosomal granules in **leucocytes**. May be the murine equivalent of **Chediak–Higashi syndrome** of man.

Bell's palsy Sudden paralysis of the muscles of one side of the face, due to impaired conduction in the lower part of the facial nerve. The cause is unknown, although there is speculation that herpes virus infection may be involved; the majority of cases recover spontaneously.

Belousov–Zhabotinsky reaction An example of chemical oscillations in **dissipative structures**, giving rise to characteristic sustained spatial patterns (concentric rings or spirals) in concentrations of reactants. The reaction is set up by mixing 0.2 M malonic acid, 0.3 M sodium bromate, 0.3 M sulphuric acid and 0.005 M ferroin (1,10-phenanthroline ferrous sulphate), and putting a thin layer (0.5–1.0 mm deep) in a Petri dish. Patterns emerge.

belt desmosome Another name for the zonula adherens or **adherens junction**.

benazepril Drug that is metabolized to active benazeprilat, a non-sulphydryl **ACE inhibitor** for treating hypertension.

Bence–Jones protein Dimers of **immunoglobulin** light chains, normally produced by **myelomas**. Bence–Jones proteins are sufficiently small to be excreted by the kidney.

benchmark concentration Statistical lower confidence limit on the concentration of a substance that produces a defined response (called the benchmark response or BMR, usually 5 or 10%) for an adverse effect compared to background, defined as 0%.

bendrofluazide *Bendroflumethiazide* **Thiazide diuretic** drug used in the treatment of oedema and hypertension.

Benedict's solution Solution used in **Benedict's test** for glucose; the qualitative solution contains 0.07 M copper sulphate, 0.67 M sodium citrate and 0.94 M sodium carbonate. The quantitative solution differs slightly, having 0.072 M copper sulphate, 0.7 M sodium carbonate, 1.29 M potassium thiocyanate and a very small amount of potassium ferrocyanide.

Benedict's test Test for glucose and other reducing disaccharides, involving the oxidation of the sugar by an alkaline copper sulphate solution (**Benedict's solution**), in the presence of sodium citrate, to give a deep red copper (I) oxide precipitate. Used in urine testing for diabetes.

benign tumour A clone of **neoplastic** cells that does not invade locally or **metastasize**, having lost **growth control** but not positional control. Usually surrounded by a fibrous capsule of compressed tissue.

benomyl A **benzimidazole** class fungicide and pesticide used on growing fruit and vegetables, although there are concerns about the health risks.

benzamidine A potent inhibitor of **serine endopeptidases** such as thrombin and trypsin

benzimidazole *N,N′-Methenyl-o-Phenylenediamine* Benzimidazole and its derivatives are used in organic synthesis and vermicides or fungicides. An example of benzimidazole class fungicides is benomyl.

benzodiazepines Class of drugs that are anxiolytics or hypnotics, widely used in medical practice as CNS depressants. Enhance the inhibitory action of **GABA** by modulating **GABA receptors**. Diazepam (Valium) is commonly used for relieving anxiety, and nitrazepam (Mogadon) for inducing hypnosis. Chlordiazepoxide (Librium) is also of this class. Are much more addictive than originally believed.

benzonatate A non-narcotic oral **antitussive** that apparently works by anaesthetizing stretch receptors in the airways.

benzopyrene Polycyclic aromatic compound. Potent mutagen and carcinogen.

benzoquinone *Cyclohexa-2,5-diene-1,4-dione* Compound (*p*-benzoquinone) used in the production of **cortisone**, as a tanning agent for leather and also used in photographic chemicals. Can be highly toxic if swallowed, inhaled or absorbed through the skin.

benzotropine mesylate *Benzotropine mesilate* Antimuscarinic drug used in treatment of **Parkinsonism**.

benzyl penicillin The first of the penicillins, but still extensively used to treat streptococcal, gonococcal and meningococcal infections. Is inactivated by bacterial penicillinases.

beri-beri A vitamin deficiency disease (deficiency of vitamin B1, thiamine), causing peripheral nerve lesions and/or heart failure.

Berk–Sharp technique *S1 mapping* A technique of genetic mapping in which **mRNA** is hybridized with **single-stranded DNA** and the non-hybridized DNA then digested with **S1 nuclease**; the residual DNA that hybridized with the messenger is then characterized by **restriction mapping**.

Bernard–Soulier syndrome Genetic deficiency in platelet membrane glycoprotein Ib (CD42); platelets aggregate normally (*cf.* **Glanzmann's thrombasthenia**) but do not stick to collagen of sub-endothelial basement membrane.

Berovin Calcium-activated photoprotein from the coelenterate *Beroe*. See **aequorin**.

beryllicosis *Berylliosis* Chronic beryllium poisoning. Causes serious and usually permanent lung damage.

Best's carmine Histochemical stain that can be used to demonstrate the presence of glycogen, which stains deep red.

beta Entry prefix is generally, but not universally, ignored for alphabetical reference; look for main portion of word.

beta-2-microglobulin *β-2-microglobulin* Immunoglobulin-like polypeptide (12 kDa, homologous with the constant region of Ig) that is found on the surfaces of most cells, associated non-covalently with Class I **histocompatibility antigens**.

beta-actinin *β-Actinin* See **capZ**.

beta-adrenoceptor blocking drugs *Beta-blockers* Group of drugs which block β-adrenoreceptors. Used in the treatment of angina, hypertension, migraine, thyrotoxicosis and anxiety states. Some are relatively unselective

(e.g. propranolol), others act primarily on β1-receptors (e.g. atenolol).

beta agonist Sympathomimetic drug that acts on β2-adrenoreceptors and has a rapid bronchodilatory effect (acting on bronchial smooth muscle) if taken by inhalation. Used especially for asthma; salmeterol is an example.

beta-alpha-beta motif *βαβ Motif* Protein **motif** comprising a beta strand–loop–helix–loop–strand arrangement, with the strands lying parallel.

beta-amylase *β-Amylase; EC 3.2.1.2* A terminal amylase that cleaves maltose units from starch, glycogen and related polysaccharides from the non-reducing ends of the chains.

beta amyloid A fragment of **amyloid precursor protein** that is produced by the action of secretases (see **presenilins**). Aggregates of beta-amyloid accumulate as plaques in the brain in **Alzheimer's disease**.

beta arch *β Arch* Protein **motif** comprising two adjacent antiparallel beta strands joined by a coil and that are part of different sheets, usually forming a **beta sandwich**.

beta-barrel *β Barrel* Protein motif in which a series of (typically **amphipathic**) beta sheets is arranged around a central pore. Example: **voltage-gated ion channel**.

beta-blockers See **beta-adrenoceptor blocking drugs**.

beta bulge *β Bulge* Protein **motif** comprising a disruption of a **beta sheet**, usually by the insertion of a single residue.

beta-CAP73 An actin-binding protein that binds only the β-isoform of actin. Has six predicted **ankyrin**-like repeats at the N-terminus.

beta carotene See **carotenes**.

beta-cells (pancreas) *β-Cells* See **B cells** of pancreas.

beta-COP *β-COP* Major component (110 kDa) of coat of non-clathrin coated vesicles derived from Golgi. Has homology with beta-**adaptin**.

betacyanin Red pigments of the **betalain** type, for example the red pigment found in beetroot.

beta-defensins Small cationic peptides with antibacterial activity. See **defensins**.

beta-emitter *β-Emitter* A radionuclide whose decay is accompanied by the emission of β particles, most commonly negatively charged electrons. Many isotopes used in biology, such as ^{3}H, ^{14}C, ^{35}S, and ^{32}P are pure β-emitters.

beta-galactosidase *β-Galactosidase; EC 3.2.1.23* Enzyme encoded by the LacZ gene and widely used as a **reporter gene**, as a variety of coloured or fluorescent compounds can be produced from appropriate substrates (typically **X-Gal**, which produces a blue colour). LacZ is incorporated in many plasmid **vectors** to allow **blue-white colour selection**.

beta-glucosidase *β-Glucosidase; EC 3.2.1.21* Enzyme catalysing the release of glucose by hydrolysis of the glycosidic link in various β-D-glucosides (compounds of the form R-β-D-glucose, where the group R may be alkyl, aryl, mono- or oligosaccharide). Favoured source: almonds, from which enzyme is known as emulsin. Defects are associated with **Gaucher's disease**.

beta-glucuronidase *β-glucuronidase;* *EC* *3.2.1.3* Enzyme that catalyses hydrolysis of a β-D-glucuronoside to D-glucuronate and the compound to which it was attached. Often used as a marker enzyme for lysosomes.

beta hairpin *β hairpin* Protein **motif** describing one possible arrangement of strands in a **beta sheet**. Strands are antiparallel and hydrogen-bonded, lying adjacent in the sheet.

beta helix *β helix; solenoid* Protein **motif** comprising a large right-handed coil (or superhelix), containing either two or three **beta sheets**.

betaine A derivative of glycine characterized by high water solubility. Can function as an osmotic agent in plant tissues. See **biogenic amines.**

beta interferon *Interferon beta-1b* Recombinant version of the endogenous biological compound, produced in *E. coli* with an inserted human fibroblast-derived gene. The recombinant form has 165 aa and is not glycosylated. Used to treat relapsing **multiple sclerosis,** although not all patients respond; proprietary name Betaseron.

beta-lactam antibiotics A large group of bactericidal antibiotics that act by inhibiting bacterial cell wall synthesis and activating enzymes that destroy the cell wall. Examples are **penicillin, ampicillin, amoxicillin.**

beta-lactamase *EC 3.5.2.6; penicillin amido-beta-lactam hydrolase* Enzyme produced by some bacteria that makes them resistant to **beta-lactam antibiotics.** Competitively inhibited by **clavulanic acid.**

betalain Nitrogen-containing red or yellow pigments functionally replacing **anthocyanins** in flowers and fruits of many Caryophyllales (ice plants, cacti, carnations). Also found in some higher fungi, where their role is obscure. Are used for food colouring, and have antioxidant and radical scavenging properties that provide protection in certain oxidative stress-related disorders

betanidin *Betanin; beetroot red* Natural pigment (**betalain**) from beetroot.

beta-oxidation *β-oxidation* The process whereby fatty acids are degraded in steps, losing two carbons as (acetyl)-CoA. Involves CoA ester formation, desaturation, hydroxylation and oxidation before each cleavage. See **omega-oxidation.**

beta-pleated sheet *β-pleated sheet* Beta secondary structure in proteins consists of two almost fully extended polypeptide chains lying side by side, linked by interchain hydrogen bonds between peptide C=O and N–H groups. When multiple chains are involved an extended sheet, the β-pleated sheet, is formed, which can consist of parallel or antiparallel sheets (where the chains run in the same or opposite directions) or mixed sheets.

beta prism *β prism* Protein **motif** comprising three antiparallel **beta sheets** arranged in a triangular, prism shape. In the orthogonal prism, strands are orthogonal to the prism access; in the aligned prism, the strands and prism axis are parallel.

beta propeller *β propeller* Protein **motif** comprising four to eight antiparallel **beta sheets** arranged like the blades of a propeller.

beta sandwich *β sandwich* Protein **motif** comprising two **beta sheets** that pack together face to face, in a layered arrangement.

beta-secretase *BACE-1, Beta-site APP cleaving enzyme 1* The major beta-secretase *in vivo,* an aspartic peptidase that generates the N-terminus of the beta-amyloid protein from amyloid precursor protein (**APP**); further cleavage is then carried out by **gamma-secretase.**

beta sheet *β sheet* See **beta-pleated sheet.**

beta strand *β strand* Region of polypeptide chain that forms part of a **beta sheet.**

beta-thymosin repeat *Beta-thymosin/WH2 domain* **Beta-thymosin** is effectively an isolated WH2 domain (**WASP-homology domain-2**), and beta-thymosin repeats are actin monomer-binding motif found in many proteins that regulate the actin cytoskeleton. See **thymosin-β4.**

beta trefoil *β trefoil* Protein **motif** consisting of three **beta hairpins** forming a triangular shape.

beta turn *β turn* Protein **motif** which consists of an abrupt 180° reversal in direction of a polypeptide chain. The turn is defined as being complete within four residues.

bevacizumab *Avastin* Monoclonal antibody directed against **VEGF,** used in treatment of colorectal carcinoma.

BFA See **brefeldin A.**

bFGF Basic **fibroblast growth factor.**

BFU-E See **burst-forming unit**-erythrocytic.

BGH 1. Bovine **growth hormone,** (bGH). **2. Brunner's gland** hyperplasia, **3.** *Blumeria graminis* f.sp. *hordei* (Bgh), a powdery mildew fungus.

BH3 1. See **Bcl-2 homology domain 3**. **2.** The borano (BH$_3$-) group.

BHK cells *Baby hamster kidney cells* A quasi-diploid established line of Syrian hamster cells, descended from a clone (Clone 13) isolated by Stoker and McPherson from an unusually rapidly growing primary culture of newborn hamster kidney tissue. Usually described as fibroblastic, although smooth muscle-like in that they express the muscle intermediate filament protein **desmin.** Widely used as a viral host, in studies of oncogenic transformation and of cell physiology.

bhlh *bHLH* Basic **helix–loop–helix**; a class of transcription factors. See Table T2.

BHT *2,6-Di-tert-butyl-4-methylphenol* Butylated hydroxytoluene, an antioxidant used in food, cosmetics, and pharmaceuticals.

Biacore Proprietary name for an instrument that uses **surface plasmon resonance** to detect the binding of a substance to the surface of a flow chamber. Using this instrument it is possible to measure the on- and off-rates for the binding of a molecule to a defined surface, e.g. the binding of an antibody to the antigen-coated surface of the flow cell or of ligand to an immobilized receptor.

Bial's orcinol test See **orcinol.**

bialaphos *Phosphinothricylalanyalanine; phosphinothricin tripeptide* A tripeptide antibiotic produced by *Streptomyces hygroscopicus,* consists of two molecules of L-alanine and one molecule of the unusual amino acid phosphinothricin (PT). Used as a selective agent in transformation experiments in plant genetic engineering and as a herbicide.

bicephalous Having two heads.

bicoid An **egg-polarity gene** in *Drosophila*, concentrated at the anterior pole of the egg, and required for subsequent anterior structures. A **maternal-effect gene**.

bicuculline From *Dicentra cucullaria* and herbs of the genus *Corydalis*. Specific blocking agent for the action of the amino acid transmitter γ-aminobutyric acid (**GABA**). See **amino acid transmitter**; **amino acid receptor** superfamily, **GABA receptor**.

Bid A pro-apoptotic Bcl-2 family member (see **Bcl-2 homology domain 3**). A substrate for the catalytic casein kinase 2 alpha subunit. DNA damage leads to **ATM**-mediated Bid phosphorylation, and this phosphorylation regulates a novel, pro-survival function of Bid important for S-phase arrest.

bifidobacteria Genus of bacteria found as a normal part of the microflora of the lower intestine and thought to assist digestive processes. Are used as a probiotic for intestinal well-being.

big brain A **neurogenic gene** of *Drosophila*, believed to encode a product involved in cell–cell communication, perhaps via **gap junctions**. Member of the **major intrinsic protein** family.

biglycan A small proteoglycan, 150–240 kDa, of the extracellular matrix. The core protein has a mass of around 42 kDa and is very similar to the core protein of **decorin** and **fibromodulin**. All three have highly conserved sequences containing 10 internal homologous repeats of around 25 amino acids with leucine-rich motifs. Biglycan has two glycosaminoglycan chains, either chondroitin sulphate or dermatan sulphate and N-linked oligosaccharides.

biguanides Drugs used in treating maturity-onset diabetes. Appear to act by increasing peripheral utilization of glucose and are of particular value in obese diabetics, e.g. metformin.

Bik *Bcl-2-interacting killer; Nbk/Bik* See **Bcl-2 homology domain 3**. BH3-only pro-apoptotic protein of bcl-2 family, targets the membrane of the endoplasmic reticulum. It is induced in human cells in response to several stress stimuli, including genotoxic stress (radiation, doxorubicin) and overexpression of **E1A** or **p53**, but not by ER stress pathways resulting from protein malfolding. Endogenous cellular BIK regulates a **BAX**, **BAK**-dependent ER pathway that contributes to mitochondrial apoptosis. Loss of Nbk/Bik is a common feature of clear-cell renal cell carcinoma.

bile pigments Pigments produced by the breakdown of haemoglobin. The main pigments are bilirubin (reddish-yellow) and its oxidation product biliverdin (green).

bile salts Sodium salts of the bile acids, a group of hydroxy steroid acids condensed with taurine or glycine, the commonest of which are the salts of taurocholic and glycocholic acids. They are powerful surfactants and are important in aiding absorption of fats from the intestine.

bilharzia *Schistosomiasis* Disease caused by the blood fluke *Schistosoma* spp., a digenean trematode. See **schistosomiasis**.

bilicyanin A blue oxidation product of **bilirubin**.

biliproteins See **phycobilins**.

bilirubin Red-brown pigment found in bile, formed by breakdown of haemoglobin.

biliverdin Green bile pigment formed by haemoglobin breakdown; can be converted into **bilirubin** by reduction.

Bim *Bcl-2 interacting mediator of death* A pro-apoptotic **Bcl-2 homology domain 3** (BH3)-only protein. Critical for eliminating most effector T-cells following an acute T-cell response; the few that survive become memory cells.

binary fission Division of a cell into two daughter cells; nuclear division precedes division of the cell body.

bindin Molecule of around 30 kDa normally sequestered in the **acrosome** of a sea urchin spermatozoon, and which through its specific **lectin**-like binding to the **vitelline membrane** of the egg confers species-specificity in fertilization.

binucleate A cell that has two nuclei.

bioaccumulation Accumulation of substances in living organisms because the rate of intake exceeds the capacity to excrete or metabolize the substance. Organisms at the top of a food chain can accumulate considerable amounts of some substances, the most notorious of which was DDT.

bioactivation Metabolic conversion of a **xenobiotic** substance to a more toxic or active derivative.

bioassay An assay for the activity or potency of a substance that involves testing its activity on living material.

bioautography The use of cells to detect, by their attachment or other reaction, the presence of a particular substance, e.g. an adhesion protein on an electrophoretic gel.

bioavailability Relative amount of a drug (or other substance) that will reach the systemic circulation when administered by a route other than direct intravenous injection.

bioblasts When Altmann first observed mitochondria he considered them to be intracellular parasites and christened them bioblasts. Not current usage.

biochip 1. A silicon chip implanted into and functioning as part of a human body. 2. An array of proteins, oligonucleotides or other molecules immobilized on a solid substratum (often a microscope slide) that can be probed with labelled reagents, mixtures or substances, etc. to identify interactions or, in the case of oligonucleotides, sequence similarity through hybridization. Increasing miniaturization allows arrays to contain thousands of individual sites. See **gene chip**.

biocidal Capable of killing living organisms.

biocompatible Capable of remaining in contact with cells or tissues without causing adverse effects. This can be simply the absence of toxicity, in the case of a culture vessel, for example, or more complex, as with materials that can be implanted into the body without exciting an inflammatory or thrombogenic response.

biodiversity The various genetic, taxonomic and ecosystem differences in the living organisms of a given area, environment, ecosystem, or indeed the whole planet.

bioengineering Rather imprecise category of activities that can range from bioreactor design, through prosthetic devices to environmental restoration. Any interface between biology and engineering that will attract students or funds.

bioethanol Ethanol produced from fermented plant material, used as a petrol additive or substitute. Chemically identical to ordinary ethanol, much of which is a by-product of petroleum refining.

biofilm A layer of bacteria, enclosed in a mucilaginous slime, attached to surfaces exposed to water or biological fluids. Multiple bacterial species may be present, as well as fungi, algae, protozoa, debris and corrosion products, and the behaviour of the bacteria in such films may be distinctly different from that exhibited in suspension culture. Plaque on teeth is a common example. See **quorum sensing**.

bioflavonoids Group of coloured phenolic pigments originally considered to be vitamins (Vitamins P, C2) but not shown to have any nutritional role. Responsible for the red/purple colours of many higher plants.

biogenic amines Amines found in both animals and plants that are frequently involved in signalling. There are several groups: ethanolamine derivatives include **choline**, **acetylcholine** and **muscarine**; catecholamines include **adrenaline**, **noradrenaline** and **dopamine**; polyamines include **spermine**; indolylalkylamines include tryptamine and **serotonin**; betaines include **carnitine**; polymethyline diamines include **cadaverine** and **putrescine**.

bioinformatics The discipline of using computers to collate and form datasets of interest to biologists. Usually used to refer to databases of DNA and protein sequences, and of mutations, disease and gene functions, in the context of genome projects.

biological oxygen demand *BOD* The oxygen required to satisfy the biological demands of contaminated water. The biological demand is from microbial flora involved in digesting organic constitutents.

bioluminescence Light produced by a living organism. The best-known system is firefly luciferase (an ATPase), which is used routinely as a sensitive ATP assay system. Many other organisms, particularly deep-sea organisms, produce light, and even leucocytes emit a small amount of light when their oxidative metabolism is stimulated. Does not really differ from **chemiluminescence**, except that the light-emitting molecule occurs naturally and is not a synthetic compound like **luminol** or lucigenin.

biomarker **1.** A biologically derived substance, the presence of which in serum may be an indication of disease. **2.** Any biological feature that is indicative of the status of the system whether that is an individual organism or an ecosystem.

biomaterials **1.** Solid materials which are produced by living organisms, such as **chitin**, **fibroin** or bone. **2.** Any materials which replace the function of living tissues or organs in humans.

biometry Statistical methods applied to biological problems. *Adj.* Biometric.

biomimetic Processes, substances, devices, or systems that imitate those found in biology.

biomimetics An area of bioengineering in which new technology is based upon mechanisms, features, methods and accomplishments found in biological systems.

biopiracy A politically charged term for the development by industrialized nations of materials native to developing countries, e.g. medicinal plants, without adequate compensation to their country of origin.

bioprospecting Investigating living organisms with the aim of discovering materials that can be exploited for commercial gain, often without recompense for the local inhabitants (**biopiracy**).

biopterin *2-Amino-4-hydroxyl-6-(1,2-dihydroxypropyl)-pteridine* Growth factor for some protozoa; present in many tissues as the reduced form, tetrahydrobiopterin, where it acts as a coenzyme for hydroxylases. Tetrahydrobiopterin is an essential cofactor for NO production from L-arginine. Defects in biopterin synthesis can lead to hyperphenylalaninemia.

bioreactors Reaction vessel for producing a biological product by fermentation or cell culture, increasingly involving modified microorganisms or cells that produce particular substances.

bioremediation Remediation of a contaminated environment through the use of biological agents, often genetically engineered or selected for the particular task – for example, the breakdown of organic molecules or the accumulation and sequestration of a toxic material. Has proved more difficult than had perhaps been hoped.

biosecurity Methods adopted to prevent harmful effects brought about by other species. May involve the use of deliberately crippled strains of virus or microorganism that are unlikely to survive outside laboratory conditions and/or containment at different levels of stringency.

biosynthesis Synthesis by a living system (as opposed to chemical synthesis).

bioterrorism The use of disease-carrying organisms or agricultural pests as a weapon in terrorism. Fears about this threat have led some governments to impose increasingly stringent regulations on the experimental use of some organisms.

biotin *Vitamin H* A prosthetic group for carboxylase enzymes. Important in fatty acid biosynthesis and catabolism, and has found widespread use as a covalent label for macromolecules which may then be detected by high-affinity binding of labelled **avidin** or **streptavidin**. Essential **growth factor** for many cells.

biotinyl Acyl group derived from **biotin**. Addition of this moiety to another molecule, for example a protein, will allow affinity purification or labelling with **avidin**.

biotinylated Having had a **biotinyl** group added.

biotope A small habitat in a large community, e.g. a cow pat on a meadow, whose several short seral stages comprise a microsere.

bioweapon Any weapon based upon a biological agent, either a biologically derived toxin (e.g. **botulinum toxin**) or an infectious agent (bacteria or viruses).

BiP Molecular chaperone (78 kDa) found in endoplasmic reticulum and related to **hsp70** family of heat-shock proteins. Originally described as immunoglobulin heavy-chain-binding protein.

bipolar cells A class of retinal **interneurons**, named after their morphology, that receive input from the photoreceptors and send it to the **ganglion cells**. Bipolar cells are **non-spiking neurons**; their response to light is evenly graded, and shows **lateral inhibition**.

bipolar filaments Filaments that have opposite polarity at the two ends; the classic example is the **thick filament** of striated muscle.

Birbeck granules Characteristic inclusion bodies seen by electron microscopy in **histiocytes** (Langerhans cells) of patients with histiocytosis X, a group of diseases with uncertain pathogenesis.

birefringence Optical property of a material in which the refractive index is different for light polarized in one plane compared to the orthogonal plane. See **birefringent**.

birefringent Any material that has different refractive index according to the plane of polarization of the light. The effect is to rotate the plane of the refracted light so that, using crossed Nicholl prisms (polarizers set at right angles to give complete extinction), the birefringent material appears bright. The birefringence can arise through anisotropy of structure (form birefringence) or through orientation of molecules either because of mechanical stretching (stress birefringence) or because of alignment in flow (flow birefringence). A classic example, often used to demonstrate the effect, is a hair, which, because of the orientation of the keratin, shows form birefringence.

bisoprolol fumarate Cardioselective **beta blocker** (β1-adrenoreceptor selective), used for treatment of hypertension and angina.

bisphosphoglycerate mutase *EC 5.4.2.4* Enzyme responsible for production of **BPG** in the erythrocyte.

bisphosphoglycerate phosphatase *EC 3.1.3.13* Enzyme that catalyses breakdown of **BPG** in the erythrocyte.

bisphosphonates Family of drugs used to prevent and treat **osteoporosis**.

bistatin A **disintegrin** found in the venom of the puff adder, *Bitis arietans*.

Biston betularia Peppered moth; famous for the shift to the melanized form as industrial pollution turned trees black and gave the melanotic form a selective advantage – and for reversion to the lighter form following the Clean Air Act in Britain.

bithorax complex A group of **homeotic** mutations of *Drosophila* that map to the bithorax region on chromosome III. The mutations all cause the third thoracic segment to develop like the second thoracic segment to varying extents. The genes of the bithorax complex are thought to determine the differentiation of the posterior thoracic segments and the abdominal segments.

Bittner agent Earlier name, now superseded, for the mouse **mammary tumour virus**.

biuret reaction Formation of a purple colour when biuret (carbamoyl urea) or any compound with two or more peptide bonds (i.e. proteins) reacts with copper sulphate in alkaline solution. Used as a colorimetric test.

bivalent Term used of two homologous chromosomes when they are in synapsis during **meiosis**.

***bla* gene** Bacterial gene coding for beta-lactamase, important for antibiotic resistance.

black disease Toxaemic disease of sheep caused by *Clostridium novyi* (*C. oedematiens*). Infection is associated with the presence of liver damage caused by immature liver flukes.

black fever See **Rocky mountain fever**.

black membrane An artificial (phospho) lipid membrane formed by 'painting' a solution of phospholipid in organic solvent over a hole in a hydrophobic support immersed in water. Drainage of the solvent from the film produces diffraction colours until the thickness falls below the wavelength of light; it then appears to be black. The structure is an extended bimolecular leaflet.

Black Widow spider venom Potent **neurotoxin** that induces catastrophic release of **acetylcholine** from **presynaptic** terminals of cholinergic **chemical synapses**.

blackhead *Infectious entero-hepatitis; histomoniasis* An infectious disease caused by infection by the protozoon, *Histomonas meleagridis*. Mainly affects turkeys.

blackleg An acute infectious disease of cattle and sheep caused by *Clostridium chauvoei*.

blackwater fever *Haemoglobinuric fever* An acute disease of tropical regions characterized by intravascular haemolysis, haemoglobinuria and acute renal failure; classically seen in European expatriates chronically exposed to *Plasmodium falciparum*. Symptoms include fever, vomiting and passage of red or dark-brown urine.

BLAST Abbrev. for Basic Local Alignment Search Tool, a commonly used web-based resource for identifying similarities in sequences of nucleotides or amino acids. See **BLAST search**.

blast cells Cells of a proliferative compartment in a cell lineage.

BLAST search The Basic Local Alignment Search Tool (BLAST) can be used to find regions of local similarity between sequences. The program compares nucleotide or protein sequences to sequence databases and calculates the statistical significance of matches. BLAST can be used to infer functional and evolutionary relationships between sequences as well as help identify members of gene families.

blast transformation The morphological and biochemical changes in B- and T-lymphocytes on exposure to **antigen** or to a **mitogen**. The cells appear to move from G0 to G1 stage of the cell cycle. They usually enlarge and proceed to S phase and mitosis later. The process probably involves receptor cross-linking on the plasma membrane.

blastema A group of cells in an organism that will develop into a new individual by asexual reproduction, or into an organized structure during regeneration.

blasticidin *Blasticidin S* A peptidyl nucleoside antibiotic isolated from the culture broth of *Streptomyces griseochromogenes*. It inhibits protein synthesis in both prokaryotes and eukaryotes by interfering with peptide bond formation in the ribosome. Frequently used to select transfected cells carrying *bsr* or *BSD* resistance genes.

blasticidin resistance genes Three **blasticidin** resistance genes have been cloned and sequenced: an acetyl transferase gene, *bls*, from a blasticidin producer strain, and two deaminase genes, *bsr* from *Bacillus cereus* and *BSD* from *Aspergillus terreus*. The latter two (*bsr* and *BSD*) are used as dominant selectable markers for transformation experiments in mammalian and plant cells.

blastocoel *Blastocele (US)* The cavity formed within the mass of cells of the **blastula** of many animals during the later stages of cleavage.

blastocyst In mammalian development, cleavage produces a thin-walled hollow sphere, whose wall is the **trophoblast**, with the embryo proper being represented by a mass of cells at one side. The blastocyst is formed before implantation and is equivalent to the **blastula.**

blastoderm In many eggs with a large amount of yolk, cell division (cleavage) is restricted to a superficial layer of the fertilized egg (meroblastic cleavage). This layer is termed the blastoderm. In birds it is a flat disc of cells at one pole of the egg, and in insects an outer layer of cells surrounding the yolk mass.

blastoma *Pl.* blastomas or blastomata Neoplasm composed of immature **blast cells**.

blastomere One of the cells produced as the result of cell division, cleavage, in the fertilized egg.

blastopore During **gastrulation** cells on the surface of the embryo move into the interior to form the **mesoderm** and **endoderm**. The opening formed by this invagination of cells is the blastopore. It is an opening from the **archenteron**, the primitive gut, to the exterior. In some animals this opening becomes the anus, whilst in others it closes up and the anus opens at the same spot or nearby. In some animals, e.g. chick, invagination occurs without a true blastopore, and the site at which the cells move in, the **primitive streak**, may be termed a virtual blastopore.

blastula Stage of embryonic development of animals near the end of cleavage but before **gastrulation**. In animals where cleavage (cell division) involves the whole egg, the blastula usually consists of a hollow ball of cells.

Blattella germanica German cockroach.

bleb Protrusion from the surface of a cell, usually approximately hemispherical; may be filled with fluid or supported by a meshwork of microfilaments.

bleomycin Any of a group of glycopeptide antibiotics from *Streptomyces verticillus.* Blocks cell division in G2: used to synchronize the division of cells in culture and as an antiproliferative agent in oncology.

Blepharisma Genus of ciliate protozoans of the order Heterotricha.

blepharoplast Alternative name for a **basal body**. An organelle derived from the **centriole** and giving rise to the **flagella**. Found chiefly in Protozoa and Algae.

Blk *B-lymphocyte kinase* A member the **Src family** tyrosine kinases involved in B-cell maturation; sustained activation of Blk induces responses normally associated with the pre-B-cell receptor activation (See **preBCR**).

BLOC-1 *Biogenesis of lysosome-related organelles complex 1* Complex that regulates trafficking to lysosome-related organelles and includes the proteins **pallidin**, **muted** and **cappuccino**. See **dystrobrevin-binding protein 1**.

blocking antibody An antibody used in a reaction to prevent some other reaction taking place, for example one antibody competing with another for a cell surface receptor.

blood group antigens The set of cell surface antigens found chiefly, but not solely, on blood cells. More than 15 different blood group systems are recognized in humans. There may be naturally occurring antibodies without immunization, especially in the case of the **ABO** system, and matching blood groups is important for safe transfusion. In most cases the antigenic determinant resides in the carbohydrate chains of membrane glycoproteins or glycolipids. See also **Rhesus**, **Duffy**, **Kell**, **Lewis** and **MN** blood groups.

blood smear A thin air-dried film of blood on a microscope slide in which the cells are thinly spread and that can be stained to examine the blood cells (and any blood-borne parasites).

blood vessels All the vessels lined with **endothelium** through which blood circulates.

blood–brain barrier The blood vessels of the brain (and the retina) are much more impermeable to large molecules (like antibodies) than are blood vessels elsewhere in the body. This has important implications for the ability of the organism to mount an immune response in these tissues, although the basis for the difference in endothelial permeability is not well understood. Also prevents some substances from entering the brain from the blood, which has importance in drug treatment and toxicology.

Bloom's syndrome Rare human autosomal recessive defect associated with genomic instability causing short stature, immunodeficiency and increased risk of all types of cancer. Caused by mutation of *BLM* locus on chromosome 15q in the gene encoding DNA **helicase** RecQ protein-like-3.

blotting General term for the transfer of protein, RNA or DNA molecules from a relatively thick acrylamide or agarose **gel** to a paper-like membrane (usually nylon or nitrocellulose) by capillarity or an electric field, preserving the spatial arrangement. Once on the membrane, the molecules are immobilized, typically by baking or by ultraviolet irradiation, and can then be detected at high sensitivity by **hybridization** (in the case of DNA and RNA), or antibody labelling (in the case of protein). RNA blots are called **Northern** blots; DNA blots, **Southern**; protein blots, **Western**. In Northwestern blotting, protein is transferred but is probed with specific RNA. See also **dot** and **slot blots**.

blue naevus A non-malignant accumulation of highly pigmented **melanocytes** deep in the **dermis**.

blue roses By some sophisticated genetic engineering Florigene managed to produce blue roses, a long-term ambition of plantsmen. This involved using **small interfering RNA** to switch off the endogenous **dihydroflavinol reductase gene** in a red rose, thereby blocking the **cyanidin** pathway, and installing a pansy **delphinidin** gene and a new DFR gene to allow complete delphinidin synthesis and thus give a blue colour.

blue-green algae Group of prokaryotes that should now be referred to as **Cyanobacteria**.

blue–white colour selection Method for identifying bacterial clones containing plasmids with inserts. Many modern **vectors** have their **polycloning site** within a part of the *LacZ* gene encoding **beta-galactosidase**, which provides **alpha complementation** in an appropriate mutant *E. coli* strain. This means that a re-ligated (empty) vector will produce blue colonies when grown on plates containing **IPTG** and **X-Gal**, but colonies with a substantial insert in their plasmid's polycloning site are unable to produce functional beta-galactosidase, and so produce white colonies.

Bluescript *pBluescript* Proprietary plasmid, sold by Stratagene. Very widely used.

bluetongue *Malarial catarrhal fever of sheep* A non-contagious viral disease caused by an Orbivirus (a genus of the Reoviridae family) that affects sheep, wild ruminants and, rarely, cattle, goats and carnivores. Transmitted by mosquitoes.

Bluetongue virus Reovirus that causes serious disease (**bluetongue**) of sheep and milder disease in cattle and pigs. Transmitted by biting flies.

blunt end End of double-stranded DNA that has been cut at the same site on both strands by a **restriction enzyme** that does not produce **sticky ends**.

Blym The human *Blym-1* transforming gene is activated in a Burkitt's lymphoma cell line and is the human homologue of chicken *Blym-1*, the transforming gene detected by transfection of chicken B-cell lymphoma DNA. See also Table O1.

B-lymphocyte *B-cells* See **lymphocyte**.

BM-40 See **osteonectin**.

BMAP-27, BMAP-28 Bovine **cathelicidins** of 26 and 27 amino acids, respectively.

B_{max} Amount of drug required to saturate a population of receptors and a measure of the number of receptors present in the sample. Usually derived from **Scatchard plot** of binding data. Analogous to V_{max} in enzyme kinetics.

BME **1.** Beta-mercaptoethanol (bME). **2.** Bone marrow edema (BME). **3.** Engineered Arabidopsis lines that exhibit **beta-glucuronidase** expression in the micropylar end of the seed, named Blue Micropylar End, BME lines.

Bmf See **Bcl-2 homology domain 3**. A key molecule for **HDAC** inhibitors (FK228 and CBHA)-mediated enhancing effect on ionizing radiation-induced cell death.

BMPs *Bone morphogenetic proteins* Multifunctional cytokines, members of the **TGFβ** superfamily. Activity of BMPs is regulated by BMP-binding proteins **noggin** and **chordin**. Receptors are serine–threonine kinase receptors (Types I and II) that link with **smad proteins**. *Drosophila* decapentaplegic (Dpp) is a homologue of mammalian BMPs. BMP2 is involved in regulating bone formation, BMP4 acts during development as a regulator of mesodermal induction and is overexpressed in fibrodysplasia ossificans. Follistatin inhibits BMP function in early *Xenopus* embryos.

BMR Basal metabolic rate, the number of calories required to maintain basic body functions at rest and in a thermally neutral zone.

BN-PAGE *Blue-Native Polyacrylamide Gel Electrophoresis* Modification of **polyacrylamide gel electrophoresis** (PAGE) in which the binding of Coomassie Blue (Kenacid Blue) to non-denatured (native) protein complexes confers negative charge which can be used to separate the proteins electrophoretically. A second dimension run using denaturing SDS can give further discrimination.

BNLF-1 Epstein–Barr virus (EBV) encoded latent membrane protein (LMP)-1. An **oncogene** from **Epstein–Barr virus**. See also Table O1.

BNP See **brain natriuretic peptide**.

BOAA *Beta-N-oxalylamino-L-alanine* Neurotoxic amino acid found in the chickling pea (*Lathyrus sativa*) and responsible for **lathyrism**.

BODIPY *Di-Pyrromethene boron difluoride; 4,4-difluoro-4-bora-3α, 4α-diaza-s-indacene* Group of fluorescent dyes that can be conjugated to a range of molecules for use as probes in fluorescence microscopy. Some derivatives show a large fluorescent enhancement upon increasing the acidity of the solution and thus can be used in aqueous solution as fluorescent pH probes.

Bodonids Free-living kinetoplastids with an anteriorly directed dorsal flagellum, a posteriorly directed ventral flagellum and a simple feeding apparatus.

Bohr effect Decrease in oxygen affinity of **haemoglobin** when pH decreases or concentration of carbon dioxide increases.

Bollinger bodies Intracytoplasmic inclusion bodies in epithelial cells infected with fowl pox virus. They are aggregates of smaller bodies (Borrel bodies) which are the actual masses of the virus collected in the cells.

bombesin Tetradecapeptide **neurohormone** with both **paracrine** and **autocrine** effects first isolated from skin of fire-bellied toad (*Bombina bombina*); mammalian equivalent is **gastrin-releasing peptide** (GRP). Bombesin cross-reacts with GRP receptors. Both are **mitogenic** for Swiss 3T3 fibroblasts at nanomolar levels. Neuropeptides of this type are found in many tissues and at high levels in pulmonary (small cell carcinoma) and thyroid tumours.

Bombyx mori Commercial silkmoth.

bone marrow Tissue found in the centre of most bones; site of **haematopoiesis**. The most radiation-sensitive tissue of the body.

bone morphogenetic protein See **BMPs**.

bongkrekic acid An inhibitory ligand of the mitochondrial adenine nucleotide translocator, also inhibits apoptosis by preventing **PARP** cleavage and **DEVD** ase activity.

bonobo *Pygmy chimpanzee* Endangered species of chimpanzee (*Pan paniscus*) found in the Democratic Republic of the Congo, probably more closely related to man than *Pan troglodytes*.

bootstrap analysis A statistical method for testing the reliability of a dataset. It involves creating a set of pseudoreplicate datasets by re-sampling and is extensively used, for example, in phylogenetic analyses.

Bordetella pertussis A small, aerobic, **Gram-negative** bacillus, causative organism of whooping cough. Produces a variety of toxins, including a dermonecrotizing toxin, an adenyl cyclase, an **endotoxin** and **pertussis toxin**, as well as surface components such as fimbrial haemagglutinin.

Borna disease Virally induced T-cell dependent immunopathological disorder of central nervous system. There are suggestions that Borna disease virus (a broadly distributed unclassified arthropod-borne virus that infects domestic animals and man) may be associated with some psychiatric disorders.

Borrel body See **Bollinger bodies**.

Borrelia burgdorferi Spirochaete, responsible for **Lyme disease**. Can be isolated from midgut of ticks (*Ixodes*).

bortezomib *Velcade* Small molecule drug, a modified dipeptidyl boronic acid, that inhibits the chymotypsin-like activity of the 26S **proteosome**; used in treatment of **multiple myeloma**.

Bos taurus Domestic cow.

Botox® Proprietary name for **Botulinum toxin** type A, injected into the skin as a temporary treatment to make lines on the face less apparent.

botrocetin *Venom coagglutinin* Lectin (22 kDa) from *Bothrops jararaca* that induces binding of von Willebrand factor (vWF) to platelet glycoprotein Ib (GP-Ib) and causes aggregation of blood platelets. There is a single **C-type lectin** domain in the α-subunit of the dimer.

botryomycosis A granulomatous infection of horses caused by *Staphylococcus aureus*.

bottle cells The first cells to migrate inwards at the **blastopore** during amphibian **gastrulation**. The 'neck' of the bottle is at the outer surface of the embryo.

botulinolysin **Cholesterol-binding toxin** from *Clostridium botulinum*.

botulinum toxin Neurotoxin (50 kDa; seven distinct serotypes) produced by certain strains of *Clostridium botulinum*. The bacterium produces the toxin as a complex with a haemagglutinin that prevents toxin inactivation in the gut. Proteolysis in the body results in cleavage into two fragments A and B; B binds to gangliosides and may stimulate the endocytosis of fragment A. The extremely high toxicity is due to the blockade of cholinergic synapses through inhibition of vesicle exocytosis. See **Botox, botulinus toxin C2, C3**, **synaptobrevin, tetanus toxin**.

botulinus toxin C2 Binary toxin with binding subunit (100 kDa) and enzymatic subunit (50 kDa) that ADP-ribosylates monomeric G-actin and blocks the formation of microfilaments. Produced by C and D strains of *Clostridium botulinum*.

botulinus toxin C3 *Exoenzyme C3* Toxin (24 kDa) produced by C and D strains of *Clostridium botulinum*. An ADP-ribosyl transferase that inactivates **Rho**. Needs to be injected into cells, and is a laboratory tool rather than a true toxin.

botulism Severe and often fatal poisoning due to eating food contaminated by *Clostridium botulinum* (see **botulinus toxin**).

Bouin's solution Bright yellow picric acid-based fixative that also contains formaldehyde and acetic acid. It has the advantage that specimens can be stored indefinitely and it generally preserves nuclear morphology quite well.

boutonnière deformity A deformity in which a finger is bent down at the middle joint and bent back up at the end joint as the result of a buttonhole-shaped tear in a tendon.

boutons Small swellings in the terminal region of an axon or along the length of an axon (*boutons en passant*) where it makes contact with, for example, a muscle fibre. Synaptic vesicles are clustered in the boutons.

Bowman–Birk peptidase inhibitors Family of **serine peptidase** inhibitors found in seeds of leguminous plants and cereals.

box Casual term for a DNA sequence that is a characteristic feature of regions that bind regulatory proteins e.g. **homeobox**, **TATA box** and **CAAT box**.

Boyden chamber Simple chamber used to test for chemotaxis, especially of leucocytes. Consists of two compartments separated by a millipore filter (3–8-μm pore size); chemotactic factor is placed in one compartment and the gradient develops across the thickness of the filter (*ca*. 150 μm). Cell movement into the filter is measured after an incubation period of less than the time taken for the gradient to decay. See also **checkerboard assay**.

bozozok *boz* Zebra fish gene that encodes the **homeodomain** protein Bozozok/Dharma; expression is activated through beta-**catenin** signaling. Mutations cause defects in axial mesoderm and anterior neurectoderm, and affect organizer formation.

BPG *2,3-Biphosphoglycerate; 2,3-BPG* A highly anionic organic phosphate, present in red blood cells at about the same molar ratio as haemoglobin. It binds to deoxyhaemoglobin and diminishes the oxygen affinity of haemoglobin thus allowing unloading of oxygen in capillaries. The concentration of BPG is regulated by **bisphosphoglycerate mutase**, which catalyses production from 3-phospho-D-glyceroyl phosphate, and bisphosphoglycerate phosphatase, which hydrolyses it to 3-phosphoglycerate and orthophosphate.

brachydactyly Feature of a number of congenital abnormalities in which the fingers and toes are short as a result of premature closure of the epiphyses. Type A1 is an autosomal dominant.

Brachydanio rerio *Danio rerio* See **zebra fish**.

brachyury Mouse gene encoding a transcription factor, one of the **T-box genes**. Product of the gene is important in tissue specification, morphogenesis and organogenesis. Mouse mutant has a short tail.

bracoviruses See **polydnaviruses**.

Bradford assay *Bradford method* Very commonly used assay for protein concentration based upon the absorbance shift in Coomassie Brilliant Blue G-250 (CBBG) when bound to arginine and aromatic residues. Standardization of the assay with a protein of comparable arginine content is important, and for some proteins, such as collagen, the results are very inaccurate.

bradycardia Condition in which the heart beats unusually slowly. Opposite of tachycardia.

bradykinin Vasoactive nonapeptide (RPPGFSPFR) formed by action of peptidases on kininogens. Very similar to **kallidin** (which has the same sequence but with an additional N-terminal lysine). Bradykinin is a very potent vasodilator and increases permeability of postcapillary venules; it acts on endothelial cells to activate phospholipase A2. It is also spasmogenic for some smooth muscles and will cause pain.

brain-derived growth factor *BDGF* See **brain-derived neurotrophic factor**.

brain-derived neurotrophic factor *BDNF; BDGF* Small basic protein (28 kDa) originally purified from pig brain; a member of the family of **neurotrophins** that also includes **nerve growth factor** and **neurotrophin-3**. In contrast to NGF, BDNF is predominantly (though not exclusively) localized in the central nervous system. It supports the survival of primary **sensory neurons** originating from the **neural crest** and ectodermal **placodes** that are not responsive to NGF. **Huntingtin** upregulates transcription of BDNF and vesicular transport of BDNF along microtubules.

brain regions The central nervous system of mammals is complex and the terminology often confusing. In development, the brain is generated from the most anterior portion of the neural tube and there are three main regions; the fore-, mid- and hindbrain. The lumen of the embryonic nervous system persists in the adult as the cerebral ventricles, filled with cerebrospinal fluid, which are connected to the central canal of the spinal cord. The forebrain develops to produce the cerebral hemispheres and basal ganglia and the diencephalons, which forms the thalamus and hypothalamus. The cerebrum consists of two hemispheres connected by the corpus callosum, the outer part being greatly expanded in man with the increased surface being thrown into fold (ridges are gyri, valleys are sulci). The outer layer (cerebral cortex) is responsible for so-called higher-order functions such as memory, consciousness and abstract thought; the deeper layers (basal ganglia) include the caudate nucleus and putamen (collectively the striatum), amygdaloid nucleus and hippocampus. The hypothalamus controls endocrine function (hunger, thirst, emotion, behaviour, sleep); the thalamus coordinates sensory input and pain perception. The midbrain is relatively small and develops to form corpora quadrigemina and the cerebral peduncle. The hindbrain develops into two regions, the more anterior being the metencephalon, and the region nearest to the spinal cord being the myelencephalon. The metencephalon contains the cerebellum, responsible for sensory input and coordination of voluntary muscles, and the pons. The myelencephalon contains the medulla oblongata, which regulates blood pressure, heart rate and other basic involuntary functions, and, dorsally, the choroid plexus.

Branchiostoma Cephalochordates, of which *Branchiostoma* is one genus (*Amphioxus* is the obsolete generic name), are the most vertebrate-like of the invertebrates, with a dorsal hollow nerve cord, notochord, postanal tail, and pharyngeal gill slits used for filter feeding.

Brassica napus Oilseed rape (canola in US). Source of edible oil (see **erucic acid**).

brassinolide See **brassinosteroids**.

brassinosteroids Plant hormones (polyoxygenated steroids) that have pronounced plant growth-regulatory effects. Are released by mature cells in response to root environmental, pest or disease stress. Commonest form is brassinolide.

brca1, brca2 *Breast cancer 1 gene Brca1* was the first gene to be linked to familial breast cancer; *brca2* was identified shortly thereafter. Mutations in *brca1* and *brca2* are associated with inherited predisposition to breast–ovarian carcinoma syndrome, although there are clearly other predisposing genes. The nuclear proteins BRCA1 and BRCA2 are involved in a large multiprotein tumour-suppressor complex but have also been associated with a range of other activities. See **BRCT, BARD**.

BRCT The BRCT domain (found and named for the C-terminal domain of BRCA1) is found predominantly in proteins involved in cell cycle checkpoint functions responsive to DNA damage. The domain is an approximately 100-amino acid tandem repeat, which appears to act as a phospho-protein binding domain.

Brdu *Bromodeoxyuridine* See **BUdR**.

bread mould *US bread mold* Favourite mould for classroom study; grows 'spontaneously' on damp bread and can be any one of a number of common fungi, though often stated to be *Rhizopus nigricans*, which has a cottony growth pattern. Bluish-green to green moulds are usually *Penicillium* or *Aspergillus*. Black to brown-black moulds can be *Aspergillus niger, Alternaria alternata, Cladosporium herbarum, Cladosporium sphaerospermum,* or *Stachybotrys chartarum* (a highly toxic mould). Reddish or pink moulds are usually species of *Fusarium*, orange mould is likely to be *Neurospora* spp.

breakpoint cluster region See **bcr**.

brefeldin A A macrocyclic lactone synthesized from palmitic acid by several fungi including *Penicillium brefeldianum*. It was initially described as an antiviral antibiotic, but it was later found to inhibit protein secretion at an early stage, probably blocking secretion in a pre-Golgi compartment. Binds to the **ARF1/GDP/Sec7** complex and blocks **GEF** activity at an early stage of the reaction, prior to guanine nucleotide release. A valuable tool for studying membrane traffic and the control of organelle structure.

brevetoxins Lipophilic 10- and 11-ring polyether toxins from the dinoflagellate *Ptychodiscus brevis* (formerly *Gymnodiunium breve*); responsible for neurotoxic shellfish poisoning. Bind to the sodium channel of nerve and muscle making it hyper-excitable by altering the critical threshold level of depolarization required to generate an action potential.

Brevibacterium Genus of Gram-positive aerobic coryneform bacteria.

bride of sevenless *Boss* In *Drosophila* eye development, the ligand for the **sevenless** tyrosine kinase receptor. Boss is expressed by the central R8 cell. It is unusual as a ligand for a tyrosine-receptor kinase in that it is on the surface of another cell and has, in addition to a large extracellular domain, seven transmembrane segments and a C-terminal cytoplasmic tail.

Bright *B cell regulator of immunoglobulin heavy chain transcription* Transcription factor that binds A+T-rich sequences in the intronic enhancer regions of the murine heavy chain locus and 5′-flanking sequences of some variable heavy chain promoters. Binds as a dimer.

Bright's disease An obsolete name for acute and chronic nephritis.

bright-field microscopy Optical **microscopy**, in which absorption, to a great extent, and diffraction, to a minor extent, give rise to the image, as opposed to **phase contrast** or **interference** methods of microscopy.

Brilliant Cresyl Blue *Brilliant Blue C* Dye used in staining of bone marrow smears.

Bristletails Small wingless insects of the Order Archaeognatha that resemble silverfish (Thysanura). Are able to jump distances of several centimetres by flexing their abdomens.

Broca's area The left inferior convolution of the frontal lobe of the brain; damage to this region leads to impairment of speech production (Broca's aphasia), though not comprehension.

bromelain *Formerly EC 3.4.22.4* Cysteine endopeptidases (thiol protease) from pineapple. EC 3.4.22.32 is stem bromelain that is distinct from EC 3.4.22.33, the fruit bromelain.

bromodomain Domain found in a variety of mammalian, invertebrate and yeast DNA-binding proteins, involved in binding to acetyllysines on histone tails. It has been suggested that the domain is involved in deciphering the 'histone code'.

bromophenol blue Dye used as pH indicator: changes from yellow to blue in the range 3.0–4.6.

bromophenol red Dye used as pH indicator: changes from yellow to red in range 5.2–6.8.

bromothymol blue Dye used as a pH indicator. Changes from yellow to blue in the pH 6.0–7.6 range.

Bromoviruses Plant viruses with a genome of three linear, positive-sense ssRNA molecules. Named originally after brome grass.

bronchiectasis Abnormal destruction and dilation of the large airways. Can be congenital or acquired, probably as a result of recurrent inflammation.

bronchodilators Substances that dilate airways by relaxing the smooth muscle of the bronchial wall, thus relieving breathlessness caused acutely by asthma or chronically by obstructive pulmonary disease. Often administered by inhalation. Examples include **sympathomimetic drugs** (salbuterol), **antimuscarinic drugs**, and xanthines (theophylline).

bronchus-associated lymphoid tissue *BALT* Subset of mucosal-associated lymphoid tissue found as lymphoid nodules in the lamina propria of the bronchus. Some mammalian species have it constitutively present (e.g. rabbits) but in others (e.g. humans) it can develop in response to infection.

brown adipose tissue Highly vascularized adipose tissue found in restricted locations in the body (in the interscapular region in the rat, for example). In hibernating animals and neonates, brown adipose tissue is important for regulating body temperature via non-shivering thermogenesis.

brown fat cells Brown fat is specialized for heat production and the **adipocytes** have many mitochondria in which an inner-membrane protein can act as an uncoupler of **oxidative phosphorylation** allowing rapid thermogenesis. See **brown adipose tissue**.

Brownian motion Random motion of small objects as a result of intermolecular collisions. First described by the nineteenth-century microscopist, Brown.

Brownian ratchet Mechanism proposed to explain protein translocation across membranes and force generation by polymerising actin filaments. Relies upon asymmetry of cis and trans sides of the membrane or biased thermal motion as a result of polymerization. Still a hypothesis.

Brucella Genus of Gram-negative aerobic bacteria which occur as intracellular parasites or pathogens in man and other animals. *Brucella abortus* is responsible for spontaneous abortion in cattle and causes undulent fever (brucellosis), a persistent recurrent acute fever, in man.

Brunner's gland *Duodenal gland* Small, branched, coiled tubular glands in the submucosa of the first part of the duodenum. Their secretion of alkaline mucus helps neutralize gastric acid from the stomach.

brush border The densely packed **microvilli** on the apical surface of, for example, intestinal epithelial cells.

Bruton's disease Sex-linked recessive **agammaglobulinaemia** caused by a deficiency in **B-lymphocyte** function. See **btk**.

Bruton's tyrosine kinase See **btk**.

Brx Dbl family member that modulates estrogen receptor activity. May thus integrate cytoplasmic signalling mediated by **rho** (for which it is a **GEF**) and nuclear receptors.

Bryophyta Plant phylum that includes mosses and liverworts.

bryostatin General name for a group of compounds (complex lactones) isolated from bryozoans; activate Protein Kinase C (**PKC**), though after longer-term exposure cells down-regulate their PKC.

Bryozoa *Polyzoa* Phylum of invertebrates (moss animals) that are mainly marine and usually colonial.

BSE Bovine **spongiform encephalopathy**. A transmissible encephalopathy that affected large numbers of cattle in the UK during the 1990s and is widely believed to have arisen through consumption by cattle of feedstuff containing sheep tissues from animals with scrapie, although the agent is clearly different from that of scrapie. A link with 'new variant CJD' is strongly suspected.

BTB/POZ domain *Bric-a-brac Tramtrack Broad/Pox virus and Zinc-finger complex; POZ/BTB* Structurally well-conserved domain (also known as the BTB domain) involved in protein–protein interactions and important in a wide range of cellular functions, including transcriptional regulation, cytoskeleton dynamics, ion channel assembly and gating, and targeting proteins for ubiquitination. Originally identified as a conserved motif present in the *Drosophila* bric-a-brac, tramtrack and broad complex transcription regulators and in many pox virus and zinc-finger proteins.

btk *Bruton's tyrosine kinase* **Tyrosine kinase** of **Tec** family, defective in Bruton's agammaglobulinaemia. Mutations in btk lead to B-cell immunodeficiencies XLA in humans, Xid in mice. Overexpression of btk enhances calcium influx following B-cell antigen receptor cross-linking. **Sab** selectively binds the SH3 domain of btk. Btk interacts with membrane through **PH domain**, and **SHIP**, by reducing PIP3 levels, regulates this association.

budding A type of cell division in fungi and in protozoa in which one of the daughter cells develops as a smaller protrusion from the other. Usually the position of the budding cell is defined by polarity in the mother cell. In some protozoa, the budded daughter may lie within the cytoplasm of the other daughter.

budesonide Synthetic corticosteroid with strong **glucocorticoid** and weak **mineralocorticoid** activity, used for treatment of **Crohn's disease**.

BUdR *Bromo-deoxyuridine* The deoxynucleoside of 5-bromouracil, an analogue of thymidine that induces point mutations because of its tendency to tautomerization: in the enol form it pairs with G instead of A. It is used as a mutagen, and also as a marker for DNA synthesis (The incorporation of BUdR can be recognized because the staining pattern differs: an even more sensitive method uses a monoclonal antibody staining procedure.)

buffer A system that acts to minimize the change in concentration of a specific chemical species in solution against addition or depletion of this species. pH buffers: weak acids or weak bases in aqueous solution. The working range is given by pKa ±1. Metal ion buffers: a metal ion chelator e.g. **EDTA**, partially saturated by the metal ion acts, as a buffer for the metal ion.

buffy coat Thin, yellow-white layer of leucocytes on top of the mass of red cells when whole blood is centrifuged.

bufotenine *3-(2-(Dimethylamino)ethyl)-1H-indol-5-ol; mappine* An indole alkaloid with hallucinogenic effects, isolated from *Piptadenia* spp. (Mimosidae); first isolated from skin glands of toad (*Bufo sp.*).

bulla *Pl.* bulli A blister or bleb. A circumscribed elevation above the skin containing clear fluid; larger than a vesicle.

bulldog calf A lethal form of **achondroplasia** (dwarfism) occurring in Dexter cattle.

bulliform cell An enlarged epidermal cell type found in longitudinal rows in the leaves of some grasses. May be responsible for rolling and unrolling of leaves in response to changes of water status.

bullous pemphigoid Form of pemphigoid (which also affects mucous membranes), in which blisters (bulli) form on the skin. Patients have circulating antibody (usually IgG) to **basement membrane** of **stratified epithelium**, although the antibody titre does not correlate with the severity of the disease.

bumetanide A potent **loop diuretic**, used in the treatment of oedema, that inhibits sodium reabsorption in the ascending limb of the loop of Henle. Reabsorption of chloride in the ascending limb is also blocked, and bumetanide is somewhat more chloruretic than natriuretic. Potassium excretion is also increased by bumetanide.

bundle of His A bundle of small, specialized conducting muscle fibres in the mammalian heart that is responsible for transmitting electrical impulses from atrium to ventricle.

bungarotoxins Toxins found in the venom of *Bungarus multicinctus*. α-bungarotoxin is a polypeptide toxin (74 residues). A powerful antagonist of **acetylcholine,** it causes virtually irreversible block of the vertebrate neuromuscular junction by binding (as a monomer) to each of the a-subunits of the postsynaptic nicotinic acetycholine receptors (nAChR) Has been much used in identifying, quantifying and localizing these receptors on muscle cells. Will also bind some neuronal nAChR. β-bungarotoxin is a a two-chain phospholipase A2 neurotoxin that acts at the presynaptic site of motor nerve terminals and blocks transmitter release. Subunit A (120 residues) is structurally homologous to other PLA2s, the B-subunit (60 residues) has homology with Kunitz-type serine peptidase inhibitors and **dendrotoxins**. Binds to subtype of voltage-sensitive potassium channels. κ-bungarotoxin (bungarotoxin 3.1; Toxin F, neuronal bungarotoxin) is a polypeptide (66 residues) also from the venom of *Bungarus multicinctus*. Has considerable homology with α-bungarotoxin. Functional toxin is a homodimer. Potent antagonist for a subset of neuronal nAChR, but is much less active against muscle receptors.

Bungarus multicinctus Formosan snake (banded krait). See **bungarotoxins**.

Bunyaviridae Single-stranded RNA (ssRNA) enveloped viruses infecting vertebrates and arthropods. Genome consists of negative sense RNA molecules. Virion are spherical or oval, 90–100 nm in diameter. Some genera contain organisms causing serious disease, e.g. viral haemorrhagic fever.

bupivacaine A powerful local anaesthetic.

buproprion hydrochloride *Wellbutrin* Aminoketone antidepressant, unrelated to **tricyclic antidepressants** and **SSRIs**. Used to assist in smoking cessation.

Burkitt's lymphoma Malignant tumour of **lymphoblasts** derived from B-lymphocytes. Most commonly affects children in tropical Africa: both **Epstein–Barr virus** and immunosuppression due to malarial infection are involved.

burr cells Triangular, helmet-shaped cells found in blood, usually indicative of disorders of small blood vessels.

Bursa of Fabricius A **lymphoid tissue** found at the junction of the cloaca and the gut of birds, giving rise to the so-called B-lymphocyte series.

bursicon Insect hormone produced by neurosecretory cells of the brain. Affects many post-ecdysal processes such as cuticular tanning.

burst-forming unit *BFU-E* A bone marrow **stem cell** lineage detected in culture by its mitotic response to **erythropoietin** and subsequent erythrocytic differentiation, in about 12 mitotic cycles, into erythrocytes.

butobarbitone *5-Butyl-5-ethylbarbituric acid; Butomet; Soneryl* A barbiturate hypnotic and sedative used for seizure disorders, as a short-term sleep aid (for insomnia) and for tension relief before surgical procedures.

butyric acid *CH3.CH2.CH2.COOH* Acid from which butyrate ion is derived. Smells of rancid butter, hence the name.

butyrophilin *BTN* Integral membrane glycoprotein of the immunoglobulin superfamily (59 kDa) of mammary secretory epithelium. Secreted in association with milk-fat globule membrane. The extracellular domain of butyrophilin has features that suggest it may have a receptor function by binding to xanthine dehydrogenase/oxidase or may act as a structural protein. The BTN gene family codes for seven proteins (BTN, BTN2A1, BTN2A2, BTN2A3, BTN3A1, BTN3A2, BTN3A3).

butyrylcholinesterase *EC 3.1.1.8; BChE; Choline esterase II; Pseudocholinesterase* A family of enzymes produced mainly in the liver that hydrolyse esters, e.g. **procaine** and **suxamethonium**. Occurs as a number of variants, dependent on four alleles, with variable degrees of cholinesterase function. The nature of the variants determines sensitivity to suxamethonium; the extreme is pseudocholinesterase deficiency. Is known to metabolize cocaine in humans.

BY2 cells Line of tobacco (*Nicotiana tabacum*) cells that will grow in suspension culture.

byssinosis Asthma-like respiratory disease in people exposed to dust from vegetable fibres (cotton, jute, flax).

bystander effect Phenomenon in which mammalian cells irradiated in culture produce damage-response signals which are communicated to their unirradiated neighbours. The mechanism is unclear, but the implications are important in radiation biology.

bystander help Lymphokine-mediated non-specific help by T-lymphocytes, stimulated by one antigen, to lymphocytes stimulated by other antigens.

bZip See **basic leucine zipper**.

C

c- Prefix used to denote the normal cellular form of, for example, a gene such as *src* (*c-src*) that is also found as a viral gene (*v-src*).

C1 First component of **complement**; actually three subcomponents, C1q, C1r and C1s, that form a complex in the presence of calcium ions. C1q, the recognition subunit, has an unusual structure of collagen-like triple helices forming a stalk for its Ig-binding globular heads. Upon binding to immune complexes, the C1 complex becomes an active endopeptidase that cleaves and activates C4 and C2.

C1–C9 Proteins of the mammalian **complement** system. See also under individual numbered components.

C2 Second component of **complement**. A beta-2-globulin.

C2-kinin A kinin-like fragment generated from **complement** C2; causes vasodilation and increased vascular permeability. Distinct from **bradykinin**.

C3 Third component of **complement**, present in plasma at around 0.5–1 mg/ml. Both classical and alternate pathways converge at C3, which is cleaved to yield C3a, an **anaphylotoxin**, and C3b, which acts as an opsonin and is bound by **CR1**; C3b in turn can be proteolytically cleaved to iC3b (ligand for **CR3**) and C3dg by C3b-inactivator. C3b complexed with factor B (to form C3bBb) will cleave C3 to give more C3b, although the C3bBb complex is unstable unless bound to **properdin** and a carbohydrate-rich surface. The C3b–C4b2a complex and C3bBb are both C5 convertases (cleave **C5**). Cobra venom factor is homologous with C3b, but the complex of cobra venom factor, properdin, and factor Bb is insensitive to C3b-inactivator.

C3 plants Plants that fix CO_2 in photosynthesis by the **Calvin–Benson cycle**. The enzyme responsible for CO_2 fixation is **ribulose bisphosphate carboxylase**, whose products are compounds containing three carbon atoms. C3 plants are typical of temperate climates. **Photorespiration** in these plants is high.

C3G Guanine nucleotide exchange factor (**GEF**) that activates **Rap1**. C3G is involved in signalling from **crk** to **JNK**

C4 Fourth component of **complement**, although the third to be activated in the classical pathway. Becomes activated by cleavage (by C1) to C4b, which complexes with C2a to act as a C3 convertase, generating C3a and C3b. The C4b2a3b complex acts on C5 to continue the cascade.

C4 plants Plants found principally in hot climates whose initial fixation of CO_2 in photosynthesis is by the **HSK pathway**. The enzyme responsible is **PEP carboxylase**, whose products contain four carbon atoms. Subsequently the CO_2 is released and re-fixed by the **Calvin–Benson cycle**. The presence of the HSK pathway permits efficient photosynthesis at high light intensities and low CO_2 concentrations. Most species of this type have little or no **photorespiration**.

C5 Fifth component of **complement**, which is cleaved by C5-convertase to form C5a, a 74-residue anaphylotoxin and potent chemotactic factor for leucocytes, and C5b. C5a rapidly loses a terminal arginine to form C5a desarg, which retains chemotactic but not anaphylotoxin activity. C5b combines with C6, C7, C8 and C9 to form a membranolytic complex.

C6, C7 Sixth and seventh components of the **complement** cascade. Contain EGF-like motifs. See **C5** and **C9**.

C8 Eighth component of **complement**: three peptide chains, α, β and γ.

C9 Ninth component of **complement**. Complexed with C5b, 6, 7, 8 it forms a potent membranolytic complex (sometimes referred to as the membrane attack complex, MAC). Membranes that have bound the complex have toroidal 'pores'; a single pore may be enough to cause lysis.

C57BL Inbred strain of black mice.

c127 cells A non-tumorigenic mouse epithelial cell line, derived from mammary gland, widely used in *in vitro* transformation assays due to its normal morphological appearance and its very low levels of spontaneous transformation.

CA1 Region of the **hippocampus**.

CAAT box Nucleotide sequence in many eukaryotic promoters usually about 75 bp upstream of the start of transcription. Binds **NF-1**.

cachectin Protein produced by macrophages that is responsible for the wasting (cachexia) associated with some tumours. Now known to be identical to **tumour necrosis factor** alpha (TNFα). Has three 17-kDa subunits, all derived from a single highly conserved gene.

cachexia Marked wasting and emaciation of the body that may occur in some patients with cancer or other serious diseases such as AIDS.

Caco cells *CACO* Cell line derived from a primary colonic carcinoma of a 72-year-old male Caucasian. Epithelial morphology.

CAD *EC 3.1.21.1* Caspase-activated DNase (40 kDa), the primary nuclease responsible for oligonucleosomal DNA fragmentation during apoptosis. See **DFF**.

CADASIL *Cerebral autosomal dominant arteriopathy with subcortical infarcts and leucoencephalopathy* Hereditary adult-onset condition, causing stroke and dementia, mapped to Chr 19 and thought to be a defect in **Notch**-3, though how a defect in this signalling pathway leads to the pathological effect remains unclear.

cadaverine *1,5-Pentanediamine* Substance formed by microbial action in decaying meat and fish by decarboxylation of lysine. The smell can be imagined. Like many of the other diamines (e.g. **putrescine**), has effects on cell proliferation and differentiation.

cadherins Integral membrane proteins involved in calcium-dependent cell adhesion. There are three types, named after their distributions: N-cadherin (neural), E-cadherin (epithelial) (equivalent to uvomorulin and LCAM) and P-cadherin (placental). Formed of a 600 amino acid extracellular domain, containing four repeats believed to contain the Ca^{2+} binding sites, a transmembrane domain and a 150 amino acid intracellular domain.

Caenorhabditis elegans Nematode much used in lineage studies since the number of nuclei is determined (there are 959 cells), and the nervous system is relatively simple. One of the first organisms to have its complete genome sequenced. The organism can be maintained axenically, and there are mutants in behaviour, in muscle proteins and in other features. Sperm are amoeboid and move by an unknown mechanism which does not seem to depend upon actin or tubulin.

caerulin Amphibian peptide hormone related to **gastrin** and **cholecystokinin**. A decapeptide that is one of the main constituents of the skin secretion of *Xenopus laevis*.

caesium chloride *cesium chloride (US)* Salt that yields aqueous solutions of high density. When equilibrium has been established between sedimentation and diffusion during ultracentrifugation, a linear density gradient is established in which macromolecules such as DNA band at a position corresponding to their own buoyant density.

caffeine A **xanthine** derivative that elevates cAMP levels in cells by inhibiting phosphodiesterases.

caged-ATP A derivative of ATP that is not biologically active until a photosensitive bond has been cleaved.

Cairns mechanism A mechanism for the replication of a double-stranded circular DNA molecule. Replication is initiated at a fixed point and proceeds either uni- or bidirectionally.

Cajal bodies *Coiled bodies* Spherical structures (0.1–2.0 μm), varying in number from 1 to 5, found in the nucleus of proliferative cells like tumour cells, or metabolically active cells like neurons. Contain factors required for splicing, ribosome biogenesis and transcription, and are dynamic structures.

calbindin Vitamin D-induced calcium-binding protein (28 kDa) found in primate striate cortex and other neuronal tissues. Contains an **EF-hand** motif.

calcein A calcium-chelating agent that fluoresces brightly in the presence of bound calcium. The acetomethoxy derivative can be transported into live cells and the reagent is useful as a viability test and for short-term marking of cells.

calcicludine Polypeptide toxin (60 residues) from *Dendroaspis angusticeps*. Blocks most high-threshold calcium channels (L-, N- or P-type). Structurally homologous to Kunitz-type serine peptidase inhibitors and **dendrotoxins**.

calciferol *Vitamin D* See Table V1.

calcimedins **Annexins**; see Table A3.

calcineurin *EC 3.1.3.16* Calmodulin-stimulated protein phosphatase, the major calmodulin-binding protein in brain. Enzymatic activity is inhibited by binding of **immunophilin**–ligand complex (immunophilin alone does not bind), and therefore may play a part in the mechanism of action of **ciclosporin A** and FK506. Thought also to be involved in the control of sperm motility.

calcinosis Deposits of calcium salts, primarily hydroxyapatite crystals or amorphous calcium phosphate, in various tissues of the body.

calciosome A membrane compartment proposed to contain the intracellular calcium store released in response to hormonal activity, and thought to be distinct from the ER. Now discredited.

calcipressins *Adapt 78; DSCR 1; MCIP 1* A family of endogenous **calcineurin** inhibitors. Calcipressin 1 is encoded by *DSCR1*, a gene on human chromosome 21 with seven exons; exons 1–4 are alternative first exons (isoforms 1–4). Calcipressin 1 isoform 1 interacts with calcineurin A and inhibits **NFAT**-mediated transcriptional activation. Calcipressin 1 is a phosphoprotein that increases its capacity to inhibit calcineurin when phosphorylated at the FLISPP motif, and this phosphorylation also controls the half-life of calcipressin 1 by accelerating its degradation.

calcisepline Polypeptide toxin (60 residues) from *Dendroaspis polylepis*. Specific blocker of some L-type calcium channels that will cause relaxation of smooth muscle and inhibition of cardiac muscle but has no effect on skeletal muscle.

calcitonin A polypeptide hormone produced by C-cells of the thyroid that causes a reduction of calcium ions in the blood.

calcitonin family peptides *ADM; CGRP1; CGRP2; amylin; calcitonin* Family of small (32–51 residues), highly homologous peptides that act through seven-transmembrane G-protein-coupled receptors. **Adrenomedullin** (ADM; 51 residues) is a potent vasodilator and has receptors on astrocytes; **amylin** (37 residues) is thought to regulate gastric emptying and carbohydrate metabolism; calcitonin (32 residues) is involved in control of bone metabolism; **calcitonin-gene-related peptides** 1 and 2 (37 residues) regulate neuromuscular junctions, antigen presentation, vascular tone and sensory neurotransmission. Receptors are themselves regulated by **RAMPs**.

calcitonin gene-related peptide *CGRP* Neuropeptide of 37 amino acids with structural homology to salmon calcitonin. Co-localizes with **substance P** in neurons. Intracerebral administration of CGRP leads to a rise in noradrenergic sympathetic outflow, a rise in blood pressure, and a fall in gastric secretion. A family of related peptides exists (**calcitonin family peptides**). See **RAMPs**.

calcitriol *1α, 25-Dihydroxyvitamin D3* The form of vitamin D3 that is biologically active in intestinal transport and calcium resorption by bone.

calcium antagonist or blocker Any compound that blocks or inhibits transmembrane calcium ion movement. Most are potent vasodilators and some are anti-arrhythmic. Examples are nifedipine, verapamil and diltiazem.

calcium ATPase Usually used of the calcium-pumping ATPase present in high concentration as an integral membrane protein of the **sarcoplasmic reticulum** of muscle. This pump lowers the cytoplasmic calcium level and causes contraction to stop. Normal function of the pump seems to require a local phospholipid environment from which cholesterol is excluded.

calcium-binding proteins There are two main groups of calcium-binding proteins; those that are similar to **calmodulin** and are called **EF-hand** proteins, and those that bind calcium and phospholipid (e.g. **lipocortin**) and are grouped under the generic name of **annexins**. Many other proteins will bind calcium, although the binding site usually has considerable homology with the calcium-binding domains of calmodulin.

calcium channel Membrane channel that is specific for calcium. Probably the best characterized is the voltage-gated channel of the sarcoplasmic reticulum which is ryanodine-sensitive. See **voltage-sensitive calcium channels**, **ryanodine receptor**.

calcium current Inflow of calcium ions through specific **calcium channels**. Critically important in release of transmitter substance from presynaptic terminals.

calcium-dependent regulator protein *CDRP* Early name for **calmodulin**.

calcium phosphate precipitation Technique used for introducing DNA or chromosomes into cells; co-precipitation with calcium phosphate facilitates the uptake of DNA or chromosomes.

calcium pump A **transport protein** responsible for moving calcium out of the cytoplasm. See **calcium ATPase**.

calcivirus Any of a group of related viruses of the genus *Picornaviridae*, many of which have cup-shaped surface depressions. Calciviruses affecting humans are a major cause of gastrointestinal upsets, and a calcivirus has been used for control of rabbit populations.

calcofluor A fluorochrome, extensively used in microscopy, that also exhibits antifungal activity and a high affinity for yeast cell-wall chitin.

calcyclin **Prolactin** receptor-associated protein, one of a family of small (around 10-kDa) calcium-binding proteins containing the **EF-hand** motif, originally isolated from Erlich ascites tumour cells, but human and rat forms now identified. Regulated through the cell cycle. Binds to annexin II (p36) and to glyceraldehyde-3-phosphate dehydrogenase.

calcyphosin *Thyroid protein p24* Calcium-binding protein that contains an **EF-hand** motif.

caldesmon Protein originally isolated from smooth muscle (h-caldesmon; 120–150 kDa on gels but 88.7 kDa from sequence) also found in non-muscle cells (l-caldesmon; 70–80 kDa on gels but 58.8 kDa from sequence). Normally dimeric, binds to F-actin blocking the myosin-binding site. Calcium–calmodulin binding to caldesmon causes its release from actin, though phosphorylation of caldesmon may also affect the link with actin. Caldesmon can block the effect of **gelsolin** on F-actin and will dissociate actin–gelsolin complexes and actin–profilin complexes.

calelectrins Membrane-associated proteins (70 and 32 kDa) of the **annexin** family. Originally from *Torpedo*, but subsequently found in bovine liver. May regulate exocytosis.

calexcitin *cp20* Calcium and GTP-binding protein (20 kDa) that interacts with and activates the ryanodine receptor; it also activates the calcium ATPase. Calexcitin is a high-affinity substrate for PKCα and once phosphorylated moves to membrane, where it inactivates voltage-dependent K$^+$ channels. Has two EF-hand motifs, and some similarities with **ARFs** (ADP-ribosylating factors).

calgizzarin *S100 calcium-binding protein A11* Calcium-binding protein of the **S100** family, originally isolated from chicken gizzard, implicated in the regulation of cytoskeletal function through its calcium-dependent interaction with **annexin I**.

calgranulins *p8, p14; MRP-8, MRP-14* Calcium-binding myeloid-associated proteins (8 and 14 kDa) expressed at high levels in neutrophils and monocytes but lost during differentiation to macrophages. Related to migration inhibitory factor (MIF). Associated with **keratinocyte** cytoskeleton. Part of the **S100** family.

calicin Actin-binding protein (67 kDa) that contains three **kelch** repeats. Found in the post-acrosomal calyx region of vertebrate sperm, which has a cytoskeletal function.

calitoxin Small toxic peptide (46 residues) from sea anemone *Calliactis parasitica* that acts on neuronal sodium channels.

Calliactis parasitica Sea anemone (an anthozoan **coelenterate**) that lives commensally with hermit crabs. See **calitoxin**.

callose A plant cell wall polysaccharide (a β-(1-3)-**glucan**) found in phloem **sieve plates**, wounded tissue, pollen tubes, cotton fibres and certain other specialized cells.

callus 1. *Bot.* undifferentiated plant tissue produced at wound edge. Callus tissue can be grown *in vitro* and induced to differentiate by varying the ratio of the hormones **auxin** and **cytokinin** in the medium. **2.** *Path.* Mass of new bony trabeculae and cartilaginous tissue formed by **osteoblasts** early in the healing of a bone fracture.

calmidazolium *Compound R24571* Inhibitor of calmodulin-regulated enzymes; also blocks sodium channel and voltage-gated calcium channel.

calmodulin *CDRP* Ubiquitous and highly conserved calcium-binding protein (17 kDa) with four **EF-hand** binding sites for calcium (three in yeast). Ancestor of **troponin** C, **leiotonin** C and **parvalbumin**.

calnexin Calcium-binding lectin-like protein (67 kDa, 592 residues) of endoplasmic reticulum that couples glycosylation of newly synthesized proteins with their folding. Calnexin and **calreticulin** act together as chaperones for newly synthesized proteins and prevent ubiquitinylation and proteosomal degradation. Can be phosphorylated by **casein kinase** II.

calpactins Calcium-binding proteins from cytoplasm. Calpactin II is identical to **lipocortin**, and is one of the major targets for phosphorylation by pp60src. See **annexin**.

calpain *Formerly EC 3.4.22.17* Calcium-activated cytoplasmic endopeptidases with broad specificity, containing the **EF-hand** motif. Calpain I (EC 3.4.22.52) is activated by micromolar calcium, Calpain II (EC 3.4.22.53) by millimolar calcium. Calpain has two subunits, the larger (80 kDa) has four domains, one homologous with **papain**, one with **calmodulin**; the smaller (30 kDa) has one domain homologous with calmodulin. First isolated from erythrocytes but now described from other cells.

calpastatin Cytoplasmic inhibitor (76 kDa) of calcium-activated peptidase, **calpain**. It consists of four repetitive sequences of 120–140 amino acid residues (domains I, II, III and IV) and an N-terminal non-homologous sequence.

calpeptin *N-benzyloxycarbonyl-L-leucyl-norleucinal* A cell-permeable **calpain** inhibitor.

calphobindins **Annexins** V and VI (35 kDa) found in placenta (see Table A3). Have substantial sequence homology with **lipocortin** and may function like **calelectrin**.

calphostin C One of a group of compounds isolated from *Cladosporium cladosporioides* that will inhibit **protein kinase C** with some specificity, though inhibits other classes of kinases if present in high concentration.

calponin Calcium and calmodulin-binding **troponin-T**-like protein (34 kDa) isolated from chicken gizzard and bovine aortic smooth muscle. Interacts with F-actin and tropomyosin in a calcium-sensitive manner and acts as a regulator of smooth muscle contraction (inhibits when not phosphorylated). Distinct from **caldesmon** and **myosin light-chain kinase,** but has some antigenic cross-reactivity with cardiac Troponin-T.

calponin homology domain *CH domain* A superfamily of actin-binding domains found in both cytoskeletal proteins and signal transduction proteins. Included are the actinin-type actin-binding domain (including **spectrin, fimbrin, ABP-280**) and the calponin-type. Usually present in pairs, although **calponin** itself has only one.

calregulin See **calreticulin.**

calreticulin *Calregulin; high-affinity calcium-binding protein; HACBP* **Calcium-binding protein** (46.6 kDa) of the **endoplasmic reticulum**. Acts as a chaperone for newly synthesized proteins, possibly in conjunction with **calnexin**. May be more selective in the proteins with which it associates than calnexin. Does not contain any EF-hand motifs.

calretinin Neuronal protein (29 kDa) of the **calmodulin** family isolated from chick retina. Has 58% sequence homology with **calbindin**, the intestinal cell isoform. Contains an **EF-hand** motif.

calsequestrin Protein (44 kDa) found in the cisternae of sarcoplasmic reticulum: sequesters calcium.

calspectin Non-erythroid **spectrin.**

calspermin High-affinity calcium/calmodulin-binding protein found in post-meiotic male germ cells. Represents the C-terminal 169 amino acids of **protein kinase IV** and is produced from a promoter located in an intron of the protein kinase IV gene. Lacks the kinase domain.

caltractin Calcium-binding (EF-hand) protein (20 kDa) from *Chlamydomonas reinhardtii*. Major component of the contractile striated rootlet system that links basal bodies to the nucleus in *Chlamydomonas*. Part of the calmodulin/troponin C family, it has sequence homology with 20 kDa calcium-binding proteins (**centrins**) found in other basal body-associated structures and with the cdc31 gene product associated with spindle pole body duplication in the yeast *S. cerevisiae.*

caltrin Inhibitor of calcium ion transport found in bovine seminal plasma (47 amino acids, MW 5411; on gels, 10 kDa approx.) and that resembles seminal antibacterial protein (confusingly called plasmin, though not related to the peptidase).

caltropin A calcium-binding protein (11 kDa) isolated from smooth muscle, a dimer under native conditions. Interacts with **calponin** in a calcium-dependent fashion in the ratio of 2 mol of dimer:1 mol of calponin.

calumenin A calcium-binding protein localized in the endoplasmic reticulum (ER) and involved in protein folding and sorting. Calumenin is a member of the **EF-hand** superfamily in the ER and Golgi apparatus named **CERC**. The *CALU* gene encodes a deduced 315-amino acid protein containing 6 EF-hand motifs, 1 potential N-glycosylation site, and a C-terminal ER retention signal. The human and mouse CALU proteins are 98% identical. CALU mRNA is ubiquitously expressed in human tissues and maps to 7q32.

calvarium One of the bones that makes up the vault of the skull (in humans these are the frontal, two parietals, the occipital and two temporals). Calvaria are often used in organ culture to investigate bone catabolism or synthesis.

calvasculin *p9Ka* An actin-bundling protein, member of the family of S100-related calcium-binding proteins, located on cytoskeletal elements of cultured mammary cells in a pattern which is identical to actin filaments stained with **phalloidin**. May be involved in the progression and metastasis of human colorectal neoplastic cells.

Calvé's disease *Juvenile osteochondrosis of spine; Scheuermann's disease* Aseptic necrosis of a vertebral body, usually in children.

Calvin–Benson cycle *Calvin cycle* Metabolic pathway responsible for photosynthetic CO_2 fixation in plants and bacteria. The enzyme that fixes CO_2 is **ribulose bisphosphate carboxylase** (RuDP carboxylase). The cycle is the only photosynthetic pathway in **C3 plants** and the secondary pathway in **C4 plants**. The enzymes of the pathway are present in the stroma of the chloroplast.

calyculin A Toxin from marine sponge, *Discodermia calyx*; potent tumour promoter and an inhibitor of protein phosphatases of types 1 and 2a.

CAM See **crassulacean acid metabolism** or **cell-adhesion molecule.**

CaM kinase I *EC 2.7.1.123; calmodulin-dependent kinase I; CamKI* Calcium-regulated kinase (37–42 kDa) that is known to phosphorylate **synapsins**, **CREB** and **CFTR**. Widely distributed. See **CaM Kinase II.**

CaM kinase II *EC 2.7.1.37; calcium/calmodulin-dependent kinase II; CaMKII* A multisubstrate calcium-sensitive kinase composed of four homologous subunits, $\alpha, \beta, \gamma, \delta$ all encoded by different genes. The heteromultimeric holoenzyme (500–600 (kDa) has 10 or 12 subunits, with the ratio of α, β, γ, and δ reflecting that present in the cell. Ca^{2+}/calmodulin-dependent protein kinase II (CaMKII) has been implicated in various neuronal functions, including **synaptic plasticity**. It is highly concentrated in the postsynaptic region and undergoes autophosphorylation at several sites in a manner that depends on the frequency and duration of Ca^{2+} spikes. Constitutively active CaMKII produces dendritic exocytosis in the absence of calcium stimulus. CaMKII activation is the primary event leading to inactivation of both CSF (cytostatic factor) and MPF (maturation-promoting factor; **cyclin**) in mammalian eggs.

cambium 1. *Bot.* **Meristematic** plant tissue, commonly present as a thin layer which forms new cells on both sides. Located either in vascular tissue (vascular cambium), forming xylem on one side and phloem on the other, or in cork (cork cambium or phellogen). **2.** *Anat.* Inner region of the **periosteum** from which **osteoblasts** differentiate.

camera lucida Attachment for a microscope that permits both a view of the object and, simultaneously, of the viewer's hand and drawing implement, thus facilitating accurate drawing of the object of interest.

cAMP See **cyclic AMP.**

cAMP- and cGMP-dependent protein kinase phosphorylation site Both cAMP- and cGMP-dependent protein kinases phosphorylate exposed serine or threonine residues near at least two consecutive N-terminal basic residues, with a consensus pattern: $RK_{(2)}$-x-**ST**.

camptothecin Cytotoxic plant alkaloid originally isolated from *Camptotheca acuminata* (Cancer tree, Tree of Life). Anti-cancer drug inhibits DNA **topoisomerase** I.

Campylobacter Genus of Gram-negative microaerophilic motile bacteria with a single flagellum at one or both poles. Found in reproductive and intestinal tracts of mammals. Common cause of food poisoning and can also cause opportunistic infections, particularly in immunocompromised patients.

canal cell 1. One of the short-lived cells present in the central cavity of the neck of the archegonium in mosses. 2. Cells of Schlemm's canal, irregular space or spaces in the sclerocorneal region of the eye that receive aqueous humour from the anterior chamber of the eye.

canaliculi In bone, channels that run through the calcified matrix between lacunae containing **osteocytes**. In liver, small channels between **hepatocytes** through which bile flows to the bile duct and thence to the intestinal lumen.

canavanine *L-canavanine* One of the non-protein amino acids (also known as secondary metabolites or antimetabolites) that exist in plants especially legumes and their seeds. L-Canavanine is an L-arginine antimetabolite shown to be a selective inhibitor of inducible **nitric oxide synthase** (iNOS).

cancellous bone *Trabecular bone* Adult bone consisting of mineralized, regularly-ordered parallel collagen fibres more loosely organized than the lamellar bone of the shaft of adult long bones. Found in the end of long bones.

cancer Originally descriptive of breast carcinoma, now a general term for diseases caused by any type of malignant tumour.

cancer susceptibility gene See **tumour-suppressor** gene.

candesartan A selective AT1-subtype **angiotensin II receptor antagonist**.

Candida albicans A dimorphic fungus that is an opportunistic pathogen of humans (causing candidiasis).

candidiasis Infection by *Candida albicans*, common on mucous membranes ('Thrush'), but in immunosuppressed patients can opportunistically infect many tissues.

canicola fever Human disease caused by infection by *Leptospira canicola*, the natural host of which is the dog.

canine distemper Paramyxovirus infection of dogs with secondary bacterial complications. Ferrets, foxes and mink are also susceptible.

Canis familiaris Dog.

canker 1. A plant disease in which there are well-defined necrotic lesions of a main root, stem or branch in which the tissues outside the xylem disintegrate. 2. Chronic eczema of the ear of dogs, often caused by mites.

cannabinoid Group of compounds, all derivatives of 2-(2-isopropyl-5-methylphenyl)-5-pentylresorcinol, found in cannabis. Most important members of the group are cannabidiol, cannabidol and various tetra-hydrocannabinols (THCs). Bind to the **cannabinoid receptors** and mimic actions of endogenous agonists **anandamide** and palmitoyl ethanolamine.

cannabinoid receptors *CB1, CB2* Seven membrane-spanning G-protein-coupled receptors for **cannabinoids** (and endogenous agonists such as anandamide). CB1 receptors are mostly found in brain and may mediate the psychotropic activities; the CB2 receptors are more peripheral and found extensively in the immune system.

cannula A tube inserted into the body for the injection or removal of fluids or gases.

canola Variety of oilseed rape (*Brassica napus*) grown extensively to produce oil for human consumption.

canonical Classical, archetypal or prototypic. For example, the canonical polyadenylation sequence is AATAAA.

canonical pathway A regular, orthodox or accepted pathway, very frequently applied, for example, to the **wnt** signalling pathway.

cantharadin *Cantharidin* A pharmaceutical product obtained from the dried elytra of the Spanish fly (blister beetles), although can now be synthesized. Has potent vesicant properties and has been used topically in the treatment of warts and **molluscum contagiosum** but is considered dangerously toxic. Inhibits **protein phosphatase** 2A.

CAP 1. **Catabolite gene-activator protein**. 2. **Cyclase associated protein**. 3. See **CAP proteins**. 4. See **cap-binding protein**.

cap-binding protein Protein (24 kDa) with affinity for cap structure at 5′-end of mRNA that probably assists, together with other initiation factors, in binding the mRNA to the 40S ribosomal subunit. Translation of mRNA *in vitro* is faster if it has a cap-binding protein.

CAP proteins CAP1 is **FADD**; CAP2 is hyperphosphorylated FADD; CAP3 is an intermediate of procaspase-8 processing, CAP4 is pro-FLICE; CAP5, CAP6 are cleaved prodomains of FLICE (**caspase 8**).

CAP-18 *CAP18* Lipopolysaccharide-binding protein (18 kDa) isolated from rabbit neutrophils. May mediate interaction of antimicrobial **protegrins** with surfaces of Gram-negative bacteria. Rabbit **cathelicidin**.

capacitance flicker Brief closings of an **ion channel** during its open phases, observed during **patch clamp**; or rapid transition of an ion channel between open and closed states such that the individual channel openings cannot be distinguished properly due to the limited bandwidth of the patch clamp amplifier.

capacitation A process occurring in mammalian sperm after exposure to secretions in the female genital tract. Surface changes take place probably involved with the **acrosome** which are necessary before the sperm can fertilize an egg.

capecitabine A fluoropyrimidine carbamate with antineoplastic activity. It is an orally available systemic prodrug converted in the liver by a 60-kDa carboxylesterase into 5′-deoxy-5-fluorouridine (5′-DFUR), which is converted by cytidine deaminase to **5-fluorouracil**.

CapG See **gCap39**.

capillary The small blood vessels that link arterioles with venules. Lumen may be formed within a single endothelial cell, and have a diameter less than that of an erythrocyte, which must deform to pass through. Blood flow through capillaries can be regulated by precapillary sphincters, and each capillary probably only carries blood for part of the time.

capillary electrophoresis High resolution electrophoretic method for separating compounds, such as peptides, by electro-osmotic flow through long ($\sim$100 cm), narrow ($<$100 μm) silica columns. With sensitive detectors, attomolar concentrations can be detected.

capnine *2-Amino-3-hydroxy-15-methylhexadecane-1-sulfonic acid* Sulphonolipid, a sulphonic acid derivative of ceramide, isolated from the envelope of the Cytophaga/Flexibacter group of **Gram-negative** bacteria. The acetylated form of capnine seems to be necessary for gliding motility.

capnophilic Organisms that grow best at concentrations of carbon dioxide higher than in air.

capping 1. Movement of cross-linked cell-surface material to the posterior region of a moving cell or to the perinuclear region. **2.** The intracellular accumulation of intermediate filament protein in the pericentriolar region following microtubule disruption by colchicine. **3.** The blocking of further addition of subunits by binding of a cap protein to the free end of a linear polymer such as actin. See also **cap-binding protein**.

capping protein *Cap32/34* Capping protein from the Z-disc of muscle, **CapZ**.

cappuccino 1. Actin-nucleation factor that regulates the onset of ooplasmic streaming in *Drosophila*. Like spire, a maternal-effect locus that participates in pattern formation in both the anteroposterior and dorsoventral axes of the early embryo. **2.** Product of a gene mutated in mouse **Hermansky–Pudlak syndrome** model, co-assembles with **pallidin** and **muted** proteins in the **BLOC-1** complex. Maps to mouse chromosome 5 and syntenic human chromosome 4p.

capsaicin *8-Methyl-N-vanillyl-6-nonenamide* Molecule in chilli peppers that makes them hot and will stimulate release of neurogenic peptides (Substance P, neurokinins) from sensory neurons. Acts on the **vanilloid receptor-1** (VR1). Can be used to desensitize nociceptors to which it binds – and which may eventually be killed.

capsazepine Competitive **capsaicin** antagonist.

Capsicum Genus that includes red peppers, pimentoes and green peppers. See **capsaicin**.

capsid A protein coat that covers the nucleoprotein core or nucleic acid of a virion. Commonly shows icosahedral symmetry and may itself be enclosed in an envelope (as in the **Togaviridae**). The capsid is built up of subunits (some integer multiple of 60, the number required to give strict icosahedral symmetry) that self-assemble in a pattern typical of a particular virus. The subunits are often packed, in smaller capsids, into five- or six-membered rings (pentamers or hexamers) that constitute the morphological unit (capsomere). The packing of subunits is not perfectly symmetrical in most cases, and some units may have strained interactions and are said to have quasi-equivalence of bonding to adjacent units.

capsomeres See **capsid**.

capsule 1. *Bact.* Thick gel-like material attached to the wall of Gram-positive or Gram-negative bacteria, giving colonies a 'smooth' appearance. May contribute to pathogenicity by inhibiting phagocytosis. Mostly composed of very hydrophilic acidic polysaccharide, but considerable diversity exists. **2.** *Path.* Cellular response in invertebrate animals to a foreign body too large to be phagocytosed. A multicellular aggregate of **haemocytes** or **coelomocytes** isolates the foreign object. In some insects the capsule is apparently acellular and composed of **melanin**. **3.** *Anat.* Dense connective tissue sheath surrounding an organ.

captopril **ACE inhibitor** used to treat hypertension.

capZ *Capping protein; cap32/34* Microfilament capping protein (32–36 kDa) originally found in *Dictyostelium* and *Acanthamoeba* and that binds to the barbed ends of thin filaments in the **Z-disc** of striated muscle. Widely distributed in vertebrate cells, though in non-muscle cells is predominantly in the nucleus. Identical to β-actinin. Some isoforms may bind to the pointed ends of microfilaments.

Carassius auratus The goldfish, one of the carp family.

carbachol *Carbamoyl choline* Parasympathomimetic drug formed by substituting the acetyl of acetylcholine with a carbamoyl group; acts on both **muscarinic** and **nicotinic acetylcholine receptors** and is not hydrolysed by **acetylcholine esterase**.

carbamazepine Anticonvulsant drug used to treat many forms of epilepsy, trigeminal neuralgia, phantom-limb pain and manic-depressive illness resistant to lithium.

carbamoyl Acyl group, –CO–NH$_2$.

carbamoylcholine See **carbachol**.

carbamyl *Carbamoyl* Obsolete: use carbamoyl.

carbendazim *Methyl benzimidazol-2-ylcarbamate* A systemic benzimidazole fungicide used extensively in plant disease control but there are concerns about its endocrine disrupting effects. Probably works as a fungicide by interfering with spindle formation at mitosis.

carbenicillin A semisynthetic penicillin-class antibiotic that interferes with final cell wall synthesis of susceptible bacteria.

carbimazole Prodrug that is converted by the body into the active agent, methimazole, which inhibits **thyroxine** synthesis by preventing peroxidase enzyme from coupling and iodinating the tyrosine residues on **thyroglobulin**; used in the treatment of thyrotoxicosis.

carbohydrase Old name for **glycosidase**.

carbohydrates Very abundant compounds with the general formula $C_n(H_2O)_n$. The smallest are monosaccharides like glucose; polysaccharides (e.g. starch, cellulose, glycogen) can be large and indeterminate in length.

carbomer Homopolymer of acrylic acid, cross-linked with an allyl ether. It is used as an emulsion stabilizer as well as an aqueous viscosity-increasing agent in many cosmetics.

carbon replica A surface replica of a specimen for examination in the EM. The specimen is coated with a structureless carbon film by vacuum deposition, and the film and specimen are subsequently removed by dissolving in an appropriate solvent.

carbonic anhydrase *EC 4.2.1.1; carbonate dehydratase* Enzyme that catalyses reversible hydration of carbon dioxide to carbonic acid. Intracellular enzyme of the erythrocyte, essential for the effective transport of carbon dioxide from the tissues to the lungs. Zinc is a co-factor.

carbonyl group Bivalent =C=O group found in aldehydes, ketones and carboxylic acids.

carboxy terminus See **c-terminus**.

carboxyglutamate γ-*Carboxyglutamate* An amino acid found in some proteins, particularly those that bind calcium. Formed by post-translational carboxylation of glutamate.

carboxyhaemoglobin *Carboxyhemoglobin* (US) Haemoglobin co-ordinated with carbon monoxide. The affinity of haemoglobin for CO is higher than for O_2, and binding is almost irreversible, hence the toxicity of CO.

carboxyl-terminal Src kinase homologous kinase *CHK; Csk homologous kinase* A negative regulatory kinase of the Src tyrosine kinase family; generally believed to inactivate Src-family tyrosine kinases (SFKs) by phosphorylating their consensus C-terminal regulatory tyrosine. Has been reported to regulate the expression of the chemokine receptor, CXCR4. CHK overexpression in neuroblastoma and astrocytoma cells inhibits their growth and proliferation and loss of CHK expression is associated with human brain tumours.

carboxypeptidase Enzymes (particularly of pancreas) that remove the C-terminal amino acid from a protein or peptide. Carboxypeptidase A, (EC 3.4.17.1) will remove any amino acid; carboxypeptidase B (EC 3.4.17.2) is specific for terminal lysine or arginine.

carboxysome Inclusion body (polyhedral body; 90–500 nm in diameter) found in some Cyanobacteria and autotrophic bacteria; contains **ribulose bisphosphate carboxylase** (RUBISCO) and is involved in carbon dioxide fixation.

carcinoembryonic antigen *CEA* Antigen found in blood of patients suffering from cancer of colon and some other diseases, and that is otherwise normally found in fetal gut tissue.

carcinogen An agent capable of initiating development of malignant tumours. May be a chemical, a form of electromagnetic radiation or an inert solid body.

carcinogenesis The generation of cancer from normal cells, correctly the formation of a **carcinoma** from epithelial cells but often used synonymously with **transformation**, **tumorigenesis**.

carcinoid Intestinal tumour arising from specialized cells with paracrine functions (APUD cells), also known as argentaffinoma. The primary tumour is commonly in the appendix, where it is clinically benign; hepatic secondaries may release large amounts of vasoactive amines such as serotonin to the systemic circulation.

carcinoma *Pl.* carcinomata Malignant neoplasia of an epithelial cell – by far the commonest type of tumour. Those arising from glandular tissue are often called **adenocarcinomas**. Carcinoma cells tend to be irregular with increased basophilic staining of the cytoplasm, and have an increased nuclear/cytoplasmic ratio and polymorphic nuclei.

carcinoma *in situ* Carcinoma that has not invaded or metastasized and remains in the tissue site of origin.

carcinomatosis *Carcinosis* Cancer widely disseminated throughout the body and thus highly malignant.

carcinosarcoma A mixed tumour with features of both carcinoma and sarcoma.

cardiac cell Strictly speaking, any cell of or derived from the cardium of the heart; however, often used loosely of heart cells.

cardiac glycoside Specific blockers of the **Na^+/K^+ pump** especially of heart muscle, e.g. **strophanthin**.

cardiac jelly Gelatinous extracellular material that lies between endocardium and myocardium in the embryo.

cardiac muscle Specialized striated but involuntary muscle, able to contract and expand indefinitely (for a lifetime anyway), found only in the walls of the heart. Responsible for the pumping activity of the vertebrate heart. The individual muscle cells are joined through a junctional complex known as the **intercalated disc,** and are not fused together into multinucleate structures as they are in **skeletal muscle**.

cardioblast An embryonic mesodermal cell which will differentiate into heart tissue.

cardiolipin A diphosphatidyl glycerol, purified from beef heart, that is also found in the membrane of *Treponema pallidum* and is the antigen detected by the Wasserman test for syphilis.

cardiotocograph Recording of the heart rate of a foetus.

cardiotoxins *Cobramine A; cobramine B; cobra cytotoxin; gamma toxin; membrane-active polypeptide* Active component of cobra (*Naja* spp.) venom; basic polypeptides of 57–62 amino acids with four disulfide bonds and a molecular weight of less than 7000. Cause skeletal and cardiac muscle contracture, interfere with neuromuscular and ganglionic transmission, depolarize nerve, muscle and blood cell membranes, thus causing hemolysis.

cardiotrophin-1 *CT-1* Cytokine (201 amino acids) belonging to the **IL-6 cytokine** family. Binds to hepatocyte cell lines and induces synthesis of various **acute phase proteins**, is a potent cardiac survival factor and supports long-term survival of spinal motoneurons.

cardiovirus Genus of viruses belonging to the Family **Picornaviridae**, isolated mostly from rodents, cause encephalitis and myocarditis.

carditis Inflammation of the heart, including pericarditis, myocarditis and endocarditis, according to whether the enveloping outer membrane, the muscle or the inner lining is affected.

carnitine β-*Hydroxy-β-trimethyl-aminobutyric acid* Compound that transports long-chain fatty acids across the inner mitochondrial membrane in the form of acyl-carnitine. Sometimes referred to as Vitamin Bt or Vitamin B7. See Table V1.

carnitine palmitoyltransferase system *CPT system* System that regulates fatty acid oxidation/ketogenesis in the liver and is itself switched off by malonyl CoA.

carnosine β-*Ala-His* Dipeptide found at millimolar concentration in vertebrate muscle.

Carnoy Fixative containing ethanol, chloroform and acetic acid. Better for nuclear structure than for cytoplasm.

carotenes Hydrocarbon **carotenoids** usually with 9 conjugated double bonds. Beta-carotene is the precursor of Vitamin A, each molecule giving rise to two Vitamin A molecules.

carotenoids Accessory lipophilic photosynthetic pigments in plants and bacteria, including **carotenes** and **xanthophylls**; red, orange or yellow, with broad absorption peaks at 450–480 nm. Act as secondary pigments of the **light-harvesting system**, passing energy to **chlorophyll** and as protective agents, preventing photoxidation of chlorophyll. Found in chloroplasts and also in plastids in some non-photosynthetic tissues, e.g. carrot root.

carotid body cell Cells derived from the neural crest, involved in sensing pH and oxygen tension of the blood.

carrageenan *Carrageenin* Sulphated cell wall polysaccharide found in certain red algae. Contains repeating sulphated disaccharides of galactose and (sometimes) anhydrogalactose. It is used commercially as an emulsifier and thickener in foods and is also used to induce an inflammatory lesion when injected into experimental animals (probably activates **complement**).

carrier 1. In human genetics, a person heterozygous for a recessive disorder. **2.** A non-radioactive compound added to a tracer quantity of the same compound, which is radio-labelled. **3.** A molecule or molecular system which brings about the transport of a solute across a cell membrane either by active transport or facilitated diffusion. **4.** An organism harbouring a parasite but showing no symptoms of disease, especially if infectious to others. **5.** A more-or-less inert material used either as a diluent or vehicle for the active ingredient of, for example, a fungicide, or as a support for, for example, cells in a bioreactor.

carrier protein 1. Protein to which a specific ligand or hapten has been conjugated, used for raising an antibody. **2.** Unlabelled protein added into an assay system at relatively high concentrations which distributes in the same manner as the labelled protein analyte that is present in very low concentrations. **3.** Protein added to prevent non-specific interaction of reagents with surfaces, sample components and each other; albumin is often used for this purpose. **4.** Protein found in cell membranes, which facilitates transport of small molecules across the membrane.

CART The cocaine- and amphetamine-regulated transcript (CART) peptide: a recently characterized hypothalamic neuropeptide implicated in the control of appetite.

cartilage **Connective tissue** dominated by **extracellular matrix** containing **collagen** type II and large amounts of **proteoglycan**, particularly **chondroitin sulphate**. Cartilage is more flexible and compressible than bone and often serves as an early skeletal framework, becoming mineralized as the animal ages. Cartilage is produced by **chondrocytes** that come to lie in small lacunae surrounded by the matrix they have secreted.

caryotype See **karyotype**.

Cas *p130Cas* Protein encoded by *Crkas* gene (**Crk**-associated protein). Adaptor molecule with SH3 domain, multiple YXXP motifs and proline-rich region. Involved in induction of cell migration and apparently contributes to tumour invasiveness. Highly homologous to p105HEF.

Cas-family proteins *Crk-associated substrate family* Large multidomain molecules that transmit signals as intermediaries through interactions with signalling molecules

such as **FAK** and other tyrosine kinases, as well as tyrosine phosphatases. After Cas is tyrosine-phosphorylated, it acts as a docking protein for binding SH2 domains of Src-family kinases. See **p130cas**. Members include **Human enhancer of filamentation 1** (HEF1), Src-interacting protein (Sin)/Efs, Cas-L (Crk-associated substrate lymphocyte type), Nedd9.

casamino acid Acid hydrolysed **casein** used as a nitrogen source in some culture media.

casein Group of proteins isolated from milk. α_s and β-caseins are amphipathic polypeptides of around 200 amino acids with substantial hydrophobic C-terminal domains that associate to give micellar polymers in divalent cation-rich medium. κ-casein is a glycoprotein rather different from α- and β-casein.

casein kinase *CKII* Casein kinase II is thought to regulate a broad range of transcription factors in which it binds the **basic leucine zipper** (bZIP) DNA-binding domains. CKII is present in brain and has been associated with process of **long-term potentiation** by phosphorylating proteins important for neuronal plasticity.

casein kinase II phosphorylation site Casein kinase II phosphorylates exposed Ser or sometimes Thr residues, provided that an acidic residue is present three residues from the phosphate acceptor site. Consensus pattern: (S/T)-x-x-(D/E).

caseous necrosis *Caseation* The development of a necrotic centre (with a cheesy appearance) in a tuberculous lesion.

CASK A member of the membrane-associated guanylate kinase (MAGUK) family of proteins. CASK is present in the nervous system where it binds to presynaptic **neurexin**. The *Drosophila* homologue of CASK is CAKI or CAMGUK.

Casparian band Region of plant cell wall specialized to act as a seal to prevent back-leakage of secreted material (analogous to **tight junction** between epithelial cells). Found particularly where root parenchymal cells secrete solutes into xylem vessels.

caspases Family of peptidases involved in processing of **IL-1β** and in **apoptosis**. See Table C1.

caspofungin *Cancidas* See **echinocandins**.

CASPR1 *Contactin-associated protein-1* One of the transmembrane adhesion/signalling proteins, part of the **neurexin** superfamily, involved in the structural definition of both neurone-to-neurone and bidirectional neurone–glia communication. Vertebrate homologue of *Drosophila* neurexin IV, also named **paranodin**.

cassette A pre-existing structure into which an insert can be moved. Fashionably used to refer to certain vectors. See **cassette mechanism**.

cassette mechanism Term used for genes such as the a and α-genes that determine **mating-type** in yeast; either one or the other is active. In this **gene conversion** process, a double-stranded **nuclease** makes a cut at a specific point in the MAT (mating-type) locus; the old gene is replaced with a copy of a silent gene from one or other flanking region, and the new copy becomes active. As the process involves replacing one ready-made construct with another in an active 'slot', it is called a cassette mechanism.

TABLE C1. Human caspases

Name	Synonyms	Substrate	
Caspase-1[a]	ICE	pro-IL-1β	
Caspase-2	Ich-1$_L$		I
Caspase-3	CPP32, Yama, apopain	PARP, PKCδ, actin, Gas2, PAK2, procaspases 6, 9, U1-SnRNP	E
Caspase-4[a]	Tx/Ich-2, ICE$_{rel}$-II	pro-ICE	
Caspase-5[a]	ICE$_{rel}$-III, Ty		
Caspase-6	Mch2	Lamins A, C, B1	E
Caspase-7	Mch3, CMH-1, ICE-LAP3	PARP	E
Caspase-8	Mch5, MACH, FLICE	Procaspases 3, 4, 7, 9	I
Caspase-9	Mch6, ICE-LAP6		I
Caspase-10	Mch4	Procaspases-3, 7	I
Caspase-11[a]		Procaspases-1, 3 and pro-IL-1beta	I
Caspase-12[a]		Procaspase 3	
Caspase-13			b
Caspase-14		Involved in differentiation of skin	

I, 'Initiator' caspases – activate effector caspases; E, 'Effector' caspases – cleave, degrade or activate other proteins.
[a] Involved in inflammation.
[b] May be bovine not human.

castanospermine Alkaloid inhibitor of **α-glucosidase I** of which the effect is to leave N-linked oligosaccharides in their 'high-mannose', unmodified state.

Castleman's disease Disease characterized by lymph node swelling, hypergammaglobulinaemia, increased levels of **acute phase proteins** and increased numbers of platelets. Probably caused by excess **IL-6** production.

CAT See **chloramphenicol acetyltransferase**.

catabolin Protein, later shown to be interleukin-1 (**IL-1**), that stimulates the breakdown of connective tissue extracellular matrix.

catabolism The sum of all degradative processes, the opposite of anabolism.

catabolite Product of catabolism, the breakdown of complex molecules into simpler ones.

catabolite gene–activator protein *CAP* Protein from *E. coli* that regulates the expression of genes for the use of alternative carbon sources if glucose is not available. For example, CAP and the **lactose repressor** protein act together to enable lactose to be utilized. If glucose is present CAP will not bind to DNA, in the absence of glucose cAMP levels rise, CAP binds cAMP, undergoes a conformational change and is then capable of binding to DNA and promoting transcription of derepressed genes.

catabolite repression Inducible enzyme systems in some microorganisms (such as the **lactose operon**) that are repressed when a more favoured carbon source, such as glucose, is available. Repression in *E. coli* is partially relieved if cAMP is bound to the cAMP-catabolite activator protein (cAMP receptor protein; CRP) that binds to DNA upstream of the repressed operon concerned. Catabolite repression (of the respiratory system) is seen in yeast in high glucose concentrations, though the mechanism is different.

catalase *EC 1.11.1.6* Tetrameric haem enzyme (245 kDa) that breaks down hydrogen peroxide.

catalytic antibody *Abenzyme; abzyme* Antibody raised against a transition-state analogue (e.g. a phosphate analogue of a carboxylic acid–ester transition state) that can then catalyse the analogous chemical reaction, though not as effectively as a true enzyme.

catalytic RNA Species of RNA that catalyse cleavage or transesterification of the phosphodiester linkage. Operates in the self-splicing of group I and group II introns and in the maturation of various tRNA species.

cataplerosis The net loss due to consumption or degradation of intermediates in a biochemical cycle; the opposite of anaplerosis.

cataract Opacity of the lens of the eye.

catch muscle See **adductor muscle**.

catechol oxidase *EC 1.10.3.1; polyphenol oxidase* Any of a family of oxidoreductases that are involved in catalysing oxidation of mono- and ortho-diphenols to ortho-diquinones. Most are found in plants and contain copper; they are responsible for the brownish discolouration of, for example, potatoes and apples when cut open.

catecholamine A type of **biogenic amine** derived from tyramine, characterized as alkylamino derivatives of *o*-dihydroxybenzene. Catecholamines include **adrenaline**, **noradrenaline** and **dopamine**, with roles as **hormones** and **neurotransmitters**.

catenate Two or more circular DNA molecules where one or more circles run through the enclosed space of another, like links in a chain.

catenins Proteins associated with the cytoplasmic domain of **uvomorulin** and presumably involved in linking to the cytoskeleton. Alpha-catenin (102 kDa) is a **vinculin**-related protein involved in adherens junction-mediated intercellular adhesion. Beta-catenin (88 kDa) is a central component of the cadherin cell-adhesion complex and binds E-cadherin and N-cadherin. Co-immunoprecipitates with the adenomatous **polyposis coli** (APC) tumour-suppressor protein, interacts with Tcf and Lef transcription factors and is an essential member of the Wingless-**Wnt** signal transduction pathway. Ubiquitination of beta-catenin is greatly reduced in wnt-expressing cells. Gamma-catenin (80 kDa) (**plakoglobin**), associates with N-cadherin and E-cadherin and is a major component of desmosomes.

cathelicidin An extensive group of mammalian cationic antimicrobial peptides. Generally stored in secretory granules of neutrophils, although may be found elsewhere. They are synthesized as precursor propeptides of 16–26 kDa that contain a structurally varied C-terminal cationic region corresponding to the antimicrobial peptide, joined to a conserved cathelin-like propiece of ~100 amino acid residues. The mature peptide is liberated by limited proteolysis by elastase or proteinase 3. Examples are LL-37/hCAP18, CAP-18, SMAP-29, BMAP-27 and BMAP-28, the protegrin PG-1 and indolicidin.

cathelin Protein (11 kDa) isolated from porcine neutrophils that was originally described as being a cysteine-peptidase inhibitor though subsequent reports have suggested that this is due to contamination with **PLCPI**. Cathelin-like sequences are found upstream of several antimicrobial peptides (protegrins) and are expressed in the propeptides. See **cathelicidins**.

cathepsins Intracellular proteolytic enzymes of animal tissues, such as cathepsin-B (EC 3.4.22.1), a lysosomal cysteine endopeptidase; cathepsin-C, dipeptidyl peptidase (EC 3.4.14.1); cathepsin-D (EC 3.4.23.5) that has pepsin-like specificity; cathepsin-G (EC 3.4.23.5), similar to chymotrypsin; cathepsin-H that possesses aminopeptidase activity; cathepsin-N that attacks N-terminal peptides of collagen and so on.

cationic proteins Proteins of azurophil granules of neutrophils, rich in arginine. A chymotrypsin-like peptidase found in azurophil granules is also very cationic as is cathepsin-G and neutrophil elastase. Eosinophil cationic protein (21 kDa) is particularly important because it damages **schistosomula** *in vitro*.

cationized ferritin Ferritin, treated with dimethyl propanediamine and used to show, in the electron microscope, the distribution of negative charge on the surface of a cell. The amount of cationic ferritin binding is very approximately related to the surface charge.

caudate nucleus The most frontal of the **basal ganglia** in the brain. Damage to caudate neurons is characteristic of **Huntington's chorea** and other motor disorders.

Caulobacter Genus of Gram-negative aerobic bacteria that have a stalk or holdfast. Found in soil and freshwater.

caveola *Pl.* caveolae Small invagination of the plasma membrane characteristic of many mammalian cells and associated with endocytosis. The membrane of caveolae contains integral membrane proteins **caveolins** (21–24 kDa) that interact with heterotrimeric G-proteins. Caveolar membranes are enriched in cholesterol and sphingolipids and may be the efflux route for newly synthesized lipids. **Clathrin** is not associated with caveolae.

caveolin scaffolding peptide-1 *CSP-1* Peptide corresponding to the **caveolin**-1 scaffolding domain, the region that is essential for caveolin interaction with signalling molecules. The peptide will block the ability of noradrenaline and histamine to induce changes in the internal calcium ion concentrations of vascular smooth muscle cells by inhibiting the activation of phospholipase C-beta3 and MAPK.

caveolins Integral membrane proteins that are essential structural components of caveolae and serve as a scaffolding onto which signalling molecules, particularly G-protein-coupled receptors, are assembled and also act as negative regulators of signal transduction. There are three closely related human isoforms, caveolin-1, -2 and -3 that are differentially expressed in various cell types.

Cavia porcellas Guinea pig.

C banding *Centromeric banding* Method of defining chromosome structure by staining with **Giemsa** and looking at the **banding pattern** in the heterochromatin of the **centromeric** regions. Giemsa banding (G banding) of the whole chromosome gives higher resolution. Q banding is done with quinacrine.

CBC Abbrev. for complete blood count, a measure of the numbers of erythrocytes, leucocytes and platelets per unit volume of blood.

CBHA 1. A synthesized histone deacetylase (**HDAC**) inhibitor, m-carboxycinnamic acid bis-hydroxamide. **2.** Cellobiohydrolase A (CbhA).

cbl Oncogene of a murine retrovirus that induces lymphomas and leukaemias. Protein has N-terminal transforming region with phosphotyrosine-binding (PTB) domain and C-terminal region with **RING finger motif**, large proline-rich region and a leucine zipper. Cbl serves as a substrate for receptor and non-receptor tyrosine kinases, and also as a multidomain adaptor protein, binding to **Grb-2**, **crk** and **p85** of PI-3-kinase. Acts as a negative regulator of tyrosine kinase signalling through its **E3 ubiquitin ligase** activity. The transforming mutant v-Cbl lacks the ubiquitin ligase RING finger domain. See also Table O1.

Cbl The ubiquitin ligases c-Cbl and Cbl-b play a crucial role in receptor downregulation by mediating multiple mono-ubiquitination of receptors and promoting their sorting for lysosomal degradation. Their function is modulated through interactions with regulatory proteins including **CIN85** and **PIX**.

CBP See **CREB-binding protein**.

CC10 See **uteroglobin**.

CC10, CC16 See **Clara cell secretory protein**.

CCAAT box Consensus sequence for RNA polymerase, found at about –80 bases relative to the transcription start site. Less well conserved than the **TATA box**.

CCCP m-*Chloro-carbonylcyanide-phenylhydrazine* An **uncoupling agent** that dissipates proton gradients across membranes.

CCK See **cholecystokinin**.

CCR5 *CKR5; CD195; CKR-5; CCCKR5; CMKBR5; CC-CKR-5* A member of the beta-chemokine receptor family, predicted to be a seven-transmembrane protein similar to G-protein-coupled receptors, present in various cells, especially macrophages, monocytes and T-cells. It is the co-receptor that HIV uses to gain entry into macrophages, and defective alleles of this gene have been associated with HIV infection resistance. Ligands of this receptor include monocyte chemoattractant protein 2 (MCP-2), macrophage inflammatory protein-1 alpha (MIP-1 alpha), macrophage inflammatory protein-1 beta (MIP-1 beta) and regulated on activation normal T-expressed and secreted protein (RANTES).

CCVs See **clathrin-coated vesicles**.

CD antigens See table of 'cluster of differentiation' antigens (Table C2).

***cdc* genes** Cell division cycle genes, of which many have now been defined, especially in yeasts. See **cyclin**. The cyclin-dependent kinases are also known as cdc2 kinases.

cdc42 *Cell division cycle 42; FDG-1* Member of the **rho** GTPase family that regulates signalling pathways controlling diverse cellular functions including cell morphology, migration, endocytosis and cell cycle progression. Thought to induce filopodium formation by regulating actin polymerization at the cell cortex by activating the **Arp2/3 complex** in concert with **WASP** family proteins. The product of oncogene *Dbl* catalyses the dissociation of GDP from this protein. Alternative splicing gives rise to at least two transcript variants.

TABLE C2. The first 200 human CD antigens

Cluster designation	Main cellular expression of antigen	Other Names	Apparent MW (kDa)	Comment
CD1	Cortical thymocytes (strong), Langerhans Cells, B-cell subset, dendritic cells	T6, Ly-38	43–49	Associated with β2microglobulin. Similar to MHC class I, probably have role in presentation
CD2	T-cells, thymocytes, NK-cells	E Rosette receptor, leucocyte function antigen-2 (LFA-2)	50	Interacts with CD58. Ig superfamily
CD3	Thymocytes, mature T-cells	T-cell receptor complex (CD3γ, CD3δ, CD3ϵ)	20, 20, 25	Complex of several proteins
CD4	T-helper/inducer cells, monocytes, macrophages	Receptor for MHC class II and HIV antigens	55	Ig superfamily member. Binds MHC class II and is an important accessory molecule
CD5	Thymocytes, T-cells, B-cell subset	T1	67	Co-precipitates with TCR. CD72R
CD6	Thymocytes, T-cells, neurons	T12	100–130	Extracellular domains similar to CD5. Ligand CD166
CD7	Majority of T-cells	gp40	40	Unknown function
CD8a, b	T-Cytotoxic/suppressor cells	MHC class 1 Co-receptor (with TCR); Ly-3	32–34, 30–32	Co-receptor in antigen recognition
CD9	Pre-B cells, monocytes, platelets	MRP-1	22–27	Major component of platelet surface, TM4 superfamily. Role in adhesion
CD10	Lymphoid progenitor cells, granulocytes	CALLA (common acute lymphoblastic leukaemia antigen); neutral endopeptidase (NEP)	100	Zinc-binding metalloproteinase/enkephalinase
CD11a	Leucocytes	leucocyte function antigen (LFA-1) Integrin α_L subunit	180	Associates with CD18. Binds ICAM-1,2,3
CD11b	Granulocytes, monocytes, NK-cells	Integrin α_M subunit, Mac-1; CR3, C3biR	170	Associates with CD18, upregulated in inflammation
CD11c	Granulocytes, monocytes, NK-cells, B-cell subset, T-cell subset	Integrin α_X subunit, p150/95; CR4	150	Associates with CD18, upregulated in inflammation
CD11d	PBLs, splenic macrophages, foam cells	Integrin α_D subunit	125	Binds ICAM-3 but not ICAM-1 or VCAM
CDw12	Granulocytes, monocytes	–	(90–120)	A phosphoprotein, function unknown

TABLE C2. (Continued)

Cluster designation	Main cellular expression of antigen	Other Names	Apparent MW (kDa)	Comment
CD13	Myeloid cells and various tissue cells	Aminopeptidase N	150–170	May inactivate small signalling peptides
CD14	Monocytes, some granulocytes and macrophages	Mo2	53–55	GPI-linked, LPS receptor
CD15	Granulocytes, monocytes	Lewis X	carbohydrate	Present on many proteins
CD15s	Neutrophils	sialyl-Lewis x (sLe-x)	(pentasaccharide)	Ligand for CD62 structures
CD16a	Macrophage, NK-cells (neutrophils)	Transmembrane form, FcγRIIIA/FcγRIIIB	50–60	Low-affinity receptor for aggregated IgG, structural similarities with CD64, CDw32
CD16b	Granulocyte form only	GPI-linked form, FcγRIIIb	48	
CDw17	Granulocytes, monocytes, platelets	Lactosylceramide	Not determined	
CD18	Leucocytes; platelets negative	Integrin β_2 subunit	95	Associates with CD11a,b,c,d
CD19	Pan B-cell except plasma cells	B$_4$	95	Involved in regulation of B-cell proliferation
CD20	Pan B-cell, T-subset	Ly-44; B1	33–37	May be B-cell calcium channel
CD21	Mature B-cells, follicular dendritic cells	C3d/EBV-receptor (CR2)	145, 110	Forms complexes with CD19 and CD81
CD22	Mature B-cells, hairy cell leucaemia cells	BLCAM	α 71 β 93	Heterodimeric (α and β subunits). IgSF. Reduces B-cell activation threshold. Interacts with Shp-1
CD23	Activated B-cells, activated macrophages, eosinophils, platelets	IgE Fc low-affinity receptor (FcεRII)	45–49	Has C-type lectin domain. Involved in regulation of IgE production
CD24	B-cells, granulocytes, normal epithelium	Heat-stable antigen (HSA)	35–45	GPI-linked sialoglycoprotein
CD25	Activated T-cells, B-cells and macrophages	IL-2 Receptor α; Tac	55	Associates with CD122 to form high-affinity receptor
CD26	Activated T-cells and B-cells, macrophages	Dipeptidyl peptidase IV	110	T-cell co-stimulation, adhesion, proteolytic function
CD27	Thymocyte subset, mature T-cells, EBV-transformed B-cells	–	50–55	Member of the NGFR superfamily. CD70 receptor
CD28	T-cell subset, thymocytes NK-cells	Tp44	44	Binds CD80 (B7) and CD86. Important co-stimulator of T-cell activation
CD29	Ubiquitous (but not on erythrocytes)	Integrin β1 subunit; gpIIaa	130	Associates with CD29, 49, 51
CD30	Activated T- and B-cells, Reed–Sternberg cells	Ki-1 antigen	105–120	NGFR Superfamily. Binds CD153 and co-stimulates T-cell proliferation
CD31	Platelets, monocytes, macrophages, granulocytes, B-cells, some T-cells, endothelial cells	PECAM-1 (platelet endothelial cell-adhesion molecule-1)	130–140	Homotypic adhesion and heterotypic interaction with integrin αvβ3

TABLE C2. (Continued)

Cluster designation	Main cellular expression of antigen	Other Names	Apparent MW (kDa)	Comment
CD32	Monocytes, granulocytes, B-cells, eosinophils	Fcγ Receptor II	40	Low-affinity receptor for aggregated IgG
CD33	Myeloid progenitor cells, monocytes	gp67	67	May mediate cell–cell adhesion or myelopoiesis
CD34	Haematopoietic precursor cells, capillary endothelial cells	Sgp90; mucosialin	105–120	Sialylated forms are ligand for L- and E-selectin
CD35	Granulocytes (basophils negative), monocytes, B-cells, erythrocytes, some NK-cells, follicular dendritic cells	Complement Receptor 1 (CR1), C3b Receptor	250	Binds complement C3b and C4b, enhances FcR-mediated phagocytosis
CD36	Monocytes, macrophages, platelets, B-cells (weakly), adipocytes	Platelet GPIV (IIIb)	88	Binds collagen, LDL, *P. falciparum* infected rbcs. Interacts with src kinases fyn, lyn and yes
CD37	Mature B-cells; T-cells and myeloid (weakly)	gp 52–40	40–52	TM4 superfamily (4 membrane-spanning domains)
CD38	Plasma cells, thymocytes, activated T-cells	T10	45	Cyclic ADP-ribose hydrolase/ADP-ribosyl cyclase
CD39	Activated B- and NK-cells, EBV-transformed B-cells, endothelium	gp80	78	Mediates B-cell homotypic adhesion. Has ecto-apyrase activity
CD40	B-cells, monocytes (weakly), some epithelia	gp50	48	NGFR superfamily, binds CD154 and essential for secondary immune response
CD41	Platelets, megakaryocytes	GPIIb, integrin αIIb	125	Associates with CD61 (β3 Integrin). Important for adhesion, defective in Glanzmann's thrombasthenia
CD42a	Platelets, megakaryocytes	GPIX	22	Defective in Bernard–Soulier syndrome
CD42b	Platelets, megakaryocytes	GPIba	145	Complexed with CD42a, binds von Willebrand factor
CD42c	Platelets, megakaryocytes	GPIbb	24	Complexed with CD42a
CD42d	Platelets, megakaryocytes	GPV	82	Complexed with CD42a–c
CD43	Leucocytes but not resting B-cells	Leucosialin, Sialophorin	115 and 135	Anti-adhesion molecule
CD44	Leucocytes, erythrocytes, platelets (weakly), brain cells	Pgp-1, H-CAM, extracellular matrix receptor III	80–95 and 130	Hermes antigen; multiple splice variants known
CD45	Pan leucocyte	Leucocyte common antigen (L-CA); T200	180–240	Protein tyrosine phosphatase, essential for signalling through TCR. Variants CD45-RO, RA, RB, RC
CD46	Haematopoietic and non-haematopoietic cells (not erythrocytes)	MCP (membrane co-factor protein)	56/66	Regulator of complement activation; binds C3b and C4b. Receptor for measles virus and *Strep. pyogenes*

TABLE C2. (Continued)

Cluster designation	Main cellular expression of antigen	Other Names	Apparent MW (kDa)	Comment
CD47	All cell types	Rh group associated; Integrin-associated protein (IAP)	47–52	Associates with β3 integrins, binds thrombospondin. Has role in leucocyte–endothelial interactions
CD47R	Less intense than CD47			Formerly CDw149
CD48	Many leucocytes, not granulocytes, platelets or rbcs	Blast-1	45	Sequence similarities with CD2, CD58, CD152
CD49a	Activated T-cells, endothelium	Very late antigen-1 (VLA-1); β1 Integrins	210	Associates with CD29. Binds laminin and collagen
CD49b	B-cells, monocytes, platelets	VLA-2	165	Mediates Mg^{2+}-dependent adhesion of platelets to collagen
CD49c	B-cells	VLA-3	125	Binds various ligands (laminin, fibronectin, collagen)
CD49d	Thymocytes, B-cells	VLA-4	150	Binds VCAM-1 and CS-1 domain of fibronectin
CD49e	Thymocytes, activated T-cells	VLA-5	135	Fibronectin receptor, enhances Fcγ-mediated phagocytosis
CD49f	Memory T-cells, thymocytes, monocytes	VLA-6	125	Laminin receptor on platelets, monocytes and T-cells
CD50	Leucocytes (platelets and erythrocytes negative)	ICAM-3	110–120	Ligand for LFA-1. May also have signalling function
CD51	Platelets, endothelial cells, various other cells	VNRα (Vitronectin receptor)	125, 24	α_V integrin chain (associates with CD61)
CD52	Leucocytes (not plasma cells, platelets or rbcs)	CAMPATH-1	21–28	GPI-anchored
CD53	Pan leucocyte	OX44	35–42	TM4 superfamily; possibly has signalling function
CD54	Endothelial cells, many activated cell types	ICAM-1	90–110	Binds CD11a/b/c/CD18 (β2 integrins)
CD55	Many haematopoietic cells and cells in contact with serum	DAF (decay accelerating factor)	60–70	GPI-anchored and transmembrane forms. Defective in paroxysmal nocturnal haemoglobinuria. Interacts with CD97
CD56	NK-cells, T-subset, neurons	NKH1, isoform of NCAM	175–185	Function unclear, probably adhesion
CD57	NK-cells, T-cells, B-cell subsets	HNK-1	110	Binds L- and P-selectin. Function obscure
CD58	Many haematopoietic and non-haematopoietic cells	LFA-3	55–70	Binds to CD2
CD59	Many haematopoietic and non-haematopoietic cells	p18, gP18, MAC inhibitor, MACIF	19	Protective against complement lysis. GPI-linked
CD60	Platelets, T-cell subset	UM4D4	carbohydrate	Subdivided into CD60a, b, c
CD61	Platelets, megakaryocytes, monocytes, macrophages, endothelium	Integrin β3, GPIIIa, Vitronectin receptor	105	Associates with CD41 or CD51
CD62E	Endothelium	E-Selectin, ELAM-1, LECAM-2	107–115	Binds sialyl-Lewis X (CD15s)

TABLE C2. (Continued)

Cluster designation	Main cellular expression of antigen	Other Names	Apparent MW (kDa)	Comment
CD62L	B- and T-cells, monocytes, NK-cells	L-Selectin, LECAM-1, LAM-1	74, 95	Mel-14 antigen, leucocyte homing
CD62P	Platelets, activated endothelial cells	P-Selectin, GMP-140, PADGEM	140	Mediates neutrophil rolling
CD63	Activated platelets, monocytes, macrophages	LAMP-3; Platelet activation antigen, granulophysin	53	TM4 superfamily
CD64	Monocytes, macrophages	High-affinity Fcγ Receptor 1, FCγR1	72	Will bind monomeric IgG
CD65	Granulocytes, monocytes	Fucoganglioside	Not confirmed	
CD66a	Neutrophil lineage cells	Biliary glycoprotein (BGP-1)	160–180	Adhesion
CD66b	Granulocytes	CEA gene member 6 (CGM6)	95–100	Previously CD67. Adhesion, neutrophil activation
CD66c	Neutrophils, colon carcinoma	Non-specific cross reacting antigen (NCA)	90	Adhesion
CD66d	Neutrophils	CEA gene member 1 (CGM1)	35	
CD66e	Adult colon, epithelia, colon carcinoma	Carcinoembryonic antigen (CEA)	180–200	
CD66f	Placental syncytiotrophoblasts	Pregnancy-specific glycoprotein (PSG)	54–72	Protects foetus against maternal immune system
CD68	Monocytes, macrophages, granulocytes, large lymphocytes	Macrosialin	110	Same family as CD107a and b (Lamp1 and 2)
CD69	Activated T- and B-cells, activated macrophages, NK-cells	gp34/28, AIM (activation inducer molecule)	28, 32	May function in signalling
CD70	Activated T- and B-cells, Reed–Sternberg cells, macrophages (weakly)	Ki-24, CD27L	50, 70, 90, 160	TNF superfamily. Co-stimulator of T- and B-cell activation
CD71	Activated T- and B-cells, macrophages, proliferating cells	T9, Transferrin receptor	95	Found as homodimer
CD72	Pan B-cell	LyB in mouse	42	Expressed as homodimer. Ligand for CD5. C-type lectin
CD73	B- and T-cell subsets	ecto-5'-nucleotidase, Lymphocyte vascular adhesion protein 2	69	GPI-linked
CD74	B-cells, macrophages, monocytes	gp41/35/33, Ii	41, 35, 33	MHC class II associated invariant chain
CD75	B-subset, T-subset	CD75 and CD75s; sialoglycan	Not confirmed	Now known not to be ligand for CD22
CD77	Activated B-cells, follicular centre	BLA, Gb3, b^k	Not confirmed	Burkitt's lymphoma-associated antigen
CD79a	B-cell specific	Igα	21	Forms heterodimer with CD79b associated with membrane Ig. B-cell receptor
CD79b	B-cell specific	B29, BCR, Igβ	23	Interacts with CD79a
CD80	B-cell subset *in vivo*, most activated B-cells *in vitro*	B7, B7-1	60	CD28 and CD152 ligand

TABLE C2. (Continued)

Cluster designation	Main cellular expression of antigen	Other Names	Apparent MW (kDa)	Comment
CD81	B-cells (broad expression including lymphocytes), T-cells, neutrophils	TAPA-1 (target of antiproliferative antibody)	26	Has a role in early T-cell development
CD82	Broad expression on leucocytes (weak), platelets, not erythrocytes	R2	50–53	Can transduce signals, will suppress metastasis
CD83	Marker for circulating dendritic cells, activated B- and T-cells, germinal centre cells	HB15	43	Function unknown
CDw84	Platelets and monocytes (strong), B-cells		73	Function unknown
CD85	Circulating B-cells (weak), monocytes (strong)		110	Inhibits NK-cells
CD86	Circulating monocytes, germinal centre cells, activated B-cells	B7-2	80	Ig superfamily. Structurally related to CD80. Has role in co-stimulation by interaction with CD28
CD87	Granulocytes, monocytes, macrophage, activated T-cells	UPa-R (urokinase plasminogen activator receptor)	39–66	May be important in extravasation
CD88	Polymorphonuclear leucocytes, mast cells, macrophage, smooth muscle	C5a receptor	40	Seven-membrane-spanning receptor
CD89	Neutrophils, monocytes, macrophage, T- and B-cell subpopulation	$Fc\alpha R$, IgA receptor	55–75	Ig superfamily member
CD90	$CD34^+$, subset on bone marrow, cord blood, foetal liver, neurons	Human Thy-1	25–35	Associated with fyn. GPI-anchored
CD91	Monocytes and some non-haemopoietic cell lines	$\alpha 2$ macroglobin receptor, LDL-receptor-related protein	600	May have a clearance function
CDw92	Neutrophils, monocytes, endothelial cells, platelets		70	Function unknown
CD93	Neutrophils, monocytes, endothelial cells		120	Function unknown
CD94	NK-cells, α/β, γ/δ, T-cell subsets	KP43	43	Role in recognition of MHC class I. C-type lectin
CD95	Variety of cell lines including myeloid and T–lymphoblastoid	Apo-1, FAS	45	NGFR superfamily. Antibody to Fas kills by apoptosis
CD96	Activated T-cells	TACTILE -(T-cell activation increased late expression)	160	Role in adhesion of activated T- and NK-cells
CD97	Activated T- and B-cells, granulocytes, monocytes	GR1, BL-KDD/F12	74, 80, 90	Function unknown but has seven-transmembrane domains
CD98	T-cells and B-cells (weak), monocytes (strong), most human cell lines	4F2	80, 40	Found as heterodimer with glycosylated heavy chain and non-glycosylated light chain

TABLE C2. (Continued)

Cluster designation	Main cellular expression of antigen	Other Names	Apparent MW (kDa)	Comment
CD99	Peripheral blood lymphocytes, thymocytes	MIC2, E2	32	Involved in T-activation and adhesion
CD99R	T-cells, NK-cells, myeloid cells, some leukaemias	MIC2, E2	32	Isoform of CD99
CD100	Broad expression on haemopoietic cells		150	A semaphorin. Involved in adhesion and activation
CD101	Granulocytes, macrophage, activated T–cells	V7, p126	120	Role in TCR/CD3 co-stimulation of T-cells
CD102	Resting lymphocytes, monocytes, vascular endothelial cells (strongest)	ICAM-2	55–65	Ligand for CD11a/CD18. Constitutively expressed
CD103	Intraepithelial lymphocytes	α_E integrin, α_6, HML-1	150, 25	Binds E-cadherin
CD104	Epithelia, Schwann cells, some tumour cells	β4 integrin chain, beta4	220	Interacts with CD49f
CD105	Endothelial cells, bone marrow cell subset, *in vitro* activated macrophage	Endoglin, receptor for TGF β1 and β3	95	Homodimer
CD106	Endothelial cells	VCAM-1, INCAM-110	110	Ligand for VLA4 (CD49d)
CD107a	Activated platelets	LAMP 1 (lysosomal-associated membrane protein)	110	Ligand for galaptin
CD107b	Activated platelets	LAMP 2	120	Ligand for galaptin
CD108	Activated T-cells in spleen, some stromal cells	GR2	80	Function unknown. GPI-linked
CD109	Activated T-cells, platelets, endothelial cells	Platelet activation factor, GR56	170	Function unknown. GPI-linked
CD110	Megakaryocytes, platelets	Thrombopoietin receptor	82–84	
CD111	Stem cell subset, neutrophils, macrophages	Nectin-1	64–72	Ig superfamily
CD112	Monocytes, neutrophils, various	Nectin-2	64–72	Ig superfamily
CD113	Testis, placenta	Nectin-3	83	Ig superfamily
CD114	Granulocytes, monocytes	G-CSF receptor	95, 139	Involved in myeloid differentiation
CD115	Monocytes, macrophage, placenta	M-CSFR (macrophage colony-stimulating factor receptor)	150	Encoded by c-fms proto-oncogene, a receptor tyrosine kinase
CD116	Monocytes, neutrophils, eosinophils, fibroblasts, endothelial cells	GM-CSF R α chain (granulocyte, macrophage colony-stimulating factor receptor)	75–85	Shares β subunit with IL3 and IL5 receptors
CD117	Bone marrow progenitor cells	Stem cell factor receptor, (SCF-R), c-KIT	145	Receptor tyrosine kinase
CD118	Epithelial cells	IFNα, β receptor; gp190	190	
CDw119	Macrophage, monocyte, B-cells, epithelial cells	IFNγ R (interferon γ receptor);	90–100	Interacts with JAK.
CD120a	Most cell types, higher levels on epithelial cell lines	TNFαR (tumour necrosis factor α receptor) Type I	55	NGFR superfamily

TABLE C2. (Continued)

Cluster designation	Main cellular expression of antigen	Other Names	Apparent MW (kDa)	Comment
CD120b	Most cell types, higher levels on myeloid cell lines	TNFα-R Type II	75–85	NGFR superfamily
CD121a	T-cells, thymocytes, fibroblasts, endothelial cells	IL-1R (interleukin-1 receptor) Type 1	80	
CD121b	B-cells, macrophages, monocytes	IL-1R, Type II	68	
CD122	NK cells, resting T-cells subpopulation, some B-cell lines.	IL-2Rβ	75	Associates with CD25.
CDw123	Bone marrow stem cells, granulocytes, monocytes, megakaryocytes	IL-3Rα	70	
CD124	Mature B- and T-cells, haemopoietic precursor cells	IL-4R, IL-13 receptor	140	
CDw125	Eosinophils and basophils	IL-5Rα chain	60	Associates with CDw131
CD126	Activated B-cells and plasma cells (strong), most leucocytes (weak)	IL-6Rα chain	80 (α subunit)	gp130 (CDw130) is the β subunit
CD127	Bone marrow lymphoid precursors, Pro-B-cells, mature T-cells, monocytes	IL-7Rα chain	65–75	CD132 is other subunit
CDw128	Neutrophils, basophils, T-cell subset	IL-8R	58–67	Re-assigned to CD181, 182
CD129	Not assigned	–	–	
CD130	Activated B-cells and plasma cells (strong), most leucocytes (weak), endothelial cells	gp130, Common β subunit for IL-6, IL11, LIF, OSM, LNTF, CT-1	130	Associates with CD126
CDw131	Myeloid cells, early B-cells	Common β chain for IL-3, IL-5, GM-CSF-R	95–120	
CD132	Broad	Common γ chain for IL-2, IL-4, IL-7, IL-9, IL-15 receptor	64	
CD133	Haematopoietic stem cells subset, epithelia	Prominin-like-1	120	
CD134	Activated T-subset	OX40	48–50	TNF-R superfamily
CD135	Progenitor cell subset	STK-1, Flt3, Flk2	130–150	Involved in growth and differentiation of primitive haematopoietic cells. A receptor tyrosine kinase
CDw136	Macrophages, epithelia	Macrophage-stimulating factor receptor	180	A receptor tyrosine kinase
CDw137	T-cell subset	4-1BB	30	Co-precipitates with lck.
CD138	Plasma cells, epithelial cells	syndecan-1	80–150	ECM receptor, possibly a co-receptor for FGF
CD139	Germinal centre B-cells		228	
CD140a, b	Fibroblasts, smooth muscle, glia, chondrocytes	PDGF-Rα,β	180	Receptor tyrosine kinases

TABLE C2. (Continued)

Cluster designation	Main cellular expression of antigen	Other Names	Apparent MW (kDa)	Comment
CD141	Endothelial cells	Thrombomodulin	100	C-type lectin
CD142	Activated monocytes and endothelial cells	Tissue factor	45	
CD143	Endothelial cell subsets	Angiotensin-converting enzyme (ACE)	170	
CD144	Endothelial cells	VE-cadherin	130	Cadherin-5
CDw145	Endothelial cells, some stromal cells		90, 110	
CD146	Endothelium, activated T-cells	MUC18	118	
CD147	Endothelium, monocytes, T-cell subset, platelets, erythrocytes	neurothelin, basigin, EMMPRIN	28	Extracellular matrix metalloproteinase inducer
CD148	Granulocytes, monocytes, dendritic cells, T-cells	HPTP-η/DEP-1	240–260	PTPase. May have role in growth regulation of epithelial cells
CD150	T- and B-cells, thymocytes	IPO-3/SLAM	75–95	Structural features similar to CD2, CD48, CD58
CD151	Platelets, endothelium	PETA-3	32	Member of TM4 superfamily
CD152	Activated T-cells and B-cells	CTLA-4	33	Negative regulator of T-cell activation. Similar to CD28, binds CD80 and CD86
CD153	Activated T-cells, neutrophils, macrophages	CD30 ligand	26	Member of TNF superfamily.
CD154	Activated T-cells	CD40 ligand	32–39	Member of TNF superfamily
CD155	Monocytes, macrophages, thymocytes	Poliovirus receptor	80–90	
CD156a	Monocytes, neutrophils	ADAM-8	69	Cell-surface metallopeptidase
CD156b	Broad	TACE, ADAM-17	100	Cell-surface metallopeptidase
CD156c	Lymphoid organs, blood leucocytes	ADAM-10	98	Cell-surface metallopeptidase, cleaves Notch, TNF αAPP.
CD157	Monocytes, neutrophils, endothelium	BST-1	42–45	ADP-ribosyl-cADP-ribose hydrolase
CD158a	NK-cells, T-cell subset	p58.1, P50.1	50–58	Inhibits NK-cell activity
CD158b	NK-cells, T-cell subset	p58.2, P50.2	50–58	Inhibits NK-cell activity
CD159a, c	T-subset, NK	NKG2A, C	43, 40	Heterodimerizes with CD94
CD160	NK-subset, T-subset		27	Involved in co-stimulation
CD161	NK-cells, T-cell subset	NKR-P1 (NK receptor P1 family)	25	Has C-type lectin domain.
CD162	T-cells, monocytes, granulocytes, B-cell subset	PSGL-1 (P-selectin glycoprotein ligand 1)	120	Mediates neutrophil and T-cell tethering and rolling
CD163	Monocytes	KiM4	130	Scavenger receptor
CD164	T-cells, monocytes, granulocytes, B-cell subset	MGC-2	80	Involved in haematopoietic cell/stroma interaction.

TABLE C2. (Continued)

Cluster designation	Main cellular expression of antigen	Other Names	Apparent MW (kDa)	Comment
CD165	Thymocytes, thymic epithelium	AD2/gp37	42	Adhesion function
CD166	Activated lymphocytes, endothelium, fibroblasts, neurons	ALCAM, CD6 ligand	62	Ig superfamily. May have role in axonal guidance.
CD167a	Epithelia, myoblasts		120	Receptor tyrosine kinase, collagen adhesion
CD168	Monocytes, T-subset	RHAMM	84–88	Adhesion and migration
CD169	Tissue macrophage subset	Sialoadhesin	185	Adhesion.
CD170	Macrophage subset, neutrophils	Siglec, CD33-like2	140	Adhesion
CD171	Various	L1	200–210	Binds CD9, CD24, CD56, CD142, CD166
CD172a	Monocytes, T-subset	SIRPα	110	Forms complex with CD47
CD172b	Monocytes, dendritic cells	SIRPβ	50	Negative regulator of receptor tyrosine kinase signals
CD173	Erythrocytes, platelets	Blood group H type 2	Carbohydrate	
CD174	Stem cell subset, epithelia	Lewis Y	Carbohydrate	
CD175	Stem cell subset		Carbohydrate	
CD176	Stem cell subset	Thomson Friedrenreich Ag	Carbohydrate	
CD177	Neutrophil subset	NB1	56–62	
CD178	Activated T-, testis	Fas ligand	38–42	CD95 ligand
CD179a, b	B-cell progenitors		16–18, 22	
CD180	B-subset, monocytes, dendritic cells		95–105	Toll-like receptor
CD181	Neutrophils, basophils, NK-cells, T-subset, monocytes	CXCR1, IL8RA	39	Chemokine receptor
CD182	Neutrophils, basophils, NK-cells, T-subset, monocytes	CXCR2, IL8-RB	40	
CD183	Eosinophils, activated T- and NK-cells	CXCR3	40	IP-10 receptor, involved in T-cell recruitment in inflammation.
CD184	B-cells, T-subset, dendritic cells	CXCR4, fusin	45	HIV-1 co-receptor
CD185	B-cells	CXCR5	45	May have regulatory function in Burkitt's lymphoma
CDw186	Activated T-cells	CXCR6	40	
CD191	T-cells, monocytes	CCR1, MIP-1αR	39	Receptor for C–C type chemokines
CD192	Activated NK-cells and monocytes, T-subset, B-cells	CCR2	40	
CD193	Eosinophils, T-subset	CCR3	45	Binds eotaxin, various other chemokines; co-receptor for HIV-1
CD195	Monocytes, T-subset	CCR5	45	MIP receptor

TABLE C2. (Continued)

Cluster designation	Main cellular expression of antigen	Other Names	Apparent MW (kDa)	Comment
CD196	T-subset, B- and dendritic cells	CCR6	45	Binds MIP-3α
CD197	T-subset, dendritic cell subset	CCR7	45	
CDw198	T esp. Th2 cells, NK-cells, monocytes	CCR8	43	Involved in allergic inflammation; co-receptor for HIV-1
CDw199	Subset of T-memory cells	CCR9	43	Alternative co-receptor for HIV-1
CD200	Thymocytes, endothelium, B, activated T	OX-2	45–50	Inhibitor of immune response

The molecular weight quoted is the apparent molecular weight from behaviour on reducing gels.
As of 2005, the CD number set reached CD339. This table was updated and cross-checked against the complete list (CD1–CD339) produced (as a poster) by eBioscience (www.ebioscience.com), which also produces a list of mouse CD numbers. Other cells may express the antigen; this depends partly upon whether it has been investigated and the level of expression. The list above should not be taken as definitive.

CDEP Human protein (1045 amino acids) containing the **ezrin**-like domain of the band 4.1 superfamily, **Dbl homology** (DH) and **pleckstrin homology** (PH) domains. CDEP mRNA is expressed not only in the differentiated chondrocytes but also in various foetal and adult tissues.

cdk See **cyclin-dependent kinases**.

cDNA *Complementary DNA* Viral **reverse transcriptase** can be used to synthesize DNA that is complementary to RNA (e.g. an isolated mRNA). The cDNA can be used, for example, as a probe to locate the gene or can be cloned in the double-stranded form.

CDP See **cytidine 5′diphosphate**.

CDR3 Abbrev. for **complementarity determining region** 3.

CDRP *Calcium-dependent regulator protein* Early name for **calmodulin**.

cdt1 *Cdc10-dependent transcript 1* A licensing protein that is recruited first to the origin of DNA replication, followed by cell division cycle 6 (Cdc6) and minichromosome maintenance proteins (Mcms). *Drosophila* homologue is double-parked (Dup). Cdt1 is present in cells in G1 phase where it is required for initiation of replication but once replication has been initiated Cdt1 is either exported out of the nucleus or degraded, thereby preventing another round of replication. Also inhibited by **geminin**.

CEA See **carcinoembryonic antigen**.

C-EBP *C/EBP; CCAAT-enhancer binding protein* A family of transcription factors with a number of closely related members, particularly implicated in **adipocyte** differentiation; play a key role in both the development and maintenance of metabolically important processes.

cecropins **Antimicrobial peptides**, 30–40 residues, act by permeabilizing the membranes of bacteria. Although originally isolated from haemolymph of *Hyalophora cecropia* (silkmoth) pupae, are now found in several other species of endopterygote insects. A structural and functional equivalent, cecropin P1, has been isolated from the pig

intestine, although subsequently it appears that this is probably the product of an intestinal nematode parasite, *Ascaris suum*. Synthetic analogues are being tested for potential therapeutic use.

ced mutant Genes identified in *Caenorhabditis elegans* after studies of developmental mutations in which cells did not die when expected. *ced* ('cell death') genes are thus thought to be involved in the pathways which control **apoptosis**.

celecoxib *Celebrex* Non-steroidal anti-inflammatory drug that acts by selective inhibition of cycloxygenase 2 (**COX-2**).

C. elegans See *Caenorhabditis elegans*.

celiac See **coeliac**.

cell An autonomous self-replicating unit (in principle) that may constitute an organism (in the case of unicellular organisms) or be a subunit of multicellular organisms in which individual cells may be more or less specialized (differentiated) for particular functions. All living organisms are composed of one or more cells. Implicit in this definition is that viruses are not living organisms – and since they cannot exist independently, this seems reasonable.

cell adhesion See **adhesins**, **cadherins**, **cell-adhesion molecules** (CAMs), **contact sites A**, **DLVO theory**, **integrins**, **sorting out**, **uvomorulin** and various specialized junctions (**adherens junctions**, **desmosomes**, **focal adhesions**, **gap junction** and **zonula occludens**).

cell-adhesion molecule *CAM* Although this could mean any molecule involved in cellular adhesive phenomena, it has acquired a more restricted sense; CAMs are molecules on the surface of animal tissue cells, antibodies (or Fab fragments) against which specifically inhibit some form of intercellular adhesion. Examples are LCAM (Liver Cell-Adhesion Molecule) and NCAM (Neural Cell-Adhesion Molecule), both named from the tissues in which they were first detected, although they actually have a wider tissue distribution.

cell aggregation Adhesion between cells mediated by cellular adhesion mechanisms rather than by a cross-linking or agglutinating agent.

cell bank Collection of cells stored by freezing that can be used to re-initiate cell cultures or, potentially, in the case of stem cells, to restore normal function to the organism from which they were removed.

cell behaviour General term for activities of whole cells such as movement, adhesion and proliferation, by analogy with animal behaviour.

cell body Used in reference to **neurons**; the main part of the cell around the nucleus excluding long processes such as **axons** and **dendrites**.

cell centre **Microtubule-organizing centre** (MTOC) of the cell, the pericentriolar region.

cell coat Imprecise term occasionally used for the **glycocalyx**.

cell culture General term referring to the maintenance of cell strains or lines in the laboratory. See Table C3.

cell cycle The sequence of events between mitotic divisions. The cycle is conventionally divided into G0, G1, (G standing for gap), S (synthesis phase during which the DNA is replicated), G2 and M (mitosis). Cells that will not divide again are considered to be in G0, and the transition from G0 to G1 is thought to commit the cell to completing the cycle and dividing.

cell death Cells die (non-accidentally) either when they have completed a fixed number of division cycles (around 60; the **Hayflick limit**) or at some earlier stage when programmed to do so, as in digit separation in vertebrate limb morphogenesis. Whether this is due to an accumulation of errors or a programmed limit is unclear; some transformed cells have undoubtedly escaped the limit. See **apoptosis**.

TABLE C3. Common cell lines

Name	Species	Tissue of origin	Cell type	Comment
3T3	Mouse	Whole embryo	Fib	Swiss or Balb/c types; very density dependent
A9	Mouse	From L929	Fib	HGPRT negative
B16	Mouse	Melanoma	Mel	High and low metastatic variants (F1 and F10)
BHK21	Hamster	Baby kidney	Fib	Syrian hamster. Usually C13 (clone 13)
BSC-1	Monkey	Kidney	Fib	Derived from African Green monkey
CHO	Hamster	Ovary	Epi	Chinese hamster ovary
Daudi	Human	Burkitt's lymphoma	Lym	
Don	Hamster	Lung	Fib	Chinese hamster
EAT	Mouse	Ascites tumour	Fib	
GH1	Rat	Pituitary tumour	Fib	Secrete growth hormone
HeLa	Human	Cervical carcinoma	Epi	Established line
HEp2	Human	Laryngeal carcinoma	Epi	
HL60	Human	–	Myl	Will differentiate to granulocytes or macrophages
J 1 1 1	Human	Monocytic leukaemia	Myl	
L1210	Mouse	Ascites fluid	Lym	Grows in suspension; DBA/2 mouse
L929	Mouse	Connective tissue	Fib	Clone of L cell
MCIM	Mouse	Sarcoma	Fib	Methylcholanthrene-induced
MDCK	Dog	Kidney	Epi	Madin–Darby canine kidney
MOPC31C	Mouse	Plasmacytoma	Lym	Grows in suspension; secretes IgG
MRC5	Human	Embryonic lung	Fib	Diploid, susceptible to virus infection
P388D1	Mouse	–	Lym	Grows in suspension
PC12	Rat	Adrenal	Neur	Phaeochromocytoma; can be induced to produce neurites
PtK1	Potoroo	Female kidney	Epi	Small number of large chromosomes
PtK2	Potoroo	Male kidney	Epi	Cells stay flat during mitosis
Raji	Human	Burkitt's lymphoma	Lym	Grows in suspension. EB virus undetectable
S180	Mouse	Sarcoma	Fib	Invasive; maintained *in vivo*
SV40-3T3	Mouse	From 3T3	Fib	Transformed by SV40 virus
U937	Human	Monocytic leukaemia	Myl	Will differentiate to macrophages
Vero	Monkey	Kidney	Fib	Often used in virus studies
W138	Human	Embryonic lung	Fib	Diploid, finite division potential
WRC-256	Rat	Carcinoma	–	Walker carcinoma; many variants

Although there are a great many cell lines available through the cell culture repositories and from trade suppliers, there are a few 'classic' lines that will be met fairly frequently. Many of these well-known lines are listed here, but the table is not comprehensive.
Epi, epithelial; Fib, fibroblastic; Lym, lymphocytic; Mel, melanin-containing; Myl, myeloid; Neur, neural.

cell division The separation of one cell into two daughter cells, involving both nuclear division (**mitosis**) and subsequent cytoplasmic division (**cytokinesis**).

cell electrophoresis Method for estimating the surface charge of a cell by looking at its rate of movement in an electrical field; almost all eukaryotic cells have a net negative surface charge. Measurement is complicated by the streaming potential at the wall of the chamber itself and by the fact that the cell is surrounded by a layer of fluid (see **double layer**). The electrical potential measured (the zeta potential) is actually some distance away from the plasma membrane. One of the more useful modifications is to systematically vary the pH of the suspension fluid to determine the pK of the charged groups responsible (mostly carboxyl groups of sialic acid).

cell fate Of an embryonic parent (progenitor) cell or cell type, the range and distribution of differentiated tissues formed by its daughter cells. For example, cells of the **neural crest** differentiate to form (among other things) cells of the peripheral nervous system.

cell fractionation Strictly, this should mean the separation of homogeneous sets from a heterogeneous population of cells (by a method such as **flow cytometry**); however, the term is more frequently used to mean subcellular fractionation (i.e. the separation of different parts of the cell by differential centrifugation) to give nuclear, mitochondrial, microsomal and soluble fractions.

cell-free system Any system in which a normal cellular reaction is reconstituted in the absence of cells – for example, *in vitro* translation systems that will synthesize protein from mRNA using a lysate of rabbit reticulocytes or wheat germ. Mostly superseded by systems in which pure enzymes are used, as in many molecular biological procedures.

cell fusion Fusion of two previously separate cells occurs naturally in fertilization and in the formation of vertebrate skeletal muscle, but can be induced artificially by the use of **Sendai virus** or fusogens such as polyethylene glycol. Fusion may be restricted to cytoplasm or nuclei may fuse as well. A cell formed by the fusion of dissimilar cells is often referred to as a **heterokaryon**.

cell growth Usually used to mean increase in the size of a population of cells, though strictly should be reserved for an increase in cytoplasmic volume of an individual cell.

cell junctions Specialized junctions between cells. See **adherens junctions**, **desmosomes**, **tight junctions**, **gap junctions**.

cell line A cell line is a permanently established cell culture that will proliferate indefinitely, given appropriate fresh medium and space. Lines differ from cell strains in that they have escaped the **Hayflick limit** and become immortalized. Some species, particularly rodents, give rise to lines relatively easily, whereas other species do not. No cell lines have been produced from avian tissues, and the establishment of cell lines from human tissue is difficult. Many cell biologists would consider that a cell line is by definition already abnormal and that it is on the way towards becoming the culture equivalent of a neoplastic cell.

cell lineage The lineage of a cell relates to its derivation from the undifferentiated tissues of the embryo. Committed embryonic progenitors give rise to a range of differentiated cells: in principle it should be possible to trace the ancestry (lineage) of any adult cell.

cell locomotion Movement of a cell from one place to another.

cell-mediated immunity Immune response that involves effector T-lymphocytes and not the production of humoral antibody. Responsible for **allograft** rejection, delayed **hypersensitivity,** and in defence against viral infection and intracellular protozoan parasites.

cell membrane Rather imprecise term usually intended to mean **plasma membrane**.

cell migration Implies movement of a population of cells from one place to another – as in the movement of neural crest cells during morphogenesis.

cell movement A more general term than cell locomotion, this can include shape-change, cytoplasmic streaming, etc.

cellobiose Reducing **disaccharide** composed of two D-glucose moieties β-1,4 linked. The disaccharide subunit of cellulose, though not found as a free compound *in vivo*.

cellobiohydrolase A *CbhA* A multimodular protein from *Clostridium thermocellum*; has an N-terminal carbohydrate-binding domain, an immunoglobulin-like domain, a glycoside hydrolase family 9 domain and a **dockerin domain**. Degrades polymeric carbohydrates such as cellulose.

cell plate Region in which the new cell wall forms after the division of a plant cell. In the plane of the equator of the spindle a disc-like structure, the **phragmoplast**, forms, into which are inserted pole-derived microtubules. Golgi-derived vesicles containing pectin come together and fuse at the plate, which develops from the centre outwards and eventually fuses with the plasma membrane thereby separating the daughter cells.

cell polarity 1. In epithelial cells the differentiation of apical and basal specializations. In many epithelia, the apical and basolateral regions of plasma membrane differ in lipid and protein composition and are isolated from one another by tight junctions. The apical membrane may, for example, be the only region where secretory vesicles fuse or have a particular ionic pumping system. 2. A motile cell must have some internal polarity in order to move in one direction at a time: a region in which protrusion will occur (the front) must be defined. Locomotory polarity may be associated with the pericentriolar **microtubule-organizing centre** and can be perturbed by drugs that interfere with microtubule dynamics.

cell proliferation Increase in cell number by division.

cell recognition Interaction between cells that is possibly dependent upon specific adhesion. Since the mechanism is not entirely clear in most cases, the term should be used with caution.

cell renewal Replacement of cells, for example those in the skin, by the proliferative activity of basal stem cells.

cell sap Casual term effectively equivalent to the term 'cytosol'.

cell signalling Release by one cell of substances that transmit information to other cells.

cell sorting The process or processes whereby mixed populations of cells, e.g. in a reaggregate, separate out into two or more populations that usually occupy different parts of the same aggregate or separate into different aggregates. Cell sorting probably takes place in the development of certain organs. See **differential adhesion, flow cytometry**.

cell strain Cells adapted to culture but with finite division potential. See **cell line**.

cell streaming See **cyclosis**.

cell-surface marker Any molecule characteristic of the plasma membrane of a cell or in some cases of a specific cell type. 5′-**nucleotidase** and Na$^+$/K$^+$ ATPase are often used as plasma membrane markers, but such enzymic markers have largely been superseded by antibodies directed against specific cell-surface molecules that are characteristic of particular cell types.

cell synchronization A process of obtaining (either by selection or by imposition of a reversible blockade) a population of growing cells that are to a greater or lesser extent in phase with each other in the cycle of growth and division.

cell trafficking The movement of leucocytes through tissues – for example, the movement of lymphocytes from skin to lymph nodes, or from blood into tissues and back into the circulation.

cell tropism A tropism is the orientation of an organism or part of an organism (e.g. leaf or stem) that involves turning or curving by differential growth. Should be distinguished from a tactic response, which involves movement of the whole cell or organism in response to a gradient of some sort. See **chemotropism**; do not confuse with **chemotrophism**.

cellubrevin Protein involved in regulating vesicle fusion. Has 60% sequence identity with **synaptobrevin** (VAMP-2), and is a target for tetanus toxin.

cellular engineering The use of techniques for constructing replacement or additional or experimental parts of cells and tissues both for fundamental investigation and as prosthetic devices. Often involves the interfacing of cells and non-living structures.

cellular immunity Immune response that involves enhanced activity by phagocytic cells and does not imply lymphocyte involvement. Since the term is easily confused with **cell-mediated immunity,** its use in this sense should be avoided.

cellular retinoic acid-binding protein *CRABP* One of the superfamily of lipid-binding proteins (**fatty acid-binding proteins**) which occur in invertebrates and vertebrates, and that acts as an initial receptor for the putative **morphogen**, **retinoic acid**. There are two isoforms, CRABPI and II, with distinct distribution and functions.

cellular slime mould See **Acrasidae**.

cellulases Enzymes that break down **cellulose** and are involved in cell wall breakdown in higher plants, especially during abscission. Produced in large amounts by certain fungi and bacteria. Degradation of cellulose **microfibrils** requires the concerted action of several cellulases.

cellulitis Inflammation of the subcutaneous connective tissues (dermis), mostly affecting face or limbs. *Streptococcus pyogenes* is commonly the causative agent. Also known as erysipelas.

cellulose A straight chain polysaccharide composed of β (1-4)-linked glucose subunits. A major component of plant cell walls where it is found as microfibrils laid down in orthogonal layers.

cellulose synthase Family of processive glycosyl synthetases involved in production of **cellulose**; plant cellulose synthase (CesA) proteins are integral membrane proteins, ~1000 amino acids in length. The sequences all have considerable similarity, but differ from genes found in *Acetobacter* and *Agrobacterium* sp. In *Arabidopsis* there are also several cellulose synthase-like (*Csl*) genes. Cellulose synthase (GDP-forming; EC 2.4.1.29) uses GTP-glucose; a similar enzyme, EC 2.4.1.12 (UDP-forming; cellulose synthase) utilizes UDP-glucose.

cellulosome A large multienzyme complex used by many anaerobic bacteria for the efficient degradation of plant cell wall polysaccharides such as cellulose. Consists of **scaffoldins** containing 9 **cohesin** domains and a cellulose-binding domain and at least 14 different enzymatic subunits, each containing a conserved duplicated sequence (or **dockerin** domain) that is responsible for holding the complex on to the bacterial cell surface through a calcium-mediated protein–protein interaction between the dockerin module from the cellulosomal scaffold and a cohesin (Coh) module of cell-surface proteins located within the proteoglycan layer.

cell wall Extracellular material serving a structural role. In plants, the primary wall is pectin-rich and the secondary wall is mostly composed of **cellulose**. In bacteria, the cell wall structure is complex; the walls of **Gram-positive** and **Gram-negative** bacteria are distinctly different. Removal of the wall leaves a **protoplast** or **spheroplast**.

cenocyte See **coenocyte**.

CENP *Centromere-associated proteins* A large group of proteins associated with the **centromere**. CENP-A is a histone H3 variant, and CENP-A nucleosomes directly recruit a proximal CENP-A nucleosome associated complex (NAC) comprising three other proteins, CENP-M, CENP-N and CENP-T, along with CENP-U(50), CENP-C and CENP-H. CENP-K, CENP-L, CENP-O, CENP-P, CENP-Q, CENP-R and CENP-S assemble on the CENP-A NAC. CENP-E is a kinesin-related microtubule motor protein that is essential for chromosome movement during mitosis. CENP-F is a 367-kDa microtubule-binding protein. See **CENP antigens**.

CENP antigens Proteins (centromere-associated proteins) of the **kinetochore** (CENP-A, 27 kDa; CENP-B, 80 kDa; CENP-C, 140 kDa; CENP-D, 50 kDa) that react strongly with antibodies from **CREST** sera. See **CENP**.

centractin *Arp1, actin-RPV* Vertebrate actin-related protein; 42 kDa, with 54% sequence homology with muscle actin and 69% similarity with cytoplasmic actin. More similar to conventional actin than to **act-2**; associated with the vertebrate **centrosome**. Generally the core structure of the protein is highly conserved and differences are in surface loops. Forms part of the **dynactin** complex, an activator of **dynein**-driven vesicle movement.

central lymphoid tissue See **lymphoid tissue**.

central nervous system *CNS* In vertebrates, the brain and spinal cord. In invertebrates, the CNS is composed of the segmental ganglia of the ventral nerve cord together with the fused ganglia or brain at the anterior end.

centrifugation The process of separating fractions of systems in a centrifuge. The most basic separation is to sediment a pellet at the bottom of the tube, leaving a supernatant at a given centrifugal force. In this case sedimentation is determined by size and density of the particles in the system, amongst other factors. Density may be used as a basis for sedimentation in **density gradient** centrifugation. At very high *g* values molecules may be separated, i.e. **ultracentrifugation**. In continuous centrifugation the supernatant is removed continuously as it is formed.

centrins Acidic calcium-binding phosphoproteins (20 kDa) involved in the duplication of centrosomes in higher eukaryotes, homologous to **caltractin**. Also found in striated flagellar roots of various algae and basal bodies of human sperm. A mutation in centrin causes genomic instability via increased chromosome loss in *Chlamydomonas reinhardtii*.

centriolar region See **pericentriolar region** or **centrosome**.

centriole Organelle of animal cells that is made up of two orthogonally arranged cylinders each with nine microtubule triplets composing the wall. Almost identical to **basal body** of cilium. The pericentriolar material, but not the centriole itself, is the major **microtubule-organizing centre** of the cell. Centrioles divide prior to mitosis, and the daughter centrioles and their associated pericentriolar material come to lie at the poles of the spindle.

centroblast Stage of B-lymphocyte differentiation after antigen exposure and activation. Centroblasts are rapidly proliferating B-lymphocytes with little or no surface immunoglobulin. These cells undergo somatic mutation and class-switching of their immunoglobulin genes.

centrocyte Non-proliferating progeny of **centroblasts** that re-express surface immunoglobulin and are thought to be positively or negatively selected by their affinity for antigen.

centrolecithal Type of egg in which the yolk is in the centre

centromere The region in eukaryote chromosomes where daughter chromatids are joined together. The **kinetochore**, to which the spindle chromosomes are attached, lies adjacent to the centromere. The centromere region generally consists of large arrays of repetitive DNA. See **CENP**.

centrophilin *Nuclear mitotic apparatus protein; NuMA* A microtubule-binding protein (236 kDa) identified by the production of monoclonal antibodies raised against isolated centromeres. In mitotic cells centrophilin is not restricted to the centromeres but is a major antigen of the spindle polar bodies.

centrosome The **microtubule**-organizing centre which, in animal cells, surrounds the **centriole**, and which will divide to organize the two poles of the **mitotic spindle**. By directing the assembly of a cell's skeleton, this organelle controls division, motility and shape.

centrosphere Alternative (rare) name for **centrosome**.

Centruroides margaritatus Scorpion. See **margaratoxin**.

Centruroides noxius Scorpion. See **noxiustoxin**.

Cepaea Genus of land snails. Two species, *C. hortensis* and *C. nemoralis*, have been much studied as convenient examples of polymorphism in colour and banding pattern.

cephaloridine Semisynthetic derivative of **Cephalosporin** C; an antibiotic used for treatment of severe sepsis.

cephalosporins Group of broad-spectrum beta-lactam (tetracyclic triterpene) antibiotics isolated from culture filtrates of the fungus *Cephalosporium* sp. Inhibit bacterial cell wall synthesis in a similar way to penicillins and are effective against Gram-positive bacteria. Various modifications have been made to produce new variants such as **cephaloridine**, cefalexin and cefprozil.

cephazolin A semisynthetic **cephalosporin** for parenteral use.

ceramide An N-acyl **sphingosine**, the lipid moiety of **glycosphingolipids**.

cerberus *DAND4* Cysteine knot superfamily protein, functioning as secreted-type **BMP** antagonist, consequently an antagonist of **Nodal**, **Wnt** and BMP signalling. Cerberus function is required in the leading edge of the anterior dorsal-endoderm of the *Xenopus* embryo for correct induction and patterning of the neuroectoderm. Cerberus-short only inhibits **nodal**.

CERC An acronym for members of the **EF-hand** superfamily found in endoplasmic reticulum and Golgi: Cab-45, Erc-55, reticulocalbin and **calumenin**.

cercidosome Specialized organelle of trypanosomes, site of terminal oxidative metabolism.

cerebello-pontine angle Region at the base of the brain where the facial nerve arises. Tumours in this region can cause a range of problems, and this is the most common location of intracranial epidermoid.

cerebellum Part of the vertebrate hindbrain, concerned primarily with somatic motor function, the control of muscle tone and the maintenance of balance. Important model for cell migration in developing mammalian brain owing to well-studied migratory pathway of the **granule cell** and to the existence of the neurological mutant mouse **weaver** in which granule cell migration fails.

cerebroside Glycolipid found in brain (11% of dry matter). **sphingosine** core with fatty amide or hydroxy fatty amide and a single monosaccharide on the alcohol group (either glucose or galactose).

cerebrotendinous xanthomatosis *CTX* An autosomal recessive lipid-storage disorder caused by a deficiency of the mitochondrial sterol 27-hydroxylase, CYP27.

cereolysin Cytolytic (haemolytic) toxin released by *Bacillus cereus*. Inactivated by oxygen, reactivated by thiol reduction (hence thiol-activated cytolysin). Binds to cholesterol in the plasma membrane and rearrangement of the toxin–cholesterol complexes in the membrane leads to altered permeability.

Cernunnos A protein involved in the process of repairing double-strand breaks in DNA (**NHEJ**). Defects in cernunnos lead to a severe immunodeficiency condition associated with microcephaly and other developmental defects in humans. Cernunnos physically interacts with the XRCC4/DNA–LigaseIV complex. NHEJ is required for the generation of diversity in the immune system.

ceruloplasmin A blue copper-containing **dehydrogenase** protein (135 kDa) found in serum (200–500 µg/ml). Apparently involved in copper detoxification and storage, and possibly also in mopping up excess oxygen radicals or **superoxide** anions.

cesium chloride *US* See **caesium chloride**.

cetuximab A recombinant, human/mouse chimeric monoclonal antibody that binds specifically to the extracellular domain of the human epidermal growth factor receptor (EGFR); used to treat advanced colorectal carcinoma.

CFSE *Carboxyfluoroscein succinimidyl ester* Fluorochrome that readily enters cells, where it is de-esterified by non-specific esterases and remains trapped. Used as a vital dye for cell tracking and lineage studies.

CFTR See **cystic fibrosis transmembrane conductance regulator**.

CFU-E **Colony-forming unit** for cell lines of **erythrocytes**.

CG island *CpG island* Region of genomic DNA rich in CG dinucleotides. Frequently found near genes that are transcribed at high level.

cGMP See **cyclic GMP**.

CGRP See **calcitonin gene-related peptide**.

chaconine Toxic trisaccharide glycoalkaloid (alpha-chaconine) found in potatoes (*Solanum tuberosum*), where it is produced in increased amounts as a response to stress (and has insecticidal and fungicidal effect). Major alkaloid may be beta-chaconine, a disaccharide breakdown product of alpha-chaconine. Related to **solanine**. Imparts an unpleasantly bitter taste. Has cytotoxic effects that have been investigated for tumour therapy, and it has been reported that alpha-chaconine induces apoptosis of HT-29 cells through inhibition of **ERK** and, in turn, activation of **caspase-3**. Also has anticholine esterase activity. See **solanidine**.

chaeotropic *Chaotropic* An agent that causes chaos, usually in the sense of disrupting or denaturing macromolecules. For example, iodide is often used in protein chemistry to break up and randomize **micelles**. In molecular biology, guanidium isothiocyanate is used to provide a denaturing environment in which RNA can be extracted intact without exposure to **RNAases**.

chaetoglobosins Chaetoglobosin J is a fungal metabolite related to cytochalasins that will inhibit elongation at the barbed end of an actin microfilament. Chaetoglobosin A is produced by *Chaetomium globosum*. Chaetoglobosin K is a plant growth inhibitor and toxin from the fungus *Diplodia macrospora*.

Chaetognatha Phylum of hermaphrodite marine coelomata with the body divided into three distinct regions: head, trunk and tail. Arrow-worms.

Chagas disease South American trypanosomiasis caused by *Trypanosoma cruzi* and transmitted by blood-sucking reduviid bugs such as *Rhodnius*.

chalaza **1.** The basal part of a plant ovule where nucellus and integument are joined. **2.** One of two spirally twisted cords of dense albumen connecting the yolk to the shell membrane in a bird's egg.

chalcone Intermediate in biosynthesis of flavanones, **flavones** and **anthocyanidins** by plants. Synthetic derivatives have been shown to have anti-inflammatory properties.

chalone Cell-released tissue-specific inhibitor of cell proliferation thought to be responsible for regulating the size of a population of cells. Their existence has been doubted but **myostatin** appears to be an example.

Chang liver cells Derived from non-malignant human tissue. Extensively used in virology and biochemistry. Cells are epithelial in morphology and grow to high density.

channel gating See **gating current**.

channel protein A protein that facilitates the diffusion of molecules/ions across lipid membranes by forming a hydrophilic pore. Most frequently multimeric with the pore formed by subunit interactions.

channel-forming ionophore An **ionophore** that makes an amphipathic pore with hydrophobic exterior and hydrophilic interior. Most known types are cation selective.

channelopathy Any disease that arises because of a defect in an ion channel. Channelopathies are known that involve the ion channels for potassium, sodium, chloride and calcium. There are also channelopathies involving the acetylcholine receptor, the glycine receptor and other receptors. Each channelopathy can play a role in a number of different diseases. For example, the calcium channelopathies include familial hemiplegic migraine, malignant hyperthermia, episodic ataxia type 2, spinocerebellar ataxia type 6, hypokalemic periodic paralysis type I, central core disease, congenital night blindness and stationary night blindness. Other examples are **Long QT syndrome**, **Andersen's syndrome**.

Chaos chaos Giant multinucleate freshwater amoeba (up to 5 mm long) much used for studies on the mechanism of cell locomotion.

chaperones Cytoplasmic proteins of both prokaryotes and eukaryotes (and organelles such as mitochondria) that bind to nascent or unfolded polypeptides and ensure correct folding or transport. Chaperone proteins do not covalently bind to their targets and do not form part of the finished product. Heat-shock proteins are an important subset of chaperones. Three major families are recognized: the **chaperonins** (groEL and hsp60), the **hsp70** family and the **hsp90** family. Outside these major families are other proteins with similar functions, including **nucleoplasmin**, secB and T-cell receptor-associated protein.

chaperonins Subset of **chaperone** proteins found in prokaryotes, mitochondria and plastids – major example is prokaryotic GroEL (the eukaryotic equivalent of which is hsp60).

CHAPS *3-((3-Cholamidopropyl) dimethylammonio)-1-propane sulphonate* Zwitterionic detergent used for membrane solubilization.

Chara See **Characean algae**.

Characean algae *Charophyceae* Class of filamentous green algae exemplified by the genus *Chara*, in which the mitotic spindle is not surrounded by a nuclear envelope. Probably the closest relatives, among the algae, to higher plants. The giant internodal cells (up to 5 cm long) exhibit dramatic **cyclosis** and have been much used for studies on ion transport and cytoplasmic streaming.

Charcot–Marie–Tooth disease A clinically and genetically heterogeneous group of hereditary motor and sensory neuropathies affecting peripheral nerves, mostly inherited as autosomal dominants, that are the most common inherited disorders of the peripheral nervous system. Type 1A, the commonest form, is caused by duplication of, or mutation in, the gene encoding peripheral myelin protein-22. Other forms are caused by mutations in a range of different genes, such as **periaxin**.

chartins Microtubule-associated proteins (**MAPs**) of 64, 67 and 80 kDa, distinct from **tau protein**. Isolated from neuroblastoma cells. They are regulated by **nerve growth factor** (NGF) and may influence microtubule distribution.

charybdotoxin Peptide (37 residues) isolated from *Leiurus quinquestriatus hebraeus* (scorpion) venom that is a selective blocker of high conductance Ca^{2+}-activated K^+ channel.

CHD proteins *Chromo-ATPase/helicase-DNA-binding proteins* Proteins that regulate ATP-dependent nucleosome assembly and mobilization through their conserved double **chromodomains**.

checkerboard assay Variant of the Boyden chamber assay for leucocyte chemotaxis introduced by Zigmond. By testing different concentrations of putative chemotactic factor in non-gradient conditions, it is possible to calculate the enhancement of movement expected due simply to **chemokinesis** and to compare this with the distances moved in positive and negative gradients. Good experimental design thus allows chemotaxis to be distinguished from chemokinesis.

checkpoint Any stage in the **cell cycle** at which the cycle can be halted and entry into the next phase postponed. Two major checkpoints are at the G1/S and G2/M boundaries. These are the points at which **cdc** proteins act.

checkpoint kinases *CHK1, CHK2* Kinases that act downstream of **ATM** in response to detection of DNA damage. Chk 1 is an essential gene for normal cell division, and Chk 2 has been found to be mutated in many human tumours.

Chediak–Higashi syndrome Autosomal recessive disorder, caused by mutation in the **lysosomal trafficking regulator** gene, characterized by the presence of giant lysosomal vesicles in phagocytes and in consequence poor bactericidal function. Some perturbation of microtubule dynamics seems to be involved. Reported from humans, albino Hereford cattle, mink, beige mice and killer whales.

chelation Binding of a metal ion by a larger molecule such as EDTA or protein (iron in haem is held as a chelate). The binding is strong but reversible, and chelating agents can be used to buffer the free concentration of the ion in question.

chelerythrine An alkaloid from the greater celandine (*Chelidonium majus*) and named from the red colour of its salts. Chelerythrine chloride inhibits **PKC**.

chelicerae In Arachnida, a pair of preoral appendages armed with fangs.

chelicerata Arthropod subphylum that includes spiders, scorpions and horseshoe crabs.

chemical potential The work required (in $J\ mol^{-1}$) to bring a molecule from a standard state (usually infinitely separated in a vacuum) to a specified concentration. More usually employed as chemical potential difference, the work required to bring one mole of a substance from a solution at one concentration to another at a different concentration, $Dm = RT.\ln (c_2/c_1)$. This definition is useful in studies of **active transport**; note that, for charged molecules, the electrical potential difference must also be considered (see **electrochemical potential**).

chemical synapse A nerve–nerve or nerve–muscle junction where the signal is transmitted by release from one membrane of a chemical transmitter that binds to a receptor in the second membrane. Importantly, signals only pass in one direction.

chemiluminescence Light emitted as a reaction proceeds. Becoming used increasingly to assay ATP (using firefly **luciferase**) and the production of toxic oxygen species by activated phagocytes (using **luminol** or **lucigenin** as bystander substrates that release light when oxidized). See also **bioluminescence**.

chemiosmosis A theoretical mechanism (proposed by Mitchell) to explain energy transduction in the mitochondrion. As a general mechanism it is the coupling of one enzyme-catalysed reaction to another using the transmembrane flow of an intermediate species, e.g. cytochrome oxidase pumps protons across the mitochondrial inner membrane, and ATP synthesis is 'driven' by re-entry of protons through the ATP-synthesizing protein complex. The alternative model is production of a chemical intermediate species, but no compound capable of coupling these reactions has ever been identified, and chemiosmosis is generally accepted as being the correct model.

chemiosmotic hypothesis See **chemiosmosis**.

chemoattractant A substance that elicits accumulation of cells.

chemoattraction Non-committal description of cellular response to a diffusible chemical – not necessarily by a tactic response. Term preferable to 'chemotaxis' when the mechanism is unknown.

chemoautotroph *Chemotroph* Chemotrophic **autotroph**. Organism in which energy is obtained from endogenous light-independent reactions involving inorganic molecules.

chemodynesis Induction of cytoplasmic streaming in plant cells by chemicals rather than by light (photodynesis).

chemokine receptors *CCR* Chemokine receptors are G-protein-linked **serpentine receptors** that, in addition to binding **chemokines**, are used as co-receptors for the binding of immunodeficiency viruses (**HIV**, **SIV**, **FIV**) to leucocytes. CXCR4 is a co-receptor for T-tropic viruses, CCR5 for macrophage-tropic (M-tropic) viruses. Individuals deficient in particular CCRs seem to be resistant to HIV-1 infection. See Table C4 (chemokines and chemokine receptors).

chemokines Small secreted proteins that stimulate **chemotaxis** of leucocytes. Chemokines can be subdivided into classes on the basis of conserved cysteine residues. The α-chemokines (IL-8, NAP-2, Gro-α, Gro-γ, ENA-78 and GCP-2) have conserved CXC motif and are mainly chemotactic for neutrophils; the β-chemokines (MCP-1-5, MIP-1a, MIP-1b, eotaxin, RANTES) have adjacent cysteines (C–C) and attract monocytes, eosinophils or basophils;

TABLE C4. Chemokines and Chemokine Receptors

Chemokine	Synonyms	Attracts	Receptor	Produced by
CXC Family	α-chemokines			
(ELR)-positive				
CXCL1	NAP-1, MONAP, MDNCF, NAF, LAI, GCP, GRO1, GROα, MGSA-a, NAP-3	Neutrophils	CXCR1, CXCR2	Many cells, including monocytes, lymphocytes, fibroblasts, endothelial cells, mesangial cells
CXCL2	MIP2-α, GROβ, GRO2, MGSA-β	Neutrophils	CXCR2	Activated monocytes, fibroblasts, epithelial and endothelial cells
CXCL3	MIP-2β, GROγ	Neutrophils, basophils, endothelial cells	CXCR2	
CXCL5	SCYB5, ENA-78	Neutrophils, endothelial cells	CXCR2	
CXCL6	SCYB6, GCP-2, CKA-3		CXCR1, CXCR2	
CXCL7	PBP, CTAP-III, NAP-2, β-TG, low-affinity platelet factor 4	Fibroblasts (βTG), neutrophils (NAP-2)	CXCR2	Platelets
CXCL8	IL-8	Neutrophils	CXCR1, CXCR2	
(ELR)-negative				
CXCL4	Oncostatin A, platelet factor 4 (PF-4)	Neutrophils	CXCR3-B	Aggregated platelets, activated T-cells
CXCL10	Interferon-inducible cytokine, γIP-10, IP-10, CRG-2, C7	Monocytes	CXCR3	Keratinocytes, monocytes, T-cells, endothelial cells and fibroblasts
CXCL12	PBSF, stromal cell-derived factor, SDF-1α	Primordial germ cells	CXCR4, CXCR7	Fibroblasts, bone marrow stromal cells
CXCL13	BCA, B-lymphocyte chemoattractant (BLC)	B-cells	CXCR5 (CD185)	
CXCL14	MIP-2γ, BRAK, BMAC	Neutrophils, myeloid cells	?	Breast and kidney
CC Family	β-chemokines			
CCL1	I-309, TCA3	Monocytes	CCR8	T-cells, mast cells
CCL2	MCP1, MCAF, JE, LDCF, GDCF, HC14, MARC	Monocytes, basophils	CCR2	Monocytes, T-cells, fibroblasts, endothelial cells, smooth muscle, some tumours. Upregulated by IFNγ
CCL3	MIP-1α LD78β, pAT464, GOS19	Eosinophils	CCR1, CCR5	T-cells, B-cells, Langerhans cells, neutrophils, macrophages
CCL4	MIP-1β, HIMAP	Memory T-cells	CCR1, CCR5	ACT-2, pAT744, hH400, hSISα, G26, HC21, MAD-5,
CCL5	RANTES, sisδ	Monocytes, memory T, eosinophils	CCR1, CCR2, CCR3	T-cells, macrophages
CCL7	MCP3	Monocytes, T-cells, NK-cells, eosinophils, dendritic cells	CCR1, CCR2, CCR3	Fibroblasts
CCL8	MCP2	Monocytes	CCR3, CCR5	
CCL11	Eotaxin	Eosinophils	CCR3	
CCL13	MCP-4	monocytes	CCR2, CCR3	
CCL14	HCC-1	T-cells, monocytes, basophils, eosinophils	CCR1, CCR5	

TABLE C4. (Continued)

Chemokine	Synonyms	Attracts	Receptor	Produced by
CCL15	HCC-2, Lkn-1, MIP-1δ	T-cells, monocytes, eosinophils, basophils, granulocytes	CCR1	
CCL16	HCC-4, LEC, LCC-1	T-cells, monocytes, eosinophils, basophils	CCR1, CCR2	
CCL17	TARC	T-cells, dendritic cells, basophils	CCR4	Activated T-cells
CCL18	DC-CK1		?	
CCL19	MIP-3b, ELC	T-cells, dendritic cells	CCR7	
CCL20	MIP-3a, LARC	T-cells, B–cells	CCR6	
CCL21	6Ckine, SLC	T-cells, dendritic cells	CCR7	
CCL22	MDC, monocyte-derived chemokine	T-cells, dendritic cells, basophils	CCR4	Monocytes
CCL23	MPIF-1, CKb8	T-cells, monocytes, granulocytes	CCR1	
CCL24	Eotaxin-2	Eosinophils, basophils, mast cells	CCR3	
CCL25	TECK	T-cells, IgA$^+$ plasma cells	CCR9	
CCL27	CTACK	T–cells	CCR10	
CCL28	MEC	T–cells	CCR3, CCR10	
C Family	γ chemokines			
XCL1	Lymphotactin, SCM-1α	Lymphocytes	XCR1	
XCL2	SCM-1β	Lymphocytes	XCL1	
CX3C Family	δ chemokines			
CX$_3$CL1	Fractalkine, neurotactin	Monocytes, lymphocytes, neutrophils	CX$_3$CR1	Endothelial cells, brain

Source: Douglas, M.R., Morrison, K.E., Salmon, M. and Buckley, C.D. (2002). *Expert Reviews in Molecular Medicine* (www-ermm.cbcu.cam.ac.uk /02005318h.htm).

the γ-chemokines have only one cysteine pair and are chemotactic for lymphocytes (lymphotactin). The δ-chemokines are structurally rather different, being membrane-anchored, have a C-X-X-X-C motif and are restricted (so far) to brain (**neurotactin**). Human genes for the α-chemokines are on Chr 4 and 10, for β-chemokines are on Chr 17, for lymphotactin are on Chr 1 and for neurotactin are on Chr16. The receptors are **G-protein-coupled**.

chemokinesis A response by a motile cell to a soluble chemical that involves an increase or decrease in speed (positive or negative **orthokinesis**) or of frequency of movement, or a change in the frequency or magnitude of turning behaviour (**klinokinesis**).

chemolithotroph Alternative name for a **chemoautotroph**.

chemoreceptor A cell or group of cells specialized for responding to chemical substances in the environment.

chemorepellant Opposite of **chemoattractant**.

chemostat Apparatus for maintaining a bacterial population in the exponential phase of growth by regulating the input of a rate-limiting nutrient and the removal of medium and cells.

chemosynthesis Synthesis of organic compounds by an organism using energy derived from oxidation of inorganic molecules rather than light (see **chemotrophy** and **photosynthesis**).

chemotactic See **chemotaxis**.

chemotaxis A response of motile cells or organisms in which the direction of movement is affected by the gradient of a diffusible substance. Differs from chemokinesis in that the gradient alters probability of motion in one direction only, rather than rate or frequency of random motion.

chemotherapy Treatment of a disease with drugs that are designed to kill the causative organism or, in the case of tumours, the abnormal cells.

chemotrophy Systems of metabolism in which energy is derived from endogenous chemical reactions rather than from food or light energy, e.g. in deep-sea hot-spring organisms.

chemotropism Growth or possibly bending of an organism in response to an external chemical gradient. Sometimes used in error when the terms **chemotaxis** or **chemokinesis** should have been used.

CHF See **chick heart fibroblasts**.

chi-squared χ-*squared* Common statistical test to determine whether the observed values of a variable are significantly different from those expected on the basis of a null hypothesis.

chiasma *Pl.* chiasmata Junction points between non-sister **chromatids** at the first **diplotene** of **meiosis**, the consequence of a **crossing-over** event between maternal and paternally derived **chromatids**. A chiasma also serves a mechanical function and is essential for normal equatorial alignment at meiotic **metaphase** I in many species. Frequency of chiasmata is very variable between species.

chick heart fibroblasts The cells that emigrate from an explant of embryonic chick heart maintained in culture. Often considered to be archetypal normal cells, and were used as the 'normal' reference cells in the original studies of contact inhibition of locomotion by Abercrombie.

Chilomonas Genus of small cryptomonad flagellate protists, 20–40 µm long.

chimera Organism composed of two genetically distinct types of cells. Can be formed by the fusion of two early blastula stage embryos, or by the reconstitution of the bone marrow in an irradiated recipient, or by somatic segregation. Since female mammals have one or the other X chromosome more or less randomly inactivated, they could also be considered chimeric.

Chinese hamster ovary cells *CHO cells* Cell line that is often used for growing viruses or for transfection; see also Table C3.

CHIP *C-terminus of Hsp70-interacting protein* A co-chaperone and ubiquitin ligase that interacts with **hsp70** through an amino-terminal tetratricopeptide repeat (**TPR motif**). Important in protection against physiologic stress.

chiral stationary phase *CSP* Stationary phase for liquid chromatography that has a chiral molecule attached and has differential binding characteristics for enantiomers of molecules passed through the column. Can therefore be used to separate racemic mixtures.

Chironomus Genus of flies (midges). Larvae live in freshwater and have been much studied because of the giant **polytene chromosomes** in the salivary glands; **haemolymph** contains haemoglobin in solution.

Chiroptera Order of flying mammals, mainly insect or fruit-eating and nocturnal. Bats.

chitin Polymer (β-1,4 linked) of *N*-acetyl-D-glucosamine, extensively cross-linked; the major structural component of arthropod exoskeletons and fungal cell walls. Widely distributed in plants and fungi.

chitinase *EC 3.2.1.14* Enzyme that catalyses the hydrolysis of 1,4-β-linkages in **chitin**.

chitosan A polymer of 1,4-β-D-glucosamine and *N*-acetyl-D-glucosamine found in the cell wall of some fungi. Can be manufactured by de-acetylation of chitin (from crustacean shells) and is sold as a health supplement, although of doubtful value. Is also used as a plant growth enhancer and boosts the ability of plants to defend themselves against fungal infections.

chitosome Membrane-bound vesicular organelle (40–70 nm diameter) found in many fungi. Contains chitin sythetase, which produces chitin microfibrils that are released and incorporated into the cell wall.

chk 1. See **chokh mutant**. 2. **Checkpoint kinases**, CHK1 and CHK2. 3. **Choline kinases**. 4. **Carboxyl-terminal Src kinase homologous kinase**.

Chlamydia Genus of minute prokaryotes that replicate in cytoplasmic vacuoles within susceptible eukaryotic cells. Genome about one-third that of *E. coli*. *C. trachomatis* causes trachoma in man; *C. psittaci* causes economically important diseases of poultry.

Chlamydomonas A genus of unicellular green algae, usually flagellated. Easily grown in the laboratory, the algae have often been used in studies on flagellar function – a range of paralysed flagellar (pf) mutants have been isolated and studied extensively.

chlamydospore A fungal cell that becomes thick-walled, separates from the parent mycelium and functions as a resting spore. Often darkly pigmented.

chloracne A type of acne caused by exposure to halogenated hydrocarbons either through the skin or by ingestion or inhalation. The characteristic large comedones progress to severe inflammation and scarring.

chloragosome Cytoplasmic granule of unknown function found in the coelomocytes of annelids.

chloramphenicol An antibiotic from *Streptomycetes venezuelae* that inhibits protein synthesis in prokaryotes and in mitochondria and chloroplasts by acting on the 50S ribosomal subunit. It is relatively toxic, but has a wide spectrum of activity against Gram-positive and Gram-negative cocci and bacilli (including anaerobes), Rickettsia, Mycoplasma and Chlamydia.

chloramphenicol acetyltransferase *CAT; EC 2.3.1.28* Enzyme that inactivates the antibiotic **chloramphenicol** by acetylation. Widely used as a **reporter gene**.

chlordiazepoxide *TN Librium* A mild tranquillizer.

Chlorella Genus of green unicellular algae extensively used in studies of photosynthesis.

chlorenchyma Form of **parenchyma** tissue active in photosynthesis, in which the cells contain many **chloroplasts**; found especially in leaf **mesophyll**.

chlorhexidine Antiseptic and disinfectant that is used for dressing minor skin wounds or burns.

chloride channel **Ion channels** selective for chloride ions. Various types including **ligand-gated** Cl-channels at synapses (the **GABA**- and **glycine**-activated channels), as well as **voltage-gated** Cl-channels found in a variety of plant and animal cells. See also **CFTR, MDR**.

chloride current Flow of chloride ions through chloride-selective **ion channels**.

chlorocruorin A green respiratory pigment found in some Polychaeta. The prosthetic group is similar, but not identical, to reduced haematin.

chlorophyll The photosynthetic pigments of higher plants but closely related to bacteriochlorophylls. Magnesium complexes of tetrapyrolles.

Chlorophyta *Green algae* Division of algae containing photosynthetic pigments similar to those in higher plants and having a green colour. Includes unicellular forms, filaments and leaf-like thalluses (e.g. *Ulva*). Some members form **coenobia**, and the **Characean algae** have branched filaments.

chloroplast Photosynthetic organelle of higher plants. Lens-shaped and rather variable in size but ~5 µm long. Surrounded by a double membrane and contains circular DNA (though not enough to code for all proteins in the chloroplast). Like the mitochondrion, it is semiautonomous. It resembles a cyanobacterium from which, on the endosymbiont hypothesis, it might be derived. The photosynthetic pigment, chlorophyll, is associated with the membrane of vesicles (thylakoids) that are stacked to form grana.

chloroquine Antimalarial drug that has the interesting property of increasing the pH within the **lysosome** when added to intact cells in culture. Chloroquine resistance seems to be due to enhanced ABC transporter activity (*Plasmodium falciparum* Chloroquine Resistance Transporter) that pumps chloroquine from the cell. Also used to treat some autoimmune diseases, such as rheumatoid arthritis and SLE.

chlorosis Yellowing or bleaching of plant tissues due to the loss of chlorophyll or failure of chlorophyll synthesis. Symptomatic of many plant diseases, also of deficiencies of light or certain nutrients.

chlorosome Elongated membranous vesicles attached to the plasma membrane of green photosynthetic bacteria; contain the light-harvesting antenna complexes of bacteria in the subOrder Chlorobiineae. Pigments include bacteriochlorophylls and carotenoids.

chlorothiazide One of the **thiazide** class of diuretics that affect the distal renal tubular mechanism of electrolyte reabsorption; used in cases of oedema and hypertension.

chlorotic A yellow or pale green colour (**chlorosis**) in plant tissues that are normally green as a result of low chlorophyll levels. May be a consequence of nutrient deficiencies, low light conditions or viral disease.

chlorpromazine Neuroleptic aliphatic phenothiazine, thought to act primarily as dopamine antagonist, but also antagonist at α-adrenergic, H1 histamine, muscarinic and serotonin receptors. Used clinically as an antiemetic. Has been shown to alter fibroblast behaviour.

chlorpropamide A **sulphonylurea** drug used in treating Type II diabetes.

chlortetracycline *Aureomycin* A **tetracycline antibiotic** used to treat a wide variety of bacterial and fungal infections. Bright yellow in colour.

chlorthalidone Diuretic of the **thiazide** type that has a long duration of action.

CHO cells See **Chinese hamster ovary cells** and Table C3.

choanocytes Cells that line the radial canals of sponges. Have long flagella that are responsible for generating the feeding current.

choanoflagellates A group of flagellate protozoa that may be ancestors of sponges. Each has a single flagellum surrounded by a ring of microvilli, forming a cylindrical or conical collar, that capture prey drawn in by the water current generated by the flagellum. Usually sessile, being attached at the opposite pole to the flagellum, and in some cases are colonial. Resemble the **choanocytes** of sponges.

chokh mutant *CHK* Mutant zebra fish that lacks eyes from the earliest stages in development; the chk gene encodes the homeodomain-containing transcription factor, Rx3.

cholangiocarcinoma Malignant tumours derived from the epithelium of the biliary duct system of the liver (**cholangiocytes**). Most (90%) are adenocarcinomas, and the remainder are squamous cell tumours.

cholangiocyte Epithelial cells that line the bile ducts. Form an important transporting epithelium actively involved in the absorption and secretion of water, ions and solutes. Can give rise to **cholangiocarcinomas**.

cholate The sodium salt of cholic acid that has strong detergent properties and can replace membrane lipids to generate soluble complexes of membrane proteins.

cholecalcin See **calbindin**.

cholecystitis Inflammatory condition of the wall of the gallbladder caused by *Salmonella typhi*.

cholecystokinin *CCK; pancreozymin* Polypeptide hormone (33 residues) secreted by the duodenum. Stimulates secretion of digestive enzymes by the pancreas and contraction of the gall bladder. The C-terminal octapeptide is found in some dorsal root ganglion neurons, where it presumably acts as a peptide neurotransmitter.

cholera toxin A multimeric protein toxin from *Cholera vibrio*. The toxic A subunit (27 kDa) activates adenyl cyclase irreversibly by **ADP-ribosylation** of a Gs protein. The B subunit (57 kDa) has five identical monomers, binds to GM1 ganglioside and facilitates passage of the A subunit across the cell membrane.

cholestasis Any condition in which excretion of bile from the liver is blocked.

cholesterol The major **sterol** of higher animals. An important component of cell membranes, especially of the plasma (outer) membrane, most notably of the **myelin** sheath. Transported in the esterified form via plasma lipoproteins.

cholesterol-binding toxins Family of 50–60-kDa pore-forming toxins from various genera of bacteria including *Streptococcus*, *Listeria*, *Bacillus* and *Clostridium*. Apparently bind to cholesterol and oligomerize to form a pore: as a result, cause cell lysis and are lethal. See **Streptolysin O**. Other examples include Pneumolysin from *S. pneumoniae*, Cereolysin O from *Bacillus cereus*, Thuringolysin O from *B. thuringiensis*, Tetanolysin from *Clostridium tetani*, Botulinolysin from *C. botulinum*, Perfringolysin O from *C. perfringens*, Listeriolysin O from *L. monocytogenes*.

choline Choline is esterified in the head group of phospholipids (phosphatidyl choline and sphingomyelin) and acetylated in the neurotransmitter acetylcholine. Otherwise a biological source of methyl groups.

choline kinases *CHKα, CHKβ; EC 2.7.1.32* Enzymes that catalyse the phosphorylation of choline to phosphocholine in the biosynthesis of phosphatidylcholine (PC).

cholinergic neurons Neurons in which acetylcholine is the neurotransmitter.

chondro- Prefix: cartilage related/associated.

chondroblast Embryonic cartilage-producing cell.

chondrocyte Differentiated cell responsible for secretion of extracellular matrix of **cartilage**.

chondroitin sulphates Major components of the extracellular matrix and connective tissue of animals. They are repeating polymers of glucuronic acid and sulphated *N*-acetyl glucosamine residues that are highly hydrophilic and anionic. Found in association with proteins. 'Chondroitin' is sold as a dietary supplement in the hope that it will relieve osteoarthritis, but the evidence for efficacy is minimal.

chondromalacia Damage to the cartilage at the back of the patella (kneecap).

chondronectin A 180-kDa homotrimeric protein (subunits 55 kDa) isolated from chick serum that specifically favours attachment of **chondrocytes** to Type II **collagen** if present with the appropriate cartilage **proteoglycan**. Structurally and chemically distinct from fibronectin and laminin.

CHORD domain *Cysteine and histidine rich domain* A zinc-binding 60 amino acid domain with uniquely spaced cysteine and histidine residues, highly conserved from plants to mammals. See **melusin**.

chordae tendineae Tendinous cords acting to restrain movement of the flaps of the valves of the heart.

chordamesoderm Embryonic mesoderm that gives rise to the **notochord**.

chordin An extracellular regulator of **BMP**; in embryogenesis, a dorsally expressed gene that is involved in mesodermal patterning and neural induction. In adults, the differential distribution and regulation of chordin in normal and osteoarthritic cartilage and chondrocytes suggests an involvement in osteoarthritis. Also found in other tissues.

chordotonal organs Insect sense organs that are sensitive to pressure, vibrations and sound.

chorea Involuntary repetitive jerky movements of the body. Seen in a number of neurological diseases, including Sydenham's chorea (or Saint Vitus's dance) and **Huntington's chorea**.

chorioallantoic membrane 1. Protective membrane around the eggs of insects and fishes. 2. Extra-embryonic membrane surrounding the embryo of amniote vertebrates. The outer epithelial layer of the chorion is derived from the trophoblast, by the apposition of the **allantois** to the inner face of the **chorion**. The chorioallantoic membrane is highly vascularized, and is used experimentally as a site upon which to place pieces of tissue in order to test their invasive capacity.

choriocarcinoma Malignant tumour of trophoblast.

chorion 1. Protective membrane around the eggs of insects and fishes. 2. Extra-embryonic membrane surrounding the embryo of amniote vertebrates. The outer epithelial layer of the chorion is derived from the trophoblast. see **chorioallantoic membrane**.

chorionic gonadotrophin *Chorionic gonadotropin* A glycoprotein hormone (244 amino acids with a molecular mass of 36.7 kDa) synthesized in the placenta that promotes the maintenance of the **corpus luteum** during the beginning of pregnancy, causing it to secrete **progesterone**; will also stimulate **Leydig cells** to synthesize testosterone. Heterodimeric, the alpha subunit is identical to that of **luteinizing hormone** (LH), **follicle-stimulating hormone**

(FSH) and **thyroid-stimulating hormone** (TSH), and the beta subunit is unique.

chorionic villus sampling *CVS* Method for diagnosing human foetal abnormalities in the sixth to tenth weeks of gestation. Small pieces of foetally derived chorionic villi are removed for chromosomal analysis and, increasingly, for DNA-based testing for disease-associate alleles.

choroid Middle layer of the vertebrate eye, between the **retina** and sclera. Well-vascularized and also pigmented to throw light back onto the retina (the tapetum is an iridescent layer in the choroid of some eyes). Not to be confused with the **choroid plexus**, a highly vascularized region of the roof of the ventricles of the vertebrate brain that secretes cerebrospinal fluid.

choroid plexus Mass of highly branched blood vessels in margin of cerebral ventricles that produces cerebrospinal fluid.

chp-1 Protein, 63% homologous to **melusin** but without the calcium-binding region, expressed in all tissues tested, including muscles, and highly conserved from invertebrates to human. Has a **CHORD domain**.

Christmas disease Congenital deficiency of blood-clotting factor IX (first described in the Christmas issue of *British Medical Journal*, 1952). Inherited in similar sex-linked way to classical haemophilia.

chromaffin cell See **granins**.

chromaffin tissue Tissue in medulla of adrenal gland containing two populations of cells, one producing adrenaline, the other noradrenaline. The **catecholamine** is associated with carrier proteins (chromogranins) in membrane vesicles (chromaffin granules).

chromagen Any substance that can give rise to a coloured product when appropriately modified by, for example, an enzyme or the products of enzymic activity.

chromatic aberration When using white light through a lens system, it is inevitable that different wavelengths (colours) are brought to a focus at slightly different points. As a consequence, there are chromatic aberrations in the image; good microscope objectives are therefore corrected for this at two wavelengths (achromats) or at three wavelengths (apochromats), as well as for **spherical aberration**.

chromatid Single chromosome containing only one DNA duplex. Two daughter chromatids become visible at mitotic metaphase, though they are present throughout G2.

chromatin Stainable material of interphase nucleus consisting of nucleic acid and associated histone protein packed into **nucleosomes**. Euchromatin is loosely packed and accessible to RNA polymerases, whereas heterochromatin is highly condensed and probably transcriptionally inactive.

chromatin assembly factor *CAF-1* Heterotrimeric protein that couples DNA replication to histone deposition *in vitro* but is not essential for yeast cell proliferation. Depletion of CAF-1 in human cell lines demonstrates that CAF-1 is, however, required for efficient progression through S-phase. Has been used as a marker of the proliferative state because the expression of both CAF-1 large subunits, p150 and p60, is massively downregulated during quiescence in several cell lines.

chromatin body Barr body; condensed X chromosome in female mammalian cell.

chromatography Techniques for separating molecules based on differential absorption and elution. Term for separation methods involving flow of a fluid carrier over a non-mobile absorbing phase.

chromatophores 1. Pigment-containing cells of the dermis, particularly in teleosts and amphibians. By controlling the intracellular distribution of pigment granules, the animal can blend with the background. **melanocytes** and **melanophores** are melanin-containing chromatophores. **2.** Term occasionally used for chloroplasts in the chromophyte algae.

chromobindin See **annexin**.

chromoblast An embryonic cell that will differentiate into a chromatophore.

chromocentre Condensed heterochromatic region of a chromosome that stains particularly strongly, although in the polytene chromosomes of *Drosophila* the chromocentre is of under-replicated heterochromatin and stains lightly.

chromodomains Modules implicated in the recognition of lysine-methylated histone tails and nucleic acids. CHD (for chromo-ATPase/helicase-DNA-binding) proteins regulate ATP-dependent nucleosome assembly and mobilization through their conserved double chromodomains. See **chromoshadow domain**.

chromogranins See **chromaffin tissue** and **granins**.

chromomere Granular region of condensed **chromatin**. Used of chromosomes at leptotene and zygotene stages of meiosis, of the condensed regions at the base of loops on lampbrush chromosomes and of condensed bands in polytene chromosomes of Diptera.

chromophore The part of a visibly coloured molecule responsible for light absorption over a range of wavelengths, thus giving rise to the colour. By extension, the term may be applied to UV- or IR-absorbing parts of molecules. Do not confuse with **chromatophores**.

chromoplast Plant chromatophore filled with red/orange or yellow carotenoid pigment. Responsible for colour of carrots and of many petals.

chromoshadow domain CSD Domain that mediates dimerization, transcription repression and interaction with multiple nuclear proteins. Related to the **chromodomain**, and found in proteins that also have a classical chromodomain. Chromodomain-containing proteins can be divided into two classes depending on the presence, e.g. in *Drosophila* Heterochromatin protein 1 (**HP1**), or absence, for example in *Drosophila* **Polycomb** (Pc), of the chromoshadow domain.

chromosome The DNA of eukaryotes is subdivided into chromosomes, presumably for convenience of handling, each of which has a long length of DNA associated with various proteins. The chromosomes become more tightly packed at mitosis and become aligned on the **metaphase plate**. Each chromosome has a characteristic length and banding pattern. See **C banding**, **G banding**.

chromosome condensation The tight packing of DNA into **chromosomes** in metaphase, in preparation for **nuclear division**.

chromosome painting See **fluorescence** *in situ* **hybridization**.

chromosome segregation The orderly separation of one copy of each chromosome into each daughter cell at **mitosis**.

chromosome synapsis The close apposition of homologous chromosomes before cell division or permanently in giant **polytene chromosomes**.

chromosome translocation The fusion of part of one chromosome onto part of another. Largely sporadic and random, there are some translocations at 'hot-spots' that occur often enough to be clinically significant. See **Philadelphia chromosome**. See **spectral karyotyping**.

chromosome walking A procedure to find and sequence a gene whose approximate position in a chromosome is known by classical genetic linkage studies. Starting with the known sequence of a gene shown by classical genetics to be near to the novel gene, new clones are picked from a **genomic library** by **hybridization** with a short probe generated from the appropriate end of the known sequence. The new clones are then sequenced, new probes generated and the process repeated until the gene of interest is reached.

chronaxie The shortest time required for excitation of a nerve when the electrical stimulus is twice the threshold intensity required to elicit a response if applied over a prolonged period.

chronic Persistent, long-lasting (as opposed to **acute**). Chronic **inflammation** is generally a response to a persistent antigenic stimulus.

chronic granulomatous disease CGD Disease, usually fatal in childhood, in which the production of hydrogen peroxide by **phagocytes** does not occur. Catalase-negative bacteria are not killed. and there is no luminol-enhanced **chemiluminescence** when the cells are tested. The absence of the oxygen-dependent killing mechanism is not itself fatal, but seriously compromises the primary defence system. At least three separate lesions can cause the syndrome, the commonest being an X-linked defect in plasma membrane cytochrome (see **p91 phox**).

chronic lymphocytic leukaemia See **leukaemia**.

chronic myelogenous leukaemia See **leukaemia**.

chrysolaminarin A modified **laminarin** found as a food reserve in **Chrysophyceae**.

Chrysophyceae A class of eukaryotic algae in the division Heterokontophyta. Golden-brown in colour due to high levels of the **xanthophyll**, fucoxanthin. Mostly found in fresh water, and are single-celled or colonial. Also called Chrysomonadida by protozoologists.

CHUK *Conserved helix-loop-helix ubiquitous kinase* Kinase responsible for phosphorylation of **IkappaB,** thus triggering degradation of IkB and allowing **NFkB** to move to the nucleus.

Churg–Strauss syndrome ANCA-associated vasculitis in which there is no complement consumption and no deposition of immune complexes. Shares many of the clinical and pathological features of **polyarteritis nodosa** but distinguished by eosinophilia. Affects predominantly small vessels.

CHX Abbrev. for (1) **Cycloheximide**, (2) **Chlorhexidine**.

chylomicron Colloidal fat globule found in blood or lymph; used to transport fat from the intestine to the liver or to adipose tissue. Has a very low density, and a low protein and high triacylglyceride content.

chymase *EC 3.4.21.39, mast cell protease* Cathepsin G-like, serine proteinase, found particularly in mast cells, that plays an important role in generating angiotensin II in response to injury of vascular tissues and converts big endothelin 1 to the 31-amino-acid length peptide **endothelin 1**.

chymosin *EC 3.4.23.4* Peptidase from the abomasum (fourth stomach) of calf that has properties similar to **pepsin**. Will cleave casein to paracasein and is used in cheesemaking.

chymostatin Low-molecular weight peptide–fatty acid compound of microbial origin that inhibits **chymotrypsins** and **papain**.

chymotrypsin *EC 3.4.21.1* Serine peptidase from pancreas. Preferentially hydrolyses Phe, Tyr or Trp peptide and ester bonds. The chymotrypsin family of peptidases are all endopeptidases.

chytrid Member of the **Chytridiomycota**.

chytridiomycota The only major group of true (chitin-walled) fungi that produce zoospores. Common as saprotrophs, facultative parasites and obligate parasites in moist soil and freshwater habitats. Chemotaxis of the rumen chytrid *Neocallimastix frontalis* has been extensively studied. *Batrachochytrium dendrobatidis* is capable of causing sporadic deaths in some amphibian populations and 100% mortality in others.

ciboulot *cib* G-actin-binding protein (14.4 kDa) from *Drosophila*; plays a major role in axonal growth during brain metamorphosis. Has three thymosin-like (WH2) repeats and binds to G-actin in the same way as **profilin**, and the complex can add to the barbed but not the pointed end of an F-actin filament.

cicatrization Contraction of fibrous tissue, formed at a wound site, by **fibroblasts**, thereby reducing the size of the wound but causing tissue distortion and disfigurement. Once thought to be due to contraction of collagen, but now known to be due to cellular activity.

ciclosporin *Cyclosporine* BAN for cyclosporine, a cyclic undecapeptide isolated from *Tolypocladium inflatum*, peptide drug used as an immunosuppressive to prevent transplant rejection. Acts selectively on the production of helper T-cells but can cause renal damage; the long-term consequences of suppressing immune function are not yet clear. See also **cyclophilin**.

CIG *Cold insoluble globulin* Obsolete synonym for **fibronectin**.

ciguatera Food poisoning caused by eating seafood containing natural **ciguatoxins**.

ciguatoxin A large, heat-stable, polyether toxin produced by certain strains of the dinoflagellate *Gambierdiscus toxicus*. The toxin can accumulate in large, predatory fish and is responsible for the poisoning syndrome known as ciguatera; activates neuromuscular sodium channels.

ciliary body Tissue that includes the group of muscles that act on the eye lens to produce accommodation and the arterial circle of the iris. The inner ciliary epithelium is continuous with the **pigmented retinal epithelium**, the outer ciliary epithelium secretes the aqueous humour.

ciliary ganglion Neural crest-derived ganglion acting as relay between parasympathetic neurons of the oculomotor nucleus in the midbrain and the muscles regulating the diameter of the pupil of the eye.

ciliary neurotrophic factor *CNTF* Neurotrophin originally characterized as a survival factor for chick ciliary neurons *in vitro*. Subsequently shown to promote the survival of a variety of other neuronal cell types and to promote the differentiation of bipotential **O-2A progenitor** cells to **type-2 astrocytes** *in vitro*. Developmental expression and regional distribution studies show that, unlike NGF, CNTF is not a target-derived neurotrophic factor. Now considered to be one of the **IL-6 cytokine family** since it acts through a receptor containing gp130.

ciliata Class of Protozoa all of which have cilia at some stage of the life cycle and that usually have a meganucleus.

cilium *Pl.* cilia Motile appendage of eukaryotic cells that contains an **axoneme**, a bundle of microtubules arranged in a characteristic fashion with nine outer doublets and a central pair ('9 + 2' arrangement). Active sliding of doublets relative to one another generates curvature, and the asymmetric stroke of the cilium drives fluid in one direction (or the cell in the other direction).

cimetidine *Tagamet* A generic H2 receptor antagonist used in treating peptic ulcers; reduces gastric acid output.

CIN85 *Cbl-interacting protein of 85 kDa* A multiadaptor protein involved in different cellular functions including the downregulation of activated receptor tyrosine kinases and survival of neuronal cells. **Cbl** (ubiquitin ligase) binds to the activated receptor, further binding of the CIN85/**endophilin** complex leads to clathrin-mediated internalization. Has three Src homology 3 (SH3) domains, a proline-rich region (PRR) and a coiled-coil domain.

CINC Rat homologue of **melanoma growth-stimulatory activity** protein.

cinchocaine *Dibucaine* Long-lasting local anaesthetic.

cingulin Rod-shaped dimeric protein (108 kDa subunit) found in cytoplasmic domain of vertebrate tight junctions. Contains globular and coiled-coil domains, and interacts *in vitro* with several tight junction and cytoskeletal proteins, including the PDZ protein **ZO-1**. GEF-H1/Lfc, a guanine nucleotide exchange factor for RhoA, directly interacts with cingulin.

CIP/KIP A family (p21, p27, p57) of cyclin-dependent kinase inhibitors (CKI) targeting CDK2.

ciprofloxacin Synthetic quinolone antibiotic with broad spectrum of action.

circadian rhythm Regular cycle of behaviour with a period of ~24 h. In most animals the endogenous periodicity, which may be of longer or shorter duration, is entrained to 24 h by environmental cues. See **periodic** and **timeless**.

circular dichroism *CD* Differential absorption of right-hand and left-hand circularly polarized light resulting from molecular asymmetry involving a chromophore group. CD is used to study the conformation of proteins in solution.

circular DNA DNA arranged as a closed circle. This brings serious topological problems for replication that are solved with **DNA topoisomerase**. Characteristic of prokaryotes, but also found in mitochondria, chloroplasts and some viral genomes.

cirrhosis Irreversible condition affecting the whole liver involving loss of parenchymal cells, inflammation, disruption of the normal tissue architecture and, eventually, hepatic failure.

cirrus *Pl.* cirri Large motor organelle of hypotrich ciliates: formed from fused **cilia**.

CIS *Cytokine-inducible immediate early gene* Gene activated by cytokine signals, the product of which may inhibit the signalling pathway. One of the **SOCS** family.

cis-activation Activation of a gene by an activator located on the same chromosome, i.e. not by a diffusible product.

cis-dominance When a gene or promoter affects only gene activity in the DNA duplex molecule in which it is placed the effect is referred to as cis, as opposed to trans effects when a gene or promoter on one DNA molecule can affect genes on another DNA molecule. cis-dominance is seen only when the appropriate pair or set of genes are all cis to each other.

cis-golgi See **Golgi apparatus**.

cisplatin cis-*Diammineplatinum dichloride* Cytotoxic drug used in tumour chemotherapy. Binds to DNA and forms platinum–nitrogen bonds with adjacent guanines.

cis-regulatory modules Short stretches of DNA that help regulate gene expression in higher eukaryotes. They may be up to 1 megabase away from the genes they regulate, and can be located upstream, downstream and even within their target genes.

cisternae Membrane-bounded saccules of the smooth and rough endoplasmic reticulum and **Golgi apparatus**. Operationally might almost be considered as an extra-cytoplasmic compartment, since substances in the cisternal space will eventually be released to the exterior.

cis–trans test The complementation test with two or more interacting genes placed in cis and in trans relationships to each other. A double mutant genome is used in the cis test, made from the two single mutant genomes used in the trans test by recombination. If the wild-type phenotype is restored by both cis and trans arrangements, it is concluded that the two mutations are in different genes and hence that the phenotype is determined by more than one gene. If the trans test is negative and the cis positive, this means that the two mutations are in the same gene. If both tests are negative, then at least one of the mutations must be dominant. Thus the double test provides a means of fine mapping of genes.

cistron A genetic element defined by means of the cis–trans **complementation** test for functional allelism; broadly equivalent to the sequence of DNA that codes for one polypeptide chain, including adjacent control regions.

citric acid cycle Also known as **tricarboxylic acid cycle** or Krebs cycle.

Citrobacter freundii A Gram-negative, facultatively anaerobic, rod-shaped member of the family Enterobacteriaceae; often the cause of significant opportunistic infections. *C. freundii* has also been associated with neonatal meningitis.

citrulline *2-Amino-5-ureiodovaleric acid* An α-amino acid not found in proteins. L-citrulline is an intermediate in the urea cycle.

CJD See **Creutzfeldt–Jacob disease**.

CLA Abbrev. usually for conjugated linoleic acid, but occasionally for cutaneous lymphocyte-associated antigen.

cladistics A method of classifying organisms into groups (taxa) based on 'recency of common descent'. Members of a clade possess shared derived characteristics.

cladogenesis See **anagenesis**.

cladogram A branching diagram (dendrogram) showing the relationships between groups of organisms determined by the methods of **cladistics**.

clamp connection In many **basidiomycete** fungi a short lateral branch of a binucleate cell develops. This is the developing clamp connection. One of the nuclei migrates into it. Both nuclei then undergo simultaneous mitosis so that one end of the cell contains two daughter nuclei from each of the parental nuclei. The nucleus in the branch and the two nuclei are separated off from the centre of the cell by **septa**. A single nucleus remains in the central region. The clamp connection then extends towards and fuses with the central section so that a binucleate cell is reformed.

clans of peptidases In the **MEROPS** peptidase database, 'clans' are composed of peptidases that can be grouped on the basis of evolutionary relationship. Many clans contain peptidases of only a single family (e.g. serine peptidases in Clan SB which contains subtilisin Carlsberg, S8 and sedolisin, S53), but others, designated 'P', have members with mixed catalytic sites (e.g. Clan PC contains both cysteine and serine peptidases).

Clara cell secretory protein *CC16; secretoglobulin 1A1* **Clara cell** secretory protein (CCSP) is an abundant 16-kDa homodimeric protein secreted by non-ciliated secretory epithelial cells in the lung. It has an important protective role against the intrapulmonary inflammatory process. Clara cell 10-kDa protein (CC10) also has anti-inflammatory properties. Both proteins are markers of lung irritation.

Clara cells The main epithelial cell type in small airways; play an important physiological role in surfactant production, protection against environmental agents, and regulation of inflammatory and immune responses in the respiratory system.

class-switching Phenomenon that occurs during the maturation of an immune response in which, for example, B-cells cease making IgM and begin making IgG that has the same antigen-specificity. Switching between other immunoglobulin classes can occur.

clastogen Substance that causes chromosome breakage.

clathrin Protein composed of three heavy chains (180 kDa) and three light chains (34 and 36 kDa) that forms the basketwork of 'triskelions' around a **coated vesicle**. There are two genes for light chains, each of which can generate two distinct transcripts by tissue-specific alternative splicing.

clathrin adaptor proteins *HA1 and HA2 adaptors* Family of proteins that bind to clathrin and promote its assembly into vesicle coats. Different **adaptor** proteins are associated with coated vesicles of Golgi or plasma membrane origin.

clathrin-coated vesicles *CCVs* Class of coated vesicles important in the receptor-mediated endocytosis. They also mediate the transport of cargo from the trans-Golgi network to the endosomal/lysosomal compartment. See **coatamer**.

claudication Fatigue and pain in the legs causing limping; usually due to poor blood flow in the calf muscles. See **intermittent claudication**.

claudins Integral membrane proteins involved in tight junction structure in epithelial and endothelial cells, a family of 24 members displaying organ- and tissue-specific patterns of expression. See **paracellin-1**.

clavulanic acid Compound isolated from *Streptomyces clavuligerusa* that acts as a competitive inhibitor of the beta-lactamases that confer resistance to beta-lactam antibiotics. Often used in combination with **amoxicillin**.

cleavage The early divisions of the fertilized egg to form blastomeres. The cleavage pattern is radial in some phyla, spiral in others.

Cleland's reagent Dithiothreitol.

CLI See **clusterin**.

climacteric A particular stage of fruit ripening, characterized by a surge of respiratory activity and usually coinciding with full ripeness and flavour in the fruit. Its appearance is hastened by ethylene at low concentrations.

clioquinol *Iodochlorhydroxyquinoline* Compound used as an internal amoebicide, and topically for fungal skin infections.

CLIP-170 *Cytoplasmic linker protein-170* Phosphorylation-regulated microtubule-binding protein that accumulates towards the plus ends of cytoplasmic microtubules; isolated from HeLa cells. At the onset of mitosis, CLIP-170 localizes to kinetochores.

CLL Chronic lymphocytic **leukaemia**.

CLN2 A yeast cyclin. See **FAR1**.

CLN3 The function of the CLN3 protein (battenin) is still unknown, although it is evolutionarily conserved, localizes in lysosomes and/or mitochondria and is mutated in **Batten's disease**. The *Schizosaccharomyces pombe* homologue of CLN3 regulates vacuole homeostasis.

cloche Zebra fish mutation that affects differentiation of endothelial and haematopoietic cells and probably acts upstream of **flk-1**.

clock gene A gene with a level of expression that varies cyclically and which might therefore be involved in the generation of biological rhythms. Examples: *period*, *timeless*.

clomipramine Tricyclic antidepressant; used to treat depression and obsessive-compulsive disorder.

clonal deletion One of the two main hypotheses advanced to explain the absence of autoimmune responses: now generally accepted. Clonal deletion is the programmed death of inappropriately stimulated, autoreactive, clones of T-cells.

clonal selection The process whereby one or more clones, i.e. cells expressing a particular gene sequence, are selected by naturally occurring processes from a mixed population. Usually the clonal selection is for general expansion by mitosis, particularly with reference to **B-lymphocytes,** where selection with subsequent expansion of clones occurs as a result of antigenic stimulation only of those lymphocytes bearing the appropriate receptors.

clone A propagating population of organisms, either single-cell or multicellular, derived from a single progenitor cell. Such organisms should be genetically identical, though mutation events may change this.

clonidine Centrally acting antihypertensive agent that works by inhibiting activity of the sympathoexcitatory neurons that regulate arterial pressure. Considered to be a mixed agonist acting on both alpha-2-adrenoreceptors and **imidazoline receptors** (I1-R). Moxonidine and rilmenidine are similar, but more specific for I1 receptors.

cloning The process whereby clones are established. The term covers various manipulations for isolating and establishing clones: in simple systems, single cells may be isolated without precise knowledge of their genotype; in other systems (see **gene cloning**) the technique requires partial or complete selection of chosen genotypes; in plants the term refers to natural or artificial vegetative propagation.

cloning vector A plasmid **vector** that can be used to transfer DNA from one cell type to another. Cloning vectors are usually designed to have convenient restriction sites that can be cut to generate sticky ends to which the DNA that is to be cloned can be ligated easily.

Clonorchis sinensis Chinese liver fluke. Can infect man if inadequately cooked fish is eaten, and can cause biliary obstruction as a result of liver infestation.

clopidogrel *TN Plavix* An inhibitor of ADP-induced platelet aggregation that works by direct inhibition of ADP binding to its receptor and of the subsequent ADP-mediated activation of the glycoprotein GPIIb/IIIa complex. Used as an antithrombotic to reduce risk of strokes or heart attacks.

Clostridium Genus of Gram-positive anaerobic spore-forming bacilli commonly found in soil. Many species produce exotoxins of great potency, the best known being *C. botulinum* and *C. tetani*. Among the toxins produced by *C. perfringens* are **perfringolysin** (theta toxin), an alpha-toxin (phospholipase C), beta, epsilon and iota-toxins (act on vascular endothelium to cause increased vascular permeability), delta-toxin (a haemolysin) and kappa toxin (a collagenase).

Clostridium difficile toxins *C. difficile* enterotoxin A (308 kDa) is secreted, enters eukaryotic cells by **receptor-mediated endocytosis,** and once in the cytoplasm glucosylates small G-proteins of the rho family, thereby inactivating them and leading to the loss of actin-filament bundles. Cytotoxin B (270 kDa) is similar.

clotrimazole Broad-spectrum antifungal drug used to treat oral candidiasis. Probably fungistatic rather than fungicidal at the concentrations normally used.

cloxacillin A semisynthetic penicillin which is resistant to staphylococcal penicillinase and therefore used to treat penicillin-resistant staphylococcal infections.

clozapine An atypical neuroleptic drug used as a sedative and to treat schizophrenia in patients unresponsive to conventional antipsychotic drugs.

CLSM Abbrev. for confocal laser scanning microscopy.

cluster of differentiation (CD) antigens See Table C2.

cluster of orthologous group *COG* Classification system for genes of orthologous sequence, a derivative of the Bidirectional Best BLAST Hits (BDBH) system. A COG consists of genes from different species that are genome-specific best hits as identified by sequence similarity searches using BLAST. The database has some limitations: the same COG number may be assigned to genes of similar yet distinct functions, the same gene may be assigned multiple COG numbers and the classification does not provide any indication of likely function.

clusterin *Complement-associated protein SP-40; complement cytolysis inhibitor (CLI); apolipoprotein J; sulphated glycoprotein 2 (SGP-2); dimeric acid glycoprotein (DAG); glycoprotein III (GpIII)* Vertebrate glycoprotein of uncertain function. Secreted as a 400 amino acid peptide, then cleaved to form two 200 amino acid peptides that are linked by a disulphide bridge. Clusterin is differentially regulated in several patho-physiological processes and invariably induced during apoptosis.

clustering Of **acetylcholine receptors**: aggregation of the receptors in developing **myotubes** in the vicinity of the presynaptic terminal, induced by nerve contact. See **agrin**.

c-maf One of a family of **bZip** transcription factors expressed during development of various organs and tissues, involved in a variety of developmental and cellular differentiation processes and in oncogenesis.

CMC 1. Carboxymethylcellulose. **2.** Critical micelle concentration. **3.** Cell-mediated cytotoxicity: term applied to the response in which T-lymphocytes that react to antigen are stimulated to develop into clones of specific T-effector cells.

CMD1 Chick homologue of **myoD**.

CML 1. Cell-mediated lympholysis: the process of target cell lysis by T-cells. **2.** Chronic myeloid **leukaemia**.

CMV See **cytomegalovirus**.

c-myc tag **Epitope tag** (EQKLISEEDL) derived from the c-myc protein.

CNFs See **cytotoxic necrotizing factors**.

Cnidaria *Coelenterata* Diverse phylum of diploblastic animals with radial or biradial symmetry that includes Classes Hydrozoa (freshwater polyps, small jellyfish), Scyphozoa (large jellyfish) and Anthozoa (sea anemones and stony corals). They possess a single cavity in the body, the enteron, which has a mouth but no anus. Characteristically the ectoderm has specialized stinging cells (cnidoblasts) containing **nematocysts**.

cnidoblast Developing form of **cnidocyte**.

cnidocyst See **nematocyst**.

cnidocyte Ectodermal cell of Cnidaria (coelenterates) specialized for defence or capturing prey. Each cell has a **nematocyst** that can be replaced once discharged.

CNP 1. C-type natriuretic peptide, the major **natriuretic peptide** in the brain, an important regulator of skeletal growth and that may have a role as a neuromodulator. **2.** 2,3-Cyclic nucleotide 3-phosphodiesterase, see **CNPase**.

CNPase *CNP; 2′,3′-cyclic nucleotide 3′-phosphodiesterase; EC 3.1.4.37* A marker enzyme for oligodendrocytes. There are two isoforms, CNP1 and CNP2, both found abundantly in myelinating cells and at much lower levels in non-myelinating cells. CNP2 is localized specifically to mitochondria in non-myelinating cells.

CNQX *6-Cyano-7-nitro-quinoxaline-2,3-dion* **AMPA receptor** antagonist that does not affect **NMDA receptors**. See **glutamate receptors**.

CNTs *Concentrative nucleoside transporters* Sodium-dependent nucleoside transporters that operate in conjunction with the equilibrative (sodium-independent) transporters (**ENTs**), although they belong to structurally unrelated protein families. There are three isoforms in humans, CNT1, CNT2, CNT3, differentially distributed between tissues.

coacervate Colloidal aggregation containing a mixture of organic compounds. One theory of the evolution of life is that the formation of coacervates in the primaeval soup was a step towards the development of cells.

coactosin *CLP* Actin-binding protein (17 kDa) originally isolated from *Dictyostelium* but since found in humans (coactosin-like protein). Binds 5-lipoxygenase and F-actin through different sites. Has homology to the ADF/**cofilin** family.

coagulation factor Group of plasma proteins, many of which contain **EGF**-like domains. See Table F1.

coagulation factor XI A plasma serine **peptidase** with an **apple domain**. See Table F1.

coat protein complex II *COPII* The coat of **coated vesicles** consists of the essential proteins Sec23p, Sec24p, Sec13p, Sec31p, Sar1p and Sec16p. It has been suggested that Sec24p and its two non-essential homologues Sfb2p and Sfb3p serve in cargo selection. **Sar1p** is involved in generating membrane curvature and vesicle formation. COPII coat proteins are required for direct capture of cargo and SNARE proteins into transport vesicles.

coatamer *COPI* A protein complex involving COPI that surrounds non-clathrin-**coated vesicles** which mediate transport in the early secretory pathway within the Golgi and between the Golgi and ER. Another class of non-clathrin-coated vesicles, COPII vesicles, bud at the ER and mediate the transport of cargo from the ER to the Golgi.

coated pit First stage in the formation of a **coated vesicle**.

coated vesicle Vesicle formed as an invagination of the plasma membrane (a coated pit), and that is surrounded by a basket of **clathrin**. Associated with receptor-mediated pinocytosis and receptor recycling.

cobalamin Vitamin B12. See Table V1.

cobra venom factor See **C3**.

cobratoxin α-*Cobratoxin* Polypeptide toxin (71 residues) from *Naja kaouthia*. One of the **alpha-neurotoxins** (curaremimetics), it binds to nicotinic **acetylcholine receptors** with high affinity.

cocaine Drug of abuse and psychostimulant that acts to increase extra-neuronal **dopamine** in the midbrain by binding to the dopamine uptake transporter and hence inhibiting dopamine re-uptake at the plasma membrane.

co-carcinogens Substances that, though not carcinogenic in their own right, potentiate the activity of a carcinogen. Strictly speaking, they differ from **tumour promotors** in requiring to be present concurrently with the carcinogen.

cocci Bacteria with a spherical shape.

coccidiomycosis Infection with the fungus *Coccidioides immitis*. Responsible for chronic infection of cattle, sheep, dogs, cats and certain rodents.

coccidiosis Infection of animals and birds by protozoa of the genera *Eimeria* and *Isospora*, usually affecting the intestinal epithelium and causing enteritis.

cochlear hair cell The sound-sensing cell of the inner ear. The cells have modified ciliary structures (hairs) that enable them to produce an electrical (neural) response to mechanical motion caused by the effect of sound waves on the cochlea. Frequency is detected by the position of the cell in the cochlea, and amplitude by the magnitude of the disturbance.

co-codamol An analgesic containing a mixture of codeine and paracetamol.

co-culture Growth of distinct cell types in a combined culture. In order to get some cells to grow at low (clonal) density it is sometimes helpful to grow them together with a **feeder layer** of **macrophages** or irradiated cells. The mixing of different cell types in culture is otherwise normally avoided, although it is possible that this could prove an informative approach to modelling interactions *in vivo*.

codeine Morphine-group alkaloid, the methyl derivative of morphine, widely used as an analgesic.

codominant Genes in which both alleles of a pair are fully expressed in the heterozygote as, for example, AB blood group in which both A- and B-antigens are present.

codon The coding unit of DNA that specifies the function of the corresponding messenger RNA. A triplet of bases recognized by anticodons on transfer RNA and hence specifying an amino acid to be incorporated into a protein sequence. The code is degenerate, i.e. each amino acid has more than one codon. The stop codon determines the end of a polypeptide. See Table C5.

coelenterate Animal of the Phylum **Cnidaria**. Mostly marine, diploblastic and with radial symmetry. Sea anemones and *Hydra* are well-known examples.

coelenterazine An imidazolopyrazine derivative which, when oxidized by an appropriate luciferase enzyme, produces carbon dioxide, coelenteramide and light. Luciferin is coelenterazine disulphate.

coeliac disease *Celiac disease (US)* Gluten enteropathy: atrophy of **villi** in small intestine leads to impaired absorption of nutrients. Caused by sensitivity to **gluten** (protein of wheat and rye). Sufferers have serum antibodies to gluten and show delayed hypersensitivity to gluten; the risk factor is ten times greater in HLA-B8-positive individuals.

TABLE C5. The codon assignments of the genetic code

First position (5′ end)	Second position				Third position (3′ end)
	U	C	A	G	
	Phe, F	Ser, S	Tyr, Y	Cys, C	U
U	Phe, F	Ser, S	Tyr, Y	Cys, C	C
	Leu, L	Ser, S	Stop (ochre)	Stop: (opal)/(Trp)[a]	A
	Leu, L	Ser, S	Stop (amber)	Trp, W	G
	Leu, L	Pro, P	His, H	Arg, R	U
C	Leu, L	Pro, P	His, H	Arg, R	C
	Leu, L	Pro, P	Gln, Q	Arg, R	A
	Leu, L	Pro, P	Gln, Q	Arg, R	G
	Ile, I	Thr, T	Asn, N	Ser. S	U
A	Ile, I	Thr, T	Asn, N	Ser, S	C
	Ile 1: (Met)[a]	Thr, T	Lys, K	Arg, R: (stop)[a]	A
	Met, M (start)	Thr, T	Lys, K	Arg, R: (stop)[a]	G
	Val, V	Ala, A	Asp, D	Gly, G	U
G	Val, V	Ala, A	Asp, D	Gly, G	C
	Val, V	Ala, A	Glu, E	Gly, G	A
	Val, V: (Met)[b]	Ala, A	Glu, E	Gly, G	G

[a] Unusual codons used in human mitochondria.
[b] Normally codes for valine, but can code for methionine to initiate translation from an mRNA chain.

coelom Body cavity characteristic of most multicellular animals (all **coelomates**). Arises within the embryonic mesoderm, which is thereby subdivided into **somatic mesoderm** and **splanchnic mesoderm**, and is lined by the mesodermally derived peritoneum. May be secondarily lost, and it is unclear whether it evolved once or several times.

coenobium *Pl.* coenobia Colony of cells formed by certain green algae, in which little or no specialization of the cells occurs. The cells are often embedded in a mucilaginous matrix. Examples: *Volvox, Pandorina*.

coenocyte Organism that is not subdivided into cells but has many nuclei within a mass of cytoplasm (a syncytium), as for example some fungi and algae, and the acellular slime mould *Physarum*.

coenzyme Either low-molecular weight intermediate that transfers groups between reactions (e.g. NAD), or catalytically active low-molecular weight component of an enzyme (e.g. haem). Coenzyme and apoenzyme together constitute the holoenzyme.

coenzyme A A derivative of adenosine triphosphate and pantothenic acid that can carry acyl groups (usually acetyl) as thioesters. Involved in many metabolic pathways, e.g. citric acid cycle, and in fatty acid oxidation.

coenzyme M *2-Mercaptoethane sulphonic acid; 2-sulfanylethylsulfonate* Substance is involved in the formation of methane from carbon dioxide by methanogenic bacteria. See **methyl-coenzyme M reductase**.

coenzyme Q *CoQ; ubiquinone* Hydrogen acceptor and donor in the electron transport chain (Preferred synonym: ubiquinone).

co-factor Inorganic complement of an enzyme reaction, usually a metal ion. See **coenzyme**.

cofilin Actin-severing protein (19 kDa) related to **destrin**. Binds to the side of filaments and is pH sensitive. Shares with tropomyosin a 13 amino acid F-actin-binding domain. Very similar to **ADF** (actin-depolymerizing factor).

cohesin domain *Coh domain* Domains found within **scaffoldins** that interact with **dockerin domains** and are responsible for maintaining the structural integrity and localization of **cellulosomes**.

cohesins Family of proteins involved in holding sister chromatids together. Rec8 (561 aa) is *S. pombe* member of this family

coil Protein secondary structure **motif** that does not qualify as **alpha-helix**, **beta sheet** or **beta turn**.

coiled body A ubiquitous nuclear organelle containing the mRNA splicing machinery (U1, U2, U4, U5, U6 and U7 snRNAs), the U3 and U8 snRNPs that are involved in pre-rRNA processing, together with nucleolar proteins such as **fibrillarin** and **coilin**. Ultrastructurally they appear to consist of a tangle of coiled threads and are spherical, between 0.5 and 1 μm in diameter.

coilin *p80-coilin* Protein, M(r) 80 kDa, (62.6 kDa calculated from 576-residue sequence) found in **coiled bodies**. A relatively short portion of the N-terminus seems to target the protein to the organelle.

co-isogenic A strain of animal that differs from others of the same inbred strain at only one locus.

Col-V A plasmid of *E. coli* that codes for **colicin** V, which confers resistance to complement-mediated killing for a **siderophore** to scavenge iron and for F-like pili that permit conjugation (see **sex pili**).

colcemid Methylated derivative of **colchicine**.

colchicine Alkaloid (400 Da) isolated from the Autumn crocus (*Colchicum autumnale*) that blocks microtubule assembly by binding to the **tubulin** heterodimer (but not to tubulin). As a result of interfering with microtubule reassembly will block mitosis at **metaphase**.

Colchicum Genus of crocuses. *C. autumnale*, the Autumn Crocus, is the source of **colchicine**.

cold agglutinins **Antibodies** that agglutinate particles with greater activity below 32°C. They are IgM antibodies specifically reactive with blood groups I and i in humans (precursors of the ABH and Lewis blood group substances), and **agglutinate** red blood cells on cooling, causing **Raynaud's disease** *in vivo*.

cold insoluble globulin (CIG) Name, now obsolete, originally given to fibronectin prepared from **cryoprecipitate**.

coleoptera Order of Insecta with the fore wings (elytra) thickened and chitinized. One of the largest of insect groups – which led J.B.S. Haldane to remark 'that God (if he existed) must have had an inordinate fondness for beetles'. Beetles.

coleoptericin Inducible antibacterial peptide found in the haemolymph of a tenebrionid beetle following the injection of heat-killed bacteria. Peptide A (glycine-rich, 74 residues) is active against Gram-negative bacteria; peptides B and C are isoforms of a 43-residue cysteine-rich peptide that has sequence homology with **defensins** and is active against Gram-positive bacteria. See **deptericins**, **cecropins**, **apidaecins**, **abaecins**.

coleoptile Closed hollow cylinder or sheath of leaf-like tissue surrounding and protecting the plumule (shoot axis and young leaves) in grass seedlings.

coleorhiza Closed hollow cylinder or sheath of leaf-like tissue surrounding and protecting the radicle (young root) in grass seedlings.

colicins Bacterial exotoxins (**bacteriocins**) that affect other bacteria. Colicins E2 and E3 are **AB toxins** with DNAase and RNAase activity, respectively. Most other colicins are channel-forming transmembrane peptides. Coded on plasmids which can be transferred at conjugation.

coliform Gram-negative rod-shaped bacillus. **1.** May be used loosely of any rod-shaped bacterium. **2.** Any Gram-negative enteric bacillus. **3.** More specifically, bacteria of the genera *Klebsiella* or *Escherichia*.

colipase Protein encoded by the pancreatic colipase (CLPS) gene, an essential co-factor needed by pancreatic triglyceride lipase (PNLIP) for efficient dietary lipid hydrolysis. See **procolipase** and **enterostatin**.

collagen Major structural protein (285 kDa) of extracellular matrix. An unusual protein in amino acid composition (very rich in glycine (30%), proline, **hydroxyproline**, lysine and **hydroxylysine**; no tyrosine or tryptophan), structure (a triple-helical arrangement of 95-kDa polypeptides giving a **tropocollagen** molecule, dimensions 300×0.5 nm), and resistance to peptidases. Most types are fibril-forming with a characteristic quarter-stagger overlap between molecules producing an excellent tension-resisting fibrillar structure. Type IV, characteristic of **basal lamina,** does not form fibrils. Many different types of collagen are now recognized. Some are glycosylated (glucose–galactose dimer on the hydroxylysine), and nearly all types can be cross-linked through lysine side chains. See **dermatosparaxis, Ehlers–Danlos syndrome**, scurvy.

collagenase Peptidase capable of breaking native collagen. Once the initial cleavage is made, less specific peptidases will complete the degradation. Collagenases from mammalian cells (EC 3.4.24.7) are metalloenzymes and are collagen-type specific. May be released in latent (proenzyme) form into tissues and require activation by other peptidases before they will degrade fibrillar matrix. Bacterial collagenases (EC 3.4.24.3) are used in tissue disruption for cell harvesting.

collapsin Glycoprotein (100 kDa) from chick brain that may act as a repulsive cue in development and inhibit regeneration of mature neurons. Causes the collapse of the nerve growth cone at picomolar concentrations. Has a domain with sequence homology to **fasciclin** IV and Ig-like domains.

collapsin response-mediator proteins *CRMP-1, -2, etc; TUC, Drp, Ulip, TOAD-64* Family of cytosolic phosphoproteins important in neuronal morphogenesis. They are involved in the signal transduction of semaphorins (of which **collapsin** is one) leading to growth cone collapse. CRMP-2 enhances the advance of growth cones by regulating microtubule assembly and Numb-mediated endocytosis. Rho kinase phosphorylates CRMP-2 during growth cone collapse and cancels the binding activity to the tubulin dimer. CRMP-4 co-localizes with F-actin. Highly homologous to *Caenorhabditis elegans* unc-33, which controls the guidance and outgrowth of neuronal cells. The tetrameric structure of CRMP resembles that of liver dihydropyrimidinase (DHPase), a protein that shares sequence similarity with the CRMPs, but purified brain CRMP does not hydrolyse several DHPase substrates.

collar cell See **choanocyte**.

collectin Family of collagenous **lectin**s believed to play an important part in first-line defence by binding to viruses and by opsonizing yeasts and bacteria. Contain a collagen-like region and a C-type lectin domain. Pulmonary surfactant proteins A and D (SP-A, SP-D), CL-43, serum mannan-binding protein (MBP) and conglutinin are all members of the family. Complement C1q is structurally related to the family.

collenchyma Plant tissue in which the **primary cell walls** are thickened, especially at the cell corners. Acts as a supporting tissue in growing shoots, leaves and petioles. Often arranged in cortical 'ribs', as seen prominently in celery and rhubarb petioles. **Lignin** and **secondary walls** are absent; the cells are living and able to grow.

collenocytes Stellate cells with long thin processes that ramify through the inhalant canal system of sponges.

colliculus A small mound or elevation. The superior and inferior colliculi are elevations on the dorsal surface of the midbrain.

colligative properties Properties that depend upon the numbers of molecules present in solution rather than their chemical characteristics.

collimating lens Lens that produces a non-divergent beam of light or other electromagnetic radiation. Simpler collimators involve slits. Essential in obtaining good illumination in microscopy and for many measuring instruments.

collodion Cellulose tetranitrate dissolved in a mixture of ethanol and ethoxyethane (1:7); the solution is used for coating materials and, medically, for sealing wounds and for dressings.

colonization factors The pili on enteropathogenic forms of *E. coli* facilitate adhesion of the bacteria to receptors (probably GM1 gangliosides) on gut epithelial cells, and are often referred to as colonization or adherence factors. Colonization factor antigens may be plasmid coded, are essential for pathogenicity and are strain-specific – for example, K88 (diarrhoea in piglets), CFAI and CFAII on strains causing similar disease in man.

colony-forming unit *CFU-S* Irradiated mice can have their immune systems reconstituted by the injection of bone marrow cells from a non-irradiated animal. The injected cells form colonies in the spleen (hence -S), each colony representing the progeny of a pluripotent stem cell. Operationally, therefore, the number of colony-forming units is a measure of the number of stem cells.

colony-stimulating factor Cytokines involved in the maturation of various leucocyte, macrophage, monocyte lines.

colony-stimulating factor-1 See **CSF-1**.

colostrum The first milk secreted by an animal coming into lactation. May be especially rich in maternal lymphocytes and Ig, and thus transfers immunity passively.

colour blindness The lack of one or more of the spectral colour receptors. The commonest form, Daltonism, is an inability to distinguish between red and green.

colour phase Any seasonal or abnormal variation in the colouration of an animal.

colposcope An endoscope for inspecting the vagina.

colposcopy Endoscopic examination of the vagina by means of a colposcope.

columnar cells Cells of columnar epithelium in which the area in contact with the basal lamina is less that the lateral cell–cell contact (in contrast to cuboidal and squamous epithelia).

comb plates Large, flat organelles formed by the fusion of many cilia. Vertical rows of comb plates form the motile appendages of Ctenophores.

combination therapy Treatment of a disease with two or more drugs, classically used with several antibiotics in tuberculosis and now in cancer and AIDS.

combinatorial chemistry Method by which large numbers of compounds can be made, usually utilizing solid-phase synthesis. In the simplest form, carrier beads would be treated separately so as to couple subunits A, B and C, then mixed and re-divided before adding subunits A, B and C in the three reaction mixtures. Thus beads +AA, AB, AC, BA, BB, BC, etc., would have been synthesized – though the products are mixed and deconvolution of an active mixture will be necessary to identify the active molecule. Using a relatively small number of reactions, enormous diversity can rapidly be generated. Increasingly, the term is used loosely for any procedure that generates highly diverse sets of compounds – the more recent tendency is to prefer high-speed parallel synthesis in which each reaction chamber contains only one compound.

combined immunodeficiency *Severe combined immunodeficiency syndrome (SCID)* Congenital immunodeficiency with thymic agenesis, lymphocyte depletion and hypogammaglobulinaemia: both cellular and humoral immune systems are affected, and life expectancy is low unless marrow transplantation is successful.

combining site Any region of a molecule that binds or reacts with a given compound. Especially of the region of immunoglobulin that combines with the determinant of an appropriate antigen.

Combretastatin® *Combrestatin A4* Drug originally derived from the African bush willow tree, used in cancer treatment. Acts by inhibition of tumour vascularization by binding to tubulin in endothelial cells. Synthetic variants are emerging with slightly different efficacy.

comC, comD, comE Operon involved in regulating competence for genetic transformation in *Streptococcus pneumoniae*. Com D, a transmembrane histidine kinase, binds **competence-stimulating peptide** (CSP-1 or CSP-2, encoded by *ComC1* and *ComC2*, respectively) and activates the response regulator ComE. Allelic variation in *ComD* determines the **pherotype** – whether CSP-1 or CSP-2 elicits a response.

comet assay *Single cell gel (SCG) electrophoresis* A rapid and very sensitive fluorescent microscopic method to examine DNA damage and repair at individual cell level. Single cells embedded in agarose gel on a microscope slide are lysed and DNA allowed to unwind before being subjected to an electrophoretic field. Following staining with acridine orange the image is of a distinct head, comprising intact DNA, and a tail, consisting of damaged or broken pieces of DNA. By adjusting the lysis and unwinding conditions, the sensitivity can be adjusted to preferentially identify particular types of damage.

comitin Actin-binding protein (24 kDa) originally from *Dictyostelium* but later found in mammalian cells. Increases the viscosity of F-actin solutions. Present on Golgi and vesicle membranes. Cells lacking comitin are impaired in the early steps of phagocytosis of yeast and bacteria, although uptake of latex beads is unaffected.

committed cells Cells become committed to particular pathways of differentiation (see **determination**) at different stages in embryogenesis; generally this is an irreversible event in mammalian cells, although reversing such commitment would potentially allow regeneration of tissues.

communicating junction Another name for a **gap junction**.

comorbid Term used for diseases or disorders that co-exist.

compactin *Mevastatin* Fungal metabolite produced by *Penicillium citrinum*, used as an intermediate compound for production of Pravastatin (one of the **statins**).

compaction Process that occurs during the morula stage of embryogenesis in which blastomeres increase their cell–cell contact area and develop gap junctions. In mice com-

paction occurs at the eight-cell stage, and after this the developmental fate of each cell becomes restricted.

companion cell Relatively small plant cell, with little or no vacuole, found adjacent to a phloem **sieve tube** and originating with the latter from a common mother cell. Thought to be involved in translocation of sugars in and out of the sieve tube.

compartment Conceptualized part of the body (organs, tissues, cells or fluids) considered as an independent system for purposes of modelling and assessment of distribution and clearance of a substance or in development as a clonal territory. In the insect wing, for example, there are two compartments, anterior and posterior, each containing several clones, but clones do not cross the boundary. It seems from studies with **homeotic mutants** that cells in different compartments are expressing different sets of genes. The evidence for such developmental compartments in vertebrates is sparse at present.

compartmental analysis Mathematical process leading to a model of transport of a substance in terms of **compartments** and rate constants for input and output.

competence-stimulating peptide *CSP-1, CSP-2* Pheromone that is involved in a **quorum-sensing** mechanism that regulates competence for genetic transformation in *Streptococcus pneumoniae*. Encoded by *comC* gene. Receptor system is a two-component signal transduction system, ComD–ComE (TCS12). See **comC**.

competent cells 1. Bacterial cells with enhanced ability to take up exogenous DNA and thus to be transformed. Competence can arise naturally in some bacteria (*Pneumococcus*, *Bacillus* and *Haemophilus* spp.); a similar state can be induced in *E. coli* by treatment with calcium chloride. Once competence has been induced, the cells can be stored at low temperature in cryoprotectant and used when needed. **2.** Cells capable of responding to an inducer in embryonic development.

competitive inhibition Inhibitor that occupies the active site of an enzyme or the binding site of a receptor and prevents the normal substrate or ligand from binding. At sufficiently high concentration of the normal ligand, inhibition is lost: the K_m is altered by the competitive inhibitor, but the V_{max} remains the same.

complement A heat-labile system of enzymes in plasma associated with response to injury. Activation of the complement cascade occurs through two convergent pathways. In the classical pathway the formation of antibody/antigen complexes leads to binding of C1 and the release of active esterase that activates C4 and C2, which in turn bind to the surface. The C42 complex splits C3 to produce C3b (an opsonin) and C3a (anaphylatoxin). C423b acts on C5 to release C5a (anaphylatoxin and chemotactic factor), leaving C5b, which combines with C6789 to form a cytolytic **membrane attack complex**. In the alternate pathway C3 cleavage occurs without the involvement of C142 and can be activated by IgA, endotoxin or polysaccharide-rich surfaces (e.g. yeast cell wall, zymosan). Factor B combines with C3b to form a C3 convertase that is stabilized by Factor P, generating a positive feedback loop. The alternate pathway is presumably the ancestral one upon which the sophistication of antibody recognition has been superimposed in the classical pathway. The enzymatic cascade amplifies the response, leads to the activation and recruitment of leucocytes, increases phagocytosis and induces killing directly.

It is subject to various complex feedback controls that terminate the response.

complement cytolysis inhibitor See **clusterin**.

complement fixation Binding of **complement** as a result of its interaction with immune complexes (the classical pathway) or particular surfaces (alternative pathway).

complementarity determining region *CDR* Hypervariable region within the antigen binding site of immunoglobulin molecules and T-lymphocyte antigen receptors. The sequence in this region determines which antigen (epitope) will bind. Also used to refer to the genomic sequence encoding the hypervariable regions.

complementary base pairs The crucial property of DNA is that the two strands are complementary: guanine and cytosine are complementary and pair up through their hydrogen bonds, as are adenine and thymine, which only form two hydrogen bonds (adenine and uracil in RNA).

complementation The ability of a mutant chromosome to restore normal function to a cell that has a mutation in the homologous chromosome when a hybrid or heterokaryon is formed – the explanation being that the mutations are in different cistrons, and between the two a complete set of normal information is present.

complexins Complexin I and complexin II are cytosolic proteins involved in the regulation of neurotransmitter release, competing with the chaperone protein alpha-SNAP for binding to **synaptobrevin** as well as the synaptic membrane proteins SNAP-25 and syntaxin, which together form the SNAP receptor (**SNARE**) complex. Complexin I is a marker of axosomatic (inhibitory) synapses, whereas complexin II mainly labels axodendritic and axospinous synapses, the majority of which are excitatory.

COMT *Catechol-O-methyltransferase; EC 2.1.1.6* Enzyme thought to functionally modulate dopaminergic neurons. Catalyses reaction between *S*-adenosyl methionine and catechol to form *S*-adenosyyl-L-homocysteine and guiacol, and methylates catecholamines, thereby terminating their signalling activity. There are two functional polymorphisms in humans, with high enzyme activity (COMT Val) and low enzyme activity (COMT Met) variants, although the significance is still unclear.

Con A See **concanavalin A**.

Con A binding sites See **Con A receptors**.

Con A receptors A common misuse of the term receptor. Con A binds to the mannose residues of many different glycoproteins and glycolipids, and the binding is therefore not to a specific site. It could be argued that the receptor is the Con A and cells have Con A ligands on their surfaces; certainly this would be less confusing.

conalbumin *Ovotransferrin* Non-haem iron-binding protein found in chicken plasma and egg white.

conantokins Class of small peptides (17–21 residues) from cone shells (*Conus* spp.) that inhibit **NMDA** class of glutamate receptors.

concanamycin A *Folimycin* Specific inhibitor of vacuolar H^+-ATPase isolated from *Streptomyces* sp.; inhibits perforin-based cytotoxic activity, mostly due to accelerated degradation of perforin by an increase in the pH of lytic granules. Other concanamycins have been isolated with slightly different characteristics.

concanavalin A *Con A* A **lectin** isolated from the jack bean, *Canavalia ensiformis*. See Table L1 (Lectins).

concatamer Two or more identical linear molecular units covalently linked in tandem. Especially used of nucleic acid molecules and of units in artificial polymers.

conchae Three pairs of turbinate bones (nasal conchae) protrude into the centre of the nasal cavity. Each concha is a cartilaginous or slightly ossified scroll that serves to increase the surface area of the nasal cavity.

condensation 1. Process of compression or increase in density. Chromosome condensation is a consequence of increased supercoiling that causes the chromosome to become shorter and thicker, and thus visible in the light microscope. **2.** A condensation reaction in chemistry is one in which two molecules combine to form a single larger molecule with the concomitant loss of a relatively small portion as water or similar small molecule. **3.** The product of the process, for example water droplets that have formed on a cold surface exposed to water vapour.

condensing vacuole Vacuole formed from the cis face of the **Golgi apparatus** by the fusion of smaller vacuoles. Within the condensing vacuole the contents are concentrated and may become semicrystalline (**zymogen granules** or **secretory vesicles**).

conditional mutation A mutation that is only expressed under certain environmental conditions – for example, temperature-sensitive mutants.

conditioned medium Cell-culture medium that has already been partially used by cells. Although depleted of some components, it is enriched with cell-derived material, probably including small amounts of growth factors; such cell-conditioned medium will support the growth of cells at much lower density and, mixed with some fresh medium, is therefore useful in **cloning**.

cone cell See **retinal cone**.

confluent culture A cell culture in which all the cells are in contact and thus the entire surface of the culture vessel is covered. It is also often used with the implication that the cells have also reached their maximum density, though confluence does not necessarily mean that division will cease or that the population will not increase in size.

confocal microscopy A system of (usually) **epifluorescence** light microscopy in which a fine laser beam of light is scanned over the object through the objective lens. The technique is particularly good at rejecting light from outside the plane of focus, and so produces higher effective resolution than is normally achieved.

conformational change Alteration in the shape, usually the tertiary structure of a protein, as a result of alteration in the environment (pH, temperature, ionic strength) or the binding of a ligand (to a receptor) or binding of substrate (to an enzyme).

conformer A molecule that has the same structural formula but different conformation. Some conformers are energetically more favourable than others.

congenic Organisms that differ in **genotype** at (ideally) one specified locus. Strictly speaking these are conisogenics. Thus one homozygous strain can be spoken of as being congenic to another.

congenital severe combined immunodeficiency See **SCID**.

conglutinin Protein present in serum that causes **agglutination** of antibody–antigen–complement complexes; binds C3b.

Congo red Naphthalene dye that is pH sensitive (blue-violet at pH 3, red at pH 5). Used as vital stain, also in staining for amyloid.

conidium Asexual spore of fungus, borne at the tip of a specialized **hypha** (conidiophore).

conjugate 1. Molecular species produced in a biological system by covalently linking two chemical moieties from different sources. **2.** Material produced by attaching two or more substances together, e.g. a conjugate of an antibody with a fluorochrome or enzyme for use as a probe.

conjugation Union between two gametes or between two cells leading to the transfer of genetic material. In eukaryotes the classic examples are in *Paramecium* and *Spirogyra*. Conjugation between bacteria involves an F$^+$ bacterium (with F-pili) attaching to an F$^-$; transfer of the F-plasmid then occurs through the sex pilus. In Hfr mutants the F$^-$ plasmid is integrated into the chromosome and so chromosomal material is transferred as well. Conjugation occurs in many Gram-negative bacteria (*Escherichia*, *Shigella*, *Salmonella*, *Pseudomonas* and *Streptomyces*).

Conn's syndrome Uncontrolled secretion of **aldosterone** usually by an adrenal **adenoma**.

connectin 1. Alpha-connectin is identical to **titin-1**, β-connectin to titin-2. Elastic connectin/titin molecules position the myosin filaments at the centre of a sarcomere by linking them to the Z line. **2.** *Drosophila* connectin (CON) is a cell-surface protein of the leucine-rich repeat family expressed on the surface of a subset of embryonic muscles, and on the growth cones and axons of the motoneurons that innervate these muscles. It is attached to the cell surface via a GPI linkage and mediates homotypic cell–cell adhesion *in vitro*.

connective tissue Rather general term for mesodermally derived tissue that may be more or less specialized. Cartilage and bone are specialized connective tissue, as is blood, but the term is probably better reserved for the less specialized tissue that is rich in extracellular matrix (**collagen**, **proteoglycan**, etc.) and which surrounds other more highly ordered tissues and organs. See **areolar connective tissue**.

connective tissue diseases A group of diseases, including rheumatoid arthritis, systemic lupus erythematosus, rheumatic fever, scleroderma and others, that are sometimes referred to as rheumatic diseases. They probably do not affect solely connective tissues, but the diseases are linked in various ways and have interesting immunological features, which suggest that they may be autoimmune in origin.

connective tissue-activating peptide III *CTAP III* **Cytokine**, produced from **platelet basic protein**, that acts as a **growth factor**.

connexins Generic term for proteins isolated from gap junctions. Currently 21 human genes and 20 mouse genes for connexins have been identified, each with tissue- or cell-type-specific expression. Most organs and many cell types express more than one connexin. Connexin phosphorylation has been implicated in connexin assembly, gap junction turnover and responses to tumour promoters and oncogenes. Connexin43 (Cx43), the most widely expressed and abundant gap junction protein, can be phosphorylated at several different serine and tyrosine residues. Mutations in connexin genes are associated with peripheral neuropathies, cardiovascular diseases, dermatological diseases, hereditary deafness and cataract. See **connexon**.

connexon The functional unit of gap junctions. An assembly of six membrane-spanning proteins (**connexins**) having a water-filled gap in the centre. Two connexons in juxtaposed membranes link to form a continuous pore through both membranes.

conotoxins Toxins from cone shells (*Conus* spp.). The α-conotoxins (small peptides, 13–18 residues) are competitive inhibitors of nicotinic acetylcholine receptors. The μ-conotoxins are small (22-residue) peptides that bind voltage-sensitive sodium channels in muscle, causing paralysis. The ω-conotoxins are similar in size and inhibit voltage-gated calcium channels, thereby blocking synaptic transmission.

consensus sequence Of a series of related DNA, RNA or protein sequences, the sequence that reflects the most common choice of base or amino acid at each position. Areas of particularly good agreement often represent conserved functional domains. The generation of consensus sequences has been subject to intensive mathematical analysis.

conservative substitution In a gene product, a substitution of one amino acid with another with generally similar properties (size, hydrophobicity, etc.), such that the overall functioning is likely not to be seriously affected.

CONSTANS *CO Arabidopsis* gene that promotes flowering in long days. Flowering is induced when CO messenger RNA expression coincides with the exposure of plants to light. A member of a family of 17 CO-like genes that are widely distributed in plants. Product is a nuclear zinc-finger protein.

constant region The C-terminal half of the light or the heavy chain of an immunoglobulin molecule. The amino acid sequence in this region is the same in all molecules of the same class or subclass whereas the variable region is antigen-specific.

constitutive Constantly present, whether there is demand or not. Thus some enzymes are constitutively produced, whereas others are inducible.

constitutive transport element *CTE* An RNA motif that has the ability to interact with intracellular RNA helicases and is essential for export of RNA from the nuclear compartment. Many retroviruses have such an element to ensure their unspliced mRNA is moved into the cytoplasm. See **Tap protein**.

constriction ring The equatorial ring of **microfilaments** that diminishes in diameter, probably both by contraction and disassembly, as **cytokinesis** proceeds.

contact following Behaviour shown by individual **slime mould** cells when they join a stream moving towards the aggregating centre. **Contact sites A** at front and rear of cell may be involved in *Dictyostelium*.

contact guidance Directed locomotory response of cells to an anisotropy of the environment, for example the tendency of fibroblasts to align along ridges or parallel to the alignment of collagen fibres in a stretched gel.

contact inhibition of growth/division See **density-dependent inhibition**.

contact inhibition of locomotion/movement Reaction in which the direction of motion of a cell is altered following collision with another cell. In heterologous contacts both cells may respond (mutual inhibition) or only one (non-reciprocal). Type I contact inhibition involves paralysis of the locomotory machinery, Type II is a consequence of adhesive preference for the substratum rather than the dorsal surface of the other cell.

contact inhibition of phagocytosis Phenomenon described in sheets of kidney epithelial cells that, when confluent, lose their weak phagocytic activity, probably because of a failure of adhesion of particles to the dorsal surface in the absence of ruffles.

contact sensitivity Allergic response to contact with irritant, usually a **hypersensitivity**.

contact sites A *csA* Developmentally regulated adhesion sites that appear on the ends of aggregation-competent *Dictyostelium discoideum* (see **Acrasidae**) – at the stage when the starved cells begin to come together to form the **grex**. Originally detected by the use of Fab fragments of polyclonal antibodies, raised against aggregation-competent cells and adsorbed against vegetative cells, to block adhesion in EDTA-containing medium. (Cell–cell adhesion mediated by contact sites A, unlike that mediated by contact sites B, is not divalent cation-sensitive.) The fact that a mutant deficient in csA behaves perfectly normally in culture is puzzling.

contact sites B See **contact sites A**.

contact-induced spreading The response in which contact between two **epithelial cells** leads to a stabilized contact and increased spreading of the cells so that the area covered is greater than that covered by the two cells in isolation.

contactin A 130-kDa **glycoprotein** attached to the **cytoskeleton** via its cytoplasmic domain; concentrated in areas of interneuronal contact. Its sequence contains both immunoglobulin-like domains and **fibronectin** type III repeats. Its close homology with NCAM suggests that it is a **CAM**. Like **L1**, **F11**, **neurofascin** and **TAG-1** in vertebrate nervous systems and **fasciclin II** in insects, thought to be associated with the process of selective **fasciculation**. Sometimes **GPI-anchored**.

contactin/F3 Adhesion molecule associated with neuronal–glial interactions as part of a complex that also includes the contactin-associated protein-1 (CASPR1), and the glial receptor protein tyrosine phosphatase beta.

contactinhibin Plasma membrane **glycoprotein** of 60–70 kDa isolated from human diploid fibroblasts, which when immobilized on silica beads has been reported to reversibly inhibit the growth of cultured cells.

contig DNA sequence assembled from overlapping shorter sequences to form one large contiguous sequence.

contractile ring See **constriction ring**.

contractile vacuole A specialized vacuole of eukaryotic cells, especially Protozoa, that fills with water from the cytoplasm and then discharges this externally by the opening of a permanent narrow neck or a transitory pore. Function is probably osmoregulatory.

contrapsin Trypsin inhibitor (**serpin**) from rat.

control element Generic term for a region of DNA, such as a **promoter** or **enhancer** adjacent to (or within) a gene that allows the regulation of gene expression by the binding of **transcription factor**.

control region General name for genomic DNA that, through binding of **transcription factor**s to its promoters, enhancers and repressors, modulates the expression level of nearby genes.

controlled drugs Drugs that can only be prescribed under guidelines laid down in legislation. Usually drugs that have the potential to cause addiction and dependence.

Conus Genus of gastropod molluscs, cone snails. See **conotoxins** and **conantokins**.

convulxin *Cvx* Snake venom toxin, a 72-kDa glycoprotein (**C-type lectin**) isolated from the venom of *Crotalus durissus terrificus* (South American rattlesnake), that activates platelets through the collagen receptor glycoprotein VI (GPVI)/Fc receptor gamma-chain (FcR γ-chain) complex, leading to tyrosine phosphorylation and activation of the tyrosine kinase **Syk** and PLCγ2. Native Cvx is an octamer composed of four αβ-heterodimers.

Coomassie Brilliant Blue *Coomassie Brilliant Blue G-250, Brilliant Blue R, Acid Blue 90, Kenacid Blue* Blue dye that binds non-specifically to proteins, used in **Bradford method** for protein estimation and for detecting proteins on gels. Originally developed as a dye for wool; in acid solution has an absorbance shift from 465 nm to 595 nm when it binds to protein.

Coombs test Diagnostic test to determine whether an individual's red cells are coated with autoantibodies or immune complexes. The erythrocytes are mixed with anti-human immunoglobulin, and if antibody is present the red cells will agglutinate.

cooperativity Phenomenon displayed by enzymes or receptors that have multiple binding sites. Binding of one ligand alters the affinity of the other site(s). Both positive and negative cooperativity are known; positive cooperativity gives rise to a sigmoidal binding curve. Cooperativity is often invoked to account for non-linearity of binding data, although it is by no means the only possible cause.

coordination complex Complex held together by coordinate (dipolar) bonds, covalent bonds in which the two shared electrons derive from only one of the two participants.

co-phenotrope *TN Lomotil* Mixture of diphenoxylate HCl, an opioid that reduces gut motility and atropine sulphate (antispasmodic) used to treat diarrhoea.

COPI, COPII See **coatamer**.

Coprinus Genus of fungi that have gills that autodigest once spores have been discharged, giving rise to a black inky fluid.

copy-number The number of molecules of a particular type on or in a cell or part of a cell. Usually applied to specific genes, or to plasmids within a bacterium.

copy-number polymorphisms *CNPs* A form of polymorphism that is becoming increasing recognized with the availability of high-density oligonucleotide microarrays for SNPs. Individuals differ by 11 CNPs on average, and copy-number variation has been found in 70 different genes. Copy-number polymorphism in the orthologous rat and human Fc gamma RIIIb receptor (FcγR3) genes is a determinant of susceptibility to immunologically mediated glomerulonephritis.

coracle A *Drosophila* **protein 4.1** homologue, required during embryogenesis and localized to the cytoplasmic face of the septate junction in epithelial cells, where it interacts with the transmembrane protein **neurexin**.

cord blood Blood taken postpartum from the umbilical cord.

cord factor Glycolipid (trehalose-6,6′-dimycolate) found in the cell walls of Mycobacteria (causing them to grow in serpentine cords) and important in virulence, being toxic and inducing granulomatous reactions identical to those induced by the whole organism.

cornea Transparent tissue at the front of the eye. The cornea has a thin outer squamous epithelial covering and an endothelial layer next to the aqueous humour, but is largely composed of avascular collagen laid down in orthogonal arrays with a few fibroblasts. Transparency of the cornea depends on the regularity of spacing in the collagen fibrils.

corneocyte Cell of the **stratum corneum**, the outer layer of the skin, heavily keratinized and dead.

cornification The process of keratinization of the outer layers of the skin that eventually leads to death of keratinocytes (forming **corneocytes**) and ultimately to shedding of squames.

cornified epithelium Epithelium in which the cells have accumulated keratin and died. The outer layers of vertebrate skin, hair, nails, horn and hoof are all composed of cornified cells.

corona radiata The layer of cylindrical cells that surrounds the developing mammalian ovum.

coronal section A cross-section of the brain taken effectively where the edge of a crown would touch.

Coronaviridae Family of single-stranded RNA viruses responsible for respiratory diseases. The outer envelope of the virus has club-shaped projections that radiate outwards and give a characteristic corona appearance to negatively stained virions. A coronavirus is responsible for **SARS**.

coronin Actin-binding protein (55 kDa) of *Dictyostelium discoideum*. Associated with crown-shaped cell-surface projections in growth phase cells. Accumulates at front of cells, responding to a chemotactic gradient of cAMP. Amino-terminal domain has similarity to β subunits of heterotrimeric G-proteins; C-terminal has high α-helical content.

corpora allata Insect endocrine organs, located behind the brain, that secrete **juvenile hormone**. In some species they are paired and laterally placed, in others they fuse during development to form a single median structure, the corpus allatum.

corpus callosum Band of white matter at the base of the longitudinal fissure dividing the two cerebral hemispheres of the brain.

corpus luteum *Pl.* corpora lutea Glandular body formed from the Graafian follicle in the ovary following release of the ovum. Secretes **progesterone**.

corpus striatum The basal ganglionic part of the wall of each cerebral hemisphere in the vertebrate brain.

corralling The proposed confinement of membrane proteins within a diffusion barrier, thereby limiting long-range translational diffusion rates without affecting short range properties (e.g. rotation rates).

cortactin A p80/85 protein first identified as a substrate for **src** kinase. An F-actin-binding protein that redistributes to membrane ruffles as a result of growth factor-induced **Rac1** activation. Has proline-rich and SH3 domains. Overexpression of cortactin increases cell motility and invasiveness.

cortex 1. *Bot.* Outer part of stem or root, between the vascular system and the epidermis; composed of **parenchyma**. 2. *Cytol.* Region of cytoplasm adjacent to the plasma membrane. 3. *Histol.* Outer part of organ.

cortexillin Calcium-independent homodimeric actin-binding (bundling) protein from *Dictyostelium*. Cortexillin activity is crucial for cytokinesis; cortexillin is enriched in the cleavage furrow and **dynacortin** is depleted.

cortical granules In sea urchin eggs, specialized secretory granules that fuse with the egg membrane following fertilization. Granule contents include (1) hyalin, which forms a layer immediately surrounding the egg; (2) a colloid, which raises the **fertilization membrane** by imbibing water; (3) a serine peptidase, which destroys receptors for sperm on the vitelline membrane; (4) a protein, termed vitelline delaminase, which may cleave the connection between the vitelline membrane and the oolemma; (5) a structural protein, which in the presence of H_2O_2 is polymerized onto the inner surface of the old vitelline membrane, now called the fertilization membrane; and (6) ovoperoxidase, a heme-dependent peroxidase that functions to block polyspermy by interacting with the structural protein.

cortical layer See **cortical meshwork**.

cortical meshwork Subplasmalemmal layer of tangled microfilaments anchored to the plasma membrane by their barbed ends. This meshwork contributes to the mechanical properties of the cell surface and probably restricts the access of cytoplasmic vesicles to the plasma membrane.

corticostatin Peptide of the alpha-defensin family that has antibiotic, antifungal and antiviral activity. It also inhibits corticotropin (ACTH) stimulated corticosterone production.

corticosteroids **Steroid** hormones produced in the **adrenal** cortex. Formed in response to **adrenocorticotrophin** (ACTH). Regulate both carbohydrate metabolism and salt/water balance. Glucocorticoids (e.g. cortisol, cortisone) predominantly affect the former and minerocorticoids (e.g. aldosterone) the latter.

corticotrophin *Corticotropin* See **adrenocorticotrophin**.

corticotrophin releasing factor See **corticotropin-releasing hormone** and **adrenocorticotrophin**.

corticotropin-releasing hormone-binding protein *CRH-BP* A 37-kDa secreted glycoprotein that binds both **corticotropin-releasing hormone** (CRH) and **urocortin** with high affinity and is structurally unrelated to the CRH receptors. It is an important modulator of CRH activity. CRH-BP orthologues have been identified in multiple invertebrate and vertebrate species, and it is strongly conserved throughout evolution.

corticotropin-releasing hormone *CRH; corticotropin-releasing factor (CRF); corticoliberin* Key regulator of the hypothalamic–pituitary–adrenal axis, a peptide hormone (41 residues) produced by the hypothalamus that stimulates corticotropic cells of the anterior lobe of the pituitary to produce ACTH (**adrenocorticotrophin**) and other biologically active substances (for example, β-endorphin). Also produced by both the placenta and foetal membranes at term in man. Stressors cause a release of corticotrophin releasing hormone. The CRH hormone family has at least four ligands, two receptors and a binding protein.

cortisol The major adrenal glucocorticoid; stimulates conversion of proteins to carbohydrates, raises blood sugar levels and promotes glycogen storage in the liver.

cortisone *11-Dehydroxy-cortisol* Natural **glucocorticoid** formed by 11β-hydroxysteroid dehydrogenase action on hydrocortisone; inactive until converted into **hydrocortisone** in the liver.

cortistatin *CST* Cortistatin (CST) is a neuropeptide (14 residues) with high structural homology with somatostatin. Its mRNA is restricted to **GABA**-containing cells in the cerebral cortex and hippocampus. CST modulates the electrophysiology of the hippocampus and cerebral cortex of rats; hence, it may be modulating mnemonic processes.

Corynebacteria Genus of **Gram-positive** non-motile rod-like bacteria, often with a club-shaped appearance. Most are facultative anaerobes with some similarities to **mycobacteria** and **nocardiae**. *C. diphtheriae* is the causative agent of diphtheria and produces a potent exotoxin, **diphtheria toxin**.

COS cells Simian fibroblasts (CV-1 cells) transformed by **SV40** that is deficient in the origin of replication region. Express **large T-antigen** constitutively and if transfected with a vector containing a normal SV40 origin have all the other early viral genes necessary to generate multiple copies of the vector and thus to give very high levels of expression.

cosegregation Of two genotypes, meaning that they tend to be inherited together, implying close linkage.

Cosmarium A very large genus of the Order Desmidiales of the **Chlorophyta**, with the cells deeply constricted to form two semicells. Each semicell contains a single, large chloroplast with two pyrenoids. The cells are slowly motile by mucilage secretion. Usually found in freshwater.

cosmid A type of **bacteriophage** lambda vector. Often used for construction of **genomic libraries**, because of their ability to carry relatively long pieces of DNA insert, compared with **plasmids**.

cos sites Sites on a **cosmid** vector that are required for integration into host DNA. The two cohesive ends, known as cos sites, are 12 nucleotides in length, and the chromosome circularizes by means of these complementary cohesive ends.

costa Rod-shaped intracellular organelle lying below the undulating membrane of *Trichomonas*. Generates active bending associated with local loss of **birefringence** at the bending zone, probably as a result of conformational change in the longitudinal lamellae. Major protein ∼90 kDa.

costameres Regular periodic submembranous arrays of **vinculin** in muscle cells; link sarcomeres to the membrane and are associated with links to extracellular matrix.

Cot curve Physicochemical technique for measuring the complexity (or size) of DNA. The DNA is heated to make it single stranded, then allowed to cool. The renaturation of the DNA is followed spectroscopically: larger DNA molecules take longer to reanneal.

cotinine A metabolite of nicotine that persists in the body for 2–4 days after tobacco use, and thus serum or urine levels of cotinine provide a more accurate marker of smoking or smoke inhalation than do questionnaires.

co-translational transport Process whereby a protein is moved across a membrane as it is being synthesized. This process occurs during the translation of the message at membrane-associated **ribosomes** in **rough endoplasmic reticulum** during the synthesis of secreted proteins in eukaryotic cells.

co-transport In membrane transport, describes tight coupling of the transport of one species (generally Na^+) to another (e.g. a sugar or amino acid). The transport of Na^+ from high to low concentration can provide the energy for transport of the second species up a concentration gradient. See secondary **active transport**.

co-trimoxazole A mixture of **trimethoprin**, which is antibacterial, and **sulphamethoxazole**, a sulfonamide antibiotic, used to treat pneumocystis carinii pneumonia, toxoplasmosis and nocardiasis. The combination seems to be more effective than either alone.

Coturnix coturnix japonica Japanese quail. Used extensively in developmental biology because quail nuclei can easily be distinguished from those of the chicken, and this facilitates grafting experiments for fate mapping.

cotyledon Modified leaf ('seed leaf'), found as part of the embryo in seeds, involved in either storage or absorption of food reserves. Dicotyledonous seeds contain two, monocotyledonous seeds only one. May appear above ground and show photosynthetic activity in the seedling.

Coulter counter Particle counter used for bacteria or eukaryotic cells; works by detecting change in electrical conductance as fluid containing cells is drawn through a small aperture. (The cell, a non-conducting particle, alters the effective cross-section of the conductive channel.)

coumarin *O-Hydroxycinnamic acid* Pleasant-smelling compound found in many plants and released on wilting (probably a major component of the smell of fresh hay). Has anticoagulant activity by competing with vitamin K. Coumarin derivatives have anti-inflammatory and antimetastatic properties, and inhibit xanthine oxidase and the production of 5-HETE by neutrophils and macrophages. Various derivatives have these activities, including esculentin, **esculin** (6,7-dihydroxycoumarin 6-O-D-glucoside),

fraxin, **umbelliferone** (7-hydroxy coumarin) and **scopoletin** (6-methoxy-7-hydroxy coumarin).

counter ion Ion of the opposite charge to that of an immobilized ionized molecule, for example the carboxyl residue of *N*-acetyl neuraminic acid on cell-surface glycoprotein. The consequence is to alter the composition of the environment immediately adjacent to the cell surface. See **double layer**.

counterstain Rather non-specific stain used in conjunction with another histochemical reagent of greater specificity to provide contrast and reveal more of the general structure of the tissue. Light Green is used as a counterstain in the Mallory procedure, for example.

counter-transport Transport system across a membrane in which movement of a molecule in one direction is matched by the movement of a different molecule in the opposite direction. If both are charged then no potential gradient will develop, provided equal numbers of charges are moved in each direction. The opposite of **co-transport**.

coupled transporter 1. A membrane transport system in which movement of one molecule or ion down an electrochemical gradient is used to drive the movement of another molecule upgradient. Common example is Na$^+$-coupled glucose transport, which uses the considerable gradient in sodium ion concentrations, low within the cell, to drive glucose uptake. See **symport** and **antiport**. **2.** May be used loosely of a transport system that is controlled by, for example, G-proteins, although the coupling is of a different kind.

coupling The linking of two independent processes by a common intermediate, e.g. the coupling of electron transport to oxidative phosphorylation, or the ATP–ADP conversion to transport processes.

coupling factors Proteins responsible for coupling transmembrane potentials to ATP synthesis in **chloroplasts** and **mitochondria**. Include ATP-synthesizing enzymes (F1 in mitochondrion) that can also act as ATPases.

COUP-TFs *Chicken ovalbumin upstream promoter-transcription factors* Nuclear receptor localized in **calbindin**-positive cells and not in **reelin**-positive cells. COUP-TFI plays a critical role in glial cell development and central nervous system myelination; expressed in cells of oligodendrocyte lineage. See **ARP-1**.

Cowden disease Germline mutations in **PTEN** are responsible for Cowden disease, a rare autosomal dominant multiple-**hamartoma** syndrome.

COX-1, COX-2, COX-3 Isoforms of cyclo-oxygenase, a key enzyme in prostaglandin synthesis. Aspirin and many **NSAIDs** inhibit both isoforms, but selective COX-2 inhibitors such as celecoxib have the analgesic and anti-inflammatory activity without deleterious effects on the gastric mucosa. The selective COX-2 inhibitors are not, however, entirely without side effects, and some have been withdrawn. In humans, COX-3 mRNA is expressed as an ~5.2-kb transcript and is most abundant in cerebral cortex and heart; acetaminophen (paracetamol) may work by inhibiting this cyclo-oxygenase variant.

coxib General suffix for any of a group of non-steroidal anti-inflammatory drugs, cyclo-oxygenase inhibitors, used in the treatment of arthritis.

Coxsackie viruses Species of enteroviruses of the **Picornaviridae** first isolated in Coxsackie, NY. Coxsackie A produces diffuse myositis, Coxsackie B produces focal areas of degeneration in brain and skeletal muscle. Similar to polioviruses in chemical and physical properties.

CPA Ambiguous abbrev. for (1) **Cyclophosphamide**, (2) **cyclopiazonic acid**, (3) cardiopulmonary arrest, (4) **cyproterone acetate**, (5) **cerebello-pontine angle**, (6) conditioned place aversion – a technique in animal behaviour studies, (7) **cyclopentyladenosine** and (8) carboxypeptidase A.

CPE Abbrev. most commonly for cytopathic effect. Less frequently an abbreviation used for (1) **Cytoplasmic polyadenylation element**, (2) *Clostridium perfringens* enterotoxin, (3) continuing professional education.

CpG island *CG island* Region of genomic DNA rich in the dinucleotide C–G. Methylation of the C in the dinucleotide is maintained through cell divisions, profoundly affects the degree of transcription of the nearby genes and is important in developmental regulation of gene expression. There are around 30 000 CpG islands in a typical mammalian genome, and these tend to be undermethylated and upstream of **housekeeping genes**.

CPK Abbrev. for creatine phosphokinase. See **creatine kinase**.

C polysaccharide *C substance* Polysaccharide released by pneumococci which contains galactosamine-6-phosphate and phosphoryl choline. **C-reactive protein** is so called because it will precipitate this polysaccharide through an interaction with the phosphoryl choline.

CPP32 *Caspase-3* See Table C1 (Caspases).

C-proteins Striated muscle thick filament-associated proteins (140–150 kDa) that show up in the C-zone of the A-band as 43-nm transverse stripes. Structurally related to various other myosin-binding proteins (**twitchin**, **titin**, **myosin light-chain kinase**, **skelemin**, 86-kDa protein, **projectin**, **M-protein**).

CPS *Carbamoyl phosphate synthetase* Enzyme responsible for production of carbamoyl phosphate, key substrate for pyrimidine biosynthesis (see **ATCase**). In *S. cerevisiae*, the multifunctional protein Ura2 carries out both CPSase and ATCase activities.

CPSF *Cleavage and polyadenylation specificity factor* Multiprotein complex comprising CPSF-1 (160 kDa), CPSF-2 (110 kDa), CPSF-3 (73 kDa) and CPSF-4 (30 kDa) involved in mRNA polyadenylation. Complex binds the AAUAAA conserved sequence in pre-mRNA. CPSF has also been found to be necessary for splicing of single-intron pre-mRNAs. CPSF-73 (CPSF3) and CPSF-100 belong to a superfamily of zinc-dependent beta-lactamase fold proteins with catalytic specificity for a wide range of substrates including nucleic acids.

CPT Abbrev. for (1) **carnitine palmitoyltransferase**, (2) **Camptothecin**; CPT-11 (**Irinotecan**) is a derivative of camptothecin, (3) in behavioural studies, continuous performance task or test.

CR1 Complement receptor 1 (CD35). Binds particles coated with C3b. Present on neutrophils, mononuclear phagocytes, B-lymphocytes and Langerhans' cells, and involved in the opsonic phagocytosis of bacteria and uptake of immune complexes. Also present on follicular dendritic cells and glomerular podocytes.

CR16 See **verprolins**.

CR2 *CD-21* Receptor for complement fragment C3d. Present only on B-lymphocytes, follicular dendritic cells and some B- and T-cell lines, and is the site to which the **Epstein–Barr virus** binds.

CR3 *CD11b/CD18; MAC-1* Receptor for complement fragment C3bi (iC3b), present on neutrophils and mononuclear phagocytes. A β2 **integrin**.

CR4 *CD11c/CD18* Receptor for C3dg, the **complement** fragment that remains when C3b is cleaved to C3bi. Thought to be present on monocytes, macrophages and neutrophils, but there is some disagreement at present. A β2 **integrin**.

CRABP See **cellular retinoic acid-binding protein**.

cranial nerves Any of the 10–12 paired nerves that have their origin in the brain of vertebrates (olfactory, optic, oculomotor, trochlear, trigeminal, abducens, facial, auditory/vestibulocochlear, glossopharyngeal, vagus, accessory and hypoglossal). Students often use mnemonics to remember the sequence: a polite one is 'Old Officers Often Trust The Army For A Glory Vague And Hypothetical'.

crassulacean acid metabolism *CAM* Physiological adaptation of certain succulent plants, in which CO_2 can be fixed (non-photosynthetically) at night into malic and other acids. During the day the CO_2 is regenerated and then fixed photosynthetically into the **Calvin–Benson cycle**. This adaptation permits the stomata to remain closed during the day, conserving water.

CRD domain *Carbohydrate-recognition domain* Domain found in C-type (calcium-dependent) lectins; there are 32 highly conserved residues in all C-type carbohydrate-recognition domains. Various CRDs have been structurally analysed.

cre Gene of *E. coli* **bacteriophage** P1 that mediates **site-specific recombination** at loxP sites. Now used in vertebrate transgenics: see **lox-Cre system**.

creatine kinase *Creatine phosphokinase; CPK; EC 2.7.3.2* Dimeric enzyme (82 kDa) that catalyses the formation of ATP from ADP and creatine phosphate in muscle.

creatine phosphate *Phosphocreatine* Storage compound of vertebrate muscle. See **creatine kinase**.

C-reactive protein A protein of the **pentraxin** family found in serum in various disease conditions, particularly during the acute phase of immune response. C-reactive protein is synthesized by **hepatocytes**, and its production may be triggered by **prostaglandin** E1 or parogen. It consists of five polypeptide subunits forming a molecule of total molecular weight 105 kDa. It binds to polysaccharides present in a wide range of bacterial, fungal and other cell walls or cell surfaces, and to **lecithin** and to phosphoryl- or choline-containing molecules. It is related in structure to serum **amyloid**. See also **acute phase proteins** and **C polysaccharide**.

CREB *Cyclic AMP response element-binding factor* **Basic leucine zipper** (bZip) transcription factor involved in activating genes through cAMP; binds to CRE element TGANNTCA. Phosphorylation by cAMP-dependent protein kinase (PKA) at serine-119 is required for interaction with DNA, and phosphorylation at serine-133 allows CREB to interact with CBP (**CREB-binding protein**), leading to interaction with RNA polymerase II.

CREB-binding protein *CBP* Transcriptional co-activator (265 kDa) of **CREB** and of c-Myb. Only binds the phosphorylated form of CREB.

CREM *cAMP response element modulator* A member of the large ATF/CREM/CREB transcriptional activator family (Mr 43 kDa). Has a highly conserved leucine zipper dimerization domain and a basic DNA-binding domain at its carboxyl terminus. There are multiple splice variants.

C-region The parts of the heavy or light chains of **immunoglobulin** molecules that are of constant sequence, in contrast to variable or V regions. The constancy of sequence is relative because there are several constant region genes and alleles thereof (see **allotypes**), but within one animal homozygous at the light and heavy chain constant region genes, all immunoglobulin molecules of any one class have constant sequences in their C-regions. The constant region sequences for the various different types of immunoglobulin, e.g. IgG, IgA, etc. will vary.

crenation Distortion of the erythrocyte membrane giving a spiky, echinocyte morphology. Results from ATP depletion or an excess of lipid species in the external lipid layer of the membrane.

CREST *Calcinosis; Reynauds phenomenon; oesophageal dysmotility; sclerodactyly; telangielactasia* A complex syndrome characterized by the presence of autoantibodies toward proteins of the **centromere**, largely the CENT A, B, C and D antigens.

Creutzfeldt–Jacob disease Rare fatal presenile dementia of humans, similar to **kuru** and other transmissible spongiform encephalopathies. Method of transmission unknown. Will induce a neurological disorder in goats 3–4 years after inoculation with CJD brain extract. A new variant, vCJD, has recently been recognized and associated with **bovine spongiform encephalopathy**. See **prions**.

CRF Abbrev. for (1) **corticotropin-releasing factor**, (2) chronic renal failure.

CRIB 1. Cdc42/Rac interactive binding: the CRIB consensus sequence is found in various signalling proteins (e.g. PAK kinases) and is involved in the binding of Cdc42. The CRIB motif itself is insufficient for high-affinity binding to Cdc42, but requires the sequence segment C-terminal to the CRIB motif for enhanced affinity. 2. Abbrev. for Clinical Risk Index for Babies, a widely used, risk-adjustment instrument to determine illness severity in infants.

CRIB cells A clone of MDBK cells that are resistant to bovine viral diarrhoea virus.

cribriform Perforated, sieve-like.

Cricetulus griseus Chinese hamster. See **CHO cells** and ***Mesocricetus auratus***.

cri-du-chat syndrome Syndrome produced by loss of part of the short arm of chromosome 5 in man. Results in severe congenital malformation, and affected infants produce a curious mewling sound said to resemble the cry of a cat. Deletion of multiple genes, including the telomerase reverse transcriptase gene (see **hTERT**), is responsible for the phenotype.

crinophagy Digestion of the contents of secretory granules following their fusion with lysosomes.

critical concentration Concentration of a substance at which adverse functional changes, reversible or irreversible, occur in a cell or an organ.

critical dose Dose of a substance at which adverse functional changes, reversible or irreversible, occur in a cell or an organ.

critical point drying A method for preparing specimens for the scanning electron microscope that avoids the problems of shrinkage caused by normal drying procedures. Water in the specimen is replaced by an intermediate fluid, for example liquid carbon dioxide, avoiding setting up a liquid/gas interface, and then the second fluid is allowed to vaporize by raising the temperature above the critical point (the temperature at which the liquid state no longer occurs).

crk An oncogene, identified in a chicken **sarcoma**. Encodes two alternatively spliced adaptor signalling proteins, CRKI (28 kDa) and CRKII (40 kDa), that have SH2 and SH3 domains (see **Grb-2**) and may recruit cytoplasmic proteins to associate with receptor tyrosine kinases. Both CRKI and CRKII have been shown to activate kinase signalling and anchorage-independent growth *in vitro*. See also Table O1.

CRM Abbrev. for (1) **Cis-regulatory modules**, (2) Certified reference materials, (3) Colorectal metastasis, (4) Continual reassessment method, used in risk analysis. See **CRM-1**.

CRM-1 *Chromosomal region maintenance 1; exportin 1* The main mediator, in many cell types, of export from the nucleus of cellular proteins that have a leucine-rich **nuclear export signal** (NES) and of RNAs. Inhibited by **leptomycin B**. In the nucleus, in association with RANBP3, binds co-operatively to the NES on its target protein and to Ran-GTPase in its active GTP-bound form. Docking of this complex to the nuclear pore complex is mediated through binding to **nucleoporins**.

CRMP Abbrev. for (1) Collapsin response-mediated protein, (2) a haeme-oxygenase (HO) enzyme inhibitor, chromium-mesoporphyrin (CrMP).

CRMP-2 *Collapsin Response-Mediator Protein-2* See **collapsin response-mediator proteins**.

crocin A yellow carotenoid pigment from *Gardenia jasminoides* and *Crocus sativus*; heat-stable and water-soluble but sensitive to oxidation.

Crohn's disease *IBD1* Inflammatory bowel disease that seems to have both genetic and environmental causes; not well understood but generally considered likely to be autoimmune. Mutations in the *CARD15* gene (caspase recruitment domain-containing protein 15) are associated with susceptibility to Crohn's disease in some families.

cro-protein Protein synthesized by **lambda bacteriophage** in the lytic state. The cro-protein blocks the synthesis of the lambda repressor (that is produced in the lysogenic stage and inhibits cro-protein synthesis). Production of the cro-protein in turn controls a set of genes associated with rapid virus multiplication.

cross hybridization Hybridization of a nucleic acid probe to a sequence that is similar, but not identical, to the target.

crossing-over Recombination as a result of DNA exchange between homologous chromatids in meiosis, giving rise to **chiasmata**.

crossover Protein **motif** that describes the connection between strands in a parallel **beta sheet**. In principle, can be extended to the region between adjacent parallel **alpha-helices**.

croton oil Oil from the seeds of the tropical plant *Croton tiglium* (Euphorbiaceae) causes severe skin irritation and contains a potent **tumour promoter, phorbol ester**.

crotoxin Neurotoxin of Brazilian rattlesnake (*Crotalus durissus terrificus*) venom. A molecular complex of an acidic non-enzymic protein (8400 Da), probably responsible for membrane-binding activity, and a basic protein (13 kDa) that has PLA2 activity. The two components in combination are necessary to produce high neurotoxicity; neither does so alone. Bind to specific proteins on presynaptic membranes and alter transmitter release. There are probably species-specific variants.

crown gall Gall, or tumour, found in many dicotyledonous plants, caused by the bacterium *Agrobacterium tumefaciens*.

CRP See **C-reactive protein**.

CRP-ductin *MUCLIN* Protein (224 kDa) expressed mainly by mucosal epithelial cells in the mouse, a membrane protein with a short cytoplasmic region, a transmembrane domain and a large extracellular region. May be a **trefoil factor** (TFF) receptor or a TFF-binding protein. CRP-ductin is the mouse homologue of human gp-340, a glycoprotein that agglutinates microorganisms and binds the lung mucosal collectin **surfactant protein**-D (SP-D).

CRP55 See **calreticulin**.

crumbs *Drosophila* gene involved in epithelial development. Gene product contains 26 repeats of the **EGF-like domain**.

CRY-1 *Cryptochrome-1* Blue-light receptor from *Arabidopsis*. A flavoprotein coded by *Hy4*, a gene that is part of a small family that also encodes CRY-2, another blue-light photoreceptor. Mutants lacking both CRY-1 and CRY-2 are deficient in phototropism.

cryofixation Fixation processes for microscopy, particularly scanning electron microscopy, carried out at low temperature to improve the quality of fixation. Often very low temperatures (liquid nitrogen or helium) and fast cooling (>10 000 degree/min) are used to prevent formation of ice crystals.

cryoglobulin Abnormal plasma globulin (IgG or IgM) that precipitates when serum is cooled.

cryomicroscopy Microscopy, either light or EM, of samples that have been prepared by rapid freezing and are maintained in a frozen glass-like state. Frozen sections can be prepared and examined rapidly, and are therefore important in pathological examination of biopsy material.

cryoprecipitate The precipitate that forms when plasma is frozen and then thawed; particularly rich in **fibronectin** and blood-clotting factor VIII.

cryoprotectant Substance that is used to protect from the effects of freezing, largely by preventing large ice-crystals from forming. The two commonly used for freezing cells are **DMSO** and glycerol.

crypt Deep pit that protrudes down into the connective tissue surrounding the small intestine. The epithelium at the base of the crypt is the site of stem cell proliferation, and the differentiated cells move upwards and are shed 3–5 days later at the tips of the villi.

cryptic plasmid Plasmid that does not confer a phenotype and is only detected by direct observation of the DNA. Many do, however, have open reading frames.

cryptobiont An organism that lives hidden away or with all signs of life disguised as in dormancy.

cryptochrome A light-sensitive pigment, a photoreceptor protein, thought to control the circadian rhythm. See **CRY-1**.

Cryptomonas Genus of flagellate protozoa with two slightly unequal flagella, and a large chromatophore in some species.

cryptophycins Naturally occurring peptides that are potent antimitotic agents, causing cell death at picomolar or low nanomolar concentrations. Bind to beta-tubulin (sharing a binding site with **dolastatin** 10, **hemiasterlin** and **phomopsin A**. Have potential as antitumour (cytotoxic) drugs.

Cryptosporidia Obligate parasitic apicomplexan protozoa, responsible for many cases of intestinal upset caused by drinking infected water. Sporulated oocysts, once in the intestinal tract (usually the ileum), release sporozoites that penetrate intestinal epithelial cells and divide to form merozoites, which can propagate and infect other cells. Eventually meronts differentiate into either macro- or microgametocytes; microgametocytes penetrate the macrogametocytes to form a zygote that undergoes meiotic division, producing sporozoites within a resistant oocyst that is then released in faeces, completing the cycle.

cryptozoology The study and attempted discovery of previously unknown animals whose existence is uncertain – for example, the Yeti.

crystal violet *gentian violet; methyl violet 10B* Deepest blue of the **methyl violet dyes**, used in **Gram stain**, for the metachromatic staining of amyloid, and as an enhancer for bloody fingerprints. Also a pH indicator, turning yellow below pH 1.8. Was used topically for skin infections, but as a suspected carcinogen this is no longer advisable.

crystallins Major proteins of the vertebrate lens. Range from high MW oligomeric species to low MW monomeric species. Immunological cross-reactivity suggests that the sequences of crystallin subunits are relatively highly conserved in evolution.

CSAT Monoclonal antibody defining integral membrane protein of chick fibroblasts. Originally thought to recognize a trimeric complex, now thought to recognize two different $\beta1$ integrins (with different α-chains).

CSF-1 *Colony-stimulating factor-1; MCSF* **Growth factor** for **haematopoietic stem cells**; stimulates **macrophage** production. Gene can be activated as an **oncogene** by **insertional mutagenesis** with a **retrovirus**. See **MCSF**.

CSIF *Cytokine synthesis inhibiting factor* Old name for IL-10.

Csk Protein tyrosine kinase that phosphorylates a tyrosine residue in **src-family** kinases, thereby allowing an inhibitory interaction with src kinase SH2 domain. (It is loss of the tyrosine residue phosphorylated by csk that makes v-src unregulated.)

CSP Ambiguous abbreviation (often with a number suffix) for (1) **Caveolin scaffolding protein**, (2) **Competence-stimulating peptide**, (3) **Cysteine string protein**, (4) **Caspase**, (5) **Chiral stationary phase**, (6) Circumsporozoite protein of *Plasmodium vivax*, (7) Common salivary protein, a secreted protein of the parotid gland, (8) Carotid sinus pressure, (9) **Ciclosporin**.

c-strand Abbrev. for the term 'complementary strand', used of nucleic acids.

C-subfibre The third partial microtubule associated with the A- and B-tubules of the outer axonemal doublets in the **basal body** (and in the **centriole**) to form a triplet structure.

CTAB *Cetyltrimethylammonium bromide; hexadecyltrimethylammonium bromide* Cationic detergent used for membrane solubilization.

CTAP III See **connective tissue-activating peptide III**.

CTCF *CCCTC-binding factor* A transcriptional regulator protein with 11 highly conserved zinc-finger domains; depending upon where it binds, it can act as either transcriptional activator or repressor through histone acetylation or de-acetylation. Mutations in this gene have been associated with invasive breast cancers, prostate cancers and **Wilms' tumours**.

CTD *Carboxy-terminal domain* Protein domain unique to Pol-II that contains multiple repeats of the YSPTSPS sequence. The CTD plays an important part in organizing the various protein factors that make up the RNA-processing 'factory' that regulates the processing of the 3′ end of mRNAs made by Pol-II.

Ctenophora Phylum of biradially symmetrical triploblastic coelomates. Lack nematocysts and cilia, though have comb plates (costae) arranged in eight rows. The comb jelly or sea gooseberry is the best-known example.

C-terminus The carboxy-terminal end of a polypeptide or protein, the 1-carboxy function of the c-terminal amino acid that is not linked to another amino acid by a peptide bond.

CTF *CCAAT box-binding transcription factor; TGGCA-binding proteins* Large family of vertebrate nuclear protein **transcription factors**, around 400–600 amino acids, that binds to the palindrome TGGCAnnnTGCCA in a range of cellular promoters. Includes nuclear factor-1(**NF-1**).

CTL See **cytotoxic T-cells**.

CTLA-4 *CD152* Type I transmembrane protein (20 kDa) of the immunoglobulin superfamily. Found on activated T-cells and binds to CD80 on B-cells. Resembles CD28 but acts as a negative regulator of T-cell activation. Cytoplasmic domain interacts with SH2 domain of **Shp2** (protein tyrosine phosphatase) and possibly with PI3-kinase. CTLA-4-deficient mice develop a severe lymphoproliferative disorder.

CTX Ambiguous abbreviation for (1) **Cerebrotendinous xanthomatosis**, (2) **Ciguatoxin**, (3) Cross-linked C-telopeptide of collagen Type I, a marker for bone degradation, (4) **Cyclophosphamide**, (5) Cefotaxime, a beta-lactam antibiotic, (6) **Crotoxin** (Ctx), (7) **Cardiotoxin**, **conotoxin** or **cholera toxin**.

CTX-M Bacterial genes responsible for producing extended-spectrum beta-lactamases that confer antibiotic resistance.

C-type lectins One of two classes of **lectin** produced by animal cells, the other being the **S-type**. The C-type lectins require disulphide-linked cysteines and Ca^{2+} ions in order to bind to a specific carbohydrate (cf. S-type lectins). The carbohydrate-recognition domain of C-type lectins consists of about 130 amino acids which contain 18 invariant residues in a highly conserved pattern. These invariant residues include cysteines which probably form disulphide bonds. So far, all identified C-type lectins are extracellular proteins and include both integral membrane proteins, such as the **asialoglycoprotein** receptor and soluble proteins.

C-type virus Originally C-type particles identified in mouse tumour tissue, and later shown to be oncogenic RNA viruses (**Oncovirinae**) that bud from the plasma membrane of the host cell, starting as a characteristic electron-dense crescent. Include feline leukaemia virus, murine leukaemia and sarcoma viruses.

CUB domain *Complement subcomponent C1r/C1s/ embryonic sea urchin protein Uegf/bone morphogenetic protein 1 domain* A widespread 110-amino acid module found in functionally diverse, often developmentally regulated proteins, for which an antiparallel beta-barrel topology similar to that in immunoglobulin V domains has been predicted. CUB domains have been found in the dorsoventral patterning protein **tolloid**, bone morphogenetic protein 1, a family of **spermadhesins**, **complement** subcomponents C1s/C1r and the neuronal recognition molecule A5. Acidic seminal fluid protein (aSFP) is built by a single CUB domain architecture. Not found in prokaryotes, plants and yeast.

cuboidal epithelium Epithelium in which the cells are approximately square in vertical section, with the area in contact with the basal lamina comparable to the area of lateral cell–cell contact – in contrast to **columnar** and **squamous** epithelia.

cucurbitacins Feeding stimulants, glucosides, for diabroticite beetles, including corn rootworms and cucumber beetles, but have been used as antitumour agents because of their interference with **STAT**3 signalling. In fibroblasts, interfere with **lysophosphatidic acid** signalling.

cuffing The accumulation of white cells in tissue immediately adjacent to a blood vessel in certain infections of the nervous system.

Culex pipiens Most widely distributed species of mosquito. Salivary glands have giant **polytene chromosomes**.

cullin *CDC53* Gene family that is involved in cell cycle control and which, when mutated, may contribute to tumour progression. Cullin (cdc53) is part of the SCF ubiquitin–protein ligase complex.

cultivar A subspecific rank used in classifying cultivated plants; particular cultivars have distinct properties that can be maintained by vegetative propagation or judicious crossing. Shown as 'cv Name' following the Genus and species names.

culture To grow *in vitro*.

cumulative incidence rate The cumulative incidence as a proportion of the total population.

cumulative median lethal dose Estimate of the total administered amount of a substance that is associated with the death of 50% of animals when given repeatedly at doses which are fractions of the median lethal dose.

cumulus Cells surrounding the developing ovum in mammals, fancifully thought to resemble the eponymous clouds.

cupins Superfamily of plant proteins that includes **germin**-like proteins, **vicilins** and **legumins**. Many are seed allergens.

curare Curare alkaloids are the active ingredients of arrow poisons used by South American Indians; they have muscle-relaxant properties because they block motor **endplate** transmission, acting as competitive antagonists for acetylcholine.

curarine Toxic alkaloid extracted as d-tubocurarine chloride from crude **curare**; used in anaesthesia as a muscle relaxant. Binds to the acetylcholine receptor channel and prevents it from opening to cause depolarization in the muscle.

curcumin Principal pigment of turmeric, a polyphenol, with a natural yellow colour. Appears to have some anti-inflammatory and antioxidant actions, and has been extensively used in traditional medicine.

CURL The compartment for uncoupling of receptors and ligands: internalized receptor–ligand complexes are stripped of the ligand and recycled.

Cushing's syndrome A type of hypertensive disease in man due probably to the over-secretion of **cortisol** due in turn to excessive secretion of **adrenocorticotrophic hormone** (ACTH). Adrenal tumours are the usual primary cause.

cutin Waxy hydrophobic substance deposited on the surface of plants. Composed of complex long-chain fatty esters and other fatty acid derivatives. Impregnates the outer wall of epidermal cells and also forms a separate layer, the cuticle, on the outer surface of the **epidermis**.

CV-1 cells *ATCC No CCL-70* Pseudodiploid, male African green monkey (*Cercopithecus aethiops*) cells derived from kidney. Morphology is fibroblastic. Much used for transfection and for studies on viral infection.

C value paradox Comparison of the amount of DNA present in the haploid genome of different organisms (the C value) reveals two problems: the value can differ widely between two closely related species, and there seems to be far more DNA in higher organisms than could possibly be required to code for the modest increase in complexity.

CVS Abbrev. for **chorionic villus sampling**.

cy3, cy5 Trade names for fluorescent cyanine dyes used for coupling to probes of various sorts.

cyanidin The cyanidin gene codes for an enzyme that modifies **dihydrokaempferol**, directing it into the cyanidin pigment pathway, which produces deep-red, pink and lilac-mauve hues.

Cyanobacteria *Cyanophyta* Modern term for the blue-green algae, prokaryotic cells that use chlorophyll on intracytoplasmic membranes for photosynthesis. The blue-green colour is due to the presence of **phycobilins**. Found as single cells, colonies or simple filaments. In Anabaena, in which the cells are arranged as a filament, heterocysts capable of nitrogen fixation occur at regular intervals. According to the **endosymbiont hypothesis** Cyanobacteria are the progenitors of **chloroplasts**.

cyanocobalamin Usual form of **Vitamin B12**. See Table V1

cyanogen bromide *CNBr* Agent that cleaves peptide bonds at methionine residues. The peptide fragments so generated can then, for example, be tested to locate particular activities.

Cyanophyta Blue-green alga See **Cyanobacteria**.

cyanosis Blueish appearance of skin due to insufficient oxygenation of blood in capillaries. May be natural (response to cold) or pathological (cyanide poisoning, among other things).

Cycads An ancient group of seed plants of the order Cycadales. Palm-like, but unrelated to palms. Evergreen gymnosperms.

cyclase associated protein *CAP; Srv2p* Protein from *S. cerevisiae* that has an actin-binding region and seems to function purely as a G-actin monomer sequestering protein. CAP homologues are found in many eukaryotes, including mammals.

cyclic ADP-ribose *cADPR; adenosine 5'-cyclic diphosphoribose* Second messenger synthesized by the multifunctional transmembrane ectoenzyme CD38 in various systems, particularly platelets, microsomes and sea urchin eggs. Endogenous regulator of intracellular calcium. May act by regulating **ryanodine** receptor, though other mechanisms are suggested.

cyclic AMP cAMP 3'5'-cyclic ester of AMP. The first second-messenger hormone signalling system to be characterized. Generated from ATP by the action of adenyl cyclase that is coupled to hormone receptors by **G-proteins (GTP-binding proteins)**. cAMP activates a specific (cAMP-dependent) protein kinase (PKA) and is inactivated by phosphodiesterase action, giving 5'AMP. Also functions as an extracellular morphogen for some slime moulds.

cyclic GMP *cGMP* 3'5'-cyclic ester of GMP. A second messenger generated by guanylyl cyclase. See **ANP**, **nitric oxide**.

cyclic inositol phosphates 1,2-cyclic derivatives of inositol phosphatide that are invariably formed during enzymic hydrolysis of phosphatidyl inositol species. Have been proposed as second messengers in hormone-activated pathways.

cyclic nucleotide phosphodiesterases Often casually referred to simply as phosphodiesterases. Multiple isoenzymes are known. PDE-I is calcium/calmodulin regulated and important in CNS and vasorelaxation; PDE-II is cGMP-stimulated and hydrolyses cAMP. PDE-III regulates vascular and airway dilation, platelet aggregation, cytokine production and lipolysis; PDE-IV (inhibited by rolipram) is important in control of airway smooth muscle and inflammatory mediator release, but also has a role in CNS and in regulation of gastric acid secretion. PDE-V and VI are cGMP-specific. PDE-V is involved in platelet aggregation; PDE-VI is regulated by interaction with **transducin** in photoreceptors. PDE-VII is abundant in skeletal muscle and present in heart and kidney.

cyclic phosphorylation Any process in which a phosphatide ester forms a cyclic diester by linkage to a neighbouring hydroxyl group.

cyclic photophosphorylation Process by which light energy absorbed by **photosystem I** in the chloroplast can be used to generate ATP without concomitant reduction of **NADP**$^+$ or other electron acceptors. Energized electrons are passed from PS-I to ferredoxin, and thence along a chain of electron carriers and back to the reaction centre of PS-I, generating ATP en route.

cyclin Proteins (A and B forms known) whose levels in a cell vary markedly during the **cell cycle**, rising steadily until **mitosis**, then falling abruptly to zero. As cyclins reach a threshold level, they are thought to drive cells into **G2** phase and thus towards mitosis. Cyclins combine with p34 kinase (cdc2) to form maturation-promoting factor (MPF). See also **M-phase promoting factor**, **cyclin-dependent kinase (cdk)**, **cdk inhibitors**, **cdk activating kinase**.

cyclin-dependent kinase *cdk* Family of kinases, including cdc28, cdc2 and p34cdc2, that are only active when they form a complex with cyclins. The complex is maturation-promoting factor (MPF), and its activity is necessary for cells to leave G2 and enter mitosis. Catalytic domain resembles that of cAMP-dependent kinase (PKA).

cyclin-dependent kinase activating kinase *CAK* Kinase that activates cdks by phosphorylation. CAK phosphorylates a threonine residue of several cdks, and a tyrosine on cdc2 (phosphatase cdc25 reverses this).

cyclin-dependent kinase inhibitors *CKIs* Two classes of CKIs are known in mammals: the p21CIP1/Waf1 class, which includes p27KIP1 and p57KIP2 and inhibits all G1/S **cyclin-dependent kinases** (cdks); and the p16INK4 class, which binds and inhibits only Cdk4 and Cdk6. The p21CIP1 inhibitor is transcriptionally regulated by p53-tumour suppressor and is important in G1 DNA-damage checkpoint; its expression is associated with terminally differentiating tissues. Deletion of p21CIP1 is non-lethal in mice; deletion of p27KIP1 leads to relatively normal mice but with some proliferation disorders, deletion of p57KIP2 causes fairly major developmental abnormalities similar to **Beckwith–Wiedemann syndrome**. See also **ICK1**, **Waf1**.

cyclizine hydrochloride An antiemetic antihistamine drug.

cyclo-oxygenase *COX* Enzyme complex present in most tissues that produces various prostaglandins and thromboxanes from arachidonic acid; inhibited by aspirin-like drugs, probably accounting for their anti-inflammatory effects. Three isoforms are known: **COX-1**, COX-2 and COX-3.

cyclodextrins Cyclic polymers of six, seven or eight α-1,4-linked D-glucose residues. The toroidal structure allows them to act as hydrophilic carriers of hydrophobic molecules.

cycloheximide Antibiotic (MW 281) isolated from *Streptomyces griseus*. Blocks eukaryotic (but not prokaryotic) protein synthesis by preventing initiation and elongation on 80S ribosomes. Commonly used experimentally.

cyclolysin Protein from *Bordetella pertussis* that is both an **adenylate cyclase** and a **haemolysin**.

cyclopamine Teratogenic steroidal alkaloid produced by the skunk cabbage (*Veratrum californicum*) that causes developmental defects such as cyclopia (one eye in the middle of the face). Cyclopamine blocks activation of the Hedgehog response pathway, specifically the multipass transmembrane proteins **Smoothened** (Smo) and Patched (Ptch). Cyclopamine will block the oncogenic effects of mutations of Ptch in fibroblasts (see **Gorlin syndrome**), and inhibits the growth of cells lacking Ptch function.

cyclopentyladenosine *CPA; N(6)-cyclopentyladenosine* A specific agonist of adenosine A(1) receptor.

cyclophilin Enzyme with **PPIase** activity; binds the immunosuppressive drug **ciclosporin**. See **immunophilin**.

cyclophosphamide An alkylating agent and important immunosuppressant. Acts by alkylating SH and NH_2 groups especially the N7 of guanine.

cyclopiazonic acid A mycotoxin produced by *Penicillium cyclopium* that selectively inhibits the sarcoplasmic-endoplasmic reticulum Ca^{2+}-ATPase (SERCA).

cyclosis Cyclical streaming of the cytoplasm of plant cells, conspicuous in giant internodal cells of algae such as *Chara*, in pollen tubes and in stamen hairs of *Tradescantia*. Term also used to denote cyclical movement of food vacuoles from mouth to **cytoproct** in ciliate protozoa.

cyclosporin A See **ciclosporin**.

cylindrospermopsin Hepatotoxin originally isolated from the cyanobacterium *Cylindrospermopsis raciborskii* and subsequently found in other species of cyanobacteria. Cylindrospermopsin is an alkaloid with a tricyclic guanidine moiety bridged to hydroxymethyl uracil.

CYP1A *Cytochrome P4501A* See **cytochrome P450**.

cypermethrin Synthetic pyrethroid used as an insecticide. Acts on sodium channels of neurons. Highly toxic to aquatic organisms, including fish. Usually a mixture of several isomers.

cypin Neuronal protein that decreases PSD-95 family member localization and regulates dendrite number. Cypin contains zinc-binding, **collapsin response-mediator protein** (CRMP) homology and **PSD-95**, discs large, zona occludens-1 binding domains. Binds tubulin via its CRMP homology domain to promote microtubule assembly; this interaction is blocked by **snapin**.

cypris Larval stage of Cirrepedia (barnacles) following nauplius stage.

cyproterone acetate *CPA; Cyprostat®* A steroidal anti-androgen which competes for testosterone binding. Used in treatment of prostatic carcinoma, as a 'chemical castrator' for male sex offenders, and in lower doses where testosterone reduction is required.

cyst 1. A resting stage of many prokaryotes and eukaryotes in which a cell or several cells are surrounded with a protective wall of extracellular materials. **2.** A pathological fluid-filled sac bounded by a cellular wall, often of epithelial origin, found on occasion in all species of multicellular animal. May result from a wide range of insults or be of embryological origin.

cystatin c One of the **cystatins** that is excreted in urine. The serum concentration of cystatin C correlates with the glomerular filtration rate, and has been suggested as a marker in testing for renal dysfunction.

cystatins A group of natural cysteine-peptidase inhibitors (13 kDa) widely distributed both intra- and extracellularly. See **stefin**.

cysteine *Cys: C* The only amino acid to contain a thiol (SH) group. In intracellular enzymes, the unique reactivity of this group is frequently exploited at the catalytic site. In extracellular proteins, found only as half-cysteine in disulphide bridges or fatty acylated.

cysteine endopeptidases *Thiol peptidases; EC 3.4.22* Clan of endopeptidases containing a cysteine residue at the active site that is involved in catalysis; the group includes papain and several cathepsins. Natural inhibitors are **alpha-2-macroglobulin** and **cystatins**.

cysteine peptidase *Thiol proteinase; thiol protease* Any peptidase of the subclass EC 3.4.22. All have a cysteine residue in the active site that can be irreversibly inhibited by sulphydryl reagents. Includes **cathepsins** and **papain**.

cysteine string protein *CSP* An abundant regulated secretory vesicle protein that is composed of a string of cysteine residues, a linker domain and an N-terminal J domain characteristic of the DnaJ/Hsp40 co-chaperone family. CSP associates with heterotrimeric GTP-binding proteins (G-proteins) and modulates signalling.

cystic fibrosis Generalized abnormality of exocrine gland secretion that affects pancreas (blockage of the ducts leads to cyst formation and to a shortage of digestive enzymes), bowel, biliary tree, sweat glands and lungs. The production of abnormal mucus in the lung predisposes to respiratory infection, a major problem in children with the disorder. A fairly common (1 in 2000 live births in Caucasians) **autosomal recessive** disease. See **cystic fibrosis transmembrane conductance regulator** (CFTR).

cystic fibrosis antigen *CFAG; MIF related protein 8* Now known to be MRP-8. See **calgranulins**.

cystic fibrosis transmembrane conductance regulator *CFTR* Gene believed to be defective in cystic fibrosis; a range of different mutations are known to occur, making preimplantation genetic diagnosis difficult. Gene encodes a chloride channel, homologous to a family of proteins that actively transport small solutes in an ATP-dependent manner (**ABC proteins**).

cystine The amino acid formed by linking two **cysteine** residues with a disulphide linkage between the two sulphydryl (SH) groups. The analogous compound present within proteins is termed two half cysteines.

cystinosin Lysosomal protein responsible for cystine export. Defects lead to nephropathic cystinosis, and mutations in the cystinosin (CTNS) gene are the most common cause of inherited renal **Fanconi syndrome**. Highly conserved in mammals. Yeast Ers1 is a functional orthologue.

cystinosis An autosomal recessive lysosomal storage disorder caused by a defect in the lysosomal cystine carrier **cystinosin**.

cystoblast In *Drosophila*, oogenesis is initiated when a germline stem cell produces a differentiating daughter cell called the cystoblast. The cystoblast undergoes four rounds of synchronous divisions with incomplete cytokinesis to generate a syncytial cyst of 16 interconnected cystocytes, within the cyst one of the cystocytes differentiates into an oocyte. cf **gonioblast**.

cystocyte See **cystoblast**.

cytidine Nucleoside consisting of D-ribose and the pyrimidine base cytosine.

cytidine 5' diphosphate *CDP* CDP (derived from cytidine 5' triphosphate) is important in phosphatide biosynthesis; activated choline is CDP-choline.

cytidylic acid Ribonucleotide of **cytosine**.

cytisine A natural alkaloid found in many of the Leguminosae family and is well known as the main toxic principle of the common garden Laburnum. Has very high affinity for the α4β2-nicotinic ACh receptors.

cytoband Synonym for **cytogenetic band**.

cytocalbins *Cytoskeleton-related calmodulin-binding protein* **Calmodulin**-binding proteins associated with the cytoskeleton. Term appears to be obsolete.

cytochalasins A group of fungal metabolites that inhibit the addition of G-actin to a nucleation site and therefore perturb labile microfilament arrays. Cytochalasin B inhibits at around 1 μg/ml, but at about 5 μg/ml begins to inhibit glucose transport. Cytochalasin D affects only the microfilament system, and is therefore preferable.

cytochemistry Branch of histochemistry associated with the localization of cellular components by specific staining methods, as for example the localization of acid phosphatases by the Gomori method. Immunocytochemistry involves the use of labelled antibodies as part of the staining procedure.

cytochrome oxidase *EC 1.9.3.1* Terminal enzyme of the electron transport chain that accepts electrons from (i.e. oxidizes) cytochrome C and transfers electrons to molecular oxygen.

cytochrome P450 *Cytochrome m; CYP; EC 1.14.14.1* Large group of mixed-function oxidases of the cytochrome *b* type, involved, among other things, in steroid hydroxylation reactions in the adrenal cortex. In liver they are found in the microsomal fraction and can be induced for the detoxification of foreign substances. Found in most animal cells and organelles, in plants and in microorganisms.

cytochromes Enzymes of the electron transport chain that are pigmented by virtue of their **haem** prosthetic groups. Very highly conserved in evolution.

cytogenetic band *Cytoban* A chromosomal subregion (band) visible microscopically after special staining. See **chromosome banding**.

cytogenetics The study of the chromosomal complement of cells, and of chromosomal abnormalities and their inheritance.

cytohesin-1 *B2 – 1* Guanine nucleotide exchange factor for human ADP-ribosylation factor (**ARF**) GTPases. Abundant in cells of the immune system, where it mediates PI3-kinase activation of β2 integrin (particularly LFA-1) through interaction with the cytoplasmic domain. Closely related to **ARNO** and **GRP-1**; all three have a central Sec7-like domain and C-terminal **pleckstrin** homology (PH) domain. The PH domain and C-terminal polybasic sequence are important for membrane-association and function.

cytokeratins Generic name for the intermediate filament proteins of epithelial cells.

cytokines Small proteins (in the range of 5–20 kDa) released by cells and that affect the behaviour of other cells. Not really different from hormones, but the term tends to be used as a convenient generic shorthand for **interleukins**,

lymphokines and several related signalling molecules such as **TNF** and **interferons**. Generally, growth factors would not be classified as cytokines, though TGF is an exception. Rather an imprecise term, though in very common usage. **Chemokines** are a subset of cytokines; see Table C4.

cytokinesis Process in which the cytoplasm of a cell is divided after nuclear division (mitosis) is complete.

cytokinins Class of **plant growth substances** (plant hormones) active in promoting cell division. Also involved in cell growth and differentiation and in other physiological processes. Examples: **kinetin**, **zeatin**, benzyl adenine.

cytology The study of cells. Implies the use of light or electron microscopic methods for the study of morphology.

cytolysis Cell **lysis**.

cytolysosome Membrane-bounded region of cytoplasm that is subsequently digested.

Cytomegalovirus Probably the most widespread of the Herpetoviridae group. Infected cells enlarge and have a characteristic inclusion body (composed of virus particles) in the nucleus. Causes disease only *in utero* (leading to abortion or stillbirth, or to various congenital defects), although can be opportunistic in the immunocompromised host.

cyton **1.** The region of a neuron containing the nucleus and most cellular organelles. **2.** In Cestodes (e.g. tapeworms), the syncytial epithelium that forms the tegument has the nucleated, proximal cytoplasm or cyton, sunk deep in the parenchyma. The cyton region contains Golgi complexes, mitochondria, rough ER and other organelles involved in protein synthesis and packaging. **3.** In cercaria of *Schistosoma mansoni* cercaria, an aggregate of subtegumental cells is found in a small, dorsoanterior area. These highly amorphous cell types, designated as cyton II, have a heterochromatic nucleus and a cytoplasm that is elaborated into coarse, tortuous processes.

cytonectin A 35 kDa adhesion protein, independent of divalent cations, expressed in a variety of organs and tissues, being evolutionarily conserved from human to avian species; overexpressed in Alzheimer's disease entorhinal cortex.

cytonemes Extensions produced by epithelial cells; other names include filopodia and cell feet.

cytophotometry Examination of a cell by measuring the light allowed through it following staining.

cytoplasm Substance contained within the plasma membrane excluding, in eukaryotes, the nucleus.

cytoplasmic bridges *Plasmodesmata* Thin strands of cytoplasm linking cells as in higher plants, *Volvox*, between **nurse cells** and developing eggs, and between developing sperm cells. Unlike gap junctions, allows the transfer of large macromolecules.

cytoplasmic determinants Slightly imprecise term usually applied to non-randomly distributed factors in maternal (oocyte) cytoplasm that determine the fate of blastomeres derived from this region of the egg following cleavage. Thus blastomeres containing cytoplasm derived from the apical region (animal pole) of the echinoderm egg form ectodermal tissues. Can be used even more generally of any feature in the cytoplasm that determines how a process or activity proceeds.

cytoplasmic inheritance Inheritance of parental characters through a non-chromosomal means; thus mitochondrial DNA is cytoplasmically inherited since the information is not segregated at mitosis. In a broader sense, the organization of a cell may be inherited through the continuity of structures from one generation to the next. It has often been speculated that the information for some structures may not be encoded in the genomic DNA, particularly in protozoa that have complex patterns of surface organelles. See **maternal inheritance**.

cytoplasmic polyadenylation element *CPE* Nucleotide sequence (often poly-U rich) that recruits the **cytoplasmic polyadenylation element binding protein** (CPEB) to quiescent mRNA that has been stockpiled in the egg (and had the poly-A tail truncated). CPEB recruits a protein complex, called **CPSF**, which physically interacts with both CPEB and another sequence in the mRNA, to the 3' side of the CPE, called the **nuclear polyadenylation hexanucleotide**. This particular sequence (AAUAAA) is required for cytoplasmic polyadenylation. CPSF, once bound to the mRNA, recruits the enzyme poly A polymerase to the mRNA.

cytoplasmic polyadenylation element binding factor *CPEB* A sequence-specific RNA-binding protein that controls polyadenylation-induced translation in germ cells and at postsynaptic sites of neurons. See **cytoplasmic polyadenylation element** and **maskin**.

cytoplasmic streaming Bulk flow of the cytoplasm of cells. Most conspicuous in large cells such as amoebae and the internodal cells of *Chara*, where the rate of movement may be as high as $100\,\mu m/s$. See **cyclosis**.

cytoplast Fragment of cell with nucleus removed (in **karyoplast**); usually achieved by cytochalasin B treatment followed by mild centrifugation on a step gradient.

cytoproct Cell anus: region at posterior of a ciliate where exhausted food vacuoles are expelled.

cytorrhysis Process in which a plant cell wall collapses inward following water loss due to hyperosmotic stress. *cf.* plasmolysis.

cytosine Pyrimidine base found in DNA and RNA. Pairs with guanine. Glycosylated base is **cytidine**.

cytosine arabinoside *Cytarabine* Cytotoxic drug used in oncology (particularly **AML**) and against viral infections. Blocks DNA synthesis.

cytoskeleton General term for the internal components of animal cells which give them structural strength and motility: plant cells and bacteria use an extracellular **cell wall** instead. The major components of cytoskeleton are the **microfilaments** (of **actin**), **microtubules** (of **tubulin**) and **intermediate filament** systems in cells.

cytosol That part of the cytoplasm that remains when organelles and internal membrane systems are removed.

cytosome 1. The body of a cell apart from its nucleus. **2.** A multilamellar body found in cells of the lung.

cytostome A specialized region of various protozoans in which phagocytosis is likely to occur. Often there is a clear concentration of microtubules or/and microfilaments in the region of the cytostome. In ciliates there may be a specialized arrangement of cilia around the cytostome.

cytotactin See **tenascin**.

cytotoxic necrotizing factors *CNFs* Toxins (110 kDa, monomeric) produced by some strains of *E. coli*. Induce ruffling and stress-fibre formation in fibroblasts, and block cytokinesis by acting on p21 Rho.

cytotoxic T-cells *CTL* Subset of T-lymphocytes (mostly CD8$^+$) responsible for lysing target cells and for killing virus-infected cells (in the context of Class I **histocompatibility antigens**).

cytotrophic Descriptive of any substance that promotes the growth or survival of cells. Not commonly used except in the tissue-specific case of **neurotrophic** factors.

cytotropic Having affinity for cells: not to be confused with **cytotrophic**.

cytotropism Movement of cells towards or away from other cells.

cytovillin Synonym for **ezrin**.

D

D₂O *Deuterium oxide* See **heavy water**.

DAF Abbrev. for (1) **decay accelerating factor**, (2) diaminofluorescein: see **DAF-2**. (3) DAF-16 is a **fork-head** transcription factor.

DAF-2 *4,5-Diaminofluorescein* Highly sensitive fluorescent probe for the real-time detection of nitric oxide (NO) *in vivo*. In the cell, DAF-2 reacts rapidly with NO in the presence of O_2 to form the highly fluorescent compound triazolofluorescein.

daft lamb disease *Border disease; hairy shaker disease; hypomyelinogenesis imperfecta* Uncommon disease of sheep and goats caused by a togavirus (see **Togaviridae**).

DAG See **diacylglycerol** and **phosphatidyl inositol**.

daidzein An isoflavone, the aglycone of daidzin, classified as a phytoestrogen since it is a plant-derived non-steroidal compound that possesses estrogen-like biological activity. Mainly found in legumes, such as soybeans and chickpeas, it has been found to have both weak estrogenic and weak anti-estrogenic effects.

DAL-1 *Differentially expressed in adenocarcinoma of the lung* A **protein 4.1** gene located on chromosome 18p11.3, which is lost in ~60% of non-small cell lung carcinomas, and exhibits growth-suppressing properties in lung cancer cell lines.

dalfopristin A **streptogramin** similar to **quinupristin**.

damaged DNA-binding proteins *DDBs* Proteins that are involved in the initial recognition of UV-damaged DNA and mediate recruitment of nucleotide-**excision repair** factor.

DAN *Differential screening-selected gene aberrative in neuroblastoma* A BMP-antagonist that regulates **BMP** activity spatially and temporally during patterning and partitioning of the medial otic tissue in ear development.

Dane particle A 42-nm particle, the complete infective virion of **hepatitis B**.

Danio rerio Formerly *Brachdanio rerio*, the **zebra fish**.

Danon disease Familial disease characterized by cardiomyopathy, myopathy and variable mental retardation. A **lysosomal storage disease** with characteristic intracytoplasmic vacuoles containing autophagic material and glycogen in skeletal and cardiac muscle. Defect is in **LAMP-2**, a lysosomal membrane structural protein, not an enzyme.

dansyl chloride *1-Dimethyl-amino-naphthalene-5-sulphonyl chloride* A strongly fluorescent compound that will react with the terminal amino group of a protein. After acid hydrolysis of all the other peptide bonds, the terminal amino acid is identifiable as the dansylated residue.

DAP Diabetes-related peptide. See **islet amyloid peptide**.

DAP-12 Disulphide-linked homodimeric protein (12 kDa) that interacts with **KIR**2DS2. Resembles γ-chain of FcεRI and ζ-chain of T-cell receptor. Cytoplasmic tail has an ITAM that will bind **zap-70** and **syk** and thus activate the NK-cell.

Daphnia magnus Cladoceran crustacean, the water flea.

DAPI stain *4,6-Diamidino-2-phenylindole* Fluorochrome that binds to DNA and is used biochemically for detection of DNA and to stain the nucleus in fluorescence microscopy.

DAP kinase *Death-associated protein kinase* A pro-apoptotic family of multidomain calcium/calmodulin (CaM)-dependent Ser/Thr protein kinases (160 kDa) that are phosphorylated upon activation of the Ras-extracellular signal-regulated kinase (ERK) pathway. DAP-related apoptotic kinase-2 (DRAK2) is highly expressed in lymphoid organs.

dapsone Drug related to the sulfonamides (diaminodiphenyl sulphone) that is used to treat **leprosy**. May act by inhibiting folate synthesis.

daptomycin A lipopeptide antibiotic active against Gram-positive bacteria that inhibits lipoteichoic acid synthesis as a consequence of membrane binding in the presence of Ca^{2+}.

dark current (of retina) Current caused by constant influx of sodium ions into the **rod outer segment** of retinal photoreceptors and that is blocked by light (leading to hyperpolarization). The plasma membrane sodium channel is controlled through a cascade of amplification reactions initiated by photon capture by **rhodopsin** in the disc membrane.

dark field microscopy *Dark ground microscopy* A system of microscopy in which particles are illuminated at a very low angle from the side so that the background appears dark and the objects are seen by diffracted and reflected patches of light against a dark background.

dark reaction The reactions in photosynthesis that occur after NADPH and ATP production and that take place in the stroma of the chloroplast. By means of the reaction, CO_2 is incorporated into carbohydrate.

Datura stramonium Jimson weed or thorn apple. Source of **scopolamine**.

Daudi B-lymphoblastoid cell line from Burkitt's lymphoma in 16-year-old male Black African. Have surface complement receptors and IgG, and are **EBV** marker positive.

dauer larva Semidormant stage of larval development in nematodes (for example *Caenorhabditis elegans*), triggered by a pheromone: essentially a survival strategy.

daunomycin *Daunorubicin; Cerubidine* An anthracycline cytotoxic antibiotic produced by a strain of *Streptomyces coeruleorubidus*. Forms complexes with DNA by intercalation between base pairs and inhibits **topoisomerase** II activity by stabilizing the DNA topoisomerase II complex.

daunorubicin Cytotoxic drug used in cancer chemotherapy. An anthracycline antibiotic which damages DNA by intercalating between base pairs resulting in uncoiling of the helix, ultimately inhibiting DNA synthesis and DNA-dependent RNA synthesis. May also act by inhibiting polymerase activity, affecting regulation of gene expression and generating free radicals. Cytotoxic activity is cell cycle phase non-specific, although effects are maximal in S-phase. Produced by *Streptomyces peucetius*.

DAXX A type I IFN-induced protein that suppresses STAT3-mediated transcriptional activation. Acts downstream of apoptosis signal-regulating kinase (**ASK1**).

dbl Human **oncogene** originally identified by transfection of NIH-3T3 cells with DNA from human diffuse B-cell lymphoma. Product is a guanine nucleotide exchange factor (**GEF**) for rho family members. Protein contains a domain of around 250 amino acids, the Dbl-homology, **DH domain**. See also Table O1.

Dbl family *Dbl, Dbs, Brx, Lfc, Lsc, Ect2, DRhoGEF2, Vav* Family of proteins containing DH domains and with guanine nucleotide exchange factor activity for **rho** family of small G-proteins.

DCC protein family *Deleted in colorectal cancer* Receptor family that includes *Drosophila* Frazzled, which binds **netrin** in association with one of the **UNC-5 family** of receptors.

D-cells *δ-Cells; delta cells* Cells of the pancreas; about 5% of the cells present in primate pancreas with small argentaffin-positive granules. Their function is unclear, but they may release **somatostatin**.

DCIP *2,6-Dichlorophenolindophenol* A commonly used electron-acceptor dye, which can accept electrons instead of P700. DCIP is blue in neutral solution and pink in acidic solution; the reduced form is colourless. Used in a test kit for the detection of haemoglobin E.

DCMU *3-(3,4-Dichlorophenyl)-1,1-dimethylurea; diuron* An inhibitor of photosynthetic electron chain transport, used agriculturally as a non-selective herbicide. Persists in the environment and breakdown products are also toxic.

DCPIP *2,6-Dichlorphenoleindophenole* A blue dye that, when reduced by electron addition, becomes colourless. Often used in measurements of the electron transport chain in plants.

DDAVP *Deamino-8-D-arginine vasopressin; desmopressin acetate* A synthetic analogue of **vasopressin** (antidiuretic hormone) used in management of diabetes insipidus and of bedwetting. See **desmopressin**.

DDB complex *Damaged DNA-binding complex* Protein complex consisting of a heterodimer of p127 (DDB1) and p48 (DDB2) subunits, believed to have a role in nucleotide-**excision repair**.

ddNTP General name for dideoxynucleotide triphosphates used in **dideoxy sequencing**.

DD-PCR See **differential display PCR**.

DE3 A T7-expression system developed under contract to the US Department of Energy (hence 'DE') that allows protein expression in competent bacterial cells to be put under the inducible control of IPTG. Competent cells, e.g. BL21(DE3), contain a DE3 lysogen which has the T7 RNA polymerase under the control of the lacUV5 promoter, and are an all-purpose strain for high-level protein expression and easy induction with IPTG.

deacetylase An enzyme that removes an acetyl group: one of the most active deacetylation reactions is the constant deacetylation (and reacetylation) of lysyl residues in histones (the half-life of an acetyl group may be as low as 10 minutes). Acetylation (which removes a positive charge on the lysine ε-amino group) is thought to be increased in active genes; therefore deacetylation would be important in switching off genes.

DEAD-box helicases Family of ATP-dependent DNA or RNA **helicases** with a four amino acid consensus, -D-E-A-D-, that resembles an ATP binding site. Examples: **p68**, a human nuclear protein involved in cell growth; **vasa**, a *Drosophila* protein required for specification of posterior embryonic structures.

DEAD-box proteins Include the **DEAD-box helicases**; may protect mRNA from degradation by endonucleases.

DEAE- *Diethyl-aminoethyl-* Group that is linked to cellulose or Sephadex to give a positive charge and thus to produce an anion exchange matrix for chromatography.

DEAH-box proteins One of the two major groups of RNA helicases, having a highly conserved motif (Asp-Glu-Ala-His). In *S. cerevisiae* there are seven members: Prp2p, Prp16p, Prp22p and Prp43p involved in mRNA splicing; Dhr1p and Dhr2p involved in ribosome biogenesis; a seventh of uncertain function). See **DEAD-box helicases**.

deamination (of nucleic acids) The spontaneous loss of the amino groups of cytosine (yielding uracil), methyl cytosine (yielding thymine) or of adenine (yielding hypoxanthine). It can be argued that the presence of thymine in DNA in place of the uracil of RNA stabilizes genetic information against this lesion, since repair enzymes would restore the GU base pair to GC.

death domain Conserved domain (around 80 amino acids) found in cytoplasmic portion of some **death receptors** (including the TNF receptor), essential for generating signals that often lead to apoptosis.

death-effector domain *DED* Domain at the C-terminus of FADD and N-terminus of FLICE. Interaction mediated by these domains leads to the assembly of the death-inducing signalling complex (DISC), which activates other **caspases**.

death receptors Superfamily of **tumour necrosis factor** receptors, including TNF-R1, CD95, TRAMP, that trigger apoptotic cell death through interaction of various adapter proteins (FADD, TRADD, etc.) with their cytoplasmic **death domains**. These adaptors then interact with **caspases** such as FLICE.

debridement The removal of necrotic, infected or foreign material from a wound.

Dec2 A regulatory protein, member of the basic helix–loop–helix transcription factor superfamily, for the timing system underlying **circadian rhythms**. Transcripts of *Dec2*, as well as those of its related gene *Dec1*, show striking circadian oscillation in the suprachiasmatic nucleus, and *Dec2* inhibits transcription from the Per1 (see **period**) promoter induced by Clock/Bmal1. Transcription of the *Dec2* gene is regulated by several clock molecules and a negative-feedback loop.

decapentaplegic *dpp* *Drosophila* gene product related to **TGFα**.

decapping Removal of the mRNA 5′ cap structure, a key step in mRNA turnover, catalysed by the Dcp1–Dcp2 complex. The decapping activator complex (Lsm1p–7p/Pat1p/Dhh1p) functions in deadenylation-dependent decapping of cellular mRNAs. See **GW bodies**.

decapping complex Set of proteins (DCP1, DCP2) located in **GW bodies**, involved in degradation of bulk mRNA in the 5′ to 3′ direction. After the decapping process has been carried out, further degradation is by the 5′ to 3′ exonuclease XRN1.

decay accelerating factor Plasma protein that regulates **complement** cascade by blocking the formation of the C3bBb complex (the C3 convertase of the alternate pathway). Widely distributed in tissues, but deficient in **paroxysmal nocturnal haemoglobinuria**.

decidua See **endometrium**.

deconvolution Process in digital image handling whereby a composite image is formed using information from several separate images taken at different levels (focal planes). The final image can be rotated and viewed from different angles, and has usually had noise filtered out so that the image is much clearer and sharper.

decorin A small **proteoglycan**, 90–140 kDa, of the **extracellular matrix**, so-called because it 'decorates' collagen fibres. The core protein has a mass of ~42 kDa and is very similar to the core protein of **biglycan** and **fibromodulin**. All three have highly conserved sequences containing 10 internal homologous repeats of ~25 amino acids with leucine-rich motifs. Decorin has one **glycosaminoglycan** chain, either chondroitin sulphate or dermatan sulphate and N-linked oligosaccharides.

decoy receptor In general, a non-functional receptor that competes for binding of a ligand and therefore reduces the effectiveness of the signal. Decoy receptors DcR1 and DcR2 compete with DR4 or DR5 receptors for binding to the ligand (**TRAIL**); decoy receptor 3 (DcR3) is a soluble receptor for **Fas ligand**.

decoy receptor 3 *DcR3* A soluble receptor belonging to the TNF-R superfamily; a receptor for both **Fas ligand** (FasL) and **LIGHT**.

DEDD superfamily Superfamily of RNases and DNases that contain four conserved acidic residues, DEDD, which are responsible for binding two metal ions involved in catalysis.

dedifferentiation Loss of differentiated characteristics. In plants, most cells, including the highly differentiated haploid **microspores** (immature pollen cells) of angiosperms, can lose their differentiated features and give rise to a whole plant; in animals this is less certain, and there is still controversy as to whether the undifferentiated cells of the blastema that forms at the end of an amputated amphibian limb (for example) are derived by dedifferentiation, or by proliferation of uncommitted cells. Neither is it clear whether dedifferentiation in animal cells might just be the temporary loss of phenotypic characters, with retention of the **determination** to a particular cell type.

deep cells Cells (blastomeres) in the teleost blastula that lie between the outer cell layer and the yolk syncytial layer, and are the cells from which the embryo proper is constructed during gastrulation; much studied in the fish, *Fundulus*.

defective virus A virus genetically deficient in replication, but that may nevertheless be replicated when it co-infects a host cell in the presence of a wild-type 'helper' virus. Most acute transforming **retroviruses** are defective, since their acquisition of oncogenes seems to be accompanied by deletion of essential viral genetic information.

defensins Family of small (30–35 residue) cysteine-rich cationic proteins found in vertebrate phagocytes (notably the azurophil granules of neutrophils) and active against bacteria, fungi and enveloped viruses. May constitute up to 5% of the total protein. Insect defensins have some sequence homology with the vertebrate forms.

defined medium Cell culture medium in which all components are known. In practice this means that the serum (which is normally added to culture medium for animal cells) is replaced by insulin, transferrin and possibly specific growth factors such as **platelet-derived growth factor**.

definitive erythroblast Embryonic erythroblast found in the liver; smaller than primitive erythroblasts, they lose their nucleus at the end of the maturation cycle and produce erythrocytes with adult haemoglobin.

degeneracy The coding of a single amino acid by more than one base triplet (**codon**). Of the 64 possible codons, 3 are used for stop signals, leaving 61 for only 20 amino acids. Since all codons can be assigned to amino acids, it is clear that many amino acids must be coded by several different codons, in some cases as many as 6.

degenerate PCR Polymerase chain reaction in which the primers are deliberately mixed (**degenerate primers**) to amplify a sequence that is imperfectly known (because translated from the peptide sequence and thus having uncertainty about codon usage). A powerful tool to find 'new' genes or gene families.

degenerate primer A single-stranded synthetic **oligonucleotide** designed to hybridize to DNA encoding a particular protein sequence. As the mapping of codons to amino acids is many-to-one, the oligonucleotide must be made as a mixture with several different bases at variable positions. The total number of different oligos in the resulting mixture is known as the degeneracy of the primer. Such primers are widely used in screening a **genomic library** or in degenerate **PCR**, to identify homologues of already known genes.

degenerins Products of *deg-1*, *mec-4* and *mec-10* genes in *Caenorhabditis elegans* which turn out to have homology with amiloride-sensitive sodium channels. Mutations cause neuronal degeneration, probably by disrupting ion fluxes. A related protein, the product of *unc-105*, interacts with collagen and may be a stretch-activated channel.

degradosome *RNA degradosome* Multienzyme complex in *E. coli* that contains exoribonuclease, polynucleotide phosphorylase (PNPase), endoribonuclease E (RNAase E), enolase and Rh1B (a member of the DEAD-box family of ATP-dependent RNA helicases).

degranulation Release of secretory granule contents by fusion with the plasma membrane.

dehydration Removal of water as in preparing a specimen for embedding or a histological section for clearing and mounting.

dehydrin Class of plant proteins expressed in response to water shortage, and notable for a run of seven contiguous serines.

dehydrogenase Broad class of enzymes that oxidize a substrate by transferring hydrogen to an acceptor that is usually either $NAD^+/NADP^+$ or a flavin enzyme, although other acceptors may be used.

delayed rectifier channels The potassium-selective **ion channels** of **axons**, so called because they change the potassium conductance with a delay after a voltage step. The name is used to denote any axon-like K-channel. Various roles, e.g. regulation of pacemaker potentials, generation of bursts of **action potentials** or generation of long plateaux on action potentials.

delayed-type hypersensitivity See **hypersensitivity**.

deletion mutation A mutation in which one or more (sequential) nucleotides is lost from the genome. If the number lost is not divisible by three and is in a coding region, the result is a **frame-shift mutation**.

DELLA A family of nuclear growth repressors, first identified as **gibberellin**-signalling components, that restrain the growth of plants. Later shown to mediate effects of other phytohormones. There are five distinct DELLAs encoded in the *Arabidopsis* genome: GAI (gibberellin insensitive), RGA (repressor of ga1-3), RGL1 (RGA-like 1), RGL2 and RGL3. RGA and GAI are negative regulators of gibberellin signalling. Gibberellin stimulates growth via 26S **proteasome**-dependent destruction of DELLAs, thus relieving DELLA-mediated growth restraint. See **GRAS family**.

delphinidin The delphinidin gene codes for an enzyme closely related to the **cyanidin** gene and modifies the anthocyanin **dihydrokaempferol** and directs pigment synthesis into the delphinidin pathway. Expression of this gene is responsible for the blue/violet colours of violas, delphiniums and grapes. Delphinidin, an active compound of red wine, inhibits endothelial cell apoptosis via the nitric oxide pathway and regulation of calcium homeostasis.

Delta *Dl* **Neurogenic gene** locus in *Drosophila*. Gene product contains nine repeats of the **EGF-like domain** and is one of the ligands for **Notch**.

delta chains δ-*chains* See **immunoglobulin**. The **heavy chains** of mouse and human IgD immunoglobulins.

delta sleep-inducing peptide *DSIP* A natural somnogenic peptide found in neurons, peripheral organs and plasma; induces mainly delta sleep in mammals. There are nine amino acid residues (Trp-Ala-Gly-Gly-Asp-Ala-Ser-Gly-Glu). DSIP has effects in pain, adaptation to stress and epilepsy; it also has anti-ischemic effects.

delta virus Hepatitis D virus. A defective RNA virus requiring a **helper virus**, usually hepatitis B virus, for replication. Delta virus infections may exacerbate the clinical effects of hepatitis B.

delta-endotoxin δ-*endotoxin* The toxic glycoprotein produced by sporulating **Bacillus thuringiensis** that can kill insects.

dematin *Band 4.9* Actin microfilament-bundling protein (52 kDa, but variants of similar molecular weight are reported); contains an **SH3** domain and is extensively palmitoylated; associated with membrane of erythrocytes (protein 4.9). A substrate for PKC and PKA and bundling of actin is regulated by PKA-mediated phosphorylation.

demissine Toxic glycoalkaloid from *Solanum* spp. See **chaconine**, **solanine**, **potato glycoalkaloids**.

demyelinating diseases Diseases in which the myelin sheath of nerves is destroyed and that often have an autoimmune component. Examples are **multiple sclerosis**, acute disseminated encephalomyelitis (a complication of acute viral infection), **experimental allergic encephalomyelitis**, **Guillain–Barré syndrome**.

denaturation Reversible or irreversible loss of function in proteins and nucleic acids resulting from loss of higher-order (secondary, tertiary or quaternary) structure produced by non-physiological conditions of pH, temperature, salt or organic solvents.

dendrite A long, branching outgrowth from a **neuron**, that carries electrical signals from synapses to the cell body – unlike an axon, which carries electrical signals away from the cell body. This classical definition, however, lost some weight with the discovery of axo–axonal and dendro–dendritic synapses.

dendritic cells 1. Follicular dendritic cells, found in germinal centres of spleen and lymph nodes; retain antigen for long periods. 2. Accessory (antigen-presenting) cells, positive for class II histocompatibility antigens, found in the red and white pulp of the spleen and lymph node cortex and associated with stimulating T-cell proliferation. 3. T-lymphocyte found in epidermis and other epithelial cells involved in antigen recognition expressing predominantly γδ-TCR receptors (dendritic epidermal cells: DECs). 4. **DOPA**-positive cells derived from neural crest and found in the basal part of epidermis: melanocytes distinct from (3). See also **Langerhans cells**.

dendritic spines Wineglass- or mushroom-shaped protrusions from dendrites that represent the principal site of termination of excitatory afferent neurons on interneurons, especially in the cortical regions.

dendritic tree Characteristic (tree-like) pattern of outgrowths of neuronal **dendrites**.

Dendroaspis angusticeps Snake, Eastern green mamba. See **calcicludine**.

Dendroaspis natriuretic peptide *DNP* Natriuretic **peptide**, 38 residues, found in the venom of the snake *Dendroaspis angusticeps*. Binds to atrial natriuretic peptide receptor-A but not to the ANP-receptors type B.

Dendroaspis polylepis Snake, Black mamba. See **calciseptine**.

dendrogram A branching diagram, like a family tree, reflecting similarities or affinities of some sort between species.

dendrotoxins Polypeptides (57–60 residues) isolated from *Dendroaspis* (snake) venom that are selective blockers of **voltage-gated** K^+ channels in a variety of tissues and cell types. Have sequence similarity with Kunitz-type serine peptidase inhibitors.

denervation Removal of nerve supply to a tissue, usually by cutting or crushing the **axons**.

dengue Tropical disease caused by a flavivirus (one of the **arboviruses**), transmitted by mosquitoes. A more serious complication is dengue shock syndrome, a haemorrhagic fever probably caused by an immune-complex hypersensitivity after re-exposure.

denitrifying bacteria Bacteria that break down nitrate and nitrite to gaseous nitrogen. Often found in soil and **biofilms** and may prove important in **bioremediation** since nitrate contamination of water is becoming a problem. Denitrification is not specific to any one phylogenetic group; the trait is found in about 50 genera of Proteobacteria and involves genes coding for NO_2^--reductase (*nirK* and *nirS*) and N_2O reductase (*nosZ*).

dense bodies Areas of electron density associated with the thin filaments in smooth muscle cells. Some are associated with the plasma membrane, others are cytoplasmic.

density-dependent inhibition of growth The phenomenon exhibited by most normal (**anchorage dependent**) animal cells in culture that stop dividing once a critical cell density is reached. The critical density is considerably higher for most cells than the density at which a monolayer is formed; for this reason, most cell behaviourists prefer the term 'density-dependent inhibition of growth', as this avoids any confusion with contact inhibition of locomotion – a totally different phenomenon that is contact dependent.

density gradient A column of liquid in which the density varies continually with position, usually as a consequence of variation of concentration of a solute. Such gradients may be established by progressive mixing of solutions of different density (as, for example, sucrose gradients) or by centrifuge-induced redistribution of solute (as for **caesium chloride** gradients). Density gradients are widely used for centrifugal and gravity-induced separations of cells, organelles and macromolecules. The separations may exploit density differences between particles, or primarily differences in size; in the latter case the function of the gradient is chiefly to stabilize the liquid column against mixing.

dentate nucleus Nerve cell mass, oval in shape, located in the centre of each of the cerebral hemispheres.

denticle Any small, tooth-like structure; the placoid scales of elasmobranchs.

dentin matrix protein *DMP-1* An acidic matrix protein, non-collagenous, present in the mineralized matrix of bone and dentine, that has a key role in mineralization. DMP-1 in solution can undergo oligomerization and temporarily stabilize the newly formed calcium phosphate nanoparticle precursors by sequestering them and preventing their further aggregation and precipitation. Binds to CD44 and RGD sequence-dependent **integrins**.

dentine *Dentin* Structural biocomposite of needle-like crystals of hydroxyapatite embedded in a fibrous collagen matrix, very similar to compact bone. Secreted by **odontoblasts**. Main component of teeth and ivory.

deoxycholate A bile salt formed by bacterial action from cholate; usually conjugated with glycine or taurine. The sodium salt is used as a detergent to make membrane proteins water-soluble.

deoxyglucose *2-Deoxyglucose* Analogue of glucose in which the hydroxyl on C-2 is replaced by a hydrogen atom. Since it is often taken up by cells but not further metabolized, it can be used to study glucose transport and also to inhibit glucose utilization.

deoxyhaemoglobin Haemoglobin without bound oxygen.

deoxynojirimycin Antibiotic produced by *Bacillus* spp; inhibits α-glucosidases and thus interferes with the **glycosylation** of cell surface **glycoproteins**.

deoxyribonuclease *DNAase; Dnase* An **endonuclease** with preference for DNA. Pancreatic DNAase I yields di- and oligonucleotide 5′ phosphates, pancreatic DNAase II yields 3′ phosphates. In chromatin, the sensitivity of DNA to digestion by DNAase I depends on its state of organization, transcriptionally active genes being much more sensitive than inactive genes.

deoxyribonucleoside A purine or pyrimidine base N-glycosidically linked to 2-deoxy-D-ribofuranose. The phosphate esters of nucleosides are nucleotides.

deoxyribonucleotide A **deoxyribonucleoside**, ester linked to phosphate.

deoxyribose *2-Deoxy-D-ribose* The sugar that, when linked by 3′–5′ phosphodiester bonds, forms the backbone of DNA.

depactin **Actin**-depolymerizing protein (17.6 kDa) originally isolated from echinoderm eggs. Similar to **actophorin**.

DEP domain A globular domain, about 80 residues, of unknown function present in signalling proteins including **Dishevelled**, Egl-10, and **pleckstrin**. Mammalian regulators of G-protein signalling contain these domains.

Dep-1 *CD148; HPTPe* Density-enhanced phosphatase-1 (220–250 kDa). Transmembrane protein with eight extracellular FnIII domains and a single cytoplasmic tyrosine phosphatase domain. Dep-1 is involved in signal transduction in lymphocytes and is also found in smooth muscle cells and tumour cells. When clustered, Dep-1 inhibits FcγRII-induced superoxide production. In many tumour cells Dep-1 is associated with a 64-kDa serine/threonine kinase that may regulate its activity. Has been speculated to play a role in density-dependent inhibition of growth.

dephosphorylation Removal of a phosphate group.

depolarization A positive shift in a cell's **resting potential** (that is normally negative), thus making it numerically smaller and less polarized (e.g. –90 mV to –50 mV). The opposite of **hyperpolarization**. In the case of excitable cells, the resting potential is around –70 mV and depolarization (due to sodium ion influx) to below the threshold level will lead to the generation of an **action potential**.

depolarizing muscle relaxant A drug that mimics the action of **acetylcholine** at the neuromuscular junction to produce a relatively short period of paralysis and relaxation. Used to facilitate the passage of an endotracheal tube to maintain airway patency at the start of a surgical operation. **Suxamethonium** is the only common drug of this type.

depsipeptides Polypeptides that contain ester bonds as well as peptides. Naturally occurring depsipeptides are usually cyclic; they are common metabolic products of microorganisms and often have potent antibiotic activity (examples are **actinomycin**, enniatins, **valinomycin**).

depurination (of DNA). The N-glycosidic link between purine bases and deoxyribose in DNA has an appreciable rate of spontaneous cleavage *in vivo*, a lesion that must be enzymically repaired to ensure stability of the genetic information.

derepress To activate by suppressing a repressor; not an uncommon feature in gene regulation.

derepression Anything that stops the repression of a gene, thereby allowing **expression** to occur.

derivatization *Derivatized* Modification of a molecule to change its solubility or other properties to enable analysis, for example by mass spectroscopy or chromatography, or to provide a label (e.g. fluorescent moiety) to facilitate identification and tracking.

dermal tissue Outer covering of plants that includes the **epidermis** and periderm (non-living bark). Compare **dermis**.

dermamyotome The embryonic region, derived from the dorsal portion of the **somite**, that gives rise to the **dermis** and axial musculature.

dermaseptins The dermaseptins are closely related peptides with broad-spectrum antibacterial activity produced by the skin of the South American frog, *Phyllomedusa sauvagei*. They are polycationic (Lys-rich), alpha-helical and amphipathic, a structure that is believed to enable the peptides to interact with membrane bilayers, leading to permeation and disruption of the target cell. Dermaseptin S9 (GLRSKIWLWVLLMIWQESNKFKKM) acts on both Gram-positive and Gram-negative bacteria.

dermatan sulphate Glycosaminoglycan (15–40 kDa) typical of **extracellular matrix** of skin, blood vessels and heart. Repeating units of D-glucuronic acid-*N*-acetyl-D-galactosamine or L-iduronic acid-*N*-acetyl-D-galactosamine with one or two sulphates per unit. Broken down by L-iduronidase, and accumulates intra-lysosomally in **Hurler's disease** and **Hunter syndrome**.

dermatitis Inflammation of the **dermis**, often a result of **contact sensitivity**.

dermatogen *Protoderm* In plants, the primary meristem that gives rise to epidermis, a **histogen**.

dermatomyositis One of a group of acquired muscle diseases called inflammatory myopathies. Probably autoimmune, with inflammation and weakness of muscles and often a purplish skin rash.

Dermatophagoides pteronyssinus House dust mite. Antigens extracted from mites and their faeces are a common cause of allergy to house dust in W. European countries. Major allergen is Der p I, a cysteine peptidase.

dermatophytes Fungi that can cause infections of the skin, hair, and nails due to their ability to utilize keratin. Examples are ringworm and athlete's foot.

dermatopontin *TRAMP* A 22-kDa extracellular matrix (ECM) protein that interacts with other ECM components, especially decorin, and regulates ECM formation and collagen fibrillogenesis. A molluscan homologue is a major shell matrix protein.

dermatosparaxis Recessive disorder of cattle in which a procollagen peptidase is absent. In consequence, the amino- and carboxy-terminal peptides of procollagen are not removed, the **collagen** bundles are disordered, and the dermis is fragile. Similar to **Ehlers–Danlos syndrome** in humans.

dermis Mesodermally derived **connective tissue** underlying the epithelium of the skin.

dermoid cyst Usually benign cyst, the walls of which are of dermal origin. Many ovarian tumours are dermoid cysts.

DES Abbrev. for **diethylstilbestrol**.

DeSanctis–Cacchione syndrome A variant of **xeroderma pigmentosum** in which a different DNA repair enzyme is involved. Hybrid fibroblasts formed by Sendai virus fusion of the two types show normal repair (complementation). Has also been associated with mutation in the ERCC6 gene, part of the nucleotide-**excision repair** (NER) pathway.

desert hedgehog *dhh* Product of one of the **hedgehog family** genes, a signalling molecule expressed by Schwann cells and essential for the structural and functional integrity of peripheral nerves.

desferrioxamine Iron transporter from *Streptomyces pilosus* that chelates ferric ions. Used clinically to treat acute iron poisoning.

desmid Chlorophyte **algae** that are usually freshwater-living and unicellular. Their cell wall often has elaborate ornamented shape.

desmin A protein (53 kDa) of intermediate filaments, somewhat similar to **vimentin**, but characteristic of muscle cells. Type III intermediate filament protein. Co-localizes with **synemin**, **paranemin** and **plectin** in the appropriate cell types.

desmocalmin A protein (240 kDa) isolated (1985) from bovine desmosomes that binds calcium–calmodulin and cytokeratin-type intermediate filaments. Not mentioned in recent literature.

desmocollins **Glycoproteins** of 130 and 115 kDa (desmocollins I and II) isolated from **desmosomes**. Antibody fragments directed against desmocollins block desmosome formation, and desmocollins are therefore thought to be involved in the adhesion.

desmoglein Transmembrane **glycoprotein** (165 kDa) found in desmosomes.

desmoplakins Proteins isolated from **desmosomes**. Types I (240 kDa) and II (210 kDa) are long, flexible, rod-like molecules about 100 nm long made of two polypeptide chains in parallel. Desmoplakin III is smaller (81 kDa).

desmoplasia Growth of fibrous or connective tissue – for example, the pervasive growth of dense fibrous tissue around some tumours.

desmopressin *DDAVP; 1-Deamino-8-D-arginine-vaso-pressin* A synthetic analogue of **vasopressin** that has been successfully used in the treatment of type I **von Willebrand's disease**, mild factor VIII deficiency and intrinsic platelet function defects.

desmosine Component of **elastin**, formed from four side chains of lysine and constituting a cross-linkage.

desmosome *Macula adherens junctions; spot desmosomes* Specialized cell junction characteristic of epithelia into which intermediate filaments (tonofilaments of cytokeratin) are inserted. The gap between plasma membranes is of the order of 25–30 nm, and the intercellular space has a medial band of electron-dense material. Desmosomes are particularly conspicuous in tissues (such as skin) that have to withstand mechanical stress.

desmotubule Cylindrical membrane-lined channel through a **plasmodesma**, linking the cisternae of **endoplasmic reticulum** in the two cells.

desmoyokin Desmosomal plaque protein (680 kDa) from bovine muzzle **keratinocytes**. The desmoyokin gene is identical to the human *Ahnak* gene, which is expressed ubiquitously and downregulated in neuroblastomas. Null mutant mice do not, however, show any marked phenotype.

desoxy- See under **deoxy-**.

desquamation Shedding of outer layer of skin (squames) or of cells from other epithelia.

destrin Actin-depolymerizing protein (19 kDa) from pig, apparently identical to **ADF** and similar to **cofilin**.

destruxins Cyclic **depsipeptide** fungal toxins that suppress the immune response in invertebrates.

desynapsis Separation of the paired homologous chromosomes at the **diplotene** stage of meiotic prophase I.

detergents **Amphipathic**, surface active, molecules with polar (water soluble) and non-polar (hydrophobic) domains. They bind strongly to hydrophobic molecules or molecular domains to confer water solubility. Examples include sodium dodecyl sulphate, fatty acid salts, the Triton family, octyl glucoside.

determinate cleavage A type of embryonic cleavage in which each blastomere has a predetermined fate in the later embryo (in contrast to the situation in so-called 'regulating' embryos).

determination The commitment of a cell to a particular path of differentiation, even though there may be no morphological features that reveal this determination. Generally irreversible, but in the case of **imaginal discs** of *Drosophila* that are maintained by serial passage, **transdetermination** may occur.

detoxification reactions Reactions taking place generally in the liver or kidney in order to inactivate toxins, either by degradation or else by conjugation of residues to a hydrophilic moiety to promote excretion.

deuterium oxide D_2O Heavy water, in which the hydrogen is replaced by deuterium. Will stabilize assembled microtubules.

Deuteromycetes Outmoded term for group now reclassified as **Deuteromycotina**. Includes fungi with no known sexual reproductive stages – the old Fungi Imperfecta.

Deuteromycotina *Fungi imperfecti; Deuteromycetes; deuteromycota* Fungi in which no sexual reproduction is known. Many appear to be Ascomycotina. Group includes many saprophytes, e.g. *Aspergillus, Penicillium,* and plant parasites such as *Fusarium* and *Verticillium.*

deuterostome Embryonic developmental pattern in which the mouth does not form from the blastopore but from a second opening: includes echinoderms and chordates. Contrasts with morphogenesis in protostome phyla, which include annelids, molluscs and arthropods. The two groups also differ in many aspects of early development, including the pattern of early cleavage and the stage at which blastomeres become committed in differentiation; in deuterostomes the early blastomeres are equipotent, whereas in protostomes there is earlier patterning and commitment to form particular cell lineages.

DEVD *Z-DEVD; bis-(N-CBZ-L-aspartyl-L-glutamyl-L-valyl-l-aspartic acid amide)* Derivatized peptide substrate for **caspase 3** with a benzyloxycarbonyl group (also known as BOC or Z) at the N-terminus to give improved cellular permeability. Can be conjugated with rhodamine or other fluorochromes for chromogenic assays. Fluoromethyl ketone (FMK)-derivatized peptides act as effective irreversible inhibitors.

Devoret test Test for potential carcinogens based on induction of prophage lambda in bacteria (*E. coli* K12 envA uvrB). There is a good correlation between the ability of aflatoxins and benzanthracenes to induce lambda and their carcinogenicity in rodents.

dexamethasone Steroid analogue (**glucocorticoid**), used as an anti-inflammatory drug.

dexamfetamine *Formerly dexamphetamine; TN Dexedrine* Acts directly on the brain as a stimulant but is a sedative in children and is used for treatment of Attention-Deficit Hyperactivity Disorder.

dextrans High molecular weight polysaccharides synthesized by some microorganisms. Consist of D-glucose linked by α-1,6 bonds (and a few α-1,3 and α-1,4 bonds). Dextran 75 (average molecular weight 75 kDa) has a colloid osmotic pressure similar to blood plasma, so dextran 75 solutions are used clinically as plasma expanders. They will also cause charge-shielding and, at the right concentrations, induce flocculation of red cells – a trick that is used in preparing leucocyte-rich plasma for white-cell purification in the laboratory. Cross-linked dextran is the basis for **Sephadex**. Commercially derived from strains of *Leuconostoc mesenteroides.*

dextrins Poly-D-glucosides formed by hydrolytic degradation of starch or glycogen. Chain length can be variable.

dextromethorphan *DMP* Semisynthetic drug that relieves cough by acting directly on the cough centre in the brain. Cough mixtures containing dextromethophan have been abused as recreational drugs.

DFF *DNA fragmentation factor* Complex of caspase-activated DNase (DFF40) in complex with its cognate 45-kDa inhibitor (inhibitor of **CAD**: ICAD or DFF45).

D-gene segment *Diversity gene segment* Part of the gene for the immunoglobulin **heavy chain**, it codes for part of the **hypervariable region** of the VH domain and is located between the VH and JH segments. There are probably about 20 different D segments.

DH5 Strain of *E. coli* K-12 that are disabled and non-colonizing and can therefore be used in experimental work where escape would be undesirable. It is recA⁻, highly transformable, and allows for selection by α-complementation.

DHAP Abbrev. for (1) **dihydroxyacetone phosphate,** (2) 3,4-dihydroxyacetophenone or 3,4-**dihydroxyacetophynone**, one of the constituents of a traditional Chinese herbal medicine, (3) Initials of drugs used in combined chemotherapy for lymphoma (**dexamethasone**, cytarabine (**Ara C**) and **cisplatin**).

DHAP-AT *Dihydroxyacetone phosphate acyltransferase* The enzyme involved in the first step of **plasmalogen** biosynthesis.

DHEA See **dihydroepiandrosterone**.

DH domain *Dbl-homology domain* Domain of around 250 amino acids found in **Dbl**, Vav and a family of other Dbl-family proteins (Lfc, Lsc, Ect2, Dbs, Brx). DH domains are invariably located immediately N-terminal to **PH domain**, and the membrane localization and enzymatic activity of the DH domain requires the PH domain for normal function.

DHFR See **dihydrofolate reductase**.

DHT Abbrev. for **dihydrotestosterone**.

diabetes insipidus Rare form of **diabetes** in which the kidney tubules do not reabsorb enough water. This can be because: (1) the renal tubules have defective receptors for **antidiuretic hormone** (ADH; vasopressin); (2) a class of **aquaporin** water channel in the collecting duct is defective; (3) there is inadequate ADH production by the pituitary, leading to the excessive production of dilute urine.

diabetes mellitus Relative or absolute lack of **insulin** leading to uncontrolled carbohydrate metabolism. In juvenile-onset (Type I) diabetes (which may be an autoimmune response to pancreatic B-cells) the insulin deficiency tends to be almost total, whereas in adult-onset (Type II) diabetes (**NIDDM**) there seems to be no immunological component but an association with obesity.

diabetes-related peptide See **islet amyloid peptide**.

diablo Smac/diablo An apoptogenic protein released from mitochondria. Promotes the release and activation of caspases (it is an inhibitor of apoptosis protein (**IAP**)-binding protein) but must first be processed by a mitochondrial protease and then released into the cytosol.

diabody A recombinant bispecific antibody (BsAb), constructed from heterogeneous single-chain antibodies. There is a five amino-acid linker between VH and VL domains, and the two antigen-binding sites of the diabody molecule are located at opposite ends of the molecule at a distance of about 7 nm apart and pointing away from each other. See also **triabody**.

diacetoxyscirpenol *DAS* **Trichothecene** mycotoxin produced by various species of fungi. Cytotoxic for human CFU-GM and BFU-E.

diacylglycerol *DAG* Glycerol substituted on the 1 and 2 hydroxyl groups with long-chain fatty acyl residues. DAG is a normal intermediate in the biosynthesis of phosphatidyl phospholipids and is released from them by phospholipase C activity. DAG from phosphatidyl inositol polyphosphates is important in signal transduction. Elevated levels of DAG in membranes activate protein kinase C by stabilizing its catalytically active complex with membrane-bound phosphatidyl serine and calcium.

diacytosis Rarely used term for the discharge of an empty pinocytotic vesicle from a cell.

diad Anything with two-fold symmetry. Thus a diad axis defines a plane where there is mirror symmetry between the two halves of the structure.

diakinesis The final stage of the first **prophase** of **meiosis**. The **chromosomes** condense to their greatest extent during this stage, and normally the nucleolus disappears and the fragments of the nuclear envelope disperse.

diallyl trisulfide A natural compound derived from garlic. Despite its reported lipid-lowering effects, the mechanisms of its actions are not yet clear.

dialysis Separation of molecules on the basis of size through a semipermeable membrane. Molecules with dimensions greater than the pore diameter are retained inside the dialysis bag or tubing, whereas small molecules and ions emerge in the dialysate outside the tubing.

diaminobenzidine *DAB* Artificial substrate for **peroxidase**, producing a coloured reaction product – but a potent carcinogen.

diaminobenzoic acid *DABA; 3,5-diaminobenzoic acid* Compound used in fluorimetric determination of DNA content: gives fluorescent product when heated in acid solution with aldehydes.

diaminopimelic acid *DAP; 2,6-diaminoheptanedioic acid* A diamino-carboxylic acid with two chiral centres, a constituent of the cell wall of Gram-negative bacteria, not known to occur in any other group.

Diamond–Blackfan anaemia A rare, progressive haematological disorder which presents in early childhood. It results from defective erythropoiesis and lack of nucleated erythrocytes in the bone marrow. Approximately 25% of cases of Diamond–Blackfan anemia are caused by mutation in the gene encoding ribosomal protein S19.

diapedesis Archaic term for the emigration of leucocytes across the endothelium.

diaphanous autoinhibitory domain *DAD* See **diaphanous-related formins**.

diaphanous-related formins *Drfs* Proteins that act as Rho GTPase effectors during cytoskeletal remodelling. See **formins**. The activity of DRFs is inhibited by an intramolecular interaction between their N-terminal regulatory region and a conserved C-terminal segment termed the diaphanous autoinhibitory domain (DAD).

diaphorase Any enzyme capable of catalysing oxidation of NAD or NADPH in the presence of an electron-acceptor other than oxygen – for example, methylene blue, quinones or cytochromes. Imprecise term.

diaphysis Central portion of a long (limb) bone, composed of an outer wall of heavily mineralized bone and a central yellow (fatty) marrow-filled cavity. See **epiphysis**.

diastema 1. A thin, yolk-free structure which is considered to play an essential role in the induction of the cleavage furrow. 2. A gap in a row of teeth.

diatom Algae of the division Bacillariophyta; largely unicellular and characterized by having cell walls of hydrated silica embedded in an organic matrix. The cell walls are formed in two halves that fit together like the lid and base of a pillbox (or petri dish) and often have elaborate patterns formed by pores. Diatoms are very abundant in marine and freshwater plankton. Deposits of the cell walls form diatomaceous or siliceous earths.

diauxic growth See **diauxie**.

diauxie *Diauxy* Adaptation of microorganisms to culture media that contain two different carbohydrates (e.g. glucose and lactose). In the first growth phase the sugar for which there are constitutive enzymes is utilized, then there is a brief pause while the enzyme systems for the second sugar are induced and synthesized.

diazepam Long-acting **benzodiazepine** used for treatment of anxiety and insomnia; also used preoperatively as a relaxant.

diazinon A non-systemic organophosphate insecticide used to control cockroaches, silverfish, ants and fleas in residential, non-food buildings, and for a range of agricultural pests and against fleas and ticks in veterinary practice. An acetylcholinesterase inhibitor.

diazotroph An organism that is capable of nitrogen fixation. Assemblages of diazotroph species are often important in relatively impoverished soils such as salt marshes. The capacity to fix nitrogen occurs in a phylogenetically diverse range of bacterial species.

dibutyryl cyclic AMP *dbCAMP* An analogue of cyclic AMP that shares some of the pharmacological effects of this nucleotide but is generally believed to enter cells more readily on account of its greater hydrophobicity.

dicer Gene encoding a ribonuclease that is required by the **RNA interference** and small temporal RNA (stRNA) pathways to produce the active small RNA component that represses gene expression. Gene in *C. elegans* is *dcr-1*.

dichlorobenzonitrile *2,6-Dichlorobenzonitrile* Inhibitor of cellulose biosynthesis in higher plants.

dichlorophenoxyacetic acid *2,4-Dichlorophenoxyacetic acid; 2,4-* A synthetic **auxin**, used as a selective herbicide and in media for plant tissue culture.

dichroism See **circular dichroism**.

Dick test An obsolete skin-test for immunity against the toxin of *Streptococcus pyogenes*, the organism that causes scarlet fever.

dickkopf-1 dkk-1 Gene in *Xenopus* that encodes a secreted protein (259 residues, 40 kDa) that induces the head region in the developing embryo. Member of a family of genes. Dkk-1 is a potent antagonist of **Wnt**.

diclofenac A non-steroidal anti-inflammatory drug. Responsible for the catastrophic decline in the numbers of Indian vultures (*Gyps indicus*) that, having eaten carcasses contaminated by diclofenac, die of gout.

Diconal® Proprietary combination of cyclizine hydrochloride (antiemetic antihistamine) and dipipanone hydrochloride (opioid analgesic) used for relief of moderate to severe pain.

dicotyledonous plants Plants belonging to the large subclass of Angiosperms that have two seed-leaves (cotyledons). Includes the majority of herbaceous flowering plants and most deciduous woody plants of the temperate regions.

Dictyoptera Order of insects that includes the cockroaches and mantises.

dictyosome Organelle found in plant cells and functionally equivalent to the **Golgi apparatus** of animal cells.

Dictyostelium A genus of the **Acrasidae**, the cellular slime moulds.

dictyotene Prolonged **diplotene** of meiosis: the stage at which oocyte nuclei remain during yolk production.

dicyclomine hydrochloride *TN Merbentyl* An antimuscarinic drug with antispasmodic properties used in treatment of irritable bowel syndrome.

dideoxy sequencing *Sanger dideoxy sequencing* The most popular method of DNA sequence determination (*cf.* **Maxam–Gilbert method**). Starting with single-stranded template DNA, a short complementary primer is annealed and extended by a DNA polymerase. The reaction is split into four tubes (called 'A', 'C', 'G' or 'T'), each containing a low concentration of the indicated dideoxynucleotide in addition to the normal deoxynucleotides. Dideoxynucleotides, once incorporated, block further chain extension, and so each tube accumulates a mixture of chains of lengths determined by the template sequence. The four reactions are denatured and run out on an acrylamide sequencing gel in neighbouring lanes, and the sequence read up the gel according to the order of the bands. In modern automated methods the labelling of each the dideoxynucleotides is different and the mixture of products can be separated on a column and the products identified spectrophotometrically.

dideoxynucleotides Any nucleoside triphosphate in which the hydroxyl groups on C2 and C3 of the pentose have been substituted by hydrogen. Used in **dideoxy sequencing**.

Didinium Fast-moving carnivorous protozoan (of the Phylum Ciliophora) that feeds almost exclusively on live *Paramecium*.

DIDS *4,4'-Diisothiocyano-2,2'-disulfonic acid stilbene* An irreversible anion transport inhibitor.

dieldrin One of the cyclodiene insecticides closely related to **aldrin** (which breaks down to form dieldrin). Used on fruit, soil and seed, and has been used to control tsetse flies and other vectors of tropical diseases. Binds to soil, where it has long half-life (years) in temperate regions and can bioaccumulate with toxic effects. Now generally banned and regarded as one of the most ecologically damaging toxins.

dielectric constant **Relative permittivity** when referring to the medium of a capacitor and independent of electric field strength. Under these conditions it is the ratio of the capacitance of the conductor to the capacitance it would have if the medium were replaced by a vacuum.

Diels–Alder reaction Reaction used in organic synthesis of six-membered rings.

diencephalon In vertebrate central nervous system, the most rostral part of the **brain** stem, consisting of the thalamus, hypothalamus, subthalamus and epithalamus. It is a key relay zone for transmitting information about sensation and movement, and also contains (in the hypothalamus) important control mechanisms for homeostatic integration.

diethylstilbestrol *Diethylstilboestrol* Non-steroidal compound that has estrogen activity; used therapeutically and, until banned, as a growth promoter in livestock. Causes clear-cell adenocarcinoma of the vagina and cervix in women exposed *in utero*. Administration of DES in large doses during pregnancy increases the subsequent risk of breast cancer, and increases the risk of testicular cancer in males exposed *in utero*.

differential adhesion The differential adhesion hypothesis was advanced by Steinberg to explain the mechanism by which heterotypic cells in mixed aggregates sort out into isotypic territories. Quantitative differences in homo- and heterotypic adhesion are proposed to be sufficient to account for the phenomenon without the need to postulate cell type-specific adhesion systems: fairly generally accepted, although tissue-specific **cell adhesion molecules** are now known to exist.

differential display PCR *ddPCR* Variation of the **polymerase chain reaction** used to identify differentially expressed genes. **mRNA** from two different tissue samples is reverse transcribed, then amplified using short, intentionally non-specific primers. The array of bands obtained from a series of such amplifications is run on a high-resolution gel and compared with analogous arrays from different samples. Any bands unique to single samples are considered to be differentially expressed; they can be purified from the gel and sequenced and used to clone the full-length cDNA. Similar in aim to **subtractive hybridization**. See also **differential screening**.

differential hybridization Technique used to compare gene expression levels under different conditions by comparing two **cDNA** libraries. The two libraries are transcribed into RNA and each set of library products is labelled with a different fluorochrome. The two RNA samples are pooled and used to probe a **DNA array**; spectrophotometric analysis of the binding reveals whether binding is comparable or whether the products of one library bind preferentially (differentially), which indicates a higher level of that particular RNA species in that library. Alternative labelling strategies can be used.

differential interference contrast Method of image formation in the light microscope based on the method proposed by Nomarski (though, strictly speaking, all forms of optical microscopy rely to a greater or lesser extent on differential interference). The light beam is split by a **Wollaston prism** in the condenser, to form slightly divergent beams polarized at right angles. One passes through the specimen (and is retarded if the refractive index is greater) and one through the background nearby; the two are recombined in a second Wollaston prism in the objective and interfere to form an image. The image is spuriously 'three-dimensional' – the nucleus, for example, appears to stand out above the cell (or be hollowed out) because it has a higher refractive index than the cytoplasm. The Nomarski system has the advantage that there is no phase-halo, but the contrast is low, and image formation with crowded cells is poor because the background does not differ from the specimen.

differential scanning calorimetry *DSC* Form of **thermal analysis** in which heat flows to a sample and a standard at the same temperature are compared, as the temperature is changed.

differential screening General term for techniques used to identify genes that are expressed differentially in two different conditions. See **subtractive hybridization**, **differential display PCR**.

differential stain 1. A histological stain which selectively stains some elements of a specimen more than others, by giving them either different colours, or different shades or intensities of the same colour. **2.** Some stains bind reversibly, and for best effect (greatest discrimination between different parts) the specimen is over-stained and then washed to remove stain; more will be released where binding is weakest.

differentiation Process in development of a multicellular organism by which cells become specialized for particular functions. Requires that there is selective expression of portions of the genome; the fully differentiated state may be preceded by a stage in which the cell is already programmed for differentiation but is not yet expressing the characteristic phenotype (**determination**).

differentiation antigen Any large structural macromolecule that can be detected by immune reagents and that is associated with the differentiation of a particular cell type or types. Many cells can be identified by their possession of a unique set of differentiation antigens. There is no implication that the antigens cause differentiation.

Difflugia A small amoeba (200–250 µm long) of the Phylum Sarcodina. Possesses a chitinous test (shell) that is usually covered completely with sand grains. Feeds mainly on green algae.

diffraction When a wave-train passes an obstacle, secondary waves are set up that interfere with the primary wave and give rise to bands of constructive and destructive interference. Around a point source of light, in consequence, is a series of concentric light and dark bands (coloured bands with white light) – a diffraction pattern.

diffusion coefficient *Diffusion constant* For the translational diffusion of solutes, diffusion is described by Fick's First Law, which states that the amount of a substance crossing a given area is proportional to the spatial gradient of concentration and the diffusion constant (D), which is related to molecular size and shape. A useful derived relationship is that the mean square distance moved (in three dimensions) by molecules in time t is $6Dt$.

diffusion limitation The boundary layer hypothesis: that the proliferation of cells in culture is limited by the rate at which some essential component (almost certainly a growth factor) diffuses from the bulk medium into the layer immediately adjacent to the plasma membrane. By spreading out, a cell obtains a supra-threshold level of the factor and can divide; if unable to spread (because of crowding or poor adhesion) then the cell will remain in the G0 stage of the **cell cycle**.

diffusion potential Potential arising from different rates of diffusion of ions at the interface of two dissimilar fluids; a junction potential.

DiGeorge syndrome Congenital absence of the thymus and parathyroid as a result of which the T-lymphocyte system is absent. May be caused by mutation in the TBX1 (***T-box*** 1) gene, which maps to the centre of the region of chromosome 22q11.2 often deleted in DiGeorge syndrome.

digestive vacuole Intracellular vacuole into which lysosomal enzymes are discharged and digestion of the contents occurs. More commonly referred to as a **secondary lysosome**.

digitalis General term for pharmacologically active compounds from the foxglove (*Digitalis*). The active substances are the cardiac glycosides, digoxin, digitoxin, strophanthin and **ouabain**. Causes increased force of contraction of the heart, disturbance of rhythm and reduced beat frequency. Also causes arteriolar constriction, venous dilation, nausea and visual disturbances.

digitonin See **saponin**.

diglyceride Generic term for any compound with two glyceryl residues. Formerly used for **diacyl glycerol**, though this is inappropriate.

digoxin Cardiac glycoside from foxglove (*Digitalis lanata*), used to treat congestive heart failure and supraventricular arrhythmias. Inhibits Na^+/K^+-ATPase. Aglycone is **digoxigenin**.

digoxygenin Small molecule derived from foxgloves that is used for labelling DNA or RNA probes, and subsequent detection by enzymes linked to anti-digoxygenin antibodies. Proprietary to Boehringer-Mannheim. See **digoxin**.

dihybrid The product of a cross between parents differing in two characters determined by single genes, each of which has two alleles. Heterozygous for two pairs of alleles.

dihydroepiandrosterone *DHEA* Predominant androgen secreted from the adrenal cortex, an intermediate in androgen and estrogen biosynthesis. Can be converted to sulphate (DHEA-S) the predominant plasma form, which can in turn be converted to potent androgens and estrogens. Considered to play an important immunomodulatory role, and the decline in DHEA levels with age correlates with reduced immune competence. Administration especially to postmenopausal women is claimed to bring benefit, especially in bone mineralization. Has been shown to have tumour suppressive and antiproliferative effects in rodent tumours.

dihydroflavinol reductase *DFR* Plant enzyme that modifies the precursor pigments for flower colour in all three pathways: cyanidin, delphinidin and pelargonidin. Precursor pigment molecules are colourless until modified by DFR, so any mutation that disrupts the DFR gene results in white flowers.

dihydrofolate reductase *DHFR; EC 1.5.1.3* Enzyme involved in the biosynthesis of **folic acid** that transfers hydrogen from NADP to **dihydrofolate**, yielding tetrahydrofolic acid – an essential vitamin cofactor in purine, thymidine and methionine synthesis. Inhibitors (e.g. aminopterin and amethopterin, components of **HAT medium**) can be used as antimicrobial and anticancer drugs.

dihydrokaempferol *DHK* The anthocyanin precursor for all three primary plant pigments: **cyanidin**, **pelargonidin** and **delphinidin**.

dihydropyridines Specific blockers of some types of **calcium channel**, e.g. nifedipine and nitrenidine; among the most widely used drugs for the management of cardiovascular disease. Fourth-generation drugs of this class, highly lipophilic dihydropyridines, are now available (lercanidipine, lacidipine).

dihydrotestosterone A potent metabolite of **testosterone**. Mediates many of the functional activities of testosterone (differentiation, growth-promotion) through the androgen receptor.

dihydroxyacetone *DHA* Old name for **glycerone**.

dihydroxyacetophynone *3,4-Dihydroxyacetophenone* One of the constituents (Qingxintong) of a traditional Chinese herbal medicine, derived from *Ilex pubescens*. Inhibits platelet function.

dihydroxylacetone phosphate One of the products of the reduction of 1,3-bisphosphoglycerate by NADPH in the Calvin cycle, used in the synthesis of sedoheptulose 1,7-bisphosphate and fructose 1,6-bisphosphate. Also the product of the dehydrogenation of L-glycerol-3-phosphate.

Dii *'di-i'* Name used for fluorescent derivatives of indocarbocyanine iodide that have two long alkyl chains and are membrane soluble. Used as general stains for membranes and also as specific probes for membrane fluidity measurements.

dikaryon Fungal hypha or mycelium in which there are two nuclei of different genetic constitution (and different mating type) in each cell (or hyphal segment). *Adj.* dikaryotic.

dikaryophase The period in the life cycle of an ascomycete or basidiomycete fungus in which the cells have two nuclei, i.e. between **plasmogamy** and **karyogamy**.

diltiazem HCl Calcium antagonist used in treatment of angina.

dilution cloning Cloning by diluting the cell suspension to the point at which the probability of there being more than one cell in the inoculum volume is small. Inevitably, on quite a few occasions there will not be any cells.

dimercaprol *British Anti-Lewisite; BAL* Drug used as an antidote to poisoning by heavy metals such as antimony, arsenic, bismuth, gold, mercury, thallium or lead, but not iron or cadmium. Not considered to be the drug of first choice because of toxic side effects. Originally synthesized as an antidote against the vesicant arsenical war gases (Lewisite), based on the fact that arsenic products react with SH radicals.

dimethyl formamide *DMF* Compound primarily used as a solvent in the production of polyurethane products and acrylic fibres. It is also used in the pharmaceutical industry, in the formulation of pesticides, and in the manufacture of synthetic leathers, fibres, films and surface coatings. Sometimes used as an alternative to **DMSO** for making stock solutions of compounds that are poorly water-soluble; the concentrated solution can then be diluted.

dimethyl sulphoxide See DMSO.

dinitrophenol *2,4-Dinitrophenol* Small molecule used as an uncoupler of oxidative phosphorylation. Also used after reaction with various proteins to provide a strong and specific identified **haptenic** group.

dinoflagellates Photosynthetic organisms of the order Dinoflagellida (for botanists, Dinophyceae). They are aquatic and abundant in marine plankton; two **flagella** lie in grooves in an often elaborately sculptured shell or **pellicle** that is formed from plates of cellulose deposited in membrane vesicles. The pellicle gives some dinoflagellates very bizarre shapes. Their chromosomes lack centromeres and may have little or no protein, and may perhaps be intermediate between pro- and eukaryote types; hence the group has been termed **mesokaryotic**. The nuclear membrane persists during mitosis. *Gymnodinium* and *Gonyaulax*, which causes 'red tides', produce toxins that, if accumulated by filter-feeding **molluscs**, can be fatal. Another common genus is *Peridinium*.

dioecious Flowering plants in which the sexes are separate; each plant is either male or female, and flowers have either stamens or pistils but not both.

dioxygenase Any oxidoreductase system in EC 1.13.11; catalyse reactions in which two oxygen atoms (from O_2) are added to a substrate.

diphtheria toxin An AB exotoxin (62 kDa) coded by β-corynephage of virulent *Corynebacterium diphtheriae* strains (that can produce a repressor of toxin production). The B-subunit binds to receptors on the surface of the target cell and facilitates the entry of the enzymically active A subunit (21 kDa) that ADP-ribosylates **elongation factor** 2, thereby halting mRNA translation.

diphtheria toxoid Diphtheria toxin treated with formaldehyde so as to destroy toxicity without altering its capacity to act as antigen. Used for active immunization against diphtheria.

dipicolinic acid *2,6-Pyridinedicarboxylic acid; DPA* Dipicolinic acid (DPA) and the Ca^{2+} complex of DPA (CaDPA) are major chemical components of bacterial spores. DPA is a chelator of metal ions.

dipipanone hydrochloride A strong opioid analgesic used for treatment of moderate to severe pain, often in conjunction with an antiemetic. See **Diconal**.

diplococcus Bacterial strain in which two spherical cells (cocci) are joined to form a pair like a dumbbell or figure-of-eight.

Diplococcus pneumoniae See *Streptococcus pneumoniae*, the formal name for this organism.

diplohaplontic Organisms that show an alternation of generations; a haploid phase (**gametophyte**) exists in the life cycle between meiosis and fertilization (e.g. higher plants, many algae and fungi); the products of meiosis are spores that develop as haploid individuals from which haploid gametes develop and fuse to form a diploid zygote. See **diplontic, haplontic**.

diploid A diploid cell has its **chromosomes** in homologous pairs, and thus has two copies of each autosomal genetic **locus**. The diploid number ($2n$) equals twice the **haploid** number, and is the characteristic number for most cells other than gametes.

diplonema A stage in **meiosis** (diplotene stage) at which the chromosomes are clearly visible as double structures. More commonly **diplotene**.

diplont Organisms in which only the zygote is diploid and the vegetative cells are haploid.

diplontic Organisms with a life cycle in which the products of meiosis behave directly as gametes, fusing to form a zygote from which the diploid, or sexually reproductive polyploid, adult organism will develop. *cf.* **haplontic, diplohaplontic**.

diplornavirus Proposed family of all double-stranded RNA viruses: considered taxonomically unsound by many virologists.

diplotene The final stage of the first prophase of meiosis. All four **chromatids** of a **tetrad** are fully visible, and homologous chromosomes start to move away from one another except at **chiasmata**.

Diptera Order of insects with one pair of wings, the second pair being modified into balancing organs, the halteres; the mouthparts are modified for sucking or piercing. The insects show complete metamorphosis in that they have larval, pupal and imaginal (imago, adult) stages. The order includes the flies and mosquitoes; best-known genera are ***Anopheles*** and ***Drosophila***.

diptericin An 82-mer **antimicrobial peptide**, originally isolated from the dipteran *Phormia terranovae* (Flesh fly).

diptericins Inducible glycine-rich antibacterial peptides (about 8 kDa) from Dipteran haemolymph.

dipyridamole A coronary vasodilator and anti-platelet drug used to treat angina pectoris and prevent blood clotting.

direct B-cells Lymphocytes responding to a small range of antigens by antibody production without any requirement for T-cells. The antigens include **flagellin** and pokeweed mitogen.

directed evolution 1. An approach to developing molecules with specific desirable properties in which sequential rounds of synthesis are informed by data on the properties of the previous set. This approach has been applied, for example, to generating novel peptide ligands and enzymes with improved characteristics. 2. The creationist view that biological systems are so complex that there must have been an external director (a.k.a. deity). Evidence is unavailable for this belief which also masquerades as 'Intelligent design'.

disaccharide Sugar formed from two monosaccharide units linked by a glycosidic bond. The trehalose types are formed from two non-reducing sugars, the maltose type from two reducing sugars.

DISC *Death-inducing signalling complex* Activation of caspase 8 is preceded by the formation of a death-inducing signalling complex by recruitment of cytoplasmic death-domain (DD)-containing proteins to the death domain of a receptor that has bound its appropriate ligand. The best-studied death-inducing ligand-receptor pairs are TNF/TNF receptor-1 (TNF-R1) and CD95L/CD95 (Fas, Apo-1). See **death-effector domain**.

disc gel Confusingly, nothing to do with shape; gels in which there is a discontinuity in pH, or gel concentration, or buffer composition.

discodermolide Anti-tumour drug (a poly-hydroxylated alkatetraene lactone) that, like **taxol**, promotes formation of stable bundles of microtubules and competes with taxol for binding to polymerized tubulin. Isolated from the marine sponge, *Discodermia dissoluta*.

discoidin A lectin, isolated from the cellular slime mould *Dictyostelium discoideum* (see **Acrasidae**), that has a binding site for carbohydrate residues related to galactose. The lectin, which consists of two distinct species (discoidins I and II), is synthesized as the cells differentiate from vegetative to aggregation phase, and was originally thought to be involved in intercellular adhesion, but discoidin I is now thought to be involved in adhesion to the substratum by a mechanism resembling that of fibronectin in animals.

Discomycetes A class of Ascomycete fungi. Includes the Lecanorales (lichen-forming fungi) and many saprophytic and mycorrhizal species, e.g. morels and truffles.

Dishevelled *Dvl* Drosophila protein (Dvl) involved in both wingless (Wg) and **Frizzled** (Fz) signalling pathways. See **wnt** and **axin**.

disintegrins Peptides found in the venoms of various snakes of the viper family that inhibit the function of some **integrins** of the β1 and β3 classes. They were first identified as inhibitors of platelet aggregation, and were subsequently shown to bind with high affinity to integrins and to block the interaction of integrins with **RGD**-containing proteins – e.g. they block the binding of the platelet integrin αIIbβ3 to **fibrinogen**. Disintegrins are effective inhibitors at molar concentrations 500–2000 times lower than short RGDX peptides. They are cysteine-rich peptides ranging from 45 to 84 amino acids in length, and almost all of them have a conserved -RGD- sequence on a β-turn, presumed to be the site that binds to integrins. The assumption is that their biological role in the venom is to inhibit blood-clotting. Found in many snake species, where they are called variously albolabrin, applagin, batroxostatin, bitis-tatin, echistatin, elegantin, flavoridin, halysin, kistrin, tri-flavin and trigramin.

disjunction mutant *Drosophila* mutant in which chro-mosomes are partitioned unequally between daughter cells at meiosis, as a result of non-disjunction.

dispase Trade name for a crude protease preparation used for disaggregating tissue in setting up primary cell cultures. Dispase gives less complete disaggregation than trypsin, but survival of cells may be better.

dispersion forces Forces of attraction between atoms or non-polar molecules that result from the formation of induced dipoles. Sometimes referred to as London disper-sion forces. Important in the **DLVO** theory of colloid floc-culation and thus in theories of cell adhesion.

disseminated intravascular coagulation Com-plication of **septic shock** in which endotoxin (from Gram-negative bacteria) induces systemic clotting of the blood, probably indirectly through the effect of endotoxin on neu-trophils. It may also develop in other situations where neu-trophils become systemically hyperactivated.

dissipative structure A system maintained far from chemical/thermodynamic equilibrium, having the potential to form ordered structures. See **Belousov–Zhabotinsky reaction**.

dissociation Any process by which a tissue is separated into single cells. Enzymic dissociation with trypsin or other peptidases is often used.

dissociation constant In a chemical equilibrium of form $A + B = AB$, the equilibrium concentrations (strictly, activities) of the reactants are related such that $[A][B]/[AB]$ = a constant, K_d, the dissociation constant, which (in this simplest case) has the dimensions of concen-tration. When A is H$^+$, this is the acid dissociation constant often designated K_a and expressed as pK_a ($-\log_{10} K_a$).

distemper virus Paramyxovirus of the genus Morbil-livirus. Commonest is the canine distemper virus that causes fever, vomiting and diarrhoea; variant that infects seals (Phocavirus) has caused significant mortality in recent years.

disulphide bond The –S–S– linkage. A linkage formed between the SH groups of two **cysteine** moieties either within or between peptide chains. Each cysteine then becomes a half-**cystine** residue. -S-S- linkages stabilize, but do not determine, secondary structure in proteins. They are easily disrupted by -SH groups in an exchange reaction, and are not present in cytosolic proteins (cytosol has a high concentration of **glutathione** that has a free -SH residue).

diterpene A fundamental class of natural products with about 5000 members known. The skeleton of every diter-pene contains 20 carbon atoms, although additional groups linked to the diterpene skeleton by an oxygen atom can increase the carbon atom count. Many have potent biolog-ical activity, and diterpene acids are known to have sub-stantial feeding deterrent and growth-inhibiting effects on a variety of insect groups and are known to inhibit a variety of fungi. Diterpenes also form the basis for biologically important compounds such as **retinol**, retinal and **phytol**.

dithioerythritol *DTE* Like **dithiothreitol,** is also referred to as Cleland's reagent and has the same properties.

dithiothreitol *Cleland's reagent* Used to protect sul-phydryl groups from oxidation during protein purification procedures or to reduce disulphides to sulphydryl groups.

diuretics Compounds (drugs) that produce diuresis. Usu-ally subdivided into **thiazide diuretics** and the more potent **loop diuretics**. See also **potassium-sparing diuretics**.

diurnal Occurring during the day or repeating on a daily basis. Use of **circadian rhythm** for the latter avoids ambiguity.

division septum The cell wall that forms between daughter cells at the end of mitosis in plant cells, or just before separation in bacteria.

dizygotic Twins arising as a result of the fertilization of two ova by two spermatozoa and thus genetically non-identical, in contrast to **monozygotic** twins.

D loop *Displacement loop* Structure formed when an addi-tional strand of DNA is taken up by a duplex so that one strand is displaced and sticks out like a D-shaped loop. Tends to happen in negatively supercoiled DNA, partic-ularly in mitochondrial DNA as an intermediate during recombination.

DLVO theory Theory of colloid flocculation advanced independently by Derjaguin and Landau and by Vervey and Overbeek, and subsequently applied to cell adhesion. There exist distances (primary and secondary minima) at which the forces of attraction exceed those of electrostatic repul-sion; an adhesion will thus be formed. For cells, there is quite good correlation between the calculated separations of primary and secondary minima and the cell separations in tight junctions (1–2 nm) and more general cell–cell appo-sitions (12–20 nm), respectively, although it is clear that other factors (particularly **cell adhesion molecules**) also play an important part.

dlx A gene family all containing a homeobox that is related to that of Distal-less (*Dll*), a gene expressed in the head and limbs of developing *Drosophila*. At least six different members, *DLX1–DLX6*. The DLX proteins are postulated to play a role in forebrain and craniofacial development.

DMARD *Disease Modifying Anti-Rheumatic Drug* Drug used for treating rheumatoid arthritis that does more than relieve symptoms. Examples include gold, penicillamine, sulphasalazine and chloroquine, though none are as effec-tive as would be desirable.

DMEM *Dulbecco Modified Eagle's Medium* Very com-monly used tissue culture medium for mammalian cells.

dmf 1. Dimethyl formamide. 2. 3′,4′-Dimethoxyflavone, an antagonist of PCB126 nuclear receptor **AhR**.

DM-GRASP A cell adhesion molecule of the immunoglob-ulin superfamily that mediates homophilic adhesion and neurite outgrowth *in vitro*.

dmp *DMP* Ambiguous abbrev. for (1) **dentin matrix protein** (DMP-1), (2) **dextromethorphan,** (3) 2,6-dimethylphenol, (4) 2′,6′-dimethylphenylalanine, (5) dimethylphosphate, (6) 4,6-diamino-2-mercaptopyrimidine, (7) disease management program, (8) diabetes management program.

DMSO *Dimethyl sulphoxide* Much used as a solvent for substances that do not dissolve easily in water and that are to be applied to cells (for example, cytochalasin B, formyl peptides), also as a cryoprotectant when freezing cells for storage. It is used clinically for the treatment of arthritis, although its efficacy is disputed.

DNA *Deoxyribonucleic acid* The genetic material of all cells and many viruses. A polymer of **nucleotides**. The monomer consists of phosphorylated 2-deoxyribose N-glycosidically linked to one of four bases: **adenine, cytosine, guanine** or **thymine**. These are linked together by 3′,5′-phosphodiester bridges. In the Watson–Crick double-helix model, two complementary strands are wound in a right-handed helix and held together by hydrogen bonds between **complementary base pairs**. The sequence of bases encodes genetic information. Three major conformations exist: **A-DNA, B-DNA** (which corresponds to the original Watson–Crick model) and **Z-DNA**.

dnaA, etc. Genes in *E. coli* that are involved in coding for replication machinery. *dnaA* and *dnaP* produce proteins involved in replication at the chromosome origin; *dnaB, C* and *D* are involved in **primosome** formation; *dnaE* codes for subunits of polymerase II; *dnaF* for ribonucleotide reductase; *dnaG* codes for primase; *dnaH, Q, X* and *Z* for components of polymerase III; *dnaI* for protein involved at the replication fork; *dnaJ* and *dnaK* products (see **dnaJ, dnaK**) are necessary to ensure survival at high temperature and are also considered essential for phage lambda replication; *dnaL* and *M* are uncharacterized; *dnaT* protein interacts with *dnaC* product, *dnaW* codes adenylate kinase.

DNA adduct DNA that has been modified by the covalent addition of another moiety. Most commonly a result of exposure to pro-oxidant species such as the hydroxyl free radical with formation of 8-hydroxyguanine.

DNA annealing The reformation of double-stranded DNA from thermally denatured DNA. The rate of reassociation depends upon the degree of repetition, and is slowest for unique sequences (this is the basis of the Cot value; see **Cot curve**).

DNAase *DNAse* See **deoxyribonuclease**.

DNA-binding domain Domain in a protein that is responsible for binding to DNA, usually with sequence-specificity. Thus many transcription factors would be expected to have such domains. Classic examples are in **zinc-finger proteins**, the **Helix–turn–helix** proteins and the **leucine zipper** proteins.

DNA-binding proteins Proteins that interact with DNA, typically to pack or modify the DNA (e.g. histones), or to regulate gene expression, transcription factors. Among those proteins that recognize specific DNA sequences, there are several characteristic conserved motifs (**DNA-binding domains**) believed to be essential for specificity.

DNA chips *DNA array; DNA microarray; gene array; biochip* A microarray of DNA or polynucleotides on a solid support (often microscope-slide sized) that can be probed with labelled RNA or DNA which will hybridize. Chips can be designed to look for expression of particular genes or for **SNPs**. Increasing miniaturization allows thousands of individual samples to be placed on a single chip.

DNA fingerprinting See **restriction fragment length polymorphism**.

DNA footprinting Technique for identifying the recognition site of DNA-binding proteins; see **footprinting**.

DNA glycosylase *DNA glycosidase* Class of enzymes involved in **DNA repair**. They recognize altered bases in DNA and catalyse their removal by cleaving the glycosidic bond between the base and the deoxyribose sugar. At least 20 such enzymes occur in cells.

DNA gyrase A type II **topoisomerase** of *E. coli* that is essential for DNA replication. This enzyme can induce or relax **supercoiling**.

DNA helicase *Unwindase* An enzyme that uses the hydrolysis of ATP to unwind the DNA helix at the **replication fork**, to allow the resulting single strands to be copied. Two molecules of ATP are required for each nucleotide pair of the duplex. Found in both prokaryotes and eukaryotes.

DNA hybridization See **hybridization**.

DNA iteron Repeated DNA sequence found near the **origin of replication** of some plasmids.

dnaJ DnaJ is a heat-shock-induced protein that interacts with the chaperone Hsp70-like **dnaK** protein and **GrpE** to disassemble a protein complex at the origins of replication of phage lambda and several plasmids. Participates actively in response to hyperosmotic and heat shock by preventing the aggregation of stress-denatured proteins and by disaggregating proteins, also in an autonomous, dnaK-independent fashion. Unfolded proteins bind initially to dnaJ, which triggers more complex interactions required for efficient folding.

dnaK Bacterial molecular chaperone of the Hsp70 family. Interacts with **DnaJ** and **GrpE** in stress responses and in refolding of misfolded proteins. DnaK is itself a weak ATPase; ATP hydrolysis by dnaK is stimulated by its interaction with another co-chaperone, dnaJ, and release of ADP is stimulated by GrpE.

DNA ladder Term often applied to the molecular weight (base pair) standards run in parallel with DNA samples on an electrophoretic gel. The 'rungs' represent different sizes of polynucleotides and calibrate the gel. See also **DNA laddering**.

DNA laddering The pattern of DNA fragmentation seen on a gel when DNA from apoptotic cells is examined. The DNA is fragmented into multiples of the 180-bp nucleosomal unit by an endonuclease, the DNase I family member, DNase gamma (endoG) (see **DFF**).

DNA library See **genomic library**.

DNA ligase Enzyme involved in DNA replication. The DNA ligase of *E. coli* seals nicks in one strand of double-stranded DNA, a reaction required for linking precursor fragments during discontinuous synthesis on the lagging strand. Nicks are breaks in the phosphodiester linkage that leave a free 3′-OH and 5′-phosphate. The ligase from phage T4 has the additional property of joining two DNA molecules having completely base-paired ends. DNA ligases are crucial in joining DNA molecules and preparing radioactive probes (by nick translation) in recombinant DNA technology.

DNA ligase D *LigD* A large polyfunctional enzyme that performs end-remodelling and end-sealing reactions during non-homologous end joining (**NHEJ**) in bacteria.

DNA markers 1. Genetic markers, for example polymorphisms such as **SNPs** and short tandem repeats that allow relationships to be established, genetic diversity to be estimated etc. **2.** Defined-length oligonucleotides used as size markers in gel electrophoresis (see **DNA ladder**).

DNA methylation Process by which methyl groups are added to certain nucleotides in genomic DNA. This affects gene expression, as methylated DNA is not easily transcribed. The degree of methylation is passed on to daughter strands at mitosis by maintenance DNA methylases. Accordingly, DNA methylation is thought to play an important developmental role in sequentially restricting the transcribable genes available to distinct cell lineages. In bacteria, methylation plays an important role in the restriction systems, as **restriction enzymes** cannot cut sequences with certain specific methylations.

DNA polymerase alpha-primase complex *pol-prim* DNA polymerase alpha (pol alpha; p180–p68) forms a four-subunit complex with DNA primase (prim-2; p58–p48) and is the only enzyme able to start DNA synthesis *de novo*. The major role of the DNA polymerase alpha–primase complex (pol-prim) is in the initiation of DNA replication at chromosomal origins and in the discontinuous synthesis of **Okazaki fragments** on the lagging strand of the replication fork.

DNA polymerases *EC 2.7.7.7* Enzymes involved in template-directed synthesis of DNA from deoxyribonucleotide triphosphates. I, II and III are known in *E. coli*; III appears to be most important in genome replication, and I is important for its ability to edit out unpaired bases at the end of growing strands. Animal cells have α, β- and γ-polymerases, with α-apparently responsible for replication of nuclear DNA, and γ for replication of mitochondrial DNA. All these function with a DNA strand as template. Retroviruses possess a unique DNA polymerase (**reverse transcriptase**) that uses an RNA template.

DNA primase *RNA primase; EC 2.7.7.6* Enzymes (RNA polymerases) that catalyse the synthesis of short (∼10 bases) RNA primers on single-stranded (ss) DNA templates that are used by DNA polymerase to initiate the synthesis of **Okazaki fragments** on the lagging strand. Bacterial primases have three functional domains in the protein: an N-terminal 12-kDa fragment contains a zinc-binding motif; a central fragment of 37 kDa has conserved sequence motifs that are characteristic of primases, including the so-called 'RNA polymerase (RNAP)-basic' motif; and a C-terminal domain of ∼150 residues interacts with the replicative helicase, DnaB, at the replication fork. Eukaryotic DNA primase is a heterodimer of large (p60) and small (p50) subunits that show little homology with prokaryotic primases.

DNA probes A short sequence of DNA that has been labelled isotopically or chemically and that can be used to detect a complementary nucleotide sequence. A diversity of labelling methods has been developed, and DNA probes are increasingly being used for example in detection and identification of pathogens and in testing for genetic abnormalities in chromosomal DNA.

DNA profiling Forensic tool to compare samples of DNA. Generally preferred to the term 'DNA fingerprinting'.

DNA rearrangement Wholesale movement of sequences from one position to another in DNA, such as occur somatically, for example in the generation of antibody diversity.

DNA renaturation See **DNA annealing**.

DNA repair Enzymic correction of errors in DNA structure and sequence that protects genetic information against environmental damage and replication errors.

DNA replication The process whereby a copy of a DNA molecule is made, and thus the genetic information it contains is duplicated. The parental double-stranded DNA molecule is replicated semiconservatively, i.e. each copy contains one of the original strands paired with a newly synthesized strand that is complementary in terms of AT and GC base-pairing. Though in this sense conceptually simple, it is mechanistically a complex process involving a number of enzymes.

DNase 1 *Deoxyribonuclease* A 33-kDa actin-binding protein which also hydrolyses DNA. Binds to G-actin and to the pointed end of actin filaments with high affinity. Has been a useful experimental tool, but the physiological relevance of the actin binding is unclear.

DNA sequence analysis Determination of the nucleotide sequence of a length of DNA. Typically, this is performed by cloning the DNA of interest, so that enough can be prepared to allow the sequence to be determined, usually by the Sanger **dideoxy sequencing** method or the **Maxam–Gilbert method**. The resulting reactions are then run on a large sequencing gel, capable of resolving single nucleotide differences in chain length. Recently, **PCR**-based methods have obviated the need to clone the DNA under some conditions, and automated DNA sequencing using column chromatographic separation has become widely available.

DNA synthesis The linking together of nucleotides (as deoxyribonucleotide triphosphates) to form DNA. *In vivo*, most synthesis is **DNA replication**, but incorporation of precursors also occurs in repair. In the special case of retroviruses, DNA synthesis is directed by an RNA template (see **reverse transcriptase**).

DNA topoisomerase *EC.5.99.1.2* An enzyme capable of altering the degree of supercoiling of double-stranded DNA molecules. Various topoisomerases can increase or relax supercoiling, convert single-stranded rings to intertwined double-stranded rings, tie and untie knots in single-stranded and duplex rings, catenate and decatenate duplex rings. Topoisomerase II of *E. coli* = gyrase.

DNA transfection A technique originally developed to allow viral infection of animal cells by uptake of purified viral DNA rather than by intact virus particles. The term, a hybrid between transformation and infection, is now generally used to describe applications of same methodology to introduction of other kinds of genes or gene fragments into cells as DNA, such as activated oncogenes from tumours into tissue culture cells.

DNA tumour virus Virus with DNA genome that can cause tumours in animals. Examples are **Papovaviridae**, **Adenoviridae** and **Epstein–Barr virus**.

DNA vaccine Vaccine in which the active principle is a DNA sequence that will be transiently expressed in host cells and generate antigens to stimulate an immune response. Such vaccines are considered to have great potential, especially in diseases for which it has been difficult to develop conventional vaccines (e.g. AIDS, malaria).

DNA virus A virus in which the nucleic acid is double- or single-stranded DNA (rather than RNA). Major groups of double-stranded DNA viruses are papovaviruses, adenoviruses, herpesviruses, large bacteriophages, and poxviruses: of single-stranded, parvoviruses and coliphages φX174 and M13.

DNS *5-(Dimethylamino)naphth-1-ylsulfonyl* Abbrev. for dansyl. See **dansyl chloride**.

docetaxel *Taxotere™* **Taxane** extracted from the English yew (*Taxus baccata*), slightly more potent than paclitaxel (**taxol**).

dockerin domain *Doc* Domain that binds the cellulose-degrading enzymes of the **cellulosome** to **cohesin** domains in the proteins of the bacterial cell wall. The structure contains a conserved fold of 42 residues and 2 calcium-binding sites with sequence similarity to the **EF-hand** motif.

docking protein See **signal recognition particle receptor**.

docosahexaenoic acid Any straight-chain fatty acid with 22 carbon atoms and 6 double bonds. The all-Z isomer is found in fish oils. It is a major omega-3 fatty acid in human brain, synapses, retina and other neural tissues. See **protectin D1**.

dodo A **peptidyl prolyl isomerase** that facilitates the degradation of the transcription factor CF2, which regulates expression of the **rhomboid** gene in *Drosophila* follicle cells. This chain of events is required to establish the dorsal/ventral polarity of the developing oocyte. Degradation is probably facilitated by isomerizing the prolyl peptide bond after MAPK-catalysed phosphorylation of the protein.

dok proteins A family of proteins that are 'downstream of tyrosine kinases' and are docking molecules characterized by an amino-terminal pleckstrin homology (PH) domain, a central putative phosphotyrosine-binding (PTB) domain and numerous potential sites of tyrosine phosphorylation.

dolastatins Family of peptides isolated from the marine nudibranch mollusc *Dolabella auricularia* (sea hare). Dolastatin 10 is a potent antimitotic pentapeptide that inhibits microtubule assembly and is being tested in cancer therapy. Variants are being generated. See **phomopsin A**. Dolastatin 11, a depsipeptide, binds to actin and stabilizes F-actin *in vitro*, like **phalloidin** and **jasplakinolide** although acting at a different site. See **doliculide**.

dolichol Terpenoids with 13–24 isoprene units and a terminal phosphorylated hydroxyl group. Function as transmembrane carriers for glycosyl units in the biosynthesis of glycoproteins and glycolipids.

doliculide An actin-binding macrocyclic depsipeptide that stimulates actin assembly, extracted from the sea hare (*Dolabella auricularia*); competes with **phalloidin** for binding and has the same effects as **jasplakinolide**.

domain Used to describe a part of a molecule or structure that shares common physicochemical features, e.g. hydrophobic, polar, globular, α-helical domains or properties, e.g. DNA-binding domain, ATP-binding domain.

dominant allele Any allele that has an effect on the phenotype when present as single copy (when heterozygous).

In change of function mutations the product has novel function; in **dominant negative** mutations the product forms inactive complexes with the product of the normal allele and renders it non-functional.

dominant negative A mutation which is capable of exerting an effect even when only one copy is present, as in a **heterozygote**. Usually explained as a mutation that disrupts one subunit of a multimeric protein, thus making the whole complex dysfunctional. Alternatively, the mutated protein may compete with the normal protein produced by the other allele so that the overall activity is reduced below a critical threshold level and function is abnormal. The latter explanation only holds if the normal product of two genes is necessary for normal function (haploinsufficiency), although experimental overexpression of an inactive form may reduce function by more than 50%.

domoic acid *DA* Toxin, a tricarboxylic acid similar to the **glutamate receptor** agonist **kainic acid**, originally isolated from the macroscopic red alga *Chondria armata*, used as an antihelminthic in a traditional medicine. Acts preferentially upon a subclass of ionotropic glutamate receptors found in nervous tissue. Has been identified as the cause of amnesic shellfish poisoning, the source of the toxin being the diatom *Pseudo-nitzschia* (previously *Nitzschia*) *pungens* forma *multiseries*.

donepezil hydrochloride *TN Aricept* Acetylcholine esterase inhibitor used to treat symptoms of mild to moderate dementia.

Donnan equilibrium An equilibrium established between a charged, immobile colloid (such as clay, ion exchange resin or cytoplasm) and a solution of electrolyte. Characteristics: ions of like charge to the colloid tend to be excluded; ions of opposite charge tend to be attracted; the colloid compartment is electrically polarized relative to the solution in the same direction as the colloid charges (a Donnan potential); and the osmotic pressure is higher in the colloid compartment.

donor splice junction The junction between an **exon** and an **intron** at the 5′ end of the intron. When the intron is removed during **processing** of **hnRNA,** the donor junction is spliced to the acceptor junction at the 3′ end of the intron.

DOPA *L-DOPA; levodopa; 3-hydroxytyrosine* Precursor of the neurotransmitter dopamine, made from L-tyrosine by tyrosine 3-mono-oxygenase and used as a treatment for **Parkinsonism**.

dopamine A **catecholamine neurotransmitter** and **hormone** (153 Da), formed by decarboxylation of dihydroxyphenylalanine (DOPA). A precursor of **adrenaline** and **noradrenaline**. See **dopamine receptors**.

dopamine receptors Family of **G-protein-coupled** receptors for **dopamine**. Fall into two classes, D1-like (D1, D3, D4; activators of adenylyl cyclase) and D2-like (D2 and D5; inhibitors of adenylyl cyclase). The prototypic ligand for the D1-like receptors is the benzazepine SCH23390; D2-like receptors efficiently bind spiperone and haloperidol but only weakly bind SCH23390. Most antipsychotic drugs are dopamine receptor antagonists, and most neuroleptics were developed as D2-receptor antagonists.

dopaminergic neurons Neurons for which **dopamine** is the neurotransmitter. See **dopamine receptors**.

doppel *Dpl* Prion-like protein related to the **prion** protein (PrP). Doppel is encoded by the gene locus, *PRND*, located on the same chromosomal region of the PrP(C) coding gene. May oppose the neuroprotective effects of PrP and be neurotoxic.

dormin See **abscisic acid**.

dorsal dl *Drosophila* polarity gene; homologue of the *rel* proto-oncogene. See **tube**, **pelle** and **toll**.

dorsal horn Region in the grey matter of the spinal cord, consisting of five zones (laminae I–V) where nociceptive information begins to be processed in the central nervous system. The dorsal (posterior) horn receives sensory input from the skin, from striated muscles or joints, or from blood vessels and internal organs.

dorsal root ganglion Nodule on a dorsal root of the spinal cord that contains cell bodies of afferent spinal nerve neurons leading into the dorsal part of the cord. Dorsal root ganglia from chick embryos are a classic source of neurites for cell culture.

dorsalin-1 Protein that stimulates **neural crest** differentiation, **neural crest** growth, bone growth and wound healing.

dosage compensation Genetic mechanisms that allow genes to be expressed at a similar level irrespective of the number of copies at which they are present. Usually invoked for genes that lie on sex chromosomes and are thus present in different copy numbers in males and females.

dose–response curve Graph of the relation between dose (concentration of substance introduced into the system) and the effect (enzyme activity, membrane potential, mortality, etc.) that is being measured. Standard dose–response curves are similar to receptor-binding curves (sigmoidal), but exceptions are found when there is cooperativity or dual-response modes. A dose–response curve with a standard slope has a **Hill coefficient** of 1.0

dot blot Method for detecting a specific protein or message. A spot of solution is dotted onto nitrocellulose paper, a specific antibody or probe is allowed to bind and the presence of bound antibody/probe then shown by using a peroxidase-coupled second antibody, as in **Western blot** or by other visualization methods. See also **slot blot**.

double helix Conformation of a DNA molecule, like a ladder twisted into a helix.

double layer The zone adjacent to a charged particle in which the potential falls effectively to zero. An excess or deficiency of electrons on the surface (charge; not to be confused with the transmembrane potential) leads to an equivalent excess of ions of the opposite charge in the surrounding fluid. For most cells (which have negative charges) there will be an excess of cations immediately adjacent to the plasma membrane, and at physiological ionic strength the double layer is likely to be around 2–3 nm thick.

double minute Gene encoding a **p53** inhibitor; murine gene is *Mdm2*.

double minute chromosome Small, paired extra-chromosomal bodies comprising circular DNA, associated with many tumours where there may be multiple copies.

double mutant Organism in which there are two mutations, often necessary if the effect is to be seen because two parallel systems need to be affected, although in some cases one mutation can mask (suppress) the effect of the other.

double-strand break Serious form of damage to DNA in which both strands are cleaved.

double-stranded RNA *dsRNA* Form of RNA in which there is base-pairing between complementary regions producing regions of duplex structure similar to that of DNA. Found as the genetic material in some viruses. dsRNA introduced into cells can cause **RNA interference**, the sequence-specific degradation of mRNA.

doublecortin *DCX* A developmentally expressed neuronal microtubule-associated protein. Mutations in the human doublecortin gene result in abnormal neuronal migration, epilepsy and mental retardation. A family of doublecortin-like proteins (including the retinitis pigmentosum 1 (*RP1*) gene product and doublecortin-like kinase 1 (Dclk)) has been identified, all with doublecortin-like (DCX) domains (usually in tandem). The domain is a microtubule-binding module and is involved in protein–protein interactions. Phylogenetically ancient with homologues in invertebrates and unicellular organisms. Doublecortin is widely used as a marker for newly generated neurons.

doublet microtubules Microtubules of the axoneme. The outer nine sets are often referred to as doublet microtubules, although only one (the A tubule) is complete and has 13 protofilaments. The B-tubule has only 10 or 11 protofilaments and shares the remainder with the A tubule. A and B tubules differ in their stability and in the other proteins attached periodically to them; it is the **dynein** affixed to the A tubule attaching and detaching from the B tubule of the adjacent doublet that generates sliding movement in the **axoneme**.

doubling time The time taken for a cell to complete the cell cycle.

Dounce homogenizer An apparatus, usually made of glass, with a tightly fitting plunger (pestle) in a glass tube. Tissue is homogenized by shear-forces generated by rotation and gentle reciprocation of the plunger; by varying the clearance between pestle and wall the particle size in the homogenate can be altered.

Down syndrome critical region *DSCR1* Region of chromosome 21. DSCR1 encodes **calcipressin** 1.

Down's syndrome Formerly referred to as mongolism, most frequently a consequence of trisomy of chromosome 21. Common (1 in 700 live births); incidence increases with maternal age. The cause is usually non-disjunction at meiosis but occasionally a translocation of fused chromosomes 21 and 14. See **Down syndrome critical region**.

downregulation Reduction in the responsiveness of a cell to a stimulus following first exposure, often by a reduction in the number of receptors expressed on the surface (as a consequence of reduced recycling). Tends to be used imprecisely.

downstream 1. Portions of DNA or RNA that are more remote from the initiation sites and that will therefore be translated or transcribed later. **2.** Shorthand term for things that happen at a late stage in a sequence of reactions, for example in a signalling cascade.

doxapram hydrochloride Centrally acting respiratory stimulant, occasionally used in the treatment of severe respiratory failure.

doxazosin An **alpha blocker** used to treat hypertension and prostatic hyperplasia.

doxorubicin Cytotoxic antibiotic from *Streptomyces peucetius*. Blocks **topoisomerase** and **reverse transcriptase** by intercalating into the DNA. Has been used in clinical oncology.

doxycyclin *Doxycycline* Broad-spectrum **tetracycline antibiotic** used for treatment of chronic bronchitis, brucellosis, chlamydial, rickettsial and mycoplasma infections. Also used prophylactically for malaria in regions where the parasite is resistant to chloroquine and/or pyrimethamine-sulfadoxine.

DP-1, DP-2 1. G-protein-coupled receptors for prostaglandin D(2) with opposite effects. DP1 activation tends to ameliorate the pathology in asthma; DP2 is preferentially expressed on type 2 lymphocytes, eosinophils and basophils, and is thought to be important in the promotion of Th2-related inflammation. **2.** Cell cycle-regulating transcription factors (DP-1 and DP-2) exist in humans, and there are additional isoforms, DP-1alpha, 278 aa; DP-1beta, 357aa. Form a heterodimer with **E2F** and regulate progression through the cycle. **3.** Gene of pneumococcal bacteriophage (Dp-1) encoding **Pal amidase**.

DPIP *DCPIP* The dye dichlorophenol-indophenol (DCPIP), often used as an indicator for the activity of the electron transport system during the light-dependent reactions of photosynthesis, becoming colourless when reduced.

dpm Abbrev. for (1) disintegrations per minute, a measure of radioactivity, (2) defects per million, in manufacturing processes, (3) a range of other non-biological things.

Dpp Protein product of the *Drosophila* gene *decapentaplegic*, related to TGF.

DPPC Dipalmitoyl-phosphatidylcholine.

DRAK See **DAP kinase**.

DRB Multiallelic locus in the class II **MHC** DR region encoding β-chains. Over a hundred alleles have been reported at the DRB locus in humans, which is more polymorphic than most of these loci. Other loci are DQ and DP.

drebin A developmentally regulated actin-binding brain protein which in the chicken has characteristic changes in expression related to developmental stage. Contains a single **ADF-H domain**. Binds F-actin. Same class as yeast ABP-1. Not in current usage.

Drickamer motif Either of the two highly conserved patterns of invariant amino acids found in the carbohydrate-recognition domain of C-type and S-type lectins, as described by Drickamer. Usage obsolete, although motifs still recognized.

drosha The RNase III enzyme that is the major nuclease involved in initiation of **microRNA** (miRNA) processing in the nucleus.

drosocin A cationic 19 amino acid **antimicrobial peptide** (GKPRPYSPRPTSHPRPIRV) secreted by *Drosophila* in response to bacterial infection. The peptide is glycosylated at Thr11, which appears to be important for its potent antimicrobial activity. Has sequence homology with **apidaecin** Ib.

drosomycin **Antimicrobial peptide**, 44 residues with four intramolecular disulphide bridges, product of a **toll**-dependent immunity gene in *Drosophila* and other Diptera. Has antifungal activity.

Drosophila A genus of small American flies, **Diptera**. The best-known species is *D. melanogaster*, often called the fruit fly, but more correctly termed the vinegar fly. First investigated by T.H. Morgan and his group, it has been extensively used in genetic studies. More recently it has been used for studies of embryonic development.

drosulphakinins *Drosophila* homologues of the **gastrin** family of peptide hormones.

drug A substance that has a physiological action and is used for the treatment of disease or the alleviation of pain. Many are also used for recreation and, depending upon societal convention, this may be classed as drug abuse or acceptable behaviour (e.g. alcohol in many cultures). Prolonged use can often lead to progressive addiction.

drug delivery The process of getting a drug into the appropriate compartment of the body – orally, if the drug can be absorbed through the gut, or by inhalation, injection or by transdermal diffusion. If the target is superficial then topical application may be suitable, and in this case absorption into deeper tissues may be undesirable.

druse crystal *Drusen* Crystals found in plant tissue, often composed of calcium oxalate, formed within the central vacuoles of parenchyma cells (idioblasts) in the cortex region just outside the phloem.

drusen See **druse crystal**.

DSB Abbrev. for (1) **double-strand breaks,** (2) 2,6-di-sec-butyl phenol, an anaesthetic, (3) 3-*O*-(3′,3′-dimethyl-succinyl)-betulinic acid, a small molecule inhibitor of the proteolytic cleavage of HIV Gag-protein, (4) the disulphide bond formation protein gene (*dsb*).

DSE See **serum response element**.

D-TACC Drosophila *TACC Drosophila* TACC (D-TACC) is concentrated at centrosomes, interacts with microtubules and is essential for normal spindle function in the early embryo. The C-terminal region of D-TACC interacts, possibly indirectly, with microtubules and is related to the mammalian transforming, acidic, coiled-coil-containing (**TACC**) family of proteins.

DTH Abbrev. for delayed-type **hypersensitivity**.

DTLET *Tyr-D-Thr-Gly-Phe-Leu-Thr; deltakephalin* Synthetic agonist of the G-protein-coupled delta **opioid receptor**. Inhibits the release of gonadotropin-releasing hormone (GnRH) from hypothalamic fragments containing the arcuate nucleus and the median eminence.

DTNB *Ellman's reagent* Abbrev. for: 5,5′-dithiobis-(2-nitrobenzoic acid), a cell impermeable dithiol-oxidizing agent. Reacts with sulphydryl group on proteins and releases 5-sulphido-2-nitrobenzoic acid which absorbs strongly at 412 nm.

dual recognition hypothesis An outmoded hypothesis that is known to be incorrect now that the structure of the T-cell receptor is known. The proposal was that viral (and some chemical) antigens were recognized in association with **histocompatibility antigens** by separate receptors on the T-cell. The generation of cytotoxic T-cells was by association with class I MHC antigens, of T-helper cells by association with class II MHC antigens. See **altered-self hypothesis**.

Duchenne muscular dystrophy A sex-linked hereditary disease confined to young males and to females with **Turner's syndrome**. It is characterized by degeneration and **necrosis** of skeletal muscle fibres, which are replaced by fat and fibrous tissue. The incidence of this disorder is about 1 in 4000 male births, and of these one-third are estimated to be new mutational events. See **dystrophin**.

ductin 1. Name for the 16-kDa transmembrane subunit of the **V-type ATPase**, reflecting a (controversial) view that it may be a multifunctional transmembrane pore protein, also involved (for example) in gap junction formation. See also **connexin. 2.** See **CRP-ductin**.

dudulin Mouse orthologue of **STEAP**.

Duffy **Blood group system**. Single gene locus.

dunce dnc *Drosophila* mutant that is deficient in short-term memory. Gene codes for cAMP-phosphodiesterase and mutation leads to elevated cAMP levels that in turn particularly affect the delayed rectifier potassium currents in neurons of brain centres associated with acquisition and retention. The effect of the mutation does also alter nerve terminal growth and synaptic plasticity. Comparable behavioural defects are associated with **rutabaga** (*rut*).

Dunn chamber *Dunn chemotaxis chamber* A special circular cell-counting chamber slide. Cells are cultured on coverslips that are then inverted onto the slide, and a temporally stable gradient of potential chemoattractant is established across the annulus that separates inner and outer circular wells. Allows directed behaviour of slow-moving cells to be observed with time-lapse methodology. A more sophisticated version of the **orientation chamber** originally developed by Zigmond.

duplicon A duplicated section of DNA. Duplicated segments of genomic DNA can allow evolution of gene function and can be associated with chromosome rearrangements. Increasing numbers of such segmental duplications are being recognized as a result of genome-sequencing studies.

Dupuytren's contracture Fibroma-like lesion of the palm of the hand that causes flexion contracture. Heritable and commoner in men. Never metastasizes.

dura mater Outermost of the three meningeal membranes covering the brain and spinal cord.

durotaxis The apparent preference of moving cells for a stiff substratum has been dubbed 'durotaxis', although it is a consequence of the physical properties of the substratum rather than being based on gradient perception.

dust cells Rare and trivial name for alveolar macrophages.

dyad Generally, any two entities regarded as some kind of unit. More specifically, half of a tetrad group of chromosomes that moves to one pole at the first meiotic division.

dyad symmetry element See **serum response element**.

dye coupling Measure of intercellular communication, usually through **gap junctions**. If a fluorescent dye (e.g. **lucifer yellow**) injected into one cell is seen to pass into a neighbouring cell, the presence of junctions at least able to pass solutes of that size can be inferred between the two. See also **electrical coupling**.

dynacortin Actin-binding protein (80-kDa dimer) from *Dictyostelium* that cross-links actin filaments into parallel arrays. Is excluded from the cleavage furrow (see **cortexillin**).

dynactin Dynein activator complex that stimulates vesicle transport. Includes dynactin (160 kDa) and polypeptides of 62, 50, 45, 37 and 32, the 45-kDa (possibly **actin-RPV**) being the most abundant. All the subunits co-sediment with antibody to dynactin 160, and the complex behaves as a stable 20S multiprotein assembly. See **centractin**.

dynamic equilibrium An equilibrium state in which forward and reverse reaction rates are exactly balanced: changes in either rate will shift the relative concentrations of reactants or products. Most methods for analysing rate constants, etc., assume that equilibrium has been reached, but the dynamic nature of this position means that this may be a transient situation and may change, as is desirable in systems that must adapt to altered circumstances.

dynamin A protein isolated from microtubule preparations and shown to cause ATP-mediated microtubule sliding towards the plus ends. A GTP-binding protein with classical G-protein motifs and with very high homology to the **Mx protein** involved in interferon-induced virus resistance. There are tissue-specific and developmentally regulated forms of dynamin in *Drosophila*. Associated with endocytic sorting of proteins.

dynamitin *p50; Jnm1p* Subunit of **dynactin**. Overexpression will interfere with the **dynein**–dynactin interaction involved in vesicle transport, and this has been used experimentally as a method for demonstrating the importance of such systems. **Immunophilins** link to dynein indirectly via dynamitin.

dynein Large multimeric protein (600–800 kDa) with ATPase activity; constitutes the side arms of the outer microtubule doublets in the ciliary axoneme and is responsible for the sliding. Probably (together with **kinesin**) involved in microtubule-associated movement elsewhere. Cytoplasmic dynein is MAP-Ic.

dynorphin Opiate peptide derived from the hypothalamic precursor pro-dynorphin (that also contains the neoendorphin sequences). Contains the pentapeptide leu-**enkephalin** sequence. Its binding affinity is greater for the κ-type than for the μ-type **opioid receptor**.

dyrks A conserved family of protein kinases that autophosphorylate a tyrosine residue in their activation loop by an intramolecular mechanism and phosphorylate exogenous substrates on serine/threonine residues. Also have nuclear targeting signal, putative leucine zipper and a very conserved 13-histidine repeat sequence. Rat gene *dyrk* is a homologue of *Drosophila* minibrain (*mnb*), a gene involved in postembryonic neurogenesis. Human homologue maps to Chromosome 21 and may be involved in pathogenesis of certain phenotypes of **Down's syndrome**.

dyschromatosis symmetrica hereditaria *DSH* An autosomal dominant disorder with high penetrance, characterized by the presence of hyperpigmented and hypopigmented macules mostly on the dorsal aspects of the extremities. Genetic studies have identified mutations in the **ADAR** gene.

dyscrasia Illness as a result of abnormal material in the blood.

dysferlin Protein that associates with the plasma membrane in primary fibroblasts, skeletal and cardiac muscles. It is the product of the Limb Girdle Muscular Dystrophy type 2 locus, and has been shown to be necessary for efficient, calcium-sensitive, membrane resealing. Mutations in the dysferlin gene (*DYSF*) underlie two main muscle diseases: Limb Girdle Muscular Dystrophy (LGMD) 2B and Miyoshi myopathy (MM). A related molecule, **myoferlin**, is found in myoblasts.

dysgenic System of breeding or selection that is genetically deleterious or disadvantageous.

dyskinesia Impaired ability to control voluntary movement, a common side effect of prolonged **levodopa** use and also a side effect of some antipsychotic drugs.

dyskinetoplasty Absence of an organized **kinetoplast** (and of kinetoplast DNA) from a flagellate protozoan cell.

dyspareunia Pain during sexual intercourse.

dysplasia Literally, 'wrong growth'. Usually used to denote early stage of carcinogenesis, marked by abnormal epithelial morphology.

dystrobrevin-binding protein 1 *DTNBP1; dysbindin* The gene encoding this protein, located on chromosome 6p, has been suggested as a potential susceptibility gene for schizophrenia. Dysbindin is widely expressed in the human brain and binds (through **dystrobrevin**) to the dystrophin-associated protein complex (DPC), which appears to be involved in signal transduction pathways. Has also been shown to be a component of **BLOC-1**. See **Hermansky–Pudlak syndrome**.

dystrobrevins A family of widely expressed **dystrophin-associated proteins** that comprises alpha and beta isoforms

and displays significant sequence homology with several protein-binding domains of the dystrophin C-terminal region. Dystrobrevin interacts with **kinesin** and may play a role in the transport and targeting of components of the dystrophin-associated protein complex to specific sites in the cell.

dystroglycan *156DAG* Complex composed of two proteins, α- and β-dystroglycans (formerly known as 156DAG and 43DAG/A3a, respectively), derived from a single precursor by proteolytic cleavage. β-Dystroglycan is a transmembrane protein that associates with **dystrophin** in the cytoplasm and α-dystroglycan, an extracellular glycoprotein, on the exterior face. α-dystroglycan binds to **dystrophin**, thus linking actin through dystrophin and β-dystroglycan to the extracellular matrix. Also associates with **dystrophin**. Dystrophin deficiency leads to a deficiency in the appearance of these proteins on the sarcolemma, even though they are not themselves defective.

dystrophic epidermolysis bullosa *DEB* A group of diseases associated in all cases with mutations of the gene coding for type VII collagen. See **epidermolysis bullosa**.

dystrophin Protein (400 kDa) from skeletal muscle that is missing in **Duchenne muscular dystrophy**. Its exact role is not yet clear, though it seems to be associated with the cytoplasmic face of the sarcolemma and T-tubules and may form part of the membrane cytoskeleton. There are sequence homologies with non-muscle α-actinin and with spectrin.

dystrophin-associated protein complex *DPC* A group of proteins (**syntrophin**, **dystrobrevin** and **dystroglycan** isoforms) believed to provide a molecular link between the actin cytoskeleton and the extracellular matrix in muscle cells, thereby sustaining sarcolemmal integrity during muscle contraction. Some of these functions are mediated by the **sarcoglycan** subcomplex. Also important for the clustering and anchoring of signalling proteins and ion and water channels.

E

E1A Oncogene from an **Adenovirus**. Product inhibits the transcriptional activation by both p73alpha and p73beta. Interacts with the *Rb* **tumour-suppressor** gene product, in a manner similar to **SV40** large T-antigen.

E1B Oncogene from an **Adenovirus**. Interacts with the *p53* tumour-suppressor gene product and is anti-apoptotic. The 19-kDa protein encoded within the adenovirus E1B gene is essential for transformation by adenovirus and for proper regulation of viral early gene transcription.

E1E2-ATPase A class of plasma membrane-localized ion-motive pumps that includes **sodium–potassium ATPase**. The phosphoenzyme has two conformational states, E1 and E2, and ion exchange is inhibited by oligomycin, orthovanadate and **ouabain**.

E1 enzymes *Ubiquitin-activating enzymes* Enzymes responsible for activating **ubiquitin** (UB) as the first step in ubiquitinylation. The E1 enzyme hydrolyses ATP and adenylates the C-terminus of UB, and then forms a thioester bond between the C-terminus of UB and the active site cysteine of E1. The thioester-linked UB is then transferred to the UB-conjugating enzyme, E2, in an ATP-dependent reaction.

E2 enzymes *Ubiquitin-conjugating enzymes* Family of different E2 enzymes (>30 in humans), which are broadly grouped into four classes, all of which have a core catalytic domain. E2 enzymes receive **ubiquitin** from **E1 enzymes** and in turn pass it to **E3 ligases**.

E2F Family of transcription factors originally identified through their role in transcriptional activation of the adenovirus E2 promoter, subsequently found to bind to promoters for various genes involved in the G1 and S phases of the cell cycle. E2F forms heterodimers with **DP-1** to produce an active transcriptional complex. E2F family members are regulated by interaction with **retinoblastoma** (Rb) proteins.

E3 ligase *E3 ubiquitin ligase* Enzyme that catalyses transfer of **ubiquitin** from the E2 enzyme to a lysine residue on a substrate protein. There are at least four classes of E3 ligases: **HECT**-type, RING-type, PHD-type, and U-box containing.

E5 Oncogene from a **papillomavirus**. Encodes a small protein that binds and blocks the 16-kDa **proteolipid** of the **V-type ATPase**, producing abnormal intravesicular processing of growth factor receptors. Is also thought to modulate growth factor receptor function, leading to a stimulation of growth factor signal transduction pathways.

E6 Oncogene from a **papillomavirus**. Encodes a 16-kDa protein. E6 and E7 proteins drive cell proliferation through their association with PDZ domain proteins and Rb (retinoblastoma), and contribute to neoplastic progression, whereas E6-mediated p53 degradation prevents the normal repair of chance mutations in the cellular genome.

E7 Oncogene from a **papillomavirus**. Interacts with the *Rb* **tumour-suppressor** gene product, in a manner similar to **SV40** large T-antigen. See **E6**.

E-64 Trans-*epoxysuccinyl-L-leucylamido-(4-guanidino)butane* A broad-spectrum cysteine peptidase inhibitor.

EA-rosettes See **E-rosettes**. A test for the presence of Fc receptors.

EAA See **excitatory amino acid**.

EaA cells Insect cell line derived from haemocytes of the salt marsh caterpillar *Estigmene acrea*. An alternative line for baculovirus expression. See **Sf9 cells**.

EAC-rosettes Rosettes (see **E-rosettes** formed from erythrocytes (E) coated with antibody (A) and complement (C). A test for C3b or C3bi receptors (**CR1** or **CR3**). The rosettes form more easily than E or EA-rosettes.

Eadie–Hofstee plot Linear transformation of enzyme kinetic data in which the velocity of reaction (v) is plotted on the ordinate, v/S on the abscissa, S being the initial substrate concentration. The intercept on the ordinate is V_{max}, the slope is $-K_m$. Preferable to the **Lineweaver–Burke plot**.

EAP-300 A developmentally regulated embryonal protein (300 kDa) that has been shown previously to be expressed by radial glia in various regions of the CNS. Has significant homology with **paranemin**. EAP-300 is a phosphoprotein, and is fatty acylated. May not always be associated with intermediate filaments.

early antigens Virus-coded cell surface antigens that appear soon after the infection of a cell by virus, but before virus replication has begun. See **early gene**.

early gene Genes that are expressed soon after viral infection of a host cell.

early region Part of a viral genome in which **early genes**, that are transcribed and expressed early during infection of a cell, are clustered.

EAST 1. EGFR-Associated protein with SH3 and TAM domains. An EGF receptor substrate that binds actin. Enriched at focal adhesions and in some cells (MDCK) at cell–cell contacts. 2. *Drosophila* EAST protein associates with an interior non-chromosomal compartment of the interphase nucleus and has a nucleoskeletal role.

EB1 A member of the RP/EB family of proteins that binds to plus end of microtubules in interphase cells and during mitosis is associated with the centrosomes and spindle microtubules. Also binds to the APC protein (see adenomatous **polyposis coli**) and associates with components of the **dynactin** complex and the intermediate chain of cytoplasmic **dynein**.

Ebola virus Filovirus that causes severe fever and bleeding. Outbreaks have so far been mostly confined to Africa.

E box The CANNTG sequence motif, found in numerous promoters and enhancers; the binding site for the basic-helix-loop-helix transcription factors.

EBP50 *ERM-binding phosphoprotein 50 kDa; Na^+/H^+ exchanger regulatory factor* Associates with **ezrin**, and with the **cystic fibrosis transmembrane conductance regulator** (CFTR) linking them to the cortical actin cytoskeleton. EBP50 has two PDZ domains; CFTR binds with high affinity to the first and **Yes-associated protein** (YAP65) binds with high affinity to the second.

EBV See **Epstein–Barr virus**.

EC$_{50}$ Effective concentration; concentration at which the substance concerned produces a specified effect in 50% of the organisms treated.

EC cells 1. **Embryonal carcinoma cells**. 2. Endocrine cells.

E classification Classification of enzymes based on the recommendations of the Committee on Enzyme Nomenclature of the International Union of Biochemistry. The first number indicates the broad type of enzyme (1 = oxidoreductase; 2 = transferase; 3 = hydrolase; 4 = lyase; 5 = isomerase; 6 = ligase (synthetase)). The second and third numbers indicate subsidiary groupings, and the last number, which is unique, is assigned arbitrarily in numerical order by the Committee.

EC number See **E classification** for enzymes

eccrine Type of gland in which the secretory product is excreted from the cells.

ecdysis Moulting of the outer layers of the integument, as in arthropods. Regulated by **ecdysone** and **juvenile hormone**.

ecdysone Family of **steroid hormones** found in insects, crustaceans and plants. In insects, α-ecdysone stimulates moulting (ecdysis). The steadily maturing character of the moults is affected by steadily decreasing levels of **juvenile hormone**. β-Ecdysone (ecdysterone) has a slightly different structure and is also found widely. Phytoecdysones are synthesized by some plants.

ECG *Electrocardiograph; electrocardiogram* A recording of the electrical activity of the heart.

Echinacea North American perennial plants of the genus *Echinacea* (e.g. purple coneflower). Herbal remedies prepared from these plants are extensively used and are claimed to boost the immune system.

echinocandins A class of antifungal agents, large lipopeptide molecules, that act on the fungal cell wall by way of non-competitive inhibition of the synthesis of 1,3-beta-glucans. Examples include Caspofungin, Micafungin. Used in clinical treatment of candidiasis and aspergillosis.

echinocytes Erythrocytes that have shrunk (in hypertonic medium) so that the surface is spiky.

Echinodermata Phylum of exclusively marine animals. The phylum is divided into five classes: the Asteroidea (starfish), the Echinoidea (sea urchins), the Ophiuroidea (brittle stars and basket stars), the Holothuroidea (the sea cucumbers) and the Crinoidea (sea lilies and feather stars).

Echinoidea Class of echinoderms (Echinodermata), commonly known as sea urchins.

echinoidin A multimeric lectin (subunits 147 kDa) from the coelomic fluid of the sea urchin *Anthocidaris crassispina*. The C-terminal sequence is highly homologous to C-terminal carbohydrate-recognition portions of rat liver mannose-binding protein and several other hepatic lectins.

Echinosphaerium Previously *Actinosphaerium*. A **Heliozoan** protozoan. The organisms are multinucleate and have a starburst of radiating **axopodia**, the microtubules of which have been much studied.

echistatin **Disintegrin** found in the venom of the saw-scaled viper, *Echis carinatus*.

Echiuroidea A phylum of sedentary marine worm-like animals.

Echoviruses A group of human **Picornaviruses**. Echo is derived from 'enteric cytopathic human orphan', where orphan implies that they are not associated with any disease, though some are now known to cause aseptic meningitis or other disorders.

ECL 1. Electrochemiluminescence: production of light during an electrochemical reaction, now being applied to various bioassay systems. 2. Enhanced chemiluminescence. Method for enhancing detection of proteins on blots. Involves the use of luminol that is oxidized by peroxidase-coupled antibody used to detect the protein of interest, and the light produced is then detected on film.

eclampsia Rare condition in which one or more convulsions occur during or immediately after pregnancy, probably as a result of impaired cerebral blood flow. Few cases of pre-eclampsia culminate in eclampsia.

eclosion Emergence of an insect from its old cuticle at a moult, particularly from pupa to adult, but also from the egg.

ecm Common abbreviation for **extracellular matrix**.

E. coli See *Escherichia coli*.

Eco RI Probably the most commonly used type II **restriction endonuclease** isolated from *E. coli*. It cuts the sequence GAATTC between G and A, thus generating 5′ **sticky ends**.

Eco RII Type II **restriction endonuclease** isolated from *E. coli*. It cuts the sequence CC(T/A)GG in front of the first C, giving 5′ **sticky ends**.

ecotropic virus **Retrovirus** which can only replicate in its original host species, *cf.* **amphotropic**.

Ecstasy Colloquial name for 3,4 methylenedioxy-methamphetamine (MDMA), a synthetic drug similar to methamphetamine (a stimulant) and mescaline (a hallucinogen).

ectoderm The outer of the three germ layers of the embryo (the other two being mesoderm and endoderm). Ectoderm gives rise to epidermis and neural tissue.

ectoenzyme Enzyme that is secreted from a cell or located on the outer surface of the plasma membrane and therefore able to act on extracellular substrates.

ectomycorrhiza *Ectotrophic mycorrhiza* Mycorrhiza with a well-developed layer of fungal mycelium on the outside of the root interconnected with hyphae both within the root cortex and also ramifying through the soil. Fungus is often a basidiomycete. See **endomycorrhiza**, **vesicular-arbuscular mycorrhiza**.

ectopic Misplaced, not in the normal location.

ectoplasm Granule-free cytoplasm of amoeba lying immediately below the plasma membrane.

ectoplasmic tube contraction Model for amoeboid movement in which it was proposed that protrusion of a pseudopod is brought about by contraction of the subplasmalemmal region everywhere else in the cell, thus squeezing the central cytoplasm forwards. See **frontal zone contraction theory**.

ectromelia Congenital absence or gross shortening of long bones of limb or limbs.

ED1 1. Antibody extensively used to identify rat monocytes/macrophages; marker antigen is a single-chain 90–110-kDa glycosylated protein, mostly on lysosomal membranes. Similar to human CD68. **2.** Gene associated with X-linked hypohidrotic ectodermal dysplasia. The putative protein is predicted to have a single transmembrane domain, and shows similarity to two separate domains of the tumour-necrosis factor receptor (TNF-R) family.

ED$_{50}$ Median effective dose, the dose that produces a response in 50% of individuals or 50% of the maximal response.

edaphic Type of physical or chemical property of soil that influences plants growing on that soil.

edema See **oedema**.

Ediacara Extensive family of ancient (600–540 million years old), preCambrian soft-bodied animals, fossils of which were first described from the Ediacaran Hills of South Australia. There are various bizarre forms and considerable uncertainty about the taxonomic position of many; some appear to be of extinct phyla.

editosome Multiprotein complex (27S) involved in **RNA processing**.

Edman degradation See **Edman reagent**.

Edman reagent *Phenyl isothiocyanate* The classic method for sequence determination of peptides using sequential cleavage of the N-terminal residue after reaction with Edman reagent. The N-terminal amino acid is removed as a phenylthiohydantoin derivative.

EDRF Endothelium-derived relaxation factor; see **nitric oxide**.

EDTA *Ethylenediaminetetraacetic acid.* Often used as the disodium salt. Chelator of divalent cations; $\log_{10} K_{app}$ for calcium at pH 7 is 7.27 (5.37 for magnesium) See **EGTA**.

EDTA-light chain Myosin light chains (18 kDa) from scallop muscle (two per pair of heavy chains), easily extracted by calcium chelation. Although the EDTA-light chains do not bind calcium they confer calcium sensitivity on the myosin heavy chains.

Edwards syndrome Complex of abnormalities caused by trisomy 18.

EEA Abbrev. for (1) early endosome antigen 1 (EEA-1), (2) end-to-end anastomosis, a surgical procedure, (3) *Euonymus europaeus* agglutinin, a alpha-galactophilic lectin that binds the sugar moiety alpha Gal (1,3) beta Gal (1,4) GlcNAc, particularly on endothelial cells.

EEG Electroencephalograph; electroencephalogram Record of electrical activity of the brain obtained using external electrodes.

EF-1 See **elongation factor**.

E-face In **freeze fracture** the plasma membrane cleaves between the acyl tails of membrane phospholipids, leaving a monolayer on each half of the specimen. The E-face is the inner face of the outer lipid monolayer. From within the cell this is the view that you would have of the outer half of the plasma membrane if the inner layer could be removed. The complementary surface is the P-face (the inner surface of the inner leaflet of the bilayer). E stands for ectoplasmic, P for protoplasmic – not terms that are in common usage!

EF-hand A very common calcium-binding motif. A 12-amino acid loop with a 12 amino acid α-helix at either end, providing octahedral coordination for the calcium ion. Members of the family include **aequorin**, **alpha-actinin**, **calbindin**, **calcineurin**, **calcyphosin**, **calmodulin**, **calpain**, **calcyclin**, **diacylglycerol** kinase, **fimbrin**, **myosin** regulatory light chains, **oncomodulin**, **osteonectin**, **spectrin**, **troponin** C.

efferent Leading away from something. The opposite of **afferent**.

eflornithine An enzyme inhibitor, used to slow hair growth and also in the treatment of trypanosomiasis.

EF-Tu See **elongation factor**.

EGF See **epidermal growth factor**, **EGF-like domain**.

EGF receptor *HER-1* Receptor tyrosine kinase encoded by *c-erbB1*. Member of the Type I family of growth factor receptors that also includes TGFα receptor, heregulin receptor.

EGF-like domain Region of 30–40 amino acids containing 6 cysteines found originally in EGF, and also in a range of proteins involved in cell signalling. Examples: **TGFα**, **amphiregulin**, **urokinase**, **tissue-plasminogen activator**, **complement** C6–C9, **fibronectin**, **laminin** (each subunit at least 13 times), **nidogen**, **selectins**. It is also found in the *Drosophila* gene products: **Notch** (36 times) **Delta**, **Slit**, **Crumbs**, **Serrate**.

EGFP Enhanced **green fluorescent protein**, often used as a reporter gene.

egg-polarity gene A gene whose product distribution in the egg determines the anterior-posterior axis of subsequent development. Best characterized in *Drosophila*: see *bicoid*, **maternal-effect gene**.

eglin C A peptidase inhibitor (70 amino acids) from leech (*Hirudo medicinalis*) but now available as a recombinant protein. A member of the potato chymotrypsin inhibitor family of serine peptidase inhibitors. In particular, inhibits neutrophil elastase and cathepsin-G.

Egr-1 Early growth response 1 (*Egr-1*) gene encodes an immediate-early response transcription factor, one of a family of C2H2-type zinc-finger proteins; upregulated after a variety of stresses.

EGS See **external guide sequence**.

EGTA *Ethyleneglycol-bis (2-aminoethyl) N,N,N,N,-tetraacetic acid)* Like **EDTA** a chelator of divalent cations but with a higher affinity for calcium (log K_{app} 6.68 at pH 7) than magnesium (log K_{app} 1.61 at pH 7). Will also bind other divalent cations. Note: the 'apparent association constant', K_{app}, is used because protons compete for binding and the association constant varies according to pH. Thus, EGTA has $\log_{10} K_{app}$ for calcium of 2.7 at pH 5, 10.23 at pH 9.

EH domain *Eps15-homology domain.* A highly conserved motif comprising approximately 100 residues that is found in many species ranging from yeast to mammals. EH domain proteins are involved in regulation of the actin cytoskeleton, signal transduction, transcriptional regulation and control of the endocytic pathway. EH domains bind to proteins that contain the tripeptide asparagine-proline-phenylalanine (NPF).

Ehlers–Danlos syndrome The classical Ehlers–Danlos Syndome (Types I and II) is characterized by loose-jointedness and fragile, bruisable skin, and is due to defects in genes for collagen alpha-1(V), alpha-2(V) or alpha-1(I). EDS III is a benign form of classic EDS. Ehlers–Danlos syndrome type IV is an autosomal dominant disorder in which there is a defect in the gene for type III collagen. In Type VI there is a defect in the gene for lysyl hydroxylase. Other forms are also recognized with defects in various aspects of collagenous connective tissue production. See **dermatosparaxis**.

Ehringhaus compensator Device used in **interference** or **polarization microscopy** to reduce the brightness of the object to zero in order to measure the phase retardation (optical path difference). The compensator consists of a birefringent crystal plate that can be tilted. An alternative to **Senarmont compensation** and has the advantage that it can be applied to retardations of more than one wavelength.

Ehrlichia Genus of rickettsia that are the cause of emerging and serious tick-borne human zoonoses, and the cause of serious and fatal infections in companion animals and livestock.

EHS cells *Englebreth-Holm-Swarm sarcoma cells* Line of mouse cells that produce large amounts of basement membrane-type extracellular matrix (ecm), rich in **laminin**, collagen type IV, **nidogen** and heparan sulphate. Often used as a source of these ecm molecules.

eicosanoids Useful generic term for compounds derived from arachidonic acid. Includes **leukotrienes**, **prostacyclin**, **prostaglandins** and **thromboxanes**.

eIF-1 One of the components of the eukaryotic initiation F-complex. eIF-1 is a low-molecular weight factor critical for stringent AUG selection and is recruited to the 43 S complex in the multifactor complex (MFC) with **eIF-2**, **eIF-3** and **eIF-5** via multiple interactions with the MFC constituents.

eIF-2 Part of the eukaryotic initiation complex. EIF-2 consists of three subunits: alpha, beta, and gamma. As initiation proceeds, eIF-2 forms a ternary complex with Met-tRNAi and GTP. EIF-2B is a guanine nucleotide exchange factor that acts to restore EIF-2 to its GTP-bound form. See **eIF-1**, **eIF-3**, **eIF-4F**, **eIF-5**.

eIF-3 Part of the eukaryotic initiation complex. EIF-3 is a multisubunit factor that contains at least eight distinct polypeptides. It plays a role in recycling of ribosomal subunits to the site of transcription initiation by promoting the dissociation of non-translating ribosomal subunits. See **eIF-1**, **eIF-2**, **eIF-4F**, **eIF-5**.

eIF-4F A trimeric peptide initiation factor complex that associates with the $5'$ cap of mRNA and is important in translation. It is composed of eIF-4A, eIF-4E and eIF-4G. See **eIF-1**, **eIF-2**, **eIF-3** and **eIF-5**.

eIF-5 A eukaryotic initiation factor that interacts with the 40S initiation complex and promotes the hydrolysis of bound GTP and subsequent release of eIF-3 and eIF-3 from the 40S subunit. The 40S subunit is then able to interact with the 60S ribosomal subunit to form the functional 80S initiation complex. See **eIF-1**, **eIF-2**, **eIF-3** and **eIF-4F**.

Eimeria Coccidian protozoan. All coccidians are intracellular parasites of various vertebrates and invertebrates. *Eimeria tenella* infects chick intestinal epithelial cells and is of veterinary importance. The trophozoites invade host cells and proliferate as merozoites by schizogony, which can then infect adjacent cells if released. Merozoites differentiate to male or female gamonts that fuse to form a zygote which undergoes division to form eight zoites that are retained within a zygocyst. If the zygocyst is ingested by a new host, the zoites emerge and reinfect the host as trophozoites.

Eisenberg algorithm An algorithm for calculating a **hydropathy plot**.

ektacytometry Method in which cells (usually erythrocytes) are exposed to increasing shear-stress and the laser diffraction pattern through the suspension is recorded; it goes from circular to elliptical as shear increases. From these measurements, a deformability index for the cells can be derived.

elaioplast Unpigmented type of **plastid** modified as an oil-storage organelle.

ELAM-1 *CD62E; E-selectin* One of the **selectin** family; upregulated on endothelial cells at sites of inflammation and partly responsible for trapping of neutrophils. The C-type lectin domain binds sialylated Lewis X and a particular glycoform of ESL-1 that is present on myeloid cells.

elastase *Pancreatic elastase; EC 3.4.21.36 (formerly EC 3.4.4.7)* Serine endopeptidase that will digest **elastin**, **collagen** Type IV and a range of other proteins; inhibited by alpha-1-protease inhibitor of plasma. A range of elastases are known; the pancreatic elastase is perhaps the commonest; that present in neutrophil granules differs, as does that from macrophages.

elasticoviscous Alternate form of the commoner term viscoelastic.

elastin Glycoprotein (70 kDa) randomly coiled and cross-linked to form elastic fibres that are found in connective tissue. Like collagen, the amino acid composition is unusual with 30% of residues being glycine and with a high proline content. Cross-linking depends upon formation of **desmosine** from four lysine side groups. The mechanical properties of elastin are poorer in old animals.

elastonectin **Elastin**-binding protein (120 kDa) found in extracellular matrix, produced by skin fibroblasts.

ELAV proteins *Embryonic lethal abnormal visual proteins* RNA-binding proteins that regulate mRNA stability. The ELAV family of RNA-binding proteins is highly conserved in vertebrates and in humans, there are four members: HuR is expressed in all proliferating cells, whereas Hel-N1, HuC and HuD are expressed in terminally differentiated neurons. See **AREs**.

electrical coupling Of two physically touching cells, denoting the presence of a **junction** that allows the passage of electrical current. Usually tested by impaling both cells with **microelectrodes**, injecting a current into one, and looking for a change in potential in the other. Usually taken as an indication of coupling by **gap junctions** or **electrical synapses**: see also **dye coupling**. Electrical coupling is not confined to excitable cells; many embryonic and adult **epithelia** are coupled, possibly to allow **metabolic cooperation**.

electrical synapse A connection between two electrically excitable cells, such as neurons or muscle cells, via arrays of **gap junctions**. This allows **electrical coupling** of the cells, and so an action potential in one cell moves directly into the other, without the 1-ms delay inherent in **chemical synapses**. Electrical synapses do not allow modulation of their connection, and so only occur in neuronal circuits where speed of conduction is paramount (e.g. the crayfish escape reflex). A few electrical synapses are rectifying, implying a more specialized property than a simple gap junction.

electrochemical potential Defined as the work done in bringing 1 mole of an ion from a standard state (infinitely separated) to a specified concentration and electrical potential. Measured in joules/mole. More commonly used to measure the electrochemical potential difference between two points (e.g. either side of a cell membrane), thus sidestepping the rather abstract concept of a standard state. If the molecule is uncharged or the electrical potential difference between two points is zero, the electrochemical potential reduces to the **chemical potential** difference of the species. At equilibrium, the electrochemical potential difference (by definition) is zero; the situation can then be described by the **Nernst equation**.

electrochemiluminescence *ECL, electrogenerated chemiluminescence* A light-emitting chemiluminescent reaction that is preceded by an electrochemical reaction. This has the advantage, for assay systems, that the time and location of the light emission can be controlled. Thus it is possible to arrange that the electrochemical reaction will only occur if the components are physically adjacent, if, for example, a receptor linked to a magnetic bead has bound labelled ligand from solution. The beads are magnetically captured and electrically stimulated, and the light emission is proportional to the binding of ligand. A common label is Ruthenium (II) tri-bipyridine, NHS ester.

electrodynamic forces London–Van der Waals forces: see **DLVO theory**.

electrofocusing Any technique whereby chemical species are concentrated using an applied electric field. See **isoelectric focusing**.

electrogenic pump Ion pump that generates net charge flow as a result of its activity. The sodium–potassium exchange pump transports two potassium ions inward across the cell membrane for each three sodium ions transported outward. This produces a net outward current that contributes to the internal negativity of the cell.

electron microprobe A technique of elemental analysis in the electron microscope based on spectral analysis of the scattered X-ray emission from the specimen induced by the electron beam. Using this technique it is possible to obtain quantitative data on, for example, the calcium concentration in different parts of a cell, but it is necessary to use ultra-thin frozen sections.

electron microscopy Any form of microscopy in which the interactions of electrons with the specimens are used to provide information about the fine structure of that specimen. In transmission electron microscopy (TEM), the diffraction and adsorption of electrons as the electron beam passes normally through the specimen is imaged to provide information on the specimen. In scanning electron microscopy (SEM), an electron beam falls at a non-normal angle on the specimen and the image is derived from the scattered and reflected electrons. Secondary X-rays generated by the interaction of electrons with various elements in the specimen may be used for **electron microprobe** analysis.

electron paramagnetic resonance *EPR; electron spin resonance; ESR* Form of spectroscopy in which the absorption of microwave energy by a specimen in a strong magnetic field is used to study atoms or molecules with unpaired electrons.

electron transport chain A series of compounds that transfer electrons to an eventual donor with concomitant energy conversion. One of the best studied is in the mitochondrial inner membrane, which takes NADH (from the **tricarboxylic acid cycle**) or FADH and transfers electrons, via **ubiquinone**, cytochromes and various other compounds, to oxygen. Other electron transport chains are involved in **photosynthesis**.

electrophoresis Separation of molecules based on their mobility in an electric field. High-resolution techniques normally use a gel support for the fluid phase. Examples of gels used are starch, acrylamide, agarose or mixtures of acrylamide and agarose. Frictional resistance produced by the support causes size, rather than charge alone, to become the major determinant of separation. The electrolyte may be continuous (a single buffer) or discontinuous, where a sample is stacked by means of a buffer discontinuity, before it enters the running gel/running buffer. The gel may be a single concentration or gradient in which pore size decreases with migration distance. In **SDS** gel electrophoresis of proteins or electrophoresis of polynucleotides, mobility depends primarily on size and is used to determine molecular weight. In pulse-field electrophoresis, two fields are applied alternately at right angles to each other to minimize diffusion-mediated spread of large linear polymers. See also **electrofocusing**, **pulse-field electrophoresis**

electrophoretogram Result of a zone electrophoresis separation or the analytical record of such a separation.

electroplax A stack of specialized muscle fibres found in electric eels, arranged in series. The fibres have lost the ability to contract; instead they generate extremely high voltages (*ca.* 500 V) in response to nervous stimulation. They contain asymmetrically distributed **sodium–potassium ATPases**, **acetylcholine receptors** and **sodium gates** at extraordinarily high concentrations.

electroporation Method for temporarily permeabilizing cell membranes so as to facilitate the entry of large or hydrophilic molecules (as in **transfection**). A brief (*ca.* 1-ms) electric pulse is given with potential gradients of about 700 V/cm.

electroretinogram Record of electrical activity in the retina made with external electrodes.

electrospray mass spectroscopy Method of mass spectroscopy in which the sample is introduced as a fine spray from a highly charged needle so that each droplet has a strong charge. Solvent rapidly evaporates from the droplets, leaving the free macromolecule. Beginning to be widely used because of its capacity to identify a wide range of compounds.

electrostatic forces Like charges in close proximity produce forces of repulsion between them. Consequently, if two surfaces bear appreciable and approximately equal densities of charged groups on their surfaces, appreciable forces of repulsion may occur between them. The range of these forces is determined in the main by the ionic strength of the intervening medium, forces being of minimal range at high ionic strength. The forces are effective over approximately twice the **double layer** thickness. See **DLVO theory**.

elegantin See **disintegrin**.

eleidin Clear substance found in stratum lucidum of skin, a keratin precursor.

elementary bodies 1. Inclusion bodies within cells, often of virus particles, although this term is more common in the older literature. **2.** Infectious extracellular form of *Chlamydia*, consisting of electron-dense nuclear material and a few ribosomes surrounded by a rigid trilaminar wall. Once taken up by cells, these reorganize into reticulate bodies.

elephantiasis *Lymphatic filariasis* Enlargement of the limbs, or of the scrotum, due to thickening of skin and blockage of lymphatic vessels by filarial nematode parasites, esp. *Brugia malayia* and *Wuchereria bancrofti*.

eleutherobin Tricyclic compound (a diterpene glycoside) that, like **taxol**, will stabilize microtubule bundles by competing for the paclitaxel binding site. Originally isolated from a marine soft coral, *Eleutherobia aurea*, also found in *Erythropodium caribaeorum*, an encrusting coral found in South Florida and the Caribbean. Synthetic routes for producing the compound have been devised.

eleutherosides Class of compounds, lignan glycosides, with anti-inflammatory and immunostimulatory activity isolated from the roots of *Eleutherococcus senticosus* (Siberian ginseng, a distant relative of Asian ginseng) and other medicinal herbs. Eleutherosides B and E have been most extensively studied.

ELF-2 *Eph ligand family 2* A transmembrane ligand (**ephrin**) for an **Eph-related receptor tyrosine kinases**. It shows closest homology to the other known transmembrane ligands in the family, **ELK-L**/LERK-2/Cek5-L: ELF-2 binds to three closely related Eph family receptors, Elk, Cek10 (apparent orthologue of Sek-4 and HEK2) and Cek5 (apparent orthologue of Nuk/Sek-3).

elicitor In general, any compound that induces a response in a system; more specifically, a substance that induces the formation of **phytoalexins** in higher plants. May be exogenous (often produced by potentially pathogenic microorganisms) or endogenous (possibly cell-wall degradation products).

elimination Disappearance of a substance from an organism, or a part of the organism, by processes of metabolism, secretion or excretion. Rates of elimination are important in toxicology and pharmacology.

ELISA *Enzyme-linked immuno-sorbent assay* A very sensitive technique for the detection of small amounts of protein or other antigenic substances. The basis of the method is the binding of the antigen by an antibody that is linked to the surface of a plate. Formation of an immune complex is detected by use of peroxidase coupled to antibody, the peroxidase being used to generate an amplifying colour reaction. Various ways of carrying out the assay are possible: if the aim is to detect antibody production from a myeloma

clone, for example, then the antigen may be bound to the plate, and the formation of the antibody/antigen complex may be detected using peroxidase coupled to an anti-Ig antibody.

elixophyllin Proprietary name for elixir (syrup) containing **theophylline** as its active ingredient.

Elk proteins *Eph-like kinases* Family of cell surface receptor tyrosine kinases restricted to brain and testis. Not to be confused with **Elk-1**.

Elk-1 Gene-regulating protein found in lung and testis. Binds to DNA at purine-rich sites. Substrate for **MAP kinases**; once phosphorylated forms complex with other transcription factors, binds to the serum response element (**SRE**) and induces transcription of *fos*.

Elk-L *LERK-2/Cek5-L* Membrane-anchored ligand (38 kDa) for **EPH-related receptor tyrosine kinase**, an **ephrin**. Becomes tyrosine phosphorylated once bound to receptor (Nuk).

ellipsosome Membrane-bounded compartment containing cytochrome-like pigment and found in the retinal cones of some fish.

Ellmans reagent *5,5'-Dithio-bis(2-nitrobenzoic acid); DNT* Reagent used to estimate the number of free sulphydryl groups.

elongation factor *EF* Peptidyl transferase components of ribosomes that catalyse formation of the acyl bond between the incoming amino acid residue and the peptide chain. There are three classes of elongation factor: EF1a (EF-Tu in prokaryotes) binds GTP and aminoacyl-tRNA, delivering it to the A site of ribosomes. EF-1b (EF-Ts) helps in regeneration of GTP-EF-1a. EF-2 (EF-G) binds GTP and peptidyl-tRNA and translocates it from the A site to the P site. Diphtheria toxin inhibits protein synthesis in eukaryotes by adding an ADP-ribosyl group to a modified histidine residue (diphthamide) in elongation factor II.

elutriation Separation of particles on the basis of their differential sedimentation rate.

EMA 1. Epithelial membrane antigen; see **episialin**. **2.** E2F-binding site modulating activity: transcriptional repressor (272 residues, 34 kDa) that has some similarity with E2F but lacks the activation domain at the carboxy terminus.

Embden–Meyerhof pathway *Glycolysis; Embden–Meyerhof–Parnas pathway* The main pathway for anaerobic degradation of carbohydrate. Starch or glycogen is hydrolysed to glucose-1-phosphate and then through a series of intermediates, yielding two ATP molecules per glucose, and producing either pyruvate (which feeds into the **tricarboxylic acid cycle**) or lactate.

embedding Tissue is embedded in wax or plastic in order to prepare sections for microscopic examination. The embedding medium provides mechanical support.

embolic gastrulation Gastrulation by invagination of part of the blastocyst wall, rather than overgrowth of the epiblast as happens with, for example, birds

embolus A clot formed by platelets or leucocytes that blocks a blood vessel.

embryo The developmental stages of an animal (or in some cases a plant) during which the developing tissue is effectively isolated from the environment by, for example, egg membranes, fetal membranes and various structures in plants.

embryo sac The female gametophyte in flowering plants (angiosperms) that develops within the ovule (megaspore) contained within an ovary at the base of the pistil of the flower. There are usually eight (haploid) cells in the female gametophyte: one egg, two synergids flanking the egg, two polar nuclei in the centre of the embryo sac, and three antipodal cells at the opposite end of the embryo sac from the egg.

embryogenesis The processes leading to the development of an embryo from egg to completion of the embryonic stage.

embryoid In plants, an embryo-like structure that may subsequently grow into a plantlet; in animals, aggregates of cells (embryoid body), derived from embryonic stem cells, that will exhibit some differentiation *in vitro*.

embryonal carcinoma cells Pluripotent cells of ectodermal origin, derived from **teratocarcinomas**.

embryonic induction The induction of differentiation in one tissue as a result of proximity to another tissue arising, for example, during gastrulation. One of the best-known examples is the induction of the neural tube in the ectoderm by the underlying chordamesoderm. Although the information to form the tube is present in the competent determined ectoderm, it must be elicited by the inducing tissue. In some cases, it is known that cell–cell contact between epithelium and mesenchyme is necessary.

embryonic stem cell *ES cell* Totipotent cell cultured from early embryo. Have the advantage that following modification *in vitro* they can be used to produce chimeric embryos and thus transgenic animals.

emerin A 254 aa type II integral membrane protein found in most cells and that forms part of a nuclear protein complex consisting of the **barrier-to-autointegration factor** (BAF), the nuclear lamina, nuclear actin and other associated proteins; apparently links A-type **lamins** to the inner nuclear envelope. Emerin is defective in some forms of X-linked **Emery–Dreifuss muscular dystrophy** (X-EDMD).

Emerson enhancement effect The effect on the rate of photosynthesis (in plants and algae) of illuminating simultaneously with far red light ($\lambda > 680$ nm) and light of shorter wavelength ($\lambda < 680$ nm). The effect is more than additive, and provides evidence for the existence of the two photosystems I and II.

Emery–Dreifuss muscular dystrophy Form of X-linked muscular dystrophy, a degenerative myopathy characterized by weakness and atrophy of muscle without involvement of the nervous system. Can arise from defects either in **emerin** gene at Xq28, or in nuclear **lamin** A.

emesis Vomiting. Antiemetic drugs are often used to reduce the nausea that is a side effect of chemotherapy.

emetic Having the power to cause vomiting.

emetine An alkaloid derived from ipecac root (*Cephaelis ipecacuanha*), used in the treatment of amoebiasis and as an emetic. Inhibits protein synthesis at the translation stage by blocking translocation of peptidyl-tRNA from the A-site to the P-site on the ribosome. Emetine-resistant CHO cell lines have been extensively studied.

EMG Abbreviation for electromyogram, a measure of electrical activity of muscles at rest and during contraction.

emperipolesis Phenomenon in which lymphocytes are apparently phagocytosed by macrophages (histiocytes) in the lymph node; associated with massive lymphadenopathy, an inflammatory disorder of obscure aetiology. In the early literature lymphocytes and leucocytes were described as entering the cytoplasm of endothelial cells during their extravasation in postcapillary venules, a process also termed emperipolesis, although this was a misapprehension and leucocytes were shown by ultrastructural studies to be moving between, not through, the endothelial cells.

emphysema Pulmonary emphysema is associated with chronic bronchitis and may be caused by excessive leucocyte clastase activity in the alveolar walls (possibly as a result of the inactivation of α1-antiprotease by active oxygen species released by leucocytes in inflammation).

EMT *Epithelial-mesenchymal transition; epithelio-mesenchymal transition.* Abbreviation for epithelial–mesenchymal transition, an essential process associated with tumour progression and metastasis in breast carcinoma in which epithelial cells alter their morphology, acquire mesenchymal markers, become detached from the epithelial sheet and are increasingly motile. The process is associated with stimulation by various growth factors such as IGF-1 and TGFβ-1.

enalapril **ACE inhibitor** used to treat hypertension and, in conjunction with diuretics, for treatment of heart failure.

enamelysin *Matrix metallopeptidase-20. MMP-20* Metallopeptidase with substrate specificity for amelogenin. See Table M1.

enantiomer Either of a pair of stereoisomers of a compound that has chirality.

enaptin *Nesprin-1* Enaptin belongs to a family of giant proteins that associate with the F-actin cytoskeleton as well as the nuclear membrane. The human enaptin gene spreads over 515 kb and gives rise to several splicing isoforms (**Nesprin-1**, Myne-1, Syne-1, CPG2). The longest assembled cDNA encompasses 27 669 bp and predicts a 1014-kDa protein.

Enc-1 *Ectoderm-Neural Cortex-1* Putative oncogene highly overexpressed in Group I vs Group II parathyroid adenomas. Product is an actin-binding protein of the **kelch** family that is an early and highly specific marker of neural induction in vertebrates. Also expressed in adipose tissue, where it appears to play a regulatory role early in adipocyte differentiation.

encapsidate To envelop a virus in a protein shell (the **capsid**).

encephalitis lethargica *Sleepy sickness; von Economo's disease.* An acute virally induced inflammation of the brain; characterized by fever and sleep disturbances and followed by various persisting forms of nervous disorder (e.g. Parkinsonism) or by changes in character. It emerged as a new infectious disease near the end of the First World War, but by 1940 new encephalitis lethargica cases had almost entirely disappeared. Probably only of historical interest.

encephalisation The increased development of the head region, and in particular the brain, in the course of the evolution of an organism.

encephalopsin *Panopsin; opsin3* An extraretinal photoreceptor molecule, primarily found in brain, that may play a role in non-visual photic processes such as the entrainment of circadian rhythm or the regulation of pineal melatonin production. Has highest homology to vertebrate retinal and pineal opsins. Encephalopsin is highly expressed in the pre-optic area and paraventricular nucleus of the hypothalamus, and is enriched in selected regions of the cerebral cortex, cerebellar Purkinje cells, a subset of striatal neurons, selected thalamic nuclei, and a subset of interneurons in the ventral horn of the spinal cord. See **opsin subfamilies**.

end plate potential Depolarization of the sarcolemma as a result of acetylcholine release from the motoneuron causing an influx of sodium ions. The end-plate potential is the sum of quantal **miniature end plate potentials**. Development of the end-plate potential is blocked by curare.

End3p Yeast actin-regulatory protein identified as being defective in the endocytosis-deficient mutant, end3Delta. Localization of **Sla1p** at the cell cortex is dependent on the **EH domain**-containing protein End3p. Forms an actin cytoskeleton-regulatory complex with **Pan1p** and Sla1p that is regulated by the serine/threonine kinase **prk1p**.

endarteritis Chronic inflammation of the arterial **intima**, often a late result of syphilis

endergonic An endergonic reaction requires the input of energy.

endiphilin An src homology 3 (**SH3**) domain-containing protein that is a major *in vitro* binding partner for **synaptojanin**. There is a high concentration of endophilin in synaptic terminals where it co-localizes with synaptojanin/amphiphysin I and II.

endocannabinoids Endogenous metabolites capable of activating the **cannabinoid receptors** (CB1 and CB2). Anandamide (arachidonylethanolamide) and 2-arachidonyl glycerol (2-AG) are the main endocannabinoids, although both bind to both receptors.

endocarditis Inflammation of the membrane lining the heart, that over the valves being particularly susceptible. May be caused by viral or bacterial infection, or indirectly as a response to rheumatic fever, scarlet fever or tonsillitis.

endochondral Term for anything situated within, or occurring within, cartilage. Endochondral bone formation occurs on a cartilage scaffold

endocrine gland Gland that secretes directly into blood and not through a duct. Examples are pituitary, thyroid, parathyroid and adrenal glands, ovary and testis, placenta, and B-cells of pancreas.

endocyte In *Dictyostelium*, cannibalistic phagosomes within the zygote giant cell that contain ingested amoebae; in *Hydra*, engulfed nurse cells found early in development. Sometimes loosely (inaccurately) used of endodermally derived cells.

endocytosis Uptake of material into a cell by the formation of a membrane-bound vesicle.

endocytotic vesicle See **endocytosis**.

endoderm A germ layer lying remote from the surface of the embryo that gives rise to internal tissues such as gut. Contrast **mesoderm** and **ectoderm**.

endodermis Single layer of cells surrounding the central stele (vascular tissue) in roots. The radial and transverse walls contain the hydrophobic **Casparian band**, which prevents water flow in or out of the stele through the **apoplast**. Also present in some stems.

endogenous Product or activity arising in the body or cell, as opposed to agents coming from outside.

endogenous pyrogen Fever-producing substance released by leucocytes (and Kupffer cells in particular) that acts on the hypothalamic thermoregulatory centre. Now known to be **interleukin-1**.

endoglin *CD105* Homodimeric glycoprotein (180 kDa) with TGF-binding activity, expressed on endothelial cells and pre-B cells.

endoglycosidase Enzyme of the subclass EC 3.2 that has the ability to hydrolyse non-terminal glycosidic bonds in oligosaccharides or polysaccharides. Endoglycosidases F and H are often used as tools to determine the role of carbohydrate moieties on glycoproteins. Endo-F, the product of *Flavobacterium meningosepticum*, cleaves glycans of high mannose and complex type at the link to asparagine in the protein; Endo H is from *Streptomyces* spp. and is an endo-β-*N*-acetyl-glucosaminidase.

endolithic Growing within rock. Such habitats are important in arid areas (e.g. Arctic, Antarctic), and a range of organisms exploit them, including algae, fungi and cyanobacteria. Bacteria within deep rock strata have been described, and the whole field is of interest to astrobiologists.

endolyn-78 Glycoprotein (78 kDa) present in membranes of endosomes and lysosomes but relatively scarce in other membranes.

endometrioma *adenomyoma* Tumour of the endometrium consisting of glandular elements and a cellular connective tissue.

endometrium Mucous membrane that lines the uterus and thickens during the menstrual cycle ready for implantation of the embryo. If implantation does not occur, the endometrium returns to its previous state and the excess tissue is shed at menstruation. If implantation does occur, the endometrium becomes the decidua and is not shed until after parturition.

endomitosis Chromosome replication without mitosis, leading to polyploidy. Many rounds of endomitosis give rise to the giant **polytene chromosomes** of Dipteran salivary glands, though in this case the daughter chromosomes remain synapsed.

endomorphins Endogenous peptides (endomorphin-1, YPWF-NH2; endomorphin-2, YPFF-NH2) with high selective affinity for μ-opiate receptor.

endomycorrhiza *Vesicular-arbuscular mycorrhiza; endotrophic mycorrhiza; Pl.* mycorrhizae Plant–fungal symbiotic association, common to many plant genera, in which fungi (usually zygomycota) penetrate roots and come to lie in close association with root cells. More common than **ectomycorrhiza**.

endomysium Connective tissue sheath surrounding individual muscle fibres.

endoneurium Connective tissue sheath surrounding individual nerve fibres in a nerve bundle.

endonexin *annexin IV* Calcium-dependent membrane-binding protein (an **annexin**) located on the endoplasmic reticulum of fibroblasts. Isolated protein will bind to liposomes if 1–10 μM calcium is present, but not if the liposomes contain sphingomyelin or cholesterol. An analogous calcium-dependent membrane-binding protein, **synexin**, codistributes with endonexin and binds particularly to phosphatidyl serine. Another of the same class is p36, a component of brush-border membrane, a target for the *src* **gene** tyrosine kinase, and that binds phosphatidyl serine or phosphatidyl inositol.

endonuclease One of a large group of enzymes that cleave nucleic acids at positions within the chain. Some act on both RNA and DNA (e.g. S1 nuclease, EC 3.1.30.1, which is specific for single-stranded molecules). **Ribonucleases** such as pancreatic, T1, etc. are specific for RNA, **Deoxyribonucleases** for DNA. Bacterial **restriction endonucleases** are crucial in recombinant DNA technology for their ability to cleave double-stranded DNA at highly specific sites.

endopeptidase An enzyme that cleaves protein at positions within the chain. Formally, the enzymes are peptidyl-peptide hydrolases, often referred to as **proteinases** or **proteolytic enzymes**.

endophilin Family of proteins involved in regulating clathrin-mediated endocytosis. Phosphorylation of endophilin by Rho-kinase inhibits the binding to **CIN85**, a key step in the internalization of ligand-activated receptors such as EGF-R. Endophilins have been proposed to have an enzymatic activity (a lysophosphatidic acid acyl transferase or LPAAT activity).

endoplasm Inner, granule-rich cytoplasm of amoeba.

endoplasmic reticulum *ER* Membrane system that ramifies through the cytoplasm. The membranes of the ER are separated by 50–200 nm and the **cisternal** space thus enclosed constitutes a separate compartment. The Golgi region is composed of flattened sacs of membrane that together with ER and lysosomes constitute the GERL system. See also **smooth ER, rough ER**.

endoplasmin *Tumour rejection antigen gp96; TRA1* Most abundant protein in microsomal preparations from mammalian cells (100-fold more concentrated in ER than elsewhere). A glycoprotein (100 kDa) with calcium-binding properties. Same as GRP (**glucose-related protein**). A member of the **hsp90** family of **heat-shock proteins**.

endorphins A family of peptide hormones that bind to **opioid receptors**. Released in response to neurotransmitters and rapidly inactivated by peptidases. Physiological responses to endorphins include analgesia and sedation.

endosmosis Movement of water into a cell as a result of greater internal osmotic pressure. The **water potential** within the vascular sap of a plant cell must be lower than that in the bathing medium or sap of a neighbouring cell.

endosome 1. Endocytotic vesicle derived from the plasma membrane. More specifically, an acidic non-lysosomal compartment in which receptor–ligand complexes dissociate. **2.** A chromatinic body near the centre of a vesicular nucleus in some protozoa.

endosperm Tissue present in the seeds of angiosperms, external to and surrounding the **embryo**, which it provides with nourishment in the form of **starch** or other food reserves. Formed by the division of the **endosperm mother cell** after fertilization; may be absorbed by the embryo prior to seed maturation, or may persist in the mature seed.

endosperm mother cell Cell of the higher plant embryo sac. Contains two polar nuclei, and fuses with the sperm cell from the pollen grain. Gives rise to the **endosperm**.

endospore 1. An asexual spore formed within a cell. **2.** Inner part of the wall of a fungal spore.

endosymbiont hypothesis The hypothesis that semiautonomous organelles such as mitochondria and chloroplasts were originally endosymbiotic bacteria or cyanobacteria. The arguments are convincing, and although the hypothesis cannot be proven it is widely accepted.

endosymbiotic bacteria Bacteria that establish a symbiotic relationship within a eukaryotic cell, e.g. the nitrogen-fixing bacteria of legume root nodules. See also **endosymbiont hypothesis**.

endothelin converting enzyme *ECE* ECE-1 is an integral membrane protein belonging to the family of metalloproteinases that also includes ECE-2, neprilysin (**endopeptidase 24.11**) and Kell blood group protein. The catalytic site is in the large extracellular domain and contains a conserved zinc-binding motif.

endothelin receptor There are thought to be two G-protein-coupled receptors for endothelin, ET(A) (427 residues) and ET(B) (427 residues in human), present on vascular smooth muscle cells mediating vasoconstriction, and on endothelium mediating **nitric oxide** release. ET(A) binds ET-1 preferentially, whereas ET(B) binds ET-1, ET-2 and ET-3 with equal affinity.

endothelins *ET-1, ET-2, ET-3* Group of peptide hormones (all 21 residues) released by endothelial cells. All have two disulphide bridges that hold them in a conical spiral shape. They are the most potent vasoconstrictor hormones known. Structurally related to the snake venom **sarafotoxins**. Pre-pro-endothelin-1 (203 residues) is cleaved to the biologically inactive big endothelin-1 (92 residues) by **endothelin converting enzyme**, which will further cleave big endothelin to form active endothelin-1. ET-1, the predominant form, is produced by endothelial cells, ET-2 and ET-3 by various tissues. In addition to their vasoconstrictive properties, endothelins have **inotropic** and **mitogenic** properties, influence salt and water balance, alter central and peripheral sympathetic activity and stimulate the **renin–angiotensin**–aldosterone system. Though ET-1 acting through **endothelin receptor**(A) is vasoconstrictive, it acts through ET(B) to induce the release of **nitric oxide,** which is a vasodilator.

endothelioma A tumour, usually benign, derived from the epithelial lining of blood vessels or lymph channels (endothelium).

endothelium Simple, generally **squamous,** epithelium lining blood vessels, lymphatics and other fluid-filled cavities (such as the anterior chamber of the eye). Mesodermally derived, unlike most epithelia. Modified in areas where there is lymphocyte traffic (see **high endothelial venule**).

endothelium-derived relaxation factor *EDRF* See **nitric oxide**.

endotherm An animal that is able to maintain a body temperature above ambient by generating heat internally. In contrast to poikilothermic (so-called cold-blooded) animals.

endothermic Process or reaction that absorbs heat and thus requires a source of external energy in order to proceed.

endotoxin Heat-stable polysaccharide-like toxin bound to a bacterial cell. The term is used more specifically to refer to lipopolysaccharide (LPS) of the outer membrane of Gram-negative bacteria. There are three parts to the molecule; the **Lipid A** (six fatty acid chains linked to two glucosamine residues), the core oligosaccharide (branched chain of ten sugars), and a variable length polysaccharide side-chain (up to 40 sugar units in smooth forms) that can be removed without affecting the toxicity (rough LPS). Some endotoxin is probably released into the medium, and endotoxin is responsible for many of the virulent effects of Gram-negative bacteria.

endotrophic mycorrhiza *Endomycorrhiza* A mycorrhiza in which the fungal hyphae grow between and within the cells of the root cortex and connect with hyphae ramifying though the soil but which do not form a thick mantle on the surface of the root. Vesicular-arbuscular mycorrhizas and the mycorrhizas of orchids and of the Ericaceae are endotrophic.

endovanilloids Endogenous ligands of the transient receptor potential vanilloid type 1 (TRPV1) channels, one of the thermo-sensitive **TRP channels**. Include *N*-arachidonoyl-dopamine, *N*-oleoyl-dopamine and **anandamide**. See **vanilloid receptor**.

endplate The area of sarcolemma immediately below the synaptic region of the motor neuron in a neuromuscular junction.

endrin A pesticide; isomeric with **dieldrin**.

enduracidin See **ramoplanin**.

engrailed *en Drosophila* gene that controls segmental polarity. It is the archetype for one of three subfamilies of **homeobox**-containing genes.

enhancement effect Property of higher plant photosynthesis, discovered by Robert Emerson. The **quantum yield** of red light (less than 680 nm) and far red light (700 nm), when shone simultaneously on a plant, is greater than the sum of the yields of the light of the two wavelengths separately. This effect provides evidence for the cooperative interaction of two **photosystems** in photosynthesis.

enhancer A DNA **control element** frequently found 5′ to the start site of a gene and which, when bound by a specific transcription factor, enhances the levels of expression of the gene, although it is not sufficient alone to cause expression. Distinguished from a **promoter**, which is sufficient alone to cause expression of the gene when bound; in practice, the two terms merge.

enhancer trap Technique for mapping gene expression patterns, classically in *Drosophila*. A **transposon** element carrying a **reporter gene** (usually **beta-galactosidase**), linked to a very weak **promoter**, is induced to jump within the genome. If the P-element reinserts within the sphere of influence of promoters and **enhancers** of some (random) gene, then the reporter gene is also expressed in a similar tissue-specific manner. Usually, many lines of flies carrying such random insertions are studied; if a line shows 'interesting' patterns of expression, it can be possible to clone the gene of interest.

enkephalins Natural **opiate** pentapeptides isolated originally from pig brain. Leu-enkephalin (YGGFL) and Met-enkephalin (YGGFM) bind particularly strongly to δ-type opiate receptors.

enolate The anion formed from an **enol**.

enols Tautomeric form of some ketones; any organic compounds with a hydroxyl group attached to a carbon that is linked to another carbon by a double bond. Loss of a proton generates an enolate anion.

entactin *Nidogen* See **nidogen**, the more commonly used name.

Entamoeba Single-celled eukaryotes that parasitize all classes of vertebrates, a few invertebrates and possibly other unicellular eukaryotes. All species have a simple life cycle consisting of an infective cyst stage and a multiplying trophozoite stage. *Entamoeba histolytica* is the only species that affects man, and is the third leading cause of morbidity and mortality due to parasitic disease in humans. Non-pathogenic entamoebiasis is now recognized as being due to infection with another species, *E. dispar*.

enteral Within the intestine, or by way of the intestine.

enteric Relating to the intestine.

enteric-coated Drug tablet or capsule that is coated in a substance that prevents it from releasing its contents until it has passed through the stomach and into the intestine.

Enterobacter Genus of enteropathic bacilli of the Klebsiella group. Not to be confused with the Family **Enterobacteria,** of which they are members.

Enterobacteriaceae A large family of Gram-negative bacilli that inhabit the large intestine of mammals. Commonest is *E. coli*; most are harmless commensals but others can cause intestinal disease (*Salmonella, Shigella*).

enterobactin Alternative name for **enterochelin**.

enterochelin *Enterobactin* Iron-binding compound (**siderophore**) of *E. coli* and *Salmonella* spp. A cyclic trimer of 2,3-dihydroxybenzoylserine.

enterochromaffin cells *Kulchitsky cells; EC cells* Neuroendocrine cells found in the epithelia lining the lumen of the gastrointestinal tract. They produce and contain about 90% of the body's store of serotonin. See **enterochromaffin-like cells**.

enterochromaffin-like cells *ECL-cells* Distinct type of neuroendocrine cell found in the gastric mucosa underlying the epithelium, particularly in the acid-secreting regions of the stomach. Synthesize and secrete histamine in response to stimulation by the hormones gastrin and pituitary adenylyl cyclase-activating peptide. Unlike **enterochromaffin cells** do not produce serotonin.

Enterococcus Gram-positive cocci that occur singly, in pairs, or in short chains. They are facultative anaerobes and live mostly in the digestive tract. Most enterococcal infections of humans are due to *E. faecalis*.

enterocyte Epithelial cell of the intestinal wall.

enterocytes Cells of the intestinal epithelium.

enteron The body cavity of Cnidaria, corresponding to the **archenteron** of a gastrula.

enterostatin The N-terminal pentapeptide (VPDPR) cleaved from **procolipase**, suppresses fat intake after peripheral and central administration. Enterostatin alters 5-HT release in the brain, and 5-HT1-B receptor antagonists block the anorectic response to enterostatin. Release of enterostatin varies in a circadian fashion in some animals.

enterotoxins Group of bacterial **exotoxins** produced by enterobacteria and that act on the intestinal mucosa. By perturbing ion- and water-transport systems they induce diarrhoea. **Cholera toxin** is the best-known example.

enterovirus A genus of **Picornaviridae** that preferentially replicate in the mammalian intestinal tract. It includes the **polioviruses** and **Coxsackie viruses**.

Entner–Doudoroff pathway Metabolic pathway for degradation of glucose in a wide variety of bacteria. Differs from the Embden–Meyerhof pathway, although end result is similar.

entopic Developed or located in the normal anatomical location (opposite of ectopic).

ENTs *Equilibrative nucleoside transporters* Members of a family of integral membrane proteins with 11 transmembrane domains. These sodium-independent nucleoside transporters are widely distributed in eukaryotes; typical inhibitors of mammalian ENTs are nitrobenzylmercaptopurine ribonucleoside, dilazep, and dipyridamole. Adenosine flux across cardiomyocyte membranes occurs mainly via equilibrative nucleoside transporters. PMAT (ENT4) is a Na^+-independent and membrane potential-sensitive transporter that transports monoamine neurotransmitters and the neurotoxin 1-methyl-4-phenylpyridinium (MPP^+) and may be a polyspecific organic cation transporter. See **CNTs** (concentrative nucleoside transporters).

env Retroviral gene encoding viral envelope glycoproteins.

envelope 1. Lipoprotein outer layer of some viruses – derived from plasma membrane of the host cell. **2.** In bacteriology, the plasma membrane and cell wall complex of a bacterium.

envoplakin Component protein (210 kDa) of transglutaminase cross-linked protein layer (cornified envelope) deposited under the plasma membrane of keratinocytes in outer layer of skin. Has sequence homology with **desmoplakin**, bullous pemphigoid antigen 1, and **plectin**.

enzootic A disease prevalent in animals in a certain area; the veterinary equivalent of an endemic disease in man.

enzyme induction An increase in enzyme secretion in response to an environmental signal. The classic example is the induction of β-galactosidase in *E. coli*.

eosin *Tetrabromofluorescein* A red dye used extensively in histology, for example in the standard H & E (haematoxylin/eosin) stain used in routine pathology.

eosinophil Polymorphonuclear leucocyte (granulocyte) of the myeloid series, of which the granules stain red with eosin. Phagocytic, particularly associated with helminth infections and with hypersensitivity.

eosinophil cationic protein Arginine-rich protein (21 kDa) in granules of eosinophils that damages schisto-

somula *in vitro*. Not the same as the MBP (major basic protein) of the granules.

eosinophil chemotactic peptide *ECF; ECF of anaphylaxis; ECF-C* Tetrapeptides (of which two are identified: VGSE and AGSE) released by mast cells and that are said to both attract and activate eosinophils.

eosinophilia Condition in which there are unusually large numbers of **eosinophils** in the circulation, usually a consequence of helminth parasites or allergy.

eosinophilic 1. Having affinity for the red dye **eosin**. **2.** Inflammatory lesion characterized by large numbers of **eosinophils**.

eosinophilopoietin Small (1500-D) peptide, possibly released by T-lymphocytes, that regulates **eosinophil** development in the bone marrow. Probably interleukin 5.

eotaxin Chemokine with specificity for eosinophils. Human eotaxin is an 8.3-kDa protein containing 74 amino acid residues. See Table C4.

epalons Class of neuroactive steroids that are positive allosteric modulators of GABA via a neurosteroid site on the $GABA_A$ receptor/Cl^- ion channel complex. Name derived from epiallopregnanolone, an endogenous metabolite of progesterone. Have anxiolytic, anticonvulsant and sedative-hypnotic properties.

ependymal cells Cells that line cavities in the central nervous system – considered to be a type of glial cells.

Eph-related receptor tyrosine kinases *EphB2 (formerly Nuk/Cek5/Sek3)* The largest known family of receptor tyrosine kinases, implicated in the control of axonal navigation and fasciculation and in vascular assembly. Efficient activation of EPH receptors generally requires that their ligands (**ephrins**) be anchored to the cell surface, through either a transmembrane (TM) region or a glycosyl phosphatidylinositol (GPI) group. Challenging cells that express the TM ligands Elk-L or Htk-L with the clustered ectodomain of Nuk induces phosphorylation of the ligands on tyrosine, thus there is bidirectional cell signalling.

ephedra *Ma huang* Ephedra is a naturally occurring plant-derived substance in which the principal active ingredient is **ephedrine**. Ephedrine alkaloids are found naturally in a number of plants, including the ephedra species. The US FDA advises that ephedra supplements present an unreasonable risk of illness or injury.

ephedrine (1R, 2S)-*1-phenyl-1-hydroxy-2-methylaminopropane* Alkaloid from plants of genus *Ephedra*. Structural analogue of epinephrine (**adrenaline**) the effects of which it mimics.

Ephemeroptera An order of insects (Mayflies) in which the adult life is very short and the mouthparts are reduced and functionless; the immature stages are active aquatic forms.

Ephestia kuhniella Mediterranean flour moth. Easily maintained in the laboratory.

ephrins Ligands of the **Eph-related receptor tyrosine kinases**. Ephrin B1 (formerly Elk-L/Lerk2) is a ligand for the EphB2 receptor (formerly Nuk/Cek5/Sek3). Ligands are transmembrane molecules on other cells that respond to being bound by the receptor by becoming tyrosine phosphorylated: both receptor and ligand are altered by the binding interaction and so signalling is bidirectional.

epi- Prefix indicating something on, above or near. Epi-illumination is from above, epithelia cover (are on top of) other tissues.

epibenthos Organisms living on the floor of a sea or lake.

epiblast The outer germinal layer of a metazoan embryo that gives rise to the ectoderm.

epiboly The process in early embryonic development in which a monolayer of dividing cells (blastoderm) spreads over the surface of a large yolk-filled egg (e.g. those of teleosts, reptiles and birds).

epicatechin Flavonoid found in chocolate, thought to be beneficial for blood-vessel function.

epichromosomal Genetic material, for example an adenovirus used as a vector, that does not become integrated into the host chromosomes but proliferates in tandem. Use of such vectors avoids the risk of activating host genes in an inappropriate fashion.

epicotyl The first shoot of a plant embryo or seedling, above the point of insertion of the cotyledon(s). Can be relatively long in some seedlings showing **etiolation**.

epidermal cell 1. Cell of epidermis in animals. **2.** Plant cell on the surface of a leaf or other young plant tissue, where bark is absent. The exposed surface is covered with a layer of **cutin**.

epidermal growth factor *EGF* A mitogenic polypeptide (6 kDa) initially isolated from male mouse submaxillary gland. The name refers to the early bioassay, but EGF is active on a variety of cell types, especially but not exclusively epithelial. A family of similar growth factors is now recognized. Human equivalent was originally named **urogastrone** owing to its hormone activity.

epidermal hair Single-celled hairs on the surfaces of plants, particularly the leaves. Brittle silicified hairs on nettles break off in the skin of an animal that comes into contact with them, and are responsible for the sting.

epidermis Outer epithelial layer of a plant or animal. May be a single layer that produces an extracellular material (as, for example, the cuticle of arthropods), or a complex stratified squamous epithelium, as in the case of many vertebrate species.

epidermolysis bullosa *EB* A very rare genetic condition in which the skin and internal body linings blister at the slightest knock or rub. EB is subdivided into (1) **epidermolysis bullosa simplex** (EBS; intraepidermal skin separation), (2) **junctional epidermolysis bullosa** (JEB; skin separation in lamina lucida or central basal membrane zone (BMZ)), (3) **dystrophic epidermolysis bullosa** (DEB; sublamina densa BMZ separation), (4) **hemidesmosomal epidermolysis bullosa** (HEB), which produces blistering at the hemidesmosomal level in the most superior aspect of the BMZ. See also **epidermolysis bullosa acquisita**.

epidermolysis bullosa acquisita *EBA* A chronic autoimmune subepidermal blistering disease of the skin and mucous membranes. Immunologically, EBA is characterized by the presence of IgG autoantibodies against the non-collagenous (NC1) domain of type VII collagen, the major component of anchoring fibrils that connect the basement membrane to dermal structures. See **epidermolysis bullosa**.

epidermolysis bullosa simplex *EBS* A collection of keratin disorders characterized by intraepidermal blistering with relatively mild internal involvement. The severity varies and the more severe EBS subtypes include Koebner, Dowling–Meara, and Weber–Cockayne forms. Most cases of EBS are associated with mutations of the genes coding for keratins 5 and 14 (which combine to form intermediate filaments in basal keratinocytes). See **epidermolysis bullosa**.

epididymis Convoluted tubule connecting the vas efferens, which comes from the seminiferous tubules of the mammalian testis, to the vas deferens. Maturation and storage of sperm occur in the epididymis.

epifluorescence Method of fluorescence microscopy in which the excitatory light is transmitted through the objective onto the specimen rather than through the specimen; only reflected excitatory light needs to be filtered out rather than transmitted light, which would be of much higher intensity.

epigenesis The theory that development is a process of gradual increase in complexity, as opposed to the preformationist view, which supposed that mere increase in size was sufficient to produce adult from embryo.

epigenetics The study of mechanisms involved in the production of phenotypic complexity in morphogenesis. According to the epigenetic view of differentiation, the cell makes a series of choices (some of which may have no obvious phenotypic expression, and are spoken of as **determination** events) that lead to the eventual differentiated state. Thus, selective gene repression or derepression at an early stage in differentiation will have wide-ranging consequences in restricting the possible fate of the cell. See **epigenomics**.

epigenomics Epigenetic effects are mediated by either chemical modifications of the DNA itself (e.g. methylation) or by modifications of proteins that are closely associated with DNA (chromatin structure). Many of these epigenetic changes alter the whole genome and need to be considered holistically, rather than in isolation: epigenomics is the analysis of genome-wide consequences of epigenetic modifications. The Human Epigenome Project is a joint effort by an international collaboration that aims to identify, catalogue and interpret genome-wide DNA methylation patterns of all human genes in all major tissues.

epiglycanin Very extensively glycosylated transmembrane glycoprotein found in TA3 Ha mouse mammary carcinoma cells, and which may mask histocompatibility antigens. Functionally analogous to **episialin,** but there is no sequence homology in the protein.

epiligrin *Laminin 5* Major glycoprotein of epidermal basement membrane, consisting of three disulphide-bonded subunits of 170, 145 and 135 kDa. Epiligrin (**kalinin**) is the major ligand for $\alpha 3/\beta 1$ integrin, is particularly prominent in the lamina lucida of the skin, and is absent in patients with lethal **junctional epidermolysis bullosa**. Now generally termed laminin 5.

epilithic Descriptor for an organism that grows on the exposed surface of a rock.

epimer Diastereomeric monosaccharides that have opposite configurations of a hydroxyl group at only one position, e.g. D-glucose and D-mannose.

epimorphosis Pattern of regeneration in which proliferation precedes the development of a new part. Opposite of **morphallaxis**.

epinasty Asymmetrical growth of a leaf or stem that causes curvature of the structure.

epinemin Intermediate filament-associated protein (44.5-kDa monomer) associated with **vimentin** in non-neural cells. Not mentioned in recent literature.

epinephrine Synonym for **adrenaline**.

epiphysis 1. Region at the end(s) of long (limb) bones that is ossified separately and only becomes united with the main portion (diaphysis) of the bone once maturity is reached and no further growth in height will occur. The outer region of compact bone is relatively thin compared to that in the diaphysis, and the remainder is composed of trabecular bone with red marrow. **2.** In Echinoidea, one of the ossicles of Aristotle's lantern. **3.** The pineal body (epiphysis cerebri). *Adj.* epiphysial.

epiphytotic A widespread outbreak of plant disease; the botanical analogue of an epidemic.

epiplakin Giant epithelial protein (>700kDa) of the **plakin** family of cytolinker proteins. It is an atypical family member consisting entirely of plakin repeat domains without any of the other domains commonly shared by plakins.

episialin *Polymorphic epithelial mucin, PEM; epithelial membrane antigen, EMA* Heavily glycosylated membrane glycoprotein. Encoded by the MUC-1 gene; has a molecular weight of around 300 kDa, more than half of which is O-linked glycan. There is a 69-residue cytoplasmic domain, and the extracellular domain may extend hundreds of nanometres beyond the plasma membrane; the increased expression in carcinoma cells may reduce the adhesion and mask antigenic properties of the cells. Similar functions are ascribed to **ASGP**, **epiglycanin** and **leukosialin**.

episome Piece of hereditary material that can exist as free, autonomously replicating DNA or be attached to and integrated into the chromosome of the cell, in which case it replicates along with the chromosome. Examples of episomes are many **bacteriophages**, such as lambda and the male sex factor of *E. coli*.

epistasis Non-reciprocal interaction of non-allelic genes – for example, when the expression of one gene masks the expression of another. Thus a gene that blocks development of an organ will mask the effects of genes that would modify the form of that organ had it developed.

epistasy See **epistasis**.

epitectin Mucin-like glycoprotein found on surface of human tumour cells (also known as CA antigen) but not non-tumourigenic cell lines. It is present on the surface of some specialized cells (sweat glands, Type II pneumocytes from lung, bladder epithelium) and may therefore be a normal **differentiation antigen** Also present in normal urine.

epithelial membrane antigen *EMA* See **episialin**.

epithelial-mesenchymal transition See **EMT**.

epithelioid cells In a general sense, a cell that has an appearance that is similar to that of epithelial cells; used specifically of the very flattened macrophages found in granulomas (e.g. in tubercular lesions).

epithelium One of the simplest types of tissues. A sheet of cells, one or several layers thick, organized above a basal lamina (see **basement membrane**), and often specialized for mechanical protection or **active transport**. Examples include skin, and the lining of lungs, gut and blood vessels.

epitope That part of an antigenic molecule to which the T-cell receptor responds; a site on a large molecule against which an antibody will be produced and to which it will bind. See also **agretope**.

epitope library *Phage display library* Large collection (hundreds of millions) of peptides each encoded by a randomly mutated piece of DNA in a phage genome and expressed on the surface of that bacteriophage, sometimes as an N-terminal extension of a coat protein. Particular phages can be selected by a binding assay and, since the peptide has its encoding DNA associated with it, sequencing is straightforward.

epitope mapping The identification and definition of the epitope recognized by an antibody. Various methods can be used including synthetic peptides (where the sequence of the protein is known), phage display libraries (see **epitope library**), protein footprinting (using monoclonal antibody to protect the protein from proteolytic degradation), isolation and characterization of the peptide bound to MHC, or expression cloning. Since some epitopes may involve glycosylation or other post-translational modifications to proteins, the process is not necessarily straightforward.

epitope tag Short peptide sequence that constitutes an **epitope** for an existing antibody. Widely used in molecular biology to tag transgenic proteins (as a translational fusion product) to follow their expression and fate by immunocytochemistry or Western blotting, but without having to raise antibodies against the specific protein. Example: **myc tag**. See also **flag tagging**.

epitrichium *periderm* The outer layer of the epidermis in an embryo or foetus, usually shed postpartum. *Adj.* epitrichial.

epizootic The veterinary equivalent of an epidemic disease.

eplin *Epithelial protein lost in neoplasm* Protein that inhibits depolymerization of F-actin and is downregulated in neoplastic cells. Eplin increases the number and size of actin stress-fibres and inhibits membrane ruffling induced by Rac. EPLIN has at least two actin-binding sites and a central **LIM domain**.

EPO See **erythropoietin**.

epothilones Compounds (epothilone A and B) isolated from the myxobacterium *Sorangium cellulosum* Str 90. Cytotoxic to tumour cells as a result of inducing microtubule assembly and stabilization in a manner similar but not identical to that of **paclitaxel**.

Eps15 An adaptor protein that is involved in epidermal growth factor (EGF) receptor endocytosis and trafficking. It is phosphorylated by EGF receptor tyrosine kinase, and the two may be brought together by ubiquitin (c-**cbl** ubiquitin-ligase activity is required for recruitment and co-localization of EGFR and Eps15 in the endosomal compartment). See **EH domain**.

EPSC Abbreviation for excitatory postsynaptic currents.

epsins Endocytic proteins with an epsin N-terminal homology (ENTH) domain that binds phosphoinositides, and a poorly structured C-terminal region that interacts with **ubiquitin** and the endocytic machinery, including **clathrin** and endocytic scaffolding proteins. Epsin 1 is an integral component of clathrin coats forming at the cell surface. A range of epsin-like proteins has been identified.

Epstein–Barr virus Species of Herpetoviridae that binds **CR2** and that causes infective mononucleosis and, in the presence of other factors, tumours such as **Burkitt's lymphoma** and nasopharyngeal carcinoma.

equatorial plate Region of the mitotic spindle where chromosomes are aligned at metaphase: as its name suggests, it lies midway between the poles of the spindle.

equilibrium constant *Equilibrium dissociation constant; dissociation constant* The ratio of the reverse and forward rate constants for a reaction of the type $A + B = AB$. At equilibrium the equilibrium constant (K) equals the product of the concentrations of reactants divided by the concentration of product, and has dimensions of concentration. $K =$ (concentration A × concentration B)/(concentration AB). The affinity (association) constant is the reciprocal of the equilibrium constant.

equilibrium dialysis Technique used to measure the binding of a small molecule ligand to a larger binding partner. The macromolecule is contained within a dialysis chamber and the diffusible ligand added to the exterior: once equilibrium is reached, an excess of ligand inside the dialysis chamber is evidence of binding and it is possible to calculate the binding affinity from a measurement of the concentrations of ligand and that of the binding macromolecule.

equinatoxins Small peptide toxins (19 kDa) from *Actinia equina*. Form cation-selective pores and are cytolytic.

equivalence The situation where two interacting molecular species are present in concentrations just sufficient to produce occupation of all binding sites. Only used to describe high avidity interactions, especially the antibody/antigen interaction.

ERAB *Endoplasmic reticulum associated binding protein* Protein (262 residues, 27 kDa) in the ER that binds amyloid β (Aβ). Found ubiquitously, but more extensively in liver, heart and brain. Overexpressed in brain of patients with **Alzheimer's disease,** and it may be the complex between Aβ and ERAB that is cytotoxic. Sequence has similarities with short-chain alcohol dehydrogenases, hydroxysteroid dehydrogenases and acetoacyl CoA reductases.

erabutotoxins Curaremimetic polypeptide toxins (62 residues) from venom of *Laticauda semifasciata*. Bind to nicotinic acetylcholine receptors.

erb Two oncogenes, *erbA* and *erbB*, associated with erythroblastosis virus (an acute transforming retrovirus). The cellular homologue of *erbB* is the structural gene for the cell surface receptor for epidermal growth factor, and that of *erbA* is a steroid hormone receptor.

ERBIN *ErbB2 receptor-interacting protein* Protein that binds to **p0071** and ErbB2; co-localized with **PAPIN** on the lateral membrane of epithelial cells.

ergastic substances Metabolically inert products of photosynthesis, such as starch grains and fat globules.

ergocalciferol Synonym for **calciferol**.

ergodic System or process in which the final state is independent of the initial state.

ergosterol *Provitamin D2* A sterol found in ergot, yeasts and other fungi. Most important precursor of vitamin D2 (ercalciol), into which it is converted by the action of ultraviolet light on the skin.

ergot Fungal (*Claviceps purpurea*) infection of rye (*Secale cornutum*); the so-called ergot that replaces the grain of the rye is a dark, purplish sclerotium, from which the sexual stage of the life cycle will form after over-wintering. The fungus produces a mycotoxin (see **ergotamine tartrate**) that contaminates rye flour and causes **ergotism**. Various bizarre behavioural symptoms caused by ergot poisoning were ascribed, in the Middle Ages, to witchcraft. A serious toxin for domestic animals that are fed on cereals, particularly rye.

ergotamine tartrate Ergotamine is a natural alkaloid of the **ergot** of rye (*Secale cornutum*). An α-adrenergic blocking agent with a direct stimulating effect on the smooth muscle of peripheral and cranial blood vessels; also causes depression of vasomotor centres. Ergotamine has serotonin antagonist properties. Used to relieve migraine by constricting cranial arteries.

ergotism A condition due to eating the grains of cereals which are infected by the ergot fungus *Claviceps purpurea*, characterized by extreme vasoconstriction leading to gangrene and convulsions. Formerly known as St Anthony's Fire.

ERKs See **MAP kinases**.

erlotinib *Tarceva* A tyrosine kinase inhibitor used to treat non-small cell lung cancer. Primary target is the EGF receptor.

ERM 1. Ezrin/radixin/moesin (ERM) proteins, involved in linking plasma membrane proteins to the cortical actin meshwork. See **FERM domain**. 2. A member of the Ets family of transcription factors, (Erm). 3. Genes (*ermA*, *ermB*, etc.) coding for dimethyl- or methyltransferases that confer drug-resistance on many pathogenic bacteria. The transferases show absolute specificity for nucleotide A2058 in 23 S rRNA; monomethylation at A2058 confers resistance to a subset of the macrolide, lincosamide, and **streptogramin** B (MLS(B)) group of antibiotics, dimethylation at A2058 confers high resistance to all MLS(B) and ketolide drugs.

E-rosettes The clustering of sheep erythrocytes (= E) around a leucocyte or other cell. E-rosette formation is used as a marker for T-lymphocytes of humans and most mammals; in this case E are untreated, compared with other rosette tests such as EA, where E have antibody bound to their surface.

error-prone repair See **SOS system**.

erucic acid *(Z)-docos-13-enoic acid* Trivial name for 22:1 fatty acid. Found in rapeseed (canola) oil.

ERV Abbrev. for (1) endogenous retroviruses, RNA viruses that have become integrated into the genome of the host, (2) expiratory reserve volume.

Erwinia chrysanthemi Phytopathogenic bacterium that causes soft-rot. Virulence factors include pectinases coded by *pelB, pelC, pelD, pelE, ogl kduI* and *kdgT* that degrade the cell walls of the plant being attacked.

Eryf1 See **erythroid transcription factor**.

erysipelas A diffuse and spreading inflammation of skin and subcutaneous cellular tissue, particularly of the face, neck, forearm and hands. Possibly associated with an allergic reaction to products of the causative organism, *Streptococcus pyogenes*.

erythema nodosum Eruption of pink or red nodules, usually on the lower limbs, as a result of infection with any of a range of bacteria, viruses or fungi. Often associated with **IBD,** and in some parts of the world is commonly associated with lepromatous leprosy.

crythritol A polyol found in various fungi and algae, probably as a storage carbohydrate. Used medicinally to dilate blood vessels.

erythroblast Rather non-committal name for a nucleated cell of the bone marrow that gives rise to erythrocytes. See also **normoblasts, BFU-E, CFU-E, primitive** and **definitive erythroblasts**.

erythroblastosis fetalis Severe haemolytic disease of the neonate as a result of transplacental passage of maternal antibodies mainly directed against Rhesus blood group antigens.

erythrocyte A red blood cell.

erythrocyte ghost The membrane and cytoskeletal elements of the erythrocyte devoid of cytoplasmic contents, but preserving the original morphology (see Table E1).

erythrogenic toxin Toxin produced by strains of *Streptococcus pyogenes* responsible for scarlet fever. Three antigenic variants of the toxin are known. It is a small protein that is complexed with hyaluronic acid and can intensify the effects of other toxins such as **endotoxin** and **streptolysin O**.

erythroid cell Cell that will give rise to erythrocytes.

erythroid Kruppel-like factor *EKLF* Red-cell-specific transcriptional activator essential for establishing high levels of adult beta-globin expression.

erythroid transcription factor *Eryf1; GF-1; NF-E1* **Transcription factor** that binds to regulatory regions of genes expressed in erythroid cells.

erythroleukacmic cell Abnormal precursor (virally transformed) of mouse erythrocytes that can be grown in culture and induced to differentiate by treatment with, for example, DMSO. See **Friend murine erythroleukaemia cells**.

erythromycin General name for a variety of wide-spectrum macrolide antibiotics isolated from *Streptomyces erythreus*. Inhibit protein synthesis by binding to the prokaryotic 50S ribosomal subunit and preventing translocation. A variety of proteins will confer resistance to erythromycin – by degrading the antibiotic, by enhancing its export from the cell, or by causing modification to the RNA so that its affinity for erythromycin is reduced.

erythrophores **Chromatophores** that have red pigment.

erythropoiesis Process of production of erythrocytes in the marrow in adult mammals. A pluripotent stem cell (CFU)

TABLE E1. Erythrocyte membrane proteins

Band number after Steck[a]	MW (kDa)	Other name or function
1	240	Spectrin α
2	220	Spectrin β
2.1	200	Ankyrin; links band 3 to spectrin
3	93	Anion transporter
4.1	82	Links spectrin to glycophorin
4.2	72	Pallidin – stabilizes link between ankyrin and band 3
4.5	46	Glucose transporter
4.9	48	*Dematin* bundles microfilaments
5	43	Actin; forms short oligomers. involved in gelation of spectrin and band 4.1
6	35	Glyceraldehyde 3-phosphate dehydrogenase
7	32	Stomatin (band 7.2): missing in hereditary stomatocytosis

The mammalian erythrocyte ghost consists of a lipid bilayer linked to a cytoskeletal network. The proteins of the ghost vary across species, but there are some common patterns. Components are identified as far as possible by comparison with the proteins of the human erythrocyte ghost, after electrophoretic separation on SDS polyacrylamide gel, and numbered according to the Steck classification (*Journal of Cell Biology* (1974), 62: 1–29).
[a] These bands are visible when the gel is stained with a typical 'protein' dye, e.g. Coomassie brilliant blue. Other bands are only detected when stained for carbohydrate with the Periodic Acid/Schiff reagent (PAS). Four bands are characterized: PAS1, PAS2, PAS3 and PAS4. Of these, PAS1 and PAS2 are the glycoprotein glycophorin (55 kDa) in different oligomeric states. PAS3 and PAS4 are minor components.

produces, by a series of divisions, committed stem cells (**BFU-Es**) which give rise to **CFU-Es**, cells that will divide only a few more times to produce mature erythrocytes. Each stem cell product can give rise to 211 mature red cells.

erythropoietin Glycoprotein (46 kDa) produced in the kidney and that regulates the production of red blood cells in the marrow. Higher concentrations are required to stimulate **BFU-Es** than **CFU-Es** to produce erythrocytes. Recombinant EPO is now being used therapeutically in patients.

erythropterin A red pterine pigment deposited in the epidermal cells or the cavities of the scales and setae of many insects.

ES cells See **embryonic stem cells**.

Escherichia coli *E. coli* The archetypal bacterium for biochemists, used very extensively in experimental work. A rod-shaped Gram-negative bacillus (0.5×3–$5 \ \mu m$) abundant in the large intestine (colon) of mammals. Normally non-pathogenic, but the *E. coli* O157 strain, common in the intestines of cattle, has recently caused a number of deaths.

Escherichia coli haemolysin α-*hemolysin; HlyA* Exotoxin of the **RTX family** of bacterial cytolysins. Synthesized as an inactive 110-kDa precursor that is activated by fatty acid acylation by accessory protein HlyC. Product of many *E. coli* strains responsible for non-intestinal infections.

esculin A **coumarin** derivative extracted from the bark of flowering ash (*Fraxinus ornus*). Esculin is used in the manufacturing of pharmaceuticals and in diagnostic microbiology: Group D Streptococci hydrolyse esculin to esculetin and dextrose. Esculetin reacts with an iron salt such as ferric citrate to form a blackish-brown coloured complex.

E-selectin See **selectin**.

eserine *Physostigmine* An alkaloid that has anticholinesterase activity, isolated from the Calabar bean (*Physostigma venenosum*), used in treatment of glaucoma.

E-site Site on the ribosome that binds deacylated tRNA after it leaves the **P-site** and prior to it leaving the ribosome. See also **A-site**.

ESL Abbrev. for (1) endothelial surface layer, (2) E-**selectin** ligand-1 (ESL-1).

espins Small actin-bundling proteins that are highly enriched in the microvilli of certain chemosensory and mechanosensory cells, often with other actin-bundling proteins. The *Drosophila* homologue is known as forked. Two espin isoforms are expressed, 30 kDa and 110 kDa. Mutations in espin are associated with DFNB36 (autosomal recessive neurosensory deafness 36).

essential amino acids Those amino acids that cannot be synthesized by an organism and must therefore be present in the diet. The term is often applied anthropocentrically to those amino acids required by humans (Ileu, Leu, Lys, Met, Phe, Thr, Try and Val), though rats need two more (Arg and His).

essential fatty acids The three fatty acids required for growth in mammals – arachidonic, linolenic and linoleic acids. Only linoleic acid needs to be supplied in the diet; the other two can be made from it.

EST See **expressed sequence tag**.

established cell line See **cell line**.

esterase An enzyme that catalyses the hydrolysis of organic esters to release an alcohol or thiol and acid. The term could be applied to enzymes that hydrolyse carboxylate, phosphate and sulphate esters, but is more often restricted to the first class of substrate.

estradiol *Oestradiol; follicular hormone* US name and **BAN** for oestradiol. Female sex hormone (272 Da) synthesized mainly in the ovary, but also in the placenta, testis and, possibly, adrenal cortex. A potent **estrogen**. Synthetic form is used in hormone replacement therapy.

estrogen *Oestrogen* (UK, outmoded) A type of hormone that induces oestrus (heat) in female animals. It controls changes in the uterus that precede ovulation, and is responsible for development of secondary sexual characteristics in pubescent girls. Some tumours are sensitive to estrogens. See **estradiol**.

ET See **endothelin**.

etanercept *Enbrel* A dimeric fusion protein consisting of the extracellular ligand-binding portion of the human 75-kDa (p75) tumour necrosis factor receptor (**TNF receptor**) linked to the Fc portion of human IgG1. Used in treatment of rheumatoid arthritis.

ethacrynic acid A loop diuretic (acts on the ascending limb of the loop of Henle and on the proximal and distal tubules); inhibits reabsorption of a much greater proportion of filtered sodium than do most other diuretic agents. Used in the treatment of oedema and oliguria due to renal failure. Has been shown to inhibit signalling by NF-kappaB.

ethidium bromide A dye that intercalates into DNA and, to some extent, RNA. Intercalation into linear DNA is easier than into circular DNA, and the addition of ethidium bromide to DNA prior to ultracentrifugation on a caesium chloride gradient was used to separate nuclear and mitochondrial or plasmid DNA for analytical purposes. Because it intercalates less into the circular DNA, the density remains higher.

ethylene *ethene* A gas at room temperature, the simplest alkene hydrocarbon. Acts as a plant growth substance (phytohormone, plant hormone) involved in promoting growth, **epinasty**, fruit ripening, senescence and breaking of dormancy. Its action is closely linked with that of **auxin**.

etiolation Growth habit adopted by germinating seedlings in the dark. Involves rapid extension of shoot and/or hypocotyls, and suppression of chlorophyll formation and leaf growth.

etioplast Form of **plastid** present in plants grown in the dark. Lacks chlorophyll, but contains chlorophyll precursors and can develop into a functional chloroplast in the light.

etoposide *VP16* Semisynthetic lignan derivative synthesized from **podophyllotoxin**. Used as an anti-tumour drug; works by inhibiting topoisomerase II.

ets An **oncogene** found in E26 transforming **retrovirus** of chickens. Encodes a nuclear protein that regulates the initiation of transcription from a range of cellular and viral promoter and enhancer elements. There is some interaction between ets protein and **AP-1**. See Table O1.

ETS domain DNA-binding domain, formed of three **alpha-helices**. Named after the DNA-binding domain of the human ETS-1 **transcription factor**.

ets-1 Helix–turn–helix **transcription factor** of the tryptophan cluster class.

Eubacteria A major subdivision of the prokaryotes (includes all except **Archaebacteria**). Most Gram-positive bacteria, cyanobacteria, mycoplasmas, enterobacteria, pseudomonads and chloroplasts are Eubacteria. The cytoplasmic membrane contains ester-linked lipids, there is **peptidoglycan** in the cell wall (if present) and no **introns** have been discovered.

Eucaryote See **Eukaryote**.

euchromatin The chromosomal regions that are diffuse during interphase and condensed at the time of nuclear division. They show what is considered to be the normal pattern of staining (eu = true) as opposed to **heterochromatin**.

Eudorina Simple multicellular alga of the Order Volvocida – often quoted as illustrating the path to multicellularity. Small spherical or ovoid colonies of between 4 and

64 flagellated cells coexist within a gelatinous envelope. *Pandorina* and *Volvox* are similar, though more complex.

Euglena *Euglena gracilis* and *E. viridis* are phytoflagellate protozoa of the algal order Euglenophyta (zoological order Euglenida). An elongate cell with two **flagella**, one emerging from a pocket at the anterior end, the organism exhibits positive **phototaxis**, determined by a photoreceptive spot on the basal part of the flagellum shaft being shielded by a carotenoid-containing stigma (eyespot) in the wall of the pocket.

euglenoid movement A type of movement shown typically by *Euglena*, which swims with a single flagellum moving in a screw-like fashion but also shows writhing of the body caused by sub-pellicular contractions of cytoplasmic filament networks which can produce coordinated peristalsis-like movements in some cases.

Eukaryote Organism whose cells have (1) Chromosomes with nucleosomal structure and separated from the cytoplasm by a two-membrane nuclear envelope; and (2) Compartmentalization of a function in distinct cytoplasmic organelles. Contrast **Prokaryotes** (bacteria and cyanobacteria).

eumelanin Form of **melanin** found in animals – particularly in skin and hair and in pigmented retinal epithelium of the eye. See **phaeomelanin**.

Eumycetozoida Order of Protozoa, includes true slime moulds (not the cellular slime moulds).

Eumycota Division of fungi having defined cell walls and forming hyphae. The other main group is the **Myxomycota**.

eupeptide bond Peptide bond formed between α-carboxyl group of one amino acid and N-2 of another; the common peptide bond of proteins.

euploid A cell or an individual with a complete set of chromosomes.

euploidy Polyploidy in which the chromosome number is an integer multiple of the starting number.

Euplotes Genus of free-living hypotrich Protozoa. Do not have cilia, but may have undulating membranes for propulsion.

euryhaline Descriptive term for marine organisms that will tolerate a wide variation in salinity.

eurytopic Descriptive term for organisms that are able to survive in a wide range of environmental conditions.

eutectic A mixture of two elements that has a lower melting point than either alone.

eutely Phenomenon exibited by a few phyla, notably nematodes, where all individuals have the same number of cells (or nuclei in a coenobium).

Evans blue A diazo dye that binds to albumin and is commonly used to estimate blood volume (knowing the amount of dye injected and the concentration in a sample taken after the dye has distributed throughout the blood, the total volume is easily calculated). Can also be used to demonstrate sites where there is leakage of plasma protein from blood vessels, for example in a site of inflammation.

even-skipped Eve A **pair-rule gene** of *Drosophila*.

evening primrose oil An oil obtained from the seeds of *Oenothera biennis*, rich in gamma linolenic acid, the precursor for prostaglandin synthesis. Remarkable claims are made for the therapeutic value of this oil as a dietary supplement.

everninomicins Class of oligosaccharide antibiotics with high activity against Gram-positive organisms; identified many years ago but being reinvestigated for use in treating antibiotic-resistant *S. aureus*. Several variants have been isolated from the fermentation broth of *Micromonospora carbonacea* var *africana*. Binds to the same site in the large 50S ribosomal subunit as **avilamycin**; resistance to this class of drugs can arise through amino acid substitution in ribosomal protein L16 or mutations in the peptidyltransferase domain of 23S ribosomal RNA.

Evi-1 *Ecotropic viral integration site 1* A gene thought to play an important role in development and could be involved in organogenesis, cell migration, cell growth and differentiation. Implicated in acute and chronic myelogenous leukaemia and **myelodysplastic syndrome**, although the mechanism by which *Evi1* induces leukaemia is not known. Product has 1051 amino acids and is a 145-kDa DNA-binding protein. It contains two domains of seven and three sets of repeats of the **zinc finger** motif, a repression domain between the two sets of zinc fingers, and an acidic domain at the C-terminal end.

Ewing's sarcoma Sarcoma that develops in bone marrow.

exaptation Phenomenon in which a character or organ is used for a purpose other than that for which it first evolved – for example, the use of wings by penguins for swimming rather than flying.

excision repair Mechanism for the repair of environmental damage to one strand of **DNA** (loss of **purines** due to thermal fluctuations, formation of pyrimidine dimers by UV irradiation). The site of damage is recognized, excised by an **endonuclease**, the correct sequence is copied from the complementary strand by a **polymerase** and the ends of this correct sequence are joined to the rest of the strand by a **ligase**. The term is sometimes restricted to bacterial systems where the polymerase also acts as endonuclease.

excitable cell A cell in which the membrane response to **depolarizations** is non-linear, causing amplification and propagation of the depolarization (an **action potential**). Apart from neurons and muscle cells, electrical excitability can be observed in fertilized eggs, some plants and glandular tissue. Excitable cells contain **voltage-gated ion channels**.

excitation–contraction coupling Name given to the chain of processes coupling excitation of a muscle by the arrival of a nervous impulse at the **motor end plate** to the contraction of the filaments of the **sarcomere**. The crucial link is the release of calcium from the sarcoplasmic reticulum, and an analogy is often drawn between this and **stimulus–secretion coupling**, which also involves calcium release into the cytoplasm.

excitatory amino acid EAA The naturally occurring amino acids L-glutamate and L-aspartate and their synthetic analogues, notably **kainate**, **quisqualate** and **NMDA**. They have the properties of excitatory neurotransmitters in the CNS, may be involved in long-term potentiation, and can act as **excitotoxins**. At least three classes of EAA receptors have been identified; the agonists of the N-type receptor are L-aspartate, NMDA and ibotenate; the agonists of the Q-type receptor are L-glutamate and quisqualate; agonists of the K-type are L-glutamate and kainate. All three receptor types are found widely in the CNS, and particularly the telencephalon. N- and Q-type receptors tend to occur together and may interact; their distribution is complementary to the K-type receptors. The ion fluxes through the Q and K receptors are relatively brief, whereas the flux through the N-type is longer, and carries a significant amount of calcium. Additionally, the N-type receptor is blockaded by magnesium near the resting potential, and thus shows **voltage-gated ion channel** properties, leading to a regenerative response; this is why N-type receptors have been linked to long-term potentiation. Invertebrate glutamate receptors may not have the same properties as those described above.

excitatory synapse A synapse (either **chemical** or **electrical**) in which an action potential in the presynaptic cell increases the probability of an action potential occurring in the postsynaptic cell. See **inhibitory synapse**.

excitotoxin Class of substances that damage neurons through paroxysmal overactivity. The best-known excitotoxins are the **excitatory amino acids**, which can produce lesions in the CNS similar to those of **Huntington's chorea** or **Alzheimer's disease**. Excitotoxicity is thought to contribute to neuronal cell death associated with stroke.

exendin Group of peptide hormones (39 residues), related to the **glucagon** family, found in the saliva of Gila monsters (*Heloderma suspectum*, *H. horridum*). Helospectin is exendin-1; helodermin is exendin-2.

exergonic A biochemical reaction in which there is a release of energy – a negative change in free energy that can be used to produce work. Such reactions will proceed spontaneously.

exfoliatin Epidermolytic toxin produced by some strains of *Staphylococcus aureus*; causes detachment of outer layer of skin by disrupting desmosomes of the stratum granulosum.

exine External part of pollen wall that is often elaborately sculptured in a fashion characteristic of the plant species. Contains **sporopollenin**. The term is also used for the outer part of a spore wall.

exobiology The study of putative living systems that, statistically, are likely to exist elsewhere in the universe.

exocrine Exocrine glands release their secreted products into ducts that open onto epithelial surfaces. See **endocrine**.

exocytosis Release of material from the cell by fusion of a membrane-bounded vesicle with the plasma membrane.

exocytotic vesicle Vesicle, for example a secretory vesicle or **zymogen granule**, that can fuse with the plasma membrane to release its contents.

exoenzyme 1. An enzyme attached to the outer surface of a cell (an ectoenzyme) or released from the cell into the extracellular space. 2. An enzyme that only cleaves the terminal residue from a polymer (in contrast to an endoenzyme).

exogen A dicotyledonous plant that grows by means of a peripheral cambial layer.

exoglycosidase Hydrolytic enzymes that cleave glycosidic bonds of terminal sugar moieties. Sequential exoglycosidase cleavage is used to sequence carbohydrates. Can

also be used to refer to glycosidases that act on the exterior of a cell although the prefix ecto- is less ambiguous.

exogonic Exothermic reaction that has negative ΔG; releases energy. Same as exergonic.

exon The sequences of the RNA **primary transcript** (or the DNA that encodes them) that exit the nucleus as part of a **messenger RNA** molecule. In the primary transcript, neighbouring exons are separated by introns.

exon shuffling Process by which the evolution of proteins with multifunctional domains could be accelerated. If exons each encoded individual functional domains, then **introns** would allow their recombination to form new functional proteins with minimal risk of damage to the sequences encoding the functional parts.

exon skipping Probably the commonest cause of alternative splicing in which, during RNA processing, one or more exons are omitted and the message is abbreviated accordingly. A recent estimate is that more than 1200 human genes exhibit exon skipping.

exon trapping Technique for identifying regions of a genomic DNA fragment that are part of an expressed gene. The genomic sequence is cloned into an intron, flanked by two exons, in a specialized exon trapping vector, and the construct expressed through a strong promoter. If the genomic fragment contains an exon, it will be spliced into the resulting mRNA, changing its size and allowing its detection.

exonuclease Enzyme that digests the ends of a piece of DNA (*cf*. **endonuclease**). The nature of the digestion is usually specified (e.g. 5′ or 3′ exonuclease).

exonuclease III *Exo III* Enzyme that degrades DNA from one end. Used to prepare deletions in cloned DNA, or for **DNA footprinting**.

exopeptidase Peptide hydrolases of the class EC 3.4 that cleave the N- or C-terminal amino acid from a peptide.

exosome 1. Antigen-presenting vesicle secreted by some professional **antigen-presenting cells**. Exosomes are membrane-bounded and enriched in MHC Class I and II proteins. 2. The exosome (nuclear exosome) complex is involved in multiple RNA processing and degradation pathways and contains $3′ \rightarrow 5′$ exoribonucleases. It is found in eukaryotes and archaea. 3. A DNA fragment taken up by a cell that does not become integrated with host DNA but nevertheless replicates.

exothermic Process or reaction in which heat is produced – the opposite of endothermic.

exotoxins Toxins released from Gram-positive and Gram-negative bacteria, as opposed to **endotoxins** that form part of the cell wall. Examples are **cholera**, **pertussis** and **diphtheria toxins**. Usually specific and highly toxic. See Table E2.

exp6 *Exportin 6* A nuclear export receptor (**exportin**) that is specific for profilin-actin complexes and responsible for maintaining the nucleus actin-free.

expansins A superfamily of plant cell wall-loosening proteins that has been divided into four distinct families, designated alpha-expansin, beta-expansin, expansin-like A and expansin-like B. They facilitate cell expansion by selectively weakening the cell wall, although it appears that this may be achieved through a non-enzymatic mechanism.

TABLE E2. Exotoxins

Name	Source	Target/mode of action
Aerolysin	*Aeromonas hydrophila*	Pore forming
α-Toxin	*Clostridium perfringens*	Phospholipase C
α-Toxin	*Staphylococcus aureus*	Pore-forming
Anthrax toxin	*Bacillus anthracis*	Three components, one a soluble adenyl cyclase
Bacteriocins	Plasmid in *E. coli*	Colicin E2 is a DNase, colicin E3 an RNase.
Botulinum toxins	*Clostridium botulinum*	Inhibits acetylcholine release
Botulinolysin	*Clostridium botulinum*	Cholesterol binding
Cereolysin	*Bacillus cereus.*	Cholesterol binding
Cholera toxin	*Vibrio cholerae*	ADP-ribosylation of G_s
Diphtheria toxin	*Corynebacterium diphtheriae*	ADP-ribosylation of EF-2
δ−Toxin	*Clostridium perfringens*	Binds to cholesterol
β, ε and ι toxins	*Clostridium perfringens*	Increase vascular permeability
Enterotoxin	*Staphyloccus aureus*	Neurotoxic
Enterotoxin	*Pseudomonas aeruginosa*	Causes diarrhoea
Erythrogenic toxin	*Streptococcus pyogenes*	Skin hypersensitivity
Exfoliatin	*Staphyloccus aureus*	Disrupts desmosomes
Haemolysins α, β, χ, δ	*Staphyloccus aureus*	β is a sphingomyelinase C, γ is haemolytic, δ is a surfactant
Haemolysin	*Serratia marcescens*	Pore forming; different method to RTX toxins
Haemolysin	*Pseudomonas aeruginosa*	Toxic for macrophages
Heat-labile toxin	*Bordetella pertussis*	Dermonecrotic
Heat-labile toxin	*Escherichia coli*	Similar to cholera toxin
Heat-stable enterotoxin	*Escherichia coli*	Analogue of guanylin
Kanagawa haemolysin	*Vibrio haemolytica*	Haemolytic, cardiotoxic
κ-Toxin	*Clostridium perfringens*	Collagenase
Leucocidin/alpha-toxin	*Staphyloccus aureus* and *Pseudomonas aeruginosa*	Lyses neutrophils and macrophages
Listeriolysin O	*Listeria monocytogenes*	Cholesterol-binding
Perfringolysin (theta toxin)	*Clostridium perfringens*	Cholesterol-binding
Pertussis toxin	*Bordetella pertussis*	ADP-ribosylates G_i
Pneumolysin	*Streptococcus pneumoniae*	Binds to cholesterol
RTX family	Various Gram-negative bacteria	Calcium-dependent pore-forming toxins
Shiga toxin/verotoxin	*Shigella dysenteriae*	Blocks eukaryotic protein synthesis
Stable toxin	*Escherichia coli*	Activates guanylate cyclase
Streptolysin D	*Streptococcus pyogenes*	Binds cholesterol
Streptolysin S	*Streptococcus pyogenes*	Membranolytic
Subtilysin	*Bacillus subtilis*	Haemolytic surfactant
Tetanolysin	*Clostridium tetani*	Binds cholesterol
Tetanus toxin	*Clostridium tetani*	Inhibits glycine release at synapse
Thuringolysin	*Bacillus thuringiensis*	Binds cholesterol
Toxin A	*Pseudomonas aeruginosa*	ADP-ribosylates EF-2

G_s, G_i: see GTP-binding proteins EF-2, elongation factor 2.

experimental allergic encephalomyelitis *EAE* An autoimmune disease that can be induced in various experimental animals by the injection of homogenized brain or spinal cord in **Freund's adjuvant**. The antigen appears to be a basic protein present in myelin, and the response is characterized by focal areas of lymphocyte and macrophage infiltration into the brain, associated with demyelination and destruction of the blood–brain barrier. Sometimes used as a model for demyelinating diseases, although whether this is entirely justifiable is not clear.

exportins Family of proteins within the **karyopherin** superfamily (which includes **importins**). Importins and exportins are both regulated by the small GTPase Ran, which is thought to be highly enriched in the nucleus in its GTP-bound form. Exportins interact with their substrates (proteins with **nuclear export signals**) in the nucleus in the presence of Ran-GTP and release them after GTP hydrolysis in the cytoplasm, causing disassembly of the export complex. Exportin 1 is the same as **CRM-1**. **Leptomycin B** inhibits export by binding to exportins and preventing the binding interaction with the nuclear export signal.

expressed sequence tag *EST* DNA sequence derived by sequencing an end of a random cDNA clone from a library of interest. Usually, tens of thousands of such ESTs are generated as part of **genome projects**. These ESTs provide a rapid way of identifying cDNAs of interest, based on their sequence tag; they can then be purchased cheaply, obviating the need to screen a library.

expression cloning Method of **gene cloning** based on **transfection** of a large number of cells with **cDNAs** in an **expression vector** (e.g. a cDNA library), then screening for a functional property (e.g. binding of a radiolabelled hormone to identify receptors, or induction of transforming activity for putative oncogenes).

expression profiling Shorthand for gene expression profiling, a procedure usually carried out using DNA microarrays to obtain an insight into tissue- and developmental-specific expression of genes and the response of gene expression to environmental stimuli.

expression vector A **vector** that results in the **expression** of inserted DNA sequences when propagated in a suitable host cell, i.e. the protein coded for by the DNA is synthesized by the host's system.

extensin Glycoprotein of the plant cell wall, characterized by its high hydroxyproline content. Carbohydrate side-chains are composed of simple galactose residues and oligosaccharides containing 1–4 arabinose residues. Part of a larger class of **hydroxyproline-rich glycoproteins**. Function uncertain.

external guide sequence *EGS* RNA oligonucleotides termed external guide sequences serve as an RNA catalyst or ribozyme by directing bound mRNA to the ubiquitous cellular enzyme **RNAse P**. By virtue of their complementarity, they are mRNA-specific.

external transcribed spacers *ETS1; ETS2* See **internal transcribed spacers**.

extinction coefficient Outmoded term for **absorption coefficient**.

extracellular matrix *ecm; ECM* Any material produced by cells and secreted into the surrounding medium, but usually applied to the non-cellular portion of animal tissues. The ecm of connective tissue is particularly extensive, and the properties of the ecm determine the properties of the tissue. In broad terms, there are three major components: fibrous elements (particularly **collagen**, **elastin**, or **reticulin**), link proteins (e.g. **fibronectin**, **laminin**) and space-filling molecules (usually **glycosaminoglycans**). The matrix may be mineralized to resist compression (as in bone) or dominated by tension-resisting fibres (as in tendon). The basal lamina of epithelial cells is another commonly encountered ecm. Although ecm is produced by cells, it has recently become clear that the ecm can influence the behaviour of cells quite markedly – an important factor to consider when growing cells *in vitro*: removing cells from their normal environment can have far-reaching effects.

extrachromosomal element Any heritable element not associated with the chromosome(s). It is usually a **plasmid** or the DNA of organelles such as mitochondria and chloroplasts.

extreme halophiles Organisms that can withstand extreme salt concentrations such as those found in salt lakes (and preserved food). There are six genera of archaeobacteria that are classified as extreme halophiles. They are aerobic chemoheterotrophs which require complex energy and carbon sources (particularly proteins and amino acids), and their minimum osmotic requirements exceed 1.5-M NaCl (8% wt/vol) although they will tolerate levels of up to 36% wt/vol).

extremophile An organism that requires an extreme environment in which to flourish – examples are **thermophiles** and **halophiles**.

extrinsic pathway Initiation of blood clotting as a result of factors released from damaged tissue, as opposed to contact with a foreign surface (the intrinsic pathway). Tissue **thromboplastin** (Factor III) in conjunction with Factor VII (proconvertin) will activate Factor X that, once activated, converts prothrombin to thrombin.

exudate cells Leucocytes that enter tissues (exude from the blood vessels) during an inflammatory response. See also **peritoneal exudate**.

eyepiece graticule *Micrometer eyepiece; (US) ocular micrometer* Grid incorporated in the eyepiece for measuring objects under the microscope. May have any sort of scale, and there are special types used in particle-size analysis consisting of a rectangular grid for selecting the particles and a series of graded circles for use in sizing. Need to be calibrated against a scale (micrometer slide) for each objective.

eyespot *Stigma* An orange or red spot found near photoreceptive areas in motile cells of many algae, phytoflagellates and protozoa. Are assumed to help detect the direction of light in phototaxis. Pigments are carotenoids. See *Euglena*.

ezrin Microfilament-bundling protein (80 kDa) from the core of microvilli. Phosphorylated following stimulation of cells.

F1 See **neuromodulin**.

F1F0 ATPase *F-type ATPase; ATP synthase; F-ATPase*
Multisubunit proton-transporting ATPase, related to the
V-type ATPase. Found in the inner membrane of **mitochondria** and **chloroplasts** and in bacterial **plasma membranes**. Normally driven in reverse by **chemiosmosis** to
make **ATP**, and so also known as ATP synthase.

F1 hybrid First filial generation, the product of crossing
two dissimilar parents. If the parents are sufficiently dissimilar the hybrid may be sterile (for example, in the crossing
of horse and donkey to produce a mule), and the term F1
hybrid generally refers to such sterile hybrids which may,
however, show desirable hybrid vigour.

F11 Neural cell recognition molecule with **immunoglobulin** type C domains and **fibronectin** type III repeats. Its
cDNA sequence is almost identical to **contactin** except that
while F11 is probably attached to the membrane via phosphatidyl inositol and lacks a cytoplasmic domain, contactin
is attached to the cytoskeleton. Like contactin, **neurofascin**,
TAG-1 and **fasciclin II**, F11 is thought to be associated
with the process of **fasciculation**.

Fab Fragment of immunoglobulin prepared by papain treatment. Fab fragments (45 kDa) consist of one light chain
linked through a disulphide bond to a portion of the heavy
chain and contain one antigen-binding site. They can be
considered as univalent antibodies.

Fab$_{(2)}$ The fragment (90 kDa) of an immunoglobulin produced by pepsin treatment. These fragments have two
antigen-combining sites and contain two light chains and
two variable region heavy chains plus one constant region
domain in each heavy chain. The fragment is divalent but
lacks the complement-fixing (Fc) domain.

FABP See **fatty acid-binding protein**.

Fabry disease *Angiokeratoma* X-linked **storage disease**
due to mutation in alpha-galactosidase (EC 3.2.1.22; also
known as ceramide trihexosidase). The enzyme deficiency
leads to inadequate breakdown of lipids, which accumulate
to harmful levels in the eyes, kidneys, autonomic nervous
system and cardiovascular system.

F-actin Filamentous **actin**.

facilitated diffusion *Passive transport* A process by
which substances are conveyed across cell membranes
faster than would be possible by diffusion alone. This is
generally achieved by proteins that provide a hydrophilic
environment for polar molecules throughout their passage
through the **plasma membrane**, acting as either shuttles or
pores. See **symport**, **antiport**, **uniport**.

facilitation Greater effectiveness of synaptic transmission by successive presynaptic impulses, usually due to
increased transmitter release.

facilitator neuron A neuron whose firing enhances the
effect of a second neuron on a third. This allows the effects
of neuronal activity to be modulated.

FACS See **flow cytometry**.

Factors I–XII Blood clotting factors, especially from
humans. These factors form a cascade in which the activation of the first factor leads to enzymic attack on the next
factor and so on, finally resulting in blood clotting. See
Table F1.

facultative heterochromatin That **heterochromatin**
which is condensed in some cells and not in others, presumably representing stable differences in the activity of genes
in different cells. The best-known example results from the
random inactivation of one of the pair of X chromosomes
in the cells of female mammals (**Lyonization**).

FAD *Flavin adenine dinucleotide* A prosthetic group of
many flavin enzymes. See **flavin nucleotides**.

FADD *Fas-associated via death domain* Adaptor protein
(208 aa) that links **death receptors** to **caspases** in the signalling pathway that leads to apoptotic cell death.

FAK See **focal adhesion kinase**.

TABLE F1. Blood clotting factors

Factor	Name	M_r (kDa)	Function
	Fibrinogen	340	Cleaved to form fibrin
II	Prothrombin	70	Converted to thrombin by Factor X
III	Thromboplastin	–	Lipoprotein which acts with VII to activate X
IV	Calcium ions	–	Needed at various stages
V	Proaccelerin	–	Product accelerin, promotes thrombin production
VII	Proconvertin	–	Activated by trauma to tissue
VIII	Antihaemophilic factor	$>10^3$	Acts with IXa to activate X
IX	Christmas factor	55	See VIII
X	Stuart factor	55	When activated converts II to thrombin
XI	Thromboplastin antecedent	124	Converts IX to active form
XII	Hagemann factor	76	Activated by surface contact
XIII	Fibrin-stabilizing factor	350	Transglutaminase which cross-links fibrin

Falconization Trade name for the treatment of polystyrene to make it appropriate for use in cell culture by increasing its wettability and thus the ability of proteins and then cells to adhere. The main commercial process was probably corona discharge in air or other gas mixtures at low pressure. Treatment of polystyrene with sulphuric acid will produce the same effect. Now superseded by other methods.

false-positive A positive result in an assay that is due to something other than the effect of interest or to some other random factor – for example, inhibition of growth by general toxicity rather than inhibition of a growth-regulating pathway. To avoid such false-positives, assays are often designed to give a positive read-out if the inhibitor works.

familial hypercholesterolaemia Excess of **cholesterol** in plasma as a result of a defect in the recycling process in which LDL (**low-density lipoprotein**) is taken up into **coated vesicles**. Various defects in the pathway lead to different forms of the disease, some dominant, others recessive.

famotidine Drug that blocks histamine H2 receptors and is used for treatment of gastric ulcers.

Fanconi syndrome Transport disease (recessive defect) in which the renal reabsorption of several substances (phosphate, glucose, amino acids) is impaired.

Fanconi's anaemia Defect in thymine-dimer excision from DNA predisposing to development of leukaemia.

far Western blot Form of Western blot in which protein/protein interactions are studied. Proteins are run on a gel and transferred to a membrane as in a normal **Western blot**. The proteins are then allowed to renature, incubated with a candidate protein and the blot washed. Areas of the blot where the protein has adhered are then detected with an antibody.

FAR1 Yeast gene, induced by a factor, so called because it is a 'Factor Arrest' gene. Product is a cyclin-dependent protein kinase inhibitor that causes cells to arrest in G1 phase, by interacting with the G1 cyclin, CLN2.

Farber's disease Lipogranulomatosis caused by deficiency of acid **ceramide**-degrading enzyme (EC 3.5.1.23), a storage disease.

Farmer's lung Type III **hypersensitivity** response to *Micropolyspora faeni*, a thermophilic bacterium found in mouldy hay. Conveniently afflicts Joe Grundy in BBC Radio 4's *The Archers*.

farnesyl transferase *EC 2.5.1.21* Enzyme that adds a farnesyl group to certain intracellular proteins. See **farnesylation**.

farnesylation The farnesyl group is the linear grouping of three isoprene units. It is specifically attached to proteins that contain the C-terminal motif CAAX by cleavage and addition to the SH group of C; the free carboxylate group is also methylated. Believed to act as a membrane attachment device. See also **polyisoprenylation**.

Farr-type assay Method of radioimmunoassay in which free antigen remains soluble and antibody–antigen complexes are precipitated.

Fas *Fas antigen; Fas receptor* Cell surface transmembrane protein (35 kDa) that mediates **apoptosis**. Has structural

homology with **TNF** receptor and **NGF** receptor. May play a part in negative selection of autoreactive T-cells in the thymus. See **Fas ligand**.

Fas ligand *FasL* Ligand for **Fas** (human 281 aa, 32 kDa); a type II transmembrane protein that belongs to the tumour-necrosis factor family. Has the ability to induce trimerization of Fas by virtue of having three receptor-binding sites; binding induces apoptosis. FasL is expressed in activated splenocytes and thymocytes. A soluble 26-kDa form of FasL is cleaved from the cell surface by matrix metallopeptidases.

fascicle Literally, a bundle. In particular, this is used to describe the tendency of **neurites** to grow together (**fasciculation**).

fasciclins Cell adhesion molecules of the **immunoglobulin superfamily** found in the central nervous system of insects. Involved with **fasciculation** of axons and probably in pathfinding during morphogenesis of the nervous system. The sequence of fasciclin II shows that it shares structural motifs with a variety of vertebrate **CAMs**. See **contactin**, **F11**, **neurofascin**, **TAG-1**, *Drosophila* **neuroglian**.

fascicular cambium Form of **cambium** present in the vascular bundles of higher plants.

fasciculation Tendency of developing **neurites** to grow along existing neurites and hence form bundles or **fascicles**. Selective fasciculation in developing vertebrate nervous systems is thought to involve the **axon**-associated **cell adhesion molecules L1**, **F11**, **contactin**, **neurofascin** and transient axonal glycoprotein **TAG-1**, and the **fasciclins** in insect nervous systems.

fascin Actin filament-bundling protein (58 kDa), the actin-binding ability being regulated by phosphorylation; originally identified from sea-urchin eggs.

FAST *Forkhead activin signal transducer* DNA-binding component (60 kDa) of the **ARF** complex, binds to **activin** response element in *mix* gene promoter. A **winged-helix transcription factor** of the **forkhead** family.

FASTA A bioinformatics resource, available on the web, that compares a protein sequence to another protein sequence or to a protein database, or a DNA sequence to another DNA sequence or a DNA library.

fat cell See **adipocyte**.

fat droplets Microaggregates of (mainly) triglycerides visible within cells.

fate map Diagram of an early embryo (usually a **blastula**) showing which tissues the cells in each region will give rise to (i.e. their developmental fate). Fate maps are normally constructed by labelling small groups of cells in the blastula with vital dyes and seeing which tissues are stained when the embryo develops.

fats A term largely applied to storage lipids in animal tissues. The primary components are triglyceride esters of long-chain fatty acids.

fatty acid-binding proteins *FABP* Group of small cytosolic proteins that bind fatty acids and facilitate their intracellular transport. At least eight different types of human FABP occur, each with a specific tissue distribution and possibly with a distinct function.

fatty acid synthase *FAS; EC 2.3.1.85* In animal tissues, a complex multifunctional enzyme consisting of two identical monomers. The FAS monomer (approximately 270 kDa) contains six catalytic activities; from the N-terminus, the order is beta-ketoacyl synthase (KS), acetyl/malonyl transacylase (AT/MT), beta-hydroxyacyl dehydratase (DH), enoyl reductase (ER), beta-ketoacyl reductase (KR), acyl carrier protein (ACP) and thioesterase (TE).

fatty acids Chemically, R-COOH where R is an aliphatic moiety. The common fatty acids of biological origin are linear chains with an even number of carbon atoms. Free fatty acids are present in living tissues at low concentrations. The esterified forms are important both as energy storage molecules and as structural molecules. See **triglycerides**, **phospholipids**.

fatty streak Superficial fatty patch in the artery wall caused by the accumulation of cholesterol and cholesterol oleate in distended **foam cells**.

fava bean The broad bean, *Vicia fava*. See **favism**.

favism Haemolytic anaemia induced, in individuals who are glucose 6-phosphate dehydrogenase-deficient, by eating fava beans (from *Vicia fava*).

F-box Motif of approximately 50 amino acids that functions as a site of protein–protein interaction. **F-box proteins** were first characterized as components of the **SCF complex,** but have subsequently been found in a number of other eukaryotic regulatory proteins.

F-box proteins Adapter proteins that are involved in associating proteins with the ubiquitin-driven proteolytic system. The F-box is a motif originally identified within *Neurospora crassa* negative regulator sulphur controller-2 but subsequently found in a wide variety of proteins including many cell cycle regulatory proteins, though various F-box proteins probably also play a part in regulation of transcription, signal transduction and development. See **SCF complexes**.

Fc That portion of an immunoglobulin molecule (fragment crystallizable) that binds to a cell when the antigen-binding sites (**Fab**) of the antibody are occupied or the antibody is aggregated; the Fc portion is also important in **complement** activation. The Fc fragment can be separated from the Fab portions by pepsin. Fc moieties from different antibody classes and subclasses have different properties.

Fc receptors Receptors for the **Fc** portion of immunoglobulins. FcγRI (CD64) is the receptor for IgG1, as are FcγRII-A (CD32), FcγRII-B2 (CD32), FcγRII-B1 (CD32) and FcγRIII (CD16). FcεRI binds IgE, FcαRI binds IgA. The distribution on cells varies and the consequences of receptor occupancy differ according to the subtype. FcγRII-B1 and FcγRII-B2 are inhibitory and have **ITIM** motifs in their cytoplasmic domains.

fccp *Carbonylcyanide-p-trifluoromethoxyphenylhydrazone* A potent **uncoupling agent** of mitochondrial oxidative phosphorylation.

FCS See **fetal calf serum**.

FDA *Food and Drug Authority* American regulatory authority responsible for assuring the safety of prescription drugs. FDA approval is essential for a new drug to be launched on the market.

feedback regulation Control mechanism that uses the consequences of a process to regulate the rate at which the process occurs: if, for example, the products of a reaction inhibit the reaction from proceeding (or slow down the rate of the reaction), then there is negative feedback – something that is very common in metabolic pathways. Positive feedback is liable to lead to exponential increase and may be explosively dangerous in some cases. Other examples are the action of voltage-gated **sodium channels** in generating action potentials, and the activation of blood clotting **factors V and VIII** by **thrombin**. Without damping, feedback can lead to resonance (hunting) and oscillation in the system.

feeder layer In order to culture some cells, particularly at low or clonal density, it is necessary to use a layer of less fastidious cells to condition the medium. Often the cells of the feeder layer are irradiated or otherwise treated so that they will not proliferate. In some cases the feeder layer may be producing growth factors or cytokines.

feline immunodeficiency virus *FIV* Widespread lentivirus (retrovirus) that causes an immunodeficiency in domestic cats. The immunodeficiency may be due to failure to generate an IL-12-dependent type I response. CXC-R4 seems to be the surface receptor for viral binding – CD4 is not required, in contrast with HIV infection.

Felis catus Cat that is said to be domesticated.

Femara See **letrozole**.

fenfluramine *TN Pondimin* Drug, now withdrawn, formerly used in conjunction with phentermine to treat obesity.

fentanyl A powerful opioid analgesic resembling morphine in its action but approximately $80\times$ more potent as an analgesic. Agonist for mu **opiate receptors**.

Fenton's reagent A solution of hydrogen peroxide and an iron catalyst, used to oxidize contaminants or waste waters. Organics targeted for treatment include chlorinated solvents, munitions, pesticides, petroleum hydrocarbons, wood preservatives, PCBs and phenolics.

FERM domain Conserved domain of about 150 residues found in Protein 4.1 (F), **ezrin** (E), **radixin** (R) and **moesin** (M) and a number of other cytoskeletal-associated proteins that link the cytoskeleton to proteins of the plasma membrane.

fermentation Breakdown of organic substances, especially by microorganisms such as bacteria and yeasts, yielding incompletely oxidized products. Some forms can take place in the absence of oxygen, in which case **ATP** is generated in reaction pathways in which organic compounds act as both donors and acceptors of electrons. Historically, the production of ethyl alcohol or acetic acid from glucose. Also applied to anaerobic **glycolysis** as in **lactate** formation in muscle.

ferredoxins Low molecular-weight iron–sulphur proteins that transfer electrons from one enzyme system to another without themselves having enzyme activity.

ferrichromes *siderochromes* Ligands for iron binding secreted by microorganisms to sequester and transport iron.

ferritin An iron-storage protein of mammals, found in liver, spleen and bone marrow. Morphologically, a shell of apoferritin (protein) with a core of ferrous hydroxide/phosphate. It is much used as an electron-dense label in electron microscopy.

ferroportin *Slc40a1 (solute-carrier family 40, member 1)* A transmembrane iron export protein expressed in macrophages and duodenal enterocytes. Heterozygous mutations in the ferroportin gene result in an autosomal dominant form of iron overload disorder, type IV haemochromatosis. See **hepcidin**.

fertilin *PH-30* Membrane protein found in the head region of sperm that appears to participate in sperm adhesion to the egg membrane. It is a heterodimer; the beta subunit has a region of homology to the disintegrin family of integrin ligands and the alpha subunit has a region of homology to viral fusion peptides. Both subunits have been identified as type I membrane glycoproteins.

fertilisin Old (obsolete) term for an agglutinin from sea-urchin egg, originally described by Lillie in 1913. Similar substances have been postulated to play a role in mammalian sperm–egg interactions. See **bindin**, **fertilin**, **speract**.

fertilization The essential process in sexual reproduction, involving the union of two specialized **haploid** cells (the gametes) to give a diploid cell (the zygote), which then develops to form a new organism.

fertilization membrane Membrane formed on the inner surface of the **vitelline membrane** of the sea-urchin egg following entry of the sperm. Following sperm penetration a wave of depolarization spreads over the egg surface and intracellular calcium levels rise, triggering the fusion of **cortical granules** that contain a complex variety of molecules which between them inhibit polyspermy.

ferulic acid Phenolic compound present in the plant cell wall that may be involved in cross-linking polysaccharide.

fes *fps* An oncogene, identified in avian and feline **sarcomas**. The *fps/fes* proto-oncogene is abundantly expressed in myeloid cells and the Fps/Fes cytoplasmic protein tyrosine kinase is implicated in signalling downstream from haematopoietic cytokines, including interleukin-3 (IL-3), granulocyte-macrophage colony-stimulating factor (GMCSF) and erythropoietin (EPO). Four somatic mutations in sequences encoding the Fps/Fes kinase domain have been identified in human colorectal carcinomas. See also Table O1.

fesselin A proline-rich actin-binding protein, similar to **synaptopodin**, but isolated from turkey-gizzard smooth muscle.

fetal calf serum *FCS, foetal calf serum* Expensive component of standard culture media for many types of animal tissue cells.

fetuin *α2-HS-glycoprotein* An α-globulin constituting up to 45% of the total protein in **fetal calf serum**. Very carbohydrate-rich, and a growth factor for many cells. Protein portion has **cystatin** features.

Feulgen reaction Specific staining procedure for DNA: mild acid hydrolysis makes the aldehyde group of deoxyribose available to react with Schiff's reagent to give a purple colour.

fexofenadine *TN Allegra* Antihistamine drug, a histamine H1-receptor antagonist, used to treat hay fever.

FFA Abbrev. usually for (1) free fatty acids, but occasionally for (2) fundus fluorescein angiography, an ophthalmological technique used to examine the vasculature of the retina.

F-factor **Plasmid** that confers the ability to conjugate (i.e. fertility) on bacterial cells and carries the *tra* genes (transfer genes); first described in *E. coli.*

FGD1 A GEF (GDP/GTP exchange factor) specific for **cdc42**, determined by positional cloning to be the locus responsible for faciogenital dysplasia (**Aarskog–Scott syndrome**). Though similar to **frabin** does not have an actin-binding domain but does have a proline-rich domain.

FGF See **fibroblast growth factor**.

***fgf* oncogenes** K-fgf/hst; fgf-5, fgf-3 Oncogenes encoding growth factors of the fibroblast growth factor (FGF) family that transform cells through an autocrine mechanism. See also Table O1.

fgr *Feline sarcoma oncogene homologue* **Oncogene** identified in a feline **sarcoma**, encoding a member of the *c-src* gene family of cytoplasmic tyrosine kinases. Expression of the *c-fgr* gene is activated in human B-lymphocytes following infection with Epstein–Barr virus. Regulates cell migration through effects on a signalling pathway involving **FAK**/Pyk2 and leading to activation of Rac and the Rho inhibitor p190RhoGAP. See also Table O1.

FIAU *2′-Deoxy-2′-fluoro-5-iodo-1-beta-D-arabinofuranosyluracil* Nucleoside analogue often used in labelled form (with ^3H, ^{14}C, ^{125}I or ^{18}F) in imaging studies to locate the site of expression of viral thymidine kinase.

fibrates A group of lipid-lowering drugs that reduce plasma triglycerides and increase breakdown of LDL-cholesterol.

fibre cell Greatly elongated type of plant cell with very thick lignified wall. Usually dead at maturity, this cell type is specialized for the provision of mechanical strength. Fibre cells and **sclereids** together make up the tissue known as **sclerenchyma**.

fibrillar centres Location of the nucleolar ribosomal chromatin at telophase: as the nucleolus becomes active, the ribosomal chromatin and associated ribonucleoprotein transcripts compose the more peripherally located dense fibrillar component.

fibrillarin Highly conserved nucleolar protein (34–36 kDa) that associates with U3-snoRNP and is found in the **coiled body** of the nucleolus. The N-terminus contains a glycine and arginine-rich domain (GAR domain). Yeast homologue is NOP1. Expression of fibrillarin (and **nucleolin**) is greater in rapidly proliferating cells and in the early stages of lymphocyte activation. Autoantibodies to fibrillarin are found in some patients with scleroderma, systematic sclerosis, **CREST** syndrome and other connective tissue diseases.

fibrillation 1. In certain neurological diseases, twitching of individual muscle fibres or bundles of fibres. 2. Uncoordinated contraction of regions of the heart, as in atrial fibrillation or ventricular fibrillation.

fibrillin Widely distributed connective tissue protein (350 kDa) associated with microfibrils (10 nm diameter). See **tropoelastin**. Mutations in the fibrillin-1 gene are responsible for **Marfan syndrome**.

fibrin Monomeric fibrin (323 kDa) is produced from **fibrinogen** by proteolytic removal of the highly charged (aspartate- and glutamate-rich) **fibrinopeptides** by thrombin, in the presence of calcium ions. The monomer readily polymerizes to form long insoluble fibres (23 nm periodicity; half-staggered) that are stabilized by covalent cross-linking (by Factor XIII, plasma transglutaminase). The fibrin gel acts as a haemostatic plug

fibrinogen Soluble plasma protein (340 kDa; 46 nm long), composed of six peptide chains (two each of Aα, Bβ and γ) and present at about 2–3 mg/ml. See **fibrin**.

fibrinolysis Solubilization of fibrin in blood clots, chiefly by the proteolytic action of **plasmin**.

fibrinopeptides Very negatively charged peptide fragments cleaved from **fibrinogen** by thrombin. Two peptides (A and B) are produced from each fibrinogen molecule.

fibroblast Resident cell of connective tissue, mesodermally derived, that secretes fibrillar procollagen, fibronectin and collagenase.

fibroblast growth factor *FGF; aFGF; bFGF; HBGF* Also known as heparin-binding growth factor (HBGF). Acidic FGF (aFGF, HBGF 1) and basic FGF (bFGF, HBGF 2) are the two founder members of a family of structurally related **growth factors** for mesodermal and neuroectodermal cells. Both aFGF and bFGF lack a signal sequence and the pathway of release is unclear. In addition to their growth promoting activity, FGFs play an important part in developmental signalling.

fibroblast growth factor receptor Family of **receptor tyrosine kinases** for **fibroblast growth factor**.

fibroblastic Many types of cultured cell become fibroblast-like in appearance; this does not mean that they are **fibroblasts**.

fibroid *Leiomyoma* A benign tumour of smooth muscle origin, usually in the uterus or occasionally in the gastrointestinal tract.

fibroin Structural protein of silk, one of the first to be studied with X-ray diffraction. It has a repeat sequence GSGAGA and is unusual in that it consists almost entirely of stacked antiparallel **beta-pleated sheets**.

fibromodulin A small proteoglycan, around 60 kDa, of the extracellular matrix. The core protein has a mass of around 42 kDa and is very similar to the core protein of **biglycan** and **decorin**. All three have highly conserved sequences containing 10 internal homologous repeats of around 25 amino acids with leucine-rich motifs. Fibromodulin has four keratan sulphate chains attached to N-linked oligosaccharides.

fibromuscular dysplasia *FMD* A non-inflammatory, non-atherosclerotic disease of the small- to medium-sized blood vessels that primarily affects young females in their second to fourth decades of life. One of the most common treatable causes of secondary hypertension.

fibronectin Glycoprotein of high molecular weight (two chains each of 250 kDa linked by disulphide bonds) that occurs in insoluble fibrillar form in extracellular matrix of animal tissues and in soluble form in plasma, the latter previously known as cold-insoluble globulin. The various slightly different forms of fibronectin appear to be generated by tissue-specific differential splicing of fibronectin mRNA, transcribed from a single gene. Fibronectins have multiple domains that confer the ability to interact with many extracellular substances, such as collagen, fibrin and heparin, and also with specific membrane receptors on

responsive cells. Notable is the **RGD** domain recognized by **integrins** and two repeats of the **EGF-like domain**. The fibronectin type III domain (FnIII), about 90 amino acids long, of which there are 15–17 per molecule, is a common motif in many cell surface proteins. Interaction of a cell's fibronectin receptors (members of the **integrin** family) with fibronectin adsorbed to a surface results in adhesion and spreading of the cell.

fibrosarcoma Malignant tumour derived from connective tissue fibroblast.

fibrosis Deposition of avascular collagen-rich matrix (**fibrous tissue**) in a wound, usually as a consequence of slow fibrinolysis or extensive tissue damage as in sites of chronic inflammation.

fibrous lamina Alternative name for the **nuclear lamina**, the region lying just inside the inner nuclear membrane.

fibrous plaque Thickened area of arterial **intima** with accumulation of smooth muscle cells and fibrous tissue (collagen, etc.) produced by the fat-laden smooth muscle cells. Below the thickening may be free extracellular lipid and debris, which, if much necrosis is also present, is referred to as an **atheroma**.

fibrous tissue Although most connective tissue has fibrillar elements, the term usually refers to tissue laid down at a wound site – well-vascularized at first (**granulation tissue**) but later avascular and dominated by collagen-rich extracellular matrix, forming a scar. Excessive contraction and hyperplasia leads to formation of a **keloid**.

fibulin Family of calcium-binding, cysteine-rich glycoproteins found in the extracellular matrix and in plasma. Alternative splicing generates three forms of fibulin with 566, 601 and 683 amino acids, respectively. All three forms have three repeated motifs near the N-terminus, with the bulk of the remaining chain formed of nine EGF-like repeats. Fibulin was originally described as a cytoplasmic protein, but this identification was based on fortuitous binding to the integrin β1 subunit. Conservation of the fibulin-1 gene throughout metazoan evolution includes fibulin-1C and fibulin-1D alternate splice variants.

ficin *Ficain; EC 3.4.22.3* Cysteine endopeptidase that selectively cleaves at Lys-, Ala-, Tyr-, Gly-, Asn-, Leu- or Val-. Similar to papain. Commercial ficin is purified from the latex of the fig tree, *Ficus glabrata* or *Ficus carica*.

Fick's law Equation that describes the process of diffusion. The flux is proportional to the concentration gradient, times the diffusion constant for the molecule in that particular medium. $J = -D \, dC/dx$

ficolin *L-Ficolin/P35; H-ficolin/Hakata antigen; M-ficolin* Family of molecules that interact with the **MASP** complex and activate the lectin complement pathway. The ficolin polypeptide chain consists of a short N-terminal domain, a middle collagen-like domain that forms a triple helix, and a C-terminal fibrinogen-like domain. Structurally very similar to **collectins**, **conglutinin**, and **surfactant proteins A and D**. Ficolins have been identified in mammals, including man, rodents and pigs, with tissue-specific distributions. They have also been found in ascidians. L- and H-ficolin are found in serum, M-ficolin in leucocytes and lung.

Ficoll Synthetic branched copolymer of sucrose and epichlorhydrin. Ficoll solutions have high viscosity and low osmotic pressures. Often used for preparing density gradients for cell separations (sometimes in conjunction with Hypaque for leucocyte separation).

Ficoll-Paque Proprietary name for premixed Ficoll and diatrizoate (Hypaque) with a density of 1.077 g/cm³ used as a cushion for separating lymphocytes (which do not pass through the Ficoll-Paque layer) from other blood cells in a one-step centrifugation method.

field ion microscope Type of microscopy in which the specimen is 'illuminated' with ions, often gallium ions, that are focused electrostatically. The ions remove components of the specimen, lower atomic masses first. These are imaged and provide information on elemental distribution with a resolution of perhaps 30 nm.

filaggrins Basic protein components of **keratohyalin granules** of the suprabasal cells of the skin. Family of intermediate filament-associated cationic proteins found in mammalian epidermis. Bundle cytokeratin filaments. Various sizes in different species (16 kDa bovine, 26 kDa mouse, 35 kDa man, 45 kDa rat). Produced by dephosphorylation and subsequent proteolysis of **profilaggrin**.

filamentous phage *Inovirus* Single-stranded DNA **bacteriophage**s of the genus *Inoviridae*. Examples that infect *E. coli* are M13, f1.

filaments See **thick filaments, thin filaments, intermediate filaments**, and **microfilaments**.

filamin A protein that binds to **F-actin**, cross-linking it to form an isotropic network; the binding does not require Ca²⁺. It was originally isolated from smooth muscle and is a homodimer 2 × 250 kDa. Similar to actin-binding protein (ABP) from leucocytes.

Filaria Genus of nematode worms causing elephantiasis and filariasis. Transmitted by insects.

filensin Protein (100 kDa) of the intermediate filament family found in lens fibre cells. Binds to **vimentin** and coassembles with **phakinin** to form the lens-specific intermediate filament system referred to as beaded-chain filaments. Filensin differs from other intermediate filament proteins in having a rather short central rod domain and will not, on its own, assemble to form intermediate filaments.

filiform papillae Curved tapering cone-shaped body on the tongue of rodents, of which the epithelial cell columns have been investigated in detail.

filipin Polyene antibiotic from *Streptomyces filipinensis*. Polymers of filipin associated with cholesterol in the cell membrane form pores which lead to cytolysis (as does **amphotericin B**).

filopodium *Pl.* filopodia A thin protrusion from a cell, usually supported by microfilaments; may be functionally the linear equivalent of the leading lamella.

Filoviridae Family of single-stranded RNA viruses, similar in some respect to rhabdoviruses. Marburg and Ebola viruses are the only two members of the family at present. Filovirus infections seem to cause intrinsic activation of the clotting cascade, leading to haemorrhagic complications and high mortality. Morphologically, virions are very long filaments (up to 14 μm, 70 nm thick), sometimes branched. The RNA is contained within a nucleocapsid that is surrounded by a cell-derived envelope.

filovirus Virus of family **Filoviridae**.

fimbria *Pl.* fimbriae See **pilus**.

fimbrillin Major subunit protein of bacterial **pili** (fimbriae). Binds to fibronectin and **statherin**. Coded by *Fim* genes. In *Porphyromonas* (*Bacteroides*) *gingivalis*, fimbrillins are around 43 kDa; in *E. coli*, around 17 kDa. Important as virulence factors.

fimbrin *Plastin* Actin-binding protein (68 kDa) from the core of epithelial brush-border **microvilli**. Contains the **EF-hand** motif.

finasteride Drug that inhibits 5-alpha-reductase, the enzyme that converts testosterone to dihydrotestosterone in the prostate. Used to treat benign prostatic hyperplasia and prostatic carcinoma.

fingerprinting The basic principle of the technique is to digest a large molecule with a sequence-specific hydrolase to produce moderate-sized fragments that can then be run on an electrophoresis gel. Provided the hydrolase only cleaves at specific sites (e.g. between particular amino acids or bases), the fragments should be characteristic of that molecule. The technique can be used to distinguish strains of virus or to differentiate between similar but non-identical proteins (peptide mapping). Not to be confused with **footprinting**.

Firmicutes Group of Gram-positive bacteria, term generally now applied only to those with low G + C levels in their DNA. They can be cocci or rod-shaped forms. Are further subdivided into the anaerobic Clostridia, the Bacilli (which are obligate or facultative aerobes) and the **Mollicutes**, (mycoplasmas).

FISH analysis See **fluorescence** *in situ* **hybridization**.

fish louse Crustacean (copepod) ectoparasites of fish that are found on the skin and in the gills, a major problem in fish farming. *Argulus americanus* is the commonest.

fission yeast See *Schizosaccharomyces pombe*.

FITC *Fluorescein isothiocyanate* FITC is used as a reagent to conjugate **fluorescein** to protein. FITC-labelled antibodies are extensively used for fluorescence microscopy: the fluorophore, when illuminated with UV, emits a yellow-green light.

FIV See **feline immunodeficiency virus**.

fixation Any chemical or physical treatment of cellular material that tends to result in its insolubilization, thus making it suitable for various types of processing for microscopy, such as **embedding** or staining. Typically, fixation involves protein denaturation.

FK506 Immunosuppressive drug (tacrolimus) that acts in a very similar way to **cyclosporin**, binding to an **immunophilin** and affecting calcineurin-mediated activation of the transcription factor **NFAT** in T-cells.

FKBP *FK506-binding protein* A family of small intracellular proteins (around 11 kDa) that bind the immunosuppressive drug FK506 (tacrolimus), thus are **immunophilins**. Like **cyclophilin** have peptidyl prolyl isomerase activity, but are not structurally similar.

FKHR *Forkhead homologue in rhabdomyosarcoma; forkhead receptor; Foxo1* A transcription factor, a member of the hepatocyte nuclear factor 3/forkhead homeotic gene family, a nuclear hormone receptor (NR) intermediary protein. FKHR interacts with both steroid and non-steroid NRs and can act as either a coactivator or co-repressor, depending on the receptor type.

flag tagging Molecular biology technique in which the gene encoding a protein of interest is mutagenized to include an **epitope** for which there is a good antibody. The fate of the protein in a transfected cell or transgenic organism can then be followed easily. Popular tags include the **myc**, **green fluorescent protein** or haemagglutinin epitopes. See also **epitope tag**.

flagellin Subunit protein (40 kDa) of the bacterial flagellum.

flagellum *Pl.* flagella Long, thin projection from a cell, used in movement. In eukaryotes, flagella (like **cilia**) have a characteristic axial '9 + 2' microtubular array (**axoneme**) and bends are generated along the length of the flagellum by restricted sliding of the nine outer doublets. In prokaryotes, the flagellum is made of polymerized **flagellin** and is rotated by the basal motor.

flame cells Specialized excretory cells found in Platyhelminthes (flatworms). The basal nucleated cell body has a distal cylindrical extension that surrounds an extracellular cavity lined by cilia. Mode of action unclear.

flanking sequence Short DNA sequences bordering a **transcription unit**. Often these do not code for proteins.

FLAP *5-Lipoxygenase-activating protein* Activator of the enzyme responsible for the production of 5-HPETE from arachidonic acid, the first step in **leukotriene** synthesis.

flare streaming Phenomenon described in isolated cytoplasm of giant amoeba when the medium contains Ca^{2+} and ATP. A loop of cytoplasm flows outward and then returns to the main mass – the appearance is reminiscent of flares around the eclipsed sun.

flat revertant Variant of a malignant-transformed animal tissue cell in which the characteristic high **saturation density** and piled-up morphology have reverted to the flatter morphology associated with non-transformed cells.

flavan *2,3-Dihydro-2-phenylbenzopyran* Parent ring compound on which flavanols, flavanones, flavones, flavonols and flavonoids are based. Should be distinguished from flavin, which shares the yellow colour but not the structure.

flavin Group of variously substituted derivatives of 7,8-dimethylisoalloxazine. Yellow coloured. The flavin group is found in FAD, FADH and flavoproteins. Not to be confused with **flavan** and **flavones**.

flavin adenine dinucleotide See **flavin nucleotides**.

flavin nucleotides General term for flavin adenine dinucleotide (FAD) or flavin mononucleotide (FMN). Act as prosthetic groups (covalently linked cofactors) for flavin enzymes.

Flaviviridae Family of enveloped RNA viruses with spherical particles 40–50 nm in diameter. Only genus is *Flavivirus*. Cause dengue haemorrhagic fever, Japanese encephalitis, tick-borne encephalitis, West Nile and yellow fever (the latter being the source of the name).

flavodoxin Electron-transfer proteins, widely distributed in anaerobic bacteria, photosynthetic bacteria and cyanobacteria, that contain flavin mononucleotide as the prosthetic group. In *E. coli*, flavodoxin is reduced by the FAD-containing protein NADPH:ferredoxin (flavodoxin) oxidoreductase; flavodoxins serve as electron donors in the reductive activation of anaerobic ribonucleotide reductase, biotin synthase, pyruvate formate lyase and cobalamin-dependent methionine synthase. Can substitute functionally for **ferredoxin**.

flavone *2-Phenylchromen-4-one; 2-phenyl-4H-1-benzopyran-4-one* Specifically the compound and more generally a group of hydroxylated derivatives. Flavone glycosides occur widely as yellow pigments in angiosperms.

flavonoids A large group of secondary metabolites (glycosides) of bryophytes and vascular plants. Some are pigments found in the vacuole, others may be phytoalexins. Term can also be used more generally for any flavone, isoflavone or their derivatives.

flavoproteins Enzymes or proteins that have a **flavin nucleotide** as a coenzyme or prosthetic group. Oxidoreductases or electron carriers in the terminal portion of the electron transport chain.

flavoridin See **disintegrin**.

flehmen *Flehmen reaction* Term used in animal behaviour to describe a response in some mammals, particularly cats, that takes the form of a grimace in which air is sucked in, allowing trace quantities of chemicals (including pheromones) to be detected by an accessory olfactory organ in the roof of the mouth.

Flemming-without-acetic An excellent cytoplasmic fixative that contains chromic acid and osmium tetroxide.

FLICE Caspase-8 See Table C1.

FliG Protein component of the rotor of the bacterial flagellum (about 25 copies per flagellum), the proposed site of torque generation.

FLIP *FLICE-inhibitory protein* Family of proteins that inhibit the **caspase** FLICE and thus protect cells from apoptotic death. Viral FLIPs (v-FLIPs) contain two **death-effector domains** that interact with **FADD** and have been shown to be produced by various herpes viruses and molluscipox virus.

flip–flop A term used to describe the coordinated transfer of two phospholipid molecules from opposite sides of a lipid bilayer membrane. Now used to mean the passage of a phospholipid species from one lamella of a lipid bilayer membrane to the other.

flippase See **flp-frp recombinase**.

FLIPR *Fluorescence imaging plate reader* Machine for fluorescence imaging using a laser that is capable of illuminating a 96-well plate and thus a means of simultaneously reading each well, thus enabling rapid measurements on a large number of samples. Used in high-throughput screening.

flk-1 *VEGFR-2; KDR* One of the receptors for **VEGF**, binds VEGF-121 and VEGF-C. See **flt-1** and **cloche**.

florigen Hypothetical plant growth substance (hormone) postulated to induce flowering. Existence not proven: it has been suggested that it might be an **oligosaccharin** and, more recently, that **CONSTANS** protein may be florigen.

flotillins *Flotillin-1 = reggie-2; flotillin-2 = reggie-1* Integral membrane protein markers of detergent-resistant lipid microdomains (**caveolae**) involved in the scaffolding of large heteromeric complexes that signal across the plasma membrane. Carry an evolutionarily conserved domain called the **prohibitin** homology (PHB) domain, and reggies/flotillins have been included within the SPFH (stomatin–prohibitin–flotillin–HflC/K) protein superfamily. Reggie/flotillin homologues are highly conserved among metazoans but are absent in plants, fungi and bacteria.

flow cytometry Slightly imprecise but common term for the use of the fluorescence activated cell sorter (FACS). Cells are labelled with fluorescent dye and then passed, in suspending medium, through a narrow dropping nozzle so that each cell is in a small droplet. A laser-based detector system is used to excite fluorescence, and droplets with positively fluorescent cells are given an electric charge. Charged and uncharged droplets are separated as they fall between charged plates and so collect in different tubes. The machine can be used either as an analytical tool, counting the number of labelled cells in a population, or to separate the cells for subsequent growth of the selected population. Further sophistication can be built into the system by using a second laser system at right angles to the first to look at a second fluorescent label or to gauge cell size on the basis of light scatter. The great strength of the system is that it looks at large numbers of individual cells and makes possible the separation of populations with, for example, particular surface properties.

flow-mediated dilation Dilation of blood vessels in response to increase flow that is mediated by endothelium-derived factors such as nitric oxide. Normally expressed as the percentage maximum change in vessel diameter from baseline, and assessed using Doppler methods following release of a temporary restriction of flow in the brachial artery by a cuff.

flp-frp recombinase (Pronounced: 'flip-furp') Yeast system for DNA rearrangement. In the presence of 'flippase', a stretch of DNA flanked by matching frp sites is excised and the ends rejoined. An example of a **cassette mechanism**.

FLRF-amide *Phe-Leu-Arg-Phe-NH$_2$* A tetrapeptide **neurotransmitter** found in invertebrates that is a member of a diverse family of RF-amide peptides; all members of which share the same C-terminal RF-amide sequence. See also **FMRF-amide**.

flt Receptors for **VEGF** isoforms. Flt-1 is VEGF receptor-1; flt-3 (Flk2; STK-1; CD135) has a ligand of 24 kDa similar to c-kit ligand and MCSF; flt-4 (VEGFR-3) binds VEGF-C and is mainly restricted to lymphatic endothelium during development. All are receptor tyrosine kinases and are involved in regulation of endothelial or haematopoietic cell development.

fluconazole *TN Diflucan* Triazole antifungal drug used for treatment of candidiasis, athlete's foot and cryptococcal menigitis.

fluctuation analysis Method used to determine (for example) how many ion channels contribute to the transmembrane current. On the assumption that each channel is either open or shut, the noise in the recorded current can be considered to arise from the statistical fluctuation in the number of channels open, and the magnitude of the fluctuation gives an estimate of the conductance of a single channel.

fluctuation test Test devised by Luria and Delbruck to determine whether genetic variation in a bacterial population arises spontaneously or adaptively. In the original version the statistical variance in the number of bacteriophage-resistant cells in separate cultures of bacteriophage-sensitive cells was compared with variance in replicate samples from bulk culture. The greater variance in the isolated populations indicates that mutation occurs spontaneously before challenge with phage. (The proportion of resistant cells depends upon when after isolation the mutation arises – which will be very different in separate populations).

fludarabine *2-Fluoro-ara-AMP* A fluorinated nucleotide analogue of the antiviral agent **vidarabine**, that is relatively resistant to deamination by adenosine deaminase. It is normally administered as the phosphate which is rapidly dephosphorylated to 2-fluoro-ara-A and then phosphorylated intracellularly by deoxycytidine kinase to the active triphosphate, 2-fluoro-ara-ATP. This acts by inhibiting DNA polymerase alpha, ribonucleotide reductase and DNA primase.

fludrocortisone acetate **Mineralocorticoid** used in replacement therapy for adrenocortical insufficiency (Addison's disease). Has **glucocorticoid** properties in addition to the mineralocorticoid properties.

fluid bilayer model Generally accepted model for membranes in cells. In its original form, the model held that proteins floated in a sea of phospholipids arranged as a bilayer with a central hydrophobic domain. Although it is now recognized that some proteins are restrained by interactions with cytoskeletal elements and that the phospholipid annulus around a protein may contain only specific types of lipid, the model is still considered broadly correct.

fluorescein Fluorophore commonly used in microscopy. Fluorescein diacetate can be used as a vital stain or can be conjugated to proteins (particularly antibodies) using isothiocyanate (**FITC**). Excitation is at 365 nm and the emitted light is green-yellow (450–490 nm). The emission spectrum is pH-sensitive, and fluorescein can therefore be used to measure pH in intracellular compartments.

fluorescence The emission of one or more photons by a molecule or atom activated by the absorption of a quantum of electromagnetic radiation. Typically the emission, which is of longer wavelength than the excitatory radiation, occurs within 10^{-8} s: phosphorescence is a phenomenon with a longer or much longer delay in re-radiation. Note that gamma rays, X-rays, UV, visible light and IR radiations may all stimulate fluorescence.

fluorescence activated cell sorter *FACS* See **flow cytometry**.

fluorescence energy transfer *Fluorescence resonant energy transfer; FRET* Transfer of energy from one fluorochrome to another. The emission wavelength of the fluorochrome excited by the incident light must approximately match the excitation wavelength of the second fluorochrome. If light at the second emission wavelength is detected, it implies that the two fluorochromes were physically within a few nanometres. Used as a technique to probe protein or cell interactions.

fluorescence *in situ* hybridization *FISH; chromosome painting* Technique of directly mapping the position of a gene or DNA clone within a genome by *in situ* hybridization to metaphase spreads, in which condensed chromosomes are distinguishable by light microscopy. The DNA probe is labelled with a fluorophore and the hybridization sites visualized as spots of light by **epifluorescence**. Frequently, several probes can be used at one time to mark specific chromosomes with different coloured fluorophores (chromosome painting).

fluorescence microscopy Any type of microscopy in which intrinsic or applied reagents are visualized. Intrinsic fluorescence is often referred to as autofluorescence. The applied reagents typically include fluorescently labelled proteins that are reactive with sites in the specimen. In particular, fluorescently labelled antibodies are widely used to detect particular antigens in biological specimens.

fluorescence recovery after photobleaching *FRAP* Many **fluorochromes** are bleached by exposure to exciting light. If, for example, the cell surface is labelled with a fluorescent probe and an area is bleached by laser illumination, then the bleached patch that starts off as a dark area will gradually recover fluorescence. The recovery is due to the repopulation of the area by unbleached molecules, and diffusion of bleached molecules to other areas. The rate and extent of recovery provide a measure of the fluidity of the membrane and the proportion of labelled molecules that are free to exchange with adjacent areas. The technique is usually applied to cell surface fluidity or viscosity measurements, but is also applicable to other structures.

fluorescence speckle microscopy Technique used to study the dynamics of protein assemblages such as microtubules using low levels of fluorescently labelled protomers *in vivo*. The polymer incorporates some labelled protomer and develops a speckled appearance; individual speckles can be tracked in time and space.

fluoride The fluoride ion F^-. Low levels of fluoride in drinking water markedly decrease the incidence of dental caries, probably because bacterial metabolism is much more sensitive to low fluoride levels. It has been claimed that fluoridation of drinking water, despite vehement protests by a minority of people, has been one of the most successful public health measures ever taken.

fluorite objective Microscope objective corrected for **spherical** and **chromatic aberration** at two wavelengths. Better than an ordinary objective corrected at one wavelength, but inferior to (and much cheaper than) a planapochromatic objective.

fluorochromes Those molecules that are fluorescent when appropriately excited; fluorochromes such as fluorescein or tetramethyl rhodamine are usually used in their isothiocyanate forms (**FITC, TRITC**).

fluorography A method used to visualize substances present in gels, blots, etc. that involves incorporating fluor or scintillant into the gel which produces light when excited by radioactively labelled molecules that are being separated.

fluoroquinolones A group of broad-spectrum antibiotics related to **nalidixic acid**. Common proprietary names are ciprofloxacin, ofloxacin or levofloxacin. Often used to treat gonorrhoea, but resistance is emerging.

fluorouracil *5-Fluorouracil* A cytotoxic antimetabolite used in the treatment of solid tumours, especially colorectal and breast carcinoma.

fluoxetine An antidepressant drug of the **SSRI** class, best known by its proprietary name, Prozac.

fluticasone proprionate A synthetic **glucocorticoid** for treatment of asthma by inhalation.

flying-spot microscope A type of light microscope in which the object is scanned in two dimensions by a light spot formed by a cathode-ray tube. Transmitted energy is collected by a photomultiplier and an image, suitable for electronic analysis, is reconstructed using the timing circuits driving the cathode-ray tube. Analogous to the scanning electron microscope.

FMD Abbrev. for (1) Foot-and-mouth disease, (2) **Flow-mediated dilation**, (3) **Fibromuscular dysplasia**.

f-met-leu-phe *Formyl-methionyl-leucyl-phenylalanine; fMLP* See **formyl peptides**.

fMLP See **formyl peptides**.

FMN *Flavine adenine nucleotide* See **flavin nucleotides**.

Fmoc Abbreviation for the fluorenylmethyloxycarbonyl group. Fmoc chemistry is important in peptide synthesis: the bulky Fmoc group protects the amino group of the Fmoc-amino acid that is being added to the growing peptide; once the peptide bond has been formed the terminal amino group is deprotected by treatment with a mild base, piperidine, and the next Fmoc-amino acid residue can be added.

FMRF-amide *Phe-Met-Arg-Phe-NH₂* A tetrapeptide **neurotransmitter**, a member of the same family of RF-amide peptides as **FLRF-amide**, sharing the same C-terminal RF-amide sequence.

fMRI *Functional magnetic resonance imaging* A technique that provides high resolution, non-invasive imaging of neural activity detected by a blood oxygen level-dependent signal based on the increase in blood flow to the local vasculature that accompanies neural activity in the brain.

fms An **oncogene**, identified in a feline **sarcoma**. The product of *c-fms* is the colony-stimulating factor-1 (CSF-1)-receptor tyrosine kinase. See also Table O1.

Fnr 1. Fumarate and nitrate reductase regulatory protein that activates a number of operons in *E. coli* during anaerobic growth and is a transcriptional regulator. 2. Ferredoxin-NADP⁺ reductase (FNR); an FAD-containing enzyme that catalyses electron transfer between NADP(H) and ferredoxin.

foam cells Lipid-laden macrophages and, to a lesser extent smooth muscle cells, found in **fatty streaks** on the arterial wall.

focal adhesion kinase *FAK* Protein kinase which is found at **focal adhesions** and is thought to mediate the adhesion or spreading processes.

focal adhesions Areas of close apposition, and thus presumably anchorage points, of the plasma membrane of a fibroblast (for example) to the substratum over which it is moving. Usually $1 \times 0.2\ \mu m$ with the long axis parallel to the direction of movement; always associated with a cytoplasmic microfilament bundle that is attached via several proteins to the plasma membrane at an area of high protein concentration (this is noticeably electron dense in electron micrographs). Focal adhesions tend to be characteristic of slow-moving cells.

focus Group of (frequently **neoplastic**) cells, identifiable by distinctive morphology or histology.

fodrin Tetrameric protein (α 240 kDa, β 235 kDa) found in brain: an isoform of **spectrin**.

folate *Tetrahydrofolate* Molecule that acts as a carrier of one-carbon units in intermediary metabolism. It contains residues of *p*-aminobenzoate, **glutamate** and a substituted **pteridine**. The latter cannot be synthesized by mammals, which must obtain tetrahydrofolate as a vitamin or from intestinal microorganisms. One-carbon units are carried at three different levels of oxidation, as methyl-, methylene- or formimino- groups. Important biosyntheses dependent on tetrahydrofolate include those of **methionine**, **thymine** and **purines**. Analogues of dihydrofolate, such as **aminopterin** and **methotrexate** block the action of tetrahydrofolate by inhibiting its regeneration from dihydrofolate.

folic acid Pteridine derivative that is abundant in liver and green plants and is a growth factor for some bacteria. The biochemically active form is tetrahydrofolate (see **folate**).

follicle **1.** Generally a small sac or vesicle. **2.** *Bot.* A kind of fruit formed from a single carpel that splits to release its seeds. **3.** *Zool.* Examples include: hair follicle, an invagination of the epidermis into the dermis surrounding the hair root; ovarian follicle, an oocyte surrounded by one or more layers of **ovarian granulosa** cells. As the ovarian follicle develops, a cavity forms and it is then termed a Graafian follicle.

follicle-stimulating hormone *FSH; follitropin* Pituitary hormone that is an acidic glycoprotein. It induces development of ovarian follicles and stimulates the release of estrogens.

follicular dendritic cells Cells of uncertain (mesenchymal or haematopoietic) lineage found in germinal centres. These cells present native antigens to potential memory cells, and only B-cells with high-affinity B-cell receptors (BCR) bind. These bound B-lymphocytes survive, whereas non-binding B-cells undergo apoptotic cell death.

follistatin Originally identified as an **activin**-binding protein, follistatin inhibits **BMP** activity in early *Xenopus* development.

footprinting A technique used to identify the binding site of, for example, a protein on a nucleic acid sequence. The basic principle is to carry out a very limited hydrolysis of the DNA with or without the protein complexed, and then to compare the digestion products. If a cleavage site is masked by the bound protein then the pattern of fragments when protein is present will be different, and it is possible to work out, by a series of such procedures, exactly where the protein binds.

Foraminifera Group of rhizopod Protozoa that secrete a test (shell) and have slender pseudopods that extend beyond the test and unite to form networks. *Allogromia* is a genus within this group. Extensive remains of Foraminiferan tests are found in sedimentary rocks from the Ordovician to the present.

foreign body giant cell Syncytium formed by the fusion of macrophages in response to an indigestible particle too large to be phagocytosed (e.g. talc, silica or asbestos fibres). There may be as many as 100 nuclei randomly distributed: similar cells but with the nuclei more peripherally located (**Langhans cells**) are found at the centre of tuberculous lesions.

forkhead *Drosophila* homeotic gene. The forkhead gene family of transcription factors belong to the winged-helix class of DNA-binding proteins. More than 40 members of the family have been identified and are involved in embryonic development, tumorigenesis and tissue-specific gene expression. The DNA-binding domain is of 100 amino acids and is referred to as the forkhead domain.

formaldehyde Commonly used fixative and antibacterial agent. As a fixative it is cheap and tends to cause less denaturation of proteins than does glutaraldehyde, particularly if used in a well-buffered solution (buffered formalin, formal saline). Old formaldehyde solutions usually contain cross-linking contaminants, and it is therefore often preferable to used a formaldehyde-generating agent such as paraformaldehyde. Formalin fumes, particularly in conjunction with HCl vapour, are potently carcinogenic.

formins Family of conserved proteins, present in all eukaryotes, that regulate actin dynamics by accelerating nucleation rate, altering filament barbed-end elongation/depolymerization rates and antagonizing capping protein binding. The highly conserved formin homology 2 (FH2) domain and its neighbouring formin homology 1 (FH1) domain, which are surrounded by regulatory domains, cooperate in rapidly assembling profilin-actin into long filaments. Formins also seem to coordinate microfilaments/microtubule interactions and bind to microtubules through a region distinct from FH1 and FH2. Mammalian formins are mDia1 and mDia2, yeast formins Bni1p and Cdc12p. Formins were originally identified as a set of protein isoforms encoded by alternatively spliced products of the *ld* (limb deformity) locus of the mouse. Mutations in *ld* lead to disruption in pattern formation, small size, fusion of distal bones and digits of limbs, and renal aplasia. See **diaphanous-related formins**.

formyl peptides Informal term for small peptides with a formylated N-terminal methionine and usually a hydrophobic amino acid at the carboxy-terminal end (f-met-leu-phe is the most commonly used). These peptides stimulate the motor and secretory activities of leucocytes, particularly neutrophils and monocytes, that have a specific receptor (about 60 kDa) of high affinity (K_d approximately 10^{-8} M). Leucocytes show chemotaxis towards formyl peptides, but the term chemotactic peptides understates the range of activities the molecules will trigger. Thought to be synthetic analogues of bacterial signal sequences, although this is unproven. The leucocytes of many animals (e.g. pig, cow, chicken) do not respond.

fornix Any arch-like structure. **1.** A major nerve-fibre tract that connects the hippocampus to the septal nuclei and mammillary bodies in the brain. **2.** Pocket-like dilations of the vagina on either side of the cervix (vaginal fornix).

forskolin *Colforsin* Diterpene from the roots of *Coleus forskohlii* that stimulates adenylate cyclase and is often used in conjunction with inhibitors of phosphodiesterase to artificially increase intracellular levels of cAMP.

Forssman antigen A glycolipid **heterophile** antigen present on tissue cells of many species. It was first described for sheep red cells, and is not present on human, rabbit, rat, porcine or bovine cells.

forward scatter Scattering of electromagnetic waves (light, radio, etc.) by particles significantly larger than the wavelength, in a direction that is within 90° of the direction of propagation of the incident beam. In **flow cytometry** the forward scatter is roughly proportional to the diameter of the cell and orthogonal scatter (side scatter) is proportional to the granularity – thus neutrophil granulocytes have higher side scatter than agranular lymphocytes. Dead cells have lower forward scatter and higher side scatter than living cells.

fos An **oncogene**, *v-fos*, carried by the Finkel–Biskis–Jenkins and Finkel–Biskis–Reilly murine osteogenic sarcoma retroviruses. The normal *c-fos* gene encodes the protein Fos that dimerizes with **Jun**, via a **leucine zipper**, to form the AP-1 **transcription factor**. See also Table O1.

fosmids An F-factor **cosmid** used as a bacterially propagated phagemid vector system for cloning genomic inserts approximately 40–50 kb in size. Largely superseded by **bacterial artificial chromosome** and P1 bacteriophage (PAC) vector systems.

founder cell Cell that gives rise to tissue by clonal expansion. For most mammalian tissues there are considerably more than two founder cells, as can be determined by forming chimaeras from genetically distinguishable embryos, but single founder cells have been found for the intestine and germ line in *C. elegans*.

four helix bundle Common protein **motif** in which four **alpha-helices** bundle closely together to form a hydrophobic core.

Fourier analysis Loosely, the use of Fourier transformations to convert a time-based signal to a frequency spectrum and back, allowing any periodic property of the signal to be identified.

fovea Small pit or depression on the surface of a structure or organ. The fovea centralis is the most cone-rich region of the retina, with maximum acuity and colour sensitivity.

foxo *Forkhead box, subgroup 'O'* Subfamily of **forkhead** transcription factors (foxhead other) that are negatively regulated by protein kinase B (PKB) in response to signalling by insulin and insulin-like growth factor in *C. elegans* (Foxo homologue, Daf-16) and mammals. dFOXO, the *Drosophila* homologue has also been described. Activation of FOXO transcription factors causes cell death or cell-cycle arrest, and FOXO factors have been implicated in stress resistance and longevity.

FoxP2 Highly conserved transcription factor of the **forkhead** family. Mutation to FOXP2 seems to result in brain defects during embryonic development that result in disruption of neural pathways essential for human speech and has other effects. The early finding of defects in FoxP2 in a family with speech dysfunction led to exaggerated claims that this was a gene for speech.

FOXP3 Mutation in the mouse *Foxp3* gene causes a phenotype called **'scurfy'**. See **polyendocrine syndrome**.

FPLC Fast protein liquid chromatography. Chromatographic method for protein purification that is much less commonly used now that recombinant proteins can be purified by affinity methods.

FPPs *Long filament-like plant proteins* Family of possible **lamin** functional homologues identified in *Arabidopsis*, tomato and rice. There are four novel unique sequence motifs and two clusters of long coiled-coil domains separated by a non-coiled-coil linker. The *Arabidopsis* homologue of the FPP family binds in a yeast two-hybrid assay to MAF1, a nuclear envelope-associated plant protein.

fps See *fes*.

fra-1 *Fos-related antigen-1* Related to **fos**, a member of the activator protein 1 (AP-1) family of transcription factors and, due to the lack of transactivation domain, Fra-1 can suppress activation of AP-1.

frabin *FGD1-related F-actin-binding protein* Actin filament-binding protein from rat brain (766 aa, 86 kDa) that, when overexpressed, induces microspike formation. Has a single actin-binding domain, a Dbl homology domain (**DH domain**), two plekstrin homology (**PH**) domains and a cysteine-rich domain. Domain structure is similar to that of GEF (GDP/GTP exchange factor) that is specific for **cdc42** (FGD1).

fractalkine Membrane-bound **chemokine** with CX3C motif. Chemokine domain (76 amino acids) is bound to membrane through mucin-like stalk (241 amino acids) or can be released as a 95-kDa glycoprotein. Highly expressed on activated endothelial cells, and is both an adhesion molecule and an attractant for T-cells and monocytes.

fraction I protein See ribulose bisphosphate carboxylase/oxidase (**RUBISCO**).

fractionation A term used to describe any method for separating and purifying biological molecules. See also **cell fractionation**.

fragile X syndrome Most frequent cause of mental retardation. There is an expanded **trinucleotide repeat**, CGG, in the *fra*(X) gene.

fragilysin *EC 3.4.24.74* Metallopeptidases of the M10 family, produced by some (~10%) pathogenic strains of *Bacteroides fragilis* that can cause diarrhoea. The fragilysin enterotoxin acts by proteolytically damaging the intestinal epithelium.

fragmentin-2 See **granzyme** B.

fragmin 1. An actin-binding protein (42 kDa) from the slime mould *Physarum polycephalum* that has calcium-sensitive severing and capping properties. **2.** Trade name for dalteparin sodium, a sterile, low molecular-weight heparin used to treat deep vein thrombosis.

frame-shift mutation Insertion or deletion of a number of bases not divisible by three in an open reading frame in a DNA sequence. Such mutations usually result in the generation, downstream, of nonsense, chain-termination codons.

Frankenstein food *Frankenfood* A foodstuff made or derived from plants or animals that have been genetically modified by methods other than conventional breeding techniques. Indicative of the level of the debate over GM crops.

Frankia Genus of **Actinomycetales** capable of nitrogen fixation, both independently and in symbiotic association with roots of certain non-leguminous plants, notably alder.

FRAP See **fluorescence recovery after photobleaching**.

Frat Frat1, Frat3 The proto-oncogene *Frat1* was originally identified as a common site of proviral insertion in transplanted tumours of Moloney murine leukaemia virus (M-MuLV)-infected Emu-Pim1 transgenic mice. More recently, FRAT/GBP (GSK-3β binding protein) family members have been recognized as critical components of the **Wnt** signal transduction pathway. *Frat1* is expressed in various neural and epithelial tissues. A second mouse *Frat* gene, designated *Frat3* has also been identified. The Frat1 and Frat3 proteins are structurally and functionally very similar; both are capable of inducing a secondary axis in *Xenopus* embryos.

frataxin Product of the *X25* gene: deficiency leads to **Friedreich's ataxia**.

fraxin *7-Hydroxy-6-methoxycoumarin 8-glucoside* Coumarinic glucoside from *Fraxinus excelsior* that has anti-inflammatory and antimetastatic properties, the former probably because of its inhibitory effect on 5-**HETE** production.

frazzled Drosophila orthologue of **DCC**, binds **netrin** and may present it to other netrin receptors of the **UNC-5** family.

free energy *Gibbs free energy, G* A thermodynamic term used to describe the energy that may be extracted from a system at constant temperature and pressure. In biological systems the most important relationship is: $\Delta G = -RT \ln(K_{eq})$, where K_{eq} is an equilibrium constant.

free radical Highly reactive and usually short-lived molecular fragment with one or more unpaired electrons.

freeze cleavage See **freeze fracture**.

freeze drying Method commonly adopted to produce a dry and stable form of biological material that has not been seriously denatured. By freezing the specimen, often with liquid nitrogen, and then subliming water from the specimen under vacuum, proteins are left in reasonably native form, and can usually be re-hydrated to an active state. Since the freeze-dried material will store without refrigeration for long periods, it is a convenient method for holding back-up or reference material, or for the distribution of antibiotics, vaccines, etc.

freeze etching If a **freeze fractured** specimen is left for any length of time before shadowing, then water will sublime off from the specimen, etching (lowering) those surfaces that are not protected by a lipid bilayer. Some etching will take place following any freeze cleavage process; in deep etching the ice surface is substantially lowered to reveal considerable detail of, for example, cytoplasmic filament systems.

freeze fracture Method of specimen preparation for the electron microscope in which rapidly frozen tissue is cracked so as to produce a fracture plane through the specimen. The surface of the fracture plane is then shadowed by heavy metal vapour and strengthened by a carbon film, and the underlying specimen is digested away, leaving a replica that can be picked up on a grid and examined in the transmission electron microscope. The great advantage of the method is that the fracture plane tends to pass along the centre of lipid bilayers, and it is therefore possible to get en face views of membranes that reveal the pattern of integral membrane proteins. The **E-face** is the outer lamella of the plasma membrane viewed as if from within the cell, the P-face is the inner lamella viewed from outside the cell. Fracture planes also often pass along lines of weakness, such as the interface between cytoplasm and membrane, so that outer and inner membrane surfaces can be viewed. Further information about the structure can be revealed by **freeze etching**. Extremely rapid freezing followed by deep etching has allowed the structure of the cytoplasm to be studied without the artefacts that might be introduced by fixation.

French flag problem The French flag (tricolour) is used to illustrate a problem in the determination of pattern in a tissue, that of specifying three sharp bands of cells with discrete properties that do not have blurred edges, using, for example, a gradient of a diffusible morphogen.

French press *French pressure cell* Hydraulic pressure system used to force a suspension of cells or organelles at very high pressure (140 MPa) through a small orifice; the shearing forces and abrupt pressure drop causes disruption of membrane-bound organelles.

frenulum The membrane attaching the foreskin to the glans and shaft of the penis.

frequenin *Neuronal calcium sensor; NCS-1* Synaptic calcium-binding protein, originally found in *Drosophila*, involved in the regulation of neurotransmission in the central and peripheral nervous systems from insects to vertebrates. Highly conserved and homologous to **recoverin** and visinin.

Freund's adjuvant A water-in-oil emulsion used experimentally for stimulating a vigorous immune response to an antigen (that is in the aqueous phase). Complete Freund's adjuvant contains heat-killed tubercle bacilli; these are omitted from Freund's incomplete adjuvant. Unsuitable for use in humans because it elicits a severe granulomatous reaction.

Friedreich's ataxia Autosomal-recessive disorder caused by trinucleotide (GAA) repeats which, unlike those in **Huntington's chorea** and **fragile X syndrome**, are within an intron of the gene 25 that codes for **frataxin**, the protein deficient in the disease. The intra-intronic repeat may interfere with hnRNA processing and thus lead to a deficiency in frataxin production.

Friend helper virus Mouse (lymphoid) leukaemia virus present in stocks of Friend virus, that was believed at one time to assist its replication. Molecular cloning of Friend virus has since shown that it is non-defective.

Friend murine erythroleukaemia cells Lines of mouse erythroblasts transformed by the Friend virus, that can be induced to differentiate terminally, producing haemoglobin, by various agents such as dimethyl sulphoxide.

Friend murine leukaemia virus Murine leukaemia virus isolated by Charlotte Friend in 1956 whilst attempting to transmit the Erlich ascites tumour by cell-free extracts. Causes an unusual erythroblastosis-like leukaemia, in which anaemia is accompanied by large numbers of nucleated red cells in blood. Does not carry a host-derived oncogene, but seems to induce tumours by proviral insertion into specific regions of host genome.

Friend spleen focus-forming virus Defective virus found in certain strains of **Friend helper virus** detected by its ability to form foci in spleens of mice and believed to be responsible in those strains for the production of a leukaemia associated with polycythaemia rather than anaemia.

fringe *Fng* Protein that regulates the location-specific expression of the **Notch** ligands **serrate protein** and **Delta** protein in the developing *Drosophila* wing. Fringe appears to be a glycosyl transferase that adds GlcNAc to O-linked fucose residues on EGF modules of Notch as it is made, thereby increasing its ability to bind Delta (but not Serrate).

frizzled *fz Drosophila* tissue-polarity gene encoding a serpentine receptor that responds to a polarity signal. Downstream signalling seems to involve JNK/SAPK-like kinases, **Rho factor** A and the product of the gene *dishevelled* (*dsh*).

frontal zone contraction theory Model proposed to account for the movement of giant amoebae in which cytoplasmic contraction at the front of the leading pseudopod (fountain zone) pulls viscoelastic cytoplasm forward in the centre of the cell and forms a tube of more rigid cytoplasm immediately below the plasma membrane behind the active region. The peripheral contracted cytoplasm relaxes into a weaker gel at the rear and is pulled forward in its turn. Contrasts with the **ectoplasmic tube contraction** model.

frozen stock Because cell lines tend to change their properties with continuous rounds of subculturing, it is common practice to keep stocks of cells frozen (either in liquid nitrogen or at −70°C) and to keep returning to this stock so that experiments are all carried out on cells of comparable passage number. The method also allows strains to be stored for long periods. Cells are usually frozen down in the presence of a cryoprotectant such as DMSO or glycerol. The method is also extensively used for storing semen for artificial insemination.

FRT Abbreviation for Flp recombinase target: see **flp-frp recombinase**.

fructose A six-carbon sugar (hexose) abundant in plants. Fructose has its reducing group (carbonyl) at C2 and thus is a ketose, in contrast to glucose, which has its carbonyl at C1 and thus is an aldose. Sucrose, common table sugar, is the non-reducing disaccharide formed by an α-linkage from C1 of glucose to C2 of fructose (latter in furanose form). Fructose is a component of polysaccharides such as inulin, levan.

frustule The cell wall, largely composed of silica, of a diatom (Bacillariophyceae) consisting of two halves, the hypotheca fitting inside the epitheca, rather like the two halves of a Petri dish. The frustule is covered in delicate markings and intricate designs and is valuable for testing a microscope's resolving power.

FSH See **follicle-stimulating hormone**.

F-spondin The F-spondins are a family of extracellular matrix molecules united by two conserved domains, FS1 and FS2, at the amino terminus plus a variable number of thrombospondin repeats at the carboxy terminus. Currently, characterized members include a single gene in *Drosophila* and multiple genes in vertebrates. The vertebrate genes are expressed in the midline of the developing embryo, primarily in the floor plate of the neural tube.

FTIR *Fourier transform infrared spectroscopy* An analytical technique used to identify (generally) organic materials. This technique is based upon the absorption of various infrared light wavelengths by the material of interest.

ftsZ Filamentous temperature-sensitive protein involved in cell division in bacteria. It is, like tubulin, a GTP-binding protein (43 kDa), with GTPase activity. Also found in mitochondria and chloroplasts. Assembly, bundling and stability of FtsZ protofilaments is important for the formation and functioning of the cytokinetic Z-ring during bacterial division.

ftz See *fushi tarazu*.

fuchsin Synthetic rosaniline dye. Used as a red dye (in **Schiff's reagent**) and as an antifungal agent.

fucose L-fucose (6-deoxy-L-galactose) is found as a constituent of *N*-glycan chains of glycoproteins; it is the only common L-form of sugar involved. D-fucose is usually encountered as a synthetic galactose analogue.

fucosyl transferase An enzyme catalysing the transfer of fucosyl residues from the nucleotide sugar GDP-fucose.

fucoxanthin **Carotenoid** pigment of certain brown algae (**Phaeophyta**) and bacteria: absorbs at 500–580 nm.

Fucus Genus of brown algae common on shore-line of Northern seas.

Fugu rubripes Japanese Puffer fish. Notorious for the poison (**tetrodotoxin**) found in lethal amounts in the poison gland (which must be removed before the fish can safely be eaten) and at low levels elsewhere. Also of interest and utility because of the very low levels of repetitive DNA found in the genome.

Fujian flu A virulent strain of influenza first described in the province of Fujian, China.

fumagillin Naturally secreted antibiotic from *Aspergillus fumigatus* that inhibits endothelial cell proliferation by binding to **methionine aminopeptidase 2** and is therefore antiangiogenic. Originally used against fungal (*Nosema apis*) infections in honeybees.

fumarate A dicarboxylic acid intermediate in the Krebs cycle (**tricarboxylic acid cycle**). Can be derived from aspartate, phenylalanine and tyrosine for input to the Krebs cycle.

functional cloning *Expression cloning* Strategy for cloning a desired gene that is based on some property (antigenicity, ligand binding, etc.) of the expressed gene. For example, a cDNA library could be produced in a eukaryotic **expression vector**, and transfected into a large number of cells. To identify a particular transport protein, a radiolabelled substrate could be added and cells containing the protein of interest identified by radiography. The plasmid could then be recovered and the genes sequence determined.

Fundulus heteroclitus The killifish. A teleost much used for the study of early embryonic development because the egg and embryo are transparent.

fura-2 Cell-permeable fluorescent indicator that exhibits a spectral shift when it chelates calcium ions, the acetoxymethyl ester (fura-2 AM) is hydrolysed within the cell by non-specific esterases and trapped in the cytosol. See **quin2**.

furan One of a class of heterocyclic aromatic compounds characterized by five-membered ring structure consisting of four CH_2 groups and one oxygen atom. The simplest furan compound is furan itself; a clear, volatile and mildly toxic liquid.

furanone Class of compounds with **furan**-like rings of four carbon and one oxygen atom. A wide variety of furanones with chemical structures similar to the *N*-acylhomoserine lactones are produced in nature. Butenolides (2(5*H*)-furanones) have been isolated from *Streptomyces* species and are also produced by marine algae, by sponges, fungi and ascidians. Some furanones are insect sex pheromones, others are important artificial flavouring compounds or are produced during cooking or fermentation. They occur naturally in pineapples or strawberries, and constitute flavouring compounds in cheese and wine. Ascorbic acid is a furanone. Naturally occurring furanones may play a role in inhibiting bacterial infections and biofilm formation by interfering with **quorum sensing**.

furanose Any monosaccharide with a furanoid ring of four carbons and one oxygen.

furfural *Furfuraldehyde* A viscous, colourless liquid that has a pleasant aromatic odour; turns dark brown or black upon exposure to air. It is the aldehyde of pyromucic acid, a derivative of **furan**. Used as a solvent and a feedstock in industrial-scale organic synthesis. Also used as a fungicide and nematicide.

furin *EC 3.4.21.75* **Subtilisin**-like eukaryotic endopeptidase (a **kexin**) with substrate specificity for consensus sequence Arg-X-Lys/Arg-Arg at the cleavage site. Furin is known to activate the haemagglutinin of fowl plague virus and will cleave the HIV envelope glycoprotein (gp160) into two portions, gp120 and gp41, a necessary step in making the virus fusion-competent.

furosemide *Frusemid* Potent diuretic that increases the excretion of sodium, potassium and chloride ions, and inhibits their resorption in the proximal and distal renal tubules.

furunculosis Disease of fish caused by *Aeromonas salmonicida*. Major problem in fish farms.

fusarium Any fungus of the genus *Fusarium*, especially those that cause serious disease in plants. See *Fusarium mycotoxins*.

Fusarium mycotoxins Important fungal mycotoxin contaminants of various food products. Include zearalenone, **diacetoxyscirpenol**, T-2 toxin, neosolaniol monoacetate, deoxynivalenol, nivalenol, fumonisin B1, fumonisin B2, moniliformin, fusarenon-X, HT-2 tioxin and beta-zearalenol.

fused gene family The 'fused' gene family comprises **profilaggrin**, **trichohyalin**, **repetin**, **hornerin**, the profilaggrin-related protein and a protein encoded by *c1orf10*. Functionally, these proteins are associated with keratin intermediate filaments and partially cross-linked to the cell envelope.

fushi tarazu ftz (Japanese for 'too few segments'); a **pair-rule gene** of *Drosophila*.

fusidic acid A steroid antibiotic related to **cephalosporin P**, isolated from fermentation broth of the fungus *Fusidium coccineum*. Blocks translocation of the elongation factor G (EF-G) from the ribosome during bacterial protein synthesis. Active particularly against Gram-positive organisms and used clinically for skin and eye infections.

fusiform Tapered at both ends, like a spindle – though the current rarity of spindles makes this a somewhat unhelpful description.

fusin Lymphocyte surface protein originally described as being an essential cofactor for HIV bound to CD4 to fuse with and enter the cell, later shown to be a chemokine receptor (CXCR4 in the case of lymphotropic virus strains, CCR5 for myelotropic strains). Since **FIV** will infect CD4⁻ cells, it is possible that the chemokine receptor is the original binding site and CD4 the co-receptor, rather than the converse.

fusion protein Protein formed by expression of a hybrid gene made by combining two gene sequences. Typically this is accomplished by cloning a cDNA into an **expression vector** in-frame with an existing gene, perhaps encoding, e.g. beta-galactosidase. See **GST fusion protein**.

fusome A membrane and cytoskeletal organelle specific to the germline in *Drosophila*; spheroid throughout stem cells and **gonialblasts**, but branches extensively throughout interconnected secondary spermatogonia.

futile cycles Any sequence of enzyme-catalysed reactions in which the forward and reverse processes (catalysed by different enzymes) are constitutively active. Frequently used to describe the cycle of phosphorylation and dephosphorylation of phosphatidyl inositol derivatives in cell membranes.

FVB Inbred strain of mice used extensively in transgenic research because of its defined background, good reproductive performance and prominent pronuclei, which facilitate microinjection of genomic material. Derived from Swiss mice and named because of sensitivity to Friend leukaemia virus B.

FW Abbrev. used variously, often for fresh weight or freshwater.

Fx 1. A homodimeric NADP(H)-binding protein of 68 kDa, first identified in human erythrocytes and identified as the human homologue of the murine protein P35B, a tumour rejection antigen. The enzyme responsible for the last step of GDP-L-fucose synthesis from GDP-D-mannose in prokaryotic and eukaryotic cells. **2.** Very small protein (5 kDa) from platelets that binds to G-actin rendering it assembly-incompetent. Now known to be identical to **thymosin $\beta4$**, and this usage is obsolete.

fyn A non-receptor **tyrosine kinase**, related to **src** and implicated in both brain development and adult brain function. Plays a role in T-cell signal transduction in concert with Lck, and excess Fyn activity in the brain is associated with conditions such as Alzheimer's and Parkinson's diseases.

FYVE A phosphatidylinositol 3-phosphate-binding **zinc finger** domain, a 65-residue module with two zinc-binding centres. The FYVE domain is found in eukaryotic proteins that are involved in membrane trafficking and phosphoinositide metabolism.

G

G0 Phase of the cell cycle in which non-proliferating cells are considered to exist. Entry into **G1** phase is a prelude to S-phase and eventually division.

G1 Phase of the eukaryotic **cell cycle** between the end of cell division and the start of DNA synthesis, **S phase**. G stands for gap.

G2 Phase of the eukaryotic **cell cycle** between the end of DNA synthesis and the start of cell division.

gab-1 *Grb2-associated binder-1* Protein (77 kDa, 694 residues) that binds to **Grb-2** and has homology with IRS-1 (insulin receptor substrate-1), particularly in the **PH domain** in the N-terminal region. Gab-1 is a substrate for the EGF-receptor and may integrate signals from different receptors into the control of cellular responses: overexpression of Gab-1 makes cells more responsive to limiting amounts of growth factors. Gab-1 is found in most human tissues.

GABA *Gamma-aminobutyric acid* Fast inhibitory **neurotransmitter** in the mammalian **central nervous system**; prevalent in higher regions of the **neuraxis**. Also mediates peripheral inhibition in crustaceans and in the leech *Hirudo medicinalis*.

GABA receptor Member of a family of receptors for neurotransmitters that includes those for **glycine receptor** and the **nicotinic acetylcholine receptor**. Opened by γ-amino butyric acid (**GABA**). There are two main classes; GABA$_A$ and GABA$_C$ receptors are **ionotropic**, GABA$_B$ receptor is **metabotropic**. **1.** GABA$_A$ receptor (*ca.* 250 kDa) is a ligand-gated chloride channel specifically blocked by **biculculine** and **picrotoxin**, a hetero-oligomer with (probably) five subunits, generally two pairs of α and β subunits and a γ subunit, but there are multiple isoform and splicing variants of these and some additional tissue-specific isoforms. The α chains (53 kDa) are needed for binding of **benzodiazepine**, though the site is probably shared with the γ subunit, and the β chains (58 kDa) bind GABA. The subunits are thought to form a tight group with the chloride channel in the centre. There is considerable protein sequence similarity between GABA$_A$ receptor and the nicotinic acetylcholine receptor. Properties of the receptor can be modified by phosphorylation. Insect GABA receptor resembles vertebrate GABA$_A$ but does not bind bicuculline and has significantly different pharmacological profiles. **2.** The GABA$_B$ receptor (80 kDa) is a G-protein-coupled receptor found in the brain and differs from the GABA$_A$ receptor both in agonist specificity (**baclofen** is a specific agonist) and its effects on cells. It is negatively coupled to adenylate cyclase through a Go-protein and thus acts indirectly on N-type calcium channels. Inhibitory effects mediated through this receptor are due to a reduction in catecholamine release. Has sequence similarity with metabotropic glutamate receptors. **3.** The GABA$_C$ receptor resembles GABA$_A$ but is restricted to the retina.

gabapentin An anticonvulsant drug used in the treatment of epilepsy. TN Neurontin.

G-actin Globular **actin**, the protomer for the assembly of **F-actin**.

gadd45 Nuclear protein induced by growth arrest and DNA damage. Level is highest in G1 phase of cell cycle and gadd associates with cyclin-dependent kinase inhibitor p21Cip1 and with **PCNA**.

gadolinium *Gd* Lanthanide element. The trivalent ion blocks current through T-type voltage-gated calcium channels and stretch-activated ion channels in a concentration-dependent manner and is used as an investigative tool.

GAG See **glycosaminoglycan**.

gag-protein *Group-specific antigen* The protein of the nucleocapsid shell around the RNA of a retrovirus.

GAI *Gibberellic acid insensitive* See **DELLA** and **GRAS family**.

gain of function mutation Gene mutation that results in higher than normal levels of activity of the gene product, for example by deletion of a regulatory phosphorylation site on the protein. Examples are **oncogenic** mutations in genes involved in growth control.

GAL promoter Inducible promoter region of the yeast operon that encodes, among other things, the enzyme **beta-galactosidase**. Extensively used, as is the prokaryotic analogue, because the blue colour generated by the action of the enzyme on **X-gal** is a convenient marker of colonies containing the vector. The transcription factor that binds to the promoter (**GAL4**) is much used in **yeast two-hybrid screening**.

GAL4 Yeast **transcription factor** that binds the UASG promoter domain. Often used in reporter gene constructs and in **yeast two-hybrid screening**.

GAL4 enhancer trap Form of **enhancer trap** (classically in *Drosophila*) in which the **reporter gene** is the yeast **transcription factor** GAL4. The advantage of this system is that such enhancer traps can drive expression of any **transgene** under control of the UAS **promoter** recognized by GAL4. This thus provides a technique for cell-specific expression of transgenes in an intact organism.

galactan Polymer of galactose, may be branched or unbranched.

galactocerebroside *GalC* Surface antigen characteristic of newly differentiated **oligodendrocytes**. GalC antibody is used to identify this glial cell type in cultures of rat optic nerve and brain.

galactosaemia Inborn disorder in which the enzyme galactose-1-phosphate uridyl transferase, which converts galactose-1-phosphate into glucose-1-phosphate, is absent. Excess galactose-1-phosphate accumulates in the blood and a variety of problems result.

galactosamine *2-Amino-2-deoxygalactopyranose* An amino sugar. *N*-acetyl galactosamine is a common component of some glycolipids, chondroitin sulphate and dermatan sulphate.

galactose Hexose identical to glucose except that orientation of -H and -OH on carbon 4 are exchanged. A component of **cerebrosides** and **gangliosides**, and glycoproteins. **Lactose**, the disaccharide of milk, consists of galactose joined to glucose by a β-glycosidic link.

galactose binding protein A bacterial periplasmic protein, most studied in *E. coli*, that acts both as a sensory element in the detection of galactose as a chemotactic signal, and in the uptake of the sugar.

galactosyl transferase *EC 2.4.1.96* Enzyme catalysing the transfer of galactose units from the sugar nucleotide, uridine diphospho-galactose (UDP-galactose) to an acceptor, commonly *N*-acetyl-glucosamine in a glycan chain, forming a glycosidic bond involving C1 of galactose. EC 2.4.1.96 is sn-glycerol-3-phosphate 1-galactosyltransferase; EC 2.4.1.137 is sn-glycerol-3-phosphate 2-α-galactosyltransferase.

galanin Neuropeptide (29 amino acids) isolated from the upper small intestine of pig but subsequently found throughout the central and peripheral nervous system. Regulates gut motility and the activity of endocrine pancreas.

galantamine *Galanthamine* Tertiary amine compound originally derived from flowers (daffodils and snowdrops), now synthesized. A specific, competitive and reversible **acetylcholine esterase inhibitor** shown to have mild cognitive and global benefits for patients with Alzheimer's disease.

galaptins Soluble **lectins** of around 130–140 residues secreted by vertebrates. Developmentally regulated; seem to be important in differentiation of tissues. Larger, related lectin is known as MAC-2 antigen, CBP-35 or IgE-binding protein. Term seems to have fallen into disuse: see **galectin**.

galectin-1 One of the galectin family, galectin-1 mediates cell–cell and cell–substratum adhesion and plays a role in immune regulation Recombinant galectin-1 will induce apoptosis in T-cells. Occurs as a homodimer, which is cell-surface associated.

galectin-3 *Formerly IgE-binding protein, Mac-2; ε-BP* One of the **galectin** family of β-galactoside binding proteins (30 kDa) that has growth-regulatory and immunomodulatory properties. Galectin-3, usually considered proinflammatory, can also act as an immunomodulator by inducing apoptosis in T-cells. Occurs as a homodimer, which is cell-surface associated. Part of the AGE-receptor (**RAGE**) complex. Aberrant expression of galectin-3 is involved in various aspects of tumour progression and p53-induced apoptosis is associated with transcriptional repression of Gal-3.

galectins Family of conserved S-type beta-galactoside-binding lectins that bind to cell surface glycoconjugates. Have been identified in a large variety of metazoan phyla, and are involved in many biological processes such as morphogenesis, control of cell death, immunological response and cancer. Specific galectins are found on the surface of human and murine neoplastic cells and have been implicated in tumorigenesis and metastasis.

gallamine A muscle relaxant used in anaesthesia. The actions of gallamine triethiodide are similar to those of **tubocurarine**.

gallic acid *3,4,5-Trihydroxybenzoic acid* Phenolic acid, commonly found in flowering plants, usually esterified with tannins. Reported to have antifungal and antiviral properties, to act as an antioxidant and helps to protect cells against oxidative damage. Gallic acid is said to be cytotoxic for cancer cells, but not normal cells. Also used for making dyes and inks.

Galliformes Order of birds that includes chickens, peacocks, grouse, pheasants and turkeys, all of which have gizzards.

GALT See **gut-associated lymphoid tissue**.

galvanotaxis The directed movement of cells induced by an applied voltage. This movement is almost always directed towards the cathode, occurs at fields around 100 mV/mm, and is argued to be involved in cell guidance during morphogenesis, and in the repair of wounds. The term galvanotropism is used for neurons, since the cell body remains stationary and the neurites grow towards the cathode. Note that these processes involve cell locomotion, and are distinct from **cell electrophoresis**.

galvanotropism See **galvanotaxis**.

gamete Specialized haploid cell produced by meiosis and involved in sexual reproduction. Male gametes are usually small and motile (spermatozoa), whereas female gametes (oocytes) are larger and non-motile.

gametocyte A cell that divides to produce gametes. Also the sexual reproductive stage of the malaria parasite that develops within erythrocytes.

gametogenesis Process leading to the production of **gametes**.

gametophyte Haploid stage of life cycle of plants; the major vegetative stage for simple plants like liverworts.

gamma-delta cells *γδ-T-cells* Lineage of T-cells possessing the γδ form of the T-cell receptor. Appear early in development and constitute around 5% of mature T-cells in peripheral lymphoid organs. May be the predominant form at epithelial surfaces. Most have neither CD4 nor CD8.

gamma-glutamyltransferase *γ-Glutamyl transpeptidase; GGT; EC2.3.2.2* Heterodimeric highly glycosylated enzyme attached to external surface of cell membrane; transfers γ-L-glutamyl residue (usually from glutathione) to the amino group of an amino acid. Elevated plasma levels are used as a diagnostic marker of hepatic disorder and pancreatitis. Type member of the threonine (T) peptidase family with Thr391 at the active site. The *E. coli* enzyme (non-glycosylated) is soluble and the enzyme is localized in the periplasmic space.

gamma-secretase A multiprotein complex – four components (**presenilin**, **nicastrin**, **APH-1** and **PEN-2**), are necessary and sufficient for gamma-secretase activity. Gamma-secretase is responsible for the intramembranous cleavage of the amyloid precursor protein (APP) and other type I transmembrane proteins such as **Notch** and **E-cadherin**. Mutations in presenilin, the catalytic core of this complex, cause Alzheimer's Disease.

gamma-toxin Bicomponent toxins produced by *Staphylococcus aureus,* forming a protein family with **leucocidins** and **Staphylococcal alpha toxin**. Two active toxins (AB and CB) can be formed, combining one of the class-S components, HlgA or HlgC, with the class-F component

HlgB. Gamma-haemolysins form cation-selective pores (hetero-oligomers formed by three or four copies of each component) with marked similarities to those formed by alpha-toxin in terms of conductance, non-linearity of the current-voltage curve, and channel stability in the open state.

gammopathy *Monoclonal gammopathy; monoclonal gammopathy of unknown significance (MGUS)* The presence of serum or urine **M-protein (5)** in asymptomatic, apparently healthy persons. Many cases seem to be benign, but some (up to 25%) progress to a B-cell malignancy or myeloma.

gamocyte Rarely used term for a phase of the life cycle of protozoa such as *plasmodium* that develops from the trophozoite and later gives rise to gametes.

gamone A pheromone released by a gamete or hypha and that is attractive to another appropriate gamete or hypha in sexual reproduction. The glycoprotein gamone 1 is produced by mating type-I cells of the ciliate *Blepharisma japonicum*. See **sirenin**.

ganciclovir Antiviral nucleoside analogue: 9-((1,3-dihydroxy-2-propoxy)methyl)-guanine. Used in treatment of cytomegalovirus infections.

ganglion A physical cluster of **neurons**. In vertebrates, the ganglia are appendages to the central nervous system; in invertebrates, the majority of neurons are organized as separate ganglia.

ganglion cell A type of **interneuron** that conveys information from the retinal **bipolar**, horizontal and **amacrine cells** to the brain.

ganglioside A **glycosphingolipid** that contains one or more residues of *N*-acetyl or other neuraminic acid derivatives. Gangliosides are found in highest concentration in cells of the nervous system, where they can constitute as much as 5% of the lipid.

gangliosidoses Diseases, such as **Tay–Sachs**, caused by inherited deficiency in enzymes necessary for the breakdown of gangliosides. Cause gross pathological changes in the nervous system, with devastating neurological symptoms.

gankyrin An oncoprotein overexpressed early in hepatocarcinogenesis and in hepatocellular carcinomas. Gankyrin regulates the phosphorylation of the **retinoblastoma** protein (pRb) by CDK4 and enhances the ubiquitylation of **p53** by the RING ubiquitin ligase MDM2.

GAP See **GTPase-activating protein**.

gap **gene** **Segmentation genes** involved in specifying relatively coarse subdivisions of the embryo. They are expressed sequentially in development between **egg-polarity genes** and **pair-rule genes**. In *Drosophila*, there are at least three such genes, e.g. *Kruppel, knirps*.

gap junction A junction between two cells consisting of many pores that allow the passage of molecules up to about 900 Da. Each pore is formed by a hexagonal array (connexon) of six transmembrane proteins (**connexins**) in each plasma membrane; when mated together the pores open, allowing communication and the interchange of metabolites between cells. **Electrical synapses** are gap junctions, and **metabolic cooperation** depends upon the formation of gap junctions.

GAP-43 *Growth-associated protein-43* A palmitoylated neuronal protein, found only in the nervous system, that regulates the response of neurons to axonal guidance signals.

gar2 1. Nucleolar protein containing a glycine- and arginine-rich (GAR) domain from *Schizosaccharomyces pombe*. Required for 18S rRNA and 40S ribosomal subunit accumulation and assembly of the pre-ribosomal particles. The functional homologue of NSR1 from *S. cerevisiae*, and structurally related to **nucleolin** from vertebrates. **2.** Suppressor mutations (gar1 and gar2) that act semidominantly to restore gibberellin responsiveness to **gibberellin**-insensitive (gai) mutant of *Arabidopsis*.

gargoyle cells Fibroblasts with large deposits of mucopolysaccharide, commonly found in storage diseases such as **Hurler's disease**.

GARP A large family of plant DNA-binding proteins that may be needed for a variety of key cellular functions, including regulation of transcription, phosphotransfer signalling and differentiation.

GARPs *Glutamic-acid-rich proteins* Glutamic-acid-rich proteins found in rod photoreceptors. Two soluble forms GARP1 (130 kDa) and GARP2 (62 kDa) exist, as does a large cytoplasmic domain (GARP′ part) of β-subunit of cGMP-gated channel. May act as multivalent proteins that hold together the **transducisome** – they interact with phosphodiesterase, guanylate cyclase and retina-specific ABC protein.

GASC1 *Gene amplified in squamous-cell carcinoma 1; JMJD2C* Gene that belongs to the JMJD2 subfamily of the jumonji family (that contains two **jumonji domains**). Product is a histone trimethyl demethylase.

gas **genes** *Growth arrest-specific genes* Genes that cause cellular quiescence. Gas1 protein shows high structural similarity to the glial cell-derived neurotrophic factor (**GDNF**) family receptors alpha, which mediate GDNF responses through the receptor tyrosine kinase Ret. The human Gas2-related gene encodes two alternatively spliced mRNA species that encode proteins of 36 kDa (GAR22alpha) and 73 kDa (GAR22beta), both of which contain a calponin homology actin-binding domain and a Gas2-related microtubule-binding domain. Gas3/peripheral myelin protein 22 (PMP22) is a component of peripheral nerve myelin, and mutations affecting the gas3/PMP22 gene are responsible for a group of peripheral neuropathies in humans. The gas5 gene is a non-protein-coding multiple small nucleolar RNA. The protein product of the growth arrest-specific gene 6 (Gas6) is a secreted ligand for tyrosine kinase receptors.

GAS motif *IFN-gamma activation site* DNA motif in the FcεRI promoter and in various interferon-gamma regulated genes.

gas vacuole A prokaryotic cellular organelle consisting of cylindrical vesicles around 75 × 300 nm, often in clusters. The wall of the gas vacuole, which is permeable to gases but not to water, is formed from a monolayer of a single protein. Gas vacuoles are found mainly in planktonic cyanobacteria and their prime function is to make the bacterium buoyant.

gastric inhibitory polypeptide Peptide hormone (43 amino acids) that stimulates insulin release and inhibits the release of gastric acid and pepsin. See **GIP**.

gastrin A group of peptide **hormones** secreted by the mucosal gut lining of some mammals in response to mechanical stress or high pH. They stimulate secretion of protons and pancreatic enzymes. Several different gastrins have been identified; human gastrin I has 16 amino acids (2116 Da). Gastrin is competitively inhibited by **cholecystokinin**.

gastrin-releasing peptide *GRP* A regulatory peptide (27 amino acids) thought to be the mammalian equivalent of **bombesin**. It elicits gastrin release, causes bronchoconstriction and vasodilation in the respiratory tract, and stimulates the growth and mitogenesis of cells in culture.

gastrocoel See **archenteron**.

gastroparesis Disorder in which paralysis of the stomach muscles delays the passage of food through the stomach. Often associated with Type 1 diabetes.

Gastropoda Class of the Phylum Mollusca; snails, slugs, limpets and conches.

gastrula Embryonic stage of an animal when **gastrulation** occurs; follows **blastula** stage.

gastrulation During embryonic development of most animals a complex and coordinated series of cellular movements occurs at the end of **cleavage**. The details of these movements, gastrulation, vary from species to species, but usually result in the formation of the three primary germ layers: **ectoderm**, **mesoderm** and **endoderm**.

GATA-1, GATA-3 Members of a family of zinc-finger transcription factors involved in vertebrate embryonic development. GATA-1 abnormalities lead to dyserythropoeitic anaemia and thrombocytopenia in mice and human. GATA-3 is essential for development of the parathyroid, auditory system and kidneys, though it may have other functions as well. Haploinsufficiency in GATA-3 in humans leads to HDR syndrome (hypoparathyroidism, sensorineural deafness and renal anomaly syndrome); homozygous knockout of the gene in mice leads to block of T-cell development and defects in the CNS but no HDR-type anomalies.

gated ion channel Transmembrane proteins of excitable cells that allow a flux of ions to pass only under defined circumstances. Channels may be either **voltage-gated**, such as the **sodium channel** of neurons, or **ligand-gated,** such as the **acetylcholine receptor** of cholinergic synapses. Channels tend to be relatively ion-specific and allow fluxes of typically 1000 ions to pass in around 1 ms; they are thus much faster at moving ions across a membrane than transport **ATPases**.

gating currents Small currents in the membrane just prior to the increase in ionic permeability, due to the movement of charged particles within the membrane. So called because they open the 'gates' for current flow through ion channels.

Gaucher's disease Familial autosomal-recessive defect of glucocerebrosidase (**beta-glucosidase**), most common in Ashkenazi Jews. Associated with hepatosplenomegaly (enlargement of liver and spleen) and, in severe early onset forms of the disease, with neurological dysfunction.

gavage Forced feeding through a tube that passes through the nose, pharynx and oesophagus into the stomach.

gax One of the family of growth arrest genes (*gas* **genes** and *gadd* genes), a homeobox gene restricted to cardiovascular tissue and downregulated by mitogens.

G banding *Giemsa banding* Spreads of metaphase chromosomes, treated briefly with protease then stained with **Giemsa**, produce characteristic **banding patterns** that allow identification of the separate chromosomes. The deeply staining G bands do not coincide with the pattern of quinacrine bands (Q bands).

GBL *Gamma-butyrolactone* A colourless liquid widely used as an industrial solvent and precursor of pyrrolidones, and as a health additive. It is a precursor of GHB (gamma hydroxybutyrate), which has become notorious as a 'date rape' drug.

GC box DNA-binding motif (GGGCG) recognized by the mammalian **transcription factor** Sp1.

gCAP39 *MCP; CapG* A ubiquitous gelsolin-family actin modulating protein (40 kDa) that binds to the barbed ends of microfilaments. Has considerable sequence homology with **gelsolin** and, like gelsolin, responds to calcium and to phosphoinositides. Widely distributed in mammalian cells, the only gelsolin-related actin-binding protein that localizes constitutively to both nucleus and cytoplasm and apparently lacks a nuclear export signal. Is also secreted into plasma, though does not have a signal sequence. Not the same as **capZ**, though function is similar.

GCN3 Alpha subunit of the translation initiation factor eIF2B, the guanine nucleotide exchange factor for eIF2; activity subsequently regulated by phosphorylated eIF2; first identified as a positive regulator of GCN4 expression.

GCN4 A basic leucine zipper protein, the primary regulator of the transcriptional response to amino acid starvation in *S. cerevisiae*. Gcn4 is regulated at the level of transcription, translation and protein stability, with its half-life ranging from approximately 2 minutes under growth in rich medium to 10 minutes under amino acid starvation conditions.

GCP Abbrev. for good clinical practice.

GDF 1. **GDI displacement factor**. 2. **Growth (and) differentiation factor**.

GDGF *Glioma-derived growth factor* Growth factor originally derived from glioma cells. GDGF-I was subsequently shown to be a homodimer of polypeptides immunologically similar to the A-chain of **PDGF,** and GDGF-II is predominantly a heterodimer containing one peptide similar to A-chain and one similar to B-chain of PDGF with some homodimers of B-chain-like peptides. Several human malignant glioma cell lines are stimulated by bacterial lipopolysaccharide to produce a high molecular weight (>200 kDa) growth activity for BALB 3T3 cells that does not cross-react with anti-PDGF antibodies. Now known to be glial-derived neurotrophic factor. See **GDNF**.

GDI GTP dissociation inhibitor: protein that inhibits GDP/GTP exchange on **ras**-like GTPases thereby maintaining them in the active GTP-bound form. GDIs specific for particular families of GTPases have been found.

GDI displacement factor *GDF* Membrane-associated protein (*ca.* 25 kDa) that is involved in **Rab** binding to membrane and dissociating from **GDI**. Yip3/PRA1 protein displays GDF activity.

gDNA Genomic DNA.

GDNF *Glial-(cell line) derived neurotrophic factor; astrocyte-derived trophic factor 1* A disulphide-linked homodimeric neurotrophic factor (monomers of 15 kDa), structurally related to Artemin, Neurturin and Persephin; belongs to the cysteine-knot superfamily of growth factors. Is specific for midbrain dopamine neurons and has been tested for effects in Parkinson's disease. Binds to GDNFR-alpha and mediates activation of the Ret protein tyrosine kinase receptor. Formerly glioma-derived growth factor (GDGF).

GDP Guanosine diphosphate. Phosphorylation gives **GTP**.

Gea-1 GEF (guanine nucleotide exchange factor) for **ARF** from *S. cerevisiae*. Contains **Sec7**-like domain. Similar to **ARNO** and **cytohesin**.

GEA1, GEA6 Proteins from *Arabidopsis thaliana* that are homologous to the early methionine-labelled (Em) proteins of wheat.

Gea-2 Yeast guanine nucleotide exchange factor for **ARF**. Contains **Sec7**-like domain.

GEFs *Guanine nucleotide exchange factors.* Family of proteins that facilitate the exchange of bound GDP for GTP on small G-proteins such as ras and rho and thus activate them. Act in the opposite way to **GTPase-activating proteins** (GAPs) which promote the hydrolysis of bound GTP, thereby switching the G-protein to the inactive form. Family includes **cytohesin**, **ARNO**, Gea-1 and 2, kalirin and yeast **Sec7**.

gel Jelly-like material formed by the coagulation of a colloidal liquid. Many gels have a fibrous matrix and fluid-filled interstices: gels are viscoelastic rather than simply viscous, and can resist some mechanical stress without deformation. Examples are the gels formed by large molecules such as collagen (and gelatin), agarose, acrylamide and starch.

gel electrophoresis Electrophoresis using a gel supporting-phase. Most commonly applied to systems where the gel is based on polyacrylamide. See **electrophoresis**.

gel filtration An important method for separating molecules according to molecular size by percolating the solution through beads of solvent-permeated polymer that has pores of similar size to the solvent molecules. Unlike a continuous filter, which separates flow according to molecular size, separation is achieved because molecules that can enter the beads take a longer path (i.e. are retarded) than those that cannot. Typical gels for protein separation are made from polyacrylamide, or from flexible (Sephadex) or rigid (agarose, Sepharose) sugar polymers. The size separation range is determined by the degree of cross-linking of the gel.

gel mobility shift *Gel retardation assay* Technique for studying DNA–protein interactions. For example, to study the levels of a particular transcription factor, nuclear protein extracts are incubated with a radiolabelled DNA fragment containing the transcription factor's binding site. When the DNA gel is run, the amount of radiolabel that runs more slowly than the free DNA is directly proportional to the amount of transcription factor present.

gel retardation assay *Mobility shift assay* Test for interaction between molecules by looking for a change in gel electrophoretic mobility. For example, to assay for levels of a **transcription factor**, cell extracts are incubated with a radiolabelled oligonucleotide corresponding to the recognition sequence of the transcription factor, and run on an agarose gel. Most of the radiolabel will run quickly through the gel, but any radioactivity that is retarded is presumably caused by DNA/transcription factor interaction.

gelactins Filament gelation proteins that have not reappeared in the recent literature.

gelatin Heat-denatured collagen.

gelatinous lesion A small area of oedema in the arterial intima, possibly a precursor of a **fibrous plaque**.

geldanamycin A naturally occurring benzoquinone ansamycin antibiotic produced by *Streptomyces hygroscopicus*. Geldanamycin binds to heat-shock protein 90 (**HSP90**) and inhibits essential ATPase activity of Hsp90, causing inactivation, destabilization and degradation of Hsp90 client proteins. Has antitumour activity, inhibits pp60src tyrosine kinase and *c-myc* gene expression in murine lymphoblastoma cells, and has been shown to have a range of other effects.

gelsolin Actin-binding protein (90 kDa) that nucleates actin polymerization, but at high calcium ion concentrations (10^{-6} M) causes severing of filaments.

gemcitabine An antitumour agent with radiosensitizing properties used to treat non-small cell lung cancer, pancreatic, bladder and breast cancer. It is a deoxycytidine analogue, related to cytarabine, and primarily kills cells at S-phase and blocks progression through the G1/S-phase boundary. Gemcitabine is a prodrug and is metabolized intracellularly to the active diphosphate (dFdCDP) and triphosphate (dFdCTP) nucleosides.

gemfibrozil Drug that lowers plasma lipoprotein levels. Activates **PPARα**.

geminin Protein that inhibits pre-replication complex assembly by binding to **Cdt1** and preventing it from recruiting the minichromosome maintenance proteins to chromatin. Also directly interacts with Six3 and Hox homeodomain proteins during embryogenesis and inhibits their functions.

geminivirus A large family of plant viruses with circular, single-stranded DNA genomes that replicate through double-stranded intermediates. Have recently emerged as leading plant pathogens that cause severe crop losses worldwide, infecting a broad range of plant species.

gemma *Pl.* gemmae **1.** Small multicellular structure involved in vegetative reproduction in algae, pteridophytes and bryophytes. **2.** Same as chlamydospore. **3.** A bud that will give rise to a new individual.

gene Originally defined as the physical unit of heredity, but the meaning has changed with increasing knowledge. It is probably best defined as the unit of inheritance that occupies a specific locus on a chromosome, the existence of which can be confirmed by the occurrence of different allelic forms. Given the occurrence of **split genes**, it might be re-defined as the set of DNA sequences (**exons**) that are required to produce a single polypeptide.

gene amplification Selective replication of DNA sequence within a cell, producing multiple extra copies of that sequence. The best-known example occurs during the maturation of the oocyte of *Xenopus*, where the set (normally 500 copies) of ribosomal RNA genes is replicated some 4000 times to give about 2 million copies.

gene annotation The process of adding comments, cross-links, information regarding alternative splicing, pseudogenes, promoter regions, the chromosomal location, homologies, etc to gene entries in databases.

gene chip An array of oligonucleotides immobilized on a surface that can be used to screen an RNA sample (after reverse transcription) and thus a method for rapidly determining which genes are being expressed in the cell or tissue from which the RNA originated. There are two alternatives for the immobilized oligonucleotides: either a random set of defined sequences or known probes for genes (probes for cytokines, adhesion molecules, etc.). In both cases the position on the chip defines the sequence that hybridizes.

gene cloning The insertion of a DNA sequence into a **vector** that can then be propagated in a host organism, generating a large number of copies of the sequence.

gene conversion A phenomenon in which alleles are segregated in a 3:1 not 2:2 ratio in meiosis. May be a result of **DNA polymerase** switching templates and copying from the other homologous sequence, or a result of mismatch repair (nucleotides being removed from one strand and replaced by repair synthesis using the other strand as template).

gene dosage Number of copies of a particular **gene** locus in the genome; in most cases either one or two. An excess of copies can cause problems, as is the case with trisomy 21 (Down's syndrome).

gene duplication A class of DNA rearrangement that generates a supernumerary copy of a gene in the genome. This would allow each gene to evolve independently to produce distinct functions. Such a set of evolutionarily related genes can be called a 'gene family'.

gene expression The full use of the information in a gene via **transcription** and **translation** leading (usually) to production of a protein and hence the appearance of the **phenotype** determined by that gene. Gene expression is assumed to be controlled at various points in the sequence leading to protein synthesis and this control is thought to be the major determinant of cellular **differentiation** in eukaryotes.

gene family *Multigene family* A set of genes coding for diverse proteins which, by virtue of their high degree of sequence similarity, are believed to have evolved from a single ancestral gene. An example is the immunoglobulin family where the characteristic features of the constant-domains are found in various cell surface receptors.

gene neighbour method Computational method of trying to deduce protein interactions based upon chromosomal location: if two genes are always found adjacent in several genomes then the assumption is that they are likely to be involved in a common function. Though obviously relevant in prokaryotes where there are operons, the principle seems to extend to eukaryotes. See **Rosetta Stone method** and **phylogenetic profile**.

gene regulatory protein Any protein that interacts with DNA sequences of a gene and controls its transcription.

gene therapy Treatment of a disease caused by malfunction of a gene, by stable **transfection** of the cells of the organism with the normal gene.

gene transfer General term for the insertion of foreign genes into a cell or organism. Synonymous with **transfection**.

general insertion protein *GIP* Protein involved in the insertion of cytochrome c1 into the mitochondrial matrix.

generation time Time taken for a cell population to double in numbers, and thus equivalent to the average length of the cell cycle.

genetic code Relationship between the sequence of bases in **nucleic acid** and the order of amino acids in the polypeptide synthesized from it. A sequence of three nucleic acid **bases** (a triplet) acts as a 'codeword' (**codon**) for one amino acid. See Table C5.

genetic drift Random change in allele frequency within a population. If the population is isolated, and the process continues for long enough, may lead to speciation.

genetic engineering General term covering the use of various experimental techniques to produce molecules of DNA containing new genes or novel combinations of genes, usually for insertion into a host cell for cloning.

genetic linkage The term refers to the fact that certain genes tend to be inherited together, because they are on the same chromosome. Thus parental combinations of characters are found more frequently in offspring than are non-parental. Linkage is measured by the percentage recombination between loci, unlinked genes showing 50% recombination. See **linkage equilibrium**, **linkage disequilibrium**.

genetic load *Genetic burden* In general terms, the decrease in fitness of a population (as a result of selection acting on phenotypes) due to deleterious mutations in the population gene pool. More specifically, the average number of **recessive lethal mutations**, in the **heterozygous** state, estimated to be present in the genome of an individual in a population.

genetic locus The position of a gene in a linkage map or on a chromosome.

genetic recombination Formation of new combinations of alleles in offspring (viruses, cells or organisms) as a result of exchange of DNA sequences between molecules. It occurs naturally, as in **crossing-over** between homologous chromosomes in meiosis or experimentally, as a result of **genetic engineering** techniques.

genetic transformation Genetic change brought about by the introduction of exogenous DNA into a cell. See **transformation**, **germ-line transformation**, **transfection**.

genetically modified organisms *GMO* An organism that has had novel genes added using the techniques of genetic engineering rather than by convention breeding and selection methods.

geneticin *Antibiotic G418* Used as a selection agent in transfection. Toxic to bacteria, yeast and mammalian cells. Vector has geneticin-resistance gene from bacteria so that positive transfectants can be selected.

genistein *4,5,7-Trihydroxyisoflavone* Phytoestrogen from soy that is an inhibitor of protein tyrosine kinases. Competes at ATP-binding site and will inhibit other kinases to some extent. A potent vasorelaxant, probably through effects on **CFTR** and **NKCC1**.

genome The total set of genes carried by an individual or cell.

genome project Coordinated programme to completely sequence the genomic DNA of an organism. Usually, genomic sequencing is combined with several associated ventures: the **physical mapping** of the genome (to allow the genome to be sequenced); the sequencing of **expressed sequence tags** (to aid in the identification of transcribed sequences in the genomic DNA sequence); for non-human organisms, a programme of systematic mutagenesis and genetic mapping of the mutants (to infer function of novel genes by reverse genetics); and an overarching computer database resource, to manage and give access to the data.

genomic clone Clone of a portion of DNA obtained from genomic DNA rather than of DNA produced by reverse transcription from mRNA. It will therefore contain introns as well as exons, but may not necessarily be expressed or complete.

genomic imprinting Parent-specific expression or repression of genes or chromosomes in offspring. There are an increasing number of recognized chromosomal imprinting events in pathological conditions, e.g. preferential transmission of paternal or maternal predisposition to diabetes or atopy, preferential retention of paternal alleles in **rhabdomyosarcoma**, **osteosarcoma**, **retinoblastoma** and **Wilms' tumour**, preferential translocation to the paternal chromosome 9 of a portion of maternal chromosome 22 to form the **Philadelphia chromosome** of chronic myeloid leukaemia.

genomic instability Abnormally high rates (possibly accelerating rates) of genetic change occurring serially and spontaneously in cell populations, as they continue to proliferate. A potentially serious feature of rapidly proliferating tumours. May well be a consequence of mutation or dysfunction of normal repair mechanisms.

genomic island Mobile genetic elements of length from 10 to 200 kb that have been integrated into an organism's genome. They allow lateral transfer of genes between a range of phylogenetically distinct organisms and can therefore facilitate the rapid spread of a trait. Can sometimes be recognized because of atypical G + C content.

genomic library Type of DNA library in which the cloned DNA is from an organism's genomic DNA. As genome sizes are relatively large compared to individual **cDNAs**, a different set of **vectors** is usually employed in addition to **plasmid** and **phage**; see **bacterial** and **yeast artificial chromosomes**, **cosmid**.

genotoxin Toxin that works by causing damage to a gene or interfering with its function.

genotype The genetic constitution of an organism or cell, as distinct from its expressed features or **phenotype**.

gentamicin A group of aminoglycoside antibiotics produced by *Micromonospora* spp. Members include the closely related gentamicins C1, C2 and C1a, together with gentamycin A. They inhibit protein synthesis on 70S ribosomes by binding to the 23S core protein of the small subunit, that is responsible for binding mRNA. Mode of action similar to that of **kanamycin**, **neomycin**, paromomycin, spectinomycin and streptomycin. Active against strains of the bacterium *Pseudomonas aeruginosa*.

geotaxis See **gravitaxis**. The prefix gravi- is preferable since the gravitational fields used as cues need not necessarily be the Earth's.

geotropism See **gravitropism**.

gephyrin Peripheral membrane protein of the cytoplasmic face of the glycinergic synapses in the spinal cord. Appears at developing postsynaptic sites before the **glycine receptor** and may therefore be important in clustering. Thought to interact with the glycine receptors and also with microtubules.

geranyl Prenyl group ((2E)-3,7-dimethyl-2,6-octadien-1-yl). Intermediate in cholesterol synthesis and in production of **geranyl-geranyl** group. Can be post-translationally added to proteins – but geranyl-geranyl prenylation is more common.

geranyl-geranyl (2E,6E,10E)-3,7,11,15-Tetramethyl-2,6,10,14-hexadecatetraen-1-yl group Prenyl group post-translationally added by **geranyl transferase** to some cytoplasmic proteins generally at CAAX motif (where X is usually leucine), serves to associate them with membranes. See **farnesylation**.

geranyl transferase *Farnesyl pyrophosphate synthetase; EC 2.5.1.10* Enzyme responsible for the post-translational transfer of geranyl-geranyl residue to protein.

geranylation The geranoyl group is a linear sequence of 2 isoprenyl residues. The term geranyl-geranyl is used for the common unit of four residues. See also **polyisoprenylation**.

GERL The Golgi–endoplasmic reticulum–lysosome system. See individual entries for each of these membranous compartments of the trans-Golgi network.

germ cell Cell specialized to produce **haploid** gametes. The germ cell line is often formed very early in embryonic development.

germ cell nuclear factor *GCNF; NR6A1* A nuclear orphan receptor that functions as a transcriptional repressor and is transiently expressed in mammalian carcinoma cells during retinoic acid (RA) induced neuronal differentiation.

germ layers The main divisions of tissue types in multicellular organisms. Diploblastic organisms (e.g. coelenterates) have two layers, ectoderm and endoderm; triploblastic organisms (all higher animal groups) have mesoderm between these two layers. Germ layers become distinguishable during late blastula/early gastrula stages of embryogenesis, and each gives rise to a characteristic set of tissues – the ectoderm to external epithelia and to the nervous system for example – although some tissues contain elements derived from two layers.

germ-line therapy Gene therapy that introduces a new gene into the reproductive cells of the body and can therefore be inherited. Generally considered an inappropriate and potentially risky procedure at present. See **germ-line transformation**.

germ-line transformation Microinjection of foreign DNA into an early embryo, so that it becomes incorporated into the **germ-line** of the individual, and thus stably inherited in subsequent generations of **transgenic** organisms. Typically, the DNA would be a **reporter gene** or **cDNA** in a **vector** such as a **transposon** that might also carry a visible **marker gene** (such as eye or coat colour), so that successful transformation could readily be detected.

germin A protein marker of the onset of growth in germinating wheat, later shown to be an extracellular matrix protein and more recently to be an oxalate oxidase. See **germin-like proteins**.

germin-like proteins *GLPs* Class of proteins with sequence and structural similarity to the cereal **germins** but mostly without oxalate oxidase activity. Germins and germin-like proteins are developmentally regulated glycoproteins characterized by a beta-barrel core structure, a signal peptide, and are associated with the cell wall. GLPs are found in organisms ranging from myxomycetes, bryophytes and pteridophytes to gymnosperms and angiosperms and have diverse activities.

germinal centre An aggregation of lymphocytes, mainly B-cells and blast forms, that develops from a primary follicle in response to antigenic stimulation. **Antigen-presenting cells** are also conspicuously present. May be sites at which B-memory cells are produced with receptors, which recognize antigens in the complexes.

Gerstmann–Straussler–Scheinker   syndrome A familial **spongiform encephalopathy**. Transgenic mice with a mutant form of the **PrP** gene from patients with this syndrome develop degenerative brain disease that is similar, but not identical, to that caused by **scrapie**.

GF-1 See **erythroid transcription factor**.

GFAP See **glial fibrillary acidic protein**.

GFP See **green fluorescent protein**.

GFRalpha *Glial cell line-derived neurotrophic factor (GDNF)-family receptor alpha; GFRA1, 2, 3, 4* Family of GPI-linked receptors for **GDNF**, **neurturin**, **artemin** and **persephin**; GFRalpha-1 is the receptor for GDNF, GFRalpha-2 for neurturin, GFRalpha-3 for artemin and GFRalpha-4 for persephin. Forms a heterodimeric (alpha/beta) receptor complex with the tyrosine kinase **ret**, the latter being the beta subunit.

GGT Abbreviation for **1. Gamma-glutamyltransferase**. **2.** Galactosyl-hydroxylysine glucosyltransferase.

ghosts See **erythrocyte ghosts**.

ghrelin Peptide hormone (28 residues), endogenous ligand for G-protein-coupled growth hormone secretagogue (GHS) receptor. Mainly synthesized by epithelial cells lining the fundus of the stomach; levels rise in fasting animals and apparently signal the necessity for increased metabolic efficiency.

GI 1. GI: Common abbreviation for gastrointestinal. **2**. Gi: See **GTP-binding protein**. **3**. Glycemic index.

giant axons Extraordinarily large unmyelinated axons found in invertebrates. Some, like the squid giant axon, can approach 1 mm in diameter. Large axons have high conduction speeds; the giant axons are invariably involved in panic or escape responses, and may (e.g. crayfish) have **electrical synapses** to further increase speed. Vertebrate axons with high conduction velocities are much narrower: they have a **myelin sheath**, allowing **saltatory conduction**.

Giardia Genus of flagellate protozoans, found as intestinal parasites of vertebrates. The human intestinal parasite is *Giardia lamblia*. The cells have a large disc or 'sucker' on their anterior ventral surfaces, by which they attach to the intestinal mucosa. The attachment of the disc is very strong and can prevent peristaltic clearing. This can result in acute or chronic diarrhoea, especially in children. The disease is termed Giardiasis or Lambliasis.

giardin Group of proteins, of 29–38 kDa, found in the ventral discs of *Giardia lamblia*.

gibberellic acids Diterpenoid compounds with **gibberellin** activity in plants. At least 70 related gibberellic acids have been described and designated as a series GA1, GA2, etc. Gibberellic acid-regulated responses are mediated by **DELLA**-domain-containing proteins including GAI, RGA and RGL1–3.

gibberellin Plant growth substance (phytohormone) involved in promotion of stem elongation, mobilization of food reserves in seeds and other processes. Its absence results in the dwarfism of some plant varieties. Chemically, all known gibberellins are **gibberellic acids**.

Giemsa A Romanovsky-type stain that is often used to stain blood films that are suspected to contain protozoan parasites. Contains both basic and acidic dyes and will therefore differentiate acid and basic granules in granulocytes.

Ginkgo biloba Ornamental tree originally native to China. Sole surviving member of the family Ginkgoales. Source of various bioactive compounds.

GINS complex *Go ichi ni san complex* Protein complex that allows the MCM (minichromosome maintenance) helicase to interact with key regulatory proteins in the large replisome progression complexes (RPCs) that are assembled at eukaryotic DNA replication forks during initiation of S phase. See **minichromosome maintenance proteins**.

GIP Acronym with too many meanings: **1. Gastric inhibitory polypeptide**. **2. Glucose-dependent insulinotropic polypeptide**. **3. General insertion protein**. **4.** Gip or Gip Light-regulated inhibitory G-protein from washed microvilli of the photoreceptors of *Octopus*. **5.** GTPase inhibitory protein. Protein that interferes with the binding of **GAP** to ras, or enhances nucleotide exchange (a **GEF**). **6.** See **gip2**.

gip2 The *gip2* oncogene encodes GTPase-deficient alpha-subunits of Gs or Gi-2 proteins and has been identified in tumours of the ovary and adrenal cortex. It will induce neoplastic transformation of Rat-1 cells but not NIH 3T3 cells and appears to be a tissue-selective oncogene.

GIRKs *G-protein-gated inward rectifying potassium channels* A gene family, the gene products of which are thought to form functional channels through the assembly of heteromeric subunits. A point mutation in the *GIRK2* gene is the cause of the neurological and reproductive defects observed in the **weaver** (wv) mutant mouse.

GLA Abbrev. for (1) gamma linolenic acid, (2) protein found in bone (bone Gla protein or osteocalcin) and matrix (matrix Gla protein), important in calcium metabolism and skeletal development.

gland Organ specialized for secretion by the infolding of an epithelial sheet. The secretory epithelial cells may either be arranged as an acinus with a duct or as a tubule. Glands from which release occurs to a free epithelial surface are exocrine; those that release product to the circulatory system are endocrine glands.

glandular fever Self-limiting disorder of lymphoid tissue caused by infection with **Epstein–Barr virus** (infectious mononucleosis). Characterized by the appearance of many large lymphoblasts in the circulation.

Glanzmann's thrombasthenia Platelet dysfunction in which aggregation is deficient. A specific glycoprotein complex (IIb/IIIa) is absent from the plasma membrane: this seems to be the fibronectin/fibrinogen receptor and is a β3-**integrin**. See Table I.1.

glaucoma An eye condition in which the intraocular pressure rises, causing damage to optic nerve fibres. Major cause of blindness after the age of 45.

GLD-1 *Germline defective-1* **STAR protein** from *C. elegans*, necessary for germ-line development. A translational repressor that acts through regulatory elements in the 3' untranslated region of the sex-determining gene *tra-2*.

Gleevec TN for imatinib mesylate, a protein tyrosine kinase inhibitor, used to treat some forms of chronic myeloid leukaemia.

GLGF Gly-Leu-Gly-Phe motif (loop) found in **PDZ** domain (also called DHR or GLGF domain) of diverse membrane-associated proteins.

gliadin Group of proline-rich proteins found in cereal seeds and constituting the major storage protein. Associate with **glutenin** to form **gluten**.

glial cells Specialized cells that surround **neurons**, providing mechanical and physical support, and electrical insulation between neurons.

glial fibrillary acidic protein *GFAP* Member of the family of **intermediate filament** proteins, characteristic of **astrocytes**. Mutations in the gene cause **Alexander disease**.

glial filaments Intermediate filaments of glial cells, made of **glial fibrillary acidic protein**.

glibenclamide *Glyburide* **Sulphonylurea** that acts via sulphonyl urea receptor (**SUR**) to regulate inwardly rectifying K⁺-ATP channels (Kir6.1) of pancreatic islet cells thereby increasing insulin release.

glicentin Peptide fragment cleaved from **glucagon** by pro-hormone convertase.

gliding motility Mode of cell motility exhibited by, for example, gregarines. There are no obvious motile appendages, little actin is detectable, and the motor mechanism is poorly understood.

glioblastoma Highly malignant brain tumour derived from glial cells. See **gliomas**.

gliomas Neuroectodermal tumours of neuroglial origin: include astrocytomas, oligodendroglioma, and ependymoma derived from **astrocytes**, **oligodendrocytes** and **ependymal cells**, respectively. All infiltrate the adjacent brain tissue, but they do not metastasize.

gliostatin Cytokine (dimeric, subunits 50 kDa) very similar to PD-ECGF. Neurotrophic for cortical neurons and inhibits proliferation of **astrocytes** (stimulates differentiation). May play a part in aberrant neovascularization of rheumatoid synovium.

glipizide A **sulphonylurea** drug used in treatment of non-insulin dependent (Type 2) diabetes mellitus.

globin The polypeptide moiety of haemoglobin. In the adult human, the haemoglobin molecule has two α (141 residues) and two β (146 residues) globin chains.

globoside *Cytolipin K* Major neutral glycosphingolipid found in kidney and erythrocytes.

globular protein Any protein that adopts a compact morphology is termed globular. Generally applied to proteins in free solution, but may also be used for compact folded proteins within membranes.

globus pallidus See **basal ganglia**.

glomerulonephritis Inflammatory response in the kidney glomerulus that often arises because immune complexes cannot pass through the basement membrane of the fenestrated epithelium where plasma filtration occurs. Circulating neutrophils are trapped on the accumulated adhesive immune complexes (that also activate complement). Immune complex tends to be irregularly distributed in contrast to the picture in **Goodpasture's syndrome**.

Glossina morsitans Tsetse fly (name is onomatopoeic), vector of African **trypanosomiasis**.

GLP-1 1. **Glucagon-like peptide 1**. 2. **Germin-like protein**-1. 3. A **GARP** homologue found in the protozoan *Giardia lamblia*, an important transcriptional activator. **4.** One of the glutaredoxin-like proteins, a subgroup of glutaredoxins with a serine replacing the second cysteine in the CxxC-motif of the active site.

glucagon A polypeptide hormone (3485 Da) secreted by the **A cells** of the Islets of Langerhans in the pancreas in response to a fall in blood sugar levels. Induces **hyperglycaemia**. A family of structurally related peptides includes **glucagon-like peptides 1** and 2 (encoded by the same gene); **gastric inhibitory polypeptide**; **secretin**; **vasoactive intestinal peptide**; **growth hormone-releasing factor**; **PACAP**; **exendins**. See also Table H2.

glucagon-like peptide 1 *GLP-1* A 36-amino acid hormone produced by intestinal L cells that acts to augment insulin secretion. Glucagon-like peptide 2 is coded by the same gene. See **exendin**.

glucans Glucose-containing polysaccharides, including **cellulose**, **callose**, **laminarin**, **starch** and **glycogen**.

glucocorticoids Steroid hormones (both natural and synthetic) that promote **gluconeogenesis** and the formation of glycogen at the expense of lipid and protein synthesis. They also have important anti-inflammatory activity. Type compound is hydrocortisone (Cortisol); other common examples are cortisone, prednisone, prednisolone, dexamethasone, betamethasone; see also Table H3.

glucomannan Hemicellulosic plant cell-wall polysaccharide containing glucose and mannose linked by β (1-4)-glycosidic bonds. May contain some side-chains of galactose, in which case it may be termed galactoglucomannan. A major polysaccharide of gymnosperm wood (softwood).

gluconeogenesis Synthesis of glucose from non-carbohydrate precursors, such as pyruvate, amino acids and glycerol. Takes place largely in liver, and serves to maintain blood glucose under conditions of starvation or intense exercise.

glucosamine Amino sugar (2-amino-2-deoxyglucose); component of **chitin**, **heparan sulphate**, **chondroitin sulphate** and many complex polysaccharides. Usually found as β-D-*N*-acetylglucosamine.

glucosaminoglycan See **glycosaminoglycan**.

glucose *Dextrose* Six-carbon sugar (aldohexose) widely distributed in plants and animals. Breakdown of glucose (**glycolysis**) is a major energy source for metabolic processes. In green plants, glucose is a major product of photosynthesis, and is stored as the polymer **starch**. In animals it is obtained chiefly from dietary di- and polysaccharides, but also by **gluconeogenesis**, and is stored as **glycogen**. Storage polymer in microorganisms is **dextran**.

glucose transporter *GLUT-1 etc* Generic name for any protein that transports glucose. In bacteria these may be **ABC proteins**, in mammals they belong to a family of 12-transmembrane integral transporters. GLUT-4 is found in muscle and is insulin-responsive. Defects in GLUT-4 may be responsible for some forms of **diabetes**.

glucose-1-phosphate Product of glycogen breakdown by phosphorylase. Converted to glucose-6-phosphate by phosphoglucomutase.

glucose-6-phosphate Phosphomonoester of glucose that is formed by transfer of phosphate from ATP, catalysed by the enzyme **hexokinase**. It is an intermediate both of the glycolytic pathway (next converted to fructose-6-phosphate), and of the NADPH-generating **pentose phosphate pathway**.

glucose-6-phosphate dehydrogenase *EC: 1.1.1.49* Ubiquitous enzyme, present in bacteria and all eukaryotic cell types, that catalyses the conversion of D-glucose-6-phosphate to D-glucono-1,5-lactone 6-phosphate, using NADP as a cofactor and generating NADPH. The first step in the pentose pathway. See **hexose monophosphate shunt**.

glucose-dependent insulinotropic polypeptide *GIP* A 42-residue peptide that stimulates insulin release from pancreatic β cells, like GLP-1 (**glucagon-like peptide 1**). An **incretin**.

glucose-related protein *GRP* One of the stress-related proteins: identical to **endoplasmin**.

glucosinolate A natural pesticide (glucoside) found mainly in brassicas such as broccoli and Brussels sprouts, responsible for the bitter or sharp taste of many common foods such as mustard, horseradish, cabbage and Brussels sprouts. About 120 different glucosinolates are known to occur naturally in plants.

glucosylation Transfer of glucose residues, usually from the nucleotide-sugar derivative UDPG. Enzymic glucosylation to generate the glucosyl-galactosyl disaccharide on the hydroxylysine of collagen is a normal process. A recent theory suggests that glucosylation of certain long-lived proteins by a non-enzymic reaction with free glucose may contribute to ageing. See **AGE**.

glucuronic acid *GA; GlcA* Uronic acid formed by oxidation of OH group of glucose in position 6. D-glucuronic acid is widely distributed in plants and animals as a subunit of various oligosaccharides.

glucuronoxylan **Hemicellulosic** plant cell-wall polysaccharide containing glucuronic acid and xylose as its main constituents. Has a β (1-4)-xylan backbone, with 4-*O*-methylglucuronic acid side-chains. Arabinose and acetyl side-chains may also be present. Major polysaccharide of angiosperm wood (hardwood).

Glufosinate A non-selective chemical herbicide that can be used only on crops tolerant to it – notoriously, those genetically engineered for such resistance. It is a natural compound isolated from two species of Streptomyces that inhibits the activity of glutamine synthetase, which is necessary for the production of glutamine and for ammonia detoxification.

GLUT1–GLUT4 Human **glucose transporters**. GLUT1 is ubiquitously expressed with particularly high levels in human erythrocytes and in the endothelial cells lining the blood vessels of the brain. GLUT3 is expressed primarily in neurons and, together, GLUT1 and GLUT3 allow glucose to cross the blood–brain barrier and enter neurons. GLUT2 is a low-affinity (high K_m) glucose transporter present in liver, intestine, kidney and pancreatic cells. The GLUT4 isoform is the major insulin-responsive transporter that is predominantly restricted to striated muscle and adipose tissue. GLUTX1 has been identified and appears to be important in early blastocyst development. See **GLUT5**.

GLUT5 A fructose transporter, expressed in insulin-sensitive tissues (skeletal muscle and adipocytes) of humans and rodents. High fructose diets lead to upregulation of the transporter.

glutamate Major fast excitatory **neurotransmitter** in the mammalian **central nervous system**. See **glutamate receptor**. Also the excitatory neuromuscular transmitter in arthropod skeletal muscles.

glutamate pyruvate transaminase *Glutamic pyruvic transaminase; alanine transferase; GPT; EC: 2.6.1.2* Enzyme that transfers an amino group from glutamate to pyruvate. Plasma levels are often assayed as an indicator of liver function. In the brain, it is an effective neuroprotectant against **glutamate** excitotoxicity.

glutamate receptor See **amino acid receptor** superfamily. Glutamate receptors are implicated in many important brain functions including **long-term potentiation** (LTP). At least four major glutamate-gated **ion channel** subtypes are at present distinguished on pharmacological grounds, named after their most selective agonists: *N*-methyl-D-aspartate (NMDA), implicated in memory and learning, neuronal cell death, ischaemia and epilepsy; kainic acid (KA); quisqualate/AMPA, and L-2-amino-4-phosphobutyrate (APB). A fifth subtype (ACPD), trans-1-amino-cyclopentane 1,3 dicarboxylate, is a G-protein-coupled receptor.

glutamate receptor-interacting proteins *GRIPs* A novel family of proteins that interact with non-NMDA **glutamate receptors** and, in particular, interact selectively with the C-termini of **AMPA** glutamate receptors. GRIPs appear to be involved in the targeting of membrane proteins that may have a critical role in neurotransmitter receptor functions, such as the formation and maintenance of excitatory synapses.

glutamic acid *Glu; E; 147 D* One of the 20 α-amino acids commonly found in proteins. Plays a central role in amino acid metabolism, acting as precursor of **glutamine**, **proline** and **arginine**. Also acts as amino group donor in synthesis by transamination of alanine from pyruvate, and aspartic acid from oxaloacetate. Glutamate is also a neurotransmitter; the product of its decarboxylation is the inhibitory neurotransmitter **GABA**.

glutamine *Gln; Q; 146 D* One of the 20 amino acids commonly found (and directly coded for) in proteins. It is the amide at the γ-carboxyl of the amino acid **glutamate**. Glutamine can participate in covalent cross-linking reactions between proteins, by forming peptide-like bonds by a transamidation reaction with lysine residues. This reaction, catalysed by clotting factor XIII stabilizes the aggregates of fibrin formed during blood clotting. Media for culture of animal cells contain some 10 times more glutamine than other amino acids, the excess presumably acting as a carbon source.

glutaraldehyde A dialdehyde used as a fixative, especially for electron microscopy. By its interaction with amino groups (and others) it forms cross-links between proteins.

glutathione The tripeptide γ-glutamylcysteinylglycine. It contains an unusual peptide linkage between the γ-carboxyl group of the glutamate side-chain and the amine group of cysteine. The concentration of glutathione in animal cells is around 5 mM and its sulphydryl group is kept largely in the reduced state. This allows it to act as a sulphydryl buffer, reducing any disulphide bonds formed within cytoplasmic proteins to cysteines. Hence few, if any, cytoplasmic proteins contain disulphide bonds. Glutathione is also important as a cofactor for the enzyme **glutathione peroxidase**, in the uptake of amino acids and participates in **leukotriene** synthesis.

glutathione peroxidase *EC 1.11.1.9* A detoxifying enzyme that eliminates hydrogen peroxide and organic peroxides. Glutathione is an essential cofactor for the enzyme and its reaction involves the oxidation of glutathione (GSH) to glutathione disulphide (GSSG). The GSSG is then reduced to GSH by glutathione reductase. Glutathione peroxidase, (GPX), has a selenocysteine residue in its active site. Three forms of the enzyme exist: cytoplasmic GPX, plasma GPX and phospholipid hydroperoxide GPX.

glutathione reductase *EC1.6.4.2* An FAD-containing enzyme, a dimer of 50-kDa subunits. It catalyses the NADP-dependent reduction of glutathione disulphide (GSSG) to glutathione (GSH). This is an essential reaction that maintains a GSH:GSSG ratio in the cytoplasm of around 500:1.

glutathione S-transferase *GST; EC 2.5.1.18* Enzyme that will couple **glutathione** to a xenobiotic as the first step in removal. Now very commonly used as a fusion with a gene of interest that is being expressed in a bacterial system. The fusion construct can be purified easily from lysate by passage down a glutathione affinity column, the purified construct then being eluted with glutathione. The GST can then be cleaved proteolytically from the protein of interest though often the complete fusion protein can be used.

glutathione synthase *GSS; EC6.3.2.3* Enzyme that catalyses synthesis of **glutathione** from ATP, γ-L-glutamyl-L-cysteine and glycine. GSS deficiency is inherited autosomal recessively, and patients with this disease can be divided into three groups, according to their clinical phenotype.

glutelin Group of proteins found in seeds of cereals such as wheat.

gluten Protein-rich fraction from cereal grains, especially wheat. When hydrated forms a sticky mass responsible for the mechanical properties of bread dough. **Glutelins** and **gliadin** form a substantial component.

glutenin A **glutelin** found in endosperm of wheatgrain. Component of gluten and through its tendency to polymerize contributes to properties of dough.

glycaemic *US glycemic* Relating to the concentration of glucose in the blood.

glycemic index *GI* A ranking of foods based on their overall effect on blood glucose levels. Slowly absorbed foods have a low GI rating, whilst foods that are more quickly absorbed will have a higher rating.

glyceraldehyde-3-phosphate Three-carbon intermediate of the glycolytic pathway formed by the cleavage of fructose 1,6-bisphosphate, catalysed by the enzyme **aldolase**. Also involved in reversible interchange between **glycolysis** and the **pentose phosphate pathway**.

glyceraldehyde-3-phosphate dehydrogenase *GAPD; GAPDH; G3PD; EC 1.2.1.12* Glycolytic enzyme that catalyses the reversible oxidative phosphorylation of glyceraldehyde-3-phosphate. Has been shown to interact with various elements of the **cytoskeleton**, and with the trinucleotide repeat in the **huntingtin** gene.

glycerination Permeabilization of the plasma membrane of cells by incubating in aqueous glycerol at low temperature. The technique was first applied to muscle, which once glycerinated, can be made to contract by adding exogenous ATP and calcium.

glycerol A metabolic intermediate, but primarily of interest as the central structural component of the major classes of biological lipids, triglycerides and phosphatidyl phospholipids. Also used as a **cryoprotectant**.

glycerone *1,3-Dihydroxy-2-propanone; dihydroxyacetone; DH* The simplest ketose with no chiral centres. Used in the cosmetics industry as a tanning substance (Coppertone) and also in fungicides.

glyceryl trinitrate Short-acting vasodilator that reduces venous return to the heart. Used in the treatment of angina pectoris.

glycine *Gly; G; 75.1 D* The simplest amino acid. It is a common residue in proteins, especially collagen and elastin, and is not optically active. It is also a major inhibitory **neurotransmitter** in spinal cord and brainstem of vertebrate **central nervous system**.

glycine receptor Chloride channel-forming receptor. One of a family of **neurotransmitter** receptors with fast intrinsic **ion channels**. See **amino acid receptors**.

glycocalyx The region, seen by electron microscopy, external to the outer dense line of the **plasma membrane** that appears to be rich in glycosidic compounds such as proteoglycans and glycoproteins. Since these molecules are often integral membrane proteins and may be denatured by the processes of fixation for electron microscopy, it might be better to avoid the term or to refer to membrane glycoproteins or to proteoglycans associated with the cell surface.

glycocholate N-*cholyl-glycine* Anion of the bile acid, glycholic acid. Usually found in bile as the sodium salt. Has powerful detergent properties.

glycoconjugate Any biological macromolecule containing a carbohydrate moiety – thus a generic term to cover **glycolipids**, **glycoproteins** and **proteoglycans**.

glycogen Branched polymer of D-glucose (mostly α (1-4)-linked, but some α (1-6) at branch points). Size range very variable, up to 105 glucose units. Major short-term storage polymer of animal cells, and is particularly abundant in the liver and to a lesser extent in muscle. In the electron microscope, glycogen has a characteristic 'asterisk' (*) appearance.

glycogenin In eukaryotes, a self-glucosylating protein that primes glycogen granule synthesis. Glycogen is bound to glycogenin to form proteoglycogen with the branched polysaccharide joined to the protein through the C-chain, a maltosaccharide considered to be 13 glucose units long.

glycolic acid Hydroxyacetic acid; found in young plants and green fruits. Glycolate is formed from ribulose-1,5-bisphosphate in a seemingly wasteful side reaction of photosynthesis, known as **photorespiration**.

glycolipid Oligosaccharides covalently attached to lipid as in the glycosphingolipids (GSL) found in plasma membranes of all animal and some plant cells. The lipid part of GSLs is sphingosine in which the amino group is acylated by a fatty chain, forming a **ceramide**. Most of the oligosaccharide chains belong to one of four series, the ganglio-, globo-, lacto-type 1 and lacto-type 2 series. Blood group antigens are GSLs.

glycolysis The conversion of a monosaccharide (generally glucose) to pyruvate via the glycolytic pathway (i.e. the **Embden–Meyerhof pathway**) in the cytosol. Generates ATP without consuming oxygen and is thus anaerobic.

glycomics The study and analysis of the carbohydrate moieties of complex macromolecules (glycolipids and glycoproteins) found within an organism, by analogy with proteomics.

glycophorins A class of abundant transmembrane glycoproteins of the human erythrocyte. The major component is a 131-residue peptide chain that is highly O-glycosylated and is rich in terminal sialic acid. The peptide chain carries the **MN blood group antigens** at its N-terminus.

glycoprotein Proteins with covalently attached sugar units, either bonded via the OH group of serine or threonine (O-glycosylated) or through the amide NH2 of asparagine (N-glycosylated). Includes most secreted proteins (serum albumin is the major exception) and proteins exposed at the outer surface of the plasma membrane. Sugar residues found include: mannose, N-acetyl glucosamine, N-acetyl galactosamine, galactose, fucose and sialic acid.

glycosaminoglycan attachment site In **proteoglycans** a number of **glycosaminoglycan** chains are attached to a core protein through a xyloside residue which is linked to a serine residue at the consensus pattern S-G-X-G.

glycosaminoglycans *Formerly mucopolysaccharides* Polysaccharide side-chains of **proteoglycans** made up of repeating disaccharide units (more than 100) of amino sugars, at least one of which has a negatively charged side-group carboxylate or sulphate. Commonest are hyaluronate (D-glucuronic acid-N-acetyl-D-glucosamine: MW up to 10 million), chondroitin sulphate (D-glucuronic acid-N-acetyl-D-galactosamine-4 or -6-sulphate), dermatan sulphate (D-glucuronic acid- or L-iduronic acid-N-acetyl-D-galactosamine), keratan sulphate (D-galactose-N-acetyl-D-glucosamine-sulphate), and heparan sulphate (D-glucuronic acid- or L-iduronic acid-N-acetyl-D-glucosamine). Glycosaminoglycan side-chains (with the exception of hyaluronate) are covalently attached to a core protein at about every 12 amino acid residues to produce a proteoglycan; these proteoglycans are then non-covalently attached by link proteins to hyaluronate, forming an enormous hydrated space-filling polymer found in **extracellular matrix**. The extent of sulphation is variable and the structure allows tremendous diversity.

glycosidase *Glycosylase* General and imprecise term for an enzyme that degrades linkage between sugar subunits of a polysaccharide. Any of the EC 3.2 class of hydrolases that cleave glycosidic bonds. They may distinguish between α and β links, for example, but are not very substrate-specific. See **endoglycosidase**.

glycosidic bond Bond between anomeric carbon of a sugar and the group to which it is attached (which may be in another sugar or in protein or lipid).

glycosome Microbody containing glycolytic enzymes, found in protozoa of the Kinetoplastida (e.g. trypanosomes).

glycosphingolipids Ceramide derivatives containing more than one sugar residue. If sialic acid is present, these are called **gangliosides**.

glycosyl phosphatidyl inositol *GPI* See **GPI-anchor**.

glycosyl transferase Class of ectoenzymes that catalyse the transfer of a sugar (monosaccharide) unit from a sugar nucleotide derivative to a sugar or amino acid acceptor. Various sugars, including galactose, glucose, N-acetylglucosamine, N-acetylneuraminic acid, mannose and fucose may be transferred.

glycosylation The process of adding sugar units, as in the addition of glycan chains to proteins.

glycylcyclines **Tetracycline** derivatives that are effective in all situations in which tetracycline was once used.

glyoxal *Ethanedial; OHC-CHO* Dialdehyde, soluble in water where it exists as the dihydrate. Readily polymerizes on standing.

glyoxisome Organelle found in plant cells, containing the enzymes of the **glyoxylate cycle**. Also contains catalase and enzymes for **beta-oxidation** of fatty acids. Together with the **peroxisome** makes up the class of organelles known as **microbodies**.

glyoxylate cycle Metabolic pathway present in bacteria and in the **glyoxisome** of plants, in which two acetyl-CoA molecules are converted to a 4-carbon dicarboxylic acid, initially succinate. Includes two enzymes not found elsewhere, isocitrate lyase and malate synthase. Permits net synthesis of carbohydrates from lipid, and hence is prominent in those seeds in which lipid is the principal food reserve.

glyoxylate shunt See **glyoxalate cycle** and **glyoxysome**.

glyphosate A broad-spectrum, non-selective systemic herbicide that is effective in killing all plant types, including grasses, perennials and woody plants. It is inactivated when it comes into contact with soil and, although an organophosphate, is not a cholinesterase inhibitor.

glypiation See **GPI-anchor**.

glypican-3 Membrane glycoprotein thought to bind **insulin-like growth factor-2** (IGF-2). Gene encodes cell surface heparan-sulfate proteoglycans, and is frequently upregulated in hepatocellular carcinoma. Defective in some cases of **Simpson-Golabi-Behmel syndrome**.

GM-CSF *Granulocyte-macrophage colony-stimulating factor* A cytokine that stimulates the formation of granulocyte or macrophage colonies from myeloid stem cells isolated from bone marrow.

Gm-types Genetically determined allotypic antigens found on IgG of some individuals.

GM130 A subfamily of Golgi-associated proteins, involved in the docking and fusion of **coatamer** (COP I) coated vesicles to the Golgi membranes; it also regulates the fragmentation and subsequent reassembly of the Golgi complex during mitosis. At the cis-side of the Golgi, the Rab1 GTPase binds directly to each of three coiled-coil proteins: p115, GM130 and **giantin**; GM130 and giantin bind to the acidic domain of p115 and stimulate p115 binding to Rab1. Other members of the GM130 subfamily, **golgins**, include rat GM130 and human golgin-95.

GMO Genetically modified organism.

gnotobiotic Organism or environment completely or almost completely depleted of all organisms or all other organisms. Animals that are SPF (specific pathogen free) are gnotobiotic.

GNRP *Guanine nucleotide releasing protein.* Originally defined as a protein that facilitated binding of GTP by ras-like GTPases, resulting in activation of the Ras signal. Examples include **sos**, RasGRP and **C3G**. They are exchange factors (**GEFs**) although the term releasing protein seems to persist.

Go A specific class (o for other) of signal-transducing heterotrimeric **GTP-binding proteins** (G-proteins) expressed in high levels in mammalian brain. Other G-proteins are involved in stimulation (Gs) or inhibition (Gi).

goblet cell 1. Cell of the epithelial lining of small intestine that secretes mucus and has a very well-developed Golgi apparatus. 2. Cell type characteristic of larval lepidopteran midgut, containing a potent H^+-**ATPase**, and thought to be involved in maintenance of ion and pH gradients.

Goldberg–Hogness box See **TATA box**.

Golden rice Rice that has been genetically engineered by the insertion of genes that produce high levels of beta-carotene, which is converted to Vitamin A within the body, and should help to prevent blindness due to vitamin deficiency. The carotene makes the rice yellow in colour, hence the name.

Goldmann equation *Goldmann constant field equation* Equation that describes the electrical potential across a membrane in terms of the distributions and relative permeabilities of the main permeant ions (typically sodium, potassium and chloride). Assumes that the electrical field across the membrane is constant and that there are no **active transport** processes; but nonetheless gives a reasonable approximation to real membranes.

Golgi apparatus Also known as the Golgi body, Golgi vesicles; in plants, the **dictyosome**; in flagellate protozoa, the parabasal body. Intracellular stack of membrane-bounded vesicles in which glycosylation and packaging of secreted proteins takes place; part of the **GERL** complex. Vesicles from endoplasmic reticulum fuse with the cis-Golgi region (the inner concave face) and progress through the vesicular stack to the trans-Golgi, whence they move towards the plasma membrane or lysosomes.

golgins Coiled-coil proteins associated with the Golgi apparatus, that are believed to be involved in the tethering of vesicles, the stacking of cisternae and linkage with the cytoskeleton. Many are peripheral membrane proteins recruited by GTPases (particularly **rab**). Relationships between golgins from different species are unclear because although they share structural features, their sequences are not well conserved.

golli proteins Proteins that have a broad distribution in the nervous and immune systems. The **myelin basic protein** (MBP) gene produces two families of proteins, the classic MBPs, important for myelination of the CNS, and the golli proteins. Oligodendrocytes (OL) express golli mRNA primarily during intermediate stages of differentiation; expression is low in proliferating OL progenitors as well as in terminally mature OLs. Golli proteins are not targeted to myelin *in vitro* and *in vivo*, in contrast to the classic MBPs, and are postulated to have a role in immune processes as well as in the development of neurons and oligodendrocytes.

Gomori procedure Cytochemical staining procedure used to localize acid phosphatases. Depends upon the production of phosphate ions from organic phosphoesters such as β-glycerophosphate. The phosphate in the presence of lead ions causes the formation of a precipitate of lead salt that is converted to the brown sulphide of lead by the action of yellow ammonium sulphide.

gonadotrophin-releasing hormones *GnRH* Peptide hormones produced in the hypothalamus that act on the pituitary to stimulate production of **luteinizing hormone** and **follicle-stimulating hormone**. Decapeptides: consensus QHXSXXXXPG (amidated).

gonadotrophins *Gonadotropins* Group of glycoprotein hormones from the anterior lobe of the pituitary gland. They stimulate growth of the gonads and the secretion of sex hormones. Examples: **follicle-stimulating hormone**, **luteinizing hormone**, **chorionic gonadotrophin**.

gonialblast In *Drosophila* spermatogenesis, the germline stem cell divides and one daughter becomes a gonialblast while the other remains a stem cell. Each gonialblast executes four divisions as secondary spermatogonia, which exit the mitotic cycle and enter a meiotic and differentiation program as a clone of 16 spermatocytes. Cf **cystoblast**.

gonioscope An instrument used to measure and inspect the anterior chamber of the eye, the region between the cornea and the iris.

gonosome Collective name for the reproductive zooids of a colonial animal such as a hydroid.

Gonyaulax Genus of **dinoflagellates**. Responsible for red tides and associated shellfish poisoning due to **saxitoxin**. Some species are bioluminescent.

Goodpasture's syndrome Disease in which there is accumulation of a very uniform layer of autoantibodies to components of basement membrane on the kidney glomerular basement membrane.

gooseberry Gsb A **segment-polarity gene** of *Drosophila*. Contains the **paired box domain**.

goosecoid A homeodomain transcription factor expressed in the dorsal lip of the *Xenopus* blastopore; may be a key factor is specifying these cells as the organizer of the embryo (**Spemann's organizer**). Name derived from *Drosophila* homedomain proteins to which its homoeodomain is similar: **gooseberry** and **bicoid**.

Gorlin syndrome *Naevoid basal cell carcinoma* A rare autosomal dominant disorder in which the patched/hedgehog/smoothened signalling pathway has been implicated; probably due to mutations in the *PTCH* gene. Oncogenic forms of **Smoothened** (Smo) have been isolated from human basal cell carcinomas.

GOS19-1 Human **macrophage inflammatory protein 1**α.

Gossypium Genus of plants that includes cotton.

gout Recurrent acute arthritis of peripheral joints caused by the accumulation of monosodium urate crytals. Usually due to overproduction of uric acid but may be a result of under-excretion. The problems partly arise because neutrophils release lysosomal enzymes as a result of damage to the phagosome membrane by ingested crystals: **colchicine** acts to reduce the attack by inhibiting lysosome–phagosome fusion.

gp130 Signalling subunit of the IL-6-family receptors. Degradation of gp130 is regulated through a phosphorylation-dephosphorylation mechanism in which protein phosphatase-2A is crucially involved; gp130 is a potential therapeutic target in cancers.

Gp-Ib *Glycoprotein Ib* Integral protein of platelets that binds to von Willebrand factor and is involved in thrombus formation. Disulphide-linked heterodimer (α 68 kDa; β 22 kDa) deficient in **Bernard–Soulier syndrome**.

gp100 Melanocyte/melanoma differentiation antigen, one of the **melanoma-associated antigens** often used in diagnostic testing.

gp41 Envelope glycoprotein of **HIV** (41 kDa), encoded by HIV *env* gene. The N-terminal part of gp41 is thought to mediate fusion between viral and cellular membranes.

GPI-anchor *Glypiation* Common modification of the C-terminus of membrane-attached proteins in which a phosphatidyl inositol moiety is linked through glucosamine and mannose to a phosphoryl ethanolamine residue that is linked to the C-terminal amino acid of the protein by its amino group. Glypiation is the sole means of attachment of such proteins to the membrane. The name comes from the addition of glycosyl phosphatidyl inositol (PI).

G-protein 1. See **GTP-binding proteins**. 2. The spike glycoprotein of vesicular stomatitis virus. This has been an important protein for investigation of membrane transport in eukaryotic cells.

G-protein-coupled receptor *GPCR* Cell surface receptors that are coupled to heterotrimeric **G-proteins** (GTP-binding proteins). All G-protein-coupled receptors seem to have seven membrane-spanning domains (are **serpentine receptors**), and have been divided into two subclasses: those in which the binding site is in the extracellular domain, e.g. receptors for glycoprotein hormones, such as **thyroid-stimulating hormone** (TSH) and **follicle-stimulating hormone** (FSH); and those in which the

ligand-binding site is likely to be in the plane of the seven transmembrane domains, e.g. **rhodopsin** and receptors for small **neurotransmitters** and **hormones,** e.g. **muscarinic acetylcholine receptor**.

GPT See **glutamate pyruvate transaminase**.

Graafian follicle Final stage in the differentiation of follicles in the mammalian ovary. Consists of a spherical fluid-filled blister on the surface of the ovary that bursts at ovulation to release the oocyte.

gradient perception Problem faced by a cell that is to respond directionally to a gradient of, for example, a diffusible attractant chemical. In a spatial mechanism, the cell would compare receptor occupancy at different sites on the cell surface; a temporal mechanism would involve comparison of concentrations at different times, the cell moving randomly between readings. In pseudospatial sensing, the cell would detect the gradient as a consequence of positive feedback to protrusive activity if receptor occupancy increased with time as the protrusion moved up-gradient. Few cell types have been unambiguously shown to detect gradients.

graft-versus-host response *GVH* When a graft of lymphocytes or a graft containing lymphocytes is carried out into an animal these may, if appropriately mismatched at MHC Class I to their host, produce lymphocyte clones that will react by a variety of processes against the host and cause damage.

Gram-negative bacteria Bacteria with thin **peptidoglycan** walls bounded by an outer membrane containing **endotoxin** (lipopolysaccharide). See **Gram stain**.

Gram-positive bacteria Bacteria with thick cell walls containing **teichoic** and **lipoteichoic acid** complexed to the **peptidoglycan**. See **Gram stain**.

Gram stain A heat-fixed bacterial smear is stained with crystal violet (methyl violet), treated with 3% iodine/potassium iodide solution, washed with alcohol and counterstained. The method differentiates bacteria into two main classes, **Gram-positive** and **Gram-negative**. Certain bacteria, notably **mycobacteria**, that have walls with high lipid content show acid-fast staining, the stain resists decolouration in strong acid.

Gramicidin A A linear peptide of alternate D- and L-amino acids that acts as a cation ionophore in lipid bilayer membranes. It is proposed that two molecules form a membrane-spanning helix containing a pore lined with polar residues.

Grammostola spatulata Chilean pink tarantula. See **grammotoxin**.

grammotoxin ω-*Grammotoxin SIA* Toxin (peptide, 36 residues) from spider, *Grammostola spatulata*, that inhibits non-L-type voltage-regulated calcium channels, thus resembling ω-**conotoxins** and ω-**agatoxins**. See **VSCC**.

graniferous Term for grain-producing plants.

granins *Chromogranins; secretogranin* Family of related acidic proteins (400–600 residues) found in many endocrine cell secretory vesicles. Secretogranin 1 = chromogranin B; secretogranin 2 = chromogranin C.

granular component of nucleolus Area of nucleolus that appears granular in the electron microscope and contains 15-nm diameter particles that are maturing ribosomes. In contrast to the pale-staining and **fibrillar** areas.

granulation tissue Highly vascularized tissue that replaces the initial fibrin clot in a wound. Vascularization is by ingrowth of capillary endothelium from the surrounding vasculature. The tissue is also rich in fibroblasts (that will eventually produce the **fibrous tissue**) and leucocytes.

granule cell Type of neuron found in the cerebellum.

granulocyte Leucocyte with conspicuous cytoplasmic granules. In humans, the granulocytes are also classified as polymorphonuclear leucocytes and are subdivided, according to the staining properties of the granules, into **eosinophils**, **basophils** and **neutrophils** (using a **Romanovsky-type stain**); some invertebrate blood cells are also referred to, not very helpfully, as granulocytes.

granulocyte/macrophage colony-stimulating factor See **GM-CSF**.

granulocytopenia Low granulocyte numbers in circulating blood.

granuloma Chronic inflammatory lesion characterized by large numbers of cells of various types (macrophages, lymphocytes, fibroblasts, giant cells), some degrading and some repairing the tissues.

granulopoiesis The production of **granulocytes** in the bone marrow.

granum *Pl.* grana Stack of **thylakoids** in the chloroplast, containing the **light-harvesting system** and the enzymes responsible for the **light-dependent reactions** of photosynthesis.

granzymes Family of serine endopeptidases found in cytotoxic T-cells and NK-cells and involved in **perforin**-dependent cell killing. Granzyme A is EC 3.4.21.78; granzyme B is EC 3.4.21.79.

GRAS *Gonadotropin-releasing hormone (GnRH) receptor activating sequence* A composite regulatory element that interacts with multiple classes of transcription factors including **Smads**, **AP-1** and a **forkhead** DNA-binding protein. But see **GRAS family**.

GRAS family Family of plant regulatory proteins named after **GAI** (gibberellin insensitive), RGA (repressor of ga1-3) and SCR (SCARECROW), the first three of its members isolated. SCR plays a significant role in the radial patterning of both roots and shoots. The **DELLA** proteins are a subfamily of the GRAS family. The *Arabidopsis* genome encodes at least 33 GRAS protein family members, although the roles of only a few are currently known. But see **GRAS**.

GRASP *Golgi reassembly stacking protein, Golgi membrane-associated protein p59* GRASP55, is structurally related to GRASP65, an *N*-ethylmaleimide-sensitive membrane protein; both play a role in stacking of the Golgi cisternae *in vitro*. GRASP55 and **golgin**-45 form a rab2 effector complex on medial-Golgi essential for normal protein transport and Golgi structure.

Graves' disease Autoimmune disease characterized by goitre, exophthalmia and **thyrotoxicosis**, caused by antibodies to the thyrotropin receptor resulting in constitutive activation of the receptor and increased levels of thyroid hormone. In Caucasians, is associated with HLA-B8 and DR3. Variations in several genes, including the CTLA4 gene, the vitamin D receptor gene and the vitamin D binding protein gene may contribute to susceptibility to the disease.

gravitaxis Directed locomotory response to gravity.

gravitropism Directional growth of a plant organ in response to a gravitational field – roots grow downwards, shoots grow upwards. Achieved by differential growth on the sides of the root or shoot. A gravitation field is thought to be sensed by sedimentation of **statoliths** (starch grains) in root caps.

GRB-2 *Growth factor receptor bound protein 2* Protein that links the cytoplasmic domain of growth factor receptor to **sos** and to **shc** through its SH2 and SH3 domains, and thus is important in the assembly of the signalling complex. Homologous to yeast sem-5 and *Drosophila* drk.

Greek key Protein **motif** in which four **beta strands** (in one or two **beta sheets**) form a twisted arrangement similar to a pattern often seen on Greek vases.

green algae See **Chlorophyta**.

green fluorescent protein GFP Protein from luminous jellyfish, *Aequorea victoria*. Excited by blue light (as produced from **aequorin** luminescence), it emits green light. The gene has been cloned and mutagenized to give brighter fluorescence and different colour variants. These have become valuable transgenic tools in **flag tagging** and as **reporter genes** and **cell lineage** markers.

gregarine movement Peculiar gliding movement shown by gregarines (Protozoa), the mechanism of which is poorly understood.

gremlin A **BMP** antagonist expressed in osteoblasts that opposes BMP effects on osteoblastic differentiation and function *in vitro*. Gremlin transgenics show bone fractures and reduced bone mineral density by 20–30%, compared with controls. Other BMP antagonists are **noggin**, chordin, cerberus and **DAN**.

grex The multicellular aggregate formed by cellular slime moulds (**Acrasidae**): the slug-like grex migrates, showing positive phototaxis and negative gravitaxis, until culmination (the formation of a fruiting body) takes place. Coordination of the activities of the hundreds of thousands of individual amoebae that compose the grex may involve pulses of cyclic AMP in *Dictyostelium discoideum*, a species in which cAMP is the chemotactic factor for aggregation.

grey crescent A region near the equator of the surface in the fertilized egg of various amphibia, often of greyish colour, that appears to contain special morphogenetic properties.

GRF See **growth hormone-releasing factor**.

Grim Regulator of apoptosis in *Drosophila*. Like **reaper** and **hid** effect is blocked by caspase inhibitors and *Drosophila* homologues of mammalian **IAPs**.

GRIP *Glutamate receptor-interacting protein* Protein (120 kDa, 1112 residues) found in postsynaptic terminal, contains 7 **PDZ** domains and is involved in the clustering of **AMPA** receptors with which it interacts through PDZ domains 4 and 5.

griseofulvin Polyketide antibiotic from *Penicillium griseofulvum*. Used therapeutically as an antifungal. Blocks microtubule assembly and thus mitosis.

GRK The family of G-protein receptor kinases that phosphorylate agonist-occupied G-protein-coupled receptors. GRK-1 is rhodopsin kinase; many others are known. See **arrestins**.

gRNA *Guide RNA* Small RNA molecules (60–80 nucleotides) that are found in the **editosome**. Guide RNAs are complementary to edited portions of the mature mRNA and contain poly-U tails that donate the Us added during editing.

gro See **melanoma growth-stimulatory activity** protein.

groEL See **chaperonins**.

groucho *Gro* An adaptor molecule that acts as a co-repressor for some negative regulators in *Drosophila* development but not others (does not act on repressor regions of **even-skipped**, **kruppel** or **knirps** transcription factors).

ground meristem Partly differentiated meristematic tissue (primary meristem) derived from the apical meristem and giving rise to relatively unspecialized plant tissues (ground tissues) such as **parenchyma** and **collenchyma**.

ground tissue Plant tissues other than those of the vascular system and the **dermal tissues**. Composed of relatively undifferentiated cells.

ground-glass cells Hepatitis B infected liver cells. The hepatitis B surface antigen appears as fine granules either diffusely spread through out the cytoplasm or concentrated in the cytoplasm peripheral to the sinusoid space and when stained by **orcein** gives the 'ground glass' appearance to the cells.

growth (and) differentiation factor *GDF* Family of proteins involved in growth and differentiation. GDF-3 has been reported to be implicated in testis carcinoma and deposition of adipose tissue; GDF-5 (also known as bone morphogenetic protein-14 and cartilage-derived morphogenetic protein-1) is important in bone repair; GDF6 is also known as BMP13; GDF7 is also known as BMP12; GDF-8 is **myostatin**; GDF-9 plays an important role in somatic cell differentiation and the formation of primordial follicles in hamster ovary; GDF-15 is the murine orthologue of macrophage inhibitory cytokine-1 (MIC-1).

growth cone A specialized region at the tip of a growing **neurite** that is responsible for sensing the local environment and moving towards the neuron's target cell. Growth cones are hand-shaped, with several long **filopodia** that differentially adhere to surfaces in the embryo. Growth cones can be sensitive to several guidance cues, for example, surface adhesiveness, growth factors, neurotransmitters and electric fields (**galvanotropism**).

growth cone collapse Loss of motile activity and cessation of advance by **growth cones**. There are specific molecules that inhibit the motility of particular growth cones, and that are important in establishing correct pathways in developing nervous systems.

growth control When applied to cells usually means control of growth of the population, i.e. of the rate of division rather than of the size of an individual cell.

growth factors Polypeptide hormones that regulate the division of cells, for example EGF (**epidermal growth factor**), PDGF (**platelet-derived growth factor**), FGF (**fibroblast growth factor**). Insulin and somatomedin are also growth factors; the status of NGF (**nerve growth factor**) is more uncertain. Perturbation of growth factor production or of the response to growth factor is important in neoplastic transformation.

growth hormone *Somatotropin* Polypeptide (191 amino acids) produced by anterior pituitary that stimulates liver to produce **somatomedins** 1 and 2.

growth hormone-releasing factor *GRF* See **growth-hormone-releasing hormone**.

growth substances See **plant growth substances**.

growth-associated proteins *GAPs* Group of developmentally regulated polypeptides thought to be critical for the formation of neural circuitry. The acidic membrane phosphoprotein GAP-43 is synthesized and transported down regenerating and developing axons; pp46 localized in growth cone membranes during embryogenesis; B-50 in mature presynaptic membranes in the regulation of phosphotidylinositol turnover and F1 in the **hippocampus** during **long-term potentiation**, are now all known to be the same protein.

growth-hormone-releasing hormone *Growth hormone-releasing factor; GRF; Growth hormone-regulating hormone (GHRH), somatocrinin; somatoliberin* Hypothalamic hormone that induces the release of growth hormone (**somatotropin**). Release is inhibited by **somatostatin**. See Table H2.

growth/differentiation factors Members of the TGFβ family of growth factors. See also **midkine**.

GRP See **glucose-related protein** or **gastrin-releasing peptide**.

GRP-1 Guanine nucleotide exchange factor with **PH domain** that binds PtdIns(3,4,5)P3 and also has domain with homology to yeast Sec7. See **cytohesin**, **ARNO**, Gea-1 and Gea-2, other members of the family.

grp78 *BiP* Glucose-regulated protein 78 (GRP78/**BiP**), an endoplasmic reticulum-specific chaperone.

GrpE Bacterial chaperone of **Hsp70** family that works in conjunction with **dnaJ** and dnaK. Dimeric GrpE is the co-chaperone for dnaK, and acts as a nucleotide exchange factor, stimulating the rate of ADP release from dnaK 5000-fold.

Gs See **GTP-binding protein**.

GSK *Glycogen synthase kinase* GSK3 is a serine–threonine protein kinase that plays an important part in various intracellular signalling pathways including the control of glycogen metabolism and protein synthesis by insulin, the modulation of **AP-1** and **CREB**, the specification of cell fate in *Drosophila* and the regulation of dorsoventral patterning in *Xenopus*. GSK3 itself is regulated by **PKB**. GSK3 is identical to Tau protein kinase I and may therefore be important in the phosphorylation of **tau** that is known to occur in the neurofibrillary tangles of Alzheimer's disease.

GSS Abbrev. for (1) **Glutathione synthetase**, (2) **Gerstmann–Straussler–Scheinker** disease, (3) Global Symptom Score, (4) Glaucoma Staging System, for classifying glaucomatous visual field defects.

GST See **glutathione S-transferase**.

GST-fusion protein One way of purifying proteins expressed by a cloned gene is to insert the gene of interest into a vector in-frame with a gene encoding glutathione S-transferase; this is then expressed in a cell line to produce a (chimeric) **fusion protein**. This can be purified on a glutathione affinity column by virtue of the affinity of GST for glutathione, eluted with free glutathione and the GST can subsequently be proteolytically cleaved off and removed by an iteration of the affinity purification – though this is not always necessary.

GST-P **Glutathione S-transferase** P gene that is strongly and specifically expressed during chemical hepatocarcinogenesis. The promoter region has a silencer (GPS1) to which various transactivators bind.

GTBP *G/T binding protein* Protein (160 kDa) important in mismatch recognition in human cells (see **mismatch repair**). The heterodimer of GTBP with hMSH2 (one of the **MSH** family) binds mismatches (G paired with T) as a first step in excision repair. Absence of either protein predisposes to tumours. GTBP gene is located near that for hMSH2 and GTBP can be considered one of the MSH family.

GTP *Guanosine 5′-triphosphate* Like **ATP** a source of phosphorylating potential, but is separately synthesized and takes part in a limited, distinct set of energy-requiring processes. Synthesis is by a substrate-linked phosphorylation involving succinyl coenzyme A, part of the tricarboxylic acid cycle. GTP is required in protein synthesis, the assembly of **microtubules**, and for the activation of regulatory G-proteins (**GTP-binding proteins**).

GTP-binding proteins *G-proteins* There are two main classes of G-proteins; the heterotrimeric G-proteins that associate with receptors of the seven-transmembrane domain superfamily and are involved in signal transduction, and the small cytoplasmic G-proteins. The Gα subunit (39–52 kDa) of the heterotrimeric G-proteins dissociates from the βγ subunits (β, 35–36 kDa; γ, 6–10 kDa) when GTP is bound, and in this state will interact with various second messenger systems, either inhibiting (Gi) or stimulating (Gs). The Gα subunit has slow GTPase activity and once the GTP is hydrolysed it reassociates with the βγ subunits. There is less diversity among the βγ subunits, but they may have direct activating effects in their own right. Most βγ subunits are post-translationally modified by myristoylation or isoprenylation that may alter their association with membranes. Stimulatory G-proteins are permanently activated by **cholera toxin**, inhibitory ones by **pertussis toxin**. **Transducin** was one of the first of the heterotrimeric G-proteins to be identified. The small G-proteins are a diverse group of monomeric GTPases that include **ras**, **rab**, **rac** and **rho** and that play an important part in regulating many intracellular processes including cytoskeletal organization and secretion. Their GTPase activity is regulated by activators (GAPs) and inhibitors (GIPs) that determine the duration of the active state. See also **GEFs**, **ras-like GTPases**, Table G1.

GTPase-activating protein *GAP* Originally purified as a 125-kDa protein from bovine brain (1044 amino acids); stimulates the GTPase activity of ras-p21 and thereby switches it to the inactive form. GAP may itself be regulated by phospholipids and by phosphorylation on a tyrosine residue by growth factor receptors (PDGF-R, EGF-R). The **neurofibromatosis** type 1 gene (NF1) codes for a protein homologous to GAP. GAP has both **SH2** and **SH3** domains. Another example is sar-1 (from yeast).

guaiacol *2-Methoxy-phenol* Guiacol is a constituent of guiacum resin and occurs in creosote. Colourless guaiacol can be oxidized to the red-brown tetraguaiacol product by H_2O_2 in the presence of peroxidase.

guanidinium chloride *Guanidine hydrochloride* Chloride salt of guanidinium $(C(NH_2)_3)^+$, a powerful chaotropic agent that is used to denature proteins.

guanine *2-Amino 6-hydroxypurine* One of the constituent bases of nucleic acids, nucleosides and nucleotides.

guanosine *9-β-D-Ribofuranosyl guanine* The nucleoside formed by linking ribose to guanine.

guanosine 5′-triphosphate See **GTP**.

guanylate cyclase Enzyme catalysing the synthesis of guanosine 3′,5′- cyclic monophosphate from guanosine 5′-triphosphate (cyclic GMP; cGMP) is used as a **second messenger** in heart muscle and **photoreceptor** cells). The plasma membrane form of guanylate cyclase is an integral membrane protein with an extracellular receptor for peptide hormones, a transmembrane domain, a **protein kinase**-like domain, and a guanylate cyclase domain. Examples: sea urchin receptors for **speract** and **resact**; and **atrial natriuretic peptide** receptors. Two soluble forms of guanylate cyclase are also known, heterodimers of highly related subunits, 70–80 kDa.

guanylin Peptide occurring in vertebrate gut that elevates the **second messenger** cyclic GMP in a variety of tissues (including gut) via a membrane **guanylate cyclase**. The receptor is also the target for the *E. coli* heat-stable enterotoxin.

guard cell Plant cells occurring in pairs in the **epidermis**, flanking each **stoma**. Changes in **turgor** in the guard cells cause the stoma to open and close.

Guarnieri body Acidophilic inclusion body found in cells infected with **vaccinia** virus; composed of viral particles and proteins, it is the location of virus replication and assembly.

guidance See **contact guidance**.

guide RNA *gRNA* Small RNA molecules that hybridize to specific mRNAs and direct their **RNA editing**.

Guillain–Barré syndrome *Landry-G-B syndrome* Acute infective polyneuritis, associated sometimes with preceding *Campylobacter jejuni* infection, in which there is cell-mediated immunity to a component of myelin; the disease may be autoimmune in origin. Some cases may be caused by mutation in the peripheral myelin protein (PMP22; *gas*-3; growth arrest-specific-3 gene). Causes a temporary paralysis, particularly of the extremities.

gurken TGFα-like ligand for the EGF-receptor **torpedo**, important for dorsoventral patterning in the *Drosophila* embryo. The gurken-torpedo signal is the first to specify dorsal fate, and in turn induces the production of the ligand spatzle that binds to toll and activates the transcription factor dorsal. The gurken–EGF–R interaction also has a role in the specification of anterior–posterior polarity of the egg.

GUS *EC:3.2.1.31* Glucuronidase; widely used as a **reporter gene**.

gustducin Taste-cell specific **GTP-binding protein**. Has a novel Gα subunit that resembles **transducin** more than any other Gα.

TABLE G1. The properties of heterotrimeric G-protein subunits

Alpha subunits:

α	MW (kDa)	Signal detector	Effector	Toxin	Comments
α_s	44–45	β-Adrenergic, glucagon and many other receptors	Activates adenylate cyclase	CT	Stimulatory. At least four splice variants known, and relative concentration varies from one cell type to another.
α_{i1}	40.4	α-Adrenergic, muscarinic cholinergic, opiate and many other receptors	Inhibits adenylate cyclase and PLC (phosphoinositide hydrolysis).	PT	Inhibitory. The sequence for $G_i\alpha_1$, and $G_i\alpha_2$ were derived from two different cDNA clones. The relative importance of the two forms, in the cell, is not yet clear
α_{i2}	40.5		Inhibits adenylate cyclase, PLC, involved in regulating K^+-channels, Ca^{2+} channels.	PT	
α_{i3}	40.5		Probably same as $\alpha_{i1,2}$	PT	
α_{olf}	44	Seven-membrane spanning	Activates adenylate cyclase	CT	In olfactory neurons
α_{gust}	40.3	Taste sensors	Unclear	Both	Component of Gustducin, restricted to tongue.
α_z	40.9	Unknown	Inhibits adenylate cyclase	None	In brain and neuronal cells
α_q	42.2		Activates PLCβ (PI hydrolysis)	None	Widely expressed except in T-cells.
α	MW (kDa)	Signal detector	Effector	Toxin	Comments
α_{11-14}	42–45		Activation PLC β	None	Widely distributed though with variations between tissues
α_o	39	Unknown	Inhibits neuronal calcium channels	PT	'o' stands for other G-protein. It was first detected in large amounts in brain. Splice variants known.
α_{t1}	40	Rhodopsin	cGMP-PDE	Both	Subunit of Transducin, found in the rod cells of the retina
α_{t2}	40.4	–	cGMP-PDE	Both	Found in the cone cells of the retina: probably the α subunit of the cone's analogue of Transducin

Beta and gamma subunits

β	MW (kDa)	Tissue distribution
β1	37	Beta subunit in transducin
β2	38	Widely distributed
β3	37	Wide, particularly cone cells of retina
β4	37	Wide, especially brain, eye, lung, heart and testis
γ1	8.4	Rod cells of retina
γ2	7.8	Brain
γ3	8.3	Brain and retina
γ4		Unknown
γ5	7.3	Wide
γ7	7.5	Brain

Based partly upon Watson, S. and Arkinstall, S. (1994). G-*Protein-Linked Receptor Facts Book*. Academic Press.

gut-associated lymphoid tissue *GALT* Peripheral lymphoid organ consisting of lymphoid tissue associated with the gut (Peyer's patches, tonsils, mesenteric lymph nodes and the appendix).

GW bodies *P bodies* Small, generally spherical, cytoplasmic domains that vary in number and size in various mammalian cell types and appear to be the site for mRNA processing, storage and degradation. Several proteins co-locate in these processing granules, including **GW182** and **argonaute** 1 and 2 which are associated with **siRNA** operations. May well be equivalent to maternal granules in eggs, which are accumulations of long-lived maternal mRNA.

GW182 RNA-binding protein initially shown to associate with a specific subset of mRNAs and to reside within dis-crete cytoplasmic foci named **GW bodies**. Originally identified as the target of an autoantibody in serum from a patient with a sensory ataxic polyneuropathy. GW182 has multiple (~60) glycine(G)-tryptophan(W) repeats and an RNA recognition motif (RRM).

Gymnospermae One of the two major division of seed-bearing vascular plants of which the most common members are conifers. See **Angiospermae**.

gyrins Family of **tetraspan vesicle membrane proteins**, in mammals the members are **synaptogyrins** 1–4.

gyrus Any of the ridge-like folds of the cerebral cortex.

H

H2 antigen An antigen of the H2 region of the **major histocompatibility complex** of mice. Divided into class I and class II antigens.

H2 blocker Antagonist of the histamine type 2 (H2) receptor. Drugs of this type block gastric acid secretion and are therefore clinically useful in treating duodenal ulcers.

H2 complex Mouse equivalent of the human MHC (**major histocompatibility complex**) system, a set of genetic loci coding for class I and class II MHC antigens and for complement components. See **histocompatibility antigen**.

H5N1 Strain of avian influenza that it is feared may give rise to a strain that can become highly infectious (transmissible) in man and potentially cause a major pandemic. The fears are partly based on the fact that a similar avian flu strain gave rise to the 1918 flu pandemic. The strain type derives from the main antigenic determinants of influenza A and B viruses, the haemagglutinin (H or HA, of which there are 16 subtypes) and neuraminidase (N or NA, of which there are 9 subtypes) both of which are transmembrane glycoproteins of the viral envelope.

H19 Mouse gene that has a differentially methylated domain (DMD); this domain is a methylation-sensitive insulator that blocks access of the **IGF2** gene to shared enhancers on the maternal allele and inactivates H19 expression on the methylated paternal allele. The imprinted H19 gene produces a non-coding RNA of unknown function.

H89 Isoquinoline **PKA** inhibitor, also reported to inhibit **Rho-kinases (ROCKs)**.

H400 See **macrophage inflammatory protein 1 β**.

habituated culture A plant tissue culture that can grow independently of exogenously added auxin.

HACBP *High-affinity Ca²⁺-binding protein* See **calreticulin**.

haem *Heme (US)* Compounds of iron complexed in a porphyrin (tetrapyrrole) ring that differ in side chain composition. Haems are the prosthetic groups of **cytochromes** and are found in most oxygen carrier proteins.

haem oxygenases *HO-1 (HSP32); HO-2, HO-3* Vascular enzymes (three isoforms known, HO-1 being an inducible heat-shock/stress protein) that metabolize haem to form carbon monoxide, iron and biliverdin. Increased haem-derived CO inhibits nitric oxide synthase. Under situations of oxidative stress, heat stress, ischaemia/reperfusion injury or endotoxaemia, HO-1 has been shown to be induced and to elicit a protective effect.

haemagglutination Agglutination of red blood cells, often used to test for the presence of antibodies directed against red-cell surface antigens or carbohydrate-binding proteins or viruses in a solution. Requires that the agglutinin has at least two binding sites.

haemagglutinin Substance that will bring about the **agglutination** of erythrocytes.

haemangioblast Earliest mesodermal precursor of both blood and vascular endothelial cells. Described in embryonic yolk sac blood-islands of birds.

Haemanthus katherinae The African blood lily, chiefly known because of classic time-lapse studies done on mitosis in endosperm cells.

haematochrome A red or orange pigment (β-carotene) accumulated in the cells of some green algae , probably for protective purposes. Algae, for example *Trentepohlia*, have haematochrome and are often photosynthetic symbionts in lichens, hence the bright red colouration of these lichens. Haematochrome in *Chlamydomonas nivalis* is responsible for 'red snow'.

haematocrit Relative volume of blood occupied by erythrocytes. An average figure for humans is 45 ml per cent, i.e. a packed red-cell volume of 45 ml in 100 ml of blood.

haematopoiesis Production of blood cells involving both proliferation and differentiation from stem cells. In adult mammals, usually occurs in bone marrow.

haematopoietic cell kinase *Hck* A protein tyrosine kinase of the **src family** found in lymphoid and myeloid cells and that is bound to B-cell receptors in unstimulated B-cells. Deletion of hck and src or hck and fgr leads to severe developmental anomalies and impaired immunity in mice.

haematopoietic stem cell Cell that gives rise to distinct daughter cells, one a replica of the stem cell, one a cell that will further proliferate and differentiate into a mature blood cell. Pluripotent stem cells can give rise to all lineages, committed stem cells (derived from the pluripotent stem cell) only to some.

haematoxylin Basophilic stain that gives a blue colour (to the nucleus of a cell for example), commonly used in conjunction with eosin, which stains the cytoplasm pink/red. Various modifications of haematoxylin have been developed. The histopathologist's 'H & E' is haematoxylin and eosin.

haemochromatosis A disease in which **haemosiderin** is deposited in excess in the organs of the body, giving rise to cirrhosis of the liver, enlargement of the spleen, diabetes (bronzed diabetes) and skin pigmentation. Classic haemochromatosis (HFE) is an autosomal recessive disorder, most often caused by mutation in a gene designated *HFE*; haemochromatosis type 2 (HFE2), is also autosomal recessive. HFE2A is caused by mutation in the *HJV* (haemojuvelin) gene; HFE2B is caused by mutation in the gene encoding **hepcidin antimicrobial peptide**. Haemochromatosis type 3, also an autosomal recessive disorder, is caused by mutation in the gene encoding transferrin receptor-2. Haemochromatosis type 4 is an autosomal dominant disorder caused by mutation in the **ferroportin** gene.

haemocoel The body cavity derived from the blood vessels that replaces the coelom in arthropods and molluscs. It is a secondary cavity, not generated from the **archenteron**.

haemocyanin *Hemocyanin (US)* Blue, oxygen-transporting, copper-containing protein found in the blood of molluscs and Crustacea. A very large protein with 20–40 subunits and molecular weight of 2–8 million Da, and having a characteristic cuboidal appearance under the electron microscope. Prior to the introduction of immunogold techniques, it was used for electron-microscopic localization by coupling to antibody. Keyhole limpet haemocyanin (KLH) is widely used as a carrier in the production of antibodies.

haemocytes Blood cells associated with a haemocoel, particularly those of insects and Crustacea. Despite the name they are more leucocyte-like, being phagocytic and involved in defence and clotting of haemolymph, and not in the transport of oxygen.

haemocytoblast *Hemocytoblast* A pluripotential stem cell of the haematopoietic tissue from which monopotential stem cells of various lineages arise.

haemocytometer A special glass slide used for counting blood cells, etc. under the microscope. A grid of lines is engraved on the bottom of a shallow rectangular trough and a stiff coverslip is placed over the trough so that the grid demarcates known volumes. Cell suspension can then be introduced and the number of cells per square counted. Various grid patterns are available (Neubauer and Fuchs-Rosenthal being common). A similar device on a smaller scale (Helber cell) is used for bacterial counting.

haemoglobin *Hemoglobin (US)* Four-subunit globular oxygen-carrying protein of vertebrates and some invertebrates. There are two α- and two β-chains (very similar to myoglobin) in adult humans; the haem moiety (an iron-containing substituted porphyrin) is firmly held in a non-polar crevice in each peptide chain.

haemoglobinopathies Disorders due to abnormalities in the haemoglobin molecule, the best known being sickle-cell anaemia in which there is a single amino acid substitution (valine for glutamate) in position 6 of the β-chain. In other cases one of the globin chains is synthesized at a slower rate, despite being normal in structure. See also **thalassaemia**.

haemolymph Circulating body fluid of invertebrates such as insects that have a **haemocoel** – sinuses and spaces between organs – rather than a closed circulatory system. Cells in the haemolymph are usually referred to as **haemocytes**. Unlike vertebrate blood cells, haemocytes do not have an oxygen-carrying function and subclasses are phagocytic with an immune function, thus resembling the granulocytes of vertebrates.

haemolysins Bacterial **exotoxins** that lyse erythrocytes.

haemolysis Leakage of haemoglobin from erythrocytes due to membrane damage.

haemolytic anaemia *Hemolytic anemia (US)* **Anaemia** resulting from reduced red-cell survival time, because of either an intrinsic defect in the erythrocyte (hereditary **spherocytosis** or ellipsocytosis, enzyme defects, **haemoglobinopathies**) or an extrinsic damaging agent – for example, autoantibody (autoimmune haemolytic anaemia), isoantibody, parasitic invasion of the cells (**malaria**), bacterial or chemical haemolysins, mechanical damage to erythrocytes.

haemolytic plaque assay A method used to detect individual cells secreting antibody *in vitro*. Sheep red cells (treated so as to bind the antibody) are mixed with the cell suspension to be assayed in a thin layer of agarose and incubated. Cells that are producing antibody are revealed, when complement is added, by an area of haemolysis surrounding them.

haemonectin A 60-kDa protein found in the bone marrow matrix of rabbits specifically aiding adhesion of granulocyte-lineage cells. The sequence shows 60–70% similarity with that of **fetuin** from other mammal species.

haemopexin Single-chain haem-binding plasma β1-glycoprotein (57 kDa); unlike **haptoglobin**, does not bind haemoglobin. Present at around 1 mg/ml in plasma. Responsible for transporting haem groups to the liver for breakdown. Structurally related to **vitronectin** and some **collagenases**.

haemophilia Sex-linked congenital deficiency of blood-clotting system, usually of factor VIII.

Haemophilus influenzae Bacterium sometimes associated with influenza virus infections, causes pneumonia and meningitis.

haemorrhagic *Hemorrhagic (US)* Related to or causing haemorrhage (bleeding).

haemosiderin A mammalian iron-storage protein related to **ferritin** but less abundant.

haemostasis Arrest of bleeding through blood clotting and contraction of the blood vessels.

Hagemann factor Plasma β-globulin (110 kDa), blood-clotting factor XII, which is activated by contact with surfaces to form Factor XIIa that in turn activates Factor XI. Factor XIIa also generates **plasmin** from plasminogen and **kallikrein** from prekallikrein. Both plasmin and kallikrein activate the complement cascade. Hagemann factor is important both in clotting and activation of the inflammatory process.

Hailey–Hailey disease *Familial chronic benign pemphigus* A blistering disease of the skin apparently due to a defect in epidermal cell junctions, even though apparently normal desmosomes and adherens junctions can be assembled. Transmitted as an autosomal dominant and caused by mutation in the ATP2C1 (calcium-transporting ATPase).

hair cells 1. Cells found in the epithelial lining of the labyrinth of the inner ear. The hairs are **stereovilli** up to 25 μm long and restrict the plane in which deformation of the apical membrane of the cell can be brought about by movement of fluid or by sound. Movement of the single **stereocilium** transduces mechanical movements into electrical receptor potentials. **2.** *Bot.* Many plant surfaces are covered with fine hairs (*Tradescantia* stamens are a common source); the hairs are made up of thin-walled cells that are convenient for studying cytoplasmic streaming and for observing mitosis.

hairpin *Alpha hairpin; α-hairpin; beta hairpin; β-hairpin* Protein **motif** formed by two adjacent regions of a polypeptide chain that lie antiparallel and alongside each other. Depending on whether the polypeptide is in **alpha-helix** or **beta strand** configuration, can be described as alpha hairpin or beta hairpin, respectively.

hairy One of the *Drosophila* **pair-rule** genes that encodes a bHLH transcription factor of the **Hairy/Enhancer of Split/Deadpan** (HES) family of proteins. *Drosophila* HES family proteins are key repressors in the developmental processes of segmentation, neurogenesis and sex determination. All have: (1) a highly conserved **bHLH domain**, (2) an adjacent Orange domain, which confers specificity among family members, and (3) a C-terminal tetrapeptide motif, WRPW, which has been shown to be necessary and sufficient for the recruitment of the co-repressor, **Groucho**.

hairy cell leukaemia *HCL* Clinically associated with severe T-cell dysfunction possibly as a result of defects in the responsiveness to activation, although there is also a very restricted repertoire of the T-cell receptor-beta family.

hairy/enhancer of split/deadpan Family of basic helix–loop–helix (bHLH) proteins that function as transcriptional repressors. See **hairy**.

Hakata antigen See **ficolin**.

half-life $t_{1/2}$ The period over which the activity or concentration of a specified chemical or element falls to half its original activity or concentration. Typically applied to the half-life of radioactive atoms, but also applicable to any other situation where the population is of molecules of diminishing concentration or activity.

Haller cell In the sinus system of the head there are air cells, air-filled cavities. In some cases these may invade the medial floor of the orbit, causing narrowing of the infundibulum and obstruction of the maxillary and anterior ethmoid sinuses. These enlarged cavities are Haller cells, and are frequently present in association with sinus disease. In medical terminology, a pneumatized infraorbital ethmoid cell; not a cell in the cell-biological sense.

halobacteria Bacteria that live in conditions of high salinity (**halophiles**).

Halobacterium halobium Photosynthetic (halophilic) bacterium that has patches of purple membrane containing the pigment **bacteriorhodopsin**.

haloperidol Drug of the butyrophenone class, all of which have antipsychotic actions. Haloperidol is used in the treatment of schizophrenia.

halophile Literally, salt-loving: organism that tolerates saline conditions, in extreme cases in concentrations considerably in excess of those found in normal sea water such as salt lakes. Some Archaebacteria (e.g. *Halobacterium halobium*) are notable for their ability to survive extremes of salinity (**extreme halophiles**). Adj. halophilic.

halophyte Plant that grows in or tolerates salt-rich environments.

halorhodopsin Light-driven chloride ion pump of halobacteria, a retinylidene protein very similar to **bacteriorhodopsin**.

halothane *2-Bromo-2-chloro-1,1,1-trifluorethane* Widely used volatile anaesthetic given by inhalation.

halysin See **disintegrin**.

hamartin Protein encoded by tumour-suppressor gene *TSC1*; interacts with the **tuberin** protein . See **tuberous sclerosis**.

hamartoma Tumour-like but non-neoplastic overgrowth of tissue that is disordered in structure. Examples are haemangiomas (that include the vascular naevus or birthmark) and the pigmented naevus (mole).

hammerhead ribozyme A **ribozyme** in which there are three helical regions radiating from a central core.

Hand–Schuller–Christian disease A chronic disorder which is histologically a type of **Langerhans cell histiocytosis**, with onset between 3 and 5 years. Granulomatous histiocytic lesions are seen in bone and visceral tissues; probably an autoimmune dysfunction.

hanging-drop preparation A preparation in which the specimen is suspended in a drop of medium on a coverslip that is inverted over a cavity ground into a microscope slide and then sealed at the edges to prevent evaporation.

Hanks BSS *HBSS* Balanced salt solution made up according to the recipe given originally by Hanks. Phosphate-buffered to pH 7.0–7.2. Suitable for mammalian and avian cells in temporary culture, though is not a growth medium and will not sustain prolonged survival. Usually contains phenol red as an indicator.

Hansen's disease Leprosy.

Hansenula wingei *Pichia canadensis* Species of budding yeast, evolutionarily distant from *S. cerevisiae*, often used for studies on mating type.

Hantavirus Hantaviruses are responsible for haemorrhagic fevers and exist in various serotypes with different pathogenicity for human beings, varying from asymptomatic infection to highly fatal disease. Human infections arise from inhalation of aerosolized excreta of persistently infected rodents.

haplodiploid A reproductive system in which females are diploid and males are haploid, as in insect species such as bees.

haploid Describes a nucleus, cell or organism possessing a single set of unpaired **chromosomes**. **Gametes** are haploid.

haploinsufficiency Situation in which a single copy of a gene is insufficient to allow normal functioning. The heterozygote is therefore affected.

haplontic Organisms in which meiosis occurs in the zygote, giving rise to four haploid cells (e.g. many algae and protozoa), only the zygote is diploid and this may form a resistant spore. Cf **diplontic**, **diplohaplontic**.

haplotype The set, made up of one **allele** of each gene, comprising the **genotype**. Also used to refer to the set of alleles on one **chromosome** or a part of a chromosome, i.e. one set of alleles of linked genes. Its main current usage is in connection with the linked genes of the **major histocompatibility complex**.

hapten Could be considered an isolated **epitope**: although a hapten (by definition) has an antibody directed against it, the hapten alone will not induce an immune response if injected into an animal, it must be conjugated to a carrier (usually a protein). The hapten constitutes a single antigenic determinant; perhaps the best-known example is dinitrophenol (DNP), which can be conjugated to BSA and against which anti-DNP antibodies are produced (antibodies to the BSA can be adsorbed out). Because the hapten is monovalent, immune complex formation will be blocked if the soluble hapten is present as well as the hapten–carrier conjugate (assuming there is more than one hapten per carrier, an immune precipitate can be formed). Competitive inhibition by the soluble small molecule is sometimes referred to as haptenic inhibition, and this term has carried over into lectin-mediated haemagglutination, where monosaccharides are added to try to block haemagglutination: the blocking sugar defines the specificity of the lectin.

haptenic inhibition See **hapten**.

haptoglobin Acid α2-plasma glycoprotein that binds to oxyhaemoglobin that is free in the plasma, and the complex is then removed in the liver. Tetrameric (2α-, 2β-subunits): the existence of two different α-chains in humans means that haptoglobins can exist in three variants in heterozygotes.

haptonema Filament extending between the paired flagella of certain unicellular algae (haptophytes). Supported by six or seven microtubules (not in an axoneme-like array) and apparently used for capturing prey, in a manner analogous to the axopodia of **heliozoa**.

haptotaxis Strictly speaking, a directed response of cells in a gradient of adhesion, but often loosely applied to situations where an adhesion gradient is thought to exist and local trapping of cells seems to occur.

ha-ras Harvey-*ras* oncogene.

Hardy–Weinberg law Mathematical formula that gives the relationship between gene frequencies and genotype frequencies in a population. If genotypes are AA, Aa, aa and the frequency of alleles A,a are respectively p,q, then $p + q = 1$ and AA:Aa:aa is p2:2pq:q2. Deviation from this equilibrium distribution suggests adverse survival characteristics of organisms with one of the alleles.

harmonin Actin-bundling and PDZ domain-containing scaffold protein. mutations in the harmonin gene are the cause of Usher syndrome type 1C (USH1C), a rare, autosomal recessive syndrome of congenital deafness and progressive blindness.

Harris–Benedict equation Equation for calculation of basal metabolic rate (BMR).

Hartnup disease Amino acid transport defect that can be caused by mutations in the SLC6A19 (solute carrier family-6) gene; leads to excessive loss of monoamino monocarboxylic acids (cystine, lysine, ornithine, arginine) in the urine and poor absorption in the gut. See **iminoglycinuria**.

Harvey sarcoma virus See *ras*.

Hashimoto's thyroiditis Autoimmune disease in which there is destruction of the thyroid by autoantibodies usually directed against thyroglobulin and a lipoprotein of thyroid cell endosomes.

Hassell's corpuscle *Thymic corpuscle* Spherical or ovoid bodies 20–$50\,\mu m$ in diameter in the medulla of the thymus, composed of flattened, concentrically arranged whorls of keratinized or hyaline cells surrounding dead cells in the core.

HA-tag An epitope tag often added to recombinant proteins to aid purification or localization. The tag is amino acid residues 97–115 of the influenza virus haemagglutinin (HA) protein (KAFSNCYPYDV-PDYASLRS).

HAT medium A selective growth medium for animal tissue cells that contains hypoxanthine, the folate antagonist aminopterin (or amethopterin) and thymine. Used for selection of hybrid somatic cell lines, as in the production of monoclonal antibodies. In HAT medium, cells are forced to use these exogenous bases, via the salvage pathways, as their sole source of purines and pyrimidines. Parental cells

lacking enzymes such as **HGPRT** or **thymidine kinase** (TK) can be eliminated whilst hybrids grow.

Hatch–Slack–Kortschak pathway *Hatch–Slack pathway* Metabolic pathway responsible for primary CO_2 fixation in **C4 plant** photosynthesis. The enzymes that are found in **mesophyll** chloroplasts include **PEP carboxylase** that adds CO_2 to phosphoenolpyruvate to give the four-carbon compound, oxaloacetate. Four-carbon compounds are transferred to bundle-sheath chloroplasts, where the CO_2 is liberated and re-fixed by the **Calvin–Benson cycle**. The HSK pathway permits efficient photosynthesis under conditions of high light intensity and low CO_2 concentration, avoiding the non-productive effects of photorespiration.

haustorium A projection from a cell or tissue of a fungus or higher plant that penetrates another plant and absorbs nutrients from it. In fungi it is a **hyphal** projection that penetrates into the cytoplasm of a host plant cell; in parasitic angiosperms, it is a modified root.

Haversian canals Small channels found in compact bone, They run along the length of the long bone and provide major blood-vessel supply to the osteocytes. Each Haversian canal is surrounded by concentric layers of bone, the whole forming an osteon.

Hayflick limit See **cell death**, **cell line**.

HB-EGF *Heparin-binding epidermal growth factor* HB-EGF, like **amphiregulin**, has a long N-terminal extension that seems to confer the ability to bind to heparin and also to other connective tissue macromolecules (glycosaminoglycans) and cell surface molecules such as CD44. Because it is immobilized the effective local concentration may be much higher and the effects may differ from those of soluble growth factors.

HBGF *Heparin-binding growth factor* See **fibroblast growth factor**.

HBV Abbrev. for **hepatitis B** virus.

HCG *hCG; human chorionic gonadotrophin* See **chorionic gonadotrophin**.

H-chain Heavy chain of immunoglobulin; see **IgG**, **IgM**, etc.

Hck See **haematopoietic cell kinase**.

HCP 1. See **histidine-rich calcium-binding protein**. 2. Health care provider.

HCT Abbrev. for (1) haematocrit (Hct), (2) human colon cancer cell line, of which there are many variants (HCT-15, HCT-116, etc.), (3) hematopoietic cell transplantation.

HCV Hepatitis C virus.

HCV core protein Hepatitis C core protein, the first 191 amino acids of the viral precursor polyprotein that is co-translationally inserted into the membrane of the endoplasmic reticulum; a viral structural protein that is considered to influence multiple cellular processes, blocks the activity of caspase-activated DNase and thus inhibits apoptotic cell death.

HDAC See **histone deacetylases**.

HDF Abbrev. for (1) haemodiafiltration, an alternative to conventional haemodialysis combining convection and diffusion, (2) Human dermal fibroblasts.

HDL High-density **lipoprotein**.

hdm2 *Human double minute 2* A ubiquitin **E3 ligase** that is the principal negative regulator of **p53**; inhibits p53 transcriptional activity and subjects it to degradation by an E3 ligase activity.

HDR syndrome *Hypoparathyroidism, sensorineural deafness and renal anomaly syndrome* See **GATA-1**.

Heaf test A commonly used tuberculin test in which tuberculin is injected intradermally with a multiple puncture apparatus. A positive reaction indicates the presence of T-cell reactivity to mycobacterial products.

heart muscle See **cardiac muscle**.

heat-shock factor See **heat-shock proteins**.

heat-shock proteins *Hsp* Families of proteins conserved through pro- and eukaryotes, induced in cells as a result of a variety of environmental stresses, though some hsps are constitutively expressed. Some serve to stabilize proteins in abnormal configurations, play a role in folding and unfolding of proteins and the assembly of oligomeric complexes, and may act as **chaperonins**. **Hsp90** complexes with inactive steroid hormone receptor and is displaced upon ligand binding. Four major subclasses are recognized: hsp90, **hsp70**, hsp60 and small hsps. Hsps have been suggested to act as major immunogens in many infections.

heavy chain In general, the larger polypeptide in a multimeric protein. Thus the immunoglobulin heavy chain is of 50 kDa, the light chain of 22 kDa, whereas in myosin the heavy chain is very much larger (220 kDa) than the light chains (~20 kDa).

heavy water *Deuterium oxide; D_2O* Most commonly used by cell biologists to stabilize **microtubules**.

HECT *Homologous to E6-AP C-Terminus* Domain that characterizes a class of ubiquitin E3 ligases that have a direct role in catalysis during ubiquitination. See **E3 ligase**.

hedgehog An autoproteolytic secreted protein morphogen that activates an essential cellular pathway (Hedgehog Signalling Pathway) required during the embryogenesis of various organisms ranging from *Drosphila* to mammals. The pathway plays multiple roles in the development of the anterior craniofacial skeleton and in development of other systems. In adults, hedgehog signalling is implicated in the maintenance of stem cell niches in the brain, renewal of the gut epithelium and differentiation of haematopoietic cells. Aberrant activation of the **hedgehog signalling complex** in the adult has been associated with a number of tumour types. See **hedgehog family**, **sonic hedgehog**, **smoothened**.

hedgehog family Family of genes that includes *hedgehog* (*Hh*), *sonic hedgehog* (*Shh*), *Indian hedgehog* (*Ihh*), and *desert hedgehog* (*Dhh*). The gene products are signalling molecules regulating multiple functions during organ development and in adult tissues. Altered hedgehog signalling has been implicated in disturbed organ development as well as in different degenerative and neoplastic human diseases. Receptors are of the **Patched** family (PTCH1 and PTCH2) that transduces signals to GLI1, GLI2 and GLI3. GLI family transcription factors then activate transcription of Hedgehog target genes, such as FOXE1 and FOXM1 encoding Forkhead-box transcription factors.

hedgehog signalling complex *HSC* Complex that involves, at a minimum, the kinesin-related protein Costal2

(Cos2), the protein kinase Fused (Fu) and the transcription factor Cubitus interruptus (Ci). See **hedgehog**, **sonic hedgehog**, **smoothened**.

Heidenhain's Azan A particularly beautiful but extremely time-consuming trichrome staining method that, properly carried out, results in chromatin, erythrocytes and neuroglia being stained red, mucus blue, collagen sharp blue, and cytoplasmic granules red, yellow or blue. Now very rarely used.

Heidenhain's iron haematoxylin One of many haematoxylin-based staining solutions and one that is particularly good for photography or automatic image processing because of the intensity of black staining that can be achieved. Requires differentiation in iron alum and thus the intensity of staining can be adjusted according to the specimen. Sections stained with Heidenhain are usually counterstained with, for example, eosin or orange-G.

Heidenhain's Susa Good general-purpose histological fixative, but has the disadvantage of containing mercuric chloride.

HEK-293 cells Human embryonic kidney cells transformed by sheared adenovirus type 5 DNA; were first described in 1977. There is speculation that these cells may have originated from a rare neuronal cell in the kidney, since they stain strongly and specifically with antibodies to several neurofilament proteins.

HeLa cells An established line of human epithelial cells derived from a cervical carcinoma (said to be from Henrietta Lacks).

Helber cell See **haemocytometer**.

Helianthus annuus The sunflower.

helicase See **DNA helicase**.

Helicobacter pylori S-shaped or curved Gram-negative bacteria ($0.5–0.9 \times 3.0\mu$ m), non-spore forming, can be flagellate; found in human stomach. Was originally named *Campylobacter pyloridis*. Infection with *H. pylori* is now considered to be a major predisposing cause of gastric ulcers and antibiotic therapy is increasingly used.

helicoidal cell wall Type of plant cell wall in which each wall layer contains parallel **microfibrils**, but in which the orientation of the microfibrils changes by a fixed angle from one layer to the next. Gives a characteristic 'herringbone' pattern in transmission electron microscopy. A similar architecture of fibrillar material is seen in some insect exoskeletons.

Heliozoa Amoeboid **Protozoa**, Order Heliozoida. They are generally free-floating, spherical cells with many straight, slender microtubule-supported **pseudopods** radiating from the cell body like a sunburst. These modified pseudopods are termed **axopodia**. Genera include *Actinophrys* and *Echinosphaerium*.

Helisoma trivolvis Pulmonate mollusc whose relatively simple nervous system contains large identifiable cells and is consequently, like *Hirudo* and *Aplysia*, a favourite preparation for studying neural mechanisms at the cellular level; and in particular for studying isolated neurons in culture.

helix-coil transition See **random coil**.

helix-destabilizing proteins *Single-stranded binding proteins* Proteins involved in DNA replication. They bind cooperatively to single-stranded DNA, preventing the reformation of the duplex and extending the DNA backbone, thus making the exposed bases more accessible for base pairing.

helix–loop–helix *hlh* A motif associated with **transcription factors**, allowing them to recognize and bind to specific DNA sequences. Two **alpha-helices** are separated by a loop. Examples: myoblast MyoD1, c-myc, *Drosophila* genes *daughterless, hairy, twist, scute, achaete, asense*. Not the same as **helix–turn–helix**.

helix–turn–helix A motif associated with **transcription factors**, allowing them to bind to and recognize specific DNA sequences. Two **amphipathic** alpha-helices are separated by a short sequence with a **beta-pleated sheet**. One helix lies across the major groove of the DNA, while the recognition helix enters the major groove and interacts with specific bases. An example in *Drosophila* is the **homeotic** gene *fushi tarazu*, the product of which binds to the sequence TCAATTAAATGA. Not the same as **helix–loop–helix**.

Heloderma horridum horridum Mexican beaded lizard. See **exendin**.

helospectin See **exendin**.

helothermin *Helothermine* Protein toxin (25 kDa) from venom of *Heloderma horridum horridum*. Probably acts by inhibiting the ryanodine-sensitive calcium release channel of sarcoplasmic reticulum.

helper factor A group of factors apparently produced by helper T-lymphocytes that act specifically or non-specifically to transfer T-cell help to other classes of lymphocytes. The existence of specific T-cell helper factor is uncertain.

helper T-cell See **T-helper cells**.

helper virus A virus that will allow the replication of a co-infecting defective virus by producing the necessary protein.

hema-, hemo- US form of UK English **haema-, haemo-** which is used throughout for headwords.

heme See **haem**.

hemiacetal An acetal, an organic compound of general formula $R^1R^2C(OR^3)OR^4$, in which R^4 is hydrogen.

hemiasterlin A tripeptide that will cause depolymerization of microtubules; binds to beta-tubulin. See **phomopsin A**.

hemicellulose Class of plant cell wall polysaccharide that cannot be extracted from the wall by hot water or chelating agents, but can be extracted by aqueous alkali. Includes **xylan, glucuronoxylan, arabinoxylan, arabinogalactan** II, **glucomannan, xyloglucan** and galactomannan. Part of the cell wall matrix.

hemicentin *Hemicentin-1* = *fibulin-6* A fibulin family extracellular matrix protein (a member of the immunoglobulin superfamily). Mutation (A16,263G) in the hemicentin-1 gene produces a non-conservative substitution of arginine for glutamine at codon 5345 which has been implicated in familial age-related maculopathy (**AMD**).

hemicyst Blebs or blisters formed by a confluent monolayer of epithelial cells in culture as a result of active fluid accumulation between cell sheet and substratum.

hemidesmosomal epidermolysis bullosa *HEB* There are two subtypes of this rare disease. The first arises from a disorder of the protein **plectin** (HD1) and is associated with muscular dystrophy. The second arises from a defect of the α6β4 integrin receptor and is associated with pyloric atresia. Each disease shows intraepidermal blistering at the most basal aspect of the lower cell layer. See **epidermolysis bullosa**.

hemidesmosome Specialized junction between an epithelial cell and its basal lamina. Although morphologically similar to half a desmosome (into which intermediate cytokeratin filaments are also inserted) different proteins are involved.

hemimetabolous Of an insect, a species without any marked change in body-plan from larval to adult, apart from the development of wings. Examples: grasshoppers and crickets (*cf.* holometabolous).

hemiparesis Weakness or paralysis that is confined to one side of the body.

Hemiptera An order of Insecta with two pairs of wings and mouthparts adapted for piercing and sucking. Many feed on plant juices. Includes bugs, cicadas, aphids, plant lice, scale insects, leaf hoppers and cochineal insects.

hemizygote Nucleus, cell or organism that has only one of a normally **diploid** set of genes. In mammals, the male is hemizygous for the **X chromosome**.

hemojuvelin *HJV* A glycosylphosphatidylinositol-linked protein that undergoes a partial autocatalytic cleavage during its intracellular processing. HJV co-immunoprecipitates with neogenin, a receptor involved in a variety of cellular signalling processes. The majority of characterized cases of juvenile **haemochromatosis** involve mutations in HJV/hemojuvelin, yet the function of this gene remains unknown.

Henderson–Hasselbach equation Equation of the form: $pH = pK_a - \lg([HA]/[A^-])$ used for the calculation of the pH of solutions where the ratio $[HA]/[A^-]$ is known and HA and A^- are respectively the protonated and deprotonated forms of an acid.

Henoch–Schonlein purpura An inflammatory disorder, thought to be an immunoglobulin A (IgA)-mediated autoimmune phenomenon, characterized by a generalized vasculitis involving the small vessels of the skin. It is the most common vasculitis in children. In adults, the predominant feature may be glomerulonephritis.

Hensen's node *Primitive knot* Thickening of the avian blastoderm at the cephalic end of the primitive streak. Presumptive notochord cells become concentrated in this region. May well be a source of retinoic acid that is acting as a morphogen in the developing embryo.

Hep2 cells Line established from human laryngeal carcinoma in a 56-year-old Caucasian male. Extensively used in viral studies.

heparan sulphate *Glycosaminoglycan* Constituent of membrane-associated **proteoglycans**. The heparan sulphate-binding domain of **NCAM** is proposed to augment NCAM–NCAM interactions, suggesting that cell–cell bonds mediated by NCAM may involve interactions between multiple ligands. The putative heparin-binding site on NCAM is a 28-amino acid peptide shown to bind both heparin and retinal cells, as well as to inhibit retinal cell adhesion to NCAM. This strengthens the argument that this site contributes directly to NCAM-mediated cell–cell adhesion.

heparanase *HSE1; HPA; EC 3.2.1.-* Lysosomal enzyme, an endoglycosidase, that will cleave heparan sulphate into characteristic large molecular weight fragments. Degradation of extracellular matrix heparan sulphate by heparanase is a key step in the extravasation of tumour cells and migrating leucocytes, and also in processes such as angiogenesis, wound healing and smooth muscle proliferation.

heparin Sulphated mucopolysaccharide, found in granules of mast cells, that inhibits the action of **thrombin** on **fibrinogen** by potentiating antithrombins, thereby interfering with the blood-clotting cascade. Platelet factor IV will neutralize heparin.

heparin-binding growth factor *HBGF* See **fibroblast growth factor**.

hepatitis A Small (27 nm in diameter), single-stranded RNA virus with some resemblance to **enteroviruses** such as polio. Causes 'infectious hepatitis'.

hepatitis B Virion (**Dane particle**), 42 nm in diameter, with an outer sheath enclosing inner 27-nm core particle containing the circular viral DNA. Aggregates of the envelope proteins are found in plasma and are referred to as hepatitis B surface antigen (HBsAg; previously called Australia antigen). Causes 'serum hepatitis'; virus can persist for long periods (and in asymptomatic carriers); association of integrated virus with hepatocellular carcinoma is now well established.

hepatitis C *HCV* An enveloped RNA virus responsible for a fairly high proportion of cases of hepatitis of the non-A, non-B type, but not all. It is a member of the **Flaviviridae**, with a particle size of ~50 nm in diameter and a positive-sense RNA genome of 9600 nucleotides. Chronic HCV infection causes a malignant tumour of the liver (hepatocellular carcinoma), although the mechanism is unknown.

hepatitis D *Delta virus* A small circular RNA (single-stranded, negative sense) virus that is replication-defective and cannot propagate in the absence of another virus. In humans, hepatitis D virus infection only occurs in the presence of **hepatitis B** infection.

hepatitis E virus *HEV* Virus responsible for enterically transmitted non-A, non-B hepatitis worldwide. A spherical, non-enveloped, single-stranded RNA virus.

hepatitis non-A, non-B Hepatitis caused by a virus that is neither hepatitis A nor B and has no antigenic cross-reaction with either; often but not always, **hepatitis C virus**.

hepatitis viruses See **hepatitis A, B, C, D, E, non-A non-B**. Hepatitis F is technically a non-existent virus, but an infection common in the Far East seems to be caused by a new virus which is neither hepatitis B nor C and is considered to be hepatitis F. Hepatitis G (HGV) is an RNA virus and another member of the Flavivirus family similar to hepatitis C. Hepatitis H will be the next one – and there is already a candidate non-A, non-E virus.

hepatocarcinoma *Hepatocellular carcinoma* Malignant tumour derived from hepatocytes. Associated with **hepatitis B** in 80–90% of cases.

hepatocyte Epithelial cell of liver. Often considered the paradigm for an unspecialized animal cell. Blood is directly exposed to hepatocytes through fenestrated endothelium, and hepatocytes have receptors for sub-terminal *N*-acetylgalactosamine residues on asialoglycoproteins of plasma.

hepatocyte growth factor *HGF* Polypeptide mitogen originally shown to cause cell division in hepatocytes. In the liver, the main sources of HGF are non-parenchymal cells. It is now clear that HGF is a mitogen for a number of cell types and it is found in many cells outside the liver, including platelets. HGF is synthesized as a single-chain precursor that is proteolytically cleaved to give a heavy chain (70 kDa) and a light chain (30 kDa) linked by a single disulphide bond. It contains multiple copies of the **kringle** domain. However, both the single-chain precursor and the two-chain forms of HGF are biologically active, and HGF is generally isolated as a mixture of the two forms. HGF also alters cell motility, and is now known to be identical to **scatter factor**.

hepatoma Carcinoma derived from liver cells: better term is hepatocarcinoma.

hepatopancreas Digestive gland of crustaceans with functions approximately analogous to liver and pancreas of vertebrates, enzyme secretion, food absorption and storage.

hepatosplenomegaly Enlargement of liver (hepatomegaly) and spleen (splenomegaly).

hepcidin antimicrobial peptide *HAMP; LEAP1* A hormone peptide produced in the liver that regulates duodenal iron absorption and iron trafficking in the reticuloendothelial system; also has antimicrobial properties. Hepcidin inhibits the cellular efflux of iron by binding to and inducing the degradation of ferroportin, the sole iron exporter in iron-transporting cells. The 84-amino acid protein has a 24-residue N-terminal signal sequence and a penta-arginyl proteolysis site followed by the active C-terminal 25-amino acid peptide. The active 25-amino acid peptide is cleaved from a 84-amino acid precursor and contains a unique 17-residue stretch with 8 cysteines forming 4 disulphide bridges. HAMP is most active against Gram-positive bacteria, but also inhibits the growth of certain yeast and Gram-negative species.

HEPES *4-(2-Hydroxyethyl)-1-piperazine-ethane-sulphonic acid* Very commonly used buffer for tissue culture medium. Its pK_a of 7.5 makes it ideal for most cell culture work. Since related compounds are molluscicides, it may be unsuitable for some invertebrate cultures. One of the series of zwitterionic buffers described by Good.

HepG2 cells Cell line derived from hepatic carcinoma. Epithelial in morphology; produce a variety of proteins such as prothrombin, alpha-fetoprotein, C3 activator and fibrinogen.

hephaestin A transmembrane ferroxidase that acts in conjunction with **ferroportin** and has been implicated in duodenal iron export. Mutations in the murine hephaestin gene produce microcytic, hypochromic anaemia that is refractory to oral iron therapy. Hephaestin shares ~50% sequence identity with the plasma multicopper ferroxidase **ceruloplasmin**.

heptad repeats Tandem repeat sequence in which a group of seven amino acids occurs many times in a protein sequence. Most coiled-coil sequences contain heptad repeats.

HER *HER-2, HER-3, HER-4* Family of receptors (HER-2, erbB2; HER-3, erbB3; HER-4, erbB4) of the EGF-receptor family of receptor **tyrosine kinases**. Ligands are **neuregulins**. Overexpression of HER-2, human homologue of erbB2, correlates with poor prognosis in breast carcinoma.

herbimycin A Tyrosine kinase inhibitor from *Streptomyces hygroscopicus*.

herceptin *Trastuzumab* Humanized monoclonal antibody that blocks EGF-mediated cell proliferation by binding to the EGF-receptor (**HER-2**; c-erbB2). An IgG1 kappa that contains human framework regions with the complementarity-determining regions of a murine antibody. HER2 protein overexpression is observed in 25–30% of primary breast cancers, and these are the only ones that can be treated with this approach.

herculin Product of the muscle regulatory gene *Myf-6*. Also known as MRF4 (muscle regulatory factor-4).

hereditary angio-oedema Condition in which there seems to be uncontrolled production of **C2-kinin** because of a deficiency in C1-inhibitor levels.

hereditary haemochromatosis *HHC* An autosomal recessive disorder of iron metabolism, usually due to missense mutation in the **HFE** gene. Iron accumulates and symptoms of iron toxicity on various organ systems begin to appear once the load exceeds 20 g.

hereditary spastic paraplegias *HSPs* A group of neurodegenerative disorders characterized by lower-extremity spasticity and weakness; most commonly caused by mutations in the **spastin** gene, although mutations in at least 20 different genes are known.

heregulins *HRG1, HRG-alpha, beta; Neu differentiation factor; neuregulin-1* Soluble secreted growth factors of the **EGF**-family that are ligands for the ErbB3/ErbB4 cell surface receptors and are involved in cell proliferation, metastasis, survival and differentiation in normal and malignant tissues. HRG-alpha is the only HRG1 isoform expressed in the mouse mammary gland; some HRG1 splice variants are translocated to the nucleus. HRG-alpha and HRG-beta mRNA levels are higher in gastric ulcer tissue than in normal gastric mucosa and may be involved in the COX-2-dependent ulcer repair process. See **ARIA**.

hERG *Human ether-a-go-go related gene* Gene encoding the pore-forming subunit of cardiac IK$_r$, the channel that is one of the two responsible for repolarization of the cardiac action potential (the outward rectifying delayed current). Mutations are associated with **long QT2 syndrome**.

Hermansky–Pudlak syndrome *HPS* A genetically heterogeneous disorder characterized by oculocutaneous albinism, prolonged bleeding and pulmonary fibrosis due to abnormal vesicle trafficking to lysosomes and related organelles, such as melanosomes and platelet dense granules. In mice, at least 16 loci are associated with HPS, including sandy (*sdy*). The *sdy* mutant mouse expresses no **dysbindin** protein owing to a deletion in the gene *Dtnbp1*, and mutation of the human orthologue *DTNBP1* causes a novel form of HPS called HPS-7. See **BLOC-1**.

Herpesviridae A group of large DNA viruses: *Herpes simplex* causes cold sores and genital herpes; *Varicella-zoster* (human herpes virus 3; HHV-3) causes chicken-pox and shingles; cytomegalovirus causes congenital abnormalities and is an opportunistic pathogen; **Epstein–Barr virus** (EBV) causes glandular fever. *Herpes simplex* type

2 and EBV are associated with human tumours (cervical carcinoma for the former and **Burkitt's lymphoma** and nasopharyngeal carcinoma in the case of EBV). *Varicella* establishes a lifelong latent infection of sensory neurons in human dorsal root ganglia and has a tendency to resurgence if the immune system is suppressed (causing shingles).

Herring bodies Granules within axons in the posterior lobe of the pituitary gland. Contain neurosecretory hormones.

Hers disease *Glycogen storage disease VI* Autosomal recessive glycogen storage disease in which there is a deficiency of liver phosphorylase. Affected individuals show mild to moderate hypoglycaemia, mild ketosis, growth retardation, and prominent hepatomegaly. Other glycogen storage diseases are known, and there is some confusion in the numbering system.

HES proteins See **hairy**.

HETE *Hydroxytetraeicosenoic acid* A family of hydroxyeicosenoic acid (C20) derivatives of arachidonic acid produced by the action of lipoxygenase. Potent pharmacological agents with diverse actions. See also **HPETE**.

hetero- Prefix denoting different or varied.

heteroagglutination **1.** The adhesion of spermatozoa to one another by the action of a substance produced by the ova of another species. **2.** The adhesion of erythrocytes to one another when blood of different groups is mixed. *cf.* iso-agglutination. See **agglutinin**.

heterochromatin The chromosomal regions that are condensed during interphase and at the time of nuclear division. They show what is considered an abnormal pattern of staining as opposed to **euchromatin**. Can be subdivided into constitutive regions (present in all cells) and facultative heterochromatin (present in some cells only). The inactive X chromosome of female mammals is an example of facultative heterochromatin.

heterochrony Lack of synchronization.

heteroclitic antibody An antibody that was produced in response to an antigen but turns out to have higher affinity for an antigen that was not present in the original immunization, presumably because of molecular mimicry.

heterocyst Specialized cell type found at regular intervals along the filaments of certain **Cyanobacteria**; site of nitrogen fixation.

heterodimer A dimer in which the two subunits are different. One of the best-known examples is tubulin that is found as an α-tubulin/β-tubulin dimer. Heterodimers are relatively common, and it may be that the arrangement has the advantage that, for example, several different binding subunits may interact with a conserved signalling subunit.

heteroduplex DNA in which the two strands are different, either of different heritable origin, formed *in vitro* by annealing similar strands with some complementary sequences, or formed of mRNA and the corresponding DNA strand.

heterogamy **1.** Situation in which gametes of different sizes are produced by different mating types or sexes. **2.** The condition in a flowering plant species of having two or more types of flowers.

heterogenous nuclear RNA *hnRNA* Originally identified as a class of RNA, found in the nucleus but not in the nucleolus, which is rapidly labelled and with a very wide range of sizes, 2–40 kbp. It represents the primary transcripts of **RNA polymerase** II and includes precursors of all **messenger RNAs** from which introns are removed by splicing.

heterokaryon Cell that contains two or more genetically different nuclei. Found naturally in many fungi and produced experimentally by cell fusion techniques, e.g. **hybridoma**. See **heterokaryosis**.

heterokaryosis The coexistence of many dissimilar nuclei in, for example, the multinucleate cells of arbuscular mycorrhizal fungi (Glomeromycota). The opposite state is homokaryosis.

Heteroloboseans A phenotypically diverse group of heterotrophic amoeboflagellates that have paddle-shaped mitochondrial cristae, eruptive pseudopodia and a flagellar apparatus consisting of parallel basal bodies (e.g. *Acrasis, Percolomonas, Tetramitus* and *Psalteriomonas*).

heterologous Derived from the tissues or DNA of a different species. *cf* **autologous, homologous**.

heterolysosome Secondary lysosome formed by fusion of a lysosome with another intracellular vesicle.

heterophile antibody An antibody raised against an antigen from one species that also reacts against antigens from other species. Also used of systems such as the **Forssman antigen** where antibody against antigens from a variety of species is present without immunization.

heteroplasmy *Heteroplasia* The occurrence of a tissue in the wrong place in an organism, as a result of inappropriate cellular differentiation.

heterosis Hybrid vigour, the superiority of a heterozygotic organism over the homozygote.

heterospecific Ab Artificially produced antibody in which the two antigen-binding sites are for different antigens.

heterospory Condition in vascular plants where the spores are of two different sizes, the smaller producing male prothalli, the larger female prothalli.

heterothallic Situation in some fungi and algae in which there are two mating types, and the individual thallus is self-sterile even if hermaphrodite.

heterotroph An organism that requires carbon compounds from other plant or animal sources and cannot synthesize them itself – not an **autotroph**.

heterotypic Of different types. Thus heterotypic adhesion would be between dissimilar cells, in contrast to homotypic adhesion between cells of the same type.

heterozygosity index Measure of the number of gene loci for which an individual is heterozygous.

heterozygote Nucleus, cell or organism with different **alleles** of one or more specific genes. A heterozygous organism will produce unlike gametes, and thus will not breed true.

heuristic Proceeding by trial and error rather than according to a planned route.

HEV Abbrev. for **High endothelial venule**. See also **hevein**.

Hevea brasiliensis Rubber tree, source of most natural rubber. Member of the Family Euphorbiaceae.

hevein *Hev b 6.02* Protein found in ***Hevea brasiliensis***, has a **CRD domain**, which is known to bind chitin and GlcNAc-containing oligosaccharides. The antifungal activity of hevein-like proteins has been associated with their chitin-binding activities. Hevein is a major IgE-binding allergen in natural rubber latex (as used in gloves).

hexamine *Hexamethylenetetramine* A condensation product of methanal with ammonia, a crystalline substance with antiseptic and diuretic properties. It is sometimes used as a smoke-free fuel in camping stoves.

hexitol Sugar alcohol with six carbon atoms. Natural examples are sorbitol, mannitol.

hexokinase *EC 2.7.1.1* Enzyme that catalyses the transfer of phosphate from ATP to glucose to form glucose-6-phosphate, the first reaction in the metabolism of glucose via the glycolytic pathway (**glycolysis**).

hexon Subunit of a hexameric structure or with hexameric symmetry, in particular the arrangement of most of the capsomers of **Adenoviridae**, one capsomer surrounded by six others to form the hexon.

hexosaminidase *EC 3.2.1.52* Enzyme involved in the metabolism of **gangliosides**. Deficient in **Tay–Sachs disease**.

hexose Monosaccharide containing six carbon atoms, e.g. **glucose, galactose, mannose**.

hexose monophosphate shunt *Pentose phosphate pathway* The main metabolic pathway in activated neutrophils, rendering them relatively insensitive to inhibitors of oxidative phosphorylation. Congenital deficiency of the first enzyme in the shunt, glucose-6-phosphate dehydrogenase (EC 1.1.1.49), produces a sensitivity to infection similar to that seen in **chronic granulomatous disease**.

Heymann nephritis Rat model of human membranous nephropathy; an autoimmune disease in which the antigen is **megalin**.

HFE Protein that binds to the transferrin receptor and reduces its affinity for iron-bound transferrin. Mutations in HFE are the most common cause of **hereditary haemochromatosis**.

Hfr High frequency, in the sense of bacterial **conjugation**.

HGF See **hepatocyte growth factor**.

HGPRT *Hypoxanthine-guanine phosphoribosyl transferase* Enzyme that catalyses the first step in the pathway for salvage of the purines hypoxanthine and guanine. The phosphoribosyl moiety is transferred from an activated precursor, 5-phosphoribosyl 1-pyrophosphate. Since animal cells can synthesize purines *de novo*, HGPRT-mutants can be selected by their resistance to toxic purine analogues. A genetic lesion in HGPRT in humans underlies the **Lesch–Nyhan syndrome**. See **HAT medium**.

HHH syndrome *Hyperornithinaemia-hyperammonaemia-homocitrullinuria syndrome* A very rare autosomal recessive metabolic disorder caused by a deficiency of the mitochondrial ornithine transporter, one of the urea cycle components. Caused by mutation in the *SLC25A15* gene on chromosome 13 (13q14).

Hib Bacterium of the genus *Haemophilus* that can cause meningitis in young children.

hib27 A protein found in the serum of hibernating mice during non-hibernation. Shares some sequence homology with the globular domain of **adiponectin**.

hid *Head involution defective Drosophila* gene that is involved in positive regulation of apoptosis (like *reaper* and *grim*).

hidden Markov models Graphical models, originally developed for the study of machine learning and speech recognition, that have been applied to analysis of gene sequences and phylogenetics. They describe a probability distribution over an infinite number of sequences.

HIF-1 *Hypoxia-inducible factor-1* A transcription factor involved in responses to environmental stress (particularly, but not exclusively, hypoxia) by regulating the expression of genes that are involved in glucose supply, growth, metabolism, redox reactions and blood supply. Also essential for myeloid cell activation in response to inflammatory stimuli.

high-density lipoproteins *HDL* Involved in cholesterol transport in serum. See **lipoproteins**.

high endothelial venule *HEV* Venules in which the endothelial cells are cuboidal rather than squamous. Found particularly in lymph nodes where there is considerable extravasation of lymphocytes as part of normal traffic. Morphologically similar endothelium is found associated with some chronic inflammatory lesions. Express particular adhesion molecules accounting for preferential lymphocyte adhesion.

high mannose oligosaccharide A subset of the *N*-glycan chains that are added post-translationally to certain asparagine residues of secreted or membrane proteins in eukaryotic cells; contain five to nine mannose residues but lack the sialic acid-terminated antennae of the so-called complex type.

high-energy bond Chemical bonds that release more than 25 kJ/mol on hydrolysis: their importance is that the energy can be used to transfer the hydrolysed residue to another compound. The risk in using the term is that students may think the bond itself is different in some way, whereas it is the compound that matters. Hydrolysis of creatine phosphate yields 42.7 kJ/mol; of phosphoenolpyruvate, 53.2 kJ/mol; of ATP to ADP, 30.5 kJ/mol. The latter is important because it shows that, energetically, the hydrolysis of creatine phosphate will suffice to reconstitute ATP – hence the use of creatine phosphate in muscle.

high-mobility group proteins Family of small, non-**histone**, nuclear proteins. Some appear to be involved in controlling transcription.

high-voltage electron microscopy *HVEM* HVEM has two advantages: the increased voltage shortens the wavelength of the electrons (and therefore increases resolving power); more importantly for the biologist, the penetrating power of the beam is increased and it becomes possible to look at thicker specimens. Thus it is possible, by using stereoscopic views (obtained with a tilting stage), to get a three-dimensional picture of the interior of a cell.

Hill coefficient A measure of cooperativity in a binding process. A Hill coefficient of 1 indicates independent binding, a value of greater than 1 shows positive cooperativity – binding of one ligand facilitates binding of subsequent ligands at other sites on the multimeric receptor complex. Worked out originally for the binding of oxygen to haemoglobin (Hill coefficient of 2.8).

Hill plot Graphical method for analysing binding of a ligand to a macromolecule with multiple binding sites. See **Hill coefficient**.

Hill reaction Reaction, first demonstrated by Robert Hill in 1939, in which illuminated chloroplasts evolve oxygen when incubated in the presence of an artificial electron acceptor (e.g. ferricyanide). The reaction is a property of **photosystem II**.

Hind II First type II **restriction endonuclease** identified by Hamilton Smith in 1970. Isolated from *Haemophilus influenzae* Rd, it cleaves the sequence 'GTPyPuAc' between the unspecified pyrimidine (Py) and purine (Pu), generating 'blunt ends'.

Hind III Commonly used type II **restriction endonuclease** isolated from *Haemophilus influenzae* Rd; it cleaves the sequence 'AAGCTT' between the two As, generating 'sticky ends'.

hinge region Flexible region of a polypeptide chain – for example, as in immunoglobulins between Fab and Fc regions, and in myosin the S2 portion of heavy meromyosin.

hippocalcin Calcium-binding protein related to **recoverin**. Found exclusively in pyramidal cells of the hippocampus.

hippocampus Area of mammalian brain known since the 1950s to be important for long-term memory storage in humans and other mammals. The hippocampus is essential for initial storing of long-term memory for a period of days to weeks before the memory trace is consolidated elsewhere. It is also the site of long-term synaptic plasticity (see **long-term potentiation**), which is exhibited by defined synaptic pathways in the hippocampus.

Hirano bodies Paracrystalline inclusions, composed of microfilament-associated proteins including actin, alpha-actinin and vinculin, found in the brain of patients with neurodegenerative disorders.

Hirschsprung's disease *HSCR* See **Waardenburg's syndrome**. Aganglionic megacolon, a congenital malformation caused by absence of ganglion cells in myenteric and submucosal neural plexuses of gut. In some cases the defect is due to mutation in RET receptor tyrosine kinase, in others to mutations in **endothelin**-3, endothelin-beta receptor, **GDGF**. When associated with Waardenburg's syndrome, the defect is due to mutation in Sox10.

hirudin An anticoagulant present in the saliva of the leech, *Hirudo medicinalis*, that prevents blood clotting by inhibiting the action of thrombin on fibrinogen.

Hirudo medicinalis The medicinal leech. The **central nervous system** of this annelid contains a relatively small number of large, identifiable cells. This has made the leech, like the molluscs *Aplysia* and *Helisoma*, a chosen preparation for studying nervous system mechanisms at the cellular level. Related species of leeches are the organisms of choice for cellular and molecular genetic studies of early development, since the early embryos also contain identifiable cells.

his tag *Histidine tag* **Epitope tag** based on a short stretch (*ca.* 6) of histidine residues. In addition to detection with antipolyhistidine antibodies, such tags permit easy protein purification on nickel-based affinity columns.

hisactophilin Actin-binding protein (13.5 kDa) from *Dictyostelium discoideum*. Promotes F-actin polymerization and binds to microfilament bundles, but is very pH-sensitive as a result of having 31 histidine residues of a total of 118. Structure (though not sequence) very similar to **FGF** and **IL1**.

hispid flagella Eukaryotic flagella with two rows of stiff protrusions (mastigonemes) at right angles to the long axis of the shaft. In hispid flagella, the normal relationship between the direction of flagellar wave-propagation and the direction of movement is reversed; a proximal to distal wave pulls the organism forward.

histamine Formed by decarboxylation of histidine. Potent pharmacological agent acting through **histamine receptors** in smooth muscle and in secretory systems. Stored in mast cells and released by antigen (see **hypersensitivity**). Responsible for the early symptoms of anaphylaxis. Also present in some venoms.

histamine H2-receptor antagonists Class of drugs that block the type of histamine receptors found in stomach and thereby inhibit production of gastric acid. Used for treatment of gastric and duodenal ulcers. Examples: cimetidine (Tagamet), famotidine (Pepcid), ranitidine (Zantac).

histamine receptors There are two main classes of receptors, H1 and H2. The former are found mostly in the skin, nose and airways, and are targeted by antihistamine drugs that are used for allergic responses. **H2-receptor antagonists** block receptors found mostly in the stomach.

histatins Family of small histidine-rich cationic proteins (ranging from seven to eight amino acids in length) secreted into saliva by parotid and submandibular glands. Have potent antifungal activity, particularly against *Candida albicans*, though the mechanism of their fungicidal activity is unclear. They counteract dietary tannins by causing them to precipitate, and are produced in larger amounts in animals with a diet that is tannin-rich.

histidine *His; H; 155 Da* An amino acid with an imidazole side chain with a pK_a of 6–7. Acts as a proton donor or acceptor, and has high potential reactivity and diversity of chemical function. Forms part of the catalytic site of many enzymes. See Table A2.

histidine-rich protein II A pH-sensitive actin-binding protein from *Plasmodium falciparum*; also binds phosphatidylinositol 4,5-bisphosphate. Similar to **hisactophilin**.

histidine-rich calcium-binding protein *HCP* Major protein (165 kDa) found in the lumen of **sarcoplasmic reticulum**, where it may play a role in sequestering calcium ions. Highly acidic, with multiple repeats of highly conserved domains, believed to be responsible for calcium binding. Binds low-density lipoprotein with high affinity.

histiocytes Long-lived resident **macrophages** found within tissues.

histoblasts Population of small diploid epithelial cells in Dipteran larvae that do not form typical **imaginal discs**, yet resemble them in some ways.

histochemistry Study of the chemical composition of tissues by means of specific staining reactions.

histocompatibility If tissues of two organisms are histocompatible, then grafts between the organisms will not be rejected. If, however, major histocompatibility antigens are different, then an immune response will be mounted against the foreign tissue.

histocompatibility antigen A set of plasmalemmal glycoproteins that are crucial for T-cell recognition of antigens, particularly the HLA system in humans and the H2 system in mice. There are two classes of histocompatibility antigens. **1.** Class I histocompatibility antigens are composed of two glycosylated subunits, a heavy chain of 44 kDa and β2-microglobulin (12 kDa). The heavy chain may be coded by K, D or L genes of mouse H2 and A, B or C genes of human HLA complex. Class I antigens are important in T-cell killing and are recognized in conjunction with foreign cell surface antigens (MHC restriction). **2.** Class II antigens are heterodimeric histocompatibility antigens composed of α- (32-kDa) and β- (28-kDa) chains, and are found mostly on B-lymphocytes, macrophages and accessory cells. The response of T-helper cells requires that the foreign antigen is presented in conjunction with the appropriate class II antigens. (Murine H2 Ia antigens and human HLA-DR antigens are class II).

histogen One of three plant meristems (dermatogen, periblem, plerome) at the shoot or root tip that give rise exclusively to particular tissues in that organ (epidermis, cortex, stele and pith). The concept of discrete histogens is now considered outmoded.

histogenesis The process of formation of a tissue, involving differentiation, morphogenesis, and other processes such as angiogenesis, growth control, cellular infiltration, etc.

histone acetylation Modification of histone by the addition of acetyl groups to N-terminal lysine residues (by **histone acetyl transferases** (HATs)). Effect is to reduce the affinity between histones and DNA making it easier for RNA polymerase and transcription factors to access the promoter region. In most cases, histone acetylation enhances transcription while histone deacetylation (by **HDACs**) represses transcription. Acetylation, particularly of histone H4, has also been proposed to play an important role in replication-dependent nucleosome assembly.

histone acetyltransferase Enzyme that catalyses addition of an acetyl group to the N-terminal lysine of histone. See **histone acetylation**.

histone deacetylases *HDACs; EC 3.5.1.* A family of 11 enzymes (isoforms) that are involved in the control of gene expression. Inhibitors of HDAC are being tested as antitumour drugs.

histones Proteins found in the nuclei of all eukaryotic cells, where they are complexed to DNA in **chromatin** and **chromosomes**. They are of relatively low molecular weight and are basic, having a very high arginine/lysine content. They are highly conserved and can be grouped into five major classes. Two copies of H2A, H2B, H3 and H4 bind to about 200 base pairs of DNA to form the repeating structure of chromatin, the **nucleosome**, with H1 binding to the linker sequence. They may act as non-specific repressors of gene transcription. See **histone acetylation** and Table H1.

TABLE H1. Classes of histones

Class	Average M$_r$ (kDa)	Arginine (%)	Lysine (%)
H1	23	1.5	29
H2A	14	8	11
H2B	14	5	16
H3	15	13.5	10
H4	11	14	11

histoplasmin Filtrate of a mycelial culture of *Histoplasma capsulatum*, the causative agent of histoplasmosis. Histoplasmin is intradermally injected in a skin test for the disease.

histoplasmosis A disease of animals and man due to infection by the soil fungus *Histoplasma capsulatum*. Affects the lungs in man.

histotope A site on an MHC class I or class II antigen (see **histocompatibility antigen**) recognized by a T-cell.

HIT strain *Escherichia coli* HIT-1 has a mutation in the Na$^+$/H$^+$ antiporter gene (*NhaB*). This strain is unable to utilize serine as a carbon source because an active NhaB is required to maintain the electrochemical potential of Na$^+$, which drives the major serine uptake process via the Na$^+$/serine carrier.

HIV *Human immunodeficiency virus* Previously known as HTLV-III, human lymphotrophic virus type III, and also referred to as LAV, lymphadenopathy-associated virus; the retrovirus that causes acquired immunodeficiency syndrome (**AIDS**) in humans, by killing CD4$^+$-lymphocytes (T-helper cells). There are multiple forms of the virus.

HL60 cells Human promyelocytic cell line that can be induced to differentiate into neutrophil- or eosinophil-like cells by various treatments. Have some, but not all, of the features of normal blood cells.

HLA *Human leucocyte antigen* Refers to the **histocompatibility antigens** found in humans.

HMEC Abbrev. for (1) human microvascular endothelial cells, (2) (rarely) human mammary epithelial cells.

HMG See **high-mobility group proteins**.

HMG-CoA reductase *3-Hydroxy-3-methylglutaryl-CoA reductase; EC 1.1.1.34* Integral membrane protein (97 kDa) of ER and peroxisomes that catalyses reaction between hydroxy-methyl-glutaryl-CoA and two molecules of NADPH to produce mevalonate, a key starting material for the synthesis of cholesterol, other sterols and, for example, geranylgeraniol groups for post-translational modification of proteins. When sterols and non-sterol end products of mevalonate metabolism accumulate in cells, the enzyme is ubiquitinated and rapidly degraded. Because the enzyme is rate-limiting in synthesis, it has made a good target for cholesterol-lowering drugs (see **fibrates**).

HMM *Heavy meromyosin* Soluble tryptic fragment of myosin that retains the ATPase activity and that will bind to F-actin to produce a characteristic **arrowhead** pattern (unless ATP is present, in which case it detaches). Papain cleavage of HMM yields S1 and S2 subfragments, the former having the ATPase activity.

hMSH2 *Human Mut S homologue 2* Human mismatch repair gene (the other major one being human mutL homologue 1 (*hMLH1*)). Microsatellite instability, due to defective mismatch repair genes, is responsible for progression of some colorectal carcinomas, and hereditary non-polyposis colon cancer is caused by germline mutations in the DNA mismatch repair genes. See also **Muir–Torre syndrome**.

HNF *Hepatocyte nuclear factor* Family of transcription factors enriched in liver. Heterozygous mutations in hepatocyte nuclear factor (HNF)-1α and HNF-1β result in maturity-onset diabetes of the young.

hnRNA See **heterogeneous nuclear RNA**.

HNT 1. Human neurotrophin. See **IgLON** and **neurotrophin**. **2.** Line of human neuroteratoma cells (hNT).

Hodgkin's disease *Hodgkin's lymphoma* A human lymphoma that appears to originate in a particular lymph node and later spreads to the spleen, liver and bone marrow. Giant cells, the Sternberg–Reed cells, with mirror-image nuclei are diagnostic. Immunological depletion, caused perhaps by the excessive growth of neoplastic histiocytes, occurs. Four types of the disease are recognized, depending on the relative predominance of various neoplastic derivatives of the lymphoid series. Pyrexia is often a feature of the disease. Death often results from generalized immunological inability to respond to infections.

hodological An approach to understanding brain function by the mapping of projections or pathways in the brain, rather than looking at the cellular architecture of different regions. The term is also used in philosophy (linkage of ideas) and in geography (study of paths).

Hoechst 33258 dye A fluorescent dye that is a specific stain for DNA and can therefore be used to visualize chromosomes and to monitor animal cell cultures for contamination by microorganisms such as mycoplasma.

Hogness box See **TATA box**.

holandry Inheritance of characters borne on the male chromosome and therefore only expressed in the male.

Holliday junction A four-stranded DNA junction, the cross-over region at meiosis that results in homologous genetic recombination, first proposed by Holliday in 1964 as a structural intermediate in a mechanistic model. The actual conformation of a DNA cross-over was speculated to be a four-way junction with separate DNA helices, or with stacked helices in either a parallel or an antiparallel orientation of the helices. Crystal structures have confirmed the antiparallel stacked-X conformation and that it can switch to the open-X conformation in order to slide.

holoblastic Eggs that exhibit total cleavage, for example those of sea urchins and mammals.

holocentric Description of a chromosome in which the centromere is diffuse rather than discrete.

holocrine Form of secretion in which the whole cell is shed from the gland, usually after becoming packed with the main secretory substance. In mammals, sebaceous glands are one of the few examples.

holoenzyme The complete enzyme complex composed of the protein portion (**apoenzyme**) and cofactor or coenzyme.

Hololena curta Funnel web spider. See **hololena toxin**.

hololena toxin Toxin from the spider *Hololena curta* that irreversibly blocks insect presynaptic calcium channels. A heterodimer with subunits of 7 and 9 kDa.

holometabolous Of an insect, a species with a marked change in body plan from larva to pupa to adult.. Examples: flies and wasps (*cf.* hemimetabolous).

Holothuria Class of the Phylum Echinodermata. Sea cucumbers are holothurians.

holotrophic Feeding behaviour in which the whole organism (food particle) is ingested.

homeobox *Homoeobox* Conserved DNA sequence originally detected by DNA–DNA **hybridization** in many of the genes that give rise to **homeotic mutants** and segmentation mutants in *Drosophila*. The homeobox consists of about 180 nucleotides coding for a sequence of 60 amino acids in a protein, sometimes termed the homeodomain, of which about 80–90% are identical in the various homeodomains identified from *Drosophila*. Homeoboxes have also been detected in the genomes of vertebrates, with about 75% amino acid homology, and a similar sequence has been found in the *MAT* gene of yeast. The homeobox codes for a protein domain that is involved in binding to DNA. Three subfamilies of homeobox-containing proteins can be identified, based on the archetypal *Drosophila* genes *engrailed*, *antennapedia* and *paired*. Interestingly, linear order within genome maps to order of expression in embryo. This may be required for the **transcriptional silencing** of certain homeotic genes (see **Polycomb**).

homeodomain See **homeobox**.

homeostasis *Homoeostasis* The tendency towards a relatively constant state. A variety of homoeostatic mechanisms operate to keep the properties of the internal environment of organisms within fairly well-defined limits.

homeotic gene Gene containing a **homeobox**, the level of expression of which is set during embryogenesis in response to positional cues, and which then directs the later formation of tissues and appendages appropriate to that part of the organism. Mutation of these genes leads to inappropriate expression of characteristics normally associated with another part of the organism (**homeotic mutants**).

homeotic mutant *Homoeotic mutant* A **mutant** in which one body part, organ or tissue, is transformed into another part normally associated with another **segment**. Examples are the **antennapedia** and **bithorax complex** mutants of *Drosophila*.

Homer Protein (28 kDa) found in postsynaptic terminal, contains a single unusual **PDZ** domain and is involved in the clustering of a subset of metabotropic glutamate receptors, mGluR1α and mGluR5. Expression level elevated by synaptic activity. See also **GRIP** and **PSD-proteins**.

homocysteine An amino acid, the oxidized form of cysteine, that occurs as a by-product of the metabolism of protein and an important intermediate in methionine metabolism. Epidemiological studies have shown that an elevated level of plasma homocysteine (strongly influenced by dietary and genetic factors) is associated with a higher risk of coronary heart disease, stroke and peripheral vascular disease (though this does not imply causality).

homocystinuria Recessive condition in which the enzyme (cystathione synthetase) that converts homocysteine and serine into cystathione, a precursor of cysteine, is missing. Deficiency of this enzyme has widespread consequences in connective tissue, circulation and nervous system.

homogentisic acid *2,5-Dihydroxyphenylacetic acid* An intermediate in the metabolism of tyrosine and phenylalanine. See **alkaptonuria**.

homograft Outmoded term for a graft from one of an individual species to another. Includes **allogeneic** grafts (allografts) between genetically dissimilar individuals, and syngeneic grafts between identical individual (e.g. twins).

homokaryosis State in which all nuclei within a multinucleate cell (syncytium) are identical. See **heterokaryosis**.

homologous 1. Derived from the tissues or DNA of a member of the same species. *cf.* **heterologous**, **autologous**, **homologous recombination**. **2.** Of genes, similar in sequence, cf. **analogous**.

homologous chromosomes Chromosomes that are identical with respect to genetic loci, and that tend to pair or **synapse** during mitosis. The two chromosomes (maternal and paternal) of each of the pairs occurring in the nuclei of diploid cells are homologous.

homologous recombination Genetic recombination involving exchange of homologous loci. Important technique in the generation of null alleles (**knockouts**) in **transgenic** mice.

homoplasious A similarity that can be explained by a shared way of life – thus wings of birds and bats evolved separately but are similar because of the requirements for flight.

Homoptera A possibly paraphyletic group of insects, often included within the Hemiptera, that contains aphids, leafhoppers, cicadas and scale insects.

homosporous A species which produces only one type of a spore.

homozygote Nucleus, cell or organism with identical **alleles** of one or more specific genes.

Hoodia gordonii Plant traditionally used in South Africa for its appetite-suppressant properties; only reported active constituent is an oxypregnane steroidal glycoside, P57AS3 (P57).

hook 1. Basal portion of bacterial flagellum, to which is distally attached the **flagellin** filament. Proximally the hook is attached to the rotating spindle of the motor. In some bacteria (Myxobacteria) the rotation of the hook itself (without an attached flagellum) may directly cause forward gliding movement. **2.** *Drosophila* gene encoding a large homodimeric protein involved in endocytosis of the **bride of sevenless/sevenless** receptor/ligand complexes from the R7 photoreceptor cell.

Hook proteins 1. A family of cytosolic coiled-coil proteins proposed to function in linking organelles to microtubules. See **Hook-related proteins**. Human Hook3 binds to Golgi membranes *in vitro* and was enriched in the *cis*-Golgi *in vivo*. **2.** See **AT hook**. **3.** See **bacterial flagellum**.

Hook-related proteins *HkRP* Family of proteins from which **Hook proteins** are derived. There are several conserved domains, including a unique C-terminal HkRP domain. The central region of each protein is comprised of an extensive coiled-coil domain, and the N-terminus contains a putative microtubule-binding domain.

hookworms Parasitic strongylid nematodes with hook-like organs on the mouth for attachment to the host. The commonest hookworm infection of man is by the genus *Ankylostoma*, which penetrates the bare feet and can induce a form of anaemia.

Hopp Wood An algorithm for calculating a **hydropathy plot**.

Hordeum vulgare Cultivated barley, very important for the brewing industry.

horizontal transmission Transmission of a disease between individuals of the same generation, as opposed to vertical transmission from mother to offspring.

hormesis A dose–response phenomenon in which low doses stimulate, but high doses cause inhibition, The dose–response curve may be J-shaped or an inverted U-shape, the latter being observed, for example, with the effect of chemotactic peptides on neutrophil migration.

hormone A substance secreted by specialized cells that affects the metabolism or behaviour of other cells possessing functional receptors for the hormone. Hormones may be hydrophilic, like **insulin**, in which case the receptors are on the cell surface, or lipophilic, like the **steroids**, where the receptor can be intracellular. See Tables H2 and H3.

hormone-sensitive lipase Class of enzymes involved in the mobilization of fatty acids from triglyceride stores in adipocytes.

hornerin A protein of the fused-type S100 family (see **fused gene family**). Hornerin shares structural features, expression profiles and intracellular localization with **profilaggrin**, indicating possible involvement of hornerin in cornification (keratinization) of epidermal cells.

horseradish peroxidase A large enzyme, frequently used in conjunction with diaminobenzidine as an intracellular marker to identify cells both at light- and electron-microscopic levels.

host-range mutant A mutant of phage or animal virus that grows normally in one of its host cells, but has lost the ability to grow in cells of a second host type.

TABLE H2. Polypeptide hormones and growth factors of vertebrates

Name	M_r	Residues	Source	Actions
Polypeptide hormones				
Adrenocorticotrophin (ACTH)[a]	4.5 kDa	39	Anterior pituitary ACTH–MSH family	Stimulates glucocortoid production from adrenals
Amylin (Islet-activating polypeptide)		37	Co-secreted with insulin by pancreatic β-cells	Moderates action of insulin
Angiotensin II[a]		8	Formed from angiotensinogen, *via* angiotensin I	Acts on adrenal gland to stimulate aldosterone release, elevates blood pressure, mitogen for vascular smooth muscle
Atrial natriuretic peptide (ANP, ANF)		21–28	Atria of heart, brain, adrenal glands	Acts on kidney to produce natriuresis and diuresis, relaxes vascular smooth muscle, inhibits catecholamine release from adrenal medulla
Bombesin[a]		14	Skin, gut P-cells, nerves	Acts on CNS, gut smooth muscle, pancreas, pituitary kidney, heart; mitogenic *in vitro*
Bradykinin[a]		9	Formed in plasma	Dilates blood vessels, increases capillary permeability
Caerulein	1352 Da	10	Amphibian skin	Hypotensive, stimulates gastric secretion
Calcitonin[a]	4.5 kDa	32	Thyroid, parathyroid	Opposes parathyroid hormone, hypocalcaemic, hypophosphataemic
Corticotrophin-releasing factor (CRH)[a]		41	Hypothalamus	Stimulates release of corticotrophin
Endorphins[a]		15–31	Pituitary, brain	Opiates
Endothelins ET-1, ET-2, ET-3		21	Endothelium	Vasoconstrictors, also inotropic and mitogenic

TABLE H2. (Continued)

Name	M$_r$	Residues	Source	Actions
Enkephalins[a]		5	Adrenal medulla, brain, gut	Opiates
Exendins (helospectin, helodermin)		39		Glucagon-related, from Gila monster
Follicle-stimulating hormone (FSH)	30 kDa		Anterior pituitary	Acts on gonads
Gastric inhibitory peptide (GIP)		43	Gut	Inhibits gastric acid secretion, stimulates intestinal secretion, stimulates insulin and glucagon secretion
Gastrin[a]		17 or 34	Gut G-cells	Stimulates acid secretion, muscle contraction
Ghrelin		28	Stomach	Ligand for growth hormone secretagogue receptor Stimulates appetite
Glucagon (HGF)[a]	3550 Da	29	α−cells of pancreas	Hyperglycaemic
Glucagon-like peptide-1 (an incretin)		36, 37	Intestinal epithelial cells	Insulinotropic
Gonadotrophin-releasing hormones		10	Hypothalamus	Stimulate release of LH and FSH
Growth hormone (somatotropin, GH)[a]	21.5 kDa	191	Pituitary	Regulates organismal growth
Growth hormone-releasing factor (somatoliberin)	44 kDa		Hypothalamus	Stimulates release of growth hormone
Hepcidin antimicrobial peptide		84	Liver	Regulates duodenal iron absorption and trafficking
Human chorionic gonadotrophin (hCG)	30 kDa		Anterior pituitary, placenta	Maintains corpus luteum during pregnancy
Hypothalamic thyrotropic hormone[a]releasing factor (TRF, TRH)	362 Da	3	Hypothalamus	Stimulates release of thyrotropin from anterior pituitary
Kallidin		10	Formed in plasma	Dilates blood vessels, increases capillary permeability
Kisspeptins		54		Regulate reproduction via the hypothalamic-releasing hormones
Lipotropins[a]		50–90	Anterior pituitary	Stimulate lipid breakdown
Luteinizing hormone (LH)	30 kDa		Anterior pituitary	Acts on gonads
Motilin[a]		22	Gut enterochromaffin-2	Increases contractile response of stomach muscle cells
Neurotensin[a]		13	Gut, hypothalamus	Effects on smooth muscle tone; may also be a neurotransmitter
Oxytocin[a]	1007 Da	8	Posterior pituitary	Uterine contraction, lactation
Pancreastatin		49	Pancreas and gut	Inhibits insulin release
Pancreatic polypeptide[a]		36	PP-cells of pancreas;	Released on feeding; alters gut muscle tone, gut mucosa
Pancreozymin-cholecystokinin (PZ-CCK)[a]		8, 33 or 39	Gall bladder, brain, gut I-cells	Secretion of enzymes and electrolytes, secretion of insulin, glucagon, pancreatic polypeptide, contraction of gall bladder
Parathyroid hormone (PTH)	9.5 kDa	84	Parathyroid glands	Raises kidney cAMP and blood Ca by stimulating bone release
Peptide YY		36	Gut	Inhibits pancreatic secretion and release of cholecystokinin

TABLE H2. (Continued)

Name	M_r	Residues	Source	Actions
Placental lactogen		191	Placenta	Promotes lactation
Prolactin	23 kDa	199	Pituitary	Promote mammary growth, lactation
Renin	40 kDa	gp	Kidney	Cleaves angiotensinogen to angiotensin I in plasma
Secretin[a]		27	Gut S-cells	Stimulates alkali secretion from pancreas: decreases gastric acid secretion
Somatostatin (SRIF)[a]		14	Hypothalamus, D-cells of pancreas	Inhibits release of several hormones, including growth hormone
Substance P[a]		11	Gut enterochromaffin-1 cells, nerves	Contracts gut musculature, decreases blood pressure
Thyroid-stimulating hormone (TSH)	30 kDa		Anterior pituitary	Stimulates release of thyroid hormones
Vasoactive intestinal peptide (VIP)[a]		28	Lung, gut H-cells, nervous tissue	Vasodilation, bronchodilation, stimulates insulin, glucagon, prolactin secretion
Vasopressin (antidiuretic hormone, ADH)[a]		8	Posterior pituitary	Antidiuretic vasopressor
Growth factors				
Activin	26 kDa	115		TGF family; 2 inhibin β-subunits
Amphiregulin	35–45 kDa	84		Heparin-binding growth factor
Amphoterin	30 kDa	215		Enhances nerve growth cone migration
Brain-derived neurotrophic factor (BDNF)	28 kDa	247	Neurotrophin-family	Neurotrophic in CNS
Epidermal growth factor (urogastrone, EGF)		53	Mouse submaxillary gland, urine	Stimulates epidermal growth, formation of GI tract
Fibroblast growth factors (FGFs)	130 kDa	59 (a), 288 (b)	Brain, pituitary	Stimulates proliferation of fibroblasts, adrenal cells, chondrocytes, endothelia
Glial-derived neurotrophic factor (GDNF; formerly Glioma-derived growth factor)	30 Da	211	Glial cells	Specific for midbrain dopamine neurons
GMCSF (MG1, CSFalpha)	23 kDa	141	Many tissues	Promotes granulocyte or macrophage colony formation
G-CSF (murine; CSF ß in human)	25/30 kDa	42	Many tissues	Promotes differentiation in myeloleukaemic cells
Heregulins (Neuregulin)	44 kDa	640		Growth factors of the EGF family
Inhibin		426	TGFβ family	Inhibits FSH synthesis and secretion
Insulin	6 kDa	21 + 30	β−cells of pancreas	Hypoglycaemic, growth factor
Insulin-like growth factor I (IGF I, somatomedin C)		70	Liver, kidney	Growth factor, released into plasma
Insulin-like growth factor II (IGF II, multiplication-stimulating activity, MSA)		67	Cultured hepatocytes	Mitogen for several cell types
MCSF (CSF-1)	40–70 kDa	–	Mouse tissues	Promotes macrophage differentiation

TABLE H2. (Continued)

Name	M_r	Residues	Source	Actions
MultiCSF (IL3)	23–30 kDa	–	Mouse primed T-cells	Proliferation of various haematopoeitic stem cells
Midkine	13 kDa			TGFβ family
Nerve growth factor (NGFβ)	130 kDa	118	Salivary gland, snake venom	Tropic and trophic effects, mainly on sensory and sympathetic neurons Peripheral nerve targets
Platelet-derived growth factor (PDGF)	3 kDa		Platelets	Stimulate fibroblast proliferation, wound healing
Somatomedin A		50–80	Liver, kidney	Stimulates growth of peripheral nervous system
Stem cell factor (Steel factor)	35 kDa (dimer)			Haematopoietic growth factor
Transforming growth factor (TGF)α	5–20 kDa		Carcinomata	Acts via EGF receptor
TGF	2 × 25 kDa	112	Tumour cells	Multifunctional role in tissue damage. Promotes or inhibits proliferation depending on cell type
Thrombopoietin	19 kDa			Regulates production of platelets
Vascular endothelial growth factor (VEGF)	45 kDa (dimer)	165		Important in angiogenesis

[a] Neuropeptide. Other neuropeptides include calcitonin gene-related peptide, neurokinin, neuromedin, neuropeptide Y, proctolin and carnosine.

TABLE H3. Steroid hormones

Name	M_r (Da)	Distribution	Actions
Female sex hormones			
β–Estradiol	272	Ovary, placenta, testis	Estrogen
Progesterone	314	Corpus luteum, adrenal, testis, placenta	Regulates pregnancy, menstrual cycle
Male sex hormones			
Testosterone, dihydrotestosterone	288	Testes	Male secondary sexual characteristics, anabolic effects
Glucocorticoids			
Cortisol	362	Adrenal cortex	Gluconeogenesis, anti-inflammatory
Cortisone	360	Adrenal cortex	Glucocorticoid, anti-inflammatory
Corticosterone	346	Adrenal cortex	
Mineralocorticoids			
Aldosterone	360	Adrenal cortex	Stimulates resorption of Na^+ from kidneys
Thyroid hormones			
Thyroxine (T4)	777	Thyroid	Controls basal metabolic rate
Tri-iodothyronine (T3)	651	Thyroid	
Other			
Vitamin D (calciferol)	384	skin (+ sunlight)	Calcium and phosphate metabolism
Ecdysone	481	–	Moulting hormone of insects and nematodes

host-versus-graft reaction The normal lymphocyte-mediated reactions of a host against allogeneic or xenogeneic cells acquired as a graft or otherwise, which lead to damage or/and destruction of the grafted cells. The opposite of **graft-versus-host response**. The common basis of graft rejection.

hotspot A region of DNA that is particularly prone to mutation or transposition.

housekeeping genes Genes which code for proteins or RNAs that are important for all cells and are thus constitutively active. Term used by contrast with luxury proteins, those that are only produced by differentiated cells.

housekeeping proteins Those sets of proteins involved in the basic functioning of a cell or the set of cells in an organism, e.g. enzymes involved in synthesis and processing of DNA, RNA, proteins or the major metabolic pathways. As opposed to **luxury proteins**.

***hox* genes** Homeobox-containing genes of vertebrates.

HP1 *Heterochromatin protein 1* Protein that co-localizes with Arp6 in pericentric heterochromatin. Binds to methylated histone H3 (methylated at lysine 9) through a single **chromodomain** and will also homodimerize. More recently has been shown also to participate in gene regulation in euchromatin.

HPA axis The hypothalamo–pituitary–adrenal neuroendocrine system, traditionally considered as the body's 'stress system', and which ultimately controls levels of cortisol and other important stress-related hormones.

HPETE *5-HPETE; 5-hydroperoxy-eicosatetraenoic acid* Intermediate in leucotriene synthesis. Generated from arachidonic acid by 5-lipoxygenase; the starting point for the formation of 5-HETE or of leucotriene A4. Other related enzymes add the hydroperoxy group in different positions.

hpf Abbrev. for (1) high power field (of a microscope), (2) hours postfertilization, (3) human plasma fibrinogen.

HPLC *High pressure liquid chromatography* Chromatographic method in which the sample is forced at high pressure through a tightly packed column of finely divided particles that present a very large surface area. Because HPLC gives good separation very rapidly (but is expensive), manufacturers tend to speak of 'high performance liquid chromatography' as an encouragement to purchasers.

HPV Human **Papillomavirus**; causes warts.

HRG1 See **heregulins**, **ARIA**.

HRGP *Hydroxyproline-rich glycoprotein* Class of plant glycoproteins and proteoglycans rich in hydroxyproline, that includes **AGP**, **extensin** and certain **lectins**. Found in the cell wall and are produced in response to injury.

Hrk *Harakiri* A pro-apoptotic **Bcl-2 homology domain 3** (BH3) peptide that has been found in brain tissues of AIDS patients, particularly in macrophages.

HS1 Leucocyte-specific homologue of **cortactin**, regulates F-actin *in vitro* and is phosphorylated in response to T-cell receptor ligation.

HSA Abbrev. for (1) Human serum **albumin**, (2) A surface glycoprotein of *Streptococcus gordonii* that binds sialic acid moieties on platelet membrane glycoprotein Ibα (Hsa).

HSK pathway See **Hatch–Slack–Kortschak pathway**.

hsp60 See **heat-shock protein**.

hsp70 Widely distributed group of conserved **heat-shock proteins** of average weight 70 kDa. Possess ATP-binding domains and may be involved in protein folding or export.

hsp90 Widely distributed group of conserved **heat-shock proteins** of average weight 90 kDa. Exact function unknown, but are found associated with **steroid receptors** and **tyrosine kinase** oncogene products. May also bind actin and tubulin.

HSR Abbrev. for (1) hypersensitivity reaction, (2) heat-shock response, (3) homogeneously staining region (of a chromosome).

hst K-fgf/hst Human **oncogene** that encodes a member of the **FGF** family.

HT1080 Line of human fibrosarcoma cells.

hTERT *Human telomerase reverse transcriptase* Telomerase is a ribonucleoprotein which *in vitro* recognizes a single-stranded G-rich telomere primer and adds multiple telomeric repeats to its 3-prime end by using an RNA template. TERT is the catalytic subunit of telomerase; it has similarity to reverse transcriptases and may represent a universal subunit for telomerases. Ectopic expression of telomerase in normal human cells extends their replicative lifespan although does not necessarily transform them. Overexpression is associated with some tumour cells, although it may not be the only change involved in progression.

HTG Hypertriglyceridaemia, a rare but well-known cause of acute pancreatitis and a major independent risk factor for coronary heart disease.

Htk-L Membrane-anchored ligand (37 kDa) for **EPH** class receptor tyrosine kinases. Becomes tyrosine phosphorylated once bound to receptor (Nuk).

HTLV-1 *Human T-cell leukaemia/lymphoma virus type I* A retrovirus causing leukaemia and sometimes a mild immunodeficiency. In addition to *gag*, *pol* and *env*, the virus carries a coding sequence *pX* that does not seem to have a normal genomic homologue and is not a conventional oncogene. The protein product of the pX region is a short-lived nuclear protein (around 40 kDa). T-cells transformed with HTLV-1 continue to proliferate and are independent of **interleukin-2**.

HTLV-II *Human T-cell leukaemia/lymphoma virus type II* Originally isolated from a T-cell line from a patient with hairy cell leukaemia. It has only partial homology with **HTLV-I**, and its pathological potential is uncertain.

HTLV-III See **HIV**.

HTRF *Homogeneous time-resolved fluorescence* Assay methodology increasingly being used in high-throughput screening. An absorbing fluorochrome is coupled to one component and the emitting fluorochrome with slow-release characteristics is coupled to the other; if the two components are in proximity (because they bind), then **fluorescence energy transfer** between emitting and absorbing fluorochromes gives a signal which is analysed in a **time-resolved fluorescence** system. No separation step is required – all the reagents are mixed together and inhibitors of binding will reduce the output signal.

HuH7 A line of human hepatoblastoma cells.

Human enhancer of filamentation 1 *HEF1* A multifunctional docking protein of the **Cas-family**, participates in integrin and growth factor signalling pathways. Consists of two isoforms, p105 and p115, the larger resulting from Ser/Thr phosphorylation of p105HEF1. Can stimulate the formation of neurite-like processes

Human Genome Diversity Project The plan to analyse and compare variations in DNA samples from hundreds of different ethnic groups, in order to understand the origins and migrations of human populations and the genetic basis of differing susceptibility to disease. Opposition from some groups has caused the project considerable difficulty.

humanized antibody Usually a mouse monoclonal antibody directed against a target of particular therapeutic value that has been modified to have all regions except the antigen-binding portion substituted by human immunoglobulin domains. The procedure should make the antibody minimally antigenic when administered to patients, though some anti-idiotype antibodies may still be made to the antigen-binding site.

humic acid A complex mixture of partially decomposed and transformed organic materials. There are several subclasses of humic acids, (tannins, lignins, fulvic acids) and their properties are complex.

humoral immune responses Immune responses mediated by antibody.

Humulin Trade name for human insulin made by recombinant methods.

hunchback hb Key regulatory gene in the early segmentation gene hierarchy of *Drosophila*. Codes for a transcription factor of the Cys2-His2 **zinc finger** type.

Hunter syndrome Recessive **mucopolysaccharidoses**, X-linked, in which dermatan and heparan sulphates are not degraded. Because two lysosomal enzymes (heparan sulphate sulphatase and α-iduronidase) are involved in the breakdown of these glycosaminoglycans, fibroblasts from Hunter's syndrome will complement the fibroblasts from **Hurler's disease** patients in culture; by recapture of lysosomal enzymes from the medium, both types of cells in mixed culture become competent to digest glycosaminoglycans.

huntingtin Protein product of the IT15 gene that has variable numbers of polyglutamine repeats in **Huntington's chorea**. The IT15 gene is widely expressed and required for normal development. The polyglutamine repeats (44 in the commonest form of the disease) increase the interaction of huntingtin with **huntingtin-associated protein-1** (HAP-1) which is enriched in the brain and may be associated with pathology.

huntingtin-associated protein-1 *Hap1* Protein that interacts with **huntingtin** and is expressed more abundantly in the hypothalamus than in other brain regions. Lack of Hap1 in mice leads to early postnatal death. Hap1 is also involved in intracellular trafficking of the GABA(A) receptor, and fasting upregulates the expression of Hap1 in the rodent hypothalamus.

Huntington's chorea *Huntington's disease* Mature-onset disease characterized by progressive loss of neuronal functioning. Caused by unstable amplification of a trinucleotide $(CAG)_n$ repeat within the coding region of a gene encoding a 348 kDa, widely expressed product (**huntingtin**).

HuP genes Human equivalents of the *Pax* genes.

Hurler's disease Autosomal mucopolysaccharidosis, a recessive storage disease in which α-iduronidase is absent, leading to accumulation of heparan and dermatan sulphates. Extensive deposits of mucopolysaccharide are found in **gargoyle cells** and in neurons. See **Hunter syndrome**.

Hurler–Scheie syndrome Although clinically distinct diseases, fibroblasts from patients with Hurler's and with Scheie syndrome do not cross-complement in culture, indicating that the enzyme defect is the same. Hurler syndrome is at the severe end of the spectrum, Scheie at the mild end, and Hurler–Scheie syndrome is intermediate in phenotypic expression.

Hurthle cells See **Askenazy cells**.

HUT-78 cells Human T-cell lymphoma from patient with **Sezary syndrome**. Have features of mature T-cells of helper/inducer phenotype and release IL-2.

HUVEC *Human umbilical vein endothelial cells* A convenient source of human endothelial cells is those that line the large vein in the umbilical cord which is usually discarded together with the placenta after childbirth. The cells can be removed as a fairly pure suspension by mild enzymatic treatment of the vein followed by some mechanical distraction and will grow relatively easily in culture, retaining their differentiated characteristics for several passages.

HVEM *High-voltage electron microscope* See **high-voltage electron microscopy**.

hyaline Clear, transparent, granule-free – as, for example, hyaline cartilage and the hyaline zone at the front of a moving amoeba.

hyaluronic acid Polymer composed of repeating dimeric units of glucuronic acid and *N*-acetyl-glucosamine. May be of extremely high molecular weight (up to several million Daltons), and forms the core of complex proteoglycan aggregates found in extracellular matrix.

hyaluronidase *EC 3.2.1.35* Enzyme that degrades **hyaluronic acid**; found in lysosomes.

hybrid Ab Artificially produced antibody made by fusing **hybridomas** producing two different antibodies; the hybrid cells produce three different antibodies, only one of which is a **heterophile antibody**. Can also be prepared chemically from two antibodies.

hybrid cells Any cell type containing components from one or more genomes, other than zygotes and their derivatives. Hybrid cells may be formed by **cell fusion** or by **transfection**. See **heterokaryon**.

hybrid dysgenesis Genetic phenomenon, in which two strains of organism produce offspring at anomalously low rates. Example: the P–M system of *Drosophila*, in which P-strain males (containing multiple **P-elements**) mated to M-strain females produce sterile hybrids.

hybridization (of nucleic acids). Technique in which single-stranded **nucleic acids** are allowed to interact so that complexes, or hybrids, are formed by molecules with sufficiently similar, complementary sequences. By this means the degree of sequence identity can be assessed and specific sequences detected. The hybridization can be carried out in solution or with one component immobilized on a gel or, most commonly, nitrocellulose paper. Hybrids are detected by various means: by visualization in the electron microscope, by radioactively labelling one component and removing non-complexed DNA, or by washing or digestion with an enzyme that attacks single-stranded nucleic acids and finally estimating the radioactivity bound. Hybridizations are done in all combinations: DNA–DNA (DNA can be rendered single-stranded by heat denaturation), DNA–RNA or RNA–RNA. *In situ* hybridizations involve hybridizing a labelled nucleic acid (often labelled with a fluorescent dye) to suitably prepared cells or histological sections. This is used particularly to look for specific **transcription** or localization of genes to specific chromosomes (**FISH analysis**).

hybridoma A cell hybrid in which a tumour cell forms one of the original source cells. In practice, confined to hybrids between T- or B-lymphocytes and appropriate **myeloma** cell lines.

hydatid cyst Large cyst in the viscera of sheep cattle or man following ingestion of eggs of the tapeworm *Echinococcus granulosus*, a cestode. In the normal host the eggs develop into intestinal worms, but in abnormal hosts the larvae penetrate through the wall of the intestine and migrate to liver and other organs. The cyst contains many scolices that can form further cysts if liberated: it is thus a metastasizing parasite.

hydatiform mole Abnormal conceptus in which an embryo is absent and there is excessive proliferation of placental villi. In most cases the tissue is diploid XX with both X chromosomes being of paternal origin, and thus it is thought that it may arise by fertilization of a dead ovum. May occasionally become invasive, though the metastases regress following removal of the mole. Not to be confused with **hydatid cyst**.

Hydra Genus of freshwater coelenterates (**Cnidaria**). They are small, solitary, and only exist in the polyp form, which is a radially symmetrical cylinder that is attached to the substratum at one end and has a mouth surrounded by tentacles at the other. They have considerable powers of regeneration and have been used in studies on positional information in morphogenesis.

hydralazine *1-Hydrazinophthalazine monohydrochloride* Antihypertensive drug that exerts a peripheral vasodilating effect through a direct relaxation of vascular smooth muscle. Hydralazine interferes with the calcium movements within the vascular smooth muscle.

hydrargyrism Poisoning by mercury and its compounds. See **Minamata disease**.

hydraulic motor By altering the internal osmotic pressure within a cell, water will enter and a considerable expansion of the compartment will occur. This has been used as a motor device in plants (turgor pressure), in eversion of **nematocysts** and possibly in the production of other cellular protrusions.

hydrocodone Hydrocodone bitartrate, an opioid analgesic and antitussive.

hydrocortisone *Cortisol; 17-hydroxy-corticosterone* Moderately potent **corticosteroid** with both mineralocorticoid and glucocorticoid activity, produced by cells of the zona reticularis in the adrenal gland. Has potent anti-inflammatory effects.

hydrogel A gel in which water forms the bulk phase.

hydrogen peroxide H_2O_2 Hydrogen peroxide is produced by vertebrate phagocytes and is used in bacterial killing (the **myeloperoxidase**-halide system).

hydrogenosome Organelle found in certain anaerobic trichomonad and some ciliate protozoa: contains hydrogenase and produces hydrogen from glycolysis.

hydrolase One of a class of enzymes (EC Class 3) catalysing hydrolysis of a variety of bonds, such as esters, glycosides, peptides.

hydrolytic enzymes See **hydrolase**.

hydronium The hydrated proton (H^+) formed by acids in water.

hydropathy plot Approximate way of deducing the higher order structure of a protein, based on the principle that 20–30 consistently hydrophobic residues are necessary to make a membrane-spanning α-helix. For each amino acid residue a weighted average of the hydrophobicity of the residue and its immediate neighbours is calculated, and the results are graphically displayed as a hydropathy plot, with hydrophobic domains plotted as positive numbers. There are several formulae for the calculation that differ in the calculation of the moving average, the window size and the scoring system for hydrophobicity of individual residues (e.g. Kyte–Doolittle; Hopp Wood and Eisenberg).

hydrophilic group A polar group or one that can take part in hydrogen bond formation, e.g. OH, COOH, NH_2. Confers water solubility, or in lipids and macromolecules causes part of the structure to make close contact with the aqueous phase.

hydrophobic bonding Interaction driven by the exclusion of non-polar residues from water. It is an important determinant of protein conformation and of lipid structures, and is considered to be a consequence of maximizing polar interactions rather than a positive interaction between apolar residues.

hydrophobins Class of surface-active proteins produced by filamentous fungi that have a role in hyphal development and in sporulation. Make fungal structures, such as spores, hydrophobic.

hydropic Generally used as an adjective to describe one of the early signs of cellular degeneration in response to injury, the accumulation of water in the cell, causing swelling.

hydroquinone Compound that decreases the formation of melanin in the skin, often used topically to reduce pigmentation. Inhibits enzymatic oxidation of tyrosine to 3,4-dihydroxyphenylalanine (DOPA).

hydroxyapatite The calcium phosphate mineral, $Ca_{10}(PO_4)_6(OH)_2$, found both in rocks of non-organic origin and as a component of bone and dentine. Used as column packing for chromatography, particularly for separating double-stranded DNA from mixtures containing single-stranded DNA.

hydroxylysine Post-translationally hydroxylated lysine is found in **collagen** and commonly has galactose and then glucose added sequentially by glycosyl-transferases. The extent of glycosylation varies with the collagen type.

hydroxyproline Specific proline residues on the amino side of a glycine residue in collagen become hydroxylated at C4, before the polypeptides become helical, by the activity of prolyl hydroxylase. This enzyme has a ferrous ion at the active site, and a reducing agent such as ascorbate is necessary to maintain the iron in the ferrous state. The presence of hydroxyproline is essential to produce stable triple-helical tropocollagen, hence the problems caused by ascorbate deficiency in scurvy. This unusual amino acid is also present in considerable amounts in the major glycoprotein of primary plant cell walls (see **HRGP**).

hydroxyproline-rich glycoprotein See HRGP.

hydroxytetraecosanoic acid See HETE.

hydroxytryptamine *5-Hydroxytryptamine, 5HT* See **serotonin**.

hydroxyurea Inhibitor of DNA synthesis (but not repair), used as an antimetabolite in treating cancer.

hydrozoa A class of **Cnidaria**, in which there is alternation of generations; the hydroid phase is usually colonial and produces the free-swimming medusoid phase by budding.

hygromycin B Aminoglycoside antibiotic from *Streptomyces hygroscopicus* that is toxic for both pro- and eukaryotic cells. Inhibits peptide chain elongation by yeast polysomes by preventing elongation factor EF-2-dependent translocation. Used as a selection agent in transfection – the hygromycin-B-phosphotransferase gene, *hph*, in the vector confers resistance.

Hymenolepis Genus of cestode parasites of the gut of mammals. The immunological response to *H. diminuta* infection has been extensively studied.

hyoscine *scopolamine* An alkaloid, present in *Datura meteloides*, that has a sedative effect on the central nervous system.

hyoscyamine The active component of belladonna extract, an anticholinergic (antimuscarinic) drug from *Atropa belladona*; used as the sulphate for relief of gut spasm. When racemized following isolation is known as **atropine**.

hyper- Prefix denoting more than, bigger than.

hyperaccumulator General term for an organism that accumulates high levels of a substance from the environment. Usually used in reference to plants that absorb large amounts of heavy metals from the soil.

hyperaemia An excess of blood in a region of the body.

hypercapnia Excess of carbon dioxide in the lungs or the blood, generally caused by hypoventilation.

hypercholesterolaemia *Hypercholesterolemia (US)* High serum levels of cholesterol. Can in some cases be caused by a defect in lipoprotein metabolism or, for example, defects in the **low-density lipoprotein** receptor (familial hypercholesterolaemia).

hyperchromasia Histopathological term for cells in which the nuclei appear to be smudged or opaque and are more darkly stained. Usually an indication of precancerous behaviour.

hyperexcitability 1. Can be used in the context of behaviour. **2.** Physiologically, a state in which the threshold for neuronal firing is reduced and there are spontaneous action potentials. This may exhibit itself as hyperalgesia (increased sensitivity to pain), and there are clinical conditions in which this maybe due to autoimmune dysfunction.

hyperforin *HF* A natural phloroglucinol purified from *Hypericum perforatum* (St John's Wort). Often used as an antidepressant, apparently working by blocking serotonin re-uptake (see **SSRIs**). Found to be inhibitory to *Staphylococcus aureus* multiresistant to conventional antibiotics, as well as to other Gram-positive bacteria, and reported to promote apoptosis of B-cell chronic lymphocytic leukaemia cells.

hypergammaglobulinaemia *Hypergammaglobulinemia (US)* A clinical condition in which the concentration of immunoglobulins in the blood exceeds normal limits. May result from continuous antigenic stimulation in chronic infections, from autoimmune diseases or from abnormal proliferation of B-cells as in **Waldenstrom's macroglobulinaemia** or in **myelomatosis**.

hyperglycaemia *Hyperglycemia (US)* An excess of plasma glucose that can arise through a deficiency in insulin production. See **diabetes mellitus, diabetes insipidus**.

hyperglycemic Having an excess of glucose in the blood, usually >11 mmol/l; often a symptom of untreated diabetes.

hyperimmune serum Serum prepared from animals that have recently received repeated injections or applications of a chosen antigen; thus the serum should contain a very high concentration of polyclonal antibodies against that antigen.

hyperkalaemia An excessive level of potassium in the blood. Most cases of hyperkalaemia are caused by disorders that reduce potassium excretion by the kidney, because of either dysfunction or a deficiency of aldosterone (as in **Addison's disease**), the hormone that regulates potassium excretion. See **pseudohypoaldosteronism**.

hyperlipidaemia *Hyperlipidemia (US)* Elevated levels of serum low-density lipoprotein, correlated with increased risk of cardiovascular disease. See **hypercholesterolaemia**.

hyperlipoproteinaemia The same as **hyperlipidaemia**.

hypermastigote Large, multiflagellate symbiotic protozoa found in the gut of termites and wood-eating cockroaches. Most bizarre example of the group is *Mixotricha paradoxical*, which actually has few flagella and is propelled by spirochaetes (bacteria) that are attached to special bracket-like regions of the cell wall.

hyperosmotic Of a liquid, having a higher osmotic pressure (usually than the physiological level).

hyperplasia Increase in the size of a tissue as a result of enhanced cell division. Once the stimulus (wound healing, mechanical stress, hormonal overproduction) is removed the division rate returns to normal (whereas in neoplasia proliferation continues in the absence of a stimulus).

hyperpolarization A negative shift in a cell's **resting potential** (which is normally negative), thus making it numerically larger – i.e. more polarized. The opposite of **depolarization**.

hypersensitive response *Bot.* An active response of plant cells to pathogenic attack in which the cell undergoes rapid necrosis and dies. Associated with the production of **phytoalexins, lignin** and sometimes **callose**. The response is thought to prevent a potential pathogen from spreading through the tissues. See **hypersensitivity**.

hypersensitivity *Immunol.* A state of excessive and potentially damaging immune responsiveness as a result of previous exposure to antigen. If the hypersensitivity is of the immediate type (antibody-mediated), then the response occurs in minutes; in delayed hypersensitivity the response takes much longer (about 24 h) and is mediated by primed T-cells. Hypersensitivity responses are not simply divisible into the two types, and it is now more common to subdivide immediate responses into types I, II and III; the delayed response being of type IV. Type I responses involve antigen reacting with IgE fixed to cells (usually mast cells) and are characterized by histamine release; anaphylactic responses and urticaria are of this type. In type II responses, circulating antibody reacts with cell surface or cell-bound antigen, and if complement fixation occurs, cytolysis may follow. In type III reactions, immune complexes are formed in solution and lead to damage (serum sickness, glomerulonephritis, **Arthus reaction**). Delayed-type responses of type IV involve primed lymphocytes reacting with antigen and lead to formation of a lymphocyte-macrophage granuloma without involvement of circulating antibody.

hypertelorism The progressive attainment of disproportionate size, either of a body part or of the whole organism.

hyperthermophile Members of the Archaea that live and thrive in temperatures above 60°C, sometimes above 100°C (*cf.* thermophiles, which have a tolerance ceiling of about 60°C).

hyperthyroidism *Graves's disease* Excess production of thyroid hormone caused by autoantibodies that bind to the **thyroid-stimulating hormone** receptor and induce secretion of **thyroxine** by raising cAMP levels in the thyroid cells.

hypertonic Of a fluid, sufficiently concentrated to cause osmotic shrinkage of cells immersed in it. Note that a mildly **hyperosmotic** solution is not necessarily hypertonic for viable cells that are capable of regulating their volumes by **active transport**. See **hypotonic, isotonic**.

hypertrophy Increase in size of a tissue or organ as a result of cell growth, rather than an increase of cell number (**hyperplasia**), though often both processes occur.

hypervariable region Those regions of the heavy or light chains of immunoglobulins in which there is considerable sequence diversity within that set of immunoglobulins in a single individual. These regions specify the antigen affinity of each antibody.

hypha Filament of fungal tissue that may or may not be separated into a file of cells by cross-walls (septa). It is the main growth form of filamentous fungi, and is characterized by growth at the tip followed by lateral branching.

hypoblast The innermost germinal layer in the embryo of a metazoan animal that gives rise to the endoderm and sometimes also to the mesoderm.

hypocotyl Part of the axis of a plant embryo or seedling between the point of insertion of the **cotyledon(s)** and the top of the radicle (root). In some **etiolated** seedlings, the hypocotyl is greatly extended.

hypocretin See **orexin**.

hypogammaglobulinaemia *Hypogammaglobulinemia (US)* Syndromes in humans and other vertebrates in which the immunoglobulin level is depressed below the normal range. Congenital, chronic and transient types are known.

hypogeal *Hypogaeous* Descriptive of (1) an organism that lives below the ground surface, (2) a plant that germinates with the cotyledons remaining in the soil.

hyponatraemia A decreased level of sodium in the blood, a relatively uncommon situation but potentially dangerous. Usually due to a disturbance in thirst or in water acquisition, in ADH, aldosterone or renal sodium transport.

hypostome 1. The raised oral cone in some Cnidaria. **2.** The labrum in insects. **3.** The lower lip or fold forming the posterior margin of the mouth in Crustacea. **4.** The lower lip formed by the fusion of the pedipalpal coxae in some Acarina.

hypotonic Of a fluid, having a concentration that will cause osmotic shrinkage of cells immersed in it. Not necessarily hypo-osmotic.

hypovolaemia *Hypovolemia (US)* Abnormal reduction of the volume of blood or blood plasma leading to a form of clinical shock in which blood pressure falls.

hypoxanthine Purine base present in inosine monophosphate (IMP) from which adenosine monophosphate (AMP) and guanosine monophosphate (GMP) are made. The product of deamination of adenine, 6-hydroxypurine.

hypoxanthine-guanine phosphoribosyl transferase *HGPRT* See **HGPRT**.

hypoxia Condition in which there is an abnormally low level of oxygen in the blood and tissues.

Hypoxia-inducible transcription factors *HIFs* See **HIF-1** as an example of one of these factors.

H-zone Central portion of the A-band of the **sarcomere**, the region that is not penetrated by thin (actin) filaments when the muscle is only partially contracted. The **M-line** is in the centre of the H-zone.

I

I-309 *TCA3* Small **cytokine** secreted by activated T-lymphocytes that is chemotactic for monocytes.

IAA See **indole acetic acid**.

Ia antigen Antigens coded for by the I-region of the MHC complex. The majority of, if not all, such antigens are class II molecules composed of α and β polypeptide chains.

IAP *Inhibitor of apoptosis* One of a family of evolutionarily conserved genes that inhibit apoptosis. Human X-linked IAP directly inhibits two **caspases**, caspase-3 and caspase-7. Produced by various viruses to allow survival of host cell whilst virus replicates.

IAPP See **islet amyloid peptide**.

iatrogenic Descriptor for a disease caused by attempts at therapy.

I-band The isotropic band of the sarcomere of striated muscle, where only thin filaments are found. Unlike the A-band, the I-band can vary in width depending upon the state of contraction of the muscle when fixed.

IBD **1.** Inflammatory bowel disease. General term that covers two different inflammatory disorders of the bowel, **Crohn's disease** and **ulcerative colitis**. **2.** Identity by descent. Two genes at a locus have identity by descent (IBD) if they were both inherited from a common ancestor. Important in linkage studies for disease susceptibility.

iberiotoxin Peptide toxin (37 residues) from the scorpion *Leiurus quinquestriatus* var. *hebraeus* that is selective for the high-conductance calcium-activated K^+ channel. Similar (and highly homologous in sequence) to **charybdotoxin** but more selective.

IBMX *Isobutylmethylxanthine* Often used as an inhibitor of cAMP-phosphodiesterase as a way of raising intracellular cAMP levels for experimental purposes.

ibotenate *Ibotenic acid* An **excitotoxin** from *Amanita* sp. that acts on the **NMDA receptor**.

ibuprofen *2-(4-Isobutylphenyl)proprionic acid* Non-steroidal anti-inflammatory drug (NSAID) that inhibits cycloxygenases I and II (**COX-1, COX-2**). Related NSAIDs are ketoprofen, flurbiprofen and naproxen.

iC3b Inactivated C3b (C3bi). See **complement**.

IC_{50} Concentration of an inhibitor at which 50% inhibition of the response is seen; should only be used of *in vitro* test systems. Needs to be used with caution, because ED_{50} (for example) is the concentration that causes an effect in 50% of tests – not necessarily the same thing.

ICAD *DFF45* Inhibitor of caspase-activated DNase, 45 kDa. See **DFF**.

ICAM *Intercellular adhesion molecule* ICAM-1 is the glycoprotein ligand (80–155 kDa) for LFA-1 (CD11a/CD18: β2-integrin) and to a lesser extent Mac-1 (CD11b/CD18). ICAM-1 is expressed on the luminal surface of endothelial cells and is upregulated in response to IL-1 or TNF treatment. ICAM-1 is also expressed on various haematopoietic cells and is upregulated on activated T- and B-cells. It is a member of the immunoglobulin superfamily and has five C2 domains. Not only is it an important ligand for leucocyte adhesion; it is also the site to which rhinovirus binds and to which *Plasmodium falciparum*-infected erythrocytes adhere. ICAM-2 (55–68 kDa) is constitutively expressed on endothelium and on resting lymphocytes and monocytes, and is not upregulated by inflammatory cytokines. It has only two Ig superfamily domains. CD50 has now been identified as ICAM-3 (120 kDa) and plays a role in the early stages of the immune response; it is also a ligand for LFA-1. Cross-linking of ICAM-3 seems to induce an increase in intracellular calcium and triggers activation of tyrosine kinases.

iccosomes Immune-complex-coated bodies, formed when antigen is injected into an immune animal and found in follicular **dendritic cells**. May serve as a reservoir of antigen to maintain B-cell memory.

ICE See **interleukin 1-converting enzyme**.

I-cell disease *Mucolipidosis II* A human disease in which the lysosomes lack hydrolases but high concentrations of these enzymes are found in the extracellular fluids. Caused by mutation in the *GNPTAB* gene (GlcNAc-phosphotransferase; EC 2.7.8.15) that catalyses the initial step in the addition of the lysosome recognition marker (mannose-6-phosphate) to lysosomal hydrolases so that they are not directed into the lysosomes but are released.

ice nucleation proteins Proteins produced by some Gram-negative bacteria that promote the nucleation of ice, apparently by aligning water molecules along repeated domains of 48 amino acids. Consist of 16-residue repeats containing the conserved octamer AGYGSTxT. Now finding commercial use in snow-making at ski resorts.

ICF Abbrev. for (1) International Classification of Functioning, Disability and Health, a conceptual scheme defined by the WHO, (2) intracortical facilitation, (3) see **ICF syndrome**.

ICF syndrome Immunodeficiency, centromeric instability and facial anomalies syndrome, a rare autosomal-recessive disease caused by mutations in the DNA methyltransferase gene *DNMT3B*.

ichnoviruses See **polydnaviruses**.

Icilin A synthetic **TRP channel** super-agonist that is nearly 200-fold more potent than menthol. Icilin induces sensations of intense cold when applied orally in humans.

ICK1 Inhibitor of cyclin-dependent kinase identified in *Arabidopsis thaliana*. Has some limited similarity with mammalian p27Kip1 kinase inhibitor. Inhibits plant cdc2 kinase but not human p34cdc2.

ICM Abbrev. for (1) iodinated contrast medium, (2) inner cell mass of the mammalian blastocyst, (3) intermediate cell mass (ICM), a major site of zebrafish primitive haematopoiesis.

ICP Abbrev. for (1) Intracranial pressure, (2) **Inductively coupled plasma**.

IcsA Actin-nucleating protein in the outer membrane of virulent strains of *Shigella flexneri*. Like **ActA**, IcsA is responsible for unipolar assembly of an F-actin bundle that pushes the bacterium through the cytoplasm. Activation of the CDC42 effector N-WASP by the IcsA protein promotes actin nucleation by **Arp2/3** complex.

ICSH Abbrev. for interstitial cell-stimulating hormone. See **luteinizing hormone**.

ICSI Intracytoplasmic sperm injection, a method used for *in vitro* fertilization.

ictal Relating to the period during a sudden attack, such as seizure (ictus) or stroke.

ictus A stroke or sudden attack.

ID$_{50}$ The dose of an infectious organism required to produce infection in 50% of subjects.

I-domain Binding domain of around 200 amino acids in the N-terminal part of a subunit of **integrins**. The I-domain has intrinsic ligand-binding activity that is divalent cation dependent.

Id proteins *Id1–Id4; inhibitor of DNA-binding proteins* Proteins that function as dominant-negative regulators of basic helix–loop–helix transcription factors and themselves contain a helix–loop–helix motif. Block transcription by forming a non-functional heterodimer with the HLH transcription factors, interaction being mediated through the HLH domains. Regulate tissue-specific transcription within several cell lineages and play a part in cell growth, senescence, differentiation and angiogenesis.

idazoxan Imidazoline alpha-2-adrenoreceptor antagonist, also binds to **imidazoline receptors**.

idioblast See **sclereid**.

idiogram *Karyogram* Picture or diagram of the chromosome complement of a cell that is arranged to show the general morphology including relative sizes, positions of centromeres, etc. Often prepared as a photo-montage.

idiomorphs Alleles, responsible for defining mating types in self-incompatible ascomycetes that have marked sequence dissimilarities. Although in any given species one idiomorph is distinct from the other, each individual idiomorph is highly conserved in DNA sequence within families and at the level of protein function among families.

idiopathic Applied to disease of unknown origin or peculiar to the individual.

idiophase In a growing culture of microorganisms, the productive phase following the trophophase or growth phase.

idiotope An antigenic determinant (**epitope**) unique to a single clone of cells and located in the variable region of the immunoglobulin product of that clone or to the T-cell receptor. The idiotope forms part of the antigen-binding site. Any single immunoglobulin may have more than one idiotope. Idiotopes are also associated with the antigen-binding sites of T-cell receptors and are the epitopes to which an anti-**idiotype** antibody or T-cell binds.

idiotype The antigenic specificities defined by the unique sequences (**idiotopes**) of the antigen-combining site. Thus anti-idiotype antibodies combine with those specific sequences, may block immunological reactions, and may resemble the epitope to which the first antibody reacts.

IDL Abbrev. for **intermediate-density lipoproteins**.

idoxuridine Analogue of thymidine that inhibits the replication of DNA viruses. Used in the treatment of herpes simplex and varicella zoster.

iduronic acid α-L-*Iduronic acid* A uronic acid, derived from the sugar idose and bearing one terminal carboxyl group. With *N*-acetyl-galactosamine-4-sulphate, a component of dermatan sulphate. See **iduronidase**.

iduronidase α-*1-Iduronidase; EC 3.2.1.76* An enzyme (653 amino acids in human) that hydrolyses the bonds between iduronic acid and *N*-acetylgalactosamine-4-sulphate; a lysosomal enzyme absent in **Hurler's disease**.

IEF Abbrev. for **isoelectric focusing**

IgA Major class of immunoglobulin of external secretions in mammals, also found in serum. In secretions, found as a dimer (400 kDa) joined by a short J-chain and linked to a secretory piece or transport piece. In serum, found as a monomer (170 kDa). IgAs are the main means of providing local immunity against infections in the gut or respiratory tract and may act by reducing the binding between an IgA-coated microorganism and a host epithelial cell. Present in human colostrum but not transferred across the placenta. Have α heavy chains.

IgD This immunoglobulin (184 kDa) is present at a low level (3–400 μg/ml) but is a major immunoglobulin on the surface of B-lymphocytes, where it may play a role in antigen recognition. Its structure resembles that of IgG, but the heavy chains are of the δ type.

IgE Class of immunoglobulin (188 kDa) associated with immediate-type **hypersensitivity** reactions and helminth infections. Present in very low amounts in serum, and mostly bound to mast cells and basophils that have an IgE-specific Fc receptor (FcεR). IgE has a high carbohydrate content and is also present in external secretions. Heavy chain of ε-type.

IGF1 See **insulin-like growth factor** 1.

IGF2 **Insulin-like growth factor** 2.

IgG *7S antibody* The classical immunoglobulin class also known as 7S IgG (150 kDa). Composed of two identical light and two identical heavy chains, the constant region sequence of the heavy chains being of the γ type. The molecule can be described in another way as being composed of two **Fab** and an **Fc** fragment. The Fabs include the antigen-combining sites; the Fc region consists of the remaining constant sequence domains of the heavy chains and contains cell-binding and complement-binding sites. IgGs act on pathogens by agglutinating them, by opsonizing them, by activating complement-mediated reactions against cellular pathogens and by neutralizing toxins. In some mammals, including man, they can pass across the placenta to the foetus as maternal antibodies, unlike other Ig classes. In humans, four main subclasses are known; IgG2 differs from the rest in not being transferred across the placenta and IgG4 does not fix complement. IgG is present at 8–16 mg/ml in serum.

IGIF *Interferon-inducing factor* Obsolete name for **IL-18**. Originally described as a protein that augments NK-cell activity in spleen cells. Precursor has 192 amino acid residues, mature protein has 157 amino acids.

IgLON A family of cell adhesion molecules, comprising OPCML, HNT, LSAMP (LAMP, limbic system-associated membrane protein) and NEGR1; two of its members have been proposed as tumour suppressors.

IgM *Macroglobulin* An IgM molecule (970 kDa) is built up from five IgG-type monomers joined together, with the assistance of J-chains, to form a cyclic pentamer. IgM binds complement and a single IgM molecule bound to a cell surface can lyse that cell. IgM is usually produced first in an immune response before IgG. The human red cell isoantibodies are IgM antibodies. Heavy chain (μchain) is rather larger than the heavy chains of other immunoglobulins.

IGS Abbrev. for (1) intergenic spacer, most commonly applied to the region between ribosomal genes in eukaryotes and prokaryotes, (2) implantable gastric stimulation, a method used to try to treat obesity.

IgSF Abbrev. for **Immunoglobulin superfamily**.

IgX An immunoglobulin class found in Amphibia.

IHC Abbrev. for immunohistochemistry. See **immunocytochemistry**.

IHF Abbrev. for **integration host factor**.

IkappaB *IκB* Protein that inhibits **NFκB** by binding to the p65 subunit. It is thought to prevent NFκB from entering the nucleus. Two forms have been identified IκBα (37 kDa) and IκBβ (43 kDa).

IKK *IkappaB kinase* Protein kinases that phosphorylate the NF-kappaB inhibitor **IkappaB**.

IL-1, IL-2, etc. See **interleukins** and individual interleukin-*n* entries.

IL-6 cytokine family Family of cytokines that all act through receptors sharing a common gp130 subunit. Includes **interleukin-6, interleukin-11, ciliary neurotrophic factor, LIF, oncostatin M** and **cardiotrophin-1**.

ILK *Integrin-linked kinase* A multifunctional, cytoplasmic, serine/threonine kinase implicated in regulating processes such as cell proliferation, survival, migration and invasion.

imaginal disc Epithelial infoldings in the larvae of holometabolous insects (e.g. Lepidoptera, Diptera) that rapidly develop into adult appendages (legs, antennae, wings, etc.) during metamorphosis from larval to adult form. By implanting discs into the haemocoele of an adult insect their differentiation can be blocked, though their **determination** remains unchanged except occasionally **transdetermination** occurs. The hierarchy of transdetermination has been studied in great detail in *Drosophila*.

imidazoles *Glyoxalines* Heterocyclic compounds produced by substitution in a five-membered ring containing two nitrogen atoms on either side of a carbon atom. Benzimidazoles are formed by the condensation of orthodiamines with organic acids and contain a condensed benzene nucleus. A group of chemically related drugs, active against fungi, a range of bacteria and in some cases as anthelminthics. Examples are ketoconazole, tiabdazole.

imidazoline receptors Receptors for **imidazolines** such as **agmatine** and **idazoxan** that modulate sympathetic outflow in the brainstem. The I1-receptor is elevated on platelets and in brains of patients with depression. Also expressed on vascular smooth muscle where they may mediate inhibition of proliferation. Candidate protein for

I1-receptor is 85 kDa and upregulated by idazoxan and agmatine. I2 receptors have been proposed to have a role in modulating pain during acute inflammation.

imine Any organic compound with one or more imino groups.

imino acid Organic acid containing an imino group in place of two hydrogen atoms. Also applied to cyclic alkylamino derivatives of aliphatic carboxylic acids, such as proline although these should properly be referred to as azacycloalkane carboxylic acids.

iminoglycinuria A defect in amino acid transport leading to abnormal excretion of glycine, proline and hydroxyproline in the urine: more seriously, absorption in the intestine may be inadequate. See **Hartnup disease**.

immaturin A heat-labile soluble cytoplasmic protein with a molecular mass of about 10 kDa that represses mating activity in *Paramecium caudatum*.

immediate-early gene *IEG* Class of genes whose expression is low or undetectable in quiescent cells, but whose transcription is activated within minutes after extracellular stimulation such as addition of a **growth factor**. *c-fos* and *c-myc* proto-oncogenes were among the first IEGs to be identified. Many IEGs encode **transcription factors** and therefore have a regulatory function.

immobilized pH gradient *IPG* Immobilized pH gradient (IPG) strips (about 3 mm wide and 180 mm long) are used in 2-D gel electrophoresis as the medium for the first dimension (isoelectric focusing) step. Commercially available strips with various pH ranges and linear or non-linear pH gradients are available; by using narrow-range strips, resolution can be improved in particular areas of the protein 'spectrum'.

immortalization Escape from the normal limitation on growth of a finite number of division cycles (the Hayflick limit), by variants in animal cell cultures, and cells in some tumours. Immortalization in culture may be spontaneous, as happens particularly readily in mouse cells, or induced by mutagens or by **transfection** of certain oncogenes.

immotile cilia syndrome Congenital defect in **dynein** (either absent or inactive) that leads to male sterility and poor bronchial function. Interestingly, non-ciliated cells show altered locomotion and 50% of patients have **Kartagener's syndrome**.

immune complex Multimolecular antibody–antigen complexes that may be soluble or insoluble depending upon their size and whether or not complement is present. Immune complexes can be filtered from plasma in the kidney, and the deposition of the complexes gives rise to **glomerulonephritis** – probably because of the trapping of neutrophils via their Fc receptors.

immune complex diseases Diseases characterized by the presence of immune complexes in body fluids. Hypersensitivity of the **Arthus** type and serum sickness are examples.

immune deficiency diseases Those diseases in which immune reactions are suppressed or reduced. Reasons may include congenital absence of B and/or T-lymphocytes, or viral killing of helper lymphocytes (see **HIV**).

immune response Alteration in the reactivity of an organism's immune system in response to an antigen; in vertebrates, this may involve antibody production, induction of cell-mediated immunity, complement activation or development of immunological tolerance.

immune response gene See **I-region**.

immune surveillance See **immunological surveillance**.

immunity A state in which the body responds specifically to antigen and/or in which a protective response is mounted against a pathogenic agent. May be innate or may be induced by infection or vaccination, or by the passive transfer of antibodies or immunocompetent cells. Use of the term should not be restricted to vertebrates, and invertebrate immune responses, though less sophisticated than the mammalian system, are nevertheless important for resisting pathogens.

immunoadsorbent Any insoluble material, e.g. cellulose, with either an antigen or an antibody bound to it and that will bind its corresponding antibody or antigen, thus removing it from a solution.

immunoblotting Techniques, such as the **Western blot**, in which very small amounts of protein are transferred from gels to nitrocellulose sheets by electrophoresis and then detected by their antibody binding, usually in combination with peroxidase- or radioactively-labelled IgG. An accurate technique for the specific recognition of very small amounts of protein. See also **blotting**.

immunoconglutinin *IK* Antibodies that react with complement components or their breakdown products. Usually directed against C3b or C4. High levels of IgG immunoconglutinins are found in plasma from patients with **systemic lupus erythematosus**.

immunocytochemistry Techniques for staining cells or tissues using antibodies against the appropriate antigen. Although in principle the first antibody could be labelled, it is more common (and improves the visualization) to use a second antibody directed against the first (an anti-IgG). This second antibody is conjugated with fluorochromes or appropriate enzymes for colorimetric reactions, or with gold beads (for electron microscopy), or with the biotin–avidin system, so that the location of the primary antibody, and thus the antigen, can be recognized.

immunodeficient See **immune deficiency diseases**.

immunoelectrophoresis Any form of **electrophoresis** in which the molecules separated by electrophoresis are recognized by precipitation with an antibody.

immunofluorescence A test or technique in which one or other component of an immunological reaction is made fluorescent by coupling with a **fluorochrome** such as fluorescein, phycoerythrin or rhodamine so that the occurrence of the reaction can be detected as a fluorescing antigen–antibody complex. Used in microscopy to localize small amounts of antigen or specific antibody.

immunogenicity The property of being able to evoke an immune response within an organism. Immunogenicity depends partly upon the size of the substance in question and partly upon how unlike host molecules it is. Highly conserved proteins tend to have rather low immunogenicity.

immunoglobulin superfamily *IgSF* A large group of proteins with immunoglobulin-like domains. Most are involved with cell surface recognition events. Sequence homology suggests that Igs, MHC molecules, some **cell adhesion** molecules and cytokine receptors share close homology, and thus belong to a **multigene family**.

immunoglobulins See **IgA, IgD, IgE, IgG** and **IgM**.

immunological memory The systems responsible for the situation where reactions to a second or subsequent exposure to an antigen are more extensive than those seen on first exposure (but see also **immunological tolerance**). The memory is best explained by clonal expansion and persistence of such clones following the first exposure to antigen.

immunological network The concept, due to Jerne, that the entire specific immune system within an animal is made up of a series of interacting molecules and cell surface receptors, based on the idea that every antibody-combining site carries its own marker antigens or **idiotypes** and that these in turn may be recognized by another set of antibody-combining sites and so on.

immunological surveillance The hypothesis that lymphocyte traffic ensures that all or nearly all parts of the vertebrate body are surveyed by visiting lymphocytes in order to detect any altered-self material, e.g. mutant cells.

immunological tolerance Specific unresponsiveness to antigen. Self-tolerance is a process occurring normally early in life due to suppression of self-reactive lymphocyte clones. Tolerance to foreign antigens can be induced in adult life by exposure to antigens under conditions in which specific clones are suppressed. Note that tolerance is not the same as immunological unresponsiveness, since the latter may be very non-specific – as in immunodeficiency states.

immunomodulation Modification of the immune response – either activating or, more commonly, suppressing.

immunophilin Generic term for intracellular protein that binds immunosuppressive drugs such as **ciclosporin, FK506, rapamycin**. Both **cyclophilin** and the receptor for FK506 are peptidyl prolyl *cis–trans*-isomerases (rotamases). Immunophilins are thought to interact with **calcineurin**.

immunoprecipitation The precipitation of a multivalent antigen by a bivalent antibody, resulting in the formation of a large complex. The antibody and antigen must be soluble. Precipitation usually occurs when there is near equivalence between antibody and antigen concentrations.

immunoregulation The various processes by which antibodies may regulate immune responses. At a simple level, secreted antibody neutralizes the antigen with which it reacts thus preventing further antigenic stimulation of the antibody-producing clone. At a more complex level, anti-idiotype antibodies can be shown to develop against the first antibodies in some cases, and perhaps further anti-idiotype antibodies against them. This is the main concept of the **immunological network** theory.

immunosuppression This occurs when T- and/or B-clones of lymphocytes are depleted in size or suppressed in their reactivity, expansion or differentiation. It may arise from activation of specific or non-specific T-suppressor lymphocytes of either T- or B-clones, or by drugs that have generalized effects on most or all T- or B-lymphocytes.

Ciclosporin and FK506 act on T-cells, as does antilymphocyte serum; alkylating agents such as cyclophosphamide are less specific in their action and damage DNA replication, while base analogues interfering with guanine metabolism act in a similar way. See **immunophilin**.

immunotoxins Any toxin that is conjugated to either an immunoglobulin or Fab fragment directed against a specified antigen. Thus if the antigen is borne by a particular type of cell, such as a tumour cell, the toxin may be targeted at the specified cell by the immunological reaction.

impermeable cell junctions See **zonula occludens**.

impetigo Contagious skin disease caused by staphylococci or streptococci.

importins Proteins that bind **nuclear localization signals** (NLS) on proteins destined for the nucleus and that, in conjunction with **ran** and pp15, are involved in transport. Importins α and β form a heterodimer in the cytoplasm that corresponds to the 'NLS-receptor': both have multiple repeated arm domains. Importin α binds importin β through an NLS-like region on importin β; importin β binds ranGTP and also several nucleoporins. Since importins α and β recycle from the nucleus to the cytoplasm at different rates, it seems that there is a dissociation step in the transport cycle.

imposex The superimposition of male sexual characteristics onto female gastropods, caused by pollutants such as the antifouling agent, tributyltin.

imprinting See **genomic imprinting**.

in ovo In the egg.

in silico Term used (rather jokingly) for experiments performed using a computer database (i.e. on a silicon chip). It is now possible to match a small sequence of nucleotides with a full-length gene by running a search on a database; often it is then possible to order the appropriate cDNA in a vector ready for use. Derived by analogy with *in vitro* and *in vivo*.

in situ Literally, in place. Used particularly in the context of *in situ* **hybridization**.

in situ **hybridization** **1.** Technique for revealing patterns of gene expression in a tissue. The tissue is fixed and prepared (usually by sectioning) on a slide, and a labelled DNA or RNA probe is hybridized to the sample. It binds only to complementary mRNA sequences, and so only stains cells that are expressing the gene in question. **2.** Use of the technique for identifying the position of a gene on a chromosome by hybridizing a probe to spread chromosomes. Widely used in *Drosophila* salivary gland polytene chromosome 'squashes', but now also possible with a variety of eukaryotic nuclei. 'FISH' analysis is the use of **fluorescence** *in situ* **hybridization**.

in vitro Literally, in glass; general term for cells in culture as opposed to in a multicellular organism (*in vivo*). Modern tissue culture techniques rarely use glass vessels, but the term persists.

in vitro **fertilization** *IVF* Fertilization of an ovum outside the body (in culture). Usually several eggs are used and the zygotes produced by sperm–egg fusion are allowed to undergo several cell divisions *in vitro* before the embryo of choice is implanted into the uterus. The opportunity therefore exists for preimplantation genetic diagnosis (PGD)

to identify embryos with abnormal genetic characteristics or select female embryos to avoid X-linked disorders. More recently intracytoplasmic sperm injection (with a micropipette) has been used, rather than relying upon normal sperm–egg fusion, especially if the donor's sperm cell count is low.

in vivo Literally, in life; used of cells in their natural multicellular environment or of experiments on intact organisms rather than on isolated cells in culture (*in vitro*).

inactivation For example, of **voltage-gated** sodium channels: process by which sodium channels that have been activated or opened by **depolarization** subsequently close during the depolarization. Distinguished from activation by its slower kinetics.

inbred strain Any strain of animal or plant obtained by a breeding strategy that tends to lead to homozygosity. Such breeding strategies include brother–sister mating and backcrossing of offspring with parents. See also **congenic**.

inclusion bodies Nuclear or cytoplasmic structures with characteristic staining properties, usually found at the site of virus multiplication. Semicrystalline arrays of **virions**, **capsids** or other viral components.

incomplete dominance *Co-dominance* Condition in which members of the F1 generation are intermediate in character between the two parents; the heterozygote is phenotypically different from both parents.

incomplete metamorphosis In insects, a pattern of development in which there is a more or less gradual change from the immature to the mature state and there is no pupal stage. Usually the young resemble the parents, but lack wings and mature sexual organs.

incretins Hormones, for example glucagon-like peptide-1 and glucose-dependent insulinotropic polypeptide, that stimulate insulin secretion from pancreatic beta-cells. They are produced in the intestine and stimulate the insulin release required after eating. An incretin mimetic has reached the market in the US for the treatment of Type II diabetes.

indels Insertion/deletion points in a DNA sequence. Indel polymorphisms are relatively common.

index case The first or original case of a disease. The term is used in the epidemiology of infectious disease, and in genetics it is synonymous with the proband or propositus.

Indian hedgehog *Ihh* One of the **hedgehog family**, actively involved in endochondral bone formation. Brachydactyly type A1 (BDA1) is caused by mutations in the Indian hedgehog gene.

indicator species A species whose presence or absence indicates particular conditions in a habitat, or one that is particularly susceptible to the effects of some environmental factor. Patterns of gene expression in such species may provide sensitive indicators of low levels of pollutants.

indinavir *TN Crixivan* A **protease inhibitor** used in the treatment of HIV.

indirect immunofluorescence A method of **immunofluorescence** staining in which the first antibody that is directed against the antigen to be localized is used unlabelled, and the location of this first antibody is then detected by use of a fluorescently labelled anti-IgG (against IgGs of the species in which the first antibody was raised).

The advantage is that there is some amplification, and a well-characterized goat anti-rabbit IgG antibody can, for example, be used against a scarce specific antibody raised in rabbits. The same technique can be used for ultrastructural localization of the first antibody by substituting peroxidase or gold-labelled second antibody.

indole acetic acid *IAA* The most common naturally occurring **auxin**. Promotes growth in excised plant organs, induces adventitious roots, inhibits axillary bud growth, regulates gravitropism.

indolicidin Bovine **cathelicidin**, 13 amino acids.

indometacin *Indomethacin* Non-steroidal anti-inflammatory drug that blocks the production of **arachidonic acid** metabolites by inhibiting **cyclo-oxygenase**. Indometacin is the British Approved Name (BAN) for this drug.

indomethacin See **indometacin**.

inducer cells Cells that induce other nearby cells to differentiate in specified pathways. Perhaps the distinction should be made, as of old, between those cells that evoke a predetermined pathway of differentiation in the target cells and those cells that can actually induce new and unexpected differentiations.

induction See **embryonic induction** or **enzyme induction**.

inductively coupled plasma *ICP* A very high temperature (7000–8000 K) excitation source that efficiently desolvates, vaporizes, excites and ionizes atoms. ICP sources are used to excite atoms for atomic-emission spectroscopy (ICP-AES) and to ionize atoms for mass spectrometry (ICP/MS).

infarction Death of tissue as a result of loss of blood supply, often as a result of thrombotic occlusion of vessels.

infectious hepatitis See **hepatitis A**.

infectious mononucleosis See **glandular fever**.

inferior colliculus An auditory centre in the midbrain located below the superior colliculus, above the trochlear nerve and at the base of the projection of the medial and lateral geniculate nuclei. It is responsive to interaural delay and amplitude differences, and may provide a spatio-topic map of the auditory environment.

inflammation Response to injury. **Acute inflammation** is dominated by vascular changes and by neutrophil leucocytes in the early stages, and mononuclear phagocytes later on. Leucocytes adhere locally and emigrate into the tissue between the endothelial cells lining the postcapillary venules. Plasma exudation from vessels may lead to tissue swelling, but the early vascular changes are independent of and not essential for the later cellular response. In chronic inflammation, where the stimulus is persistent, the characteristic cells are **macrophages** and **lymphocytes**.

infliximab *Remicade* MAb against TNFα used for the treatment of inflammatory disease.

influenza virus Member of the **Orthomyxoviridae** that causes influenza in humans. There are three types of influenza virus: Type A causes the worldwide epidemics (pandemics) of influenza and can infect other mammals and birds, Type B only affects humans, and Type C causes only a mild infection. Types A and B virus evolve continuously,

resulting in changes in the antigenicity of their spike proteins, preventing the development of prolonged immunity to infection. The spike proteins, external haemagglutinin (HA) and **neuraminidase**, have been studied as models of membrane glycoproteins. See **H5N1**.

informosome Cytoplasmic complex of mRNA and non-ribosomal protein. May protect the message from degradation during passage from nucleus to cytoplasm.

infusoria A collective term for various minute aquatic organisms found in freshwater pond water, probably some of the first specimens ever observed under a microscope. Strictly speaking, the term relates to ciliate protozoa.

ING1 A tumour-suppressor gene that has similar biological functions to **p53**.

inhibin Polypeptide hormone secreted by the hypophysis that selectively suppresses the secretion of pituitary FSH (**follicle-stimulating hormone**). The molecule has two subunits (14 and 18 kDa) and is a product of the gene family that includes **TGFβ**. There are two forms, αβA and αβB, the β subunit being shared with **activin**. Inhibin is now, on the basis of gene-knockout experiments, considered to be a tumour suppressor, the key gene being that for inhibin-α.

inhibition constant K_i The equilibrium dissociation constant for the reaction between enzyme and inhibitor. $K_i=[E]\times[I]/[EI]$ where $[E]$, $[I]$, and $[EI]$ are concentrations of enzyme, inhibitor and the complex, respectively.

inhibitory postsynaptic potential *ipsp* The change in membrane voltage of a postsynaptic neuron which makes it more difficult for an action potential to be produced and reduces the firing rate of the neuron. A consequence of activation of inhibitory neurotransmitter receptors such as **GABA receptors** and **glycine receptors**.

inhibitory synapse A synapse in which an action potential in the presynaptic cell reduces the probability of an action potential occurring in the postsynaptic cell. The most common inhibitory neurotransmitter is **GABA**; this opens channels in the postsynaptic cell which tend to stabilize its **resting potential**, thus rendering it less likely to fire. See **excitatory synapse**.

initial cell Actively dividing plant cell in a **meristem**. At each division, one daughter cell remains in the meristem as a new initial cell and the other is added to the growing plant body. The animal equivalent is a stem cell (potentially confusing terminology in plants).

initiation codon *Start codon* The **codon** 5'-AUG in mRNA, at which polypeptide synthesis is started. It is recognized by formylmethionyl-tRNA in bacteria and by methionyl-tRNA in eukaryotes.

initiation complex Complex between mRNA, 30S ribosomal subunit and formyl-methionyl-tRNA that requires GTP and **initiation factors** to function.

initiation factors *IFs* The set of catalytic proteins required, in addition to mRNA and ribosomes, for protein synthesis to begin. In bacteria, three distinct proteins have been identified: IF-1 (8 kDa), IF-2 (75 kDa) and IF-3 (30 kDa). At least six to eight proteins have been identified in eukaryotes. IF-1 and -2 enhance the binding of initiator tRNA to the **initiation complex**.

INK4 *P16/INK4* Tumour-suppressor gene that generates several transcript variants (p15, p16, p18, p19), two of which encode structurally related isoforms known to function as inhibitors of CDK4 kinase (cyclin-dependent kinase 4). p16 has been found to be mutated more frequently than p53 in various cancer cells.

inner cell mass A group of cells found in the mammalian blastocyst that give rise to the embryo and are potentially capable of forming all tissues, embryonic and extraembryonic, except the trophoblast.

inner sheath The material that encases the two central microtubules of the ciliary axoneme.

innexin Protein components of **gap junctions** in protostomes like *Drosophila*, *C. elegans* and *Hirudo*, distantly related to the vertebrate **pannexins** and encoded by a multigene family. Highly conserved.

inoculum Cells added to start a culture or, in the case of viruses, viruses added to infect a culture of cells. Also a term for biological material injected into an animal to induce immunity (a vaccine).

inosine The 'fifth base' of nucleic acids. Important because it fails to form specific pair bonds with the other bases. In **transfer RNAs**, this property is used in the **anticodon** to allow matching of a single tRNA to several codons. **PCR** performed with primers containing inosine tolerates a limited degree of mismatch between primer and template, useful when trying to clone homologous protein by using degenerate primers.

inositol A cyclic hexahydric alcohol with six possible isomers. The biologically active form is myo-inositol.

inositol 3-kinase See **PI-3-kinase**.

inositol phosphates Virtually all of the possible phosphorylated states of inositol have been reported to occur in living tissues. The hexaphosphate, **phytic acid**, is abundant in many plant tissues and is a powerful calcium chelator. See **PI-3-kinase**.

inositol phosphoglycans *IPGs* A family of putative second messengers of insulin, released outside cells by hydrolysis of membrane-bound glycosylphosphatidylinositols. There are two subfamilies, IPG-A and IPG-P; IPG-A inhibits PKA from bovine heart, decreases phosphoenolpyruvate carboxykinase mRNA levels in rat hepatoma cells and stimulates lipogenesis and inhibits leptin release in rat adipocytes. IPG-P stimulates bovine heart pyruvate dehydrogenase phosphatase.

inositol trisphosphate *IP3, InsP3* Inositol 1,4,5 trisphosphate is important as a second messenger. It is released from the membrane phospholipid phosphatidyl inositol bisphosphate by the action of a specific phospholipase C enzyme (PLCγ) and binds to and activates a calcium channel in the endoplasmic reticulum.

inotropic Altering rate of heartbeat. Example: adrenaline has a positive inotropic effect and increases rate of beating.

INR Abbrev. for **international normalized ratio**.

insect defensins See **defensins**.

Insecta A class of Arthropoda, mainly terrestrial, whose members breathe by tracheae. The head is distinct from the thorax and bears one pair of antennae; there are three pairs of similar legs attached to the thorax, which may also bear wings. May be the inheritors of the Earth, given their success so far.

insertin Protein (30 kDa) from chicken gizzard smooth muscle. Binds to the barbed ends of actin filaments and apparently allows insertion of further monomers. Highly homologous to amino acids 962–1292 of **tensin,** and probably formed from tensin by proteolysis.

insertion sequence *IS elements* Mobile nucleotide sequences that occur naturally in the genomes of bacterial populations. When inserted into bacterial DNA, they inactivate the gene concerned; when they are removed the gene regains its activity. Closely related to transposons and range in size from a few hundred to a few thousand bases, but are usually less than 1500 bases.

insertional mutagenesis Generally, mutagenesis of DNA by the insertion of one or more bases. Specific examples: (1) oncogenesis by insertion of a retrovirus adjacent to a cellular proto-oncogene, (2) a strategy of mutagenesis with **transposons**. After a round of transposition, progeny are screened by **PCR**, with transposon- and gene-specific primers, for the proximity of the transposon sequence to the gene of interest. As PCR can only produce products up to 1–2 kb, a large fraction of progeny identified as positive by PCR will have a transposon close enough to the gene to inactivate or otherwise alter its pattern of expression.

insertosome See **insertion sequence**.

inside-out patch A variant of the **patch clamp** technique, in which a disc of plasma membrane covers the tip of the electrode with the inner face of the plasma membrane facing outward to the bath.

inside-out vesicle *IOV* Mechanical disruption of cell membranes gives rise to small closed vesicles surrounded by a bilayer membrane. These may be right-side out (ROV) or IOV if the topography is inverted.

insig-1, insig-2 *Insulin-induced gene-1, Insulin-induced gene-2* Genes coding for proteins required for sterol-mediated inhibition of the proteolytic processing of sterol regulatory element-binding proteins (**SREBPs**) to their active nuclear forms. The sequences of human insig-1 and -2 are 59% identical. Both are predicted to contain six transmembrane helices.

instructive theory Theory of antibody production, now considered untenable, in which antigen acted as template for the production of specific antibody, as opposed to the **clonal selection** theory in which pre-existing variation occurs and appropriate clones are selectively expanded.

insufflation The action of blowing gas, air, vapour or powder into a body cavity, usually the lungs.

insulin A polypeptide hormone (bovine insulin; 5780 Da) found in both vertebrates and invertebrates. Secreted by the **B-cells** of the pancreas in response to high blood sugar levels, it induces hypoglycaemia. Defective secretion of insulin is the cause of **diabetes mellitus**. Insulin is also a mitogen, has sequence homologies with other **growth factors** and is a frequent addition to cell culture media for demanding cell types.

insulin C-peptide *C-peptide; connecting peptide* A single chain of 31 amino acids (MW 3020) connecting the A and B chains of insulin in the proinsulin molecule. Has no known physiological function but, because C-peptide has a longer half-life in plasma before being cleared than does insulin (two to five times longer), plasma C-peptide concentrations reflect pancreatic insulin secretion more reliably than the level of insulin itself and laboratory tests for C-peptide levels are routinely used in assessing diabetics.

insulin sensitizer Category of drug (thiazolidinediones (TZDs), e.g. rosiglitazone) that enhance insulin action in muscle, fat and other tissues. Require the presence of insulin in order to work.

insulin-like growth factor *IGF* IGFs I and II are polypeptides with considerable sequence similarity to insulin. They are capable of eliciting the same biological responses, including mitogenesis in cell culture. On the cell surface there are two types of IGF receptor, one of which closely resembles the insulin receptor (which is also present). IGF I = somatomedin A = somatomedin C; IGF II = MSA (multiplication-stimulating activity). IGF-1 is released from the liver in response to growth hormone.

***int* oncogenes** Oncogenes first identified as targets for insertional activation by the mouse mammary tumour virus (MMTV) in mammary carcinomas. *Int1, Int2* and *Int3* are unrelated genes; the similarity in nomenclature is simply because they are targets for MMTV insertion mutation (insertion sites 1, 2 and 3). Int1 is a homologue of *Drosophila* **wingless** and a member of the **Wnt** family, the *Int2* product is fibroblast growth factor-3, the *Int3* product is **Notch4**. See also Table O1.

int-1 Now renamed **Wnt-1**.

Integra® Artificial skin composed of bovine collagen and shark-cartilage glycosaminoglycan with a silicone outer layer; used to temporarily close a wound.

integral membrane protein A protein that is firmly anchored in a membrane (unlike a peripheral membrane protein). Most is known about the integral proteins of the plasma membrane, where important examples include hormone receptors, ion channels and transport proteins. An integral protein need not cross the entire membrane; those that do are referred to as transmembrane proteins.

integrase protein Enzyme of the bacteriophage lambda (λ) that catalyses the integration of phage DNA into the host DNA.

integration Incorporation of the genetic material of a virus into the host genome.

integration host factor *IHF* An architectural protein which assists formation of high-order protein–DNA complexes such as those found in replication and long-distance transcription regulation in bacteria. It is a heterodimeric protein, with ~90 residue subunits with very similar three-dimensional structures, that binds to a specific 35-bp site.

integrator gene In the Britten and Davidson model for the coordinate expression of unlinked genes in eukaryotes, sensor elements respond to changing conditions by switching on appropriate integrator genes, which then produce transcription factors that activate appropriate subsets of structural genes.

integrins Superfamily of cell surface proteins that are involved in binding to extracellular matrix components in some cases. Most are heterodimeric with a β subunit of 95 kDa that is conserved through the superfamily and a more variable α subunit of 150–170 kDa. The first examples described were **fibronectin** and **vitronectin** receptors of fibroblasts, which bind to an RGD (Arg-Gly-Asp) sequence in the ligand protein, though the context of the RGD seems important and there is also a divalent cation dependence. Subsequently the platelet IIb/IIIa surface glycoprotein (fibronectin and fibrinogen receptor) and the LFA-1 class of leucocyte surface protein were recognized as integrins, together with the VLA surface protein. The requirement for the RGD sequence in the ligand does not seem to be invariable. See Table I1.

integron Class of DNA element composed of a DNA integrase gene adjacent to a recombination site, at which one or more genes can be found inserted. Frequently, antibiotic resistance genes are found inserted in integron sites in samples of resistant Gram-positive bacteria.

inteins Intervening sequences (introns) in proteins, by analogy with those in mRNA, the polypeptide chain is removed by proteolytic processing and splicing to produce the mature protein. Some intein polypeptides are site-specific endonucleases as well as protein-splicing catalysts. Endonuclease action results in insertion of the intein nucleic acid in a target site. About 90% of inteins contain domains whose amino acid sequences are about 34% similar to those of the HO site-specific endonuclease involved in the **cassette mechanism**.

intelligent design The creation of the universe by a rational agent (deity) rather than by random processes; a matter of faith rather than a scientific hypothesis. Creationists have attempted to force the teaching of 'intelligent design' in science lessons in schools.

interactome The complete set of molecular interactions going on within cells. Unlike the genome or, to a lesser extent the proteome, the interactome has a temporal aspect – not all interactions take place at one time or even in one cell's lifetime.

intercalated disc An electron-dense junctional complex, at the end-to-end contacts of cardiac muscle cells, that contains **gap junctions** and **desmosomes**. Most of the disc is formed of a convoluted type of **adherens junction** into which the actin filaments of the terminal sarcomeres insert (they are therefore equivalent to half Z-bands); desmosomes are also present. The lateral portion of the stepped disc contains gap junctions that couple the cells electrically and thus coordinate the contraction.

intercalation Insertion into a pre-existing structure – for example, nucleotide sequences into DNA (or RNA), molecules into structures such as membranes.

intercellular Between cells: can be used either in the sense of connections between cells (as in intercellular junctions) or as an antonym for intracellular.

intercellular adhesion molecule See **ICAM**.

interdigitating cells Cells found particularly in thymus-dependent regions of lymph nodes; they have dendritic morphology and **accessory cell** function.

interference diffraction patterns The patterns arising from the recombination of beams of light or other waves after they have been split and one set of rays have undergone a phase retardation relative to the other. Such patterns formed by simple objects give information on the correctness of the focus and the presence or absence of optical defects.

TABLE I1. Vertebrate integrins

Subunits		Ligand	Binding site	Synonyms[a]
β_1	α_1	Collagens, laminin		VLA-1
	α_2	Collagens, laminin	-DGEA-	VLA-2, platelet glycoprotein Ia/IIa
	α_3	Collagens, fibronectin, laminin	-RGD-?	VLA-, chick integrin
	α_4	Fibronectin (alternatively spliced domain), VCAM-1	-EILDV-	VLA-4
	α_5	Fibronectin (RGD)	-RGD-	VLA-5 'fibronectin receptor', platelet glycoprotein Ic/IIa
	α_6	Laminin		VLA-6, platelet glycoprotein Ic/IIa
	α_7	Laminin		
	α_8	Fibronectin		
	α_9	Tenascin C, osteopontin		
	α_V	Vitronectin, fibronectin	-RGD-	
β_2	α_L	ICAM-1, ICAM-2		LFA-1, CD11a
	α_M	C3bi, fibrinogen, Factor X, ICAM-1		Mac-1, Mo-1, CR3, CD11b
	α_X	Fibrinogen, C3bi?	-GPRP-	gp150,95, CR4, CD11c
	α_D	VCAM-1		
β_3	α_{IIb}	Fibronectin, fibrinogen, vitronectin, thrombospondin	-RGD-	Platelet glycoprotein IIb/IIIa
		von Willebrand Factor	-KQAGDV-	
	α_V	Collagen, fibronectin, fibrinogen, osteopontin, thrombospondin, vitronectin, von Willebrand Factor	-RGD-	'Vitronectin receptor'
β_4	α_6	Laminin-5 > laminin-1		
β_5	α_V	Vitronectin	-RGD-	
β_6	α_V	Fibronectin	-RGD-	
β_7	α_4	Fibronectin (alternatively spliced domain), VCAM-1	-EILDV-(in fibronectin only)	LPAM-2
	α_{IEL}	Expressed on intraepithelial lymphocytes		Does not bind MAdCAM-1, VCAM-1 or fibronectin
β_8	α_V	Vitronectin		

[a] CD numbers are only given for the β_2 family of integrins.

interference microscopy Although all image formation depends on interference, the term is generally restricted to systems in which contrast comes from the recombination of a reference beam with light that has been retarded by passing through the object. Because the phase retardation is a consequence of the difference in refractive index between specimen and medium, and because the refractive increment is almost the same for all biological molecules, it is possible to measure the amount of dry mass per unit area of the specimen by measuring the phase retardation. Quantification of the phase retardation is usually done by using a compensator to reduce the bright object to darkness (see **Senarmont** and **Ehringhaus compensators**). Two major optical systems have been used – the **Jamin–Lebedeff** system and the **Mach–Zehnder** system. These instruments are often referred to as interferometers, since they are designed for measuring phase retardation. Although their use has passed out of fashion, it may be that they will be employed more frequently in future in conjunction with image-analysing systems.

interference reflection microscopy An optical technique for detecting the topography of the side of a cell in contact with a planar substratum and for providing information on the separation of the plasmalemma from the substratum. Interference between the reflections from the substratum–medium interface and the reflections from the plasmalemma–medium interface generate the image.

interferon-stimulated response element *ISRE* A conserved **response element** present upstream of some, but not all, IFN-alpha/beta-responsive genes.

interferon-regulatory factor-1 *IRF-1* Transcription factor. Deletion of the IRF-1 gene in mice leads to severe deficiency in NK-cell function. Regulates gene for IL-15.

interferons *IFN* A family of glycoproteins produced in mammals that prevent virus multiplication in cells. IFN-α is made by leucocytes and IFN-β by fibroblasts after viral infection. IFN-γ is produced by immune cells after antigen stimulation. IFN-α and -β are also known as type

I interferons. IFN-γ, a type II interferon, is more usually classed as a cytokine.

intergenic suppression The situation where a primary gene and the gene that suppresses it do not lie in the same chromosomal locus. Compare **intragenic suppression**.

interleukin A variety of substances produced by leucocytes (though not necessarily exclusively) and that function during inflammatory responses. (This is the definition recommended by the IUIS-WHO Nomenclature Committee.) Interleukins are of the larger class of T-cell products, **lymphokines**. Now more frequently considere as **cytokines**. New interleukins are being discovered all the time, and there appears to be no end in sight. Generally speaking, the information given for the individual members relates to the human form – which is often, but not always, similar to mouse equivalent.

interleukin-1 *IL-1; IL-1α (LAF; MCF); IL-1β (IFNβ -inducing factor, OAF, catabolin)* Protein (17 kDa: 152 amino acids) secreted by **macrophages** or **accessory cells** involved in the activation of both T- and B-lymphocytes in response to antigens or mitogens, as well as affecting a wide range of other cell types. At least two IL-1 genes are active; α and β forms of IL-1 are recognized. There is an endogenous receptor antagonist, IL-1ra that binds to the receptor but does not elicit effects. IL-1α, IL-1β and IL-1ra are remarkably different in sequence though similar in binding properties. See also **catabolin**, **endogenous pyrogen**.

interleukin-1-converting enzyme *ICE; caspase-1; EC 3.4.22.36* Cytoplasmic cysteine endopeptidase (**caspase**) that is uniquely responsible for cleaving proIL-1β (31 or 33 kDa) into mature IL-1β (17.5 kDa); the active cytokine is then released by a non-standard mechanism (there is no signal sequence and it does not pass through the Golgi). The enzyme seems to be composed of two non-identical subunits derived from a single proenzyme. The ICE gene has some homology with the *ced-9* gene of *Caenorhabditis elegans*, the product of which is a **caspase** involved in mediating cell death by apoptosis.

interleukin-2 *IL-2; T-cell growth factor (TCGF); thymocyte-stimulating factor (TSF)* Cytokine (17 kDa) released by activated T-cells that causes activation, stimulates and sustains growth of other T-cells independently of the antigen. Blocking production or release of IL-2 would block the production of an immune response.

interleukin-3 *IL-3; HCGF; MCGF; multi-SCF* Product (15–17 kDa) of mitogen-activated T-cells: **colony-stimulating factor** for bone-marrow stem cells and **mast cells**. Species-specific; the human form is ineffective in mouse systems and *vice versa*.

interleukin-4 *IL-4, B-cell stimulating factor, BSF-1* Cytokine, 13–14 kDa, provides B-cell help, stimulates IgG1 and IgE production, regulates T-helper subsets. An alternatively spliced variant lacking exon 2 is an antagonist of the complete molecule. Species-specific; the human form is ineffective in mouse systems and *vice versa*.

interleukin-5 *IL-5* A B-cell growth and differentiation factor (homodimeric, 30–40 kDa); also stimulates eosinophil precursor proliferation and differentiation.

interleukin-6 *IL-6; BCSF; IFN-β2* Cytokine that is co-induced with interferon from fibroblasts, a B-cell differentiation factor, a hybridoma growth factor, an inducer of acute phase proteins and a colony-stimulating factor acting on mouse bone marrow.

interleukin-7 *IL-7; lymphopoietin 1* Single-chain cytokine (17 kDa) originally described as a pre-B-cell growth factor but now known to have effects on a range of other cells, including T-cells. Produced by monocytes, T-cells and NK-cells.

interleukin-8 *IL-8; neutrophil-activating protein, NAP-1, CXCL8* One of the first **chemokines** to be isolated; one of the C–X–C family (8 kDa). Secreted by a variety of cells, and potently chemokinetic and chemotactic for neutrophils and basophils but not monocytes. Receptor is G-protein coupled.

interleukin-9 *IL-9; mast cell growth factor (MCGF)* Cytokine (14 kDa) produced by T-cells, particularly when mitogen-stimulated, that stimulates the proliferation of erythroid precursor cells (**BFU-E**). May act synergistically with **erythropoietin**. Receptor belongs to haematopoietic receptor superfamily. Together with IL-3, promotes mast cell growth.

interleukin-10 *IL-10; CSIF; TGIF* **Cytokine**, 18 kDa, homodimeric, produced by Th2 helper T-cells, some B-cells and LPS-activated monocytes. Regulates cytokine production by a range of other cells and is an inhibitor of immune responses.

interleukin-11 *IL-11; megakaryocyte CSF* **Pleiotropic** cytokine originally isolated from primate bone marrow **stromal cell** line. Stimulates T-cell-dependent B-cell maturation, megakaryopoiesis, various stages of myeloid differentiation. Receptor shares gp130 subunit with other members of **IL-6 cytokine family**.

interleukin-12 *IL-12; NK-stimulatory factor; cytotoxic lymphocyte maturation factor* Heterodimeric cytokine (35 and 40 kDa: the heterodimer often referred to as IL-12/p70) that enhances the lytic activity of **NK**-cells, induces **interferon**-γ production, stimulates the proliferation of activated T-cells and NK-cells. Is secreted by human B-lymphoblastoid cells (NC-37). May play a role in controlling immunoglobulin isotype selection, and is known to inhibit IgE production.

interleukin-13 *IL-13; NC30; p600* Cytokine (12.4 kDa) with anti-inflammatory activity. Produced by activated T-cells; inhibits IL-6 production by monocytes and also the production of other proinflammatory cytokines such as TNFα, IL-1, IL-8. Stimulates B-cells and the production of IL-1ra. Gene is located in cluster of genes on human chromosome 5q that also has IL-4 gene.

interleukin-14 *IL-14; high molecular weight B-cell growth factor; HMW-BCGF* Cytokine (53 kDa) produced by T-cells that enhances proliferation of activated B-cells and inhibits immunoglobulin synthesis. Unrelated to other cytokines but has homology with complement factor Bb.

interleukin-15 *IL-15; IL-T* Cytokine of the IL-2 family that has effects very similar to IL-2 but in addition is potently chemotactic for lymphocytes. Levels are elevated in the rheumatoid joint. Receptor shares β and γ subunits with IL-2 receptor, but has unique α-subunit.

interleukin-16 *IL-16; lymphocyte chemoattractant factor (LCF)* Secreted from CD8+ cells and will induce migratory responses in CD4+ cells (lymphocytes, monocytes and eosinophils). May bind to CC-CKR-5 and contribute to the blocking of **HIV** internalization.

interleukin-17 *IL-17; CTLA-8* Proinflammatory T-cell product (17 kDa) that acts on receptors on a range of cells to activate NFκB. Induces expression of IL-6, IL-8 and ICAM-1 in fibroblasts and enhances T-cell proliferation stimulated by suboptimal levels of **PHA**. Receptor is a type I transmembrane protein, though a soluble form is also found, and has no homology with other known sequences.

interleukin-18 *Interferon-gamma-inducing factor; IGIF; IIF* First isolated from liver of mice during toxic shock; has sequence homology with IL-1β and IL-1ra and has also been designated IL-1γ. A proinflammatory cytokine that induces interferon-γ production and has a role in angiogenesis.

interleukin-19 Melanoma-differentiation associated protein-like protein. One of the IL-10 family of cytokines; produced by activated monocytes and B-cells; induces IL-6 and TNF production by monocytes, IL-4, IL-5, IL-10 and IL-13 production by activated T-cells; shares receptor with IL-20 and IL-24.

interleukin-20 Member of IL-10 family of cytokines; produced by monocytes and keratinocytes; an autocrine factor for keratinocyte function, differentiation and proliferation.

interleukin-21 An IL-2-like cytokine, produced by activated Th cells; signals through a specific IL-21R and the IL-2R gamma-chain. Influences the function of T-cells, NK-cells and B-cells. NB: In some early reports **IL-22** was mistakenly named IL-21.

interleukin-22 *IL-10-related T-cell-derived inducible factor; IL-TIF* A novel human cytokine, 25 kDa, distantly related to IL-10, produced by activated T-cells. IL-22 is a ligand for CRF2-4, a member of the class II cytokine receptor family that, together with a new member of the interferon receptor family (IL-22R), enables IL-22 signalling. Cell lines have been identified that respond to IL-22 by activation of STATs 1, 3 and 5, but were unresponsive to IL-10. Unlike IL-10, IL-22 does not inhibit the production of proinflammatory cytokines by monocytes in response to LPS, but it has modest inhibitory effects on IL-4 production from Th2 T-cells.

interleukin-23 *IL-23* A heterodimer of IL-12p40 and a different subunit, p19. Proinflammatory, involved in stimulating proliferation of memory T-cells and induction of IL-17. May play an important role during the early effector phase in immune-mediated demyelination of the peripheral nerve.

interleukin-24 *IL-24; ST16, melanoma differentiation-associated gene-7 (MDA-7); FISP* Cytokine (35 kDa) of the IL-10 family. Involved in megakaryocyte differentiation. Induces IL-6 and TNF production by monocytes.

interleukin-25 *IL-25; IL-17E; SF20* A T-helper-2 (Th2)-produced cytokine of the IL-17 family that has a role in allergic inflammation. Upregulates IgG and IgE production, eosinophil levels and inflammatory response. Effects are mediated through induction of IL-4, IL-5 and IL-13. A disulphide-linked homodimer with subunits of 17 kDa.

interleukin-26 *IL-26; AK155.1* Cytokine (homodimer of two 17.7-kDa monomers) of the IL-10 family that induces secretion of IL-8 and IL-10. Originally cloned from herpes virus saimiri (HVS)-transformed T-cells and named AK155.1. The IL-26 receptor complex is highly specific for IL-26, although the individual subunits of the IL-26 receptor complex are components in receptor complexes for other class II cytokines.

interleukin-27 *IL-27* A heterodimeric member of the IL-6/IL-12 family of long type I cytokines, composed of EBI3 (EBV-induced gene 3; a 34-kDa glycoprotein that is related to the p40 subunit of IL-12 and IL-23) and p28 (a 28-kDa glycoprotein related to the p35 chain of IL-12). Has both anti- and proinflammatory properties, and is expressed by monocytes, endothelial cells and dendritic cells.

interleukin-28A *Interferon λ2* Ligand for class II cytokine receptor, distantly related to members of the IL-10 family and type I IFN family. Like interferon I, has antiviral activity and upregulates MHC class I antigen expression. IL-28A, IL-28B and IL-29 all signal through the same heterodimeric receptor complex, which is composed of the IL-10 receptor β (IL-10 Rβ) and a novel IL-28 receptor β (IFN-λR). Mouse IL-28A is 20 kDa.

interleukin-28B *IL-28B; IFNλ3* See **interleukin-28A** to which it is very similar.

interleukin-29 *IL-29; IFNλ1* Similar to **interleukin-28A**.

interleukin-30 *IL-30; p28* A subunit of **interleukin-27**.

interleukin-31 *IL-31* A member of the alpha-helical family of cytokines, 24 kDa. Signals through a receptor composed of IL-31 receptor-A and **oncostatin M**-receptor constitutively expressed by keratinocytes. IL-31 is involved in allergic skin reactions and possibly pruritus (itch).

interleukin-32 *IL-32; NK4* An inflammatory cytokine that induces TNFα, IL-8 and MIP-2 production. Apparently unrelated to any other cytokines but activates typical cytokine signal pathways. At least four isoforms (αβγδ) are reported.

intermediate filaments A class of cytoplasmic filaments of animal cells so named originally because their diameter (nominally 10 nm) in muscle cells was intermediate between thick and thin filaments. Unlike microfilaments and microtubules, the protein subunits of intermediate filaments show considerable diversity and tissue specificity. See **cytokeratins**, **desmin**, **glial fibrillary acidic protein**, **neurofilament** proteins, **nestin** and **vimentin**; see also Table I2.

intermediate-density lipoproteins *IDL* Class of plasma lipoproteins formed by the degradation of very low-density lipoproteins, constituted essentially of triacylglycerols and cholesterol esters. Are approximately 25–35 nm in diameter, and are rapidly cleared by receptor-mediated endocytosis. Reportedly the lipoprotein fraction most closely associated with aortic atherosclerosis in haemodialysis patients.

intermedin *Adrenomedullin-2* Originally identified in fish, vertebrate forms have subsequently been identified. Like **adrenomedullin** a vasodilatory peptide.

intermembrane space Region between the two membranes of mitochondria and chloroplasts. On the **endosymbiont hypothesis**, this space would represent the original phagosome.

intermittent claudication The most prominent symptom of peripheral artery disease, in which inadequate blood supply to the leg muscles leads to pain when walking.

TABLE I2. Intermediate filaments and sequence-related proteins

Name of protein	Sequence homology group	M_r (kDa)	Cell type
Cytokeratins: epithelial keratins (~20), trichocytic (hair) keratins (~13)	I (acidic)	40–60	Epithelial
	II (neutral-basic)	50–70	
Vimentin	III	53	Many, especially mesenchymal
Desmin	III	52	Muscle
Glial fibrillary acidic protein	III	51	Glial cells. Astrocytes
Peripherin	III	57–58	Co-expressed with neurofilaments in periperal neurons
Neurofilament polypeptides (L, M and H)	IV	57–150	Neurons (vertebrates)
		60–120	Neurons (invertebrates)
α-Internexin		68	CNS neurons
Synemin (three alternatively spliced isoforms)		230	Smooth muscle, co-localized with desmin
Syncoilin		53	Striated and cardiac muscle; associates with α-dystrobrevin
Nuclear lamins	V	60–70	All eukaryotic cells
Nestin	VI	200	Developing rat brain

internal bias Applied to the motile behaviour of crawling cells that, in the short term, show **persistence** and do not behave as true random walkers. Any intrinsic regulation of the random motile behaviour of the cell could be considered as internal bias.

internal membranes General term for intracellular membrane systems such as endoplasmic reticulum. Not particularly helpful, but has the advantage of being non-committal.

internal transcribed spacers *ITS1; ITS2* Eukaryotic ribosomal RNA genes (rDNA) are found as parts of repeat units arranged in tandem arrays, located at nucleolar organizing regions. Each repeat unit consists of a transcribed region (having genes for 18S, 5.8S and 26S rRNAs and the external transcribed spacers, i.e. ETS1 and ETS2) and a non-transcribed spacer (NTS) region. In the transcribed region, internal transcribed spacers (ITS) are found on either side of 5.8S rRNA gene and are described as ITS1 and ITS2. The ITS region is highly conserved intraspecifically but variable between different species, so is often used in taxonomy.

internalin Surface proteins (InlA, InlB) that mediate entry of *Listeria monocytogenes* into epithelial cells that express E-cadherin or L-CAM. There appears to be an diverse internalin multigene family (Inl C, D, E, F) although not all the products are involved in bacterial entry into cells.

international normalized ratio *INR* A system established by the World Health Organization (WHO) for reporting the results of blood coagulation (clotting) tests. It is not a test but a mathematical calculation that corrects for the variability in **prothrombin time** (PT) results attributable to the variable sensitivities (**international sensitivity index**, ISI) of the **thromboplastin reagents** used by laboratories.

International Sensitivity Index *ISI* Value assigned to each batch of thromboplastin reagent after comparing each batch to a 'working reference' reagent preparation. This 'working reference' has been calibrated against internationally accepted standard reference preparations which have an ISI value of 1. The more sensitive the thromboplastin reagent the longer the resulting **prothrombin time** (PT); its ISI will be less than 1. See **international normalized ratio**.

interneurons Neurons that connect only with other neurons and not with either sensory cells or muscles. They are thus involved in the intermediate processing of signals.

internexin α-*Internexin* Neuronal intermediate filament protein (68 kDa). Subunit of type IV filaments found in neurons of CNS.

interphase The stage of the cell or nucleus when it is not in mitosis, hence comprising most of the cell cycle.

interstitial cell-stimulating hormone *ICSH* See **luteinizing hormone**.

interstitial cells 1. Cells lying between but distinct from other cells in a tissue, a good example being the interstitial cells in *Hydra* that serve as stem cells. 2. Cells lying between the testis tubules of vertebrates and that are responsible for the secretion of testosterone.

intervening sequence Alternative but uncommon name for an **intron**.

intestinal calcium-binding protein *ICaBP* Calcium-binding proteins containing the **EF-hand** motif, induced by vitamin D3.

intestinal epithelium The endodermally derived epithelium of the intestine varies considerably, but the absorptive epithelium of small intestine is usually implied. The apical surfaces of these cells have **microvilli** (which increases the absorptive surface and probably also provides a larger surface area for enzyme activity). The lateral subapical regions have well-developed intercellular **junctions**.

intima Inner layer of blood vessel wall, comprising an endothelial monolayer on the luminal face with a subcellular elastic extracellular matrix containing a few smooth muscle cells. Below the intima is the **media**, then the **adventitia**. The term may be applied to other organs.

intimin Bacterial protein (**adhesin**) located in the outer membrane that mediates adhesion between bacterium and mammalian cell. Intimin binds to the **translocated intimin receptor**, a bacterially produced protein that is transferred to the mammalian host cell. Intimins are important in pathology of attaching and effacing pathogens such as enterohaemorrhagic *E. coli* O157:H7. Intimin has four Ig-like domains and a C-terminal lectin-like domain that is similar to that in **invasin**.

intine Inner layer of the wall of a pollen grain, resembling a **primary cell wall** in structure and composition. Also the name for the inner wall layer of a **spore**.

intragenic suppression The situation where a primary gene and the mutated gene that suppresses it lie within the same locus.

intramembranous particles *IMP* Particles (or complementary pits) seen in **freeze fractured** membranes. The cleavage plane is through the centre of the bilayer and the particles are usually assumed to represent **integral membrane proteins** (or polymers of such proteins).

intrinsic factor A mucoprotein normally secreted by the epithelium of the stomach and that binds vitamin B12; the intrinsic factor/B12 complex is selectively absorbed by the distal ileum, though only the vitamin is taken into the cell.

intrinsic pathway See **extrinsic pathway**.

intron *Intervening sequence* A non-coding sequence of DNA within a gene (cf. **exon**) that is transcribed into **hnRNA** but is then removed by RNA splicing in the nucleus, leaving a mature mRNA that is then translated in the cytoplasm. Introns are poorly conserved and of variable length, but the regions at the ends are self-complementary, allowing a hairpin structure to form naturally in the hnRNA; this is the cue for removal by RNA splicing. Introns are thought to play an important role in allowing rapid evolution of proteins by **exon shuffling**. Genes may contain as many as 80 introns.

intussusception The insertion of new material into the thickness of an existing cell wall or other structure.

inulin A polysaccharide of variable molecular weight (around 5 kDa) that is a polymer of fructofuranose. Widely used as a marker of extracellular space, an indicator of blood volume in insects (by measuring the dilution of the radiolabel) and in food for diabetics.

invariant chain Polypeptide chain of invariant sequence that masks the peptide-binding groove of MHC class II molecules inside antigen-presenting cells until antigen is encountered in the lysosomal compartment.

invasins Proteins produced by bacterial cells that promote bacterial penetration into mammalian cells. The invasin produced by *Yersinia pseudotuberculosis* seems to bind to the fibronectin receptor (α5-β1 integrin) at a site close to the fibronectin-binding site, though the invasin does not have an RGD sequence.

invasion A term that should be used with caution; although most cell biologists would follow Abercrombie in meaning the movement of one cell type into a territory normally occupied by a different cell type (see **invasion index**), some pathologists might not agree.

invasion index An index devised by Abercrombie and Heaysman as a means to estimate the invasiveness of cells *in vitro*. The index is derived from measurements on confronted explants of the cells and embryonic chick heart fibroblasts growing in tissue culture: it is the ratio of the estimated movement, had the cells not been hindered, and the actual movement in the zone in which collision occurs.

inverse agonist *Reverse antagonist* Any ligand that binds to receptors and reduces the proportion in the active form. Has the opposite effects to an agonist, and may actually reduce the background level of activity. Not the same as a **partial agonist**.

inversion heterozygote Individual in which one chromosome contains an inversion whereas the homologous chromosome does not.

invertase 1. ('sucrase') Enzyme catalysing the hydrolysis of sucrose to glucose and fructose, so-called because the sugar solution changes from dextro-rotatory to laevo-rotatory during the course of the reaction. 2. Generally, a name for an enzyme that catalyses certain molecular rearrangements. DNA invertases are a class of **resolvase**.

involucrin Marker protein for **keratinocyte** differentiation first appearing in the upper spinous layer of the **epidermis**. Together with **trichohyalin** forms the scaffold for the cell envelope.

involuntary muscle A muscle that is not under conscious control – as, for example, the heart.

involution 1. Restoration of the normal size of an organ. 2. Infolding of the edges of a sheet of cells, as in some developmental processes, notably gastrulation.

ion channel A transmembrane pore that presents a hydrophilic channel for ions to cross a lipid bilayer down their electrochemical gradients. Some degree of ion specificity is usually observed, and typically a million ions per second may flow. Channels may be permanently open, like the potassium leak channel, or they may be **voltage-gated**, like the **sodium channel**, or **ligand-gated**, like the acetylcholine receptor.

ion exchange chromatography Separation of molecules by absorption and desorption from charged polymers. An important technique for protein purification. For small molecules the support is usually polystyrene, but for macromolecules, cellulose, acrylamide or agarose supports give less non-specific absorption and denaturation. Typical charged residues are CM (carboxymethyl) or DEAE (diethylaminoethyl).

ion-selective electrode An electrode half-cell, with a semipermeable membrane that is permeable only to a single ion. The electrical potential measured between this and a reference half-cell (e.g. a calomel electrode) is thus the **Nernst potential** for the ion. Given that the solution filling the ion-selective electrode is known, the activity (rather than concentration) of the ion in the unknown solution can be measured. Commercial ion-selective electrodes frequently use a hydrophobic membrane containing an **ionophore**, such as **valinomycin** (for potassium) or **monensin** (for sodium). A pH electrode is made with a thin membrane of pH-sensitive (i.e. proton permeable) glass.

ionic coupling The same as **electrical coupling**.

ionizing radiation Radiation capable of ionizing, either directly or indirectly, the substances it passes through. α and β radiation are far more effective at producing ionization (and therefore are more likely to cause tissue or cell damage) than γ radiation or neutrons.

ionomycin Antibiotic (a diacidic polyether) that acts as a potent and selective calcium ionophore; more effective than **A23187**. Ionomycin binds Ca^{2+} in the 7.0–9.5 pH range. Ionomycin induces apoptosis in immature B-cell lines, e.g. in Burkitt's lymphoma cells, and in cultured embryonic rat cortical neurons.

ionophore A molecule that allows ions to cross lipid bilayers. There are two classes: carriers and channels. Carriers, like **valinomycin**, form cage-like structures around specific ions, diffusing freely through the hydrophobic regions of the bilayer. Channels, like **gramicidin**, form continuous aqueous pores through the bilayer, allowing ions to diffuse through. See **ion channels** and Table I3.

ionotropic receptors Ligand-gated ion channels involved in fast inhibitory or excitatory neurotransmission. Examples include the **nicotinic acetylcholine receptors** of the neuromuscular junction, some **glutamate receptors** in the CNS and **GABA receptors**. Cf. **metabotropic receptors**.

iontophoresis Movement of ions as a result of an applied electric field. For example, the delivery of a charged molecule from the end of a micropipette without hydraulic flow.

TABLE I3. Ionophores

	M_r (Da)	Ion selectivity	Comments
Neutral			
CA 1001	685	Ca	Selective for Ca
Cryptate 211	288	$Li > Na > K \approx Rb \approx Cs$; $Ca > Sr \approx Ba$	Amino ether; one of a substantial family
Enniatin A	681	$K > Rb \approx Na > Cs >> Li$	Cyclic hexadepsipeptide
Enniatin B	639	$Rb > K > Cs > Na >> Li$; $Ca > Ba > Sr > Mg$	Cyclic hexadepsipeptide
Monactin	750	$NH_4 > K > Rb > Cs > Na > Ba$	Macrotetralide
Narasin	765	$K > Na$	4-methylsalinomycin
Non-actin	736	$NH_4 > K \approx Rb > Cs > Na$	Macrotetralide, product of Actinomyces strains.
Salinomycin	773	$K > Na$	Polyether antibiotic used as coccidiostat
Valinomycin	1110	$Rb > K > Cs > Ag >> NH_4 > Na > Li$	Depsipeptide, uncoupler
Carboxylic			
A23187	523	$Li > Na > K$; $Mn > Ca > Mg > Sr > Ba$	Predominantly selective for divalent cations
Ionomycin	747	$Mn > Ca > Mg >> Sr > Ba$	Diacidic polyether
Monensin	670	$Na >> K > Rb > Li > Cs$	Blocks transport through Golgi
Nigericin	724	$K > Rb > Na > Cs >> Li$	
X-537A (lasalocid)	590	$Cs > Rb \approx K > Na > Li$; $Ba > Sr > Ca > Mg$	Macrotetralide
Channel-forming			
Alamethicin	–	$K > Rb > Cs > Na$	Peptide; voltage dependent
Gramicidin A	≈ 1700	$H > Cs \approx Rb > NH_4 > K > Na > Li$	Peptide; works as a dimer
Monazomycin	≈ 1422	$Cs > Rb > K > Na > Li$	Polyene-like; voltage sensitive
Miscellaneous			
Nystatin	926	Cation selective	Activates Na/K ATPase
Palytoxin	2670	Na	Converts Na/K ATPase into channel; highly toxic

Many uncouplers such as FCCP (carbonyl cyanide-trifluoro-methoxyphenylhydrazone) also act as ionophores. Amphotericin and filipin may be anion-specific ionophores.

IP-10 *Interferon-inducible protein-10; CXCL10* T-cell chemokine, IFN-gamma-inducible protein of 10 kDa. Signals through the CXCR3 receptor and selectively chemoattracts Th1-lymphocytes and monocytes, and inhibits cytokine-stimulated hematopoietic progenitor cell proliferation. Additionally, it is angiostatic and mitogenic for vascular smooth muscle cells.

IP3 See **inositol trisphosphate**.

IPG Abbrev. for (1) **Inositol phosphoglycans**, (2) **Immobilized pH gradient**.

IPNS *Isopenicillin N synthase; EC1.21.3.1* Non-haem iron-dependent oxidase (38 kDa) that catalyses the formation of isopenicillin N from alpha-aminoadipylcysteinyl-valine (ACV).

ipomeamarone *Ipomoeamarone* One of the furanoterpenoids produced in sweet potato (*Ipomoea batatas*) that has become infected with the black rot fungus, *Ceratocystis fumbriata*. A phytoalexin, hepatotoxic.

iporin Iporin, which is similar to KIAA0375 is a rab1-interacting protein. Contains a SH3 domain and two polyproline stretches, which are known to play a role in protein/protein interactions. In addition, iporin encloses a RUN domain, which seems to be a major part of the rab1-binding domain (R1BD). Iporin is ubiquitously expressed and immunofluorescence staining displays a cytosolic punctual distribution. Also interacts with another rab1-interacting partner, the **GM130** protein.

ipratropium bromide An antimuscarinic drug used as a bronchodilator in treating chronic bronchitis.

IPSP Abbrev. for **inhibitory postsynaptic potential**.

***ipt* oncogene** Plant oncogene from *Agrobacterium tumefaciens* that encodes **isopentenyl transferase** under the control of the transcriptional regulator Ros.

IPTG *Isopropyl β-D-thiogalactoside* Used to trigger gene expression that is under the control of **gal promoter**, particularly used in expression systems for producing protein.

IQ motif Motif, a basic amphiphilic helix, usually present in one to seven copies that is a crucial determinant for **calmodulin** binding. Found, for example in the C-terminal region of voltage-gated calcium channel subunits and in myosin molecules as the binding sites for light chains, often calmodulin.

IQGAPs A conserved group of Rho family G-proteins that typically present as multiple isoforms and that have been implicated in cytokinesis. Have a **calponin homology domain** in the amino-terminal region that confers actin-binding capacity.

I-region **1.** The inducible gene region of the genome of *E. coli* involved in the **lactose operon**. **2.** Region of the murine genome coding for products involved in many aspects of the immune response (from immune region). Most of the products are class II histocompatibility antigens.

I-region-associated antigens Class II major histocompatibility (MHC) antigens.

Ir genes Immune response genes, located within the MHC of vertebrates. Originally recognized as controlling the level of immune response to various synthetic polypeptides, they are now also recognized as mapping within the regions controlling T-cell help and suppression (**I-region**).

Ir-associated antigens Antigens coded for by Ir (immune response) genes or antigens coded for by the genome close to the **Ir genes**. See also **Ia antigens**.

IRAK *IL-1 receptor-associated kinase* Part of the kinase cascade that eventually leads to NFκB translocation to the nucleus and altered gene expression. Has homology with **pelle**. IRAK associates with the IL-1R once **interleukin-1** (IL-1) binds.

IRFs Interferon-regulatory factors (IRF) 1 and 2 are DNA-binding proteins that control interferon (IFN) gene expression. IRF1 functions as an activator for IFN and IFN-inducible genes, whereas IRF2 represses the action of IRF1. Expression of the two regulatory genes is itself IFN-inducible. The IRF-3 gene encodes a 50-kDa protein that binds specifically to the IFN-stimulated response element (ISRE) but not to the IRF-1-binding site.

iridocyte *Iridoid* A reflecting cell found in some eyes and responsible for giving an iridescent appearance. See **tapetum**.

iridoviruses A group of non-occluded viruses of insects; the crystalline array of virus particles in the cytoplasm of epidermal cells gives infected insects an iridescent appearance.

Irinotecan *CPT-11* An antineoplastic agent, a derivative of **camptothecin**, that inhibits **topoisomerase I**.

IRL-2500 N-*(3,5-dimethylbenzoyl)*-N-*methyl-3-(4-phenyl)-(D)-phenylalanyl-(L)-tryptophan* **Endothelin** ET(B) receptor antagonist. A range of such antagonists have been developed, including ETA receptor antagonist FR-139317.

irs-1 *Insulin receptor substrate-1* Multisite docking protein (180 kDa) that is phosphorylated by insulin and **IGF-1** receptors following ligand binding. The tyrosine-phosphorylated form of irs-1 will interact with the SH2 domains of **p85** of PI-3-kinase, **Grb-2** and **PTP-2**.

Is element See **insertion sequence**.

ischaemia *Ischemia (US)* Inadequate blood flow leading to hypoxia in the tissue.

ISCOMS *Immunostimulatory complexe* Small cage-like structures that make it possible to present viral proteins to the immune system in an array, much as they would appear on the virus. Produced by mixing the viral protein with Quill A, a substance isolated from the Amazonian oak, in the presence of detergent. ISCOMS are being used successfully in vaccines.

islet amyloid peptide *IAPP* Peptide of 37 amino acids that selectively inhibits insulin-stimulated glucose uptake in muscle. Structurally related to **calcitonin gene-related peptide**.

islet cells Cells of the **Islets of Langerhans** within the pancreas. See **A-cells**, **B-cells**, **D-cells**.

Islets of Langerhans Groups of cells found within the pancreas: A-cells and B-cells secrete **insulin** and **glucagon**. See also **D-cells**.

iso-agglutination **1.** The agglutination of spermatozoa by the action of a substance produced by the ova of the same species. **2.** The adhesion of erythrocytes to one another within the same blood group. *cf.* **heteroagglutination**.

isoantibody Antibody made in response to antigen from another individual of the same species.

isochores Long stretches of GC- or AT-rich sequences of DNA associated with R and G chromosome bands, respectively.

isocitrate An intermediate in the **tricarboxylic acid cycle** (citric acid cycle).

isodesmosine A rare amino acid found in **elastin**, formed by condensation of four molecules of lysine into a pyridinium ring. Isodesmosine and **desmosine** are involved in the intramolecular cross-links between elastin chains.

isoelectric focusing **Electrophoresis** in a stabilized pH gradient. High-resolution method for separating molecules, especially proteins, that carry both positive and negative charges. Molecules migrate to the pH corresponding to their **isoelectric point**. The gradient is produced by electrophoresis of amphiphiles, heterogenous molecules giving a continuum of isoelectric points. Resolution is determined by the number of amphiphile species and the evenness of distribution of their isoelectric points.

isoelectric point The pH at which a protein carries no net charge. Below the isoelectric point proteins carry a net positive charge, above it a net negative charge. Due to a preponderance of weakly acid residues in almost all proteins, they are nearly all negatively charged at neutral pH. The isoelectric point is of significance in protein purification because it is the pH at which solubility is often minimal and at which mobility in an **isoelectric focusing** system is zero (and therefore the point at which the protein will accumulate).

isoenzymes Variants of enzymes that catalyse the same reaction, but owing to differences in amino acid sequence can be distinguished by techniques such as electrophoresis or isoelectric focusing. Different tissues often have different isoenzymes. The sequence differences generally confer different enzyme kinetic parameters that can sometimes be interpreted as fine-tuning to the specific requirements of the cell types in which a particular isoenzyme is found.

isoflavone A phytoestrogen found especially in soy beans and claimed to have nutritional benefits.

isoform A protein having the same function and similar (or identical) sequence, but the product of a different gene and (usually) tissue-specific. Rather stronger in implication than 'homologous'.

isogamy Situation in which the two gametes (isogametes) that fuse to form the zygote are morphologically identical.

isohaemagglutinins Natural antibodies that react against normal antigens of other members of the same species.

isokont Descriptor for flagella of equal length or identical morphology.

isoleucine *Ileu; I; 131 Da* Hydrophobic amino acid. See Table A2.

isolog *Isologue* Term that has been used to indicate similarity rather than identity between genetic sequences; unclear whether this will become standard usage. *cf.* **homologue**, paralogue, etc.

isomastigote A protozoan having two or four flagella of equal length at one pole.

isomers Alternative stereochemical forms of molecules containing the same atoms.

isometric tension Tension generated in a muscle without contraction occurring: cross-bridges are being reformed with the same site on the thin filament and the tension (in striated muscle) is proportional to the overlap between thick and thin filaments.

isoniazid *Isonicotinyl hydrazine* Bactericidal antibiotic used in combination therapy for tuberculosis, inhibits the synthesis of **mycolic acids**.

isopentenyl transferase *EC 2.5.1.-; Trans-zeatin-producing protein* Enzyme involved in biosynthesis of the plant growth hormone **cytokinin**. Overexpression of the isopentenyltransferase gene (ipt) from the Ti-plasmid of *Agrobacterium tumefaciens* increases cytokinin levels and leads to the generation of shoots from transformed plant cells. See *ipt* **oncogene**.

isopeptide bond A peptide bond other than a **eupeptide bond**, e.g. bond formed between β-carboxyl of aspartic acid or with the ε-amino group of lysine.

isoprenaline See **isoproterenol**.

isoprenoid Large family of molecules that include **carotenoids**, phytoids, prenols, steroids, **terpenoids** and tocopherols. May form only a portion of a molecule being attached to a non-isoprenoid portion. Isoprenoids are synthesized from diphosphate of isopentyl alcohol (isopentenyl diphosphate).

isoprenylation See **prenylation**.

isoprostanes Class of **prostaglandin**-like compounds that are produced by peroxidation of lipoproteins; thought to play a causative role in atherogenesis. 8-Isoprostane is considered a marker of oxidative stress.

isoproterenol *Isoprenaline: isopropyl-noradrenaline* Synthetic β-adrenergic agonist; causes peripheral vasodilation, bronchodilation and increased cardiac output.

isopycnic Having equal density: thus in equilibrium density gradient centrifugation a particle (molecule) will cease to move when it reaches a level at which it is isopycnic with the medium.

isosbestic Wavelength at which the absorption coefficients of equimolar solutions of two different substances are identical.

isosmotic Having the same osmotic pressure.

isosorbide mononitrate A **nitrate** drug used in treatment of angina and congestive heart failure.

isothiocyanate Any of a group of sulphur-containing compounds, some of which are produced by cabbages and other cruciferous vegetables, and act as herbicides or fungicides. Allyl isothiocyanate is also called mustard oil. General formula, -N=C=S.

isotonic Of a fluid, having a concentration that will not cause osmotic volume changes of cells immersed in it. Note that an isotonic solution is not necessarily isosmotic. See **hypotonic**, **hypertonic**.

isotonic contraction Contraction of a muscle, the tension remaining constant. Since the contractile force is proportional to the overlap of the filaments, and the overlap is varying, the numbers of active cross-bridges must be changing.

isotropic environments Environments in which the properties are the same at all points and there are no vectorial or axial cues.

isotype 1. Applied to a set of macromolecules sharing some features in common. In immunology isotype describes the class, subclass, light-chain type and subtype of an immunoglobulin. **2.** Antigenic determinant that is uniquely present in individuals of a single species. **3.** A conventionalized method for the graphical display of statistical data.

isotype switching The switch of immunoglobulin isotype that occurs, for example, when the immune response progresses (IgM to IgG). The switch from IgM to IgG involves only the constant region of the heavy chains (from μ to γ), the light chain and variable regions of the heavy chain remaining the same, and involves the switch regions, upstream (on the 5′ side) of the constant region genes, at which recombination occurs. Similarly, IgM and IgD with the same variable region of the heavy chain, but with different heavy chain constant regions (μ and δ), seem to coexist on the surface of some lymphocytes.

isotypic variation Variability of antigens common to all members of a species, for example the five classes of immunoglobulins found in humans. See **idiotype** and **allotype**.

isozyme See **isoenzyme**.

ISRE See **interferon-stimulated response element**.

ITAM *Immunoreceptor tyrosine-based activation motif* When phosphorylated binds **zap70** and **syk** and initiates T-cell activation. Contrast with **ITIMs**.

ITIM *Immunoreceptor tyrosine-based inhibitory motif* Phosphorylation of the ITIM motif, found in the cytoplasmic tail of some inhibitory receptors (**KIRs**) that bind MHC class I, leads to the recruitment and activation of a protein tyrosine phosphatase.

Ito cells Hepatic stellate cells that become activated in liver fibrosis due to intoxication or hepatotoxic compounds such as carbon tetrachloride. Activation is associated with expression of a sodium/calcium exchanger.

ITS See **internal transcribed spacers**.

IUPAC Abbrev. for the International Union of Pure and Applied Chemistry.

ivermectin A broad-spectrum semisynthetic anthelminthic used in the treatment of human infection with parasitic nematodes (e.g. *Strongylus* and ***Onchocerca***) and extensively in veterinary practice. Ivermectin is derived from the avermectins, a class of highly active broad-spectrum antiparasitic agents isolated from the fermentation products of *Streptomyces avermitilis*.

IVET In vivo *expression technology* A method extensively used in functional genomics. It was the first practical strategy described for selecting bacterial genes expressed preferentially during infection of an animal host; genes are detected because they are highly expressed in host tissues or cell culture infection models, but poorly expressed on laboratory media.

IVS Abbrev. for (1) interventricular septum, the wall separating the two ventricles of the heart, (2) nomenclature for description of sequence variations in intronic nucleotides; the number of the preceding exon, a plus sign and the position in the intron – e.g. 'IVS2+1G>T' denotes the G to T substitution at nucleotide +1 of intron 2.

J

J 1. The joule, SI unit of energy. **2.** Used in the single letter code for amino acids to represent trimethyl lysine, e.g. in calmodulin.

J774.2 cells Mouse (Balb/c) monocyte/macrophage cells with surface receptors for IgG and complement.

JAB *JAK-binding protein* Cytokine-inducible inhibitor of JAKs, probably a negative regulator of cytokine signalling. See **SOCS**.

Jacobson's organ See **vomeronasal organ**.

JAK *Janus kinase* Family of intracellular tyrosine kinases (120–140 kDa) that associate with cytokine receptors (particularly but not exclusively interferon receptors) and are involved in the signalling cascade. JAK is so-called either from Janus kinase (Janus was the gatekeeper of heaven) or 'just another kinase'. JAK has neither SH2 nor SH3 domains.

Jamin–Lebedeff system **Interference microscopy** in which object and reference beams are split and later recombined by birefringent calcite plates, but pass through the same optical components (in contrast to the **Mach–Zehnder system**).

janiemycin See **ramoplanin**.

jasmonate *Jasmonic acid* Any of several organic compounds that occur in plants and are thought to control processes such as growth and fruit ripening, and to aid the plant's defences against disease and insect attack.

jasmonic acid See **jasmonate**.

jasplakinolide A cyclic peptide (~710 Da) isolated from the marine sponge, *Jaspis johnstoni*. Behaves like **phalloidin** in stabilizing F-actin and binds to the same or similar site, but permeates into cells readily and so can be used to assay actin-dependent processes in living cells. Although it stabilizes actin filaments *in vitro*, *in vivo* it can disrupt actin filaments and induce polymerization of monomeric actin into amorphous masses.

jaundice Yellowing of the skin (and whites of eyes) by bilirubin, a bile pigment. Frequently because of a liver problem.

J-chain *J-piece* Polypeptide chain (15 kDa) found in IgA and in IgM that joins heavy chains (H chains) to each other to form dimers (in IgA) and pentamers (in IgM). Disulphide bonds are formed between the J-chain and H chains near the Fc ends of the heavy chains. Despite the similar name, it is not identical with the **J region** or coded for by the **J gene**.

JC virus A human **retrovirus** similar to **polyoma** virus, but which has only recently been found associated with any human cancer. Member of the **Papoviridae**. The causative agent of progressive multifocal leukoencephalopathy in immune-compromised patients

JE See **monocyte chemotactic and activating factor**.

jelly roll Complex protein topology in which four **Greek key** motifs form an eight-stranded **beta sandwich**. So called because the overall structure resembles a swiss (or jelly, USA) roll.

jervine Steroidal alkaloid produced by the skunk cabbage (*Veratrum californicum*); has similar effects to **cyclopamine**.

J gene Gene(s) coding for the Joining segment of polypeptide chain which links the V (variable regions) to the C (constant) regions of both light and heavy chains of immunoglobulins. During lymphoid development the DNA is rearranged so that the V genes are linked to the J region sequences.

JH1 domain Domain in **JAKs** that is probably the binding site for the **SH2** domain of **SOCS**.

Jijoye cells Human lymphoblastic cell line, CD23 positive. Model for B-lymphocytes. Derived from **Burkitt's lymphoma**.

jimpy Mouse mutant with reduced life span due to a recessive sex-linked defect in **PLP** gene. Has a severe CNS **myelin** deficiency associated with complex abnormalities affecting all glial populations. Jimpy-J4, the most severe of the jimpy mutants, has virtually no PLP protein. See **Pelizaeus–Merzbacher disease**.

JM109 Strain of competent *E. coli* K12 cells said to be an ideal host for many molecular biology applications.

JNK *Jun kinase; c-jun N-terminal kinase; stress-activated protein kinase; SAPK* Family of kinases involved in intracellular signalling cascades. JNKs are distantly related to **ERKs** and are activated by dual phosphorylation on tyrosine and threonine residues. In addition to **c-jun,** will also phosphorylate p53. JNK1, 46 kDa; JNK2 55 kDa. JNK3 (464 residues) is mainly found in neurons and may play a part in regulation of apoptosis.

Jnm1p See **dynamitin**.

Job's syndrome Thought to be due to a defect in neutrophil chemotaxis which predisposes to infection by staphylococci, often without the normal signs of inflammation (cold abscesses). At one time all patients described were female, with red hair and elevated plasma IgE levels, but this is no longer the case.

J region The polypeptide chains coded for by **J genes**.

jumonji domain The *jumonji* (*jmj*) gene was identified by a mouse gene trap approach and has essential roles in the development of multiple tissues. The Jmj protein has a DNA-binding domain, **ARID**, and two conserved jmj domains (jmjN and jmjC).

jumping gene Populist term for **transposon**.

jun Oncogene from an avian **sarcoma** virus. Protein product, jun, dimerizes with **fos** via a **zipper** motif to form the **transcription factor** AP1. See Table O1.

junction potential Potential difference at the boundary between dissimilar solutions; arises from differences in diffusion constants between ions.

junctional basal lamina Specialized region of the **extracellular matrix** surrounding a muscle cell, at the **neuromuscular junction**. May be responsible for localization of **acetylcholine receptors** in the synaptic region, and also binds acetylcholinesterase to this region

junctional epidermolysis bullosa *JEB* A collection of diseases characterized by blistering within the lamina lucida. Primary subtypes include a lethal subtype termed Herlitz or JEB letalis (a severe defect in laminin 5), a nonlethal subtype termed JEB mitis, and a generalized benign type termed generalized atrophic benign EB (GABEB). Mutations in genes coding for laminin 5 subunits (α3 chain, laminin β3 chain, laminin γ2 chain), collagen XVII (BP180), α6 integrin and β4 integrin have been demonstrated. See **epidermolysis bullosa**.

junctions See **adherens junction, desmosome, gap junction, zonula occludens**.

junk DNA Genomic DNA that, as yet, does not have any known function.

Jurkat cells Human T-lymphocyte line much used for studies of IL-2 production *in vitro*. Although a convenient model system they are not identical to real T-cells, particularly in their activation behaviour.

juvenile hormone *JH* A hormone found in insects which affects the balance between mature and juvenile attributes of certain tissues at each moult. In particular, the **imaginal discs** of many larval insects only develop into adult wings, sexual organs or limbs when blood juvenile hormone levels fall below a threshold level. There is a complex interaction between juvenile hormone and **ecdysone**. Synthetic analogues of JH include farnesol and methoprene, which have been tested for insecticide potential (known, with diflubenzuron, as Insect Growth Regulators, IGRs; see also **chitin**).

juxtacrine activation Activation of target cells by membrane-anchored growth factors; also used for activation of leucocytes by **PAF** bound to endothelial cell surface.

juxtaposition The state of being close to something else (juxtaposed). From Latin, *juxta*, close or near.

K

K562 cells A line of blast cells established from a female Caucasian patient with chronic myelogenous leukaemia in terminal blast crisis. They are highly undifferentiated and of the granulocytic series. Recent studies have shown the K562 blasts are multipotential, haematopoietic malignant cells that spontaneously differentiate into recognizable progenitors of the erythrocyte, granulocyte and monocytic series.

K_a 1. Acid **dissociation constant**. Often encountered as pK_a (i.e. $-\log_{10}K_a$). 2. Association constant (K_{ass}). The equilibrium constant for association, the reciprocal of K_d, with dimensions of litres/mole. Better to use K_d, thereby removing any ambiguity.

KaiABC A cluster of genes, originally identified in the cyanobacterium *Synechococcus* PCC 7942, that is essential in maintaining circadian rhythmicity in cyanobacteria. They do not share any homology with any known eukaryotic clock genes.

kainate An agonist for the K-type **excitatory amino acid** receptor. It can act as an **excitotoxin** producing symptoms similar to those of **Huntington's chorea**, and is also used as an anthelminthic drug. Originally isolated from the alga *Digenea simplex*. The receptor is an amino acid-gated **ion channel**, one of several types gated by the transmitter.

kainic acid See **kainate**.

kairomone A subclass of **pheromone**, defined as an interspecific secretion which benefits the receiver. See **allomone**.

KAL A glycosylated peripheral membrane protein with an apparent molecular weight of approximately 100 kDa. Proteolytically processed on the cell membrane to yield a 45-kDa diffusible component. See **Kallmann syndrome**.

kala-azar *Visceral leishmaniasis* Disease caused by infection with the protozoan *Leishmania donovani*, spread by the bite of infected sand flies; characterized by enlargement of the liver and spleen, anaemia, wasting and fever.

kalinin *Laminin 5; epiligrin; nicein; BM600* Protein that provides adhesion between epidermal keratinocytes and dermis. Localizes to anchoring filaments of basement membrane; 400–440 kDa with fragments of 165, 155, 130 and 105 when disulphide bonds are reduced. Forms an asymmetric 170-nm long rod with two globules at one end, one at the other.

kaliotoxin *KTX* Toxin from the scorpion *Androctonus mauretanicus m.* (peptide, 4 kDa) that blocks some potassium channels. Closely related to **charybdotoxin** and **agitoxins**.

kalirin *P-CIP10* A neuronal Rho-GEF; a cytoplasmic protein with spectrin-like and guanine exchange factor (**GEF**) domains that interacts with peptidylglycine alpha-amidating mono-oxygenase (peptide-processing enzyme). One of the **Dbl family**. May link secretory pathway to cytosolic regulatory pathways.

kallidin Decapeptide (lysyl-bradykinin; amino acid sequence KRPPGFSPFR) produced in kidney. Like bradykinin, an inflammatory mediator (a **kinin**); causes dilation of renal blood vessels and increased water excretion.

kallikrein Plasma serine endopeptidases normally present as inactive prekallikreins which are activated by **Hagemann factor**. Act on **kininogens** to produce **kinins**. Contain an **apple domain**. Three forms are recognized, plasma kallikrein (EC 3.4.21.34), tissue kallikrein (EC 3.4.21.35), and prostate-specific antigen (PSA; EC 3.4.21.77).

Kallmann syndrome A syndrome characterized by hypogonadotropic hypogonadism and anosmia, caused by a defect of migration and targeting of gonadotropin-releasing hormone-secreting neurons and olfactory axons during embryonic development. The gene responsible for the X-linked form of the disease encodes a 680-amino acid protein, **KAL**, which displays the unusual combination of a protease inhibitor domain with fibronectin type III repeats.

Kanagawa haemolysin See Table E2.

kanamycin An **aminoglycoside antibiotic** obtained from a bacterium (*Streptomyces kanamyceticus*) found in soil. Used in the treatment of Gram-negative bacterial infections.

K antigen Capsular polysaccharide antigens of Gram-negative bacteria, often used to define strain types (**serotypes**), e.g. *E. coli* K12.

Kaposi's sarcoma A sarcoma of spindle cells mixed with angiomatous tissue. Usually classed as an angioblastic tumour. A fairly frequent concomitant to **HIV** infection or long-term immunosuppression.

kappa chain *K-Light chains* See **L-chain**.

kappa particle **Gram-negative** bacterial endosymbiont of *Paramecium* spp., (*Caedobacter taeniospiralis*), that confers the 'killer' trait; infected *Paramecium* are resistant to the toxin liberated by infected forms. Killing activity is associated with the **induction** of defective phage in the endosymbiont, leading to the release of R-bodies, coded for by the phage genome and apparently of mis-assembled phage-coat protein.

kappa toxin *κ-Toxin* Exotoxin produced by *Clostridium*; a collagenase that presumably aids tissue infiltration.

kaptin *2E4* An actin-associated protein from human blood platelets also found in lamellipodia and the tips of the stereocilia of the inner ear. Binds F-actin.

Kar3 A **kinesin-14** motor of *Saccharomyces cerevisiae* required for mitosis and karyogamy. Like **NCD**, differs from kinesin in that it moves towards the minus end of the microtubule (like cytoplasmic dynein). Kar3 forms a heterodimer with either Cik1 or Vik1, both of which are non-catalytic polypeptides. Kar3Cik1 depolymerizes microtubules from the plus end and promotes minus-end-directed microtubule gliding. The interaction with Vik1 is distinct and also important in regulating interactions with microtubules.

KARP-1 *Ku86 Autoantigen-Related Protein-1* Leucine zipper protein expressed from the human Ku86 autoantigen locus that appears to play a role in mammalian DNA double-strand break repair as a regulator of the DNA-dependent protein kinase complex. KARP-1 gene expression is significantly upregulated following exposure of cells to DNA damage.

Kartagener's syndrome *Situs inversus* Condition in which the normal left/right asymmetry of the viscera is reversed. Associated with a **dynein** defect (dynein is absent or dysfunctional in some cases) and with **immotile cilia syndrome**.

karyogamy The formation of a zygote by fusion of two gametic nuclei, usually immediately after cytoplasmic fusion, although in some fungi there may be a prolonged binucleate stage (dikaryophase). See **plasmogamy**.

karyokinesis *Mitosis* Division of the nucleus, whereas cytokinesis is the division of the whole cell.

karyopherins Superfamily of proteins that are nucleo-cytoplasmic shuttling receptors (**importins** and **exportins**), which bind to transport signals on the cargoes, and by means of interactions with **nuclear pore complex** (NPC) proteins (nucleoporins), direct cargo translocation through the NPC.

karyoplast A nucleus isolated from a eukaryotic cell surrounded by a very thin layer of cytoplasm and a plasma membrane. The remainder of the cell is a cytoplast.

karyorrhexis Degeneration of the nucleus of a cell. There is contraction of the chromatin into small pieces, with obliteration of the nuclear boundary.

karyosomes Concentrated irregular clumps of chromatin within the interphase nucleus.

karyotype The complete set of chromosomes of a cell or organism. Used especially for the display prepared from photographs of mitotic chromosomes arranged in homologous pairs.

kassinin A tachykinin (**neuromedin**) dodecapeptide of amphibian origin that binds preferentially to the mammalian **tachykinin**-2 (NK-2) receptor.

katanin Microtubule-severing enzyme, similar to **spastin**. Consists of a subunit termed P60 that breaks the lattice of the microtubule and another subunit termed P80, the functions of which are not well understood.

katanosin B Katanosin B and plusbacin A(3) are naturally occurring cyclic depsipeptide antibiotics from strains of *Cytophaga* and *Pseudomonas*, respectively, containing a lactone linkage. They show strong antibacterial activity against methicillin-resistant *Staphylococcus aureus* and VanA-type vancomycin-resistant enterococci. Antibacterial activity is due to blocking of transglycosylation in bacterial cell wall peptidoglycan synthesis via a mechanism differing from that of vancomycin.

Katayama disease A disease due to infection with *Schistosoma japonicum*. Characterized by urticaria, painful enlargement of liver and spleen, bronchitis, diarrhoea, loss of appetite and fever.

Kawasaki disease Acute inflammatory disease with systemic angiitis, most commonly occurring in infants and young children. Cause uncertain.

kazal proteins Family of serine **peptidase inhibitors**. Includes seminal **acrosin** inhibitors, pancreatic secretory trypsin inhibitor (PSTI), Bdellin B-3 from leech.

KB cell A line of cells originally described as 'epidermoid carcinoma established from the mouth of a Caucasian man in 1954'; however, probably a **HeLa** subclone.

KC Mouse homologue of **melanoma growth-stimulatory activity** protein.

K$_{cat}$ *Catalytic constant* Catalytic constant of an enzyme, also referred to as the turnover number. Represents the number of reactions catalysed per unit time by each active site.

K cells See **killer cells**.

K$_d$ An equilibrium constant for dissociation. Thus, for the reaction: $A + B = C$, at equilibrium $K_d = [A][B]/[C]$. Dimension: moles per litre in this case. K_d is the reciprocal of K_a. In general the concept of K_d is more readily understood than that of K_a; for example, in considering the conversion of A to C by the binding of ligand B, the $K_d = [B]$ when $[A] = [C]$. Thus K_d is equal to the ligand concentration which produces half-maximal conversion (response).

KDEL Single letter code for the C-terminal amino acid consensus, in animals and many plants, for proteins targeted to the **endoplasmic reticulum**. Other variants in some plants and other Phyla include HDEL, DDEL, ADEL and SDEL.

KDEL receptor A Golgi/intermediate compartment-located integral membrane protein that carries out the retrieval of escaped ER proteins bearing a C-terminal **KDEL retention signal**. Expression of an activated form of src will cause the KDEL receptor to relocate from the Golgi apparatus to the endoplasmic reticulum.

KDR See **VEGF**.

kelch domain See **kelch proteins**.

kelch proteins The kelch family of proteins is defined by a 50 amino-acid repeat (kelch domain) that has been shown to associate with actin. *Drosophila* kelch has four protein domains, two of which are found in kelch-family proteins and in numerous non-kelch proteins. In *Drosophila*, kelch is required to maintain ring canal organization during oogenesis.

Kell Blood group system. The K-antigen is relatively uncommon (9%) but after the **Rhesus** antigens is the next most likely cause of haemolytic disease of the newborn.

keloid A bulging scar, the result of excess collagen production. Tendency to produce keloids seems to be heritable (particularly in Black Africans) and is associated in some cases with low plasma fibronectin levels.

Kenacid blue See **Coomassie Brilliant Blue**.

kendrin See **pericentrin**.

K$_{eq}$ The equilibrium constant for a reversible reaction. $K_{eq} = [AB]/[A][B]$.

keratan sulphate See **glycosaminoglycans**

keratinizing epithelium An epithelium such as vertebrate epidermis in which a keratin-rich layer is formed from intracellular **cytokeratins** as the outermost cells die.

keratinocyte Skin cell, of the keratinized layer of epidermis: its characteristic **intermediate filament** protein is **cytokeratin**.

keratinocyte growth factor KG A growth factor structurally related to **fibroblast growth factor**.

keratins Group of highly insoluble fibrous proteins (of high α-helical content) which are found as constituents of the outer layer of vertebrate skin and of skin-related structures such as hair, wool, hoof and horn, claws, beaks and feathers. Extracellular keratins are derived from **cytokeratins**, a large and diverse group of **intermediate filament** proteins.

keratitis Inflammation of the cornea, associated with herpes virus I infection and with congenital syphilis.

keratohyalin granules Granules found in living cells of **keratinizing epithelia** and which contribute to the **keratin** content of the dead cornified cells. Some, but not all, contain sulphur-rich keratin.

keratomileusis Surgical reshaping of the cornea to correct defective vision.

keratoses *Actinic keratoses* Benign but precancerous lesions of skin associated with ultraviolet irradiation.

kernicterus A form of brain damage in newborn children caused by excessive jaundice, usually as a complication of haemolytic disease of the newborn.. Unconjugated bilirubin crosses the blood–brain barrier and causes necrosis of neurons.

ketoacidosis Form of acidosis in which there is excess production of **ketone bodies** by the ketogenic pathway. Frequently a complication of diabetes, hence the sweet smell said to be characteristic of the breath of diabetics.

ketoconazole Powerful imidazole antifungal drug used in treating resistant candidiasis, gastrointestinal infections and infections of skin and nails.

ketogenesis Production of **ketone bodies**. Occurs in mitochondria, mostly in liver.

ketoglutarate α-*Ketoglutarate* Intermediate of the **tricarboxylic acid cycle**, also formed by deamination of **glutamate**.

ketone body Acetoacetate, β-hydroxybutyrate or acetone. None of these are bodies as defined by morphologists! Ketone bodies accumulate in the body following starvation, in diabetes mellitus, and in some disorders of carbohydrate metabolism.

ketosis Metabolic production of abnormal amounts of ketones. Often a consequence of diabetes mellitus.

kettin A large (472-kDa) actin-binding protein of the **connectin/titin** family with many immunoglobulin-like (Ig) repeats (31 in *C.elegans*), which is associated with the thin filaments in invertebrate muscles. Appears to be an important regulator of myofibrillar organization and provides mechanical stability to the myofibrils during contraction. See **projectin**.

kexins Family of subtilisin-like peptidases, of which **furin** is another example. All members are from eukaryotes although there are various kexin-like peptidases from prokaryotes.

keyhole limpet haemocyanin *KLH* A **haemocyanin** from the keyhole limpet (*Megathura crenulata*). Widely used as a carrier in the production of antibodies; it is chemically coupled to the immunogenic peptide or protein before injection.

KGF See **keratinocyte growth factor**.

Ki-67 Human Ki-67 protein is strictly associated with cell proliferation. During interphase, the antigen can be exclusively detected within the nucleus, whereas in mitosis most of the protein is relocated to the surface of the chromosomes. The fraction of Ki-67-positive tumour cells (the Ki-67 labelling index) is often correlated with the clinical course of the disease.

Ki antigen *PA28gamma* Component of the **proteosome** activator protein **PA28** or 11S regulator. Antibody against Ki antigen labels the nucleus but not the nucleoli while in the cytoplasm it labels two different classes of structures identified as microtubular-like extensions and inclusion bodies that are most likely autophagosomes. Patients with **systemic lupus erythematosus** (SLE) produce autoantibodies against a variety of nuclear antigens including Ki antigen.

KIAA genes Uncharacterized human genes isolated at the Kazusa DNA Research Institute, Japan.

KIF Abbrev. for (1) **Keratin** intermediate filaments, (2) **Kinesin superfamily proteins** (KIFs), (3) Ki antigen (KiF), (4) A *Cannabis sativa* preparation smoked in the Rif mountains of northern Morocco (kif).

killer cells *K cells* **1.** Mammalian cells which can lyse antibody-coated target cells. They have a receptor for the Fc portion of IgG, and are probably of the mononuclear phagocyte lineage, though some may be lymphocytes. Not to be confused with **cytotoxic T-cells** (CTL), which recognize targets by other means and are clearly a subset of T-lymphocytes: this confusion exists in the early literature. **2.** (NK-cells) Natural killer cells are CD3-negative large granular lymphocytes, mediating cytolytic reactions that do not require expression of Class I or II major **histocompatibility antigens** on the target cell. **3.** (LAK cells) Lymphokine-activated killer cells are NK-cells activated by **interleukin-2**.

killer plasmid These plasmids are found in some strains of *Kluyveromyces marxianus* where the cells contain multiple cytoplasmic copies of dsDNA plasmids. Such cells secrete a glycoprotein toxin. The plasmids and the killer function can be transferred to yeast.

kilobase *kb; kbp* One thousand base pairs of DNA. Strictly, should probably be kbp (kilobase pairs), but usually truncated.

kinase Widely used abbreviation for phosphokinase, an enzyme catalysing transfer of phosphate from ATP to a second substrate usually specified in less abbreviated name, e.g. creatine phosphokinase (**creatine kinase**), **protein kinase**. Serine/threonine kinases phosphorylate on serine or threonine residues, tyrosine kinases on tyrosines.

kinectin Integral membrane protein (160 kDa) of the endoplasmic reticulum and probably other membrane compartments; binds to **kinesin** and is the membrane anchor for kinesin-driven vesicle movement. Kinectin has extensive α-helical coiled-coil regions and, like the myosin tail with which it has sequence and structural similarities, may form a very long molecule, possibly 100 nm in length when fully extended.

kinesin Cytoplasmic protein (110 kDa) that is responsible for moving vesicles and particles towards the distal (plus) end of microtubules. Differs from cytoplasmic **dynein** (MAP1C) in the direction in which it moves and its relative insensitivity to vanadate. It has two heavy chains and two light chains. A large number of related gene products are believed to be motor proteins active in mitosis. See **kinesin-14 proteins**.

kinesin-14 proteins *Formerly C-terminal motor proteins* Kinesin-like motor proteins that have a C-terminal motor domain, differing from that of other kinesin proteins, and unlike conventional kinesins move towards the minus end of microtubules. There are at least four members of the group (**Ncd**, **KAR3**, CgCHO2, AtKCBP).

kinesin superfamily proteins *KIFs* Motor proteins that transport membranous organelles and macromolecules along microtubules. Their roles in transport in axons and dendrites have been studied extensively, but KIFs are also used in intracellular transport in general. See **kinesin-14 proteins**.

kinesis Alteration in the movement of a cell, without any directional bias. Thus speed may increase or decrease (**orthokinesis**), or there may be an alteration in turning behaviour (**klinokinesis**). See **chemokinesis**.

kinetin *6-Furfurylaminopurine* A **cytokinin** used as a component of plant tissue culture media. Obtained by heat-treatment of DNA, and does not occur naturally in plants.

kinetochore Multilayered structure, a pair of which develop on the mitotic chromosome, adjacent to the **centromere**, and to which spindle microtubules attach – but not at the end normally associated with a **microtubule-organizing centre**.

kinetodesma Longitudinally oriented cytoplasmic fibrils associated with, and always on the right of, the **kinetosomes** of ciliates.

kinetoplast Mass of mitochondrial DNA, usually adjacent to the flagellar **basal body**, in flagellate protozoa.

kinetosome **Basal body** of cilium: used mostly of ciliates.

kinety A row of **kinetosomes** and associated **kinetodesmata** in a ciliate protozoan.

kingdom In modern taxonomy, any of the six major groupings used in the classification of living organisms: Monera, Archaea, Protista, Plantae, Animalia, Fungi. Many alternative schemata have been proposed, and Monera are often considered to include Archaea, making only five kingdoms.

kininogen Inactive precursor in plasma from which **kinin** is produced by proteolytic cleavage.

kinins Inflammatory mediators that cause dilation of blood vessels and altered vascular permeability. Kinins are small peptides produced from **kininogen** by **kallikrein**, and are broken down by kininases. Act on phospholipase and increase arachidonic acid release and thus prostaglandin (PGE2) production. See **bradykinin**, **kallidin**, **C2-kinin**.

KIR *Killer cell inhibitory receptor; killer cell immunoglobulin-like receptor* Set of receptors on killer cells. Killer cell immunoglobulin-like receptors are distinct from the CD94/NKG2 KIR (CD94 and NKG2 are C-type lectins) that is HLA-E specific. KIR3D binds to HLA-B, KIR2D to HLA-C. Together these receptors confer tolerance to self by an active signalling in the killer cell.

Kirby–Bauer test Agar disc-diffusion test used to test antibiotics. Bacteria under test are distributed throughout the agar and discs of paper impregnated with different antibiotics are placed on the surface: the extent of the area of growth inhibition around each test disc is a measure of the susceptibility of the organism to the antibiotic.

Kirsten sarcoma virus A murine sarcoma-inducing retrovirus, generated by passaging a murine erythoblastosis virus in newborn rats. Source of the *Ki-ras* oncogene.

KiSS1 A metastasis-suppressor gene that inhibits metastasis of human melanomas and breast carcinomas. The human *KiSS1* gene encodes a C-terminally amidated active peptide (145 residues, **kisspeptin**) that is the ligand of a G-protein-coupled receptor.

kisspeptins Peptides (54 amino acids), derived from the *KiSS* gene, which signal through the G protein-coupled receptor 54 (GPR54), and have recently been shown to be critical regulators of reproduction via the hypothalamic **gonadotrophin-releasing hormone** (GnRH) system. Mutations of the GPR54 gene are linked to absence of puberty onset and hypogonadotrophic hypogonadism in humans. *KiSS-1* was originally identified as a metastasis-suppressor gene.

kistrin Naturally occurring inhibitor (68 residue peptide) of platelet aggregation found in the venom of Malayan pit viper *Agkistrodon rhodostoma*. Kistrin has an RGD site that competes for the platelet IIb/IIIa **integrin** and is therefore one of the **disintegrins**.

kit *Mast cell growth factor* An **oncogene**, identified in feline **sarcoma**, encoding a tyrosine **protein kinase** that is the receptor for **stem cell factor**. See **W locus** and Table O1.

Klebs–Löffler bacillus A rod-shaped Gram-positive bacterium (*Corynebacterium diphtheriae*) that is responsible for diphtheria in humans and similar diseases in other animals.

Klebsiella Genus of Gram-negative bacteria, non-motile and rod-like, associated with respiratory, intestinal and urinogenital tracts of mammals. *K. pneumoniae* is associated with pneumonia in humans.

Klein–Waardenburg syndrome See **Waardenburg's syndrome**.

Klenow fragment Larger part of the bacterial DNA polymerase I (76 kDa) that remains after treatment with **subtilisin**; retains some but not all exonuclease and polymerase activity.

Klinefelter's syndrome Human genetic abnormality in which the individual, phenotypically apparently male, has three sex chromosomes (XXY).

klinokinesis Kinesis in which the frequency or magnitude of turning behaviour is altered. Bacterial chemotaxis can be considered as an adaptive klinokinesis; the probability of turning is a function of the change in concentration of the substance eliciting the response.

K_m *Michaelis constant* A kinetic parameter used to characterize an enzyme; defined as the concentration of substrate that permits half-maximal rate of reaction. An analogous constant K_a is used to describe binding reactions, in which case it is the concentration at which half the receptors are occupied.

km-fibres Bundles of microtubules running longitudinally below and to one side of the bases of cilia in a **kinety**.

knirps *Kni* A *Drosophila* *gap* **gene**, asymmetric distribution of which is essential for normal expression of striped patterns of **pair-rule genes** and thus abdominal segmentation. Encodes a steroid/thyroid orphan receptor-type **transcription factor**.

knock-in A transgenic animal to which a new gene has been added (and often another similar gene deleted) rather than eliminated (knocked out).

knockout *Gene knockout* Informal term for the generation of a mutant organism in which the function of a particular gene has been completely eliminated (a null allele). See also **homologous recombination, transposon**.

knockout mouse Transgenic mouse in which a particular gene has been deleted. Often shows disappointingly little phenotypic change – usually because there are alternative mechanisms or because the right challenge is not being made (some genes are probably unnecessary for the survival of a well-fed laboratory mouse in very well-regulated surroundings).

Koch's postulates The criteria, first advanced by Robert Koch in 1890, by which the causative agent of a disease can be unambiguously identified. For an organism to be accepted as the causative agent, (1) it must be present in all cases, (2) it must be isolatable in pure culture, (3) inoculation with the pure isolated organism should cause the disease, and (4) the organism should be observable in the experimentally infected host.

Kohler illumination The recommended type of optical microscope illumination in which the image of the lamp filament is focused in the lower focal plane of the sub-stage condenser. As opposed to collimated illumination, in which the light-emitting surface is imaged in the object. Collimated illumination requires even intensity across the light-emitting surface, but is preferable for certain types of microscopy. Kohler illumination gives even illumination on the object even if there are irregularities in the brightness of the light-emitting surface.

koilocytes Large cells with clear cytoplasm and pyknotic nuclei with inconspicuous nucleoli. Koilocytosis is induced by human papilloma virus infection of the superficial epithelial cells of the uterine cervix.

koilocytosis Histological appearance of cells that cells have halos around the nuclei, often caused by papilloma virus infection. See **koilocytes**. *Adj.* koilocytotic.

Kostmann's syndrome Autosomal recessive disease characterized by profound **neutropenia**. It appears that bone marrow precursor cells fail to respond to the endogenous (normal) levels of functional G-CSF, though they will respond to pharmacologic doses of G-CSF and the G-CSF receptors seem normal. Defect may be in intracellular signalling downstream of the receptor. Autosomal dominant or sporadic congenital neutropenia is associated with mutation in the neutrophil elastase gene, but this is not the case with the autosomal recessive form (Kostmann's syndrome).

Kozak consensus Consensus for **translational** start site of an **mRNA**. Although the trinucleotide ATG (coding for methionine) is generally considered to be the start site, statistical analysis of a large number of mRNAs revealed several conserved residues around this sequence. In eukaryotes, RNNMTGG; in prokaryotes, MAYCATG.

KRAB *Kruppel associated box* Subset of **zinc finger**-type transcription factors.

k-ras Kirsten-*ras*; see also **kirtsen sarcoma virus**.

Krebs cycle Tricarboxylic acid cycle or citric acid cycle.

kringle Triple-looped, disulphide-linked protein domains, found in some serine peptidases and other plasma proteins, including plasminogen (five copies), tissue-plasminogen activator (two copies), thrombin (two copies), hepatocyte growth factor (four copies), apolipoprotein A (38 copies). Resemble the eponymous Scandinavian pastry.

Kruppel *Kr Gap* **gene** of *Drosophila*, encoding a zinc-finger **transcription factor**.

Kruppel-like factors *KLF* A subfamily of transcription factors characterized by the presence of a conserved DNA-binding domain comprising three **kruppel**-like zinc fingers. Various tissue-specific forms have been identified (**erythroid Kruppel-like factor** (KLF1), gut-enriched Kruppel-like factor (GKLF)). KLF2 regulates T-cell quiescence and survival and is essential for T-cell trafficking; KLF8 mediates cell-cycle progression downstream of **focal adhesion kinase** (FAK) by upregulating **cyclin** D1.

KTX See **kaliotoxin**.

Ku A heterodimeric protein comprised of 70- and 80-kDa subunits which participates in the non-homologous end joining (**NHEJ**) repair pathway for rejoining DNA double-strand breaks.

Kunitz domain Domain that characterizes a family of serine protease inhibitors that inhibit peptidases of the S1 family, mostly confined to metazoa. The type example for this family is **aprotinin** (bovine pancreatic trypsin inhibitor); other examples include trypstatin, a rat mast cell inhibitor of trypsin, and tissue factor pathway inhibitor precursor.

Kuppfer cell Specialized macrophage of the liver sinusoids; responsible for the removal of particulate matter from the circulating blood (particularly old erythrocytes).

Kurloff cells Cells found in the blood and organs of guinea pigs that contain large secretory granules but are of unknown function.

kuru Degenerative disease of the central nervous system found in members of the Fore tribe of New Guinea: a **spongiform encephalopathy**.

kwashiokor Form of severe malnutrition of children in the tropics. Generally considered to be due to protein deficiency, though it could be due to deficiency in a single essential amino acid. Contrast with **marasmus**.

kymograph A recording instrument consisting of a rotating drum holding paper on which a stylus traces a continuous record of some variable output: largely of historical interest.

kynurenine pathway The major route of L-tryptophan catabolism, activated by IFNγ and IFNα, resulting in the production of nicotinamide adenine dinucleotide and other neuroactive intermediates, in particular **quinolinic acid**.

kyotorphin An endogenous analgesic opioid dipeptide, (L-Tyr-L-Arg).

kyphosis A deformity of the spine that results in an exaggerated 'round-back'.

Kyte–Doolittle An algorithm for calculating a **hydropathy plot**.

L

L1 *NgCAM* Neural adhesion molecule with six immunoglobulin-type C2 domains and fibronectin type III repeats, making it another member of the **immunoglobulin superfamily** with binding domains similar to **fibronectin**. The purified molecule, immobilized on a culture dish, is a potent substrate for neurite outgrowth. See also **neuroglian** and **NCAM**.

L2 Abbrev. for (1) second lumbar vertebra, (2) inner lipoyl domain (L2) of the dihydrolipoyl acetyltransferase, (3) rat lung epithelial L2 cells, (4) reovirus lambda2 core spike (L2) gene, (5) *Toxocara canis* L2 excretory/secretory antigen.

L32 Ribosomal protein that forms part of the 50S ribosomal subunit, found in both prokaryotes and eukaryotes. Ribosomal protein L32 of yeast binds to and regulates the splicing and the translation of the transcript of its own gene.

L929 Line of fibroblasts originally cloned from areolar and adipose tissue of an adult C3H mouse.

lac operon See **lactose operon**.

laccase *EC.1.10.3.2; urishiol oxidase* A group of multicopper proteins of low specificity. Act on both O- and P-quinols, and often also on aminophenols and phenylenediamine. First found in the lac (lacquer) tree but now known to be widespread.

lactacystin A microbial metabolite isolated from *Streptomyces* that is widely used as a selective and irreversible inhibitor of the 20S **proteosome**.

lactadherin *MFG-E8; milk-fat globule E8; BA46* Mucinassociated glycoprotein (46 kDa) found in milk; a phosphatidylserine-binding glycoprotein secreted by macrophages. Has been used as a tumour marker because it is expressed in most breast cancer cells. Has protective effects for suckling young because it binds to rotavirus and inhibits replication.

lactalbumin Milk protein fraction containing β-lactoglobulin and α-lactalbumin. α-lactalbumin is the regulatory subunit of lactose synthetase: thought to be related to lysozyme C.

lactate *2-Hydroxypropionic acid* Important as the terminal product of anaerobic glycolysis. Accumulation of lactate in tissues is responsible for the so-called oxygen debt.

lactate dehydrogenase *LDH; EC 1.1.2.3* The enzyme that catalyses the formation and removal of lactate according to the equation: $pyruvate + NADH = lactate + NAD$. The appearance of LDH in the medium is often used as an indication of cell death and the release of cytoplasmic constituents.

Lactobacillus Genus of Gram-positive anaerobic or facultatively aerobic bacilli; the product of their glucose fermentation is lactate. Important in production of cheese, yoghurt, sauerkraut and silage.

lactoferrin Iron-binding protein of very high affinity (K_d 10^{-19} at pH 6.4, 26-fold greater than that of **transferrin**), found in milk and in the specific granules of neutrophil leucocytes.

lactoglobulin A globular protein (18.4 kDa) present in most milk (except human); constitutes 50–60% of bovine whey protein.

lactoperoxidase *EC 1.11.1.7* Peroxidase enzyme from milk that finds an important use in generating active iodine as a non-permeant radiolabel for membrane proteins.

lactose *4-O-β-D galactopyranosyl-β-D glucose* The major sugar in human and bovine milk. Conversion of lactose to lactic acid by *Lactobacilli*, etc. is important in the production of yoghurt and cheese.

lactose carrier protein The best-known example is the product of the *lacY* gene, coded for in the **lactose operon** and responsible for the uptake of lactose by *E. coli*.

lactose operon Group of adjacent and coordinately controlled genes concerned with the metabolism of lactose in *E. coli*. The lac operon was the first example of a group of genes under the control of an **operator** region to which a **lactose repressor** binds. When the bacteria are transferred to lactose-containing medium, allolactose (which forms by transglycosylation when lactose is present in the cell) binds to the repressor, inhibits the binding of the repressor to the operator, and allows transcription of mRNA for enzymes involved in galactose metabolism and transport across the membrane (beta-galactosidase, galactoside permease, and thiogalactoside transacetylase). *LacZ* codes for beta-galactosidase, *lacY* for the permease, *lacA* for the transacetylase.

lactose repressor Protein (tetramer of 37-kDa subunits) that normally binds with very high affinity to the **operator** region of the **lactose operon** and inhibits transcription of the downstream genes by blocking access of the polymerase to the promoter region. When the lactose repressor binds allolactose, its binding to the operator is reduced and the gene set is derepressed.

lacuna *Pl.* lacunae Small cavity or depression, for example, the space in bone where an **osteoblast** is found.

LacZ *E. coli* gene encoding **beta-galactosidase**. Part of the **lac operon**.

LAD syndrome *Leucocyte adhesion deficiency syndrome* See **LFA-1**.

laddering Apoptotic cells show a regular pattern of oligonucleotide sizes on electrophoretic gels; the ladder-like arrangement is a consequence of the cleavage of the DNA strand between **nucleosome** beads by **endonucleases** as part of the process by which cell death occurs. See **apoptosis**.

laidlomycin A polyether ionophore, much more effective for monovalent cations than divalents. Used in bovine feed as an antibiotic.

LAL test See *Limulus polyphemus*.

LAMB2 *Laminin beta2, formerly s-laminin* An important protein constituent (1797 aa) of certain kidney and muscle basement membranes. Defects in this gene are responsible for **Pierson syndrome**, an autosomal recessive disorder, usually fatal in early infancy.

lambda bacteriophage *λ-phage* Bacterial DNA **virus**, first isolated from *E. coli*. Its structure is similar to that of the **T-even phages**. It shows a **lytic** cycle and a **lysogenic** cycle, and studies on the control of these alternative cycles have been very important for our understanding of the regulation of gene **transcription**. It is used as a cloning vector, accommodating fragments of DNA up to 15 kbp long. For larger pieces, the **cosmid** vector was constructed from its ends.

lambda chain See **L-chain**.

lamellar phase See **phospholipid bilayer**.

lamellipodium Flattened projection from the surface of a cell, often associated with locomotion of fibroblasts.

lamina Flat sheet; as in **basal lamina**.

lamina propria Fibrous layer of connective tissue underlying the basal lamina (**basement membrane**) of an epithelium. May contain smooth muscle cells and lymphoid tissue in addition to fibroblasts and extracellular matrix.

laminarin Storage polysaccharide of Laminaria and other brown algae; made up of β (1–3)-glucan with some β (1–6) linkages.

laminin Link proteins of basal lamina; consist of an A chain (400 kDa) and two B chains (200 kDa). Each subunit contains at least 12 repeats of the **EGF-like domain**. The first laminin studied was from mouse **EHS cells**, but it is now clear that different forms of laminin occur. In laminin from placenta the A chain is replaced with merosin, in laminin found near the neuromuscular junction the B1 chain is replaced by laminin B2 (**LAMB2**, formerly s-laminin (synapse laminin)). Laminin induces adhesion and spreading of many cell types and promotes the outgrowth of neurites in culture. The three chains A, B1 and B2 have been renamed alpha, beta and gamma, respectively.

lamins Proteins that form the nuclear lamina, a polymeric structure intercalated between chromatin and the inner nuclear envelope. Lamins A and C (70 and 60 kDa, respectively) have C-terminal sequences homologous to the head and tail domains of **keratins**; their peptide maps are similar, and significantly different from that of lamin B (67 kDa), although there are some common epitopes.

lamotrigine An anticonvulsant drug, a phenyltriazine, used to treat partial seizures. Thought to act at voltage-sensitive sodium channels to stabilize neuronal membranes and inhibit the release of excitatory amino acid neurotransmitters (e.g. glutamate, aspartate) that may play a role in the generation and spread of epileptic seizures.

LAMP *Limbic system-associated membrane protein* 1. An adhesion molecule (**IgLON** family member) involved in specifying regional identity during development, enriched in the neuropil of limbic brain regions in mammals. Can promote outgrowth of limbic axons. An immunoglobulin (Ig) superfamily member with three Ig domains and a glycosylphosphatidylinositol anchor. **2.** See **LAMP-1**, a lysosome-associated membrane protein.

LAMP-1 *Lysosomal-associated membrane protein 1* Heavily glycosylated protein of lysosomal and plasma membrane. Depending on extent of glycosylation may have molecular weight between 90 and 140 kDa; function may be to protect membrane from attack by lysosomal enzymes. Also known as LEP100 (lysosomes, endosomes and plasma membrane, 100 kDa) and LGP120.

LAMP-2 *Lysosomal-associated membrane protein 2* Highly glycosylated transmembrane protein (410 residues) of lysosomal membrane with a large intraluminal head region with two internally homologous domains. Defects in LAMP-2 cause **Danon disease** but the function of the protein is still unclear though it may regulate fusion of lysosomes with primary autophagocytic vacuoles.

lampbrush chromosomes Large chromosomes (as long as 1 μm), actually meiotic **bivalents**, seen during prophase of the extended meiosis in the oocytes of some Amphibia. Segments of DNA form loops in pairs along the sides of the sister chromosomes, giving them a brush-like appearance. These loops are not permanent structures but are formed by the unwinding of **chromomeres** and represent sites of very active RNA synthesis.

Landry–Guillain–Barré syndrome See **Guillain–Barré syndrome**.

Langendorff perfused heart Classic pharmacological organ preparation in which a rodent heart is maintained *in vitro* by perfusion of the aorta with oxygenated fluid so that the fluid passes into the coronary arteries. Can be used to study metabolism of cardiac muscle.

Langerhans See **Islets of Langerhans** and **Langerhans cells**.

Langerhans cell histiocytosis *Histiocytosis X* A set of clinical syndromes including, in decreasing severity: Letterer–Siwe disease, **Hand–Schuller–Christian disease,** and eosinophilic granuloma of bone. The classification and causes of these diseases are confused at present.

Langerhans cells Cells of dendritic appearance, strongly MHC Class II positive and weakly phagocytic, found in the basal layers of the epidermis where they serve as **accessory cells**, responsible for **antigen processing** and **antigen presentation**. Having been exposed to antigen they migrate to the lymph nodes. Are derived from bone marrow and are immature **dendritic cells**. Part of the immune surveillance system, their location means that they are readily exposed to antigens that penetrate the dermal barrier.

Langhans' giant cells Multinucleate cells formed by fusion of epithelioid macrophages and associated with the central part of early tubercular lesions. Similar to **foreign body giant cells**, but with the nuclei peripherally located.

Langmuir trough A device for studying the properties of lipid monolayers at an air/water interface. A moveable barrier connected to a balance allows measurement of surface pressure.

Langmuir–Blodgett film In biophysics, an ordered monolayer of molecules produced on the surface of water. An **amphipathic** molecule is floated at low concentration on the surface of the water and steadily compressed into an ordered surface by moving a barrier across the surface (usually done in a **Langmuir trough**).

lansoprazole Proton-pump inhibitor drug used in the treatment of gastric ulcers.

lanthanum *La* Lanthanum salts are used as a negative stain in electron microscopy, and as calcium-channel blockers.

lanthionine A non-protein amino acid composed of two alanine residues cross-linked on their carbon atoms by a thioether linkage. Found in human hair, lactalbumin and feathers, also in bacterial cell walls. Are a component of gene encoded peptide antibiotics, **lantibiotics**.

lanthiopeptin See **ancovenin**.

lantibiotics Lanthionine-containing peptide antibiotics. The prototype lantibiotic **nisin** inhibits peptidoglycan synthesis and forms pores through specific interaction with the cell wall precursor lipid II. The flexible amphiphilic type A lantibiotics (e.g. nisin and epidermin) act primarily by pore formation in the bacterial membrane. The rather rigid and globular type-B lantibiotics inhibit enzyme functions through interaction with the respective substrates: **mersacidin** and actagardine inhibit the cell wall biosynthesis by complexing lipid II, whereas the cinnamycin-like peptides inhibit phospholipases by binding phosphoethanolamine. Other examples are **ancovenin**, **subtilin**.

LAPF *Lysosome-associated apoptosis-inducing protein* One of the **phafins**.

La protein Protein (45 kDa) transiently bound to unprocessed cellular precursor RNAs that have been produced by polymerase III; stabilizes nascent pre-tRNAs from nuclease degradation, influences the pathway of pre-tRNA maturation, and assists correct folding of certain pre-tRNAs. Mainly located in the nucleus. An autoantigen in patients suffering from **Sjogren's syndrome**, **systemic lupus erythematosus** and neonatal lupus.

LA-PF4 See **connective tissue-activating peptide III**.

LAR *Leucocyte antigen-related protein* LAR is the prototype for a family of transmembrane protein tyrosine phosphatases with extracellular domains composed of Ig and fibronectin type III (FnIII) domains and two cytoplasmic catalytic domains, one active, one inactive. LAR-family phosphatases (LAR, PTPdelta, PTPsigma) play a role in axon guidance, mammary gland development, regulation of insulin action and glucose homeostasis. See **liprins**.

LARGE The *Large* gene glycosyltransferase is required for normal glycosylation of **dystroglycan**, and defects in either Large or dystroglycan cause abnormal neuronal migration. There are mutations in Large in both the myodystrophy mouse and congenital muscular dystrophy type 1D (MDC1D).

large T-antigen See **T-antigen**.

large-cell lymphoma Highly malignant group of tumours arising from transformed **lymphocytes** or **myeloid** precursors. Cell of origin often obscure.

Lariam Proprietary name for mefloquine, an antimalarial drug related to quinine; appears to interfere with the transport of haemoglobin and other substances from the erythrocytes to the food vacuoles of the malaria parasite.

Laron dwarfism Human growth defect in which cells do not respond to growth hormone; the defect is due to mutation in the growth hormone receptor gene.

Las17p *S. cerevisiae* homologue of human Wiskott–Aldrich syndrome protein (**WASP**); interacts with the **Arp2/3** complex.

LASEK Laser-assisted epithelial keratomileusis (or keratectomy), the surgical reshaping of the surface of the cornea using a laser.

LASIK Laser-assisted *in situ* keratomileusis, surgical reshaping of tissue immediately below the corneal surface with the aim of improving vision.

Lassa fever See **Lassa virus**.

Lassa virus Virulent and highly transmissible member of the **Arenaviridae** whose normal host is a rodent (*Mastomys natalensis*); first recorded from Nigeria.

LAT adapter *Linker for activation of T-cells* A transmembrane adapter (36 and 38 kDa isoforms) essential for the transmission of T-cell receptor (TCR)-mediated signalling. The intracytoplasmic domain contains nine tyrosines which, when phosphorylated by ZAP-70 upon receptor aggregation, recruit SH2 domain-containing cytosolic enzymes and adapters. LAT is tyrosine phosphorylated in human platelets in response to collagen, collagen-related peptide (CRP) and FcγRII-A cross-linking.

latanoprost Prostaglandin analogue used to reduce intraocular pressure caused by glaucoma. Acts by increasing efflux of aqueous fluid.

late gene Gene expressed relatively late after infection of a host cell by a virus, usually structural proteins for the viral coat.

latency 1. In electrophysiology, the time between onset of a stimulus and peak of the ensuing **action potential**. 2. Of an infection, a period in which the infection is present in the host without producing overt symptoms.

latent membrane protein *LMP* Epstein–Barr virus (EBV)-encoded latent membrane protein (LMP-1) is a potential target for immunotherapy of some proportion of Hodgkin's disease cases, nasopharyngeal carcinomas, EBV-associated natural killer (NK)/T-lymphomas and chronic active EBV infection (CAEBV). LMP-1 upregulates anti-apoptotic genes, including *bcl-2*.

latent virus Virus integrated within host genome but inactive: may be reactivated by stress such as ultraviolet irradiation.

lateral diffusion Diffusion in two dimensions, usually referring to movement in the plane of the membrane, such as the motion of fluorescently labelled lipids or proteins measured by the technique of **fluorescent recovery after photobleaching** (FRAP).

lateral inhibition A simple form of information processing. The classic example is found in the eye, whereby ganglion cells are stimulated if photoreceptors in a well-defined field are illuminated, but their response is inhibited if neighbouring photoreceptors are excited (an 'on field/off surround' cell), or *vice versa* an 'off field/on surround' cell. The effect of lateral inhibition is to produce edge- or boundary-sensitive cells, and to reduce the amount of information that is sent to higher centres; a form of peripheral processing.

latex Milky fluid that exudes from cells and vessels (laticifers) when many plants are cut. It is a watery solution containing many different substances, including terpenoids (which form rubber), alkaloids (e.g. opium alkaloids), sugar, starch, etc.

lathyrism Disorder of collagen cross-linking as a result of copper sequestration by nitriles. (Lysyl oxidase is a copper-containing metalloenzyme.) In animals, caused by eating chickling peas (*Lathyrus sativus*) the toxic component of which is the neurotoxic amino acid beta-*N*-oxalylamino-L-alanine (BOAA).

Laticauda semifasciata Sea snake. See **erabutotoxins**.

Latrodectus Genus of spiders, Black Widows. See **alpha-latrotoxin**.

latroinsectotoxin α-*Latroinsectotoxin* See **latrotoxin**.

latrotoxin α-*Latrotoxin;* α-*LTx* Major toxin from *Latrodectus* spp. (1401 residues). Causes release of neurotransmitters from all synapses. An insect-specific toxin, α-latroinsectotoxin, is also present in the venom and has substantial homology and similar mode of action to α-latrotoxin.

latrunculin A macrolide inhibitor of actin polymerization (binds G-actin rendering it assembly incompetent), now often used in preference to cytochalasin because of greater potency. Latrunculins A and B were isolated from the Red Sea sponge *Negombata magnifica* (Demospongiae, Latrunculiidae), although chemical syntheses have been developed. The compound is found within vesicles in archaeocytes and choanocytes and may be for defensive purposes.

laulimalide A natural product from marine sponges; a microtubule-stabilizing agent that binds to tubulin at a site distinct from that of the **taxanes**.

lavage Washing out of a cavity (e.g. peritoneal cavity) in order to remove loosely adherent cells.

lazy leucocyte syndrome A rare human complaint in which neutrophils display poor locomotion towards sites of infection. Thought to be due to a defect in the cytoplasmic actomyosin system leading to impaired movement and reduced deformability leading to difficulty in release from the bone marrow and emigration into tissues, but sufficiently rare that this has not been explored in detail.

LB medium A growth medium for *E. coli*, typically contains 10 g of tryptone and 5 g of yeast extract per litre. LB medium was originally developed by Bertani to optimize *Shigella* growth and plaque formation. The abbreviation, according to Bertani, was intended to stand for lysogeny broth.

Lbc Oncoprotein of Dbl-like family. See **Lfc**.

LBP *Lipopolysaccharide-binding protein* Serum protein that binds lipopolysaccharide; levels rise in patients with severe Gram-negative sepsis and it may function as a recognition molecule in conjunction with CD14.

LC$_{50}$ Concentration of substance that is lethal to 50% of individuals tested.

LCAT *EC 2.3.1.43; lecithin:cholesterol acyltransferase* An enzyme bound to high-density **lipoproteins** (HDLs) and low-density lipoproteins in the plasma that catalyses the formation of cholesterol esters in lipoproteins. There are two autosomal recessive disorders caused by mutations of the *LCAT* gene; in familial LCAT deficiency (complete LCAT deficiency) all activity is lost, in fish eye disease (partial LCAT deficiency) there is a defect in LCAT activity towards HDL only.

L cells Cell line established by Earle in 1940 from mouse connective tissue. L929 cells are a subclone of this original line.

L-chain *Light chain* Although **light chains** are found in many multimeric proteins, L-chain usually refers to the light chains of immunoglobulins. These are of 22 kDa and of one of two types, kappa (κ) or lambda (λ). A single immunoglobulin has identical light chains (2κ or 2λ). Light chains have one variable and one constant region. There are **isotype** variants of both κ and λ.

LD$_{50}$ That dose of a compound which causes death in 50% of the organisms to which it has been administered. Routine use of LD$_{50}$ tests is now being replaced by more sensitive (and less wasteful) methods.

LDH See **lactate dehydrogenase**.

LDL See **low-density lipoprotein**.

LDL receptor-related protein *LRP* Low-density lipoprotein receptor-related protein, one of the three main **apolipoprotein** E-recognizing endocytic receptors involved in the clearance of triglyceride-rich lipoproteins from plasma. Also affects the processing of **amyloid precursor protein** (APP) and amyloid beta (Aβ) protein production as well as mediates the clearance of Aβ from the brain.

LE body A globular mass of nuclear material that stains with haematoxylin; associated with lesions of **systemic lupus erythematosus**.

LE cell Phagocyte that has ingested nuclear material of another cell: characteristic of **systemic lupus erythematosus**.

leader peptide See **leader sequence**.

leader sequence In the regulation of gene expression for enzymes concerned with amino acid synthesis in prokaryotes, the leader sequence codes for the leader peptide that contains several residues of the amino acid being regulated. Transcription is closely linked to translation, and if translation is retarded by limited supply of aminoacyl tRNA for the specific amino acid, the mode of transcription of the leader sequence permits full transcription of the operon genes; otherwise complete transcription of the leader sequence prematurely terminates transcription of the **operon**.

leading lamella Anterior region of a crawling cell, such as a fibroblast, from which most cytoplasmic granules are excluded.

leaf spot Any of several plant diseases characterized by the appearance of dark spots on the leaves.

leaky mutation Mutation in which sub-normal function exists, for example if a mutation leads to instability in a protein rather than its complete absence, or there is reduced expression of a gene.

Leber's disease *Hereditary optic atrophy; hereditary optic neuritis* A rare inherited disorder of the eye that is characterized by the relatively slow, painless, progressive loss of vision. The cause is a mutation in mitochondrial DNA which affects the mitochondrial respiratory chain, and it is therefore maternally inherited.

LECAM *CD62L; LAM-1; L-selectin; MEL-14 antigen; leu-8* Leucocyte-endothelial cell adhesion molecule (37 kDa polypeptide), a **selectin**, expressed on most haematopoietic cells and on mature monocytes, eosinophils and neutrophils. Important for initial adhesion step (margination) and is then rapidly lost from cell surface through activity of a membrane-associated sheddase, a metallopeptidase. Important function is in lymphocyte homing to **high endothelial venule** in peripheral lymph nodes, Peyer's patches and areas of inflammation. Has N-terminal **C-type lectin** domain and binds particularly to carbohydrates on CD34, CD162, GlyCam and MAdCAM.

lecithin Phospholipids of egg yolk (usually hen's eggs). A mixture of phosphatidyl choline and phosphatidyl ethanolamine but usually refers to phosphatidyl choline.

lecithinase See **phospholipases**.

lectin Proteins obtained particularly from the seeds of leguminous plants, but also from many other plant and animal sources, that have binding sites for specific mono- or oligosaccharides. Named originally for the ability of some to selectively agglutinate human red blood cells of particular blood groups. Lectins such as **concanavalin A** and **wheat germ agglutinin** are widely used as analytical and preparative agents in the study of glycoproteins. They are classified according to the carbohydrate-recognition domain (CRD) of which there are two main types, **S-type lectins** and **C-type lectins**. See **selectin** and Table L1.

leghaemoglobin Form of haemoglobin found in the nitrogen-fixing root nodules of legumes. Binds oxygen, and thus protect the nitrogen-fixing enzyme, **nitrogenase**, which is oxygen-sensitive.

Legionella Genus of Gram-negative asporogenous bacteria. Most species are pathogenic in humans, causing pneumonia-like disease – e.g. Legionnaires' disease, named after an outbreak in Philadelphia amongst members of an American Legion reunion.

Legionnaires' disease See *Legionella*.

legumins Plant storage proteins that occur in the seeds of many dicot families; a compound similar to legumin is also produced in monocots. These molecules are usually polymers with molecular weights around 300–400 kDa. Typically have two subunits, one acidic and one basic, and their quaternary structure involves six acidic and six basic polypeptides linked by disulphide bonds. See **cupins**.

Leidig cells See **Leydig cells**.

leiomyoma Benign tumour of smooth muscle in which parallel arrays of smooth muscle cells form bundles which are arranged in a whorled pattern. The amount of fibrous connective tissue is very variable. Leiomyoma of the uterus (fibroid) is the commonest form.

TABLE L1. Lectins

Source	Abbreviation	Sugar specificity
Some common plant-derived lectins		
Bandieraea simplicifolia	BSL1	α-D-gal > α-D-GalNAc
Concanavalla ensiformis (Jack bean)	ConA	α-D-Man > α-D-Glc > α-D-GlcNAc
Dolichos biflorus	DBA	α-D-GalNAc
Lens culinaris (lentil)	LCA	α-D-Man > α-D-Glc > α-D-GlcNAc
Phaseolus vulgaris (red kidney bean)	PHA	β-D-Gal (1-4)-D-GlcNAc
Arachis hypogaea (peanut)	PNA	β-D-Gal (1-3)-D-GalNAc
Pisum sativum (garden pea)	PSA	α-D-Man > α-D-Glc
Ricinus communis (castor bean)	RCA1	β-D-Gal > α-D-Gal
Sophora japonica	SJA	β-D-GalNAc > β-D-Gal > α-D-Gal
Glycine max (soybean)	SBA	α-D-GalNAc > β-D-GalNAc
Ulex europaeus (common gorse)	UEA1	α-L-fucosyl
Triticum vulgaris (wheat germ)	WGA	β-D-GlcNAc(1-4) GlcNAc > β-D-GlcNAc(1-4)-β-D-GlcNAc

Animal lectins	
Subfamily	Example
C-type Lectins	
Endocytic receptors	macrophage mannose receptor
Collectins	tetranectin
Selectins	ELAM-1
Lymphocyte lectins	CD69
Proteoglycans	versican core protein
Miscellaneous	endothelial cell scavenger receptor
Viral lectins	gp22–24 of vaccinia virus
Snake venom	botrocetin
Invertebrate lectins	echinoidin
S-type Lectins	
Galectins 1–10	

Gal, galactose; GalNAc, galactosamine; Glc, glucose; GlcNAc, glucosamine; Man, mannose.
[a]Many plant-derived lectins have been characterized, and only a small sample of commonly encountered ones are listed above.

leiotonin Smooth-muscle analogue (homologue?) of **troponin**. Two subunits, leiotonins A and C, the latter similar in size, and homologous, to **calmodulin** and **troponin** C. Name apparently obsolete.

Leishman stain Romanovsky-type stain; a mixture of basic and acid dyes used to stain blood smears and that differentially stains various classes of leucocytes.

leishmaniasis Disease caused by protozoan parasites of the genus *Leishmania*. The parasite lives intracellularly in macrophages. Various forms of the disease are known, depending upon the species of parasite: in particular visceral leishmaniasis (**kala-azar**) and mucocutaneous leishmaniasis.

Leiurus quinquestriatus hebraeus Scorpion. see **scyllatoxin** and **charybdotoxin**.

LEM domain The LEM (lamina-associated polypeptide-**emerin**-MAN1) domain is a motif shared by a group of **lamin**-interacting proteins in the inner nuclear membrane (INM) and in the nucleoplasm. The LEM domain mediates binding to a DNA-cross-linking protein, **barrier-to-autointegration factor** (BAF).

lenticel A small patch of the periderm in which intercellular spaces are present allowing some gas exchange between the internal tissues of the stem and the atmosphere.

lentigo Relatively common pigmented lesion of the skin in which melanocytes replace the basal layer of the epidermis.

Lentivirinae Subfamily of non-oncogenic retroviruses that cause 'slow diseases' that are characterized by horizontal transmission, long incubation periods and chronic progressive phases. Visna virus is in this group, and there are similarities between visna and equine infectious anaemia virus.

lentoid Spherical cluster of retinal cells, formed by aggregation *in vitro*, that has a core of lens-like cells inside which accumulate proteins characteristic of normal lens. The cells concerned derive from retinal glial cells.

Lepidoptera Order of insects that comprises butterflies and moths.

Lepore haemoglobin Variant haemoglobin in a rare form of **thalassemia**: there is a composite $\delta - \beta$ chain as a result of an unequal **crossing-over** event. The composite chain is functional but synthesized at reduced rate.

leprosy *Hansen's disease* Disease caused by *Mycobacterium leprae*, an obligate intracellular parasite that survives lysosomal enzyme attack by possessing a waxy coat. Leprosy is a chronic disease associated with depressed cellular (but not humoral) immunity; the bacterium requires a lower temperature than 37°C and thrives particularly in peripheral Schwann cells and macrophages. Only humans and the nine-banded armadillo are susceptible.

leptin Product (16 kDa) of the *ob* (obesity) locus. Found in plasma of mouse and man: reduces food uptake and increases energy expenditure. Mutations in mouse lead to obesity, but this seems only very rarely to be the case in man.

leptin receptor Receptor for **leptin**. A single membrane-spanning peptide containing an approximately 300-amino acid intracellular domain. Highly expressed in the hypothalamus, the site of appetite regulation in the brain; downstream regulates a K^+ATP channel similar to that involved in regulation of insulin release (see **SUR**). Most closely related to the gp130 receptors (see **IL-6 cytokine family**). Three alternatively spliced versions, one a soluble form, have been identified. Mutations in the receptor are responsible for the obese phenotype of the *fa/fa* rat.

leptine Toxic glycoalkaloid from *Solanum* sp. (esp. wild potato). Aglycone is acetylleptinidine; different trisaccharides are added to produce Leptine I and II. See **chaconine**.

leptinine Glycoalkaloid from Solanaceae. Similar to **chaconine**. Aglycone is leptinidine.

leptinotarsins β-*Leptinotarsins; leptinotoxin* Toxic proteins (45–47 kDa) present in the haemolymph of potato beetles (*Leptinotarsa* spp.) Causes release of neurotransmitters from synapses of insect and vertebrates by inducing calcium entry, possibly by acting as a calcium-channel agonist.

leptinotoxin See **leptinotarsins**.

leptokurtic Description of the shape of a distribution curve; a normal random variable has a kurtosis of three irrespective of its mean or standard deviation; if the kurtosis is greater than three, it is said to be leptokurtic. The probability density function (distribution) is peaked but with extensive tails. The opposite (a kurtosis less than three) is said to be platykurtic.

leptomycin B Antibiotic first identified in *Streptomyces* sp. as an unsaturated, branched-chain fatty acid with antifungal activity. Has a range of inhibitory activities, but the key one is inhibition of the export of proteins from the nucleus. Binds to **exportin-1** in its conserved central region at a critical cysteine residue and prevents formation of the complex between the exportin and the nuclear export signal of cargo proteins.

leptonema See **leptotene**.

Leptospira Genus of **spirochaete** bacteria that cause a mild chronic infection in rats and many domestic animals. The bacteria are excreted continuously in the urine and contact with infected urine or water can result in infection of humans via cuts or breaks in the skin. Infection causes leptospirosis or Weil's disease, a type of jaundice that is an occupational hazard for sewerage and farm workers.

leptospirosis Weil's disease, caused by infection with *Leptospira*.

leptotene Classical term for the first stage of **prophase** I of **meiosis**, during which the chromosomes condense and become visible.

Lepus Generic name for the hare, the common European brown hare is *Lepus europaeus*.

Lesch–Nyhan syndrome A sex-linked recessive inherited disease in humans that results from mutation in the gene for the purine salvage enzyme **HGPRT**, located on the X chromosome. Results in severe mental retardation and distressing behavioural abnormalities, such as compulsive self-mutilation.

lethal mutation Mutation that eventually results in the death of an organism carrying the mutation.

letrozole *TN Femara* A non-steroidal competitive inhibitor of the aromatase enzyme system that inhibits the conversion of androgens to estrogens. Used as a treatment for hormone-dependent breast cancer in postmenopausal women.

LETS *Large extracellular transformation/trypsin-sensitive protein* Original name (now obsolete) for a cell surface protein that was altered on transformation *in vitro*: now known to be **fibronectin**.

Leu-enkephalin A natural peptide neurotransmitter; see **enkephalins**.

Leu-phyllolitorin A bombesin-related peptide, Leu^8-phyllolitorin increases branching of developing airways and augments thymidine incorporation in cultured lung buds. Central injection of Leu^8-phyllolitorin has been shown to produce hypothermia in animals exposed to a cold environment. (Another form, with Phe in the 8-position is also known.)

leucine *leu; L; 2-amino-4-methylpentanoic acid; 131 Da* The most abundant amino acid found in proteins. Confers hydrophobicity and has a structural rather than a chemical role. See Table A2.

leucine aminopeptidase *EC 3.4.11.1* An **exopeptidase** that removes neutral amino acid residues from the N-terminus of proteins.

leucine zipper Motif found in certain **DNA-binding proteins**. In a region of around 35 amino acids, every seventh is a leucine. This facilitates dimerization of two such proteins to form a functional **transcription factor**. Examples of proteins containing leucine zippers are products of the **proto-oncogenes** *myc, fos* and *jun*. See also **AP-1**.

leucine-responsive regulatory proteins *LRP* Family of bacterial transcriptional regulators that control a large variety of genes, including those coding for cell appendages and other potential virulence factors.

leucine-rich repeat *LRR* Short motif (around 24 residues) with 5–7 leucines generally at positions 2, 5, 7, 12, 21, 24. Forms an amphipathic region and is probably involved in protein–protein interactions.

leucinopine *Dicarboxypropyl leucine* An analogue of **nopaline** found in crown gall tumours (induced by *Agrobacterium tumefasciens*) that do not synthesize octopine or nopaline.

leucistic Lacking pigmentation in the skin but not fully albinotic, eyes being pigmented.

leucocidin Exotoxins from staphylococcal and streptococcal species of bacteria that cause leucocyte killing or lysis. There are two subunits, S and F, each inactive alone but synergistically form a pore. Myeloid but not lymphoid cells are affected.

leucocyte *Leukocyte (US)* Generic term for a white blood cell. See **basophil**, **eosinophil**, **lymphocyte**, **monocyte**, **neutrophil**.

leucocytosis An excess of **leucocytes** in the circulation.

leucopenia An abnormally low count of circulating **leucocytes**.

leucoplast Colourless **plastid** that may be an **etioplast** or a storage plastid (**amyloplast, elaioplast** or **proteinoplast**).

leukaemia *Leukemia (US)* Malignant neoplasia of **leucocytes**. Several different types are recognized according to the stem cell that has been affected, and several virus-induced leukaemias are known (e.g. that caused by feline leukaemia virus). Both acute and chronic forms occur. **1.** Acute lymphoblastic leukaemia (ALL): neoplastic proliferation of white cell precursors in which the blood has large numbers of primitive lymphocytes (high nuclear/cytoplasmic ratio characteristic of dividing cells and few specific surface antigens expressed); tends to be common in the young. **2.** Acute myeloblastic leukaemia (AML): more common in adults; the proliferating cells are of the **myeloid** haematopoietic series and the cells appearing in the blood are primitive **granulocytes** or **monocytes**. **3.** Chronic lymphocytic leukaemia (CLL): neoplastic disease of middle or old age, characterized by excessive numbers of circulating lymphocytes of normal, mature appearance, usually B-lymphocytes; presumably a neoplastic transformation of lymphoid stem cells. **4.** Chronic myelogenous leukaemia (CML): neoplasia of myeloid stem cells, commonest in middle-aged or elderly people, characterized by excessive numbers of circulating leucocytes, most commonly neutrophils (or precursors) but occasionally eosinophils or basophils.

leukaemia inhibitory factor *LIF* Polypeptide **growth factor** or **cytokine** with wide range of activities. Regulates growth and differentiation of primordial germ cells and embryonic stem cells, but has effects on peripheral neurons, osteoblasts, adipocytes and various cells of the myeloid lineage. Given to adult animals, induces weight loss, behavioural disorders and bone abnormalities. Many of the effects of LIF *in vitro* can be mimicked by **interleukin-6**, **oncostatin M** and **ciliary neurotrophic factor**, all of which interact indirectly with gp130, a shared transducer subunit.

leukemia See **leukaemia**.

leuko- See **leuco-**.

leucosialin *CD43; sialophorin* Widely distributed membrane-associated mucin, the major sialoglycoprotein of thymocytes and mature T-cells. Transmembrane protein with extensive O-linked glycosylation (75–85 oligosaccharides on the 239 residue extracellular domain). Extends at least 45 nm beyond plasma membrane. Similar but not homologous to **episialin**.

leucosis The correct term for an excess of leucocytes in the circulation and other parts of the body, preferable in place of the term leucocytosis.

leucosulfakinin *LSK* Cockroach **peptide hormones**, that affect gut motility. Related to **gastrin**.

leucotrienes *LTA4, LTB4, LTC4, LTD4 , LTE4* A family of hydroxyeicosatetraenoic (HETE) acid derivatives. LTA4 and LTB4 are modified lipids; leucotrienes C, D and E have the lipid conjugated to glutathione (LTC4) or cysteine (LTD4, LTE4) to form the peptidyl leucotrienes. A mixture of the latter (LTC4, LTD4, LTE4) constitute SRS-A, the **slow reacting substance of anaphylaxis** that has potent bronchoconstrictive effects. LTB4 is a potent neutrophil chemotactic factor.

leupeptin Family of modified-tripeptide peptidase inhibitors. Commonest is *N*-acetyl-Leu-Leu-argininal.

levaquin *Levofloxacin* A synthetic broad-spectrum antibacterial agent, a chiral fluorinated carboxyquinolone, used for oral and intravenous administration.

levodopa The L-amino acid precursor of dopamine that increases dopamine levels in the basal ganglia thereby improving mobility in Parkinsonism. Usually given together with a dopa-decarboxylase inhibitor (carbidopa) to prevent peripheral effects of dopamine.

levonorgestrel A synthetic and biologically active progestin that exhibits no significant estrogenic activity. Used in oral contraceptives, particularly as an emergency contraceptive taken after intercourse.

levothyroxine *L-thyroxine sodium* Preparation of thyroid hormone (thyroxine) given when normal levels are deficient (hypothyroidism).

Lewis blood group A pair of blood group activities associated with the A, B, H substances. Lewis *Lea* is a separate gene, whereas Leb arises from the combined activity of the enzymes specified by *Lea* and *H* genes.

Lewy body Hyaline eosinophilic concentrically laminated inclusions found in the **substantia nigra** and locus ceruleus of patients with **Parkinsonism** and Lewy body dementia.

LexA *E. coli* **repressor** of the **SOS** system for response to DNA damage.

Leydig cell Interstitial cells of the mammalian testis, involved in synthesis of testosterone.

LFA-1 *CD11a/CD18; lymphocyte function-related antigen-1* Heterodimeric lymphocyte plasma membrane protein (αL 180 kDa, β 95 kDa) that binds **ICAM-1**, particularly involved in cytotoxic T-cell killing. One of the **integrin** superfamily of adhesion molecules. Deficiency of LFA-1 in leucocyte adhesion deficiency (LAD) syndrome leads to severe impairment of normal defences and poor survival prospects. The related surface adhesion molecules (sometimes referred to as the LFA-1 class of adhesion molecules) are Mac-1 (αM 170 kDa, β 95 kDa; CD11b/CD18) and p150,95 (αX 150 kDa, β 95 kDa; CD11c/CD18); they are also defective in severe forms of LAD because the β-subunit, which is apparently common to all three, is missing. Mac-1 (also known as Mo-1 in earlier literature) is the complement C3bi receptor (CR3) and is present on mononuclear phagocytes and on neutrophils; p150,95 is less well characterized, but is particularly abundant on macrophages.

LFA-3 *Lymphocyte function-related antigen-3* Ligand for the CD2 adhesion receptor that is expressed on cytolytic T-cells. LFA-3 is expressed on endothelial cells at low levels. The CD2/LFA-3 complex is an adhesion mechanism distinct from the **LFA-1/ICAM-1** system, and binding of erythrocyte LFA-3 to T-lymphocyte CD2 is the basis of **E-rosetting**.

Lfc Oncoprotein of the Dbl-related family. Contains a Dbl-homology domain in tandem with a **PH domain** and is similar to Lsc, Lbc, **Tiam-1** and Dbl. Has rho-**GEF** activity.

L-forms Bacteria lacking cell walls, a phenomenon usually induced by inhibition of cell wall synthesis, sometimes by mutation.

LH See **luteinizing hormone**.

LHRF See **luteinizing hormone-releasing factor**.

Li-Fraumeni syndrome A clinically and genetically heterogeneous predisposition to tumours caused by mutation in the **p53** gene.

library See **genomic library**.

lichen A large group of symbiotic associations between fungi and green and occasionally blue-green algae. Several genera of algae and of fungi are involved and the associations are so stable and of such varied but distinct types that the lichens have been classified into genera and species. A variety of incompatibility phenomena are often manifested between individual lichens. Confined to terrestrial habitats and often used as indicators of pollution status of the environment.

lichen planus Rare skin disorder in which there is marked hyperkeratosis and extensive infiltration of lymphocytes into the lower epidermis.

Liddle's disease Hereditable (autosomal dominant) form of salt-sensitive human hypertension caused by mutation in the beta or gamma subunit of the multisubunit epithelial **sodium channel** (ENaC).

lidocaine *Lignocaine* Commonly used local anaesthetic.

LIF See **leukaemia-inhibitory factor**.

ligand Any molecule that binds to another; in normal usage a soluble molecule such as a hormone or neurotransmitter, that binds to a receptor. The decision as to which is the ligand and which the receptor is often a little arbitrary when the broader sense of receptor is used (where there is no implication of transduction of signal). In these cases it is probably a good rule to consider the ligand to be the smaller of the two, thus in a lectin–sugar interaction, the sugar would be the ligand (even though it is attached to a much larger molecule, recognition is of the saccharide).

ligand-gated ion channel A transmembrane **ion channel** whose permeability is increased by the binding of a specific **ligand**, typically a neurotransmitter at a **chemical synapse**. The permeability change is often drastic; such channels let through effectively no ions when shut, but allow passage at up to 10^7 ions/s when a ligand is bound. The receptors for both **acetylcholine** and **GABA** have been found to share considerable sequence homology, implying that there may be a family of structurally related ligand-gated ion channels.

ligand-induced endocytosis The formation of coated pits and then **coated vesicles** as a consequence of the interaction of ligand with receptors, which then interact with **clathrin** and associated proteins (coatomers) on the cytoplasmic face of the plasma membrane and come together to form a pit. Not all coated vesicle uptake of receptors requires receptor occupancy.

ligase amplification reaction *LAR* Method for detecting small quantities of a target DNA, with utility similar to **PCR**. It relies on DNA ligase to join adjacent synthetic oligonucleotides after they have bound the target DNA. Their small size means that they are destabilized by single base mismatches, and so form a sensitive test for the presence of mutations in the target sequence.

ligases *Synthetases* Major class of **enzymes** that catalyse the linking together of two molecules (category 6 in the **E classification**), e.g. DNA ligases that link two fragments of DNA by forming a **phosphodiester** bond.

ligatin A filamentous, baseplate protein (10-kDa monomer) that binds and localizes peripheral glycoproteins to the external cell surface and within endosomes (a trafficking receptor for phosphoglycoproteins). In hippocampal neurons, long-lasting downregulation of ligatin mRNA levels occurs following glutamate receptor activation.

LIGHT A 29-kDa type II transmembrane protein produced by activated T-cells, a member of the tumour-necrosis factor (TNF) superfamily that forms a membrane-anchored homotrimeric complex; function is mediated by at least two receptors, including lymphotoxin receptor (LTR) and herpes simplex virus entry mediator. Name apparently based on 'homologous to lymphotoxin, exhibits inducible expression, competes with herpesvirus glycoprotein D for herpes virus entry mediator on T-cells' and justifiably is abbreviated.

light chain Non-specific term used of the smaller subunits of several multimeric proteins, for example **immunoglobulin**, **myosin**, **dynein**, **clathrin**. See also **L-chain**.

Light Green A stain often used for counterstaining cytoplasm following iron haematoxylin; a component of Masson's trichrome stain.

light microscopy In contrast to electron microscopy. See **bright field**, **phase contrast**, **interference**, **interference contrast**, **interference reflection**, **dark field**, **confocal** and **fluorescence microscopy**. See also Table L2.

light scattering Particles suspended in a solution will cause scattering of light, and the extent of the scattering is related to the size and shape of the particles (in a somewhat complex relationship).

light-dependent reaction The reaction taking place in the chloroplast in which the absorption of a photon leads to the formation of ATP and NADPH.

light-harvesting system Set of photosynthetic pigment molecules that absorb light and channel the energy to the photosynthetic **reaction centre**, where the light reactions of **photosynthesis** occur. In higher plants, contains **chlorophyll** and **carotenoids**, and is present in two slightly different forms in **photosystems I and II**.

Lightcycler Trade name for a **real-time PCR** machine used to quantify specific DNA sequences, or, after reverse transcription, mRNA levels. It uses fluorescent detection during the PCR reaction to quantify the DNA at each cycle.

TABLE L2. Types of light microscopy

	Method	Physical parameter detected
A. With axial illumination		
Without spatial filtration[a]		
I	Bright field	Absorption by specimen (May be operated in Visible, UV or IR regions of the spectrum and in quantitative microspectrophotometric modes)
II	Interference	
	Transmitted	Path difference arising in specimen, qualitative or quantitative
	Reflected (interference reflection = IRM)	Path difference in films 1–10 wavelengths thick next to substrate. For cell contacts
III	Fluorescence	Natural fluorescence or that of probes applied to system
IV	Dark field	Refractive index discontinuities revealed by scattered light
V	Polarization	Birefringent and/or dichroic properties
With spatial filtration		
VI	Confocal scanning microscopy	Contrast and resolution enhanced by selection of light paths modified by the object at the back focal plane of the objective. Bright-field or fluorescence modes. Usually combined with video processing of the image.
VII	Phase contrast	Path differences revealed as contrast differences non-quantitatively and non-regularly, using phase plate at back focal plane
VIII.	Differential interference contrast (DIC) = Nomarski	Path difference gradients revealed as contrast or colour differences
IX	Out-of-focus phase contrast	Path differences revealed as diffraction patterns.
B. With anaxial illumination: with spatial filtration		
X	Hoffman modulation contrast	Path differences
XI	Single side-band edge enhancement (SSEE microscopy)	Path differences from first-order diffractions

Nearly all systems can be run in the epi- (incident) illumination mode. Video (image) processing can enhance contrast and resolution in images by the application of simple algorithms to expand the grey scale, reduce noise and subtract background. More complex processing is possible, including the extraction of further information by Fourier transforms.
[a]Spatial filtration – this is the application of methods to remove those ray paths which have not interacted with the object. This is done at the back focal plane of the objective. It can also be applied to select or remove ray paths that have interacted in some specified way with the object.

lignin Complex polymer of phenylpropanoid subunits, laid down in the walls of plant cells such as **xylem** vessels and **sclerenchyma**. Imparts considerable strength to the wall, and also protects it against degradation by microorganisms. It is also laid down as a defence reaction against pathogenic attack, as part of the **hypersensitive response** of plants.

lignocaine *Lidocaine* Local anaesthetic agent that stabilizes the neuronal membrane and prevents the initiation and transmission of nerve impulses; also used to treat ventricular arrhythmias of the heart.

LIM domain Domain found in proteins required for developmental decisions. Contain 60-residue conserved, cysteine-rich repeats. Named after first three genes in group: Lin-11 (*Caenorhabditis elegans* – required for asymmetric division of blast cells), IsI-1 (mammalian insulin gene-binding enhancer protein), mec-3 (*C. elegans* – required for differentiation of a set of sensory neurons).

LIM kinases *LIMK-1, LIMK-2* Actin-binding kinases that phosphorylate members of the ADF/cofilin family of actin binding and filament severing proteins. Are activated by **PAK**.

limatin *Dematin* homologue from human retina.

limb-bud The limbs of vertebrates start as outpushings of mesenchyme surrounded by a simple epithelium. The distal region is referred to as the progress zone. There has been extensive study of positional information within the limb-bud that determines, for example, the proximal–distal pattern of bone development and the anterior–posterior specification of digits.

limbic system Those regions of the central nervous system responsible for autonomic functions and emotions. Includes hippocampus, amygdaloid nucleus and portions of the midbrain.

limbus A distinct edge or border, for example that is between cornea and sclera in the eye.

LIME A transmembrane adaptor (**TRAP**) required for B-cell receptor (BCR)-mediated B-cell activation. LIME is expressed in mouse splenic B-cells. Upon BCR cross-linking, LIME is tyrosine phosphorylated by **Lyn** and associated with Lyn, **Grb2**, **PLC-gamma2** and **PI3K**.

limit of resolution See **resolving power**.

limitin *Interferon (IFN)-Zeta* An interferon-like cytokine (shows some sequence homology with IFN-alpha and IFN-beta), suppresses B lymphopoiesis through ligation of the interferon-alpha/beta (IFN-alpha/beta) receptor, activation of **Tyk2** and the upregulation and nuclear translocation of **Daxx**.

limonene The major component of the oil extracted from citrus rind; D-limonene is characteristic of oranges, L-limonene of lemons. Used for a wide variety of purposes as a solvent and additive to cleaning solutions.

Limulus polyphemus Now renamed *Xiphosura*, though *Limulus* is still in common usage as a name. The king crab or horseshoe crab, found on the Atlantic coast of North America. It is more closely related to the arachnids than the Crustacea, and horseshoe crabs are the only surviving representatives of the subclass Xiphosura. Its compound eyes have been widely used in studies on visual systems, but it is probably better known from the Limulus-amoebocyte lysate (LAL) test; LAL is very sensitive to small amounts

of **endotoxin**, clotting rapidly to form a gel, and the test is used clinically to test for septicaemia.

lincomycin Antibiotic (macrolide) active against Gram-positive bacteria from *Streptomyces lincolnensis*. Acts by blocking protein synthesis by binding to the 50S subunit of the ribosome and blocking peptidyl transferase reaction. Clindamycin, a semisynthetic derivative of lincomycin, is used as an antimalarial drug.

lindane *Benzene hexachloride; gamma-hexachlorocyclo-hexane* A neurotoxic and probably carcinogenic organochlorine pesticide used for treating head lice and scabies and as a general pesticide.

linear dichroism See **circular dichroism**.

Lineweaver–Burke plot A plot of $1/v$ against $1/S$ for an enzyme-catalysed reaction, where v is the initial rate and S the substrate concentration. Using the equation $1/v = 1/V_{max}(1 + K_m/S)$, the parameters V_{max} and K_m can be determined. The equation overweights the contribution of the least accurate points, and other methods of analysis are preferred; see **Eadie–Hofstee plot**.

linezolid Antibacterial drugs of the oxazolidinone class, chemically unrelated to currently used antibiotics and with enhanced activity against Gram-positive bacteria, including enterococci that are resistant to vancomycin.

lining epithelium An epithelium lining a duct, cavity or vessel and that is not particularly specialized for secretion or as a mechanical barrier. Not a precise classification.

linkage Tendency for certain genes tend to be inherited together because they are on the same chromosome. Thus parental combinations of characters are found more frequently in offspring than are non-parental. Linkage is measured by the percentage recombination between loci.

linkage disequilibrium The occurrence of some genes together, more often than would be expected. Thus, in the HLA system of **histocompatibility antigens**, HLA-A1 is commonly associated with B8 and DR3, and A2 with B7 and DR2, presumably because the combination confers some selective advantage.

linkage equilibrium Situation that should exist in a population undisturbed by selection, migration, etc., in which all possible combinations of linked genes should be present at equal frequency. The situation is no more common than are such undisturbed populations.

linoleic acid An essential **fatty acid** (9, 12, octadeca-dienoic acid); occurs as a glyceride component in many fats and oils.

linolenic acid An 18-carbon fatty acid with three double bonds (9, 12, 15, octadecatrienoic acid) and α- and γ-isomers. Essential dietary component for mammals. See **fatty acids**.

lipaemia Presence in the blood of an abnormally large amount of lipid.

lipases Enzymes that break down mono-, di- or triglycerides to release fatty acids and glycerol. Calcium ions are usually required. Triglyceride lipases (EC 3.1.1.3) hydrolyse the ester bond of triglycerides.

lipid A The lipid associated with polysaccharide in the **lipopolysaccharide** (LPS) of Gram-negative bacterial cell walls.

lipid bilayer See **phospholipid bilayer**.

lipidoses Storage diseases in which the missing enzyme is one that degrades sphingolipids (sphingomyelin, ceramides, gangliosides). In **Tay–Sachs disease** the lesion is in hexosiminidase A, an enzyme that degrades ganglioside Gm2; in **Gaucher's disease**, glucocerebrosidase; in **Niemann-Pick** disease, sphingomyelinase.

lipids Biological molecules soluble in apolar solvents but only very slightly soluble in water. They are a heterogenous group (being defined only on the basis of solubility) and include fats, waxes and terpenes. See Table L3.

lipoamide The functional form of lipoic acid in which the carboxyl group is attached to protein by an amide linkage to a lysine amino group.

lipoamide dehydrogenase *EC.1.8.1.4* An enzyme that regenerates **lipoamide** from the reduced form dihydrolipoamide.

lipocalins Superfamily of carrier proteins that transport small, hydrophobic molecules, such as retinol, porphyrins, odorants. Characterized by two orthogonally stranded **beta sheets**. Examples: alpha-1-microglobulin, **purpurin**, **orosomucoid**. Neutrophil-gelatinase-associated lipocalin (NGAL) has been used as a biomarker in the detection of acute renal failure in children after cardiac surgery.

lipocortin The name given to calcium-binding protein believed to be secreted by macrophages that acts as an inhibitor of phospholipase A2 enzymes and has a possible role in mediating the anti-inflammatory effects of steroids. Lipocortins are identified as proteins of the **annexin** class and their extracellular role is in some doubt.

lipocyte Liver cell that stores lipid.

lipofectamine *Lipofect amine; lipofectin* Proprietary liposome preparation, formulated from cationic lipids, for lipid-mediated **transfection** of cultured cells.

lipotuscin Brown pigment characteristic of ageing. Found in lysosomes, it is the product of peroxidation of unsaturated fatty acids and is symptomatic, perhaps, of membrane damage rather than being deleterious in its own right.

lipoic acid *Thioctic acid; 1,2-dithiolane-3-valeric acid* Regarded as a coenzyme in the oxoglutarate dehydrogenase complex of the **tricarboxylic acid cycle**. Involved generally in oxidative decarboxylations of α-keto acids. A growth factor for some organisms.

TABLE L3. Lipids

(i) Fatty acid. These are the most important feature of the majority of biological lipids. They occur free in trace quantities and are important metabolic intermediates. They are esterified in the majority of biological lipids. Compounds are included either because they are common components of biological lipids or are used in synthetic 'model' analogues of these lipids. General formula R-COOH. Branched chain compounds are widespread but are not found in mammalian lipids. All the examples given are straight-chain compounds.

Saturated fatty acids

Number of carbon atoms	Name	M_r(Da)
2	Acetic	60
3	Propionic	74.1
4	Butyric	88.1
5	Valeric	102.1
6	Hexanoic (caproic)	116.2
7	Heptanic	130.2
8	Octanoic (caprylic)	144.2
9	Nonanoic (pelargonic)	158.2
10	Decanoic (capric)	172.2
11	Undecanoic	186.3
12	Lauric	200.3
13	Tridecanoic	214.4
14	Myristic	228.4
15	Pentadecanoic	242.4
16	Palmitic	256.4
17	Margaric	270.7
18	Stearic	284.5
20	Eicosanoic (Arachidic)	312.5
21	Docosanoic (Behenic)	340.6

Unsaturated fatty acids

Designation	Name	M_r(D)
Mono-unsaturated acids*		
16:1 (*cis* 9)	Palmitoleic	254.2
18:1 (*cis* 9)	Oleic	282.5
18:1 (*trans* 9)	Elaidic	282.5
18:1 (*cis* 11)	*cis*-Vaccenic	282.5
18:1 (*trans* 11)	*Trans*-Vaccenic	282.5
Poly-unsaturated acids* (all *cis* double bonds)		
18:2 (9, 12)	Linoleic	280.4
18:3 (9, 12, 15)	α-Linolenic	278.4
18:3 (6, 9, 12)	γ-Linolenic	278.4
20:4 (5, 8, 11, 14)	Arachidonic (eicosenoic)	304.5
22:6 (4, 7, 10, 13, 16, 19)	Dodecosahexaenoic acid	328.6

Number of carbon atoms: number of double bonds (configuration and position of bonds).

TABLE L3. (Continued)

(ii) Acyl glycerols Glycerol esters of fatty acids. Acyl glycerols are the parent compounds of many structural and storage lipids. Diglycerides (DG) may be considered as the parent compounds of the major family of phosphatidyl phospholipids. Triglycerides (TG) are important storage lipids.

Diglycerides present as trace components of membranes. They are important metabolites and second messengers in signal–response coupling.

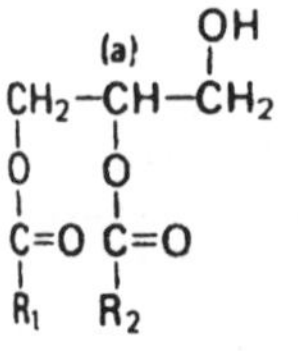

(a) $CH_2-CH-CH_2$ with OH
$O \quad O$
$C=O \quad C=O$
$R_1 \quad R_2$

i.e.

OH

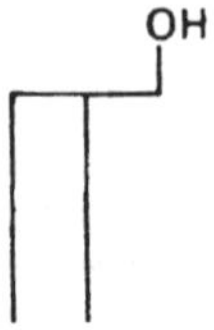

(a) This carbon is asymmetric. See below under phosphatidic acid.

(iii) Sphingolipids. Important and widespread classes of phospholipids and glycolipids.
The parent alcohol is **sphingosine**:
where X = OH and the primary amino-group is free.

sphingosine is normally substituted at X and Y. When Y is a long-chain unsaturated fatty acyl group the derivative is a **ceramide**. When the **ceramide** carries uncharged sugars as the X substituent this is a **cerebroside** and where the sugars include sialic acid it is a **ganglioside**.

(iv) Phospholipids. In animal cell membranes the major class of phospholipids are the phosphatidyl phospholipids for which phosphatidic acid can be considered as the simplest example. These are diacylglycerol (DG) derivatives and in most cases DG is the immediate metabolic precursor.
Outline structure (see diglyceride):

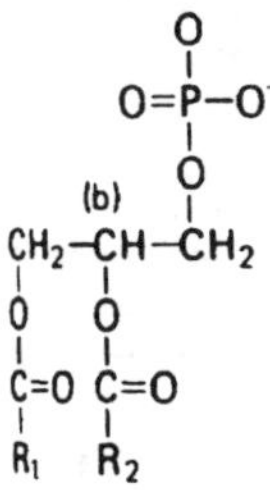

O
$O=P-O^-$ (H) pK_a ~6.5
O
(b) $CH_2-CH-CH_2$
$O \quad O$
$C=O \quad C=O$
$R_1 \quad R_2$

i.e.

O^-
$^-O-P=O$
O

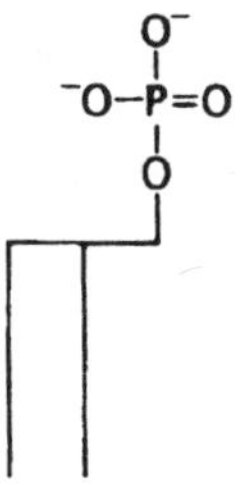

(b) As in diglycerides, this carbon atom is asymmetric. The biologically important configuration is *syn*. R_1 is usually saturated and R_2 is unsaturated in animal cell membranes.

lipolysis The breakdown of fat into fatty acids and glycerol.

lipomodulin The name originally given to **lipocortin** from neutrophils.

lipophilic Having an affinity for lipids, and thus hydrophobic.

lipophorin A family of high-density lipoproteins (6–700 kDa) from insect **haemolymph** that transport diacyl glycerols. The molecule comprises heavy (250 kDa) and light (85 kDa) subunits, the remainder of the molecular weight being accounted for by the high lipid content (40–50%, depending on insect species). Lipophorin forms large aggregates during the haemolymph clotting process.

lipopolysaccharide *LPS* The major constituents of the cell walls of Gram-negative bacteria. Highly immunogenic and stimulates the production of endogenous pyrogen **interleukin-1** and **tumour-necrosis factor** (TNF). Has three parts: variable polysaccharide (O antigen) side chains, core polysaccharides and **lipid A**.

lipoproteins An important class of serum proteins in which a lipid core with a surface coat of phospholipid monolayer is packaged with specific proteins (**apolipoproteins**). Classified according to density: chylomicrons, large low-density particles; very low-density (VLDL); low-density (LDL) and high-density (HDL) species. Important in lipid transport, especially cholesterol transport.

liposomes Artificially formed single-layer or multilayer spherical lipid bilayer structures. Made from solutions of lipids, etc. in organic solvents dispersed in aqueous media. Under appropriate conditions, liposomes form spontaneously. Often used as models of the plasma membrane. May also be used experimentally and therapeutically for delivering drugs etc. to cells, since liposomes can fuse with a plasma membrane and deliver their contents to the interior of the cell (see **lipofectamine**). Vary in size from submicron diameters to (in a few record-breaking cases) centimetres.

lipoteichoic acid Compounds formed from **teichoic acid** linked to glycolipid and found in the walls of most Gram-positive bacteria. The lipoteichoic acid of streptococci may function as an **adhesin**.

lipotropin *LPH; lipotropic hormone; adipokinetic hormone* Polypeptide hormone (β form: 9894 Da, 91 residues; γ form has only residues 1–58 of β) from the pituitary hypophysis, of particular interest because it is the precursor of **endorphins**, which are released by proteolysis. Promotes lipolysis and acts through the adenylyl cyclase system. Part of the **ACTH** group of hormones.

lipovitellin The predominant lipoprotein found in the yolk of egg-laying animals, both vertebrate and invertebrate; involved in lipid and metal storage; it is formed from the precursor **vitellogenin**.

lipoxins *LX* Bioactive **eicosanoids**, generated by **lipoxygenases** acting on arachidonic acid, that activate human monocytes and inhibit neutrophils. LXA4 is rapidly converted by monocytes to inactive products. In human myeloid cells, lipoxin LXA4 actions are mediated by interaction with a G-protein-coupled receptor.

lipoxygenase *5-Lipoxygenase; 5-LO; EC 1.13.11.34* Enzyme that catalyses the addition of a hydroperoxy group to the 5-position of arachidonic acid, the first step in **leucotriene** synthesis.

liprins Family of proteins that interact with **LAR**-family phosphatases. The C-terminal portion of liprins binds to the membrane-distal phosphatase domain of LAR and the N-terminal region may be involved in dimerization. Some liprins are widely distributed, others are more tissue specific. May affect LAR distribution in the cell, in particular bringing LAR to regions of contact between cell and extracellular matrix.

lisinopril Angiotensin-converting enzyme inhibitor **ACE inhibitor** used in treatment of heart failure and hypertension. Mode of action primarily through suppression of the rennin–angiotensin–aldosterone system.

Listeria monocytogenes Rod-shaped Gram-positive bacterium. Wide spread and can grow over an unusually wide range of temperatures (0–45°C). Normally **saprophytic**, but is an opportunistic parasite in that it can survive within cells (particularly leucocytes) and can be transmitted transplacentally. It has caused a number of serious outbreaks of food poisoning with a high mortality rate in recent years.

listeriolysin O **Cholesterol-binding toxin** from *Listeria monocytogenes*.

lithium The lightest alkali metal, although it has the largest hydrated cation. Important as an antidepressant, and is thought to act by inhibiting the regeneration of **inositol** from **IP3** and thus reducing the efficiency of the **phosphatidyl inositol** signalling pathways. Lithium salts (carbonate, citrate) are used for treatment and prevention of mania, manic-depressive illness and recurrent depression; lithium succinate has anti-inflammatory and antifungal activity.

lithotroph Cell or organism that depends upon inorganic compounds as electron donors for energy production.

litorin A peptide that mimics **bombesin** in its mitogenic effects, and has a carboxy-terminal octapeptide in common with bombesin.

liver cells Usually implies **hepatocytes**, even though other cell types are found in the liver (**Kupffer cells,** for example). Hepatocytes are relatively unspecialized epithelial cells, and are the biochemist's 'typical animal cell'.

LKB1 LKB1 is a 50-kDa serine/threonine kinase that was originally discovered as the product of the gene mutated in the autosomal dominant human disorder, Peutz–Jeghers syndrome (PJS). People with PJS develop benign polyps in the gastrointestinal tract, but also have a 15-fold increased risk of developing malignant tumours in other tissues.

LKR Abbrev. for (1) lysine-alpha-ketoglutarate reductase, EC 4.6.1.10, (2) LKR-13, a lung adenocarcinoma cell line derived from *K-ras* (LA1) mice, (3) leucokinin receptor (Lkr) from *Drosophila*.

LL-37 *hCAP18* Human **cathelicidin** with homology to rabbit CAP18, a neutrophil-derived peptide that shares its antimicrobial and LPS-neutralizing activity with SLPI but, unlike SLPI, does not function as an elastase inhibitor. Is also present in lymphocytes and monocytes, and in skin and other epithelia, where the synthesis is upregulated in inflammatory states.

LMM *Light meromyosin* The rod-like portion of the myosin heavy chain (predominantly α-helical) that is involved in lateral interactions with other LMM to form the thick filament of striated muscle, and is separated from heavy meromyosin (HMM) by cleavage with trypsin.

LMP 1. **Latent membrane protein**. 2. Low-molecular weight proteins (LMP), components of the **proteosome**.

L-myc Relative of the *myc* **proto-oncogene** overexpressed in lung **carcinoma**.

LNCaP cells A line of androgen-sensitive prostate cancer cells derived from a lymph node metastatic lesion of human prostatic adenocarcinoma; widely used in the study of prostate cancer.

lobopodia Hemispherical protrusions from the front of a moving tissue cell.

local circuit theory A generally accepted model for neuronal conduction, by which depolarization of a small region of a neuronal plasma membrane produces transmembrane currents in the neighbouring regions, tending to depolarize them. As the **sodium channels** are **voltage-gated**, the depolarization causes further channels to open, thus propagating the action potential.

lock and key models Specific recognition in biological systems might be mediated through interactions that depend upon very precise steric matching between receptor and ligand, or between enzyme and substrate. The commonly used analogy is between lock and key, and implies a precise, sterically determined interaction.

locomotion Term used by some authors to distinguish movement of cells from place-to-place from movements such as flattening, shape-change, **cytokinesis**, etc.

locus *Pl.* loci The site in a linkage map or on a chromosome where the gene for a particular trait is located. Any one of the alleles of a gene may be present at this site.

locus coeruleus A dense cluster of neurons in the dorsorostral pons region of the brain. It is the major source of norepinephrine with neuronal projections throughout most of the CNS, including the cerebral cortex, hippocampus, thalamus, midbrain, brainstem, cerebellum and spinal cord. Considered to be a key brain centre for anxiety and fear.

locus control region *Control region; LCR* Region of DNA which contains the **promoters** and **enhancers** that regulate the expression of a particular gene. Often taken to be a single region 0–2 kb upstream of the transcriptional **start site**, although there are probably few genes where things are that simple.

Lod score Logarithm of the odds score: a statistical test for the probability that there is linkage. For non-X-linked genetic disorders, a Lod score of +3 (1000:1) is usually taken to indicate linkage.

Loeffler's medium Suboptimal coagulated-serum medium used to culture *Corynebacterium diphtheriae* in diagnostic bacteriology.

Loeffler's syndrome Acute but mild and self-limiting eosinophilic pneumonia.

Lofexidine A non-opiate drug, a centrally acting alpha-2 adrenergic agonist, which is used for the management of withdrawal symptoms in patients undergoing opiate detoxification.

log-normal distribution Distribution function $F(y)$, in which the logarithm of a quantity has a Gaussian (normal) distribution.

logP $log K_o$ The octanol–water **partition coefficient**, serves as a quantitative indicator of lipophilicity, an important parameter in pharmacokinetics and for predicting distribution of a chemical in the environment.

Loligo Squid. Source of giant axons for electrophysiologists.

lomasome Membranous structure, often containing internal membranes, located between the plasma membrane and cell wall of plant cells. Included in the more general term, **paramural body**.

Long QT *Long QT syndrome* Genetic channelopathy in which a defect in a potassium channel increases the risk of sudden heart failure; probably responsible for most cases of Sudden Adult Death syndrome. The QT interval is defined from the characteristics of the electrocardiogram, and represents the time taken for electrical activation and inactivation of the ventricles.

long-term potentiation *LTP* Increase in the strength of transmission at a **synapse** with repetitive use that lasts for more than a few minutes. As a form of long-term **synaptic plasticity,** it is important as a possible cellular basis of learning and memory storage. It has been studied most extensively at excitatory synapses onto principal neurons of the **hippocampus,** where it was first demonstrated. Selective inhibition of **NMDA receptor** channels has been shown to block LTP and also spatial learning.

long-terminal repeat *LTR* Identical DNA sequences, several hundred nucleotides long, found at either end of **transposons** and the proviral DNA, formed by reverse transcription of **retroviral** RNA. They are thought to have an essential role in integrating the transposon or provirus into the host DNA. LTRs have inverted repeats – that is, sequences close to either end are identical when read in opposite directions. In proviruses, the upstream LTR acts as a promoter and enhancer and the downstream LTR as a polyadenylation site.

loop diuretics Drugs that inhibit resorption from the loop of Henle in the renal tubule and thus increase the rate of fluid excretion. They are powerful diuretics, and include frusemide, bumetanide and ethacrynic acid.

lophocytes Cells found beneath the dermal membrane of a few species of sponges. Have been postulated to constitute a primitive nervous system, though this is uncertain.

Lophophorates Group of minor protostome coelomate phyla. Includes Bryozoa, Phoronida and Brachiopoda.

lophotrichous Descriptor for a cell with flagella arranged as a tuft at one end.

loratadine *TN Clarityn* Non-sedating antihistamine used to relieve symptoms of hay fever and urticaria.

lorazepam Short-acting **benzodiazepine** for short-term treatment of anxiety and insomnia.

lorica Shell or test secreted by a protozoan; often vase-shaped.

loricrin Major protein of the **keratinocyte** cell envelope. Mutations in the gene for loricrin are responsible for the inherited skin diseases Vohwinkel's syndrome and progressive symmetric erythrokeratoderma.

loss of heterozygosity *LOH* Situation in which heterozygosity is lost in some tissues, probably as a result of mitotic recombination so that one daughter cell obtains two identical alleles and the other gets two mutant alleles or a major deletion on the homologous chromosome so that only one allele remains. It is fairly common in various tumours, and indicates tumour progression.

loss-of-function mutation Mutation that causes the loss of function in a protein or system. Much more common than gain of function mutations, since there are more ways to disrupt things than there are to add additional functionality.

Lou Gehrig's disease See **amyotrophic lateral sclerosis**.

LOV domains *Light-oxygen-voltage-sensing domains* Photosensitive domains of the blue-light-sensitive protein YtvA from *Bacillus subtilis* also found in **phototropin** from higher plants.

lovastatin *Mevinolin; 6α-methylcompactin* Fungal metabolite isolated from cultures of *Aspergillus terreus* that inhibits hydroxymethylglutaryl-CoA reductase (HMG-CoA reductase) and is used as an antihypercholesterolemic drug.

low-density lipoprotein *LDL* See **lipoprotein**.

low-density lipoprotein receptor *LDL receptor* A cell surface protein that mediates the endocytosis of LDL by cells. Genetic defects in LDL receptors lead to abnormal serum levels of LDL and hypercholesterolaemia.

low-affinity platelet factor IV See **connective tissue-activating peptide III**.

lowest-observed-adverse-effect level *LOAEL* Lowest concentration or amount of a substance (dose) that causes an adverse effect on morphology, functional capacity, growth, development or life span of a target organism. Not an absolute level since it depends upon the ability to observe an effect, which is, in turn, dependent upon analytical methodology.

lowest-observed-effect level *LOEL* Lowest concentration or amount of a substance (dose) that has an observable effect under defined conditions. As with the **lowest-observed-adverse-effect level**, depends upon the sensitivity of the observational technique.

Lowry assay One of the most commonly used assays for protein content – the paper describing it is said to be the most frequently cited in the biological literature. Depends upon the interaction of Folin–Ciocalteau reagent with tyrosine or phenylalanine. Proteins that are deficient in these amino acids (e.g. collagen) will be underestimated, and for this reason the Bradford assay is often preferred.

lox Site in **bacteriophage** P1 DNA that is recognized by the **cre** recombinase. Now used in vertebrate transgenics: see **lox-Cre system**.

lox-Cre system Site-specific recombination system from *E. coli* **bacteriophage** P1. Now used in transgenic animals to produce conditional mutants. If two **lox** sites are introduced into a transgene, the intervening DNA is spliced out if active **cre** recombinase is expressed.

LPA *LysoPA* See **lysophosphatidic acid**.

LPL *Lipoprotein lipase; EC 3.1.1.34* The major enzyme for hydrolysis of circulating triglyceride-rich lipoproteins. Familial lipoprotein lipase deficiency is characterized by increased plasma triglyceride levels caused by delayed clearance of chylomicrons from the plasma after digestion of dietary fat.

LPS See **lipopolysaccharide**.

L-ring Outermost ring of the basal part of the bacterial flagellum in Gram-negative bacteria. It may serve as a bush to anchor the flagellum relative to the lipopolysaccharide layer.

LRP 1. **Leucine-responsive regulatory proteins**. 2. **LDL receptor-related protein**.

LRR See **leucine-rich repeat**.

Lsc One of the Dbl-like oncoproteins. See **Lfc**.

LSD See **lysergic acid diethylamide**.

LST8 A 34-kDa **WD-repeat protein** that is part of the **raptor–mTOR** complex (**TORC**). Has a scaffolding role.

LTA4, LTB4, LTC4, etc. See **leucotrienes**.

LTP See **long-term potentiation**.

L-type channels A class of **voltage-sensitive calcium channels**. L-type channels are found in neurons, neuroendocrine cells, and smooth, cardiac and striated muscle; they are involved in neurotransmitter release at some synapses and inactivate relatively slowly. They are activated at membrane potentials more positive than $-30\,$mV. The long-lasting properties and possible role in **long-term potentiation** are the reason for them being designated L-type. They are insensitive to ω-**conotoxin** but inhibited by dihydropyridines, benzodiazepines and phenylalkylamines.

Lubrol TN for a family of non-ionic detergents. Lubrol 12A9 is 2-dodecoxyethanol; Lubrol TSC 5110 is trioctadecyl citrate. Lubrol PX and WX are often used for isolation of integral membrane proteins with retention of their enzymic activity.

lucifer yellow Bright yellow fluorescent molecule (similar to fluorescein), widely used by **microinjection** in developmental biology and neuroscience to study the outline of cells, in **cell lineage** studies, or as an indicator of **dye coupling** between cells.

luciferase *EC 1.13.12.7* Enzyme from firefly tails that catalyses the production of light in the reaction between luciferin and ATP. Used by the male firefly for producing light to attract females, and used in the laboratory in a **chemiluminescence** bioassay for ATP.

luciferase reporter system **Reporter genes** that are based on firefly **luciferase** gene offer luminescent detection of reporter activity. As few biological processes emit light, this assay has a very low background.

luciferins Substrates for the enzyme **luciferase** that catalyse an oxidative reaction leading to photon emission (bioluminescence).

lucigenin Compound used as a bystander substrate in assaying the **metabolic burst** of leucocytes by **chemiluminescence**. When oxidized by **superoxide** it emits light.

Lucké carcinoma A renal carcinoma, caused by a herpesvirus, in frogs; it aroused interest because its abnormal growth appears to be dependent on a restricted temperature range. Nuclei from these cells give rise to normal frogs if transplanted into enucleated eggs, giving support to the **epigenesis** theories of neoplasia.

Lugol's iodine An aqueous solution of iodine and potassium iodide in water. Historically, used as a disinfectant and also as a reagent for demonstrating the presence of starch.

lumefantrine *Benflumetol* A synthetic antimalarial drug developed in the 1970s by the Academy of Military Medical Sciences in Beijing, China. Frequently used in combination with Artemether, a drug related to **artemisinin**; lumefantrine has a half-life of about 3–6 days in patients but is slow acting, whereas Artemether shows rapid antimalarial activity. The combination is distributed as Riamet™.

lumen A cavity or space within a tube or sac.

lumican Isoform (37 kDa) of corneal keratan sulphate **proteoglycan** also found in arterial wall and many other tissues.

lumicolchicine A derivative of **colchicine** produced by exposure to ultraviolet light and that does not inhibit tubulin polymerization, although it has many of the non-specific effects of colchicine.

luminol *o-Aminophthaloyl hydrazide* Compound used as a bystander substrate in assaying the **metabolic burst** of leucocytes by **chemiluminescence**. When oxidized by the myeloperoxidase/hydrogen peroxide system, it emits light. Also used forensically to detect traces of blood.

luminometer Laboratory instrument used to measure light emission, for example in measurement of ATP levels by interaction with luciferase, which produces light.

lumirhodopsin Altered form of **rhodopsin** produced as a result of illumination.

lumisome Subcellular membrane-enclosed vesicle that is the site of bioluminescence in some marine **coelenterates**.

lunula Generally, a crescent-shaped mark, more specifically, the white crescent lying at the root of the nail.

lupus erythematosus Skin disease in which there are red scaly patches, especially over the nose and cheeks. May be a symptom of **systemic lupus erythematosus**.

lusitropic Anything that relaxes the heart, the opposite of **inotropic**.

luteinizing hormone *LH; lutropin* A glycoprotein hormone (26 kDa) and **gonadotrophin**. Made up of an α-chain (96 amino acids) identical to other gonadotrophins and a hormone-specific β-chain. Acts with **follicle-stimulating hormone** to stimulate sex hormone release.

luteinizing hormone-releasing factor *LHRF* A decapeptide releasing hormone (1182 Da) that stimulates release of **luteinizing hormone**.

lutropin Synonym for **luteinizing hormone**.

luxury protein A term sometimes used to describe those proteins that are produced specifically for the function of differentiated cells and are not required for general cell maintenance (the so-called 'housekeeping' proteins).

LXR LXRalpha and LXRbeta, are differentially expressed ligand-activated transcription factors that induce genes controlling cholesterol homeostasis and lipogenesis. Expressed in liver, intestine and adrenal gland, and can complex with retinoid-X receptor (RXR). Ligand may be oxysterol metabolites of cholesterol (particularly 22(R)-hydroxycholesterol).

lyases Enzymes of the EC Class 4 (see **E classification**) that catalyse the non-hydrolytic removal of a group from a substrate with the resulting formation of a double bond; or the reverse reaction, in which case the enzyme is acting as a synthetase. Include decarboxylases, aldolases and dehydratases.

Lyb antigen Surface antigens of mouse B-lymphocytes. Nomenclature largely superseded by use of **CD antigen** markers.

lycopene A linear, unsaturated hydrocarbon **carotenoid** (536 Da); the major red pigment in some fruit. Extracted and used as an antioxidant.

Lyme disease Disease caused by *Borrelia burgdorferi*, a tick-borne spirochaete.

lymph Fluid found in the lymphatic vessels, which drain tissues of the fluid that filters across the blood vessel walls from blood. Lymph carries lymphocytes that have entered the lymph nodes from the blood.

lymph node *Lymph gland* Small organ made up of a loose meshwork of reticular tissue in which are enmeshed large numbers of lymphocytes, macrophages and accessory cells. Recirculating lymphocytes leave the blood through the specialized high-endothelial venules of the lymph node and pass through the node before being returned to the blood through the lymphatic system. Because the lymph nodes act as drainage points for tissue fluids, they are also regions in which foreign antigens present in the tissue fluid are most likely to begin to elicit an immune response.

lymphadenitis Inflammation of **lymph nodes**.

lymphadenopathy Pathological disorder of **lymph nodes**. Lymphadenopathy-associated virus (LAV) was the name given to **HIV** by the Pasteur Institute group.

lymphangioma An uncommon benign neoplasm or hamartoma arising from lymphatic vessels and consisting of lymphatic channels.

lymphoblast Often referred to as a blast cell. Unlike other usages of the suffix –blast, a lymphoblast is a further differentiation of a lymphocyte (T- or B-) occasioned by an antigenic stimulus. The lymphoblast usually develops by enlargement of a lymphocyte, active re-entry to the S phase of the cell cycle, mitogenesis, and production of much mRNA and ribosomes.

lymphocyte White cells of the blood that are derived from stem cells of the lymphoid series. Two main classes, T- and B-lymphocytes (T-cells and **B-cells**), are recognized; the latter are responsible (when activated) for production of antibody, the former are subdivided into subsets (**helper, suppressor, cytotoxic T-cells**) and are responsible both for cell-mediated immunity and for stimulating B-lymphocytes.

lymphocyte activation *Lymphocyte transformation* The change in morphology and behaviour of lymphocytes exposed to a mitogen or to an antigen to which they have been primed. The result is the production of **lymphoblasts**, cells that are actively engaged in protein synthesis and divide to form effector populations. Should not be confused with transformation of the type associated with oncogenic viruses, and 'activation' is therefore perhaps a better term.

lymphocyte transformation See **lymphocyte activation**.

lymphocytic leukaemia See **leukaemia**.

lymphoid cell Cells derived from stem cells of the lymphoid lineage: large and small lymphocytes, plasma cells.

lymphoid tissue Tissue that is particularly rich in lymphocytes (and accessory cells such as macrophages and reticular cells), particularly the **lymph nodes**, spleen, **thymus, Peyer's patches**, pharyngeal tonsils, adenoids and (in birds) the **Bursa of Fabricius**. See also **BALT, GALT, MALT, SALT** (bronchus-, gut-, mucosal- and skin-associated lymphoid tissues).

lymphokine Substance produced by a leucocyte that acts upon another cell. Examples are **interleukins, interferon**γ, lymphotoxin (tumour-necrosis factor β), granulocyte–monocyte colony-stimulating factor (**GM-CSF**). The term is becoming less common and **cytokine**, a more general term, is taking over. Cytokines include lymphokines.

lymphoma Malignant neoplastic disorder of lymphoreticular tissue that produces a distinct tumour mass, not a leukaemia (in which the cells are circulating). Includes tumours derived both from the lymphoid lineage and from mononuclear phagocytes; lymphomas arise commonly (but not invariably) in lymph nodes, spleen or other areas rich in lymphoid tissue. Lymphomas are subclassified as **Hodgkin's disease** or non-Hodgkin's lymphomas (e.g. **Burkitt's lymphoma**, large-cell lymphoma, histiocytic lymphoma).

lymphotoxin Cytotoxic product of T-cells: the term is usually restricted to **tumour-necrosis factor** β which is also known as lymphotoxin.

lymphotropic Having an affinity for lymphocytes – as, for example, some forms of HIV.

lyn Non-receptor tyrosine **kinase**, related to **src**. Plays a critical role in B-cell development and intracellular signalling. Lyn-deficient mice exhibit splenomegaly, elevated serum IgM, production of autoantibody and, later, glomerulonephritis.

Lyon hypothesis Hypothesis, first advanced by Lyon, concerning the random inactivation of one of the two X chromosomes of the cells of female mammals. In consequence females are chimaeric for the products of the X chromosomes, a situation that has been exploited in female Black Americans (who are heterotypic for isozymes of glucose-6-phosphate dehydrogenase) as a means to confirm the monoclonal origin of papillomas and of atherosclerotic plaques.

Lyonization See **Lyon hypothesis**.

lyophilic Characteristic of a material that readily forms a colloidal suspension. Molecules of the solvent form a shell around the particles; if the solvent is water then 'hydrophilic'.

lyophilization Now generally restricted to mean freeze drying, removal of water by sublimation under vacuum.

lyotropic series A listing of anions and cations in order of their effect on protein solubility (tendency to cause salting out). Essentially a competition between the protein and the ion for water molecules for hydration.

lysergic acid diethylamide *Lysergide; LSD* Hallucinogenic compound which causes effects that resemble schizophrenia.

lysigeny A mechanism for the formation of **aerenchyma** in which cells die to create the gas space. Lysigenous aerenchyma is found in many important crop species, including barley, wheat, rice and maize. See **schizogeny**.

lysine *Lys; K; 146 Da* Amino acid; the only carrier of a side-chain primary amino group in proteins. Has important structural and chemical roles in proteins. See Table A2.

lysis Rupture of cell membranes and loss of cytoplasm.

lysogenic bacteriophage Bacteriophage that can take part in a lysogenic or lytic cycle in its bacterial host. See **lysogeny**.

lysogenic conversion See **lysogeny**.

lysogeny The ability of some phages to survive in a bacterium as a result of the integration of their DNA into the host **chromosome**. The integrated DNA is termed a prophage. A regulator gene produces a **repressor protein** that suppresses the lytic activity of the phage, but various environmental factors, such as ultraviolet irradiation, may prevent synthesis of the repressor, leading to normal phage development and lysis of the bacterium. The best example of this is the **lambda bacteriophage**.

lysophosphatides Monoacyl derivatives of diacyl phospholipids that are present in membranes as a result of cyclic deacylation and reacylation of phospholipids. Membranolytic in high concentrations and fusogenic at concentrations that are just sublytic. May have important modulatory roles. See **lysophosphatidic acid**.

lysophosphatidic acid *LPA* A signalling phospholipid that will, for example, induce neurite retraction and the formation of retraction fibres in young cortical neurons by actin rearrangement. Also implicated in various inflammatory signalling cascades. The lysophosphatidic acid receptor-1 (LPA 1) is constitutively localized in the nucleus of mammalian cells.

lysophospholipase D *EC 3.1.4.39* Enzyme found in mammalian plasma and serum that hydrolyses lysoPE and lysoPC with different fatty acyl groups to the corresponding lysophosphoric acids. See **autotaxin**.

lysosomal diseases Diseases (also known as storage diseases) in which a deficiency of a particular lysosomal enzyme leads to accumulation of the undigested substrate for that enzyme within cells. Not immediately fatal, but within a few years lead to serious neurological and skeletal disorders and eventually to death. See the following diseases or syndromes: **Hurler's**, **Hunter's**, **San Fillipo**, **Gaucher's**, **Niemann-Pick**, **Pompe's**, **Tay–Sachs**.

lysosomal enzymes A range of degradative enzymes, most of which operate best at acid pH. The best-known marker enzymes are **acid phosphatase** and **beta-glucuronidase**, but many others are known.

lysosomal trafficking regulator LYST Gene that is mutated in **Chediak–Higashi syndrome**. Product in human is a 2186-amino acid polypeptide with homology to orphan proteins identified in *S. cerevisiae* (e.g. CDC4), and is apparently required for sorting endosomal resident proteins into late multivesicular endosomes by a mechanism involving microtubules.

lysosome Membrane-bounded cytoplasmic organelle containing a variety of hydrolytic enzymes that can be released into a phagosome or to the exterior. Release of lysosomal enzymes in a dead cell leads to autolysis (and is the reason for hanging game, to tenderize the muscle), but it is misleading to refer to lysosomes as 'suicide bags', since this is certainly not their normal function. Part of the GERL complex or trans-Golgi network. Secondary lysosomes are phagocytic vesicles with which primary lysosomes have fused. They often contain undigested material.

lysosome-associated membrane glycoproteins *LAMP-1, LAMP-2* Group of lysosome-specific integral membrane glycoproteins. Long luminal domain, short transmembrane domain, very short cytoplasmic tail. Function not yet clear. See **LAMP-1**, **LAMP-2**).

lysosome–phagosome fusion A process that occurs after the internalization of a primary phagosome. Fusion of the membranes leads to the release of lysosomal enzymes into the phagosome. Some species of intracellular parasite evade immune responses by interfering with this process.

lysosomotropic Having affinity for and thus accumulating in **lysosomes**.

lysozyme *EC 3.2.1.17; muramidase* Glycosidase that hydrolyses the bond between *N*-acetyl-muramic acid and *N*-acetyl-glucosamine, thus cleaving an important polymer of the cell wall of many bacteria. Present in tears, saliva and in the **lysosomes** of phagocytic cells, it is an important antibacterial defence, particularly against Gram-positive bacteria.

lysyl oxidase *EC 1.4.3.13* Extracellular enzyme that deaminates lysine and hydroxylysine residues in collagen or elastin to form aldehydes, these then interact with each other or with other lysyl side chains to form cross-links.

Lyt antigen A set of plasmalemmal surface glycoproteins on mouse T-lymphocytes. Possession of Lyt 1 partly defines a T-helper cell and of Lyt 2,3 suppressor and cytotoxic cells. Formerly known as Ly antigens; see also **Lyb antigen**, largely superseded by **CD antigen** nomenclature.

lytic Of, relating to, causing lysis.

lytic complex The large (2000-kDa) cytolytic complex formed from complement C5b6789. See **complement**.

lytic infection The normal cycle of infection of a cell by a **virus** or **bacteriophage**, in which mature virus or phage particles are produced, and the cell is then lysed.

lytic vacuole Vacuole found in plant cells that contains hydrolytic enzymes, analogous to the **lysosome** of animal cells but differing in morphology, function, enzyme content and mode of origin.

LZTR-1 A **BTB/kelch** protein, deleted in the majority of **DiGeorge syndrome** patients, believed to act as a transcriptional regulator.

M

m119 See **Mig**.

Mab See **monoclonal** antibody.

MAC *Membrane attack complex* See **complement** and **C9**.

Mac-1 *CR3; CD11b/CD18* αMβ2 **integrin** of leucocytes.

Macaca mulatta Rhesus monkey.

Macardle's disease *Glycogen storage disease type 5* A disorder caused by a phosphorylase B deficiency that leads to abnormal glycogen storage in the muscle and causing muscle cramping and stiffness following exercise.

MacConkey's agar Agar-based medium used for isolation of bacteria from faeces, etc. Contains lactose and neutral red as an indicator so that lactose-fermenting bacteria will produce red-pink colonies.

macedocin An anticlostridial bacteriocin **lantibiotic** (2.7 kDa) produced by *Streptococcus macedonicus*, a natural cheese isolate. Inhibits a broad spectrum of lactic acid bacteria, as well as several food spoilage and pathogenic bacteria.

Mach–Zehnder system Interferometric system in which the original light beam is divided by a semitransparent mirror: object and reference beams pass through separate optical systems and are recombined by a second semitransparent mirror. Interference fringes are displaced if the optical path difference for the reference beam is greater, and this can be compensated with a wedge-shaped auxiliary object. The position of the wedge allows the phase retardation of the object to be measured. The Mach–Zehnder system was used in a microscope designed by Leitz.

Machupo virus A member of the **Arenaviridae** that may cause a severe haemorrhagic fever in humans. The natural hosts are rodents and transmission from human to human is not common.

macrocytes Abnormally large red blood cells, numerous in **pernicious anaemia**.

macroglobulin Globulin such as IgM that has a high molecular weight: 400 kDa in the case of **IgM**, 725 kDa in the case of **alpha-2-macroglobulin**.

Macrolepidoptera A rather non-specific group, the larger butterflies and moths.

macrolides A group of antibiotics produced by various strains of *Streptomyces* that have a complex macrocyclic structure. They inhibit protein synthesis by blocking the 50S ribosomal subunit. Include erythromycin, carbomycin. Used clinically as broad-spectrum antibiotics, particularly against Gram-positive bacteria.

macromolecule Non-specific and rather imprecise term covering proteins, nucleic acids and carbohydrates, but probably not phospholipids.

macronucleus The larger nucleus (or sometimes nuclei) in ciliate protozoans. Derived from the micronucleus by a process of formation of **polytene chromosomes**. The DNA in the macronucleus is actively transcribed but the macronucleus degenerates before conjugation.

macrophage Relatively long-lived phagocytic cell of mammalian tissues, derived from blood **monocyte**. Macrophages from different sites have distinctly different properties. Main types are peritoneal and alveolar macrophages, tissue macrophages (**histiocytes**), **Kuppfer cells** of the liver and **osteoclasts**. In response to foreign materials may become stimulated or activated. Macrophages play an important role in killing of some bacteria, protozoa and tumour cells, release substances that stimulate other cells of the immune system, and are involved in **antigen presentation**. May further differentiate within chronic inflammatory lesions to epithelioid cells or may fuse to form **foreign body giant cells** or **Langhans' giant cells**.

macrophage colony-stimulating factor See **MCSF**.

macrophage inflammatory protein 1 *MIP1* Cytokine now recognized to exist in two forms, MIP1α and MIP1β (SIS-α, TY-5, L2G25B, 464.1, GOS 19-1). Small **cytokine** (monokine) with inflammatory and chemokinetic properties.

macrophage inflammatory protein 2 *MIP2* **Cytokine** that is **chemotactic** for **neutrophils**.

macrophage inhibition factor *MIF; macrophage migration inhibition factor* Substance that has had a complex history, partly because of doubts concerning the assay systems that were used to define the phenomenon: originally considered to be a mixture of lymphokines (including a 14-kDa glycoprotein) produced by activated T-lymphocytes that reduced macrophage mobility and probably increased macrophage–macrophage adhesion. More recently it has been suggested that MIF is a protein secreted by anterior pituitary cells in response to LPS stimulation but having the same biological activity.

macrophage-stimulating protein *MSP* A serum protein belonging to the plasminogen-related growth factor family and the ligand for **ron** receptor tyrosine kinase. A pleiotropic growth factor, it also inhibits NO production by macrophages in response to IFN-gamma and LPS-induced production of various inflammatory mediators.

macrosialin *CD68* Mouse homologue of CD68. Transmembrane **scavenger receptor** of the mucin-like class that includes **LAMP**-1 and -2. Expressed specifically on macrophages and related cells, binds oxidized **LDL**. Predominantly in lysosomal and endosomal membranes and, to a lesser extent, on cell surface.

MACS 1. Membrane-anchored C peptides; peptides derived from the C-terminal heptad repeat domain of HIV-1 gp41. 2. Medium-chain acyl-CoA synthetases; enzymes that catalyse the ligation of medium-chain fatty acids with CoA. 3. Magnetic-activated cell separation or magnetic-assisted cell sorting. 4. Membrane-associated adenylyl cyclases (mACs). 5. Mammalian artificial chromosomes (MACs).

macula adherens Spot desmosome: see **desmosome**.

macula densa A group of close-packed epithelial cells in the distal convoluted tubule of the kidney that control renin release in response to changes in the sodium concentration of the fluid passing through the tubule.

macule A spot: only commonly met in the construct 'immaculate', meaning unspotted (but see **macula adherens**).

MAD Basic **helix–loop–helix** leucine zipper transcription factor (28 kDa) involved in **Dpp** signalling (Dpp is TGFβ-related). MAD (mothers against dpp) contains a specific DNA-binding activity that activates an enhancer in a *Drosophila* wing-patterning gene, *vg* (vestigial). See also **SMAD**.

Madin–Darby canine kidney *MDCK* Line of canine epithelial cells that grow readily in culture and form confluent monolayers with relatively low trans-monolayer permeability (varies between clones). Often used as a general model for epithelial cells.

MAF *Macrophage activating factor* Imprecise term for a **lymphokine** that activates macrophages. Main example is interferon gamma.

MAG *Myelin-associated glycoprotein* A cell adhesion molecule of the **immunoglobulin superfamily** that plays a part in regulating neurite outgrowth. See **nogo** and **oligodendrocyte–myelin glycoprotein**.

magainins Cationic peptides of about 20 amino acid residues with antimicrobial activity, originally found in amphibian skin. Probably have membrane insertion and lytic properties. Sequence related to **melittin**.

MAGE antigens Melanoma-associated antigens, a superfamily of proteins associated with tumours and part of the larger 'cancer/testis antigen' family. The MAGE superfamily includes five families: MAGE-A, MAGE-B, MAGE-C, MAGE-D and **necdin**. Are thought to have some promise as vaccine targets.

MAGI-1 Membrane-associated guanylate kinase protein, a component of epithelial tight junctions both in Madin–Darby canine kidney cells and in intestinal epithelium.

magnesium *Mg* An essential divalent cation. The major biological role is as the chelated ion in ATP and presumably other triphosphonucleotides. The Mg^{2+}/ATP complex is the sole biologically active form of ATP. The other essential role of Mg^{2+} is as the central ion of chlorophyll. Cellular concentration is less than 5 mM, serum concentration approx. 1 mM.

magnetic resonance imaging *MRI* Non-invasive imaging technique that uses nuclear magnetic resonance to look at intact tissues in the body. Particularly valuable for studies on brain and soft tissues.

magnetoencephalography Measurement of the weak magnetic signals that are produced by the brain.

magnetoresistive Term for a metal or semiconductor that exhibits a change in electrical resistance when subjected to a magnetic field.

magnetosome Enveloped compartment in magnetotactic bacteria containing magnetite particles. By using this organelle to detect the vertical component of the Earth's magnetic field, the bacteria swim towards the bottom of the sea.

magnetotaxis Tactic response to magnetic field; in magnetotactic bacteria the Earth's magnetic field is used as a guide to 'up' and 'down' in deep sediment.

magnocellular neuron A neuron in the magnocellular region of the brain. Perhaps the first class of neuron from the central nervous system shown to be sensitive to **nerve growth factor** (which had previously been thought only to act at the periphery).

MAGUKs Membrane-associated guanylate kinase family of proteins of which **CASK** is a member.

maitotoxin *MTX* Toxin from the dinoflagellate, *Gambierdiscus toxicus*. Activates L-type voltage-sensitive calcium channels (**VSCC**) and mobilizes intracellular calcium stores.

major histocompatibility antigen *MHC antigen* A set of plasmalemmal glycoprotein antigens involved in rapid (e.g. 7 days in the mouse) graft rejection and other immune phenomena. The minor histocompatibility antigens are involved in much slower rejection phenomena. The major antigens show remarkable polymorphism and occur as class I and class II types in mammals; birds may have a class III molecule as well. See **histocompatibility antigens**, **MHC restriction**.

major histocompatibility complex *MHC* The set of gene loci specifying major histocompatibility antigens, e.g. HLA in man, H-2 in mice, RLA in rabbits, RT-1 in rats, DLA in dogs, SLA in pigs, etc.

major intrinsic protein *MIP* Family of structurally related proteins with six transmembrane segments, associated with gap junctions or vacuoles. MIP is found in lens fibre gap junctions. Other members: nodulin-26 (soybean), tonoplast intrinsic protein (TIP) found in plant storage vacuoles, *Drosophila* neurogenic protein 'big brain'.

major urinary protein *MUP-1* A pheromone-carrying protein of the **lipocalin** family.

majusculamide C Analogue of the natural antitumour agent **dolastatin**.

malabsorption syndrome A variety of conditions in which digestion and absorption in the small intestine are impaired. Multiple causes, including lymphoma, amyloid and other infiltrations, **Crohn's disease**, gluten-sensitive enteropathy, and the sprue syndrome in which the villi atrophy for unknown reasons.

malacia General pathological term for softening of any organ or tissue.

malaria In humans, the set of diseases caused by infection by protozoans of the genus *Plasmodium*. *P. vivax* causes the tertian type, *P. malariae* the quartan type and *P. falciparum* the quotidian or irregular type of disease, the names referring to the frequency of fevers. The fevers occur when the merozoites are released from the erythrocytes. The organisms are transmitted by species of the *Anopheles* mosquito, and are becoming increasingly resistant to antimalarial drugs.

Malassez cells Cells found in the periodontal ligament as 'epithelial rests of Malassez'. Malassez cells retain the major characteristics of epithelial cells throughout their differentiation from the root sheath epithelium into the rests of Malassez.

malate The ion from malic acid, a component of the citric acid cycle.

MALDI-TOF Common abbreviation for matrix-assisted laser-desorption ionization, a method for generating molecular ions, combined with time of flight (TOF) mass spectroscopy. An increasingly used method for analysing biological samples.

maleate The ion from maleic acid, often used in biological buffers.

malignant As applied to tumours, means that the primary tumour has the capacity to show **metastatic spread** (metastasize). Implies loss of both **growth control** and positional control.

malignant hyperthermia Channelopathy affecting calcium channels, a rare but often fatal genetic condition during anaesthesia.

Mallory's one-step stain A modified trichrome method that can give good tissue differentiation using a relatively brief procedure.

malonate The ion from malonic acid, $HOOC \cdot CH_2COOH$. Malonate is a competitive inhibitor for succinate dehydrogenase in the **tricarboxylic acid cycle**. Malonyl-SCoA is an important precursor for fatty acid synthesis.

Malpighian tubule Blind-ending tubule opening into the lower intestine of insects and responsible for fluid excretion, the arthropod equivalent of the kidney.

maltase EC 3.2.1.20 Enzyme that hydrolyses maltose (and the glucose trimer maltotriose) to glucose during the enzymic breakdown of starch.

maltose Disaccharide intermediate of the breakdown of starch, glucose-α(1-4)-glucose. Fermentable substrate in brewing.

maltose-binding protein Protein of the bacterial (*E. coli*) surface that links with **MCP**-II and is involved in the chemotactic response to maltose; probably derived from a similar protein that links with a transmembrane transport system.

MAML proteins *Mastermind-like proteins* A family of three co-transcriptional regulators essential for **Notch** signalling. They have distinct tissue-specific distributions; MAML1 interacts with **MEF2C**.

mammalian expression vector In molecular biology, a **vector** that will produce large amounts of eukaryotic protein (taxonomy notwithstanding, not necessarily a protein from a mammal).

mammamodulin Protein (52–55 kDa) that was reportedly expressed by hormone-independent mammary tumour cells and that affected morphology, motility, growth and hormone receptor expression. There have been no recent reports or entries in databases, and the term apparently only survives in dictionary entries remarkably like the one in the 3rd Edition.

mammary gland Milk-producing gland of female mammals. An adapted sweat gland, it is made up of milk-producing alveolar cells surrounded by contractile myoepithelial cells, together with considerable numbers of fat cells. Milk production is hormonally controlled.

mammary tumour virus *Bittner agent* **Retrovirus** that induces mammary **carcinoma** in mice. Isolated from highly inbred strains that had very high incidence of the tumours, after the discovery that the disease was transmitted in milk by nursing mothers. Endogenous provirus is present in germ line of all inbred mice. Transcription of the provirus is regulated by a viral promoter that increases transcription in response to glucocorticoid hormones. May transform by proviral insertion activating the cellular *int-1* oncogene.

mammary-derived growth inhibitor *MDGI* Heart-type **Fatty acid-binding protein** (H-FABP) that inhibits proliferation of mammary carcinoma cells. Overexpression or knockout of the gene in mice causes no overt phenotypic change.

Manduca sexta A species of Lepidopteran insect, also known as the tobacco hornworm moth. The caterpillars, which are very large, are used in studies of ion transport, moulting and as a system for transgenic gene expression (see **baculovirus**).

manganese *Mn* An essential trace element. Present in cells at concentrations of around 0.01 mM. Activates a wide range of enzymes, e.g. **pyruvate carboxylase** and one family of **superoxide dismutases**. Resembles **magnesium** and may replace it in many enzymes when it can modify substrate specificities. The addition of manganese salts to buffer solutions will often make cells very adhesive.

mannan Mannose-containing polysaccharides found in plants as storage material, in association with cellulose as hemicellulose. In yeasts, a wall constituent.

mannan-binding lectin *MBL; formerly mannan-binding protein, MBP* Plasma protein (32 kDa) structurally related to **complement** C1 that binds specific carbohydrates (**mannans**) on the surface of various microorganisms, including bacteria, yeasts, parasitic protozoa and viruses; activates the complement cascade through **MASP** and promotes phagocytosis. Deficiency is associated with frequent infections in childhood.

mannitol *D-Mannitol* Hexitol related to D-mannose. Found in plants, particularly fungi and seaweeds.

mannopeptimycins Mannopeptimycins alpha, beta, gamma, delta and epsilon are cyclic glycopeptide antibiotics produced by *Streptomyces hygroscopicus* LL-AC98. Mannopeptimycin epsilon is effective in treating infection due to methicillin-resistant staphylococci and vancomycin-resistant enterococci. Inhibit cell wall biosynthesis through lipid II binding.

mannoprotein Glycoproteins in which there are many mannose residues. Mannoproteins are a major component of the cell wall of *S. cerevisiae*, comprising about 40% of its mass. Mannoproteins are believed to be determinants of cell wall permeability, and some are essential for developmental events such as mating and transition to hyphal growth.

mannose *D-Mannose* Hexose identical to D-glucose except that the orientation of the -H and -OH on carbon 2 are interchanged (i.e. the 2-epimer of glucose). Found as constituent of polysaccharides and glycoproteins.

mannose-6-phosphate Mannose derivative, formed by phosphorylation in the **Golgi complex**, of certain mannose residues on *N*-glycan chains of lysosomal enzymes. Believed to function as targeting signal that causes entry of these enzymes to the lysosomes. The receptor (215 kDa) is enriched in specialized pre-lysosomes.

mannosidases Enzymes catalysing hydrolysis of the glycosidic bond between mannose residues and a variety of hydroxyl-containing groups. Alpha-mannosidases (EC 3.2.1.113) in rough endoplasmic reticulum and *cis*-Golgi are responsible for removing four mannose residues during the synthesis of the complex type N-linked glycan chains of glycoproteins.

mannosidosis Autosomal-recessive **storage disease** in which there is a defect in lysosomal alpha or beta-mannosidase.

mantle-cell lymphoma *MCL* A subtype of B-cell non-Hodgkin lymphoma that behaves aggressively and is incurable.

Mantoux test Test for tuberculin reactivity in which tuberculin PPD (purified protein derivative) is injected intracutaneously. The injection site is examined after 2–3 days; a positive reaction, indicating current or previous infection with *Mycobacterium tuberculosis* (in an uninoculated individual), is an oedomatous and reddened area caused by T-cell reactivity.

MAP kinase *Mitogen-activated protein kinase; externally regulated kinase, ERK* Serine–threonine kinases that are activated when quiescent cells are treated with mitogens and that therefore potentially transmit the signal for entry into cell cycle. One target is transcription factor p62TCF. MAP kinase itself can be phosphorylated by MAP kinase kinase, and this may in turn be controlled by **raf**-1. Confusingly, also phosphorylate **microtubule-associated proteins**.

MAP1C Microtubule-associated protein (two heavy chains of 410 kDa associated with six or seven light chains of about 50–70 kDa), now considered to be the two-headed cytoplasmic equivalent of ciliary dynein and to be responsible for retrograde transport (transport towards the centrosome).

MAPKK MAP **kinase** kinase, phosphorylates MAPK on both the tyrosine and threonine residues, and in that order. It is itself a substrate for phosphorylation by **MAPKKK**.

MAPKKK *MAP kinase kinase kinase* Family of kinases that regulate the activity of the kinases that phosphorylate MAP kinases. Lie at the top of signalling cascades.

MAPs *Microtubule-associated proteins* May form part of the electron-lucent zone around a microtubule. MAP1A and 1B (approximately 350 kDa), from brain, form projections from microtubules; MAP2A and 2B (270 kDa) are also from brain microtubules and form projections. MAP3 (180 kDa) and MAP4 (220–240 kDa) have been described as co-purifying with MAPs 1 and 2. Microtubule-associated protein 2 is associated with axonal processes of vertebrate nerve cells, and its presence distinguishes them from dendritic processes. **MAP1C** is in a separate class, being a motor molecule.

MAPTAM *1,2-Bis-(2-amino-5-methylphenoxy)ethane tetraacetic acid tetraacetoxymethyl ester* Compound which readily enters cells, where it is converted to 5-methyl **BAPTA**, an indicator of calcium concentration.

marasmus Wasting of the body due to a deficiency in energy-giving food, as opposed to protein deficiency (kwashiorkor).

Marburg virus A **filovirus** that causes Marburg disease, a severe haemorrhagic fever developed in many people who work with African green monkeys.

MARCKS *Myristoylated alanine-rich protein kinase C-substrate* Membrane-associated (through the myristoyl residue) calmodulin- and actin-binding protein that cycles between membrane and cytoplasm depending on its phosphorylation state and on the presence of calcium/calmodulin. As a result of phosphorylation by protein kinase C, MARCKS is displaced from the membrane; dephosphorylation leads to reassociation. Implicated in macrophage activation, neurosecretion and growth factor-dependent mitogenesis.

Marek's disease Infectious cancer of the lymphoid system (lymphomatosis) in chickens, caused by a contagious herpes virus. An effective vaccine is now available.

Marfan syndrome Dominant disorder of connective tissue in which limbs are excessively long and loose-jointed. Due to mutation in the fibrillin-1 gene that causes collagen fibril-assembly disorder that can be mimicked in mice by aminonitriles that interfere with cross-linking.

margaratoxin Charybdotoxin-related peptide toxin (39 residues) from scorpion *Centruroides margaritatus*. Blocks mammalian voltage-gated potassium channels in neural tissues and lymphocytes. Very similar to **noxiustoxin** and **kaliotoxin**.

marginal band A bundle of equatorially located microtubules that stabilize the biconvex shape of platelets and avian erythrocytes. They are unusual in that they do not derive from the centrosomal **MTOC**.

margination Adhesion of leucocytes to the endothelial lining of blood vessels, particularly postcapillary venules; often, but not always, a prelude to leaving the circulation and entering the tissues.

mariner Group of **transposons** with broad phylogenetic distribution (arthropods, nematodes, planaria, humans). Mariner elements consist of a transposase gene flanked by short inverted repeats.

marker gene Gene that confers some readily detectable phenotype on cells carrying the gene, either in culture or in transgenic or chimaeric organisms. Gene could be an enzymic **reporter gene**, a selectable marker conferring antibiotic resistance, or a cell membrane protein with a characteristic **epitope**.

marker rescue The restoration of gene function by replacing a defective gene with a normal one by recombination. The most common application is in co-infection of cells with a mutant phage that is unable to replicate and a wild-type phage; recombination between the phages repairs the replication defect and the recombinant derivative of the mutant phage can reproduce – it has been rescued.

Markov process A **stochastic** process in which the probability of an event in the future is not affected by the past history of events.

Maroteaux–Lamy syndrome Mucopolysaccharidosis type VI; deficiency of the lysosomal enzyme arylsulphatase B; resembles **Hurler's disease** in some respects.

MARPS *Microtubule repetitive proteins MARP-1 and MARP-2* Heat-stable high molecular weight proteins (*ca.* 320 kDa) with many 38-residue repeats. Isolated from membrane skeleton of ***Trypanosoma*** *brucei*; probably stabilize microtubules; each repeat has tubulin-binding activity.

mas mas1 Oncogene isolated from DNA of a human epidermoid carcinoma cell line. The *mas1* gene product contains seven potential transmembrane domains and is a G-protein-coupled receptor for the peptide angiotensin 1–7 that has opposite effects to **angiotensin** II and has been shown to stimulate the proliferation of multipotential and differentiated progenitor cells in cultured bone marrow and human cord blood. See Table O1.

masked messenger RNA Long-lived and stable mRNA found originally in the oocytes of echinoderms and constituting a store of maternal information for protein synthesis that is unmasked (derepressed) during the early stages of morphogenesis. In these early stages the rate of cell division is so rapid that transcription from the embryonic genome cannot occur. Undoubtedly not restricted to oocytes, and the term can be applied to any mRNA which is present in inactive form.

maskin Protein in oocytes that binds eukaryotic translation initiation factor 4E (eIF4E) and prevents formation of the eIF4F initiation complex. When the oocytes are stimulated to re-enter the meiotic divisions (maturation), CPEB promotes cytoplasmic polyadenylation of the formerly quiescent mRNA, the newly elongated poly(A) tail becomes bound by poly(A)-binding protein (PABP), which in turn binds eIF4G and helps it displace maskin from eIF4E, thereby inducing translation. See **cytoplasmic polyadenylation element-binding factor** (CPEB).

MASP *MBL-associated serine protease; EC 3.4.21.104* Peptidases (*ca.* 76 kDa) in the **complement** system, activated by the binding of mannan to **mannan-binding lectin** (MBL), (formerly mannose-binding protein, MBP). MASP-1 and MASP-2 are similar to the **C1q**-associated proteases, C1r and C1s. All four have a C1r/C1s-like domain, an **EGF-like domain** and a second C1r/C1s-like domain, two complement control protein (CCP) domains and a serine protease domain. MASP-1 and MASP-2 are probably responsible for the proteolytic cleavage and activation of C2 and C4.

maspin Serpin (42 kDa) expressed in normal mammary epithelium but reduced in mammary tumours.

mass spectrometry *MS* Widely used method for determining the relative mass and abundance of ionized molecules or fragments of molecules (generated, for example, by **MALDI**) that are accelerated *in vacuo* by electrical means, and their trajectories within the beam of charged particles modified by the application of magnetic fields; particles are separated on the basis of mass/charge ratio. The spectrum of peaks on the detector system indicates the composition of the mixture and is diagnostic of the starting material. See **MALDI-TOF** and **electrospray mass spectroscopy**.

Masson's trichrome stain Trichrome stains are used particularly for connective tissue. Masson's trichrome method uses haemalum, acid fuchsin and methyl blue, and has the effect of staining nuclei blue-black, cytoplasm red and collagen blue.

mast cell Resident cell of connective tissue that contains many granules rich in **histamine** and heparan sulphate. Release of histamine from mast cells is responsible for the immediate reddening of the skin in a weal-and-flare response. Very similar to **basophils,** and possibly derived from the same stem cells. Two types of mast cells are now recognized; those from connective tissue, and a distinct set of mucosal mast cells. The activities of the latter are T-cell dependent.

mastigonemes Lateral projections from eukaryotic flagella. May be stiff and alter the hydrodynamics of flagellar propulsion, or flexible and alter the effective diameter of the flagellum (flimmer filaments).

mastocytoma Neoplastic **mast cells**.

mastoparans Basic peptides from wasp venoms. Analogous to **melittin** in honeybee venom, they can act as **phospholipase** A2 activators, but their relevance to the toxic action of the venoms is not known.

maternal antibody Any antibody transferred from a mammalian mother transplacentally into the foetus. See **IgG**.

maternal inheritance Inheritance through the maternal cell line, e.g. through the oocyte and eggs. Mitochondrial genes are maternally inherited, and various other non-Mendelian forms of inheritance may also appear as maternal inheritance.

maternal mRNA Messenger RNA found in oocytes and early embryos that is derived from the maternal genome during oogenesis. See **masked messenger RNA**.

maternal-effect gene Gene, usually required for early embryonic development, whose product is secreted into the egg by the mother. The phenotype is thus determined by the mother's, rather than the egg's, genotype, cf. **zygotic-effect gene**. See also **egg-polarity gene**.

mating-type genes Genes that, in *S. cerevisiae*, specify into which of the two mating types (a and α) a particular cell falls. Only unlike mating-type haploids will fuse. The interest derives from the way in which mating type is switched: the existing gene is removed and a new gene, derived from a (silent) master copy elsewhere in the genome, is spliced in. Later this gene will in its turn be replaced by a new copy of the old gene, also derived from a silent 'master'. The a and α genes code for pheromones that affect cells of the opposite mating type. Similar mating-type genes are known from other yeasts, and the switching mechanism (**cassette mechanism**) may be used more generally.

Matrigel Proprietary name for gel-forming matrix material derived from **EHS cells**. Cells grown on Matrigel often show morphological characteristics distinct from those seen on a solid tissue culture substratum, and are probably in a more normal environment both chemically and physically.

matrilysin *Matrix metallopeptidase 7; EC 3.4.24.23* Matrix metalloproteinase that is thought to play a part in cellular invasion of tissues by digesting extracellular matrix. Matrilysin-2 is MMP-26. See Table M1.

matriosome See **striosome**.

matrix Ground substance in which things are embedded or that fills a space (as, for example, the space within the mitochondrion). Most common usage is for a loose meshwork within which cells are embedded (e.g. extracellular matrix), although it may also be used of filters or absorbent material.

TABLE M1. Matrix metallopeptidases

Enzyme	Designation	Main substrate
Interstitial collagenase	MMP-1	Fibrillar collagens
Gelatinase A	MMP-2	Progelatinase A
Stromelysin-1	MMP-3	Non-fibrillar collagen, Ln, Fn
Matrilysin	MMP-7	Ln, Fn, non-fibrillar collagen
Neutrophil collagenase	MMP-8	Fibrillar collagens I, II and III
Gelatinase B	MMP-9	Interleukin-8 precursor, metastasis-suppressor KiSS1 precursor
Stromelysin-2	MMP-10	Ln, Fn, non-fibrillar collagen
Stromelysin-3	MMP-11	Serpin
Metalloelastase	MMP-12	Elastin
Collagenase-3	MMP-13	Fibrillar collagens
MT-MMP	MMP-14	Pro-gelatinase-A
Membrane-type matrix metallopeptidase-3	MMP-16	Pro-gelatinase-A
Collagenase 4 (*Xenopus*)	MMP-18	α-1-proteinase inhibitor, type I collagen α-1
RASI-1, RASI-6	MMP-19	Aggrecan core protein
Enamelysin	MMP-20	Amelogenin
Xenopus MMP-21	MMP-21	?
MMP-27 (*Homo sapiens*)	MMP-22	?
MIFR protein (*Homo sapiens*)	MMP-23B	Synthetic
Matrilysin-2	MMP-26	Fibrinogen, vitronectin, matrilysin, α-1-proteinase inhibitor
Epilysin	MMP-28	?

For more detailed information consult the MEROPS peptidase database, http://merops.sanger.ac.uk/ Fn, fibronectin; Ln, laminin.

matrix attachment region *MAR; scaffold/matrix attachment regions* Regions of chromatin sequence that binds directly to nuclear matrix. Found in intergenic DNA, especially flanking the 5′ ends of genes or clusters of genes. Many have a sequence motif called the MAR/SAR recognition signature sequence that binds to matrix attachment region-binding proteins.

matrix metallopeptidases *MMP; matrix metalloproteinases* Proteolytic enzymes that degrade extracellular matrix. Include collagenases and elastases. Inhibitors are predicted to have benefits in arthritis and metastasis though this remains to be proven. See Table M1.

matrix proteins Proteins of the outer layer of the cell wall of Gram-negative bacteria.

matrix-associated region-binding proteins *MARBP* Group of plant coiled-coil nuclear proteins that may serve as functional homologues of **lamins**.

maturation-promoting factor *MPF* See **cyclin**.

Mauthner neuron Large neuron in the **mesencephalon** of fishes and amphibians. A rare example of an individually identifiable neuron in a vertebrate nervous system.

Max Transcription factor: forms homodimers which then interact with CACGTG motif of DNA repressively, but will form heterodimers with **Myc protein** that bind the same motif with greater affinity and activate the downstream gene.

Maxam–Gilbert method A method of DNA sequencing, based on the controlled degradation of a DNA fragment in a set of independent, nucleotide-specific reactions. The resulting fragments have characteristic sizes, depending on the sequence of the template, that can be resolved on a sequencing gel. Although no longer the main protocol, Maxam–Gilbert sequencing still has advantages, e.g. for oligonucleotides or covalently modified DNA. See also **dideoxy sequencing**.

maximum tolerable concentration *MTC; LC_0* Highest concentration of a substance in an environmental medium that does not cause death of test organisms or species.

maximum tolerable dose *MTD; LD_0* Highest amount of a substance that, when introduced into the body, does not kill test animals.

maximum tolerable exposure level *MTEL* Maximum amount (dose) or concentration of a substance to which an organism can be exposed without leading to an adverse effect after prolonged time of exposure.

maximum tolerated dose *MTD* Highest dose of a substance that, on the basis of subchronic testing, is expected to produce limited toxicity when administered chronically for the duration of the test period.

maxiprep Slang, denoting a large-scale purification of **plasmid** from a bacterial culture. Usually used to describe preparations from 100 to 500 ml culture. See also **miniprep**, **midiprep**, **megaprep**.

mayven An actin-binding protein expressed predominantly in the CNS, has six **kelch** repeats. Thought to have a role in the formation of processes from oligodendrocyte precursor (O2-A) cells. Associates with the SH3 domain of **fyn**. See **actinfilin**.

M-band Central region of the **A-band** of the **sarcomere** in striated muscle.

MBC Abbrev. variously for (1) Metastatic breast cancer, (2) Minimum (minimal) bactericidal concentration, (3) **Carbendazim**, methyl benzimidazol-2-ylcarbamate, (4) Microbial biomass carbon.

MBF *Mlu1-binding factor* Protein complex from the fission yeast *Schizosaccharomyces pombe* that contains the proteins Res1p and Res2p and binds to the Mlu1 cell cycle box (**MCB**) element in DNA, activating the transcription of genes required for S phase.

MBL See **mannan-binding lectin**.

MBP See **myelin basic protein**.

MCAF See **monocyte chemotactic and activating factor**.

McArdle's disease *Glycogen storage disease type V* Glycogen storage disease in which the defective enzyme is muscle glycogen phosphorylase (EC 2.4.1.1).

mcb 1. Metaplastic carcinoma of the breast, a rare form of cancer containing mixture of epithelial and mesenchymal elements in variable combinations. **2.** Monochlorobimane. **3.** Multicolour banding (of chromosomes). **4.** MluI cell-cycle boxes (MCB; see **MBF**). **5.** Monochlorobenzene.

M cells Cells found amongst the other cells of the cuboidal surface epithelium of **gut-associated lymphoid tissue**; have a complex folded surface.

MCF7 Line of human breast adenocarcinoma cells derived from a metastatic site.

MCH Abbrev. for (1) **Melanin-concentrating hormone**, (2) Mean erythrocyte haemoglobin content, (3) Methacholine (MCh).

M-channels Voltage-sensitive **potassium channels** inactivated by **acetylcholine**. ACh acting at **muscarinic acetylcholine receptors** produces an internal messenger that turns off this class of K-channel. A mechanism for regulating the sensitivity of cells to synaptic input.

MCHC Mean corpuscular haemoglobin concentration; the amount of haemoglobin per erythrocyte, a standard measure in haematology.

MCI *Mild cognitive impairment* Slight impairment of memory that may be a precursor of dementia.

MCL A software package (Markov CLustering) using an efficient algorithm for detecting protein families on the basis of sequence data.

MCM Abbrev. for (1) **Minichromosome maintenance proteins**, (2) Modified Chee's medium, (3) Microglia-conditioned medium, (4) **Methylmalonyl-Coenzyme A mutase**. See also **MCM-41**.

MCM-41 *Mobile crystalline material-41* A silicate produced by a templating mechanism that has arrays of non-intersecting hexagonal channels. By changing the length of the template molecule, the width of the channels can be controlled to be within 2–10 nm. The walls of the channels are amorphous SiO_2; used as a selective adsorption agent and molecular sieve.

MCP-1 See **monocyte chemoattractant protein-1**.

MCPA *4-Chloro-2-methylphenoxyacetic acid* Herbicide. Current view is that MCPA is not genotoxic *in vivo*, which is consistent with its lack of carcinogenicity in rats and mice.

MCPs *Methyl-accepting chemotaxis proteins* Proteins of the inner cytoplasmic face of the bacterial plasma membrane with which the receptors of the outer face interact. Four different MCPs are known in *E. coli*, each with a separate set of receptors. Can be methylated at various sites; methylation is part of the **adaptation** to the signal. Although important intermediate signal integration sites, they are not directly connected to the flagellar motor system used in chemotaxis.

MCS See **polycloning site**.

MCSF *CSF-1; macrophage colony-stimulating factor* A 40–76-kDa glycoprotein that plays an important role in the activation and proliferation of **microglial cells** both *in vitro* and in injured neural tissue. It is important in adipocyte hyperplasia, in osteoclast differentiation, and in one of the early events in atherosclerosis – monocyte to macrophage differentiation in the arterial intima. The receptor for MCSF (**c-fms**) is expressed on the pluripotent precursor and mature osteoclasts and macrophages. Mutation in MCSF leads to osteopetrosis because of osteoclast deficiency.

M-current Flow of potassium ions through **M-channels**.

MCV Abbrev. for (1) Mean cell volume or mean corpuscular volume, a standard parameter in haematological analyses, (2) Measles-containing vaccine, (3) **Molluscum contagiosum virus**.

MDCK cells Madin–Darby canine kidney cells.

MDGI See **mammary-derived growth inhibitor**.

Mdm2 Oncoprotein that inhibits **p53** by binding to the transcriptional activator domain of p53, preventing it from regulating its target genes. Mdm2 expression is activated by p53, and Mdm2 promotes degradation of p53 by the proteasome – a complex feedback regulation.

Mdr See **multidrug transporter**.

mdr-TB Multidrug-resistant tuberculosis.

Mdx mouse Mouse mutant deficient in **dystrophin** and thus a good model system for **Duchenne muscular dystrophy**.

ME See **myalgic encephalomyelitis**.

mean residence time *MRT* In pharmacokinetics, the average time a drug molecule remains in the body or an organ after rapid intravenous injection.

measles virus **Paramyxovirus** that causes the childhood disease measles and is responsible for subacute sclerosing panencephalitis.

mechanoreceptor A sense organ or cell specialized to respond to mechanical stimulation.

meclozine HCl An antihistamine drug used for treatment of motion sickness.

media Avascular middle layer of the artery wall, composed of alternating layers of elastic fibres and smooth muscle cells.

median effective concentration EC_{50} Concentration of a substance in an environmental medium expected, on the basis of statistical calculation, to produce a certain effect under a defined set of conditions.

median effective dose ED_{50} Dose of a chemical or physical agent (radiation) that is expected to produce a certain effect or to produce a half-maximal effect in a biological system under a defined set of conditions.

median lethal concentration LC_{50} Concentration of a substance expected to kill 50% of organisms under a defined set of conditions. Estimation of LC_{50} was a regulatory requirement in drug testing until superceded by more sensitive and less wasteful measurements.

median lethal dose LD_{50} Dose of a chemical or physical agent (radiation) expected to kill 50% of organisms.

median lethal time TL_{50} Average time interval during which 50% of a given population may be expected to die following acute administration of a chemical or physical agent (radiation) at a given concentration or dose under a defined set of conditions.

mediastinal Term to describe the mesentery-like membrane which separates the pleural cavities of the two sides in higher vertebrates; also, in mammals, applied to a mass of fibrous tissue that represents an internal prolongation of the capsule of the testis.

medicinal leech The freshwater leech (*Hirudo medicinalis*) formerly used for bloodletting. The large size and simple nervous system have made it a favourite animal for some neurophysiological studies.

medium Shorthand for culture medium or growth medium, the nutrient solution in which cells or organs are grown.

medroxyprogesterone Synthetic progestogen used for the treatment of menstrual disorders and, when given as a deep intramuscular injection, as a depot (long-lasting) contraceptive.

medulla oblongata Region of the brain where the spinal cord tapers into the brain stem. Neurons in this region regulate some very basic functions such as respiration.

MEF Most commonly, mouse embryonic fibroblasts, but see **MEF2**.

MEF2 *Myocyte-enhancer factor 2* Group of transcription factors of the **MADS** superfamily. MEF2C is expressed in skeletal muscle, spleen, brain and various myeloid cells. In monocytic cells, MEF2C has its transactivation activity enhanced by **LPS** acting through the MAP kinase **p38**. The result of activation is increased *c-jun* transcription. MEF-2A is a calcium-regulated transcription factor that promotes cell survival during nervous system development.

mefloquine See **Lariam**.

megakaryocyte Giant polyploid cell of bone marrow that gives rise to 3000–4000 platelets.

megalin *gp330* Epithelial endocytic receptor that internalizes various ligands, including apolipoproteins E and B. Major antigen against which there is autoimmune response in **Heymann nephritis**.

megaloblast An embryonic cell that will give rise to erythroblasts by mitotic division within the blood vessels.

megaloblastic anaemia *Pernicious anaemia* A heterogeneous group of disorders that have common morphological characteristics. Erythrocytes are larger than normal, neutrophils can be hypersegmented and megakaryocytes are abnormal. **Megaloblasts** are found in the bone marrow and macrocytes are often found in the peripheral blood. The usual cause is a deficiency in vitamin B12 or folic acid.

megaprep Slang, denoting a medium-scale purification of **plasmid** from a bacterial culture. Usually used to describe preparations from over 500 ml culture. See also **miniprep**, **midiprep**, **maxiprep**.

megaspore Haploid spore, produced by a plant sporophyte, that develops into a female gametophyte.

meiocytes Cell that will undergo meiosis; a little-used term.

meiosis A specialized form of nuclear division in which there are two successive nuclear divisions (meiosis I and II) without any chromosome replication between them. Each division can be divided into four phases similar to those of mitosis (pro-, meta-, ana- and telophase). Meiosis reduces the starting number of $4n$ chromosomes in the parent cell to n in each of the four daughter cells. Each cell receives only one of each homologous chromosome pair, with the maternal and paternal chromosomes being distributed randomly between the cells. This is vital for the segregation of genes. During the prophase of meiosis I (classically divided into the stages **Leptotene**, **Zygotene**, **Pachytene**, **Diplotene** and **Diakinesis**), homologous chromosomes pair to form **bivalents**, thus allowing **crossing over** – the physical exchange of chromatid segments. This results in the **recombination** of genes. Meiosis occurs during the formation of **gametes** in animals, which are thus haploid, and fertilization gives a diploid egg. In plants meiosis leads to the formation of the spore by the sporophyte generation.

meiospore Haploid spore formed after meiotic division.

meiotic spindle The meiotic equivalent of the **mitotic spindle**.

mek *MAPK/ERK kinase* Mitogen-activated protein kinase (Mek1, 45 kDa; Mek2, 46 kDa).

MEKK Mitogen-activated protein/ERK kinase kinases, a family of enzymes that phosphorylate mitogen-activated protein **MAP kinases** (meks).

Mel-14 Antibody that reacts with L-selectin (CD62L). Blocks lymphocyte binding to **HEV** both *in vitro* and *in vivo*.

melanin Pigments largely of animal origin. High molecular-weight polymers of indole quinone. Colours include black/brown, yellow, red and violet. Found in feathers, cuttle ink, human skin, hair and eyes, in cellular immune responses and in wound healing in arthropods.

melanin-concentrating hormone *MCH* A hypothalamic cyclic neuropeptide with key central and peripheral actions on the regulation of energy balance. Two G-protein-coupled receptors (MCHR1 and -2) are known. Pharmacological antagonism at MCHR1 in rodents diminishes food intake and results in significant and sustained weight loss in fat tissues, particularly in obese animals. MCHR1 antagonists have been shown to have anxiolytic and antidepressant properties.

melanocortin A group of pituitary peptide hormones that include **adrenocorticotrophin** (ACTH) and the alpha, beta and gamma melanocyte-stimulating hormones (MSH).

melanocyte Cells that synthesize melanin pigments. The pigments are stored in **melanosomes (chromatophores)** that can be redistributed in the cytoplasm to change pigment patterns in fish and reptiles.

melanocyte-stimulating hormone *MSH* A releasing hormone produced in the mammalian hypophysis and related structures in lower vertebrates. Made up of α-MSH (1665 Da), the same as amino acids 1–13 of **ACTH** and β-MSH (18 amino acids in pig, 22 in humans). Causes darkening of the skin by expansion of the **melanophores,** but its role in mammals is unclear.

melanoma Neoplasia derived from **melanocytes**; benign forms are moles, but often are highly malignant. Generally the cells contain melanin granules, and for this reason they have been used in studies on metastasis because the secondary tumours are easily located in lung.

melanoma growth-stimulatory activity *MGSA, neutrophil-activating protein 3, NAP-3, gro, KC, N51, CINC, CXCL1* **Chemokine** of the C–X–C subfamily. Potent **mitogen**. Activates (and is chemotactic for) neutrophils.

melanoma-associated antigens *MAA* There are five known MAA: melan-A, tyrosinase, gp100, **MAGE-1** and **MAGE-3**.

melanophore Cell type found in skin of lower vertebrates (amphibian skin, fish scales) that contains granules of the black pigment **melanin**. The granules can be rapidly redeployed between a dispersed state (which darkens the skin) and concentration at the centre (which lightens it). One of a family of pigmented or light-diffracting, coloured cells, known collectively as **chromatophores**.

melanopsin *Opsin 4* A photopigment (534 aa), initially cloned from cultured photosensitive dermal melanophores derived from *Xenopus laevis* embryos, found in a subpopulation of mammalian retinal ganglion cells that are intrinsically photosensitive. Important for photoentrainment of behaviour; more closely related to opsin proteins found in invertebrates. Gq-coupled. See **opsin subfamilies**.

melanosome Membrane-bounded organelle found in **melanocytes**; when **melanin** synthesis is active its internal structure is characteristic, containing melanofilaments that have a periodicity of around 9 nm and are arranged in parallel arrays. Mature melanosomes, in which the filamentous structure is masked by the dense accumulation of melanin, are transferred to keratinocytes in the skin. Also found in **pigmented retinal epithelium** and in some cells of the connective tissue.

melatonin N-*acetyl 5-methoxytryptamine* A hormone secreted by the pineal gland. In lower vertebrates, causes aggregation of pigment in melanophores and thus lightens skin. In humans, believed to play a role in establishment of circadian rhythms and purported to help overcome the effects of jet lag.

Meleagris gallopavo The turkey.

melibiose *6-O-α-D-galactopyranosyl-D-glucose* Disaccharide of galactose and glucose. Found in exudates and nectar of some plants; used as a pheromone by the desert locust and some other insects. Added to a variety of cosmetic preparations, although no evidence exists for any beneficial effects.

melioidosis *Whitmore's disease* A fatal infectious disease, predominantly in the tropics, caused by the bacterium *Burkholderia* (formerly *Pseudomonas*) *pseudomallei*. Affects lymph nodes and viscera; clinically and pathologically similar to glanders. Has been seen as potential bioterror agent.

melittin The major component of bee venom, responsible for the pain of the sting. A 26-amino acid peptide that has a hydrophobic and a positively charged region. Can lyse cell membranes and activate phospholipase A2 enzymes; it has a very high affinity for **calmodulin**, but the biological relevance of this is unclear.

mellitose See **raffinose**.

meloxicam Non-steroidal anti-inflammatory drug used for the short-term treatment of acute osteoarthritis and for the long-term treatment of rheumatoid arthritis.

melting curve Graph of the melting (denaturation and strand separation, detected by change in absorbance) of DNA as a function of temperature, from which can be calculated the temperature (T_m) at which half the molecules have undergone thermal denaturation. Melting-curve analysis can distinguish products of the same length but different GC/AT ratio, and products with the same length and GC content but differing in their GC distribution along the sequence will have very different melting curves. Small sequence differences may even be observable.

melusin A muscle-specific protein required for heart hypertrophy in response to mechanical overload. The N-terminal portion has a tandemly repeated **CHORD domain**. Melusin also has a C-terminal calcium-binding stretch of 30 acidic amino acid residues which is absent in **chp-1**.

memantine *1-Amino-3,5, dimethyl adamantane* Relatively low-affinity non-competitive **NMDA** antagonist used in the treatment of **Parkinsonism** and some other brain disorders.

membrane Generally, a sheet or skin. In cell biology the term is usually taken to mean a modified lipid bilayer with integral and peripheral proteins, as forms the plasma membrane. Because this usage is so general, it is advisable to avoid other uses where possible, particularly in histology or ultrastructure.

membrane attack complex See **complement**.

membrane capacitance The electrical capacitance of a membrane. Plasma membranes are excellent insulators and dielectrics: capacitance is the measure of the quantity of charge that must be moved across unit area of the membrane to produce unit change in membrane potential, and is measured in Farads. Most plasma membranes have a capacitance around $1\,\mu F/cm^2$.

membrane depolarization See **depolarization**.

membrane fluidity Biological membranes are viscous, two-dimensional fluids within their physiological temperature range.

membrane fracture See **freeze fracture**.

membrane potential More correctly, transmembrane potential difference: the electrical potential difference across a plasma membrane. See **resting potential**, **action potential**.

membrane protein A protein with regions permanently attached to a membrane (peripheral membrane protein) or inserted into a membrane (integral membrane protein). Insertion into a membrane implies hydrophobic domains in the protein. All **transport proteins** are integral membrane proteins.

membrane recycling The process whereby membrane is internalized, fuses with an internal membranous compartment and is then re-incorporated into the plasma membrane. In cells that are actively secreting by an exocrine method (in which secretory granules fuse with the plasma membrane), it is obviously essential to have some way of reducing the area of the plasma membrane. The membrane can then be used to form new secretory vesicles. The converse is true for phagocytic cells.

membrane transport The transfer of a substance from one side of a plasma membrane to the other, in a specific direction and at a rate faster than diffusion alone. See **active transport**.

membrane vesicles Closed unilamellar shells formed from membranes either in physiological transport processes or when membranes are mechanically disrupted. They form spontaneously when membrane is broken because the free ends of a lipid bilayer are highly unstable.

membrane zippering See **zippering**.

memory cells Cells of the immune system that 'remember' the first encounter with an antigen and facilitate the more rapid secondary response when the antigen is encountered on a subsequent occasion. The long-lasting immune memory is humoral and resides in B-lymphocytes, although it appears that persistence of the antigen may be essential. T-cell memory is shorter.

MEN1 See **menin**.

menadione *Vitamin K3; 2-methyl-1,4-naphthalene dione* Synthetic naphthoquinone derivative with properties similar to those of vitamin K. See also Table V1.

menaquinone *Methyl-napthoquinone* Class of naphthaquinones (vitamins K) produced by bacteria in the intestine and important in blood clotting.

Mendelian inheritance Inheritance of characters according to the classical laws formulated by Gregor Mendel, which give the classic ratios of segregation in the F2 generation. In sexually reproducing organisms, any process of heredity explicable in terms of chromosomal segregation, independent assortment and homologous exchange.

menin Nuclear protein (610 aa) that binds to DNA through its nuclear localization signals, product of the *MEN1* gene. Defects in this gene lead to the autosomal-dominant disease multiple endocrine neoplasia type 1, and *MEN1* is thought to be a tumour suppressor gene. The biochemical function(s) of menin are currently unknown.

meninges Three layers of tissue surrounding the brain: See **dura mater**, **pia mater** and arachnoid layer.

meningitis Inflammation of the meninges of the brain and spinal cord. It can be caused by viral infections, by lymphocytic infiltrations and by various bacteria, but the most serious form is due to infection by *Neisseria meningitidis*, with rapidly fatal consequences in up to 70% of untreated cases.

meningococcus *Neisseria meningitidis*; Gram-negative non-motile pyogenic coccus that is responsible for epidemic bacterial **meningitis**.

Menkes disease *(incorrectly) Menke's disease* X-linked human defect in copper metabolism caused by mutation in the gene encoding the alpha polypeptide of Cu^{2+}-transporting ATPase resulting in an inability to absorb copper from the gut. Copper is required in synthesis of various elements of connective tissue, so the disease is serious and is lethal at an early age.

menthol A camphor compound extracted from common (wild) mint (*Mentha arvensis*). The L-isomer is the chief constituent of peppermint oil. It is used as an antiseptic and local analgesic, and as a flavouring in chewing gum, toothpaste, cigarettes, etc.

mepacrine *Quinacrine* Prophylactic antimalarial drug.

meprins *Endopeptidase 2* Members of the astacin family of metalloendopeptidases, capable of hydrolysing a wide variety of peptide and protein substrates. Meprin A (EC 3.4.24.18) is responsible for degrading a variety of signalling peptides, such as bradykinin and TGFα. There are two subunits: Meprin A contains both alpha and beta subunits, Meprin B contains only beta subunits.

meprobamate Anxiolytic drug, used in motion sickness and alcoholism.

mepyramine Antihistamine used topically to treat insect bites and nettle stings. Appears to be an **inverse agonist** of the histamine H1-receptor (reduces G-protein availability for other non-related receptors associated with the same signalling pathway).

merbromin Green iridescent crystalline compound that forms a red solution in water, used as an antiseptic.

mercaptans *Thiols* Any of a class of organic compounds with a sulphydryl group bonded to a carbon atom. Low molecular-weight mercaptans have very disagreeable odours: methyl mercaptan is produced as a decay product of animal and vegetable matter; allyl mercaptan is released when onions are cut; butanethiol (butyl mercaptan) derivatives are present in skunk secretions.

mercaptoethanol A pungent water-soluble thiol, not of biological origin. Used in biochemistry to cleave disulphide bonds in proteins, or to protect sulphydryl groups from oxidation.

Mercurochrome® Proprietary name for **merbromin**.

meristem Group of actively dividing plant cells, found as apical meristems at the tips of roots and shoots and as lateral meristems in vascular tissue (vascular **cambium**) and in cork tissue (**phellogen**). Also found in young leaves, and at the bases of internodes in grasses. Consists of small, non-photosynthetic cells, with primary walls and relatively little vacuole.

Merkel cell carcinoma Rare and highly malignant skin tumour that arises from neuroendocrine cells with features of epithelial differentiation.

merlin *Schwannomin* Product of the **neurofibromatosis** 2 (*NF2*) tumour-suppressor gene, a diverged member of the **protein 4.1** family of membrane-associated proteins. Merlin is defective or absent in schwannomas and meningiomas, and 80% of the merlin mutants studied significantly altered cell adhesion.

MeroCaM Calcium-sensitive fluorophore that can be used to measure calcium levels within live cells.

merocrine Commonest mode of secretion in which a secretory vesicle fuses with the plasma membrane and releases its contents to the exterior.

merogony 1. Development of a portion only of an ovum. 2. **Schizogony** resulting in the production of merozoites.

meromyosin Fragments of myosin formed by trypsin digestion. Heavy meromyosin (HMM) has the hinge region and ATPase activity, light meromyosin (LMM) is mostly α-helical and is the portion normally laterally associated with other LMM to form the thick filament itself.

meront The asexual developmental stage of certain protozoa, especially non-sporozoa, that gives rise to merozoites.

MEROPS Database of peptidases classified into 'Families' on the basis of their catalytic site (see Table P1) or 'Clans of peptidases' on the basis of evolutionary relationships. The database provides sequence data as well as links to structural information, and also lists inhibitors. It can be accessed at http://merops.sanger.ac.uk/ (Rawlings, N.D., Morton, F.R. and Barrett, A.J. (2006) MEROPS: the peptidase database. *Nucleic Acids Res.*, 34: D270–D272). It provides far more information than this dictionary.

merosin See **laminin**.

merotomy Partial cutting: used in reference to experiments in which protozoa are enucleated and the behaviour of the residual cytoplasm is studied.

merozoite Stage in the life cycle of the malaria parasite (*Plasmodium*); formed during the asexual division of the schizont. Merozoites are released and invade other cells.

merozygote A bacterium that is in part haploid and in part diploid because it has acquired exogenous genetic material, e.g. during transduction or conjugation.

mersacidin An antibiotic peptide produced by *Bacillus* sp. strain HIL Y-85,54728 that belongs to the group of lantibiotics. Its activity *in vivo* against methicillin-resistant *Staphylococcus aureus* strains compares with that of the glycopeptide antibiotic vancomycin, although the target appears to be different. Unlike type A lantibiotics, mersacidin does not form pores in the cytoplasmic membrane but rather inhibits cell wall biosynthesis.

mesangial cells Cells found within the glomerular lobules of mammalian kidney, where they serve as structural supports, may regulate blood flow, are phagocytic and may act as **accessory cells**, presenting antigen in immune responses.

mescaline *3,4,5-Trimethoxy-β-phenethylamine* A hallucinogenic drug derived from the Mexican cactus, peyote (*Lophophora williamsii*).

mesencephalon Region of the brain below the thalamus and above the pons developed from the middle of the three cerebral vesicles of the embryonic nervous system. Includes the superior and inferior colliculi, and cerebral peduncles.

mesenchyme Embryonic tissue of mesodermal origin. *Adj.* mesenchymal.

mesocotyl The tubular, white, stem-like tissue that connects the seed and the base of the coleoptile, the first internode of the stem. The mesocotyl pushes the coleoptile of a seedling above the soil surface.

Mesocricetus auratus Syrian golden hamster. See *Cricetulus griseus*.

mesoderm Middle of the three **germ layers**; gives rise to the musculoskeletal, vascular and urinogenital systems, and to connective tissue (including that of dermis); it also contributes to some glands.

mesokaryotic Description applied to the unusual nuclei (dinokaryons) of Dinoflagellates in which the chromosomes lack histone and are attached to the nuclear membrane. This arrangement was once considered to be an intermediate between the nucleoid region of prokaryotes and the true nuclei of eukaryotes, but this is no longer considered appropriate and the term should probably be allowed to disappear into historical oblivion.

mesomere 1. A **blastomere** that is intermediate in size between a micromere and a macromere. 2. The middle zone of the mesoderm from which excretory tissue will develop. 3. The median series of bones supporting the pectoral fin of sarcopterygian fish.

mesopelagic Descriptive term for the intermediate regions between the surface and depths of the ocean.

mesophase *Smectic mesophase* Arrangement of phospholipids in water where the liquid-crystalline phospholipids form multilayered parallel-plate structures, each layer being a bilayer with the layers separated by aqueous medium.

mesophile Organism that thrives at moderate temperatures (say, 20–40°C).

mesophyll Tissue found in the interior of leaves, made up of photosynthetic (parenchyma) cells, also called **chlorenchyma** cells. Consists of relatively large, highly vacuolated cells with many **chloroplasts**. Includes **palisade parenchyma** and spongy mesophyll.

mesosecrin Glycoprotein (46 kDa) secreted by mesothelial cells (including endothelium) and in culture; forms a fine coating on the substratum. Cloning revealed it to be identical to plasminogen activator inhibitor-1 (PAI-1), and the term has become obsolete.

mesosome Invagination of the plasma membrane in some bacterial cells, sometimes with additional membranous lamellae inside. May have respiratory or photosynthetic functions.

mesothelin A cell surface glycoprotein, a differentiation antigen that is strongly expressed in normal mesothelial cells, mesotheliomas, non-mucinous ovarian carcinomas and some other malignancies. First described as the antigenic target of the monoclonal antibody K1.

mesothelioma Malignant tumour of the **mesothelium**, usually of lung; frequently caused by exposure to asbestos fibres – particularly those of crocidolite, the fibres of which are thin and straight and penetrate to the deep layers of the lung. Because of their shape, the fibres puncture the macrophage phagosome and are released, leading to a chronic inflammatory state that is thought to contribute to development of the tumour.

mesothelium Simple squamous epithelium of mesodermal origin. It lines the peritoneal, pericardial and pleural cavities, and the synovial space of joints. The cells may be phagocytic.

mesotherapy The practice of microinjecting conventional or homeopathic medication and/or vitamins into the middle layer of skin. Used in cosmetic surgery.

mesotocin Peptide hormone secreted by posterior lobe of pituitary; structure and function similar to **oxytocin**.

messenger RNA *mRNA* Single-stranded RNA molecule that specifies the amino acid sequence of one or more polypeptide chains. This information is translated during protein synthesis when ribosomes bind to the mRNA. In prokaryotes, mRNA is the primary transcript from a DNA sequence and protein synthesis starts while the mRNA is still being synthesized. Prokaryote mRNAs are usually very short-lived (average $t_{1/2}$ 5 min.). In contrast, in eukaryotes the primary transcripts (**HnRNA**) are synthesized in the nucleus and they are extensively processed to give the mRNA that is exported to the cytoplasm where protein synthesis takes place. This processing includes the addition of a $5'$-$5'$-linked 7-methyl-guanylate 'cap' at the $5'$ end and a sequence of adenylate groups at the $3'$ end, the **poly-A tail**, as well as the removal of any **introns** and the splicing together of **exons**; only 10% of HnRNA leaves the nucleus. Eukaryote mRNAs are comparatively long-lived, with $t_{1/2}$ ranging from 30 min to 24 h.

met Transforming gene from a chemically transformed human osteosarcoma-derived cell line; the beta-subunit of the *c-Met* proto-oncogene product is the cell surface receptor tyrosine kinase for **hepatocyte growth factor**. Overexpressed in a significant percentage of human cancers. See also Table O1.

met repressor-operator complex Repressor protein, 104 residues, product of the *metJ* gene, which regulates methionine biosynthesis in *E. coli*. Dimeric molecules bind to adjacent sites 8 bp apart on the DNA; sequence recognition is by interaction between antiparallel beta-strands of protein and the major groove of the B-form DNA duplex.

met-enkephalin *YGGFM* See **enkephalins**.

meta-analysis The analysis of multiple data sets to produce a single statistical analysis – for example, using published clinical trial data from multiple trials. There has been controversy over the validity of the technique, but with appropriate reduction in the degrees of freedom it is generally thought to be acceptable.

metabolic burst *Respiratory burst* Response of phagocytes to particles (particularly if **opsonized**) and to agonists such as **formyl peptides** and **phorbol esters**; an enhanced uptake of oxygen leads to the production, by an NADH-dependent system, of hydrogen peroxide, superoxide anions and hydroxyl radicals (reactive oxygen species, ROS), all of which play a part in bactericidal activity. Defects in the metabolic burst, as in **chronic granulomatous disease**, predispose to infection, particularly with **catalase**-positive bacteria, and are usually fatal in childhood.

metabolic cooperation Transfer between tissue cells in contact of low molecular-weight metabolites such as nucleotides and amino acids. Transfer is via channels constituted by the **connexons** of gap junctions, and does not involve exchange with the extracellular medium. First observed in cultures of animal cells in which radiolabelled purines were transferred from wild-type cells to mutants unable to utilize exogenous purines.

metabolic coupling The same as **metabolic cooperation**.

metabolic half-life *Metabolic half-time* The time required for 50% of a substance to be metabolized in the body.

metabolic syndrome A group of metabolic risk factors in one person. They include abdominal obesity, high levels of triglycerides, low HDL-cholesterol and high LDL-cholesterol, elevated blood pressure, insulin resistance, high fibrinogen or plasminogen activator inhibitor-1 levels in the blood and elevated C-reactive protein levels. Metabolic syndrome is associated with a high risk of coronary heart disease.

metabolism Sum of the chemical changes that occur in living organisms.

metabolome Study of information gleaned from endogenous metabolic profiling; the quantitative analysis of all the low molecular-weight molecules present in cells in a particular physiological or developmental state. Unlike the genome it has a temporal component, and the complexity of the data is considerable. Metabonomics is the study of dynamic changes in the metabolome.

metabonomics The study of dynamic changes in the **metabolome**.

metabotropic Type of neurotransmitter receptor which affects cell activity but not through changes in ion channel properties (*cf.* **ionotropic**).

metacentric Descriptive of a chromosome that has its centromere (**kinetochore**) at or near the middle of the chromosome, as opposed to **acrocentric,** where the centromere is near one end.

metachromasia *Metachromatic staining* The situation where a stain when applied to cells or tissues gives a colour different from that of the stain solution.

metachromatic See **metachromasia**.

metachronal rhythm See **metachronism**.

metachronism *Metachronal rhythm* Type of synchrony found in the beating of cilia. A metachronal process is one that happens at a later time, and the synchronization is such that the active stroke of an adjacent cilium is slightly delayed so as to minimize the hydrodynamic interference; coordination is by visco-mechanical coupling. Different patterns of metachronal synchronization are recognized: in symplectic m, the wave of activity in the field passes in the same direction as the active stroke of the individual cilium; in antiplectic m, the opposite is true. In dexioplectic and laeoplectic m, the wave of activity in the field is normal to the beat axis. Symplectic and antiplectic m are considered orthoplectic, the other forms diaplectic.

metacyclic The final developmental stage of a trypanosome in the tsetse fly, the infective form of the parasite that is transmitted by the tsetse fly when it takes a blood meal from a mammalian host.

metafemales Human females in which there are four X chromosomes in addition to 44 autosomes.

metagon RNA particle found in *Paramecium*, where it behaves as mRNA, and that can behave like a virus if ingested by the protozoan *Didinium*. Metagon RNA hybridizes specifically with DNA from paramecia bearing an M gene.

metalloenzyme An enzyme that contains a bound metal ion as part of its structure. The metal may be required for enzymic activity, either participating directly in catalysis or stabilizing the active conformation of the protein.

metallopeptidases *Metalloproteases; metalloproteinases* The most diverse group of peptidases, with more than 30 families identified. A divalent cation, usually zinc, activates the water molecule and is held in place by amino acid ligands, usually three in number.

metalloprotein A protein that contains a bound metal ion as part of its structure.

metalloproteinase *Metalloendopeptidase, EC 3.4.24; metallocarboxypeptidase, EC 3.4.17* Proteolytic enzymes in which a divalent cation is part of the catalytic mechanism. See **matrix metallopeptidases**.

metallothioneins Small, cysteine-rich metal-binding proteins found in the cytoplasm of many eukaryotes. Synthesis can be induced by heavy metals such as zinc, cadmium, copper and mercury, and metallothioneins probably serve a protective function. Metallothionein gene promoters are used in studies of gene expression.

metamere Unit of **segmentation** or metamerism.

metameric *Chem.*: having two or more constitutional isomers. *Biol.*: having a segmented body form (metamerism).

metamorphosis Change of body form, for example in the development of the adult frog from the tadpole or the butterfly from the caterpillar.

metaphase Classically, the second phase of **mitosis** or one of the divisions of **meiosis**. In this phase the **chromosomes** are well condensed and aligned along the **metaphase plate,** making it an ideal time to examine the chromosomes in a cytological preparation (metaphase spread) that flattens the nuclei.

metaphase plate The plane of the **spindle** approximately equidistant from the two poles along which the chromosomes are lined up during **mitosis** or **meiosis**. Also termed the equator.

metaplasia Change from one differentiated phenotype to another – for example, the change of simple or transitional epithelium to a stratified squamous form as a result of chronic damage.

metaraminol A potent **sympathomimetic** amine that increases both systolic and diastolic blood pressure and also produces vasoconstriction. Given by injection as metaraminol bitartrate.

metastasis Development of secondary tumour(s) at a site remote from the primary; a hallmark of malignant cells.

metastatic spread Process of development of secondary tumours. Involves local invasion (in most cases), passive transport, lodgement and proliferation at a remote site.

metavinculin Splice variant of **vinculin** found in smooth and cardiac muscle; has an additional exon 19 that encodes 68 amino acids. In cardiac muscle, connects microfilaments to the intercalated disc. Mutation (R975W), in the alternatively spliced exon 19, is associated with hypertrophic cardiomyopathy.

metaxolone *Skelaxin®; 5-[(3,5-dimethylphenoxy)methyl]-2-oxazolidinone.* A skeletal muscle relaxant. Mode of action unclear, but does not act directly on contractile systems.

metchnikowin A proline-rich **antimicrobial peptide**, 26-residues, from *Drosophila* with antibacterial and antifungal properties. Induction of metchnikowin gene expression can be mediated either by the **TOLL** pathway or by the *imd* gene product.

metformin A biguanide drug used in treatment of Type II diabetes. Metformin lowers the rate of gluconeogenesis in the presence of insulin and is considered an insulin sensitizer.

methacholine A synthetic choline ester that is a nonselective **muscarinic receptor** agonist. Used in the 'methacholine challenge test' to diagnose bronchial hyperreactivity in asthma.

methadone A synthetic opioid analgesic with properties similar to morphine but less marked. Used in treatment of opioid addiction, although can itself be habit-forming.

methaemoglobin An oxidized form of haemoglobin containing ferric iron that is produced by the action of oxidizing poisons. Non-functional.

Methanobacterium A genus of strictly anaerobic bacteria that reduce CO_2 using molecular hydrogen, H_2, to give methane. They show a number of features that distinguish them from other bacteria, and are now classified as a separate group, the **Archaebacteria**. Methanobacteria are found in the anaerobic sediment at the bottom of ponds and marshes (hence marsh gas as the common name for methane), and as part of the microflora of the rumen in cattle and other herbivorous mammals.

methanochondrion A structure of involuted plasma membrane found in many methanogenic bacteria and thought to be an organelle of methane formation.

methanotroph Bacteria (Methylococcaceae and Methylocystaceae) that can use methane as their sole source of energy. Can be found in soil at landfill sites and in lake sediments where methane is being generated.

methicillin *Meticilli* Synthetic penicillinase-resistant betalactam antibiotic. Has a rather narrow spectrum of effectiveness, but has become the defining antibiotic for **MRSA**.

methicillin-resistant *Staphylococcus aureus* *MRSA* An increasing problem, particularly in hospitals, suggesting that bacteria are beginning to win the arms race against antibiotics. Many are now also resistant to the newer antibiotics, such as **vancomycin**.

methimazole See **carbimazole**.

methionine *Met; M; 149 Da* Contains the SCH_3 group, which can act as a methyl donor (see **S-adenosyl methionine**). Common in proteins, but at low frequency. The met-x linkage is subject to specific cleavage by cyanogen bromide. See also **formyl peptides** and Table A2.

methionine aminopeptidase 2 *EC 3.4.11.18* An intracellular metallopeptidase. Inhibition by **fumagillin** blocks endothelial cell proliferation, and knockout of the gene causes an early gastrulation defect; targeted deletion specifically in the hemangioblast lineage results in abnormal vascular development and embryonic lethality.

methionine puddle A term used to describe a region of a protein surface composed of a cluster of methionine side chains. Proposed as the active hydrophobic site of calmodulin and also of signal recognition particle (SRP). The concept is of a highly fluid hydrophobic patch.

methisazone N-*methylisatin; β-thiosemi-carbazone* Drug that specifically blocks the translation of late viral mRNA in poxvirus infection and was used prophylactically for smallpox.

methotrexate Analogue of dihydrofolate. Inhibits **dihydrofolate reductase** and kills rapidly-growing cells. Therapeutic agent for leukaemias, but has a low therapeutic ratio.

methyl- -*CH* Specific reference to the methyl group is made when macromolecules are modified after synthesis by enzymic addition of methyl groups. The group is transferred to nucleic acids and proteins. See also **methyltransferase** and **DNA methylation**.

methyl violet dyes *Gentian violet* Mixtures of tetramethyl, pentamethyl and hexamethy pararosanilin (the latter being **crystal violet**). All are deep blue except at very acid pH, and increasing methylation makes the colour darker. Metachromatic, and used to demonstrate presence of **amyloid**. Gentian violet is an imprecise synonym.

methyl-accepting chemotaxis proteins See MCPs.

methyl-coenzyme M reductase *MCR; EC 2.8.4.1* The enzyme responsible for microbial formation of methane. It is a hexamer composed of two alpha, two beta and two gamma subunits with two identical nickel porphinoid active sites.

methyl-CpG-binding protein 2 *MeCP2* Originally thought to be a global transcriptional repressor, but now thought to have a role in regulating neuronal activity-dependent expression of specific genes such as BDNF. X-linked gene is defective in **Rett syndrome**.

methylcholanthrene *3-Methylcholanthrene* Carcinogenic polycyclic hydrocarbon. One of many such substances formed during incomplete combustion of organic material.

methyldopa An antihypertensive drug, preferred in pregnant patients.

methylene blue *Swiss blue; Basic Blue 9; tetramethythionine chloride* Water-soluble dye that can be reduced to a colourless form and can be oxidized by atmospheric oxygen. Used as a stain in bacteriology and histology.

methylmalonyl-coenzyme A mutase *MCM; EC 5.4.99.2* A mitochondrial adenosylcobalamin-requiring enzyme that catalyses the rearrangement of methylmalonyl-CoA to succinyl CoA. Defects are associated with methylmalonic acidaemia, a disorder of organic acid metabolism that is frequently fatal.

methylotroph Yeasts, like *Hansenula polymorpha*, that utilize methanol as an energy source.

methylphenidate *TN Ritalin* **Dexamfetamine**-like drug that stimulates the central nervous system and is used in attention-deficit hyperactivity disorder (ADHD). Mode of action not completely clear, but is thought to block the re-uptake of norepinephrine and dopamine into the presynaptic neuron and increase the release of these monoamines into the extraneuronal space.

methylprednisolone Corticosteroid (**glucocorticoid**) used to treat inflammatory diseases.

methyltransferases *EC 2.1.1* Enzymes that transfer a methyl group from *S*-adenosyl methionine to a substrate. Most commonly encountered in bacterial chemotaxis, where the methyl-accepting chemotaxis proteins (**MCPs**) become methylated in the course of adaptation.

methylxanthines Naturally occurring purine alkaloids such as theobromine, theophylline and caffeine (trimethylxanthine). They inhibit cAMP-phosphodiesterase and thus cause an increase in the intracellular cAMP concentration. Found in various beverages, coffee, tea, etc.

methysergide A synthetic **ergot** alkaloid that is a 5-HT receptor antagonist, occasionally used to prevent migraine but can cause adverse effects including retroperitoneal fibrosis.

metJ Gene for the repressor of methionine biosynthetic operon. Protein product (12 kDa) binds to DNA.

metoprolol tartrate Cardioselective β1-adrenoreceptor blocking agent (**beta-blocker**) used for treatment and prevention of heart arrhythmia and hypertension.

metorphamide Amidated **opioid** octapeptide from bovine brain. Derived by proteolytic cleavage from proenkephalin.

metrizoate *3-Acetimido-5-(N-methyl-acetamido)-triiodobenzoate* The sodium salt of metrizoate is used to produce solutions with high densities suitable for cell-density gradient separations.

metronidazole *2-Methyl-5-nitro-1H-imidazole-1-ethanol* Antiprotozoal and antibacterial drug. The mechanism of action is unknown.

metyrapone *Metopirone* Drug that reduces endogenous cortisol and corticosterone production by inhibiting the 11-β-hydroxylation reaction in the adrenal cortex. Used to suppress cortisol output in some tumours of the adrenal gland, and as a diagnostic drug for testing hypothalamic–pituitary ACTH function.

mevalonic acid Key intermediate in polyprenyl biosynthesis and thus cholesterol synthesis. Derived from hydroxymethylglutaryl-CoA (HMG-CoA), a reaction inhibited by **statins**.

mevinolin Intermediate in terpene synthesis; an analogue of **compactin**, a fungal metabolite that is used to lower plasma **low-density lipoprotein** levels. It acts as an inhibitor of **HMG-CoA-reductase**, the rate-controlling enzyme in cholesterol biosynthesis.

MG132 *Carbobenzoxy-L-leucyl-L-leucyl-L-leucinal* A potent and reversible inhibitor of the **proteosome** system at nanomolar levels; at higher concentrations also activates c-Jun N-terminal kinase (**JNK1**) and inhibits NFκB activation ($IC_{50} = 3 \mu M$). Prevents β-secretase cleavage.

MGP *Matrix Gla protein* One of the family of vitamin-K dependent, gamma-carboxyglutamic acid (Gla)-containing proteins; the calcification inhibitor produced in cartilage and in arterial walls. Comparable in function to bone Gla protein (Bgp or **osteocalcin**).

MGSA See **melanoma growth-stimulatory activity** protein.

MHC See **major histocompatibility complex**.

MHC restriction Restriction on interaction between cells of the immune system because of the requirement to recognize foreign antigen in association with MHC antigens (major histocompatibility antigens). Thus, cytotoxic T-cells will only kill virally infected cells that have the same class I antigens as themselves, whereas helper T-cells respond to foreign antigen associated with class II antigens.

MHCP 1. Methylhydroxy chalcone polymer, an active component of cinnamon that will increase glucose metabolism of adipocytes roughly 20-fold *in vitro*. **2.** MHC–Peptide Interaction Database (MHCP), a curated database for sequence-structure–function information on MHC–peptide interactions.

mianserin Antidepressant drug similar to the **tricyclic antidepressants** but with fewer antimuscarinic effects. Works by preventing the reabsorption of noradrenaline into neurons, and is a relatively selective 5-HT2 receptor antagonist.

MIB-1 Nuclear antigen Ki-67, linked to proliferative activity and prognosis in a variety of tumours.

mibolerone A synthetic **androgen**.

micelle One of the possible ways in which amphipathic molecules may be arranged; a spherical structure in which all the hydrophobic portions of the molecules are inwardly directed, leaving the hydrophilic portions in contact with the surrounding aqueous phase. The converse arrangement will be found if the major phase is hydrophobic.

Michaelis constant See K_m and **Michaelis–Menten equation**.

Michaelis–Menten equation Equation derived from a simple kinetic model of enzyme action that successfully accounts for the hyperbolic (adsorption–isotherm) relationship between substrate concentration S and reaction rate V. The equation is $V = V_{max} \, S/(S + K_m)$, where K_m is the **Michaelis constant** and V_{max} is the maximum rate approached by very high substrate concentrations.

microaerophiles Organisms that grow well at low oxygen concentrations but are unable to survive normal oxygen concentrations. Examples are *Borrelia burgdorferi*, *Helicobacter pylori*, *Lactobacillus* spp. and *Treponema pallidum*.

microangiopathic haemolytic anaemia *Microangiopathic hemolytic anemia (US)* Consequence of **disseminated intravascular coagulation** (DIC): fragments of red blood cells, damaged by being forced through a fibrin meshwork, are found in the circulation.

microarray A small membrane or glass slide containing samples of biological material (DNA, protein, etc.) arranged in a regular pattern and used as an analytical tool. Arrays are becoming progressively smaller as analytical techniques improve.

microbicide Substance that kills microbes; rather an imprecise term.

microbody See **peroxisome**.

microcarrier Microcarriers are small, solid (or in some cases immiscible liquid) spheres on which cells may be grown in suspension culture. They provide a means of obtaining large yields of cells in small volumes. The cells must exhibit anchorage dependence of growth, and the

dimensions of the carrier bead may be important in controlling growth rate. The term is imprecise and has other potential meanings.

microcentrum Obsolete name for the pericentriolar region.

microchimaerism The presence of two genetically distinct populations of cells in an individual or in an organ, one population being at a low concentration. May arise through transfer of cells between mother and foetus, between twins *in utero*, or as a result of blood transfusions and transplantation.

microcinematography The making of films using a microscope and cine camera.

Micrococcus Genus of Gram-positive aerobic bacteria, cells around 1–2 μm in diameter. *M. lysodeikticus* (now *M. luteus*) was commonly used as the source of the bacterial cell wall suspension on which lysozyme activity was measured by a decrease in turbidity.

microcolliculi Broad swellings (0.5 μm) on the dorsal surface of a moving epidermal cell in culture that move rearward as the cell moves forward (as do ruffles on fibroblasts).

microcystins Toxic, seven-amino acid cyclic peptides that inhibit liver function. The amino acid composition of the individual microcystins or **nodularins** may vary, but the novel hydrophobic amino acid 3-amino-9-methoxy-10-phenyl-2,6,8 trimethyl deca,4,6 dienoic acid (ADDA) is essential to its pharmacological activity. Potent inhibitors of the serine/threonine protein phosphatases; bind to the same site as **okadaic acid**.

microcytes Abnormally small red blood cells, found in some types of anaemia.

microdialysis **Dialysis** on a small scale, giving microlitre range samples. Used, for example, in studies of *in vivo* release of transmitters in brain tissue.

microelectrode An electrode with tip dimensions small enough (less than 1 μm) to allow non-destructive puncturing of the plasma membrane. This allows the intracellular recording of **resting** and **action potentials**, the measurement of intracellular ion and pH levels (using **ion-selective electrodes**), or **microinjection**. Microelectrodes are generally pulled from glass capillaries and filled with conducting solutions of potassium chloride or potassium acetate to maximize conductivity near the tip. Electrical contact, if required, is usually made with a silver chloride-coated silver wire.

microfibril Basic structural unit of the plant **cell wall**, made of **cellulose** in higher plants and most algae, **chitin** in some fungi, and **mannan** or **xylan** in a few algae. Higher plant microfibrils are about 10 nm in diameter and extremely long in relation to their width. The cellulose molecules are oriented parallel to the long axis of the microfibril in a paracrystalline array, which provides great tensile strength. The microfibrils are held in place by the wall matrix, and their orientation is closely controlled by the protoplast.

microfilament Cytoplasmic filament, 5–7 nm thick, of **F-actin** that can be decorated with **HMM**; may be laterally associated with other proteins (tropomyosin, α-actinin) in some cases, and may be anchored to the membrane. Microfilaments are conspicuous in **adherens junctions**.

microglial cell Small glial cells of mesodermal origin, with scanty cytoplasm and small spiny processes. Distributed throughout grey and white matter. Derive from monocytes and invade neural tissue just before birth; capable of enlarging to become macrophages.

microglobulin Any small globular plasma protein. See **beta-2-microglobulin**.

microinjection The insertion of a substance into a cell through a **microelectrode**. Typical applications include the injection of drugs, histochemical markers (such as **horseradish peroxidase** or lucifer yellow) and RNA or DNA in molecular biological studies. To extrude the substances through the very fine electrode tips, either hydrostatic pressure (pressure injection) or electric currents (ionophoresis) is employed.

microkeratome Instrument used in eye surgery to make incisions of a predetermined depth in the cornea.

micromere One of the small cells that are formed in the upper or animal hemisphere of a fertilized egg during **holoblastic** cleavage.

micronucleus The smaller nucleus in ciliate protozoans, fully active in inheritance and passed after meiosis to conjugating pairs. Gives rise to the macronucleus or macronuclei. Genes in the micronucleus are not actively transcribed.

microperoxidase Part of a cytochrome c molecule that retains haem group and has **peroxidase** activity.

microperoxisome Small **peroxisomes** of 150–250 nm in diameter found in most cells.

micropinocytosis Pinocytosis of small vesicles (around 100 nm in diameter). Not blocked by **cytochalasins**.

microplate reader Analytical instrument for spectroscopic analysis of multi-well plates, commonly with 96 wells but increasingly having more (but smaller) wells per plate to economize on reagents. Different instruments will analyse radioactivity, optical absorbance, fluorescence or luminescence.

micropore filters *Millipore filter* Filters made of a meshwork of cellulose acetate or nitrate and with defined pore size. They can be autoclaved, and the smaller pore sizes (0.22, 0.45 μm) are used for sterilizing heat-labile solutions by filtering out microorganisms. Larger pore-size filters are used in setting up **Boyden chambers**. They are about 150 μm thick, and should be distinguished from **Nucleopore filters**. 'Millipore' is a trade name for micropore filters.

microprobe See **electron microprobe**.

micropyle 1. Small hole or aperture in the protective tissue surrounding a plant ovule through which the pollen tube enters at fertilization. Develops into a small hole in the seed coat through which, in many cases, water enters at germination. 2. Perforation in the shell (chorion) of an insect egg through which the sperm enters at fertilization.

microRNA *miRNA* A collection of hundreds of snippets of non-coding RNA, each typically no more than 22 nucleotides in length. Pattern may be disease-specific, as in breast carcinoma, and they may well be acting as **small interfering RNAs**.

microsatellites Short sequences of di- or trinucleotide repeats of very variable length distributed widely throughout the genome. Using **PCR** primers to the unique sequences upstream and downstream of a microsatellite their location and polymorphism can be determined, and the technique is extensively used in investigating genetic associations with disease.

microscopic polyangiitis Necrotizing **vasculitis**, with few or no immune deposits, affecting small vessels (capillaries, venules, arterioles). Now differentiated from **polyarteritis nodosa**.

microsequencing Term generally (though rather imprecisely) used for the sequencing of very small amounts of protein – often as a prelude to producing an oligonucleotide probe, screening a cDNA library and cloning.

microsleep An episode lasting a few seconds during which external stimuli are not perceived; associated with narcolepsy or sleep deprivation.

microsomal fraction See **microsomes**.

microsomes Heterogenous set of vesicles, 20–200 nm in diameter, formed from the **endoplasmic reticulum** when cells are disrupted.

microspikes Projections from the leading edge of some cells, particularly (but not exclusively) nerve **growth cones**. They are usually about 100 nm diameter and 5–10 μm long, and are supported by loosely bundled microfilaments. They are referred to by some authors as filopodia. Functionally, a sort of linear version of a ruffle on a leading lamella.

microspore 1. A haploid spore produced by a plant sporophyte that develops into a male gametophyte. In seed plants, it corresponds to the developing pollen grain at the uninucleate stage. 2. The smaller of the spores of a heterosporous species. 3. See **Microsporidia**.

Microsporidia Obligate intracellular parasites, considered to be extremely reduced fungi, ranging in size from 1.5 to 2.0 μm. There are eight genera of microsporidia that can infect humans: the more common are *Encephalitozoon* spp., *Septata intestinalis* and *Enterocytozoon bieneusi*; the less common are *Brachiola* spp., *Microsporidium* spp., *Nosema* spp., *Pleistophora* spp., *Trachipleistophora* spp. and *Vittaforma* spp. Microsporidia multiply extensively with the host cell cytoplasm; the life cycle includes repeated divisions by binary fission (merogony) or multiple fission (schizogony) and spore production (sporogony), and spores can then infect other cells.

microtome A device used for cutting sections from an embedded specimen, for either light- or electron-microscopy.

microtrabecular network Complex network arrangement seen using the high-voltage electron microscope to look at the cytoplasm of cells prepared by very rapid freezing. The suggestion was that most cytoplasmic proteins are in fact loosely associated with one another in this fibrillar network and are separate from the aqueous phase that contains only small molecules in true solution. If it exists, then it must certainly be very labile in cells where there is cytoplasmic flow and rapid organelle movement. Now considered artefactual by most microscopists.

microtubule Cytoplasmic tubule of 25 nm outside diameter, with a 5-nm thick wall. Made of tubulin heterodimers packed in a three-start helix (or of 13 protofilaments, looked at another way), and associated with various other proteins (**MAPs**, **dynein**, **kinesin**). Microtubules of the ciliary

axoneme are more permanent than cytoplasmic and spindle microtubules.

microtubule-organizing centres *MTOC*   Rather amorphous region of cytoplasm from which microtubules radiate. The pattern and number of microtubules is determined by the MTOC. The **pericentriolar region** is the major MTOC in animal cells; the basal body of a cilium is another example. Activity of MTOCs can be regulated, but the mechanism is unclear.

microtubule-associated proteins See **MAPs**.

Microtus Genus of voles and meadow mice.

microvillus *Pl.* microvilli Projection from the apical surface of an epithelial cell that is supported by a central core of microfilaments associated with bundling proteins such as **villin** and **fimbrin**. In the intestinal **brush border** the microvilli presumably increase absorptive surface area, whereas the stereovilli (**stereocilia**) of the cochlea have a distinct mechanical role in sensory transduction.

Microviridae A diverse group of single-stranded DNA bacteriophages, also known as ϕX phage group or isometric ssDNA phages.

Mid1p *S. pombe* homologue of **anillin**. Mid2p is another anillin homologue but is not orthologous with Mid1p, and influences **septin** ring organization at the site of cell division. Overproduction of Mid2p depolarizes cell growth and affects the organization of both the septin and actin cytoskeletons.

midbody Dense structure formed during **cytokinesis** at the cleavage furrow. It consists of remnants of **spindle fibres** and other amorphous material, and disappears before cell division is completed.

middle lamella First part of the plant **cell wall** to be formed, laid down in the **phragmoplast** during cell division as the **cell plate**. Subsequently makes up the central part of the double cell wall that separates two adjacent cells, cementing together the two primary walls. Rich in **pectin** and relatively poor in **cellulose**.

midecamycin Macrolide antibiotic that inhibits staphylococcal enterotoxin B-induced mRNA expression of the Th2 cytokines IL-4 and IL-5 in peripheral blood mononuclear cells from patients with atopic dermatitis.

midiprep Slang term denoting a medium-scale purification of **plasmid** from a bacterial culture. Usually used to describe preparations from 10 ml to 100 ml culture. See also **miniprep**, **maxiprep**, **megaprep**.

midkine Heparin-binding growth factor (13 kDa) of the TGF-β superfamily; has 50% sequence identity with heparin-binding growth-associated molecule (HB-GAM). Structurally unrelated to fibroblast growth factor (FGF). Midkine was originally described as associated with tooth morphogenesis induced by epithelial–mesenchyme interactions. **Nucleolin** binds midkine.

MIF See **macrophage inhibition factor** or **migration inhibitory factor**.

mifepristone A synthetic compound that inhibits the action of **progesterone**, used to induce abortion up to the twentieth week of pregnancy.

Mig *m*119 Mouse protein induced by IFNγ (interferon γ). See also **cytokines**.

migfilin A **LIM domain**-containing protein that localizes to cell–matrix adhesions, associates with actin filaments and is essential for cell shape modulation. Migfilin interacts with the cell–matrix adhesion protein Mig-2 (mitogen inducible gene-2) and **filamin** through its C- and N-terminal domains, respectively.

migration inhibitory factor Factor that inhibits macrophage movement. Originally defined on the basis of inhibition of emigration of mononuclear cells from capillary (haematocrit) tubes; more recently a 13-kDa protein with migration inhibitory activity has been isolated. See **macrophage inhibition factor**.

mil v-raf-1 Avian homologue of *v-raf-1*, encoding a serine/threonine **protein kinase**. The **Raf/MEK/ERK** cascade is a highly conserved signal transduction cascade. See Table O1.

miliary tuberculosis A form of tuberculosis in which small lesions (about 2 mm in diameter, described as 'milletlike') are found in various organs of the body, especially in the meninges and in the lungs. Relatively uncommon, and only seen in 1–2% of cases of TB.

Miller unit Beta-galactosidase (β-gal) assays using *o*-nitrophenyl-β-D-galactoside (ONPG) as a substrate are referred to as 'Miller' assays, and a standardized amount of β-Gal activity is a 'Miller unit' (after Jeffrey Miller, who published the first protocol in 1972).

Millipore filter Trade name for a well-known brand of **micropore filters**.

MIM *Missing in metastasis* A **WASP homology domain-2**-containing protein (37 kDa) that binds ATP-actin monomers more tightly than ADP-actin monomers; interacts directly with the SH3 domain of **cortactin**. MIM appears to regulate cell motility by modulating different **Arp2/3** activators.

mimivirus An exceptionally large double-stranded DNA virus found in amoebae. The particle size of 400 nm makes it dimensionally comparable to mycoplasma. The genome is 1.2 Mb in size and there are 1262 putative open reading frames, only 10% of which resemble proteins of known function.

Mimosa pudica The 'sensitive plant', the leaflets of which fold inwards very rapidly when touched. A more vigorous stimulus causes the whole leaf to droop, and the stimulus can be transmitted to neighbouring leaves.

mimosine *2 Amino-3-(3-hydroxy-4-oxo-1H-pyridin-1-yl)-propanoic acid* A non-protein amino acid found in leaves, pods and seeds of tropical legumes of the genus *Leucaena*. It is toxic to animals consuming the plant, and is an extremely effective inhibitor of DNA replication in mammalian cells; it appears to prevent the formation of replication forks when delivered to mammalian cells approaching the G/S boundary. It is an iron/zinc chelator, and may inhibit iron-dependent ribonucleotide reductase and the transcription of the cytoplasmic serine hydroxymethyltransferase gene (SHMT). Inhibition of serine hydroxymethyltransferase is moderated by a zinc-responsive unit located in front of the SHMT gene.

mimotope Compound that mimics the structure of a conformational **epitope** and that will elicit an identical antibody response (whereas a mimetic would not have the same antigenicity). Mostly used of peptides from **phage display libraries**; potentially useful as vaccines.

Minamata disease Classic and infamous case of poisoning by methylmercury in humans who ate fish and shellfish contaminated by waste water (in 1956) from a chemical plant in Minamata, South West Kyushu, Japan. The disease is characterized by sensory and motor disturbances and foetal damage in pregnant women.

mineralocorticoid Natural or synthetic corticosteroid that acts on water and electrolyte balance by promoting retention of sodium ions and excretion of potassium ions in the kidney. Aldosterone is the most potent natural example, and is produced in the outer layer of the adrenal cortex.

minichromosome maintenance proteins *MCM proteins* Family of conserved proteins involved in the initiation and elongation of DNA replication forks in archaea and eukaryotes. MCM proteins can form hexameric complexes that possess ATP-dependent DNA unwinding activity. In eukaryotes MCM2-7 is a heterohexameric **helicase**, composed of six related subunits that assembles at the DNA replication forks in Archaea. MCM-2 is often used as a marker for proliferation competence; MCM-4 and other MCMs are phosphorylated during the cell cycle, at least in part by cyclin-dependent kinases. MCM-7 is upregulated in a variety of tumours, including neuroblastoma and prostate, cervical and hypopharyngeal carcinomas.

miniature end plate potential *MEPP* Small fluctuations (typically 0.5 mV) in the resting potential of postsynaptic cells. They are the same shape as, but much smaller than, the **end plate potentials** caused by stimulation of the presynaptic cell. MEPPs are considered as evidence for the quantal release of **neurotransmitters** at **chemical synapses**, a single MEPP resulting from the release of the contents of a single synaptic vesicle.

minicell Spherical fragment of a bacterium produced by abnormal fission and not containing a bacterial chromosome.

minichromosome **1.** Certain viruses complex with the histones of the host eukaryote cells they have infected to form a chromatin structure resembling a small chromosome. **2.** A plasmid that contains a chromosomal **origin of replication**. **3.** See **minichromosome maintenance proteins**.

minimal medium *Minimal essential medium; MEM* The simplest tissue culture medium that will support the proliferation of normal cells.

minimyosin Form of **myosin** isolated from *Acanthamoeba*; only 180 kDa, but capable of binding to actin.

miniprep Slang term denoting a small-scale purification of **plasmid** from a bacterial culture. Usually used to describe preparations from 1 ml to 10 ml culture. See also **midiprep, maxiprep, megaprep**.

minisatellite *Variable number tandem repeat; VNTR* Class of highly repetitive **satellite DNA**, comprising variable (typically 10–20) repeats of short (e.g. 64 bases) DNA sequences. The high level of polymorphism of such minisatellites make them very useful in genomic mapping.

minisegregant cells Human cells with small amounts of DNA and few chromosomes; obtained experimentally by perturbing cell division. Can readily be fused with whole cells.

minK Widely expressed protein (15 kDa) that forms **potassium channels** by aggregation with other membrane proteins. A variety of channels have been shown to have minK associated with them, and minK seems to be important in regulating structure and activity of the channel.

minoxidil Peripheral vasodilator used to treat severe hypertension, although it can cause side effects, and used topically to stimulate the regrowth of hair in cases of male-pattern baldness.

minute mutant A class of recessive lethal mutants of *Drosophila*. The heterozygotes grow more slowly, are smaller and less fertile than the wild-type flies. There are about 40 loci that produce minute mutants.

MIP See **major intrinsic protein** or **macrophage inflammatory protein**. Macrophage inflammatory protein is known to have various subclasses, MIP-1α, MIP-1β, MIP-2. See Table C4 (chemokines).

MIP-90 *Microtubule-interacting protein-90* Protein that associates with microtubules and actin filaments in different cell domains. Co-localizes with actin at the leading edge of fibroblasts.

miracidium The ciliated first stage larva of a trematode, 150–180 μm in length by 70–80 μm in width, the free-living stage responsible for infecting the intermediate host; in the case of *Schistosoma* sp., the aquatic snail.

miranda *Drosophila* gene, the product of which (830 residues) co-localizes with **prospero** in mitotic neuroblasts and apparently directs prospero exclusively to the ganglion mother cell (GMC) during the asymmetric division that gives rise to another neuroblast and the GMC. Anchors prospero selectively to the basal side of the cell cortex during mitosis and releases it after cytokinesis. Miranda has two **leucine zipper** motifs and eight consensus sites for PKC phosphorylation.

mirtazapine Antidepressant that is thought to enhance central noradrenergic and serotonergic activity, although the exact mechanism of action is unclear.

mismatch repair A DNA repair system that detects and replaces wrongly paired, mismatched bases in newly replicated DNA. *E. coli* has a mismatch correction enzyme coded for by three *mut* genes, *mutH*, *mutL* and *mutS*, which is directed to the newly synthesized strand and removes a segment of that strand including the incorrect nucleotide. The gap is then filled by DNA polymerase.

misoprostol Synthetic **prostaglandin** analogue used chiefly in treating and preventing gastric ulcers, especially those caused by anti-inflammatory drugs. It is rapidly de-esterified to its free acid, which is responsible for its clinical activity.

missense mutation A mutation that alters a **codon** for a particular amino acid to one specifying a different amino acid.

mitochondrial diseases Illnesses, frequently neurological, which can be ascribed to defects in mitochondrial function. If the defect is in the mitochondrial rather than the nuclear genome, unusual patterns of inheritance can be observed.

Mitochondrial Eve Purportedly the most recent common matrilineal ancestor of all living humans, since all mitochondria are supposed to derive from the egg, although there has been some recent doubt cast upon this. 'Mitochondrial Eve' was a member of a population that existed about 150 000 years ago and produced multiple lineages, but all others have disappeared.

mitochondrial trifunctional protein *TFP* Protein, composed of four hydroacyl-CoA dehydrogenase-alpha (HADHA) and four hydroacyl-CoA dehydrogenase-beta (HADHB) subunits, that catalyses the last three steps in the long-chain fatty acid beta-oxidation pathway in mitochondria. TFP deficiency leads to a wide clinical spectrum of disease, ranging from severe neonatal/infantile cardiomyopathy and early death to mild chronic progressive sensorimotor polyneuropathy with episodic rhabdomyolysis.

mitochondrion *Pl.* mitochondria Highly pleiomorphic organelle of eukaryotic cells that varies from short, rod-like structures present in high number to long-branched structures. Contains DNA and **mitoribosomes**. Has a double membrane, and the inner membrane may contain numerous folds (cristae). The inner fluid phase has most of the enzymes of the **tricarboxylic acid cycle** and some of the urea cycle. The inner membrane contains the components of the **electron transport chain**. Major function is to regenerate ATP by oxidative phosphorylation (see **chemiosmotic hypothesis**).

mitogen-activated kinases *MEKs; MAPKs* See **MAP kinases**.

mitogenesis The process of stimulating transit through the cell cycle, especially as applied to lymphocytes. Concanavalin A is a mitogen for T-lymphocytes; the best mitogen for B-lymphocytes is Cowan strain *Staphylococcus aureus*.

mitogenic Causing re-entry of cells into the cell cycle, not just into mitosis.

mitogillin Toxin (149 residues) produced by *Aspergillus restrictus*. One of the aspergillins; see **alpha-sarcin**.

mitomycin C Aziridine antibiotic isolated from *Streptomyces caespitosus*. Inhibits DNA synthesis by cross-linking the strands, and is used as an antineoplastic agent. Most active in late G1 and early S phase. Mitomycin-treated cells are sometimes used as feeder layers.

mitoplasts Isolated mitochondria without their outer membranes. They have finger-like processes and retain the capacity for oxidative phosphorylation.

mitoribosomes Mitochondrial ribosomes; these more closely resemble prokaryotic ribosomes than cytoplasmic ribosomes of the cells in which they are found, though they are even smaller and have fewer proteins than bacterial ribosomes.

mitosis The usual process of nuclear division in the somatic cells of **eukaryotes**. Mitosis is classically divided into four stages. The chromosomes are actually replicated prior to mitosis during the **S phase** of the **cell cycle**. During the first stage, prophase, the chromosomes condense and become visible as double strands (each strand being termed a chromatid) and the **nuclear envelope** breaks down. At the same time, the mitotic **spindle** forms by the polymerization of **microtubules** and the chromosomes are attached to spindle fibres at their kinetochores. In metaphase, the chromosomes align in a central plane perpendicular to the long axis of the spindle. This is termed the metaphase plate. During anaphase, the paired chromatids are apparently pulled to opposite poles of the spindle by means of the spindle fibre microtubules attached to the kinetochore, though the actual mechanism for this movement is still controversial. This separation of chromatids is completed during telophase, when they can be regarded as chromosomes proper. The chromosomes now lengthen and become

diffuse, and new nuclear envelopes form round the two sets of chromosomes. This is usually followed by cell division or cytokinesis in which the cytoplasm is also divided to give two daughter cells. Mitosis ensures that each daughter cell has a **diploid** set of chromosomes that is identical to that of the parent cell.

mitotic apparatus See **spindle**.

mitotic death Cells fatally damaged by ionizing radiation may not die until the next **mitosis**, at which point the radiation damage to the DNA becomes evident, particularly when there is fragmentation of chromosomes.

mitotic index The fraction of cells in a sample that are in mitosis. It is a measure of the relative length of the mitotic phase of the **cell cycle**.

mitotic recombination *Somatic crossing-over* **Crossing-over** can occur between **homologous chromosomes** during mitosis, but is very rare because the chromosomes do not normally pair. When it occurs, it can lead to new combinations of previously linked genes. Although infrequent, mitotic recombination has been utilized for genetic analysis in *Aspergillus* and in studies on developmental **compartments** in *Drosophila*, where the frequency of mitotic recombination can be increased by X-irradiation.

mitotic segregation See **mitotic recombination**.

mitotic shake-off method A method of collecting cells in **mitosis**, so that the chromosomes can be examined and the karyotype determined. Many cultured cells round up during mitosis and so become less firmly attached to the culture substratum. Cells in mitosis thus can be removed into suspension by gentle shaking of the culture vessel, leaving the non-mitotic cells still attached. The number of cells that are in mitosis is usually increased by using a drug, such as **colcemid**, that blocks mitosis at **metaphase**.

mitotic spindle See **spindle** and **mitosis**.

mitrocomin Calcium-activated photoprotein from coelenterate *Halistaura*. See **aequorin**.

mix Homeobox genes (*mix1, mix2*), expressed in prospective mesoderm and endoderm after mid-blastula stage, that respond to TGFβ (**transforming growth factor-β**) superfamily signals, including **activin**, TGF-β, Vg-1 and BMP-4, but not to non-TGF inducers.

mixed lymphocyte reaction *MLR* Reaction of mitogenesis produced in T-lymphocytes when allogeneic (i.e. mixed) lymphocytes are brought together, provided they are mismatched in histocompatibility loci. Once used as a test for possible graft compatibility in human grafting. It is now known that a negative reaction is a poor predictor of graft acceptance.

mixotroph An organism that combines two or more fundamental methods of nutrition.

MKK *MAPK kinase* Protein kinase that phosphorylates **MAP kinase** on serine or threonine residues.

ML-7 *1-(5-Iodonaphthalene-1-sulfonyl)homopiperazine* A specific inhibitor of **myosin light-chain kinase** ($K_i=300$ nM).

MLCK See **myosin light-chain kinase**.

MLEE *Multilocus enzyme electrophoresis* Form of two-dimensional **electrophoresis** used to distinguish **polymorphisms** between strains or populations.

M-line Central part of the **A-band** of striated muscle (and of the **M-band**): contains M-line protein (**myomesin**, 165 kDa), **creatine kinase** (40 kDa) and glycogen phosphorylase b (90 kDa). Involved in controlling the spacing between thick filaments.

MLKs *Mixed lineage kinases, MAP3K9, 10, 11* A family of serine/threonine kinases with the features of mitogen-activated protein kinase kinase kinases (MAPKKKs). The MLK family consists of two subgroups, one containing the highly related MLK1 (MAP3K9), MLK2/MST (MAP3K10) and MLK3/SPRK/PTK (MAP3K11) kinases. MLKs regulate signalling by the **JNK** and **p38** MAPK pathways.

MLST *Multilocus sequence typing* A nucleotide sequence-based approach for the unambiguous characterization of isolates of bacteria and other organisms using the sequences of internal fragments of (usually) seven housekeeping genes. It is based upon the concept of multilocus enzyme electrophoresis (**MLEE**) adapted so that alleles at each locus are defined directly, by nucleotide sequencing, rather than indirectly from the electrophoretic mobility of their gene products.

MluI cell-cycle boxes *MCB elements* A hexameric nucleotide sequence found near the start site of yeast genes expressed at G1/S that may play a role in controlling entry into the cell division cycle; cut by the type II restriction enzyme MluI.

MLV Abbrev. for (1) **Murine leukaemia virus,** and, less commonly, for (2) Modified-live virus, (3) Multilamellar liposomes or multilamellar vesicles, (4) Mechanical lung ventilation.

MMR Abbrev. for (1) Measles, mumps and rubella vaccine, a combined vaccine given in infancy and mistakenly hypothesized to be associated with autism, (2) **Mismatch repair**.

MMTV Mouse **mammary tumour virus**.

MN blood group antigens A pair of blood group antigens governed by genes that segregate independently of the ABO locus. The alleles are co-dominant and there are three types: MM, NN and MN. **Glycophorin** has M or N activity, and this is associated with oligosaccharides attached to the amino terminal portion of the molecule. M-type glycophorin differs from N-type in amino acid residues 1 and 5, although the antigenic determinants are associated with the carbohydrate side chains.

mnemiopsin Calcium-activated photoprotein from the coelenterate *Mnemiopsis*. See **aequorin**.

MNNG N-*Methyl*-N′-*nitro*-N-*nitrosoguanidine* A potent mutagen and carcinogen, an alkylating agent used experimentally to induce mutations.

Mnt Protein that interacts with **Max** and functions as a transcriptional repressor. Not a member of the **Myc** or **Mad** families. Binds to same site on Max as does **Sin3**.

Mo-1 αMβ2 **integrin** of leucocytes (CD11b/CD18).

mobile genetic elements See **transposons**.

mobile ion carrier See **ionophore**.

modafinil A stimulant drug that enhances wakefulness, although the mode of action is unknown and the pharmacological profile differs from that of sympathomimetic amines. Used in the treatment of narcolepsy.

modification enzyme Enzyme that introduces minor bases into DNA or RNA or that alters bases already incorporated. Serves to alter the sequence so that **restriction enzymes** will not damage the strand.

modulation See **neuromodulation**.

moesin *Membrane-organizing extension spike protein* Isolated from placenta, a member of the **FERM domain** (protein 4.1, **ezrin**, **radixin**, moesin) protein family, all members of which have been shown to serve as cytoskeletal adaptor molecules. **Pasin-2** has been shown to be identical to moesin.

molecular clock This term has two separate uses. In one sense it means the rate of fixation of mutations in DNA and thus times the rate of genetic diversification. In the second sense it means a biological system capable of maintaining a timing rhythm or pulse. All such clocks are thought to be entrained by a natural oscillator such as the **diurnal** rhythm.

Mollicutes Subgroup of the **Firmicutes**, mycoplasmas. Although they lack cell walls and so do not respond to Gram staining, they are considered Gram-positive and lack the second membrane found in Gram-negative bacteria.

molluscan catch muscle Muscle responsible for holding close the two halves of the shell of bivalves. Specialized to maintain tension with low expenditure of ATP. Rich in **paramyosin**.

molluscum bodies Intracellular inclusions of poxviruses found in cells of human epidermis; harmless, but contagious. Associated with skin lesions (**molluscum contagiosum**).

molluscum contagiosum virus Poxvirus that causes a benign viral disease of the skin. The virus is large, 240–320 nm in diameter. See **molluscum bodies**.

Moloney murine leukaemia virus *Moloney murine leukemia virus (US)* Replication-competent retrovirus (Oncovirinae) that causes leukaemia in mice, isolated by Moloney from cell-free extracts made from a transplantable mouse sarcoma.

Moloney murine sarcoma virus Replication-defective retrovirus, source of the **oncogene** *v-mos,* responsible for inducing fibrosarcomas *in vivo* and transforming cells in culture.

Moloney test Obsolete skin test for immunity to diphtheria in which active toxin was injected into one site and toxoid into another. This was to control for pseudopositive reactions to the toxin.

MOLT-4 cells Suspension culture derived from human male with acute lymphoblastic leukaemia. A stable T-cell leukaemia line.

molybdopterin Trivial name for the pterin cofactor common to molybdenum- and tungsten-containing enzymes.

MOM *Mitochondrial outer membrane* It is used particularly for mitochondrial outer membrane proteins, in conjunction with the molecular mass in kiloDaltons, e.g. MOM19 and MOM72, proteins of 19 and 72 kDa, respectively.

mometasone furoate Potent topical steroid used for treatment of psoriasis, allergic dermatitis and allergic rhinitis.

Mona *Gads* A molecular adapter protein that links the adaptors **Slp-76** and **LAT** upon T-cell receptor activation and, in platelets, collagen receptor activation. Platelet activation by thrombin results in rapid induction of Mona expression.

monastrol Membrane-permeable drug that reversibly inhibits activity of mitotic **kinesin** (Eg5). Treatment of cells with monastrol results in the formation of monopolar spindles.

monellin Basic non-glycosylated heterodimeric protein (44- and 50-residue protomers) that has intensely sweet taste.

monensin A sodium ionophore (671 Da) from *Streptomyces cinnamonensis*. Has antibiotic properties and is used as a feed additive in chickens. Also used in **ion-selective electrodes**.

Monera Kingdom that contains all prokaryotic organisms (bacteria and cyanobacteria) in the Five Kingdom scheme.

mongolism See **Down's syndrome**.

moniliasis Infection with the fungus *Monilia* (*Candida*). See **candidiasis**.

monoamine neurotransmitters See **biogenic amines**.

monoamine oxidase *MAO; EC 1.4.3.4* Enzyme catalysing breakdown of several **biogenic amines**, such as serotonin, adrenaline, noradrenaline, dopamine. See **monoamine-oxidase inhibitors**.

monoamine-oxidase inhibitors *MOAIs* Class of drugs used to treat very severe depressive illness. Now rarely used except in patients unresponsive to **tricyclic antidepressants** or **SSRIs,** because of their dangerous interactions with foods containing tyramine and with **sympathomimetic** drugs.

monocentric chromosome Chromosome with a single **centromere**, i.e. most chromosomes.

monocistronic RNA A **messenger RNA** that gives a single polypeptide chain when **translated**. All **Eukaryote** mRNAs are monocistronic, but some bacterial mRNAs are polycistronic, especially those transcribed from **operons**.

monoclonal Used of a cell line whether within the body or in culture to indicate that it has a single clonal origin. Monoclonal antibodies are produced by a single clone of **hybridoma** cells and are therefore a single species of antibody molecule.

monocotyledonous *Monocot* Plants in which the developing plant has only one **cotyledon**. Grasses are perhaps the commonest examples of the Class (which also contains palms, lilies and orchids).

monocyte Mononuclear phagocyte circulating in blood that will later emigrate into tissue and differentiate into a **macrophage**.

monocyte chemoattractant protein-1 See **monocyte chemotactic and activating factor**.

monocyte chemotactic and activating factor *MCAF; JE; monocyte chemoattractant protein-1; MCP-1;*

CCL2 **Cytokine** of the C–C subfamily, co-induced with **interleukin-8** on stimulation of endothelial cells, fibroblasts or monocytes, that activates and is **chemotactic** for **monocytes**. A **chemokine**.

monokines Soluble factors, derived from macrophages, that act on other cells (e.g. interleukin-1). Term becoming unusual – all monokines are **cytokines,** and that term is more commonly used.

monolayer A single layer of any molecule, but most commonly applied to polar lipids. Can be formed at an air/water interface in experimental systems. The term should not be used to describe one layer of a lipid bilayer, for which the term 'leaflet' is generally used. See also **monolayering of cells**.

monolayering of cells Tendency of animal tissue cells growing on solid surfaces to cover the surface with a complete layer only one cell thick, before growing on top of each other. This non-random distribution is generated by contact inhibition of locomotion, a phenomenon in which colliding cells change direction rather than move over one another. Of the theories why some (but by no means all) types of cells stop growing when a monolayer is formed, present evidence favours limitation by supply of growth factors from the medium, rather than any inhibitory effect of contact on growth.

mononuclear cells Cells of the blood other than erythrocytes and **polymorphonuclear leucocytes** (neutrophils, basophils and eosinophils).

mononuclear phagocytes Monocytes and their differentiated products, macrophages. 'Mononuclear cells' are leucocytes other than polymorphonuclear cells, and include lymphocytes.

monopodial Adjective describing an **amoeba** that has only one **pseudopod** (as opposed to polypodial forms).

monosaccharide A simple sugar that cannot be hydrolysed to smaller units. Empirical formula is $(CH_2O)_n$; range in size from trioses ($n = 3$) to heptoses ($n = 7$).

monosome **1.** A single ribosome attached to a strand of mRNA. **2.** A ribosome that has dissociated from a polysome. **3.** Chromosome in an aneuploid set that does not have a homologue.

monosomy Situation in a normally **diploid** cell or organism in which one or more of the **homologous chromosome** pairs is represented by only one chromosome of the pair. For example, sex determination in grasshoppers depends on the fact that females are XX and males XO; that is, males have only one sex chromosome and are monosomic for the X chromosome.

monotremes *Monotremata* The single order of the Prototherian mammals. Examples are the spiny anteater (*Tachyglossus*) and the duck-billed platypus (*Ornithorhynchus*).

montelukast **Leukotriene** receptor antagonist used to treat asthma.

MOPC Cell lines (e.g. MOPC-21, MOPC-315) derived from murine myeloma (plasmacytoma) originally induced by intraperitoneal injections of mineral oil in highly inbred BALB/CJ mice.

MOPS *Morpholino-propane sulphonic acid* A 'biological' buffer; a synthetic zwitterionic compound, with a pK_a of 7.2, that is non-toxic and has a low temperature coefficient. Widely used in biochemical studies, largely as a replacement for phosphate buffers.

Moraxella *Moraxella catarrhalis* is a Gram-negative, aerobic, oxidase-positive diplococcus that was described for the first time in 1896. The organism has also been known as *Micrococcus catarrhalis*, *Neisseria catarrhalis* and *Branhamella catarrhalis*. It is commensal in the upper respiratory tract, a common cause of otitis media and sinusitis, and an occasional cause of laryngitis.

Morbilli virus Genus of viruses (of the **Paramyxoviridae**). Type species is **measles virus**; other species include canine distemper virus (CDV) and the related seal virus (phocine distemper virus (PDV)).

morphallaxis Regenerative process in which part of an organism is transformed directly into a new organism without replication at the cut surface.

morphiceptin *Tyr-Pro-Phe-Pro-NH*$_2$ A tetrapeptide amide fragment (Tyr-Pro-Phe-Pro-NH$_2$) of a milk protein, α-casein, that is a mu-selective **opioid receptor** ligand. May occur naturally.

morphine An **opioid** alkaloid, isolated from opium, with a complex ring structure. It is a powerful analgesic with important medical uses, but is highly addictive. Functions by occupying the receptor sites for the natural neurotransmitter peptides, **endorphins** and **enkephalins**, but is stable to the peptidases that inactivate these compounds.

morphogen Diffusible substance that carries information relating, for example, to position in the embryo and thus determines the differentiation that cells perceiving this information will undergo.

morphogenesis The process of shape formation: the processes that are responsible for producing the complex shapes of adults from the simple ball of cells that derives from division of the fertilized egg.

morphogenetic movements Movements of cells or of groups of cells in the course of development. Thus the invagination of cells in gastrulation is one of the most dramatic of morphogenetic movements; another much-studied example is the migration of neural crest cells.

morphometry Method that involves measurement of shape. A variety of methods exist to enable examination of, for example, the distribution of objects in a 2-D section of a cell and then to use this to predict the shapes and the distribution of these objects in three dimensions.

Morquio-Brailsford disease A rare storage disease characterized by dwarfism, kyphosis and skeletal defects in the hip joint. The defect is in degradation of keratan sulphate with deposits in tissues of glycosaminoglycans due to *N*-acetylgalactosamine-6-sulphatase deficiency. Keratin sulphate is excreted in large amounts in urine. Occurs in two forms, depending on which gene is mutated: in type A there is a deficiency of galactosamine-6-sulphate sulphatase; in type B a deficiency of β-galactosidase. Inheritance is autosomal recessive.

mortalin *75-kD glucose-regulated protein; GRP75; MOT-1; MOT-2* Member of the **HSP70** family – functions as a mortality marker in fibroblasts. Mortal fibroblasts have MOT-1 uniformly distributed in the cytoplasm; immortalized fibroblasts have MOT-2 in a juxtanuclear concentration. Associates preferentially with duplicated centrosomes, and promotes dissociation of **p53** from centrosomes through physical interaction.

morula Stage of development in holoblastic embryos. The morula stage is usually likened to a spherical raspberry, a cluster of blastomeres without a cavity.

mos *Moloney murine sarcoma viral oncogene homologue* An **oncogene**, identified in mouse **sarcoma**, encoding a serine/threonine **protein kinase**. Normal *c-mos* is expressed only in the germ cells of both testis and ovary. Overexpression of *c-Mos* proto-oncogene product stimulates activity of **jun**. See Table O1.

mosaic egg At one time a distinction was drawn between those organisms in which the egg seemed to have a firmly committed fate map built in, and 'regulating' embryos. In the former, after the first cleavage one blastomere was committed to produce one set of tissues, the other blastomere a different set, and removal of one blastomere led to the production of an incomplete embryo. This was particularly obvious in mollusc development, where one blastomere had the polar lobe material. This early differentiation (or **determination**) of blastomeres for particular fates was in distinction to regulating embryos in which the removal of one blastomere did not matter, the other blastomere(s) compensating and producing a full set of tissues. The distinction is, however, only based upon the timing of differentiative events, and within a few divisions the regulating embryo also becomes a mosaic of determined cells.

mosaicism A condition in which an individual is composed of a mixture of cells that are karyotypically or genotypically distinct.

motif A small structural element that is recognizable in several proteins, e.g. **alpha-helix**.

motilin Peptide (22 residues) found in duodenum, pituitary and pineal that stimulates intestinal motility. Apparently unrelated to other hormones.

motogen Term proposed for substances that stimulate cell motility – by analogy with those that stimulate cell division (**mitogens**). **Scatter factor** (hepatocyte growth factor) is an example, though it seems likely that factors may be motogens for some cells and mitogens for others, and may be motogens, mitogens or both depending upon the local conditions in which the cell is operating.

motoneuron A **neuron** that connects functionally to a **muscle fibre**.

motor end plate Synonym for **neuromuscular junction**.

motor neuron Synonym for **motoneuron**.

motor neuron disease *Amyotrophic lateral sclerosis; Lou Gehrig's disease* Degenerative disease of unknown cause that affects predominantly motor neurons of the spinal cord, cranial nerve nuclei and motor cortex. See **amyotrophic lateral sclerosis**.

motor protein Proteins that bind ATP and are able to move on a suitable substrate with concomitant ATP hydrolysis. Most eukaryotic motor proteins move by binding to a specific site on either actin filaments (myosin) or on microtubules (dynein, kinesin). They are normally elongated molecules with two active binding sites, although some kinesin analogues have a single site. The distal end of the molecule normally binds adaptor proteins that enable them to make stable interactions with membranous vesicles or with filamentous structures, which then constitute the cargo to be moved along the substrate filament.

Mott cells Plasma cells containing large eosinophilic inclusions; found in the brain in cases of African **trypanosomiasis**.

Mounier–Kuhn syndrome Tracheobronchomegaly, a rare disorder of the lower respiratory tract. May be a connective tissue disease.

moxonidine See **clonidine**.

MPAC Carboxybutoxy ether of **mycophenolic acid** (MPA).

MPF See **cyclin**.

M-phase Mitotic phase of cell cycle of eukaryotic cells, as distinct from the remainder, which is known as interphase (and that can be further subdivided as G1, S and G2). Beginning of M is signalled by separation of centrioles, where present, and by the condensation of chromatin into chromosomes. M-phase ends with the establishment of nuclear membranes around the two daughter nuclei, normally followed immediately by cell division (**cytokinesis**).

M-phase-promoting factor MPF Protein whose levels rise rapidly just before, and fall away just after, **mitosis**. Thought to be a trigger for mitosis. Now known to be the **Cyclin B2-Cdc**2 complex.

MPN *Most probable number* A term used in estimating the extent of bacterial contamination of a sample. In practice, a minimum of three dilutions and three, five or ten tubes per dilution are used. After incubation, the pattern of positive and negative tubes is noted and a standardized MPN Table is consulted to determine the most probable number of organisms (causing the positive results) per unit volume of the original sample.

M-protein 1. Galactoside carrier in *E. coli*. 2. Cell surface antigen of *Brucella*. 3. Structural protein in the M-line of striated muscle (**myomesin**). 4. **Streptococcal M-protein**. 5. See **paraprotein**.

MPTP *1-Methyl-4-phenyl-1,2,3,6-tetrahydropyridine* Compound that causes dopaminergic neuronal degeneration, used to treat mice, thereby generating a model for Parkinson's disease ('MPTP mouse model'). Inbred mouse strains differ remarkably in their susceptibility to MPTP, indicating a genetic element.

MPV Abbrev. for (1) mean platelet volume, a determinant of platelet function, and a risk factor for atherothrombosis, (2) mouse parvovirus-1.

MRC-5 cells Cell line established from normal human male foetal lung tissue. Will double 50–60 times before showing senescence. Often used as 'normal' cells.

MRF-4 *myf-6* Member of the **myoD** family of muscle regulatory proteins.

M-ring Innermost (motor) ring of the bacterial flagellar base, located in the outer leaflet of the plasma membrane. It is this ring that is linked to the hook region (and thus to the flagellum itself) and that rotates. Composed of 16 or 17 subunits (one more or less than the **S-ring**).

mRNA See **messenger RNA**.

MRP See **calgranulins**.

MRSA See **methicillin-resistant** *Staphylococcus aureus*.

MS2 1. Type of F-specific coliphage (RNA bacteriophage). 2. Occasionally used as an abbreviation for MS-MS – mass spectroscopy (MS) followed by a second MS analysis of the fragments from the first analysis.

MSC Abbrev. for (1) mesenchymal stem cells, (2) bone marrow stromal cells – slightly confusing, because there can be bone marrow mesenchymal stem cells.

MSDS Abbrev. for (1) musculoskeletal disorders (MSDs), (2) material safety data sheet, (3) mean square displacements, of atoms, particles, etc., (4) membrane-spanning domains.

msh Multigene family coding for proteins involved in **mismatch repair**. Homologous to *S. cerevisiae MutS*. Included in family are *MSH1*, *MSH2*, *hMSH2*, *hMLH1*, *hPMS1*, *hPMS2* and probably *GTBP*.

MSH See **melanocyte-stimulating hormone**.

MSI 1. Microsatellite instability. 2. Magnetic source imaging. 3. Amphipathic antimicrobial peptides, MSI-78 and MSI-594 derived from **magainin**-2 and **melittin**, respectively.

MSM 1. Minimal salt medium. 2. Men who have sex with men – a common abbreviation in studies of sexually transmitted diseases.

MSO See *Msx*.

MSS4 1. Ubiquitous conserved protein that binds to a subgroup of **rab** proteins, a relatively slow exchange factor that forms a long-lived nucleotide-free complex with RabGTPase. See **translationally controlled tumour protein**. 2. In yeast, *MSS4* encodes a phosphatidylinositol-4-phosphate 5-kinase that synthesizes phosphatidylinositol (4,5)-bisphosphate.

Msx 1. Homeobox genes (*msx*) generally expressed in areas of cell proliferation and in association with multipotent progenitor cells. 2. L-methionine-DL-sulfoximine, a potent convulsant which metabolically and morphologically primarily affects astroglia (sometimes MSO). An inhibitor of glutamine synthetase.

MTBE *Methyl tertiary-butyl ether* Compound used as a petrol additive in place of tetra-ethyl lead, to reduce toxic emissions.

MTOC See **microtubule-organizing centre**.

mTOR *Mammalian/molecular target of rapamycin* A member of the phosphoinositide 3-kinase-related kinase (PIKK) family, a serine/threonine kinase and a central modulator of cell growth. The target of rapamycin (TOR) proteins in *S. cerevisiae*, TOR1 and TOR2, redundantly regulate growth in a rapamycin-sensitive manner. TOR2 additionally regulates polarization of the actin cytoskeleton in a rapamycin-insensitive manner. See **raptor** and **rictor**.

MTS Abbrev. for (1) **Microtubules** (MTs), (2) **Metallothioneins**, (3) Mitochondrial targeting sequence, (4) See **MTS assay**.

MTS assay *3-(4,5-dimethylthiazol-2-yl)-5-(3-carboxy-methoxyphenyl)-2-(4-sulphophenyl)-2H-tetrazolium assay* A cell viability assay. MTS is chemically reduced by cells into formazan and the production of formazan is proportional to the number of living cells, so the intensity of the produced colour is a good indication of the cell viability. Toxicity of a substance is determined using the dose–response curve to determine IC_{50}, the concentration of the test substance required to reduce the light absorbance capacity (at 492 nm) by 50%. Often used in place of the **MTT** assay.

MTT *3-(4,5-Dimethylthiazol-2-yl)-2,5-diphenyltetrazolium bromide* Compound used in a cell viability assay (MTT assay) based on the ability of a mitochondrial dehydrogenase enzyme from viable cells to cleave the tetrazolium rings of the pale yellow MTT and form a dark blue formazan crystals. Principle is the same as in the **MTS assay**.

MUC-1 An antiadhesion molecule, mucin type-1: see **episialin**.

mucilage Sticky mixture of carbohydrates in plants.

mucocyst Small membrane-bounded vesicular organelle in **pellicle** of ciliate protozoans that will discharge a mucus-like secretion.

mucopeptide Synonym for **peptidoglycan**.

mucopolysaccharide The polysaccharide components of proteoglycans, now more usually known as **glycosaminoglycans**.

mucopolysaccharidoses Inherited diseases in humans resulting from inability to break down glycosaminoglycans. **Hunter syndrome** and **Hurler's disease**, for example, result from defects in lysosomal enzymes needed to break down sulphated mucopolysaccharides. See **storage diseases**.

mucosal-associated lymphoid tissue *MALT* Lymphoid tissue and lymphoid aggregates associated with mucosal surfaces; includes bronchus-associated lymphoid tissue (BALT) and gut-associated lymphoid tissue (GALT).

mucous gland A type of **merocrine** gland that produces a thick (mucopolysaccharide-rich) secretion (as opposed to a **serous gland**).

mucus Viscous solution secreted by various membranes; rich in glycoprotein.

Muir–Torre syndrome *MTS* An autosomal, dominantly inherited disorder characterized by sebaceous neoplasms and visceral malignancies; the lesions are characterized by loss of **hMSH2** expression.

Muller cell Supporting cell of the neural retina. Cell body and nucleus lie in the middle of the inner nuclear region; their bases form the internal and external limiting membranes.

multilocus enzyme electrophoresis See **MLEE**.

multicopy inhibition Inhibition of translation of the transcript of a **transposase** gene by a multicopy **plasmid** with suitable inhibitory gene. The plasmid inhibits transposition events in the host bacterium.

multidrug transporter *Mdr; P-glycoprotein* Closely related family (**ABC proteins**) of integral membrane glycoproteins that export a variety of solutes from the cytoplasm.

multienzyme complex Cluster of distinct enzymes catalysing consecutive reactions of a metabolic pathway that remain physically associated through purification procedures. Multifunctional enzymes, found in eukaryotes, are a somewhat different phenomenon, since the several enzymic activities are associated with different domains of a single polypeptide.

multigene family See **gene family**.

multipain ATP-dependent protease (500 kDa) isolated from the cytoplasm of skeletal muscle. May form a complex with the 20S **proteosome** to form a 26S-like particle. Has not appeared in recent literature.

multiple cloning site See **polycloning site**.

multiple isomorphous replacement *MIP* Method of solving the phase problem in X-ray crystallography of proteins. Protein crystals consist of an array of geometrically identical unit cells arranged in three dimensions, each unit cell containing one or more identical asymmetric units. By substitution of a heavy atom at a small number of sites in each molecule, it may be possible to produce isomorphous crystals with identical geometry and molecular structure. Diffraction patterns from the unlabelled protein and two or more isomorphous derivatives can then be used to calculate the phases of the unlabelled crystal and, with the amplitude data, the molecular structure deduced.

multiple myeloma See **myeloma cell**.

multiple sclerosis Neurodegenerative disease characterized by the gradual accumulation of focal plaques of demyelination, particularly in the periventricular areas of the brain. Peripheral nerves are not affected. Onset usually in third or fourth decade, with intermittent progression over an extended period. Cause still uncertain.

multipotent cell Progenitor or precursor cell that can give rise to diverse cell types in response to appropriate environmental cues.

multipotential colony-stimulating factor See **interleukin-3**.

multivesicular body Secondary **lysosome** containing many vesicles of around 50 nm in diameter.

mulundocandin An **echinocandin**-like anti-fungal compound.

MUP 1. 4-methylumbelliferylphosphate, a fluorogenic substrate used in phosphatase assays. **2.** Motor unit potential, action potentials occurring in muscle units. **3.** **Major urinary protein**. **4.** See **Mup genes**.

***mup* genes** Cluster of genes encoding type I polyketide synthases and monofunctional enzymes that are involved in conversion of the product of the polyketide synthase into the active antibiotic, **mupirocin**.

mupirocin A polyketide-derived antibiotic from *Pseudomonas fluorescens* NCIMB10586, a mixture of pseudomonic acids (PA) that target isoleucyl-tRNA synthase.

muramic acid *3-O-α-carboxyethyl-D-glucosamine* Subunit of **peptidoglycan** (**murein**) of bacterial cell walls.

muramidase See **lysozyme**.

muramyl dipeptide Fragment of **peptidoglycan** from cell wall of mycobacteria that is used as an **adjuvant**.

murein Cross-linked **peptidoglycan** complex from the inner cell wall of all Eubacteria. Constitutes 50% of the cell wall in Gram-negative and 10% in Gram-positive organisms and comprises β (14)-linked *N*-acetyl-glucosamine and *N*-acetyl-muramic acid extensively cross-linked by peptides.

murine Pertaining to mice.

murine leukaemia virus A group of type C **Retroviridae** infecting mice and causing, in some strains, lymphatic leukaemia after a long latent period. Nearly all are replication-competent and v-onc negative. See also **Abelson leukaemia virus**, **Moloney murine leukaemia virus**, **Friend murine leukaemia virus**.

Mus musculus House mouse.

Musca domestica House fly.

muscarine Toxin (alkaloid) from the mushroom *Amanita muscaria* (Fly Agaric) that binds to **muscarinic acetylcholine receptors**.

muscarinic acetylcholine receptor Distinct from the **nicotinic acetylcholine receptor** in having no intrinsic ion channel; a seven-membrane spanning **G-protein-coupled receptor**.

muscle Tissue specialized for contraction. See also **muscle cell**, **twitch muscle**, **smooth muscle**, **catch muscle**, **cardiac muscle**.

muscle cell Cell of muscle tissue; in striated (skeletal) muscle it comprises a **syncytium** formed by the fusion of embryonic **myoblasts**, in cardiac muscle a cell linked to the others by specialized junctional complexes (**intercalated discs**), in smooth muscle a single cell with large amounts of actin and myosin capable of contracting to a small fraction of its resting length.

muscle fibre Component of a skeletal muscle comprising a single syncytial cell that contains **myofibrils**.

muscle spindle A specialized muscle fibre found in tetrapod vertebrates. A bundle of muscle fibres is innervated by sensory neurons. Stretching the muscle causes the neurons to fire; the muscle spindle thus functions as a stretch receptor.

muscular dystrophy A group of diseases characterized by progressive degeneration and/or loss of muscle fibres without nervous system involvement. All or nearly all of them have a hereditary origin, but details of the type of genetic defect and of the prognosis for the disease vary from type to type. **Duchenne muscular dystrophy** (pseudohypertrophic MD) is the most common form. It is due to a sex-linked recessive allele, and this is expressed as an absence of the protein **dystrophin**; the disease in boys shows extensive but insufficient muscle fibre reformation from **satellite cells**.

MuSK Tyrosine kinase localized on the postsynaptic surface of the neuromuscular junction. Mice lacking MuSK fail to form neuromuscular junctions. MuSK is probably involved in **agrin** signalling, though it does not interact directly with agrin.

Mustela Genus of mustelids, mink, ferret, stoat, etc.

mut Family of genes, products of which are involved in **mismatch repair**. *MutH* codes for a repair endonuclease (25 kDa) specific for unmethylated GATC, *mutL* codes for a protein (68 kDa) involved in the repair process, *mutS* (95 kDa) may be involved in the recognition step and has some ATPase activity.

mutagenicity tests See **Ames test**.

mutagens Agents that cause an increase in the rate of mutation; includes X-rays, ultraviolet irradiation (260 nm) and various chemicals.

mutarotation Change in optical rotation with time as an optical isomer in solution converts into other optical isomers.

mutation Usage usually restricted to change in the DNA sequence of an organism, which may arise in any of a variety of different ways. See **frame-shift**, **nonsense** and **missense mutation**.

mutation rate The frequency with which a particular mutation appears in a population or the frequency with which any mutation appears in the whole genome of a population. Normally the context makes the precise use clear. See **fluctuation analysis**.

muted Product of a **Hermansky–Pudlak syndrome** gene: see **BLOC-1**. Fibroblasts derived from the muted mouse strain exhibit reduced levels of **pallidin**, suggesting that the absence of the muted protein destabilizes pallidin.

mutein Protein with altered amino acid sequence, usually enough to alter properties. Uncommon usage.

Mx proteins GTPases (70–100 kDa) found in interferon-treated cells. Mx1 is found in the nucleus, and determines the resistance of mice to influenza A virus by blocking transcription of the viral RNA genome. Other Mx proteins are cytoplasmic and are related to dynamin; Mx proteins are involved in the innate antiviral response of fish.

myalgia Muscle pain.

myalgic encephalomyelitis *ME* A long-term postviral syndrome with chronic fatigue and muscle pain on exercise. Although dismissed for many years as a psychosomatic construct it has become recognized as a real phenomenon, although diagnostic markers are poor.

myasthenia gravis The characteristic feature of the disease is easy fatigue of certain voluntary muscle groups on repeated use. Muscles of the face or upper trunk are especially likely to be affected. In most (and perhaps all) cases, due to the development of autoantibodies against the **acetylcholine receptor** in neuromuscular junctions. Immunization of mice or rats with this receptor protein leads to a disease with the features of myasthenia.

myb An **oncogene**, identified in avian myeloblastosis, encoding a nuclear protein that binds the DNA sequence YAAC(G/T)G and is involved in the regulation of haematopoiesis. See Table O1.

myc A **proto-oncogene**, identified in several avian tumours, encoding a nuclear protein with a **leucine zipper** motif. See **Myc proteins** and Table O1.

Myc proteins Family of proteins involved in control of transcription; have a C-terminal basic helix–loop–helix–zipper domain. Myc-**Max** heterodimers specifically bind the sequence CACGTG with higher affinity than homodimers of either.

myc tag **Epitope tag** frequently expressed as a translational fusion with a transgenic protein of interest. As there are good antibodies to the myc epitope, this allows localization of the fusion gene product by **immunocytochemistry** or **Western blot** or its immunoaffinity purification.

mycelium Mass of **hyphae** that constitutes the vegetative part of a fungus (the conspicuous part in most cases is the fruiting body). Similar though smaller structures are found with some saprophytic bacteria such as **Nocardia**.

mycetocytes Cells containing symbiotic microorganisms, found in the **mycetome** of insect, ticks and mites.

mycetome A specialized organ in some species of insects, ticks and mites that is the site for intracellular symbionts such as bacteria, fungi and rickettsiae.

mycobacteria Bacteria with unusual cell walls that are resistant to digestion, being waxy, very hydrophobic and rich in lipid, especially esterified **mycolic acids**. Staining properties differ from those of Gram-negative and Gram-positive organisms, being acid-fast. Many are intracellular parasites, causing serious diseases such as **leprosy** and tuberculosis. Cell wall has strong immunostimulating (**adjuvant**) properties due to muramyl dipeptide (MDP).

mycolic acids Saturated fatty acids found in the cell walls of **mycobacteria**, *Nocardia* and **corynebacteria**. Chain lengths can be as high as 80, and the mycolic acids are found in waxes and in glycolipids.

mycophenolate mofetil Immunosuppressant drug used in conjunction with **ciclosporin** and corticosteroids to prevent graft rejection.

mycophenolic acid Antibacterial and antitumour compound from *Penicillium brevicompactum*. Inhibits *de novo* nucleotide synthesis.

mycoplasma Prokaryotic microorganisms lacking cell walls and therefore resistant to many antibiotics. Formerly known as pleuro-pneumonia-like organisms (PPLO). *Mycoplasma pneumoniae* is a causative agent of pneumonia in humans and some domestic animals. Troublesome contaminants of animal cell cultures, in which they may grow attached or close to cell surfaces, subtly altering properties of the cells, but escaping detection unless specifically monitored. Similar organisms, spiroplasms, cause various diseases in plants.

mycorrhiza Fungi associated with roots of higher plants: relationship is mutually beneficial and in some cases essential to survival of the higher plant. See **vesicular-arbuscular mycorrhiza, endomycorrhiza**.

mycosides Complex glycolipids found in **mycobacterial** cell wall. Non-toxic, non-immunogenic molecules that influence the form of the colony and the susceptibility of the bacteria to bacteriophages.

mycosis Any disease caused by a fungus.

mycosis fungoides A human disease in which a frequent secondary feature is fungal infection of lesions in the skin. Recognized as a tumour of T-lymphocytes that accumulate in the dermis and epidermis and cause loss of the epidermis.

myd 1. Common abbreviation for myotonic dystrophy (MyD). **2.** A gene (*myd*) that is involved in the determination of muscle cells (commitment to the myogenic lineage), not the same as MyoD. **3.** In the Large(*myd*) mouse, dystroglycan is incompletely glycosylated and thus cannot bind its extracellular ligands, causing a muscular dystrophy that is usually lethal in early adulthood. The Large(*myd*) mutation alters the composition and organization of the sarcolemma of fast-twitch skeletal muscle fibres. **4.** Myeloid differentiation factor (MyD)-88 is a key adaptor protein that plays a major role in the innate immune pathway.

mydriatic Drug that causes dilation of the pupil of the eye.

myelin The material making up the **myelin sheath** of nerve axons.

myelin basic protein Major component of the **myelin sheath** in mammalian CNS. Used as an antigen will induce **experimental allergic encephalomyelitis**, possibly a model for some neurodegenerative disorders.

myelin figures Structures that form spontaneously when bilayer-forming phospholipids (e.g. egg lecithin) are added to water. They are reminiscent of the concentric layer structure of myelin.

myelin sheath An insulating layer surrounding vertebrate peripheral **neurons** that dramatically increases the speed of conduction. It is formed by specialized **Schwann cells**, which can wrap around neurons up to 50 times. The exposed areas are called **nodes of Ranvier;** they contain very high densities of **sodium channels**, and **action potentials** jump from one node to the next without involving the intermediate axon – a process known as saltatory conduction.

myelodysplasia A group of disorders in which the bone marrow malfunctions and fails to produce normal numbers of blood cells. Various categories are recognized: refractory anaemia with or without ring sideroblasts (RA or RARS); refractory cytopenia with multilineage dysplasia (RCMD); 5q-syndrome; refractory anaemia with excess blasts (RAEB); unclassified (none of the previous types).

myelodysplastic syndrome MDS Haematological disorder that occurs mainly in the elderly as an acquired sporadic disease.

myeloid cells One of the two classes of marrow-derived blood cells; includes **megakaryocytes**, erythrocyte-precursors, **mononuclear phagocytes** and all the **polymorphonuclear leucocytes**. That all these are ultimately derived from one stem cell lineage is shown by the occurrence of the **Philadelphia chromosome** in these but not **lymphoid** cells. Most authors tend, however, to restrict the term myeloid to mononuclear phagocytes and granulocytes, and commonly distinguish a separate erythroid lineage.

myeloma cell Neoplastic **plasma cell**. The proliferating plasma cells often replace all the others within the marrow, leading to immune deficiency, and frequently there is destruction of the bone cortex. Because they are monoclonal in origin they secrete a monoclonal immunoglobulin. Bence-Jones proteins are monoclonal immunoglobulin light chains overproduced by myeloma cells and excreted in the urine. Myeloma cell lines are used for producing **hybridomas** in raising monoclonal antibodies.

myeloma proteins The immunoglobulins and Bence-Jones proteins secreted by **myeloma cells**.

myeloperoxidase *EC 1.11.1.7* A metallo-enzyme containing iron, found in the lysosomal granules of **myeloid cells**, particularly macrophages and neutrophils; responsible for generating potent bacteriocidal activity by the hydrolysis of hydrogen peroxide (produced in the **metabolic burst**) in the presence of halide ions. Deficiency of myeloperoxidase is not fatal, and the enzyme is reportedly absent entirely in chickens.

myeloplax *Megakaryocyte* A giant cell of bone marrow and other haematopoietic organs that gives rise to the blood platelets. May be multinucleated.

myf-5 Member of the **myoD** family of muscle regulatory genes/proteins.

myf-6 See **MRF-4**.

myo-inositol *Inositol* 'Muscle sugar' – a name that is only really of historical interest.

myoblast Cell that, by fusion with other myoblasts, gives rise to **myotubes** which eventually develop into skeletal muscle fibres. The term is sometimes used for all the cells recognizable as immediate precursors of skeletal muscle fibres. Alternatively, the term is reserved for those post-mitotic cells capable of fusion, others being referred to as presumptive myoblasts.

myocarditis Inflammation of heart muscle usually due to bacterial or viral infection.

myocardium Middle and thickest layer of the wall of the heart, composed of cardiac muscle.

myocilin *Trabecular meshwork glucocorticoid-inducible response protein; TIGR* A cytoskeletal protein expressed in many ocular tissues, including the **trabecular meshwork**. Mutations in *MYOC* gene encoding myocilin are responsible for primary open-angle glaucoma (POAG). Interacts with **flotillin-1**.

myoclonic Descriptive of a type of congenital tremor involving jerky spasms of muscles.

myoclonus 1. Paramyoclonus multiplex. A condition in which there occur sudden shock-like contractions of muscles, often associated with epilepsy and mental deterioration. **2.** A sudden, spasmodic contraction of a muscle.

myoD *MyoD1 MyoD* (myogenic determination) was originally described as a master regulatory gene for the determination of muscle cells, a process now thought to involve a family of genes. It is normally only expressed in myoblasts and skeletal muscle cells, but if transfected into cells will convert many differentiated cells into muscle cells. The *myoD* gene codes for the myoD protein, which then switches on the transcription of many muscle-specific genes. *MyoD* is a member of a family of genes that all code for nuclear proteins which have a basic DNA-binding motif and a **helix–loop–helix** dimerization domain (bHLH or mHLH proteins; b = basic, m = myogenic). They bind to a consensus sequence –CANNTG – and can form homodimers or heterodimers with other members of the bHLH superfamily.

myoepithelial cell *Basket cell; basal cell* Cell found between epithelium of exocrine glands (e.g. salivary, sweat, mammary, mucous) and their basement membranes, which resembles a smooth muscle cell and is thought to be contractile.

myoferlin Protein that is highly expressed in myoblasts undergoing fusion and is expressed at the site of myoblasts fusing to myotubes. Like **dysferlin,** it binds phospholipids in a calcium-sensitive manner.

myofibril Long, cylindrical organelle of striated muscle, composed of regular arrays of thick and thin filaments and constituting the contractile apparatus.

myofibroblasts Histological term for fibroblast-like cells that contain substantial arrays of actin microfilaments, myosin and other muscle proteins arranged in such a way as to suggest that they produce contractile forces. Are commonly described as occurring in granulation tissue (formed during wound healing) and in certain forms of arterial thickening where they are found in the intima. Behave in much the same way as smooth muscle cells, and have markers characteristic of these cells.

myogenesis The developmental sequence of events leading to the formation of adult muscle that occurs in the animal and in cultured cells. In vertebrate skeletal muscle, the main events are the fusion of **myoblasts** to form **myotubes** that increase in size by further fusion of myoblasts to them, the formation of **myofibrils** within their cytoplasm, and the establishment of functional **neuromuscular junctions** with **motoneurons**. At this stage they can be regarded as mature muscle fibres.

myogenin Member of the **MyoD** family of muscle regulatory genes/proteins. Related to the *myc* proto-oncogene family.

myoglobin Protein (17.5 kDa) found in red skeletal muscle. The first protein for which the **tertiary structure** was determined by **X-ray diffraction**, by J.C. Kendrew's group working on sperm whale myoglobin. A single polypeptide chain of 153 amino acids, containing a **haem** group bonded via its ferric iron to two histidine residues, that binds oxygen non-cooperatively and has a higher affinity for oxygen than **haemoglobin** at all partial pressures. In capillaries, oxygen is effectively removed from haemoglobin and diffuses into muscle fibres, where it binds to myoglobin, which acts as an oxygen store.

myomesin Protein (165 kDa) found in the **M-line** of the **sarcomere**.

myometrium Uterine **smooth muscle**.

myoneme Contractile organelle of ciliate protozoans; referred to as M-bands in *Stentor*, where they are composed of 8–10-nm tubular fibrils. The **spasmoneme** of peritrich ciliates was originally called a myoneme.

myopodin An actin-bundling protein that shuttles between nucleus and cytoplasm in response to cell stress or during differentiation – it has a nuclear localization sequence. Myopodin shows no significant homology to any known protein except **synaptopodin**. See **nuclear actin-binding proteins**.

myosin A family of motor ATPases that interact with F-actin filaments. An increasing number of different myosins are being described. Classical striated muscle myosin is myosin II. (See also **myosin light chains**, **meromyosin**.) Myosin I is a low molecular-weight (111–128 kDa) form found in protozoa (*Acanthamoeba* and *Dictyostelium*) that does not self-assemble and is found in the cytoplasm as a globular monomeric molecule that can associate with membranes and transport membrane vesicles along microfilaments.

myosin heavy chain See **myosin**: do not confuse with **heavy meromyosin,** which is a subfragment of the heavy chain of myosin II.

myosin light-chain kinase *MLCK* Calmodulin-regulated kinase of myosin II light chains: molecular weight varies according to source, and is 130 kDa in non-muscle mammalian cells. May regulate activity of myosin in some cells.

myosin light chains Small subunit proteins (17–22 kDa) of **myosin** II, all with sequence homology to **calmodulin**, but not all with calcium-binding activity: two pairs of different light chains are found per myosin. Several types are known: regulatory light chains (LC-2, DNTB-light chains) probably regulate the ATPase activity of the heavy chain directly (through the binding of calcium) or indirectly (activating when they themselves are phosphorylated by **myosin light-chain kinase**); essential light chains (LC-1, LC-3; alkali light chains) have a more subtle and apparently non-essential role. In molluscan muscle, the EDTA-light chains (similar to LC-2 from vertebrate muscle) confer calcium sensitivity on the myosin itself.

myositis Inflammation of muscle. Bacterial myositis can be caused by *Clostridium welchii* (gas gangrene). Viral myositis (epidemic myalgia) is usually due to Coxsackie B virus. Parasitic myositis can be a result of infection with the nematode worm *Trichinella*.

myostatin *Growth/differentiation factor-8a; GDF8* Member of the transforming growth factor-beta superfamily; blood protein that limits muscle growth. Mutation of the myostatin gene in mice, cattle and humans causes a massively developed skeletal muscle, characterized by muscle hypertrophy and hyperplasia.

myotilin A 57-kDa cytoskeletal protein. Its N-terminal sequence is unique, but the C-terminal half contains two Ig-like domains homologous to **titin**. Myotilin is expressed in skeletal and cardiac muscle, co-localizes and interacts with **alpha-actinin** in the sarcomeric I-bands. The human myotilin gene maps to chromosome 5q31. Defects in the myotilin gene may cause muscular dystrophy.

myotonic dystrophy An inherited human neuromuscular disease classed as an autosomal-dominant disease in which there is progressive muscle weakening and wasting. Caused by an unstable nucleotide repeat (CTG) in the $3'$ untranslated region.

myotoxins Small basic proteins (42–45 amino acids) in rattlesnake venom. Induce rapid necrosis of muscle.

myotube Elongated multinucleate cells (three or more nuclei) that contain some peripherally located **myofibrils**. They are formed *in vivo* or *in vitro* by the fusion of **myoblasts** and eventually develop into mature muscle fibres that have peripherally located nuclei and most of their cytoplasm filled with myofibrils. In fact, there is no very clear distinction between myotubes and muscle fibres proper.

myristic acid The myristoyl group is one of the less common fatty acyl residues of phospholipids in biological membranes (see Table L3), but is found as an N-terminal modification of a large number of membrane-associated proteins and some cytoplasmic proteins. It is a common modification of viral proteins. In all known examples, the myristoyl residue is attached to the amino group of N-terminal glycine. The specificity of the myristoyl transferase enzymes is extremely high with respect to the fatty acyl residue. For many proteins, the addition of the myristoyl group is essential for membrane association. There is some evidence that myristoylated proteins do not interact with free lipid bilayer, but require a specific receptor protein in the target membrane.

myristoylation Many proteins in eukaryotes are covalently attached to **myristic acid** in membranes through amide linkages formed by myristoyl-CoA:protein N-myristoyl transferase (NMT), at a glycine, with a consensus site: G-(EDRKHPFYW)-x-x-(STAGCN)-P. This allows a range of enzymes to be concentrated in specific domains within the cell.

Mytilus edulis The edible mussel, a marine bivalve mollusc. The ciliated gills are used for filter feeding, and these are utilized in studies on the **cilium** and on **metachronism**.

myxamoebae In the Myxomycetes, such as *Physarum*, each spore on germination produces two amoeboid cells, myxamoebae, which then transform into flagellated cells.

Myxobacteria Group of Gram-negative bacteria found mainly in soil. They are non-flagellated with flexible cell walls. They show a gliding motility, moving over solid surfaces leaving a layer of slime (myxo = slime). At some stage in their growth the cells of this group swarm together and form fruiting bodies and spores in a fashion similar to the **slime moulds**.

myxoedema Severe hypothyroidism usually as a result of autoimmunity to **thyroglobulin**. A variety of severe physiological problems accompany the reduction in thyroid function.

myxoma virus A poxvirus (see **Poxviridae**) that causes myxomatosis. Originally isolated from a species of wild rabbit, *Sylvilagus*, in Brazil, in which it causes a mild non-fatal disease, it was found to be 99% fatal in the European rabbit *Oryctlagus*. It causes the characteristic subcutaneous gelatinous swellings, 'myxomata', and usually kills in 2–5 days. It has been used to control rabbit populations in Australia and Britain, but there are signs that rabbits have developed immunity.

myxomycota The slime moulds. Amoeboid heterotrophic organisms, formerly classed as fungi but now as Protista. May be free-living phagocytes that feed on bacteria, or may be parasitic within plant cells. Includes Acrasiomycetes (cellular slime moulds, e.g. *Dictyostelium*) and Myxomycetes (acellular slime moulds, e.g. *Physarum*).

Myxoviridae Single-stranded RNA viruses of animals. Orthomyxoviruses include influenza viruses, Paramyxoviruses include mumps virus.

N

N51 Mouse homologue of **melanoma growth-stimulatory activity** protein.

Na⁺/H⁺ exchanger regulatory factor *NHE-RF; EBP50* See **EBP50**.

NAA See **naphthalene acetic acid**.

Nabothian cysts Benign cysts on the cervix of the womb composed of endocervical columnar cells covered by squamous metaplasia and filled with mucus.

NAC See **nascent polypeptide-associated complex**.

N-acetyl glucosamine *2-Acetamido glucose* A sugar unit found in glycoproteins and various polysaccharides such as **chitin**, bacterial **peptidoglycan** and in **hyaluronic acid**.

N-acetyl muramic acid Sugar unit of bacterial peptidoglycan, consisting of *N*-acetyl glucosamine bearing an ether-linked lactyl residue on carbon 3. The repeating unit of the cell wall polysaccharide is *N*-acetyl muramic acid linked to *N*-acetyl glucosamine via a β (1-4)-glycosidic bond, which can be cleaved by **lysozyme**.

N-acetyl neuraminic acid A nine-carbon sugar, structurally a condensation product of *N*-acetyl mannosamine and pyruvate. Also known as sialic acid, but more correctly is a member of the family of **sialic acids**. Found in **glycolipids** (especially **gangliosides**) and in **glycoproteins**, and therefore in the **plasma membrane** of animal cells, to the outer surface of which it contributes negative charge by virtue of its carboxylate group.

NAD *Nicotinamide adenine dinucleotide; NAD⁺; formerly DPN* Coenzyme in which the nicotine ring undergoes cyclic reduction to NADH and oxidation to NAD. Acts as a diffusible substrate for dehydrogenases etc. NADH⁺ is one source of reducing equivalents for the electron transport chain. NAD is of special interest as the source of ADP-ribose (see **ADP-ribosylation**).

NADP *TPN (formerly)* Analogue of **NAD**, but NADPH is used for reductive biosynthetic processes (e.g. pentose phosphate synthesis) rather than ATP generation.

Naegleria gruberi A normally amoeboid protozoan found in the soil. When it is flooded with water or a solution of low ionic strength, it transforms into a swimming form with two **flagella**.

naevus *Nevus (US)* Tumour-like but non-neoplastic **hamartoma** of skin. A vascular naevus is a localized capillary-rich area of the skin ('strawberry birthmark'; sometimes the much more extensive 'port-wine stain'). A mole (benign melanoma) is a pigmented naevus, a cluster of melanocytes containing melanin.

nagarse *Nagarase; subtilisin BPN; EC.3.4.21.62* Broad-specificity serine peptidase, from bacteria.

Nagler's reaction Standard method for identifying *Clostridium perfringens*. When the bacterium is grown on agar containing egg yolk, an opalescent halo is formed around colonies that produce α-toxin (lecithinase).

Naja kaouthia Asian cobra (one of the Elapidae). See alpha-**cobratoxin**.

nalidixic acid Synthetic antibiotic that interferes with **DNA gyrase** and inhibits prokaryotic replication. Often used in selective media.

naloxone An **alkaloid** antagonist of **morphine** and of the opiate peptides.

Namalwa cells Line of human B-lymphocytes grown in suspension and used to produce interferon (stimulated by Sendai virus infection). Derived from patient with **Burkitt's lymphoma**.

nanobacteria Nanobacteria are the smallest cell-walled bacteria, only recently discovered in human and cow blood, and commercial cell culture serum. They can produce apatite in media mimicking tissue fluids and glomerular filtrate and the aggregates produced closely resemble those found in tissue calcification and kidney stones.

nanobot Casual term for a hypothetical nano-scale robot, although it could perhaps be argued that some microorganisms are the nearest approach to such an entity.

nanopore A microscopic pore or opening, strictly speaking one between 10^{-7} and 10^{-9} m.

nanovid microscopy Technique of bright-field light microscopy using electronic contrast enhancement and maximum numerical aperture.

NAP-1 See **interleukin-8**.

NAP-3 See **melanoma growth-stimulatory activity**.

naphthalene acetic acid *NAA* A synthetic auxin, often used in plant physiology and in plant tissue culture media because it is more stable than **IAA**.

naphthylamine β-*Naphthylamine* Potent carcinogen; used in production of aniline dyes, one of the first chemicals to be associated with a tumour (bladder cancer). The compound itself is not directly carcinogenic; a metabolite produced by hydroxylation (1-hydroxy-2-aminonaphthalene) is detoxified in the liver by conjugation with glucuronic acid, but reactivated by a glucuronidase in the bladder.

napin Angiosperm 2S albumin seed-storage protein from *Brassica napus* (oilseed rape). Consists of two polypeptide chains (3.8 and 8.4 kDa) linked by two disulphide bridges. Interacts with calmodulin and has antifungal properties.

naproxen A non-steroidal anti-inflammatory drug probably working as a prostaglandin synthetase inhibitor.

napthoquinones Plant pigments derived from napthoquinone.

narasin A polyether monocarboxylic acid antibiotic produced by *Streptomyces aureofaciens*. Acts as an ionophore; used as a coccidiostat in chicken feed.

naratriptan Agonist for subclass of serotonin receptors (5HT₁ receptors), used to treat acute migraine.

NASBA Nucleic acid sequence-based amplification, a technology used for the continuous amplification of nucleic acids in a single mixture at one temperature. Normally applied to amplify single-stranded target RNA, producing RNA amplicons, although it can be modified to amplify DNA.

nascent polypeptide-associated complex *NAC* Heterodimeric complex, a peripheral component of cytoplasmic ribosomes, that interacts with nascent polypeptides as they emerge; it may also act as a transcriptional coactivator and binds to nucleic acids. NAC prevents targeting of nascent polypeptide chains that lack a **signal sequence** to the ER, the opposite function to that carried out by the **signal recognition particle**. Interacts with **alpha-taxilin**. There are two subunits, alpha- and beta-NAC, with high homology. NAC is highly conserved from yeast to humans, and mutations in NAC cause severe embryonically lethal phenotypes in mice, *Drosophila* and *C. elegans*.

nasopharyngeal carcinoma *NPC* Carcinoma, highly prevalent in Southern China, associated with infection by **Epstein–Barr virus** and probably exposure to inhaled co-carcinogens.

nastic movement Non-directional movement of part of a plant in response to external stimulus. The tips of growing shoots of plants that twine around supports show nastic movement. See **epinasty**.

natriuretic Of a substance or hormone, causing natriuresis (elimination of extra sodium in the urine). See **atrial natriuretic peptide**.

natriuretic peptides Family of four peptides all sharing significant sequence and structural homology, and all acting through guanylyl cyclase. The mammalian members are atrial natriuretic peptide (ANP), B-type natriuretic peptide (BNP), C-type natriuretic peptide (CNP) and possibly osteocrin/musclin. C-type natriuretic peptide (CNP-22) is a vasodilator and plays an important part in regulating blood pressure, renal function, volume homeostasis and long-bone growth. It is produced by endothelial and renal cells, and is considered an autocrine regulator of endothelium as well as a neuropeptide. Effects may oppose those of **atrial natriuretic peptide**. Receptor is natriuretic peptide receptor-B. See also **brain natriuretic peptide** and **dendroaspis natriuretic peptide**.

natural killer cells *NK-cells* See **killer cells**.

natural selection The hypothesis that genotype-environment interactions occurring at the phenotypic level lead to differential reproductive success of individuals and hence to modification of the gene pool of a population.

nauplius larva The typical first larval stage of Crustacea; approximately egg-shaped, unsegmented, and with three pairs of appendages and a median eye.

navicular disease Chronic osteitis of the navicular bone in the foot of the horse arising from a range of disparate and often poorly defined causes. At one time thought to be due to local ischaemia, but this is no longer considered the basic cause.

Nbk *Nbk/Bik* Pro-apoptotic **BH3**-only protein, natural-born killer; see **Bik**.

NBQX *6-Nitro-7-sulphamoylbenzo[f] quinoxaline-2,3-dione* Blocker of **AMPA receptors**.

NBT *Nitroblue tetrazolium* See **nitroblue tetrazolium reduction**.

N-cadherin See **cadherins**.

NCAM *Neural cell adhesion molecule* One of the first of the **CAMs** to be isolated from chick brain. Part of the **immunoglobulin superfamily**, as is NgCAM (neural-glial CAM). Initially defined by adhesion-blocking antiserum. Thought to be important in divalent-cation-independent intercellular adhesion of neural and some embryonic cells. See also **neuroglian**.

NCAP Abbrev. for (1) non-cell-autonomous proteins, proteins exchanged between plant cells via plasmodesmata, (2) *N*-acetyl-4-cystaminylphenol; used topically for the treatment of hyperpigmentation, (3) N-terminal residue of a protein, (4) sleep phase in which there is not a cyclic alternating pattern (CAP) in the EEG.

NCD Protein of the **kinesin-14 protein** family that differs from kinesin in that it moves towards the minus end of the microtubule (like cytoplasmic dynein). Implicated in spindle organization. See also **Kar3**.

N-chimaerin A phorbol ester/diacyl glycerol binding protein found in brain. A GTPase-activating protein for **rac**.

Nck Small adaptor protein with SH2 and SH3 domains. Similar to **Crk** and **GRB-2**. The Nck family of adaptors function to link tyrosine phosphorylation induced by extracellular signals with downstream regulators of actin dynamics.

N-CoR *Nuclear receptor co-repressor* Proteins involved in transcriptional repression by thyroid hormone and retinoic acid receptors. The binding of the co-repressor leads to the assembly of a larger complex that may, for example, contain **Sin3** and **histone deacetylases** (HDAC). RIP140, one of the family, has a role in the regulation of energy expenditure. **SMRT** and N-CoR are paralogs and possess similar molecular architectures and mechanistic strategies, but exhibit distinct molecular and biological properties.

NCP1 Family of cell adhesion molecules that includes **neurexin** IV, contactin-associated protein (**Caspr**) and **paranodin**, present at the vertebrate axo-glial synaptic junction.

NCS family Neuronal calcium sensor family of proteins. See **frequenin**, **visinin-like proteins**.

Ndk *Nucleoside diphosphate kinase; E.C.2.7.4.6* Enzyme that generates nucleoside triphosphates or their deoxy derivatives by terminal phosphotransfer from ATP or GTP.

nearest neighbour analysis Statistical method that can be used to analyse spatial distributions (e.g. of organisms in an environment) according to whether they are clustered, random or regular. It has also been used, for example, to determine the frequency of pairs of adjacent bases in DNA, revealing a deficiency of the pair CG in most eukaryotes.

nebulin Family of large matrix proteins (600–900 kDa) found in the **N-line** of the **sarcomere** of striated muscle. Consist of many (more than 200) repeats of conserved actin-binding motifs; bind to F-actin and may serve as templates for assembly of the sarcomere.

necdin A growth suppressor expressed predominantly in postmitotic neurons and implicated in terminal differentiation. Necdin-like proteins form a family within the **MAGE**

superfamily. Necdin is one of the chromosome 15 products disrupted in **Prader-Willi syndrome**.

necrosis Death of some or all cells in a tissue as a result of injury, infection or loss of blood supply. Necrosis, unlike **apoptosis** is likely to elicit an inflammatory response.

necrotizing fasciitis A bacterial infection that causes rapid decay of the fascia and soft tissue. The causative bacteria may be aerobic, anaerobic or mixed, and aggressive treatment is required to prevent major tissue loss.

nectin 1. Another name for SAM (substrate adhesion molecule), e.g. fibronectin. **2.** A protein forming the stalk of mitochondrial ATPase. **3.** Calcium-independent immunoglobulin-like adhesion molecule that interacts with other nectin molecules to cause cell–cell adhesion and interacts, through its cytoplasmic domain-associated protein **afadin**, with **catenin** which interacts in turn with the cytoplasmic domain of **cadherin**.

nedd2 *Caspase-2; Ich-1* See **caspases**.

nef **HIV** protein that is important for pathogenesis, enhances infectivity and regulates the sorting of at least two cellular transmembrane proteins, CD4 and MHC Class I. Has a proline-rich sequence that interacts with **Hck**-SH3 domain and activates kinase activity of Hck; also interacts with a serine/threonine kinase of 62 kDa. Nef itself has a sorting signal (ENTSLL) that functions as an endocytosis marker, and it has some amino acid homology with alpha-scorpion toxins that bind to potassium channels.

negative feedback Situation in which the products of a process act at an earlier stage in the process to inhibit their own formation. The term was first used widely in conjunction with electrical amplifiers where negative feedback was applied to limit distortion of the signal by the amplification mechanism. Tends to stabilize the process. In contrast to **positive feedback**.

negative regulation Negative feedback in biological systems mediated by allosteric regulatory enzymes.

negative staining Microscopic technique in which the object stands out against a dark background of stain. For electron microscopy, the sample is suspended in a solution of an electron-dense stain such as sodium phosphotungstate and then sprayed onto a support grid. The stain dries as structureless solid and fills all crevices in the sample. When examined in the electron microscope, the sample appears as a light object against a dark background. Quite fine structural detail can be observed using negative staining, and it has been used extensively to study the structure of viruses and other particulate samples.

negative-stranded RNA virus Class V **viruses** that have an RNA genome that is complementary to the mRNA, the positive strand. They also carry the virus-specific **RNA polymerase** necessary for the synthesis of the mRNA. Includes **Rhabdoviridae**, **Paramyxoviridae** and Myoviridae (e.g. the T-even phages).

NEGR1 See **IgLON**.

Negri body Acidophilic cytoplasmic inclusion (mass of **nucleocapsids**) characteristic of rabies virus infection.

Neisseria Gram-negative non-motile pyogenic cocci. Two species are serious pathogens: *N. meningitidis* (see **meningitis**) and *N. gonorrhoeae*. The latter associates specifically with urinogenital epithelium through surface **pili**. Both species seem to evade the normal consequences of attack by phagocytes.

nekton *Necton* Actively swimming aquatic organisms, in contrast to passively drifting planktonic organisms (plankton).

nemaline Fibrous or thread-like.

nemaline bodies Sarcoplasmic inclusions composed largely of alpha-actinin and actin, probably derived from the Z-disc. See **nemaline myopathies**

nemaline myopathies Muscle disorders of variable severity and age of onset, with characteristic nemaline bodies in the sarcoplasm. Nemaline myopathy (NM) is a rare autosomal dominant skeletal muscle myopathy caused by mutation in the alpha-actin-1 gene and characterized by severe muscle weakness and the subsequent appearance of nemaline rods within the muscle fibres.

nematocyst *Cnidocyst* Stinging mechanism used for defence and prey capture by *Hydra* and other members of the Cnidaria (Coelenterata). It is located within a specialized cell, the **nematocyte,** and consists of a capsule containing a coiled tube. When the nematocyte is triggered, the wall of the capsule changes its water permeability and the inrush of water causes the tube to evert explosively, ejecting the nematocyst from the cell. The tube is commonly armed with barbs and may also contain toxin.

nematocyte *Cnidoblast* Stinging cells found in *Hydra*, used for capturing prey and for defence. There are four major types, containing different sorts of **nematocysts**: stenoteles (60%), desmonemes, holotrichous isorhizas and atrichous isorhizas. They differentiate from interstitial cells and are almost all found in the tentacles.

Nematoda A class of the phylum Aschelminthes. Unsegmented worms with an elongated rounded body pointed at both ends, a mouth and alimentary canal, and a simple nervous system. The sexes are separate and larvae resemble the adults; many species are of economic importance as pests; most are free-living but some are parasitic. Best-known example in cell biology is *Caenorhabditis elegans* , which, because of the determinate number of cells, has proved a valuable experimental organism, especially now that its genome is sequenced and the effects of many mutations analysed.

nematode sperm The nematode *Caenorhabditis elegans* has an unusual amoeboid spermatozoon that is actively motile yet appears to lack both actin and tubulin.

nematosome Cytoplasmic inclusion in some neurons.

N-end rule The N-end rule holds that the *in vivo* half-life of a protein is determined by the N-terminal residues.

neoantigen Antigen acquired after a cell has been transformed by an oncogenic virus.

neoblasts Totipotent stem cells responsible for regeneration and tissue renewal in planarians.

neoendorphin Opioid peptide (**endorphin**) cleaved from pro-dynorphin.

neointima The new intima laid down in a vessel that has been dilated by **angioplasty**; often hyperplastic and the cause of re-stenosis.

neomycin Either of two aminoglycosides (B and C) produced by *Streptomyces fradiae* that have generalized antibiotic activity. Neomycin A (Ineamine) contains 2-deoxy-1,3-diamino-inositol combined with the aminoglycoside.

neoplasia Literally new growth, usually refers to abnormal new growth, and thus means the same as **tumour**, which may be benign or malignant. Unlike **hyperplasia**, neoplastic proliferation persists even in the absence of the original stimulus.

neopterin A pteridine derivative, the D-erythro enantiomer of **biopterin**, produced by human monocytes/macrophages upon stimulation with interferon-gamma. Increased neopterin concentrations in human serum and urine indicate activation of cell-mediated (Th1-type) immune responses.

neotenin *Juvenile hormone* Insect hormone, a derivative of farnesoic acid. It is produced by the **corpora allata** and suppresses the development of adult characteristics at each moult except the last when metamorphosis to the adult form occurs.

neoteny *Paedomorphosis* The persistence in the reproductively mature adult of characters usually associated with the immature organism.

neoxanthin A **xanthophyll carotenoid** pigment, found in higher plant chloroplasts as part of the **light-harvesting system**.

NEP *Neutral endopeptidase; EC 3.4.24.11.* Cell surface zinc endopeptidase that hydrolyzes regulatory peptides such as **ANP**. Spontaneously hypertensive hamsters have elevated levels of NEP in two organs that contribute appreciably to vascular resistance, skeletal muscle and kidney.

nephelometer Instrument for measuring turbidity, in which light scattered orthogonally to the incident beam is measured. Light scattering depends upon the number and size of particles in suspension.

nephelometry Any method for estimating the concentration of cells or particles in a suspension by measuring the intensity of scattered light, often at right angles to the incident beam. See **nephelometer**.

nephron The structural and functional unit of the vertebrate kidney. It is made up of the glomerulus, Bowman's capsule and the convoluted tubule.

neprilysin *EC 3.4.24.11* Neuropeptide-degrading neutral endopeptidase, a zinc metallopeptidase similar to bacterial thermolysin with some homology to **endothelin converting enzyme**.

Nernst equation A basic equation of biophysics that describes the relationship between the equilibrium potential difference across a **semipermeable membrane**, and the equilibrium distribution of the ionic permeant species. It is described by: $E = (RT/zF). \ln([C1]/[C2])$, where E is the potential on side two relative to side one (in volts), R is the gas constant (8.314 J/Kper mol), T is the absolute temperature, z is the charge on the permeant ion, F is the Faraday constant (96 500°C/mol) and $[C1]$ and $[C2]$ are the concentrations (more correctly activities) of the ions on sides 1 and 2 of the membrane. It can be seen that this equation is a solution of the more general equation of **electrochemical potential**, for the special case of equilibrium. The equation describes the voltage generated by ion-selective electrodes, like the laboratory pH electrode, and approximates the behaviour of the resting plasma membrane (see **resting potential**).

Nernst potential See **Nernst equation** and **ion-selective electrodes**.

nerve cell See **neuron**.

nerve ending See **synapse**.

nerve growth cone See **growth cone**.

nerve growth factor *NGF* A peptide (13.26 kDa) of 118 amino acids (usually dimeric) with both chemotropic and chemotrophic properties for **sympathetic** and **sensory neurons**. Found in a variety of peripheral tissues, NGF attracts **neurites** to the tissues by chemotropism, where they form synapses. The successful neurons are then 'protected' from neuronal death by continuing supplies of NGF. It is also found at exceptionally high levels in snake venom and male mouse submaxillary salivary glands, from which it is commercially extracted. NGF was the first of a family of nerve tropic factors to be discovered. Amino acids 1–81 show homology with proinsulin. Besides its peripheral actions, NGF selectively enhances the growth of **cholinergic neurons** that project to the forebrain and that degenerate in **Alzheimer's disease**.

nerve impulse An **action potential**.

nesidioblast Precursor cell of pancreatic **B cells**.

nesprins *Nuclear envelope spectrin repeat; nesprin-1 (synaptic nuclear envelope protein-1 (syne-1); enaptin; myne; MSP-300; Ank-1)* Proteins distantly related to **spectrin**, a family of alpha-actinin type actin-binding proteins residing at the nuclear membrane. Nesprin-1 and -2 make direct connections with the actin cytoskeleton through their NH_2-terminal actin-binding domain, and with a C-terminal region in **Sun1** through a conserved C-terminal motif, PPPX. Nesprin-3, an outer nuclear membrane protein, lacks an actin-binding domain and associates with the cytoskeletal linker protein **plectin**. Nesprin-2 is also found as a large isoform (Nesprin-2 Giant; 800 kDa) in the outer nuclear membrane. Also associated with Golgi membranes.

nested PCR Variety of **polymerase chain reaction** in which specificity is improved by using two sets of primers sequentially. An initial PCR is performed with the 'outer' primer pairs, then a small aliquot is used as a template for a second round of PCR with the 'inner' primer pair.

nested primers Sets of primers for PCR so arranged that the second set to be used lie within the sequence amplified by the first set of primers and so on.

nestin *Neural stem cell protein* Large (200-kDa) intermediate filament protein found in developing rat brain. Functionally similar to other intermediate filament proteins, but the sequence is very different. Forms class VI of the **intermediate filaments**.

netrin Products of genes identified in studies of vertebrate neuronal development. Netrins are **chemotropic** for embryonic commissural neurons: netrin 1 is secreted by the floorplate, whereas netrin-2 is distributed ventrally except for the floorplate. The netrins are homologous to the product of *unc-6*, a gene identified in studies of neuronal development of *C. elegans*. Receptors include members of the Deleted in Colorectal Cancer (**DCC**) protein family (see *Drosophila* **Frazzled**), and members of the UNC-5 family.

netropsin Basic peptide antibiotic from *Streptomyces*. Binds selectively in minor groove of **B-DNA** and will induce A to B transition.

network theory 1. In general, theories about the properties of networks of which signalling pathways and the Internet are classic examples. An important property of a network is its robustness to perturbation – alternative pathways can be used to compensate for damage or deletion. **2.** In immunology, a theory proposed by Jerne in 1974 that the immune system is controlled by a network of interactions between antigen-binding sites (paratopes), each of which is capable of binding an epitope on an external antigen and also an idiotope, with a shape resembling the epitope, present on another immunoglobulin molecule.

neu erb-B2 **Oncogene**, originally identified in a **neuroblastoma**, that encodes a receptor **tyrosine kinase** of the EGF-receptor family. Ligand is **neuregulin**. See Table O1.

NeuN Antigenic marker frequently used to identify mature neurons. NeuN (neuronal nuclei) protein is a DNA-binding, neuron-specific protein present in most CNS and PNS neuronal cell types of all vertebrates tested. NeuN protein distributions are apparently restricted to neuronal nuclei, perikarya and some proximal neuronal processes in both fetal and adult brain.

neurabin *Neurabin-1; neurabin-2 (spinophilin)* Neuron-specific actin-binding protein that is enriched in dendritic spines, and tethers protein phosphatase-1 to regions of actin-rich postsynaptic density. Will cause F-actin bundling. Neurabin-1 and spinophilin appear to have distinct roles.

neural cell adhesion molecule See **NCAM**.

neural crest A group of embryonic cells that separate from the **neural plate** during neurulation and migrate to give several different lineages of adult cells: the spinal and autonomic ganglia, the **glial cells** of the peripheral nervous system, and non-neuronal cells, such as **chromaffin cells**, **melanocytes**, and some haematopoietic cells.

neural fold A crease that forms in the **neural plate** during **neurulation**.

neural induction In vertebrates, the formation of the nervous system from the **ectoderm** of the early embryo as a result of a signal from the underlying **mesoderm** of the archenteron roof; also known as primary neural induction. The mechanism of neural induction is not yet clear, but in *Xenopus* neural induction results from the combined inhibition of **BMP** receptor regulated serine/threonine kinases and activation of receptor tyrosine kinases that signal through MAPK and phosphorylate **Smad1** in the linker region, further inhibiting Smad1 transcriptional activity.

neural plate A region of embryonic ectodermal cells, called neuroectoderm, that lie directly above the **notochord**. During **neurulation**, they change shape, so as to produce an infolding of the neural plate (the neural fold) that then seals to form the neural tube.

neural retina Layer of nerve cells in the retina, embryologically part of the brain. The incoming light passes through nerve fibres and intermediary nerve cells of the neural retina before encountering the light-sensitive rods and cones at the interface between neural retina and the pigmented retinal epithelium.

neural tube The progenitor of the central nervous system. See **neural plate**, **neurulation**.

neuraminic acid See **N-acetyl neuraminic acid**.

neuraminidase *Sialidase; EC3.2.1.18* Enzyme catalysing cleavage of **neuraminic acid** residues from oligosaccharide chains of glycoproteins and glycolipids. Since these residues are usually terminal, neuraminidases are generally exo-enzymes, although an endoneuraminidase is known. For use as a laboratory reagent, common sources are from bacteria such as *Vibrio* or *Clostridium*. A neuraminidase is one of the transmembrane proteins of the envelope of influenza virus and a target for antiviral drugs.

neuraxin Protein associated with neuronal **microtubules**. Structurally related to **MAP**-1B.

neuraxis The neural axis of the body, the brain and spinal cord.

neuregulins *NRG1, NRG2 neu differentiation factor (NDF); heregulin-α* A family of growth and differentiation factors that are related to epidermal growth factor (EGF) and are ligands for the ErbB family of tyrosine kinase transmembrane receptors. They induce growth and differentiation of epithelial, glial and muscle cells in culture. Gene disruption is lethal during embryogenesis, with heart malformation and defects in Schwann cells and neural ganglia. There are three major isoforms of neuregulin 1 with distinct domain structures, although all contain an EGF-like domain. Type I NRG1, (44 kDa), is also known as **heregulin**, Type II NRG1 is also called glial growth factor-2, and Type III is sensory and motor neuron-derived factor (SMDF).

neurexin *NRX* Neurexins constitute a large family of highly variable cell surface molecules, related to alpha-**latrotoxin** receptor, laminin and agrin, that may function in synaptic transmission and/or synapse formation. Each of the known vertebrate neurexin genes encodes two major neurexin variants, alpha- and beta-neurexins, that are composed of distinct extracellular domains linked to identical intracellular sequences. Alpha-neurexins regulate presynaptic N- and P/Q-type Ca^{2+} channels. Beta-neurexin binds to presynaptic **neuroligin**. Neurexin IV is a component of *Drosophila* septate junctions: see **NCP**, **paranodin**. Binds to **coracle**.

neurite A process growing out of a neuron. As it is hard to distinguish a **dendrite** from an **axon** in culture, the term neurite is used for both.

neuroblast Cells arising by division of precursor cells in neural ectoderm (**neurectoderm**) that subsequently differentiate to become neurons.

neuroblastoma Malignant tumour derived from primitive ganglion cells. Mainly a tumour of childhood. Commonest sites are adrenal medulla and retroperitoneal tissue. The cells may partially differentiate into cells having the appearance of immature neurons.

neurocalcin A dimeric 24-kDa, calcium-sensing protein from neurons that belongs to a family with **recoverin**, visin, VILIP and **hippocalcin**. The interaction with F-actin requires calcium. Neurocalcin is myristoylated, and also binds clathrin and tubulin. Abundant in CNS.

neuroectoderm Ectoderm on the dorsal surface of the early vertebrate embryo that gives rise to the cells (neurons and glia) of the nervous system. Also called the **neural plate**.

neuroendocrine cell See **neurohormone**.

neuroepithelium See **neuroectoderm**.

neurofascin *NF155; NF186* An axonal member of the L1 subgroup of the immunoglobulin superfamily, implicated in neurite extension during embryonic development. There are two main forms of neurofascin, referred to as NF155 and NF186 on the basis of their molecular weight, and these are found in oligodendrocytes and neurons respectively. In myelinated axons of the CNS, NF186 is found within nerve cells at the nodes of Ranvier. The other form of neurofascin, NF155, is found within the oligodendrocyte, and is located on either side of the nodes of Ranvier.

neurofibrillary tangle A characteristic pathological feature of the brain of patients with **Alzheimer's disease** is the presence of tangles of coarse **neurofibrils** within large neurons of the cerebral cortex. Whether this causes neuronal degeneration or is a secondary consequence remains contentious.

neurofibrils Filaments found in neurons; not necessarily **neurofilaments** in all cases, and in the older literature 'fibrils' are composed of both microtubules and neurofilaments. Originally used by light microscopists to describe much larger fibrils seen particularly well with silver-staining methods.

neurofibromatosis Tumours of neuronal sheath. The most common genetic disease, Type 1 neurofibromatosis, is associated with the von Recklinghausen Neurofibromatosis locus, which encodes the NF 1 protein **neurofibromin** – a **GTPase-activating protein** that interacts with the **ras** proteins. See **neurofibromatosis type 2**. Other variants of the disease are recognized.

neurofibromatosis type 2 *NF2* A genetic disorder due to inactivating mutations in the *NF2* (**neurofibromin-2**) tumour-suppressor gene on chromosome 22, the product of which is **merlin**. The Type 2 form is characterized by tumours of the eighth cranial nerve (usually bilateral), meningiomas of the brain, and schwannomas of the dorsal roots of the spinal cord, and has few of the hallmarks of the peripheral (Type 1) form of **neurofibromatosis**.

neurofibromin *NF-1, NF-2* The neurofibromin-1 (*NF1*) gene product has a GTPase-activating protein domain (GRD) that interacts with the Ras protein; mutations in the gene lead to **neurofibromatosis** Type 1. Mutations in *NF-2* (product of which is **merlin**) cause **neurofibromatosis Type 2**.

neurofilament Member of the class of **intermediate filaments** found in axons of nerve cells. In vertebrates, assembled from three distinct protein subunits (NF-L, 68 kDa; NF-M, 160 kDa; NF-H, 200 kDa) These proteins, if introduced into fibroblasts, will incorporate into the vimentin filament system.

neurogenesis Differentiation of the nervous system from the **ectoderm** of the early embryo. There are major differences between neurogenesis in vertebrates and invertebrates.

neurogenic gene Best described in *Drosophila*, genes that are required to determine a neuronal fate. Examples: *Notch, Delta*.

neurogenins Family of bHLH transcription factors involved in specifying neuronal differentiation. Related to *Drosophila* atonal; neurogenin-1 is expressed in and required for specification of dopaminergic progenitor cells.

neuroglia See **glial cells**.

neuroglian Protein isolated from *Drosophila* nervous system that is a member of the **immunoglobulin superfamily,** invertebrate homologue of L1-CAM. It contains six immunoglobulin-like domains and five fibronectin-type III domains, and has strong sequence homology to mouse **NCAM** and **L1**. Two different forms of neuroglian arise by differential splicing. These have identical extracellular domains but differ in the size of the cytoplasmic domains: the long form is restricted to neurons in central and peripheral nervous systems of embryos and larvae.

neurogranin *RC3* A postsynaptic substrate (78 aa) for protein kinase C; its expression is related to dendritic spine development and postsynaptic plasticity. Binds calmodulin at low Ca^{2+} levels, and the binding is affected by PKC phosphorylation. Neurogranin apparently enhances **long-term potentiation** and learning by promoting calcium-mediated signalling; knockout mice have learning difficulties.

neurohaemal organs Organs specialized for the release of products of neurosecretory cells into the circulating blood or body fluid. Term usually applied to invertebrate organs such as the corpora cardiaca of insects.

neurohormone A hormone secreted by specialized **neurons** (neuroendocrine cells); e.g. releasing hormones.

neurokinin See **tachykinins**.

neuroleptic drugs *Anti-schizophrenic drugs; antipsychotic drugs; tranquillizers* Literally 'nerve-seizing': used of chlorpromazine-like drugs. Antagonize the effects of **dopamine**.

neuroligin-1 A postsynaptic transmembrane protein that has the presynaptic ligand **neurexin**. Neuroligin also interacts with SAP90/PSD95, a multidomain scaffolding protein thought to recruit proteins to postsynaptic sites.

neurolin A growth-associated cell surface glycoprotein from goldfish and zebrafish which has been shown to be involved in axonal pathfinding in the goldfish retina and suggested to function as a receptor for axon guidance molecules. A member of the immunoglobulin superfamily of cell adhesion proteins, neurolin consists of five N-terminal extracellular immunoglobulin (Ig)-like domains, a transmembrane and a short cytoplasmic domain.

neuromedin U *NMU* One of the **neuromedin** family of neuropeptides. Neuropeptide with potent activity on smooth muscle. First isolated from porcine spinal cord, subsequently from other species. NMU will stimulate smooth muscle, increase blood pressure, alter ion transport in gut, control local blood flow and regulate adrenocortical function. Receptors are G-protein coupled: NMU1R is abundantly expressed in peripheral tissues, NMU2R in specific regions of the brain – particularly the ventromedial hypothalamus, where levels of NMU are reduced following fasting.

neuromedins Family of neuropeptides. Four classes are recognized: **kassinin**-like, **bombesin**-like, **neurotensin**-like and **neuromedin U**.

neuromeres Alternate swellings and constrictions seen along the **neuraxis** at early stages of **neural tube** development, thought to be evidence of intrinsic segmentation in the central nervous system. Neuromeres or segments in the hindbrain region are called **rhombomeres** and have been shown to be lineage-restriction units, each constructing a defined piece of hindbrain.

neuromodulation Alteration in the effectiveness of **voltage-gated** or **ligand-gated ion channels** by changing the characteristics of current flow through the channels. The mechanism is thought to involve **second messenger** systems.

neuromodulin *GAP-43; pp46; B-50; F1; P-57* Protein associated with actively growing axons, especially in the **growth cone**. Binds **calmodulin**, is phosphorylated by **protein kinase C**.

neuromuscular junction A **chemical synapse** between a motoneuron and a muscle fibre. Also known as a motor end plate.

neuron *Neurone; nerve cell* An **excitable cell** specialized for the transmission of electrical signals over long distances. Neurons receive input from sensory cells or other neurons, and send output to muscles or other neurons. Neurons with sensory input are called 'sensory neurons'; neurons with muscle outputs are called 'motoneurons'; neurons that connect only with other neurons are called 'interneurons'. Neurons connect with each other via **synapses**. Neurons can be the longest cells known – a single **axon** can be several metres in length. Although signals are usually sent via **action potentials**, some neurons are **non-spiking**.

neuronal calcium sensor-1 See **frequenin**.

neuronal ceroid lipofuscinoses *NCLs* A range of progressive neurological disorders primarily affecting children. Although six of the causative genes have been characterized, the underlying disease pathogenesis for this family of disorders is unknown. See **Batten's disease**.

neuronal differentiation Acquisition during development of specific biochemical, physiological and morphological properties by nerve cells.

neuronal guidance See **axonal guidance**.

neuronal plasticity Ability of nerve cells to change their properties, for example by sprouting new processes, making new synapses or altering the strength of existing synapses. See **long-term potentiation** and **synaptic plasticity**.

neuronal polarity Distribution of specific functions to discrete cellular domains, e.g. axons and dendrites that have different molecular composition, morphology and ultrastructure and perform different functions.

neuropeptide AF *NPAF* A mammalian pain-modulating and anti-opiate neuropeptide (analgesic *in vivo*) of the RF-amide class (AGEGLNSQFWSLAAPQRF-NH2); receptor is an orphan G-protein-coupled receptor.

neuropeptide FF *NPFF* Neuropeptide (FLFQPQRF-amide) that, like neuropeptides AF and SF (NPAF, NPSF), is involved in pain modulation and opioid tolerance. In several mammalian species NPFF and NPAF are derived from the same gene that is expressed mainly in the CNS. Binds to a G-protein-coupled receptor.

neuropeptide SF *NPSF* Similar to **neuropeptide FF** and **neuropeptide AF**; Sequence is SLAAPQRF-amide.

neuropeptide Y *NPY; melanostatin* Peptide neurotransmitter (36 residues) found in adrenals, heart and brain. Potent stimulator of feeding and regulates secretion of gonadotrophin-releasing hormone. **Leptin** inhibits NPY

gene expression and release. Receptors are G-protein coupled.

neuropeptides Peptides with direct synaptic effects (peptide neurotransmitters) or indirect modulatory effects on the nervous system (peptide neuromodulators). See Table N1.

neurophysin Carrier protein (10 kDa, 90–97 amino acids) that transports neurohypophysial hormones along axons, from the hypothalamus to the posterior lobe of the pituitary. See also **brain**.

neuropil *Neuropile* A network of axons, dendrites and synapses within the central nervous system of vertebrates or within the brain and the central portion of segmental ganglia of arthropods.

neuropilin Neuropilin-1 (NRP1) is a receptor for two unrelated ligands with disparate activities, vascular endothelial growth factor-165 (**VEGF**165), an angiogenesis factor, and **semaphorins/collapsins**, mediators of neuronal guidance.

neuropore The anterior and posterior openings of the neural tube of the early embryo; usually close at very specific times in development.

neuropsin *Opsin 5; KLK8* Neuropsin, a secreted serine peptidase of the **chymotrypsin**-family, has a role in neuronal plasticity, and its expression has been shown to be upregulated in response to injury to the CNS. Has also been implicated in learning and memory and the type II splice form of neuropsin is only found in hominoid species (humans and apes). Neuropsin shares 25–30% amino acid identity with all known opsins. It is expressed in the eye, brain, testis and spinal cord. See **opsin subfamilies**, **neurosin**.

neurosecretory cells Cells that have properties of electrical activity, carrying impulses, and also a secretory function, releasing hormones into the bloodstream. In a sense they are behaving in the same way as any chemically signalling neuron, except that the target is the blood (and remote tissues), not another nerve or postsynaptic region.

neurosin *Kallikrein 6 (KLK6); protease M* An arginine-specific serine endopeptidase of the kallikrein family, expressed by neurons and glial cells. Expression is similar, but not identical, to that of **neuropsin**, especially following injury to the CNS.

Neurospora An Ascomycete fungus, haploid and grows as a **mycelium**. There are two mating types, and fusion of nuclei of two opposite types leads to meiosis followed by mitosis. The resulting eight nuclei generate eight ascospores, arranged linearly in an ordered fashion in a pod-like **ascus** so that the various products of meiotic division can be identified and isolated. Because of this, *Neurospora crassa* is one of the classic organisms for genetic research; studies on biochemical mutants led Beadle and Tatum to propose the seminal 'one gene-one enzyme' hypothesis.

neurosteroids Steroids synthesized in the brain that have effects on neuronal excitability. The **epalons** may regulate type A **GABA receptors** by allosteric potentiation.

neurotactin *CX3CL2* **1.** Membrane-anchored chemokine (395 amino acid residues in mouse, 397 in man) with unique CXXXC pattern (unlike α-chemokines, CXC; β-chemokines, CC; and γ-chemokines, C). Neurotactin is predominantly expressed in brain and is upregulated on capillary vessels and microglia in LPS-induced inflammation

TABLE N1. Neurotransmitters

Transmitter	Peripheral nervous system	Central nervous system
Noradrenaline	Some postganglionic sympathetic neurons	Diverse pathways especially in arousal and blood pressure control
Dopamine	Sympathetic ganglia	Diverse; perturbed in Parkinsonism and schizophrenia
Serotonin	Neurons in myenteric plexus	Distribution very similar to that of noradrenergic neurons; lysergic acid (LSD) may antagonize
Acetylcholine	Neuromuscular junctions (nmj); all postganglionic parasympathetic and most postganglionic sympathetic neurons	Widely distributed, usually excitatory; possibly antagonizes dopaminergic neurons
GABA	Inhibitory at nmj of arthropods	Inhibitory in many pathways
Glutamate	Excitatory at nmj of arthropods	Widely distributed; excitatory
Glycine	–	Diverse; particularly in grey matter of spinal cord
Aspartate	Locust nmj	–
Neuropeptides	Diverse actions in both peripheral and central nervous systems; see Table H2	
Histamine	–	Minor role
Purines	Particularly neurons controlling blood vessels	Mostly inhibitory
Octopamine	Invertebrate nmj	–
Substance P	Sensory neurons of vertebrates	Sensory neurons
RF-amides	Invertebrates	

Other substances known, or proposed to have neurotransmitter function are: adrenaline, agmatine, β-alanine, cholecystokinin, taurine, proctolin and cysteine.

and **EAE**. Unlike other chemokines, gene is on chromosome 8 (mouse) and 16q (man). Proteolytically released soluble active fragments may be generated. Chemotactic for neutrophils. 2. *Drosophila* neurotactin is a transmembrane receptor and a member of the cholinesterase-homologous protein family; is required for neurotactin-mediated cell adhesion and axon fasciculation.

neurotensin Tridecapeptide hormone (sequence: ELYEN-KPRRPYIL) of gastrointestinal tract: has general vascular and neuroendocrine actions.

neurotoxin A substance, often exquisitely toxic, that inhibits neuronal function. Neurotoxins act typically against the **sodium channel** (e.g. **TTX**) or block or enhance **synaptic transmission** (**curare**, **bungarotoxin**).

neurotransmitter A substance found in **chemical synapses** that is released from the presynaptic terminal in response to depolarization by an action potential, diffuses across the synaptic cleft, and binds a ligand-gated ion channel on the postsynaptic cell. This alters the resting potential of the postsynaptic cell, and thus its excitability. Examples: **acetylcholine**, **GABA**, **noradrenaline**, **serotonin**, **dopamine**. See Table N1.

neurotrophic Involved in the nutrition (or maintenance) of neural tissue. Classic example is **nerve growth factor**, see **neurotrophins**.

neurotrophin-3 *NT-3; hippocampal-derived neurotrophic factor or NGF-2* Member of the family of neurotrophic factors or **neurotrophins** that also includes **nerve growth factor** and **brain-derived neurotrophic factor** (BDNF)

that have about 50% amino acid sequence identity. NT-3 shows strong similarities to NGF and BDNF (including strictly conserved domains that contain six cysteine residues) but has a different pattern of neuronal specificity and regional expression.

neurotrophins Molecules with closely related structures that are known to support the survival of different classes of embryonic neurons. See **nerve growth factor** (NGF), **brain-derived neurotrophic factor** (BDNF), **neurotrophin-3** (NT-3), **GDGF** and **ciliary neurotrophic factor**.

neurotropic Having an affinity for, or growing towards, neural tissue. Rabies virus, which localizes in neurons, is referred to as neurotropic; the term can also be used to refer to chemicals. Not to be confused with **neurotrophic**.

neurotubules A term for **microtubules** in a neuron.

neurula The stage in vertebrate embryogenesis during which the neural plate closes to form the central nervous system.

neurulation The embryonic formation of the **neural tube** by closure of the **neural plate**, directed by the underlying notochord.

neutral mutation A mutation that has no selective advantage or disadvantage. Considerable controversy surrounds the question of whether such mutations can exist.

neutral protease Peptidase that is optimally active at neutral pH: may be from any of several classes of peptidases. Eukaryotic example is **calpain**; a range of bacterial enzymes were originally designated neutral proteases, e.g. neutral protease from *Staphylococcus hyicus*, hyicolysin, a metallopeptidase.

neutropenia Condition in which the number of **neutrophils** circulating in the blood is below normal. (Normal levels are $4.3–10.8 \times 10^9$ cells per litre.)

neutrophil *Neutrophil granulocyte; polymorphonuclear leucocyte; PMN or PMNL* Commonest ($2500–7500$ mm^{-3}) blood leucocyte; a short-lived phagocytic cell of the **myeloid** series, which is responsible for the primary cellular response to an acute inflammatory episode and for general tissue homeostasis by removal of damaged material. Adheres to endothelium (**margination**) and then migrates into tissue, possibly responding to chemotactic signals. Contain **specific** and **azurophil granules**.

neutrophil-activating protein See **interleukin-8**.

neutrophil-activating protein 3 *NAP3* See **melanoma growth-stimulatory activity**.

neutrophilin Neutrophil-derived platelet activator, probably a serine endopeptidase (cathepsin-G) that acts via receptor proteolysis (a tethered ligand mechanism) similar to the mechanism of thrombin activation, but distinct.

nevus See **naevus**.

Newcastle disease virus *Avian paramyxovirus-1* A paramyxovirus that causes the disease, fowl-pest, in poultry; the lethality of different strains varies and highly virulent strains can cause major mortality in all species of birds.

nexilin An actin filament-binding protein localized at cell-matrix adherens junction. Two splice variants have been isolated, b-nexilin (78 kDa) from brain and s-nexilin (72 kDa) from fibroblasts. s-nexilin co-localizes with **vinculin**, **talin**, and **paxillin** at focal contacts but unlike b-nexilin, which has two actin-binding domains, does not seem to bundle F-actin.

nexin 1. Protein (165 kDa) that links the adjacent microtubule doublets of the **axoneme**. There is a repeat at 96-nm intervals. 2. See **protease nexin-1**. 3. See **sorting nexins**.

nexus A connection or link.

Nezelof syndrome Congenital T-cell deficiency associated with thymic hypoplasia and distinct from Bruton-type agammaglobulinaemia in which the tonsillar system is absent, and from severe combined immunodeficiency (**SCID**).

NF-1 1. Nuclear factor-1; CCAAT-binding transcription factor. Family of dimeric transcription factors (74 kDa); there are four *NF1* genes (*NF1-A, -B, -C* (**CTF**/NF1) and *-X*) that give rise to multiple isoforms by alternative splicing in many tissues. Essential for adenovirus DNA replication and the transcription of many cellular genes. 2. Do not confuse with **neurofibromatosis** Type 1 (NF1).

NF-B Nuclear factor-B, more commonly **NFκ-B**.

NF-E1 See **erythroid transcription factor**.

NFAR *Nuclear factors associated with dsRNA* Evolutionarily conserved proteins; two alternatively spliced variants, NFAR-1 and -2, are generated from the single gene on chromosome 19. Associate with the **spliceosome** and may play a role in regulating gene expression in response to dsRNA-regulated signaling events in the cell.

NFAT *Nuclear factor of activated T-cells* Transcription factor involved in regulation of IL-2 and IL-4 gene transcription (in concert with other transcription factors). NFAT is cytoplasmic until dephosphorylated by **calcineurin**, a step that is inhibited by **ciclosporin** and **FK506**, then translocates to the nucleus.

NFκB *NF-kappaB* A **transcription factor** (originally found to switch on transcription of genes for the kappa class of immunoglobulins in B-lymphocytes). It is involved in activating the transcription of more than 20 genes in a variety of cells and tissues. NFκB is found in the cytoplasm in an inactive form, bound to the protein IκB. A variety of stimuli, such as tumour-necrosis factor, phorbol esters and bacterial lipopolysaccharide activate it, by releasing it from IκB, allowing it to enter the nucleus and bind to DNA. It has two subunits p50 and p65 that bind DNA as a heterodimer. The dimerization and DNA-binding activity are located in N-terminal regions of 300 amino acids that are similar to regions in the Rel and dorsal transcription factors.

NgCAM Neural–glial cell adhesion molecule (**CAM**). See **NCAM**.

NGF Abbrev. for **nerve growth factor**.

***N*-glycanase** *EC 3.5.1.52* Enzyme that cleaves asparagine-linked oligosaccharides from glycoproteins.

***N*-glycans** Oligosaccharides, based on the common core pentasaccharide Man$_3$GlcNAc$_2$, that are linked to a protein backbone via an amide bond to asparagine residues in an Asn-X-Ser/Thr motif, where X can be any amino acid, except Pro. Glycosylation with *N*-glycans occurs co-translationally; the completed oligosaccharide is transferred from the dolichol precursor to the Asn of the target glycoprotein by oligosaccharyltransferase (OST) in the cisternal space of the ER where trimming and further modification may also take place. In the Golgi, high mannose *N*-glycans can be converted to a variety of complex and hybrid forms

N-glycosylation site Exposed extracellular asparagine residues are often glycosylated (see **glycosylation**). The consensus site is Asn-X-Ser/Thr-X, where X can be any amino acid except Pro.

NHE-RF *Na$^+$-H$^+$ exchanger regulatory factor* Cytoplasmic phosphoprotein involved in **protein kinase A** (PKA) mediated regulation of ion transport. Contains two **PDZ domains** and will bind to C-terminus of **CFTR**.

NHEJ Abbreviation for non-homologous end joining, the primary mammalian DNA repair mechanism that occurs through recognition of double-strand breaks by a variety of proteins that process and rejoin DNA termini by direct ligation. Proteins known to play a role in NHEJ include the DNA-dependent protein kinase catalytic subunit (DNA-PKcs), the **Ku** heterodimer, XRCC4, and DNA ligase IV. Originally thought to be restricted to eukaryotes but now known to occur in prokaryotes (see **DNA ligase D**).

niacin *Nicotinic acid* One of the B vitamins. See Table V1.

nibrin *p95 protein of the MRE11/RAD50 complex* Protein thought to be defective in **Nijmegen breakage syndrome**. Has two modules found in cell cycle checkpoint proteins and a **forkhead**-associated domain. Identical to the p95 protein component of the double-strand break (DSB) repair complex.

nicardipine Calcium channel blocker used to treat hypertension. Inhibits the transmembrane influx of calcium ions into cardiac muscle and smooth muscle without changing serum calcium concentrations.

nicastrin Transmembrane glycoprotein (709 amino acids) that interacts with **β-amyloid precursor protein** (βAPP) and with **presenilins**. Probably involved, together with presenilins, in processing of βAPP to amyloid-β-peptide and in processing of **notch**.

nick A point in a double-stranded DNA molecule where there is no **phosphodiester bond** between adjacent nucleotides of one strand typically through damage or enzyme action.

nick translation A technique used to radioactively label DNA. *E. coli* DNA polymerase I will add a nucleotide, copying the complementary strand, to the free 3′-OH group at a **nick**, at the same time its exonuclease activity removes the 5′-terminus. The enzyme then adds a nucleotide at the new 3′-OH and removes the new 5′-terminus. In this way, one strand of the DNA is replaced starting at a nick, which effectively moves along the strand. Nick translation refers to this translation or movement and not to protein synthesis. In practice, DNA is mixed with trace amounts of **DNAase** I to generate nicks, **DNA polymerase** I and labelled nucleotides. Because the nicks are generated randomly the DNA preparation can be uniformly labelled and to a high degree of specific activity.

nicotinamide adenine dinucleotide See **NAD**.

nicotinamide adenine dinucleotide phosphate See **NADP**.

nicotine A plant alkaloid from tobacco; blocks transmission at nicotinic synapses. See **nicotinic acetylcholine receptor**.

nicotine replacement therapy Treatment to help smoking cessation in which small quantities of nicotine are administered by patches, chewing gum, etc., until the craving is lost.

nicotinic acetylcholine receptor *nAChR* Integral membrane protein of the postsynaptic membrane to which **acetylcholine** binds. The receptor contains an integral **ion channel**; as a result of binding of acetylcholine, ion channels in the subsynaptic membrane are opened. At the **neuromuscular junction**, the nicotinic acetylcholine receptor initiates muscle contraction. Currently the best-characterized ion channel protein: made of a heteropentamer of related subunits, although a homo-pentamer is functional in insects. Structural studies show that the acetylcholine-binding site and the ionic channel are part of the same macromolecular unit. The nAChR mediates rapid transduction events (1 ms), whereas receptors activating **G-protein**-coupled receptors operate on slower time scales (millisecond to second range).

nicotinic acid *Pyridine 3-carboxylic acid* A precursor of **NAD**, that is a product of the oxidation of nicotine.

NIDDM *Non-insulin dependent diabetes mellitus* Type II, maturity (adult) onset diabetes. Can usually be treated by regulating sugar intake.

nidogen *Entactin* A dumbbell-shaped 150-kDa sulphated **glycoprotein**, found in all basement membranes, consisting of three globular regions, G1–G3. G1 and G2 are connected by a thread-like structure, whereas that between G2 and G3 is rod-like. The nidogen G2 region binds to collagen IV and **perlecan**. The connecting rod has five to six **EGF-like domains** of cysteine-rich repeats, one of which has an **RGD** sequence for cellular interaction. Also interacts with **laminin** and ablation of the high-affinity nidogen-binding site of the laminin gamma1 chain is lethal. Oddly, the nidogen G2 beta-barrel domain has structural similarity to green fluorescent protein. There are two isoforms, nidogen-1 and nidogen-2, with broadly similar properties.

Niemann-Pick disease Severe lysosomal storage disease. Types A & B are caused by deficiency in **sphingomyelinase**; excess sphingomyelin is stored in 'foam' cells (macrophages) in spleen, bone marrow and lymphoid tissue. More common in Ashkenazi Jews than other groups. Niemann-Pick diseases Type C1 and D are due to a mutation in the NPC1 gene that encodes a 1278-amino acid protein with sequence similarity to the morphogen receptor 'patched', the putative sterol-sensing regions of **SREBP** and **HMG-CoA reductase** and that has a critical role in regulating of intracellular cholesterol trafficking.

***nif* genes** The complex of genes in nitrogen-fixing bacteria that code for the proteins required for **nitrogen fixation**, particularly the **nitrogenase**. Present as an operon in *Klebsiella* and carried on plasmid in *Rhizobium*.

nifedipine *BAYa1041; Nifedin; Procardia* A calcium channel blocker (346 Da) used experimentally and as a coronary vasodilator.

niflumic acid *(2,(3-Trifluoromethyl)-anilino) nicotinic acid* A rather non-specific inhibitor of chloride channels.

nigericin An ionophore capable of acting as a carrier for K^+ or Rb^+ or as an exchange carrier for H^+ with K^+. Originally used as an antibiotic. Has been used in investigating chemiosmosis and other transport systems.

NIH 3T3 cells Very widely used mouse fibroblast cell line; 3T3 cells have been derived from different mouse strains and it is therefore important to define the particular cell line. NIH strain were from the National Institute of Health in the USA.

Nijmegen breakage syndrome *NBS* Autosomal recessive chromosomal instability syndrome characterized by microencephaly, growth retardation, immunodeficiency and predisposition to tumours. Cells from patients are hypersensitive to ionizing radiation in the same way as cells from **ataxia telangiectasia**. A novel protein, **nibrin**, has been implicated in the syndrome.

NIK *NF-kappaB-inducing kinase* NIK is involved in CD3/CD28 activation of IL-2 transcription. Splenic T-cells from *aly/aly* mice (defective in NIK) have a severe impairment in IL-2 and GM-CSF but not TNF secretion in response to CD3/CD28. Apparently activates the CD28 responsive element (CD28RE) of the IL-2 promoter and strongly synergizes with c-Rel in this activity.

ninhydrin *Triketohydrindene hydrate* Pale yellow substance used to detect amino acids and proteins (compounds containing free amino or imino groups), with which it forms a deeply coloured purple-blue compound.

niosomes Multilamellate **liposomes** made from non-ionic lipids and used for drug delivery.

nisin A post-translationally modified antimicrobial peptide (34 residues), a (**lantibiotic**), that is widely used as a food preservative. Gallidermin and epidermin possess the same putative lipid II-binding motif as nisin; however, both peptides are considerably shorter (22 amino acids).

Nissl granules Discrete clumps of material seen by phase contrast microscopy in the perikaryon of some neurons, particularly motor neurons. They are basophilic and contain much RNA, and are regions very rich in rough endoplasmic reticulum. Their reaction following damage to neurons is characteristic; they disperse through the cytoplasm giving a general basophilia to the whole cell body.

Nitella Characean alga that has giant, multinucleate internodal cells. These show **cytoplasmic streaming** at rates of up to 100 μm/s and have been used as models for motile phenomena in cells, and in studies on ionic movement.

nitrates Vasodilatory drugs used in treatment of angina and congestive heart failure. Cause dilatation of large veins and thus reduce workload on heart. Example is nitroglycerin.

nitrazepam *TN Mogadon* Widely used non-barbiturate hypnotic used as a sleep inducer, although can be addicitve.

nitric oxide *Endothelium-derived relaxation factor; NO* Gas produced from L-arginine by the enzyme **nitric oxide synthase**. Acts as an intracellular and intercellular messenger in a wide range of processes, in the vascular and nervous systems. The intracellular 'receptor' is a soluble (cytoplasmic) **guanylate cyclase**. In the immune system, large amounts can be generated as a cytotoxic attack mechanism. NO signalling is phylogenetically widespread, suggesting it is an ancient mechanism.

nitric oxide synthase *NO synthase; EC 1.14.13.39* Enzyme that produces the vasorelaxant **nitric oxide** (endothelium-derived relaxation factor) from L-arginine. There are two isoforms, one constitutive and calmodulin-dependent, the other inducible (iNOS) and calcium-independent.

nitroblue tetrazolium reduction Nitroblue tetrazolium, a yellow dye, is taken up by phagocytosing neutrophils and reduced to insoluble formazan, which is deep-blue, if the **metabolic burst** is normal. Reduction does not take place in **chronic granulomatous disease**.

nitrocellulose paper Paper with a high non-specific absorbing power for biological macromolecules. Very important as a receptor in blot-transfer methods. Bands are transferred from a chromatogram or electropherogram either by blotting on nitrocellulose sheets or by electrophoretic transfer. The replica can then be used for sensitive analytical detection methods.

nitrogen fixation The incorporation of atmospheric nitrogen into ammonia by various bacteria, catalysed by **nitrogenase**. This is an essential stage in the nitrogen cycle, and is the ultimate source of all nitrogen in living organisms. In the sea, the main nitrogen fixers are **Cyanobacteria**. There are several free-living bacteria in soil that fix nitrogen including species of *Azotobacter*, *Clostridium* and *Klebsiella*. *Rhizobium* only fixes nitrogen when in symbiotic association, in root nodules, with leguminous plants. The oxygen-sensitive nitrogenase is protected by plant-produced leghaemoglobin, and the plant obtains fixed nitrogen from the bacteria. See *Frankia*.

nitrogen mustards A series of tertiary amine compounds having vesicant (blistering) properties similar to those of mustard gas. They have the general formula $RN(CH_2CH_2Cl)_2$. They can alkylate compounds such as DNA and derivatives have been used for cancer chemotherapy.

nitrogenase *EC 1.18.6.1* Enzymes found in nitrogen-fixing bacteria that reduce nitrogen to ammonia (also ethylene to acetylene). See **nitrogen fixation**.

nitrosamines These molecules contain the N-N=O group (*N*-nitrosamines): many are carcinogens or suspected carcinogens.

nitzin Protein with 42% homology to the **Band 4.1** protein superfamily, capable of linking integral membrane proteins to the cytoskeleton. Nitzin mRNA is high throughout the developing nervous system. Has not reappeared in the literature since being described in 1998.

NK-cells *Natural killer cell* See **killer cells**.

NKCC1 $Na^+-K^+-2Cl^-co$-transporter-1 Electroneutral cation-coupled chloride co-transporter found in kidney and that is regulated by **SPAK** and **OSR1** and inhibited by **thiazide diuretics**.

N-lines Regions in the sarcomere of striated muscle. The N1 line is in the I-band near the Z-disc, the N2 line is at the end of the A-band. The N-lines may represent the location of proteins such as **nebulin** that contribute to the stability of the sarcomere.

NMDA *N-methyl-D-aspartic acid* A powerful agonist for a subclass of glutamate receptors (**NMDA receptor**).

NMDA receptor Glutamate receptor subtype (see **excitatory amino acids**. NMDA channels seem to be potentiated by intracellular **arachidonic acid**.

N-methyl-D-aspartate See **NMDA**.

NMRI Abbreviation for nuclear **magnetic resonance imaging**.

N-myc Oncogene, related to *myc*, found in neuroblastomas.

NO See **nitric oxide**.

NO synthase See **nitric oxide synthase**.

no-effect level *NEL* Maximum dose (of a substance) that produces no detectable changes under defined conditions of exposure. Obviously depends on the sensitivity of the detection system.

no-observed-adverse-effect level *NOAEL* Greatest concentration or dose of a substance that causes no detectable adverse alteration of morphology, functional capacity, growth, development, or life span of the target organism under defined conditions of exposure.

no-observed-effect level *NOEL* Greatest concentration or dose of a substance that causes no detectable alteration of morphology, functional capacity, growth, development, or life span of the target organism under defined conditions of exposure and by comparison with normal untreated organisms.

Nocardia Genus of Gram-positive bacteria that form a **mycelium** that may fragment into rod- or coccoid-shaped cells. They are very common **saprophytes** in soil but some are opportunistic pathogens of humans, causing nocardiosis. This is characterized by abscesses, particularly of the jaw, which if untreated may invade the surrounding bone.

nociceptin See **orphanin FQ**.

nociception Detection of pain. See **capsaicin**.

nociceptive Class of sensory nerve ending that sends signals that cause pain in response to certain stimuli.

nociceptor Pain receptor. Many nociceptors respond to **capsaicin**.

nocistatin *NST* Neuropeptide (EQKQLQ) derived from the same precursor molecule, prepronociceptin (ppNCP), as nociceptin (NCP)/orphanin FQ, and that is involved in pain transmission. Nocistatin has been shown to antagonize several effects of nociceptin by acting on a different receptor.

nocodazole Microtubule disrupting compound that binds to the tubulin heterodimer rendering it assembly-incompetent.

Noctiluca A bioluminescent dinoflagellate. Responsible for many instances of marine phosphorescence.

NOD mice *Non-obese diabetic mice.* Have unique histocompatibility antigens; pancreatic B cells are destroyed by an autoimmune response.

nodal A protein related to transforming growth factor-beta (TFG β), which is expressed in the epiblast and visceral endoderm of the mammalian embryo. Nodal signals induce mesoderm and endoderm. See **cerberus**.

node A point in a plant stem at which one or more leaves are attached.

node of Ranvier A region of exposed neuronal plasma membrane in a myelinated axon. Nodes contain very high concentrations of **voltage-gated ion channels**, and are the site of propagation of action potentials by saltatory conduction.

nodularins Hepatotoxic cyclic peptides similar to microcystins but containing only five amino acids. The amino acid composition may vary but the hydrophobic amino acid, ADDA, is essential for activity. Like **microcystin** binds to the same site on serine/threonine phosphatases as **okadaic acid**.

nodulin Plant protein. Soybean nodulin-24 is closely related to **major intrinsic protein**.

noelins A family of extracellular proteins with proposed roles in neural and neural crest development. Are members of the **olfactomedin** family of proteins.

noggin Dorsalizing factor (26 kDa) from **Spemann's organizer** region of the amphibian embryo.

nogo *Nogo-A* A neuronal protein, expressed by oligodendrocytes, involved in diverse processes that include axonal fasciculation and apoptosis. Like **MAG** and **OMgp**, nogo can cause growth cone collapse and inhibit neurite outgrowth *in vitro*. Although Nogo, MAG and OMgp lack sequence homologies, they all bind to the Nogo receptor (NgR), a GPI-linked cell surface molecule which, in turn, binds p75 to activate RhoA.

nojirimycin Antibiotic produced by *Streptomyces* strains; inhibits α-glucosidases and prevents normal glycosylation of proteins by interfering with the early pruning down to the core carbohydrate, a step that is normally followed by addition of specific sugar residues.

Nomarski differential interference contrast See **differential interference contrast**.

non-coding DNA DNA that does not code for part of a polypeptide chain or RNA. This includes **introns** and **pseudogenes**. In eukaryotes the majority of the DNA is non-coding. Non-coding strand refers to the so-called non-sense strand, as opposed to the sense strand that is actually translated into mRNA.

non-competitive inhibitor Reversible inhibition of an enzyme by a compound that binds at a site other than the substrate-binding site.

non-cyclic photophosphorylation Process by which light energy absorbed by **photosystems I and II** in chloroplasts is used to generate ATP (and also NADPH). Involves photolysis of water by photosystem II, passage of electrons along the photosynthetic electron transport chain with concomitant phosphorylation of ADP, and reduction of $NADP^+$ using energy derived from photosystem I.

non-depolarizing muscle relaxants Group of drugs which block neuromuscular transmission by competing with acetylcholine at the receptor site. Used to produce paralysis and muscle relaxation during anaesthesia. Common example is tubocurarine.

non-disjunction Failure of homologous chromosomes or sister **chromatids** to separate at meiosis or mitosis, respectively. It results in aneuploid cells. Non-disjunction of the X chromosome in *Drosophila* allowed Bridges to confirm the theory of chromosomal inheritance.

non-equivalence Term used in cell determination for cells that will give rise to the same sorts of differentiated tissues but that have different positional values (e.g. cells of fore-limb and hind-limb buds).

non-histone chromosomal proteins Chromatin consists of DNA, **histones** and a very heterogeneous group of other proteins that include DNA polymerases, regulator proteins, etc. They are often generically referred to as non-histone proteins, or acidic proteins, to distinguish them from the basic histones.

non-ionic detergent Detergent in which the hydrophilic head group is uncharged. In practice, hydrophilicity is usually conferred by -OH groups. Examples are the polyoxyethylene *p-t*-octyl phenols known as Tritons, and octyl glucoside. Non-ionic detergents can be used to solubilize intrinsic membrane proteins with less tendency to denature them than charged detergents. They do not usually cause disassembly of structures, such as microfilaments and microtubules, that depend on protein–protein interactions.

non-Mendelian inheritance In eukaryotes, patterns of gene transmission not explicable in terms of segregation, independent assortment and linkage. May be due to **cytoplasmic inheritance**, **gene conversion**, meiotic drive, etc.

non-Newtonian fluid A fluid in which the viscosity varies depending upon the shear stress. The effect can arise because of alignment of non-spherical molecules as flow is established or because of suspended deformable particles as in blood.

non-reciprocal contact inhibition Collision behaviour between different cell types in which one cell shows contact inhibition of locomotion, and the other does not. An example is the interaction between sarcoma cells and fibroblasts (the former not being inhibited).

non-spiking neuron A neuron that can convey information without generating action potentials. As passive electrical potentials are attenuated over distances greater than the space constant for a neuron (typically 1 mm), this implies that most non spiking neurons are involved in signalling over relatively short distances. Typical examples are invertebrate stretch receptors and **interneurons** in the central nervous system.

non-steroidal anti-inflammatory drugs *NSAIDs* See **NSAIDs**.

non-transcribed spacer See **internal transcribed spacers**.

Nonidet Trade name for non-ionic detergents, usually octyl- or nonyl- phenoxy-polyethoxy-ethanols.

non-permissive cell Originally a cell of a tissue type or species that does not permit replication of a particular virus. Early stages of the virus cycle may be possible in such a cell, that in the case of tumour viruses the cell may become **transformed**. Now used in a more general sense, of agents and treatments other than viruses.

non-polar group *Hydrophobic group* Group in which the electronic charge density is essentially uniform, and that cannot therefore interact with other groups by forming hydrogen bonds, or by strong dipole–dipole interactions. In an aqueous environment, non-polar groups tend to cluster together, providing a major force for the folding of macromolecules and formation of membranes. Clusters are formed chiefly because they cause a smaller increase in water structure (decrease in entropy) than dispersed groups. (Non-polar groups interact with each other only by the relatively weak London–van der Waals forces).

non-receptor protein tyrosine kinase See **tyrosine kinase**.

nonsense codon *Nonsense triplet* The three **codons**, UAA (known as ochre), UAG (amber) and UGA (opal), that do not code for an amino acid but act as signals for the termination of protein synthesis. Any mutation that causes a base change which produces a nonsense codon results in premature termination of protein synthesis and probably a non-functional or nonsense protein. See also Table C5.

nonsense mutation Mutation in coding DNA producing a **nonsense codon** that prevents the protein from being synthesized.

nonsense strand See **non-coding DNA**.

nootropic drugs A class of drugs that act as cognitive enhancers. Have been referred to as 'smart drugs'.

nopaline *N-alpha-(1,3-dicarboxypropyl)-L-arginine* An **opine**. The gene for nopaline synthase is carried on the T-DNA of the **Ti-plasmid**.

noradrenaline *Norepinephrine; arterenol* Catecholamine neurohormone, the neurotransmitter of most of the sympathetic nervous system (of so-called adrenergic neurons): binds more strongly to α-adrenergic receptor than to β-adrenergic receptor. Stored and released from **chromaffin cells** of the adrenal medulla.

norepinephrine See **noradrenaline**.

norleucine *Ahx; Nle; 2-aminohexanoic acid* Non-protein amino acid. Formyl-norleucyl-leucyl-phenylalanine

has been used as a substitute for fMLP in studies on neutrophil chemotaxis since it is not so susceptible to oxidation.

normoblast Nucleated cell of the **myeloid cell** series found in bone marrow, that gives rise to red blood cells. See **erythroblast**.

normocyte Erythrocyte of normal size and shape.

norovirus Genus of viruses of the family Caliciviridae. Noroviruses contain a positive-strand RNA genome of ~7.5 kbp. Causes intestinal illness. **Norwalk virus** is in this genus.

Northern blot An electroblotting method in which RNA is transferred to a filter and detected by hybridization to ^{32}P-labelled RNA or DNA. See **blots**.

Northwestern blot Technique for identifying protein/RNA interactions in which protein is run on a gel, blotted and probed with a labelled RNA of interest. Interactions are detected as hot-spots on the filter. So-called because it involves both RNA (**Northern blot**) and protein (**Western blot**).

Norwalk virus Unclassified single-stranded RNA virus (**norovirus**) causing common acute infectious gastroenteritis.

nosocomial infections Hospital-acquired infections: commonest are due to *Staphylococcus aureus*, *Pseudomonas aeruginosa*, *E. coli*, *Klebsiella pneumoniae*, *Serratia marcescens* and *Proteus mirabilis*.

nosology The branch of medical science which deals with the systematic classification of diseases.

Nostoc Common genus of freshwater nitrogen-fixing **cyanobacteria** that form colonies of intertwined filaments in a gelatinous sheath.

Notch Family of large transmembrane receptor proteins (350 kDa) that mediate developmental cell-fate decisions; Notch contains 36 repeats of the **EGF-like domain**. Mammalian Notch gene mutations have been associated with leukaemia, breast cancer, stroke and dementia (see **CADASIL**). In *Drosophila* wing development, Notch receptor is activated at the dorsal/ventral boundary and is important in growth and patterning. Notch binds transmembrane ligands encoded by Ser (**serrate** protein) and Dl (**Delta** protein). **Fringe** (*fng*) is also involved in Notch signalling, encoding a pioneer protein.

notch signaling See **Notch**.

notexins Notexins Np and Ns are phospholipase A2 isoforms found in the venom of *Notexis scutatus scutatus*. Block acetylcholine release at the neuromuscular junction.

Notexis scutatus scutatus Tiger snake. Venom contains a range of toxins including **notexins**.

notochord An axial mesodermal tissue found in embryonic stages of all chordates and protochordates, often regressing as maturity is approached. Typically a rod-shaped mass of vacuolated cells immediately below the nerve cord, and may provide mechanical strength to the embryo.

notoplate Region of the **neural plate** overlying the **notochord**.

novobiocin An antibiotic obtained from *Streptomyces niveus* and other *Streptomyces* species, used clinically chiefly against staphylococci and other Gram-positive organisms. Acts as an inhibitor of prokaryotic DNA **gyrase** and eukaryotic type II **topoisomerase** enzymes, interferes with *in vitro* chromatin assembly using purified histones, DNA and **nucleoplasmin**.

Noxa See **Bcl-2 homology domain 3**.

noxiustoxin Charybdotoxin-related peptide toxin (39 residues) from scorpion *Centruroides noxius*. Blocks mammalian voltage-gated potassium channels and high-conductance calcium-activated potassium channels.

NP-40 *Nonidet P40* Non-ionic detergent useful for the isolation and purification of functional membrane proteins. Apparently no longer available commercially but an equivalent, IGEPAL CA-630 (octylphenoxy) polyethoxyethanol, is said to be chemically indistinguishable.

NPAF See **neuropeptide AF**.

NPC Abbrev. for (1) **Niemann-Pick disease** type C (NPC), a storage disease, (2) **Nuclear pore complex,** (3) Nasopharyngeal carcinoma, (4) Neural precursor cells.

NPFF See **neuropeptide FF**.

NPRAP *Neural plakophilin-related armadillo repeat protein; delta-catenin; neurojungin* Protein found in specialized adhesion plaques of the outer limiting zone of the retina. Interacts with and activates **sphingosine kinase** 1.

N-protein 1. Anti-terminator protein of the **lambda bacteriophage** and other phages that plays a key role in the early stages of infection. During the early phase only two genes, *N* and *cro* , are transcribed, by transcription of the DNA in opposite directions. N-protein binds to sites on the DNA (nut sites for N-utilization), prevents rho-dependent termination and allows transcription of the genes. 2. Name was once used for **GTP-binding proteins** (G-proteins); now obsolete, and should be avoided because of confusion with N-protein of bacteriophages.

NPSF See **neuropeptide SF**.

Nramp *Natural resistance associated macrophage protein; Nramp1, Nramp2 Nramp1* codes for an integral membrane protein (65 kDa) expressed only in macrophages/monocytes and PMNs. Localized to endosomal/lysosomal compartment and rapidly recruited to the phagosome membrane following phagocytosis. Mutations in *Nramp1* seem to abrogate the ability of macrophages to kill intracellular parasites such as *Mycobacterium tuberculosis* and to be associated with onset of rheumatoid arthritis. Nramp2 is very similar to Nramp1 but is expressed in more tissues and is known to be an iron transporter. Yeast homologues Smf1 and Smf2 transport divalent cations.

NRG 1. See **neuregulin**. 2. *Drosophila* **neuroglian** (Nrg).

NRG1 *HRG-alpha* **Neuregulin**-1alpha.

NRK cells Normal rat (*Rattus norvegicus*) kidney cell line, grow adherently and exhibit epithelial morphology.

NS1 Viral non-structural protein, NS1, a nuclear, dimeric protein that is highly expressed in infected cells and has dsRNA-binding activity. Various biochemical functions, such as ATP binding, ATPase, site-specific DNA-binding and nicking, and helicase activities, have been assigned to the protein NS1.

NSAID *Non-steroidal anti-inflammatory drug* Range of anti-inflammatory drugs that include aspirin, ibuprofen and a wide range of derivatives. Mostly act on the production of early low molecular-weight mediators of the acute inflammatory response (particularly **COX-1** etc.). Are particularly good at inhibiting swelling, but often have undesirable side effects, including gastric irritation.

NSF *N-ethyl-maleimide-sensitive factor* Homotetrameric protein (76 kDa) involved, together with three **SNAPs**, in mediating vesicle traffic between medial and trans-Golgi compartments.

NSO cells Murine **myeloma** cell line (plasmacytoma).

NT2 cells Human teratocarcinoma cell line; have properties similar to those of progenitor cells in the central nervous system (CNS). Can differentiate into all three major lineages: neurons, astrocytes and oligodendrocytes.

NT3 Neurotrophin-3; see **neurotrophins**.

NTAL *Non-T-cell activation linker* A transmembrane adaptor protein (**TRAP**) in mature B-cells that is phosphorylated following immunoreceptor engagement. Phosphorylated NTAL recruits **GRB-2**.

NTG Abbrev. for (1) Normal-tension **glaucoma**, (2) **Nitroglycerin**, (3) Non-transgenic.

Ntk Protein tyrosine kinase (56 kDa) similar to **Csk** but found in nervous tissue and T-cells. Has SH2 and SH3 domains but lacks consensus tyrosine phosphorylation and myristoylation sites of src. Ntk levels drop when T-cells are activated

NTRK1 *Neurotrophic tyrosine kinase receptor* Receptor tyrosine kinase, the high-affinity receptor for **nerve growth factor**; mutations are associated with hereditary sensory neuropathy type IV. Somatic rearrangements of NTRK1, producing chimeric oncogenes with constitutive tyrosine kinase activity (*trk* oncogene), have been detected in a consistent fraction of papillary thyroid tumours.

N-type channels A class of **voltage-sensitive calcium channels**. Restricted to neurons and neuroendocrine cells, where they are involved in regulation of neurotransmitter or neurohormone release. Require substantial depolarization to become activated, and become inactivated in a time-dependent fashion. Potently inhibited by ω-conotoxin.

nuclear actin binding protein *NAB* Nuclear protein, dimer of 34-kDa subunits. Binds actin with K_d of around 25 mM. Has not appeared in recent literature, but could be what is now termed **gCAP39** or **myopodin**.

nuclear envelope Membrane system that surrounds the nucleus of eukaryotic cells. Consists of inner and outer membranes separated by perinuclear space and perforated by nuclear pores. The term should be used in preference to 'nuclear membrane', which is potentially very confusing.

nuclear export signal *NES* A leucine-rich motif (e.g. LQLPPLERLTL, the NES in rev protein of HIV-1) in proteins destined for export from the nucleus by **exportins** such as **CRM-1**.

nuclear factor 1 *NF1* See **CTF**.

nuclear lamina A fibrous protein network lining the inner surface of the nuclear envelope. The extent to which this system also provides a scaffold within the nucleus is controversial. Proteins of the lamina are **lamins** A, B and C, which have sequence homology to proteins of **intermediate filaments**.

nuclear localization signal *NLS* In eukaryotes, peptide signal sequence that identifies a protein as being destined for the nucleus (see **importins**). Frequently the signal sequence is a collection of basic amino acids downstream of a helix-breaking proline; e.g. SV40 T (Pro-Lys-Lys-Lys-Arg-Lys-Val).

nuclear magnetic resonance *NMR* Biophysical technique which allows the spectroscopy or imaging of molecules containing at least one paramagnetic atom (e.g. ^{13}C, ^{31}P). Although non-invasive, the scale of the equipment needed to generate the radiofrequency electromagnetic and magnetic fields, and the computer power needed to analyse the results, are non-trivial. Widely used as a medical imaging technology (**fMRI**).

nuclear matrix Protein latticework filling the nucleus that anchors **DNA replication** and **transcription** complexes.

nuclear matrix constituent protein *NMCP1* Plant protein that may function as a **lamin** homologue. Identified in carrot, celery and *Arabidopsis*; contains a central domain with long α-helices exhibiting heptad repeats of apolar residues, and terminal domains that are predominantly non-helical. It also contains potential **nuclear localization sequence** motifs. NMCP1 shares a weak sequence similarity to myosin, tropomyosin and IF proteins, and fractionates with the nuclear matrix.

nuclear matrix protein 1 *NMP1* A small (36-kDa) tomato protein, identified in the cytoplasm and nucleus. This protein is unique to plants, and is highly conserved among flowering and non-flowering plants, moss, and the liverwort *Marchantia polymorpha*. It is predominantly alpha-helical with multiple stretches of short amphipathic regions that associate with the nuclear matrix, probably functioning as a plant-specific structural protein

nuclear membrane See **nuclear envelope**.

nuclear polyadenylation hexanucleotide Sequence of mRNA (AAUAAA) that is required for polyadenylation in the nucleus. It is also necessary for cytoplasmic polyadenylation. See **cytoplasmic polyadenylation element**.

nuclear pore Openings in the nuclear envelope, diameter about 10 nm, through which molecules such as nuclear proteins (synthesized in the cytoplasm) and mRNA must pass. Pores are generated by a large protein assembly.

nuclear pore complex General term for the **nuclear pore** and the associated multiprotein complex. Components include **nucleoporins**.

nuclear receptor Receptor for a diffusible signal molecule that can enter the nucleus, particularly receptors for **steroid hormones**.

nuclear RNA The nucleus contains RNA that has just been synthesized, but in addition there is some that seems not to be released, or is only released after further processing, the heterogenous nuclear RNA (**hnRNA**) and small RNA molecules associated with protein to form **snRNPs** (small nuclear ribonucleoproteins).

nuclear run-on *Nuclear run-off* Strictly different, the two terms tend to be used interchangeably. A nuclear run-on assay is intended to identify the genes that were being transcribed at a particular instant; nuclei are rapidly isolated from cells and incubated with labelled nucleotides. This gives a population of labelled RNAs that were being transcribed immediately before isolation. These can be studied directly or (more commonly) used as a probe to identify corresponding cDNAs.

nuclear transplantation Experimental approach to study of nucleocytoplasmic interactions, in which a nucleus is transferred from one cell to the cytoplasm (which may be anucleate) of a second.

nuclear transport Passage of molecules in and out of the nucleus, presumably via nuclear pores. Passage of proteins into the nucleus may depend on possession of a **nuclear location sequence** containing five consecutive positively-charged residues (PKKKRKV). See **exportins**, **importins**.

nuclease *EC3.1.33.1* An enzyme capable of cleaving the phosphodiester bonds between nucleotide subunits of nucleic acids.

nucleation A general term used in polymerization or assembly reactions where the first steps are energetically less favoured than the continuation of growth. Polymerization is much faster if a pre-formed seed is used to nucleate growth. (e.g. microtubule growth is nucleated from the **microtubule-organizing centre**, although the nature of this nucleation is not known).

nucleic acids Linear polymers of nucleotides, linked by $3',5'$ phosphodiester linkages. In DNA (deoxyribonucleic acid), the sugar group is deoxyribose and the bases of the nucleotides are adenine, guanine, thymine and cytosine. RNA (ribonucleic acid) has ribose as the sugar, and uracil replaces thymine. DNA functions as a stable repository of genetic information in the form of base sequence. RNA has a similar function in some viruses, but more usually serves as an informational intermediate (mRNA), as a transporter of amino acids (tRNA), in a structural capacity or, in some newly discovered instances, as an enzyme.

nucleocapsid The coat (**capsid**) of a virus plus the enclosed nucleic acid genome.

nucleocytoplasmic transport Transport of molecules from the nucleus to the cytoplasm. See **exportins**.

nucleoid Region of cell in a bacterium that contains the DNA.

nucleolar organizer Loop of DNA that has multiple copies of rRNA genes. See **nucleolus**.

nucleolin A major nucleolar protein (100 kDa) that functions as a shuttle protein between nucleus and cytoplasm and is also found on the cell surface. Nucleolin binds **midkine** and heparin-binding growth associated molecule (HB-GAM).

nucleolus A small, dense body (sub-organelle) within the nucleus of eukaryotic cells, visible by phase contrast and interference microscopy in live cells throughout interphase. Contains RNA and protein, and is the site of synthesis of ribosomal RNA. The nucleolus surrounds a region of one or more chromosomes (the nucleolar organizer) in which are repeated copies of the DNA coding for ribosomal RNA.

nucleophosmin *NPM* An abundant, predominantly nucleolar protein that influences numerous cellular processes. Associates specifically with the bodies of messenger RNAs as a result of the process of $3'$-end formation. Binds the tumour-suppressors p53 and p19 (Arf) and is thought to be indispensable for ribogenesis, cell proliferation and survival after DNA damage. Mutations in exon 12 of the nucleophosmin (*NPM1*) gene occur in about 60% of cases of adult acute myeloid leukaemia (AML) with normal karyotype.

nucleoplasm By analogy with cytoplasm, that part of the nuclear contents other than the nucleolus.

nucleoplasmin First protein to be described as a molecular **chaperone**; major function seems to be in assembly of **nucleosomes**.

nucleopore filter Filter of defined pore size made by etching a polycarbonate filter that has been bombarded by neutrons, the extent of etching determining the pore size. Very thin, with neat circular holes going right through the membrane; not a complex meshwork like micropore filters.

nucleoporins Proteins that make up the nuclear pore complex that regulates the traffic of proteins and nucleic acids into and out of the nucleus. Many contain *N*-acetylglucosamine residues.

nucleoproteins Structures containing both nucleic acid and protein. Examples are chromatin, ribosomes, certain virus particles.

nucleoside Purine or pyrimidine base linked glycosidically to ribose or deoxyribose, but lacking the phosphate residues that would make it a nucleotide. Ribonucleosides are adenosine, guanosine, cytidine and uridine. Deoxyribosides are deoxyadenosine, deoxyguanosine, deoxycytidine and deoxythymidine (the latter is almost universally referred to as thymidine).

nucleoskeletal DNA DNA that is proposed to exist mostly to maintain nuclear volume and not for coding protein.

nucleosome Repeating units of organization of chromatin fibres in chromosomes, consisting of around 200 base pairs, and two molecules each of the **histones** H2A, H2B, H3 and H4. Most of the DNA (around 140 bp) is believed to be wound around a core formed by the histones, the remainder joins adjacent nucleosomes, thus forming a structure reminiscent of a string of beads.

nucleotidase *5′-Nucleotidase; EC 3.1.3.5* Enzyme that cleaves the 5′ monoester linkage of nucleotides, and thus converts them to the corresponding nucleoside.

nucleotide Phosphate esters of **nucleosides**. The metabolic precursors of nucleic acids are monoesters with phosphate on carbon 5 of the pentose (known as 5′ to distinguish sugar from base numbering). However, many other structures, such as adenosine 3′5′-cyclic monophosphate (cAMP), and molecules with two or three phosphates are also known as nucleotides. See Table N2.

nucleotide-binding fold Protein motif consisting of a fold or pocket with certain conserved residues, required for the binding of nucleotides.

nucleotide-excision repair factor *NEF* Multiprotein complexes involved in nucleotide-excision repair; possesses DNA damage recognition and endodeoxynuclease activities. In yeast, Rad14 and Rad1–Rad10 form one subassembly called NEF1; the Rad4–Rad23 complex is named NEF2; Rad2 and TFIIH constitute NEF3; and the Rad7–Rad16 complex is called NEF4

nucleus The major organelle of eukaryotic cells, in which the chromosomes are separated from the cytoplasm by the **nuclear envelope**.

nucleus accumbens A region of the brain involved in functions ranging from motivation and reward to feeding and drug addiction.

TABLE N2. Nucleotides

(a) Phosphate esters of nucleosides, which are themselves conjugates between the biological bases and sugars, either ribose or 2-deoxyribose. Nucleosides are derived from the bases by the addition of a sugar in the position indicated (H). Ribonucleotides are precursors of RNA and also common metabolic intermediates and regulators; examples of the shorthand nomenclature are given.

	Adenine	Cytosine	Guanine	Uracil
Mononucleotide	AMP	CMP	GMP	UMP
Dinucleotide	ADP	CDP	GDP	UDP
Trinucleotide	ATP	CTP	GTP	UTP
Cyclic nucleotide	3′, 5′ cyclic AMP		3′, 5′ cyclic GMP	

(b) Deoxyribonucleotides, required for the synthesis of DNA, are made by the biological reduction of the corresponding ribose dinucleotides and the deoxyribonucleotides are phosphorylated to give the triphosphonucleotides. dTMP is made by methylation of dUMP, which is then phosphorylated to give dTTP.

	Adenine	Cytosine	Guanine	Thymidine
Dinucleotide	dADP	dCDP	dGDP	
Trinucleotide	dATP	dCTP	dGTP	dTTP

Nucleotides occur as part of other biological molecules, e.g. NAD is the ADP-ribose derivative of nicotinamide. Nucleotide adducts are important intermediates in anabolic processes. CDP derivatives occur in the biosynthesis of lipids. UDP and TDP derivatives are important in sugar metabolism.

nude mice Strains of athymic mice bearing the recessive allele nu/nu which are largely hairless and lack all or most of the T-cell population. Show no rejection of either allografts or xenografts; nu/nu alleles on some backgrounds have near normal numbers of T-cells.

null cell Lymphocytes lacking typical markers of T- or B-cells capable of lysing a variety of tumour or virus-infected cells without obvious antigenic stimulation, also effect **antibody-dependent cell lysis**, and in humans carry CD16 marker.

null mutant Mutation in which there is no gene product.

nullo Protein that appears to stabilize the initial accumulation of **cadherins** and **catenins** as they form a mature basal junction during the process of cellularization of the initially syncytial *Drosophila* embryo.

NuMA Nuclear mitotic apparatus protein, identical to **centrophilin**.

Numb *mNumb* Mammalian Numb (mNumb) has multiple functions and plays important roles in the regulation of neural development, including maintenance of neural progenitor cells and promotion of neuronal differentiation in the central nervous system. There are multiple splicing isoforms. The signalling pathway mediated by Numb is antagonistic to that mediated by **Notch**; both play essential roles in enabling the two daughters to adopt different fates after a wide variety of asymmetric cell divisions.

numerical aperture *N.A* For a lens, the resolving power depends upon the wavelength of light being used and inversely upon the numerical aperture. The N.A. is the product of the refractive index of the medium (1 for air, 1.5 for immersion oil) and the sine of the angle, *i*, the semi-angle of the cone formed by joining objects to the perimeter of the lens. The larger the value of N.A., the better the resolving power of the lens; most objectives have their N.A. value engraved on the barrel, and this should be quoted when describing an optical system.

nurse cells Cells accessory to egg and/or sperm formation in a wide variety of organisms. Usually thought to synthesize special substances and to export these to the developing gamete.

nutlins Small selective inhibitors (cis-imidazoline derivatives) of murine double minute-2 (Mdm2)-ligase binding to p53 that will drive tumour cells into apoptosis.

nutraceutical Food that has, or is engineered to have, therapeutic properties. **Golden rice**, engineered to provide higher Vitamin A in the diet, could be considered an example.

nutrigenomics The study of the combined effect of diet and genomic make-up on health.

NXT *p*15 An NTF2-related cofactor for **Tap protein**-mediated mRNA export in both human and invertebrate cells. Stimulates binding of a Tap-RNA complex to **nucleoporin**.

Nycodenz[R] Proprietary name for a dense, non-ionic substance, solutions of which will form density gradients for separation procedures if centrifuged. Solutions are heat-stable and can be sterilized by autoclaving. A density of 2.1 g/ml can be achieved.

nystatin A polyene antibiotic active against fungi. The name is derived from 'New York State Health Department', where it was discovered as a product of *Streptomyces noursei*. Exhibits selectivity for Na^+ and increases the activity of the Na^+-K^+ pump.

NZB, NZW mice *New Zealand Black, NZ White* Inbred strains of mice which develop spontaneous autoimmune diseases. There is evidence for an underlying retroviral aetiology, and a NZB virus has been isolated.

O

O-2A progenitor Bipotential progenitor cells in rat optic nerve that give rise initially to oligodendrocytes and then to type-2 astrocytes. Production of type-2 astrocytes from O-2A progenitor cells *in vitro* is triggered by **ciliary neurotrophic factor** (CNTF).

O-antigens Tetra- and pentasaccharide repeat units of the cell walls of Gram-negative bacteria. They are a component of **lipopolysaccharide**.

oat cell carcinoma Form of carcinoma of the lung in which the cells are small, spindle-shaped and dark-staining. May derive from argyrophilic **APUD** cells of the mucosa, and certainly tends to be associated with endocrine symptoms.

obelin Calcium-activated photoprotein in the photocyte of the colonial hydroid coelenterate, *Obelia geniculata*.

obestatin Peptide, 23 amino acids (FNAPFDVGIKLSG VQYQQHSQAL-NH$_2$), that has appetite-suppressing activity. Produced by post-translational modification of a precursor that also gives rise to **ghrelin**, an appetite stimulator.

oblique fracture A bone fracture in which the break is oblique to the long axis. Often mistaken for a spiral fracture.

obscurin Giant multidomain muscle protein (~800 kDa) that apparently plays a role in spatial positioning of contractile proteins and the structural integration and stabilization of myofibrils, especially at the stage of myosin filament incorporation and A-band assembly.

occipital lobe The posterior lobe of each cerebral hemisphere, particularly involved in the interpretation of visual images.

occludens junction Tight junction. See **zonula occludens**.

occludin A four-pass integral plasma membrane protein (~65 kDa), a functional component of the **occludens junction**. The predicted structure has two extracellular loops and N- and C-terminal cytoplasmic domains.

occlusion bodies Large proteinaceous structures (0.3 × 0.5 mm^2) formed late in the infection of cells by **baculovirus**; major protein is **polyhedrin,** embedded within which are virions.

OCD 1. Obsessive-compulsive disorder. **2.** See **ocd lesion**.

ocd lesion *Osteochondritis dissecans lesion* Lesion in joint, usually the knees or elbow, caused by a loose piece of bone and cartilage that separates from the end of the bone because of a loss of blood supply. The loose piece (joint mouse) may stay in place or fall into the joint space, making the joint unstable.

ocellus *Pl.* ocelli A simple eye or eyespot found in some invertebrates.

ochnaflavone A biflavonoid present in the human diet and a medicinal herbal product isolated from *Lonicera japonica*; inhibits angiotensin II-induced hypertrophy and

serum-induced smooth muscle cell proliferation. Also has antifungal and anti-inflammatory activities.

ochre codon The **codon** UAA, one of the three that causes termination of protein synthesis. The most frequent termination codon in *E. coli* genes.

ochre mutation Mutation that changes any codon to the **termination** codon UAA.

ochre suppressor A gene that codes for an altered tRNA so that its **anticodon** can recognize the **ochre codon** and thus allows the continuation of protein synthesis. A suppressor of an **ochre mutation** is a tRNA that is charged with the amino acid corresponding to the original codon or a neutral substitute. Ochre suppressors will also suppress **amber codons**.

ochronosis Deposition of dark brown pigment in cartilage, joint capsules and other tissues, usually as a result of **alkaptonuria**.

OCIF Osteoclastogenesis inhibitory factor. See **osteoprotegerin**.

oct Family of genes for transcription factors that act as RNA Polymerase II promoters. Protein products contain a **POU domain** and are **leucine zipper** proteins that bind to **octamer** sequences. See **Oct4, OCTN1**.

Oct4 Transcription factor required to maintain the pluripotency and self-renewal of embryonic stem cells. At the expanding blastocyst stage, Oct4 is confined to the inner cell mass; following detachment of the hypoblast and formation of the embryonic disc, Oct4 is selectively observed in the epiblast and disappears at different stages as the three germ layers are differentiated.

octamer 1. Assembly of eight histone proteins (two each of H2A, H2B, H3 and H4) that forms the core of nucleosome. **2.** Eight-base sequence motif common in eukaryotic promoters. Consensus is ATTTGCAT; binds various transcription factors.

octamer binding protein Transcription factor that binds to the **octamer** motif. Examples: mammalian proteins Oct-1, Oct-2.

octamer motif A DNA motif found in certain promoters that can produce B-cell-specific gene expression. Sequence: ATGCAAAT.

OCTN1 A proton/organic cation transporter cloned from human foetal liver. A 551-amino acid protein with 11 transmembrane domains and 1 nucleotide-binding site motif; appears to be a proton antiporter that functions for active secretion of cationic compounds across the renal epithelial brush-border membrane. It may play a role in the renal excretion of xenobiotics and their metabolites.

octopaline N-*alpha-(D-1-carboxyethyl)-L-arginine* An **opine**.

octopamine A **biogenic amine** found in both vertebrates and invertebrates (identified first in the salivary gland of

Octopus). Octopamine can have properties both of a hormone and a neurotransmitter, and acts as an adrenergic agonist.

octyl glucoside A biological detergent characterized by its ease of removal from hydrophobic proteins. Used to solubilize membrane proteins.

OD$_{600}$ Optical density at 600 nm; provides a reasonable estimate of cell numbers present in a growing bacterial or yeast culture.

ODC Abbrev. for (1) **Ornithine decarboxylase** EC 4.1.1.17, (2) Oxygen (oxyhaemoglobin) dissociation curve.

ODD Abbrev. for oppositional defiant disorder, a category of disorder exhibited by young adolescents.

ODF Abbrev. for (1) **Osteoclast differentiation factor**, (2) **Outer dense fibres**, (3) Orientation distribution function, a formal mathematical method for describing orientation of materials (fibres, crystals, etc).

odontoblasts Columnar cells derived from the dental papilla after **ameloblasts** have differentiated, and that give rise to the dentine matrix that underlies the enamel of a tooth.

odontogenic epithelial cells Epithelial layer that will give rise to teeth.

oedema *Edema (USA)* Swelling of tissue: can result from increased permeability of vascular endothelium or increased blood pressure, for example, as a result of being at high altitude without adequate acclimatization.

Oedogonium The type genus of the Oedogoniaceae; freshwater green algae with long unbranched filaments.

oestrogen *Estrogen (USA)* See **estrogen**.

OFAGE *Orthogonal field alternation gel electrophoresis* Electrophoresis in which macromolecules are electrophoresed in a gel using electric fields applied alternately at right angles to each other.

off-label Term used in the US for a drug (usually a prescription-only medicine, POM) that is used to treat a condition for which it has not been approved by the Food and Drug Administration.

O-glycans O-linked glycans are linked to protein (usually large) through the hydroxyl group of serine or threonine. O-linked glycosylation is a post-translational modification, does not require a consensus sequence and no oligosaccharide precursor is required for protein transfer. The commonest O-linked glycans contain an initial GalNAc residue.

oil body Small droplets (0.2–1.5 μm in diameter), containing mostly triacylglycerol, that are surrounded by a phospholipid/**oleosin** annulus. Found in oil-rich seeds.

okadaic acid A toxin first isolated from the sponge *Halichondria okadai*, although produced by various species of dinoflagellates; a complex lipophilic polyether. It is a potent inhibitor of serine-threonine-specific protein **phosphatases** 1 and 2A, and can act as a tumour promoter. **Microcystins** and **nodularins** bind to the same site on the phosphatase.

Okazaki fragments Short fragments of newly synthesized DNA strands produced during DNA replication. All the known **DNA polymerases** synthesize DNA in the 5′ to 3′ direction. However, as the strands separate, replication forks will be moving along one parental strand in the 3′ to 5′ direction and along the other parental strand in the 5′ to 3′ direction. On the former, the leading strand, DNA can be synthesized continuously in the 5′ to 3′ direction. On the other, the lagging strand, DNA synthesis can only occur when a stretch of single-stranded DNA has been exposed and proceeds in the direction opposite to the movement of the replication fork (still 5′ to 3′). It is thus discontinuous, and the series of fragments are then covalently linked by **ligases** to give a continuous strand. Such fragments were first observed by Okazaki using pulse-labelling with radioactive thymidine. In eukaryotes the Okazaki fragments are typically a few hundred nucleotides long, whereas in prokaryotes they may contain several thousands of nucleotides.

olanzapine Atypical antipsychotic drug (thienobenzodiazepine class) used for treatment of schizophrenia. A selective monoaminergic antagonist that is thought to act through a combination of dopamine and serotonin type 2 (5HT$_2$) antagonism.

oleic acid See **fatty acids** and Table L3.

oleosins Proteins (16–24 kDa) that form a hydrophilic shell around **oil body** inclusions in plant cells. Oleosin has three distinct domains; the N- and C-terminal domains, which are amphipathic, and a central extremely hydrophobic domain with long stretches of non-polar amino acids. May have a structural role in stabilizing the lipid body during desiccation of the seed.

oleosome Plant spherosome rich in lipid that serves as a storage granule in seeds and fruits. There are none of the enzymes characteristic of lysosomes.

Olestra$^{(TM)}$ A synthetic fat substitute, marketed as a dieting aid, that is neither digested nor absorbed in the gastrointestinal tract.

olfactomedin family A family of proteins with a conserved protein motif through which extracellular protein–protein interactions occur. Although they share sequence similarity, they have diversified roles in many important biological processes. Includes **amassin**, **myocilin**, **noelins** (which enhances neural crest generation in the chick), **pancortins**, **photomedins** and **tiarin**.

olfactory epithelium The **epithelium** lining the nose. Has the diverse G-protein-coupled receptors responsible for the sense of smell.

olfactory neuron **Sensory neuron** from the lining of the nose. They are some of the only neurons that continue to divide and differentiate throughout an organism's life.

oligoadenylate synthetases *EC 2.7.7.* The human 2′-5′ oligoadenylate synthetases (OAS) form a conserved family of interferon-induced proteins consisting of four genes: *OAS1*, *OAS2*, *OAS3*, and the 2′-5′ oligoadenylate synthetase-like gene (*OASL*). When activated by double-stranded RNA, OAS1-3 polymerizes ATP, which in turn activates a latent endoribonuclease that degrades viral and cellular RNAs.

oligodendrocyte **Neuroglial** cell of the central nervous system in vertebrates whose function is to myelinate CNS axons.

oligodendrocyte-myelin glycoprotein *OMgp; myelin-oligodendrocyte glycoprotein, (MOG)* An integral membrane protein expressed in oligodendrocytes and outer myelin lamellae; like **myelin-associated glycopro-**

tein (MAG) and **nogo**-A, binds to the nogo receptor, causes growth cone collapse and inhibits neurite outgrowth *in vitro*.

oligomycin A bacterial toxin inhibitor of oxidative phosphorylation that acts on a small subunit of the **F1-ATPase**. See **oligomycin sensitivity conferral protein**.

oligomycin sensitivity conferral protein *OSCP* The δ-subunit of the **F-type ATP synthase**, believed to link the F1 catalytic segment to the F0 proton-conduction segment. Binds the toxin **oligomycin**.

oligonucleotide Linear sequence of up to 20 nucleotides joined by phosphodiester bonds. Above this length the term polynucleotide begins to be used.

oligopeptide A peptide of a small number of component amino acids as opposed to a polypeptide. Exact size range is a matter of opinion, but peptides from 3- to about 40-member amino acids might be so described.

oligosaccharide A saccharide of a small number of component sugars, either O- or N-linked to the next sugar. Number of component sugars is not rigorously defined.

oligosaccharin An oligosaccharide derived from the plant cell wall that in small quantities induces a physiological response in a nearby cell of the same or a different plant, and thus acts as a molecular signal. Sometimes considered to be a plant hormone or plant growth substance. The best authenticated examples are involved in host–pathogen interactions and in the control of plant cell expansion.

oligotroph Organism that can grow in an environment poor in nutrients.

olomoucine *2-(2-Hydroxyethylamino)-6-benzylamino-9-methylpurine* Purine derivative that is an inhibitor of cyclin-dependent kinase **cdk5** but inactive against cdk4 and cdk6. A similar compound, **roscovitine**, is somewhat more potent.

omega fatty acid Type of fatty acid found in unsaturated fat. Two forms are found, omega-3 and omega-6. Generally regarded as beneficial in reducing risk of heart disease.

omega-oxidation Minor metabolic pathway in the ER for medium-chain-length fatty acids. The omega carbon in a fatty acid is the carbon furthest in the alkyl chain from the carboxylic acid: this carbon is progressively oxidized first to an alcohol and then to a carboxylic acid, creating a molecule with a carboxylic acid on both ends. The dicarboxylic acid product can enter the beta-oxidation pathway to be shortened at both ends of the molecule at the same time.

omentum Portion of the serosa, the thin tissue infolding from the peritoneal lining, that connects two or more folds of the alimentary canal in vertebrates.

omeprazole Non-proprietary name for gastric proton-pump (H^+/K^+ transport ATPase) inhibitor, much prescribed for gastric ulcers since it inhibits acid secretion by **oxyntic** (parietal) cells.

OMgp See **oligodendrocyte-myelin glycoprotein**.

OMIM The 'Online Mendelian Inheritance in Man' database that is a catalogue of human genes and genetic disorders, authored and edited by Dr V.A. McKusick and his colleagues at Johns Hopkins University, and developed for the World Wide Web by NCBI, the National Center for Biotechnology Information. An invaluable resource that provides much more information than the brief Dictionary entries, and well worth consulting.

ommatidium *Pl.* ommatidia Single facet of an insect compound eye. Composed of a set of photoreceptor cells, overlain by a crystalline lens.

Onchocerca Genus of filarial nematode parasites that cause river blindness.

oncocytes See **Askenazy cells**.

oncogen Synonym for **carcinogen**, an agent causing cancer.

oncogene Mutated and/or overexpressed version of a normal gene of animal cells (the **proto-oncogene**) that in a dominant fashion can release the cell from normal restraints on growth, and thus alone, or in concert with other changes, convert a cell into a tumour cell. See Table O1.

oncogenic virus A virus capable of causing cancer in animals or in humans. These include DNA viruses, ranging in size from papova viruses to herpes viruses, and the RNA-containing retroviruses. See **Oncovirinae**.

oncomodulin *Parvalbumin-beta* Calcium-binding protein containing the **EF-hand** motif. Released by macrophages and can act as a growth factor for neurons of the mature central and peripheral nervous systems. First discovered in a rat hepatoma and subsequently in outer hair cells in the organ of Corti (in the cochlea).

OncoMouse™ Registered proprietary name for transgenic mice, genetically engineered to contain the activated *v-Ha-ras* oncogene fused to a mouse zeta-globin promoter. The patenting of these animals was a matter of considerable debate. Because they are predisposed to papillomas, they have been used extensively in screening tumour promoters and non-genotoxic carcinogens and for assessing antitumour and antiproliferative agents.

oncoprotein 18 See **stathmin**.

oncostatin M Multifunctional cytokine (28 kDa) of the **IL-6 cytokine family**. Produced by activated T-cells; inhibits tumour cell growth and induces IL-6 production by endothelial cells via the tyrosine kinase p62yes .

Oncovirinae The subfamily of retroviruses (Retroviridae) that can cause tumours. They are enveloped by membrane derived from the plasma membrane of the host cell, from which they are released by budding without lysing the cell. Within each virion is a pair of single-stranded RNA molecules. Replication involves a DNA intermediate made on an RNA template by the enzyme **reverse transcriptase**. No longer considered to be an appropriate class, since the genera of the subfamily are unrelated.

Ondansetron Drug used to treat vomiting and nausea, in particular that resulting from chemotherapy or radiotherapy. It is a selective blocking agent of the serotonin $5-HT_3$ receptor type, although how this reduces nausea is not clear.

ONPG *Ortho-nitrophenyl-beta-D-galactopyranoside.* A chromogenic substrate for the enzyme β-galactosidase that produces a yellow colour (dinitrophenol) when hydrolysed.

ontogeny The total of the stages of an organism's life history. Once believed to recapitulate phylogeny, though this is an outmoded conceit.

TABLE O1. Oncogenes and tumour viruses

Acronym	Virus	Species	Tumour origin	Comments
Abl	Abelson leukaemia	Mouse	Chronic myelogenous leukaemia	TyrPK(src)
Akt	AKT8	Human		Ser/Thr kinase (PKB)
Arg	Abelson leukaemia	Mouse		TyrPK
Bcl		Human	B-cell lymphoma	Involved in apoptosis
Cbl				
Crk		Chicken	Sarcoma	Adaptor signalling proteins
Dbl			Diffuse B-cell lymphoma	Encodes a GEF
E1A	Adenovirus			Interacts with Rb product
E1B	Adenovirus			Interacts with p53
E5, 6, 7	Papilloma			E5 blocks processing of growth factor receptors, E6 and E7 drive proliferation
erbA	Erythroblastosis	Chicken		Homology to human glucocorticoid receptor
erbB	Erythroblastosis	Chicken		TyrPK EGF/TGFc receptor
ets	E26 myeloblastosis	Chicken		Nuclear
fes(fps)[a]	Snyder–Theilen sarcoma; Gardner–Arnstein sarcoma	Cat		TyrPK (src)
fgf oncogenes				
fgr	Gardner-Rasheed sarcoma	Cat		TyrPK(src)
fms	McDonough sarcoma	Cat		TyrPK CSF-1 receptor
fps(fes)[a]	Fujinami sarcoma	Chicken		TyrPK(src)
fos	FBJ osteosarcoma	Mouse		Nuclear, TR
gip2		Rat	Ovarian and adrenal	GTPase-defective Gs and Gi
hst	NVT	Human	Stomach tumour	FGF homologue
intl	NVT	Mouse	MMTV-induced carcinoma	Nuclear, TR
int2	NVT	Mouse	MMTV-induced carcinoma	FGF homologue
ipt	Agrobacterium	Plants		Isopentenyl transferase
jun	ASV17 sarcoma	Chicken		Nuclear, TR
kit	Hardy–Zuckerman 4 feline sarcoma	Cat	Sarcoma	PTK receptor for SCF
mas	NVT	Human	Epidermoid carcinoma	Potentiates response to angiotensin II ?
met	NVT	Mouse	Osteosarcoma	TyrPK GFR L?
mil (raf)	Mill Hill 2 acute leukaemia	Chicken		Ser/ThrPK
mos	Moloney sarcoma	Mouse		Ser/ThrPK
myb	Myeloblastosis	Chicken	Leukaemia	Nuclear, TR
myc	MC29 Myelocytomatosis	Chicken	Lymphomas	Nuclear TR?
N-myc	NVT	Human	Neuroblastomas	Nuclear
neu(ErbB2)	NVT	Rat	Neuroblastoma	TyrPK GFR L?
pim-1, -2, -3		Human	Prostatic carcinoma	Ser/ThrPK
ral(mil)[b]	3611 sarcoma	Mouse		Ser/ThrPK
Ha-ras	Harvey murine sarcoma	Rat	Bladder, mammary and skin carcinomas	GTP-binding

TABLE O1. (Continued)

Ki-ras	Kirsten murine sarcoma	Rat	Lung, colon carcinomas	GTP-binding
N-ras	NVT	Human	Neuroblastomas, Leukaemias	GTP-binding
ral			Small GTPase	
rel	Reticuloendotheliosis	Turkey		
ret		Human	Thyroid, multiple endocrine neoplasia	RTK for GDNF
ros	UR2	Chicken		TyrPK GFR L?
sea	S13 avian erythroblastosis homologue			RTK for hepatocyte growth factor
sis	Simian sarcoma	Monkey		One chain of PDGF
src	Rous sarcoma	Chicken		TyrPK
ski	SKV770	Chicken		Nuclear
tre		Human	Transfected 3T3	De-ubiquitinating enzyme
trk	NVT	Human	Colon carcinoma	RTK for NGF
yes	Y73, Esh sarcoma	Chicken		TyrPK(src)

[a] fps/fes are species equivalents.
[b] mil/raf are species equivalents.
GFR L?, from sequence, a growth factor receptor for unknown ligand; MMTV, mouse mammary tumour virus; NVT, isolated from non-retroviral tumour, in most cases detected by transfection of 3T3 cells; RTK, receptor tyrosine kinase; Ser/ThrPK, serine/threonine protein kinase; TyrPK, tyrosine protein kinase; TR, transcriptional regulator.

oocyst 1. In some Protozoa, the cyst formed around two conjugating gametes, generally with a thick protective wall that facilitates survival. **2.** In **Sporozoa**, the passive phase from which sporozoites are released.

oocyte The developing female gamete before maturation and release.

oocyte expression Technique whereby the cellular translational machinery of an oocyte (typically *Xenopus*) is utilized to generate functional protein from microinjected mRNA or to produce protein encoded by an introduced expression vector.

oogenesis The process of egg formation.

oogonium Female sexual structure in certain algae and fungi, containing one or more gametes. After fertilization the oogonium contains the oospore.

oomycetes Group of fungi in which the mycelium is non-septate, i.e. lacks cross-walls, and the nuclei are diploid. Sexual reproduction is oogamous.

oospore A zygote with food reserves and a thick protective wall, formed from a fertilized oosphere in some algae and the Oomycetes.

ootid An immature female gamete that will develop into an ovum.

opal codon The codon UGA, one of the three that causes **termination** of protein synthesis.

opal mutation Mutation that changes any codon to the termination codon UGA.

opal suppressor A gene that codes for an altered tRNA so that its anticodon can recognize the **opal codon** and thus allows the continuation of protein synthesis. A suppressor of an **opal mutation** is a tRNA that is charged with the amino acid corresponding to the original codon or a neutral substitute. Some eukaryote cells normally synthesize opal suppressor tRNAs. The function of these is not clear and they usually do not prevent normal termination of protein synthesis at an opal codon.

Opalina A genus of parasitic protozoa found in the guts of frogs and toads. They look superficially like ciliates, but are classified in a separate group as they have a number of similar nuclei.

OPC 1. Oropharyngeal candidiasis, the most common opportunistic infection in immunosuppressed patients. **2.** Oral and pharyngeal cancer. **3.** Oligodendrocyte precursor cell, a lineage-restricted precursor cell that expresses most neurotransmitter receptors. **4.** Organophosphorus compounds. **5.** OPC-14523 (OPC), a compound with high affinity for sigma and 5-HT$_{1A}$ receptors that shows 'antidepressant-like' effects in animal models of depression.

OPCML Opioid-binding cell adhesion molecule of the IgLON family. May act as a tumour suppressor.

open reading frame *ORF* A possible **reading frame** of DNA which is capable of being translated into protein, i.e. is not punctuated by stop codons. (This capacity does not indicate *per se* that the ORF is translated.)

operator The site on DNA to which a specific **repressor protein** binds and prevents the initiation of transcription at the adjacent **promoter**.

operon Groups of bacterial genes with a common **promoter**, that are controlled as a unit and produce mRNA as a single, **polycistronic** messenger. An operon consists of two or more structural genes, which usually code for proteins with related metabolic functions, and associated control elements that regulate the transcription of the structural genes. The first described example was the **lac operon**.

OPG See **osteoprotegerin**.

opiates See **opioids**.

opines Carbon compounds produced by crown galls and hairy roots induced by *Agrobacterium tumefaciens* and *A. rhizogenes,* respectively. They are utilized as nutritional sources by *Agrobacterium* strains that induced the growth and are, in some cases, chemoattractants. Chemotactic activity seems to be specific for plasmids carrying the relevant opine synthase gene. **Octopaline, nopaline,** mannopine and agrocinopines A and B are examples.

opioid receptors Four distinct classes of G-protein-coupled receptors, δ, μ and κ together with an orphan receptor, the ligand for which is **orphanin FQ**.

opioids Naturally occurring basic (alkaloid) molecules with a complex fused ring structure. Group includes the opiates such as morphine, found in the opium poppy (*Papaver somniferum*) and synthetically modified derivatives. Effective pain-killers but tend to be addictive and are drugs of abuse. Weak opioids include codeine and dihydrocodeine; strong opioids include morphine, diamorphine, fentanyl and phenazocine. Have high pharmacological activity and are likely to cause addiction. See **morphine, opioid receptors**.

opportunistic pathogen *Secondary pathogen* Pathogenic organism that is often normally a commensal, but which gives rise to infection in immunocompromised hosts.

opsin General term for the apoproteins of the **rhodopsin** family.

opsin subfamilies The protein moiety of rhodopsin; opsins are G-protein-coupled receptors and many variants are known. They can be subdivided into families: (1) the vertebrate visual (transducin-coupled) and nonvisual opsin subfamily, (2) the **encephalopsin**/tmt-opsin subfamily, (3) the Gq-coupled opsin/**melanopsin** subfamily, (4) the Go-coupled opsin subfamily, (5) the **neuropsin** subfamily, (6) the **peropsin** subfamily and (7) the retinal photoisomerase subfamily.

opsonin Substance that binds to the surface of a particle and enhances the uptake of the particle by a phagocyte. Probably the most important opsonins in mammals derive from **complement** (C3b or C3bi) or immunoglobulins (which are bound through the **Fc receptor**).

opsonization Process of coating with an **opsonin**. Often done simply by incubating particles (e.g. **zymosan**) with fresh serum.

optic disc The blind spot of the eye where the optic nerve passes through the retina.

optic nerve Projection from the vertebrate retina to the **optic tectum** of the midbrain. Embryologically, a CNS tract rather than a peripheral nerve. Popular experimental preparation for studies of regeneration of **retino-tectal connections** in lower vertebrates, and also for studies of glial cell lineage in CNS.

optic tectum A region of the midbrain in which input from the optic nerve is processed. Because the retinally derived neurons of the optic nerve 'map' onto the optic tectum in a defined way, the question of how this specificity is determined has been a long-standing problem in cell biology. Although there is some evidence for adhesion gradients and for some adhesion specificity, the problem is unresolved.

optical diffraction A technique used to obtain information about repeating patterns. Diffraction of visible light can be used to calculate spacings in the object.

optical isomers Isomers (stereoisomers) differing only in the spatial arrangement of groups around a central atom. Optical isomers rotate the plane of polarized light in different directions. For most biological molecules in which the possibility of optical isomerism exists, only one of the isomers is functional – although there are some exceptions. For example, both glucose and mannose are used, and D-amino acids are found in the cell walls of some bacteria (whereas L-forms are universal in proteins).

optical tweezers *Laser tweezers; optical trap* By focusing a beam of light on a particle it is possible to trap the particle as a result of the forces due to radiation pressure (the forces involved are of the order of pico-Newtons). Laser beams exert sufficient force for it to be possible to move small organelles around under the microscope or to measure the forces that motor molecules are exerting by measuring the force needed to oppose their activity.

optimedin *Olfactomedin 3* An **olfactomedin** domain-containing protein that appears to be a downstream target regulated by Pax6 in eye development.

orange domain A functional domain for the **Hairy**/E(SPL) proteins that mediates the specificity of their biological action *in vivo*.

orbivirus Genus of **Reoviridae** that infects a wide range of vertebrates and insects.

orcein The purple dye originally extracted from lichens, formerly used as a food colouring and as a microscopical stain. Contains a variety of phenazones: hydroxy-orceins, amino-orceins and amino-orceinimines, and is the standard histological stain for elastic fibres and for **ground-glass cells** (hepatitis B infected liver cells).

orcinol *5-Methyl-1,3-benzenediol* Bial's orcinol test is used to distinguish pentoses from hexoses. Bial's reagent (an acidified alcoholic solution of orcinol with ferric chloride) will develop a green to deep blue colour with pentoses; hexoses give a muddy brown to grey colour.

orexigenic Appetite-stimulating.

orexin *Hypocretin* Orexin-A and -B are potent orexigenic peptides (also known as hypocretin-1 and -2) that are derived from the same precursor peptide. They are specifically localized in neurons located in the lateral hypothalamic area, a region classically implicated in feeding behaviour. The major endogenous molecule of orexin-A is a 33-amino acid peptide, and that of orexin-B a 28-amino acid peptide. Receptor is G-protein coupled. There is increasing evidence that other behaviours are regulated by orexin.

organ culture Culture *in vitro* of pieces of tissue (as opposed to single cells) in such a way as to maintain some normal spatial relationships between cells and some normal function. Contrast with **tissue culture**.

organelle A structurally discrete component of a cell.

organizing centre See **microtubule-organizing centre**.

organogenesis The process of formation of specific organs in a plant or animal involving morphogenesis and differentiation.

Oriental sore Skin disease caused by the flagellate protozoan, *Leishmania tropica*.

orientation chamber Chamber designed by Zigmond in which to test the ability of cells (particularly **neutrophils**) to orient in a gradient of chemoattractant. The chamber is similar to a haemocytometer, but with a depth of only *ca.* 20 μm. The gradient is set up by diffusion from one well to the other, and the orientation of cells towards the well containing chemoattractant is scored on the basis of their morphology or by filming their movement. A variation on this is the **Dunn chamber**.

origin of replication Regions of DNA that are necessary for its replication to begin, such as pBR322 ori, required for plasmid replication.

orlistat *TN Xenical* An orally administered drug that reduces the absorption of dietary fat by inhibiting the action of enzymes in the digestive system, used to treat obesity.

ornithine decarboxylase *ODC; EC 4.1.1.17* The enzyme that converts ornithine to putrescine (dibasic amine) by decarboxylation. Rate-limiting in the synthesis of the polyamines spermidine and spermine that regulate DNA synthesis.

ornithine transcarbamylase deficiency X-linked disorder, the most common cause of inherited urea cycle disorders.

orosomucoid α-*1-seromucoid;* α-*1-acid glycoprotein* Plasma protein of mammals and birds, 38% carbohydrate. In humans, a single-chain glycoprotein of 39 kDa. Increased levels are associated with inflammation, pregnancy and various diseases.

orotic acid *Orotate* Intermediate in the *de novo* synthesis of pyrimidines. Linked glycosidically to ribose 5'-phosphate, orotate forms the pyrimidine nucleotide orotidylate that, on decarboxylation at position 5 of the pyrimidine ring, yields the major nucleotide uridylate (uridine 5'-phosphate).

orphan drug Drug for a small patient population, given special status and financial support by governmental or charitable agencies to enable it to be produced; given drug development costs it is commercially uneconomic without such support.

orphan receptor A receptor for which no ligand has been identified.

orphanin FQ *Nociceptin* Endogenous ligand for the fourth opioid receptor (opioid-like orphan receptor), a 17-amino acid peptide resembling **dynorphin**. Produced as a preproprotein which is proteolytically processed. The receptor is widely distributed in the brain, but the function of orphanin FQ remains unclear.

orthodromic Conduction of impulses in the normal direction down a nerve fibre, from the cell body towards the distal presynaptic region, as opposed to the antidromic direction.

orthogonal arrays Arrays that are at (approximately) right angles to one another. Confluent fibroblasts often become organized into such arrays; other examples are the packing of collagen fibres in the cornea, and cellulose fibrils in the plant cell wall.

orthograde transport Axonal transport from the cell body of the neuron towards the synaptic terminal. Opposite of retrograde transport and dependent on a different mechanochemical protein **kinesin** interacting with microtubules.

orthokinesis Kinesis in which the speed or frequency of movement is increased (positive orthokinesis) or decreased.

orthologous genes Genes related by common phylogenetic descent and usually with a similar organization; contrast with **paralogous genes**.

Orthomyxoviridae Class V **viruses**. The genome consists of a single negative strand of RNA that is present as several separate segments each of which acts as a template for a single mRNA. The **nucleocapsid** is helical and has a viral-specific RNA polymerase for the synthesis of the mRNAs. They leave cells by budding out of the plasma membrane and are thus enveloped. They usually have two classes of spike protein in the envelope. One has **haemagglutinin** activity and the other acts as a **neuraminidase,** and both are important in the invasion of cells by the virus. The major viruses of this group are the **influenza viruses**.

Orthopoxviridae Genus of double-stranded DNA viruses (250–390×200–260 nm^2) that preferentially infect epithelial cells. Includes variola (smallpox) and **vaccinia**.

orthotropism A tropism in which a plant part (e.g. a shoot or root) becomes aligned directly towards (positive-) or away from (negative-) the source of the orientating stimulus. Most seedling shoots are negatively orthogravitropic and positively orthophototropic.

Oryctolagus cuniculus Rabbit.

Oryza sativa Rice.

oryzalin A dinitroaniline (sulfonamide) herbicide; a selective pre-emergence surface-applied herbicide used for control of annual grasses and broadleaf weeds. Thought to be relatively non-hazardous.

oscar See **oskar**.

oscillator Something that changes regularly or cyclically. Examples: oscillator neurons, which generate regular breathing or locomotory rhythms; slime moulds, which secrete cyclic AMP in regular pulses.

Oscillatoria princeps Large cyanobacterium that exhibits gliding movements, possibly involving the activity of helically arranged cytoplasmic fibrils of 6–9 nm diameter.

oscillin Soluble protein (oligomeric: 33 kDa subunits) from mammalian sperm that is involved with the oscillations in calcium concentration that occur in the egg following fertilization. Has sequence similarity with prokaryote hexose phosphate isomerases

OSCP See **oligomycin sensitivity conferral protein**.

oseltamivir An antiviral drug (Tamiflu$^{(R)}$) used to treat certain types of influenza. Inhibits viral neuraminidase.

oskar *Os* An **egg-polarity gene** in *Drosophila*, concentrated at the posterior pole of the egg, and required for subsequent posterior structures. A **maternal-effect gene**.

osmiophilic Having an affinity for **osmium tetroxide**.

osmium tetroxide *OsO4* Used as a post-fixative/stain in electron microscopy. Membranes in particular are osmiophilic, i.e. bind osmium tetroxide.

osmole The amount of a solute, dissolved in water, that produces a solution with the same osmotic pressure as would be produced by dissolving one mole of an ideal non-ionized solute. Sea water is approximately 1000 milliosmolar, mammalian isotonic saline is about 290 milliosmolar.

osmoreceptors Cells specialized to react to osmotic changes in their environment, and in mammals involved in the regulation of secretion of antidiuretic hormone by the neurohypophysis.

osmoregulation Processes by which a cell regulates its internal osmotic pressure. These may include water transport, ion accumulation or loss, synthesis of osmotically active substances such as glycerol in the alga *Dunaliella*, activation of membrane ATPases, etc.

osmosis The movement of solvent through a membrane impermeable to solute, in order to balance the chemical potential due to the concentration differences on each side of the membrane. Frequently misused in the popular press.

osmotic pressure See **osmosis**. The pressure required to prevent osmotic flow across a semipermeable membrane separating two solutions of different solute concentration. Equal to the pressure that can be set up by osmotic flow in this system.

osmotic shock Passage of solvent from a hypotonic solution into a membrane-bound structure due to osmosis, causing rupture of the membrane. A method of lysing cells or organelles.

OSR1 *Oxidative stress-responsive kinase-1* See **SPAK** and **WNK**.

Osteichthyes The bony fishes; characterized by a bony skeleton and a swim bladder. The largest class of fishes with representatives in marine, estuarine and freshwater habitats in tropical, temperate and polar regions.

osteoarthritis Disease of joints due to mechanical trauma. There is major disturbance in homeostasis of extracellular matrix with cartilage degradation (involving **matrix metalloproteinases**) and loss of normal joint function. Unlike **rheumatoid arthritis** there is not an autoimmune element, and though there is inflammation it is generally considered to be secondary rather than causative.

osteoblast Mesodermal cell that gives rise to bone.

osteocalcin *Bone γ-carboxyglutamic acid protein: BGP* Polypeptide of 60 residues formed from a 76–77 amino acid precursor, and found in the extracellular matrix of bone. Binds **hydroxyapatite**. Has limited homology of its leader sequence with that of other vitamin K-dependent proteins such as prothrombin, Factors IX and X, and Protein C.

osteoclast Large multinucleate cell formed from differentiated **macrophage**, responsible for breakdown of bone.

osteoclast differentiation factor *ODF; RANKL (receptor activator of NF-kappaB ligand)* A member of the tumour-necrosis factor (TNF) superfamily, a key cytokine involved in the differentiation of the immune system and the regulation of immunity as well as in bone metabolism. In particular, RANKL-deficient mice show defects in the early differentiation of T-lymphocytes, suggesting that RANKL is a novel regulator of early thymocyte development. RANKL-RANK signalling is essential for osteoclast development and plays a major role in pathological bone destruction. **Osteoprotegerin** and anti-RANKL antibody act as a specific inhibitor of RANKL and are therapeutically applicable to osteoporosis and rheumatoid arthritis.

osteocyte **Osteoblast** that is embedded in bony tissue and which is relatively inactive.

osteodystrophia fibrosa *Osteofibrosis* A disease of the skeletal system in animals in which excessive amounts of calcium and phosphorus are withdrawn from the bones and replaced by fibrous tissue.

osteogenesis Production of bone.

osteogenesis imperfecta Heterogenous group of human genetic disorders that affect connective tissue in bone, cartilage and tendon. Bones are very brittle and fracture-prone. Type I is a dominantly inherited, generalized connective tissue disorder caused by mutation in either of the collagen Type1A genes (*COL1A1* and *COL1A2*); other types also involve mutations in one or the other of these genes but have phenotypic differences.

osteogenin *Bone morphogenetic protein-3 (BMP-3)* Bone-inducing protein (less than 50 kDa) associated with extracellular matrix. Binds heparin. See **bone morphogenetic protein**.

osteoid Uncalcified bone matrix, the product of osteoblasts. Consists mainly of collagen, but has **osteonectin** present.

osteoma Benign tumour of bone.

osteomalacia Softening of bone caused by vitamin D deficiency: adult equivalent of rickets.

osteomyelitis Inflammation of bone marrow caused by infection.

osteonectin *Basement membrane protein BM-40; SPARC* Calcium-binding protein of bone, containing the **EF-hand** motif. Binds to both collagen and **hydroxyapatite**.

osteopetrosis The formation of abnormally dense bone, as opposed to **osteoporosis**.

osteophytes Small dense areas seen in X-ray pictures of bone around margin of joints with osteoarthritis. Sometimes larger, more porotic, protrusions are also seen.

osteopontin *Bone sialoprotein 1; secreted phosphoprotein 1(SPP-1); nephropontin; uropontin* Bone-specific sialoprotein (40 kDa) that links cells and the hydroxyapatite of mineralized matrix; has **RGD** sequence. Found mostly in calcified bone, produced by osteoblasts to which it binds via αvβ3 integrin; also bind CD44. Synthesis is stimulated by **calcitriol** (Vitamin D3).

osteoporosis Loss of bony tissue; associated with low levels of **estrogen** in older women.

osteoprotegerin *OPG* A soluble decoy receptor of **osteoclast differentiation factor**/RANKL that inhibits both differentiation and function of osteoclasts, and thus acts as an inhibitor of bone destruction. OPG-deficient (OPG$^{-/-}$) mice exhibited severe osteoporosis, and deficiency of OPG in human has been shown to result in juvenile **Paget's disease**.

osteosarcoma Malignant tumour of bone (probably neoplasia of **osteocytes**).

ostium Generally, a mouth-like aperture. In sponges, an inhalant opening on the surface; in arthropods, an aperture in the wall of the heart by which blood enters the heart from the pericardial cavity; in mammals, the internal aperture of a Fallopian tube.

OTC drugs 'Over the counter' – drugs for which a doctor's prescription is not required.

ouabain *Strophanthin G* A plant alkaloid from *Strophantus gratus* that specifically binds to and inhibits the **sodium–potassium ATPase**. Related to **digitalis**.

Ouchterlony assay Immunological test for antigen–antibody reactions in which diffusion of soluble antigen and antibody in a gel leads to precipitation of an antigen–antibody complex, visible usually as a whitish band. The system has the advantage that, because of radial diffusion of the reagents, a very wide range of ratios of antigen to antibody concentration develop; thus it is likely that precipitation will occur somewhere in the gel even when no care is taken with quantitation of the system.

outer dense fibres *ODF* Fibres found in the sperm tail; include Odf1, which utilizes its leucine zipper to associate with Odf2, another major ODF protein; Spag4, which localizes to the interface between ODF and axonemal microtubule doublets; and Spag5. Rat Spag5 has close sequence similarity with human **astrin**, a microtubule-binding spindle protein.

outron Found at the 5′ end of pre-mRNAs that are to be trans-spliced: contains an intron-like sequence, followed by a splice acceptor.

outside-out patch A variant of **patch clamp** technique, in which a disc of plasma membrane covers the tip of the electrode, with the outer face of the plasma membrane facing outward towards the bath.

oval cells Stem cells of the adult liver that can differentiate into both hepatocytes and bile duct epithelial cells. In normal adult liver, oval cells are quiescent; they exist in low numbers and proliferate following severe, prolonged liver trauma. There is some evidence implicating oval cells in the development of hepatocellular carcinoma.

ovalbumin A major protein constituent of egg white. A phosphoprotein of 386 amino acids (44 kDa) with one N-linked oligosaccharide chain. Synthesis is stimulated by estrogen. The gene, of which there is only one in the chicken genome, has eight exons and is of 7.8 kbp; it was one of the first genes to be studied in this sort of detail.

ovalocytosis Hereditary disorder of erythrocytes relatively common in areas where malaria is endemic. Not only are the erythrocytes more rigid, but there is also a mutation in **band III**, the anion transporter.

ovarian follicle In mammals, the group of cells around the primary **oocyte** proliferate and form a surrounding noncellular layer. A space opens up in the follicle cells, and the whole structure is then the ovarian (Graafian) follicle.

ovarian granulosa cells During oogenesis in mammals, the ovarian (Graafian) follicle, in which the developing ovum lies, is lined with follicle cells; the peripheral follicle cells form the stratum granulosum or ovarian granulosa.

overlap index A measure of the extent to which a population of cells in culture forms multilayers. The predicted amount of overlapping is calculated knowing the cell density, the projected area of the nucleus (usually),

and assuming a Poisson distribution. The actual overlap is measured on fixed and stained preparations, and the ratio of actual/predicted is derived. A value of 1 implies a random distribution with no constraint on overlapping; normal fibroblasts may have values as low as 0.05. Although a useful measure it does not unambiguously indicate the reason for the effect, which may be **contact inhibition of locomotion** or differential adhesion of cells between substratum and other cells.

overlapping In cell locomotion, the situation in which the **leading lamella** of one cell moves actively over the dorsal surface of another cell – should be distinguished from **underlapping**.

overlapping genes Different genes whose nucleotide coding sequences overlap to some extent. The common nucleotide sequence is read in two or three different reading frames thus specifying different polypeptides.

oviduct The tubular tract in female animals through which eggs are discharged either to the exterior or, in mammals, to the uterus.

Ovis aries Domestic sheep (e.g. Dolly).

ovocleidin A major protein of the avian eggshell calcified layer.

ovomucoid Egg-white protein produced in tubular gland cells in the epithelium of the chicken oviduct in response to progesterone or estrogen.

ovum *Pl.* ova An egg cell.

owl-eye cells Enlarged cells infected with **Cytomegalovirus** that contain large inclusion bodies surrounded by a halo, hence the name.

oxa1 Translocase involved in the insertion of proteins into the inner mitochondrial membrane. Similar to **Albino3** and **YidC**.

oxacladiellanes *1,4-oxacladiellanes* Class of compounds that includes **sarcodictyin A** and **eleutherobin** that stabilize microtubules, and the valvidones and eleuthrosides that are anti-inflammatory.

oxalic acid Occurs in plants, and is toxic to higher animals by virtue of its calcium-binding properties; it causes the precipitation of calcium oxalate in the kidneys, prevents calcium uptake in the gut, and is not metabolized.

oxaloacetate Metabolic intermediate. Couples with acetyl CoA to form citrate, i.e. the entry point of the **tricarboxylic acid cycle**. Formed from aspartic acid by transamination.

oxidation Occurs when a compound donates electrons to an oxidizing agent. Also, combination with oxygen or removal of hydrogen in reactions where there is no overt passage of electrons from one species to another.

oxidation-reduction potential See **redox potential**.

oxidative metabolism Respiration in the biochemical sense.

oxidative phosphorylation The phosphorylation of ATP coupled to the **respiratory chain**.

oxidative stress A condition of increased oxidant production in animal cells; reactive oxygen species including superoxide, hydroxyl radicals and singlet oxygen are potently bacteriostatic but also potentially very damaging.

oxidoreductase Any oxidase that uses molecular oxygen as the electron acceptor.

oxybutinin Antimuscarinic drug used to treat urinary incontinence by acting on musculature of the bladder wall.

oxycodone Opioid analgesic derived from **thebaine**.

oxygen electrode A sensitive method to detect oxygen consumption; involves a PTFE (Teflon) membrane that is permeable to oxygen; underneath this membrane is a compartment containing a saturated KCl solution and two electrodes; a platinum cathode and a silver anode. A fixed polarizing voltage is applied between the electrodes and the resulting current ($\sim$1 μA) is proportional to oxygen concentration.

oxygen radical Any oxygen species that carries an unpaired electron (except free oxygen). Examples are ·OH, the hydroxyl radical and O_2^-, the superoxide anion. These radicals are very powerful oxidizing agents and cause structural damage to proteins and nucleic acids. They mediate the damaging effects of ionizing radiation.

oxygen-dependent killing One of the most important bactericidal mechanisms of mammalian phagocytes involves the production of various toxic oxygen species (hydrogen peroxide, superoxide, singlet oxygen, hydroxyl radicals) through the **metabolic burst**. Although anaerobic killing is possible, the oxygen-dependent mechanism is crucial for normal resistance to infection, and a defect in this system is usually fatal within the first decade of life (**chronic granulomatous disease**). See **myeloperoxidase**, **chemiluminescence**.

oxygenases *Subclass EC 1.13* Enzymes catalysing the incorporation of the oxygen of molecular oxygen into organic substrates; they differ from those in EC 1.14 in that a second hydrogen donor is not required. Sub-subclasses are EC 1.13.11, when two atoms of oxygen are incorporated (dioxygenases); EC 1.13.12, when only one oxygen atom is used (mono-oxygenases); and EC 1.13.99, for other cases. Both types are used by bacteria in degradation of aromatic compounds. Dioxygenases all contain iron, e.g. tryp-2,3 dioxygenase. Examples of mono-oxygenases are the enzymes that hydroxylate proline and lysine of collagen, using α-ketoglutarate.

oxylipin Secreted, hormone-like lipogenic molecules derived from linolenic acid. Examples include signalling compounds such as **jasmonates**, antimicrobial and antifungal compounds such as leaf aldehydes or divinyl ethers, and a plant-specific blend of volatiles including leaf alcohols.

oxyntic cell *Parietal cell* Cell of the gastric epithelium that secretes hydrochloric acid.

oxyntomodulin A 37 amino acid peptide that contains the 29 amino acid sequence of **glucagon** followed by an 8 amino acid carboxy-terminal extension. Produced by cleavage of proglucagon by prohormone convertases. Oxyntomodulin inhibits meal-stimulated gastric acid secretion in rodents, and has been shown to inhibit food intake following icv administration in rats. No specific receptor has been identified, although it does act on glucagon and **GLP-1** receptors.

oxyphil cells See **Askenazy cells**.

oxysome Multimolecular array that acts as a unit in **oxidative phosphorylation**.

oxytetracycline A broad-spectrum **tetracycline**-class antibiotic from *Streptomyces rimosus*, active against a wide range of Gram-negative and Gram-positive organisms.

oxytocin A peptide hormone (1007 Da) from hypothalamus: transported to the posterior lobe of the pituitary (see **neurophysin**). Induces smooth muscle contraction in uterus and mammary glands. Related to **vasopressin**.

Oxyuranus scutelatus scutelatus The Australian taipan snake. See **taicatoxin**.

P

p Prefix, usually of a number, indicating a protein (sometimes a phosphoprotein), e.g. **p53**, or a region of a chromosome (see **p-region**. The entries are organized in strict numerical order, irrespective of whether the 'p' is upper or lower case.

p0071 *Plakophilin-4; PKP4* A member of the armadillo protein family, is most closely related to p120(ctn) (**catenin**) and the **plakophilins**. Depending upon the cell type, may be localized in adherens junctions and desmosomes. See **PAPIN, ERBIN**. The C-terminal fragment of **presenilin** 1 directly binds to p0071.

P1, P2, etc. Often used as an abbreviation for postnatal day 1, day 2. etc.

p14 *p14(ARF)* One of two functionally and structurally different proteins (p16(INK4a) and p14(ARF)) that are encoded by the gene *INK4a/ARF* located at 9p21. Both are cyclin-dependent kinase (cdk) inhibitors and important cell cycle regulators. Notice that ARF here means alternative reading frame, not **ADP-ribosylation factor**.

p14 **gene** One of a family of eggshell proteins found in *Schistosoma mansoni*. The *p14* gene is expressed only in vitelline cells of mature female worms in response to a male stimulus.

p14arf Human equivalent of **p19arf**.

p15 1. Binding partner of the export receptor **TAP**. **2.** One of three tumour suppressors p15 (INK4b), encoded by the *INK4/ARF* locus (the others are ARF and p16 (INK4a)).

p16 *INK4a* See **p15**.

p17 HIV-1 p17 is a viral matrix (structural) protein that also acts as a cytokine that promotes proliferation, proinflammatory cytokine release and HIV-1 replication in preactivated, but not on resting, human T-cells.

p18 1. Ink4c. See **Ink4**. **2.** A haplo-insufficient tumour suppressor and a key factor for ATM/ATR-mediated **p53** activation. **3.** A surface protein produced by the fish pathogen *Flavobacterium psychrophilum*. **4.** N-terminal-truncated version of **Bax** (p18 Bax.). **5.** An antimicrobial peptide, P18 (KWKLFKKIPKFLHLAKKF-NH2) designed from a **cecropin** A-**magainin** 2 hybrid.

p19 1. INK4D. Expression of the gene encoding p19 (INK4D) is induced by the hormonal form of vitamin D and by retinoids. **2.** A line of murine embryonal carcinoma stem cells (P19 cells). **3.** One subunit of the cytokine **IL-23**, the other subunit being the p40 subunit of IL-12. **4.** See **p19arf**.

p19arf A tumour suppressor that blocks the ubiquitin-ligase activity of mdm2 from targeting and inactivating **p53**. Absence of p19 therefore leads to decreased p53 activity and facilitates the formation of tumours. The CDK inhibitor Cdkn2a (p16INK4a) also controls the p53 pathway by generating an alternative transcript that encodes Cdkn2a (p14ARF) in humans or Cdkn2a (p19ARF) in mice

p20 1. p20 CCAAT enhancer-binding protein beta (C/EBPbeta), a truncated C/EBPbeta isoform. **2.** *Bacillus* *thuringiensis* helper (chaperone) protein P20 is important in production of vegetative insecticidal proteins (VIPs). **3.** *Xenopus* p20 protein influences the stability of MeCP2 (**methyl-CpG-binding protein 2**), mutations in which cause **Rett syndrome**. **4.** A cardiac heat-shock protein, p20. **5.** *E. coli* p20 is a thioredoxin-dependent thiol peroxidase. **6.** Cdk5/p20.

p21 *Waf1; Cip1; CDKN1A* **Cyclin-dependent kinase** (CDK) inhibitor (21 kDa) that inhibits multiple cdks; transcriptionally regulated by **p53**. See **p21-activated kinase** (PAK).

p21-activated kinases *PAKs; PAK-1, 2, 3; p21-associated kinases* Serine/threonine protein kinases of the **STE20** subfamily. Form an activated complex with GTP-bound ras-like proteins (P21, cdc2 and Rac1). Activated, autophosphorylated PAK acts on a variety of targets, shows highly specific binding to the SH3 domains of phospholipase Cγ and of adapter protein **Nck** and regulates various morphological and cytoskeletal changes. **Caspase**-mediated cleavage of 62-kDa PAK2 generates a constitutively active catalytic fragment (34 kDa) and induces apoptosis in **Jurkat** cells. Its activity is apparently essential for the formation of apoptotic bodies.

p22 phox *CYBA* See **p91 phox**.

p23 1. p23 and p25 are members of a family of putative cargo receptors important for vesicular trafficking between Golgi complex and ER. **2.** Co-chaperone (p23) that forms a complex with the chaperone **Hsp90**. **3. Pancreatitis-associated protein I**.

P24 A family of small and abundant transmembrane proteins of the secretory pathway.

p25 1. A transmembrane protein (p25) that has a role in localization of protein tyrosine phosphatase TC48 to the ER. p23 and p25 are members of a family of putative cargo receptors involved in vesicular trafficking between Golgi complex and ER. **2.** Potent proteolytic fragment of **p35** (cdk5). **3.** P25 and P28 proteins are essential for *Plasmodium* parasites to infect mosquitoes. **4.** P25 is one of three polypeptide components of the fibroin synthesized in the larval silk gland of silkworm.

p26 1. An abundantly expressed small heat-shock protein from *Artemia* (brine shrimp).

p27kip1 A member of the Cip1/Kip1 family of cyclin-dependent kinase inhibitors.

p28 1. See **p25** definition 3. **2.** One of the subunits of **IL-27**. **3.** The poxviral **RING** protein p28 is a virulence factor whose molecular function is unknown. **4.** Subunit (p28) of eukaryotic initiation factor 3 (eIF3k). **5.** Replicase proteins p28 and p65 of mouse hepatitis virus.

p29 1. See **BAG-1**. **2.** Papain-like protease, p29 from the N-terminal portions of the *Cryphonectria* hypovirus 1 (CHV1)-EP713-encoded open reading frame that shares similarities with the potyvirus-encoded suppressor of RNA-silencing HC-Pro and can suppress RNA-silencing in the natural host, the chestnut blight fungus *Cryphonectria parasitica*. **3.** Subline (P29) of the mouse Lewis lung cancer line. **4.** Type IV collagen-binding protein (p29) of ML-SN2 (murine *fyn* cDNA-transfected clone). **5.** Adhesin from *Mycoplasma fermentans*. **6.** Stable and active **calpain** cleavage product of the cdk5 neuronal activator p39 (*cf.* **p25**). **7.** Antigen (P29) localized in the dense granules of *Toxoplasma gondii*. **8.** Estrogen-receptor-related protein, p29. **9. Synaptogyrin** (p29) is a synaptic vesicle protein that is uniformly distributed in the nervous system.

p30 **1.** HTLV-I p30 interferes with **toll-like receptor** (TLR)-4 signalling. **2.** A major surface protein, P30, from *Toxoplasma gondii*. **3.** Terminal organelle protein (P30) of *Mycoplasma pneumoniae* that has a role in gliding motility that is distinct from its requirement for adherence. **4.** Serine–arginine-rich protein p30 directs alternative splicing of glucocorticoid receptor pre-mRNA to glucocorticoid receptor-beta in neutrophils.

P32 **1.** Beta-emitting isotope of phosphorus, extensively used for labelling biomolecules (^{32}P). **2.** Protein from *Spiroplasma citri* (32 kDa deduced from the electrophoretic mobility). The P32-encoding gene (714 bp) is carried by a large plasmid of 35.3 kbp present in transmissible strains and missing in non-transmissible strains. **3.** A cofactor of splicing factor ASF/SF-2. **4.** Antigenic protein of African swine fever virus. **5.** A mitochondrial matrix protein that binds the capsid of Rubella virus. **6.** Almost any protein of about 32 kDa that does not have another name.

p33 **1.** Tomato bushy stunt virus replication protein p33, also found in other tombus viruses. **2.** A tumour suppressor p33 (ING1b). **3.** Complement and kininogen-binding protein gC1qR/p33 (gC1qR), distinct from **calreticulin**.

p34 kinase *cdc2* A cyclin-dependent kinase. See **cyclin**.

p35 **1.** A neuronal activator of **cyclin-dependent kinase** 5 (Cdk5). **2. IL-12** p35, a subunit of IL-12 p70. **3.** *Baculovirus* caspase inhibitor, p35 protein. **4.** P35 surface antigen of *Toxoplasma gondii*.

p36 **1.** See **endonexin**. **2.** A member of a family of secreted proteins distributed throughout the genus *Mycobacterium*. The central domain of these proteins contains several amino acid PGLTS repeats, which differ considerably between species. P36, also called exported repetitive protein (Erp) in *M. tuberculosis*, has been shown to be associated with virulence. **3.** One of the immunodominant sperm antigens identified by antibodies eluted from the spermatozoa of infertile men.

p37 **1.** A major structural protein of African swine fever virus, p37 protein. **2.** A membrane lipoprotein of *Mycoplasma hyorhinis*. **3.** The p37 AUF1 isoform (similarly the p40 isoform) has AU-rich element (**ARE**)-mRNA-destabilizing activity. **4.** One of two RNA-binding proteins (p34 and p37) from *Trypanosoma brucei*. **5.** A 37-kDa platelet-agglutinating protein, p37, probably prethrombin-2. **6.** *Vaccinia* virus major envelope protein P37. **7.** The 37-kDa protein (P37) of *Borrelia burgdorferi* elicits an early IgM response in **Lyme disease** patients.

p38 *p38MAPK* Serine/threonine protein kinase activated by MAP kinase kinase (MKK6b) that acts in the signalling cascade downstream of various inflammatory cytokines such as IL-1 and TNFa. Homologous to the yeast HOG protein. See **MEF2**.

p39 **1.** Neuronal activator of cdk5, like **p35**. See **p29**. **2.** Bacterioferritin (BFR) or P39 proteins of *Brucella* spp. that are T dominant antigens.

p40 **1.** Subunit of both **IL-12** and **IL-23**. **2.** Long interspersed nucleotide element-1 (LINE-1) p40 protein expressed in childhood malignant germ cell tumours. **3.** Transport factor p40 is an associated protein of the PtdIns 5-P/PtdIns 3,5-P2-producing kinase PIK **Fyve**.

p41 **1.** The MHC II-associated chaperone molecule of **invariant chain** (inhibitory p41 Ii) suggested to regulate stability and activity of cathepsin-L in antigen-presenting cells. **2.** A putative regulatory component of the mammalian **Arp2/3 complex**, p41-Arc. **3.** Human herpesvirus 6 (HHV-6) early protein, p41. **4.** Polypeptide p41 of a Norwalk-like virus is a nucleic acid-independent nucleoside triphosphatase.

p42 One of the **MAP kinases**, p42/MAP kinase is erk2, p44MAP kinase is erk1. Often collectively referred to as p42/p44MAP kinase or p44/p42 MAPK.

p43 **1.** Former name for aminoacyl-tRNA synthetase-interacting multifunctional protein (AIMP1), an auxiliary factor associated with a macromolecular tRNA synthetase complex; also has cytokine activity and acts on endothelial and immune cells to control angiogenesis and inflammation. **2.** Mitochondrial tri-iodothyronine receptor. **3.** Subunit (p43) of telomerase from the ciliate *Euplotes aediculatus*. **4.** Human placental isoferritin is composed of a 43 kDa subunit (p43) and ferritin light chains.

p44 *p44/42 MAPK; ERK 1/2* **1.** A **MAP kinase**, also called erk1. **2.** The 44-kDa major outer membrane proteins (P44s) of *Anaplasma phagocytophilum*, the causative agent of human granulocytic anaplasmosis. **3.** Splice variant of **arrestin**, p44.

p45 **1.** Nuclear factor E2 p45-related factor 2 (Nrf2). **2. F-box protein** p45(SKP2) the substrate-specific receptor of ubiquitin-protein ligase involved in the degradation of p27(Kip1).

p46 **1.** Isoform of **Shc**, p46Shc. **2.** Isoform of cdk11. **3.** A 46-kDa glucose 6-phosphate translocase (P46), one of the components of the glucose-6-phosphatase enzyme complex of the ER. **4.** Isoform of **JNK/SAPK**, p46JNK. **5.** One of two subunits of mouse DNA primase; p46 is the catalytic subunit capable of RNA primer synthesis, the role of p54 is unclear, although it has a nuclear localization signal.

p47 **1.** Cytosolic component of the NAD(P)H oxidase in phagocytes, p47(phox). **2.** p47 GTPases (immunity-related GTPases (IRG) family) are essential, interferon-inducible resistance factors in mice that are active against a broad spectrum of intracellular pathogens. **3.** Constitutively expressed heat-shock protein, p47.

p48 **1.** One of two splice variants of visual arrestin (p44 and p48). **2.** Nucleotide excision repair protein p48 encoded by DDB2 gene. **3.** A DNA-binding subunit of the transcription factor PTF1 (pancreas-specific transcription factor 1a). **4.** Surface lipoprotein P48 of *Mycoplasma bovis*. **5.** One subunit (p48) of **chromatin assembly factor-1**. **6.** Norwalk virus non-structural protein p48 that may disrupt intracellular protein trafficking in infected cells. **7.** The e-subunit (p48) of mammalian initiation factor 3 (eIF3e). **8.** The most conserved subunit of mammalian DNA polymerase alpha-primase.

p49 1. A 49-kDa protein (p49/STRAP) that specifically interacts with an acidic amino acid motif (Q7IGSEDG) in the N-terminus of **GLUT4**. **2.** A member of the **p35** family, p49, that inhibits mammalian and insect **caspases**. p49 will block apoptosis triggered by treatment with **Fas ligand** (FasL), **TRAIL**, or ultraviolet radiation but not apoptosis induced by **cisplatin**. **3.** An HLA-G1-specific inhibitory receptor (p49) present in NK-cells from placenta but undetectable in peripheral blood NK-cells.

p50 See **dynamitin**.

p51 1. A **p53** homologue, p51/p63/p73L/p40/KET. **2.** Subunit, p51, of **HIV-1** reverse transcriptase. The other subunit is p66. **3.** Major antigenic 51-kDa protein in *Neorickettsia risticii*, P51.

p52 1. **IKKalpha** activates a non-canonical **NFκB** pathway in which p100 (NF-kappaB2) is processed to p52 which can enter the nucleus and induce genes important to adaptive immunity. **2.** Isoform of **Shc**.

p53 A tumour suppressor, 393-residue phosphoprotein in humans. Mutations in p53 are found in most tumours and if only one p53 gene is functional (Li–Fraumeni syndrome) there is a predisposition to tumours. p53 protein binds DNA and stimulates **p21** production: p21 interacts with cdk2 and the complex inhibits progression through the cell cycle. See **p53 family** and Table O1.

p53 family Family of transcription factors consisting of **p53**, **p63** and p73, each of which has multiple isoforms due to transcription at two separate promoters and alternative splicing.

p54 See **p56** definition 1.

p55 1. Membrane-associated guanylate kinase (**MAGUK**) protein, erythrocyte protein p55; expressed in **stereocilia** of outer hair cells in the inner ear. Has single **PDZ domain** followed by SH3, HOOK and guanylate kinase (GuK or GK) domains. **2.** TNF receptor alpha, p55. **3.** *Drosophila* orthologue (chromatin assembly factor, p55/dCAF-1) of **retinoblastoma** protein RbAp46/RbAp48.

p56 1. Viral stress-inducible murine protein, P56, that inhibits initiation of translation by binding to the 'e' subunit of eukaryotic initiation factor 3 (eIF3). P54 has similar effects. **2. Src family kinase**, p56(Lck). **3. Borna disease** virus surface glycoprotein (p56), implicated in viral entry involving receptor-mediated endocytosis. **4.** The cortical granules of the eggs of mice, rats, hamsters, cows and pigs contain a pair of proteins designated p62/p56. **5.** p56(**dok-2**) acts as a multiple docking protein downstream of receptor or non-receptor tyrosine kinases.

p57 1. See **neuromodulin**. **2.** An oxypregnane steroidal glycoside, P57AS3 (P57), the only reported active constituent from *Hoodia gordonii*. **3.** An actin-binding protein p57 is a mammalian **coronin**-like protein. **4.** See **p57/kip2**. **5.** See **p57lck**.

p57/kip2 Cyclin-dependent kinase inhibitor of the p21CIP1, p27KIP1 family. Mice deficient in this inhibitor have major developmental defects similar to those seen in **Beckwith–Wiedemann syndrome** as a result of altered cell proliferation and differentiation. Inhibits G1/S phase **cdks**. Encoded by a maternally imprinted gene in both human and mouse.

p57lck *lck* Lymphoid isoform of src family tyrosine kinase. Expressed predominantly in thymocytes and peripheral

T-cells. Associates with cytoplasmic domains of CD4 and CD8, and with the β-chain of the IL2 receptor.

p58 1. PITSLRE/CDK11(p58) cyclin-dependent protein kinase. **2.** p58/ERGIC-53 is a calcium-dependent animal lectin that acts as a cargo receptor, binding to a set of glycoproteins in the endoplasmic reticulum (ER) and transporting them to the Golgi complex. **3.** P58 (IPK) is a cellular inhibitor of the mammalian double-stranded RNA-activated protein kinase (PKR). **4.** Insulin receptor substrate p53/p58 (IRSp53) is involved in cytoskeletal dynamics. **5.** The p58 subunit of human DNA primase is important for primer initiation (see **DNA polymerase alpha-primase complex**).

p59 1. Src family kinase, p59(fyn). **2.** p59, **oligoadenylate synthetase**-like gene (OASL) product. **3.** p59(scr), protein predominantly expressed in the testis and is developmentally regulated during spermatogenesis. **4.** p59 is the human homologue of **GRASP**55. **5.** Rabbit p59-HBI (Heat-shock protein-Binding Immunophilin) or rFKBP59 (FK506-Binding Protein), a component of the steroid receptor complex.

p60 1. *Listeria monocytogenes* protein p60 that affects haemolytic activity and uptake of bacteria by macrophages. **2. Katanin** p60, a microtubule-severing protein. **3.** Type 1 TNF receptor, p60. The type 2 receptor is p80. **4. Chromatin assembly factor** 1 p60 subunit (CAF-1 p60). **5.** Tyrosine kinase p60(c-**Src**). **6. Caveolin** isoform, cav-p60.

p61 1. p61/63 **myc** nuclear proteins. **2.** Immunodominant antigen, P61, from *Nocardia brasiliensis*.

p62 1. Subunit of **dynactin**. **2.** Signalling adaptor (p62) that controls pathways that modulate cell differentiation. **3.** Polyubiquitin-binding protein p62/SQSTM1, the **sequestosome 1** (SQSTM1) gene product. **4.** Subunit of the protein kinase C zeta (PKCζ)-p62-K$_{vbeta}$ (beta-subunit of delayed rectifier K$^+$ channel) complex, a K$_v$ channel-modulating complex.

p63 1. A transcription factor of the **p53 family**; complex cross-talk between **Notch** and p63 is involved in the balance between keratinocyte self-renewal and differentiation. **2.** A **surfactant protein A**-binding protein on type II pneumocytes (CKAP4/ERGIC-63/CLIMP-63).

p64 1. A chloride channel, p64, of intracellular membranes; present in regulated secretory vesicles. See **parchorin**. **2. Interleukin** 2-receptor gamma-chain, p64.

p65 1. Protein from human T-cells that has sequences in common with **plastin**. Major target for IL-2-stimulated phosphorylation. On basis of sequence has two calcium-binding sites, a calmodulin-binding site and two actin-binding sites. **2.** RelA/p65, a subunit of **NFκB**.

p68 One of the **DEAD-box helicases**.

p70 1. See **annexin**. **2.** p70 S6 kinase. **3.** The heterodimeric form of **IL-12**, p70.

p73 One of the **p53 family** of transcription factors. Overexpression of p73 protein and aberrant expression of its particular isoforms, with very low frequency of P73 hypermethylation or mutations, were found in malignant myeloproliferations, including acute myeloblastic leukaemia. In contrast, hypermethylation and subsequent inactivation of the P73 gene are the most common findings in malignant lymphoproliferative disorders, especially acute lymphoblastic leukaemia (ALL) and non-Hodgkin's lymphomas. See **E1A** oncogene.

p80 1. Type I **interleukin-1** receptor (IL-1RI). **2. IL-12** p80. 3. p80 **coilin** is a nuclear autoantigen that strongly accumulates in **Cajal bodies**. 4. Type 2 TNF receptor, p80.

p84 1. A regulatory subunit for **PI3K**gamma. Present in human, mouse, chicken, frog and *Fugu* genomes. Broadly expressed in cells of the murine immune system. 2. **STAT**1beta.

p85 1. Regulatory subunit of phosphatidylinositol 3-kinase (see **p110**). 2. Pluronic P85; pluronic block copolymers are potent sensitizers of multidrug resistant (MDR) cancer cells.

p87 *p87(PIKAP); PI3Kgamma adapter protein* A regulatory subunit of **PI3K**gamma, functionally homologous to **p101** in many ways: binds to both p110gamma and Gβγ and mediates activation of p110gamma downstream of G-protein-coupled receptors. Highly expressed in heart.

p91 phox *CYBB* Cytochrome *b* (-245) is a heterodimer of the p91-phox beta polypeptide (CYBB) (phox for phagocyte oxidase) and a smaller p22-phox alpha polypeptide (CYBA); mutation in p91phox is responsible for the commonest form of **chronic granulomatous disease**. It is an essential component of phagocytic NADPH-oxidase, a membrane-bound enzyme complex that generates large quantities of microbicidal superoxide and other oxidants upon activation (the respiratory or **metabolic burst**).

p100 *NF-kappaB2* See **NFκB**.

p101 Regulatory subunit of PI3Kgamma that binds to Gβγ and recruits the catalytic p110gamma subunit to the plasma membrane.

p107 Protein (107 kDa) with many similarities to the **retinoblastoma** gene product. Binds to **E2F** and is found in the **cyclin**/E2F complex together with p33cdk2.

p110 Catalytic subunit of phosphatidyl inositol 3-kinase (**PI-3-kinase**) of which various isoforms (α, β and δ) are coupled to receptor tyrosine kinases. The activity of the p110 is regulated by various subunits of which there are at least seven (p85α, p85β, p55γ and their splicing variants). Upon growth factor stimulation, p110 is recruited to the membrane and activated via the interaction of SH2 domains on the regulatory subunit and phosphotyrosine motifs on the stimulated RTKs. PI3Kgamma isoform is activated by receptor-stimulated G-proteins and recruited to the membrane by **p101**.

p120 *p120ctn; p120-catenin* **Catenin** that regulates adherens junction stability in cultured cells. There is evidence that p120 affects NFκB activation and immune homeostasis, in part through regulation of **Rho** GTPases.

p125 A protein that interacts with Sec23p, is only expressed in mammals and exhibits sequence homology with phosphatidic acid-preferring phospholipase A(1). Appears to be essential for localization of **COPII**-coated vesicles to ER exit sites.

p130cas *BCAR1* Substrate for **src family** kinases. An adaptor protein that, once phosphorylated, can act as docking protein for proteins with **SH2** domains; a key mediator of focal adhesion turnover and cell migration. Related proteins are Sin/Efs and Nedd9 (a melanoma metastasis gene). Founder member of the **Cas-family proteins** that serve as docking proteins in integrin-mediated signal transduction.

p150 1. P150 replicase protein of *Rubella* virus. 2. Subunit (p150) of **dynactin** (dynactin 1) that is only present in the dynactin members of the CAP-Gly family of proteins. Also referred to as p150(Glued). 3. The hVPS34/p150 phosphatidylinositol (PtdIns) 3-kinase complex serves to regulate late endosomal phosphatidylinositol signalling that is important for protein sorting and intraluminal vesicle sequestration. 4. Leucocyte beta2 integrin, CD11c/CD18 (p150,95/CR4). 5. p150(Sal2), a vertebrate homologue of the *Drosophila* homeotic transcription factor Spalt.

p185 The oncogene product **HER-2**, a 185-kDa transmembrane glycoprotein of the epidermal growth factor (EGF) receptor family, also called p185/neu or c-erbB-2.

p300 The transcriptional coactivator p300 is a **histone acetyltransferase** (HAT) whose function is critical for regulating gene expression in mammalian cells.

P388.D1 cells Clonal derivative of P388 cells, a mouse macrophage line that produces a large amount of interleukin-1.

p400 protein 1. Inositol 1,4,5-triphosphate receptor (InsP3-R), originally described as a glycoprotein (250 kDa) closely associated with the membranes of Purkinje cells. 2. Adenovirus E1A-associated p400 is one of the **SWI2/SNF2** family of chromatin-remodelling proteins and a component of the **p53-p21** (WAF1/CIP1/sid1) pathway, regulating the p21 transcription and senescence induction programme.

P680 Form of chlorophyll that has its absorption maximum at 680 nm. See **photosystem II**.

P700 Form of chlorophyll that has its absorption maximum at 700 nm. See **photosystem I**.

PA28 *11S regulator* Proteasome activator protein composed of two homologous subunits (alpha and beta) and a separate but related protein termed **Ki antigen** or PA28gamma.

PABA p-*Aminobenzoic acid* Compound present in yeast as an intermediate in the synthesis of **folic acid**. Sometimes called vitamin BX, although is not a true vitamin. Used as an UV-blocking ingredient in sunscreen.

PAC 1. Protein phosphatase PAC-1, a key positive regulator of inflammatory cell signalling and effector functions, mediated through **Jnk** and Erk **MAP kinase** cross-talk. 2. Polyaluminum chloride, used in gel column chromatography. 3. Protein antigen c (PAc) of *Streptococcus mutans*. 4. A ligand-mimetic anti-alphaIIb/beta3 monoclonal antibody, PAC-1. 5. Photoactivated adenylyl cyclase (PAC), a blue-light photoreceptor that mediates photomovement in *Euglena gracilis*. 6. See **PAC vector**.

PAC vector P1-derived artificial chromosome vector, a type of cloning vector that will accept large DNA fragments.

PACAP *Pituitary adenylate cyclase-activating peptide* Member of the **secretin** superfamily of neuropeptides expressed in both the brain and peripheral nervous system, with neurotrophic and neurodevelopmental effects *in vivo*. Promotes the differentiation of **PC12 cells**. Receptor is G-protein coupled.

pachynema See **pachytene**.

pachytene Classical term for the third stage of prophase I of meiosis, during which the homologous chromosomes are closely paired and **crossing-over** takes place.

Pacinian corpuscles In vertebrates, receptors in the skin that are sensitive to pressure. The nerve ending is surrounded by multiple concentric layers of connective tissue that provide mechanical resistance to deformation.

packaging Of a virus, the process by which the genetic material is encapsulated by the coat proteins.

paclitaxel See **taxol**.

pactamycin Antibiotic, isolated from *Streptomyces pactum* that inhibits translation in pro- and eukaryotes by preventing release of initiation factors from the 30S initiation complex. The binding site is distinct from that of tetracycline and that of hygromycin B.

PADGEM *CD62P; P-selectin; GMP-140* One of the **selectin** family, present on megakaryocytes, activated platelets and activated endothelial cells; rapidly upregulated in platelets and endothelial cells and only transiently expressed. Mediates rolling of neutrophils, platelets and some T-cell subsets along luminal surface of vessels. Although knockout is defective in cellular infiltration into inflammatory sites, it is necessary to inhibit both P-selectin and E-selectin to see total blockade.

paedogenesis Sexual reproduction by immature or larval forms. See **neoteny**.

paedomorphosis Synonym for **neoteny**.

PAGE See **polyacrylamide gel electrophoresis**.

Paget's disease Breast carcinoma characterized by large cells with clear cytoplasm. See **Paget's disease of bone**.

Paget's disease of bone *Osteitis deformans; PDB* A chronic disease in which there is progressive enlargement and softening of bones, particularly of the skull and of the lower limbs, due to activated osteoclasts. Genetically heterogeneous; some cases (PDB2) are caused by mutation in the *TNFRSF11A* gene which encodes **RANK**, a protein essential in **osteoclast** formation. PDB type 3 is caused by mutation in the *SQSTM1* gene, the product of which is associated with the RANK pathway.

pagoda cells Ganglion cells, from the central nervous system of a leech, with a spontaneous firing pattern that can look a little like a pagoda on an oscilloscope.

PAI-1 *Plasminogen activator inhibitor-1* PAI-1 and PAI-2 are plasma **serpins** that inhibit **plasminogen**.

pair-rule gene A **segmentation gene**, expressed sequentially between **gap genes** and **segment-polarity genes**. In development of *Drosophila*, a set of about eight genes that are expressed only in alternate segments (odd or even) of the developing embryo. Loss-of-function mutants thus lack alternate segments. Examples: *even-skipped (eve), fushi tarazu (ftz), hairy*.

paired prd Developmentally regulated gene in *Drosophila* that contains the **paired box domain**.

paired box domain Conserved domain of 128 amino acids, found in several developmentally regulated proteins in *Drosophila* (e.g. paired, gooseberry, Pox), mouse and human (e.g. Pax, HuP1, HuP48).

PAL Abbreviation for **1. phenylalanine ammonia lyase**. **2.** Pyothorax-associated lymphoma, a rare B-cell non-Hodgkin lymphoma. **3.** Peptidoglycan-associated lipoprotein, a highly conserved structural outer membrane protein among Gram-negative bacteria. See **tol-pal proteins**.

4. Physical activity level, an important parameter in nutritional studies. **5.** See **PAL motif**.

PAL motif A motif in the C-terminal region of the aspartyl peptidases **gamma-secretase** and **signal peptide peptidase**. Motif contributes to the active site conformation of the enzymes.

Palade pathway The routing of protein(s) from the site of their synthesis to the final cellular or secreted position. Several different pathways are known and others suspected. Glycosylation of the proteins may provide specific 'address labels' for the proteins. Probably only of historical interest.

palindromic sequence Nucleic acid sequence that is identical to its complementary strand when each is read in the correct direction (e.g. TGGCCA). Palindromic sequences are often the recognition sites for **restriction enzymes**. Degenerate palindromes with internal mismatching can lead to loops or hairpins being formed (as in tRNA).

palisade parenchyma Tissue found in the upper layers of the leaf **mesophyll**, consisting of regularly shaped, elongated parenchyma cells, orientated perpendicular to the leaf surface, which are active in **photosynthesis**.

palladin An actin-associated proline-rich protein that binds to vasodilator-stimulated phosphoprotein (**VASP**), **alpha-actinin**, **ezrin** and **profilin**. May play a role in recruiting profilin to sites of actin dynamics.

pallidin Protein (25 kDa) encoded by a gene that is mutated in the pallid mouse strain. Has no homology to any other known protein and no recognizable functional motifs. See **BLOC-1**. Pallidin bind to **syntaxin** 13, a member of the syntaxin family of soluble *N*-ethylmaleimide-sensitive factor attachment protein receptors (**SNAREs**).

Pallister–Killian syndrome Rare disorder with multiple congenital abnormalities, seizures and mental retardation. Cause is an extra metacentric chromosome 12p only in skin fibroblasts, so that the body is a tissue-specific mosaic.

palmitic acid n-*Hexadecanoic acid* One of the most widely distributed of fatty acids. The palmitoyl residue is one of the common acyl residues of membrane phospholipids. It is also found as a thioester attached to cysteine residues on some membrane proteins. The proteins so modified are often transmembrane proteins, and the modified residue is on the cytoplasmic surface of the membrane. The specificity of the transferase for the acyl residue is not high and both stearoyl and oleoyl residues can replace the palmitoyl residue. (*cf*. **myristoylation**).

palmitoylation See **palmitic acid**.

palynology The study of the past occurrence and abundance of plant species by an analysis of pollen grains and other spores that have been preserved in peat and sedimentary deposits of known age. The characteristic morphology and toughness of pollen makes the technique possible.

palytoxin *PTX* Linear peptide (2670 Da) from corals of *Palythoa* spp. that binds to Na^+/K^+ ATPase at a site overlapping that of **ouabain** and converts it into a channel. Extremely toxic and said to be the most potent animal-derived toxin. It is a complex molecule with 64 stereocentres and a backbone of 115 contiguous carbon atoms but has been synthesized. There are suggestions that the coral is simply concentrating the toxin made by the dinoflagellate *Ostreopis siamensis*.

Pan1p A yeast actin cytoskeleton-associated protein localized in actin patches. It activates the **Arp2/3 complex**. Has multiple protein–protein interaction domains, including two **EH domains**, a coiled-coil domain, an acidic Arp2/3-activating region and a proline-rich domain.

pancortins Neuron-specific **olfactomedin**-related glycoproteins, components of the extracellular matrix of the brain. They comprise four alternatively spliced variants.

pancreastatin Peptide hormone (49 aa) that inhibits insulin release from the pancreas. Derived from **chromogranin A** by proteolytic processing in several peptide hormone-producing cells, such as pancreatic islet cells and gut endocrine cells.

pancreatic acinar cells Cells of the pancreas that secrete digestive enzymes; the archetypal secretory cell upon which much of the early work on the sequence of events in the secretory process was done.

pancreatic peptide *PP; pancreatic hormone* Peptide synthesized in **islets of Langerhans** that acts as a regulator of pancreatic and gastrointestinal functions. Produced as a larger propeptide, which is enzymatically cleaved to yield the mature active peptide 36 amino acids in length.

pancreatic triglyceride lipase *PNLIP; EC 3.1.1.3* Lipolytic enzyme that hydrolyses ester linkages of triglycerides; cofactor is **colipase**. Similar hepatic and gastric/lingual isozymes exist. Plays a key role in dietary fat absorption by hydrolysing dietary long-chain triacyl glycerol to free fatty acids and monoacylglycerols in the intestinal lumen.

pancreatitis-associated protein I *PAP-I; HIP; p23; Reg2* Protein initially characterized as being overexpressed in acute pancreatitis, now also associated with a number of inflammatory diseases, such as **Crohn's disease**.

pancreozymin See **cholecystokinin**.

pancytopenia Simultaneous decrease in the numbers of all blood cells: can be caused by aplastic anaemia, hypersplenism or tumours of the marrow.

Pandorina Colonial phytomonad in which the cells are held together in a gelatinous matrix. More complex than *Eudorina*, less complex than *Volvox*.

Paneth cells Coarsely granular secretory cells found in the basal regions of crypts in the small intestine and most abundant in the distal small intestine. Secrete alpha-defensin microbicidal peptides as mediators of innate enteric immunity.

panmictic A panmictic population is one in which there is random mating.

pannexins **Innexin** homologues found in vertebrates, pannexins are highly conserved in worms, molluscs, insects and mammals. Both innexins and pannexins are predicted to have four transmembrane regions, two extracellular loops, one intracellular loop and intracellular N and C-termini.

panniculitis Inflammation of the subcutaneous fat in any part of the body. Fat is divided into lobules by connective tissue septae that contain the blood supply supplying the lobule; disorders which disrupt the arterial supply lead to a lobular panniculitis, while venous disorders lead to a septal panniculitis.

panning Method in which cells are added to a dish with a particular surface coat or a layer of other cells and the non-adherent cells are then washed off. Those that remain are expressing particular surface adhesive properties and can be cloned, or, in the case of an expression library, the identity of the adhesion molecule can be determined.

pannus 1. Vascularized granulation tissue rich in fibroblasts, lymphocytes and macrophages, derived from synovial tissue; overgrows the bearing surface of the joint in rheumatoid arthritis and is associated with the breakdown of the articular surface. **2.** Granulation tissue that invades the cornea from the conjunctiva in response to inflammation.

pantetheinase *EC 3.5.1.* An ubiquitous enzyme that will hydrolyse pantotheine to pantothenic acid (vitamin B5) and cysteamine, a potent antioxidant. The **vanin-1** gene product is a GPI-anchored pantetheinase, an ectoenzyme.

pantetheine N-*pantothenylcysteamine* Intermediate in pathway for CoA biosynthesis and a growth factor for *Lactobacillus*. Pantethine is the disulfide dimer of pantetheine and is sold as a health food supplement.

P antigen Antigenic determinant on the surface of human red blood cells to which the Donath–Landsteiner antibody reacts. This antibody binds in the cold (a 'cold IgG'), but elutes from red cells at 37° C, is particularly associated with tertiary syphilis and its binding causes paroxysmal nocturnal haemoglobinuria. (See also **decay accelerating factor**.)

pantonematic flagella **Flagella** without **mastigonemes**; *cf.* **hispid flagella**.

pantophysin Ubiquitously expressed **synaptophysin** homologue (29 kDa) found in cells of non-neuroendocrine origin. May be a marker for small cytoplasmic transport vesicles.

pantothenic acid Vitamin of the B2 group. See Table V1.

PAP technique 1. Colloquial abbreviation for **Papanicolaou's stain**. **2.** Peroxidase–antiperoxidase method for obtaining an enhanced peroxidase reaction to indicate antibody binding to antigen. In the first stage the material, e.g. a section, is reacted with a specific antiserum (say rat) against the antigen. In the next stage a large excess of (say) rabbit anti-rat immunoglobulin is applied so that only one of the binding sites is bound to the first antibody. Then a rat antiperoxidase antiserum is bound to the second antibody's unfilled sites and finally peroxidase is added and binds to the third antiserum before the peroxidase is used to develop a colour reaction.

papain *EC 3.4.22.2* Thiol peptidase from *Carica papaya* (pawpaw). Thermostable and will act in the presence of denaturing agents. Although it will cleave a variety of peptide bonds there is greatest activity one residue towards the C-terminus from a phenylalanine.

Papanicolaou's stain *PAP stain* A complex stain for detecting malignant cells in cervical smears. Contains in separate staining stages: (a) haematoxylin, (b) Orange-G phosphotungstic acid, (c) light green, Bismarck Brown, eosin and phosphotungstic acid.

papaverine Constituent of opium that acts as a smooth muscle relaxant probably by blocking membrane calcium channels and inhibiting phosphodiesterase.

paper chromatography Separation method in which filter paper is used as the support. Not a very sensitive method, but historically important as one of the first methods available for separating natural compounds.

Papilio glaucus Swallowtail butterfly.

papilla *Pl.* papillae **1.** A projection occurring in various animal tissues and organs. **2.** A small, blunt hair on plants.

papilloma *Pl.* papillomata Benign tumour of epithelium. Warts (caused by papilloma virus) are the most familiar example and each is a clone derived from a single infected cell.

papillomaviruses Genus of **Papovaviridae**. See **papilloma**, **Shope papilloma virus**.

PAPIN *Plakophilin-related armadillo-repeat protein-interacting protein* Protein that is diffusely distributed on the plasma membrane of epithelial cells. Has six **PDZ domains** and interacts with **p0071**, a **catenin**-related protein. See **plakophilins**.

Papovaviridae Family of oncogenic **DNA viruses** including papilloma, polyoma and simian vacuolating virus (SV40). Non-enveloped small viruses that mainly infect mammals.

pappalysin-1 *EC 3.4.24.79; PAPP-A, insulin-like growth factor binding protein-4 protease; pregnancy-associated plasma protein A* A metalloendopeptidase that cleaves insulin-like growth factor (IGF) binding protein-4 (IGFBP-4), causing a dramatic reduction in its affinity for IGF I and II and is also responsible for IGF-dependent degradation of IGFBP-2. Pappalysin-2 may be a IGFBP-5 proteinase in many tissues.

papule Small, raised spot on skin (as in the rash of chickenpox).

PAR See **protease-activated receptor**.

paraben *Para-hydroxybenzoic acid* A group of preservatives with a broad spectrum of antimicrobial activity. Propyl paraben and methyl paraben are commonly used preservatives in cosmetics and in foodstuffs.

parabiosis Surgical linkage of two organisms so that their circulatory systems interconnect.

paracellin *Claudin-16* Paracellin-1 is a member of the tight junction **claudin** protein family and mutations in the paracellin-1 gene cause a human hereditary disease, familial hypomagnesaemia with hypercalciuria and nephrocalcinosis (FHHNC) with severe renal Mg^{2+} wasting. Renal tubular dysplasia is an autosomal recessively inherited disorder in Japanese black cattle that is due to deletion mutations in the claudin-16 gene.

paracentric Descriptor for a portion of a chromosome that does not include the **centromere**.

paracentric inversion Chromosomal rearrangement in which a **paracentric** portion of a chromosome (not including the centromere) has been rotated through 180 degrees and reinserted in the same chromosomal location. If no genes have been lost this may not be particularly deleterious.

Paracentrotus lividus Species of sea urchin commonly used in developmental biology.

paracetamol *Acetaminophen; 4′-hydroxyacetanilide* A non-opiate, non-salicylate analgesic and antipyretic that may work by inhibition of **COX-3**.

paracortex Mid-cortical region of lymph node; area that is particularly depleted of T-lymphocytes in thymectomized animals and is referred to as the thymus-dependent area.

paracrine Form of signalling in which the target cell is close to the signal-releasing cell. Neurotransmitters and neurohormones are usually considered to fall into this category.

paradominant Heterozygous individuals carrying a paradominant mutation are generally phenotypically normal and the trait only becomes manifest when there is loss of heterozygosity in some lineages during development, so that there is a mosaic population of cells, some of which are either homozygous or hemizygous for the mutation. A few rare disorders, particularly of skin, are thought to arise through paradominant inheritance.

parainfluenza virus Species of the **Paramyxoviridae**; there are four types: type 1 is also known as Sendai virus or Haemagglutinating virus of Japan (HVJ) and the inactivated form is used to bring about cell fusion. Types 2–4 cause mild respiratory infections in humans.

parakeratosis Condition in which there is retention of nuclei in the **stratum corneum** of the epidermis. This is a normal finding on mucous membranes, and also occurs in **psoriasis**.

paralogous genes Genes that result from duplication of existing genes and then divergence of function; contrast with **orthologous genes**.

paralogue *Paralogous gene* A gene that has a similar DNA configuration to another gene in the same organism owing to the duplication of a common ancestral gene, *cf.* **orthologue**.

Paramecium *Paramoecium* Genus of ciliate protozoans. The 'slipper animalcule' is cigar-shaped, covered in rows of cilia and about 250 μm long. Free-swimming, common in freshwater ponds: feeds on bacteria and other particles. Reproduces asexually by binary fission and sexually by conjugation involving the exchange of micronuclei. See **kappa particle**.

paramural body Membranous structure located between the plasma membrane and cell wall of plant cells. If it contains internal membranes it may be called a **lomasome**; if not, it may be termed a plasmalemmasome.

paramutation An allele-dependent transfer of epigenetic information, which results in the heritable silencing of one allele by another. In maize the *mop1* (mediator of paramutation1) gene is required for paramutation; the gene product is an RNA-dependent RNA polymerase. A similar modification of the mouse *Kit* gene in the progeny of heterozygotes with the null mutant *Kit(tm1Alf)* (a lacZ insertion) has been reported: even the homozygous wild-type offspring maintain, to a variable extent, the white spots characteristic of Kit mutant animals.

paramylon Storage polysaccharide of ***Euglena*** and related algae, present as a discrete granule in the cytoplasm and consisting of β (1-3)-glucan.

paramyosin Protein (200–220 kDa) that forms a core in the thick filaments of invertebrate muscles. The molecule is rather like the rod part of myosin and has a two-chain coiled-coil a-helical structure, 130×2 nm. Paramyosin is present in particularly high concentration in the **catch muscle** of bivalve molluscs, where it forms the almost crystalline core of the thick filaments.

Paramyxoviridae Class V viruses of vertebrates. The genome consists of a single negative strand of RNA as one piece. The helical nucleocapsid has a virus-specific RNA polymerase (transcriptase) associated with it. They are enveloped viruses: main members are **Newcastle Disease virus**, **measles virus** and the **parainfluenza viruses**.

paranemic Topological term for the joint that is made by wrapping one circle around another without cutting either circle. The two circles can always be pulled apart. See **plectonemic**.

paranemin Developmentally regulated protein (280 kDa) associated with **desmin** and **vimentin** filaments. Contains the rod domain characteristic of all cytoplasmic intermediate filament proteins. Coassembles with desmin in muscle cells. **EAP-300** and IF-associated protein (IFAPa-400) are highly homologous to paranemin, and it has significant homology with human **nestin** and frog tanabin.

paranode Region flanking the **nodes of Ranvier** in myelinated fibres where glial cells closely appose and form specialized septate-like junctions with axons. These junctions contain a *Drosophila* **neurexin** IV-related protein, Caspr/**Paranodin** (NCP1).

paranodin *CASPR* One of the **NCP** family of transmembrane neuronal cell adhesion molecules found at synaptic junctions and highly enriched in paranodal regions of myelinated axons. A vertebrate homologue of **neurexin** IV.

paraoxonase *PON1* Human serum protein (45-kDa glycoprotein) located on high density **lipoprotein** (HDL) and that has been implicated in detoxification of organophosphates and possibly preventing the oxidation of LDL. Mice lacking serum paraoxonase are susceptible to organophosphate toxicity and to atherosclerosis.

parapatric speciation Formation of a new species without geographical separation, usually assumed to arise through subdivision of an environmental niche with reproductive isolation of those members of the parent species that occupy the new niche. *cf.* **allopatric** and **sympatric speciation**.

paraprotein *M-protein* Protein found in plasma of patients with monoclonal **gammopathy**: can be composed of whole immunoglobulin molecules (IgG, IgM, IgΛ) or their constituent subunits (heavy or light chains) and derived from a single clone of cells.

parasegment In development of *Drosophila*, the genetic boundaries between developing segments are thought to lie along the middle of each visible segment. To distinguish them from the segments in everyday use, these compartments are called parasegments.

parasitaemia Infection of a host by a parasite or the level of infection by the parasite, depending upon context.

parasympathetic nervous system One of the two divisions of the vertebrate **autonomic nervous sytem**. Parasympathetic nerves emerge cranially as preganglionic fibres from oculomotor, facial, glossopharyngeal and vagus nerves, and from the sacral region of the spinal cord. Most neurons are cholinergic, and responses are mediated by **muscarinic acetylcholine receptors**. The parasympathetic system innervates, for example, the salivary glands, thoracic and abdominal viscera, bladder and genitalia. *cf.* **sympathetic nervous system**.

parathormone *Parathyrin; parathyroid hormone* A peptide hormone of 84 amino acids (9402 Da). Stimulates osteoclasts to increase blood calcium levels, the opposite effect to **calcitonin**.

parathyroid hormone See **parathormone**.

paratope In immune network theory, an **idiotope**; an antigenic site of an antibody that is responsible for that antibody binding to an antigenic determinant (**epitope**). Also used of the site on a ligand molecule to which a cell surface receptor binds.

paratyphoid Enteric fever due to infection by *Salmonella* spp. other than *S. typhi*; usually *Salmonella enterica* serovar *Paratyphi*. The disease is acquired through ingestion of heavily contaminated food and water and is similar to, but milder than, typhoid fever.

paraxial Lying along an axis; commonest use is in reference to paraxial mesoderm, the mesoderm that forms somites as opposed to the axial mesoderm that forms notochord.

parchorin Parchorin, **p64** and the related chloride intracellular channel (CLIC) proteins are thought to be auto-inserting, self-assembling intracellular anion channels involved in a wide variety of fundamental cellular events including regulated secretion, cell division and apoptosis.

Pardachirus marmoratus Red sea flatfish. See **pardaxin**.

pardaxin Polypeptide (33 residues) from toxin gland of *Pardachirus marmoratus* that forms an eight-subunit voltage-dependent pore that will induce neurotransmitter release.

parenchyma Type of unspecialized cell making up the ground tissue of plants. The cells are large and usually highly vacuolated, with thin, unlignified walls. They are often photosynthetic in which case they may be termed **chlorenchyma**.

parenteral Administration of a substance to an animal by any route other than the alimentary canal.

parenthesome Structure shaped rather like a parenthesis '(', found on either side of pores in the septum of a basidiomycete fungus. More logically called septal pore caps.

paresis Slight or incomplete paralysis.

parfocal Microscope objectives mounted in such a way that changing objectives does not cause the specimen to go out of focus.

parkin Gene found to be mutated in an unusual form of **Parkinsonism** (autosomal-recessive juvenile parkinsonism). Located on long arm of chromosome 6. Gene is large (500 kb), very active in the **substantia nigra** and codes for large protein that is an E3 ligase, an integral component of the cytoplasmic ubiquitin/proteasomal protein degradation pathway.

Parkinsonism *Parkinson's disease; paralysis agitans* Disease characterized by tremor and associated with the underproduction of L-DOPA (dihydroxyphenylalanine) by dopaminergic neurons and their death, particularly in the substantia nigra of the brain. Can be treated quite successfully in many cases by administration of L-DOPA. See **parkin**.

paroral membrane In ciliates, the cilia in the region of the mouth may be fused into a paroral membrane.

paroxetine Antidepressant drug, of the **SSRI** class, used to treat major depressive disorder, social anxiety disorder, obsessive-compulsive disorder (OCD), panic disorder (PD), generalized anxiety disorder (GAD) and post-traumatic stress disorder (PTSD).

paroxysmal cold haemoglobinuria *Donath–Landsteiner syndrome* An antibody-induced anaemia caused by a cold-reacting polyclonal immunoglobulin G (IgG) known as the Donath–Landsteiner autoantibody. Antibody binding occurs at temperatures below normal body temperature and complement-dependent lysis follows after warming. Autoantibody formation may be a result of infection with a microorganism-derived antigen that induces antibodies that cross-react with the P antigen in the erythrocyte membrane.

paroxysmal nocturnal haemoglobinuria Disease in which there is haemolysis by complement as a result of deficiency in **decay accelerating factor**.

PARP See **poly(ADP-ribose)polymerase**.

pars fibrosa *pars intermedia* Region of the posterior lobe of the pituitary body in higher vertebrates.

pars nervosa Part of the posterior lobe of the pituitary, developed from the infundibulum.

parthenocarpy Fruit formation without fertilization. Occurs spontaneously in some plants, e.g. banana, and in other plants can be induced by application of **auxin**. Results in seedless fruits.

parthenogenesis Development of an ovum without fusion of its nucleus with a male pronucleus to form a zygote.

parthenolide A sesquiterpene lactone, the primary bioactive compound in feverfew (*Tanacetum parthenium*); an inhibitor of **NFκB** with anti-inflammatory and antitumour activity.

partial agonist Agonist for a receptor population that is unable to produce a maximal response even if all the receptors are occupied.

partition coefficient Equilibrium constant for the partitioning of a molecule between hydrophobic (oil) and hydrophilic (water) phases. A measure of the affinity of the molecule for hydrophobic environments, and thus, for example, a rough guide to the ease with which a molecule will cross the plasma membrane.

parvalbumins Calcium-binding proteins (12 kDa), found in teleost and amphibian muscle, with sequence homology to **calmodulin** but only two **EF-hand** calcium-binding sites. Parvalbumin-beta is also known as **oncomodulin**.

parvins *Actopaxin; affixin; CH-ILKBP* A family of actin-binding proteins (~42 kDa) from the α-actinin superfamily. Alpha-parvin (actopaxin) and beta-parvin (affixin) are located at focal contacts, some cell–cell adhesion junctions, ruffling membranes and the nucleus. Gamma-parvin is specifically expressed in several lymphoid and mono-

cytic cell lines and directly associates with integrin-linked kinase (ILK).

Parvoviridae Class II viruses. The genome of these simple viruses is single-stranded DNA, and they have an icosahedral nucleocapsid. The autonomous parvoviruses have a negative strand DNA and include viruses of vertebrates and arthropods. The defective Adeno-associated viruses cannot replicate in the absence of helper adenoviruses and have both positive and negative-stranded genomes, but packaged in separate virions.

parvulin *Par14/PIN4* A **peptidyl prolyl cis/trans-isomerase** (PPIase), highly conserved in all metazoans and assumed to play a role in cell cycle progression and chromatin remodelling. It is predominantly localized to the nucleus.

PAS 1. See **periodic acid-Schiff reaction**. **2.** *p*-Aminosalicylic acid. **3.** Pre-autophagosomal structure, from which the autophagosome is thought to originate. **4.** See **PAS genes**.

PAS domain *Per-Arnt-Sim domain* A ubiquitous protein module with a common three-dimensional fold involved in a wide range of regulatory and sensory functions. See **period** and **ARNT** (aryl hydrocarbon receptor nuclear translocator).

PAS genes Genes essential for the biogenesis and proliferation of **peroxisomes** in yeast (*S. cerevisiae*). *PAS1* codes for a rather hydrophilic 117-kDa protein with two ATP-binding sites and similarity with some ATPases, *PAS2* codes for a 183-residue polypeptide that seems to be a member of the ubiquitin-conjugating protein family, *PAS3* codes for a 48-kDa integral membrane protein that may be part of the import machinery.

pasin Proteins bound, on the cytoplasmic face, to the sodium–potassium ATPase. Two pasins have been identified, pasin 1 (77 kDa) and pasin 2 (73 kDa). Pasin 2 is identical to **moesin**, a member of the FERM (**protein 4.1**, **ezrin**, **radixin**, moesin) protein family, all members of which have been shown to serve as cytoskeletal adaptor molecules.

passage Term that derives originally from maintenance of, for example, a parasite by serially infecting host animals, passaging the parasite each time. Subsequently also used to describe the subculture of cells in culture, and therefore not equivalent to cell division number.

passage number The number of times a culture of cells or organisms has been subcultured (passaged). With increasing **passage** number, which equates to (but is not identical to) increased numbers of cell cycles, mutations will inevitably accumulate and selection pressures will alter the characteristics of the culture.

passive immunity Immunity acquired by the transfer from another animal of antibody or sensitized lymphocytes. Passive transfer of antibody from mother to offspring is important for immune defence during the perinatal period.

passive transport The movement of a substance, usually across a plasma membrane, by a mechanism that does not require metabolic energy. See **active transport**, **transport protein**, **facilitated diffusion**, **ion channels**.

Pasteur effect Decrease in the rate of carbohydrate breakdown that occurs in yeast and other cells when switched from anaerobic to aerobic conditions. Results from a relatively slow flux of material through the biochemical pathways of respiration compared with those of fermentation.

Pasteurella pestis Old name for **Yersinia pestis**.

Patau syndrome *Trisomy 13* Set of congenital defects in man caused by presence of an extra chromosome 13. Usually fatal within days of birth.

patch clamp A specialized and powerful variant of **voltage clamp** method in which a patch electrode of relatively large tip diameter (5 μm) is pressed tightly against the plasma membrane of a cell, forming an electrically tight 'gigohm' seal. The current flowing through individual **ion channels** can then be measured. Different variants on this technique allow different surfaces of the plasma membrane to be exposed to the bathing medium: the contact just described is a 'cell-attached patch'. If the electrode is pulled away, leaving just a small disc of plasma membrane occluding the tip of the electrode, it is called an 'inside-out patch'. If suction is applied to a cell-attached patch, bursting the plasma membrane under the electrode, a 'whole-cell patch' (similar to an intracellular recording) is formed. If the electrode is withdrawn from the whole-cell patch, the membrane fragments adhering to the electrode reform a seal across the tip, forming an 'outside-out patch'.

patching Passive process in which integral membrane components become clustered following cross-linking by an external or internal polyvalent ligand. See **capping**.

pathogenic Capable of causing disease.

pathognomonic A sign or symptom that is diagnostic of a disease.

pattern formation One of the classic problems in developmental biology is the way in which complex patterns are formed from an apparently uniform field of cells. Various hypotheses have been put forward, and there is now evidence for the existence of gradients of diffusible substances (morphogens) specifying the differentiative pathway that should be followed according to the concentration of the **morphogen** around the cell.

patulin A mycotoxin (a polyketide lactone) produced by certain species of *Penicillium*, *Aspergillus* and *Byssochlamys* growing on fruit, particularly apples, pears and grapes. It is a carcinogen resistant to low pH and tolerant of high temperature. Causes breaks in DNA and inhibits aminoacyl-tRNA synthetase.

PAUP* *Phylogenetic Analysis Using Parsimony* Software package used for phylogenetic tree reconstruction, a process in which the ancestral relationships among a group of organisms are inferred from their DNA sequences.

Pauropoda A class of myriapods similar to centipedes. Mainly found in soil.

pavementing Term used to describe the **margination** of leucocytes on the endothelium near a site of damage.

pawn Mutant of Paramecium that, like the chess-piece, can only move forward and is unable to reverse to escape noxious stimuli. Defect is apparently in the voltage-sensitive calcium channel of the ciliary membrane.

Pax genes Mouse genes that contain a DNA-binding domain similar to one in the pair-rule genes of *Drosophila*. Eight *Pax* genes have been identified, and most of them are expressed in the nervous system during development. A number of mouse mutations have been found to map to *Pax* genes: for example, *undulated*, which causes distortions of the vertebral column and sternum, results from point mutations of *Pax-1* and is expressed in the sclerotome.

paxillin Cytoskeletal protein (68 kDa) that localizes, like **talin**, to focal adhesions, to dense plaques in smooth muscle and to the myotendonous and neuromuscular junctions of skeletal muscle. Binds to **vinculin**.

PAZ domain PIWI/Argonaute/Zwille domains. See **piwi** and **argonaute**.

PBMC *Peripheral blood mononuclear cells* A mixture of **monocytes** and **lymphocytes**; blood leucocytes from which **granulocytes** have been separated and removed.

PBP 1. Platelet basic protein. 2. Penicillin-binding protein 3. Nuclear receptor coactivator PBP (**peroxisome proliferator-activated receptor** (PPAR)-binding protein) functions as a coactivator for PPARs and other nuclear receptors. **4.** Pheromone-binding proteins (PBPs) located in the antennae of male moth species.

pBR322 Plasmid that is one of the most commonly used *E. coli* cloning vectors.

PC12 A rat **phaeochromocytoma** cell line from adrenal medulla. Widely used in the study of stimulus-secretion coupling and, because it differentiates to resemble sympathetic neurons on application of nerve growth factor.

PCA Abbrev. for (1) Principal component analysis; a widely used analytical technique that can be applied to a variety of complex systems, (2) Patient-controlled analgesia, (3) *p*-chloroamphetamine, (4) Primary cutaneous aspergillosis, (5) Prostate cancer antigen-1 (pca-1).

PCAF *p300/CREB-binding protein-associated factor* Histone acetyltransferase (HAT) paralogue which plays a role in the remodelling of chromatin.

PCD 1. Programmed cell death (see **apoptosis**). **2. Primary ciliary dyskinesia. 3.** Premature centromere division. **4.** Mutant mice (pcd mice) in which there is Purkinje cell degeneration, often used as a model for neurodegenerative disorders.

PCDHα *Protocadherin-alpha* One of the synaptic-specifier/adhesion molecules. See **neuroligin**, **neurexin**, **contactin**.

pcDNA Expression vectors used experimentally. pcDNA3.1$^{(+)}$ and pcDNA3.1$^{(-)}$ are 5.4-kb vectors derived from pcDNA3 and designed for high-level stable and transient expression in mammalian hosts. They contain human cytomegalovirus immediate-early (CMV) promoter, multiple cloning sites in the forward (+) and reverse (−) orientations and a neomycin-resistance gene for selection of stable cell lines.

pCEF-4 See **9E3**.

pCMBS *p-Chloromercuriphenylsulfonic acid* An organo-mercurial sulphydryl-reactive compound that inhibits water movement through **aquaporin-1**.

PCMT *Protein-L-isoaspartate (D-aspartate) O-methyltransferase; EC 2. 1.1.77* Enzyme that catalyses the methyl esterification of the free alpha-carboxyl group of abnormal L-isoaspartyl residues, which occur spontaneously in protein and peptide substrates as a consequence of molecular ageing.

pCMV Usually part of the name of an expression vector that contains the cytomegalovirus (CMV) promoter, e.g. pCMV-Tag from Stratagene.

PCNA *Proliferating cell nuclear antigen* Commonly used marker for proliferating cells, a 35-kDa protein that associates as a trimer, and as a trimer interacts with DNA polymerases δ and ε; acts as an auxiliary factor for DNA repair and replication. Transcription of PCNA is modulated by p53.

PCR See **polymerase chain reaction**.

PCR *in situ* hybridization Technique for detection of very rare mRNA or viral transcripts in a tissue. Tissue sections are subjected to **PCR**, usually in a temperature-cycling oven, before detection of the (hugely amplified) transcript.

PCTA-1 *Prostate carcinoma tumor antigen-1* A **galectin** that is highly expressed in prostate cancer and, because it is secreted, may be a useful serum marker. See **STEAP**.

PD-ECGF *Platelet-derived endothelial cell growth factor* Cytokine (471 residues), also known as thymidine phosphorylase, produced by platelets, fibroblasts and smooth muscle cells. Stimulates endothelial proliferation *in vitro* and angiogenesis *in vivo*. Also promotes survival and differentiation of neurons. **Gliostatin** is related.

PD98059 *2'-Amino-3'-methoxyflavone* A flavone derivative that is a selective mitogen-activated protein kinase kinase-1 (MEK-1) inhibitor.

PDA 1. Personal digital assistant, a small handheld computer. 2. Patent ductus arteriosus.

PDE *Phosphodiesterase* Any enzyme (in EC 3.1 class) that catalyses the hydrolysis of one of the two ester linkages in a phosphodiester. PDE-I (EC 3.1.4.1) catalyses removal of 5'-nucleotides from the 3' end of an oligonucleotide. PDE-II (EC 3.1.16.1) catalyses removal of 3'-nucleotides from the 5'-end of a nucleic acid. Often the name is used loosely when cAMP-phosphodiesterase is meant. See **cyclic nucleotide phosphodiesterases**.

PDGF See **platelet-derived growth factor**.

PDK 1. **Pyruvate dehydrogenase kinase**. 2. **Phosphoinositide-dependent kinase**-1.

P domain See **trefoil motif**.

PDX-1 *Pancreas–duodenum homeobox-1* Transcription factor that plays a central role in regulating insulin gene transcription and differentiation of insulin-producing cells.

PDZ domains Domains found in various intracellular signalling proteins associated with the plasma membrane; named for the postsynaptic density, disc-large, ZO-1 proteins in which they were first described. May mediate formation of membrane-bound macromolecular complexes, for example of receptors and channels, by homotypic interaction, also of cell–cell junctions. Usually bind to short linear C-terminal sequences in the protein with which they interact.

PEA-15 *Phosphoprotein enriched in astrocytes, 15 kDa* PEA-15 is an acidic serine-phosphorylated protein highly expressed in the CNS, where it can play a protective role against cytokine-induced apoptosis. PEA-15 is phosphorylated in astrocytes by CaMKII (or a related kinase) and by protein kinase C in response to endothelin. Phosphory-

lation of PEA-15 may control whether PEA-15 influences proliferation or apoptosis.

peanut agglutinin Lectin from *Arachis hypogaea* that binds to **glycoproteins** containing β-D-gal (1-3) D-galNAc in membranes; used to investigate differential adhesiveness in developing systems.

Pecten *Scallop* A bivalve mollusc. The adductor muscle, a **catch muscle**, has been a favourite with muscle physiologists and biochemists as well as with gourmets.

pectin Class of plant cell wall polysaccharide, soluble in hot aqueous solutions of chelating agents or in hot dilute acid. Includes polysaccharides rich in galacturonic acid, rhamnose, arabinose and galactose, e.g. the polygalacturonans, rhamnogalacturonans, and some arabinans, galactans and arabinogalactans. Prominent in the **middle lamella** and **primary cell wall**. Important for the setting of jam.

PEDF *Pigment epithelium-derived factor* A natural extracellular component of the retina, a non-inhibitory **serpin**, that is a potent inhibitor of angiogenesis, has neuronal differentiating activity and inhibits endothelial cell injury *in vitro*.

pedicels See **podocytes**.

pedin A peptide of 13 amino acids that stimulates foot formation in *Hydra*. The precursor protein, thypedin, is about 110 kDa, which contains 13 copies of the peptide; the deduced amino acid sequence of the precursor comprises 27 copies of a **beta-thymosin repeat** domain.

PEG See **polyethylene glycol**.

pegylation Covalent coupling of polyethylene glycol (PEG) to a molecule. Proteins modified in this way have improved stability, biological half-life, water solubility and immunological characteristics following injection. A PEG mass of 40–50 kDa is sufficient to increase the size of a small molecule to such an extent that it is less readily excreted through the kidneys and therefore persists in the body for longer. In addition, as they are more or less surrounded by the attached PEGs, pegylated proteins are less rapidly broken down by the body's enzymes than are unmodified proteins. *Adj.* pegylated.

PEI Polyethylenimine.

pelargonidin One of the three primary plant pigments of the anthocyanin class. See **cyanidin**, **delphinidin**.

pelB The 18-residue N-terminal leader sequence of pelB is commonly used in various vector constructs. The gene from which it comes, *pelB*, codes for pectin lyase B, one of the many virulence factors of *Erwinia chrysanthemi*, pectinases that degrade cell walls of plants.

P-element A class of *Drosophila* **transposon**, widely used as a vector for reporter genes, for efficient germ line transformation and for **enhancer trap** or **insertional mutagenesis** studies.

Pelizaeus–Merzbacher disease Dysmyelinating disease resulting from defects in PLP (proteolipid protein) gene. Mouse model is **jimpy**.

pellagra Chronic disease due to a deficiency of vitamin B3 (niacin) or of tryptophan. Often a consequence of a diet consisting predominantly of maize, where the nicotinic acid is in bound form and there is a lack of the tryptophan precursor of nicotinic acid.

pelle *Drosophila* protein kinase that is involved in the activation of **dorsal** (NFκB homologue).

pellicle The outer covering of a **protozoan**: the plasma membrane plus underlying reinforcing structures, for example the membrane-bounded spaces (alveoli) just below the plasma membrane in ciliates.

Pelomyxa Genus of giant amoebae, usually 500–800 μm but occasionally larger; multinucleated; found in fresh water.

PEM *Polymorphic epithelial mucin* See **episialin**.

pemphigus A group of dermatological diseases characterized by the production of bullae (blisters).

PEN-2 *Presenilin enhancer* One of the four components of **gamma-secretase**; triggers endoproteolysis of **presenilin**, conferring gamma-secretase activity.

Pendred's syndrome An autosomal-recessive disease characterized by goitre and congenital sensorineural deafness. The genetic defect has been linked to chromosome 7q22-31.1 and the Pendred's syndrome gene (*PDS* gene) encodes pendrin, a highly hydrophobic 780 amino acid protein with 11 transmembrane domains. Its function is unknown, although it has high sequence homology to several sulphate transporters.

pendrin See **Pendred's syndrome**.

penetrance The proportion of individuals with a specific genotype who express that character in the phenotype.

penicillamine *Dimethyl cysteine* Product of acid hydrolysis of **penicillin** that chelates heavy metals (lead, copper, mercury) and assists in their excretion in cases of poisoning. Also used in treatment of rheumatoid arthritis, although its mode of action as an antirheumatic drug is not clear.

penicillin Probably the best known of the **beta-lactam antibiotics**, derived from the mould *Penicillium notatum*. It blocks the cross-linking reaction in **peptidoglycan** synthesis and therefore destroys the bacterial cell wall, making the bacterium very susceptible to damage. See **penicillin-binding protein**.

penicillin-binding protein *PBP* Proteins that catalyse both polymerization of glycan chains (glycosyltransferases) and cross-linking of pentapeptidic bridges (transpeptidases) during the biosynthesis of the peptidoglycan bacterial cell wall. PBPs are the targets for beta-lactam antibiotics and thus play key roles in drug-resistance mechanisms. Altered penicillin-binding protein 2X (PBP 2X), for example, is essential to the development of penicillin and cephalosporin resistance in *Streptococcus pneumoniae*.

pentazocine A powerful synthetic analgesic drug of the opiate class.

pentobarbital An anticonvulsant and anaesthetic, usually used as the sodium or calcium salt.

pentosan Glycan that, when hydrolysed, yields only pentoses.

pentose phosphate pathway *Pentose shunt; hexose monophosphate pathway; phosphogluconate oxidative pathway* Alternative metabolic route to **Embden–Meyerhof pathway** for breakdown of glucose. Diverges from this when **glucose-6-phosphate** is oxidized to ribose 5-phosphate by the enzyme glucose-6-phosphate dehydrogenase. This step reduces **NADP** to NADPH, generating a source of reducing power in cells for use in reductive biosyntheses. In plants, part of the pathway functions in the formation of hexoses from CO_2 in photosynthesis. Also important as source of pentoses, e.g. for nucleic acid biosynthesis. This pathway is the main metabolic pathway in neutrophil leucocytes; congenital deficiency in the pathway produces sensitivity to infection.

pentoses Sugars (monosaccharides) with five carbon atoms. Include **ribose** and **deoxyribose** of nucleic acids and many others, such as the aldoses **arabinose** and **xylose** and the ketoses **ribulose** and **xylulose**.

pentoxifylline *Trental* A tri-substituted xanthine derivative that is used to reduce blood viscosity, and thus improve blood flow, in patients with chronic peripheral arterial disease and intermittent claudication. Mode of action unclear.

pentraxins Family of proteins that share a discoid arrangement of five non-covalently linked subunits. Includes **CRP** and **serum amyloid P**.

PEP See **phosphoenolpyruvate**.

PEP carboxylase *EC 4.1.1.31* Enzyme responsible for the primary fixation of CO_2 in **C4 plants**. Carboxylates PEP (**phosphoenolpyruvate**) to give oxaloacetate. Also important in **crassulacean acid metabolism**, since it is responsible for CO_2 fixation in the dark.

peplomers Glycoproteins of the outer viral envelope; particularly large and conspicuous in Coronavirus and responsible for the 'sunburst' appearance.

pepsin *EC 3.4.23.1, formerly EC 3.4.4.1* Aspartic peptidase (formerly 'acid protease' or 'carboxyl proteinases') from stomach of vertebrates. Cleaves preferentially between two hydrophobic amino acids (e.g. F-L, F-Y) and will attack most proteins except protamines, keratin and highly glycosylated proteins. A single-chain phosphoprotein (327 amino acids; 34.5 kDa) released from the enzymatically inactive zymogen, pepsinogen, by autocatalysis at acid pH in the presence of HCl. One of the peptides cleaved off in this process is a pepsin inhibitor and has to be further degraded to allow the pepsin to have full activity. Pepsin is the type example of peptidase family A1.

pepsinogen The inactive precursor (42.5 kDa) of **pepsin**.

pepstatin Peptide from *Streptomyces* spp. that inhibits pepsin and other aspartic peptidases, for example cathepsin-D and renin.

peptidase Alternative name for a **protease** or proteinase; there is a move to phase out those terms and use the general term 'peptidase'. Peptidases can be grouped into clans and families. Clans are groups of families for which there is evidence of common ancestry. Families are grouped by their catalytic type – aspartic, cysteine, glutamic acid, metallo, serine, threonine, and unknown-type. A recent initiative, designed to rationalize the classification, is the **MEROPS** database (Rawlings, N.D., Morton, F.R. andBarrett, A.J. (2006). MEROPS: the peptidase database. *Nucleic Acids Res.*, 34, D270–D272).

peptide bond The amide linkage between the carboxyl group of one amino acid and the amino group of proline. The linkage does not allow free rotation and can occur in *cis* or *trans* configuration; the latter is the most common in natural peptides, except for links to the amino group of praline, which are always *cis*.

peptide histidine methionine *PHM* One of the **secretin family** of neuropeptides. Human analogue (27-amino acids, N-terminal histidine and C-terminal methionine) of peptide histidine-isoleucine (PHI), generated from the same precursor as **vasoactive intestinal peptide**, prepro VIP. Has vasodilatory activity.

peptide map Peptidases (proteases) will produce fragments of a characteristic size from a protein, and this can be used as a test for the identity or otherwise of two similar-sized proteins. It is possible to produce a peptide fragment map from a single gel band.

peptide neurotransmitter Small peptides used as primary or co-transmitters in nerve cells, e.g. **FMRF-amide**, **FLRF-amide**.

peptide nucleic acid *PNA* Synthetic **nucleic acid** mimic, in which the sugar-phosphate backbone is replaced by a peptide-like polyamide. Instead of 5′ and 3′ ends, PNAs have N- and C-termini. Their resistance to both **nucleases** and **peptidases**, and their ability to bind closely to complementary DNA or RNA sequences, have made them promising candidates in **antisense** and **gene therapy** technologies.

peptide receptor Specific receptor for **peptide neurotransmitters**.

peptide YY Gut-derived peptide hormones (PYY1–36 and PYY3–36) with anorectic properties, antagonists of the neuropeptide Y2 receptor (Y2R). Inhibit both pancreatic secretion and the release of cholecystokinin.

peptidoglycan *Murein* Cross-linked polysaccharide–peptide complex of indefinite size found in the inner cell wall of all bacteria (50% of the wall in Gram-negative, 10% in Gram-positive). Consists of chains of approximately 20 residues of β (1-4)-linked *N*-acetyl glucosamine and *N*-acetyl muramic acid cross-linked by small peptides (4–10 residues).

peptidomimetics In general, any compound that mimics the properties of a peptide; in practice, often a synthetic peptide with some non-natural amino acids in the sequence that confer greater stability or resistance to degradation.

peptidyl prolyl cis/trans-isomerases *PPIases; EC 5.2.1.8* Enzymes that catalyse the *cis–trans* isomerization of prolyl bonds in oligopeptides and various folding states of proteins. Three distinct classes of PPIases have been identified: **cyclophilins**, FK506-binding proteins (**FKBPs**) and **parvulins**.

peptidyl transferase *EC 2.3.2.12* Integral enzymic activity of the large subunit of a ribosome, catalysing the formation of a **peptide bond** between the carboxy terminus of the nascent chain and the amino group of an arriving tRNA-associated amino acid.

peptidyl-arginine deiminase *PAD; EC 3.5.3.15* Enzyme responsible for formation of protein-bound **citrulline**, a major amino acid in the inner root sheath and medulla of the hair follicle. Substrate is **trichohyalin** and postsynthetic modification of trichohyalin by PAD alters its properties so that it is able to act as a rigid matrix component.

peptidyl prolyl isomerase *Peptidyl-prolyl* cis–trans-*isomerase* See **PPIase** and **immunophilin**.

peptoid Oligomer composed of N-substituted glycines, a specific subclass of **peptidomimetics** in which side chains are appended to nitrogen atoms along the molecule's backbone, rather than to the α-carbons (as they are in amino acids).

peptones Mixture of partial degradation products of proteins; often used in culture media for microorganisms.

Percoll Trademark for colloidal silica coated with polyvinylpyrrolidone that is used for density gradients. Inert, and will form a good gradient rapidly when centrifuged. Useful for the separation of cells, viruses and subcellular organelles.

perforins *PRF* Perforins 1 and 2 form tubular transmembrane complexes (16 nm in diameter) at the sites of target cell lysis by **NK-cells** and **cytotoxic T-cells**. Perforin (*PRF1*) gene alterations have been documented in 40% of patients with familial haemophagocytic lymphohistiocytosis (FHLH), a heterogeneous autosomal-recessive disorder characterized by hyperactivation of monocytes/macrophages.

perfringolysin O *Theta toxin; θ-toxin* **Cholesterol-binding toxin** from *Clostridium perfringens*. Shares with other **thiol-activated haemolysins** a highly conserved sequence (ECTGLAWEWWR) near the C-terminus.

perfusion Passage of a fluid through a compartment or chamber; in physiology, often used for the process of passing fluid through the vessels of a specific tissue or organ.

periaxin *PRX* Protein, localized to the plasma membrane of **Schwann cells** that plays an important role in the myelination of the peripheral nerve. Two isoforms exist coded by a single gene. L-periaxin (147 kDa) is localized to plasma membrane of Schwann cells, S-periaxin (16 kDa) is diffusely cytoplasmic. Both possess **PDZ domains**. Mutations in the gene for periaxin have been reported in eight families with **Charcot–Marie–Tooth disease**.

peribacteroid membrane Membrane derived from the plasma membrane of a plant cell and that surrounds the nitrogen-fixing bacteroids in legume root nodules. Has a high lipid content, and may regulate the passage of material from the plant cell cytoplasm to the symbiotic bacterial cell. The idea that it restricts **leghaemoglobin** to the peribacteroid space seems untenable since leghaemoglobin is found in the cytoplasm of some cells.

pericanicular dense bodies Electron-dense membrane-bounded cytoplasmic organelles found near the canaliculi in liver cells: lysosomes.

pericarp That part of a fruit that is produced by thickening of the ovary wall. Composed of three layers: epicarp (skin), mesocarp (often fleshy) and endocarp (membranous or stony in the case of, for example, plums).

pericentric inversion Chromosomal inversion in which the region that is inverted includes the kinetochore.

pericentrin *Pcnt* Conserved coiled-coil protein (200–220 kDa) of the pericentriolar region involved in organization of microtubules during meiosis and mitosis; concentration highest at metaphase, lowest at telophase. A similar but larger human protein, kendrin, has been identified. AKAP450 (also known as AKAP350, CG-NAP or hyperion) and pericentrin share a well-conserved 90-amino acid domain near their C-termini.

pericentriolar region Rather amorphous region of electron-dense material surrounding the centriole in animal cells: the major **microtubule-organizing centre** of the cell.

perichondrial cell Cell of the perichondrium, the fibrous connective tissue surrounding cartilage.

periclinal In botanical anatomy, periclines are planes that are parallel to the outer surface. For a dome-like surface, e.g. the tip of a growing shoot, two kinds of periclines can be distinguished: meridional (longitudinal) and latitudinal (transverse). For an organ with bilateral symmetry, longitudinal and transverse periclines can be distinguished. A periclinal cell wall is parallel to a nearby surface, usually the outer surface of the plant. Anticlines are trajectories perpendicular to periclines. Usage in geology is slightly different.

pericyte Cell associated with the walls of small blood vessels; not a smooth muscle cell nor an endothelial cell.

periderm The outer cork layer of a plant that replaces the epidermis of primary tissues. Cells have their walls impregnated with **cutin** and **suberin**.

peridinin Accessory pigment (carotenoid) found in the Dinophyceae.

peridinium *Peridium* General term for the outer wall of the fruiting body of a fungus.

perikaryon Cell body surrounding nucleus of a neuron – does not include axonal and dendritic processes.

perilipin Protein (57 kDa) found in adipocytes as a coating of lipid droplets. Perilipin phosphorylation is apparently essential for the translocation of **hormone-sensitive lipase** from the cytosol to the lipid droplet, a key event in stimulated lipolysis.

perimysium The connective tissue sheath that binds muscle fibres into bundles.

perinuclear space Gap, 10–40 nm wide, separating the two membranes of the **nuclear envelope**.

period per *Drosophila* gene regulating circadian rhythm. Expressed in **central nervous system**, **Malpighian tubules** and a number of other tissues. Per contains a structural **PAS domain**, a nuclear localization sequence and a cytoplasmic localization domain that restricts it to the cytoplasm in the absence of Tim (product of *timeless*), with which it forms a heterodimer.

periodic acid-Schiff reaction *PAS* A method for staining carbohydrates: adjacent hydroxyl groups are oxidized to form aldehydes by periodic acid (HIO_4) and these aldehyde groups react with Schiff's reagent (basic fuchsin decolourized by sulphurous acid) to give a purple colour. Used in histochemistry and in staining gels on which glycoproteins have been run.

periodontal Adjective describing the region around the teeth, i.e. gums and gingival crevice. Periodontal disease is a common consequence of inadequate phagocyte function.

peripheral lymphoid tissue Secondary lymphoid tissue, not necessarily located peripherally. See **lymphoid tissue**.

peripheral membrane protein Membrane proteins that are bound to the surface of the membrane and not integrated into the hydrophobic region. Usually soluble, and were originally thought to bind to integral proteins by ionic and other weak forces (and could therefore be removed by high ionic strength, for example). However, it is now

clear that some peripheral membrane proteins are covalently linked to molecules that are part of the membrane bilayer (see **glypiation**), and that there are others that fit the original definition but are perhaps more appropriately considered proteins of the cytoskeleton (e.g. band 4.1 and **spectrin**) or extracellular matrix (e.g. **fibronectin**).

peripherin 1. Type III intermediate filament protein (57–58 kDa) co-expressed with **neurofilament** triplet proteins. 2. Photoreceptor-specific glycoprotein found on the rim region of rod outer segment disk membranes. Thought to be essential for assembly, orientation and physical stability of outer segment discs. Predicted sequence 346 residues, highly conserved between rodents, man and cattle. Mutations in the gene (*RDS* gene: Retinal Degeneration, Slow) can cause retinitis pigmentosa or macular dystrophy.

Periplaneta Genus of insects that includes *P. americana*, the American cockroach, a favourite experimental animal but an unwelcome commensal.

periplasmic-binding proteins Transport proteins located within the **periplasmic space**. Some act as receptors for bacterial chemotaxis, interacting with **MCPs**. Their mode of action is unclear.

periplasmic space Structureless region between the plasma membrane and the cell wall of Gram-negative bacteria.

periseptal annulus Organelle associated with cell division in Gram-negative bacteria. There are two circumferential zones of cell envelope in which membranous elements of the envelope are closely associated with **murein**. The annuli appear early in division, and in the region between them (the periseptal compartment) the division septum is formed.

perithecium A flask-shaped **ascocarp** with a pore or ostiole at the top through which the ascospores are discharged.

peritoneal exudate A term most commonly used to describe the fluid drained from the peritoneal cavity some time after the injection of an irritant solution. For example, a standard method for obtaining neutrophil leucocytes is to inject intraperitoneally saline with glycogen (to activate complement) and drain off the leucocyte-rich peritoneal exudate some hours later.

peritrichous Descriptor for bacteria that have flagella distributed uniformly over the surface of the cell.

perivitelline space The space surrounding the yolk of an egg.

perlecan Proteoglycan found in all basement membranes. The core protein (*ca.* 400 kDa) is 80 nm long with five to seven variable-length globular domains, and has two or three heparan sulphate chains of about 40–60 kDa located in the N-terminal end of the molecule. May occasionally contain both heparan sulphate and chondroitin/dermatan sulphate chains. It contains domains homologous to **LDL receptor**, **laminin**, neuronal cell adhesion molecule (**NCAM**) and **epidermal growth factor** (EGF). Binds to **nidogen** and plays an important role in maintaining the structural integrity of the basement membrane. In man, nonfunctional mutations of perlecan cause a lethal chondrodysplasia, dyssegmental dysplasia, Silverman–Handmaker type (DDSH); partially functional mutations of perlecan also cause Schwartz–Jampel syndrome (SJS), characterized by myotonia and chondrodysplasia.

permease General term for a membrane protein that increases the permeability of the plasma membrane to a particular molecule, by a process not requiring metabolic energy. See **facilitated diffusion**.

permissive cells Cells of a type or species in which a particular virus can complete its replication cycle.

permissive temperature Of a temperature-sensitive mutation, a temperature at which the mutated gene product behaves normally and so the cell or organism survives as if wild-type (compare with restrictive temperature, at which the gene product takes on a mutant phenotype).

pernicious anaemia Commonest cause of megaloblastic anaemia in which there is impaired absorption of vitamin B12 because of an autoimmune response to **intrinsic factor**.

Peromyscus Genus of mice native to Central and North America.

peropsin Retinal pigment epithelium-derived rhodopsin homologue (RRH). May act as a retinal isomerase.

peroxidase A **haem** enzyme that catalyses reduction of hydrogen peroxide by a substrate that loses two hydrogen atoms. Within cells, may be localized in peroxisomes. Coloured reaction products allow detection of the enzyme with high sensitivity, so peroxidase-coupled antibodies are widely used in microscopy and **ELISA**. **Lactoperoxidase** is used in the catalytic surface labelling of cells by radioactive iodine.

peroxins Gene products of *PEX* genes, of which at least 29 are known; necessary for **peroxisome** biogenesis.

peroxisomal targeting sequence 1 *PTS1* Proteins are targeted to peroxisomes if they have a C-terminal serine–lysine–leucine (SKL) sequence. Examples include firefly (*Photinus pyralis*) luciferase and urate oxidase.

peroxisome Organelle containing peroxidase and catalase, sometimes as a large crystal. A site of oxygen utilization, but not of ATP synthesis. In plants, associated with **chloroplasts** in **photorespiration** and considered to be part of a larger group of organelles, the **microbodies**.

peroxisome biogenesis disorders *PBDs* Fatal autosomal-recessive diseases in which peroxisomes are defective or deficient. There are 12 complementation groups, most of which have been linked to specific gene mutations. Examples are Zellweger syndrome and neonatal adrenoleucodystrophy. See **peroxins**.

persephin *PSPN* Neurotrophic factor that binds to **GFRalpha-4**. Promotes the survival of multiple populations of neurons.

persistence 1. The tendency of a cell to continue moving in one direction: an internal bias on the random walk behaviour that cells exhibit in isotropic environments. **2.** Descriptive of viruses that persist in cells, animals, plants or populations for long periods, often in a non-replicating form. Persistence is achieved by such strategies as integration into host DNA, immunological suppression, or mutation into forms with slow replication.

pertussis toxin Protein complex (*ca.* 117 kDa). An **AB toxin**, the active subunit is a single polypeptide (28 kDa), the binding subunit a pentamer (two heterodimers, 23 + 11.7 kDa, 11.7 + 22 kDa and a monomer (9.3 kDa)

that binds the heterodimers). The active subunit **ADP-ribosylates**, the α-subunit of the inhibitory **GTP-binding protein** (Gi). Crucial to the pathogenicity of *Bordetella pertussis*, the causative agent of whooping cough.

pes *Pl.* pedes The foot or a foot-like part.

PEST sequence *Pro-Glu-Ser-Thr* Amino acid motif that is thought to target cytoplasmic proteins for rapid proteolytic degradation.

pET expression system System that uses the pET vector to produce large amounts of protein in a bacterial expression system. The pET vector (plasmid) contains an ampicillin resistance marker, the *lacI* gene, the T7 transcription promoter, the lac operator region 3′ to the **T7 promoter,** and a polylinker region. There are two origins of replication – one that enables the production of a single-stranded vector under appropriate conditions, and the other the conventional origin of replication. The gene for the protein of interest is cloned into the polylinker region, a bacterial host expressing T7 polymerase under lac control is transfected and the system is activated by IPTG (which displaces repressor from the lac operon of the vector and the T7 polymerase).

petechia *Pl.* petechiae Small, round, red-purple spot, not raised; caused by intradermal haemorrhage.

pethidine A synthetic opioid analgesic and hypnotic.

petite mutants A class of yeast mutants, most studied in *Saccharomyces cerevisiae*. Mutants grow slowly and rely on anaerobic respiration: mitochondria, although present, have reduced cristae and are functionally defective (termed promitochondria). There are three types of petite mutant: (1) segregational mutants that show Mendelian behaviour and result from mutations in mitochondrial genes located in the nucleus; (2) neutral petites, which are recessive genotypes and result from the complete absence of mitochondrial DNA; and (3) suppressive petites, in which most of the mitochondrial DNA is lost (60–99%), though what remains is often amplified.

Petromyzon *Lamprey* Primitive marine vertebrate (Class: Agnatha) with eel-like body and lacking true jaws. Their relatively simple nervous system has been studied in some detail.

Peutz–Jeghers syndrome (*PJS*) See **LKB1**.

PEV 1. Position-effect variegation, a heterochromatin-associated gene silencing phenomenon. **2.** Porcine enterovirus 1 (PEV-1).

PEVK domains Domains first identified in vertebrate **titin** and associated with the elasticity of the protein.

Peyer's patches Lymphoid organs located in the submucosal tissue of the mammalian gut containing very high proportions of IgA-secreting precursor cells. The patches have B- and T-dependent regions and germinal centres. A specialized epithelium lies between the patch and the intestine. Involved in gut-associated immunity.

PF 4 See **cytokine**.

PFA PFA-100 is a laboratory test designed to measure platelet function. It measures the time required for whole blood to occlude a membrane impregnated with either epinephrine or ADP. The results are reported as closure time in seconds.

P-face See **freeze fracture**.

Pfam A large collection of multiple sequence alignments and hidden Markov models covering many common protein domains and families. Can be used, for example, to view the domain organization of proteins. At the end of 2006 there were 8957 protein families in the database; the number will undoubtedly grow.

PFG Pulsed field gradient. PGF-NMR is a well-established method for the determination of diffusion coefficients, which are indicative of molecular size and shape.

PFK *6-Phosphofructo-1-kinase; EC 2.7.1.1.1* Abbreviation for phosphofructokinase, a glycolytic enzyme. There are three isoenzyme in humans – muscle (**M**), liver (**L**) and platelet (**P**) – which are encoded by different genes.

Pfr The form of **phytochrome** that absorbs light in the far red region, 730 nm, and is thus converted to **Pr**. It slowly and spontaneously converts to Pr in the dark.

P(GAL4) Synthetic **P-element** of *Drosophila melanogaster*, comprising **long-terminal repeats** flanking a mini-*white* gene to mark flies carrying the P-element by their red eye colour, **Bluescript** to allow plasmid rescue of DNA flanking the genomic insertion site and a gene encoding the yeast transcription factor **GAL4**, downstream of a weak (permissive) promoter. Although itself unable to move within the genome, P(GAL4) mobilization can be induced by crossing in a source of transposase. Patterns of expression of neighbouring genes can be detected (see **enhancer trapping**) by crossing in a reporter gene (e.g. *LacZ*, **green fluorescent protein** (*GFP*)) under the control of the UAS promoter recognized by GAL4.

PGC *Peroxisome proliferator-activated receptor gamma coactivator-1* One of a family of transcriptional coactivators involved in several aspects of energy metabolism.

PGD Preimplantation genetic diagnosis, a technique for screening embryos that have been fertilized *in vitro* for congenital defects before implantation in the uterus. The range of tests available is increasing rapidly, raising some ethical concerns.

pGEM Proprietary name for cloning vectors with multiple restriction sites, and the *lacZ* gene that codes for a peptide that enables blue/white screening and containing both the SP6 and T7 RNA polymerase promoters. Various modified forms are sold to facilitate particular cloning requirements.

pGEX Proprietary name for a family of expression vectors that have an expanded multiple cloning site that facilitates the unidirectional cloning of cDNA inserts obtained from libraries. Proteins are expressed as fusion proteins with the 26-kDa **glutathione S-transferase** (GST) that facilitates purification on Glutathione Sepharose™ 4B.

PGK Abbrev. for **phosphoglycerate kinase**.

pGLO Plasmid that encodes the gene for **GFP** and a gene for resistance to the antibiotic ampicillin. The expression of the gene for GFP can be switched on in transformed cells by adding the sugar arabinose to the medium.

P-glycoprotein See **multidrug transporter**.

PGP Abbreviation for **P-glycoprotein** (Pgp). See **multidrug transporter**; see also **PGP 9.5**.

PGP 9.5 The protein gene product 9.5 (PGP 9.5) is a member of the ubiquitin hydrolase family of proteins, is con-fined to neural and neuroendocrine cells, and is the general immunohistochemical marker for nerves.

P granules Granules found in germ cells, and germ cell precursors, in the nematode *C. elegans*. P granules are segregated during early embryogenesis into those blastomeres that eventually produce the germ line. All the known protein components of P granules contain putative RNA-binding motifs, but no specific mRNAs have been identified within P granules in the gonad.

PGs *Prostaglandins; PGA, PGB, PGD, PGE, PGF, PGG, PGH, PGI* PGA is **prostaglandin** A, etc. PGI is more commonly known as **prostacyclin**.

pH $-log_{10}[H^+]$ A logarithmic scale for the measurement of the acidity or alkalinity of an aqueous solution. Neutrality corresponds to pH 7, whereas a 1 molar solution of a strong acid would approach pH 0 and a 1 molar solution of a strong alkali would approach pH 14.

PH-30 Heterodimeric sperm surface transmembrane protein involved in sperm–egg fusion. The α-subunit has some similarities to viral fusion proteins and the β-subunit has a domain similar to soluble integrin ligands (**disintegrins**).

PHA 1. **Phytohaemagglutinin**. 2. *Pha-2* is the *C. elegans* homologue of the vertebrate homeobox gene *Hex*. 3. Polyhydroxyalkanoates; the organically produced basis for biodegradable plastics.

phaeo- *Pheo- (US)* Prefix meaning dark-coloured. The UK English form is used in this Dictionary.

phaeochromocytoma *Pheochromocytoma (US)* A normally benign neoplasia (**neuroblastoma**) of the **chromaffin tissue** of the adrenal medulla. In culture, the cells secrete enormous quantities of **catecholamines** and can be induced to form neuron-like cells on addition of (for example) cyclic AMP or nerve growth factor. Excessive production of adrenaline and noradrenaline leads to secondary hypertension, sometimes paroxysmal.

phaeomelanin One of the two chemically distinct types of melanin: the red-yellow phaeomelanin and the brown-black eumelanin. Both have been detected in human hair, epidermis and cultured melanocytes.

Phaeophyta *Pheophyta (US); brown algae* Division of algae, generally brown in colour, with multicellular, branched thalluses. Includes large seaweeds such as *Laminaria* and *Fucus*. The brown colour is due to the **xanthophylls**, fucoxanthin and lutein. Many have **laminarin** as a food reserve and alginic acid as a wall component.

phaeophytin *Pheophytin* Chlorophyll from which the metal ion (magnesium) has been removed and two protons substituted.

phafins A family of proteins that contain both the **PH domain** (pleckstrin homology) and the **FYVE** domains.

phage See **bacteriophage**.

phage display library Phage library in which the insert is expressed as a translational fusion with a phage-coat protein. This makes it particularly easy to screen with antibodies. Widely used to identify epitopes recognized by some particular antibody; commercial phage display libraries randomly encoding all possible six or seven residue peptides can be screened with the antibody, and the inserts of bound phage sequenced to build up a picture of the binding profile of the antibody.

phage integrase family See **recombinases**, **site-specific recombination**.

phage typing Bacteria may be typed by their susceptibility to a range of bacteriophages though confusion may arise if the bacteria carry plasmids encoding **restriction endonucleases**.

phagemid **Bacteriophage** whose genome contains a **plasmid** that can be excised by co-infection of the host with a helper phage. Useful as **vectors** for library production, as the library can be amplified and screened as phage, but the inserts of selected **plaques** can readily be prepared as plasmids without subcloning. An example of a commercial phagemid is λZap, from which pBluescript can be excised with helper phage.

phagocyte A cell that is capable of phagocytosis. The main mammalian phagocytes are **neutrophils** and **macrophages**.

phagocytic vesicle Membrane-bounded vesicle enclosing a particle internalized by a phagocyte. The primary phagocytic vesicle (phagosome) will subsequently fuse with **lysosomes** to form a secondary phagosome in which digestion will occur.

phagocytosis Uptake of particulate material by a cell (endocytosis). See **opsonization**, **phagocyte**.

phakinin Eye-lens-specific protein, 47 kDa, that coassembles with **filensin** (three phakinin molecules per filensin) to form beaded-chain **intermediate filaments**. Phakinin has very strong sequence homology with **cytokeratins** but lacks the rod domains that are involved in filament formation. It is effectively a tail-less intermediate filament protein.

phalangeal cells Cells of the organ of Corti (in the inner ear).

phalloidin Cyclic peptide (789 Da) from the Death Cap fungus (*Amanita phalloides*) that binds to, and stabilizes, F-actin. Fluorescent derivatives are used to stain actin in fixed and permeabilized cells, although there is some uptake by live cells.

phallotoxins Toxic compounds (bicyclic peptides) produced by *Amanita phalloides*. Bind to F-actin and inhibit depolymerization; hepatotoxicity is the primary cause of problems. **Phalloidin** is a phallotoxin.

pharate Of an insect, having its new cuticle formed beneath its present cuticle and thus ready for its next moult.

pharmacodynamics The study of how drugs affect the body: contrast with **pharmacokinetics**.

pharmacogenetics The study of genetic causes of individual variations in drug response; the term is often used interchangeably with **pharmacogenomics**.

pharmacogenomics Pharmacogenomics involves genome-wide analysis of the genetic determinants of drug efficacy and toxicity rather than individual genetic differences (polymorphisms). Often used interchangeably with **pharmacogenetics**.

pharmacokinetics The study of what the body does to drugs, in contrast to **pharmacodynamics**. The pharmacokinetics of a drug relate to the rate and extent of its uptake (absorption), its transformation as a result of metabolism, the kinetics of distribution of the drug and its metabolites in the tissues and the rate and route of elimination (excretion)

from the body. Commonly abbreviated as **ADME** (absorption, distribution, metabolism and excretion).

pharming The commercial production of substances from transgenic plants or animals for medical (pharmaceutical) use; however, be aware that this term is also applied to the covert redirection of computer users from legitimate websites to counterfeit sites in order to gain confidential information about them.

phase contrast microscopy A simple non-quantitative form of **interference microscopy** of great utility in visualizing live cells. Small differences in optical path length due to differences in refractive index and thickness of structures are visualized as differences in light intensity.

phase separation The separation of fluid phases that contain different concentrations of common components. Occurs with partially miscible solvents used in many biochemical separation methods. Also, temperature-dependent phase separation occurs with some detergent solutions. With reference to membranes, means the segregation of lipid components into 'domains' that have different chemical composition.

phase variation Alteration in the expression of surface antigens by bacteria. For example, *Salmonella* can express either of two forms of **flagellin**, H1 and H2, that are coded by different genes. Control of which form is expressed is brought about by inversion of the promoter for the *H2* gene, which if functional (non-inverted) is associated with the expression of H2 and the production of a repressor of the *H1* gene. Inversion occurs about every 1000 bacterial divisions and is under the control of another gene, *hin*, that is within the invertable sequence.

phaseolin Vacuolar storage proteins of the 7S class and the major trimeric seed-storage protein found in the bean, *Phaseolus vulgaris*.

phaseollin A **phytoalexin** produced by *Phaseolus* (bean) plants in response to pathogenic attack or other stress. Not a misprint for **phaseolin**.

phasic See **adaptation**.

phasing of nucleosomes A non-random arrangement of **nucleosomes** on DNA, in which, at certain segments of the genome, nucleosomes are positioned in the same way relative to the nucleotide sequence in all cells. Most nucleosomes are arranged randomly, but phasing has been detected in some genes.

phasmid Hybrid phage/plasmid formed by integration of plasmid containing the att site, and lambda phage, mediated by phage integrase site-specific recombination.

PHB Abbrev. for (1) poly(beta-3-hydroxybutyrate), a biodegradable polymer that can be produced by bacteria, (2) protein domain, the **prohibitin** homology (PHB) domain, characteristic of a family of proteins.

PH domain *Pleckstrin homology domain* Domain found in various intracellular signalling cascade proteins (e.g. **pleckstrin**, **tec family kinases**). Seems to be involved in interactions with phospholipids, particularly PIP3. At one stage it was suggested that they were involved in the interaction with heterotrimeric G-proteins.

PHD-type See **E3 ligase**.

phelloderm Tissue containing parenchyma-like cells, in the bark of tree roots and shoots. Produced by cell division in the **phellogen**.

phellogen **Meristematic** tissue in plants, giving rise to cork (phellem) and **phelloderm** cells. Also termed 'cork cambium'.

phenanthroline o-*Phenanthroline; 1,10-phenanthroline* A tricyclic aromatic hydrocarbon composed of three fused benzene rings (isomeric with anthracene) in which two carbon atoms are replaced by two nitrogens in the ring structure. Used as an analytical reagent for photometric determination of Fe(II). The complex of Fe(II) and phenanthroline has red colour.

phenazopyridine Analgesic used for symptomatic relief of pain from infections of lower urinary tract. It is excreted in the urine, where it exerts a topical analgesic effect on the mucosa. The precise mechanism of action is unknown.

phencyclidine *1-(1-Phenylcyclohexyl) piperidine; Angel dust; PCP* Anaesthetic and drug of a kind that can produce marked behavioural effects. Interacts with the **NMDA receptor**.

phenetics Numerical taxonomy.

phenobarbital *Phenobarbitone;   phenylethyl-barbituric acid* Barbiturate used mostly as an anticonvulsant in the treatment of epilepsy. Phenobarbital is the British Approved Name (BAN).

phenobarbitone Now **phenobarbital**.

phenocopy An environmentally produced phenotype simulating the effect of a particular genotype.

phenol red Dye used as pH indicator: changes from yellow to red in range 6.8–8.4. Very commonly used in tissue culture medium (which turns acid as it becomes exhausted) though it can interfere with luminescence assays.

phenolphthalein Indicator dye that is colourless at neutral pH and red-pink in slightly alkaline solutions. Also used as a laxative.

phenome Phenotypic equivalent of the genome, the sum of all phenotypic characters.

phenothiazines A group of antipsychotic drugs, thought to act by blocking dopaminergic transmission in the brain. Examples are **chlorpromazine** and **trifluoperazine**. Trifluoperazine binds to and inhibits **calmodulin**, and has been used experimentally to block calcium/calmodulin-controlled reactions.

phenotype The characteristics displayed by an organism under a particular set of environmental factors, regardless of the actual genotype of the organism.

phenoxyacetic acids Phenoxyacetic acid is used as a precursor in antibiotic fermentations, especially for penicillin V, and is a main skeleton of plant-growth regulators and herbicides such as 2,4-D, 2,4,5-T and **MCPA**. It is used as an intermediate for manufacturing dyes, pharmaceuticals, pesticides, fungicides.

phentermine An appetite-suppressing drug, a sympathomimetic amine, used in treating obesity. Has activity similar to the prototype drugs of this class, the amphetamines.

phenylalanine *Phe; F; 165 Da* An essential amino acid with an aromatic side chain. See Table A2.

phenylalanine ammonia lyase *PAL; EC 4.3.1.5* Enzyme involved in the synthesis of **lignin** and other phenolic compounds from phenylalanine. Used as an enzymic marker for lignification and other developmental processes in plant cells. Commonly activated as part of a plant's response to disease.

phenylbutazone Analgesic and antipyretic non-steroidal anti-inflammatory drug used in the treatment of rheumatic conditions.

phenylephrine hydrochloride An α1-adrenergic agonist (204 Da).

phenylethylamine Backbone for compounds which have important physiological functions within the body as neurotransmitters in the central nervous system and hormones in the blood circulation, as well as alkaloids found in substances such as chocolate.

phenylketonuria *PKU* Congenital absence of phenylalanine hydroxylase (an enzyme that converts phenylalanine into tyrosine). Phenylalanine accumulates in blood and seriously impairs early neuronal development. The defect can be controlled by diet and is not serious if treated in this way. Incidence highest in Caucasians.

phenylmethylsulphonyl fluoride *PMSF* Widely used as a broad-spectrum serine peptidase inhibitor. Sulphonylates the active-site histidine.

phenylthiourea *PTU; phenylthiocarbamide (PTC)* An inhibitor of **tyrosinase** and melanin synthesis, widely used in zebrafish research to suppress pigmentation in developing embryos/fry. There is a common human polymorphism in the ability to taste PTC.

phenytoin Drug widely used in the treatment of epilepsy, also used to control abnormal heart rhythms. Stabilizes the threshold against hyperexcitability, possibly by promoting sodium efflux from neurons.

pheo- Prefix meaning dark-coloured. See phaeo-.

pheomelanin See **phaeomelanin**.

pheresis Procedure in which the blood is filtered, particular elements (cells or plasma) separated and the remainder returned to the donor's circulation. In leucapheresis, white cells are removed; in plasmapheresis, plasma is removed and all cellular components returned.

pheromone A volatile hormone or behaviour modifying agent. Normally used to describe sex attractants (e.g. bombesin for the moth *Bombyx*), but includes volatile aggression stimulating agents (e.g. isoamyl acetate in honey bees).

pherotype Strain-specific variation of *Streptococcus pneumoniae* in response to the quorum-sensing pheromone **competence-stimulating peptide** (CSP). The pherotypes are determined by whether the response is to CSP-1 or CSP-2. See **comC, D, E**.

phi X-174 ϕX-174 Bacteriophage of *E. coli* with a single-stranded DNA genome and an icosahedral shell. This was the first DNA phage to be fully sequenced: the genome consists of 10 genes, some of which are **overlapping genes**.

phialidin Calcium-activated photoprotein from the coelenterate *Phialidium*. See **aequorin**.

Philadelphia chromosome Characteristic chromosomal abnormality of chronic myelogenous **leukaemia** in which a portion of chromosome 22 is translocated to chromosome 9.

philopatry Tendency for species or groups to remain in or habitually return to their native regions or territories. *Adj:* philopatric.

phlebolith *Venous calculus* A venous thrombus that has become calcified; a complication of deep venous thrombosis, most commonly found in pelvic veins.

phlebotomy The cutting of veins; a fancy name for taking blood by venepuncture, usually with a needle not a knife.

phloem Tissue-forming part of the plant vascular system, responsible for the transport of organic materials, especially sucrose, from the leaves to the rest of the plant. Consists of **sieve tubes**, **companion cells**, **fibre cells** and **parenchyma**.

phoA *EC 3.1.3.1* Alkaline phosphatase from *E. coli.*

phomopsin A Naturally occurring peptide that acts on tubulin to block microtubule assembly. Shares a binding site (distinct from the **vinca alkaloid** binding site) on beta-tubulin with **cryptophycin** 1, cryptophycin 52, **dolastatin 10** and **hemiasterlin**. Has antitumour properties.

phorbol esters Polycyclic compounds isolated from croton oil in which two hydroxyl groups on neighbouring carbon atoms are esterified to fatty acids. The commonest of these derivatives is phorbol myristoyl acetate (PMA). Potent co-carcinogens or tumour promoters, they are diacyl glycerol analogues and activate protein kinase C irreversibly.

phormicin Insect **defensin** produced by the blowfly, *Phormia terranovae.*

phosducin *PDC* Protein (33 kDa) that inhibits Gs-GTPase activity by binding the Gβγ subunit of heterotrimeric G-proteins and making them unavailable for signalling. Isolated from bovine brain, and found in retina, pineal gland and many other tissues. Activity of phosducin is inhibited if phosphorylated by a cAMP-dependent protein kinase.

phosducin-like protein *PdcL protein* A family of regulators of G-protein function expressed throughout brain and body and in a wide range of organisms, with properties similar to **phosducin**.

phosmid Cosmid-phage hybrid vector which has the phage λ origin of replication. Was developed to facilitate restriction enzyme mapping.

phosphatases Enzymes that hydrolyse phosphomonoesters. Acid phosphatases are specific for the single-charged phosphate group and alkaline phosphatases for the double-charged group. These specificities do not overlap. The phosphatases comprise a very wide range of enzymes, including broad- and narrow-specificity members. Phosphoprotein phosphatases specifically dephosphorylate a particular protein, and are essential if phosphorylation is to be used as a reversible control system; they are also specific for phosphoserine/threonine or phosphotyrosine residues within the target protein.

phosphatides The family of phospholipids based on 1, 2 diacyl 3-phosphoglyceric acid. See **phospholipids**.

phosphatidic acid *PA; diacyl glycerol 3-phosphate* The 'parent' structure for phosphatidyl phospholipids, present in low concentrations in membranes. The acyl groups are derived from long-chain fatty acids. An intermediate in the synthesis of diacyl glycerol, the immediate precursor of most of the phosphatidyl phospholipids (except phosphatidyl inositol) and of triacyl glycerols.

phosphatidyl choline *PC* The major phospholipid of most mammalian cell membranes, where the 1-acyl residue is normally saturated and the 2-acyl residue unsaturated. Choline is attached to phosphatidic acid by a phosphodiester linkage. Major synthetic route is from diacyl glycerol and CDP-choline. Forms monolayers at an air/water interface and forms bilayer structures (**liposomes**) if dispersed in aqueous medium. A zwitterion over a wide pH range. Readily hydrolysed in dilute alkali.

phosphatidyl ethanolamine *PE* A major structural phospholipid in mammalian systems. Tends to be more abundant than phosphatidyl choline in the internal membranes of the cell, and is an abundant component of prokaryotic membranes. Ethanolamine is attached to phosphatidic acid by a phosphodiester linkage. Synthesis from diacyl glycerol and CDP-ethanolamine.

phosphatidyl inositol *PI* Very important minor phospholipid in eukaryotes, involved in signal transduction processes. Contains myo-inositol linked through the 1-hydroxyl group to phosphatidic acid. The 4-phosphate (PIP) and 4,5 bisphosphate derivatives (PIP2) are formed and broken down in membranes by the action of specific kinases and phosphatases (futile cycles). Signal-sensitive phospholipase C enzymes remove the inositol moiety, in particular from 1,4,5 trisphosphate (PIP2) as inositol 1,4,5-triphosphate (InsP3: IP3). Both the diacyl glycerol and inositol phosphate products act as **second messengers**.

phosphatidyl serine *PS* An important minor species of phospholipid in membranes. Serine is attached to phosphatidic acid by a phosphodiester linkage. Synthesis is from phosphatidyl ethanolamine by exchange of ethanolamine for serine. Distribution is asymmetric, as the molecule is only present on the cytoplasmic side of cellular membranes. It is negatively charged at physiological pH and interacts with divalent cations; involved in calcium-dependent interactions of proteins with membranes (e.g. protein kinase C).

phosphocreatine *Creatine phosphate* Present in high concentration (about 20 mM) in striated muscle, and is synthesized and broken down by creatine phosphokinase to buffer ATP concentration. It acts as an immediate energy reserve for muscle.

phosphodiester bond Not a precise term. Refers to any molecule in which two parts are joined through a phosphate group. Examples are found in RNA, DNA, phospholipids, cyclic nucleotides, nucleotide diphosphates and triphosphates.

phosphodiesterase An enzyme that cleaves phosphodiesters to give a phosphomonoester and a free hydroxyl group. Examples include RNAase, DNAase, phospholipases C and D and the enzymes that convert cyclic nucleotides to the monoester forms. In casual usage the cAMP-phosphodiesterase is usually meant.

phosphodiesterase inhibitors Compounds that inhibit the breakdown of the important intracellular second messenger cyclic AMP, although strictly speaking the cAMP-phosphodiesterase (EC:3.1.4.17) is only one of a number of such enzymes. Allowing cAMP to accumulate potentiates the action of the sympathetic nervous system.

Most commonly encountered examples are the xanthines such as theophylline.

phosphoenolpyruvate *PEP* An important metabolic intermediate. The enol (less stable) form of pyruvic acid is trapped as its phosphate ester, giving the molecule a high phosphate transfer potential. Formed from 2-phosphoglycerate by the action of enolase.

phosphofructokinase *EC 2.7.1.56* The pacemaker enzyme of glycolysis. Converts fructose 6-phosphate to fructose 1,6-bisphosphate. A tetrameric allosteric enzyme that is sensitive to the ATP/ADP ratio.

phosphoglycerate The molecules 2-phosphoglycerate and 3-phosphoglycerate are intermediates in glycolysis. 3-Phosphoglycerate is the precursor for synthesis of phosphatidic acid and diacyl glycerol, hence of phosphatidyl phospholipids.

phosphoglycerate kinase *PGK; EC 2.7.2.3* X-linked enzyme that plays a key role in the glycolytic pathway catalysing the phosphorylation of 3-phospho-D-glycerate to 3-phospho-hydroxypyruvate.

phosphoinositide-dependent kinase *PDK-1; EC 2.7.1.37* A protein kinase that is critical for the activation of many downstream protein kinases in the **AGC kinase** superfamily, through phosphorylation of the activation loop site on these substrates.

phospholamban *PLN; PLB* Integral membrane homopentameric protein (52 aa) that is the endogenous regulator of the sarcoplasmic reticulum calcium Ca^{2+} ATPase (**SERCA**). Phosphorylation by protein kinase A and dephosphorylation by protein phosphatase 1 modulate the inhibitory activity of phospholamban. A mutation in the human phospholamban gene, deleting arginine 14, results in lethal, hereditary cardiomyopathy.

phospholemman *PLM; FXYD1* A small transmembrane protein that, like **phospholamban**, interacts with P-type ATPases and regulates ion transport in cardiac cells and other tissues.

phospholipase D *PLD; EC 3.1.4.4* A family of phospholipases that are found in a wide variety of organisms. Plays a central signalling function in eukaryotic cells. Glycosylphosphatidylinositol-specific phospholipase D (GPI-PLD) is abundant in serum although its function is unclear.

phospholipases Enzymes that hydrolyse ester bonds in phospholipids. They comprise two types: aliphatic esterases (phospholipase A1, A2 and B) that release fatty acids, and phosphodiesterases (types C and D) that release diacyl glycerol or phosphatidic acid, respectively. Type A2 is widely distributed in venoms and digestive secretions. Types A1, A2 and C (the latter specific for phosphatidyl inositol) are present in all mammalian tissues. Type C is also found as a highly toxic secretion product of pathogenic bacteria. Type B attacks monoacyl phospholipids and is poorly characterized. Type D is largely of plant origin. PLA2 type II (a secreted enzyme, but not the same as the type I digestive pancreatic enzyme) is probably very important in inflammation because its action can release arachidonic acid, the starting point for **eicosanoid** synthesis. Phosphatidylinositol bisphosphate-specific phospholipase C is important in generating **diacylglycerol** and **inositol trisphosphate**, both **second messengers**.

phospholipid The major structural lipid of most cellular membranes (except the chloroplast, which has galactolipids). Contains phosphate, usually as a diester. Examples include phosphatidyl phospholipids, plasmalogens and sphingomyelins. See Table L3.

phospholipid bilayer A lamellar organization of phospholipids that are packed as a bilayer with hydrophobic acyl tails inwardly directed and polar head groups on the outside surfaces. It is this bilayer that forms the basis of membranes in cells, though in most cellular membranes a very substantial proportion of the area may be occupied by integral proteins. The triple-layered appearance of membranes seen in electron microscopy is thought to arise because the **osmium tetroxide** binds to the polar regions leaving a central, unstained, hydrophobic region.

phospholipid transfer protein Cytoplasmic proteins that bind phospholipids and facilitate their transfer between cellular membranes. May also cause net transfer from the site of synthesis.

phosphomannose See **mannose-6-phosphate**.

phosphoprotein Proteins that contain phosphate groups esterified to serine, threonine or tyrosine (S, T or Y). The phosphate group usually regulates protein function.

phosphoramidite Nucleotide derivative used in oligonucleotide synthesis.

phosphorescence 1. Emission of light following absorption of radiation. Emitted light is of longer wavelength than the exciting radiation, and is a result of decay of electrons from the triplet to the ground state. Lasts longer than fluorescence (electron decay from singlet to ground state) and occurs after a longer delay. **2.** Popularly misused as a term for biological luminescence, e.g. by fireflies.

phosphorimaging Method for detecting radioactivity using 'phosphor' compounds that emit visible light when exposed to radiation; used in the same way as autoradiography (e.g. for detecting labelled bands on gels) but is much more sensitive.

phosphorylase *Glycogen phosphorylase; EC 2.4.1.1* Enzyme that catalyses the sequential removal of glycosyl residues from glycogen to yield one glucose-1-phosphate per reaction. Its activity is controlled by phosphorylation (by **phosphorylase kinase**).

phosphorylase kinase *EC 2.7.1.38* The enzyme that regulates the activity of **phosphorylase** and glycogen synthetase by addition of phosphate groups. A large and complex enzyme, itself regulated by phosphorylation. Integrates the hormonal and calcium signals in muscle.

phosphorylation of proteins Addition of phosphate groups to hydroxyl groups on proteins (to the side chains of serine, threonine or tyrosine) catalysed by a protein kinase (often specific) with ATP as phosphate donor. Activity of proteins is often regulated by phosphorylation.

phosphotransferase An enzyme that transfers a phosphate group from a donor to an acceptor. Very important in metabolism.

phosphotyrosine Strictly speaking, tyrosine phosphate, but normally refers to the phosphate ester of a protein tyrosine residue. Present in very small amounts in tissues, but believed to be important in systems that regulate growth control, and is therefore of interest in studies of malignancy.

The *src* gene product (pp60src) was one of the first kinases shown to phosphorylate at a tyrosine residue.

photoadduct Generally, any compound formed between two reacting molecules as a result of exposure to light. Most commonly used in the context of covalent modifications of DNA as a result of UV irradiation, sometimes for therapeutic reasons, as with **psoralens** in **photodynamic therapy**. See **photoaffinity labelling**.

photoaffinity labelling A technique for covalently attaching a label or marker molecule onto another molecule such as a protein. The label, which is often fluorescent or radioactive, contains a group that becomes chemically reactive when illuminated (usually with UV light) and will form a covalent linkage with an appropriate group on the molecule to be labelled: proximity is essential. The most important class of photoreactive groups used is the aryl azides, which form short-lived but highly reactive nitrenes when illuminated.

photobleaching Light-induced change in a **chromophore**, resulting in the loss of its absorption of light of a particular wavelength. A problem in fluorescence microscopy, where prolonged illumination leads to progressive fading of the emitted light because less of the exciting wavelength is being absorbed.

photodynamic therapy Therapeutic approach in which a light-sensitive prodrug is given and then the target area (usually a tumour) is illuminated to generate the active drug in the right place.

photodynesis Initiation of cytoplasmic streaming by light. Uncommon usage.

photolithography Originally a form of lithography in which light-sensitive plates or stones were exposed to a photographic image; now more widely applied to any process in which selective masking generates light patterns which cause chemical transformations on exposed areas of a photosensitive surface, an approach that has been used in semiconductor manufacture and in production of complex substrata for cell behavioural studies, especially on **contact guidance**.

photolyase *DNA photolyase* Family of ubiquitous enzymes found in bacteria, archaebacteria and eukaryotes that can repair UV-induced DNA damage. The protein (between 454 and 614 residues) is associated with two prosthetic groups, FADH and a light-harvesting cofactor, MTHF (5,10-methenyltetrahydrofolyl polyglutamate). Light is needed for the repair step.

photolysis Light-induced cleavage of a chemical bond, as in the process of photosynthesis.

photomedins **Olfactomedin family** proteins of the retina; photomedin-1 is selectively expressed in the outer segment of photoreceptor cells and photomedin-2 in all retinal neurons. Photomedins preferentially bind to chondroitin sulphate-E and heparin.

photoperiodism Events triggered by duration of illumination or pattern of light/dark cycles: often the wavelength of the illuminating light is important, as, for example, in the control of circadian rhythm in plants. See **phytochromes**.

photophosphorylation The synthesis of ATP that takes place during photosynthesis. In non-cyclic photophosphorylation the photolysis of water produces electrons that generate a **proton motive force** which is used to produce ATP, the electrons finally being used to reduce $NADP^+$ to NADPH. When the cellular ratio of reduced to non-reduced NADP is high, **cyclic photophosphorylation** occurs and the electrons pass down an electron transport system and generate additional ATP, but no NADPH.

photopigment Pigment involved in **photosynthesis** in plants. Includes **chlorophyll**, **carotenoids** and **phycobilins**.

photoreceptor A specialized cell type in a multicellular organism that is sensitive to light. This definition excludes single-celled organisms, but includes non-eye receptors, such as snake infrared (heat) detectors or photosensitive pineal gland cells. See **retinal rods**, **retinal cones**.

photorespiration Increased respiration that occurs in photosynthetic cells in the light, due to the ability of **RuDP carboxylase** to react with oxygen as well as carbon dioxide. Reduces the photosynthetic efficiency of **C3 plants**.

photosynthesis Process by which green plants, algae and some bacteria absorb light energy and use it to synthesize organic compounds (initially carbohydrates). In green plants, occurs in **chloroplasts** that contain the photosynthetic pigments. Occurs by slightly different processes in **C3** and **C4 plants**. See also **Z scheme of photosynthesis** and contrast with **chemosynthesis**.

photosynthetic bacteria Bacteria that are able to carry out **photosynthesis**. Light is absorbed by **bacteriochlorophyll** and **carotenoids**. Two principal classes are the green bacteria and the purple bacteria.

photosynthetic unit Group of photosynthetic pigment molecules (**chlorophylls** and **carotenoids**) that supply light to one **reaction centre** in **photosystems I or II**.

photosystem I Photosynthetic system in **chloroplasts** in which light of up to 700 nm is absorbed and its energy used to bring about charge separation in the **thylakoid** membrane. The electrons are passed to ferredoxin and then used to reduce $NADP^+$ to NADPH (non-cyclic electron flow) or to provide energy for the phosphorylation of ADP to ATP (cyclic photophosphorylation).

photosystem II Photosynthetic system in **chloroplasts** in which light of up to 680 nm is absorbed and its energy used to split water molecules, giving rise to a high-energy reductant, Q^-, and oxygen. The reductant is the starting point for an electron transport chain that leads to **photosystem I** and that is coupled to the phosphorylation of ADP to ATP.

phototaxis Movement of a cell or organism towards (positive phototaxis) or away from (negative phototaxis) a source of light.

phototransduction The transformation by photoreceptors (e.g. **retinal rods** and **cones**) of light energy into an electrical potential change.

phototrophic Any organism that can utilize light as a source of energy.

phototropins *phot1, phot2* Phototropins 1 and 2 function as blue-light photoreceptors for phototropism, chloroplast relocation, stomatal opening and leaf flattening in *Arabidopsis thaliana*. The photoreceptor comprises a C-terminal Ser/Thr protein kinase domain and two structurally similar flavin mononucleotide (FMN) binding **LOV domains**, LOV1 and LOV2. Homologues are present in many plants.

phototropism Movement or growth of part of an organism (e.g. a plant shoot) towards (positive phototropism) a source of light, without overall movement of the whole organism.

phox *phox47 (p47phox); phox67 (p67phox)* Components of the NADPH-oxidase system in phagocytes, the system responsible for generating an oxidative burst and thus bacterial killing. Phox47 and phox67 are cytoplasmic and only associate with the integral membrane component following activation. Both contain **SH3** domains. Deletion or mutation in either leads to **chronic granulomatous disease**.

phragmoplast Central region of mitotic spindle of a plant cell at telophase, in which vesicles gather and fuse to form the **cell plate**, apparently guided by spindle microtubules.

phragmosome In plant cells, the region of the cytoplasm in which the nucleus is located during nuclear division. Can also refer to **microbodies** associated with the developing **cell plate** after nuclear division.

phthalate Family of organic compounds used as plasticizers in PVC films. May be persistent and bioaccumulate in fatty tissue although their toxicity is disputed.

phycobilins Photosynthetic pigments found in certain algae, especially red algae (Rhodophyta) and **cyanobacteria**.

phycobiliprotein Phycobilins.

phycobilisome An accessory light energy harvesting structure in **cyanobacteria**. They have cores of allophycocyanin with radiating rods composed of discs of **phycocyanin** and **phycoerythrin**. Linker polypeptides attach the core to the **thylakoid** membranes. These structures, 20–70 nm across, contain the pigments named above that transfer light energy to chlorophyll *a*. The pigments are extracted and used as fluorochromes for labelling various probe reagents.

phycocyanin Blue **phycobilin** found in some algae, and especially in **cyanobacteria**.

phycoerythrin Red **phycobilins** found in some algae, especially red algae (**Rhodophyta**).

Phycomycetes A group of fungi possessing hyphae that are usually non-septate (without cross-walls).

phycoplast A set of microtubules oriented parallel to the plane of the new cell wall during cytokinesis and involved in wall formation in some algae.

phyllolitorin Bombesin-like peptide subfamily, originally identified from the skin of the South American frog *Phyllomedusa sauvagei*. See **Leu-phyllolitorin**. There is a precursor of 90 amino acids containing a signal peptide sequence, an amino-terminal extension peptide, the phyllolitorin peptide of nine amino acids and a carboxy-terminal extension peptide.

phylogenetic profile Computational method of trying to deduce functional interactions between proteins based upon the presence of both the proteins in some genomes and neither in other genomes: the argument is that it is unlikely their joint presence would be invariable if they did not interact. See **Rosetta Stone method** and **gene neighbour method**.

physaliphorous cells Cells of chordoma (tumour derived from notochordal remnants) that appear vacuolated because they contain large intracytoplasmic droplets of mucoid material.

Physarum A member of the Myxomycetes or acellular slime moulds. Normally exists as a multinucleate **plasmodium** that may be many centimetres across, but if starved and stimulated by light will produce spores that later germinate to produce amoeboid cells, myxamoebae, which may transform into flagellated swarm cells. Either of these cell types may fuse to produce a zygote that forms the plasmodium by synchronous nuclear division. Easily grown in the laboratory and much used for studies on cytoplasmic streaming and on the cell cycle (because they show synchronous DNA synthesis and nuclear division).

physical mapping The process of assembling genomic DNA clones that completely cover a genetic locus. In **genome projects**, this is an essential prerequisite for sequencing; in **positional cloning**, it assists in designing a strategy to identify the gene of interest. The procedure is to screen candidate clones for a series of characteristic marker sequences, based either on **satellite DNA** or on **PCR**-derived sequence-tagged sites. Clones that share particular markers are assumed to overlap in that region, and computer analysis is used to identify the smallest set of clones that completely cover the region.

physins Subgroup of the **tetraspan vesicle membrane proteins**. Mammalian physins include **synaptophysin**, **synaptoporin**, **pantophysin** and mitsugumin29.

physostigmine *Eserine* Reversible acetylcholine esterase inhibitor derived from the Calabar bean (from *Physostigma venenosum*), used as a drug in treating glaucoma and for treating anticholinergic syndrome.

phytic acid Inositol hexaphosphate, found in plant cells, especially in seeds, where it acts as a storage compound for phosphate groups.

phytoalexins Toxic compounds produced by higher plants in response to attack by pathogens and to other stresses. Sometimes referred to as plant antibiotics, but rather non-specific, having a general fungicidal and bactericidal action. Production is triggered by **elicitors**. Examples: **pisatin, phaseollin**.

phytochelatins **Metallothionein**-type peptides of plants that bind heavy metals such as cadmium, zinc, lead, mercury and copper. General form is $(\gamma\text{-glutamylcysteinyl})_n$-glycine, where *n* is from 2 to 11. Involved in the detoxification of heavy metals and the homeostasis of non-essential metals.

phytochrome Plant pigment protein that absorbs red light and then initiates physiological responses governing light-sensitive processes such as germination, growth and flowering. Exists in two forms, **Pr** and **Pfr** that are interconverted by light.

phytoestrogen A plant-derived substance with weak **estrogen**-like properties.

phytohaemagglutinin *PHA* Sometimes used as synonym for **lectins** in general, but more usually refers to lectin from seeds of the red kidney bean *Phaseolus vulgaris*. Binds to oligosaccharide containing *N*-acetyl galactosyl residues. Binds to both B- and T-lymphocytes, but acts as a **mitogen** only for T-cells.

phytohormones See **plant growth substances**.

phytol Long-chain fatty alcohol (C20) forming part of **chlorophyll**, attached to the protoporphyrin ring by an ester linkage.

phytonutrient Any of various organic substances derived from plants that are believed to have health-giving properties.

phytoremediation The use of plants to decontaminate soil, for example by absorbing pollutants such as heavy metals.

phytotoxins A diverse group of substances, often designed to inhibit herbivores or pests. See: aristolochic acids, pyrrolizidine alkaloids, beta-**carotene**, **coumarin**, the alkenylbenzenes safrole, methyleugenol and estragole, **ephedrine** alkaloids and synephrine, kavalactones, anisatin, St John's wort ingredients (**hyperforin**), cyanogenic glycosides, **picrotoxin**, **solanine** and **chaconine**, thujone and glycyrrhizinic acid.

pi protein π *protein* Polypeptide (35 kDa) that is required for the initiation of DNA replication in the R6K antibiotic-resistance plasmid, of which there are 12–18 copy equivalents in the *E. coli* chromosome.

PI-3-kinases *Phosphatidyl inositol-3-kinases; PI kinases* Lipid kinases that phosphorylate phosphatidylinositol phosphate on the 3 position. Now recognized to be key enzymes acting downstream of many receptors, particularly receptor tyrosine kinases such as PDGF-receptor (in the case of class Ia). Classical form has p85 regulatory subunit and p110 enzymatic subunit. The p85 adaptor associates with the cytoplasmic domain of various **growth factor receptors** through SH2 domains that bind to phosphotyrosine residues in the ligated (phosphorylated) receptor and with the catalytic subunit. An increasing family is being identified, some of which are regulated by **G-proteins** or calcium. Most are inhibited by wortmannin.

pia mater Innermost of the three meningeal membranes that surround the **brain**, lying between the dura mater and the arachnoid layer. Contains a plexus of small blood vessels.

Pichia canadensis See *Hansenula wingei*.

Pichia pastoris A methylotrophic yeast, capable of metabolizing methanol as the sole carbon source, that has been developed into a heterologous protein expression system.

Pick's disease Rare neurodegenerative disease similar in clinical symptoms to **Alzheimer's disease**. Affects mostly frontal and temporal lobes. Caused by mutations in the gene encoding **tau** or, in some cases, **presenilin-1**.

Picornaviridae Class IV viruses, with a single positive strand of RNA and an icosahedral capsid. There are two main classes: enteroviruses, which infect the gut and include poliovirus, and the rhinoviruses, which infect the upper respiratory tract (common cold virus, Coxsackie A and B, foot-and-mouth disease virus and hepatitis A).

picrotoxin *Cocculin* Toxic plant alkaloid, found primarily in the fruit ('fish berries') of *Anamirta cocculus*, an East Indian woody vine. Acts as a non-competitive antagonist of GABA$_A$ receptors and, since GABA is an inhibitory neurotransmitter, infusion of picrotoxin has a stimulative effect, sometimes used as an antidote for barbiturate poisoning.

Pierson syndrome *Microcoria-congenital nephrosis syndrome* Rare autosomal-recessive disease due to mutation in the gene encoding laminin beta-2 (**LAMB2**).

pigment cells Cells that contain pigment: see **melanocytes**, **chromatophores**.

pigmented retinal epithelium *PRE; retinal pigmented epithelium, RPE* Layer of unusual phagocytic epithelial cells lying below the photoreceptors of the vertebrate eye. The dorsal surface of the PRE cell is closely apposed to the ends of the rods, and as discs are shed from the rod outer segment they are internalized and digested by the PRE. Do not have **desmosomes** or **cytokeratins** in some species.

pigtail *Glypiation* One name for the covalent assembly of sugars linked to phosphatidyl inositol joined to the C-terminal residue of many proteins by a modified ethanolamine residue. Also called a greasy foot. Another term for this modification is glypiation. The function of the pigtail is to act as the sole anchor of the protein to the external surface of the lipid bilayer. The moiety is added to the protein during co-translational insertion into the ER membrane on the luminal side. The addition is synchronized with the removal of a large C-terminal polypeptide sequence that is usually hydrophobic and could itself have formed a membrane anchor. The surface proteins of many unicellular protozoa very commonly have this modification, the best known being the variable surface glycoprotein of **trypanosomes** and of **malaria** parasites. Examples are probably present in all eukaryotic plasma membranes.

PIIF *Proteinase-inhibitor-inducing factor* Factor produced by a plant in response to attack by insects. Induces the formation of a substance that inhibits the proteinase that the insect secretes to digest plant tissues. May be mobile within the plant, thus inducing inhibitor formation away from the site of original attack.

pilin 1. General term for the protein subunit of a **pilus**. 2. Protein subunit (7.2 kDa) of F-pili, **sex pili** coded for by the F-plasmid.

pilocarpine Alkaloid with muscarinic cholinomimetic activity isolated from *Pilocarpus jaborandi* (Jaborandi tree).

pilus *Pl.* pili; *fimbrium* (*Pl.* fimbria) Hair-like projection from surface of some bacteria. Involved in adhesion to surfaces (may be important in virulence), and specialized **sex-pili** are involved in conjugation with other bacteria. Major constituent is a protein, **pilin**.

pim pim-1, pim-2, pim-3 Family of oncogenes that encode serine/threonine **protein kinases**. The *pim-1* product is upregulated in prostate cancer; *pim-2* is 53% identical to *pim-1* at the amino acid level and shares substrate preference; *pim-3* is aberrantly expressed in human pancreatic cancer and phosphorylates **bad**. See also Table O1.

PIN-1 Protein of the **parvulin** family within the **peptidyl prolyl cis/trans-isomerase** (PPIase) group of proteins. Modulates the assembly, folding, activity and transport of essential cellular proteins. It is a mitotic regulator, interacting with a range of proteins that are phosphorylated prior to cell division, but phosphorylation of the PIN-1 **WW domain** (at Ser-16) by PKA abolishes the interactions between PIN-1 and its target proteins.

pinacocyte Flattened polygonal cell that lines ostia and forms the epidermis of sponges. Capable of synthesizing collagen.

pindolol A synthetic non-selective beta-adrenergic antagonist (beta-blocker) for the treatment of cardiovascular diseases such as hypertension and angina pectoris.

pinitol *D-Pinitol; 3-O-methyl-D-chiro-inositol* Naturally occurring compound found in certain plants, trees and foods, such as soy. It is claimed to have an insulin-like activity and to cause insulin sensitization, although double-blind clinical trials do not show it to have any detectable effects (neither toxic nor therapeutic).

pinocytosis Uptake of fluid-filled vesicles into cells (endocytosis). Macropinocytosis and micropinocytosis are distinct processes, the latter being energy independent and involving the formation of receptor–ligand clusters on the outside of the plasma membrane and **clathrin** on the cytoplasmic face.

pinocytotic vesicle Fluid-filled endocytotic vesicle, usually less than 150 nm in diameter. Micropinocytotic vesicles are around 70 nm in diameter.

pinosome A **pinocytotic vesicle**.

pioglitazone A thiazolidinedione glucose sensitizer used for the treatment of type II diabetes.

pioneer species Species that colonize 'bare' environments, for example after fire or pollution has destroyed the previous flora and fauna. Most attention has been paid to plants, but bacterial and animal pioneers must also exist.

PIP2 *Phosphatidyl inositol 4,5-bisphosphate* Formed by linked 'futile cycles' from **phosphatidyl inositol** via phosphatidyl inositol phosphate (PIP). Chiefly important because a ligand-activated PIP2-specific phosphodiesterase (phospholipase Cγ) breaks down PIP2 to form diacyl glycerol, which stimulates protein kinase C, and inositol 1,4,5 trisphosphate (InsP3) (which releases calcium from the endoplasmic store).

PIPES *1,4-Piperazinediethanesulfonic acid* One of the Good buffers; pK_a (20°C) = 6.8.

piroplasm Class of Protista, Phylum Apicomplexa (Sporozoa or Telosporidea), which includes the tick-transmitted parasite, Babesia.

Pisaster Echinoderm of the Class Asteroidea, a starfish.

pisatin **Phytoalexin** produced by peas.

pit Region of the plant cell wall in which the **secondary wall** is interrupted, exposing the underlying **primary cell wall**. One or more **plasmodesmata** are usually present in the primary wall, communicating with the other half of a pit pair. May be simple or bordered; in the latter case, the secondary wall overarches the pit field. Do not confuse with **coated pits**.

PITSLRE kinase family *CDK11 family kinases* A family of p34Cdc2-related protein kinases (CDKs), named according to the single amino acid code of an important regulatory region; generated by alternative splicing and promoter utilization from three duplicated and tandemly linked genes on human chromosome 1. Their function has been related to cell cycle regulation, splicing and apoptosis. Homologues are found in non-mammalian species.

pituicytes Dominant intrinsic cells of the neural lobe of the hypophysis. Have long branching processes and resemble **neuroglia**: secrete **antidiuretic hormone**.

pituitary *Hypophysis* Possibly the most important of the vertebrate endocrine glands. Located below the **brain** to which it is attached by a stalk. Has two lobes, the anterior adenohypophysis and posterior neurohypophysis. Secretes a wide range of hormones including **somatotrophin**, **follicle-stimulating hormone**, **gonadotrophins**, **thyroid-stimulating hormone**, **lipotropin** and many others. See Table H2.

Pitx2 Bicoid-type homeobox gene that is expressed asymmetrically in the left lateral plate mesoderm and may be involved in determining left-right asymmetry in mouse and chick.

pityriasis rosea A common self-limiting skin rash of young people, especially young adults. The aetiology is unclear.

piwi Family of *Drosophila* genes that plays an essential role in stem cell self-renewal, gametogenesis and RNA interference in diverse organisms ranging from *Arabidopsis* to human (*hiwi* genes). See **piwi domain**.

PIWI domain A highly conserved motif within **argonaute** proteins that has been shown to adopt an RNase H fold critical for the endonuclease cleavage activity of **RISC**.

pK_a See **association constant**.

PKA, B, etc. See **protein kinase A**, **protein kinase B**, etc.

PKI Abbrev. for (1) **Protein kinase inhibitor peptide,** (2) A cell proliferation marker protein, **pKi-67**, the antigen recognized by the Ki-67 MAb, (3) A **Her-2**/Her-1 inhibitor (PKI-166), (4) The negative log dissociation constant of a competitive inhibitor (pK_i), (5) Potato-kallikrein-inhibitors (PKI).

pKi-67 A cell proliferation marker protein that distributes to the chromosome periphery during mitosis and nucleolar heterochromatin during interphase. The antigen recognized by the Ki-67 MAb.

PKR See **protein kinase R**.

PLA2 *Phospholipase A2* See **phospholipases**.

placental calcium-binding protein *18a2; nerve growth factor-induced protein 42a; pE2-9δ; calvasculin p9k* Calcium-binding protein of placenta, uterus and vasculature containing the **EF-hand** motif.

placode Area of thickened ectoderm in the embryo from which a nerve ganglion or a sense organ will develop.

plakalbumin Fragment of ovalbumin produced by **subtilisin** cleavage: more soluble than ovalbumin itself.

plakins A family of giant cytoskeleton-binding proteins. One member is bullous pemphigoid antigen 1 (Bpag1)/dystonin, which has neuronal and muscle isoforms that have actin-binding and microtubule-binding domains at either end separated by a plakin domain and several spectrin repeats.

plakoglobin *γ-Catenin* Polypeptide (83 kDa) present at cell–cell but not cell–substratum contacts. Associated with **desmosomes** and with **adherens junctions**: soluble 7S form present in cytoplasm. See **catenins**.

plakophilins *PKP* Members of the **armadillo** family (arm-repeat) family of proteins, found in various cell types, both as an architectural component in desmosomes and dispersed in cytoplasmic particles. Mutations in PKP1 are the underlying cause of ectodermal dysplasia-skin fragility syndrome. Mutations in the plakophilin-2 gene (*PKP2*) have been found in patients with arrhythmogenic right ventricular dysplasia/cardiomyopathy (ARVC). Plakophilin-3 (PKP3) interacts with **plakoglobin, desmoplakin** and the epithelial keratin 18, and has been shown to bind all three **desmogleins, desmocollin**-3a and -3b, and possibly also desmocollin-1a and -2a. See **p0071**.

planapochromat Expensive microscope objective that is corrected for **spherical aberration** and **chromatic aberration** at three wavelengths.

plant growth substances Substances that, at low concentration, influence plant growth and differentiation. Formerly referred to as plant hormones or phytohormones, these terms are now suspect because some aspects of the 'hormone concept', notably action at a distance from the site of synthesis, do not necessarily apply in plants. Also known as 'plant growth regulators'. The major classes are **abscisic acid, auxin, cytokinin, ethylene** and **gibberellin**; others include steroid and phenol derivatives.

planula Free-living larval form of a hydrozoan cnidarian. Has an outer layer of ciliated ectoderm and an inner mass of endoderm cells, is flattened and bilaterally symmetrical.

plaque assay **1.** Assay for virus in which a dilute solution of the virus is applied to a culture dish containing a layer of the host cells; convective spread is prevented by making the medium very viscous. After incubation the 'plaques', areas in which cells have been killed (or transformed) can be recognized and the number of infective virus particles in the original suspension estimated. **2.** Assay for cells producing antibody against erythrocytes or antigen that has been bound to the erythrocytes. The cell is surrounded by a clear plaque of haemolysis. Basic principle behind the assay is the same as for the virus plaque assay.

plaque-forming cell Antibody-secreting cell detected in a **plaque assay**.

plaque-forming unit *pfu* Number of Ig-producing cells or infectious virus particles per unit volume. Of a virus-like bacteriophage λ, the number of viable viral particles, established by counting the number of plaques (see **plaque assay**) formed by serial dilution of the library. For example, a cDNA library might have a titre of 50 000 pfu/μl of library.

plasma Acellular fluid in which blood cells are suspended. Serum obtained by defibrinating plasma (plasma-derived serum) lacks platelet-released factors and is less suitable to support the growth of cells in culture.

plasma cell A terminally differentiated antibody-forming, and usually antibody-secreting, cell of the B-cell lineage.

plasma kallikrein A plasma **serine peptidase** with an **apple domain**.

plasma membrane The external, limiting **phospholipid bilayer** membrane of cells. See **transmembrane proteins**.

plasmacytoma Malignant tumour of **plasma cells**, very similar to a **myeloma** (plasmacytomas usually develop into multiple myeloma). Can easily be induced in rodents by the injection of complete **Freund's adjuvant**. Plasmacytoma cells are fused with primed lymphocytes in the production of monoclonal antibodies.

plasmal reaction Long-chain aliphatic aldehydes occurring in **plasmalogens** react with **Schiff's reagent** in the so-called plasmal reaction, to form, e.g. palmitaldehyde, stearaldehyde.

plasmalemma Archaic name for the plasma membrane of a cell (the term often included the cortical cytoplasmic region). Adjectival derivative (plasmalemmal) still current.

plasmalemmasome See **paramural body**.

plasmalogens A group of glycerol-based phospholipids in which the aliphatic side chains are not attached by ester linkages. Widespread distribution. Less easily studied than the acyl phospholipids.

plasmid *Episome* A small, independently replicating piece of cytoplasmic DNA that can be transferred from one organism to another. Linear or circular DNA molecules found in both pro- and eukaryotes capable of autonomous replication. 'Stringent' plasmids occur at low copy number in cells, 'relaxed' plasmids at high copy number, *ca.* 10–30. Plasmids can become incorporated into the genome of the host or can remain independent. An example is the **F-factor** of *E. coli*. May transfer genes and plasmids carrying antibiotic-resistant genes can spread this trait rapidly through the population. Described largely from bacteria and protozoa. Widely used in genetic engineering as vectors of genes (**cloning vectors**).

plasmid prep Generic term for the isolation of **recombinant** plasmids from liquid bacterial culture, usually by alkaline/detergent lysis, selective precipitation of other components and affinity purification of plasmid. As this is the most exciting thing most molecular biologists ever do, there is an informal shorthand for the scale of the preparation based on the size of the overnight culture: see **miniprep, midiprep, maxiprep** and **megaprep**.

plasmin *Fibrinolysin; fibrinase; thrombolysin; E.C. 3.4.21.7* Trypsin-like **serine endopeptidase** of the peptidase S1 family (human plasmin ~86 kDa) that is responsible for digesting **fibrin** in blood clots. Generated from **plasminogen** by the action of another peptidase, **plasminogen activator**. Also acts on activated **Hagemann factor** and on complement.

plasminogen Inactive precursor of **plasmin**; occurs at 200 mg/l in blood plasma. Contains multiple copies of the **kringle** domain. Human plasminogen is a 291 amino-acid glycoprotein with as many as 24 disulphide bonds.

plasminogen activator Serine peptidase that acts on **plasminogen** to generate **plasmin**. Has also been implicated in invasiveness and is produced by many normal and invasive cells. The vascular form (tPA; 55 kDa) is very similar to tissue-plasminogen activator (uPA; 70 kDa) and to **streptokinase** and **urokinase**.

plasminogen activator inhibitor-1 *PAI-1; mesosecrin* Principal tissue inhibitor of **plasminogen activator**, a **serpin**, 50 kDa. Plasminogen activator inhibitor-2 (PAI-2) is secreted by the placenta and is only present in significant amounts during pregnancy.

plasmodesma *Pl.* plasmodesmata Narrow tube of cytoplasm penetrating the plant cell wall, linking the protoplasts of two adjacent cells. A desmotubule runs down the centre of the tube, which is lined by plasma membrane.

plasmodium 1. Multinucleate mass of protoplasm bounded only by a plasma membrane; the main vegetative form of acellular slime moulds (e.g. *Physarum*). **2.** See *Plasmodium*.

Plasmodium Genus of parasitic protozoa that cause **malaria**. *Plasmodium vivax* causes the tertian type, *P. malariae* the quartan type and *P. falciparum* the quotidian or irregular type of disease. The life cycle is complex and involves an intermediate host, the female mosquito (*Anopheles*), which infects the vertebrate host when taking a blood meal. Predominant form of the organism in humans is the intracellular parasite (the **merozoite**) in the erythrocyte, where it undergoes a form of multiple cell division termed **schizogony**. As a result, the erythrocyte bursts and the progeny infect other erythrocytes. Eventually some cells develop into gametes that, when ingested by a female mosquito, will fuse in her gut to form a zygote (ookinete). Multiple cell division within the resultant oocyte, attached to the gut wall, gives rise to infective sporozoites; these migrate to the salivary glands and are ejected with the saliva the next time the mosquito takes a blood meal.

plasmogamy Fusion of cytoplasm that occurs when protoplasts or gametes fuse. In most organisms the latter is followed more or less immediately by karyogamy (fusion of nuclei); in some fungi it may result in the formation of a heterokaryon.

plasmolysis Process by which the plant cell protoplast shrinks, so that the plasma membrane becomes partly detached from the wall. Occurs in solutions of high osmotic potential, due to water moving out of the protoplast by osmosis.

plastid Type of plant cell organelle, surrounded by a double membrane and often containing elaborate internal membrane systems. Partially autonomous, containing some DNA, RNA and ribosomes, and reproducing itself by binary fission. Includes **amyloplasts**, **chloroplasts**, **chromoplasts**, **etioplasts**, **leucoplasts**, **proteinoplasts** and **elaioplasts**. Develop from **proplastids**.

plastin Microfilament-bundling protein from mammalian cells very similar to **fimbrin** with two actin-binding domains. Two forms, L-plastin (627 residues, may be identical to **acumentin**) from leucocytes and t-plastin (630 residues). See **p65**.

plastocyanin An electron-carrying protein present in chloroplasts, forming part of the electron transport chain. Contains two copper atoms per molecule. Associated with **photosystem I**.

plastoglobuli Globules found in plastids, containing principally lipid, including **plastoquinone**.

plastoquinone A **quinone** present in chloroplasts, forming part of the photosynthetic electron transport chain. Closely associated with **photosystem II**. May be stored in **plastoglobuli**.

plate count Number of bacterial colonies that grow on a nutrient agar plate under defined conditions. Used to estimate bacterial contamination of water, etc.

platelet Anucleate discoid cell (3 μm diameter) found in large numbers in blood; important for blood coagulation and for haemostasis. Platelet α-granules contain lysosomal enzymes; dense granules contain ADP (a potent platelet-aggregating factor) and **serotonin** (a vasoactive amine).

They also release **platelet-derived growth factor,** which presumably contributes to later repair processes by stimulating fibroblast proliferation.

platelet-activating factor *PAF; PAFacether; 1-0-hexadecyl-2-acetyl-sn-glycero-3-phosphorylcholine* Potent activator of many leucocyte functions, not just platelet activation.

platelet basic protein *PBP* Protein (94 residues) that is naturally processed via N-terminal cleavage to yield connective tissue activating peptide III (85 residues; CTAP-III), **beta-thromboglobulin** (81 residues) and neutrophil-activating peptide-2 (NAP-2; 70 residues).

platelet factor 3 Phospholipid associated with the platelet plasma membrane that contributes to the blood clotting cascade by forming a complex (thromboplastin) with other plasma proteins and activating **prothrombin**.

platelet factor 4 Platelet-released protein that promotes blood clotting by neutralizing **heparin**.

platelet-derived growth factor *PDGF* The major mitogen in serum for growth in culture of cells of connective tissue origin. It consists of two different but homologous polypeptides A and B ($\sim$30 000 Da) linked by disulphide bonds. Believed to play a role in wound healing. The B-chain is almost identical in sequence to p28sis, the transforming protein of simian sarcoma virus that can transform only those cells that express receptors for PDGF, suggesting that transformation is caused by **autocrine** stimulation. Receptor is a **tyrosine kinase**.

platykurtic See **leptokurtic**.

PLCPI *Porcine leucocyte cysteine protease inhibitor* Stefin-type peptidase inhibitor (103 residues, 11 kDa) that co-purifies with **cathelin**. Inhibits papain and cathepsins L and S by forming a tight complex.

PLD 1. Phospholipase D. **2.** Pegylated liposomal doxorubicin. **3.** PLD-118 is a novel, oral antifungal drug, formerly BAY 10-8888, a synthetic derivative of the naturally occurring beta-amino acid cispentacin. **4.** The Protein Ligand Database (PLD) is a publicly available web-based database that aims to provide further understanding of protein–ligand interactions.

pleckstrin Protein of 47 kDa, the major substrate for protein kinase C in platelets. Pleckstrin homology domains (**PH domains**) are found in a number of proteins.

plectin Abundant linker protein of cytomatrix (apparent 300 kDa, but 466 kDa on basis of cDNA sequence). Co-localizes with various intermediate filament proteins and may be involved in their cross-linking or anchoring. Also has an actin-binding domain comprising two **calponin** homology domains near the amino terminus.

plectonemic One of two topologically distinct forms of supercoiling of DNA: plectonemic supercoiling is found in DNA molecules freely suspended in solution; solenoidal supercoiling is a characteristic of DNA molecules wrapped around histones. Plectonemic associations of DNA molecules cannot be disrupted by deproteination, whereas paranemic joints can.

pleiomorphic Having more than one body shape during the life cycle, or having the ability to change shape or to adopt a variety of shapes.

pleiotrophin *PTN* A secreted 18-kDa heparin-binding protein that stimulates mitogenesis and angiogenesis and neurite and glial process outgrowth guidance activities *in vitro*.

pleiotropic Having multiple effects. For example, the cyclic AMP concentration in a cell will have a variety of effects because the cAMP acts to control a protein kinase that in turn affects a variety of proteins.

plesiomorphic See **apomorphic**.

Pleurobrachia Small, free-swimming marine organism, member of the Phylum Ctenophora. Roughly spherical and transparent, with most of the body made up from transparent jelly-like material. The animal has two long tentacles for catching prey, and swims by means of eight rows of **comb plates** (made of fused cilia) that run along the body.

Pleurodeles Genus of salamanders.

pleuropneumonia-like organism *PPLO* See **mycoplasma**.

plexins Transmembrane receptors of **semaphorins**. Subtype-specific functions of the majority of the nine members of the mammalian plexin family are largely unknown. Plexin-A is the receptor for semaphorin 3A; binding of which leads to local translation of RhoA in the nerve growth cone and subsequent collapse.

PLGF *Placental growth factor* Growth factor similar in activity to **VEGF**.

P-light chain *DNTB light chain* Myosin light chain that can be phosphorylated by **myosin light-chain kinase**; as a result of phosphorylation, the myosin is activated.

P-loop See **ATP-binding site**.

pluripotent stem cell Cells in a stem cell line capable of differentiating into several different final differentiated types, e.g. there may be a pluripotent stem cell line for erythrocytes, granulocytes and megakaryocytes.

pluronic Proprietary name for surfactants, based on ethylene oxide and propylene oxide that can function as antifoaming agents, wetting agents, dispersants, thickeners and emulsifiers.

plusbacin A See **katanosin B**.

pluteus Free-swimming ciliated larval stage of some echinoderms.

PM1 **1.** Particulate matter (PM) in air, with a diameter less than 1 μm. PM2.5, PM10, etc. are particles of greater diameter. **2.** Bacterial strain PM1 will rapidly and completely biodegrade the petrol-additive methyl tertiary-butyl ether (MTBE) in groundwater. **3.** A transformed CD4$^+$ T-cell clone derived from the Hut78 T-cell line. **4.** Monoclonal antibody (precursor marker 1: PM1) that labels most neuroepithelial cells in day 4 embryonic chick retinal sections. **5.** A lignin-degrading basidiomycete, strain PM1 (= CECT 2971).

PMA *TPA; tumour promotor activity* Phorbol myristate acetate, a **phorbol ester**.

PMAT *Plasma membrane monoamine transporter* See **ENTs**.

PMF **1.** **Proton motive force**. **2.** Potential of mean force (PMF). Concept used in molecular dynamics, the energy associated with the probability of being in a particular state.

PML body *Promyelocytic leukaemia body* Nuclear structure containing multimers of PML protein and a range of other nucleoproteins including the **Nijmegen breakage disease** syndrome protein (p95/Nbs1, **nibrin**) which assists in repair of double-strand breaks in DNA.

PMN *PMNL* Polymorphonuclear leucocyte: could be an **eosinophil**, **basophil** or **neutrophil**, but usually intended to mean the latter (an idle habit).

PMSF See **phenylmethylsulphonyl fluoride**.

PNA See **peptide nucleic acid**.

pneumococci Gram-positive pyogenic organisms (about 1 μm diameter), usually encapsulated, closely related to streptococci; associated with diseases of the lung.

pneumoconiosis An occupational disease of the lungs caused by excessive inhalation of dust. Various types are recognized including silicosis, asbestosis and anthrocosis.

Pneumocystis carinii Organism that commonly causes pneumonia in immunocompromised patients (e.g. with AIDS). Apparently most closely related to ustomycetous yeasts.

pneumocyte Cells that line the alveoli of the lung. Type I pneumocytes are squamous. Type II pneumocytes are smaller, roughly cuboidal cells, usually found at the alveolar septal junctions, responsible for secreting surfactant and will replicate to replace damaged type I pneumocytes.

pneumolysin Cholesterol-binding toxin from *Streptococcus pneumoniae*.

PNMT *Phenylethanolamine N-methyl transferase* Terminal enzyme in the catecholamine biosynthetic pathway; it converts noradrenaline to adrenaline.

podocalyxin A heavily glycosylated single-pass transmembrane protein (140 kDa) mainly found on the apical membrane of rat renal glomerular epithelial cells (**podocytes**) and also in endothelial, hematopoietic and tumour cells.

podocytes Cells of the visceral epithelium that closely invest the network of glomerular capillaries in the kidney. Most of the cell body is not in contact with the **basal lamina**, but is separated from it by trabeculae that branch to give rise to club-shaped protrusions, known as pedicels, interdigitating with similar processes on adjacent cells. The complex interdigitation of these cells produces thin filtration slits that seem to be bridged by a layer of material (of unknown composition) which acts as a filter for large macromolecules.

podophyllotoxin Glucoside toxin (414 Da), derived from the roots of the American Mayapple (*Podophyllum peltatum*), that binds to tubulin and prevents microtubule assembly. **Etoposide** is a derivative of podophyllotoxin.

podosomes Punctate substratum-adhesion complexes in osteoclasts. Contain **vinculin**, **talin**, **fimbrin** and **F-actin**. Podosomes form a broad ring of contacts with the underlying bone, and the enclosed area below the cell is then absorbed.

poikilocytosis Irregularity of red cell shape.

poikilotherm Organism whose body temperature varies with environment; opposite of homeotherm. Though poikilothermic animals are often referred to as cold-blooded, this is not necessarily true.

point mutation **Mutation** that causes the replacement of a single base pair with another pair.

pokeweed mitogen Any of the **lectins** derived from the pokeweed, *Phytolacca americana*, all of which will stimulate T-cells. Binds β-D-acetylglucosamine.

***pol* genes** Genes coding for **DNA polymerases,** of which there are three in *E. coli* – *polA, polB* and *polC*, coding for polymerases I, II and III, respectively. *Pol* genes in oncogenic retroviruses code for **reverse transcriptase**.

polar body In animals, each meiotic division of the oocyte leads to the formation of one large cell (the egg) and a small polar body as the other cell. Polar body formation is a consequence of the very eccentric position of the nucleus and the spindle.

polar granules Granules containing a basic protein found in insect eggs that induce the formation of and become incorporated into germ cells.

polar group Any chemical grouping in which the distribution of electrons is uneven enabling it to take part in electrostatic interactions.

polar lobe In some molluscs, a polar lobe appears as a clear protrusion close to the vegetal pole of the cell prior to the first cleavage and becomes associated with only one of the daughter cells. Removal of the first polar lobe, or of any polar lobe that forms at a subsequent mitosis, leads to defects in the embryo; it seems that the polar lobe contains special morphogenetic factors.

polar plasm Differentiated cytoplasm associated with the animal or vegetal pole of an oocyte, egg or early embryo.

polarity Literally 'having poles' (like a magnet), but used to describe cells that have one or more axes of symmetry. In epithelial cells, the polarity meant is between apical and basolateral regions; in moving cells, having a distinct front and rear. Some cells seem to show multiple axes of polarity (which will hinder forward movement).

polarization microscopy Any form of microscopy capable of detecting birefringent objects. Usually performed with a polarizing element below the stage to produce plane-polarized light, and an analyser that is set to give total extinction of the background, and thus to detect any birefringence.

pole cell A cell at or near the animal or vegetal pole of an embryo.

pole fibres Microtubules inserted into the pole regions of the mitotic spindle, each pole is the product of the division of the centrioles and constitutes a **microtubule-organizing centre**.

polehole *phl* *Drosophila* homologue of the *raf* **oncogene**.

poliovirus A member of the enterovirus group of **Picornaviridae** that causes poliomyelitis.

pollen mother cell A diploid plant cell that forms four **microspores** by meiosis; the microspores give rise to pollen grains in seed plants.

pollen tube A tubular outgrowth produced when a pollen grain germinates. In angiosperms it grows through the tissue of the stigma to the embryo sac and delivers the male gamete(s). In *Arabidopsis* growth of the pollen tube is guided by gradients of **GABA**.

polo Founding member of the family of **polo-like kinases** (Plks), identified in a *Drosophila* screen for mutants affecting spindle pole behaviour.

polo-box domain *PBD* See **polo-like kinases**.

polo-like kinases *Plks* A conserved family of serine/threonine kinases, characterized by the presence of a C-terminal domain termed the polo-box domain (PBD) in addition to the N-terminal kinase domain, with many members throughout various species. Multiple Plks are present in mammalian cells (Plk1, Plk2/Snk, Plk3/Fnk/Prk and Plk4/Sak) and *Xenopus* (Plx1-3), whereas in other species only one member has been identified, like Polo in *Drosophila*, Cdc5 in budding yeast and Plo1 in fission yeast. Required for various stages of mitosis.

poly(A) binding protein *PABP* See **maskin**.

poly-A polymerase *PAP; EC 2.7.7.19* Enzyme that polyadenylates mRNA (adds the poly(A) tail).

poly(ADP-ribose)polymerase *PARP; EC 2.4.2.30* An abundant nuclear protein activated by DNA nicks and important in DNA repair. PolyADP ribosylation, brought about by ADP-ribosyl protein ligase, is a post-transcriptional modification of proteins and **p53** is one of the proteins that can be modified in this way. PARP knockout mice show defects in fibroblast proliferation and impaired capacity to handle radiation-induced damage. One of the earliest proteins cleaved by **caspase 3** in apoptosis.

poly-A See **polyadenylic acid**.

poly-A tail Polyadenylic acid sequence of varying length found at the 3′ end of most eukaryotic mRNAs. Histone mRNAs do not have poly-A tail. The poly-A tail is added post-transcriptionally to the primary transcript as part of the nuclear processing of RNA, yielding **hnRNAs** with 60–200 adenylate residues in the tail. In the cytoplasm, the poly-A tail on mRNAs is gradually reduced in length. The function of the poly-A tail is not clear, but it is the basis of a useful technique for the isolation of eukaryotic mRNAs. The technique uses an **affinity chromatography** column with oligo(U) or oligo(dT) immobilized on a solid support. If cytoplasmic RNA is applied to such a column, poly-A-rich RNA (mRNA) will be retained.

polyacrylamide gel electrophoresis *PAGE* Analytical and separative technique in which molecules, particularly proteins, are separated by their different electrophoretic mobilities in a hydrated gel. The gel suppresses convective mixing of the fluid phase through which the electrophoresis takes place and contributes molecular sieving. Commonly carried out in the presence of the anionic detergent sodium dodecylsulphate (SDS). SDS denatures proteins so that non-covalently associating subunit polypeptides migrate independently, and by binding to the proteins confers a net negative charge roughly proportional to the chain weight. See also **SDS-PAGE**.

polyadenylic acid **Polynucleotide** chain consisting entirely of residues of adenylic acid (i.e. the base sequence is AAAA... AAAA). Polyadenylic chains of various lengths are found at the 3′ end of most eukaryotic mRNAs, the **poly-A tail**.

polyamine Polycations at physiological pH, polyamines can bind and interact with various other molecules within the cell. In particular interact with DNA, but may also modulate ion channels and act as growth factors. **Spermine** has four positive charges, **spermidine** has three. The precursor of both, **putrescine**, has two.

polyanion Macromolecule carrying many negative charges. The commonest in cell-biological systems is nucleic acid.

polyarteritis nodosa Rare necrotizing inflammation of medium-sized or small-sized arteries, without glomerulonephritis or vasculitis in arterioles, capillaries or venules. Now distinguished from **microscopic polyangiitis**. An autoimmune disease, although the origin of the immune complexes is uncertain; antineutrophil cytoplasmic antibodies (**ANCA**) are present in some, but not all, patients.

polycation Macromolecule with many positively charged groups. At physiological pH the most commonly used in cell biology is poly-L-lysine; this is often used to coat surfaces, thereby increasing the adhesion of cells (which have net negative surface charge). See also **cationized ferritin**.

polycistronic mRNA A single **mRNA** molecule that is the product of the **transcription** of several tandemly arranged genes; typically the mRNA transcribed from an **operon**.

polyclonal antibody An antibody produced by several clones of B-lymphocytes as would be the case in a whole animal. Usually refers to antibodies raised in immunized animals, whereas a **monoclonal** antibody is the product of a single clone of B-lymphocytes, usually maintained *in vitro*.

polyclonal compartment When the progeny of several cells occupy an area or volume with a defined boundary, it is referred to as a polyclonal compartment, e.g. clones lying close to the midline of the wing of *Drosophila*.

polycloning site *Multiple cloning site; MCS* Region of a phage or plasmid vector that has been engineered to contain a series of **restriction sites** that are usually unique within the entire vector. This makes it particularly easy to insert or excise (subclone) DNA fragments.

Polycomb *Pc Drosophila* gene that when mutated leads to extra sex combs on the legs of male flies, suggesting that the posterior legs have become anterior legs. There are at least 10 genes in the Polycomb group; they are thought to act by **transcriptional silencing** of **homeotic genes**.

polycystic ovary syndrome *PCOS; Stein–Leventhal syndrome* A metabolic syndrome with many other symptoms, ovarian cysts arise through incomplete follicular development or failure of ovulation. Associated with insulin resistance and consequent hyperinsulinaemia and (frequently) hyperlipidaemia and obesity. There is evidence for polymorphism at the CYP11A1 (cytochrome P450, subfamily XIA) locus.

polycystin *PC1, PC2* Polycystin-1 is a modular membrane protein with a long extracellular N-terminal portion with several ligand-binding domains, 11 transmembrane domains and an intracellular C-terminal portion (*ca.* 200 aa) with several phosphorylation sites. Polycystin-1 (*PKD1* gene product) may act as a mechanosensor, receiving signals from the primary cilia, neighbouring cells and extracellular matrix and transducing them into cellular responses that regulate proliferation, adhesion and differentiation

of renal tubules and kidney morphogenesis. Mutations in polycystin-1 (PC1) underlie most cases of autosomal-dominant polycystic kidney disease (ADPKD). Polycystin-2 (PC2), the *PKD2* gene product, and the related protein polycystin-L, function as Ca^{2+}-permeable, non-selective cation channels in different expression systems.

polycythaemia Increase in the haemoglobin content of the blood, because of either a reduction in plasma volume or an increase in red cell numbers. The latter may be a result of abnormal proliferation of red cell precursors (polycythaemia vera, Vaquez–Osler disease). Most cases of polycythaemia vera are associated with a somatic mutation in the *JAK2* gene.

polydnaviruses *PDVs* Viruses that have been described in thousands of parasitoid wasp species; have a segmented DNA genome in viral particles and an integrated form that persists as a provirus in the wasp genome. Two genera of PDVs phylogenetically unrelated exist, the bracoviruses (BVs) and the ichnoviruses (IVs), associated with braconid and ichneumonid wasps, respectively.

polyelectrolyte An ion with multiple charged groups.

polyendocrine syndrome Autoimmune disorder (the antigen to which the response is mounted is in the **B cells** of the pancreas) in which there is involvement of several organ systems. Autoimmune polyendocrinopathy syndrome type I is caused by mutation in the autoimmune regulator gene (*AIRE*), the product of which is probably a transcription factor. Polyendocrinopathy, immune dysfunction and diarrhoea, X-linked (XPID) are caused by mutation in *FOXP3*.

polyene Any organic compound containing two or more carbon–carbon double bonds.

polyene antibiotics Group of structurally related antibiotics produced by *Streptomyces* spp. Interact with sterols in eukaryotic membranes. Examples are **amphotericin B** and **nystatin**.

polyethylene glycol *PEG* A hydrophilic polymer that interacts with cell membranes and promotes fusion of cells to produce viable hybrids. Often used in producing **hybridomas**. See **pegylation**.

polyethylenimine *PEIs* Synthetic polymers with a high cationic charge density which function as transfection reagents based on their ability to compact DNA or RNA into complexes.

polygalacturonan Plant cell wall polysaccharide consisting predominantly of galacturonic acid. May also contain some rhamnose, arabinose and galactose. Those with significant amounts of rhamnose are termed **rhamnogalacturonans**. Found in the **pectin** fraction of the wall.

polygalacturonase Enzyme that degrades **polygalacturonan** by hydrolysis of the glycosidic bonds that link galacturonic acid residues. Important in fruit ripening and in fungal and bacterial attack on plants.

polygenic Something that is controlled or caused by the action of many genes. Thus many of the major non-infectious diseases (for example, arthritis, cardiovascular disease, asthma, diabetes) are likely to be caused by the interaction of many genes; no single gene mutation is responsible, rather the coincidence of polymorphic variants that together contribute risk factors that predispose an individual to the disease.

polyhedrin Major protein (28 kDa) forming the crystalline matrix of viral polyhedral bodies (occlusion bodies) that form within baculovirus-infected cells.

polyisoprenylation See **geranylation**.

polylysine A polycationic polymer of **lysine**, it carries multiple positive charges and is used to mediate adhesion of living cells to synthetic culture substrates, or of fixed cells to glass slides (for observation by fluorescence microscopy, for example).

polymer A macromolecule made of repeating (monomer) units or **protomers**.

polymerase chain reaction *PCR* The first practical system for *in vitro* amplification of DNA, and as such one of the most important recent developments in molecular biology. Two synthetic oligonucleotide primers, which are complementary to two regions of the target DNA (one for each strand) to be amplified, are added to the target DNA (that need not be pure), in the presence of excess deoxynucleotides and **Taq polymerase**, a heat-stable DNA polymerase. In a series (typically 30) of temperature cycles, the target DNA is repeatedly denatured (around 90°C), annealed to the primers (typically at $50 - 60°C$), and a daughter strand extended from the primers (72°C). As the daughter strands themselves act as templates for subsequent cycles, DNA fragments matching both primers are amplified exponentially, rather than linearly. The original DNA need thus be neither pure nor abundant, and the PCR reaction has accordingly become widely used not only in research but also in clinical diagnostics and forensic science.

polymerization The process of polymer formation. In many cases this requires **nucleation** and will only occur above a certain critical concentration.

polymorphic epithelial mucin See **episialin**.

polymorphism 1. The existence, in a population, of two or more alleles of a gene, where the frequency of the rarer alleles is greater than can be explained by recurrent mutation alone (typically greater than 1%). HLA alleles of the **major histocompatibility complex** are very polymorphic. 2. The differentiation of various parts of the units of colonial animals into different types of unit specialized for different purposes, e.g. as in the colonial hydroid *Obelia*.

polymorphonuclear leucocyte *PMNL; PMN* Mammalian blood leucocyte (granulocyte) of myeloid series in distinction to mononuclear leucocytes: see **neutrophil**, **eosinophil**, **basophil**.

polymyxins Group of peptide antibiotics produced by *Bacillus* spp. Molecular weights are around 1000–1200 Da and the molecules are cyclic. Act against many Gram-negative bacteria, working apparently by increasing membrane permeability.

polynucleotide Linear sequences of **nucleotides**, in which the 5′-linked phosphate on one sugar group is linked to the 3′ position on the adjacent sugars. In the polynucleotide DNA the sugar is **deoxyribose,** and in RNA it is **ribose**. They may be double-stranded or single-stranded, with varying amounts of internal folding.

polyol Any polyhydric alcohol. Common examples are inositol, mannitol and sorbitol.

polyomavirus A DNA tumour virus with very small genome (of the **Papovaviridae**). Polyoma was isolated from mice, in which it causes no obvious disease, but when injected at high titre into baby rodents (including mice) it causes tumours of a wide variety of histological types (hence polyoma). *In vitro*, infected mouse cells are permissive for virus replication, and thus are killed, whilst hamster cells undergo **abortive infection**, and at a low frequency become transformed.

polyp 1. Growth, usually benign, protruding from a mucous membrane. 2. The sessile stage of the Cnidarian (**coelenterate**) life cycle; the cylindrical body is attached to the substratum at its lower end and has a mouth surrounded by tentacles bearing **nematocysts** at the upper end; *Hydra* and the feeding-polyps of the colonial *Obelia* are examples.

polypeptide Chains of α-**amino acids** joined by peptide bonds. Distinction between peptides, oligopeptides and polypeptides is arbitrarily by length; a polypeptide is perhaps more than 10 residues.

polypeptide antibiotics Bactericidal antibiotics (**bacitracin**, colistin, **Polymyxin B**) with activity against Gram-negative aerobic bacilli including *Pseudomonas aeruginosa*. Polymyxin B and colistin are not active against *Proteus* sp. and have no activity against Gram-positive organisms. Both act by disrupting the bacterial cell membrane.

polyphenism An adaptation in which a genome is associated with discrete alternative phenotypes in different environments – for example, the solitary and migratory forms of the locust.

polyploid Of a nucleus, cell or organism that has more than two **haploid** sets of **chromosomes**. A cell with three haploid sets (*3n*) is termed triploid, with four sets (*4n*) tetraploid, and so on.

polypodial Adjective describing an amoeba with several pseudopods.

polyposis coli *Adenomatous polyposis coli; familial adenomatous polyposis; FAP* Hereditary disorder (Mendelian dominant) characterized by the development of hundreds of adenomatous **polyps** in the large intestine, which show a tendency to progress to malignancy. The *APC* gene has also been implicated in a chromosome 5 gastric and pancreatic cancer.

polyprotein Protein that, after synthesis, is cleaved to produce several functionally distinct polypeptides. Some viruses produce such proteins, and some polypeptide hormones seem to be cleaved from a single precursor polyprotein (**pro-opiomelanocortin**, for example).

polyribosome Functional unit of protein synthesis consisting of several **ribosomes** attached along the length of a single molecule of mRNA.

polysaccharide Polymers of (arbitrarily) more than about 10 monosaccharide residues linked glycosidically in branched or unbranched chains.

polysialic acid *PSA* Potential regulator of cell–cell interactions. Polysialic acid chains in **glycoproteins** may have negative regulatory effects on cell–cell contact. Thus the low PSA form of **NCAM** is thought to promote cell–cell contact and enhance **fasciculation** whereas NCAM with a high PSA content is thought to prevent close membrane–membrane apposition.

polysome See **polyribosome**.

polysomy Situation in which all chromosomes are present, and some are present in greater than the diploid number – for example, trisomy 21 (**Down's syndrome**).

polyspermy Penetration of more than one spermatozoon into an ovum at time of fertilization. Occurs as normal event in very yolky eggs (e.g. bird), but then only one sperm fuses with egg nucleus. Many other eggs have mechanisms to block polyspermy.

Polysphondylium A genus of **Acrasidae**, the cellular slime moulds.

polytene chromosomes Giant **chromosomes** produced by the successive replication of homologous pairs of chromosomes, joined together (synapsed) without chromosome separation or nuclear division. They thus consist of many (up to 1000) identical chromosomes (strictly chromatids) running parallel and in strict register. The chromosomes remain visible during interphase and are found in some ciliates, ovule cells in angiosperms, and in larval dipteran tissue. The best known polytene chromosomes are those of the salivary gland of the larvae of *Drosophila melanogaster*, which appear as a series of dense bands interspersed by light interbands, in a pattern characteristic for each chromosome. The bands, of which there are about 5000 in *D. melanogaster*, contain most of the DNA (*ca.* 95%) of the chromosomes, and each band roughly represents one gene. The banding pattern of polytene chromosomes provides a visible map to compare with the linkage map determined by genetic studies. Some segments of polytene chromosome show chromosome **puffs**, areas of high transcription.

Polytron Proprietary name for a tissue homogenizer.

polyuridylic acid Homopolymer of uridylic acid. Historically was used as an artificial mRNA in cell-free **translation** systems, where it coded for polyphenylalanine; thus began the deciphering of the genetic code.

polyvinylidene fluoride *PVDF* A polymer that is very non-reactive and can be used in contact with biological materials and in medical applications. Also used in sensors.

polyvinylpyrrolidone *PVP* Polymer used to bind phenols in plant homogenates, and hence to protect other molecules, especially enzymes, from inactivation by phenols. Also occasionally used to produce viscous media for gradient centrifugation.

POMC See **pro-opiomelanocortin**.

POMC/CART neurons See **arcuate nucleus**.

Pompe's disease Severe glycogen **storage disease** caused by deficiency in α (1-4)-glucosidase, the lysosomal enzyme responsible for glycogen hydrolysis. Even though the non-lysosomal glycogenolytic system is normal, glycogen still accumulates in the lysosomes.

Ponceau red *Ponceau S; Fast Ponceau 2B* Dye used to stain proteins.

ponticulin Developmentally regulated 17-kDa transmembrane glycoprotein from *Dictyostelium* that regulates actin binding and nucleation. Preferentially located at actin-rich regions such as sites of cell adhesion. Analogue found in human neutrophils.

POP2 *PGK promoter-directed over production-2; CAF1* The yeast POP2 protein (Pop2p) is a component of a global transcription regulatory complex and is required for gene expression of many genes in *Saccharomyces cerevisiae*. It is a nuclease (RNase) of the **DEDD superfamily**, a subunit of the Ccr4-Not complex that mediates 3′ to 5′ mRNA deadenylation involved in mRNA turnover.

population diffusion coefficient Coefficient that describes the tendency of a population of motile cells to diffuse through the environment. Its use presupposes that the cells move in a random walk. Can also be applied to populations of free-living motile organisms.

porins *Outer membrane channel proteins* Transmembrane matrix proteins (37 kDa) found in the outer membranes of Gram-positive bacteria. Associate as trimers to form channels (1 nm in diameter, *ca.* 105 per bacterium) through which hydrophilic molecules of up to 600 Da can pass. Similar porins are also found in outer mitochondrial membranes (VDAC, voltage-dependent anion-selective channel). Multiple genes coding for VDAC homologues have been discovered in eukaryotic genomes, but their function is unclear.

porocyte Type of cell in asconoid sponges (small, simple sponges with a tube-shaped body) through which water enters the spongocoel.

porphyria Any of a group of disorders in which there is excessive excretion of porphyrins or their precursors.

porphyrins Pigments derived from porphin: all are chelates with metals (Fe, Mg, Co, Zn, Cu, Ni). Constituents of haemoglobin, chlorophyll, cytochromes.

Porphyromonas gingivalis collagenase *Proteinase C; prtC gene product* Peptidase with 'collagenase' activity from a Gram-negative anaerobe associated with periodontal lesions (strain W83 has been fully sequenced). Catalytic site of unknown character, hence in U-family of peptidases. Inhibited by EDTA and thiol blocking agents.

POSH *Plenty of SH3s* A scaffold protein for the Jun N-terminal kinase (**JNK**) signal transduction pathway.

position effect Effect on the expression of a gene depending upon its position relative to other genes on the chromosome. Moving (transposing) a gene from an inactive region to an active region can alter expression markedly – sometimes with unfortunate consequences, as with the **Philadelphia chromosome** abnormality that leads to CML.

positional cloning Identification of a gene based on its location in the genome. Typically, this will result from **linkage** analysis based on a mutation in the target gene, followed by a **chromosome walk** from the nearest known sequence.

positional information The instructions that are interpreted by cells to determine their differentiation in respect of their position relative to other parts of the organism, e.g. digit formation in the limb-bud of vertebrates.

positive control Mechanism for gene regulation that requires that a regulatory protein must interact with some region of the gene before transcription can be activated.

positive feedback See **feedback**.

positive-strand RNA viruses Class IV and VI viruses that have a single-stranded RNA **genome** that can act as mRNA (plus strand) and in which the virus RNA is itself infectious. Includes **Picornaviridae**, **Togaviridae** and **Retroviridae**.

post-transcriptional gene silencing Inactivation of a gene by destruction of the mRNA, similar to quelling in fungi and **RNA interference** (RNAi) in animals.

post-translational modification Changes that occur to proteins after peptide bond formation has occurred. Examples include glycosylation, acylation, limited proteolysis, phosphorylation, isoprenylation.

postcapillary venule That portion of the blood circulation immediately downstream of the capillary network; the region having the lowest wall-shear stress and the most common site of leucocytic margination and endothelial transmigration (diapedesis).

postsynaptic cell In a chemical **synapse**, the cell that receives a signal (binds neurotransmitter) from the presynaptic cell and responds with depolarization. In an electrical synapse, the postsynaptic cell would just be downstream, but since many electrical synapses are **rectifying**, one of the two cells involved will always be postsynaptic.

postsynaptic potential In a synapse, a change in the **resting potential** of a postsynaptic cell following stimulation of the presynaptic cell. For example, in a cholinergic synapse, the release of acetylcholine from the presynaptic cell causes channels to open in the postsynaptic cell. Each channel opening causes a small depolarization, known as a **miniature end plate potential** (mepp); these sum to produce an excitatory postsynaptic potential.

postsynaptic protein *43-kDa postsynaptic protein* A peripheral membrane protein closely associated with the cytoplasmic portion of the **nicotinic acetylcholine receptor** and thought to help anchor them in the postsynaptic membrane. Highly conserved. Probably **rapsyn**.

potassium channel Ion channel selective for potassium ions. There are diverse types with different functions, for example: **delayed rectifier channels**, **M-channels**, A-channels, inward rectifier channels, Ca-dependent K^+ channels.

potassium-sparing diuretics Class of mild diuretics that act on the kidney to promote excretion of water without loss of potassium ions. Example is amiloride.

potato blight Destructive disease of the potato caused by either of the parasitic fungi *Alternaria solani* (early blight) or *Phytophthora infestans* (late blight). Late blight was responsible for the Irish potato famine of the 1840s.

potato glycoalkaloids Potatoes, and other members of the Solanaceae, upregulate their production of range of glycoalkaloids in response to stress. These compounds are a defence mechanism against pathogens such as viruses, bacteria, fungi and insects. The major alkaloids are **chaconine** and **solanine**, but a range of others are known, including: leptine I and II, leptine I and II, commersonine, demissine, alpha-solamargine, alpha-solasonine, beta-solamarine, alpha-solamarine and alpha-tomatine. All are likely to be toxic to mammals and many act by inhibition of acetylcholine esterase.

potato lectin Lectin from the potato, *Solanum tuberosum*. Binds to *N*-acetyl glucosaminyl residues.

potency In toxicology, an expression of relative toxicity of an agent as compared to a given or implied standard or reference.

potentiation **1.** Increase in quantal release at a synapse following repetitive stimulation. Whereas **facilitation** at

synapses lasts a few hundred milliseconds, potentiation may last minutes to hours. **2.** Phenomenon in which a substance or physical agent, at a concentration or dose that does not itself have an effect, enhances the effect or response to another substance or physical agent. Sometimes referred to as 'priming'.

potocytosis Transport of small molecules across membrane using **caveolae** rather than **coated vesicles**.

POU domain A conserved protein domain of around 150 amino acids, composed of a 20 amino-acid **homeobox** domain and a larger POU-specific domain, and so is the target of some **transcription factors**. Named POU (Pit-Oct-Unc) after three such proteins: Pit-1, which regulates expression of certain pituitary genes; Oct-1 and 2, which bind an octamer sequence in the promoters of histone H2A and some immunoglobulin genes; and Unc-86, involved in nematode sensory neuron development.

Poxviridae Class I viruses with double-stranded DNA genome that codes for more than 30 polypeptides. They are the largest viruses and their shell is complex, consisting of many layers, and includes lipids and enzymes, amongst which is a DNA-dependent RNA polymerase. Uniquely among the DNA viruses they multiply in the cytoplasm of the cell, establishing what is virtually a second nucleus. The most important poxviruses are **vaccinia**, **variola** (smallpox) and **myxoma virus**.

POZ domain *Poxvirus zinc-finger domain* A family of transcription factors, characterized by the presence of a protein–protein interaction domain called the POZ or BTB domain at their N-terminus and zinc fingers at their C-terminus that play a role in the control of growth arrest and differentiation in several types of mesenchymal cell. See **BTB/POZ domain**.

PP1 See **protein phosphatase-1**.

pp46 See **neuromodulin**.

pp60src The phosphoprotein (60 kDa) encoded by the *src* oncogene. A protein tyrosine kinase; see **src family**.

PPAR *Peroxisome proliferator-activated receptors* PPARα stimulates β-oxidative degradation of fatty acids, PPARγ promotes lipid storage by regulating adipocyte differentiation. Are implicated in metabolic disorders predisposing to atherosclerosis and inflammation. PPARα-deficient mice show prolonged response to inflammatory stimuli. PPARα is activated by gemfibrozil and other fibrate drugs.

PPD *Purified protein derivative* Protein purified from the culture supernatant of tubercle bacteria (*Mycobacterium tuberculosis*) and used as a test antigen in **Heaf** and **Mantoux tests**.

PPI See **proton-pump inhibitor**.

PPIase *Peptidyl-prolyl* cis–trans-*isomerase* Enzymes that accelerate protein folding by catalysing *cis–trans* isomerizations. **Immunophilins** are PPIases though their enzymic activity may not be essential for their immunosuppressive effects.

PPLO Pleuropneumonia-like organisms. See **mycoplasma**.

P-proteins *Phloem-specific proteins; PP1, PP2* Proteins found in large amounts in phloem sieve tubes. There are two major proteins: PP1, the phloem filament protein (appears as thin strands when seen in the electron microscope), which contains structural motifs in common with cysteine proteinase inhibitors, and PP2, which has lectin activity and RNA-binding properties. PP2 is widely distributed through the vascular plants (even in the absence of PP1), and though it has conserved sequence motifs is polymorphic in size between species.

Pr The form of **phytochrome** that absorbs light in the red region (660 nm) and is thus converted to **Pfr**. In the dark, the equilibrium between Pr and Pfr favours Pr, which is therefore more abundant.

Prader–Willi syndrome Syndrome in which there is an absence of paternal chromosome 15q11q13. Short stature, obesity and mild mental retardation are features of the syndrome. Uniparental disomy leads to differences between this and **Angelman syndrome,** where it is the equivalent maternal region that is deleted. See **imprinting**.

prazosin Antagonist of α-adrenergic receptors.

pRb Protein product of the *retinoblastoma* gene.

PRC Abbrev. for (1) phase response curve: the phase shifts produced in an oscillator by stimuli applied at different initial phase states of that oscillator, (2) bacterial protease (**Prc protease**), (3) progesterone receptor isoform, PRc, (4) plasma renin concentration, (5) see **PRC-barrel domain**.

Prc protease *EC 3.4.21.102; tail-specific protease (tsp)* The *E. coli* protease Prc (Tsp) exhibits specificity *in vitro* for proteins with non-polar carboxyl termini, and may be involved in protection of the bacterium from thermal and osmotic stresses.

PRC-barrel domain Superfamily of protein domains, the PRC-barrels, approximately 80 residues long, widely represented in bacteria, archaea and plants. This domain is also present at the carboxyl terminus of the pan-bacterial protein RimM, which is involved in ribosomal maturation and processing of 16S rRNA. Prototype is the PRC-barrel identified in the H subunit of the purple bacterial photosynthetic reaction centre (PRC-H).

pre-eclampsia Condition during pregnancy in which there is oedema, high blood pressure and albuminuria; may cause **eclampsia** unless treated.

pre-pro-protein A preprotein is a form that contains a signal sequence that specifies its insertion into or through membranes. A proprotein is one that is inactive; the full function is only present when an inhibitory sequence has been removed by proteolysis. A pre-pro-protein has both sequences still present. Pre-pro-proteins usually only accumulate as products of *in vitro* protein synthesis.

pre-prophase band Band of microtubules 1–3 μm wide that appears just below the plasma membrane of a plant cell before the start of mitosis. The position of the pre-prophase band determines the plane of cytokinesis and of the cell plate that will eventually separate the two cells.

preBCR *PreB-cell antigen receptor* Receptor on immature B-cells that contains the immunoglobulin mu heavy chain (Ig mu) and signals to the preB-cell that heavy chain rearrangement has been successful, a process termed heavy chain selection. In association with the **B-cell receptor** signals through tyrosine kinases including **Blk**.

precipitin Any antibody that forms a precipitating complex (a precipitin line) with an appropriate multivalent antigen. The term is now outmoded.

precocial Descriptive of those birds in which the hatchlings are independent from the outset, able to move around and feed independently, as opposed to altricial species, where the young require an extended period of parental attention before becoming independent.

prednisolone *1,4-Pregnadiene-11β, 17α, 21-triol-3,20-dione* Steroid with glucocorticoid action, very similar to **prednisone**. An effective anti-inflammatory drug but with serious side effects.

prednisone *1,4Pregnadiene-17α, 21-diol-3,11,20-trione* Synthetic steroid that acts as a glucocorticoid, with powerful anti-inflammatory and antiallergic activity.

p-region The centromere divides each chromosome into two regions: the smaller one, which is the p-region, and the bigger one, the q region. Loci are identified, for example as 1p35 (chromosome 1, p-region, subregion 35).

preleptonema Rarely used term that designates an extra stage in the prophase of meiosis I. Usually included in **leptotene**.

prenylation Post-translational addition of prenyl groups to a protein. Farnesyl, geranyl or geranyl–geranyl groups may be added. Consequence is usually to promote membrane association.

pre-prophase Rarely used term to designate an extra stage of mitosis, normally included as part of prophase.

presecretory granules Vesicles near the maturation face of the Golgi. Also known as Golgi-condensing vacuoles.

presenilins *PS1, PS2* Multipass transmembrane proteins, PS1 and PS2, found in Golgi. Mutations in genes for PS1 are associated with 25% of early onset **Alzheimer's disease** and altered amyloid β protein (β-**amyloid precursor protein**) processing. PS1 is a functional homologue of SEL-12, a protein found in *C. elegans* that facilitates signalling mediated by **Notch**, and the expression of PS1 seems to be essential for the spatio-temporal expression of *Notch1* and *Dll1* (*Delta-like gene 1*) during embryogenesis. PS1 and PS2 are also similar to *C. elegans Spe-1*, a gene involved in protein trafficking in the Golgi during spermatogenesis. See **gamma-secretase**.

prespore cells Cells in the rear portion of the migrating slug (grex) of a cellular slime mould, which will later differentiate into spore cells. Can be recognized as having different proteins by immunocytochemical methods. See also **Acrasidae**.

prestalk cells Cells at the front of the migrating grex of cellular slime moulds that will form the stalk upon which the **sorocarp** containing the spores is borne. See **prespore cells**.

presynaptic cell In a chemical synapse, the cell that releases neurotransmitter that will stimulate the **postsynaptic cell**. In an electrically synapsed system, the cell that has the first action potential, but since electrical synapses are usually rectifying, one of the two cells involved is always presynaptic.

presynaptic receptor Receptors located on presynaptic terminals at **synapses**.

prezygonema Rarely used term to designate an extra stage in the prophase of meiosis I. Usually included within **zygotene**.

PRH See **proline-rich homeodomain**.

Pribnow box See **promoter**.

prickle cell Large, flattened polygonal cells of the stratum germinosum of the epidermis (just above the basal stem cells) that appear in the light microscope to have fine spines projecting from their surfaces; these terminate in desmosomes that link the cells together, and have many tonofilaments of **cytokeratin** within them.

primaquine An 8-aminoquinoline drug, used prophylactically and therapeutically to treat malaria. Affects the mitochondria of the *exo*-erythrocytic stages (see *Plasmodium*), but the mechanism is not understood. Currently the most effective drug at preventing spread of all four species of human malaria.

primary cell culture Of animal cells, the cells taken from a tissue source and their progeny grown in culture before subdivision and transfer to a subculture. See also Table C3.

primary cell wall A plant cell wall that is still able to expand, permitting cell growth. Growth is normally prevented when a **secondary wall** has formed. Primary cell walls contain more **pectin** than secondary walls, and no lignin is present until a secondary wall has formed on top of them.

primary ciliary dyskinesia *Immotile cilia syndrome* Disorder in which ciliary function is abnormal as a result of a defect in **dynein**, although this may arise in several different ways. Can result in **Kartagener's syndrome** and **situs inversus**.

primary immune response The immune response to the first challenge by a particular antigen. Usually less extensive than the secondary immune response, being slower and shorter lived with smaller amounts of lower affinity antibody being produced.

primary lymphoid tissue See **lymphoid tissues**.

primary lysosome A **lysosome** before it has fused with a vesicle or vacuole.

primary meristem Synonym for an **apical meristem**.

primary oocyte The enlarging ovum before maturity is reached, as opposed to the secondary oocyte or polar body.

primary spermatocyte A stage in the differentiation of the male germ cells. Spermatogonia differentiate into primary spermatocytes, showing a considerable increase in size in doing so; primary spermatocytes divide into secondary spermatocytes.

primary transcript RNA transcript immediately after transcription in the nucleus, before **RNA splicing** or polyadenylation to form the mature **mRNA**.

primary tumour The mass of tumour cells at the original site of the neoplastic event – from the primary tumour **metastasis** will lead to the establishment of secondary tumours.

primase *EC 2.7.7*. The enzyme that polymerizes nucleotide triphosphates to form oligoribonucleotides in a 5′ to 3′ direction. The enzyme synthesizes the RNA for RNA–DNA sequences that later become **Okazaki fragment**s and also RNA primers for some types of phage using an sDNA template.

primer extension Technique for finding the **transcriptional** start site of a gene. **mRNA**s cannot be relied on to be complete at the 5′ end, so a labelled antisense oligonucleotide **primer** is designed to complement the putative mRNA near its 5′ end, and used to prime a **reverse transcription** reaction. The products are run on a sequencing gel and the lengths of products allow the putative start sites to be deduced.

priming Treatment that does not in itself elicit a response from a system but that induces a greater capacity to respond to a second stimulus. See **potentiation**.

primitive erythroblast Large cell with euchromatic nucleus found in mammalian embryos. In the mouse, the cells are located in the yolk sac and are responsible for early production of erythrocytes with foetal haemoglobin.

primitive streak Thickened elongated region of cells in early mammalian and avian embryos that marks the location of embryonic axis. Hensen's node is at the end of the primitive streak until the cellular movements of gastrulation cause it to regress caudally.

primordial germ cells Germ cells at the earliest stage of development. Since germ cells may originate in the embryo at some distance from the gonads, they then have to migrate to the gonadal primordia – a process that may involve chemotaxis or, more probably, random movement with trapping.

primosome Complex of proteins involved in the synthesis of the RNA primer sequences used in DNA replication. Main components are **primase** and **DNA helicase**, which move as a unit with the **replication fork**.

P-ring One of the bushes at the base of the flagellum of Gram-negative bacteria, anchoring it in the peptidoglycan layer of the cells wall. Lies below the **L-ring**.

prions Suggested, and generally accepted, as the causative agents of several infectious diseases (transmissible **spongiform encephalopathies**) such as **scrapie** (in sheep), kuru and **Creutzfeldt–Jakob disease** in man. Prions (proteinaceous-infective particles) apparently contain no nucleic acid. The 27-kDa protein of scrapie is related to a normal cell protein and may possibly cause its overproduction. See **PrP, Gerstmann–Straussler–Scheinker syndrome**.

pristane *2,6,10,14-Tetramethylpentadecane* Extracted from shark liver. Will induce a lupus-like syndrome in non-autoimmune mice, and a form of experimental arthritis.

PRK *PKC-related kinase; PKN* Serine/threonine kinases (120 kDa). PRK-1 is found in hippocampus and is activated by phospholipids and arachidonic acid, binds to rho GTP and possibly regulates cytoskeletal changes. PRK-2 (PKN-2; PAK-2) was isolated from U937 cells and foetal brain, and is activated by cardiolipin and acidic phospholipids. Binds **rac** and **rho** which will activate its kinase activity and interacts with SH3 domain of **Nck** and PLCγ.

PRK1 See **protein kinase N**.

Prk1p One of the yeast actin-regulating serine/threonine kinases that is localized to cortical actin patches, which may be sites of endocytosis. Involved in regulating the **Sla1p/End3p/ Pan1p** complex. See **Ark1p**.

PRL 1. Phosphatases found in regenerating liver (PRL)-1, PRL-2 and PRL-3 (also known as PTP4A1, PTP4A2 and PTP4A3, respectively). A family of **protein tyrosine phosphatases** (PTPs) modified by farnesylation. **2. Prolactin.**

PRND Gene for **doppel**.

prodrug Compound that is pharmacologically inactive (or relatively inactive) but is metabolized to the active form of the drug once in the body.

proenzyme Enzyme that does not have full (or any) function until an inhibitory sequence has been removed by limited proteolysis. See also **zymogen**.

pro-opiomelanocortin Polyprotein produced by the anterior pituitary that is cleaved to yield **adrenocorticotrophin**, α, β and γ-t**melanocyte-stimulating hormones**, lipotropic hormones, β-**endorphin** and other fragments.

proacrosin *EC 3.4.21.10* Proenzyme (zymogen) of **acrosin**, activated by zona pellucida glycoproteins into beta-acrosin during the acrosome reaction. Activation involves the removal of a C-terminal segment rich in proline residues and the cleavage of the Arg23–Val24 bond, leading to the formation of the light and heavy chains.

proadrenomedullin-20 *PAMP* Post-translational processing of proadrenomedullin generates two biologically active peptides, **adrenomedullin** (AM) and proadrenomedullin N-terminal 20 peptide (PAMP). PAMP has both angiogenic potential and antimicrobial capability. Binds to the G-protein-coupled receptor MrgX2, one of the family of *Mas-related* gene products or sensory neuron-specific G-protein-coupled receptors.

proanthocyanidins *Oligomeric proanthocyanidins; OPCs* Class of flavonoid complexes found in grape seeds and skin; act as antioxidants, they are the main precursors of the blue-violet and red pigments in plants. See **anthocyanidin**.

proband First patient to present with a disorder, usually heritable, and from whom the descent can be traced.

probe General term for a piece of DNA or RNA corresponding to a gene or sequence of interest that has been labelled either radioactively or with some other detectable molecule, such as **biotin**, **digoxygenin** or **fluorescein**. As stretches of DNA or RNA with complementary sequences will hybridize, a probe will label viral **plaques**, bacterial colonies or bands on a gel that contains the gene of interest. See also **Northern blots, Southern blots**.

probenecid *4-[(dipropylamino)sulfonyl] benzoic acid* Drug used in treating gout; promotes excretion of uric acid by inhibiting the tubular reabsorption of urate (and a number of other compounds such as penicillin, aspirin, indometacin, sulfonamides and sulfonylureas).

procaine Organic base (234 Da). Procaine butyrate, borate and hydrochloride (novocaine) are used as local anaesthetics.

procambium Plant **meristem** that gives rise to the primary vascular system.

Procardia Proprietary name for **nifedipine**.

procaryote See **prokaryote**.

procentriole The forming centriole composed of microtubules. Multiple procentrioles are present in some cells as a structure called the blepharoplast.

procolipase Precurspor of **colipase**, the protein cofactor for pancreatic lipase. The N-terminal pentapeptide is cleaved off as **enterostatin**. Procolipase (–/–) knockout mice have a severely reduced fat digestion and fat uptake.

procollagen Triple-helical trimer of collagen molecules in which the terminal extension peptides are linked by disulphide bridges; the terminal peptides are later removed by specific proteases (**procollagen peptidases**) to produce a **tropocollagen** molecule.

procollagen peptidases *EC 3.4.24.19* Enzyme (100 kDa) of the peptidase family M12 (astacin family) that removes the terminal extension peptides of **procollagen**; deficiency of these enzymes leads to **dermatosparaxis** or **Ehlers–Danlos syndrome**. Activity is increased by Ca^{2+} and by an enhancer glycoprotein.

proctolin Bioactive neuropeptide (RYLPT) that modulates interneuronal and neuromuscular synaptic transmission in a wide variety of arthropods; orphan G-protein-coupled receptor CG6986 of *Drosophila* is probably the proctolin receptor.

prodigiosin See *Serratia marcescens*.

prodromal A prodromal sign is an early indication of a disease, often before classical symptoms appear.

profilaggrin A major protein component of the **keratohyalin granules** of mammalian epidermis. It contains 10–12 tandemly repeated **filaggrin** units and is processed into the intermediate filament-associated protein filaggrin by specific dephosphorylation and proteolysis during terminal differentiation of the epidermal cells. One of the *fused* gene family.

profilin Actin-binding protein (15 kDa) that forms a complex with G-actin, rendering it incompetent to nucleate F-actin formation. The profilin/G-actin complex seems to interact with inositol phospholipids that may regulate the availability of nucleation-competent G-actin.

proflavine A synthetic acridine dye, deep orange in colour, used as a powerful antiseptic during WWII to dress wounds. Proflavine is mutagenic and intercalates between base pairs in DNA.

progenitor cell In development, a 'parent' cell that gives rise to a distinct **cell lineage** by a series of cell divisions.

progeria Accelerated ageing syndrome in which most of the characteristic stages of human senescence are compressed into less than a decade. Hutchinson–Gilford progeria syndrome is caused by mutation in the *lamin* A gene.

progesterone *Luteohormone* Hormone (314 Da) produced in the **corpus luteum**, as an antagonist of **estrogens**. Promotes proliferation of uterine mucosa and the implantation of the blastocyst; prevents further follicular development. See Table H3.

programmed cell death A form of cell death, best documented in development, in which activation of the death mechanism requires protein synthesis. Morphologically, the cell appears to die by **apoptosis** though this is not necessarily the case. Presumably requires some form of genetic code that determines that certain cells are to die at specific stages and specific sites during development. Classic example is the death of cells in the spaces between the developing digits of vertebrates, thus dividing them.

progress zone An undifferentiated population of mesenchyme cells beneath the **apical ectodermal ridge** of the chick limb-bud from which the successive parts of the limb are laid down in a proximo-distal sequence.

prohibitin A highly conserved protein with multiple functions in the nucleus and the mitochondria. Prohibitin is involved in mitochondrial biogenesis and function and is a potential tumour suppressor that represses the activity of E2F transcription factors while enhancing p53-mediated transcription. See **PHB domain**.

prohormone A protein hormone before processing to remove parts of its sequence and thus make it active.

proinsulin **Insulin** precursor produced in the pancreas that is enzymically processed to generate active insulin.

projectin Projectin and kettin are **titin**-like proteins mainly responsible for the high passive stiffness of insect indirect flight muscles. Projectin is very large (1 MDa) and has its amino-terminus embedded in the Z-bands of the sarcomere with an adjacent elastic region, possibly the **PEVK**-like domain. A member of the functionally and structurally heterogeneous family of **myosin light-chain kinases**, its location is different in synchronous and asynchronous muscles.

Prokaryotes Organisms, bacteria and cyanobacteria, characterized by the possession of a simple naked DNA chromosome, occasionally two such chromosomes, usually of circular structure, without a nuclear membrane and possessing a very small range of organelles, generally only a plasma membrane and ribosomes.

prokineticins *PK1 (endocrine gland vascular endothelial growth factor); PK2 (Bv8)* Small, cysteine-rich secreted proteins that regulate diverse biological processes, including gastrointestinal motility, angiogenesis and circadian rhythms. Two closely related G-protein-coupled receptors (PKR1 and PKR2) mediate signal transduction of prokineticins.

prolactin Pituitary lactogenic hormone (23 kDa), synthesized on ER-bound ribosomes as preprolactin that has an N-terminal signal peptide that is cleaved from the mature form. The conversion of preprolactin to prolactin has been much used as an assay for membrane insertion.

prolactin-releasing peptide *PrRP* Originally reported to act in the anterior lobe of the pituitary gland to stimulate **prolactin** release but subsequently shown to have various other effects. In the central nervous system, PrRP inhibits food intake, stimulates sympathetic tone and activates stress hormone secretion. Receptor is G-protein-coupled receptor 10 (GPR10).

prolamellar body The disorganized membrane aggregations in chloroplasts that have been deprived of light (**etioplasts**).

prolamine proteins Plant storage proteins that form granules within specialized areas of ER (prolamine protein bodies; PPBs). Prolamine mRNA specifically locates to restricted areas of ER, apparently through a microfilament-based mechanism, and prolamines are retained, even though they lack a lumenal retention signal, by BiP-mediated folding and aggregation into PPBs.

proliferating cell nuclear antigen *PCNA; cyclin* Commonly used marker for proliferating cells, a 35-kDa protein that associates as a trimer and as a trimer interacts with DNA polymerases δ and ε; acts as an auxiliary factor for DNA repair and replication. Transcription of PCNA is modulated by p53. The sequence of PCNA is well conserved between plants and animals and there are homologues in prokaryotes.

proliferative unit (of epidermis). The basal layer of the mammalian epidermis contains cells that undergo repeated divisions. The cells outwards from a particular basal cell are often derived from this cell or a nearby one so that columns of cells exist running outwards from the stem cell in the basal layer from which they were derived. Such columns of cells are referred to as proliferative units.

proliferin *PLF; mitogen-regulated protein* A hormone, related to **prolactin**, associated with the induction of cell division that is triggered by serum. A family of proliferin genes has been recognized.

proline *pro; P; 115 Da* One of the 20 amino acids directly coded for in proteins. Structure differs from all the others, in that its side chain is bonded to the nitrogen of the α-amino group, as well as the α-carbon. This makes the amino group a secondary amine and so proline is described as an imino acid. Has strong influence on secondary structure of proteins and is much more abundant in collagens than in other proteins, occurring especially in the sequence glycine-proline-**hydroxyproline**. A proline-rich region seems to characterize the binding site of **SH3** domains. See Table A2.

proline-rich homeodomain *PRH* The proline-rich homeodomain protein (PRH/Hex) is an important transcription factor involved in the control of cell proliferation and differentiation and in the regulation of multiple processes in embryonic development.

proline-rich polypeptide *PRP; colostrinin* A peptide isolated from ovine colostrum that has a regulatory effect on the immune response. A nonapeptide fragment (VESYV-PLFP) seems to have full activity.

proline-rich proteins *PRPs* Proline-rich proteins include collagens, complement 1q and salivary PRPs. Salivary PRPs may protect against dietary tannins. Protein domains that bind proline-rich motifs (PRMs) are frequently involved in signalling events. See **proline-rich homeodomain**.

prolyl hydroxylase *EC 1.14.11.2* See **hydroxyproline**.

promastigote Stage in the life cycle of certain trypanosomatid protozoa (e.g. *Leishmania*) that resembles the typical adult form of members of the genus *Leptomonas*. The cell is elongate or pear-shaped with a central nucleus and at the anterior end a **kinetoplast** and a basal body from which arises a single long, slender flagellum.

prometaphase Rarely used term that designates an extra stage in mitosis, starting with the breakdown of the nuclear envelope. Usually lumped in with **metaphase**.

promethazine Antihistamine drug with sedative properties. Used for treating hay fever, urticaria, as an antiemetic and to relieve insomnia.

prominin A stem cell marker, prominin-1 (CD133), a pentaspan membrane protein (115/120 kDa) found on membrane protrusions of the apical surface of neuroepithelial cells and on haematopoietic stem cells. Other members of the prominin family are being identified, as well as splice variants of prominin-1.

promoter A region of DNA to which RNA polymerase binds before initiating the transcription of DNA into RNA. The nucleotide at which transcription starts is designated +1 and nucleotides are numbered from this, with negative numbers indicating upstream nucleotides and positive downstream nucleotides. Most bacterial promoters contain two **consensus sequences** that seem to be essential for the binding of the polymerase. The first, the Pribnow box, is at about −10 and has the consensus sequence 5′-TATAAT-3′. The second, the −35 sequence, is centred about −35 and has the consensus sequence 5′-TTGACA-3′. Most factors that regulate gene transcription do so by binding at or near the promoter and affecting the initiation of transcription. Much less is known about eukaryote promoters; each of the three RNA polymerases has a different promoter. RNA polymerase I recognizes a single promoter for the precursor of rRNA. RNA polymerase II, which transcribes all genes coding for polypeptides, recognizes many thousands of promoters. Most have the Goldberg–Hogness or **TATA box** that is centred about position −25 and has the consensus sequence 5′-TATAAAA-3′. Several promoters have a CAAT box around −90 with the consensus sequence 5′-GGCCAATCT-3′. There is increasing evidence that all promoters for genes for 'housekeeping' proteins contain multiple copies of a GC-rich element that includes the sequence 5′-GGGCGG-3′. Transcription by polymerase II is also affected by more distant elements known as enhancers. RNA polymerase III synthesizes 5S ribosomal RNA, all tRNAs and a number of small RNAs. The promoter for RNA polymerase III is located within the gene either as a single sequence, as in the 5S RNA gene, or as two blocks, as in all tRNA genes.

promoter insertion Activation of a gene by the nearby **integration** of a virus. The **long-terminal repeat** acts as a promoter for the host gene. A form of **insertional mutagenesis**.

promyelocytes Cells of the bone marrow that derive from myeloblasts and will give rise to myelocytes; precursors of **myeloid cells** and neutrophil granulocytes.

pronase Mixture of proteolytic enzymes from *Streptomyces griseus*. At least four enzymes are present, including trypsin and chymotrypsin-like peptidases.

pronucleus Haploid nucleus resulting from meiosis. In animals, the female pronucleus is the nucleus of the ovum before fusion with the male pronucleus. The male pronucleus is the sperm nucleus after it has entered the ovum at fertilization but before fusion with the female pronucleus. In plants, the pronuclei are the two male nuclei found in the pollen tube.

propantheline bromide Antimuscarinic drug that blocks parasympathetic innervation and thereby reduces secretion and mobility of the stomach and intestine.

properdin *Factor P* Component of the alternative pathway for **complement** activation: complexes with C3b and stabilizes the alternative pathway C3 convertase (C3bBbP) that cleaves C3.

prophage The genome of a **lysogenic** bacteriophage when it is integrated into the chromosome of the host bacterium. The prophage is replicated as part of the host chromosome.

prophase Classical term for the first phase of mitosis or of one of the divisions of meiosis. During this phase the chromosomes condense and become visible.

prophylaxis Preventative action that will, for example, prevent infection; thus vaccination is a prophylactic treatment.

propidium iodide Used as a fluorescent stain for DNA and also for detecting dead cells which are permeable and therefore stain.

Propionibacterium Genus of bacteria that will ferment glucose to propionic acid or acetic acid.

proplastid Small, colourless **plastid** precursor, capable of division. It can develop into a chloroplast or other form of plastid and has little internal structure. Found in cambial and other young cells.

propolis A resinous substance collected by bees from the leaf buds and bark of trees, especially poplar and conifer trees and used, along with beeswax, to construct the combs. Has some antibiotic properties and is used as a nutritional supplement although the health benefits are unproven.

propranolol Potent adrenergic antagonist acting at β1 and β2 adrenergic receptors.

proprietary name Trade name of a drug – for example, Zantac is the proprietary name for ranitidine. With a few exceptions, drugs are described in the Dictionary only under their generic names.

proprioception Sensory awareness of body position, often unconscious but essential for motor control. *Adj.* proprioceptive.

prorenin Inactive precursor of **renin**.

Prosite Searchable database of conserved protein domains. Useful in inferring likely function of novel proteins.

prosome Raspberry-shaped ribonucleoprotein particle (19S) composed of small cytoplasmic RNA (15%) and heat-shock proteins, thought to be involved in post-transcriptional repression of mRNA translation: found in both nucleus and cytoplasm.

prospero *Drosophila* gene, product of which is asymmetrically distributed in the division of neural stem cells (neuroblasts) and is not present in one daughter (pluripotent neuroblast) but is retained in the ganglion mother cell (which has more restricted developmental potential). Prospero protein, once released from interaction with *miranda*, translocates to the nucleus and causes differential gene expression.

prospherosome Proposed stage in the development of **spherosomes** in plant cells. There is an accumulation of lipid in the prospherosome that is mobilized at a later stage.

prostacyclin *PGI2* Unstable **prostaglandin** released by mast cells and endothelium, a potent inhibitor of platelet aggregation; also causes vasodilation and increased vascular permeability. Release enhanced by **bradykinin**.

prostaglandins *PGs* Group of compounds derived from arachidonic acid by the action of **cyclo-oxygenases** that produces cyclic endoperoxides (PGG2 and PGH2) that can give rise to **prostacyclin** or **thromboxanes** as well as prostaglandins. Were originally purified from prostate (hence the name), but are now known to be ubiquitous in tissues. PGs have a variety of important roles in regulating cellular activities, especially in the inflammatory response, where they may act as vasodilators in the vascular system, cause vasoconstriction or vasodilation together with bronchodilation in the lung, and act as hyperalgesics. Prostaglandins are rapidly degraded in the lungs, and will not therefore persist in the circulation. Prostaglandin E2 (PGE2) acts on **adenylate cyclase** to enhance the production of **cyclic AMP**.

prostanoids Collective term for **prostaglandins, prostacyclins** and **thromboxanes**: slightly narrower than **eicosanoids**.

prosthetic group A tightly bound non-polypeptide structure required for the activity of an enzyme or other protein, e.g. the **haem** of **haemoglobin**.

protamine Highly basic (arginine-rich) protein that replaces **histone** in sperm heads, enabling DNA to pack in an extremely compacted form, e.g. clupein, iridin (4 kDa). See also **transition proteins**.

protease See also **peptidase**, the preferred modern term. The term was normally reserved for endopeptidases that have very broad specificity and would cleave most proteins into small fragments. These are usually the digestive enzymes, e.g. **trypsin**, **pepsin**, etc., or enzymes of plant origin (e.g. ficin, papain) or bacterial origin (e.g. pronase, proteinase K). Proteases are widely used for peptide mapping and for structural studies. See Table P1.

protease inhibitors Any inhibitor of an enzyme (peptidase) that breaks down proteins, but commonly used as shorthand for drugs, such as indinavir, which inhibit the action of the protease involved in producing mature virus particles and are used in combination therapy for AIDS.

protease M *Neurosin* See **neurosin**.

protease nexin-1 *PN-1; glial-derived neurite-promoting factor; glia-derived nexin (GDN); serpinE2* A serine protease inhibitor (**serpin**), 44 kDa. Can inhibit thrombin, plasmin and plasminogen activators, but when associated with glycosaminoglycans its activity is mainly directed towards thrombin. PN-1 has been shown to be a neuroprotective factor in a number of assay systems.

protease-activated receptor *PAR1, 2, 3* PAR1 is the human **thrombin** receptor, PAR2 is a possible trypsin receptor and PAR3 is similar to PAR1. All are G-protein-coupled receptors activated by cleavage of part of the extracellular domain at K38/T39: the cleaved fragment then acts as a ligand and stimulates phosphoinositide hydrolysis. Found on platelets and megakaryocytes.

protectin *CD59* Membrane-bound (GPI-anchored) complement regulatory glycoprotein, which protects cells from bystander attack by autologous complement by preventing the formation of the membrane attack complex. See **protectin D1**.

protectin D1 *PD1; neuroprotectin D1; 10,17S-docosatriene* A member of a family of bioactive products generated from docosahexaenoic acid. PD1 blocks T-cell migration *in vivo*, inhibits TNFα and interferon-gamma secretion and promotes apoptosis mediated by raft clustering. See **protectin**.

protegrin-1 Porcine **cathelicidin** of 18 amino acids.

protegrins Family of **cathelin**-associated antimicrobial peptides found in mammalian leucocytes. Will kill a range of bacteria, including Gram-positives, and are active against multidrug-resistant strains.

protein A linear polymer of amino acids joined by peptide bonds in a specific sequence.

Protein 4.1 *Band 4.1; 4.1R, HGMW-approved symbol EPB41; 4.1G, HGMW-approved symbol EPB41L2* An abundant protein (80 kDa phosphoprotein) of the human erythrocyte, in which it stabilizes the **spectrin**/actin cytoskeleton. Multiple protein 4.1 isoforms are generated by alternative pre-mRNA splicing, differential use of two translation initiation sites, and post-translational modifications. Mutations in 4.1R can cause hereditary elliptocytosis. Protein 4.1 can be found in nucleoplasm and centrosomes at interphase, in the mitotic spindle during mitosis, in perichromatin during telophase, as well

TABLE P1. Peptidases

Family	Feature	Inhibitors	Examples
Aspartic (formerly acid- or carboxyl proteinases)	Two Asp at active site	Pepstatin	Pepsin
Cysteine	CysSH at active site	Iodoacetate	Papain, caspase-1, cathepsin-B, bromelain
Glutamic	Glu136 at active site	1,2-epoxy-3-(p-nitrophenoxy)propane	Scytalidoglutamic peptidase, endopeptidases from fungi
Metallo	Metal ion, often zinc	o-phenanthroline, EDTA	See Table M1
Serine	Serine at active site	Organic phosphate esters (DFP, PMSF)	Trypsin, chymotrypsin, thrombin, plasmin, elastase, subtilisin
Threonine	Thr at active site	L-azaserine	γ-Glutamyltransferase
Unknown	Unknown catalytic site	Often unknown	Collagenase from *Porphyromonas gingivalis*

Proteolytic enzymes, now properly referred to as peptidases, the older names (proteases; proteinases) being deprecated, can be divided into 'mechanistic' sets (families) according to their mode of action. Most inhibitors tend to be specific for one set alone (the important exception being the plasma inhibitor **alpha-2-macroglobulin**). Alternatively, peptidases can be classified simply according to whether they act on terminal amino acids (exopeptidases; aminopeptidases act at the N-terminal, carboxypeptidases at the C-terminus) or on peptide bonds within the chain (endopeptidases). An important web-based classification scheme has recently been developed and should be consulted for more information (the **MEROPS** peptidase database, http://merops.sanger.ac.uk/. This also lists inhibitors. See also **clans of peptidases**.

as in the midbody during cytokinesis. Erythrocyte 4.1 (4.1R) is encoded by a large, complexly spliced gene located on human chromosome 1p32–p33. A second 4.1 gene, 4.1G, maps to human chromosome 6q23 and is widely expressed among human tissues. 4.1N, a neuronal homologue of 4.1R, is expressed in almost all central and peripheral neurons of the body.

protein 4.2 A major erythrocyte membrane skeletal protein, playing an important role in maintaining the integrity and stability of the membrane. It is a transglutaminase-like molecule, but has no enzymatic cross-linking activity.

protein A Protein obtained from *Staphylococcus aureus* that binds immunoglobulin molecules without interfering with their binding to antigen. Widely used in purification of immunoglobulins and in antigen detection, e.g. by **immunoprecipitation**. A very effective B-cell mitogen.

protein B Cell surface protein of group B streptococci that, like **protein A**, will bind Fc region of immunoglobulin – but preferentially IgA.

protein C Vitamin K-dependent glycoprotein (62 kDa) that is the zymogen of a serine endopeptidase (activated protein C; EC 3.4.21.69) found in plasma. Activated protein C in combination with **protein S** will hydrolyse blood-clotting factors Va and VIIIa, thereby inhibiting blood coagulation.

protein engineering Normally means the use of recombinant DNA technology to produce proteins with desired modifications in the primary sequence. See **site-specific mutagenesis**.

protein G Protein from Group C *Streptococci* that binds the Fc portion of IgG; less species-specific than **protein A**.

protein kinase Enzyme catalysing transfer of phosphate from **ATP** to hydroxyl side chains on proteins, causing changes in function. Most phosphates on proteins of animal cells are on **serine** residues, less on **threonine**, with a very small amount on **tyrosine** residues. Tyrosine kinases phosphorylate proteins on tyrosine; serine/threonine kinases on serine or threonine. See **protein kinase A, C, G**, etc.

protein kinase IV A calcium/calmodulin-dependent protein kinase (53 kDa) found in brain, T-cells and postmeiotic male germ cells. Present in nucleus where it phosphorylates and activates **CREB** and CREM-tau. See **calspermin** and **reticalmin**. May be important in preventing apoptosis during T-cell development and during activation of T-cells in response to mitogens.

protein kinase A *PKA; cAMP-dependent protein kinase; EC 2.7.11.1* A family of serine/threonine protein kinases whose activity is dependent on the level of **cyclic AMP** (cAMP) in the cell. Consists of two regulatory and two catalytic subunits; cAMP binds to the regulatory subunits causing conformational change and activation of the catalytic subunits. An important regulatory enzyme having pleiotropic effects because of the diversity of substrates.

protein kinase B *AKT kinases; EC 2.7.1.37* See **AKT**. Mitogen-activated kinases downstream from cell surface receptors that activate **PI-3-kinase**. A major phosphorylation target is glycogen synthase kinase-3 (GSK-3).

protein kinase C *PKC; EC 2.7.11.1* Family of protein serine/threonine kinases activated by phospholipids that play an important part in intracellular signalling. The classical PKCs (α, β1, β2, γ) are also calcium dependent and can be activated by diacyl glycerol, one of the products of phospholipase C, or, non-physiologically, by phorbol esters. A growing set of non-classical calcium-independent isoforms is known. The catalytic domain is highly conserved and specific properties are conferred by a variety of regulatory domains, including a pseudo-substrate region which is displaced upon activation. The specific physiological substrates for these enzymes are not yet well defined.

protein kinase C phosphorylation site Protein kinase C tends to phosphorylate serine or threonine residues near a C-terminal basic residue, with the consensus pattern: [ST-x-RK].

protein kinase G *PKG* cGMP-dependent protein kinase, member of the **AGC kinase** family that regulates **p21-activated kinase**.

protein kinase inhibitor peptide *PKI* An endogenous thermostable peptide (8 kDa) that modulates cAMP-dependent protein kinase function. Distinct PKI isoforms (PKIα, PKIβ, PKIγ have been identified, and each isoform is expressed in the brain.

protein kinase N *PKN; PRK1* Serine/threonine kinase (100 kDa) probably regulated by **rho**-dependent phosphorylation. Kinase activity resides in the C-terminal region (which has high homology with the catalytic domain of PKC). The N-terminal region, through which regulation occurs, has a polybasic region and a **leucine zipper** domain. PKN has some sequence homology with **rhophilin**.

protein kinase R *PKR* RNA-activated protein kinase, a serine/threonine, dsRNA-dependent protein kinase, 68 kDa in human cells; expression is induced by interferon. A member of a family of evolutionary conserved dsRNA-binding molecules that includes *E. coli* RNase III, *Drosophila* staufen and *Xenopus* 4F.1. Interaction with dsRNA structures of greater than 35 bp causes PKR to autophosphorylate, subsequently to catalyse the phosphorylation of substrate targets and cause inhibition of protein synthesis in the cell. Contains an N-terminal RNA-binding domain and a C-terminal kinase domain. Plays a key role in interferon-mediated host defense against viral infection and is implicated in cellular transformation and apoptosis.

protein phosphatases *PP1, PP2A* Protein phosphatase types 1 (PP1) and 2A (PP2A) represent two major families of serine/threonine protein phosphatases implicated in the regulation of many cellular processes, including cell growth and apoptosis in mammalian cells. Both types are oligomeric complexes comprising a catalytic structure (PP1c or PP2AC) containing the enzymatic activity and at least one more interacting subunit.

protein S Single-chain glycoprotein (69 kDa) that promotes binding of **protein C** to platelets; a vitamin K-dependent cofactor.

protein sequencing There are two major methods, Edman degradation and mass spectroscopy, the latter becoming more popular as instrumentation becomes more readily available. In the Edman degradation method, peptides of no more than about 50 residues are affixed to a solid support and the N-terminal sequences sequentially reacted with **Edman reagent**, removed and identified. Larger proteins can be proteolytically fragmented in order to make analysis possible. In mass spectroscopic methods, the protein is proteolytically cleaved and small peptides are identified by their characteristic mass, often using **electrospray**

methods to introduce them into the spectrometer. With sufficient data, and by comparison with database information on proteins of known sequence, it is possible to work out the full sequence using overlapping peptides from different proteolytic digestions.

protein tyrosine phosphatase *PTP* A phosphatase that specifically cleaves the phosphate from a tyrosine residue in a protein, thus reversing the action of a **tyrosine kinase**. Examples include CD45, **shp**, **dep**, **lar**.

protein Z Major protein (43 kDa) of barley endosperm, structurally similar to a **serpin**.

protein zero *Pf* The major **glycoprotein** of peripheral nerve **myelin**, an integral transmembrane protein (28 kDa), synthesized by **Schwann cells**.

proteinase-inhibitor-inducing factor See **PIIF**.

proteinoid droplets Membrane-bounded droplets supposed to have been formed in 'primaeval soup' as an early stages in the evolution of cells.

proteinoplast *Proteoplast* Form of **plastid** adapted as a protein storage organelle; the protein may be crystalline.

proteoglycan A high molecular weight complex of protein and polysaccharide, characteristic of structural tissues of vertebrates, such as bone and **cartilage**, but also present on cell surfaces. Important in determining viscoelastic properties of joints and other structures subject to mechanical deformation. **Glycosaminoglycans** (GAGs), the polysaccharide units in proteoglycans, are polymers of acidic disaccharides containing derivatives of the amino sugars glucosamine or galactosamine.

proteoheparan sulphate A **proteoglycan** containing as its **glycosaminoglycan** heparan sulphate whose constituent *N*-acetyl glucosamine is often sulphated. Hence highly negatively charged. **Syndecan** is one example.

proteolipid Obsolete term for hydrophobic integral membrane proteins.

proteolipid protein *PLP* Highly conserved membrane protein (30 kDa) that accounts for about half of the protein content of adult CNS myelin. Cellular function obscure but mutations lethal, e.g. **jimpy** mouse and **Pelizaeus–Merzbacher disease** of man.

proteolysis Cleavage of proteins by peptidases (proteases). Limited proteolysis occurs where proteins are functionally modified (activated in the case of zymogens) by highly specific peptidases.

proteolytic enzyme See **peptidase** and **protease**.

proteome All the proteins encoded by the genome of an organism. Though all proteins are coded for within the genome not all are expressed in every cell, and their differential expression, temporally and spatially, is key to understanding how cells and organisms work. Alternative splicing and post-translational modification add further complexity.

proteomics The study of **proteomes**, by analogy with genomics.

proteosome *Proteasome* The 20S proteosome has 28 protein subunits arranged as an ($\alpha1$–$\alpha7$, $\beta1$–$\beta7$)$_2$ complex in four stacked rings. The interior of the complex has the active sites. The β-type subunits are synthesized as proproteins and are proteolytically cleaved before assembly. Nomenclature is complex: see Table P2.

TABLE P2. Proteasome subunits

20S core subunits

	Synonyms (*S. cerevisiae*)	Human eq	Size (kDa)
$\alpha1$	C7/PRS2	iota	24
$\alpha2$	Y7	C3	27
$\alpha3$	Y13	C9	29
$\alpha4$	PRE6	C6	28
$\alpha5$	PUP2	zeta	29
$\alpha6$	PRE5	C2	26
$\alpha7$	C1/PRS1	C8	31
$\beta1$	PRE3	LMP2	16
$\beta2$	PUP1	MECL1	25
$\beta3$	PUP3	C10	23
$\beta4$	C11/PRE1	C7	22
$\beta5$	PRE2	LMP7	23
$\beta6$	C5/PRS3	C5	25
$\beta7$	PRE4	N3/beta	26

19S cap subunits

	Synonyms (yeast)	Human orthologue	Size (kDa)
Rpt1	Yta3/Cim5	S7/Mss1	52
Rpt2	Yta5	S4	49
Rpt3	Yta2/Ynt1	S6/Tbp7	48
Rpt4	Sug2/Crl13	S10b	49
Rpt5	Yta1	S6/Tbp1	48
Rpt6	Sug1/Cim3	S8/Trip1	50
Rpn1	Hrd2/Nas1	S2-Trap2	109
Rpn2	Sen3	S1	104
Rpn3	Sun2	S3	60
Rpn4	Son1/Ufd5		60
Rpn5			52
Rpn6	NAS5	S9	50
Rpn7		S10	49
Rpn8	NAS3	S12	38
Rpn9	RPN8		46
Rpn10	Mcb1/Sun1	S5a	30
Rpn11	Mpr1	Poh1	34
Rpn12	Nin1	S14	32

The 26S proteasome consists of a 20S protease core that is capped at one or both ends by the 19S regulatory particle. The 20Score particle has two copies each of seven differentα- and seven different β-subunits arranged into four stacked rings ($\alpha_7\beta_7\beta_7\alpha_7$). In addition there are a number of proteasome-interacting proteins which further regulate its function. Based upon http://biochemie.web.med.uni-muenchen.de/feldmann/proteasome_units.html.

Proteus 1. Genus of highly motile Gram-negative bacteria. They are found largely in soil, but are also found in the intestine of humans. They are opportunistic pathogens; *P. mirabilis* is a major cause of urinary tract infections. 2. An urodele amphibian. It is a cave dweller and is blind, has external gills and lacks any pigment.

prothallus Independent gametophyte phase of horsetail or fern.

prothrombin Inactive precursor of **thrombin**, found in blood plasma.

prothrombin time *PT* The time until a fibrin clot forms, measured in seconds, after a specific volume of thromboplastin reagent is added to the sample of citrated blood plasma. Used in calculating the **international normalized ratio** (INR).

protirelin See **thyrotropic-releasing hormone**.

Protista The kingdom of eukaryotic unicellular organisms. It includes the **Protozoa**, unicellular eukaryotic algae and some fungi (myxomycetes, acrasiales and oomycetes).

proto-oncogene The normal, cellular equivalent of an **oncogene**; thus usually a gene involved in the signalling or regulation of cell growth. In general, cellular proto-oncogenes are prefixed with a 'c', rather than their abnormal viral counterparts, which are prefixed with a 'v', e.g. *c-myc* and *v-myc*.

protochlorophyllide Precursor of chlorophyll, found in **proplastids** and **etioplasts**. Lacks the phytol side chain of chlorophyll.

protofilaments One way of viewing microtubule structure is to consider it to be built of (usually) 13 protofilaments arranged in parallel.

protolignin An immature form of **lignin** that can be extracted from the plant cell wall with ethanol or dioxane.

protolysosome Primary lysosome that has not been involved in fusion with another vesicle or in digestive activity.

protomers Subunits from which a larger structure is built. Thus the tubulin **heterodimer** is the protomer for microtubule assembly, G-actin the protomer for F-actin. Because it avoids the difficulty that arises with, for example, dimers that serve as subunits for assembly, it is a useful term that deserves wider currency.

proton ATPase H^+-*ATPase* An ion pump that actively transports hydrogen ions across lipid bilayers in exchange for ATP. Major groups are the **F-type ATPases**, which run in reverse to synthesize ATP in bacterial, mitochondrial and chloroplast membranes (ATP synthase); and the **V-type ATPases,** found in intracellular vesicles with an acidic lumen and on certain epithelial cells (e.g. kidney intercalated cells). Gastric H^+/K^+ ATPase is a proton ATPase.

proton motive force *PMF* The proton gradient across a prokaryote or mitochondrial membrane that provides the coupling between oxidation and ATP synthesis. In bacteria, used to drive the flagellar motor. See **chemiosmosis**.

proton-pump inhibitors *PPIs* Drugs that inhibit the secretion of gastric acid by acting on the cellular proton ATPase; used as a short-term treatment for gastric and duodenal ulcers. An example is omeprazole.

protonophore Ionophore that carries protons. Many **uncoupling agents** are protonophores.

protophloem Primary phloem, the first phloem to be produced; characteristically matures while the organ is elongating.

protoplast A bacterial cell deprived of its cell wall, for example by growth in an isotonic medium in the presence of antibiotics that block synthesis of the wall **peptidoglycan**. Alternatively, a plant cell similarly deprived by enzymic treatment.

protoporphyrin Porphyrin ring structure lacking metal ions. The most abundant is protoporphyrin IX, the immediate precursor of **haem**.

protostome Invertebrate phylum in which the mouth forms from the embryonic blastopore. Major protostome phyla are Annelida, Mollusca and Arthropoda. See **deuterostome**.

prototroph An organism able to grow in unsupplemented medium. *cf.* **auxotroph**.

protoxylem The first-formed primary xylem with narrow tracheary elements with annular or helical thickening. Becomes stretched and crushed as the organ elongates.

Protozoa A very diverse group comprising some 50 000 eukaryotic organisms that consist of one cell. Because most of them are motile and heterotrophic, the Protozoa were originally regarded as a phylum of the animal kingdom. However, it is now clear that they have only one common characteristic, that they are not multicellular, and Protozoa are now usually classed as a Subkingdom of the Kingdom **Protista**. On this classification the Protozoa are grouped into several phyla, the main ones being the Sarcomastigophora (flagellates, heliozoans and amoeboid-like protozoa), the Ciliophora (ciliates) and the Apicomplexa (sporozoan parasites such as *Plasmodium*).

provacuoles In plant cells, provacuoles are budded directly from the rough endoplasmic reticulum and fuse with other provacuoles to form vacuoles. Since vacuoles may contain hydrolytic enzymes, it is possible to consider them as analogues of primary lysosomes in animal cells.

provirus The genome of a virus when it is integrated into the host cell DNA. In the case of the retroviruses, their RNA genome has first to be transcribed to DNA by **reverse transcriptase**. The genes of the provirus may be transcribed and expressed, or the provirus may be maintained in a latent condition. The integration of the **oncogenic** viruses, such as **Papovaviridae** and retroviruses, can lead to cell **transformation**.

Prozac *Fluoxetine hydrochloride* Proprietary name for a selective serotonin re-uptake inhibitor (SSRI) used as an antidepressant. Inhibiting the re-uptake of serotonin (5-HT) increases the levels of serotonin in the central nervous system.

prozone Prozone phenomena occur in immunological reactions when the concentrations of antibody or other active immune agent are so high that the optimum concentration for maximal reaction with antigen is exceeded. Immunological phenomena in the prozone region may show partial or total inhibition.

prozymogen granule *Condensing vacuole* Stage in the development of a mature **secretory vesicle** (zymogen granule).

PrP PrPc is a normal protein anchored to the outer surface of neurons and, to a lesser extent, the surfaces of other cells, including lymphocytes. The **prion** thought to be responsible for **scrapie** and other **spongiform encephalopathies** is hypothesized to be a modified form of PrPc, PrPSc. See **ure2p** and **doppel**.

PRP 1. Platelet-rich plasma. 2. **Proline-rich polypeptide**. 3. **Proline-rich proteins**. 4. **Prion** proteins PrPc, etc. 5. See **Prp3, Prp19**.

Prp19 *Precursor RNA processing-19* An essential splicing factor and a member of the U-box family of E3 ubiquitin ligases. The Prp19-associated complex consists of at least eight protein components and is involved in **spliceosome** activation by specifying the interaction of U5 and U6 with pre-mRNA. See **Prp3**; do not confuse with **PrP** (prion protein) or **PRPs** (proline-rich proteins).

Prp3 **Spliceosome** protein involved in Precursor RNA processing. Prp3 is a U4/U6-associated splicing factor. See Prp19-associated complex consists of at least eight protein components and is involved in spliceosome activation by specifying the interaction of U5 and U6 with pre-mRNA. Prp19 is an essential splicing factor and a member of the U-box family of E3 ubiquitin ligases.

pruritus Severe and persistent itching, a characteristic of many skin diseases.

PS2 Abbrev. for (1) **Presenilin**-2, (2) **Trefoil factor** 1 (pS2/TFF1).

PSA *Prostate-specific antigen; semenogelase; EC 3.4.21.77* Antigen in serum that seems to be a relatively reliable marker for prostatic hyperplasia/carcinoma. Antigen is a serine endopeptidase of the **kallikrein** family.

PSCA A cell surface antigen associated with prostatic carcinoma; a member of the Thy-1/Ly-6 family of GPI-anchored surface proteins, expressed primarily in basal cells of normal prostate tissue, suggesting that it is a potential stem cell marker. Expression is upregulated in cancer epithelia and is detected in 80% of prostate cancer. See **STEAP**.

PSD-proteins *Postsynaptic density proteins* Family containing **PDZ** domains and apparently responsible for the clustering of receptors. The PSD-95 protein is responsible for the clustering of NMDA receptors and K$^+$ channels, a similar protein is responsible for the formation of synaptic complexes. See **postsynaptic protein**.

P-selectin See **selectins**.

pseudogene Non-functional DNA sequences that are very similar to the sequences of known genes. Examples are those found in the β-like globin gene cluster. Some probably result from gene duplications that become non-functional because of the loss of promoters, accumulation of stop codons, mutations that prevent correct processing, etc. Some pseudogenes contain a **poly-A tail** suggesting that a mRNA, at some point, was copied into DNA that was then integrated into the genome.

pseudohypoaldosteronism *Gordon hyperkalaemia-hypertension* **Hyperkalaemia** despite normal renal glomerular filtration, hypertension and correction of physiological abnormalities by thiazide diuretics. Caused by a defect in WNK1 or WNK4, kinases that are apparently activated by hyperosmotic stress. See **SPAK**.

pseudoknots A topological structure, composed of non-nested double-stranded stems connected by single-stranded loops into which RNA can fold. mRNA pseudoknots have a stimulatory function in programmed ribosomal frame-shifting.

Pseudomonas Genus of Gram-negative bacteria. They are rod-shaped and motile, possessing one or more polar **flagella**. Several species produce characteristic water-soluble fluorescent pigments. They are found in soil and water. *P. syringae* is a plant pathogen causing leaf spot and wilt. *P. aeruginosa*, normally a soil bacterium, is an opportunistic pathogen of humans who are immunocompromised. It can infect the wounds of victims with severe burns, causing the formation of blue pus.

Pseudonaja textilis textilis Australian brown snake. See **textilotoxin**.

pseudopod Blunt-ended projection from a cell – usually applied to cells that have an amoeboid pattern of movement.

pseudopterosins Class of natural compounds (diterpene-pentose glycosides) isolated from the soft coral *Pseudopterogorgonia elisabethae*, and that interfere with arachidonic acid metabolism. Have anti-inflammatory and analgesic properties.

pseudospatial gradient sensing Mechanism for sensing a gradient of a diffusible chemical in which the cell sends protrusions out at random; upgradient protrusions are stabilized by positive feedback (because receptor occupancy is rising with time) and others are transitory because of adaptation. Possibly the mechanism by which neutrophils sense chemotactic gradients.

pseudouridine *5-β-D-Ribofuranosyluracil* Unusual nucleotide found in some tRNA: glycosidic bond is associated with position 5′ of **uracil**, not position 1′.

pseudovirus Virus-like particle composed of a viral coat protein enclosing an unrelated DNA sequence. Pseudoviruses are potentially useful as a means of delivering DNA into cells for therapeutic purposes or to induce antigen production and thus act as a vaccine.

PSGL-1 *P-selectin glycoprotein ligand; CD162* An extended mucin-like transmembrane glycoprotein expressed as a homodimer on neutrophils, monocytes and most lymphocytes. See **PADGEM**.

Psilotum nudum *Whisk fern* A spore-producing vascular plant, an epiphyte occasionally found as a terrestrial plant in rocky crevices in sandy soils. Related to ferns but may not be as primitive as formerly thought.

P-site The peptidyl-tRNA binding site on the **ribosome**, the one to which the growing chain is attached; the incoming **aminoacyl tRNA** attaches to the A-site.

PSM A cell surface antigen expressed in prostate cancer, a type II transmembrane protein with hydrolase activity and 85% identity to a rat neuropeptidase, also expressed in the small intestine and the brain. May have a role in neuropeptide catabolism in the brain. See **STEAP**.

psoralens Drugs capable of forming photoadducts with nucleic acids if ultraviolet-irradiated.

psoriasis Chronic inflammatory skin disease characterized by epidermal hyperplasia. Lesions may be limited or widespread, and in the latter case the disease can be life-threatening. Unlike many chronic inflammatory conditions, seems to be T-cell mediated. There is a fairly strong genetic predisposition.

psychotomimetic Class of drugs that cause bizarre psychic effects in humans as well as marked behavioural changes in animals.

psychrophile Organism that grows best at low temperatures.

psychrophilic Growing best at a relatively low temperature. In the case of microorganisms, having a temperature optimum below 20°C.

Psyllium Common name for several members of the plant genus *Plantago*, the seeds of which are used commercially for the production of mucilage. Seed husks are an ingredient in high-fibre breakfast cereals that are claimed to have cholesterol-lowering properties.

PTB domain *Phosphotyrosine-binding domain* Domain that is present in many proteins downstream from receptors that are tyrosine-phosphorylated when they bind ligand (e.g. **shc**, **IRS-1** and **dok proteins**).

PTEN *Phosphatase and tensin homologue deleted on chromosome ten; MMAC1/PTEN; TEP-1* A **protein tyrosine phosphatase** with homology to **tensin**, is the product of a tumour suppressor gene on chromosome 10q23. Somatic mutations in PTEN occur in multiple tumours, most markedly glioblastomas. Germ line mutations in PTEN are responsible for Cowden disease (CD), a rare autosomal-dominant multiple-hamartoma syndrome. Mutated in MMAC1/PTEN (multiple advanced cancers 1/phosphatase and tensin) is a homologue.

pteridine *Pyrazino[2,3-d]pyrimidine* Nitrogen-containing compound composed of two six-membered rings (pyrazine and pyrimidine rings). Structural component of **folic acid** and **riboflavin** and parent compound of pterins such as xanthopterin, a yellow pigment found in the wings of some butterflies.

Pteridophyta Division of the plant kingdom that includes ferns, horsetails and clubmosses.

PTGS See **post-transcriptional gene silencing**.

PtK2 cells Cell line from *Potorous tridactylis* (potoroo or kangaroo rat) kidney. Often used in studies on mitosis because there are only a few large chromosomes and the cells remain flattened during mitosis.

ptosis Drooping of the upper eyelid for any one of a number of causes. May be a result of damage to the third cranial nerve, of **myasthenia gravis** or of Horner's syndrome, or simply be an isolated congenital feature.

PTP *Protein tyrosine phosphatase* A **phosphatase** that reverses the effect of a tyrosine kinase.

ptRNA *Precursor tRNA* See **RNase P**.

PTS Abbrev. for (1) **Phosphotransferase** system, (2) Platinum monosulphide (PtS).

PTX Abbrev. for (1) Paclitaxel (**taxol**), (2) **Pertussis toxin**, (3) **Pentoxifylline**, (4) Pectenotoxins, dinoflagellate toxins from *Dinophysis* sp. that accumulate in shellfish, (5) **Palytoxin**, another dinoflagellate toxin from *Ostreopsis* sp., (6) **Picrotoxin**, (7) Homeobox transcription factor, Ptx-2.

ptyalin Common name for **amylase** found in saliva.

P-type ATPase One of three major classes of ion transport **ATPases**, characterized by vanadate sensitivity and a phosphorylated intermediate. The archetype is the **sodium pump**. See **F1F0 ATPase**, **V-type ATPase**.

P-type channels A class of **voltage-sensitive calcium channels**. Found in various neurons, but particularly in Purkinje cells of the cerebellum (hence the name). Involved in induction of long-term depression. Require substantial depolarization to activate and inactivate slowly. Inhibited by **agatoxin** and a polyamine FTX, through a G-protein-coupled mechanism.

PU box Purine-rich sequence recognized by the Sp-1 transcription factor.

pUC18/19 Commonly used plasmid-cloning vectors in *E. coli*, a double-stranded circle, 2686 bp in length. pUC18 and pUC19 are identical except that they contain multiple cloning sites (MCS) arranged in opposite orientations.

pUC9 *E. coli* **phagemid vector**, derived from pUC19 and M13MP9.

PUFA *Poly-unsaturated fatty acid* Increasing the ratio of unsaturated to saturated fatty acids can alter the behaviour of cells, probably by altering physical characteristics of membranes and thus influencing the behaviour of integral membrane proteins. Whether this is true for whole organisms is a matter for debate.

puffs Expanded areas of a **polytene chromosome**. At these areas the chromatin becomes less condensed and the fibres unwind, though they remain continuous with the fibres in the chromosome axis. A puff usually involves unwinding at a single band, though they can include many bands, as in **Balbiani rings**. Puffs represent sites of active RNA transcription. The pattern of puffing observed in the larvae of *Drosophila*, in different cells, and at different times in development provides possibly the best evidence that differentiation is controlled at the level of transcription.

pulchellin A highly toxic type 2 ribosome-inactivating protein isolated from seeds of the *Abrus pulchellus tenuiflorus* plant.

pull-down assay A form of affinity purification in which a 'bait' protein (e.g. either **GST**-tagged so that it can be purified on a glutathione column or directly coupled to a matrix) is used to capture proteins (prey) with which it interacts. Often used to confirm results of **yeast two-hybrid screening** studies.

pullulan An extracellular glucan, produced from starch by *Aureobasidium pullulans*, a chain of maltotriose units linked by alpha-1,4- and alpha-1,6-glucosidic bonds. Used to produce edible films.

pullulanase *EC 3.2.1.41; pullulan-6-glucanohydrolase* A glucanase that degrades **pullulan**. It is an extracellular, cell surface-anchored lipoprotein produced by Gram-negative bacteria of the genus *Klebsiella*.

pulse-chase An experimental protocol used to determine cellular pathways, such as precursor–product relationships. A sample (organism, cell or cellular organelle) is exposed for a relatively brief time to a radioactively labelled molecule, the pulse. The labelled compound is then replaced with an excess of the unlabelled molecule, the chase (cold chase). The sample is then examined at various later times to determine the fate of radioactivity incorporated during the pulse.

pulse-field electrophoresis A method used for high-resolution electrophoretic separation of very large (megabase) fragments of DNA. Electric fields 100° apart (the angle may vary) are applied to the separation gel alternately. The continuous change of direction prevents the molecules aligning in the electric field and greatly improves resolution on the axis between the two fields.

Puma See **Bcl-2 homology domain 3**.

punctuated equilibrium A view of the evolutionary process that holds that there were long periods of stasis interrupted by relatively short periods of rapid change and speciation.

Punnett's square The checkerboard (matrix) method used to determine the types of zygotes produced by fusion of gametes from parents of defined genotype.

purine A heterocyclic compound with a fused pyrimidine/imidazole ring. Planar and aromatic in character. The parent compound for the purine bases of nucleic acids.

purinergic receptors Receptors that use purine nucleotides (e.g. ATP) as ligands.

Purkinje cell A class of **neuron** in the **cerebellum**; the only neurons that convey signals away from the cerebellum. See **Purkinje fibres**.

Purkinje fibres Specialized cardiac muscle cells that conduct electrical impulses through the heart and are involved in regulating the beat.

puromycin An antibiotic that acts as an **aminoacyl tRNA** analogue. Binds to the A-site on the **ribosome**, forms a peptide linkage with the growing chain and then causes premature termination.

purple membrane Plasma membrane of *Halobacterium* and *Halococcus* that contains a protein-bound carotenoid pigment that absorbs light and uses the energy to translocate protons from the cytoplasm to the exterior. The proton gradient then provides energy for ATP synthesis. The binding protein is called **bacteriorhodopsin** or purple membrane protein.

purpura Condition in which small spontaneous haemorrhages appear beneath the skin and the mucous membranes and form purple patches.

purpurin **1.** A secretory retinol-binding protein (20 kDa) in developing chicken retinas. **2.** A highly active photodynamic therapy sensitizer, purpurin-18. **3.** An anthraquinone constituent from madder (*Rubia tinctorum*) root, reportedly antimutagenic.

putamen See **basal ganglia**.

putrescine A dibasic amine associated with putrefying tissue. Associates strongly with DNA. Has been suggested as a growth factor for mammalian cells in culture. Metabolic precursor of the polyamines **spermine** and **spermidine**.

PVDF See **polyvinylidene fluoride**.

PVP *Polyvinyl pyrrolidone* Water-soluble white compound that when dissolved makes a very viscous solution.

pyaemia Invasion of bloodstream by pyogenic organisms.

PYK2 A calcium-dependent proline-rich **tyrosine kinase**, activated by tyrosine phosphorylation and associated with **focal adhesion** proteins. Has been linked to proliferative

and migratory responses in a variety of mesenchymal and epithelial cell types, is activated in neurones following NMDA receptor stimulation via PKC and is involved in hippocampal **long-term potentiation**.

pyknosis Contraction of nuclear contents to a deep-staining irregular mass; a sign of cell death.

pyocins **Bacteriocins** produced by bacteria of the genus *Pseudomonas*.

pyocyanin Blue-green phenazine pigment produced by *Pseudomonas aeruginosa*; has antibiotic properties.

pyogenic Causing the formation of pus, a thick yellow or greenish liquid formed at a site of infection, and that contains dead leucocytes, bacteria and tissue debris.

pyramidal cells Commonest nerve cells of the cerebral cortex.

pyranose Sugar structure in which the carbonyl carbon is condensed with a hydroxyl group (i.e. in a hemiacetal link), forming a ring of five carbons and one oxygen. Most hexoses exist in this form although in sucrose, fructose is found with the smaller (four carbon) furanose ring.

pyranoside Compound containing a pyrenoid ring of five carbons and one oxygen atom.

pyrazinamide Bactericidal drug, the pyrazine analogue of nicotinamide, used in treatment of tuberculosis. Mode of action is unclear.

pyrenoid Small body found within some chloroplasts that may contain protein. In green algae, may be involved in starch synthesis.

pyrethrum *Pyrethroid* Toxic hydrocarbon, originally from the pyrethrum daisy (*Tanacetum cinerariaefolium*) flower, that now forms the basis for a wide range of 'natural' synthetic pyrethroid insecticides.

pyridoxal phosphate The coenzyme derivative of vitamin B6. Forms **Schiff's bases** of substrate amino acids during catalysis of transamination, decarboxylation and racemization reactions.

pyrimidine *1,3 Diazine* A heterocyclic six-membered ring, planar and aromatic in character. The parent compound of the pyrimidine bases of nucleic acid.

pyrogen Substance or agent that produces fever. The major endogenous pyrogen in mammals is **interleukin-1**.

pyrogenic Causing fever. See **pyrogen**.

pyroninophilic cells Cells that stain strongly with methyl green pyronin and have bright red cytoplasm indicative of large amounts of RNA, implying very active protein synthesis. **Plasma cells** are very pyroninophilic, for example.

pyrophilous: Organisms that thrive in areas that have recently been burnt; some beetles are particularly attracted to such areas and have IR sensors; some plants are particularly adapted to rapidly colonize such areas.

pyrophosphate Two phosphate groups linked by esterification. Released in many of the synthetic steps involving nucleotide triphosphates (e.g. protein and nucleic acid elongation). Rapid cleavage by enzymes that have high substrate affinity ensures that these reactions are essentially irreversible.

pyrosequencing A DNA sequencing technique based on the detection of released pyrophosphate (PPi) during DNA synthesis. The unknown sequence is used as a template and a cascade of enzymatic reactions generate light in proportion to the number of incorporated nucleotides. Nucleotides are added separately, allowing the sequence to be deduced (only the complementary base will be incorporated, triggering pyrophosphate production and subsequent activation of the luciferase system to produce light.)

Pyrrhophyta An alternative name for **dinoflagellates**.

pyrrole ring A heterocyclic ring structure, found in many important biological pigments and structures that involve an activated metal ion, e.g. chlorophyll, haem.

pyruvate *2-Oxopropanoate;* CH_3COCOO^- Important intermediate in many metabolic pathways, particularly of glucose metabolism and the synthesis of many amino acids.

pyruvate carboxylase *EC 6.4.1.1* An enzyme that catalyses the formation of oxaloacetate from pyruvate, CO_2 and ATP in **gluconeogenesis**.

pyruvate dehydrogenase *EC 1.2.1.51* A complex multienzyme system that catalyses the conversion of (pyruvate+CoA+NAD$^+$) to (acetyl CoA + CO_2+NAD).

pyruvate dehydrogenase kinase *PDK; EC 2.7.11.2* Enzyme that inactivates the multienzyme mitochondrial **pyruvate dehydrogenase** complex by the phosphorylation of three seryl residues in the pyruvate dehydrogenase moiety, and thus has a role in the control of glucose homeostasis. Genetically and biochemically distinct PDK family isozymes have been identified in various species.

Pythium Genus of cellulose-walled fungi in the Oomycota that are best known as pathogens of crop plants, causing Pythium blight, damping-off and other seedling diseases, and that also progressively destroy the root tips of older plants. A few *Pythium* species are, however, aggressive parasites of other fungi and may be useful as biological control agents.

Q

Q10 Ratio of the velocity of reaction at one temperature and that at a temperature 10°C lower. Usually around 2 for biological reactions.

Q banding See **banding patterns** and **quinacrine**.

Q beta *Q* An RNA virus that infects *E. coli*. Genome circular, single stranded and acts both as template for replication of a complementary strand and as messenger RNA.

Q box Glutamine-rich sequence found in some transcription factors.

Q-enzyme *1, 4-α-Glucan branching enzyme; EC 2.4.1.18* Converts amylose to amylopectin. Should be qualified according to the product, e.g. glycogen branching enzyme, amylopectin branching enzyme. The latter has frequently been termed Q-enzyme.

Q fever Typhus-like illness caused by rickettsia, *Coxiella burneti*. Mainly a disease of domestic animals, but can be caught by man.

QH2-cytochrome *c* reductase Membrane-bound complex in the mitochondrial inner membrane, responsible for electron transfer from reduced coenzyme Q to cytochrome *c*. Contains cytochromes *b* and *c*1 and iron–sulphur proteins.

qmf1, qmf2, qmf3 Quail homologues of MyoD, myogenin and myf-5, respectively.

q region See **p-region**.

QSAR See **quantitative structure–activity relationship**.

Q-type channels A class of **voltage-sensitive calcium channels**. May be identical or very similar to **P-type channels**. Inhibited by neurotransmitters that act through G-protein-coupled receptors, high concentrations of ω-**conotoxin** and ω-**agatoxin**.

QT interval *QT period* The period between the start and finish of electrical activity in the ventricles of the heart during a single heartbeat. The interval is measured between the ORS complex and the end of the T-wave in the electrocardiogram.

quail Small galliform bird; *Coturnix coturnix coturnix* is the European quail. Quail embryos are often use in developmental studies because quail cells can be distinguished from chicken cells, yet the two are sufficiently closely related that it is possible to graft embryonic tissue from one to the other.

quaking Mouse **STAR proteins** (signal transduction and activation of RNA), QKI-5, QKI-6 and QKI-7, essential for embryogenesis and myelination.

quantal mitosis A controversial concept in cellular differentiation proposed by H. Holtzer and defined by him as a mitosis 'that yields daughter cells with metabolic options very different from those of the mother cell as opposed to proliferative mitoses in which the daughter cells are identical to the mother cell'. Implicit in this is the idea that the changes in cell **determination** that occur during development take place at these special quantal mitoses.

quantasome Smallest structural unit of photosynthesis, a particulate component of the **thylakoid** membrane containing chlorophyll and cytochromes.

quantitative character A character displaying continuous variation, therefore likely to be polygenic.

quantitative structure–activity relationship *QSAR* Relationship between the structure of a compound and its activity in binding or inhibiting something, based upon computed parameters of structure. Computational chemists rely upon complex computational methods to derive appropriate parameters of shape and electronic distribution.

quantum dot A small particle of semiconductor material, typically a few nanometers in diameter; the small size makes quantum mechanical effects more significant than in the (macroscopic) bulk material. Have been used for intravital staining and have the advantage that they are available in a wide range of colours.

quantum yield *Quantum requirement* The number of photons required for the formation of one oxygen molecule in photosynthesis. Varies from 8 to 14 depending on the system used to measure it.

quasi-equivalence Term used to refer to the way in which subunits pack into a quasi-crystalline array as, for example, in viral coat assembly. There is usually some strain in the packing.

quaternary structure Fourth-order level of structural organization of proteins. Tertiary structure defines the shape of single protein molecules; quaternary structure the way in which dimers or multimers are arranged.

Quellung reaction Swelling of the capsule surrounding a bacterium as a result of interaction with anticapsular antibody; consequently the capsule becomes more refractile and conspicuous.

quercetin Mutagenic flavonol pigment found in many plants. Inhibits **F-type ATPases**.

quiescence Quietness. In cells, the state of not dividing; in neurons, the state of not firing.

quiescent stem cell A stem cell that is not currently undergoing repeated cell cycles but that might be stimulated so to do later. For example, the satellite cells in the skeletal muscles of mammals are quiescent myoblasts that will proliferate after wounding and give rise to more muscle cells by fusion.

quin2 A fluorescent calcium indicator. Resembles the chelator **EGTA** in ability to bind calcium much more tightly than magnesium. Binding of calcium causes large changes in ultraviolet absorption and fluorescence.

quinacrine A fluorescent dye that intercalates into DNA helices. Chromosomes stained with quinacrine show typical banding patterns of fluorescence at specific locations, **Q bands**, that can be used to recognize chromosomes and their abnormalities.

quinapril **ACE inhibitor** used for treatment of hypertension.

quinate: NAD oxidoreductase *EC 1.1.1.24* A plant enzyme-converting hydroquinic acid (a derivative of the shikimate pathway) to quinic acid. The enzyme is activated by a calcium- and calmodulin-dependent phosphorylation.

quinine An alkaloid isolated from Cinchona bark (*Cinchona* is a genus of South American trees and shrubs). Used as an antimalarial drug. It is believed to act by raising the pH of endocytotic vesicles and inhibiting internal membrane fusion processes.

quinoline *1-Azanaphthalene; 1-benzazine* Heterocyclic aromatic organic compound used in dye, polymer and agrochemical production. It is also a preservative, disinfectant and solvent, but is generally considered fairly toxic.

quinolinic acid One of the end products of the **kynurenine pathway**; thought to play a role in the pathogenesis of several major neuroinflammatory diseases; an endogenous NMDA receptor agonist and neurotoxin.

quinolone antibiotics The quinolones and fluoroquinolones are bactericidal and inhibit the activity of DNA gyrase. The older quinolones, nalidixic acid and cinoxacin, are active only against Enterobacteriaceae, but the newer fluoroquinolones (e.g. Ciprofloxacin) have a broader spectrum of activity.

quinolones See **quinolone antibiotics**.

quinone Aromatic dicarbonyl compound derived from a dihydroxy aromatic compound. Ubiquinone (coenzyme Q) is a dimethoxy-dicarbonyl derivative of benzene involved in electron transport. Other quinones may act as tanning agents.

quinone reductase *EC:1.6.5.5* Enzymes that reduce quinones to phenols, usually using NADH or NADPH as a source of reductant.

quinupristin One of the **streptogramins**, naturally occurring antibiotics produced by streptomycetes. Quinupristin/dalfopristin is active *in vitro* against all Gram-positive organisms except *Enterococcus faecalis*; against some Gram-negative organisms, such as *Moraxella* and *Neisseria*; and against mycoplasma, *Legionella* and *Chlamydia pneumoniae*.

quisqualate An agonist of the Q-type **excitatory amino acid** receptor. See **kainate**.

quorum sensing Phenomenon whereby single bacteria are able to sense the population density of bacteria in the immediate neighbourhood. Depends upon the accumulation of species-specific signalling molecules (e.g. *N*-acyl-L-homoserine lactones), although species that commonly live together may share the same signalling molecule. Significant behavioural changes can be triggered when critical population densities are reached.

QX-222 Open-channel blocker at **nicotinic acetylcholine receptors**; a quaternary ammonium derivative of the local anaesthetic lidocaine. In mouse muscle AChR it blocks by interacting with adjacent turns of the M2 (membrane-spanning region) helix.

R

R7G See **G-protein-coupled receptors**.

R17 bacteriophage Bacteriophage with RNA genome that codes for the enzyme **RNA synthetase** and for the coat protein, a protein to which the RNA is attached and that is involved in attachment to the bacterium.

rab genes 1. One of the three main groups of *ras*-like genes specifying small GTP-binding proteins (the others are *ras* and *rho*). Rab proteins are involved in vesicular traffic and seem to control translocation from donor to acceptor membranes. **2.** Gene family in plants – 'responsive to abscisic acid': encode proteins of 15–17 kDa.

rabaptin Rabaptin-5 (117 kDa) interacts with rab5-GTP and is essential for rab-mediated endosomal fusion. Both ends of rabaptin have coiled-coil domains characteristic of vesicular transport proteins.

rabies virus Species of the **Rhabdoviridae** that causes rabies in humans. The virus infects the cells in the brain, causing a fatal encephalomyelitis. It is found all over the world, but strict quarantine regulations have excluded it from Britain and Australia. The virus infects a number of domestic and wild mammals, whose saliva is infective. Some bats and small mammals can carry the virus without showing any symptoms of disease.

rabin A human protein (50 kDa: Rabin8) and its rat equivalent (Rabin3) bind Rab8 and function as Rab8-specific nucleotide exchange factors (**GEFs**), but not Rab3A and Rab5 (although an earlier report suggested it associated with Rab3).

rabphilin Receptor (704 amino acid residues) for the small GTP-binding protein rab3A that is implicated in regulated secretion, particularly of neurotransmitters. The N-terminal region interacts with rab3A, the C-terminal domain interacts with calcium and phospholipid, and rabphilin is found in association with sites of exocytosis in neurites. There are indications that rabphilin may inhibit **GAP**-activated GTPase activity of rab3A.

rac Small GTP-binding protein involved in regulating actin cytoskeleton – the activated form of rac seems to induce membrane ruffling (whereas **rho** acts on stress-fibres). Rac may be activated by specific **GAPs** such as **bcr** and **N-chimaerin**.

RACE See **rapid amplification of DNA ends**.

racemic mixture *Racemate* A mixture containing equimolar amounts of two enantiomers (D- and L-forms) of a chiral molecule.

rachitic To do with rickets, a disease caused by vitamin D deficiency.

RACKs *Receptors for activated C kinase* Proteins, usually anchored to specific areas of the cell, which selectively bind activated protein kinase C and thus control the regions of the cell on which it acts.

rad 1. Abbrev. for radian. **2.** Unit of radiation, 1 rad = 0.01 Gy. **3.** *rad1* is a *Schizosaccharomyces pombe* checkpoint control gene important in both DNA damage-dependent and replication-dependent cycle control; various *rad* genes are of comparable function in other organisms (*Hrad1* from man, *Mrad1* from mouse, *RAD17* from *Saccharomyces cerevisiae*). **4.** See **rad proteins**.

rad proteins *rad-50, 51, 52, 54, 55, 57, 59; MRE11, XRS2* Yeast proteins coded by genes originally identified as being particularly sensitive to X-rays. Required for spontaneous and induced mitotic recombination, meiotic recombination and mating-type switching. Human homologues of many of these proteins have been identified. rad51 (37 kDa) is the functional homologue of **recA** and promotes ATP-dependent homologous pairing and DNA strand exchange; binds Rad52. See **rad23**.

rad23 Conserved protein that is important in nucleotide-**excision repair**. N-terminal domain has similarity with **ubiquitin** and links rad23 to the **proteosome**. Binds to rad14 and TFIIH and forms a stable complex with rad4.

radial cleavage Cleavage pattern, in holoblastic eggs, characteristic of the **deuterostomes**, in which the spindle axes are parallel or at right angles to the polar axis of the oocyte. *cf.* **spiral cleavage**.

radial glial cell A type of glial cell, organized as parallel fibres joining the inner and outer surfaces of the developing cortex. They are thought to play a role in **neuronal guidance** in development. See **contact guidance**.

radial spoke The structure that links the outer microtubule doublet of the ciliary axoneme with the sheath that surrounds the central pair of microtubules. The spokes are arranged periodically along the axoneme every 29 nm, have a stalk about 32 nm long and a bulbous region adjacent to the sheath. At least 23 different polypeptides are associated with the spokes (see **radial spoke protein**). Spokes are thought to restrict the sliding of doublets relative to one another; digestion of the radial spokes will allow sliding apart of the doublets.

radial spoke proteins *RSPs* Protein of the **radial spokes** of the ciliary **axoneme**. Among the 18 spoke proteins identified in *Chlamydomonas* so far, at least 12 have apparent homologues in humans, indicating that the radial spoke has been conserved throughout evolution. Many of them are predicted to contain domains associated with signal transduction, suggesting that the spoke stalk is both a scaffold for signalling molecules and itself a transducer of signals. RSP2 protein contains Ca^{2+}-dependent calmodulin-binding motifs and a GAF domain, a domain found in diverse signalling proteins for binding small ligands including cyclic nucleotides. Radial spoke protein (RSP) 3 is an A-kinase anchoring protein (**AKAP**). Axonemal protein kinase A (PKA) operates in a pathway involving the radial spokes and inner arm dynein I1 to regulate microtubule sliding.

radiation inactivation The technique of inactivating proteins in freeze-dried (lyophilized) preparations using high energy particles (e.g. electrons). One high-energy particle can apparently inactivate all of the components of a multisubunit polypeptide; the method is therefore used to determine the molecular weight of functional oligomers.

radicicol A macrocyclic antifungal compound, structurally unrelated to geldanamycin, but that competes for the same binding site on Hsp90. Inhibits Src tyrosine kinase.

radioautography See **autoradiography**.

radioimmunoassay Any system for testing antigen–antibody reactions in which use is made of radioactive labelling of antigen or antibody to detect the extent of the reaction. A standard approach is to spike the sample with a known amount of radiolabelled analyte and estimate the concentration of analyte in the sample by measuring how much radiolabel is bound. Analyte in the sample competes with the labelled material for binding.

radioisotope Form of a chemical element with unstable neutron number, so that it undergoes spontaneous nuclear disintegration. Major use in biology is to trace the fate of atoms or molecules that follow the same metabolic pathway or enzymic fate as the normal stable isotope, but that can be detected with high sensitivity by their emission of radiation. Also used to locate the position of the radioactive metabolite, as in **autoradiography**, and to measure relative rates of synthesis of compounds from radioactive precursors.

Radiolaria Subclass of the **Sarcodina**. Marine protozoans with silicaceous exoskeleton and radiating filopodia.

radixin Barbed-end capping actin-modulating protein (82 kDa) found in **adherens junctions** and in the cleavage furrow of many cells.

raf Gene for **raf**; see ***mil*** oncogene.

raf Serine/threonine protein kinase implicated in signal-response transduction pathways involving tyrosine kinases. Apparently raf-1 is downstream of ras1 in the signalling cascade.

raffinose *Mellitose* A non-reducing trisaccharide found in sugar beet and many seeds, consisting of the disaccharide **sucrose** bearing a D-galactosyl residue linked α (1–6) to its glucose group.

RAGE *Receptor of Advanced Glycation Endproducts* Member of the immunoglobulin superfamily. Interacts with a variety of advanced glycation end products (**AGE**), amyloid β peptide, **amphoterin**, and members of the **S100** family.

RAIDD *Receptor-interacting protein **RIP**-associated ICH-1/CED-3-homologous protein with a death domain* A dual-domain adaptor protein that mediates the recruitment of **caspase**-2 to **tumour-necrosis factor** receptor-1 (TNF-R1) signalling complex through **RIP kinase**. May have an additional function in cell differentiation and RAIDD-deficiency may be embryonically lethal.

Raji cell-binding test A test for the detection of soluble IgG–antigen complexes. Raji cells are a line of EBV-transformed lymphocytes with surface Fc receptors. Complexes are detected by their ability to compete with a radiolabelled aggregated IgG for binding to the cells.

Raji cells A line of **EBV**-transformed lymphocytes with surface Fc receptors that grows in suspension; derived from patient with **Burkitt's lymphoma**.

rak *FRK (fyn-related kinase)/RAK* Tyrosine kinase (54 kDa) found in nucleus, originally isolated from breast cancer cells. Has N-terminal **SH2** and **SH3** domains and has similarities with src. Binds to **Rb** protein and leads to growth suppression. Previously named GTK/Bsk/IYK (GTK, gut tyrosine kinase; Bsk, beta-cell Src homology kinase; IYK, intestinal tyrosine kinase).

ral Oncogene related to *ras*. Protein product is a multifunctional small GTPase involved in tumorigenesis and in controlling intracellular membrane trafficking. It is mainly activated by factors downstream of Ras.

raloxifene Drug that acts selectively on estrogen receptors in bones, heart and arteries but is not estrogenic in uterus and breast. Used to protect against osteoporosis.

Raman spectroscopy Method for measuring the Raman spectrum, the plot of Raman scattering of light that produces weak radiation at frequencies not present in the incident radiation. The spectrum is characteristic of the compound and independent of the wavelength of the incident light.

ramipril **ACE inhibitor** used to treat hypertension.

ramoplanin A peptide antibiotic that works by sequestration of lipid intermediates for peptidoglycan biosynthesis, making them unavailable to the late-stage peptidoglycan biosynthesis enzymes. Ramoplanin is structurally related to two cell-wall active lipodepsipeptide antibiotics, janiemycin and enduracidin, and is functionally related to members of the **lantibiotic** class of antimicrobial peptides (mersacidin, actagardine, **nisin** and epidermin) and glycopeptide antibiotics (**vancomycin** and **teicoplanin**).

RAMPs *Receptor activity modifying proteins* RAMPs are type I transmembrane proteins that determine receptor phenotype of various G-protein-coupled receptors. If the calcitonin receptor-like receptor is transported to the membrane by RAMP1 it behaves as a **calcitonin gene-related peptide** receptor; if associated with RAMP2 its glycosylation pattern is different and it acts as an **adrenomedullin**-receptor.

ramus A physically and physiologically independent individual plant.

ran Small G-protein (GTPase) required, together with **importins** α and β and pp15, for protein transport into the nucleus. The only known nucleotide exchange factor for ran is nuclear (RCC1), whereas the only known activating factor is cytoplasmic. This would provide a mechanism for vectorial transport. Ran-GTP binds importin and may cause dissociation of the transport complex. GTP-loaded Ran also induces the assembly of microtubules into aster-like and spindle-like structures in *Xenopus* egg extract.

Rana pipiens Common European frog.

ranatensin Subfamily of **bombesin**-like peptides, of which **neuromedin**-B (NMB) is the mammalian form. See **phyllolitorin**.

random amplification of polymorphic DNA See **RAPD**.

random coil A term originally invented by polymer chemists to describe a disordered tangle of a linear polymer chain with curved sections. In DNA parlance, the random coil refers to the structure that results from melting or other forms of separation of the double helix, i.e. helix–coil transition.

random priming Method of labelling a DNA probe for use in hybridization. Double-stranded DNA is denatured to form a single-stranded template. Random oligonu-cleotide primers (usually hexamers) are allowed to anneal, nucleotides and DNA polymerase added, and new DNA fragments synthesized in the presence of trace amounts of radioactive or non-radioactive label. The result is a popu-lation of short, labelled DNA molecules of indeterminate length that represent the whole length of the template DNA.

random walk A description of the path followed by a cell or particle when there is no bias in movement. The direction of movement at any instant is not influenced by the direction of travel in the preceding period. If changes of direction are very frequent then the displacement will be small (unless the speed is very great) and the object will appear to vibrate on the spot. Although the behaviour of moving cells in a uniform environment can be described as a random walk in the long term, this is not true in the short term because of **persistence**.

ranitidine *Zantac®* See **histamine H2-receptor antago-nists**.

RANK Receptor activator of NF-kappaB. Ligand is **RANKL**, q.v. See **Paget's disease of bone**.

RANKL Ligand for RANK (receptor activator of **NF-kappaB**), and the decoy receptor **osteoprotegerin** (OPG); part of the regulatory system for osteoclast development and function.

RANTES *Regulated upon activation normal T-expressed and secreted* **Cytokine** of the C–C subfamily, produced by T-cells, and chemotactic for monocytes, memory T-cells and eosinophils. Uniquely among the **chemokines**, it is downregulated when the secreting cells are activated.

RAP 1. Oncogene (*rap*) related to *ras*. Protein product, Rap1 GTPase, is a key regulator of cell adhesion and cell migration in a number of systems. (see **rap1**) **2.** N-RAP is a muscle-specific protein concentrated in myofibril precur-sors during sarcomere assembly and at intercalated disks in adult heart. **3.** See **receptor-associated protein** (RAP). **4.** *Bacillus subtilis* encodes several Rap proteins that are regulators antagonized by Phr signalling peptides that are imported into the cell. The processes regulated by many of these Rap proteins and Phr peptides are unknown. **5.** Rhoptry-associated protein 1: see **rap1**. **6.** See **RAP74**.

rap1 1. Small GTP-binding protein which seems to antag-onize **ras** activity. Has anti-mitogenic activity, seems to be involved in NGF-induced neuronal differentiation and T-cell activation. Activated by **CRK** adaptor proteins and rap-specific **GEFs**. See **tuberous sclerosis**. **2.** Rhoptry-associated protein 1 of *Plasmodium falciparum*, a potential component of an antimalarial vaccine.

RAP74 Large subunit of the transcription initiation factor (**TFIIF**).

rapamycin Immunosuppressive macrolide antibiotic with structural similarity to **FK506**; inhibits T- and B-cell pro-liferation but at a much later stage than FK 506, despite binding to the same **immunophilin**. Inhibits TOR (target of rapamycin) in the Ras/MAP kinase signalling pathway.

RAPD *Random amplification of polymorphic DNA* Vari-ant of polymerase chain reaction used to identify differ-ential gene expression. The mRNA samples are reverse transcribed, then amplified using short, intentionally non-specific primers. The array of bands obtained from a series of such amplifications is compared with analogous arrays from different samples. Any bands unique to single samples

are considered to be differentially expressed; they can be purified from the gel, and sequenced and used to clone the full-length cDNA. Similar in aim to **subtractive hybridiza-tion**. See also **differential display PCR**.

raphide crystal A needle-shaped crystal, usually of cal-cium oxalate, found in the vacuole of some plant cells.

raphidosome Rod-shaped particle found in bacterial cells near DNA-rich region.

rapid amplification of DNA ends *RACE; 3′ RACE; 5′ RACE* Techniques, based on the **polymerase chain reac-tion**, for amplifying either the 5′ end (5′ RACE) or 3′ end (3′ RACE) of a cDNA molecule, given that some of the sequence in the middle is already known. The two proce-dures differ slightly; in the more straightforward 3′ RACE, first-strand cDNA is prepared by reverse transcription with an oligo dT primer (to match the poly-A tail), from an mRNA population believed to contain the target. PCR then proceeds with a gene-specific, forward-facing primer and an oligo-dT reverse facing primer. 5′ RACE is an exam-ple of **anchored PCR**; the first-strand cDNA population is tailed with a known sequence, either by homopolymer tailing (e.g. with dA) or by ligation of a known sequence. PCR then proceeds as before, with a primer specific for the gene and one specific for the added tail.

rapsyn Cytoplasmic protein (43 kDa) that causes cluster-ing and anchorage of acetylcholine receptors in the post-synaptic membrane of the neuromuscular junction. Mutations in rapsyn can be one cause of congenital myas-thenic syndrome.

raptor 1. Binding partner of the mammalian target of rapamycin (mTOR); the raptor–mTOR complex is a key component of a nutrient-sensitive signalling pathway that regulates cell size by controlling the accumulation of cellular mass. The raptor-mTOR complex phosphorylates the rapamycin-sensitive forms of S6K1, while the dis-tinct **rictor**-mTOR complex phosphorylates the rapamycin-resistant mutants of S6K1. **2.** A bird of prey; of either the Order Falconiformes (diurnal raptors) or the Order Strigi-formes (the nocturnal raptors, owls).

ras One of a family of **oncogenes**, first identified as trans-forming genes of **Harvey** and **Kirsten** murine **sarcoma viruses** (ras from rat sarcoma because Harvey virus, though a mouse virus, obtained its transforming gene during pas-sage in a rat). Transforming protein coded is p21ras, a GTP-binding protein with GTPase activity that resembles regulatory G-proteins. See **ras-like GTPases**.

ras-like GTPases *rac, rab, ran, rad, rheb, rho, gem, kir, ric, rin, rit, Ypt* Family of small G-proteins. The **rab** sub-family is required for membrane traffic in eukaryotic cells, ral has been associated with growth factor-induced DNA synthesis and oncogenic transformation. **Ran** is highly con-served and found in the nucleus. **Rho** and **rac** are involved in cytoskeletal control. rin, ric and rit lack prenylation sequences and are well conserved between *Drosophila* and man; rin is confined to neuronal cells. Ypts are the yeast homologues, of which 11 are known. The Rad subfam-ily, rad (ras associated with diabetes), gem (immediate-early gene expressed in mitogen-stimulated T-cells) and kir (tyrosine kinase-inducible ras-like), bind calmodulin in a calcium-dependent manner via a C-terminal extension, and also have various serine phosphorylation sites so their activ-ity may be regulated by kinases including CaMKII, PKA, PKC and CKII. Rheb is a ras homologue enriched in brain.

ratio-imaging fluorescence microscopy A method of measurement of intracellular pH or intracellular calcium levels, using a fluorescent probe molecule (see **fura-2**), in which the two different excitation wavelengths are used and the emitted light levels are compared. If emission at one wavelength is sensitive to the intracellular ion level and emission at the other wavelength is not, then standardization for intracellular probe concentration, efficiency of light collection, inactivation of probe and thickness of cytoplasm can all be performed automatically.

Rauber's cells Cells from the caudolateral deep part of the avian blastoderm (Rauber's sickle) or from the thin layer of trophoblast (Rauber's layer) that covers the inner cell mass in the early mammalian embryo.

Rauwolfia serpentina *Indian snake-root* Source of various pharmacologically active compounds, including the alkaloids **reserpine**, serpentine, sarpagine and ajmalicine, that have been used in traditional Ayurvedic medicine.

RAW 264 cells Murine monocyte/macrophage line derived from ascitic tumour induced with **Abelson leukaemia virus**.

Raynaud's disease *Raynaud's phenomenon* Constriction of blood supply to digits, producing 'white finger', most often in response to cold.

Rb **Tumour-suppressor** gene encoding a nuclear protein that, if inactivated, enormously raises the chances of development of cancer, classically **retinoblastoma**, but also other **sarcomas** and **carcinomas**.

rbc Red blood cell or erythrocyte.

RBD Abbrev. for (1) REM sleep behaviour disorder, (2) ras-binding domain, (3) RBD-1 (RNA-binding domain-1), the *C. elegans* homologue of Mrd1p, required for 18S ribosomal RNA (rRNA) processing in yeast.

RBL-1 cells Rat basophilic leukaemia cell line: shows wide variation, but can be used as a model for basophils.

R body A protein structure, visible by optical microscopy, found in various bacteria and probably related to plasmid presence. Found both in free-living pseudomonads and in various bacteria endosymbiotic in *Paramecium*. Has toxic activity against *Paramecium* and confers killer characteristics on *Paramecium* that ingest bacteria containing the structure.

RBS 1. **Rutherford backscattering spectrometry.** 2. *rbs* operon, encodes the genes responsible for ribose utilization in bacteria.

RC3 See **neurogranin**.

rDNA DNA that codes for ribosomal RNA.

RDP 1. Ribosomal Database Project, a database of aligned and annotated rRNA gene sequences. (http://rdp.cme.msu.edu/). **2.** Rapid-onset dystonia-parkinsonism.

RDW 1. Red cell distribution width, a standard parameter in haematology. **2.** See **RDW rats**.

rdw rats Strain of dwarf rats in which there is a missense point mutation of the **thyroglobulin** (Tg) gene.

reaction centre The site in the chloroplast that receives the energy trapped by chlorophyll and accessory pigments, and initiates the electron transfer process.

reactive oxygen species *ROS* Oxygen-containing radical or reactive ions such as superoxide, singlet oxygen and hydroxyl radicals, the product of the **respiratory burst** in phagocytes and responsible for bacterial killing as well as incidental damage to surrounding tissue.

reading frame One of the three possible ways of reading a nucleotide sequence. As the genetic code is read in non-overlapping triplets (**codons**), there are three possible ways of **translation** of a sequence of nucleotides into a protein, each with a different starting point. For example, given the nucleotide sequence AGCAGCAGC, the three reading frames are AGC AGC AGC, GCA GCA GCA, CAG CAG CAG.

readthrough Transcription or translation that continues beyond the normal termination signals in DNA or mRNA, respectively.

reagin Reaginic antibodies; an outmoded term for **IgE**.

real-time PCR *RT-PCR* A method in which the rate of accumulation of PCR products is measured in real time using a fluorescent marker. The signal increases in direct proportion to the amount of PCR product which allows, from the kinetics, an estimation of the original concentration of the target. Rather misleadingly often called RT-PCR, leading to the risk of confusion with reverse transcriptase PCR.

reannealing Renaturation of a DNA sample that has been dissociated by heating. In reannealing, the two strands that recombine to form a double-stranded molecule are from the same source. Differences in the rate of reannealing led to the early recognition of repetitive sequences – which rapidly recombine (have low values on the **Cot curve**).

reaper *rpr* Regulator of apoptosis in *Drosophila*. Has no known homology with vertebrate proteins, but reaper-induced apoptosis is blocked by **caspase** inhibitors and human **IAPs**. See also **grim** and **hid**.

rec8 See **cohesins**.

RecA The prototype of a class of proteins playing a central role in genomic repair and recombination in all organisms. RecA (40 kDa; product of the *rec* (recombination) gene) aligns a single strand of DNA with a duplex DNA and mediates a DNA strand switch, ATP hydrolysis being required. As part of the process, a loop of single-stranded DNA (D-loop) is generated.

recapitulation The outmoded theory that the stages of development (ontogeny) recapitulated the evolutionary stages through which an organism had passed (phylogeny); thus the primitive mammalian embryo was supposed to go through fish-like and amphibian-like stages before gaining mammal-like features.

recB protein Protein (140 kDa); one subunit of the nuclease that unwinds double-stranded DNA and fragments the strands sequentially; the other subunit is recC (128 kDa).

RecBCD A bipolar DNA helicase that employs two single-stranded DNA motors of opposite polarity to drive translocation and unwinding of duplex DNA, and thus facilitates loading of **RecA** protein onto ssDNA produced by its helicase/nuclease activity. This process is essential for RecBCD-mediated homologous recombination.

receiver cell Cells in the photosynthetic tissues of plants into which the solutes from xylem are pumped.

receptor In general terms, a membrane-bound or membrane-enclosed molecule that binds to or responds to something more mobile (the ligand), with high specificity. Examples: **acetylcholine receptor**, **photoreceptors**, **nuclear receptors**.

receptor activator of NF-kappaB *RANK* See **osteoclast differentiation factor**.

receptor downregulation A phenomenon observed in many cells: following stimulation with a ligand, the number of receptors for that ligand on the cell surface diminishes because internalization exceeds replenishment. Often used very loosely, thus destroying the utility of the term.

receptor-interacting protein kinase 1 *RIP kinase* Plays a key role in TNFα-induced IκB kinase (IKK) activation and subsequent activation of transcription factor NFκB. The **death domain** (DD) of RIP interacts with the DD of **TRADD** (TNF-R1-associated death domain protein) in two different ways: one that subsequently recruits CRADD (apoptosis/inflammation) and another that recruits NFκB (survival/proliferation). Multiple isoforms (RIP1, 2, 3 and 4) have been described. See **RAIDD**.

receptor potential The transmembrane potential difference of a sensory cell. Such cells are not generally excitable, but their response to stimulation is a gradual change in their **resting potential**.

receptor tyrosine kinase Class of membrane receptors that phosphorylate tyrosine residues. Many play significant roles in development or cell division. Examples: insulin receptor family, c-ros receptor, *Drosophila* sevenless, trk family.

receptor-associated protein *RAP* A chaperone/escort protein for members of the LDL receptor family. Binds to **lipoprotein lipase** (LPL) and may play a role in the maturation of LPL.

receptor-mediated endocytosis Endocytosis of molecules by means of a specific receptor protein that normally resides in a **coated pit**, but may enter this structure after complex formation occurs. The structure then forms a coated vesicle that delivers its contents to the endosome whence it may enter the cytoplasm or the lysosomal compartment. Many bacterial toxins and viruses enter cells by this route.

receptors for activated C kinase *RACKS* Proteins, usually anchored to specific areas of the cell, which selectively bind activated protein kinase C and thus control the regions of the cell on which it acts.

receptosome Synonym for **endosome**.

recessive An **allele** or **mutation** that is only expressed phenotypically when it is present in the homozygous form. In the heterozygote it is obscured by dominant alleles.

recombinant DNA Spliced DNA formed from two or more different sources that have been cleaved by **restriction enzymes** and joined by **ligases**.

recombinant protein Protein product from a gene that has been cloned and introduced into an appropriate expression system.

recombinase Enzymes that mediate **site-specific recombination** in prokaryotes. They fall into two families, 'phage integrases' and 'resolvases'.

recombination The creation, by a process of intermolecular exchange, of chromosomes combining genetic information from different sources, typically two genomes of a given species. Site-specific, homologous, transpositional and non-homologous (illegitimate) types of recombination are known. Recombination can be intragenic, between two alleles of a gene (**cistron**), or intergenic, where there is information exchange between non-allelic genes.

recombination nodule Protein-containing assemblies of about 90 nm in diameter placed at intervals in the **synaptonemal complexes** that develop between homologous chromosomes at the zygotene stage of meiosis. Some nodules may be associated with the site of **recombination**.

recombination, site-specific See **site-specific recombination**, **recombinase**.

recon Unit of genetic **recombination**, the smallest section of a chromosome that is capable of recombination.

recoverin Calcium-binding protein containing three **EF-hand** motifs that inhibits rhodopsin kinase and prevents premature phosphorylation of rhodopsin until the opening of cGMP-gated ion channels causes a decrease in intracellular calcium levels, signalling completion of the light response. At one time was wrongly thought to be an activator of photoreceptor guanylate cyclase. Related to visinin, P26, 23-kDa protein, S-modulin, also to 21-kDa CaBP and neurocalcin.

recreational drug A drug taken for non-medical reasons. Unfortunately, such drugs are often addictive.

recruitment zone Region of cytoplasm in the rear third of a moving amoeba where endoplasm is recruited from ectoplasm.

rectifying synapse An **electrical synapse** at which current flow can only occur in one direction.

red blood cell *Erythrocyte* Cell specialized for oxygen transport, having a high concentration of **haemoglobin** in the cytoplasm (and little else). Biconcave, anucleate discs, ca. 7 μm diameter in mammals; nucleus contracted and chromatin condensed in other vertebrates.

red coral Any of various red-coloured corals of warm seas, especially *Corallium nobile*.

red drop effect Experimental observation that the photosynthetic efficiency of monochromatic light is greatly reduced above 680 nm, even though chlorophyll absorbs well up to 700 nm. Led to the discovery of the two light reactions of photosynthesis; see **photosystems I and II**.

red tide Phenomenon in which the sea appears red as a result of massive increase in the population of a dinoflagellate (*Gymnopodium*). Unfortunately the dinoflagellate contains a **saxitoxin** that is fatal to humans; shellfish that have filtered out the dinoflagellates in large numbers become poisonous. Other algal blooms do occur, though the effect may only be to asphyxiate fish by clogging gills.

redia The secondary larval stage of Trematoda.

redox potential The reducing/oxidizing power of a system measured by the potential at a hydrogen electrode.

Reed-Sternberg cell Giant histiocytic cells, a common feature of **Hodgkin's disease**.

reeler Mouse autosomal recessive mutant, deficient in **reelin** and with disruption in large areas of the brain. Reelin, a large extracellular protein, is secreted by Cajal–Retzius cells in the forebrain and by granule neurons in the cerebellum.

reelin Protein product (99 kDa) of mouse **reeler** gene, an extracellular matrix component produced by pioneer neurons and important in cortical neuronal migration.

refractile Adjective usually used in describing granules within cells that scatter (refract) light. Not to be confused with refractory.

refractory period Most commonly used in reference to the interval (typically 1 ms) after the passage of an **action potential** during which an axon is incapable of responding to another. This is caused by inactivation of the sodium channels after opening. The maximum frequency at which neurons can fire is thus limited to a few hundred Hertz. An analogous refractory period occurs in individuals of *Dictyostelium discoideum*, which are insensitive to extracellular cyclic AMP immediately after a pulse of cAMP has been secreted. The term can be applied to any system where a similar insensitive period follows stimulation.

Refsum's disease A rare recessive lipid disorder in which there is accumulation of the branched chain fatty acid, phytanic acid, because of deficient alpha-oxidation of (14)C-phytanic acid to pristanic acid, the normal mechanism of degradation. Clinical features include retinitis pigmentosa, chronic polyneuropathy and cerebellar signs. Can be caused by mutation in either the gene encoding phytanoyl-CoA hydroxylase or the gene encoding **peroxin**-7.

REG proteins *REGα = PA28α; REGβ; REGγ (Ki antigen)* Components of the 11S **proteosome** activator that stimulates peptidase activity and enhance the processing of antigens for presentation. All three proteins share substantial sequence homology. Purified REGα (27.8 kDa) forms a heptamer in solution with a 2–3 nm cone-shaped pore, but will preferentially form a heteromeric complex with REGβ.

Reg1 Islet cell mitogen, the product of the pancreatic regulating (*reg1*) gene. Expression of *reg1* inversely correlates with cell differentiation and can be modulated by glucocorticoid receptor but not gastrointestinal hormones. Reg1 is associated with pancreatic islet regeneration and recombinant rat reg1 will stimulate pancreatic **B-cell** replication *in vitro* and *in vivo*. Also expressed in developing and Alzheimer's disease-affected cerebral cortex.

Reg2 Potent Schwann cell mitogen, a secreted protein of 16 kDa. Produced by motor and sensory neurons during development, production possibly regulated by **LIF/CNTF**.

regeneration Processes of repair or replacement of missing structures.

regulatory sequence *Control element* DNA sequence to which regulatory molecules such as **promotors** or **enhancers** bind, thereby altering the expression of the adjacent gene.

regulatory T-cell *T-reg cells* Vague term for any class of T-lymphocyte not directly involved in the effector side of immunity, but involved in controlling responses and actions of other cells; especially T-helper and T-suppressor cells. Thought to control immune responses to self-antigens and pathogens. They are now considered to be CD4$^+$CD25$^+$ and distinct from CD8$^+$CD28$^-$ **suppressor T-cells**.

regulatory T-cells *T-reg* T-cells that regulate the immune response.

regulon A situation in which two or more spatially separated genes are regulated in a coordinated fashion by a common regulator molecule.

rejection Usually used of grafts. Any process leading to the destruction or detachment of a graft or other specified structure.

Rel Protein that acts as a transcription factor. It was first identified as the oncogene product of the lethal, avian retrovirus Rev-T. It has an N-terminal region of 300 amino acids that is similar to the N-terminal regions of NFκB subunits.

rel Oncogene encoded by avian reticuloendotheliosis virus; the acutely transforming member of the Rel/NF-kappaB family of transcription factors. *V-Rel* is a truncated and mutated form of *c-rel* and transforms cells by increasing the expression of genes regulated by Rel/NF-kappaB proteins; induces oncogenic transformation and inhibits apoptosis. See Table O1.

rel homology domain *RHD* Domain found in a family of eukaryotic transcription factors, including NF-kappaB, Dorsal, Relish, NFAT, among others. Phosphorylation of the domain may regulate its activity. Has two immunoglobulin-like beta-barrel subdomains that grip the DNA in the major groove.

relative excess risk *RER* Measure used in comparison of adverse reactions to drugs, or other exposures; the component of risk solely due to the exposure or drug under investigation, making allowance for risk due to background exposure that is experienced by everybody.

relative permittivity The ratio of the **permittivity** of a substance to the permittivity of vacuum, a measure of the extent to which a substance resists the flow of electric charge (units are farads/metre). Symbol ε. Normally around 10, but for ferromagnetic materials can be as high as 4000. Formerly **dielectric constant**.

relative risk **1.** Ratio of the risk of disease or death among those exposed to the risk among those unexposed. Synonym: risk ratio. **2.** Ratio of the cumulative incidence rate in the exposed to the cumulative incidence rate in the unexposed. Synonym: rate ratio.

relaxation time Time taken for a system to return to the resting or ground state or a new equilibrium state following perturbation. Often used in context of receptor systems that have a **refractory period** after responding and then relax to a competent state. Can be used more precisely to mean the time for a system to change from its original equilibrium value to 1/e of this original value.

relaxin *RLX* Polypeptide hormone (6 kDa) produced by corpus luteum and found in the blood of pregnant animals. Acts, as its name suggests, to cause muscle relaxation during parturition, but is a multifunctional endocrine and paracrine factor that is important in several organs, including the normal and diseased cardiovascular system. Among other things it regulates matrix metallopeptidases (MMPs) and thus the properties of extracellular matrix. Human relaxin has an A chain of 24 amino acids and a B chain of 29. Has structural similarity to **insulin**. See **relaxin-like factor**.

relaxin-like factor *RLF; INSL3* A critical component in the chain of events that lead to the normal positioning of the gonads in the male fetus. **Relaxin** cross-reacts with the LGR8, the RLF receptor.

relaxosome Complex multisubunit structure forming at the plasmid origin of replication which nicks supercoiled DNA.

relay cell An interneurone of the central nervous system, particularly one linking afferent and efferent neurones of a reflex arc.

RelB Gene that encodes *v-rel* reticuloendotheliosis viral oncogene homologue B; forms heterodimeric complexes with NF-kappaB1 or NF-kappaB2.

release factor A component of the specialized transport system involved in the transport of cobalamin (vitamin B12) across the wall of the intestine. Dissociates the complex between cobalamin and the extracellular cobalamin-binding glycoprotein known as **intrinsic factor**.

Relenza® Drug used in treating influenza; works by binding to the neuraminidase on the surface of the virus particles, and stops newly formed viral particles being released from infected cells. Proprietary name for **zanamivir**.

renal Associated with the kidney.

renaturation The conversion of denatured protein or DNA to its native configuration. This is rare for proteins. However, if DNA is denatured by heating, the two strands separate and if the heat-denatured DNA is then cooled slowly the double-stranded helix reforms. This renaturation is also termed reannealing.

Renilla reniformis A soft coral, the sea pansy, source of a green fluorescent protein that has some advantages over the **GFP** from *Aequorea*; extensively used as a reporter gene, often after modification.

renin *Angiotensinogenase; EC 3.4.23.15* An **aspartic peptidase** released from the walls of afferent arterioles in the kidney when blood flow is reduced, plasma sodium levels drop, or plasma volume diminishes. Catalyses splitting of **angiotensin** I from **angiotensinogen**, an α2-globulin of plasma. Renin inhibitors (e.g. Aliskiren) have potential for the clinical treatment of hypertension.

rennet A preparation containing the enzyme rennin (chymosin) that cleaves κ-casein and causes milk to curdle. Used in the production of cheese; originally prepared from the mucous membrane of the stomach of calves but now generally made by engineered bacteria.

Reoviridae Class III viruses, with a segmented double-stranded RNA genome; there are about 8–10 segments each coding for a different polypeptide and only one strand of the RNA (minus strand) acts as template for **mRNA** (plus strand). Icosahedral capsid, and the **virion** includes all the enzymes needed to synthesize mRNA. The viruses originally included in this group do not seem to cause any disease in humans, though they have been isolated from the respiratory tract and gut of patients with a variety of diseases; the name is derived from Respiratory, Enteric, Orphan viruses. Several pathogenic viruses are now classed as reoviruses including orbivirus (a tick-borne virus that causes Colorado tick fever) and **rotavirus**.

repair nucleases Class of enzymes involved in DNA repair. It includes **endonucleases** that recognize a site of damage or an incorrect base pairing and cut it out, and exonucleases that remove neighbouring nucleotides on one strand. These are then replaced by a **DNA polymerase**.

repellant guiding molecule Specific molecules that inhibit the activity of growth cones and are thought to be important in establishing axon pathways during nervous system development. See **growth cone collapse**.

reperfusion injury Damage that occurs to tissue when blood flow is restored after a period of ischaemia. The damage is caused by neutrophils that adhere to the microvasculature and release various inflammatory mediators, hydrolytic enzymes, etc. as a response to damage. Can exacerbate the effects of stroke or cardiac ischaemia; can also follow extended periods of arterial occlusion during surgical operations.

repetin Protein of 784 amino acids, product of one of the **fused gene family**. Contains EF-hands of the S100 type and internal tandem repeats typical for cell envelope precursor proteins. Repetin expression is scattered in the normal epidermis but strong in the acrosyringium, the inner hair root sheath and in the filiform papilli of the tongue.

repetitive DNA Nucleotide sequences in DNA that are present in the genome as numerous copies. Originally identified by the value on the **Cot curve** derived from kinetic studies of DNA renaturation. These sequences are not thought to code for polypeptides. One class of repetitive DNA, termed highly repetitive DNA, is found as short sequences, 5–100 nucleotides, repeated thousands of times in a single long stretch. It typically comprises 3–10% of the genomic DNA and is predominantly **satellite DNA**. Another class, which comprises 25–40% of the DNA and is termed moderately repetitive DNA, usually consists of sequences about 150–300 nucleotides in length dispersed evenly throughout the genome, and includes **Alu** sequences and **transposons**.

replica methods Methods in the preparation of specimens for transmission electron microscopy. The specimen (for example, a piece of **freeze fractured** tissue) is shadowed with metal and coated with carbon, and then the tissue is digested away. The replica is then picked up on a grid, and it is the replica that is examined in the microscope.

replica plating Technique for testing the genetic characteristics of bacterial colonies. A dilute suspension of bacteria is first spread, in a Petri dish, on agar containing a medium expected to support the growth of all bacteria – the master plate. Each bacterial cell in the suspension is expected to give rise to a colony. A sterile velvet pad, the same size as the Petri dish, is then pressed onto it, picking up a sample of each colony. The bacteria can then be 'stamped' onto new sterile Petri dishes, plates, in the identical arrangement. The media in the new plates can be made up to lack specific nutritional requirements or to contain antibiotics. Thus colonies can be identified that cannot grow without specific nutrients or that are antibiotic-resistant, and cells with mutations in particular genes can be isolated.

replicase Generic (and rather unhelpful) term for an enzyme that duplicates a polynucleotide sequence (either RNA or DNA). The term is more usefully restricted to the enzyme involved in the replication of certain viral RNA molecules.

replication Copying, but usually the production of daughter strands of nucleic acid from the parental template.

replication factor A *Replication protein A; RPA* A heterotrimeric (three subunits, p70, p32 and p14) single-stranded-DNA-binding protein which is conserved in all eukaryotes. Associates, together with the **PIK3-kinase**-like kinase (**ATM**), at sites where homologous regions of DNA interact during meiotic prophase and at breaks associated with meiotic recombination after **synapsis**.

replication factor C *RFC* A hetero-pentameric AAA+ protein complex consisting of the Rfc1, Rfc2, Rfc3, Rfc4 and Rfc5 subunits; acts as an accessory protein (clamp loader) required to load, in an ATP-dependent manner, the **proliferating cell nuclear antigen** (PCNA) onto DNA in the replication process in eukaryotes.

replication fork Point at which DNA strands are separated in preparation for **replication**. Replication forks thus move along the DNA as replication proceeds.

replicative intermediate Intermediate stage(s) in the replication of an RNA virus; a copy of the original RNA strand, or of a single-strand copy of the first replicative intermediate. Essentially an amplification strategy.

replicons Tandem regions of replication in a chromosome, each about 30 μm long, derived from an **origin of replication**. By definition, a replicon must contain an origin of replication.

replisome Complex of proteins involved in the replication (elongation) of DNA that moves along as the new complementary strand is synthesized. On this basis, a minimum content would be DNA polymerase III and a **primosome**. An RNA-replisome has been proposed as a putative ancestor of the ribosome.

reporter gene A gene that encodes an easily assayed product (e.g. **CAT**) that is coupled to the upstream sequence of another gene and transfected into cells. The reporter gene can then be used to see which factors activate response elements in the upstream region of the gene of interest.

repressor protein A protein that binds to an **operator** of a gene, preventing the **transcription** of the gene. The binding affinity of repressors for the operator may be affected by other molecules. Inducers bind to repressors and decrease their binding to the operator, while co-repressors increase the binding. The archetype of repressor proteins is the lactose repressor protein, which acts on the **lac operon** and for which the inducers are β-galactosides such as **lactose**; it is a polypeptide of 360 amino acids that is active as a tetramer. Other examples are the λ repressor protein of lambda bacteriophage that prevents the transcription of the genes required for the lytic cycle leading to **lysogeny**, and the cro-protein, also of lambda, which represses the transcription of the λ repressor protein establishing the lytic cycle. Both of these are active as dimers and have a common structural feature, the **helix–turn–helix** motif that is thought to bind to DNA with the helices fitting into adjacent major grooves.

reproduction Propagation of organisms. The act of producing new organisms. May be asexual or sexual.

reproductive cloning Cloning of an organism, by intracytoplasmic nuclear transfer into an enucleated oocyte *in vitro*, with the intention that full development should occur; in the case of mammals this means that the blastocyst must successfully implant in a surrogate mother and the foetus must come to term. In most countries this is illegal for humans, although the technique has been used in sheep and a few other species. Strictly speaking the embryo is chimeric, since only nuclear genes have been transferred and mitochondrial genes are derived from the egg into which the nucleus was transplanted.

resact Sea urchin peptide hormone (CVTGAPGCVGGRL-NH$_2$) affecting sperm motility and metabolism. Receptor is a plasma membrane **guanylate cyclase,** and resact is a chemoattractant for sperm.

resealed ghosts Membrane shells formed by lysis of erythrocytes resealed by adjusting the cation composition of the medium. Relatively impermeable, although more permeable than the original membrane.

reserpine **Alkaloid** derived from *Rauwolfia serpentina* or *R. vomitoria*; blocks the packaging of noradrenaline into presynaptic vesicles. Useful experimental tool to determine the involvement of sympathetic innervation; formerly used for the treatment of hypertension.

residual body **1.** **Secondary lysosomes** containing material that cannot be digested. **2.** The surplus cytoplasm shed by spermatids during their differentiation to spermatozoa. Usually the cytoplasm from several spermatids connected by **cytoplasmic bridges. 3.** Surplus cytoplasm containing pigment and left over after production of merozoites during schizogony of **malaria** parasites.

resilin Amorphous rubber-like protein found in insect cuticle: similar to **elastin**, though there is no fibre formation. One of the most elastic materials known, and very important in insect flight. The first exon of the resilin gene has been expressed in a bacterial system, and the cross-linked product has a resilience (recovery after deformation) greater than that of unfilled synthetic polybutadiene – a high-resilience rubber.

resiniferatoxin Potent analogue of **capsaicin** from *Euphorbia resinifera* (flowering cactus). An agonist at **vanilloid receptor-1**.

resistin A 12.5-kDa polypeptide hormone produced by adipocytes and in humans particularly by immunocompetent cells, associated with obesity and type II diabetes in animal models; believed to modulate insulin resistance in humans.

resolution Complete return to normal structure and function; used, for example, of an inflammatory lesion, or of a disease. See also **resolving power**.

resolvase See **recombinase, site-specific recombination**.

resolving power **1.** The resolution of an optical system defines the closest proximity of two objects that can be seen as two distinct regions of the image. This limit depends upon the **Numerical Aperture** (N.A.) of the optical system, the contrast step between objects, and background and the shape of the objects. The often-quoted Airy limit applies only to self-luminous discs. **2.** In genetics, the smallest map distance measurable by an experiment involving a certain number of classified recombinant progeny.

resonance energy transfer See **fluorescence energy transfer**.

resorcinol *Resorcin; 1,3-dihydroxybenzene.* Used in the production of resins and some dyes, also topically to treat acne, eczema, psoriasis and other skin disorders.

resorufin Pink fluorescent dye. A caged form of resorufin, non-fluorescent unless activated but released by irradiation with UV, coupled to G-actin and microinjected, has been used as a marker for microfilaments in the leading lamella of moving cells.

respiration Term used by physiologists to describe the process of breathing and by biochemists to describe the intracellular oxidation of substrates coupled with production of ATP and oxidized coenzymes ($NAD^+$and FAD). This form of respiration may be anaerobic as in glycolysis, or aerobic in the case of oxidations operating via the **tricarboxylic acid cycle** and the **electron transport chain**.

respiratory burst See **metabolic burst**.

respiratory chain The mitochondrial **electron transport chain**.

respiratory enzyme complex The enzymes that make up the **respiratory chain**: NADH-Q reductase, succinate-Q reductase, cytochrome reductase, cytochrome C and cytochrome oxidase.

respiratory quotient Molar ratio of carbon dioxide production to oxygen consumption.

response elements The recognition sites of certain transcription factors, e.g. **CREB**, ISRE. Most are located within 1 kbp of the transcriptional start site.

re-stenosis Re-occlusion of coronary arteries after **angioplasty** (PTCA) or after replacement with blood vessels from elsewhere. Probably due to excessive proliferation of vascular smooth muscle that inappropriately thickens the intima and narrows the lumen.

resting potential The electrical potential of the inside of a cell, relative to its surroundings. Almost all animal cells are negative inside; resting potentials are in the range –20 to –100 mV, –70 mV typical. Resting potentials reflect the action of the **sodium pump** only indirectly; they are mainly caused by the subsequent diffusion of potassium out of the cell through potassium leak channels. The resting potential is thus close to the **Nernst potential** for potassium. See **action potential**.

restriction endonucleases *Restriction enzymes* Class of bacterial enzymes that cut DNA at specific sites. In bacteria, their function is to destroy foreign DNA, such as that of **bacteriophages** (host DNA is specifically modified at these sites). Type I restriction endonucleases occur as a complex with the methylase and a polypeptide that binds to the recognition site on DNA. They are often not very specific and cut at a remote site. Type II restriction endonucleases are the classic experimental tools. They have very specific recognition and cutting sites. The recognition sites are short, 4–8 nucleotides, and are usually **palindromic sequences**. Because both strands have the same sequence running in opposite directions the enzymes make double-stranded breaks, which, if the site of cleavage is off-centre, generates fragments with short single-stranded tails; these can hybridize to the tails of other fragments and are called sticky ends. They are generally named according to the bacterium from which they were isolated (first letter of genus name and first two letters of the specific name). The bacterial strain is identified next, and multiple enzymes are given Roman numerals – for example, the two enzymes isolated from the R strain of *E. coli* are designated *Eco*RI and *Eco*RII. The more commonly used restriction endonucleases are shown in Table R1.

restriction enzyme See **restriction endonuclease**.

restriction fragment length polymorphism *RFLP; DNA fingerprinting* Technique that allows familial relationships to be established by comparing the characteristic polymorphic patterns that are obtained when certain regions of genomic DNA are amplified (typically by PCR) and cut with certain restriction enzymes. In principle, an individual can be identified unambiguously by RFLP (hence the use of RFLP in forensic analysis of blood, hair or semen). Similarly, if a polymorphism can be identified close to the locus of a genetic defect, it provides a valuable marker for tracing the inheritance of the defect.

TABLE R1. Recognition sequences of various type II restriction endonucleases

Enzyme	Bacterium from which enzyme is derived	Recognition sequence
Aha lll	*Aphanothece halophytica*	↓ TTT I AAA
Alu I	*Arthrobacter luteus*	↓ AG I CT
Ava I	*Anabaena variabilis*	↓ C PyC I GPuG
Bam Hl	*Bacillus amyloliquefaciens* H	↓ m G GA I TCC
Bst Ell	*Bacillus stearothermophilus* ET	↓ G GTNACC
Cla I	*Caryphanon latum*	↓ AT C I GAT
Dde I	*Desulfovibrio desulfuricans*	↓ C TNAG
Eco Rl	*Escherichia coli*	↓ m G AA I TTC
Eco Rll	*Escherichia coli*	↓ m CC (AT) TT
Eco RV	*Escherichia coli*	↓ GAT I ATC
Hae ll	*Haemophilus aegyptius*	↓ PuGC I CG Py
Hae 111	*Haemophilus aegyptius*	↓ m GG I CC
Hha I	*Haemophilus haemolyticus*	m ↓ GC I GC
Hin dl ll	*Haemophilus influenzae* R_d	m ↓ AA G I CTT
Hin f l	*Haemophilus influenzae* R_f	↓ G ANTC
Hpa I	*Haemophilus parainfluenzae*	↓ GTT I AAC
Hpa ll	*Haemophilus parainfluenzae*	↓ m CC I GG
Kpn I	*Klebsiella pneumoniae* OK8	↓ GGT I AC C
Mbo I	*Moraxella bovis*	↓ GA I TC
Msp I	*Moraxella* species	↓ C C I GG

TABLE R1. (Continued)

Mst I	Microcoelus species	↓
		TGC I GCA
Pst I	Providencia stuartii	↓
		CTG I CA G
Pvu I	Proteus vulgaris	↓
		CGA I T CG
Sac I	Streptomyces achromogenes	↓
		GAG I CT C
Sma I	Serratia marcescens	↓
		CCC I GGG
Xba I	Xanthomonas badrii	↓
		T CT I AGA
Xho I	Xanthomonas holcicola	↓
		C TC I GAG

↓
5′-XXX X|XXXX-3′ (↓: Cleavage site |: Axis of symmetry). Pu, Purine i.e. A or G is recognized; Py, Pyrimidine i.e. C or T is recognized; N, any base m; X, base methylated by corresponding methylase, where known, to give N6-methyladenosine or 5-methylcytosine.

restriction fragments The fragments of DNA generated by digesting DNA with a specific **restriction endonuclease**. Each of the fragments ends in a site recognized by that specific enzyme.

restriction map Map of DNA showing the position of sites recognized and cut by various **restriction endonucleases**.

restriction nucleases See **restriction endonuclease**.

restriction point (of cell cycle) A point, late in **G1**, after which the cell must, normally, proceed through to division at its standard rate. See **checkpoint**.

restriction site Any site in DNA that can be cut by a **restriction enzyme**.

restrictive temperature See **permissive temperature**.

restrictocin Toxin (149 residues) produced by *Aspergillus restrictus*. One of the aspergillins; see **alpha-sarcin**.

resveratrol Trans-*3,5,4′-trihydroxystilbene* A fungicidal phenol with antioxidant properties found in grape skins and other plant tissues, thought to be responsible for the beneficial effects of moderate red wine consumption in protecting against heart disease and marketed as a nutritional supplement.

ret *Rearranged during transfection proto-oncogene* A human **oncogene**, encoding a receptor **tyrosine kinase**. Glial cell-line-derived neurotrophic factor (**GDNF**) exerts its effect through a multicomponent receptor system consisting of GFRalpha1, RET and **NCAM**. Hereditary medullary thyroid carcinoma (MTC) is caused by autosomal dominant gain of function mutations in the *ret* proto-oncogene, and loss of function mutations of the *ret* gene are associated with Hirschsprung's disease. *Ret* mutations are also associated with multiple endocrine neoplasia. See Table O1.

retention signal The sequence of amino acids on proteins that indicates that they are to be retained in the secretory processing system, for example the **Golgi apparatus**, and not passed on and released. Can be applied to any similar situation in which sorting of macromolecules into different compartments occurs. The carboxy-terminal tetrapeptide KDEL sequence is an important retrieval sequence for retention in the mammalian ER (HDEL in yeast).

reticalmin Calcium/calmodulin-dependent protein kinase IV-like protein (35 kDa) found in rat retina, mainly in outer segment of photoreceptors and dendrites of inner plexiform layers. See also **calspermin**.

reticular fibres Fine fibres (of **reticulin**) found in extracellular matrix, particularly in lymph nodes, spleen, liver, kidneys and muscles.

reticular lamina The lower region of extracellular matrix underlying an epithelial monolayer, separated from the basal surface of the epithelial cells by the basal lamina. The reticular lamina contains fibrillar elements (collagen, elastin, etc.) and is probably secreted by fibroblasts of the underlying connective tissues. The reticular lamina and the basal lamina together form what older textbooks refer to as the **basement membrane**.

reticulin Constituent protein of **reticular fibres**: collagen type III.

reticulocyte Immature red blood cells found in the bone marrow, and in very small numbers in the circulation.

reticulocyte lysate Cell lysate produced from **reticulocytes**; used as an *in vitro* translation system.

reticuloendothelial system The phagocytic system of the body, including the fixed macrophages of tissues, liver and spleen. Rather old-fashioned term that is coming back into use; mononuclear phagocyte system is probably better when only phagocytes are meant.

reticulum cells Cells of the **reticuloendothelial system**, found particularly in lymph nodes, bone marrow and spleen. In lymph nodes they are stromal cells and probably not reticuloendothelial cells in the current sense of that term.

retina Light-sensitive layer of the eye. In vertebrates, looking from outside, there are four major cell layers: (1) the outer neural retina, which contains neurons (ganglion cells, **amacrine cells**, **bipolar cells**) as well as blood vessels; (2) the photoreceptor layer, a single layer of rods and cones; (3) the **pigmented retinal epithelium** (PRE or RPE); and (4) the choroid, composed of connective tissue, fibroblasts and including a well-vascularized layer, the chorio capillaris, underlying the basal lamina of the PRE. Behind the choroid is the sclera, a thick organ capsule. See **retinal rods**, **retinal cones**, **rhodopsin**. In molluscs (especially cephalopods such as the squid), the retina has the light-sensitive cells as the outer layer with the neural and supporting tissues below.

retinal Aldehyde of **retinoic acid** (vitamin A); complexed with opsin forms **rhodopsin**. Photosensitive component of all known visual systems. Absorption of light causes retinal to shift from the 11-*cis*-form to the all-*trans* configuration and, through a complex cascade of reactions, excites activity in the neurons synapsed with the rod cell.

retinal cone The other light-sensitive cell type of the retina that, unlike **retinal rods**, is differentially sensitive to particular wavelengths of light, and is important for colour vision. In the human eye there are three types of cones, each type sensitive to red, green or blue. Present in large numbers in the fovea.

retinal ganglion cell See **ganglion cell**.

retinal pigmented epithelial cell See **pigmented retinal epithelium** and **retina**.

retinal rod Major photoreceptor cell of vertebrate retina (about 125 million in a human eye). Columnar cells (about 40 μm long, 1 μm in diameter) having three distinct regions: a region adjacent to, and synapsed with, the neural layer of the **retina** contains the nucleus and other cytoplasmic organelles; below this is the inner segment, rich in mitochondria, which is connected through a thin neck (in which is located a **ciliary body**) to the outer segment. The outer segment largely consists of a stack of discs (membrane infoldings that are incompletely separated in cones) that are continually replenished near the inner segment and are shed from the distal end and phagocytosed by the pigmented epithelium. The membranes of the discs are rich in **rhodopsin**, the pigment that absorbs light.

retinitis pigmentosa Disease caused by overactivity of the pigmented retinal epithelial cells, leading to damage and occlusion of photoreceptors and blindness. The mutation in Retinitis pigmentosa 1 (RP-1) is in an oxygen-regulated photoreceptor protein called ORP1; in RP-12 in the gene for the homologue of *Drosophila* **crumbs**; in the X-linked form in the gene for RPGR (retinitis pigmentosa GTPase regulator); in RP-7 in the gene encoding the photoreceptor type of **peripherin**. Other forms are known.

retino-tectal connection A problem that has exercised developmental biologists is the way in which nerve fibres from the developing retina are mapped onto the tectum of the brain. There seems to be a good positioning system in operation, and a variety of mechanisms probably operate, including control of the fasciculation of fibres in the optic nerve, and some specific recognition of the correct target area by the nerve growth cone.

retinoblastoma Malignant tumour of the retina, usually arising in the inner nuclear layer of the neural retina. Retinoblastoma is unusual in being caused by an autosomal dominant mutation in some cases (about 6%), in which case it may be bilateral. The gene product of the retinoblastoma gene (pRb) is a tumour suppressor that interacts with transcription factors such as **E2F** to block transcription of growth-regulating genes. The *Rb* gene plays a role in normal development, not just that of the retina.

retinoic acid *Vitamin A* The aldehyde (**retinal**) has long been known to be involved in photoreception, but retinoic acid has other roles. There are cytoplasmic retinoic acid-binding proteins and retinoic acid response elements that regulate gene transcription. Retinoic acid is thought to be a morphogen in chick limb-bud development and in early development of the chick (see **Hensen's node**), which probably accounts for its potent teratogenic action. See also Table V1.

retraction fibres Thin projections from crawling cells associated with areas where the cell body is becoming detached from the substratum but **focal adhesions** persist. Usually contain a bundle of microfilaments that are under tension.

retrograde axonal transport The transport of vesicles from the synaptic region of an axon towards the cell body: involves the interaction of **MAP1C** (cytoplasmic dynein) with microtubules.

retrolental fibroplasia *Retinopathy of prematurity* Damage to the retina in newborn babies exposed to excessive oxygen concentrations in early designs of incubators. There is excessive proliferation of the developing retinal vasculature and scar tissue later forms in the vitreous humour; this can, by contraction, cause retinal detachment.

retromer Complex that regulates retrieval of the cation-independent mannose 6-phosphate receptor from endosomes to the trans-Golgi network. See **sorting nexins**.

retrotransposon **Transposable element** with a transpositional mechanism requiring **reverse transcriptase** in a manner reminiscent of **retroviruses**, to which they may be related. The transposon replicates by producing an RNA transcript of itself and translating this back into a DNA copy, which can then be inserted at a different site in a chromosome or in a different chromosome in the same cell.

retroviral vector See **Retroviridae**. Retroviral vectors are used in the genetic modification of cells as a means of introducing foreign DNA into the genome. For example, RVs encoding histochemical markers (**reporter genes**) are used in the study of neural cell lineage in vertebrates. RVs may contain the bacterial *lacZ* gene that encodes for the enzyme beta-galactosidase. When the retrovirally infected cells divide, they replicate the foreign DNA. Progeny of infected cells will therefore express the protein and can then be detected histochemically.

Retroviridae Viruses with a single-stranded RNA genome (Class VI). On infecting a cell the virus generates a DNA replica by action of its virally coded **reverse transcriptase**. **Oncovirinae** are one of the three subclasses of retroviruses, the others being **Lentivirinae** and Spumavirinae. See **retroviral vector**.

retrovirus See **Retroviridae**.

Rett syndrome Severe progressive genetic neurological disorder that mainly affects baby girls, causing dyspraxia and impaired learning and communication. The gene associated with the disorder is on the X chromosome, and encodes methyl-CpG-binding protein 2 (MeCP2), which regulates transcription of various other proteins. Lethal in males.

reverse genetics The technique of determining a gene's function by first sequencing it, then mutating it, and then trying to identify the nature of the change in the phenotype.

reverse passive haemagglutination If antibodies are bonded to the surface of red blood cells, haemagglutination will occur if the appropriate bi- or multivalent antigen is added in soluble or microparticulate form. Used as a test for, for example, **Hepatitis B** virus in the serum.

reverse transcriptase RNA-directed DNA polymerase. Enzyme first discovered in retroviruses that can construct double-stranded DNA molecules from the single-stranded RNA templates of their genomes. Reverse transcription now appears also to be involved in movement of certain mobile genetic elements, such as the Ty plasmid in yeast, in the replication of other viruses such as **Hepatitis B**, and possibly in the generation of mammalian **pseudogenes**.

reversion Reversion of a mutation occurs when a second mutation restores the function that was lost as a result of the first mutation. The second mutation causes a change in the DNA that either reverses the original alteration or compensates for it.

Reynold's number A constant without dimensions that relates the inertial and viscous drag that act to hinder a body moving through fluid medium. For cells the Reynold's number is very small; viscous drag is dominant and inertial resistance can be neglected.

RF *Release factor; RF-1, RF-2, RF-3, eRF-1* Proteins that are involved in the release of the nascent polypeptide from the **ribosome**. In bacteria, RF-1 (40 kDa) is specific for UAG/UAA codons and RF-2 (41 kDa) is specific for UGA/UAA. Act on the ribosomal A-site and are assisted by RF-3, which is not codon-specific. Eukaryotic equivalents (eRF-1, etc.) have also been identified.

Rf value *Retardation factor* In chromatographic separation, the ratio of the distance travelled by the substance of interest to the distance simultaneously travelled by the mobile phase: always less than 1.

RFC 1. **Replication factor C. 2.** Reduced-folate carrier, one of the major proteins mediating transmembrane folate transport.

RFLP See **restriction fragment length polymorphism**.

RGD A domain found in **fibronectin** and related proteins, recognized by **integrins**. In most cases, the consensus is -R-G-D-S- (arginine-glycine-aspartic acid-serine).

RGS proteins Regulators of G-protein signalling proteins; a large and diverse family initially identified as GTPase-activating proteins (GAPs) of heterotrimeric G-protein Gα-subunits, now recognized to have a broader range of activities. Characteristically have a RGS domain with a conserved stretch of 120 amino acids responsible for direct binding to activated G-protein alpha subunits

RH factor *Rh factor* Rhesus factor: see **rhesus blood group**.

Rhabdocoela Order of aquatic turbellaria (flatworms). Superseded as a classification.

rhabdom The photosensitive portion of each ommatidium of the compound eye of arthropods. The rhabdom is formed by a cluster of rod-like cells that are tightly packed and have, on their inner faces, densely packed microvilli (forming the rhabdomeres) orthogonally disposed with respect to the long axis of the cells composing the rhabdom and containing rhodopsin in their membranes. In the majority of insects, the rhabdomeres of the neighbouring cells are interlaced to form a fused rhabdom. The orientation of the microvilli may be important for detecting the polarity of light.

rhabdomere The inner portion of the photosensitive cells that are clustered to form the **rhabdom** in each ommatidium of the compound eyes of Arthropods.

rhabdomyosarcoma Malignant tumour (sarcoma) derived from striated muscle.

Rhabdoviridae Class V viruses with a single negative strand RNA genome and an associated virus-specific RNA polymerase. The capsid is bullet shaped and enveloped by a membrane that is formed when the virus buds out of the plasma membrane of infected cells. The budded membrane contains host lipids but only glycoproteins coded for by the virus, of which there are usually between one and three species. In the electron microscope these appear as regularly arranged spikes about 10 nm long, and are called spike glycoproteins. This group includes **rabies virus**, **vesicular stomatitis virus** and a number of plant viruses.

rhamnogalacturonan Plant cell-wall polysaccharide consisting principally of rhamnose and galacturonic acid. Present as a major part of the pectin of the primary cell wall. Two types are known: rhamnogalacturonan I (RG-I), the major component, which contains rhamnose, galacturonic acid, arabinose and galactose; and rhamnogalacturonan II (RG-II), which contains at least four different sugars in addition to galacturonic acid and rhamnose.

rhamnolipid Simple glycolipids, normally produced by *Pseudomonas aeruginosa*, composed of a fatty-acid tail with either one or two rhamnose rings at the carboxyl end of the fatty acid. They have biosurfactant properties that are exploited in bioremediation of oil and heavy-metal-contaminated soils. Rhamnolipids also have antifungal properties and are used as agricultural fungicides.

rhamnose *6-Deoxy L-mannose* A sugar found in plant glycosides.

rheotaxis A **taxis** in response to the direction of flow of a fluid.

Rhesus blood group Human blood group system with allelic red cell antigens C, D and E. The D antigen is the strongest. Red cells from a Rhesus-positive foetus cross the placenta and can sensitize a Rhesus-negative mother, especially at parturition. The mother's antibody may then, in a subsequent pregnancy, cause haemolytic disease of the newborn if the foetus is Rhesus-positive. The disease can be prevented by giving anti-D IgG during the first 72 hours after parturition to mop up D^+ red cells in the maternal circulation.

rheumatic fever Disease involving inflammation of joints and damage to heart valves that follows streptococcal infection and is believed to be autoimmune, i.e. antibodies to streptococcal components cross-react with host-tissue antigens.

rheumatoid arthritis Chronic inflammatory disease in which there is destruction of joints. Generally considered to be an autoimmune disorder in which immune complexes are formed in joints and excite an inflammatory response (complex-mediated **hypersensitivity**), although the antigen(s) responsible is unclear. Cell-mediated (Type IV) hypersensitivity also occurs, and macrophages accumulate. This in turn leads to the destruction of the synovial lining (see **pannus**). For reasons that are also unclear, the pattern of joint damage tends to be bilaterally symmetrical.

rheumatoid factor Complex of IgG and anti-IgG formed in joints in **rheumatoid arthritis**. Serum rheumatoid factors are more usually formed from IgM antibodies directed against IgG.

rhinovirus **Picornaviridae** that largely infect the upper respiratory tract. Include the common cold virus and foot-and-mouth disease virus.

Rhizobium Gram-negative bacterium that fixes nitrogen in association with roots of some higher plants, notably legumes. Forms root nodules, in which it is converted to the nitrogen-fixing **bacteroid** form. See *nif genes*.

rhizoid Portion of a cell or organism that serves as a basal anchor to the substratum.

rhizoplast Striated contractile structure attached to the basal region of the **cilium** in a variety of ciliates and flagellates. May regulate the flagellar beat pattern, and is sensitive to calcium concentration. Composed of a 20-kDa protein rather similar to **spasmin**.

Rhizopoda Phylum that includes single-celled amoebae such as *Amoeba proteus*.

rhizopodin Cytostatic drug from the culture broth of the myxobacterium, *Myxococcus stipitatus*. Causes adherently growing L929 mouse fibroblasts and PtK2 potoroo kidney cells to produce long, narrow, branched extensions in the same way as **latrunculin** and apparently disrupts the actin cytoskeleton.

rho factor ρ *factor* A hexameric ring-shaped helicase that uses the energy derived from ATP hydrolysis to dissociate RNA transcripts from the ternary elongation complex in *E. coli*. Mutations in rho may cause the RNA polymerase to read through from one **operon** to the next. Not to be confused with rho (small G-protein) in eukaryotic cells.

rho genes Genes coding for small GTP-binding proteins; implicated in actin organization and the interaction of the cytoskeleton with intracellular membranes. See also *ras* and *rab*.

rho-kinases *ROCKs* The best-characterized effectors of the small G-protein RhoA. Abnormal activation of the RhoA/ROCK pathway has been observed in major cardiovascular disorders such as atherosclerosis, re-stenosis, hypertension, pulmonary hypertension and cardiac hypertrophy.

rhodamines A group of triphenylmethane-derived dyes are referred to as rhodamines, lissamines, etc. Many are fluorescent and are used as fluorochromes in labelling proteins and membrane probes.

Rhodnius prolixus Reduviid blood-sucking bug, vector of *Trypanosoma cruzi* in South America. Much used by insect physiologists because the transition from one instar to the next is triggered by a blood meal. The blood meal, which may exceed substantially the body weight of the insect, triggers complex changes in the mechanical properties of abdominal cuticle, allowing the insect to swell, and activates excretory mechanisms. The haemolymph contains relatively few **haemocytes**, which made some of the classical parabiosis experiments much easier.

Rhodophyta *Red algae* Division of algae, many of which have branching filamentous forms and red colouration. The latter is due to the presence of **phycoerythrin**. The food reserve is floridean (starch), found outside the plastid. The walls contain sulphated galactans such as **carrageenan** and **agar**.

rhodopsin *Visual purple* Light-sensitive pigment formed from **retinal** linked through a **Schiff's base** to **opsin**: rhodopsin is an integral membrane protein found in the discs of **retinal rods** and cones, comprising some 40% of the membrane. Vertebrate opsins are proteins of 38 kDa. See also **bacteriorhodopsin** and **opsin subfamilies**.

rhodopsin kinase *RK* Kinase that mediates adaptation of photoreceptors to light and protects against light-induced injury. Phototransduction initiated by light is quenched rapidly by RK-phosphorylation of photoexcited rhodopsin. The phosphorylated rhodopsin is then bound by cytosolic **arrestin**, and thereby uncoupled from the G-protein-**transducin**. In *Drosophila* the rhodopsin kinase appears to be G-protein-coupled receptor kinase 1 (GPRK1), which is most similar to the mammalian beta-adrenergic receptor

kinases, G-protein-coupled receptor kinase 2 (GRK2) and GRK3. See **recoverin**.

Rhodospirillum rubrum A purple non-sulphur bacterium with a spiral shape; contains the pigment **bacteriochlorophyll** and under anaerobic conditions photosynthesises using organic compounds as electron donors for the reduction of carbon dioxide. The purple colour results from the presence of **carotenoids**, though the bacteria are often more red or brown.

rhombencephalon The hindbrain, comprising the metencephalon, the myelencephalon and the reticular formation. Functions include attention and sleep, autonomic functions, motor coordination, reflex movements and simple learning.

rhomboid rho *Drosophila* gene, product is the polytopic membrane protein rhomboid-1 that promotes the cleavage of the membrane-anchored TGFα-like growth factor Spitz, allowing it to activate the *Drosophila* EGF receptor. Rhomboid-1 has the sequence characteristics of a serine protease and spitz processing is blocked by specific serine protease inhibitors. See **dodo**.

rhombomere **Neuromeres** or segments in the hindbrain region that are of developmental significance. Shown to be lineage-restriction units in that cells of adjacent rhombomeres do not mix with each other. Regulatory genes have been shown to be expressed in patterns in the developing hindbrain that relate to the neuromeric or **segmentation** pattern.

rhophilin Protein (71 kDa, 643 residues) that is a **Rho** GTPase. Has sequence homology with the N-terminal region of **protein kinase N,** though possesses no catalytic activity. Has a **PDZ domain** at the carboxy terminus suggesting that it may act as an adaptor molecule.

rhoptry Electron-opaque dense body found in the apical complex of parasitic protozoa of the Phylum Apicomplexa.

rhotekin One of a group of proteins that contain a Rho-binding domain and are target peptides (effectors) for the Rho GTPases. Rhotekin is expressed at a low level in normal cells and is overexpressed in many cancer-derived cell lines, which may lead to activation of NF-kappaB.

ribavirin Ribavirin is an orally active guanoside analogue that inhibits the replication of a variety of RNA viruses. It inhibits inosine monophosphate dehydrogenase.

ribbon synapse Ultrastructurally distinct type of synapse found in a variety of sensory receptor cells such as retinal **photoreceptor** cells, **cochlear hair cells** and vestibular organ receptors, as well as in a non-sensory neuron, the retinal bipolar cell. Unlike most neurons, these cells do not use regenerative action potentials but release transmitter in response to small graded potential changes. Ribbon synapses have different exocytotic machinery from conventional synapses in containing dense bars or ribbons anchored to the presynaptic membrane covered with a layer of synaptic vesicles. The ribbons have been proposed to shuttle synaptic vesicles to exocytotic sites.

riboflavin *Vitamin B2* Ribose attached to a flavin moiety that becomes part of FAD and FMN. See also Table V1.

ribonuclease *RNAse, RNAase* Widely distributed type of enzyme that cleaves RNA. May act as endonucleases or exonucleases, depending upon the type of enzyme. Generally recognize target by tertiary structure rather than sequence. Ribonuclease E is an RNAase involved in the formation of 5S ribosomal RNA from pre-rRNA. F is stimulated by interferons and cleaves viral and host RNAs and thus inhibits protein synthesis. H specifically cleaves an RNA base-paired to a complementary DNA strand. P is an endonuclease that generate tRNAs from their precursor transcripts. T is an endonuclease that removes the terminal AMP from the 3′ CCA end of a non-aminoacylated tRNA. RNAase T1 cleaves RNA specifically at guanosine residues. RNAase III cleaves double-stranded regions of RNA molecules.

ribonucleic acid See **RNA**.

ribonucleoprotein RN Complexes of RNA and protein are involved in a wide range of cellular processes. Besides **ribosomes** (with which RNP was originally almost synonymous), in eukaryotic cells both initial RNA transcripts in the nucleus (**hnRNA**) and cytoplasmic **mRNAs** exist as complexes with specific sets of proteins. Processing (splicing) of the former is carried out by small nuclear RNPs (**snRNPs**). Other examples are the **signal recognition particle** responsible for targeting proteins to endoplasmic reticulum, and a complex involved in termination of transcription.

ribophorin Glycoproteins of the endoplasmic reticulum that interact with ribosomes whilst co-translational insertion of membrane or secreted proteins is taking place. Ribophorins may form a pore through which the nascent polypetide chain passes.

riboprobe Somewhat casual term for an RNA segment used to probe for a complementary nucleotide sequence, either in the mRNA pool or in the DNA of a cell.

ribose *D-Ribose* A monosaccharide pentose of widespread occurrence in biological molecules, e.g. RNA.

ribose-binding protein Periplasmic-binding proteins of bacteria that interact either with the ribose transport system or with the **methyl-accepting chemotaxis protein**, MCP-III (trg).

ribosomal protein Proteins present within the ribosomal subunits. In prokaryotes, there are 31 proteins in the large subunit and 21 in the small subunit. Eukaryotic subunits have 50 large-subunit and 33 small-subunit proteins.

ribosomal RNA See **RNA**.

ribosome A heterodimeric multisubunit enzyme composed of **ribonucleoprotein** and protein subunits. Interacts with aminoacylated tRNAs and mRNAs, and translates protein-coding sequences from messenger RNA. During protein elongation the nascent protein is held at the P-site (peptidyl-tRNA complex), while aminoacyl-tRNAs bearing new aminoacids are bound at the A-site. Similar ribosomes are found in all living organisms, all composed of large and small subunits, as well as in chloroplasts and mitochondria. Differences are apparent between prokaryotic and eukaryotic ribosomes.

ribotype The RNA complement of a cell, by analogy with phenotype or genotype.

ribozyme RNA with catalytic capacity, an enzyme made of nucleic acid not protein. Of particular interest because of the implications for self-replicating systems in the earliest stages of the evolution of (terrestrial) life.

ribulose 1,5-bisphosphate An intermediate in the **Calvin–Benson cycle** of photosynthesis.

ribulose bisphosphate carboxylase *RUBISCO; D-Ribulose 1,5-diphosphate carboxylase; EC 4.1.1.39.* Enzyme responsible for CO_2 fixation in photosynthesis. Carbon dioxide is combined with ribulose diphosphate to give two molecules of 3-phosphoglycerate, as part of the **Calvin–Benson cycle**. It is the sole CO_2-fixing enzyme in C3 plants, and collaborates with **PEP carboxylase** in CO_2 fixation in C4 plants. In the presence of oxygen, the products of the reaction are one molecule of phosphoglyceric acid and one molecule of phosphoglycolic acid. The latter is the initial substrate for photorespiration, and this oxygenase function occurs in C3 plants where the enzyme is not protected from ambient oxygen; in C4 plants the enzyme acts exclusively as a carboxylase since it is protected from oxygen. Also known as Fraction 1 protein, the major protein of leaves.

ribulose diphosphate carboxylase *RuDPC; RuDP carboxylase* See **ribulose bisphosphate carboxylase**, the recommended name.

ricin Highly toxic **lectin** (66 kDa) from seeds of the castor bean, *Ricinus communis*. Has toxic A subunit (32 kDa), carbohydrate-binding B subunit (34 kDa). Toxic subunit inactivates ribosomes by depurinating A4324 in the 28S rRNA fragment of the 60S RNA chain; and the binding subunit is specific for β-galactosyl residues. See **sarcin/ricin loop**.

***Ricinus communis* agglutinin** *RCA* **Lectin** (120 kDa) from castor bean, with specificity similar to **ricin**, but much less toxic.

rickets *Rachitis* Disease due to deficiency of vitamin D that leads to defective ossification and softening of bones.

Rickettsia Genus of Gram-negative bacteria responsible for a number of insect-borne diseases of man (including scrub typhus and **Rocky Mountain spotted fever**). Obligate intracellular parasites.

rictor *Rapamycin-insensitive companion of mTOR* The rictor–**mTOR** complex modulates the phosphorylation of PKCalpha and the actin cytoskeleton. Rictor shares homology with pianissimo from *D. discoideum* and STE20p from *S. pombe*. See **raptor**.

Riedel's disease An extremely rare form of chronic inflammation of the thyroid gland, which becomes enlarged and forms a hard mass of scar tissue. Aetiology unknown.

rifampicin Semisynthetic member of the **rifamycin** group of antibiotics.

rifamycin Antibiotic produced by *Streptomyces mediterranei* that acts by inhibiting prokaryotic, but not eukaryotic, DNA-dependent RNA synthesis. Blocks initiation but not elongation of transcripts.

rigor Stiffening of muscle as a result of high calcium levels and ATP depletion, so that actin–myosin links are made, but not broken.

rilmenidine See **clonidine**.

RING See **E3 ligase** and **RING finger motif**.

RING finger motif A **zinc finger** motif found in various nuclear proteins and in some receptor-associated proteins. The RING (Really Interesting New Gene) finger, or

$Cys_3HisCys_4$, family of zinc-binding proteins play important roles in differentiation, oncogenesis and signal transduction.

Ringer's solution Isotonic salt solution used for mammalian tissues; original version (for frog tissues) much modified, and the term is often used loosely to mean any physiological saline.

ringworm *Tinea* A contagious disease caused by infection of the skin by dermatophyte fungi. Ring-shaped patches of inflammation may form, hence the name. The same fungi are responsible for athlete's foot.

rINNs Recommended International Non-Proprietary Names for medicinal substances. British Approved Names (BANs) have been harmonized with rINNs.

RIPA 1. Regulator of iron proteins A (RipA), an AraC-type regulator of bacterial genes. 2. Radioimmuno-precipitation assay.

RISCs *RNA-induced silencing complexes* Complexes formed during **RNA interference**: siRNAs incorporate into RISCs and provide guide functions for sequence-specific ribonucleolytic activity. Protein components of RISCs include **argonaute** family members, nucleases, and other factors.

risk In toxicolgy, the probability that an agent will cause adverse effects in an organism, a population or an ecological system, or the expected frequency of such an event.

risk assessment An attempt to identify and quantify the **risk** involved in the use of a substance or the carrying out of a procedure. By considering the potential hazardous outcomes that might occur, precautionary measures can be taken (and litigation avoided). But faint heart never won fair lady, and excessive caution may not help discovery research!

risperidone Atypical antipsychotic drug, a psychotropic agent belonging to the chemical class of benzisoxazole derivatives, used in treatment of acute and chronic psychoses, including schizophrenia. May act through antagonism at dopamine type 2 (D2) and serotonin type 2 ($5HT_2$) receptors, although this is uncertain.

ristocetin Mixture of ristocetins A and B: isolated from actinomycete, *Nocardia lurida*. Induces platelet aggregation.

Ritalin® Proprietary name for **methylphenidate**.

ritonavir A **protease inhibitor** used in the treatment of HIV.

rituximab *Rituxan* Monclonal antibody directed against CD20. Used for treatment of non-Hodgkin's lymphoma.

rivastigmine A reversible cholinesterase inhibitor used to treat Alzheimer's disease. May work by increasing the concentration of acetylcholine in cholinergic neuronal pathways that project from the basal forebrain to the cerebral cortex and hippocampus, and that are involved in memory, attention, learning and other cognitive processes.

R-loop Structure formed when RNA hybridizes to double-stranded DNA by displacing the identical DNA strand.

RMP pathway *Ribulose monophosphate pathway; allulose phosphate pathway* A metabolic pathway used by methylotrophic bacteria for the conversion of formaldehyde to hexose sugars, etc. In the first stage, ribulose-5-phosphate is condensed with HCHO.

RNA *Ribonucleic acid* This molecular species has an informational role, a structural role and an enzymic role, and is thus used in a more versatile way than either DNA or proteins. Considered by many to be the earliest macromolecule of living systems. The structure is of ribose units joined in the 3′ and 5′ positions through a phosphodiester linkage with a purine or pyrimidine base attached to the 1′ position. All RNA species are synthesized by transcription of DNA sequences, but may involve post-transcriptional modification.

RNA editing A process responsible for changes in the final sequence of mRNA that are not coded in the DNA template. Excludes mRNA **splicing** and modifications to tRNA. Various kinds of editing are known, the commonest being cytidine (C) to uridine (U) substitution, though polyadenylation of mitochondrial mRNA and guanosine (G) insertion in paramyxoviruses are other examples of editing. The process involves guide RNA (gRNA) in some cases but not all. Though the commonest examples are in organelles, the process of editing does also occur in nuclear transcripts.

RNA interference The regulation of mRNA half-life, and thus gene expression, by the selective degradation of some messages following the binding of small interfering RNA (**siRNA**) and formation of **RISCs**. Has proved a powerful experimental tool for selective deletion of particular proteins.

RNA plasmid DsRNA found in yeasts, also known as killer factors. Their nomenclature is uncertain, and some scientists consider them viruses.

RNA polymerases Enzymes that polymerize ribonucleotides in accordance with the information present in DNA. Prokaryotes have a single enzyme for the three RNA types that is subject to stringent regulatory mechanisms. Eukaryotes have type I, which synthesizes all **rRNA** except the 5S component; type II, which synthesizes **mRNA** and **hnRNA;** and type III, which synthesizes **tRNA** and the 5S component of rRNA.

RNA primase Although a RNA polymerase, is usually referred to as **DNA primase**.

RNA primer The short RNA sequence synthesized by **DNA primase**.

RNA processing Modifications of primary RNA transcripts including splicing, cleavage, base modification, capping and the addition of **poly-A tails**. See also **RNA editing**.

RNAse protection assay *RNAase protection assay* Sensitive and quantitative alternative to **Northern blots** for the measurement of gene expression levels. Labelled antisense cRNA is transcribed from a DNA clone in an appropriate vector, hybridized with an mRNA sample, and single-stranded RNA digested away with RNAase, then run out on a gel. The amount of labelled RNA surviving is directly proportional to the amount of target mRNA present in the sample.

RNase P A ribonucleoprotein responsible for the 5′ maturation of precursor tRNAs (**ptRNAs**) in all organisms, will cleave any target mRNA that forms a ptRNA-like structure and sequence-specific complex when bound to an RNA, termed the EGS (**external guide sequence**).

RNases E, F, H, P, T, III *RNAases* See **ribonucleases**.

RNA splicing The removal of **introns** from primary RNA transcripts. See **alternative splicing**.

RNA tumour virus See **Oncovirinae**.

RNP See **ribonucleoprotein**.

Robertsonian translocation A special type of non-reciprocal translocation in chromosomes whereby the long arms of two non-homologous acrocentric chromosomes are attached to a single centromere. The short arms become attached to form a reciprocal structure that often disappears some divisions after its formation. Common Robertsonian translocations in man are confined to the acrocentric chromosomes 13, 14, 15, 21 and 22, the short arms of which contain no essential genetic material.

robo *Roundabout* Receptor for **slit**.

Rock *Rock-1, ROKβ; Rock-2, ROKα* Serine–threonine kinases, putative targets for **rho** activation. Rock-2 may act as an effector for GTP-Rho-A in inducing cytoskeletal rearrangement.

Rocky Mountain spotted fever Acute infectious tick-borne rickettsial disease caused by *Rickettsia rickettsii*.

rod cell See **retinal rod**.

rod outer segment See **retinal rod**.

rodent ulcer *Basal cell carcinoma* Carcinoma arising from the basal cells of the skin, usually affecting areas, such as the upper face, that have had much exposure to sun. Slow-growing and of low malignancy.

Rohypnol® *Flunitrazepam* Proprietary name for a powerful benzodiazepine sedative used for short-term treatment of insomnia. Until modified by the addition of a substance that turns blue upon dissolving, was used as a date rape drug.

rolipram Inhibitor of cAMP-specific phosphodiesterase IV. The effect of administration of rolipram is increased synthesis and release of norepinephrine, which enhances central noradrenergic transmission, and suppression of the production of proinflammatory cytokines and other inflammatory mediators. Thus rolipram attenuates endogenous depression and inflammation in the central nervous system.

rolling circle mechanism A mechanism of DNA replication in many viral DNAs, in bacterial F factors during mating, and of certain DNAs in gene amplification in eukaryotes. DNA synthesis starts with a cut in the + strand at the replication origin, the $5'$ end rolls out and replication starts at the $3'$ side of the cut around the intact circular DNA strand. Replication of the $5'$ end (tail) takes place by the formation of **Okazaki fragments**.

Romanovsky-type stain Composite histological stains including methylene blue, Azure A or B and eosin, sometimes with other stains. Often used to stain blood films, and allow differentiation of the various leucocyte classes. Examples are Giemsa, Wright's and Leishman's stains.

ron *Recepteur d'origine nantaise* A receptor tyrosine kinase for **macrophage-stimulating protein**, a member of the receptor family that includes the proto-oncogene product **met** and the avian oncogene product **sea**. It is a transmembrane heterodimer comprising of one alpha- and one beta-chain originating from a single-chain precursor and held together by several disulphide bonds.

root cap Tissue found at the apex of roots, overlying the root apical meristem and protecting it from friction as the root grows through the soil. Secretes a glycoprotein mucilage as a lubricant.

root hair cell Root epidermal cell, part of which projects from the root surface as a thin tube, thus increasing the root surface area and promoting absorption of water and ions.

root nodule Globular structure formed on the roots of certain plants, notably legumes and alder, by symbiotic association between the plant and a nitrogen-fixing microorganism (*Rhizobium* in the case of legumes and *Frankia* in the case of alder and a variety of other plants).

rootlet system Microtubules associated with the base of the flagellum in ciliates and flagellates. Also associated with this region is the **rhizoplast**.

ros 1. An **oncogene**, *ros*, identified in bird **sarcoma**, encoding a receptor **tyrosine kinase** that has high homology with the *Drosophila* **sevenless** protein. See Table O1. **2.** Reactive oxygen species (ROS); see **metabolic burst**. **3.** *Agrobacterium* transcriptional regulator Ros is a prokaryotic zinc-finger protein that regulates the plant *ipt* **oncogene**.

roscovitine A purine derivative that is a potent and selective inhibitor of **cyclin-dependent kinases**.

Rosetta Stone method A method in bioinformatics used to detect functional linkage between two proteins based on the principle that if two proteins A and B are found as separate sequences in some species but as a fused protein with both A and B regions in others then it is probable that the proteins are functionally associated. For example, yeast genes *Pur2* and *Pur3* encode enzymes in the purine synthesis pathway: in *C. elegans* the single Ade 5,7,8 protein has sequences highly homologous with both Pur2 and Pur3. Knowing that both are found in a single protein makes it very probable that they are involved in a linked function. See **phylogenetic profile**, **gene neighbour method**.

rosiglitazone **Insulin sensitizer** used in treatment of Type II diabetes.

rotamase *EC 5.2.1.8* Prokaryotic peptidyl prolyl *cis–trans*-isomerase, homologue of **immunophilins** but not inhibited by **ciclosporin**. Located in the periplasm.

rotamer A rotational isomer – conformationally different by rotation at a single bond.

Rotarix Proprietary name for the first vaccine for the prevention of gastroenteritis caused by **rotavirus**.

Rotavirus Genus of the Reoviridae having a double-layered capsid and 11 double-stranded RNA molecules in the genome. They have a wheel-like appearance in the electron microscope, and cause acute diarrhoeal disease in their mammalian and avian hosts.

rotenone An inhibitor of the **electron transport chain** that blocks transfer of reducing equivalents from NADH dehydrogenase to coenzyme Q. A very potent poison for fish and for insects.

rotifera *Wheel animalcules* Small, unsegmented, pseudocoelomate animals of the phylum Aschelminthes. Found in many freshwater environments and in moist soil, where they inhabit the thin films of water around soil particles. Move using cilia which may cause them to rotate, hence their name.

rough endoplasmic reticulum *RER* Membrane organelle of eukaryotes that forms sheets and tubules. Contains the receptor for the signal receptor particle and binds ribosomes engaged in translating mRNA for secreted proteins and the majority of transmembrane proteins. Also a site of membrane lipid synthesis. The membrane is very similar to the nuclear outer membrane. The lumen contains a number of proteins that possess the C-terminal signal **KDEL**.

rough microsome Small vesicles obtained by sonicating cells and that are derived from the rough endoplasmic reticulum. Have bound ribosomes and can be used to study protein synthesis.

rough strain Bacterial strains that have altered outer cell-wall carbohydrate chains, causing colonies on agar to change their appearance from smooth to dull. In Streptococci, the smooth strains are virulent whereas the rough strains are not. This is partly because the rough strains are much more readily phagocytosed.

rouleaux Cylindrical masses of red blood cells. Horse blood will spontaneously form rouleaux; in other species they can be induced by reducing the repulsion forces between erythrocytes, for example by adding dextran of appropriate molecular weight.

Rous sarcoma virus *RSV* The virus responsible for the classic first cell-free transmission of a solid tumour, the chicken **sarcoma**, first reported by Rous in 1911. An avian **C-type** oncorna virus, original source of the *src* **gene**, an oncogene.

royalisin Insect **defensin** found in honeybee royal jelly.

R plasmid *R factor; drug resistance factor* A plasmid that confers resistance to one or more antibiotics or other poisonous compounds in a bacterium.

RPMI 1640 Culture medium developed at Roswell Park Memorial Institute that utilizes a bicarbonate buffering system. Widely used for the culture of human normal and neoplastic leukocytes, and when properly supplemented will support growth of many types of cultured cells, including fresh PHA-stimulated human lymphocytes.

R point of cell cycle See **restriction point**.

RRM *RNA recognition motif* A fold that is found in many eukaryotic RNA-binding proteins. Also known as RNA-binding domain (RBD) or ribonucleoprotein domain (RNP).

rRNA *Ribosomal RNA* Structural RNA components of the ribosome. Prokaryotes have 5S and 23S species in the large subunit and a 16S species in the small subunit. Eukaryotes have a 5S, 5.8S and 28S species in the large subunit and an 18S species in the small subunit.

RS domain Arginine-serine-rich domain required for RNA binding of splicing factors such as **U2AF**. See **SR-proteins**.

RSC complex RSC, an abundant, essential chromatin-remodelling complex, related to the **SWI/SNF complex**, binds nucleosomes and naked DNA with comparable affinities. The RSC complex of *Saccharomyces cerevisiae* is closely related to the SWI/SNF complex. Both complexes are involved in remodelling chromatin structure and they share conserved components. The RSC proteins Sth1, Rsc8/Swh3, Sfh1 and Rsc6 are homologues of the SWI/SNF proteins Swi2/Snf2, Swi3, Snf5 and Swp73, respectively.

RSV See **Rous sarcoma virus**.

RT-PCR *Reverse transcriptase polymerase chain reaction; reverse transcription PCR* PCR in which the starting template is RNA, implying the need for an initial reverse transcriptase step to make a DNA template. Some thermostable polymerases have appreciable reverse transcriptase activity; however, it is more common to perform an explicit reverse transcription, inactivate the reverse transcriptase or purify the product, and proceed to a separate conventional PCR. Abbreviation ambiguous because also used sloppily for **real-time PCR**.

RTF *Resistance transfer factor* The part of a conjugative R plasmid, usually self-repressed, that specifies conjugation.

RTX family of toxins A group of related cytolysins and cytotoxins produced by Gram-negative bacteria including *E. coli*, *Proteus vulgaris* (haemolysin), *Pasteurella haemolytica* (leukotoxin) and *Bordetella pertussis* (adenylate cyclase-haemolysin). Characteristically contain a repeat domain (hence the designation, repeats in toxins) with glycine- and aspartate-rich motifs repeated within the domain. All are produced in inactive pro-form that must be post-translationally modified to generate an active toxin, and are calcium-dependent pore-forming toxins. See *Escherichia coli* **haemolysin**. *Vibrio cholerae* RTX toxin causes the depolymerization of actin stress fibers, through the unique mechanism of covalent actin cross-linking.

R-type channels A class of neuronal **voltage-sensitive calcium channels** that are unaffected by dihydropyridines, phenylalkylamines and **conotoxins**. Thought to carry much of the current, stimulated by glutamate release in response to ischaemia, that induces neuronal death.

rubber Natural rubber is a polyterpene (polyisoprene) in which the double bonds are *cis*. A related compound in which the bonds are *trans* is gutta-percha.

rubella *German measles* A mild acute viral childhood infection caused by the rubella virus, with a pink rash somewhat like that of measles. Infection during pregnancy can cause severe foetal abnormalities.

rubidium *Rb* One of the alkali earth metals, used to substitute for potassium in some ion flux experiments.

RUBISCO See **ribulose bisphosphate carboxylase**.

rubp See **ribulose bisphosphate carboxylase**.

RUDP carboxylase See **ribulose bisphosphate carboxylase**.

Ruffinini's corpuscles Ovoid encapsulated sensory nerve ending in subcutaneous tissue. Probably mechanosensors.

ruffles Projections at the leading edge of a crawling cell. In time-lapse films the active edge appears to ruffle. The protrusions are apparently supported by a microfilament meshwork, and can move centripetally over the dorsal surface of a cell in culture.

rugae Wrinkles. *Adj.* rugose or, if the wrinkles are small, rugulose.

Runx2 *Runt-related transcription factor 2* A transcription factor that is a global regulator of osteoblast differentiation, expressed several days before osteoblast genes in bone anlage. Inhibited by **twist**.

rutabaga rut *Drosophila* memory mutant; gene codes for calcium/calmodulin-responsive adenylyl cyclase; net result is elevated cAMP levels and a comparable behavioural defect to **dunce**.

ruthenium red A stain used in electron microscopy for acid mucopolysaccharides on the outer surfaces of cells.

Rutherford backscattering spectrometry *RBS* Method used to investigate quantitatively the depth profile of individual elements in a solid by detecting the energies and amount of the backscattered ions when the target is bombarded by a mono-energetic light ion beam, typically 1–2-MeV He ions.

ryanodine Drug that blocks the release of calcium from the sarcoplasmic reticulum of skeletal muscle. Ryanodine-binding proteins have also been found in the CNS. The water-soluble plant extract ryania from the powdered stem of the tropical shrub *Ryania speciosa* has been used as an insecticide. The extract contains several structurally related ryanoids, including ryanodine. See **ryanodine receptor**.

ryanodine receptor *RyR* Large transmembrane proteins of 565 kDa that form tetrameric Ca^{2+} channels which release calcium ions from the sarcoplasmic reticulum into the cytosol during muscle contraction. They are stimulated to transport Ca^{2+} into the cytosol by recognizing Ca^{2+} on the cytosolic side of the SR (calcium-induced calcium release), a positive feedback that leads to a rapid response. Isoforms (RyR_2 and RyR_3) are found in neural tissue and in muscle. Show sequence similarity with InsP3-gated calcium channels of the endoplasmic reticulum but are pharmacologically distinct. RyR_1 mutations are associated with malignant hyperthermia.

S

S1 1. Soluble fragment (102 kDa) of **heavy meromyosin** that is produced by papain cleavage: retains the ATPase, actin-binding activity and motor function, and can be used to decorate actin filaments for identification by electron microscopy. **2.** Ribosomal protein S1 plays a critical role in translation initiation and elongation in *E. coli* and is believed to stabilize mRNA on the ribosome.

S1 nuclease *EC 3.1.30.1* A single-strand specific nuclease, usually isolated from certain *Neurospora* and *Aspergillus* species, which degrades single-stranded nucleic acids and is more active against DNA than RNA. Commonly used to analyse the structure of DNA–RNA hybrids (S1 nuclease mapping), and to remove single-stranded extensions from DNA to produce blunt ends.

S2 Fibrous fragment of heavy meromyosin (**HMM**). Links the **S1** head to the light meromyosin (**LMM**) region that lies in the body of the thick filament and acts as a flexible hinge.

S6 kinase A serine/threonine **kinase**, activated by **MAP kinase**; phosphorylates ribosomal protein S6 to elevate protein production in cells stimulated by a **mitogen**. S6 kinase alpha-1 is EC 2.7.11.1; S6 kinase alpha 3 is EC 2.7.1.37.

S9 1. Postmitochondrial supernatant fraction of liver homogenate rich in drug-metabolizing enzymes (P450s), sometimes used in the **Ames test**. **2.** A ribosomal protein. **3.** See **dermaseptins**.

S100 Family of calcium-binding proteins containing an **EF-hand**. Found on glial cell surfaces. S100b has neurotrophic and mitogenic activity through the ability to cause a rise in intracellular calcium. The family includes calcyclin, MRP8 and MRP14, **calmodulin**, **troponin** and **calgranulins**.

S180 *Sarcoma 180* Highly malignant mouse sarcoma cells, often passaged in **ascites** form. Used in some of the classical studies on **contact inhibition of locomotion**.

Sab *SH3BP5* Protein (425 residues) that binds selectively to SH3 domain of **btk** and binds to and serves as a substrate for **JNK**.

saccade An eye movement in which the eyes jump from one point to another; seeing something is actually based on a series of snapshots focused on some limited elements of the whole visual field.

saccharomicins Saccharomicins A and B are heptadecaglycoside antibiotics isolated from the fermentation broth of the rare actinomycete *Saccharothrix espanaensis*. They are active both *in vitro* and *in vivo* against bacteria and yeast.

Saccharomyces Genus of Ascomycetes; yeasts. Normally haploid unicellular fungi that reproduce asexually by budding. Also have a sexual cycle in which cells of different mating types fuse to form a diploid **zygote**. Economically important in brewing and baking, and are also suitable eukaryotic cells for the processes of genetic engineering and for the analysis of, for example, cell division cycle control by selecting for mutants (see **cdc genes**). *S. cerevisiae* is

baker's yeast; *S. carlsbergensis* is now the major brewer's yeast. See also *Schizosaccharomyces pombe*.

saccharopine An intermediate in the aminoadipic pathway for the synthesis or degradation of lysine, synthesized from lysine and α-oxoglutarate in mammalian liver.

saccharopine dehydrogenase *EC 1.5.1.9* Cytoplasmic enzyme in plants that catalyses the NAD$^+$-dependent cleavage of saccharopine to L-lysine and 2-oxoglutarate. In some organisms this enzyme is found as a bifunctional polypeptide with lysine ketoglutarate reductase, the first two linked enzymes of lysine catabolism. One of the AlaDH/PNT (alanine dehydrogenase/pyridine nucleotide transhydrogenase) family of enzymes.

sad1 1. Constitutive membrane-bound component of the yeast **spindle pole body** (SPB) that interact with Kms1. See **suns**. **2.** *Arabidopsis* sad1 (supersensitive to ABA and drought) mutation increases plant sensitivity to drought stress and abscissic acid in seed germination. **3.** Mycobacterial semialdehyde dehydrogenases, Sad1 and Sad2.

S-adenosyl methionine S-*(5′-deoxyadenosine-5′)-methionine* An activated derivative of **methionine** that functions as a methyl group donor, in (for example) phospholipid methylation and bacterial **chemotaxis**.

Saethre–Chotzen syndrome *SCS* An autosomal dominant craniosynostosis syndrome (fusion of bones in the skull that are normally separate), caused by loss-of-function mutation in **twist-1**, with uni- or bilateral coronal synostosis and mild limb deformities.

safranin *Basic red 2* Safranin O is a mixture of dimethyl and trimethyl safranin used as a counterstain in histochemistry. It stains nuclei red and metachromatically stains cartilage yellow.

SAGA complex *Spt-Ada-Gcn5-acetyltransferase* A multisubunit **histone acetyltransferase** complex involved in transcriptional regulation in *S. cerevisiae*, required for RNA polymerase II-dependent transcription of several genes and that facilitates the binding of TATA-binding protein (TBP) during transcriptional activation. The two histone fold-containing core subunits, Spt7 and Ada1, apparently provide a SAGA-specific interface with the **Tafs** (TBP-associated factors). Gcn5 is a transcriptional adaptor protein that has nuclear histone acetyltransferase (HAT) activity.

sagittal section Section through the median vertical longitudinal plane of an animal.

salbutamol Sympathetomimetic drug that stimulates beta adrenoreceptors in airways and acts as a bronchodilator giving rapid relief of acute asthma attacks and alleviating the symptoms of chronic bronchitis and emphysema. TN Combivent.

salicylic acid *2-Hydroxybenzoic acid* A naturally occurring antiseptic and anti-inflammatory compound found especially in the bark of willow (*Salix*); the acetylated form is **aspirin** (acetylsalicylic acid). The ester, ethyl salicylate, is oil of wintergreen.

salmeterol Beta-2 adrenergic agonist with selectivity for bronchial smooth muscle; used by inhalation for asthma. Longer acting than **salbutamol**.

Salmonella Genus of Enterobacteriaceae; motile, Gram-negative bacteria that, if invasive, cause enteric fevers (e.g. typhoid, caused by *S. typhi*), food poisoning (usually *S. typhimurium* or *S. enteridis*, the latter notorious for contamination of poultry) and occasionally septicaemia in non-intestinal tissues.

saltatory conduction A method of neuronal transmission in vertebrate nerves, where only specialized **nodes of Ranvier** participate in excitation. This reduces the capacitance of the neuron, allowing much faster transmission. See **myelin**, **Schwann cells**.

saltatory movements Abrupt jumping movements of the sort shown by some intracellular particles. Mechanism unclear.

saltatory replication The sudden amplification of a DNA sequence to generate many copies in a tandem arrangement. Possible mechanism for the origin of **satellite DNA**.

salvage pathways Metabolic pathways that allow synthesis of important intermediates from materials that would otherwise be waste products. An experimentally important pathway is that from hypoxanthine to nucleotides. See **HGPRT**.

SAM domain *Sterile alpha motif domain* A protein interaction module (~70 amino acids) that is present in diverse signal-transducing proteins including over 40 **EPH-related receptor tyrosine kinases**, *Drosophila* bicaudal-C, a p53 from *Loligo forbesi*, serine/threonine protein kinases, cytoplasmic scaffolding and adaptor proteins, regulators of lipid metabolism and GTPases as well as members of the ETS family of transcription factors. SAM domains are known to form homo- and hetero-oligomers.

Sam68 *Src-associated in mitosis, 68 kDa* Protein that associates with and is tyrosine phosphorylated by Src in a mitosis-specific manner. Has KH RNA-binding domain, SH2 and SH3 domains. Interacts with RNA, src family kinases, grb2 and PLCγ. Inhibition of phosphorylation of Sam68 by **radicicol** will block exit from mitosis.

Sanfillipo syndrome **Lysosomal disease** in which either keratan sulphate sulphatase or *N*-acetyl-α-D-glucosaminidase is defective: cross-correction (complementation) of co-cultured fibroblasts from apparently clinically identical patients can therefore occur if a different enzyme is missing in each.

Sanger–Coulson method See **dideoxy sequencing**.

S-antigen 1. Abundant protein of the retina and pineal gland that elicits experimental autoimmune uveitis; now known to be **arrestin**. 2. Soluble heat-stable antigens (195 kDa) on the surface of *Plasmodium falciparum* that are responsible for antigenic heterogeneity.

SAP 1. See **serum amyloid P-component**. 2. **SLAM-associated protein**.

sapecin Insect **defensin** produced by the flesh fly, *Sarcophaga peregrina*.

saponin Glycosidic surfactants produced by plant cells. Used to permeabilize membranes (being less harsh than, for example, Triton X-100) and as foaming agents in some beverages.

saporin A ribosome-inactivating protein (30 kDa) from seeds of the plant *Saponaria officinalis* (soapwort). Acts as an RNA-*N*-glycosidase, specifically depurinizing the 28S RNA of ribosomes.

saprophyte Organism that feeds on complex organic materials, often the dead and decaying bodies of other organisms. Many fungi are saprophytic.

saquinavir A **protease inhibitor** drug used in the treatment of HIV.

SAR See **structure–activity analysis**.

sar1p *Secretion-associated and Ras-related protein* A small GTPase that controls the assembly of the cytosolic COPII coat that mediates export from the ER. Sar1p directly initiates membrane curvature during vesicle biogenesis; when sar1p binds GTP, membrane insertion of the N-terminal amphipathic alpha helix deforms synthetic liposomes into narrow tubules. sara2 is the human homologue.

sara2 Human homologue of **Sar1p**.

sarafotoxins Group of snake cardiotoxic venoms from *Atractaspis engaddensis* (burrowing asp). Small peptides structurally related to the **endothelins**.

sarcin/ricin loop A highly conserved sequence found in the RNA of all large ribosomal subunits. **Alpha-sarcin** and **ricin** both inactivate ribosomes by cleaving a single bond in the loop. This prevents the interaction with elongation factors and stops translation.

sarcodictyin A Tricyclic compound that, like **taxol**, will stabilize microtubule bundles.

Sarcodina Group of aquatic protozoa that includes Amoebae, Foraminifera and Radiolaria.

sarcoglycan *Adhalin* Complex of α- β- and γ-sarcoglycans. α-sarcoglycan (50DAG; A2, adhalin), β-sarcoglycan (43DAG; A3b) and γ-sarcoglycan (35DAG; A4) are all transmembrane glycoproteins that associate with dystroglycan in the **sarcolemma** (approximate molecular weights are indicated by the old names). Defects in sarcoglycans have been shown to be associated with autosomally inherited **muscular dystrophy**.

sarcoidosis Disease of unknown aetiology in which there are chronic inflammatory granulomatous lesions in lymph nodes and other organs.

sarcolemma Plasma membrane of a striated muscle fibre.

sarcolipin *SLN* A proteolipid that inhibits the cardiac sarco(endo)plasmic reticulum Ca^{2+} ATPase (SERCA2a) by direct binding and is superinhibitory if it binds as a binary complex with **phospholamban** (PLN).

sarcoma cells Cells of a malignant tumour derived from connective tissue. Often given a prefix denoting tissue of origin, e.g. osteosarcoma (from bone).

sarcoma growth factor Polypeptide released by **sarcoma** cells that promotes the growth of cells by binding to a cell surface receptor; the sarcoma cell is therefore self-sufficient and independent of normal growth control. See **growth factors**. The name is no longer commonly used.

sarcoma virus Virus that causes tumours originating from cells of connective tissue such as **fibroblasts**. See **Rous sarcoma virus** and *src* **gene**.

Sarcomastigophora Phylum of unicellular protozoa with pseudopodia or flagella or both.

sarcomere Repeating subunit from which the **myofibrils** of striated muscle are built. Has **A-** and **I-bands**, the I-band being subdivided by the **Z-disc**, and the A-band being split by the **M-line** and the H-zone.

sarcoplasm Cytoplasm of striated muscle fibre.

sarcoplasmic reticulum Endoplasmic reticulum of striated muscle, specialized for the sequestration of calcium ions that are released upon receipt of a signal relayed by the **T tubule** from the neuromuscular junction.

sarcosine *N-methyl glycine* A natural amino acid found in muscle and other tissues; has a sweet taste and is used in toothpaste. Reported to be as an endogenous antagonist of glycine transporter-1 and have some beneficial effects in treating schizophrenia.

sarcosyl *Sarkosyl; N-lauroyl-sarcosine* A mild, biodegradable anionic surfactant derived from fatty acids and **sarcosine** used in preparing solubilized fractions of biological materials.

sarin Nerve gas; inhibitor of acetylcholine esterase.

SARS *Severe Acute Respiratory Syndrome* A highly infectious viral respiratory disease caused by a **coronavirus**, first recognized in the Far East in 2003, where it caused a major epidemic with a fatality rate of 11%.

satellite cell **1.** Sparse population of mononucleate cells found in close contact with muscle fibres in vertebrate skeletal muscle. Seem normally to be inactive, but may be important in regeneration after damage. May be considered a quiescent stem cell. **2.** An alternative name for glial cell.

satellite DNA *Minisatellite; microsatellite* DNA, usually containing highly repetitive sequences, that has a base composition (and thus density) sufficiently different from that of normal DNA that it sediments as a distinct band in caesium chloride density gradients. Typically, 10% of mammalian and 50% of insect genomes are composed of satellites. As satellites are dispersed widely in the genome, they are easily detectable (with a highly repetitive probe) and are frequently polymorphic in length; they are ideal markers for **linkage** studies of disease or inheritance and for genomics.

satellite virus A term used in plant virology for a virus associated functionally, at least for the purpose of its own replication, with another virus.

Sativex® A selective **cannabinoid** preparation containing two of the main active ingredients of cannabis: tetrahydrocannabinol and cannabidiol. Developed to relieve spasticity in patients with multiple sclerosis.

saturated fatty acids In eukaryotic membranes, refers to stearic, palmitic and myristic acids, which are linear aliphatic chains with no double bonds. Prokaryotes have numerous branched-chain saturated fatty acids.

saturation density The maximal population density achieved by a cell type grown under particular *in vitro* culture conditions. Although transformed cells generally grow to a higher saturation density than normal cells, this is not necessarily the case. Many factors affect the final density

achieved by a cell population; the critical factor may be availability of surface upon which to spread, or the serum concentration in the medium. Population densities in culture never approach those found in whole organisms.

saturation of receptors Saturation, the state in which all receptors are effectively occupied all the time, can be said to occur in a simple binding equilibrium when the concentration of ligand is more than five times the K_d value, although strictly it will only be true at infinite ligand concentration.

sauvagine Peptide (40 amino acids) originally isolated from the skin of the frog, *Phyllomedusa sauvagei*, and that is closely related to **corticotrophin-releasing factor** and to urotensin I.

saxitoxin *STX* **Neurotoxin** produced by the 'red tide' dinoflagellates, particularly *Alexandrium* (formerly *Gonyaulax*) *catenella*, *A. minutum*, *A. ostenfeldii*, *A. tamarense*, *Gymnodinium catenatum* and *Pyrodinium bahamense* var. *compressum*. It accumulates in shellfish, and when ingested binds to the **sodium channel**, blocking the passage of action potentials. Its action closely resembles that of **tetrodotoxin**. The toxin was originally isolated from the clam, *Saxidomus giganteus*, and is responsible for paralytic shellfish poisoning.

SC-35 See **ASF/SF-2**.

scabies A contagious skin disease caused by the mite *Sarcoptes scabiei* which burrows in the horny layer of the skin and causes an inflammatory reaction.

scaffold/radial loop model Model for chromatin organization in eukaryotic metaphase chromosomes. Involves a non-histone protein core that is coiled and to which the linear DNA molecule has an ordered series of attachment points every 30–90 kb, with intervening DNA forming a loop that is supercoiled or folded. The 150–200-nm diameter central core contains structural maintenance of chromosomes 2 (SMC2) protein and topoisomerase II.

scaffoldin *CipA, cipB, etc* The scaffoldin subunits (~210 kDa) of the bacterial **cellulosome** function to organize and position other protein subunits into the complex. The scaffoldins (scaA (cipBc), scaB, scaC and scaD) can also serve as an attachment device for fastening the cellulosome to the cell surface and/or for its targeting to substrate.

scaffolding proteins In general, proteins that assist in the formation of large multimolecular complexes. The term is applied, for example, to proteins that maintain the clustering of particular receptors at synapses (see, *inter alia*, **caveolins**, **flotillins**), to proteins involved in assembling the viral capsid and to proteins involved in eukaryotic chromosome structure. See also **AKAP79**, **involucrin**, **titin**.

SCAMP See **secretory carrier-associated membrane proteins**.

scanning electron microscopy *SEM* Technique of electron microscopy in which the specimen is coated with heavy metal and then scanned by an electron beam. The image is built up on a monitor screen, in the same way as the raster builds a conventional television image. The resolution is not so great as with transmission electron microscopy, but preparation is easier (often followed by **critical point drying**), the depth of focus is relatively enormous, the surface of a specimen can be seen (though not the interior unless the specimen is cracked open), and the image is aesthetically pleasing.

scanning probe microscopy *SPM* Methods for visualizing surfaces at microscopic scale that rely on moving a tiny probe over a surface (usually in an *x*–*y* scan) and recording some property of interest (current, force) at each coordinate. These techniques have the ability to resolve detail down to single atoms. See also **scanning tunnelling microscopy** and **atomic force microscopy**.

scanning transmission electron microscopy *STEM* Method of electron microscopy in which image formation depends upon analysis of the pattern of energies of electrons that pass through the specimen. Has comparable resolving power to conventional transmission EM.

scanning tunnelling microscopy *STM* A form of ultra-high-resolution microscopy of a surface in which a very small current is passed through a surface and is detected by a microprobe of atomic dimensions at its tip that scans the surface by use of a piezodrive. In the simplest form, the current transferred to the probe is recorded as an indication of the contours of molecules on the surface above the local plane. In more complex forms, feedback is used to hold the probe at a constant difference and the signal in the feedback loop indicates the contours of the molecule. Capable of resolving single atoms, and known to work for non-conducting molecules as well as conducting ones.

SCAP *SREBP cleavage-activating protein* Protein involved in feedback inhibition of the sterol regulatory element-binding protein (**SREBP**) pathway. Sterols prevent movement of the SCAP/SREBP complex from the endoplasmic reticulum to the Golgi, where proteolytic cleavage of SREBPs releases the transcription factor domain and thereby activates genes for lipid biosynthesis.

scaphognathite A thin, leaf-like appendage on the second maxilla of decapod crustaceans, used to sweep water into the gill cavity.

scaphoid One of the small wrist bones (carpal) on the thumb side; the one most likely to be broken in a fall.

Scar A **WASP**-related protein, binds the p21 subunit of the **Arp2/3** complex and is an endogenous activator of actin polymerization. See **WAVE/SCAR**.

SCARECROW *SCR* See **GRAS family**.

Scatchard plot A method for analysing data for freely reversible ligand/receptor-binding interactions. The graphical plot is: ([Bound ligand]/[Free ligand]) against ([Bound ligand]); the slope gives the negative reciprocal of the binding affinity, the intercept on the *x*-axis the number of receptors (Bound/Free becomes zero at infinite ligand concentration). The Scatchard plot is preferable to the **Eadie–Hofstee plot** for binding data because it is more dependent upon the values at high ligand concentration – which will be the most reliable values. A non-linear Scatchard plot is often taken to indicate heterogeneity of receptors, although this is not the only explanation possible.

scatter factor A motility factor (**motogen**) isolated from conditioned medium in which human fibroblasts have been grown. It causes colonies of epithelial and endothelial cells, in culture, to separate into single cells that move apart, i.e. they scatter. It has been shown to be identical to human **hepatocyte growth factor**, but it is not mitogenic for all cell types.

scavenger receptor Structurally diverse family of receptors on macrophages that are involved in the uptake of modified **LDL** and have been implicated in development of atherosclerotic lesions. Six classes are recognized, with different binding preferences. Macrophage scavenger receptors class A bind a wide range of ligands, including bacteria, and it is speculated that scavenger receptors may be important in recognizing apoptotic cells.

Scenedesmus A non-motile colonial alga, of the Order Chlorococcales, consisting of two, four or eight elongated cells, often with long spines on the terminal cells. Common in ponds and as planktonic forms in rivers and lakes.

SCF See **stem cell factor**.

SCF complexes Class of E3 ubiquitin-protein ligases that are composed of a Skp1p-cdc53p-F-box protein complex and play a role in regulation of cell division. Substrate specificity for targeting for ubiquitinylation is conferred by the F-box component. See **skp** and **cullin**.

scFv *Single-chain variable fragment* A fusion of the variable regions of the immunoglobulin heavy and light chains linked together with a short peptide linker and produced by recombinant methods in bacteria. Can be selected to be specific for antigens of choice by various selection methods.

SCG10 gene A neural-specific gene (superior cervical ganglion10) that encodes a growth-associated protein expressed early in the development of neuronal derivatives of the neural crest. The 22-kDa intracellular protein product is associated with the membranous organelles that accumulate in growth cones. Recently reported to be a microtubule regulator. Product has amino acid sequence similarity with **stathmin**.

Scheie syndrome Mucopolysaccharidosis (**lysosomal disease**) in which there is a defect in α-L-iduronidase. Fibroblasts from Scheie syndrome patients do not cross-correct fibroblasts from **Hurler's disease**; Hurler's and Scheie syndromes represent phenotypes at the severe and mild ends of the clinical spectrum.

Schick test A test, introduced by Schick in 1913, to assess the degree of susceptibility or immunity of individuals to diphtheria by challenging with a small amount of diphtheria toxin injected intracutaneously into one forearm.

Schiff base The reaction of a primary amine with an aldehyde or ketone yields an imine, sometimes called a Schiff base. When an arylamine is used the Schiff base may form an intermediate in a staining reaction, e.g. for polysaccharides.

Schiff's reagent See **periodic acid–Schiff reaction**.

Schilder's disease *Myelinoclastic diffuse sclerosis; diffuse sclerosis; encephalitis periaxialis* A rare, progressive demyelinating disorder which usually begins in childhood. Symptoms may include dementia, aphasia, seizures, personality changes, poor attention, tremors, balance instability, incontinence, muscle weakness, headache, vomiting, and vision and speech impairment. The disorder is a variant of multiple sclerosis.

schistocytes Fragments of red blood cells found in the circulation.

schistosomiasis *Bilharzia* Disease caused by trematode worms (flukes). Three main species, *Schistosoma haematobium*, *S. japonicum* and *S. mansoni*, cause disease in man. Larval forms of the parasite live in freshwater snails; cercariae liberated from the snail burrow into skin, transform to the schistosomulum stage and migrate to the urinary tract

(*S. haematobium*), liver or intestine (*S. japonicum, S. mansoni*) where the adult worms develop. Eggs are shed into the urinary tract or the intestine and hatch to form miracidia, which then infect snails, completing the life cycle. Adult worms cause substantial damage to tissue and seem to resist immune damage by mechanisms that are not fully understood.

schistosomulum *Pl.* schistosomula See **schistosomiasis**.

schizocoel Coelom that is developed within the mass of mesoderm by splitting or cleavage. *cf.* **enterocoel**.

schizogeny A mechanism of **aerenchyma** formation in plants in which development results in the cell separation. Schizogeneous aerenchyma is common in wetland species like *Rumex* (dock) and is formed by cell separation, without the cells involved dying. See **lysigeny**.

schizogony The division of cells, especially of protozoans, in non-sexual stages of the life history of the organism.

Schizosaccharomyces pombe Species of fission yeast commonly used for studies on cell cycle control because there is a distinct G2 phase to the cycle. Only distantly related to the budding yeast **Saccharomyces** *cerevisiae*. A further advantage is that some mammalian introns are processed correctly.

Schultz–Charlton test Test for scarlet fever in which antitoxin to **erythrogenic toxin** of *Streptococcus pyogenes* is injected subcutaneously.

Schwann cell A specialized glial cell that wraps around vertebrate **axons** providing extremely good electrical insulation. Separated by **nodes of Ranvier** about once every millimetre, at which the axon surface is exposed to the environment. See **saltatory conduction, myelin**.

Schwannoma-derived growth factor *SDGF* A **growth factor** containing an **EGF-like domain**, mitogenic for astrocytes, Schwann cells and fibroblasts. Like **amphiregulin**, one of the EGF-family of growth factors.

Schwartzmann reaction Mis-spelling of **Shwartzman** reaction.

SCID *Severe Combined (or Congenital) Immunodeficiency Disease* An heterogeneous group of inherited disorders characterized by gross functional impairment of the immune system. In all types T-cells are deficient (T^-), but SCID can be divided into two main classes: those with B-lymphocytes (B^+ SCID) and those without (B^- SCID). NK-cells may or may not be present. The most common form of SCID is X-linked T^-, B^+, NK^- SCID caused by mutation in the IL2RG gene. Autosomal recessive SCID includes T^-, B^+, NK^- SCID caused by mutation in the JAK3 gene; T^-, B^+, NK^+ SCID is caused by mutation in the IL7R gene, the CD45 gene or the CD3D gene; T^-, B^-, NK^- SCID is caused by mutation in the ADA (**adenosine deaminase**) gene; T^-, B^-, NK^+ SCID with sensitivity to ionizing radiation caused by mutation in the Artemis gene; and T^-, B^-, NK^+ SCID is caused by mutation in the RAG1 and RAG2 genes. About half the patients with recessive autosomal form have ADA deficiency, which has, in a few cases, been corrected by gene therapy.

scinderin Protein (80 kDa) of the **gelsolin** family isolated from vertebrate neural and secretory tissue. Subcortical scinderin is redistributed into patches following stimulation of **chromaffin cells** through nicotinic receptors. Similar to **adseverin**.

scintillation counting Technique for measuring quantity of a radioactive isotope present in a sample. In biology, liquid scintillation counting is mainly used for β emitters such as ^{14}C, ^{35}S and ^{32}P ,and particularly for the low-energy β emission of ^3H. Gamma emissions are often measured by counting the scintillations that they cause in a crystal. Autoradiographic images can be enhanced by using a screen of scintillant behind the film.

scintillation proximity assay Assay system in which antibody or receptor molecule is bound to a bead that will emit light when β emission from an isotope occurs in close proximity, i.e. from a radioactively labelled ligand. Avoids the need for scintillant in order to measure the amount of bound isotope and thus the amount of antigen or ligand present.

SCIP *Oct-6; Tst-1* POU domain transcription factor expressed by promyelinating Schwann cells (where it represses expression of the myelin structural genes) and, in tissue culture, by oligodendrocyte progenitors.

scirrhous carcinoma Carcinoma having a hard structure because of excessive production of dense connective tissue.

sclereid Type of **sclerenchyma** cell that differs from the **fibre cell** by not being greatly elongated. Often occurs singly (an idioblast) or in small groups, giving rise to a gritty texture in, for instance, the pear fruit, where it is known as a 'stone cell'. May also occur in layers, e.g. in hard seed coats.

sclerenchyma Plant cell type with thick lignified walls, normally dead at maturity and specialized for structural strength. Includes **fibre cells,** which are greatly elongated, and **sclereids,** which are more isodiametric. Intermediate types exist.

scleritis An inflammatory disease that affects the conjunctiva, sclera and episclera (the connective tissue between the conjunctiva and sclera). It is associated with underlying systemic diseases in about half of the cases.

scleroderma Hardening of skin.

sclerosis Pathological hardening of tissue.

sclerotin Hard, dark-coloured cross-linked (tanned) protein found in the cuticle of insects and some other arthropods.

SCN 1. Suprachiasmatic nucleus of the hypothalamus. 2. Thiocyanate anion (SCN^-).

SCO-spondin A brain-secreted glycoprotein specifically expressed in the subcommissural organ, an ependymal differentiation located in the roof of the Sylvian aqueduct. SCO-spondin makes part of Reissner's fibre, a phylogenetically and ontogenetically conserved structure present in the central canal of the spinal cord of chordates. This secretion is a large multidomain protein probably involved in axonal growth and/or guidance. There are several conserved domains, including thrombospondin type 1 repeats (TSRs), low-density lipoprotein receptor (LDLr) type A repeats and EGF-like domains.

scoliosis Abnormal lateral curvature of the spine.

scombrotoxin Causative agent of scombroid poisoning, caused by eating foods (spoiled fish, some cheeses) with high levels of histamine and possibly other vasoactive amines and compounds produced by bacteria. Also called histamine poisoning.

scopolamine *Hyoscine* Alkaloid found in thorn apple (*Datura stramonium*). Related to atropine both in effects and structure and acts as a **muscarinic acetylcholine receptor** antagonist.

scopoletin *7-Hydroxy-6-methoxycoumarin* A naturally occurring fluorescent component of some plants that acts as a plant growth inhibitor. Said to lower blood pressure in hypertension and raise it in hypotension; also to be bacteriostatic and anti-inflammatory. An acetylcholine esterase inhibitor.

scorpion toxins Polypeptide toxins (7 kDa) with four disulphide bridges. The α-toxins are found in venom of Old World scorpions, β-toxins in those of the New World. Bind with high affinity to the voltage-sensitive **sodium channel** of nerve and muscle (α- and β-toxins bind to different sites).

scotophobin Peptide (15 residues) isolated from brains of rats trained to avoid the dark that will transfer this aversion to naive animals. The original claim was treated with considerable scepticism, but subsequent work seems to have provided validation.

Scoville units A scale developed by Scoville in 1912 to measure the hotness of chillies. Originally a subjective taste test but now refined by the use of HPLC, the unit is named in honour of its inventor. The greater the number of Scoville units, the hotter the pepper.

scrapie A chronic neurological disease of sheep and goats, similar to other **spongiform encephalopathies** and much used as a model for studying the diseases. Controversy still surrounds the nature of the transmissible agent, although the idea of slow viruses has been overtaken by Prusiner's **prion** hypothesis, which is now fairly generally accepted. Atypical forms of the disease seem to be emerging in sheep of the genotypes that are resistant to the classical form.

scRNP Small cytoplasmic ribonucleoprotein. See **small interfering RNA**.

scrofula A tuberculous infection of the skin of the neck, most often caused by *Mycobacterium tuberculosis* in adults. There may be enlargement and infection of the cervical lymph nodes. During mediaeval times, the King's touch was thought to be curative.

scruin Actin-binding protein found associated with the acrosomal process of *Limulus polyphemus* sperm. Scruin holds the microfilaments of the core process in a strained configuration so that the process is coiled. The myosin-binding sites on the microfilaments are blocked so HMM decoration is impossible, indicating that there is an unusual packing conformation; when the scruin–actin binding is released the process straightens, the conformation of the actin changes, and myosin binding is possible.

scurfy The murine X-linked lymphoproliferative disease scurfy is similar to the **Wiskott–Aldrich syndrome** in humans. Disease in scurfy (sf) mice is mediated by CD4+ T-cells, but there is general overproduction of various cytokines.

scurvy Disease caused by Vitamin C deficiency. The effects are due to a failure of the hydroxylation of proline residues in collagen synthesis and the consequent failure of fibroblasts to produce mature collagen. See **hydroxyproline**.

scutellum Part of the embryo in seeds of the Poaceae (grasses). Can be considered equivalent to the cotyledon of other monocotyledonous seeds. During germination, absorbs degraded storage material from the endosperm and transfers it to the growing axis.

scyllatoxin Toxin from the scorpion *Leiurus quinquestriatus hebraeus*. Peptide of 31 residues that specifically blocks low conductance calcium-dependent potassium channels that are also a target for **apamin**.

scyphozoa A class of Cnidaria in which the polyp stage is inconspicuous or completely absent. Jellyfish.

scytalidoglutamic peptidase *EC 3.4.23.31; scytalidopepsin A* Type peptidase of the glutamic (G1) peptidase family from the fungus *Scytalidium lignicolumin*. The Glu136 residue appears to be important at the catalytic site. Originally identified as one of the pepstatin-insensitive carboxyl proteases. Substrates include angiotensin 1 and oxidized insulin B-chain.

SD sequence *Shine–Dalgarno sequence* See **Shine–Dalgarno** region.

SDF-1 *CXCL12* A chemokine, stromal cell-derived factor-1 that controls many aspects of stem cell function, including trafficking and proliferation. The receptor is **CXCR4**. Originally known as pre-B-cell growth-stimulating factor and identical to human intercrine reduced in hepatomas (hIRH).

SDGF See **Schwannoma-derived growth factor**.

SDH Most commonly: **1. Succinate dehydrogenase.** The *SDHA*, *SDHB*, *SDHC* and *SDHD* genes encode the subunits of succinate dehydrogenase (succinate: ubiquinone oxidoreductase), a component of both the Krebs cycle and the mitochondrial respiratory chain. Less often: **2. Sorbitol dehydrogenase. 3. Saccharopine dehydrogenase. 4. Serine dehydratase. 5. Shikimate 5-dehydrogenase.**

SDS *Sodium dodecyl sulphate; sodium lauryl sulphate* Anionic detergent that at millimolar concentrations will bind to and denature proteins, forming an SDS–protein complex. The amount of SDS bound is proportional to the molecular weight of the protein, and each SDS molecule, bound by its hydrophobic domain, contributes one negative charge to the protein, thus swamping its intrinsic charge. This property is exploited in the separation of proteins by **SDS-PAGE**.

SDS-PAGE **Polyacrylamide gel electrophoresis** (PAGE) in which the charge on the proteins results from their binding of **SDS**. Since the charge is proportional to the surface area of the protein, and the resistance to movement proportional to diameter, small proteins migrate further.

sea *S13 avian erythroblastosis oncogene homologue* An **oncogene** that encodes a member of the Met/**hepatocyte growth factor**/scatter factor family of **receptor tyrosine kinases**. See Table O1.

sea hare See *Aplysia*.

seborrhoea *Seborrhea (US)* Overactivity of the sebaceous glands of the skin. *Adj.* seborrheic.

SEC 1. *Serpin–enzyme complex* Receptor that mediates catabolism of α-1-antitrypsin/elastase complexes and elevates α-1-antitrypsin synthesis. Also implicated in neutrophil chemotaxis and neurotoxicity of amyloid β-peptide.

2. Selenocysteine (Sec). **3.** Size-exclusion chromatography (SEC). **4.** The general secretory (sec) pathway in bacteria (see **sec-dependent transport**). **5.** See **sec7, sec61, sec65**.

sec7 *ySec7p* Protein from *S. cerevisiae* that plays an important part in the secretory pathway. Mutations in sec7 lead to accumulation of Golgi cisternae and loss of secretory granules. Sec7 contains a domain of around 200 amino acids that is found in several **GEF**s for ADP-ribosylation factors (**ARF**s).

sec61 A conserved protein-conducting channel (translocon) in eukaryotes (homologous to the SecY channel in eubacteria and archaea), translocates proteins across cellular membranes and integrates proteins containing hydrophobic transmembrane segments into lipid bilayers. Sec 61 is actually a heterotrimeric complex α, β and γ) and associates with the subcomplex Sec62/Sec63. Sec61 α is a multispanning membrane protein. Sec61 β kinetically facilitates co-translational translocation and interacts with the 25-kDa subunit of the signal peptidase complex (SPC25). See **sec-dependent transport**.

sec65 Gene of *S. cerevisiae* that encodes a protein very similar to the SRP19 subunit of the mammalian **signal recognition particle**.

sec-dependent transport *SecYEG* Pathway for the secretion of proteins across the inner membrane into the periplasm of Gram-negative bacteria using the general translocase SecYEG. Also used for the insertion of inner membrane proteins, in some cases in association with **YidC**. Other translocation mechanisms involve the **Tat** system and YidC. SecYEG is a trimeric complex where Y and E are related to **sec61** α and γ subunits.

second messenger In many hormone-sensitive systems the systemic hormone does not enter the target cell but binds to a receptor and indirectly affects the production of another molecule within the cell; this diffuses intracellularly to the target enzymes or intracellular receptor to produce the response. This intracellular mediator is called the second messenger. Examples include cyclic AMP, cyclic GMP, IP3 and diacylglycerol.

secondary immune response The response of the immune system to the second or subsequent occasion on which it encounters a specific antigen.

secondary lymphoid tissue See **lymphoid tissue**.

secondary lysosome Term used to describe intracellular vacuoles formed by the fusion of lysosomes with organelles (**autosomes**) or with primary phagosomes. **Residual bodies** are the remnants of secondary lysosomes containing indigestible material.

secondary phloem Phloem formed during secondary growth by the activity of a vascular **cambium** as opposed to primary phloem that is derived from procambium.

secondary product End product of plant cell metabolism, which accumulates in, or is secreted from, the cell. Includes **anthocyanins**, **alkaloids**, etc. Some are of major economic importance, e.g. as drugs. In contrast to a primary product that is involved in the vital metabolism of the plant.

secondary structure Structures produced in polypeptide chains involving interactions between amino acids within the chain. Especially alpha-helical and **beta-pleated sheet** structures. Also applies to the complex folding of

nucleic acids as, for example, the cloverleaf structure of tRNA.

secondary wall That part of the plant cell wall which is laid down on top of the **primary cell wall** after the wall has ceased to increase in surface area. Only occurs in certain cell types, e.g. tracheids, vessel elements and sclerenchyma. Differs from the primary wall both in composition and structure, and is often diagnostic for a particular cell type.

secondary xylem Xyem formed by the activity of a vascular **cambium** as a plant grows. Wood is composed largely of secondary xylem.

secretagogue Substance that induces secretion from cells; originally applied to peptides inducing gastric and pancreatic secretion.

secretases See **alpha-, beta-, gamma-secretase**.

secretin Peptide hormone of gastrointestinal tract (27 residues) found in the mucosal cells of duodenum. Stimulates pancreatic, pepsin and bile secretion, inhibits gastric acid secretion. Considerable homology with **gastric inhibitory polypeptide**, **vasoactive intestinal peptide** and **glucagon**. See **secretin family**.

secretin family Family members are **vasoactive intestinal peptide** (VIP), pituitary adenylate cyclase-activating polypeptide (**PACAP**), **secretin, glucagon, glucagon-like peptide-1** (GLP(1)), GLP(2), **gastric inhibitory peptide** (GIP), **growth-hormone-releasing hormone** (GHRH or GRF), **peptide histidine methionine** (PHM) and **helodermin**. Most of the family members are present both in central nervous system (CNS) and in various peripheral tissues.

secretion Release of synthesized product from cells. Release may be of membrane-bounded vesicles (merocrine secretion) or of vesicle content following fusion of the vesicle with the plasma membrane (apocrine secretion). In holocrine secretion whole cells are released.

secretogranins See **granins**.

secretoneurin A 33-amino acid polypeptide generated by proteolytic cleavage of **secretogranin** II, widely distributed throughout the central and peripheral nervous systems; stimulates dopamine release from striatal neurons and activates monocyte migration, thereby probably playing a role in neurogenic inflammation. Receptor is G-protein coupled.

secretor A person who secretes ABO blood group substances into mucous secretions, e.g. saliva; at least 80% of humans are secretors.

secretory carrier-associated membrane proteins *SCAMPs* Integral membrane proteins (**tetraspan vesicle membrane proteins**) of post-Golgi membranes that function as recycling carriers to the cell surface. At least four members of the family have been identified.

secretory cells Cells specialized for secretion, usually epithelial. Those that secrete proteins characteristically have well developed rough endoplasmic reticulum, whereas conspicuous smooth endoplasmic reticulum is typical of cells that secrete lipid or lipid-derived products (e.g. **steroids**).

secretory component of IgA A polypeptide chain of about 60 kDa that aids secretion of **IgA**; a portion of the IgA receptor on the plasmalemma of the inner side of the epithelial cells lining the gut, which is proteolysed when the IgA receptor complex has travelled through the cell after receptor-mediated endocytosis at the inner face, to the outer (luminal) face.

secretory proteins In eukaryotes, proteins synthesized on **rough endoplasmic reticulum** and destined for export. Nearly all proteins secreted from cells are glycosylated in the **Golgi apparatus**, although there are exceptions (**albumin**). In prokaryotes, secreted proteins may be synthesized on ribosomes associated with the plasma membrane or exported post-translation.

secretory vesicle Membrane-bounded vesicle derived from the **Golgi apparatus** and containing material that is to be released from the cell. The contents may be densely packed, often in an inactive precursor form (**zymogen**).

securin An inhibitor of the anaphase activator **separin** (separase/Esp1p) that is ubiquitinated by activated **anaphase-promoting complex**; loss of securin leads to proteolytic cleavage of **cohesin**.

sedimentation Settling of a component of a mixture under the influence of gravity or centrifugation so that the mixture separates into two or more phases or zones.

sedimentation coefficient The ratio of the velocity of sedimentation of a molecule to the centrifugal force required to produce this sedimentation. It is a constant for a particular species of molecule, and the value is given in Svedberg units (S) that, it should be noted, are non-additive.

sedimentation test A standard blood test that involves measuring the rate of settling of erythrocytes in anticoagulant treated blood. Sedimentation rates are increased in inflammation and in pregnancy.

sedoheptulose Seven-carbon sugar, whose phosphate derivatives are involved in the **pentose phosphate pathway** and the **Calvin–Benson cycle**.

seed fern *Pteridospermae* A group of ferns, mostly extinct, that flourished in the Devonian period; the only living representatives are the cycads.

segment long-spacing collagen See **SLS collagen**.

segment-polarity gene A **segmentation gene** responsible for specifying anterior–posterior polarity within individual embryonic segments. In *Drosophila* there are at least 10 such genes, for example *gooseberry*.

segmentation Organization of the body into repeating units called segments is a common feature of several phyla, e.g. arthropods and annelids, although the segments arise by very different mechanisms. Segmentation also occurs during embryonic development in vertebrates, e.g. partition of the mesoderm into **somites,** and is a feature of early **CNS** development. See **rhombomeres, neuromeres.**

segmentation gene Genes required for the establishment of **segmentation** in the embryo. In *Drosophila*, about 20 such genes are required.

segregation of chromosomes The separation of pairs of **homologous chromosomes** that occurs at meiosis so that only one chromosome from each pair is present in any single gamete.

selectin Group of cell adhesion molecules that bid to carbohydrates via a **lectin**-like domain. The name is derived from select and lectin. They are integral membrane glycoproteins with an N-terminal, **C-type lectin** domain, followed by an EGF-like domain, a variable number of repeats of the short consensus sequence of complement regulatory proteins and a single transmembrane domain. Three selectins have been identified and are distinguished by capital letters based on the source of the original identification, i.e. E-, L- and P-selectin. Examples: ELAM-1, GMP-140, LECAM. See Table S1.

TABLE S1. Selectins

Selectin	Synonyms	Cellular expression	Domain structure	Role in adhesion
E-selectin	ELAM-1	Activated endothelial cells	1 C-lectin	Binding of leucocytes in the inflammatory response
			1 EGF	
			6 CRP	
			(90 kDa)	
L-selectin	LECAM-1, LAM-1, gp90^{MEL-14}, LHR, lymphocyte homing receptor	Lymphocytes, polymorphonuclear neutrophils	1 C-lectin	Homing of lymphocytes to peripheral lymph nodes
			1 EGF	
			2 CRP	
			(115 kDa)	
P-selectin	CD62, GMP-140, PADGEM	α-granules of platelets and Weibel–Palade granules of endothelial cells.	1 C-lectin	Binding of cells to monocytes and leucocytes?
			1 EGF	
		Plasma membranes of both cells after activation	9 CRP	
			(140 kDa)	

selector genes A group of genes that determines which part of a developmental pattern cells will be allocated within a developmental segment. *Antennapedia* is an example, and the neural selector gene *cut*, a homeobox transcription factor, is required for the specification of the correct identity of external (bristle-type) sensory organs in *Drosophila*.

selenium *Se* Essential trace element that must be provided as a supplement in serum-free culture media for most animal cells. See **selenocysteine** and **selenoprotein**.

selenocysteine *Sec* An unusual amino acid of proteins, the selenium analogue of **cysteine**, in which a selenium atom replaces sulphur. Involved in the catalytic mechanism of seleno-enzymes such as formate dehydrogenase of *E. coli* and mammalian glutathione peroxidase. May be co-translationally coded by a special **opal suppressor** tRNA that recognizes certain UGA nonsense codons.

selenoprotein Protein that contains selenium in the form of **selenocysteine**.

self-antigens *Autoantigen* The antigens of an individual's body have the potential to be self-antigens for the immune system, and unless clones of immune cells reactive with self-antigens are eliminated an autoimmune response can be initiated.

self-assembly The property of forming structures from subunits (protomers) without any external source of information about the structure to be formed such as a priming structure or template.

self-cloning Any system in which inappropriate cell types or organisms are eliminated because they possess some character that allows them to die or to remove themselves from the system. Thus a transfected cell with genetic material including a drug resistance marker will be self-cloning in the presence of the drug and non-transfected cells will die.

self-incompatibility Inability of pollen grains to fertilize flowers of the same plant or its close relatives. Acts as a mechanism to ensure out-breeding within some plant species, e.g. in the case of the **S-gene complex** in Brassicas.

self-replicating Literally, replication of a system by itself without outside intervention. In practice, often taken to refer to systems that replicate without the contribution of any information from outside the system.

self-splicing Self-catalysed removal of group five **introns**, mediated by six paired conserved regions.

sem-5 Cell-signalling gene of *C. elegans* that encodes a protein (228 residues) with SH2 and SH3 domains and that acts in vulval development and sex myoblast migration.

semaphorins Family of proteins that mediate neuronal guidance by inhibiting **nerve growth cone** movement. Both transmembrane and secreted proteins are included, and many domains of the proteins are highly conserved between invertebrates and vertebrates. Most are around 750 residues with a conserved 'sema' domain of up to 500 residues extracellularly with a single immunoglobulin C2-type domain C-terminally to this. **Collapsin**, responsible for the collapse of nerve growth cones of chick sensory neurites in culture following contact with retinal axons, was one of the first semaphorins described. Receptors are **plexins**; see also **collapsin response-mediator proteins**. See Table S2.

semeiotic *Semiotic* **1.** Relating to semantics. **2.** Relating to symptomatology, the symptoms and signs of disease.

semelparity The production of offspring only once in the lifetime of the organism. *Adj.* semelparous.

TABLE S2. Semaphorins

Class	New name	Old name	Features of class
Invertebrate			
Class 1	Sema-1a	G-Sema I, D-Sema I, T-sema I, Ce-Sema I	TM domain and short cytoplasmic tail
	Sema-1b	Sema 1b	
Class 2	Sema 2a	D-Sema II, Ce-Sema II, gSemaII	Secreted and have Ig domain
Vertebrate			
Class 3	Sema3A	C-Collapsin-1 (Coll-1), H-Sema III, M-SemD, R-Sema III, Sema-Z1a	Ig domain, short basic domain, secreted
	Sema3B	M-SemaA, H-SemaA, H-Sema V	
	Sema3C	M-SemE, C-Coll-3, H-Sema E	
	Sema3D	C-Coll-2, Sema-Z2	
	Sema3E	C-Coll-5, M-Sema H	
	Sema3F	H-Sema IV, M-Sema IV, H-Sema-3F	
Class 4	Sema4A	M-SemB	Ig domain, TM domain, short cytoplasmic domain
	Sema4B	M-SemC	
	Sema4C	M-sema F	
	Sema4D	CD100. M-Sema G, C-Coll-4	
	Sema4E	Sema-Z7	
	Sema4F	M-Sema W, R-Sema W, H-Sema W	
	Sema4G		

TABLE S2. (Continued)

Class	New name	Old name	Features of class
Class 5	Sema5A	M-SemF	7 thrombospondin repeats, TM and short cytoplasmic domain
	Sema5B	M-SemG	
Class 6	Sema6A	M-Sema VIa	TM and cytoplasmic domain
	Sema6B	M-SemaVIb, R-Sema Z	
	Sema6C	M-Sema Y, R-Sema Y	
Class 7	Sema7A	H-Sema K1, H-Sema L, M-Sema L, M-Sema K1	Ig domain and GPI anchor
Viral			
Class V	SEMAVA	Vaccinia sema, Variola sema	Truncated sema domain
	SEMAVB	AHV sema	Sema domain + Ig domain

semiautonomous Of systems or processes that are not wholly independent of other systems or processes.

semiconservative replication The system of replication of DNA found in all cells in which each daughter cell receives one old strand of DNA and one strand newly synthesized at the preceding **S phase**. The existence of semiconservative replication was demonstrated by the Meselson–Stahl experiment and implied the two- or multi-strandedness of DNA.

semipermeable membrane A membrane that is selectively permeable to only one (or a few) solutes. The potential developed across a membrane permeable to only one ionic species is given by the **Nernst equation** for the species: this is the basis for the operation of **ion-selective electrodes**.

Semliki forest virus Enveloped virus of the alphavirus group of **Togaviridae**. First isolated from mosquitoes in the Semliki Forest in Uganda; not known to cause any illness. The synthesis and export of its three spike glycoproteins, via the endoplasmic reticulum and Golgi complex, have been used as a model for the synthesis and export of plasma membrane proteins.

Senarmont compensation In interference microscopy, compensation for the phase difference introduced by the object, measured by introducing a quarter-wavelength plate and rotating the analyser: the angle of rotation is proportional to the optical path difference.

Sendai virus Parainfluenza virus type 1 (Paramyxoviridae). Can cause fatal pneumonia in mice, and may cause respiratory disease in humans. The ability of ultraviolet-inactivated virus to fuse mammalian cells has been extensively used in the study of **heterokaryons** and **hybrid cell lines**.

senescent cell antigen An antigen (62 kDa) that appears on the surface of senescent erythrocytes and is immunologically cross-reactive with isolated **band III**. Seems to be recognized by an autoantibody, and the immunoglobulin-coated erythrocyte is then removed from circulation by cells such as Kuppfer cells of the liver that have Fc receptors. Intracellular cleavage of intact band III

by a calcium-activated peptidase, **calpain**, may reveal the antigen *in situ*.

senile plaque Characteristic feature of the brains of **Alzheimer's disease** patients and aged monkeys, consisting of a core of amyloid fibrils surrounded by dystrophic neurites. The principal component of amyloid fibrils in senile plaques is B/A4, a peptide of about 4 kDa that is derived from the larger **amyloid precursor protein** (APP). The B/A4 sequence is located near the C-terminus of APP.

sensitization A state of heightened responsiveness, usually referring to the state of an animal after primary challenge with an antigen. The term is frequently used in the context of **hypersensitivity**.

sensory neuron 1. A **neuron** that receives input from sensory cells. 2. Sensory cells such as cutaneous mechanoreceptors and muscle receptors.

separin *Separase; Esp1; EC 3.4.22.49* Cysteine endopeptidase ($\sim$180 kDa), related to caspases, that uniquely cleaves the Sccl subunit of **cohesin** at the metaphase to anaphase transition and thereby allows sister chromatid separation. Separin is the downstream target of the **anaphase-promoting complex** (APC). In *S. cerevisiae* separin is Esp1. See also **securin**.

Sephacryl Trade name for a covalently cross-linked allyl dextrose gel formed into beads. Used in **gel filtration** columns for separating molecules in the size range 5 kDa to 1.5 million Da.

Sephadex Trade name for a cross-linked dextran gel in bead form used for **gel filtration** columns: by varying the degree of cross-linking the effective fractionation range of the gel can be altered.

Sepharose Trade name for a gel of agarose in bead form from which charged polysaccharides have been removed. Used in **gel filtration** columns.

septate junction An intercellular junction found in invertebrate epithelia that is characterized by a ladder-like appearance in electron micrographs. Thought to provide structural strength and also a barrier to diffusion of solutes through the intercellular space. Occurs widely in transporting epithelia and is controversially considered analogous to tight junctions (**zonula occludens**).

septic shock Condition of clinical shock caused by **endotoxin** in the blood. A serious complication of severe burns and abdominal wounds, frequently fatal. Part of the problem seems to be due to increased leucocyte adhesiveness, which leads to massive sequestration of neutrophils in the lung, increased vascular permeability and acute (adult) respiratory distress syndrome.

septicaemia A potentially life-threatening infection in which many bacteria are present in the blood. Commonly referred to as blood poisoning. See **bacteraemia**.

septin Family of homologous proteins (around 40 kDa) first identified in *S. cerevisiae*, where they associated with cytokinesis and septum formation. Encoded by CDC3, CDC10, CDC11 and CDC12 genes in *S. cerevisiae*: seem to form 10-nm filaments that form a ring around the plasma membrane in the mother-bud neck. Homologous proteins, associated with cleavage furrows, are reported from *Drosophila*, amphibians and mammals.

septum Literally, a separating wall. Mainly applied to the structure composed of plasmalemmae and cell wall material formed in cell division in prokaryotes and fungi. Also applied to the sealing layers in various packages of sterile fluids, or barriers through which injections, needles, etc. may be passed.

Sequenase™ Proprietary name for a genetically engineered form of T7 DNA polymerase used in DNA sequencing.

sequence homology Strictly, refers to the situation where nucleic acid or protein sequences are similar because they have a common evolutionary origin. Often used loosely to indicate that sequences are very similar. Sequence similarity is observable; homology is a hypothesis based on observation.

sequestosome *Aggresome* Protein aggregates (inclusion bodies) composed of ubiquitin-linked proteins destined for degradation. See **sequestosome-1**.

sequestosome-1 *Ubiquitin-binding protein p62* A polyubiquitin-chain binding protein involved in ubiquitin proteasome degradation. Some cases of Paget's disease of bone are due to mutations in this gene.

sequon A consensus sequence of amino acids, as for example the tripeptide motif Asn-Xaa-Ser/Thr that is the site for N-linked glycosylation.

SER See **smooth endoplasmic reticulum**.

SERCA *Sarcoplasmic–endoplasmic reticulum Ca²⁺-ATPase* The pump responsible for reducing sarcoplasmic calcium levels following contraction of muscle. Two different Ca^{2+} pumps are involved, SERCA sequestrates calcium ions into the SR; the plasma membrane Ca^{2+} ATPase (PMCA) extrudes calcium to the exterior.

serglycin An intracellular **proteoglycan**, found particularly in the storage granules of connective tissue mast cells. The core protein consists of 153 amino acids with 24 serine–glycine repeats between amino acids 89 and 137, hence the name. The serine–glycine repeats are the linkage sites for around 15 **glycosaminoglycan** chains that are either heparin or highly sulphated chondroitin sulphate. These negatively charged chains are thought to concentrate positively charged proteases, histamine and other molecules within the storage granules.

sericin Protein found in silk. Very serine-rich: 30% of the residues are serine.

serinc *Serinc-1, -2, etc* Class of carrier proteins that incorporates a polar amino acid serine into membranes and facilitates the synthesis of two serine-derived lipids, phosphatidyl serine and sphingolipids. Serinc is a unique protein family that shows no amino acid homology to other proteins but is highly conserved among eukaryotes. The members contain 11 transmembrane domains.

serine *Ser; S: 105 Da* One of the amino acids found in proteins and that can be phosphorylated. See Table A2.

serine dehydratase *EC 4.2.1.13; SDH; SerDH* A gluconeogenic enzyme, a member of the beta-family of pyridoxal phosphate-dependent (PLP) enzymes, catalyses the deamination of L-serine and L-threonine to yield pyruvate or 2-oxobutyrate.

serine hydroxymethyltransferase *SHMT; EC 2.1.2.1* One of the alpha-class of pyridoxal phosphate enzymes, a catabolic protein (43 kDa) involved in converting serine to glycine and a key enzyme in the formation and regulation of the folate one-carbon pool. *E. coli* SHMT has little sequence similarity to the enzyme family. In primates the cytosolic (cSHMT) and mitochondrial (mSHMT) genes constitute the functional members of the gene family. See **mimosine**.

serine peptidase *Serine protease* Peptidases that share a common reaction mechanism based on formation of an acyl-enzyme intermediate on a specific active serine residue. Most are inhibited by generic serine peptidase inhibitors (**serpins**) and irreversibly inactivated by a series of organophosphorus esters, such as di-isopropylfluorophosphate (DFP). They are, however, diverse in molecular structure and catalytic mechanisms and are not homologues of each other. Examples are **trypsin**, **chymotrypsin** and the bacterial enzyme **subtilisin**.

seroconvert Shorthand term for responding to immunization with an antigen and becoming seropositive, i.e. having antibodies directed against the antigen in question.

serogroup A group of bacteria or other microorganisms that have a certain antigen in common.

seroma A sterile accumulation of serum in a local area of tissue; unlike an abscess does not contain leucocytes.

serosa 1. A serous epithelium, having **serous glands** or cells, as opposed to a mucous membrane. 2. Thin infolding of the lining of the peritoneal cavity that forms the **omentum**.

serotonin *5-Hydroxytryptamine; 5-HT* A **neurotransmitter** and **hormone** (176 Da), found in vertebrates, invertebrates and plants.

serotype The genotype of a unicellular organism as defined by antisera directed against antigenic determinants expressed on the surface.

serous gland An exocrine gland that produces a watery, protein-rich secretion, as opposed to a carbohydrate-rich mucous secretion.

serpentine receptors *Seven-spanners; seven-transmembrane receptors* Receptors in which there are seven transmembrane-spanning regions. All are G-protein coupled. A variety of alternative names have been used, some clumsier than others.

serpins Superfamily of proteins, mostly **serine peptidase** inhibitors, that includes **ovalbumin, −α-1-antitrypsin, antithrombin**.

serrate *Ser* Transmembrane ligand for **Notch**, contains 14 repeats of the **EGF-like domain**, expressed on dorsal cells of *Drosophila* wing, activates Notch on ventral cells and induces the expression of **Delta** protein. Serrate protein expression is reciprocally induced by Delta and modulated by **fringe**.

Serratia marcescens A Gram-negative bacterium that is very common in soil and water; most strains produce a characteristic pigment, prodigiosin. Opportunistic human pathogens, infecting mainly hospital patients.

Serratia marcescens haemolysin *ShlA, ShlB* A new type of haemolysin that has nothing in common with the pore-forming toxins of *E. coli* type (RTX toxins), the *Staphylococcus aureus* alpha-toxin or the thiol-activated toxin of group A beta-haemolytic streptococci (Streptolysin O). Composed of two proteins ShlB and ShlA; activation (involving a conformational change) of ShlA by ShlB requires phosphatidylethanolamine as a cofactor. ShlA not only forms pores in erythrocytes but also in fibroblasts and epithelial cells.

Sertoli cell Tall, columnar cells found in the mammalian testis, closely associated with developing spermatocytes and spermatids. Probably provide appropriate microenvironment for sperm differentiation and phagocytose degenerate sperm.

serum Fluid that is left when blood clots; the cells are enmeshed in **fibrin** and the clot retracts because of the contraction of platelets. It differs from plasma in having lost various proteins involved in clot formation (**fibrinogen, prothrombin**, various blood-clotting factors such as **Hagemann factor, Factor VIII**, etc.) and in containing various platelet-released factors, notably **platelet-derived growth factor**. For this reason serum is a better supplement for cell culture medium than is defibrinated plasma (plasma-derived serum).

serum amyloid *SAA* In secondary amyloidosis the fibrils deposited in tissues are unrelated to immunoglobulin light chains (in contrast to the situation in primary amyloidosis) and are made of amyloid A protein (AA protein). This is derived from serum amyloid A (SAA), which is the apolipoprotein of a high-density lipoprotein and an **acute phase protein**. Partial proteolysis converts SAA into the **beta-pleated sheet** configuration of the amyloid fibrils. Amyloid P protein is also found as a minor component of the fibrils (in both primary and secondary amyloidosis) and is derived from serum amyloid P, which has similarity to **C-reactive protein**. The physiological role remains obscure.

serum amyloid P-component *SAP* Precursor of amyloid component P, found in basement membrane. Member of the **pentraxin** family. See **serum amyloid**.

serum- and glucocorticoid-inducible kinases *SGK* Serine/threonine protein kinases (of which several isoforms have been identified) that are transcriptionally regulated by corticoids, serum and cell volume. Sgk1 plays an important role in the regulation of epithelial ion transport by inactivating (by phosphorylation) Nedd4-2, an E3 ubiquitin-protein ligase that targets the epithelial Na$^+$ channel (ENaC) for degradation. The excitatory amino acid transporter (EAAT)2 is the major glutamate carrier in the mammalian CNS is similarly targeted by Nedd4-2; SGK isoforms alter the turnover of the transporter and this may account for the role of SGK in facilitating memory formation of spatial learning in rats. SGKs are related to **Akt** (PKB).

serum hepatitis See **hepatitis B**.

serum requirement The amount of serum that must be added to culture medium to permit growth of an animal cell in culture. Transformed cells frequently have less-stringent serum requirements than their normal counterparts.

serum response element *SRE Nucleic acid* motif found (for example) in the *c-fos* **promoter**, which is bound by the **serum response factor**.

serum response factor *SRF; p67SRF* **Transcription factor** which interacts with Elk-1 (p62TCF) to bind the **serum response element** promoter found in many growth-related genes.

serum sickness A **hypersensitivity** response (Type III) to the injection of large amounts of antigen, as might happen when large amounts of antiserum are given in a passive immunization. The effects are caused by the presence of soluble immune complexes in the tissues.

seven-membrane-spanning receptors See **serpentine receptors**.

sevenless *sev Drosophila* gene that is required for development of the R7 cell in each **ommatidium** in the eye. Gene product is a **receptor tyrosine kinase**, related to the insulin receptor. Ligand is the product of the *bride of sevenless* gene. In the downstream signalling cascade, **son-of-sevenless** plays an important part.

severin Calcium-dependent F-actin cleaving protein (40 kDa) isolated from the slime mould *Dictyostelium discoideum* that binds irreversibly to the barbed ends of the microfilament; not, apparently, essential for movement.

sex chromatin Condensed chromatin of the inactivated X chromosome in female mammals (**Barr body**).

sex chromosome Chromosome that determines the sex of an animal. In humans, where the two sex chromosomes (X and Y) are dissimilar, the female has two X chromosomes and the male is heterogametic (XY). In birds the opposite is the case, the male being XX and the female XY; in many organisms there is only one sex chromosome, and one sex is XX, the other X0. A portion of the X and Y chromosomes is similar, and is known as the pseudoautosomal region.

sex hormone Hormone that is secreted by gonads, or that influences gonadal development. Examples are **estrogen, testosterone, gonadotrophins**.

sex pili Fine, filamentous projections (**pili**) on the surface of a bacterium that are important in conjugation. Often seem to be coded for by plasmids that confer conjugative potential on the host; in the case of the F-plasmid, the F-pili are 8–9 nm in diameter and several microns long, composed of **pilin**. Whether the pili merely serve to establish and maintain adhesive contact between the partners in conjugation or whether DNA is actually transferred through the central core of the pilus is still unresolved, although a simple adhesion role is more generally accepted.

sex-duction The transfer of genes from one bacterium to another by the process of conjugation. May involve one bacterium with an F′-plasmid, in which case the process is called F-duction.

sex-linked disorder A genetic defect, usually due to a gene on the unpaired portion of the X chromosome. Recessive X-linked alleles are fully expressed in the heterogametic sex because they can have only one copy of the gene. Thus X-linked mutant disorders are more common in human males than in females.

Sezary cells See **Sezary syndrome**.

Sezary syndrome A cutaneous T-cell lymphoma, an advanced form of **mycosis fungoides**, in which skin all over the body is reddened, itchy, peeling and painful. There may also be patches, plaques or tumours on the skin. Cancerous T-cells (Sezary cells) are found in the blood and infiltrating the skin.

Sf9 cells Insect cell line derived from *Spodoptera frugiperda* much used for production of recombinant protein. Gene is incorporated into **baculovirus** vector which is then used to infect the cells.

S-gene complex Genes coding for molecular components of the pollen–stigma recognition system in the cabbage genus (*Brassica*). The gene products govern the **self-incompatibility** response, and include a glycoprotein found on the stigma surface and a lectin on the pollen grain surface that binds to the stigma glycoprotein.

SGK See **serum- and glucocorticoid-inducible kinases**.

SGOT *Serum glutamic-oxaloacetic transaminase; aspartate transaminase; EC 2.6.1.1* Enzyme that catalyses the reversible transfer of an amine group from glutamic acid to oxaloacetic acid, forming alpha-ketoglutaric acid and aspartic acid. High levels in serum are an indication of liver or heart damage.

SGPT *Serum glutamic pyruvic transaminase.; alanine aminotransferase (ALT); EC 2.6.1.2* Old name for **alanine aminotransferase**. In routine blood testing, the SGOT/SGPT ratio is often taken as an indication of alcoholic hepatitis.

sgRNA See **subgenomic RNA**.

SH domains *Src homology domains* Domains within proteins that, from their homology with **src**, are involved in the interaction with phosphorylated tyrosine residues on other proteins (SH2 domains) or with proline-rich sections of other proteins (SH3 domains).

SH2 See **SH domains**.

SH3 See **SH domains**.

SH3BP4 *SH3-domain binding protein 4; TTP* The human gene encodes a protein with three Asn-Pro-Phe (NPF) motifs, an **SH3 domain**, a PXXP motif, a bipartite nuclear targeting signal and a tyrosine phosphorylation site. Involved in cargo-specific control of clathrin-mediated endocytosis, specifically controlling the internalization of the **transferrin** receptor (TfR). TTP interacts with endocytic proteins, including **clathrin**, **dynamin** and the TfR, and localizes selectively to TfR-containing coated pits and vesicles.

shadowing Procedure much used in electron microscopy, in which a thin layer of material, usually heavy metal or carbon, is deposited onto a surface from one side in such a way as to cast 'shadows'. Deposition is usually done by vaporizing the metal on an electrode under vacuum.

shaker *Drosophila* gene encoding a potassium channel. Related genes *shab*, *shal* and *shaw* are known in flies and humans. The mutation is so-called because the fly's legs shake under ether anaesthesia.

shc Gene family identified by the presence of **SH2** domains. Shc also has a tyrosine motif that, when phosphorylated, will bind the SH2 of the adaptor protein **grb2** and may link receptor tyrosine kinases with the ras signalling pathway. Overexpression of shc will transform fibroblasts.

shear stress response element *SSRE* Various cells can be stimulated to divide if subject to fluid shear stress. This is particularly interesting in the case of endothelial and vascular smooth muscle cells, and efforts have been made to identify the response element that activates gene expression in response to shear; in the human PDGF-A promoter there is a GC-rich region near the TATA box that is required for shear-inducible reporter gene expression. There have been suggestions, however, that the signalling is through the **ras** pathway.

shibire *Drosophila* gene that encodes **dynamin**. *Shibire* is temperature sensitive, and in affected flies synaptic vesicles are depleted at high temperatures but are restored in nerve terminals when endocytosis resumes at lower temperatures.

shiga toxin *Verotoxin* Bacterial toxin from *Shigella dysenteriae* that blocks eukaryotic protein synthesis. See **Shiga-like toxins**.

shiga-like toxins *SLT* Group of structurally related toxins that block eukaryotic protein synthesis by cleaving a single residue from the 28S rRNA subunit of ribosomes thus blocking interaction with elongation factors eEF-1 and eEF-2. Examples: **Shiga toxin**, Shiga-like toxins SLT-1 and SLT-2 of *E. coli*.

Shigella Genus of non-motile Gram-negative enterobacteria (Escherichiae group): cause dysentery. See **Shiga toxin**.

shikimate 5-dehydrogenase *SDH; SKDH; EC 1.1.1.25* A key enzyme in the aromatic amino acid biosynthesis pathway, catalyses the reversible reduction of 3-dehydroshikimate to shikimate.

shikimic acid pathway Metabolic pathway in plants and microorganisms, by which the aromatic amino acids (phenylalanine, tyrosine and tryptophan) are formed from phosphoenolpyruvate and erythrose-4-phosphate via shikimic acid. The aromatic amino acids in turn serve as precursors for the formation of lignin and other phenolic compounds in plants. Inhibitors of this pathway are used as herbicides.

shimamushi fever *Scrub typhus; flood fever; Japanese river fever; tsutsugamushi fever* An acute fever cause by *Rickettsia tsutsugamushi*, transmitted by the bite of a larval mite (chigger), *Leptotrombidium akamushi*.

Shine–Dalgarno region A poly-purine sequence found in bacterial mRNA about seven nucleotides in front of the **initiation codon**, AUG. The complete sequence is 5′-AGGAGG-3′, and almost all messengers contain at least half of this sequence. It is complementary to a highly conserved sequence at the 3′ end of 16s ribosomal RNA, 3′-UCCUCC-5′, and it is thought to be involved in the binding of the mRNA to the ribosome.

shingles Disease in adults caused by **Varicella zoster** virus (Herpetoviridae), which in children causes chicken pox. Disease arises by reactivation (usually associated with a decline in cell-mediated immunity) of latent virus that persists in spinal or cranial sensory nerve ganglia.

SHIP Lipid phosphatase containing an SH2 domain; dephosphorylates 5-inositol phosphate. Important in regulation of mast cell degranulation and cytokine signal transduction in lymphoid and myeloid cells generally. SHIP also modulates PI3-kinase signalling downstream of growth factor and insulin receptors. Negative signalling through SHIP appears to inhibit the ras pathway by competition with **grb2** and **shc** for SH2 domain binding.

ShK toxin A potassium channel-blocking polypeptide (35 residues), from the sea anemone *Stichodactyla helianthus*.

shmoo Polarized morphological form of the yeast *S. cerevisiae* that has been exposed to mating pheromone (either **alpha factor** or a-factor). The cytoskeleton and proteins involved in mating are localized to a cell surface projection; the tips of the projections from the two cells eventually fuse.

shock Condition associated with circulatory collapse, a result of blood loss, bacteraemia, an anaphylactic reaction, or emotional stress.

Shope fibroma virus Poxvirus associated with the production of benign skin tumours in rabbits.

Shope papilloma virus Papovavirus that produces **papillomas** (warts) in rabbits.

short interspersed elements *SINEs* Non-autonomous non-LTR retrotransposons that are found in most animal genomes in large numbers. They are around 300 bp long and contain a region that is homologous to a tRNA and are typified by the human **Alu** repeat.

short stop *Shot; Kakapo* Spectraplakin isoform. **Shot** isoforms are similar to **spectrin** and **dystrophin**, with an actin-binding domain followed by spectrin repeats. In short stop **plakin** repeats are inserted between the actin-binding domain and spectrin repeats of shot. Localized to **adherens junctions** of embryonic and follicular epithelia.

short tandem repeat *STR* DNA sequences repeated in tandem, widespread throughout the human genome. Can be subgrouped on the basis of the size of the repeat region: **minisatellites** (variable number of tandem repeats, VNTRs) have core repeats with 9–80 bp, **microsatellites** (short tandem repeats, STRs) contain 2–5 bp repeats. Tandem repeats are frequently used in genetic mapping, linkage analysis and human identity testing. See also **fragile X, Huntington's chorea**.

short-term exposure limit *STEL* Limit of exposure to a hazard that should not be exceeded at any time during a (working) day. A 15-min time-weighted average exposure is often used.

shot *Previously Kakapo* A *Drosophila* **plakin** family member with actin-binding and microtubule-binding domains. In *Drosophila*, it is required for a wide range of processes, including axon extension, dendrite formation, axonal terminal arborization at the neuromuscular junction, tendon cell development and adhesion of wing epithelium.

shotgun approach Casual term for any approach that analyses a large number of small data points rather than a single large one: a shotgun fires many small pellets over a wider area whereas a rifle fires a single large bullet. Generally used in the context of sequence analysis (shotgun sequencing), carried out by chopping DNA (or even the whole genome) into many small sections of random length, sequencing them all, some many times, and pasting them together using computer methods that recognize overlap, as opposed to working systematically from one end.

shotgun sequencing See **shotgun approach**.

Shp *Shp-1, Shp-2* Protein tyrosine phosphatases with SH2 domains. Are recruited to **ITIM** motif of receptor tyrosine kinases and play an important role in the control of cytokine signalling. Shp-1 is important in regulating antigen responses in T-cells, and the mouse mutant (motheaten) is immunosuppressed. Shp-2 is more ubiquitously expressed and functions downstream of a variety of growth factor receptors and has a role in cell spreading and migration; the homozygous mouse knockout is embryonic lethal.

shRNA *Short hairpin RNA In vivo* transcription from plasmids coding for short hairpin RNAs (shRNAs) is a mechanism to generate long-term expression of siRNAs. Cellular processing of shRNAs shares common features with that of naturally occurring interfering RNA.

shroom A PDZ domain-containing actin-binding protein (210 kDa) that is required for neural tube morphogenesis in mice. Shroom-related proteins (APX, APXL and KIAA1202) are found in a wide range of species and are likely to be renamed as the family grows.

shuttle flow See **cytoplasmic streaming**.

shuttle vector Cloning vector that replicates in cells of more than one organism, e.g. *E. coli* and yeast. This combination allows DNA from yeast to be grown in *E. coli* and tested directly for **complementation** in yeast. Shuttle vectors are constructed so that they have the origins of replication of the various hosts.

Shwartzman reaction Reaction that occurs when two injections of **endotoxin** are given to the same animal, particularly rabbits, 24 h apart. In the local Shwartzman reaction the first injection is given intradermally, the second intravenously, and a haemorrhagic reaction develops at the dermal site. If both injections are intravenous the result is a generalized Shwartzman reaction, often accompanied by **disseminated intravascular coagulation**. The reaction depends upon the response of platelets and neutrophils to endotoxin.

sialic acid See **neuraminic acid**.

sialidase See **neuraminidase**.

sialin Lysosomal protein responsible for sialic acid export. Defects lead to Salla disease or infantile free sialic acid storage disease, both neurodegenerative conditions.

sialoglycoprotein Glycoprotein of which the *N*- or *O*-glycan chains include residues of **neuraminic acid**.

sialophorin See **leucosialin**.

sialyl Lewis X *sLex ; CD15s* Sialylated form of CD15, the ligand for E-, P- and L-**selectins**. Expressed on neutrophils, basophils and monocytes and on some lymphocytes. Also present on some **HEV**. Deficiency in sialyl Lewis-X will cause leucocyte adhesion deficiency Type II.

sialylate To add **sialic acid** to a glycoprotein or glycolipid, usually in a terminal position.

siamois The transcriptional mediator of the dorsal **Wnt** signalling pathway, necessary for formation of the Spemann organizer and dorsoanterior development in *Xenopus*.

SIC1 S phase **cyclin-dependent kinase inhibitor** from *S. cerevisiae*. See **Skp**.

sickle cell anaemia Disease common in races of people from areas in which malaria is endemic. The cause is a point mutation in haemoglobin (valine instead of glutamic acid at position 6), and the altered haemoglobin (HbS) crystallizes readily at low oxygen tension. In consequence, erythrocytes from homozygotes change from the normal discoid shape to a sickled shape when the oxygen tension is low, and these sickled cells become trapped in capillaries or damaged in transit, leading to severe anaemia. In heterozygotes, the disadvantages of the abnormal haemoglobin are apparently outweighed by increased resistance to *Plasmodium falciparum* malaria, probably because parasitized cells tend to sickle and are then removed from circulation.

sideramines Naturally occurring iron-binding compounds, hydroxamic acids.

sideroblasts Red blood cells containing Pappenheimer bodies: small, deeply basophilic granules that contain ferric iron.

sideromycins Non-chelating antibiotic analogues produced by some enteric bacteria; interfere with the uptake of **sideramine**–ferric ion complexes.

siderophilins Family of non-haem iron chelating proteins (about 80 kDa) found in vertebrates. Examples are **lactoferrin** and **transferrin**.

siderophores Natural iron-binding compounds that chelate ferric ions (which form insoluble colloidal hydroxides at neutral pH and are then inaccessible) and are then taken up together with the metal ion. See **sideramines**.

siderosis 1. Lung disease (pneumonoconiosis) caused by the inhalation of metallic particles. An occupational disease of workers in tin, copper, lead and iron mines, and of steel grinders. 2. Excessive deposition of iron in the body tissues.

sieve plate Perforated end walls separating the component cells (sieve elements) that make up the phloem **sieve tubes** in vascular plants. The perforations permit the flow of water and dissolved organic solutes along the tube, and are lined with **callose**. The plates are readily blocked by further deposition of callose when the sieve tube is stressed or damaged.

sieve tube The structure within the phloem of higher plants that is responsible for transporting organic material (sucrose, raffinose, amino acids, etc.) from the photosynthetic tissues (e.g. leaves) to other parts of the plant. Made up of a column of cells (sieve elements) connected by **sieve plates**.

sigma factor σ *Factor* **Initiation factor** (86 kDa) that binds to *E. coli* DNA-dependent **RNA polymerase** and promotes attachment to specific initiation sites on DNA. Following attachment, the sigma factor is released.

signal peptidase complex *SPC* A multisubunit serine peptidase (protease) complex located in the ER membrane that cleaves the **signal sequence** from proteins that are destined for export. Cleavage occurs as soon as the cleavage site of the translocating polypeptide is exposed in the lumen of the ER. Mammalian signal peptidase is a complex of five different polypeptide chains; SPC12 and SPC25 have substantial cytoplasmic domains and span the membrane twice; SPC18, SPC21 and SPC22/23 are single-spanning membrane proteins mostly exposed to the lumen of the endoplasmic reticulum. See also **sec61**.

signal peptide See **signal sequence**.

signal peptide peptidase *SPP* Like **gamma-secretase**, an unusual GxGD aspartyl peptidase. A family of SPP-like proteins (SPPLs) of unknown function has been identified.

signal recognition particle *SRP* A complex between a 7SRNA and six proteins. SRP binds to the nascent polypeptide chain of eukaryotic proteins with a **signal sequence** and halts further translation until the ribosome becomes associated with the rough endoplasmic reticulum. One of the SRP proteins (srp54) binds GTP and, in association with 7SRNA and srp19, has GTPase activity.

signal recognition particle receptor *Docking protein* Receptor for the **signal recognition particle** (SRP) found in the membrane of the endoplasmic reticulum. Heterodimeric, both protomers having GTP-binding capacity, though dissimilar binding sites. Not until the complex of SRP, ribosome, message and nascent polypeptide chain binds to the SRP-receptor is the block to further chain elongation released; concurrently the SRP is released, leaving the ribosome attached to the ER membrane. **Co-translational transport** of the polypeptide delivers it into the lumen of the ER.

signal sequence A peptide present on proteins that are destined either to be secreted or to be membrane components. It is usually at the N-terminus and normally absent from the mature protein. Normally refers to the sequence (~20 amino acids) that interacts with the **signal recognition particle** and directs the ribosome to the **endoplasmic reticulum**, where co-translational insertion takes place. Could also refer to sequences that direct post-translational uptake by organelles. Signal peptides are highly hydrophobic but with some positively charged residues. The signal sequence is normally removed from the growing peptide chain by the **signal peptidase complex** located on the cisternal face of the endoplasmic reticulum. See **signal recognition particle.**

signal transduction The cascade of processes by which an extracellular signal (typically a hormone or neurotransmitter) interacts with a receptor at the cell surface, causing a change in the level of a **second messenger** (for example calcium or cyclic AMP) and ultimately effects a change in the cell's functioning (for example, triggering glucose uptake, or initiating cell division). Can also be applied to sensory signal transduction, e.g. of light at photoreceptors.

signal-response coupling See **signal transduction**.

signet-ring cell Cell (adipocyte) with a large, central, fat-filled vacuole that pushes the nucleus to one side to give an appearance reminiscent of a signet ring.

silanization *Silanisation* Modification of hydroxyl groups on silica or glass surfaces with silane coupling agents (e.g. (3-mercaptopropyl)trimethoxysilane) to give the inactive -*O*-SiR3 grouping. Silanization can neutralize surface charges, thus eliminating non-specific binding. Often used to prepare surfaces for the binding of DNA fragments in microarrays.

silanizing Conversion of active silanol (–SiOH) groups on surface of (for example) glass into less polar silyl ethers (–SiOR), thereby making the surface less adhesive. See **siliconization**.

sildenafil Any of several compounds, especially sildenafil citrate (Viagra), that increase blood flow to the penis, used in treating male impotence. A drug all too frequently advertised by e-mail spammers.

silent gene A gene that is not phenotypically expressed. This can arise because of mutation of the gene itself (e.g. a nonsense codon causing premature termination of translation), a defect in some upstream control, or a mutation that renders the product inactive in some way.

silent mutation Mutations that have no effect on **phenotype** because they do not affect the activity of the product of the gene, usually because of codon ambiguity.

siliconization Non-covalent coating of surface with a layer of silicone oil making it less adhesive or reactive. See **silanizing**.

silicosis Inflammation of the lung caused by foreign bodies (inhaled particles of silica): leads to fibrosis but unlike **asbestosis** does not predispose to neoplasia.

Simian Virus 40 See **SV40**.

Simmonds' disease *Hypopituitarism* The failure of the anterior lobe of the pituitary to produce any one or more of its six hormones (ACTH, TSH, FSH, LH, GH and prolactin).

simple epithelium An epithelial layer composed of a single layer of cells all of which are in contact with the basal lamina (see **basement membrane**). May be cuboidal, columnar, squamous or pseudo-stratified; though the last of these appears superficially to be multilayered, all cells are in contact with the basal lamina.

simple sequence length polymorphism *SSLP* Tandem repeat sequences that are of variable length in different individuals, i.e. exhibit polymorphism. Since the flanking sequences are the same in all individuals it is relatively easy to choose primers that will selectively amplify such repeat sequences, and the length variation (difference in repeat number) can be detected easily on the basis of product size.

Simpson–Golabi–Behmel syndrome *SGBS* Human disorder leading to overgrowth. Arises through mutation in **glycipan-3** gene on X chromosome – probably reducing the extent to which IGF II is bound and unavailable for growth stimulation.

simvastatin A **statin** used to lower levels of blood cholesterol.

Sin3 Component of the multiprotein complex involved in repression of transcription. Sin3 is a co-repressor and forms a complex with Rpd3 histone deacetylase; the complex then interacts with DNA-binding proteins. Yeast SIN3 is involved in the repression of a diverse range of genes. Sin3 does not itself bind to DNA.

Sindbis virus Enveloped virus of the alphavirus group of **Togaviridae**. It is thought to be an infection of birds spread by fleas, and there is little evidence that it causes any serious infection in humans. The synthesis and export of the spike proteins, via the endoplasmic reticulum and Golgi complex, have been used as a model for the synthesis and export of plasma membrane proteins.

single cell protein Protein(s) produced by single cells in culture, especially *Candida* species. Of possible commercial importance in providing food sources from biotechnological processes

single-stranded DNA *ssDNA* DNA that consists of only one chain of nucleotides rather than the two **base-pairing** strands found in DNA in the double-helix form. **Parvoviridae** have a single-stranded DNA genome. Single-stranded DNA can be produced experimentally by rapidly cooling heat-denatured DNA. Heating causes the strands to separate and rapid cooling prevents **renaturation**.

single-channel recording Variant of **patch clamp** technique in which the flow of ions through a single channel is recorded.

single-stranded conformational polymorphism *SSCP* Technique for detecting point mutations in genes by amplifying a region of genomic DNA (using asymmetric **PCR**) and running the resulting product on a high quality gel. Single base substitutions can alter the secondary structure of the fragment in the gel, producing a visible shift in its mobility.

singlet oxygen $^1O_2$ An energized but uncharged form of oxygen that is produced in the **metabolic burst** of leucocytes and can be toxic to cells. One of the so-called reactive oxygen species (ROS)

SipA Salmonella *invasion protein A* An actin-binding protein from *Salmonella enterica*, apparently stabilizes F-actin bundles by increasing polymerization and decreasing depolymerization. One of the effectors of the *Salmonella* pathogenicity island-1 (SPI-1) type 3 secretion system. SipC is also involved in the process.

sir2 *Silent information regulator-2 protein* Protein from *S. cerevisiae*, original member of the **sirtuin** family to be identified. A NAD^+-dependent protein **histone deacetylase**.

sirenin Sexual pheromone, a bicyclic sesquiterpenediol, produced by female gametes of the water moulds, *Allomyces macrogynus* and *A. arbuscula*. Male gametes respond chemotactically.

siRNA *Small interfering RNA* See **RNA interference**.

sirtuins *Silent information regulator-2 (Sir2) enzymes* Family of NAD^+-dependent protein **histone deacetylases** (HDACs) that includes **Sir2** and its mammalian orthologues: play an important role in epigenetic gene silencing, DNA recombination, cellular differentiation and metabolism, and the regulation of aging.

sis **1.** An **oncogene**, *sis*, originally identified in monkey **sarcoma**, encoding a B-chain of **PDGF**. Human *c-sis/PDGF-B* proto-oncogene has been shown to be overexpressed in a large percentage of human tumour cells. See Table O1. **2.** See **macrophage inflammatory protein 1** α and β. (SIS). **3.** Small intestinal submucosa: porcine SIS has been used as a cell-free, biocompatible biomaterial for surgical repair purposes.

sister chromatid One of the two **chromatids** making up a **bivalent**. Both are semiconservative copies of the original chromatid.

SIT *SHP2-interacting transmembrane adapter protein* Transmembrane adaptor protein (**TRAP**), a disulfide-linked homodimeric glycoprotein, that is expressed in lymphocytes. Interacts with the SH2-containing protein tyrosine phosphatase 2 (**SHP2**) via an immunoreceptor tyrosine-based inhibition motif (**ITIM**). Also interacts with **grb2**.

Site-1 protease *S1P* See **ski**.

site-directed mutagenesis See **site-specific mutagenesis**.

site-specific mutagenesis *Site-directed mutagenesis* An *in vitro* technique in which an alteration is made at a specific site in a DNA molecule, which is then reintroduced into a cell. Various techniques are used; for the cell biologist, a very powerful approach to determining which parts of a protein or nucleotide sequence are critical to function.

site-specific recombination A type of **recombination** that occurs between two specific short DNA sequences present in the same or in different molecules. An example is the integration and excision of lambda prophage.

situs inversus Condition in which the normal asymmetry of the body (in respect of the circulatory system and intestinal coiling) is reversed. Interesting because it occurs in ~50% of patients with **immotile cilia syndrome**, a disorder of ciliary **dynein**.

SIV *Simian immunodeficiency virus* Very similar to **HIV**, and used extensively as an animal model.

Sjögren's syndrome One of the so-called connective tissue diseases that also include **rheumatoid arthritis**, **systemic lupus erythematosus** and **rheumatic fever**. Characterized by inflammation of conjunctiva and cornea.

skelemin Differential splicing of a single exon in the gene encoding sarcomeric myomesin gives rise to two polypeptides, **myomesin** and skelemin. Skelemin (195 kDa) is one of a superfamily of cytoskeletal proteins that contain fibronectin type III-like motifs and immunoglobulin C2-like motifs and that regulate the organization of myosin filaments in muscle and may be involved in linking integrins to the cytoskeleton in non-muscle cells.

skeletal muscle A rather non-specific term usually applied to the striated muscle of vertebrates that is under voluntary control. The muscle fibres are syncytial and contain myofibrils, tandem arrays of **sarcomeres**.

skeletrophin *mindbomb homolog 2 (MIB2)* A RING finger-dependent ubiquitin ligase, which targets the intracellular region of **Notch** ligands. It is an actin-binding cytoskeleton-related molecule, which is induced by the overexpression of truncated human SWI1 (SMARCF1). Human SWI1 is a subunit of the chromatin-remodelling complex.

ski 1. An oncogene, *ski* (Sloan-Kettering Institute proto-oncogene), identified in avian **carcinoma**, encoding a nuclear protein. C-ski is a regulating factor for fibroblast proliferation and an important co-repressor of **Smad3**. See Table O1. **2. Subtilisin**/kexin isozyme-1 (SKI-1), otherwise known as Site-1 protease (S1P), a Golgi proteinase mediating the proteolytic activation of the precursor to sterol-regulated element-binding proteins (**SREBPs**) 1 and 2.

skin-associated lymphoid tissue *SALT* The immune system cells associated with the dermis and epidermis of the skin. Includes **Langerhans cells** and resident phagocytes although some authors would include cutaneous nerve termini containing calcitonin gene-related peptide (CGRP): release of the peptide will stimulate mast cells to release cytokines such as IL-10 and TNFα. More of a concept than a morphological entity: *cf.* **gut-, bronchus-** and **mucosal-associated lymphoid tissue**.

skl 1. The *skl* gene from *Streptococcus mitis* SK137 encodes a peptidoglycan hydrolase (Skl), an *N*-acetylmuramoyl-L-alanine amidase (EC 3.5.1.28). Skl is a unique member of the choline-binding family of proteins since it contains a cysteine, histidine-dependent amidohydrolases/peptidases (CHAP) domain. **2.** Carboxyl-terminal amino acid sequence serinelysine–leucine (SKL) is the consensus **peroxisomal targeting sequence 1** (PTS1) that directs a polypeptide to peroxisomes in plants, animals and yeasts.

Skn-1 Maternally expressed transcription factor that specifies the fate of certain blastomeres during early development of *C. elegans*. Binds DNA with high affinity as a monomer even though it has a basic region similar to basic-leucine zipper (bZIP) proteins that only bind as dimers.

skotomorphogenesis *Etiolation* Development, growth and differentiation of a plant under conditions of darkness.

skp *S-phase kinase-associated proteins; Skp1, Skp2* Component of the **SCF complex**. The SCF (Skp1-cullin-F-box complex) ubiquitin-protein ligase of *S. cerevisiae* triggers DNA replication by catalysing ubiquitinylation of the S phase cyclin-dependent kinase inhibitor, SIC1.

skyllocytosis Phagocytic process in *Allogromia* (a marine rhizopod with which the general zoologist is usually acquainted).

Sl locus *Steel locus* Mouse mutant; see **stem cell factor**.

SL1 1. Transcription factor composed of four proteins, including TATA-binding protein **TBP**, required for activity of RNA polymerase I, resembles in some respect bacterial **sigma factor**. **2.** Stem-loop structure 1 (SL1), a characteristic feature of some viral RNAs.

Sla1p In yeast, an actin-regulatory protein that, together with **End3p** and Sla2p, is important in maintaining a rapid turnover of F-actin in cortical patches. Binds both to activators of actin dynamics (**Las17p** and Pan1p) and to cargo proteins, such as the pheromone receptor Ste2p.

SLAM *Signalling lymphocyte-activation molecule; CD150* Glycosylated type I transmembrane protein (70 kDa) present on T- and B-cell surfaces that is a high-affinity self-ligand. Triggering of SLAM coactivates T- or B-cell responses. A set of SLAM family receptors, CD150 (formerly CDw150), CD244 and CD48, are expressed at different stages of hematopoiesis and, in association with **SAP** family adaptors, have crucial roles during normal immune reactions in innate and adaptive immune cells.

SLAM-associated protein *SAP* A T-cell-specific protein (128 residues) that interacts with the cytoplasmic tail of **SLAM** and blocks recruitment of **Shp2**, thus blocking activation. SAP contains an SH2 domain. Mutations in SAP have been found in patients with X-linked lymphoproliferative disease (XLP).

SLAP *Src-like adapter protein* A dimeric adapter protein containing a SH3 and a SH2 domain; interacts with **ZAP70**, **Syk**, **LAT** and T-cell receptor (TCR)-zeta chain in Jurkat T-cells.

S-layer paracrystalline array which completely covers the surfaces of many pathogenic bacteria of both archae- and eubacteria. Usually consist of a single (glyco-)protein species with molecular masses ranging from about 40 to 200 kDa that form lattices of oblique, tetragonal or hexagonal architecture, and probably represent the earliest cell wall structures. About 10 nm thick.

SLC 1. Secondary lymphoid tissue chemokine (CCL21): promotes co-clustering of T-cells and dendritic cells in lymph nodes and spleen. **2. Sodium-lithium counter-transport**. **3.** An orphan G-protein-coupled receptor, somatostatin-like receptor 1 (SLC-1), for which the ligand appears to be melanin-concentrating hormone (MCH), an orexigenic peptide.

sleeping sickness See *Trypanosoma*.

SLEEPY *SLY1* F-box gene involved in *Arabidopsis* **gibberellin** signalling. The SLY1 gene encodes an F-box subunit of a Skp1-cullin-F-box (SCF) E3 ubiquitin ligase complex that positively regulates GA signalling. See also **SNEEZY**.

slicer activity Human **Argonaute2** is apparently responsible for target RNA cleavage in RNA interference, a process referred to as Slicer activity.

sliding filament model Generally accepted model for the way in which contraction occurs in the **sarcomere** of striated muscle, by the sliding of the **thick filaments** relative to the **thin filaments**.

slime moulds Two distinct groups of fungi: the cellular slime moulds or **Acrasidae,** which include *Dictyostelium*; and the **acellular slime moulds** or Myxomycetes, which include *Physarum*.

slipped strand mispairing Mispairing of the complementary DNA strands of a single DNA double helix during replication. Slippage can be either forward (causing deletion) or backward (causing insertion). **Short tandem repeats** are thought to be particularly prone to slipped strand mispairing.

slit Family of proteins first identified in *Drosophila* and subsequently in many vertebrates. Secreted diffusible protein that acts as a chemorepellant for neurons migrating from the subventricular zone to the olfactory bulb in developing mouse brain. Slit genes are expressed in the septum (at the midline of the telencephalon and caudal to the subventricular zone) and the effect of the repulsion is to force neurons to move in the rostral migratory stream. Also acts as a repellant for axonal movement. Receptor is **robo** (roundabout).

SLK *Ste20-like kinase* A microtubule-associated kinase that can regulate actin reorganization during cell adhesion and spreading.

SLM-1 *Sam68-like mammalian protein* A member of the **STAR protein** family, and related to **SAM68** and SLM-2. Regulates splice site selection *in vivo* via a purine-rich enhancer but in contrast to the widely expressed SAM68 and rSLM-2 proteins, rSLM-1 is found primarily in brain and, to a much smaller degree, in testis.

slot blot A **dot blot** in which samples are placed on a membrane through a series of rectangular slots in a template. This is slightly advantageous because hybridization artefacts are usually circular.

slow muscle Striated muscle used for long-term activity (e.g. postural support). Depends therefore on oxidative metabolism and has many mitochondria and abundant **myoglobin**.

slow reacting substance of anaphylaxis *SRS-A* Potent bronchoconstrictor and inflammatory agent released by mast cells; an important mediator of allergic bronchial asthma. A mixture of three **leucotrienes** (LTC4 mainly, LTD4 and LTE4).

slow virus 1. Specifically, one of the **Lentivirinae**. **2.** Any virus causing a disease that has a very slow onset. Diseases such as subacute **spongiform encephalopathy**, Aleutian disease of mink, **scrapie, kuru** and **Creutzfeldt–Jacob disease** may be caused by slow viruses, although this hypothesis has fallen out of favour. See also **prion**.

SLP-76 An adapter protein required for T-cell receptor (TCR) signalling. See **TRAPS**.

SLS collagen *Segment long-spacing collagen* Abnormal packing pattern of collagen molecules formed if ATP is added to acidic collagen solutions, in which lateral aggregates of molecules are produced. Each aggregate is 300 nm long, and the molecules are all in register. If SLS-aggregates are overlapped with a quarter-stagger, the 67-nm banding pattern of normal fibrils is reconstituted.

Slug *SNAI2* **Snail**-related zinc-finger transcription factor, SLUG (SNAI2), is critical for the normal development of neural crest-derived cells, and loss-of-function SLUG mutations have been proven to cause piebaldism and **Waardenburg's syndrome** type 2.

smac *Second mitochondria-derived activator of caspases; Diablo* Smac (monomer 21 kDa: oligomeric in solution) eliminates the inhibitory effect of **IAPs** and is the mammalian functional homologue of *Drosophila* **Reaper**, Grim and Hid. May be a key regulator of apoptosis.

Smad proteins Intracellular proteins that mediate signalling from receptors for extracellular TGFβ-related factors. Smad2 is essential for embryonic mesoderm formation and establishment of anterior–posterior patterning. Smad4 is important in gastrulation. Smads1 and 5 are activated (serine/threonine phosphorylated) by **BMP** receptors, Smad2 and 3 by **activin** and TGFβ receptors. Smads activated by occupied receptors then form complexes with Smad4/DPC4 and move into the nucleus, where they regulate gene expression. Interact with **FAST2**.

small acid-soluble spore proteins *SASP* DNA-binding proteins in the spores of some bacteria, thought to stabilize the DNA in the **A-DNA** configuration, so protecting it from cleavage by enzymes or UV light.

small cell carcinoma Common malignant neoplasm of bronchus. Cells of the tumour have endocrine-like characteristics and may secrete one or more of a wide range of hormones, especially regulatory peptides like **bombesin**.

small interfering RNA *siRNA* See **RNA interference**.

small nuclear RNA *snRNA* Abundant class of RNA found in the nucleus of eukaryotes, usually including those RNAs with sedimentation coefficients of 7S or less. They are about 100–300 nucleotides long. Although 5S rRNA and tRNA are of a similar size, they are not normally regarded as snRNAs. Most are found in complexes with proteins (see **ribonucleoprotein** particles, SnRNPs), and at least some have a role in processing hnRNA.

SMARC *BRG-1-associated factors* The **SWI/SNF complex**-related, matrix-associated, actin-dependent regulators of chromatin (SMARC), also called BRG1-associated factors, are components of human SWI/SNF-like chromatin-remodelling protein complexes.

SMCT *Sodium-coupled monocarboxylate transporter; slc5a8* A Na$^+$-coupled transporter for lactate, pyruvate and short-chain fatty acids.

Smith–Watermann alignment Algorithm for detecting sequence similarities when searching a genomic database.

SMO 1. See **smoothened**. **2. Spermine oxidase**. **3.** The SMO genetic locus in strains of the fungus *Magnaporthe grisea* directs the formation of correct cell shapes in asexual spores, infection structures and asci. **4.** Sulfamethoxazole, an antibacterial sulfonamide, now used primarily in combination with **trimethoprim**.

smooth endoplasmic reticulum *SER* An internal membrane structure of the eukaryotic cell. Biochemically similar to the rough endoplasmic reticulum (RER), but lacks the ribosome-binding function. Tends to be tubular rather than sheet-like, may be separate from the RER or may be an extension of it. Abundant in cells concerned with lipid metabolism, and proliferates in hepatocytes when animals are challenged with lipophilic drugs.

smooth microsome Fraction produced by ultracentrifugation of a cellular homogenate. It consists of membrane vesicles derived largely from the **smooth endoplasmic reticulum**.

smooth muscle Muscle tissue in vertebrates made up from long tapering cells that may be anything from 20 to 500 μm long. Smooth muscle is generally involuntary, and differs from striated muscle in the much higher actin/myosin ratio, the absence of conspicuous sarcomeres and the ability to contract to a much smaller fraction of its resting length. Smooth muscle cells are found particularly in blood vessel walls, surrounding the intestine (particularly the gizzard in birds) and in the uterus. The contractile system and its control resemble those of motile tissue cells (e.g. fibroblasts, leucocytes), and antibodies against smooth muscle myosin will cross-react with myosin from tissue cells whereas antibodies against skeletal muscle myosin will not. See also **dense bodies**.

smooth strain See **rough strain**.

smoothelins Actin-binding proteins that are expressed abundantly in visceral (smoothelin-A) and vascular (smoothelin-B) smooth muscle and are often used as marker proteins for contractile smooth muscle cells. Co-localize with alpha smooth-muscle actin in stress fibres, and are thought to have a role in the contractile process. Calponin homology-associated smooth muscle (CHASM) also has a **calponin homology domain** and shares sequence similarity with the smoothelin family.

smoothened *Smo Drosophila* seven-transmembrane protein Smoothened (Smo), which is thought to convert the Gli family of transcription factors from transcriptional repressors to transcriptional activators. Protein kinase A (PKA) and casein kinase I (CKI) regulate Smo cell surface accumulation and activity in response to **hedgehog**. Smo transmits its activation signal to a microtubule-associated **Hedgehog signalling complex**. Mammalian Smoothened is expressed on the primary cilium. See **cyclopamine**.

SMRT *Silencing mediator of retinoic acid and thyroid hormone receptors* Part of a co-repressor complex but distinct from **N-CoR**.

Snail Zinc-finger transcription factor involved in epithelial–mesenchymal transition; binds conserved E-box elements in the prostaglandin dehydrogenase (PGDH) promoter to repress transcription. The snail family includes **Slug**, which suppresses several epithelial markers and adhesion molecules, including E-cadherin.

SNAP 1. *S*-nitroso-*N*-acetyl penicillamine. **2.** Soluble **NSF** attachment (accessory) protein (25 kDa), involved in the control of vesicle transport. α-and γ-SNAPs are found in a wide range of tissues. β-SNAP is a brain-specific isoform of α-SNAP. SNAPs bind, together with NSF, to **SNAREs**. See **SNIP** definition 2.

snapin Neuronal protein that binds to the CRMP (**collapsin response-mediator protein**) homology domain of **cypin** and blocks the interaction of this domain with tubulin, thereby promoting microtubule disassembly and dendrite collapse.

SNAREs Receptors for **SNAPs**. The neuronal receptor for vesicle-SNAPs (v-SNARE) is **synaptobrevin**, also known as VAMP-2. The target (t-SNARE) associated with the plasma membrane of the axonal terminal is **syntaxin**. The SNAP–SNARE complex is apparently responsible for regulating vesicle targeting: neurotoxins such as **tetanus toxin** and **botulinum toxin** selectively cleave SNAREs or SNAPs. See also **cellubrevin**.

SNEEZY *SNE* **SLEEPY** (SLY1) homologue. SNE F-box protein can replace SLY1 in the **gibberellic acid**-induced proteolysis of RGA. See **DELLA**.

SNIP 1. An enzyme (snRNA incomplete 3′ processing) involved in processing snRNA. The central 158-amino acid domain of SNIP is related to the exonuclease III (ExoIII) domain of the 3′ →5′ proofreading epsilon subunit of *E. coli* DNA polymerase III. **2.** SNAP-25-interacting protein, a hydrophilic, 145-kDa protein that comprises two predicted coiled-coil domains, two highly charged regions and two proline-rich domains. Selectively expressed in brain where it codistributes with SNAP-25 in most brain regions.

snoRNA Small nucleolar ribonucleic acid. See **fibrillarin**.

SNP *Single nucleotide polymorphism* Any of the variations in single nucleotides in a DNA sequence that contribute to human individuality. SNP-mapping is increasingly used in searching for genetic associations of disease.

snRNA See **small nuclear RNA**.

snRNP Small nuclear ribonucleoprotein. See **small nuclear RNA**.

SOC media A cell growth medium used to ensure maximum transformation efficiency, it is Super Optimal Broth with the 'B' in SOB changed to 'C', for catabolite repression, reflective of the added glucose.

SOCS *Suppressor of cytokine signalling* Family of proteins (SOCS 1–3 and CIS; 211 amino acids) rapidly induced in response to IL-6 and other cytokines, thought to act as negative feedback regulator of **JAKs** in the intracellular signalling pathway. All contain a central **SH2** domain. Similar, possibly identical, to JAB and SSI-1.

sodium channel *Sodium gate* The protein responsible for electrical excitability of neurons. A multisubunit transmembrane **ion channel**, containing an aqueous pore around 0.4 nm in diameter, with a negatively charged region internally (the 'selectivity filter') to block passage of anions. The channel is **voltage-gated**: it opens in response to a small depolarization of the cell (usually caused by an approaching action potential), by a multistep process. Around 1000 sodium ions pass in the next millisecond, before the channel spontaneously closes (an event with single-step kinetics). The channel is then refractory to further depolarizations until returned to near the **resting potential**. There are around 100 channels per μm^2 in unmyelinated axons; in myelinated axons, they are concentrated at the **nodes of Ranvier**. The sodium channel is the target of many of the deadliest **neurotoxins**.

sodium cromoglycate *Intal*™ Drug used prophylactically for allergic asthma. It apparently acts by inhibiting release of bronchoconstricting agents from the mast cells in the airways by indirectly blocking calcium ion influx.

sodium dodecyl sulphate See **SDS**.

sodium lithium counter-transport *SLC; Na/LiCT* Erythrocyte sodium lithium counter-transport, sodium-stimulated lithium efflux from lithium-loaded erythrocytes, has been observed to be abnormal in several hypertension-related diseases and is increased in patients with diabetes mellitus.

sodium pump See **sodium–potassium ATPase**.

sodium–potassium ATPase A major transport protein of the plasma membrane. A multiunit enzyme, it moves three sodium ions out of the cell, and two potassium ions in, for each ATP hydrolysed. The sodium gradient established is used for several purposes (see **facilitated diffusion**, **action potential**), while the potassium gradient is dissipated through the potassium leak channel. Must not be confused with a **sodium channel**.

sodoku *Rat-bite fever* Infection by *Streptobacillus moniliformis* or *Spirillum minus*, following the bite of a rat or contact with rat saliva; it is characterized by fever, chills, headache and muscle pain, and a diffuse rash, primarily in the extremities.

soft agar Semi-solid agar used to gelate medium for culture of animal cells. Placed in such a medium, over a denser agar layer, the cells are denied access to a solid substratum on which to spread, so that only cells that do not show **anchorage dependence** (usually transformed cells) are able to grow.

sol–gel transformation Transition between more fluid cytoplasm (**endoplasm**) and stiffer gel-like **ectoplasm** proposed as a mechanism for amoeboid locomotion: since the endoplasm cannot really be considered a simple fluid and has viscoelastic properties like a gel, the term is misleading.

solanidine The aglycone of **solanine** and **chaconine** that are produced by the action of solanidine UDP-glycosyltransferases. Solanine has a branched trisaccharide galactose-(glucose)-rhamnose moiety and chaconine a glucose-(rhamnose)-rhamnose moiety.

solanine A glycoalkaloid toxin found in species of the nightshade family (Solanaceae) that includes potatoes. It can occur naturally in any part of the plant, including the leaves, fruit and tubers. It is very toxic even in small quantities. Solanine has both fungicidal and pesticidal properties,

and it is one of the plant's natural defences. Has sedative and anticonvulsant properties, and has been used as a treatment for bronchial asthma. See **solanidine**, **chaconine**.

Solanum tuberosum The potato.

soluble tyrosine kinase *STK-1, STK-2* STK-1 is a soluble form of c-**src** underphosphorylated on C-terminal tyrosine residues; STK-2 is a 48-kDa protein tyrosine kinase molecularly and functionally related to **Csk**.

solute carrier family proteins *SLCs* Superfamily of proteins involved in the transport of molecules across membranes; 46 subclasses are recognized, each with several members. For example, solute carrier family 1 (SLC1) comprises neuronal/epithelial high affinity glutamate/neutral amino acid transporters, SLC2 family is of facilitated glucose transporters (GLUT proteins) and has 14 members.

somatic cell Usually any cell of a multicellular organism that will not contribute to the production of gametes, i.e. most cells of which an organism is made: not a **germ cell**. Notice, however, the alternative use in **somatic mesoderm**.

somatic cell genetics Method for identifying the chromosomal location of a particular gene without sexual crossing. Unstable **heterokaryons** are made between the cell of interest and another cell with identifiably different characteristics (or without the gene in question) and a series of clones isolated. By correlating retention of gene expression with the remaining chromosomes, it is possible to deduce which chromosome must carry the gene. Human–mouse heterokaryons have been extensively used in this sort of work.

somatic cell nuclear transfer Transfer of the nucleus of a somatic cell into an enucleated egg as a means of cloning.

somatic hybrid **Heterokaryon** formed between two somatic cells, usually from different species. See **somatic cell genetics**.

somatic mesoderm That portion of the embryonic mesoderm that is associated with the body wall and is seperated from the splanchnic (visceral) mesoderm by the coelomic cavity.

somatic mutation **Mutation** that occurs in the somatic tissues of an organism and that will not, therefore, be heritable since it is not present in the **germ cells**. Some neoplasia is due to somatic mutation; a more conspicuous example is the reversion of some branches of variegated shrubs to the wild-type (completely green) phenotype. Somatic mutation is probably also important in generating diversity in V-gene regions of immunoglobulins.

somatic recombination One of the mechanisms used to generate diversity in antibody production is to rearrange the DNA in B-lymphocytes during their differentiation, a process that involves cutting and splicing the immunoglobulin genes. Somatic recombination via homologous crossing-over occurs at a low frequency in *Aspergillus*, *Drosophila* and *Saccharomyces*, and in mammalian cells in culture. It may be detected through the production of homozygous patches or sectors after mitosis of cells heterozygous for suitable marker genes.

somatocrinin *Growth hormone-releasing factor* See **growth-hormone-releasing hormone**.

somatoliberin *Growth hormone-releasing factor; somatotropin-releasing factor* See **growth-hormone-releasing hormone**.

somatomedin Generic term for **insulin-like growth factors** (IGFs) produced in the liver and released in response to **somatotropin**. Somatomedins stimulate the growth of bone and muscle, and also influence calcium, phosphate, carbohydrate and lipid metabolism. Somatomedin A is IGF II, somatomedin C is IGF I. Somatomedin B is a serum factor of uncertain function. See **somatostatin**.

somatostatin Gastrointestinal and hypothalamic peptide hormone (two forms: 14 and 28 residues); found in gastric mucosa, pancreatic islets, nerves of the gastrointestinal tract, in posterior pituitary and in the central nervous system. Inhibits gastric secretion and motility· in hypothalamus/pituitary, inhibits **somatotropin** release. Somatostatin acts through five G-protein-coupled receptor subtypes (SSTR$_{1–5}$), displaying a tissue-specific distribution with multiple subtypes present on many cells. A neuropeptide, **cortistatin**, strongly resembles somatostatin. See also Table H2.

somatotrope Cell in the anterior pituitary which secretes growth hormone.

somatotrophin *Somatropin (US); growth hormone* Hormone (human growth hormone, hGH, 191 amino acids; 22 kDa) released by anterior pituitary that stimulates release of **somatomedin**, thereby causing growth. See also Table H2.

somites Segmentally arranged blocks of mesoderm lying on either side of the notochord and neural tube during development of the vertebrate embryo. Somites are formed sequentially, starting at the head. Each somite will give rise to muscle (from the myotome region), spinal column (from the sclerotome) and dermis (from dermatome).

son-of-sevenless *Drosophila* ras-GRF (GDF-releasing factor), mammalian homologues of which (sos1, sos2) play an important part in intracellular signalling. See **sos**.

sonic hedgehog *Shh* Secreted protein that is involved in organization and patterning of several vertebrate tissues during development. The zone of polarizing activity (**ZPA**) that determines anterior–posterior patterning of the limb expresses sonic hedgehog. See **hedgehog**.

sorbitol *Glucitol* The polyol (polyhydric alcohol) corresponding to glucose. Occurs naturally in some plants, is used as a growth substrate in some tests for bacteria and is sometimes used to maintain the tonicity of low ionic strength media.

sorbitol dehydrogenase *EC 1.1.1.14; SORD* A member of the superfamily of medium-chain dehydrogenases/reductases, 38 kDa, the second enzyme in the polyol pathway; oxidizes sorbitol to fructose in the presence of NAD$^+$.

sorbose A monosaccharide hexose: L-sorbose is an intermediate in the commercial synthesis of **ascorbic acid**.

soredium *Pl.* soredia Specialized asexual reproductive products of lichens consisting of algal cells surrounded by fungal hyphae; emerge from soralia on the lichen surface and are wind-dispersed.

Soret band A very strong absorption band in the blue region (414 nm) of the optical absorption spectrum of a haem protein.

Soret effect The mass diffusion of chemical species due to an imposed thermal gradient. A temperature gradient across a fluid or gaseous mixture generally leads to a net mass flux and the build-up of a concentration gradient. This effect is known as thermal diffusion or Ludwig–Soret effect. The amplitude of the concentration gradient is determined by the Soret coefficient, ST.

sorocarp Fruiting body formed by some cellular **slime moulds**; has both stalk and spore-mass.

sorting nexins *SNX* A family of proteins that characteristically have a phox-homology (PX) domain and are required for the endocytosis and the sorting of transmembrane proteins. Sorting nexin-1 is a component of the mammalian **retromer** complex. Human SNX-1, -2 and -4 have been proposed to play a role in receptor trafficking and have been shown to bind to several receptor tyrosine kinases. SNX-6 interacts with members of the TGF-beta family of receptor serine–threonine kinases. Yeast Vps5p is the orthologue of mammalian sorting nexin-1.

sorting out Phenomenon observed to occur when mixed aggregates of dissimilar embryonic cell types are formed *in vitro*. The original aggregate sorts out so that similar cells come together into homotypic domains, usually with one cell type sorting out to form a central mass that is surrounded by the other cell type. Much controversy has arisen over the years as to the underlying mechanism – whether there is specificity in the adhesive interactions (which would imply tissue-specific receptor–ligand interactions), or whether it is sufficient to suppose that there are quantitative differences in homo- and heterotypic adhesion (the **differential adhesion** hypothesis). With the exception perhaps of the main protagonists, most cell biologists consider that there are probably elements both of tissue specificity (**CAMs**) and of quantitative adhesive differences involved.

sorus A group of **sporangia** or spore cases, e.g. on the underside of fern leaves.

sos 1. Guanine nucleotide-releasing factor (155 kDa), the mammalian homologue of **son-of-sevenless**. The proline-rich region of sos binds to the **SH3** domain of **GRB-2**. Has homology with CDC-25, the yeast GTP-releasing factor for ras. A family of related proteins are now known and include vav, **C3G**, Ost, NET1, Ect2, RCC1, tiam, RalGDS and **Dbl**. **2.** See **SOS system**.

SOS system The DNA repair system also known as error-prone repair in which apurinic DNA molecules are repaired by incorporation of a base that may be the wrong base but that permits replication. RecA protein is required for this type of repair. *SOS* genes function in control of the cell cycle in prokaryotes and eukaryotes.

Southern blots See **blotting**. Originally developed by Dr Ed Southern, hence the name.

sox 1. SoxR: Redox sensory protein in *E. coli*. **2.** SOX syndrome: Sialadenitis, osteoarthritis and xerostomia syndrome. **3.** *Sox* genes: Gene family involved in many developmental processes. *Sox-2* regulates transcription of FGF-4 gene, *sox-3* is involved in neural tube closure and lens specification, *sox-9* is related to *sry* and found in mouse testis, *sox-10* is important in neural crest development.

soybean trypsin inhibitor *STI; SBTI* Single polypeptide (21 kDa; 181 amino acids) that forms a stable, stoichiometric, enzymically inactive complex with trypsin.

SP-1 1. A proinflammatory **transcription factor** of the Cys_2His_2 zinc-finger family. **2. Tunichrome** Sp-1 is a modified pentapeptide from the haemocytes of the ascidian *Styela plicata*.

spacer DNA The DNA sequence between genes. In bacteria, only a few nucleotides long. In eukaryotes, can be extensive and include **repetitive DNA**, comprising the majority of the DNA of the genome. The term is used particularly for the spacer DNA between the many tandemly repeated copies of the ribosomal RNA genes.

SPAK *Ste20/SPS-1-related proline-, alanine-rich kinase* SPAK and OSR1 (oxidative stress-responsive kinase-1) kinases interact and phosphorylate NKCC1 (Na^+-K^+-$2Cl^-$ co-transporter-1), leading to its activation. Both are phosphorylated and activated by **WNKs** (with no K (lysine) protein kinase-1).

SPARC *Secreted protein, acidic and rich in cysteine* See **osteonectin**. Overexpressed in the fibroblasts of skin biopsy specimens obtained from patients with systemic sclerosis.

sparsomycin Antibiotic, isolated from the fermentation broth of *Streptomyces sparsogenes*, that inhibits **peptidyl transferase** in both prokaryotes and eukaryotes.

spartin A frame-shift mutation in the spartin (SPG20) gene is the cause of Troyer syndrome, an autosomal recessive form of spastic paraplegia. Spartin is both cytosolic and membrane-associated and apparently binds **Eps15**.

spasmin Protein (20 kDa) that forms the **spasmoneme**. Thought to change its shape when the calcium ion concentration rises and to revert when the calcium concentration falls: the reversible shape-change is used as a motor mechanism. Contraction does not require ATP; relaxation does, probably to pump calcium ions back into the smooth endoplasmic reticulum.

spasmoneme Contractile organelle found in *Vorticella* and related ciliate protozoans. Capable of shortening faster than any actin–myosin system and of expanding actively. See **spasmin**.

spastin Product of the *SPG4* gene that is mutated in most cases of autosomal dominant **hereditary spastic paraplegia**. Protein product belongs to the family of ATPases associated with various cellular activities (AAAs) and is related to the microtubule-severing protein **katanin**.

spatial sensing Mechanism of sensing a gradient in which the signal is compared at different points on the cell surface and cell movement directed accordingly. Translocation of all or part of the cell is not required. See **temporal** and **pseudospatial gradient sensing**.

speciation Formation of new biological species. Usually considered to require isolation of a subpopulation of the ancestral species in one of the following ways: geographically (classically seen with remote islands); by occupying a different niche (adaptive radiation); through acquisition of behavioural changes that restrict mating, so that distinct genetic variations accumulate and prevent further interbreeding; or by polyploidy that makes interbreeding with the original species impossible. See **sympatric speciation**.

species A group of individuals that are capable of interbreeding with each other but not with members of other groups. Usually a result of some form of isolation; in the early stages **speciation** can be reversed if reproductive isolation is lost or the niches change and overlap. Species

are grouped into genera and divided into subspecies and varieties or cultivars. The name of the genus should be capitalized and the species name should not – thus *Homo sapiens*, genus *Homo*, species *sapiens* – and both should be italicized. This nicety is increasingly neglected.

specific activity The number of activity units (whatever is appropriate) per unit of mass, volume or molarity. Perhaps most often encountered in the context of radiochemicals, the number of microcuries per micromole.

specific granules *Secondary granules* One of the two main classes of granules found in neutrophils: contain lactoferrin, lysozyme, Vitamin B12 binding protein and elastase. Are released more readily than the **azurophil** (primary) granules which have typical lysosmal contents.

spectinomycin A bacteriostatic aminocyclitol antibiotic produced by a species of soil bacterium, *Streptomyces spectabilis*; binds to the 30S subunit of the bacterial ribosome, thus inhibiting protein synthesis. It is used in the treatment of gonococcal infections.

spectral karyotyping *SKY* A technique that allows scientists to visualize all of the human chromosomes at one time by 'painting' each pair of chromosomes in a different fluorescent color; **translocations** show up very conspicuously because the affected chromosome is multicoloured.

spectraplakins A superfamily of giant cytoskeletal linker proteins, with multiple isoforms produced from each gene. Bind actin, tubulin and intermediate filaments. Spectraplakins have in common two **calponin homology domains**, a **plakin** domain, a series of **plectin** repeats, numerous **spectrin** repeats and finally a GAS2 domain. Shot is the sole spectraplakin in *Drosophila*.

spectrin Membrane-associated dimeric protein (240 and 220 kDa) of erythrocytes. Forms a complex with **ankyrin**, **actin** and probably other components of the membrane cytoskeleton, so that there is a meshwork of proteins underlying the plasma membrane, potentially restricting the lateral mobility of integral proteins. Isoforms have been described from other tissues (**fodrin**, **TW-240/260 protein**), where they are assumed to play a similar role. Contains the **EF-hand** motif.

spectrophotometry Quantitative measurements of concentrations of reagents made by measuring the absorption of visible, ultraviolet or infrared light.

Spemann's organizer Signalling region located on the dorsal lip of the blastopore in the early embryo, essential for defining the main body axis.

speract Sea urchin peptide hormone, from the jelly coat of the eggs of the sea urchins *Strongylocentrotus purpuratus* and *Hemicentrotus pulcherrinus*, affecting sperm motility and metabolism. Receptor is a plasma membrane **guanylate cyclase**.

spermatids The haploid products of the second meiotic division in spermatogenesis. Differentiate into mature spermatozoa.

spermatocytes Cells of the male reproductive system that undergo two meiotic divisions to give haploid **spermatids**.

spermatogenesis The process whereby primordial germ cells form mature spermatozoa.

spermatogonium Plant gonad cell that undergoes repeated mitoses, leading to the production of **spermatocytes**.

Spermatophyte Division of the plant kingdom, consisting of plants that reproduce by means of seeds.

spermatozoon Mature sperm cell (male gamete).

spermidine N-*(3-aminopropyl)-1,4-butanediamine* A polybasic amine (polyamine); see **spermine**.

spermine N, N′*bis(3-aminopropyl)-1,4-butanediamine* Polybasic amine (polyamine). Found in human sperm, in ribosomes and in some viruses. Involved in nucleic acid packaging. Synthesis is regulated by **ornithine decarboxylase** which plays a key role in control of DNA replication.

spermine oxidase *SMO(PAOh1); EC 1.5.3.* An inducible enzyme that may play a direct role in the cellular response to the antitumour polyamine analogues. Oxidizes both spermine and $N(1)$-acetylspermine but not spermidine.

SPF *Specific pathogen free* Animals that have been raised in carefully controlled conditions so that they are not infected with any known pathogens. May require that they are delivered by Caesarean section and raised in strict quarantine.

S phase The phase of the cell cycle during which DNA replication takes place. S stands for synthesis.

spherical aberration Deficiency in simple lenses in which the image is sharp in the centre but out of focus at the periphery of the field, more a problem when taking photographs than when observing directly. Lenses compensated for this defect are referred to as plan-lenses (e.g. **planapochromat**).

spheroblast *Spheroplast* Bacterium or yeast that has been treated in such a way as to have lost the outer cell wall. This leads to the cell being osmotically fragile, but makes it easier to get large molecules across the plasma membrane in, for example, transfection.

spherocytosis A condition in which erythrocytes lose their biconcave shape and become spherical. It occurs as cells age, and is also found in individuals with abnormal cytoskeletal proteins, (hereditary spherocytosis, a disorder that leads to haemolytic anaemia).

spheroplast Bacterium from which the cell wall has been removed but that has not lysed.

spherosome Lysosome-like compartment in plants that derives from the endoplasmic reticulum and is a site for lipid storage.

sphingolipid Structural lipid of which the parent structure is sphingosine rather than glycerol. Synthesized in the Golgi complex.

sphingomyelin A **sphingolipid** in which the head group is phosphoryl choline. A close analogue of phosphatidyl choline. In many cells the concentration of sphingomyelin and phosphatidyl choline in the plasma membrane seems to bear a reciprocal relationship.

sphingomyelinase *Sphingomyelin phosphodiesterase-1; acid sphingomyelinase; EC 3.1.4.12* Stress is believed to activate sphingomyelinase to generate ceramide, which serves as a second messenger in initiating the apoptotic response. Deficiency in the enzyme leads to **Niemann–Pick disease** types A and B.

sphingosine Long-chain amino alcohol that bears an approximate similarity to glycerol with a hydrophobic chain attached to the 3-carbon. Forms the class of **sphingolipids** when it carries an acyl group joined by an amide link to the nitrogen. Forms **sphingomyelin** when phosphoryl choline is attached to the 1-hydroxyl group. Gives rise to the cerebroside and ganglioside classes of glycolipids when oligosaccharides are attached to the 1-hydroxyl group. Not found in the free form.

sphingosine kinase *SPHK; sphinganine kinase; EC 2.7.1.91* Key enzyme catalysing the formation of sphingosine 1-phosphate (SPP), a lipid messenger implicated in the regulation of a wide variety of important cellular events acting through intracellular, as well as extracellular, mechanisms.

SPHK See **sphingosine kinase**.

Spi-1 **1.** Proto-oncogene (*spi-1*) encoding a transcription factor (PU1) that binds to purine-rich sequences (PU boxes) expressed in haematopoietic cells. **2.** *Salmonella typhimurium* pathogenicity island 1 (SPI-1), a region that carries genes mediating invasion into intestinal epithelial cells and that induce cell death in murine macrophages. **3.** *Vaccinia* virus host-range/antiapoptosis genes, SPI-1 and SPI-2.

spin labelling The technique of introducing a grouping with an unpaired electron to act as an electron spin resonance (ESR) reporter species. This is almost invariably a nitroxide compound (–N–O) in which the nitrogen forms part of a sterically hindered ring. See **electron paramagnetic resonance**.

spina bifida A neural tube defect in which the bones of the spinal canal fail to meet and fuse during development. Varies in severity depending upon the extent of the failure: mild forms (spina bifida occulta), where only the most posterior vertebrae are affected, may be almost asymptomatic. In more serious forms (myelomeningocele and meningocele) there can be severe disability. The addition of folic acid to the diet of women of childbearing age significantly reduces the incidence of neural tube defects.

spinal cord Elongated, approximately cylindrical part of the central nervous system of vertebrates that lies in the vertebral canal, and from which the spinal nerves emerge.

spinal ganglion *Dorsal root ganglion* Enlargement of the dorsal root of the spinal cord containing cell bodies of afferent spinal neurons. Neural outgrowth from dorsal root ganglia has been studied extensively *in vitro*.

spindle See **mitosis**.

spindle fibres Microtubules of the spindle that interdigitate at the equatorial plane with microtubules of the opposite polarity derived from the opposite pole **microtubule-organizing centre**. Usually distinguished from **kinetochore** fibres, which are microtubules that link the poles with the kinetochore, although these could be included in a broader use of the term.

spindle pole body *SPB* In budding yeast, the functional equivalent of the **centrosome** that organizes both the spindle and cytoplasmic microtubules throughout the cell cycle. It is anchored to the nuclear envelope. See **sad1**.

spinner culture Method for growing large numbers of cells in suspension by continuously rotating the culture vessel.

spiral cleavage Pattern of early cleavage found in molluscs and annelids (both **mosaic eggs**). The animal pole blastomeres are rotated with respect to those of the vegetal pole. The handedness of the spiral twist shows **maternal inheritance**.

spiral fracture Fracture of a bone caused by a twisting movement, one of the commonest forms of fracture.

spire *Drosophila* protein of the Spir family (interacts with **cappuchino**). Spir proteins nucleate actin polymerization by binding four actin monomers to a cluster of four WASP-homology domain 2 (WH-2 domains) in the central region of the proteins. This mechanism is distinct from actin nucleation by the **Arp2/3** complex or by **formins**.

spirillum *Pl.* spirilla A fairly rigid helically twisted (corkscrew-shaped) bacterial cell, often, but not necessarily, a member of the genus *Spirillum*. Common examples are *Vibrio cholerae* and *Treponema pallidum*, the causative agents of cholera and syphilis, respectively.

spirochaete *US spirochete* An elongated, spirally shaped bacterium, e.g. the organism (*Treponema pallidum*) responsible for syphilis.

Spirogyra Genus of green filamentous algae found in freshwater ponds. Contain helically disposed ribbon-like chloroplasts.

spironolactone Aldosterone antagonist: diuretic; used to treat low-**renin** hypertension, and **Conn's syndrome** (in which there is overproduction of aldosterone).

Spiroplasma citri A plant-pathogenic mollicute phylogenetically related to Gram-positive bacteria. *Spiroplasma* cells are restricted to the phloem sieve tubes and are transmitted from plant to plant by the leafhopper vector *Circulifer haematoceps*.

Spirostomum Genus of large free-living ciliate protozoans with an elongated body.

splanchnic Relating to the viscera. See also **splanchnic mesoderm**.

splanchnic mesoderm That portion of the embryonic **mesoderm** that is associated with the inner (endodermal) part of the body in contrast to **somatic mesoderm** which is associated with the body wall. The two mesodermal regions are separated by the **coelom**.

splenocytes Vague term usually referring to phagocytic cells (macrophages) of the spleen.

splice variants Proteins that are related but differ in their sequence as a result of **alternative splicing**, variation in the exons that are spliced to form the mRNA. Some proteins exhibit enormous variation and a significant number of proteins have splice variants.

spliceosomes The macromolecular RNA–protein complexes involved in intron removal and exon ligation as mRNA is processed. Components include U2, U5 and U6 snRNAs and the essential spliceosomal protein Prp8. Different subclasses of spliceosome process particular classes of introns, for example U12-dependent introns are spliced by the so-called minor spliceosome.

splicing The process by which introns are removed from hnRNA to produce mature messenger RNA that contains only exons. Alternative splicing seems to occur in many proteins and by alternative exon usage a set of related proteins can be generated from one gene, often in a tissue or developmental stage-specific manner (**alternative splicing**). See **spliceosome**.

split gene See **introns**.

split ratio The fraction of the cells in a fully grown culture of animal cells that should be used to start a subsequent culture. Minimum may be dictated by inadequacies of the medium that result in poor growth of some cells at high dilution.

spokein Term formerly used for the constituent protein of the **radial spokes** of the ciliary **axoneme**. Since a number of complementary spoke mutants are known to occur in *Chlamydomonas*, and one mutant lacks 17 proteins, it seems likely that spokein is a complex mixture. The radial spokes are regularly repeating axonemal structures composed of at least 23 proteins and are required for normal axonemal motility. See **radial spoke proteins**.

spongiform encephalopathies A group of diseases characterized by long incubation and fatal progressive course with characteristic spongiform degeneration of grey matter of the cortex. The two main human diseases are **kuru** and **Creutzfeldt–Jacob disease**, there being a new variant (vCJD or nvCJD) of the latter, possibly as a result of infection with the agent of bovine spongiform encephalopathy (BSE). Diseases such as **scrapie**, mink encephalopathy and BSE are considered to be similar. There is still controversy regarding the causative agent, although general opinion now favours Prusiner's **prion** hypothesis. See also **Gerstmann–Straussler–Scheinker syndrome**.

spongin Protein that forms the basis for the extracellular matrix of silicaceous sponges and creates a meshwork that links the silica-rich spicules.

spongioblast Cell found in developing nervous system: gives rise to **astrocytes** and **oligodendrocytes**.

spongioblastoma Rare tumours of childhood and adolescence with embryonal features. Sometimes described as a class of neuroepithelial tumours, which are probably not a distinct entity but could be considered either ependymomas or neuroblastomas.

spongiocytes Lipid droplet-rich cells from the middle region of the cortex of the adrenal gland.

spongy parenchyma Tissue usually found in the lower part of the leaf **mesophyll**. Consists of irregularly shaped, photosynthetic parenchyma cells, separated by large air spaces.

spontaneous transformation Transformation of a cultured cell that occurs without the deliberate addition of a transforming agent. Cells from some species, especially rodents, are particularly prone to such spontaneous transformation.

sporadic Of a **tumour** or genetic disease, a novel occurrence without any previous family history of the disease (*cf.* inherited). Examples of diseases with both sporadic and inherited forms: **retinoblastoma**, **Wilms' tumour**.

sporangium Spore case, within which asexual spores are produced.

spore Highly resistant dehydrated form of reproductive cell produced under conditions of environmental stress. Usually have very resistant cell walls (integument) and low metabolic rate until activated. Bacterial spores may survive quite extraordinary extremes of temperature, dehydration or chemical insult. Gives rise to a new individual without fusion with another cell.

sporocarp Multicellular structure in fungi, lichens, ferns or other plants. Location of spore formation.

sporophyte Spore-producing plant generation. The dominant generation in **pteridophytes** and higher plants, and alternates with the gametophyte generation.

sporopollenin Polymer of **carotenoids**, found in the exine of the pollen wall. Extremely resistant to chemical or enzymic degradation.

sporotrichosis A fungal infection, usually of the skin, caused by *Sporothrix schenckii*. Mainly an occupational disease of farmers, gardeners and horticulturists.

Sporozoa Class of spore-forming parasitic protozoa without cilia, flagella or pseudopodia. See **Apicomplexa**.

sporozoite Infective stage of the life cycle of **Apicomplexa** such as *Plasmodium* and *Cryptosporidia*.

spot desmosome Macula adherens: see **desmosome**.

sprouting 1. Production of new processes (outgrowths) by nerve cells: e.g. by embryonic neurons undergoing primary differentiation; by adult neurons in response to nervous system damage; or by dissociated neurons redifferentiating in culture. **2.** Casual term for the germination of a seed.

squalene A 30-carbon isoprenoid lipid found in large quantities in shark liver oil and in smaller amounts in olive oil, wheat germ oil, rice bran oil and yeast. A key intermediate in the biosynthesis of **cholesterol**.

squames Flat, keratinized, dead cells shed from the outermost layer of a squamous **stratified epithelium**.

squamous epithelium An epithelium in which the cells are flattened. May be a **simple epithelium** (e.g. **endothelium**) or a **stratified epithelium** (e.g. **epidermis**).

squamous-cell carcinoma Carcinoma that develops from the squamous layer of the epithelium. Slow growing.

squid giant axon Large axons, up to 1 mm in diameter, that innervate the mantle of the squid. Because of their large size, many of the pioneering investigations of the mechanisms underlying resting and action potentials in excitable cells were done on these fibres.

squidulin *Squid calcium-binding protein (SCaBP)* Calcium-binding protein (16.8 kDa) from the optic lobe of squid, which contains the **EF-hand** motif.

SR-proteins A family of highly conserved nuclear phosphoproteins required for constitutive pre-mRNA splicing and also influence **alternative splicing**. They are characterized by one or two N-terminal RNA-binding domains and a C-terminal SR domain enriched in serine–arginine dipeptides (RS domain). SR-proteins influence splice site selection and are required at an early step in **spliceosome** assembly.

SR-type splicing factors See **SR-proteins**.

SRBC Sheep red blood cells.

src family Family of protein tyrosine kinases of which src was the first example (see *src* **gene**). Includes Fyn, Yes,

Fgr, Lyn, Hck, Lck, Blk and Yrk. All cells studied so far have at least one of these kinases which act in cellular control. Family members all have characteristic src homology (SH) domain structure, a kinase domain (SH1), SH2 and SH3 domains and a domain (SH4), which has myristoylation and membrane-localization sites. Inter-domain interactions, themselves regulated by phosphorylation (see **Csk**), regulate the activity of the kinase.

src **gene** The transforming (sarcoma-inducing) gene of Rous sarcoma virus. Protein product is pp60vsrc, a cytoplasmic protein with tyrosine-specific **protein kinase** activity that associates with the cytoplasmic face of the plasma membrane.

SRE See **serum response element**.

SREBP *Sterol regulatory element-binding proteins* Family of transcription factors that bind to steroid response elements in the promoter regions of genes involved in the metabolism of cholesterol and fatty acids.

S region The non-**MHC** gene in the midst of the H-2 **major histocompatibility complex** of the mouse genome that codes for complement component C4. Sometimes confusingly known as the gene for the type III MHC product in mice.

SRF See **serum response factor**.

S-ring The static part of the bacterial motor: a ring of 15 or 17 subunits (one less or one more than the **M-ring**), anchored to the inner surface of the cell wall.

SRP See **signal recognition particle**.

SRS-A See **slow reacting substance of anaphylaxis**.

SRTX See **sarafotoxin**.

sry *Sex-related gene on Y* Genes for a family of high-mobility group (**HMG**) proteins that bind to a subset of sequences recognized by C/EBP family of DNA-binding proteins. *Sry* itself is the primary testis-determining gene and located, as its name suggests, on the Y chromosome. The protein product, SRY (testis-determining factor), binds to DNA and causes it to bend sharply thereby affecting the expression of other genes on the Y chromosome. *Sox*-10 is related to *Sry*.

SSB 1. Single-stranded DNA-binding (SSB) protein that binds selectively to single-stranded DNA intermediates during DNA replication, recombination and repair. **2.** Single-strand breaks (SSB) in DNA. **3.** La/SSB phosphoprotein is the target antigen of autoantibodies in sera of patients with **Sjogren's syndrome** (SS) and **systemic lupus erythematosus** (SLE).

SSCP See **single-stranded conformational polymorphism**.

ssDNA phage Single-strand DNA phages such as MS2, ΦX174, as opposed to double-stranded DNA phages or RNA phages.

SSEA 1. Stage-specific embryonic antigen. 2. SseA is a key *Salmonella* virulence determinant, a small (12 kDa), basic pI protein that serves as a type III secretion system chaperone for SseB and SseD. **3.** SseA, the translation product of the *E. coli sseA* gene, is a 31 kDa protein with 3-mercaptopyruvate:cyanide sulphurtransferase activity *in vitro*.

SSH See **suppression subtractive hybridization**.

SSI-1 *STAT-induced STAT inhibitor-1* Protein induced by cytokines (through **STAT** pathway) that binds to and inhibits JAK2 and Tyk2 thereby acting as a negative feedback signal. See **SOCS**.

SSLP See **simple sequence length polymorphism**.

SSRIs *Selective serotonin re-uptake inhibitors* Antidepressant drugs that inhibit re-uptake of serotonin released in the brain, thereby prolonging its action as a neurotransmitter. Less sedative than the tricyclic antidepressants. Examples: citalopram, fluoxetine, paroxetine, sertraline and related drug venlafaxine.

stable transfection *Stable expression* When **transfecting** animal cells, a clone of cells in which the **transgene** has been physically incorporated into the genome. It thus provides stable, long-term expression although is more difficult to produce.

stachyose Digalactosyl-sucrose, a compound involved in carbohydrate transport in the phloem of many plants and also in carbohydrate storage in some seeds.

stacking gel An upper layer of weak gel at the top of a gel in which an electrophoretic separation is to be run. Because the upper stacking gel is weak, all the large molecules move through it rapidly and accumulate at the very top of the separating gel; thus they all start from the same level.

stage-specific embryonic antigen *SSEA* SSEA-1 (CD15) is a surface antigen that can be used to define immature retinal progenitor cells; others, SSEA-3, -4, etc, are similarly cell surface antigens.

staggered cut Situation when the two strands of a DNA molecule are cut in different places, slightly offset so that the two halves have sticky ends, a short segment of single-stranded DNA overhanging the duplex.

stanniocalcin *STC-1* Glycoprotein hormone secreted by the corpuscle of Stannius, an endocrine gland in teleosts. Prevents hypercalcaemia and inhibits calcium uptake through gills. Mammalian homologues (STC-1, STC-2) have been identified and are involved in various physiological processes, such as ion transport, reproduction and development.

stanol *24-Alpha-ethylcholestanol* Plant-derived sterol, similar to cholesterol but poorly absorbed from the intestine. Thought to compete with cholesterol for binding sites and, by decreasing dietary intake of cholesterol, to contribute to lowering blood levels.

stanozolol An **anabolic steroid** that has been used as a performance-enhancing drug by some athletes.

Staphylococcal alpha toxin *Leucocidin* Pore-forming exotoxin (33 kDa) secreted by *Staphylococcus aureus*. Protein (monomer) has two domains connected by flexible hinge region: oligomerizes in the plasma membrane by lateral mobility to form a hexameric oligomer (220 kDa) that has a pore ~1 nm in diameter with some anion selectivity. Osmotic lysis leads to death of the cell. Has been used to selectively permeabilize cells to small molecules.

Staphylococcal delta toxin Small peptide exotoxin (26 residues) secreted by *Staphylococcus aureus*. Binds to membranes and a range of cellular components. Very amphipathic and surface active. Properties, though not sequence, very similar to **melittin**.

staphylococcins **Bacteriocins** produced by staphylococci.

Staphylococcus Genus of non-motile Gram-positive bacteria that are found in clusters and produce important exotoxins (see Table E2). *Staphylococcus aureus* (*S. pyogenes*) is pyogenic, an opportunistic pathogen and responsible for a range of infections. It has **protein A** on the surface of the cell wall. Coagulase production correlates with virulence: hyaluronidase, lipase and **staphylokinase** are released in addition to the toxins.

staphylokinase *Streptokinase; EC 3.4.99.22* Enzyme released by *Staphylococcus aureus* that acts as a **plasminogen activator**.

STAR proteins **1.** Signal transduction and activation of RNA (STAR) family of RNA-binding proteins; evolutionarily conserved from yeast to humans and important for a number of developmental decisions. Contain a conserved KH domain as well as two conserved domains called QUA1 and QUA2. Examples include mouse **quaking** and *C. elegans* **germline defective-1** (GLD-1). **2. Steroidogenic acute regulatory protein**, StAR.

starch Storage carbohydrate of plants, consisting of **amylose** (a linear α (1-4)-glucan) and **amylopectin** (an α (1-4)-glucan with α (1-6)-branch points). Present as starch grains in plastids, especially in **amyloplasts** and **chloroplasts**.

start codon See **initiation codon**.

start site Imprecise term that can refer to either a **transcriptional** start site or a **translational** start site.

statherin A low-molecular weight (5380 Da, 43 amino acid residues) acidic tyrosine-rich phosphoprotein secreted mainly by salivary glands. Binds **fimbrillin**.

stathmin *Oncoprotein 18; Op18* Ubiquitous coiled-coil cytosolic phosphoprotein (19 kDa) that interacts with tubulin heterodimers and increases the rate of rapid (catastrophic) disassembly of microtubules. Overexpressed in some tumours and probably regulated by phosphorylation.

statins Group of lipid-lowering drugs that act by inhibiting **HMG CoA reductase**, a key enzyme in cholesterol biosynthesis. Beneficial effects in other diseases have been reported. Examples are atorvastatin (Lipitor), rosuvastatin, pravastatin, simvastatin.

statocyst An organ for the perception of gravity and thus body orientation, found in many invertebrate animals; a cavity lined with sensory cells and containing a **statolith**.

statocyte A root-tip cell containing one or more **statoliths**, involved in the detection of gravity in geotropism.

statolith **1.** *Bot.* A type of **amyloplast** found in root-tip cells of higher plants. It can sediment within the cell under the influence of gravity and is thought to be involved in the detection of gravity in geotropism. **2.** *Zool.* A sand grain or a structure of calcium carbonate or other hard-secreted substance, found in the cavity of a **statocyst**. It stimulates sensory cells lining the cavity with which it comes into contact under the influence of gravity.

STATs *Signal transducers and activators of transcription* Contain **SH2** domains that allow them to interact with phosphotyrosine residues in receptors, particularly cytokine-type receptors; they are then phosphorylated by **JAKs**, dimerize and translocate to the nucleus where they act as transcription factors. Many STATs are known; some are relatively receptor-specific, others more promiscuous. Thus a wide range of responses is possible, with some STATs being activated by several different receptors, sometimes acting synergistically with other STATs.

staufen Double-stranded RNA-binding protein, a component of RNA granules, binds to the 3′ UTR of specific mRNAs and acts in their localization. Involved in creating and maintaining cellular asymmetry in the *Drosophila* oocyte. Mammalian homologues Staufen 1 and Staufen 2 play an important role in dendritic mRNA targeting. Staufen2, a brain-specific isoform has been shown to shuttle between nucleus and cytoplasm, and can enter the nucleolus.

staurosporine Inhibitor of PKC-like protein kinases derived from *Streptomyces* sp. Has a rather broad inhibitory spectrum and cannot be used to ascribe a specific role for protein kinase C in a signalling pathway.

STE20 Evolutionarily conserved serine/threonine kinases that regulate fundamental cellular processes including the cell cycle, apoptosis and stress responses. One of the **p21-activated kinase** (PAK) family.

STEAP *Six-transmembrane epithelial antigen of the prostate* One of a small family of cell surface antigens that are expressed in prostate cancer and are potential targets for antibody-mediated therapy and diagnosis. Includes **PSM**, **PTCA-1**, **PSCA**. STEAP is expressed at the cell–cell junctions of the secretory epithelium of prostate and strongly expressed in prostate cancer cells. May be a channel or transporter protein. ORF of 339 amino acids with a predicted molecular mass of 40 kDa and with no significant homology to any known genes.

stearic acid *n-octadecanoic acid* See **fatty acids**.

steatoblasts Cells that give rise to fat cells (adipocytes).

steel factor Murine equivalent of **stem cell factor**.

stefin *Cystatin* Family of **cysteine peptidase** inhibitors. Stefin A has 98 residues and inhibits **cathepsins** D, B, H and L.

stele *Vascular cylinder* The vascular tissue (xylem and phloem) inside the cortex of roots and stems of vascular plants.

stem cell 1. Cell that gives rise to a lineage of cells. 2. More commonly used of a cell that, upon division, produces dissimilar daughters, one replacing the original stem cell, the other differentiating further (e.g. stem cells in basal layers of skin, in haematopoietic tissue and in meristems).

stem cell factor *SCF; Steel factor in mice; mast cell growth factor; c-kit ligand* Haematopoietic growth factor, 18.6 kDa from sequence; found as dimer (35 kDa protein, 53 kDa in its glycosylated form).

stem cell-derived tyrosine kinase *STK* One of the hepatocyte growth factor receptor family and the murine homologue of the human RON receptor tyrosine kinase. Expressed on macrophages. Ligand is macrophage-stimulating protein (**MSP**), a serum protein activated by the coagulation cascade.

stem-and-loop structure Term for the structure of tRNAs which has four base-paired stems and three loops (not base-paired), one of which contains the anticodon.

stenohaline Descriptive of an organism that is unable to tolerate a range of salinities.

stent A device inserted into a blood vessel to keep it open, usually a small metal coil or mesh tube. Similar devices are used in the GI tract and ureter.

Stentor Genus of large, multinucleate protozoa (up to 2 mm long) with a ring of apical cilia used in feeding on bacteria. Spirotrich ciliates. Usually attached to a surface in freshwater ponds, etc., but can relocate. Sometimes called the trumpet animalcule because of its shape.

stereocilium *Pl.* stereocilia Microfilament bundle-supported projection, several microns long, from the apical surface of sensory epithelial cells (**hair cells**) in inner ear: like a **microvillus** but larger. It is stiff and may act as a transducer directly, or merely restrict the movement of the sensory cilium (which does have an axoneme). Also described on cells of pseudo-**stratified epithelium** of the epididymal duct. Recently, stereocilia have been referred to as stereovilli – a much better and less confusing name.

stereotaxis A system by which the precise 3D coordinates of a target site in tissue are identified. May be done using imaging systems, and once the coordinates are known it is possible to deliver a focused beam onto the area from several directions – as, for example, in radiation treatment of brain tumours. Can also be used to guide the insertion of microelectrodes.

stereovillus *Pl.* stereovilli Better name for **stereocilium**.

Sternberg–Reed cells See **Hodgkin's disease**.

steroid finger motif See **steroid receptor**.

steroid hormone A group of structurally related hormones, based on the cholesterol molecule. They control sex and growth characteristics, are highly lipophilic, and are unique in that their receptors are in the nucleus rather than on the plasma membrane. Examples: **testosterone**, **estrogen**.

steroid receptor Family of nuclear **transcription factors**, most of which are receptors for hormones of the steroid family – for example, androgen, estrogen, glucocorticoid, mineralocorticoid, progesterone, retinoic acid, ecdysone, thyroid hormone; and the *Drosophila* transcription factors knirps, ultraspiracle and seven-up. This family contains a conserved domain (the 'steroid finger' motif) containing two C4-type **zinc fingers**.

steroid response element DNA sequence that is recognized and bound by a **steroid receptor**.

steroidogenic acute regulatory protein *StAR protein* A 30-kDa mitochondrial protein that mediates cholesterol transfer and promotes steroid hormone production in the ovary, testis and adrenal gland. Defective in congenital lipoid adrenal hyperplasia (lipoid CAH). The pesticide, Roundup, inhibits steroidogenesis by disrupting STAR protein expression.

sterol regulatory element-binding proteins *SREBPs* A family of membrane-bound transcription factors important in both cholesterol and fatty acid metabolism. There are three SREBPs that regulate the expression of over 30 genes. SREBPs are regulated by proteolytic cleavage (see **insig-1**, **SCAP**), rapid degradation by the ubiquitin-proteasome pathway and **sumoylation**.

sterols Molecules that have a 17-carbon steroid structure but with additional alcohol groups and side chains. Commonest example is **cholesterol**.

STI Abbrev. for a sexually transmitted infection.

sticky ends The short stretches of single-stranded DNA produced by cutting DNA with **restriction endonucleases** whose site of cleavage is not at the axis of symmetry. The cut generates two complementary sequences that will hybridize (stick) to one another or to the sequences on other DNA fragments produced by the same restriction endonuclease.

stilboestrol Synthetic estrogen that was given to pregnant women in the belief that it would prevent threatened miscarriage, subsequently shown to be a risk factor for vaginal clear cell carcinoma in women exposed *in utero*. Now occasionally used in treatment of prostate carcinoma.

Still's disease A systemic inflammatory disorder of unknown aetiology characterized by the association of a high spiking fever, an evanescent skin rash, arthritis and hyperleucocytosis. Adult-onset Still's disease has often been regarded as the adult version of systemic juvenile idiopathic arthritis.

stimulus-secretion coupling A term used to describe the events that link receipt of a stimulus with the release of materials from membrane-bounded vesicles (the analogy is with excitation–contraction coupling in the control of muscle contraction). A classical example is the link between membrane depolarization at the presynaptic terminal and the release of neurotransmitter into the synaptic cleft.

sting cells **Nematocysts** of coelenterates.

stipe A stalk, especially of fungal fruiting bodies or of large brown algae.

STK 1. Stem cell-derived tyrosine kinase. 2. Soluble tyrosine kinase (STK-1, STK-2). **3. Streptokinase.**

stochastic Random or probabilistic.

stoichiometry Ratio of the participating molecules in a reaction; in the case of an enzyme–substrate or receptor–ligand interaction, should be a small integer.

Stokes' radius Stokes' law of viscosity defines the frictional coefficient for a particle moving through a fluid, a coefficient that depends upon the viscosity of the fluid and the radius of the particle. The apparent radius of a molecule sedimenting under centrifugal force calculated from this law (the Stokes' radius) is a feature of the tertiary structure and thus informative about the molecule in question.

stoma *Pl.* stomata Pore in the epidermis of leaves and some stems, which permits gas exchange through the epidermis. Can be open or closed, depending upon the physiological state of the plant. Flanked by stomatal **guard cells**.

stone cell See **sclereid**.

stop codon See **termination codon**.

stop transfer sequence *Membrane anchor sequence* Amino acid sequence that causes cessation of the co-translational transfer of a protein across a membrane and leaves the protein embedded in the membrane. Generally consists of long sequences of hydrophobic residues.

STOPS *Stable tubulin-only proteins* A family of calcium–calmodulin-regulated microtubule-stabilizing proteins that confer cold-stability on microtubules. Subunit seems to be 145 kDa, but may bind as multimer.

storage diseases Another name for **lysosomal diseases**.

storage granules 1. Membrane-bounded vesicles containing condensed secretory materials (often in an inactive, zymogen, form). Otherwise known as zymogen granules or condensing vacuoles. **2.** Granules found in plastids, or in cytoplasm; assumed to be food reserves, often of glycogen or other carbohydrate polymer.

STR See **short tandem repeat**.

strain birefringence See **birefringent**.

stratified epithelium An epithelium composed of multiple layers of cells, only the basal layer being in contact with the basal lamina (see **basement membrane**). The basal layer is of **stem cells** that divide to produce the cells of the upper layers; in skin, these become heavily keratinized before dying and being shed as squames. Stratified epithelia usually have a mechanical/protective role.

stratum corneum Outermost layer of skin, composed of clear, dead, scale-like cells with little remaining except **keratin**.

stratum granulosum Layer of granular cells underlying the **stratum corneum** in the skin of vertebrates. The cells accumulate keratin and gradually become compressed to form the cornified cells of the outermost layer.

stratum lucidum A thin, clear layer of dead cells in vertebrate skin, lying between the **stratum corneum** and **stratum granulosum** of thick skin, as on palms of the hands and the soles of the feet.

stratum Malpighii One of the layers of the skin in vertebrates, lying between the proliferating cells of the basal layer (**stratum germinatum**) and the **stratum granulosum** where keratin deposition occurs. Also known as the prickle cell layer. Location of many of the disorders of the skin.

streptavidin Analogue of **avidin**. A tetrameric protein (4×13 kDa) isolated from *Streptomycetes avidinii* that has a high affinity for biotin. Each monomer of streptavidin binds one molecule of biotin. Used to detect biotin markers.

Streptococcal M-protein Cell wall protein of streptococci: antibody typing of the M-protein is important in identification of different strains of Group A streptococci (at least 55 serotypes are known). The M-protein confers anti-phagocytic properties on the cell and is present as hair-like **fimbriae** on the surface. M-protein is an important virulence factor, and antibodies directed against M-protein are essential for phagocytic killing of the bacteria.

streptococcal toxins Group of haemolytic **exotoxins** released by *Streptococcus* spp. α-haemolysin, 26–39 kDa (four types), forms ring-like structures in membranes (see **streptolysin O**); lipid target unclear. β-haemolysin is a hot-cold **haemolysin** with sphingomyelinase C activity. γ-haemolysin is a complex of two proteins (29 and 26 kDa) that act synergistically, rabbit erythrocytes particularly sensitive. δ-toxin is a heat-stable peptide (5 kDa) with high proportion of hydrophobic amino acids; seems to act in a detergent-like manner (*cf.* **subtilysin**), but may form hydrophilic transmembrane pores by cooperative interaction with other δ-toxin molecules. Leucocidin (Panton-Valentine leucocidin) has two components, F (fast migration on CM-cellulose column, 32 kDa) and S (slow,

38 kDa); mode of action is contentious. See also **strepto-coccus**, **streptolysins O and S**, **erythrogenic toxin**.

streptococcins Bacteriocins released by streptococci.

Streptococcus Genus of Gram-positive cocci that grow in chains. Some species (*S. pyogenes* in particular) are responsible for important diseases in humans (pharyngitis, scarlet fever, rheumatic fever): *S. pneumoniae* is the main culprit in lobar- and broncho-pneumonia. Streptococci have anti-phagocytic components (hyaluronic acid-rich capsule and **M-protein**) and release various toxins (**streptolysins O and S**, **erythrogenic toxin**) and enzymes (**streptokinase**, **streptodornase**, hyaluronidase and proteinase). α-haemolytic streptococci (viridans streptococci) produce limited haemolysis on blood agar; include *S. mutans*, *S. salivarius*, *S. pneumoniae*. β-haemolytic streptococci, of which *S. pyogenes* is the only species (though there are many serotypes), produce a broad zone of almost complete haemolysis on blood agar as a result of streptolysin O and S release. γ-streptococci are non-haemolytic (e.g. *S. faecalis*).

Streptococcus pneumoniae Pneumococcus pneumoniae; Diplococcus pneumoniae; *Fraenkel's bacillus* Gram-positive, non-motile, facultative anaerobic bacteria that occurs as pairs, hence *Diplococcus*. The leading cause of bacterial pneumonia and otitis media (middle ear infections), and an important contributor to bacterial meningitis.

streptodornase *EC 3.1.21.1* Mixture of four DNAases released by streptococci. By digesting DNA released from dead cells, the enzyme reduces the viscosity of pus and allows the organism greater motility.

streptogramins Group of antibiotics discovered some time ago but coming back into favour as a result of increasing antibiotic resistance. Consist of mixtures of two structurally distinct compounds, type A and type B, which are separately bacteriostatic but are bactericidal in appropriate ratios. These antibiotics act at the level of inhibition of translation, through binding to the bacterial ribosome. Examples include **dalfopristin**, **quinupristin**.

streptokinase Plasminogen activator released by *Streptococcus pyogenes*. Occurs in two forms, A and B.

streptolydigin Antibiotic that blocks peptide chain elongation by binding to the polymerase.

streptolysin O Oxygen-labile thiol-activated haemolysin (native toxin is 61 kDa and is cleaved to form 55-kDa fragment which retains activity). Haemolysis is inhibited by cholesterol, and only cells with cholesterol in their membranes are susceptible. Toxin aggregates are linked to cholesterol to form a channel of 30-nm diameter in the membrane, and non-osmotic lysis follows. Markedly inhibits neutrophil movement and stimulates secretion, but has little effect on monocytes.

streptolysin S *SLS* Oxygen-stable haemolysin of *Streptococcus pyogenes*. Thought to be a peptide of ∼28 residues: causes zone of β-haemolysis around streptococcal colonies on blood agar. Like complement-mediated haemolysis, it appears to act in a one-hit mechanism. Toxic to leucocytes, platelets and several cell lines. Has sequence homologies to the **bacteriocin** class of antimicrobial peptides. A nine-gene operon is required for SLS production: this includes a candidate gene for a bacteriocin prepropeptide; SagA, the likely SLS precursor and candidate genes for chemical modification of the bacteriocin propeptide and self protection;

an ABC transporter for export and maturation proteolysis of the leader peptide; and an internal terminator motif for differential transcription of structural gene and accessory gene mRNAs.

Streptomyces Genus of Gram-positive spore-forming bacteria that grow slowly in soil or water as a branching filamentous mycelium similar to that of fungi. Important as the source of many antibiotics, e.g. **streptomycin, tetracycline, chloramphenicol, macrolides**.

streptomycin A water-soluble **aminoglycoside antibiotic** derived from the bacterium *Streptomyces griseus*. Commonly used antibiotic in cell culture media: acts only on prokaryotes, and blocks transition from **initiation complex** to chain-elongating ribosome. Clinically, used mainly in the treatment of tuberculosis, usually in conjunction with other drugs to prevent the development of resistance.

streptovaricins Antibiotics of the ansamycin class, produced by various Actinomycetes, that block initiation of transcription in prokaryotes. (*cf.* **rifamycins** and **rifampicin**).

streptozotocin Methyl nitroso-urea with a 2-substituted glucose, used as an antibiotic (effective against growing Gram-positive and Gram-negative organisms) and also to induce a form of diabetes in experimental animals (rapidly induces pancreatic **B-cell** necrosis if given in high dose). By using multiple low doses in a particular strain of mice, it is possible to produce insulitis followed later by diabetes – a model for Type 1 (juvenile-onset) diabetes in humans.

stress-fibres Bundle of microfilaments and other proteins found in fibroblasts, particularly slow-moving fibroblasts cultured on rigid substrata. Shown to be contractile; have a periodicity reminiscent of the **sarcomere**. Anchored at one end to a **focal adhesion**, although sometimes seem to stretch between two focal adhesions.

stress-induced proteins Alternative and preferable name for **heat-shock proteins** of eukaryotic cells, which emphasizes that the same small group of proteins is stimulated both by heat and various other stresses.

striated border Obsolete term for the apical surface of an epithelium with microvilli.

striated muscle Muscle in which the repeating units (**sarcomeres**) of the contractile **myofibrils** are arranged in register throughout the cell, resulting in transverse or oblique striations observable at the level of the light microscope, e.g. the voluntary (skeletal) and cardiac muscle of vertebrates.

stringency *Low stringency; high stringency; stringency wash* In nucleic acid **hybridization**, the labelled **probe** is used to label matching sequences by base-pairing. Unbound probe is removed through a series of stringency washes. Low stringency washing (low temperature, high ionic strength) allows some mismatching of probe and target, and thus the detection of similar sequences at some cost in specificity. By contrast, high stringency conditions allow only closely matching sequence to remain base-paired.

striosome One of the components of the striatum region of the basal ganglia in the brain. The striatum is made of two parts, the matriosome and the striosome. Both receive input from the cortex (mostly frontal) and from dopaminergic (DA) neurons, but the striosome projects principally to DA neurons in the ventral tegmental area (VTA) and the substantia nigra (SN). Hypothesized to act as a reward predictor, allowing the DA signal to compute the difference between the expected and received reward. The matriosome projects back to the frontal lobe.

strobila *Pl.* strobilae A part or structure that buds to form a series of segments. **1.** in Scyphozoa, the sessile larval stage that produces medusoids by transverse fission. **2.** InCestoda, the segmented body consisting of proglottides. *cf* **strobilus**.

strobilus *P.* strobili **1.** The cone-like reproductive structures of most gymnosperms and some pteridophytes. **2.** An angiosperm inflorescence of cone-like appearance. Cf **strobila**.

stroma 1. The soluble, aqueous phase within the chloroplast, containing water-soluble enzymes such as those of the **Calvin–Benson cycle**. The site of the **dark reaction** of photosynthesis. **2.** Loose connective tissue with few cells.

stromal cell Resident cell of loose fibrous connective tissue. Relatively non-committal term.

stromatolites Large, rounded, multilayered fossils found in rocks dating from the Early Archaean (3.5 mya), possibly before modern cyanobacteria evolved. The layers are produced as a result of photosynthetic activity of bacterial colonies causing the precipitation of calcium carbonate that is combined with other sedimentary material trapped within the mucilage surrounding the colony. New layers of bacteria then develop on the outer surface. Modern equivalents are produced by cyanobacteria under certain circumstances, e.g. on the beaches at Shark Bay in W. Australia.

stromelysins Metallopeptidases involved in breaking down the extracellular matrix (matrix metallopeptidases, MMPs). Stromelysin-1 (EC 3.4.24.17) is MMP-3; stromelysin-2 (EC 3.4.24.22) is MMP-10; stromelysin-3 (EC 3.4.24.-) is MMP-11.

strong promoter Promoter that, when bound by a **transcription factor**, strongly activates expression of the associated gene. The term is widely used but lacks precision.

Strongylocentrotus purpuratus Common sea urchin. Echinoderms are popular tools for developmental biology because the early embryo is transparent and cell movements can easily be observed. Sea urchin eggs are available in large numbers and were a convenient material for early biochemical studies on histones and mRNA. Because they are relatively large, they have also been convenient for various electrophysiological studies.

strongyloidiasis *Strongyloidosis* Intestinal infestation of man with the nematode worm, *Strongyloides stercoralis.*

strophanthin Mixture of glycosides from *Strophanthus kombe* (a tropical liana) that has properties similar to digoxin and **ouabain**. See **digitalis**.

STRP **Short tandem repeat** polymorphism.

structural gene A gene that codes for a product (e.g. an enzyme, structural protein, tRNA), as opposed to a gene that serves a regulatory role.

structure–activity analysis *Structure–activity relationship; SAR* Study in which systematic variation in the structure of a compound is correlated with its activity, in an attempt to determine the characteristics of the (receptor) site at which it acts.

structure–activity relationship *SAR* Association between specific aspects of molecular structure and defined biological action. See also **QSAR**.

struvite A mineral composed of orthorhombic crystals of magnesium ammonium phosphate. Struvite stones in the kidney are composed of a struvite–carbonate–apatite matrix and are associated with urinary infections, particularly with urease-producing bacteria, including *Ureaplasma urealyticum* and *Proteus*.

strychnine **Alkaloid** obtained from the Indian tree *Strychnos nux-vomica*; specific blocking agent for the action of the amino acid transmitter glycine. Convulsive effects of strychnine are probably due to its blockage of inhibitory synapses onto spinal cord motoneurons.

STX 1. See **saxitoxin**. **2.** *Shiga* toxin. See **shiga-like toxin**. **3.** Sialyltransferase.

Stylonychia mytilus Large ciliate protozoan of the Order Hypotrichida which has compound cilia (cirri) that can be used for walking or swimming.

S-type lectins *Galectins* One of two classes of **lectin** produced by animal cells. The classification of animal lectins into two classes (the other being the C-type), was originally proposed by K. Drickamer. The carbohydrate-binding activity of the S-type lectins requires their cysteines to have free thiols and does not need divalent cations (*cf*. **C-type lectins**). They mostly have molecular masses in the range 14–16 kDa, and often form dimers and higher oligomers. The carbohydrate-recognition domain contains a number of critically conserved amino acids and largely binds to β-galactosides. S-type lectins occur as cytoplasmic proteins and lack a signal sequence for secretion, yet do exist extracellularly.

subacute Description of a disease that progresses more rapidly than a chronic disease and more slowly than an acute one.

subacute sclerosing panencephalitis *SSPE* Chronic progressive illness seen in children a few years after measles infection and involving demyelination of the cerebral cortex. Virus apparently persists in brain cells: usually considered a **slow virus** disease.

suberin Fatty substance, containing long-chain fatty acids and fatty esters, found in the cell walls of cork cells (phellem) in higher plants. Also found in the **Casparian band**. Renders the cell wall impervious to water.

subfragment 1 of myosin See **S1**.

subgenomic RNA *sgRNA* Copies of viral RNA from positive-strand RNA viruses that are truncated at the 5′ ends, but have the same 3′ ends, so that the 5′ end of the RNA has the start codon of a downstream gene on the complete viral genomic RNA. (Because more than one gene is encoded on the genomic RNA, and only the first open reading frame (ORF) would normally be translated, this allows the downstream genes to be expressed.)

submitochondrial particle Formed by sonicating mitochondria. Small vesicles in which the inner mitochondrial membrane is inverted to expose the innermost surface.

substance P A **vasoactive intestinal peptide** (1348 Da) found in the brain, spinal ganglia and intestine of vertebrates. Induces vasodilation, salivation and increases capillary permeability. Sequence: RPKPQFFGLM.

substantia nigra Area of darkly pigmented dopaminergic neurons in the ventral midbrain thought to control movement and damaged in **Parkinsonism**.

substrate Substance that is acted upon by an enzyme: one can also speak of a suitable substrate for maintaining a species of bacterium, the compound is one that can support cell growth.

substratum The solid surface over which a cell moves, or upon which a cell grows: should be used in this sense in preference to **substrate**, to avoid confusion.

subtilin Cationic pore-forming **lantibiotic** produced by *Bacillus subtilis*. Acts preferentially on Gram-positive microorganisms but the producer cells have immunity mediated by the four genes *spaIFEG*. SpaFEG is an **ABC transporter**-2 subfamily member, a multidrug resistance protein. SpaI is a membrane-localized lipoprotein that may be a subtilin-intercepting protein.

subtilisin *EC 3.4.21.62* Extracellular serine endopeptidase produced by *Bacillus* spp.

subtilysin Haemolytic surfactant produced by *Bacillus subtilis*; hexapeptide linked to a long-chain fatty acid.

subtractive hybridization *Subtraction cloning* Technique used to identify genes expressed differentially between two tissue samples. A large excess of **mRNA** from one sample is hybridized to cDNA from the other, and the double-stranded hybrids removed by physical means. Remaining cDNAs are those not represented as RNA in the first sample, and thus presumably expressed uniquely in the second. To improve specificity, the process is often repeated several times. See also **differential screening**.

subunits Components from which a structure is built; thus myosin has six subunits, microtubules are built of tubulin subunits. In some cases it may be more informative to speak of **protomers**.

succinate *Ethane dicarboxylic acid* Intermediate of the **tricarboxylic acid cycle** and **glyoxylate cycle**.

succinate dehydrogenase *SDH* **1.** EC 1.3.5.1. Succinate-coenzyme Q reductase; complex II in electron transport; succinate-ubiquinone oxidoreductase. An enzyme complex of four subunits located in the inner mitochondrial membrane. It has two main activities, as part of the citric acid cycle, oxidizing succinate into fumarate while passing electrons on to FAD and as complex II of the electron transport chain, which uses electrons freed from succinate, to reduce ubiquinone to ubiquinol. **2.** EC 1.3.99.1 Fumarate reductase; Fumarate dehydrogenase. Oxidizes succinate to fumarate with the involvement of an acceptor. A bacterial enzyme or degraded form of mitochondrial SDH.

succinyl CoA An intermediate product in the **tricarboxylic acid cycle**.

succinylcholine *Suxamethonium chloride* Cholinergic antagonist and therefore a skeletal muscle relaxant.

sucralose An artificial sweetener made by the selective chlorination of sucrose. Much sweeter (500–600-fold) than sucrose and non-digestible. In Europe additive code number E955.

sucrase See **invertase**.

sucrose *Table sugar* Non-reducing disaccharide, α-D-glucopyranosyl-β-D-fructofuranose.

Sudan stains Histochemical stains used for lipids.

sudden oak death A disease of oak trees caused by the fungus *Phytophthora ramorum*.

sudorific Connected with the secretion of sweat or a drug that stimulates sweating.

sugars See separate entries and Table S3.

TABLE S3. Sugars

Monosaccharides

pentoses

L-arabinose   D-ribose 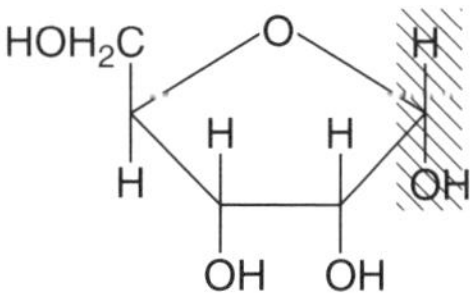   D-xylose 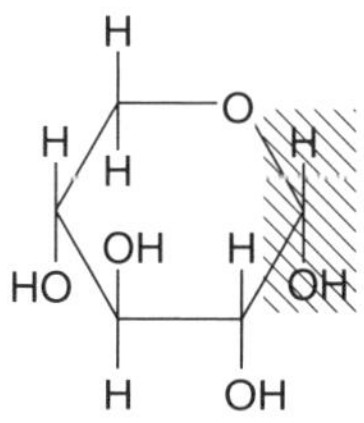   2-deoxy-D-ribose MW 134 1

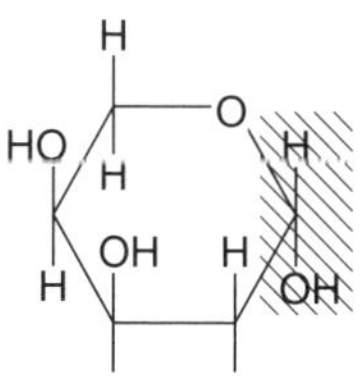

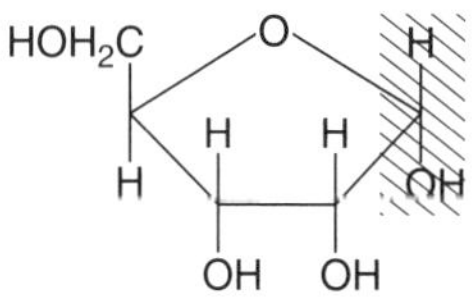

hexoses

D-fructose   D-galactose 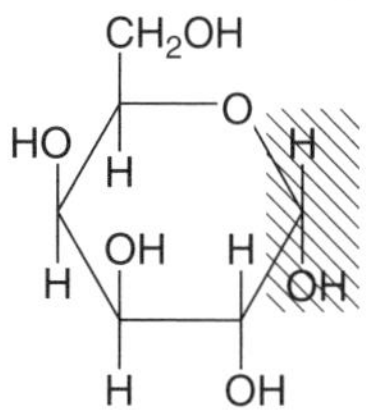   D-gulucose 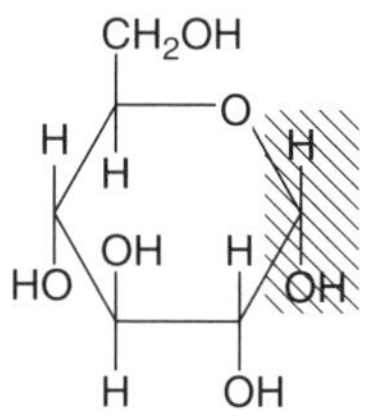   D-mannose

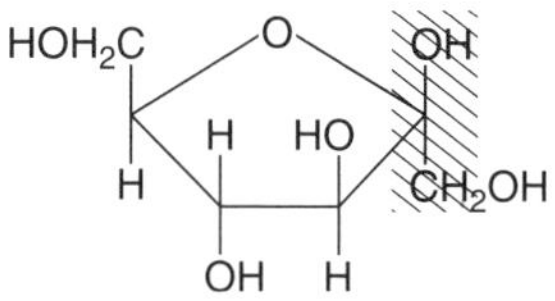

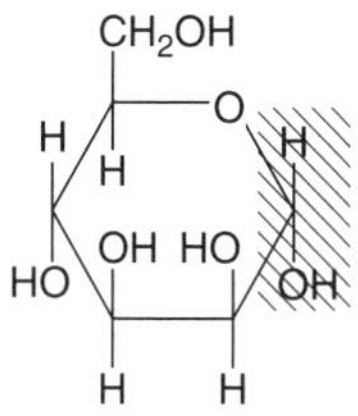

TABLE S3. (Continued)

Free amino-sugars are not found in structural oligosaccharides but *N*-acetyl aminohexoses are widely distributed. Most common are:

N-acetylgalactosamine   N-acetylglucosamine

Other common components of structural oligosaccharides are:

fucose   sialic acids (*N*-acetyl-neuraminic acid)

Hexose derivatives found in proteoglycans also include:

D-glucuronic acid   muramic acid

Sulphated derivatives of *N*-acetyl aminohexoses are also widespread and include the 4- and 6-sulphate esters of *N*-acetyl glucosamine and *N*-acetyl galactosamine.

Disaccharides and polysaccharides

These are fully specified by the residue names, sequence, bond-direction and the position numbers of the carbon atoms giving rise to the linkage. The configuration around the glycosidic carbon is also specified as alpha or beta.

The list includes only the most common compounds found in metabolic pathways and in structural molecules. The structures are presented as Haworth models and it should be noted the configuration at the carbon which carries the carbonyl oxygen is not determined unless the hydroxyl-group takes part in a glycosidic linkage, which it always does in higher oligomers. The convention for depicting glycosidic linkages is: glycosyl carbon → acceptor hydroxyl. *n* Configuration not defined in free molecule.

sulcus *Pl.* sulci A groove or furrow, e.g. on the surface of the cerebrum in mammals.

sulfa drugs *Sulpha drugs* General term for **sulfonamides**.

sulfasalazine *Sulphasalazine* Combination of aminosalicylic acid and the **sulfonamide** sulfapyridene. Used to treat inflammatory bowel disease and some cases of rheumatoid arthritis.

sulfinpyrazone *Formerly sulphinpyrazone in UK* Pyrazole compound related to phenylbutazone but without anti-inflammatory activity. Has no effect on platelet aggregation *in vitro*, but inhibits platelet adhesion and release reactions. Inhibits uric acid resorption in the proximal convoluted tubule of the kidney.

sulfonamides Synthetic bacteriostatic antibiotics derived from sulphanilamide (a red dye) with a wide spectrum of activity against most Gram-positive and many Gram-negative organisms. Sulfonamides inhibit multiplication of bacteria by acting as competitive inhibitors of p-aminobenzoic acid in the folic acid metabolism cycle. In UK formerly sulphonamides, but sulfonamide is the British Approved Name (BAN).

sulfur, sulfo- The British spelling, sulphur, sulpho- is used throughout except where the British Approved Name (BAN) for drugs is the sulf- spelling.

sulphatases *EC 3.1.6* An **esterase** in which one of the substituents of the substrate is a sulphate group. Thus IDS (iduronate-2-sulphatase; EC 3.1.6.13) is a lysosomal exo-sulphatase that is involved in the degradation of heparan sulphate and dermatan sulphate.

sulpholipids Lipids in which the polar head group contains sulphate species. Synthesized in the Golgi complex.

sulphonylurea receptor *SUR* The sulphonylurea receptor-1 (SUR-1) regulates glucose-induced insulin secretion by controlling K^+-ATP channel activity of the pancreatic beta-cell membrane. The ATP-sensitive potassium (KATP) channels in neuron and neuroendocrine cells consist of four pore-forming Kir6.2 and four regulatory sulphonylurea receptor (SUR1) subunits. SURs contain two nucleotide-binding folds (NBFs) that sense changes in the metabolic status ([ATP]/[ADP]) of the cell. Tritiated **glibenclamide** is often used to detect their tissue distribution.

sulphonylureas Group of drugs which act by augmenting insulin secretion. Used in the treatment of Type II diabetes. Examples include glibenclamide, which is the sulphonylurea most commonly used in experimental studies. See **sulphonylurea receptor**.

sulphydryl reagent Compounds that bind to SH groups. Include *p*-chlormercuribenzoate, *N*-ethyl maleimide, iodoacetamide. Very important in studies of protein structure.

sumatriptan Antagonist of vascular $5HT_1$ receptors, used for treatment of migraine.

SUMO *Small ubiquitin-related modifier proteins* See **sumoylation**.

sumoylation Post-transcriptional modification of a protein by the conjugation of SUMO (small ubiquitin-related modifier) proteins: stabilizes some proteins and may alter subcellular localization. Three different SUMO proteins are conjugated to proteins, SUMO-1, SUMO-2 and SUMO-3. The SUMO-2 and SUMO-3 genes are closely related, with 86% sequence identity, while SUMO-1 is less closely related, with about 50% sequence identity with SUMO-2 and SUMO-3. SUMO-1 conjugates as a monomer, while SUMO-2 and SUMO-3 are conjugated to proteins as higher molecular weight polymers. Targets include p53, ran-Gap; several transcription factors are regulated by sumoylation, including C/EBP proteins and c-Myb.

Suns *Sun-1, Sun-2* Integral membrane proteins of the inner nuclear membrane that share a Sun domain (Sad1/UNC-84 homology domain) and interact with **lamins** on the inner (nuclear) face of the inner nuclear membrane and with **nesprins** in the outer nuclear membrane. Originally cloned by homology with *C. elegans* protein UNC-84 that is involved in nuclear migration/positioning.

superantigen Antigens, mostly of microbial origin, which activate all those T-lymphocytes that have a T-cell receptor with a particular Vβ sequence; as a consequence superantigens activate large numbers of T-cells. Are presented on MHC class II but are not processed and, though they bind with high affinity, not in the groove of the MHC molecule where peptides are normally bound. Presentation is not MHC-restricted. *Staphylococcal* enterotoxins are the best known superantigens and stimulate $CD4^+$ T-cells in humans. The *Mls* gene product in mice can act as a self-superantigen.

supercoiling In circular DNA or closed loops of DNA, twisting of the DNA about its own axis changes the number of turns of the double helix. If twisting is in the opposite direction to the turns of the double helix, i.e. anticlockwise, the DNA strands will either have to unwind or the whole structure will twist or supercoil – termed negative supercoiling. If twisting is in the same direction as the helix, clockwise, which winds the DNA up more tightly, positive supercoiling is generated. DNA that shows no supercoiling is said to be relaxed. Supercoiling in circular DNA can be detected by electrophoresis because supercoiled DNA migrates faster than relaxed DNA. Circular DNA is commonly negatively supercoiled, and the DNA of eukaryotes largely exists as supercoils associated with protein in the **nucleosome**. The degree of supercoiling can be altered by **topoisomerases**.

superhelix A supercoil of a molecule, like DNA, that is already coiled.

superoxide *Superoxide radical* Term used interchangeably for the superoxide anion $.O_2^-$, or the weak acid HO_2^-. Superoxide is generated both by prokaryotes and eukaryotes and is an important product of the **metabolic burst** of neutrophil leucocytes. A very active oxygen species, it can cause substantial damage and may be responsible for the inactivation of plasma antiproteases that contributes to the pathogenesis of emphysema.

superoxide dismutase *SOD; EC 1.15.1.1* Any of a range of metalloenzymes that catalyse the formation of hydrogen peroxide and oxygen from superoxide, and thus protect against superoxide-induced damage. Usually have either iron or manganese as the metal cation in prokaryotes, copper or zinc in eukaryotes.

supershift Phenomenon in bandshift assays where the reduction in mobility on a gel induced by a binding interaction with a protein is enhanced by the addition of an antibody to the protein (or another interacting protein). Net result is that the mobility of the band of interest is further decreased (shifted).

supervillin *Archvillin* An actin-binding protein (205 kDa) that has homology to the **gelsolin/villin** family and is implicated in various signalling pathways, being an androgen receptor co-regulator. Also has myosin II-binding activity. Expressed in muscle in large amounts; archvillin (250 kDa) is a splice variant of supervillin found in smooth muscle.

suppression subtractive hybridization *SSH* A method developed for cDNA comparisons, based on PCR suppression by inverted terminal repeats (PS-effect). In complex mixtures, the PS-effect allows precise amplification only of molecules that are flanked by different adapters at opposing termini (asymmetrically flanked molecules). This principle is used in Suppression Subtractive Hybridization (SSH), where the molecules of interest, those unique to the test sample, are driven to the asymmetrically flanked state and selectively amplified.

suppressor factor 1. Factors released by T-suppressor cells. 2. See **suppressor mutation**, **ochre suppressor**, **opal suppressor**.

suppressor mutation Mutation that alleviates the effect of a primary mutation at a different locus. May be through almost any mechanism that can give a primary mutation, but perhaps the most interesting class are the **opal**, **amber** and **ochre supressors**, where the anticodon of the tRNA is altered so that it misreads the termination codon and inserts an amino acid, preventing premature termination of the peptide chain.

suppressor T-cells See **T-suppressor cell**.

suprachiasmatic nucleus *SCN* A region in the hypothalamus, immediately above the optic chiasm. The SCN generates a circadian rhythm of neuronal activity, which regulates many different body functions. **Melanopsin**-containing ganglion cells in the retina have a direct connection to the SCN.

SUR *Sulphonylurea receptor; SUR1, SUR2* **ABC protein** that interacts with K-ATP (Kir6.1 and Kir6.2) channels and regulates the response of the cell to glucose levels by sensing intracellular ATP concentration. The channel is formed by four SUR and four Kir subunits; the presence of both is essential for function.

suramin A polysulphonated naphthylurea that uncouples G-proteins from receptors, inhibits phospholipase D and inhibits binding of EGF, PDGF to cell surface receptors. A $P2_x$ and $P2_y$ purinergic receptor antagonist. Has been used in treatment of trypanosomiasis and more recently for treatment of hormone-refractory prostate carcinoma.

surface envelope model A way of treating the hydrodynamics of a ciliary field – by considering the whole surface of the ciliate to have an undulating surface. The undulations arise because of **metachronism**.

surface plasmon resonance Alteration in light reflectance as a result of binding of molecules to a surface from which total internal reflection is occurring. Used in the Biacore (Pharmacia Trademark) machine that detects the binding of ligand to surface-immobilized receptor or antibody.

surface potential The electrostatic potential due to surface charged groups and adsorbed ions at a surface. It is

usually measured as the zeta potential at the Helmholtz slipping plane outside the surface.

surface-active compound Usually, in biological systems, means a detergent-like molecule that is amphipathic and that will bind to the plasma membrane, or to a surface with which cells come in contact, altering its properties from hydrophobic to hydrophilic, or *vice versa*.

surfactant A **surface-active compound**; the best known example of which is the lung surfactant (see **surfactant proteins A and D**) that renders the alveolar surfaces hydrophobic and prevents the lung filling with water by capillary action. The lung surfactant is produced just at parturition, and it has often been speculated that deficiencies in surfactant metabolism might cause cot death.

surfactant proteins A and D *SP-A; SP-D* Proteins, members of the **collectin** family, found in the respiratory tract that appear to play an important role in mammalian first-line host defence. SP-A is oligomeric, consisting of 18 protomers with collagen and lectin-like domains, and recognizes glycoconjugates, lipids and protein determinants on both host cells and invading microorganisms. SP-D also occurs in the gastric mucosa at the luminal surface.

surfactin A cyclic lipopeptide biosurfactant produced by *Bacillus subtilis*. Because it can be produced from cheap feedstock by biofermentation, it is being investigated for use in bioremediation.

surrogate marker A readily measured parameter that is associated with a disease state and can be monitored as a substitute for more complex clinical signs that may be less easy to quantify.

survivin *Baculoviral IAP repeat-containing protein (BIRC-5)* Protein in tumour cells that blocks apoptosis, possibly by inhibiting **caspases**. Related to **IAPs**. Several splice variants have been described.

Sus scrofa Domestic pig.

sushi domains *Complement control protein (CCP) modules; short consensus repeats (SCR)* Domains identified in a wide variety of complement and adhesion proteins; involved in protein–protein and protein–ligand interactions. Contain four cysteines forming two disulphide bonds in a 1-3 and 2-4 pattern. See **sushi peptides**.

sushi peptides Peptides derived from the LPS-binding domains of an LPS-sensitive serine peptidase, Factor C, from the horseshoe crab (*Carcinoscorpius rotundicauda*); have potent antibacterial properties. See **sushi domains**.

suspensor cell Plant cell linking the growing embryo to the wall of the embryo sac in developing seeds.

sustentacular Something that supports or maintains. Sustentacular cells (Sertoli cells) of the testis are involved in support and possibly nutrition of developing sperm.

suxamethonium *Succinylcholine* A depolarizing neuromuscular blocking agent that is composed of two acetylcholine molecules linked by their acetyl groups, and binds to acetylcholine receptors, acting as an agonist. Unlike acetylcholine persists for long enough to cause the loss of electrical excitability.

SV2 1. An integral membrane glycoprotein of synaptic vesicles, SV2 has isoforms A, B and C and is found in vertebrate neuronal and endocrine tissues, where it is thought to regulate synaptic vesicle exocytosis. Similar proteins have been found in invertebrates. SV2 is the protein receptor for botulinum neurotoxin A. **2.** A line of immortal non-tumorigenic human lung fibroblasts, MRC-5 SV2 cells.

SV3T3 Swiss 3T3 cells transformed with **SV40**.

SV40 *Simian virus 40* A small DNA **tumour virus**, member of the **Papovaviridae**. Isolated from monkey cells, which were being used for the preparation of **poliovirus** vaccine, and originally named vacuolating agent owing to a cyto-pathic effect observed in infected cells. Found to induce tumours in newborn hamsters. In culture, transforms the cells of many non- and semipermissive species, including mouse and human. See also **T-antigen**.

S value Svedberg unit. See **sedimentation coefficient**.

Svedberg unit The unit applied to the sedimentation coefficient of a particle in a high-speed or ultracentrifuge. The unit S is calculated as follows, S = rate of sedimentation $\times 1/\rho^2 r$, where ρ is the speed of rotation in rad/s and r is the radius to a chosen point in the centrifuge tube. 1 Svedberg unit is defined as a velocity gradient of 10^{-13} s. Named after a pioneer of the ultracentrifuge. The units are non-additive: a particle formed from two 5S particles will not have a sedimentation coefficient of 10S.

swainsonine Fungal indolizidine alkaloid that is a reversible inhibitor of lysosomal alpha-**mannosidase** and of the Golgi complex alpha-mannosidase II which is involved in processing the oligosaccharide chains of glycoproteins.

SWI/SNF complex The SWI/SNF complex remodels nucleosome structure in an ATP-dependent manner. In yeast, the SWI/SNF chromatin-remodelling complex comprises 11 tightly associated polypeptides (SWI1, SWI2, SWI3, SNF5, SNF6, SNF11, SWP82, SWP73, SWP59, SWP61 and SWP29). SWP59 and SWP61 are encoded by the *ARP9* and *ARP7* genes, respectively, which encode members of the actin-related protein (ARP) family. The similarity of ARP7 and ARP9 to the **heat-shock protein** and HSC family of ATPases suggests the possibility that chromatin remodelling by SWI/SNF may involve chaperone-like activities.

Swiss 3T3 cells An immortal line of fibroblast-like cells established from whole trypsinized embryos of Swiss mice (not an inbred stock) under conditions that favour establishment of cells with low saturation density in culture.

switch regions The nucleotide sequences in heavy chain immunoglobulin genes located in the introns at the 5′ end of each CH locus concerned with DNA recombination events that lead to changes in the type of heavy chain produced by a B-cell, e.g. IgM to IgG switching. These regions are highly conserved sequences. See **isotype switching**.

syk Tyrosine kinase (72 kDa), an effector of the B-cell receptor-signalling pathway. Contains two tandem SH2 domains through which it interacts with **ITAM** motif. More widely distributed than **zap70**, and important in signalling in both myeloid and lymphoid cells.

symbiont One of the partners in a symbiotic relationship.

symbiosis Living together for mutual benefit. See **symbiont, symbiotic algae**.

symbiotic algae Algae (often *Chlorella* spp.) that live intracellularly in animal cells (e.g. endoderm of *Hydra viridis*). The relationship is complex, because lysosomes do not fuse with the vacuoles containing the algae, and the growth rates of both cells are regulated to maintain the symbiosis. There is considerable strain-specificity. The term is imprecise, since there are many other symbiotic algae (as in lichens) where the relationship is different.

sympathetic nervous system One of the two divisions of the vertebrate **autonomic nervous system** (the other being the **parasympathetic nervous system**). The sympathetic preganglionic neurons have their cell bodies in the thoracic and lumbar regions of the spinal cord, and connect to the paravertebral chain of sympathetic ganglia. Innervate heart and blood vessels, sweat glands, viscera and the adrenal medulla. Most sympathetic neurons, but not all, use noradrenaline as a postganglionic neurotransmitter.

sympathetic ophthalmia An autoimmune response in which there is inflammatory disease in the undamaged eye following perforating injury to the other eye.

sympathomimetic Mimicking the sympathetic nervous system. See **sympathomimetic drugs**.

sympathomimetic drugs Drugs that mimic the stimulation of the sympathetic nervous system to produce, for example, tachycardia and increased output from the heart (e.g. isoprenaline and dobutamine) or bronchodilatation and vasodilatation (salbutamol).

sympatric speciation Development of a new species without (geographical) isolation. Controversial. See **speciation**.

symplast The intracellular compartment of plants, consisting of the cytosol of a large number of cells connected by **plasmodesmata**. See **apoplast**.

symplectic metachronism See **metachronism**.

symplesiomorphic See **apomorphic**.

symport A mechanism of transport across a membrane in which two different molecules move in the same direction. Often, one molecule can move up an electrochemical gradient because the movement of the other molecule is more favourable (see **facilitated diffusion**). Example: the sodium/glucose co-transport. See **antiport, uniport**.

synapomorph **Apomorphic** features possessed by two or more taxa in common. Phylogenic trees are based upon identifying groups united by synapomorphies.

synapse A connection between **excitable cells**, by which an excitation is conveyed from one to the other. **1.** Chemical synapse: one in which an **action potential** causes the exocytosis of neurotransmitter from the presynaptic cell, which diffuses across the synaptic cleft and binds to **ligand-gated ion channels** on the postsynaptic cell. These ion channels then affect the resting potential of the postsynaptic cell. **2.** Electrical synapse: one in which electrical connection is made directly through the cytoplasm, via **gap junctions**. **3.** Rectifying synapse: one in which action potentials can only pass across the synapse in one direction (all chemical and some electrical synapses). **4.** Excitatory synapse: one in which the firing of the presynaptic cell increases the probability of firing of the postsynaptic cell. **5.** Inhibitory synapse: one in which the firing of the presynaptic cell reduces the probability of firing of the postsynaptic cell.

synapsid One of the so-called mammal-like reptiles of the Carboniferous, Permian and Triassic periods. Although many had characteristics in common with mammals, none of them were actually reptiles.

synapsins Family of phosphoproteins associated with synaptic vesicles and implicated in control of release. Synapsin Ia (84 kDa) and Ib (80 kDa) are alternatively spliced variants as are synapsins IIa (74 kDa) and IIb (55 kDa). Can be phosphorylated by several **protein kinases**. Thought to be involved in regulation of neurotransmitter release at **synapses**.

synapsis 1. The specific pairing of the chromatids of homologous chromosomes during **prophase** I of meiosis. It allows **crossing-over** to take place. **2.** Process that brings the ends of double-strand breaks in DNA together, prior to end joining in **NHEJ**. Synapsis results in the autophosphorylation of DNA-PKcs, which is required to make the DNA ends available for ligation.

synaptic cleft The narrow space between the presynaptic cell and the postsynaptic cell in a chemical **synapse**, across which the **neurotransmitter** diffuses.

synaptic plasticity Change in the properties of a synapse, usually in the context of learning and memory. Very few synapses provide simple 1:1 transfer of **action potentials**, and very small changes in the efficiency of a synapse (usually mediated by changes in either the pre- or postsynaptic membrane) can have profound influences on the electrical properties of a neuronal circuit. See also **neuronal plasticity**.

synaptic transmission The process of propagating a signal from one cell to another via a **synapse**.

synaptic vesicle Intracellular vesicles found in the presynaptic terminals of chemical synapses, which contain **neurotransmitter**.

synaptobrevin *v-SNARE; VAMP-2* Small integral membrane proteins (16.7 kDa) of synaptic vesicles. Two isoforms, VAMP-1 and VAMP-2 are known. Bind SNAPs and also interact with target-SNARE (syntaxin). Cleaved by clostridial toxins encoding zinc endopeptidases, such as tetanus toxin and botulinum toxin, blocking synaptic release.

synaptogenesis Formation of a **synapse**.

synaptogyrin *p29* Integral component (29 kDa) of synaptic vesicle with some similarity to synaptophysin. Has four transmembrane domains.

synaptojanin Protein of the vertebrate nerve terminal (145 kDa) that seems to participate with **dynamin** in the process of vesicle recycling. Has phosphatase activity and is a member of the inositol-5-phosphatase family. Amino-terminal region has homology with yeast Sac1 (involved in phospholipid metabolism) and C-terminal region has proline-rich sequences that probably interact with **SH3** domains of **amphiphysin** and **GRB-2**.

synaptomorphic See **apomorphic**.

synaptonemal complex Structure, identified by electron microscopy, lying between chromosomes during **synapsis**; consists of two lateral plates closely apposed to the chromosomes and connected to a central plate by filaments. It appears to act as a scaffold and is essential for **crossing-over**.

synaptophysin Abundant glycoprotein component of synaptic vesicle membranes composed of a 38-kDa subunit

that spans the membrane four times and has both its N- and C-termini located cytoplasmically. Its transmembrane organization and putative quaternary structure resemble the molecular topology of **gap junction** proteins, **connexins**.

synaptopodin An actin-associated proline-rich protein (100 kDa) found in kidney podocytes and a subset of mature telencephalic dendritic spines of neurons; a regulator of RhoA signalling and cell migration. Blocks Smurf1-mediated ubiquitination of RhoA, thereby preventing the targeting of RhoA for proteasomal degradation.

synaptoporin *SPO* Putative channel protein of synaptic vesicles, and a member of the **synaptophysin/connexin** superfamily. It has 58% amino acid identity to synaptophysin, with highly conserved transmembrane segments but a divergent cytoplasmic tail.

synaptosome A subcellular fraction prepared from tissues rich in chemical **synapses**, used in biochemical studies. Consists mainly of vesicles from presynaptic terminals.

synaptotagmin *p65* Calcium-binding synaptic vesicle protein that binds acidic phospholipids and recognizes the cytoplasmic domain of the **neurexins**. Functions as a Ca^{2+} sensor that facilitates SNARE-mediated membrane fusion.

SynCAM *Synaptic cell adhesion molecule* Molecule involved in synaptic adhesion, along with β-**neurexin**/**neuroligin**, and involved in triggering presynaptic differentiation.

synchronous cell population A culture of cells that all divide in synchrony. Particularly useful for certain studies of the cell cycle, cells can be made synchronous by depriving them of essential molecules, which are then restored. Synchronization breaks down after a few cycles, however, as individual cells have unique division rates.

syncoilin An intermediate filament-type III protein (53 kDa) found in striated and cardiac muscle; binds to α-dystrobrevin and may link the desmin-associated intermediate filament network of muscle with the dystrophin-associated protein complex (DAPC).

syncolin Microtubule-associated protein (280 kDa) found in chicken erythrocytes. Has some similarities with MAP-2, but thought to be distinct. Has not appeared in the literature since first described in 1991. Not to be confused with **syncollin**.

syncollin Protein (13 kDa) found within zymogen granules that is required for efficient regulated exocytosis. Normally exists as a doughnut-shaped homo-oligomer (possibly a hexamer) in close association with the luminal surface of the zymogen granule membrane.

syncytiotrophoblast Syncytial layer that forms the outermost foetal layer in the placenta and is thus the interface with maternal tissue. Has invasive capacity, though in a regulated manner.

syncytium An **epithelium** or tissue in which there is cytoplasmic continuity between the constituent cells.

syndecan An integral membrane proteoglycan (250–300 kDa) associated largely with epithelial cells. The core protein of 294 amino acids has an extracellular domain of 235 amino acids and a single transmembrane domain of 25 amino acids. The extracellular domain has up to three heparan sulphate and two chondroitin or dermatan sulphate chains plus an N-linked oligosaccharide. The heparan sulphate chains bind to several proteins of the extracellular matrix, including collagens, fibronectin and **tenascin**. The cytoplasmic domain is thought to interact with actin filaments. Its name is derived from the Greek *syndein*, to bind together. Ligation of *N*-syndecan (syndecan-3) by heparin-binding growth-associated molecule increases phosphorylation of c-src and **cortactin**, and *N*-syndecan may act as a neurite outgrowth receptor.

synemin An intermediate filament-associated protein (230 kDa) isolated from avian smooth muscle, but homologue also found in mammalian muscle. Co-localizes with **desmin** near myofibrillar **Z-discs**. Three synemin isoforms, of 180 kDa (H), 150 kDa (M) and 41 kDa (L), are produced by alternative splicing of the pre-mRNA, and are regulated differentially during development.

synexin Annexin VII. See Table A3.

syngamy Fusion of two haploid gametic nuclei to form the diploid nucleus of the zygote.

syngeneic Organisms that are antigenically identical; monozygotic twins or highly inbred strains of animals. Thus cells injected into a syngeneic host will not be rejected because they are histocompatible (do not differ in histocompatibility antigens).

synkaryon A somatic hybrid cell in which chromosomes from two different parental cells are enveloped in a single nucleus.

synomone See **allomone**.

synoviocytes Fibroblastic cells of the synovial membrane (lining) that produce synovial fluid and the extracellular matrix of the bearing surface of the joint.

synovium Connective tissue that forms the bearing surface of the joint and that is eroded in arthritis.

synphilin-1 A neural protein that interacts with **synuclein**, a protein which has been genetically linked to **Parkinson's disease**. Though synphilin-1 shares little similarity with other proteins in the public databases, synphilin comprises several interaction domain motifs. *In vivo* binding interaction of synphilin with synuclein was demonstrated by immunoprecipitation studies on brain extracts. Co-expression of both proteins in cultured cells results in cytoplasmic inclusions similar to the **Lewy bodies** characteristic of neurons from patients with Parkinson's disease.

syntaxins *t-SNAREs* A family of receptors for intracellular transport vesicles. Members of the syntaxin family have sizes ranging from 30 to 40 kDa; all have a hydrophobic C-terminal region that anchors the protein on the cytoplasmic surface of cellular membranes, a central conserved region and a variable N-terminal cytoplasmic domain. Syntaxin 1 is mainly expressed in brain tissue and is thought to function specifically in neurotransmitter release, whereas syntaxin 2, 3 and 4 have a wider tissue distribution. Syntaxin-1a and -4 preferentially interact with beta- and gamma-**taxilin**, respectively. Syntaxin 5 is a Golgi-localized **SNARE** protein that has been shown to be required for ER–Golgi traffic in yeast.

syntenic Syntenic genes lie on the same chromosome. Some loci are syntenic in both man and mouse, others are not.

synthetase Enzymes of Class 6 in the **E classification**; catalyse synthesis of molecules, their activity being coupled to the breakdown of a nucleotide triphosphate.

syntrophins A family of proteins that form a complex with **dystrophin**, **dystrobrevin** and diacylglycerol kinase (DGK)-zeta at the plasma membrane of muscle and nerve. Different isoforms have distinct expression patterns and are differentially affected by loss of dystrophin anchorage and denervation in human neuromuscular disease. DGK-zeta, syntrophin and Rac1 form a regulated signalling complex that controls polarized outgrowth in neuronal cells. There is some evidence that syntrophin is also an actin-binding protein. See **dystrophin-associated protein complex**.

synuclein Family of proteins (3 genes coding α-, β- and γ-synuclein) sharing structural resemblance to apolipoproteins, abundant in neuronal cytosol and particularly in presynaptic terminals. Have been implicated in various diseases, including **Alzheimer's disease**, **Parkinsonism** and breast cancer. In Alzheimers, α-synuclein is a component of plaque amyloid; in Parkinsonism an α-synuclein allele is linked to various familial cases and the protein accumulates in **Lewy bodies**. Interacts with **synphilin**.

Syp *PTP1D; Shp-2* An adaptor molecule mediating **GRB-2**/ras signalling. See **Shp**.

syphilis A contagious, sexually transmitted disease caused by infection with the spirochaete *Treponema pallidum*. Can also be vertically transmitted from mother to foetus.

syringyl alcohol *Sinapyl alcohol* A phenylpropanoid alcohol, one of the three precursors of lignin.

systemic effect An effect that is manifest throughout the organism, not just at the point of application. Thus systemic insecticides are distributed throughout the plant and many drugs act on the whole body, not just on limited local areas or specialized tissues.

systemic lupus erythematosus *SLE* Disease of humans, probably autoimmune, with antinuclear and other antibodies in plasma. Immune complex deposition in the glomerular capillaries is a particular problem. Multiple genes affect susceptibility, mostly genes that regulate the immune system.

systemin An 18 amino-acid polypeptide released from wound sites on tomato leaves caused by insects or other mechanical damage. Through the **systemin receptor,** regulates more than 20 defensive genes. Produced, by proteolytic cleavage, from a precursor prohormone, called prosystemin.

systemin receptor The tomato systemin receptor, SR160, a plasma membrane-bound, leucine-rich repeat receptor kinase (160 kDa) that signals systemic plant defence through a complex signal transduction pathway. The receptor regulates an intracellular cascade including, depolarization of the plasma membrane, the opening of ion channels, an increase in intracellular Ca^{2+}, activation of a **MAP kinase** activity and a phospholipase A2 activity. As a consequence, linolenic acid is released from plant membranes and converted to **jasmonic acid**.

systems biology Fashionable term for an integrated approach to biology in which effects on the whole organism are studied, as opposed to the reductionist approaches of modern molecular bioscience. This dictionary deliberately contains definitions that span a wide range of biological subspecialties, since molecular cell biologists are generally well aware of the importance of putting their work into context but may need a glossary for some of the more arcane terminologies.

syzygy In some parasitic protozoa, the pairing of gamonts prior to sexual fusion; in gregarines, the end-to-end attachment of the sporonts; in some crinoids, the fusion of organs or skeletal elements.

T

T7 *Bacteriophage T7* A lytic T-odd phage that infects *E. coli*, the prototype of a group of virulent phages that have an icosahedral head and short, stubby tail to which are attached six tail fibres The genome (39 937 bp) contains 56 genes that encode 59 known proteins. Genes are grouped into three classes: class I genes, which enter the host cell and are transcribed first and moderate the transition in metabolism from host to phage; class II genes, which are primarily responsible for T7 DNA replication; and class III genes, which code for particle, maturation and packaging proteins. See **T7 promoter**.

T7 promoter Promoter region comprising a highly conserved 23-bp sequence that is selective for the phage T7 RNA polymerase, used in the **pET Expression System**.

T7 RNA polymerase Polymerase (single polypeptide, 99 kDa) that catalyses the synthesis of RNA in the 5′ to 3′ direction in the presence of a DNA template containing a T7 phage promoter. Very specific for the **T7 promoter**.

TA cloning Cloning strategy for **PCR** products that relies on the tendency of **Taq polymerase** to add an extra dA at the 3′ end of newly synthesized DNA strands, thus leaving a single base 3′ overhang. Vectors are accordingly prepared with single base dT 3′ overhangs, allowing ligation of **sticky ends**.

Tacaribe complex Group of eight **Arenaviridae** isolated from bats in S. America.

TACC proteins *Transforming, acidic, coiled-coil-containing proteins* Family of proteins originally described in mammals though *Drosophila* homologue now also identified (**D-TACC**). The function of the TACC proteins is unknown, but the genes encoding the known TACC proteins are all associated with genomic regions that are rearranged in certain cancers. At least one of the mammalian TACC proteins appears to be associated with centrosomes and microtubules in human cells.

TACE *TNFα converting enzyme; ADAM17; EC 3.4.24.* A zinc metallopeptidase of the **ADAM** family involved in processing of TNFalpha and releasing the soluble cytokine from the inactive membrane-bound precursor. Also mediates the cleavage and shedding of **fractalkine** (CX3CL1).

tachykinins A group of neuropeptide hormones including **substance P**, substance K (neurokinin A) and neurokinin B in mammals, eledoisin from *Octopus* and physalaemin (amphibian). All have 10 or 11 residues with a common -FXGLM-NH2 ending. Elicit a wide range of responses from neurons, smooth muscle, endothelium, exocrine glands and cells of the immune system; effects similar in many ways to **bradykinin** and **serotonin**.

tachyphylaxis A decrease in the response to an agonist following repeated exposure. Can arise through a variety of mechanisms.

tachyzoite An asexual stage of rapid growth in the tissue phase of certain coccidial infections such as *Toxoplasma gondii*. The proliferation occurs in parasitophorous vacuoles in the infected cells.

tacrine A reversible acetylcholinesterase inhibitor used in the treatment of Alzheimer's disease. Early pathological changes in Alzheimer's disease involve, fairly selectively, cholinergic neuronal pathways that project from the basal forebrain to the cerebral cortex and hippocampus: inhibiting acetylcholine esterase increases the duration of cholinergic signals.

tacrolimus *FK506* Drug that inhibits T-cell activation and used as an immunosuppressant to prevent transplant rejection. Binds to **FKBP-12**.

TAF 1. **Tumour angiogenesis factor**. 2. **TATA-binding protein**-associated factor.

TAG-1 *Transient axonal glycoprotein* A 135-kDa surface glycoprotein that is expressed transiently on commissural and **motoneurons** in developing vertebrate nervous system. TAG-1 and **L1** have been shown to be on different segments of the same embryonic spinal axons. See **axonin** and **tax-1**

tagma *Pl.* tagmata A distinct section of an arthropod, consisting of two or more adjoining segments, e.g. the thorax of an insect.

tagmosis The grouping or fusion of somites to form definite regions (tagmata) in a metameric animal.

taicatoxin Complex oligomeric protein toxin from *Oxyuranus scutelatus scutelatus* (taipan snake). Blocks high- but not low-threshold calcium channels of heart muscle. The oligomer contains a neurotoxin-like peptide (8 kDa), a phospholipase (16 kDa) and a serine peptidase inhibitor (7 kDa).

taipoxin Heterotrimeric toxin from *Oxyuranus scutelatus scutelatus* (taipan snake). All three subunits (α,β,γ) have homology with pancreatic phospholipase A2. Blocks transmission at the neuromuscular junction.

TAK Prefix to drugs developed by Takeda Pharmaceutical Co., e.g. TAK-220, a CCR5 antagonist; TAK-599, a new cephalosporin, etc.

talin Protein (215 kDa) that binds to **vinculin**, but not to actin, and is associated with the sub-plasmalemmal cytoskeleton. The amino-terminal head consists of a **FERM domain** that binds an NPxY motif within the cytoplasmic tail of most **integrin** beta subunits.

Talon resin Proprietary name for immobilized nickel-beads. Used to purify **his-tagged** recombinant proteins.

Tamiami virus Arenavirus of the **Tacaribe complex**.

Tamiflu® Proprietary name for the antiviral drug **oseltamivir**.

tamoxifen Synthetic anti-estrogen used in chemotherapy of breast carcinoma. Probably has other effects, including inhibition of chloride channel conductance.

tamsulosin An **alpha blocker** used to treat benign prostatic hyperplasia.

tanabin See **paranemin**.

tandem repeats Copies of genes repeated one after another along a chromosome: for example, the 40S-rRNA genes in somatic cells of toads, of which there are about 500 copies.

tangential longitudinal section A section of an approximately cylindrical organ taken longitudinally along a tangent at its surface.

tannic acid Penta-(*m*-digalloyl)-glucose, or any soluble **tannin**; used in electron microscopy to enhance the contrast. Addition of tannic acid to fixatives greatly improves, for example, the image obtained of tubulin subunits in the microtubule, or the **HMM** decoration of microfilaments.

tannins Complex phenolic compounds found in the vacuoles of certain plant cells, e.g. in bark. They are strongly astringent, and are used in tanning and dyeing.

t-antigen *Small-t-antigen, little-t-antigen* The small **T-antigen** of polyoma virus and of **SV40** exert pleiotropic effects on biological processes such as DNA replication, cell cycle progression and gene expression, possibly through stimulation of NFkappaB-responsive genes.

T-antigen *Large-T-antigen* Proteins coded by viral genes that are expressed early in the replication cycle of papovaviruses such as SV40 and polyoma. Essential for normal viral replication, they are also expressed in non-permissive cells transformed by these viruses. Originally detected as tumour antigens by immunofluorescence with antisera from tumour-bearing animals. SV40 has two, large T and small t; polyoma has three, large, middle and small-t. Appear to be collectively responsible for transformation by these viruses.

Tap protein Member of the evolutionarily conserved nuclear RNA export factor (NXF) family of proteins that mediates the sequence non-specific nuclear export of cellular mRNAs as well as the sequence-specific export of retroviral mRNAs bearing the **constitutive transport element** (CTE). Contains separate domains for binding to **nucleoporins** and **NXT1** and the functional heterodimer with NXT/p15 mediates the export process.

tapasin Accessory protein required for the interaction of MHC Class I with **TAPs** thus ensuring efficient peptide binding. Tapasin is related to the immunoglobulin superfamily and has an ER retention signal.

tapetum 1. Layer of reflective tissue just behind the pigmented retinal epithelium of many vertebrate eyes. May consist of either a layer of guanine crystals or a layer of connective tissue. In bovine eyes, reflects a blue-green iridescent colour. 2. Layer of cells in the sporangium of a vascular plant that nourishes the developing spores.

TAPs Transporters associated with antigen processing: **ABC proteins** involved in transporting protein fragments across ER membranes during antigen processing. TAP is composed of TAP1 and TAP2, each containing a transmembrane domain and a nucleotide-binding domain (NBD). TAP-like proteins have also been found. See **tapasin**.

Taq polymerase A heat-stable **DNA polymerase** that is normally used in the **polymerase chain reaction**. It was isolated from the bacterium *Thermus aquaticus*.

TAR RNA *Transactivating-response RNA* RNA structure at the extreme 5′ terminus of virion RNA. HIV-2 TAR RNA domain (TAR-2) is the target for binding the **Tat-2 protein**.

target cell An erythrocyte with increased surface area to volume ratio, so called because it resembles a target with a bull's eye. Can arise as a result of liver disease, iron deficiency or **thalassaemia**.

target regulation General term for an interaction between neurons and their targets by which target-derived signals influence the differentiation of the innervating neurons.

targeting signal Peptide sequence within a protein that determines where it will be located. Thus there are targeting signals for proteins that accumulate in the nucleus, others for ER, lysosomes, etc.

Tat protein Transactivator protein from lentiviruses, notably **HIV**; sequence-specific RNA-binding protein that recognizes **TAR RNA**. Will induce endothelial cell migration and invasion *in vitro*, and rapid angiogenesis *in vivo*. Peptides from this protein are potent neurotoxins, implying a possible route for HIV-mediated toxicity.

Tat system Translocase of the inner bacterial membrane that transports proteins with firmly bound cofactors. Independent of the **SecYEG** system.

TATA box *Goldberg–Hogness box* A consensus sequence found in the promoter region of most genes transcribed by eukaryotic **RNA polymerase** II. Found about 25 nucleotides before the site of initiation of transcription, and has the consensus sequence: 5′-TATAAAA-3′. This sequence seems to be important in determining accurately the position at which transcription is initiated.

TATA-binding protein *TBP* A 30-kDa component of TFIIIB and of **SL1**, responsible for positioning the polymerase. Also involved in positioning RNA polymerase II, in which case it binds directly to the **TATA box**. Spinocerebellar ataxia type 17, a neurodegenerative disorder in man, is caused by an expanded polymorphic polyglutamine-encoding trinucleotide repeat in the gene for TBP.

tau protein *Tau factor* Protein (60–70 kDa) that co-purifies with **tubulin** through cycles of assembly and disassembly, and the first microtubule associated protein to be characterized. Tau proteins are a family made by alternative splicing of a single gene. It has tandem repeats of a tubulin-binding domain and promotes tubulin assembly. Although tau proteins are found in all cells they are major components of neurons, where they are predominantly associated with microtubules of the axon. See **MAPs** and **tauopathy**.

tauopathy Neuropathy in which there is deposition of an excess of the microtubule-binding protein **tau** in brain lesions.

taurine *2-aminoethanesulphonic acid* Compound derived from cysteine by oxidation of the sulphydryl group and decarboxylation. Present in the cytoplasm of some cells (particularly neutrophils) at high concentration.

taurocholate Major bile salt (derived from taurocholic acid) with strong detergent activity. Formed by conjugation of taurine with cholate.

tautomerase *EC 5.3.2.* An enzyme that catalyses a **tautomerism**, such as the keto–enol equilibrium – for example, oxaloacetate tautomerase, EC 5.3.2.2. However, some other enzymes may be called tautomerases, e.g. D-dopachrome decarboxylase (EC 4.1.1.84), which is also called phenylpyruvate tautomerase.

tautomerism Form of isomerism in which there are two or more arrangements usually of hydrogens bonded to oxygen. Keto–enol tautomerism is one common example. The balance between two co-existing tautomers may shift with time or as a result of changes in conditions.

tautomycin Antibiotic, isolated from *Streptomyces spiroverticillatus*, that is an inhibitor of Type 1 and Type 2a protein phosphatases.

tax-1 Axonal surface glycoprotein (135 kDa), the human homologue of rat **TAG**-1 and chicken **axonin**-1. GPI-linked to neuronal plasma membrane and is involved in adhesion. There are six Ig-like and four fibronectin III-like domains. Will support neurite outgrowth *in vitro*.

taxane Generic name for a cytotoxic drug that inhibits cell growth by stopping cell division (antimitotic or antimicrotubule agents). Best known is paclitaxel (**taxol**).

taxilins Proteins of the **syntaxin** family, implicated in intracellular vesicle traffic. Alpha-taxilin (62 kDa) interacts with the **nascent polypeptide-associated complex**. Beta- and gamma-taxilins preferentially interact with syntaxin-1a and -4, respectively.

taxis A response in which the direction of movement is affected by an environmental cue. Should be clearly distinguished from a **kinesis**.

taxol Drug isolated from Pacific yew (*Taxus brevifolis*) that stabilizes microtubules: analogous in this respect to **phalloidin**, which stabilizes microfilaments. Strictly speaking, Taxol is the proprietary name for paclitaxel; however, taxol is commonly used in the cell-biological literature pre-dating the clinical use.

Tay–Sachs disease Lysosomal disease (**lipidosis**) in which hexosaminidase A, an enzyme that degrades **ganglioside** GM2, is absent. A lethal autosomal recessive; mostly affects brain, where ganglion cells become swollen and die.

TBC domains *Tre-2/Bub2/Cdc16-domains* Domains that are predicted to encode GTPase-activating proteins (GAPs) for Rab family G-proteins. The *TRE17* gene (also referred to as *Tre-2* and *USP6*) was originally identified by virtue of its ability to transform NIH 3T3 cells and its TBC domain actually functions to bind GDP-Arf6 and promote its plasma membrane localization.

T-box genes The T-box gene family codes for transcription factors (and putative transcription factors) that share a unique DNA-binding domain, the T-domain. In all metazoans studied, from *C. elegans* to man, they are found as a small, highly conserved group of genes; mutations are associated with developmental defects. See **brachyury**.

TBP **1.** **TATA-binding protein**. **2.** Thioredoxin binding protein-2 (TBP-2).

TCA cycle See **tricarboxylic acid cycle**.

TCA3 Mouse **I-309**.

T-cell *T-lymphocyte* A class of lymphocytes, so called because they are of thymic origin and have been through thymic processing. Involved primarily in cell-mediated immune reactions and in the control of B-cell development. They bear T-cell antigen receptors (CD3) and lack Fc or C3b receptors. Major T-cell subsets are CD4$^+$ (mainly helper cells) and CD8$^+$ (mostly cytotoxic or suppressor T-cells). See **T-helper**, **T-suppressor**.

T-cell factor *TCF* Transcription factor, one of the high-mobility-group domain proteins, activated by wnt/wingless signalling and repressed by **CREB-binding protein**. Coactivator is beta-catenin. Activation of TCF in colonic epithelium and other cells leads to tumours.

T-cell growth factor See **interleukin-2**.

T-cell leukaemia/lymphoma viruses See **HTLV-I**, **HTLV-II**.

T-cell receptor The antigen-recognizing receptor on the surface of **T-cells**. Heterodimeric (disulphide-linked), one of the immunoglobulin superfamily of proteins; binds antigen in association with the **major histocompatibility complex** (MHC), leading to the activation of the cell. There are two subunits (α and β, 42–44 kDa in mouse; 50–40 kDa in humans), each with variable and constant regions, that are associated non-covalently with T3 (20–30 kDa). A second heterodimer on CD3$^+$ cells with γ (35 kDa in mice, 55 kDa in humans) and δ (45 kDa in mice, 40 kDa in humans) chains is a second T-cell antigen receptor that is not **MHC-restricted**. The $\gamma\delta$ T-cell receptors (TCRs) are formed on very early T-cells in the thymus.

TCID$_{50}$ *50% tissue culture infective dose* An assay, usually for viruses, done by using serial dilutions inoculated into several wells containing susceptible cells and calculating the titre that causes infection of 50% of the wells.

TCP-1 *T-complex polypeptide 1; chaperonin containing T-complex (CCT)* A protein complex (eight related $\sim$60-kDa proteins) that mediates folding of the nascent actin molecule. CCT is localized to the leading edge of fibroblasts and neurons.

TDG *Thymine-DNA glycosylase; EC 3.2.2.* Enzyme responsible for repair of G/T mispairings.

T-DNA **1.** DNA coding for tRNA (tDNA). **2.** Transfer DNA (T-DNA), part of the Ti-plasmid that is transferred to plant cells from *Agrobacterium tumefaciens*. Following its nuclear import, the single-stranded T-DNA is stripped of its escorting proteins, most likely converts to a double-stranded (ds) form, and integrates into the host genome. Several genes on the T-DNA have been identified, the most important for crown gall induction are concerned with indole acetic acid (IAA) synthesis and the synthesis of a cytokinin. The transfer of these genes from bacterium to plant induces cell division and explains the hormone independence of crown gall tissue. T-DNA is the most frequently used insertion element for gene tagging in *Arabidopsis thaliana*. See **activation tagging**.

TDT **1.** Transmission disequilibrium test. A test for linkage between a genetic marker and a disease susceptibility locus. **2.** Terminal deoxynucleotidyl transferase (TdT).

Tec family Family of intracellular protein **tyrosine kinases** involved in signalling. Includes **Btk**. Unlike **src family**, are not regulated by C-terminal phosphorylation. Have **PH** and **TH domains** in N-terminal region.

teichoic acid Acidic polymers (glycerol or ribitol linked by phosphodiester bridges) found in cell wall of Gram-positive bacteria. May constitute 10–50% of wall dry weight, and are cross-linked to peptidoglycan. Related to **lipoteichoic acid**.

teicoplanin Glycopeptide antibiotic used as a less toxic alternative to vancomycin for treating methicillin-resistant *S. aureus*. Actually a group of at least six glycopeptides, all with very similar activities, A2-1 to A2-5, and A3-1 being the main components. Works by inhibiting formation of cell walls in Gram-positive bacteria by interfering with formation of links in the cell wall (transglycosylation), acting on amino acyl-D-alanyl-D-alanine residues.

tektins Family of filamentous proteins (A, 55 kDa; B, 51 kDa; C, 47 kDa) associated with some microtubules in ciliary and flagellar axonemes. Have homology with some intermediate filament proteins (keratins and lamins).

telangiectasia Condition in which capillary vessels are dilated. See **ataxia telangiectasia**.

teleomorph Mycological term that describes the form of a fungus when reproducing sexually. The anamorph form describes the fungus when reproducing asexually. The holomorph is a description of the whole fungus, encompassing both forms of description.

teleost melanophores Large stellate cells found in the epidermis of fish. Cytoplasmic pigment granules (containing **melanin**) can be centrally located, or rapidly dispersed, using a microtubule-associated system. Altering the granule distribution changes the colour of the skin. **Chromatophores** containing other pigments extend the capacity to adopt protective colouration.

telethonin *T-cap* **Titin**-capping protein

teleutospore Thick-walled resting spore of rust and smut fungi. *Ustilago coicis* teleutospores can retain their ability to germinate even when maintained under dry conditions for about 5 years.

telocentric chromosome Chromosome with the centromere located at one end.

telodendria Branched termination of axons and axon collaterals, the distal ends of which are slightly enlarged to form synaptic bulbs.

telokin Acidic protein (24 kDa) found in some muscle tissues, identical to the C-terminal 155 residues of smooth muscle **myosin light-chain kinase** (MLCK) and independently expressed.

telolecithal A type of egg which is relatively large, with yolk constituting most of the volume of the cell. Cytoplasm is concentrated at one pole. Typical of sharks, reptiles and birds.

telomerase *Telomere terminal transferase; EC 2.7.7.-* A DNA polymerase with rather unusual properties that will only elongate oligonucleotides from the telomere and not other sequences. The enzyme contains an essential 159-residue RNA sequence that provides a template for the replication of the G-rich telomere sequences (so that the enzyme could in fact be considered a reverse transcriptase).

telomere The end of a chromosome.

telomeric repeat binding factors *TRF-1, TRF-2* Major protein component (33 kDa) of **telomeres**, colocalizes with telomeres in interphase and is located at chromosome ends during mitosis. Critical for the control of telomere structure and function; TRF-2 specifically recognizes TTAGGG tandem repeats at chromosomal ends and TRF2 dysfunction results in the exposure of the telomere ends and activation of ATM (ataxia telangiectasia mutated)-mediated DNA damage response.

telomestatin A telomerase inhibitor, isolated from *Streptomyces anulatus* 3533-SV4, which is known to stabilize G-quadruplex structures at 3′ single-stranded telomeric overhangs (G-tails) and rapidly dissociates TRF2 from telomeres in cancer cells. Telomestatin has a large polycyclic planar structure resembling a G-quadruplex.

telopeptides Portions of the amino acid sequence of a protein that are removed in maturation of the protein. Best-known examples are the N- and C-terminal telopeptides of procollagen that are involved in development of the quaternary structure and are then proteolytically removed by **procollagen peptidases**.

telophase The final stage of mitosis or meiosis, when chromosome separation is completed.

temazepam Short-acting **benzodiazepine** drug useful for inducing sleep, but now a **controlled drug** because of recreational abuse.

TEMED N,N,N,N-*tetramethyl-ethylenediamine* Compound used in conjunction with ammonium persulfate to accelerate polymerization of acrylamide.

temozolomide Drug of the imidazotetrazine class, that is rapidly converted at physiological pH to the reactive compound MTIC. The cytotoxicity of MTIC is primarily due to alkylation of DNA at the O6 and N7 positions of guanine. Crosses the blood–brain barrier and is used in the treatment of recurrent malignant glioma.

temperate phage A bacteriophage that integrates its DNA into that of the host (**lysogeny**) as opposed to virulent phages that lyse the host.

temperature-sensitive mutation *Ts mutation* A type of conditional mutation in organism, somatic cell or virus that makes it possible to study genes whose total inactivation would be lethal. Such ts mutations can also make possible studies of the effect of reversible switching (by temperature changes) in expression of the mutated gene. The usual mechanism of temperature sensitivity is that the mutated gene codes for a protein with a temperature-dependent conformational instability, so that it possesses normal activity at one temperature (the permissive temperature) but is inactive at a second (non-permissive) temperature.

template A structure that in some direct physical process can cause the patterning of a second structure, usually complementary to it in some sense. In current biology, almost exclusively used to refer to a nucleotide sequence that directs the synthesis of a sequence complementary to it by the rules of Watson–Crick base-pairing.

temporal sensing Mechanism of gradient sensing in which the value of some environmental property is compared with the value at some previous time, the cell having moved position between the two samplings. Initial movement is random; until the second observation is made the gradient cannot be detected. See **spatial** and **pseudospatial** sensing mechanisms. Bacterial chemotaxis (so called) is based on this mechanism.

temporomandibular joint The joint that connects the lower jawbone to the skull.

tenascin *Myotendinous antigen; cytotactin* Protein of the extracellular matrix (240-kDa subunit: usually as a hexabrachion, a six-armed hexamer of more than 1000 kDa) selectively present in mesenchyme surrounding foetal (but not adult) rat mammary glands, hair follicles and teeth. Found in the matrix surrounding mammary tumours of rat. Tenascin contaminates cell surface **fibronectin** and accounts for most of the haemagglutinating activity of extracellular matrix protein. Contains 14 repeats of the **EGF-like domain**.

tensegrity The hypothesis that cells can behave like structures in which shape results from balancing tensile and hydrostatic forces.

tensin Actin-binding component of **focal adhesions** and submembranous cytoskeleton. Has SH2 domain and can be **tyrosine phosphorylated**; interacts with PI3-kinase and JNK signalling pathways. See **PTEN**.

tenuin Sub-plasmalemmal protein (400 kDa) from **adherens junctions**, associated with membrane insertions of **microfilament** bundles, and membrane adjacent to circumferential microfilament bundles of epithelial cells. Has not reappeared in recent literature.

TEP1 *TGFα-regulated and epithelial cell-enriched phosphatase* Also termed **PTEN** or MMAC1 (mutated in multiple advanced cancers 1).

teratocarcinoma Malignant tumour (teratoma), thought to originate from primordial germ cells or misplaced blastomeres, that contains tissues derived from all three embryonic layers, e.g. bone, muscle, cartilage, nerve, toothbuds and various glands. Accompanied by undifferentiated, pluripotent epithelial cells known as embryonal carcinoma cells.

teratogen Agent capable of causing malformations in embryos. Notorious example is **thalidomide**.

teratoma See **teratocarcinoma**.

terfenadine Non-sedating antihistamine drug, a histamine H1-receptor antagonist, used in the treatment of hay fever.

terminal bar Obsolete name for **zonula occludens** (tight junction).

terminal buttons The small swellings at the end of an axon that release neurotransmitters; the presynaptic region.

terminal cisternae Regions of the **sarcoplasmic reticulum** adjacent to **T-tubules**, and from which calcium is released when striated muscle is activated.

terminal dideoxynucleotidyl transferase *TdT* An intranuclear DNA polymerase that catalyses the template-independent addition of deoxynucleotides to the 3′ hydroxyl terminus of oligonucleotide primers. Such terminal additions at the junctions of rearranging V (D) J gene segments greatly contribute to antigen-receptor diversity. TdT is expressed only on immature lymphocytes and acute lymphoblastic leukaemia cells and has been identified in several vertebrate species, where it is highly conserved.

terminal web The cytoplasmic region at the base of microvilli in intestinal epithelial cells, a region rich in microfilaments from the microvillar core and from **adherens junctions**, in myosin, and in other proteins characteristic of an actomyosin motor system.

termination codon The three codons, UAA known as **ochre**, UAG as **amber** and UGA as **opal**, that do not code for an amino acid but act as signals for the termination of protein synthesis. They are not represented by any tRNA and termination is catalysed by protein release

factors. There are two release factors in *E. coli*; RF1 recognizes UAA and UAG, RF2 recognizes UAA and UGA. Eukaryotes have a single GTP-requiring factor, eRF.

terminator DNA sequence at the end of a **transcription unit** that causes **RNA polymerase** to stop transcription.

terpene Lipid species, very abundant in plants. In principle, terpenes are polymers of isoprene units. Function in plants is not clear. In animals, **dolichol**, an important carrier species in the formation of glycoproteins, is a terpenoid. Similarly, squalene, an intermediate in the synthesis of cholesterol, is a terpene.

terpenoids A group of plant secondary metabolites related to **terpenes**, based on one to four or more isoprene (C5) units. Include many essential oils, gibberellins and carotenoids.

tertiary structure The third level of structural organization in a macromolecule. The primary structure of a protein (for example) is the amino acid sequence, the secondary structure is the folding of the peptide chain (e.g. alpha-helical or beta-pleated), the tertiary structure is the way in which the helices or sheets are folded or arranged to give the three-dimensional structure of the protein. Quaternary structure refers to the arrangement of protomers in a multimeric protein.

testa Outer covering of a seed, also called the seed coat; derived from the integument of the ovary.

testicular feminization If genetic males lack receptors for testosterone, they develop as females and are unresponsive to male hormones.

testosterone Male sex hormone (androgen) secreted by the interstitial cells of the testis of mammals and responsible for triggering the development of sperm and of many secondary sexual characteristics.

tetanolysin Thiol-activated haemolysin (50 kDa) released by the bacterium *Clostridium tetani*. Tetanolysin apparently forms water-filled pores that may vary in size, depending on the tetanolysin concentration utilized.

tetanospasmin See **tetanus toxin**.

tetanus *Lock-jaw* Disease caused by the bacterium *Clostridium tetani*, spores of which persist in soil but can proliferate anaerobically in an infected wound. Disease entirely due to the **tetanus toxin**, released by bacterial autolysis.

tetanus antitoxin Antibody to **tetanus toxin**, usually from horses hyper-immunized with *Clostridium tetani* exotoxin. Can cause serum sickness, an immune response to the horse proteins.

tetanus toxin *Tetanospasmin* Neurotoxin released by *Clostridium tetani*; becomes active when peptide cleaved proteolytically to heavy (100-kDa) and light (50-kDa) chains held together by disulphide bond. Heavy chain binds to disialogangliosides (GD2 and GD1b), and part of the peptide (the amino-terminal B-fragment) forms a pore; light chain is a zinc endopeptidase that specifically attacks **synaptobrevin**, to block neurotransmitters. See also **botulinum toxin**.

tetracaine *Amethocaine* Potent local anaesthetic.

tetracycline antibiotics A group of closely related bacteriostatic antibiotics, with similar antibacterial spectrum and toxicity. They act by binding to the 30S subunit of the bacterial ribosome and inhibiting protein synthesis. Effective against many streptococci, Gram-negative bacilli, rickettsiae, spirochaetes, Mycoplasma and *Chlamydia*. Tetracycline itself is produced by *Streptomyces aureofasciens*.

tetrad Four homologous chromatids paired together during first meiotic prophase. More generally, any group of four objects.

tetraethylammonium ion *TEA* A monovalent cation widely used in neurophysiology as a specific blocker of potassium channels. It is similar in size to the hydrated potassium ion and gets stuck (reversibly) in the channels.

tetrahydrobiopterin See **biopterin**.

tetrahydrocannabinol *THC* A **cannabinoid** and one of the more psychoactive components of cannabis.

tetrahydrogestrinone *THG* A synthetic steroid that has been used as a performance-enhancing drug by some athletes.

Tetrahymena Genus of ciliate protozoa frequently used in studies on ciliary axonemes, self-splicing RNA and telomere replication.

tetranectin A 67-kDa serum glycoprotein (*ca.* 10 mg/l), a homotrimeric protein containing a **C-type lectin**-like domain, thought to regulate proteolytic processes via its binding to plasminogen kringle 4 and indirect activation of plasminogen. Decreased plasma levels of tetranectin correlates with cancer progression.

tetraploid Nucleus, cell or organism that has four copies of the normal **haploid** chromosome set.

tetraspan vesicle membrane proteins *Tetraspanins; TVPs* Ubiquitous and abundant family of widely expressed integral membrane proteins that associate extensively with one another and with other membrane proteins to form specific membrane microdomains. They are characterized by four transmembrane regions and cytoplasmically located end domains. TVP-containing vesicles shuttle between various membranous compartments and are localized in biosynthetic and endocytotic pathways. TVPs can be grouped into three distinct families: physins, gyrins and **secretory carrier-associated membrane proteins** (SCAMPs). Examples are CD9 (Tspan29) and CD81 (Tspan28). See **tetraspanin web**.

tetraspanin web A network of **tetraspan vesicle membrane proteins** (tetraspanins) and their partner proteins (**ADAMs** and **integrins**) that facilitates cellular interactions and cell fusion.

tetrodotoxin *TTX* A potent **neurotoxin** (319 D) from the Japanese puffer fish (*Fugu rubripes*). It binds to the sodium channel, blocking the passage of action potentials. Its activity closely resembles that of **saxitoxin**.

tetrose General term for a monosaccharide with four carbon atoms.

T-even phage A group of dsDNA bacteriophages of enterobacteria including T2, T4, T6, as opposed to T-odd phage (T1, 3, 5 and 7). They are related serologically, and all have large genomes. The T-odd phages fall into three serological groups: T3 and T7 are related to each other but not to T1 or to T5, which are unrelated.

Texas red Rhodamine-type fluorophore well suited for excitation by the 568-nm spectral line of Ar–Kr mixed-gas laser or the 594-nm spectral line of the orange He–Ne laser; emits at about 620 nm, with very little spectral overlap with fluorescein; often conjugated to antibodies for use in FACS analysis.

textilinin *Txln* Kunitz-type serine peptidase inhibitor (59 amino acids) from the venom of *Pseudonaja textilis* that has effects on plasmin kinetically distinct from those of **aprotinin** (with which it has ~50% sequence homology). Two distinct forms have been isolated, with molecular weights of 6688 Da and 6692 Da. The effect is to slow fibrinolysis.

textilotoxin Protein neurotoxin (70 kDa) from venom of *Pseudonaja textilis textilis* (Australian common brown snake) that blocks neuromuscular transmission. All five subunits of the toxin have some phospholipase A2 activity. See **textilinin**.

TFG 1. See **TRK-fused gene**. 2. *Trigonella foenumgraecum* (fenugreek; TFG), leaf extracts of which exert analgesic, anti-inflammatory and antipyretic effects in various experimental models. 3. TFG1 and TFG2 encode the two larger subunits of the **TFIIF** complex. 4. Relatively frequent misprint for **TGF**.

TFIID A multicomponent transcription factor that consists of a DNA-binding subunit that recognizes the TATA element (a **TATA-binding protein**), as well as several TBP-associated factors (or TAFs). Nucleates the formation of the transcription complex.

TFIIF General transcription initiation factor which in humans consists of a heterodimer. TFIIF-alpha (RNA polymerase II-associating protein 74; RAP74; EC 2.7.11.1) is the large subunit, TFIIF-beta is smaller (26 kDa), also known as RAP30. Helps recruit RNA polymerase II to the initiation complex and promotes translation elongation.

TFIIIA Transcription factor, one of the first to be cloned and characterized. Has a crucial role in transcription of 5S ribosomal RNA. Multiple cysteine/histidine **zinc finger** motifs. Interacts with **TFIIIB**.

TFIIIB Transcription factor consisting of **TBP** and two other proteins. An initiation factor required by RNA polymerase III; TFIIIA and TFIIIC assist its binding to the appropriate DNA sequence. Effectively acts as a positioning factor for polymerase.

TFIIIC Transcription factor; large (500 kDa) complex containing at least five subunits.

TFIIX Any one of a number of accessory proteins involved in the binding of RNA Polymerase II to DNA in association with **TBP**.

TFP 1. **Trifluoperazine**. 2. **Mitochondrial trifunctional protein** (TFP). 3. Type IV pili (Tfp) of various bacteria.

TGEs *Tra-2 and GLI elements* Regions of the 3′ untranslated region of the sex-determining gene *tra-2* in *C. elegans*. Thought to be the target of **STAR proteins**.

TGF See **transforming growth factor**.

TGN Abbreviation for the **trans-Golgi network**.

TH domain *Tec homology domain* Domain characteristic of **Tec family** protein kinases, probably ligand region for the **SH3** domain that follows downstream. The N-terminal

27 residues of the TH domain are highly conserved (the Btk motif) and are followed by a proline-rich (PRR) region.

thalassaemia *Thalassemia* Hereditary blood disease in which there is abnormality of the globin portion of haemoglobin. Widespread in Mediterranean countries. Alpha-thalassaemias have mutations in the haemoglobin alpha chain, beta-thalassaemias in the beta chain.

thalidomide Sedative drug that, when taken between third and fifth weeks of pregnancy, produced a range of malformations of the foetus – in severe cases, complete absence of limbs (amelia) or much reduced limb development (phocomelia). A **teratogen**. Thalidomide inhibits TNF-alpha production and therefore has potential as an anti-inflammatory drug.

thallus Simple plant body, not differentiated into stem, root, etc. Main form of the gametophyte generation of simpler plants such as liverworts.

thapsigargin Cell-permeable inhibitor (a sesquiterpene lactone extracted from *Thapsia garganica*) of calcium ATPase of endoplasmic reticulum (**SERCA**); leads to increase in cytoplasmic calcium ions. Acts independently of InsP3. A tumour promoter.

thaumatin Protein from the African plant *Thaumatococcus daniellii*. It tastes 10^5 times sweeter than sucrose.

thebaine *Paramorphine* Minor constituent of opium, chemically similar to both morphine and codeine, but stimulatory rather than depressant. Thebaine can be converted into a variety of compounds, including **codeine**, hydrocodone, **oxycodone**, naloxone.

T-helper cells *TH cells; Th1; Th2* There are now recognized to be two subclasses of CD4 positive T-helper cells, Th1 and Th2. Th1 cells produce IL-2, IFNγ and TNFα and do not produce IL-4, IL-5 and IL-10. They are associated with cell-mediated immunity. Selective activation of Th1 cells is promoted by IFNγ and IL-12 and inhibited by IL-4 and IL-10, the products of Th2 cells. Th2 cells are involved with the humoral immune response, produce IL-4, IL-5 and IL-10 and promote antibody production; IL-4 is essential for growth and differentiation of Th2 cells. There is cross-inhibition between the two classes; if one subclass is activated it will inhibit the activity of the other so that the response is polarized.

theobromine *3,7-Dimethyl xanthine* Principal alkaloid of cacao bean; has similar properties to theophylline and caffeine.

theophylline *1,3-Dimethylxanthine* Inhibits cAMP **phosphodiesterase**, and is often used in conjunction with exogenous dibutyryl cyclic AMP to raise cellular cAMP levels. Other, less potent, methylxanthines are caffeine, theobromine and aminophylline.

therapeutic cloning Production of a cloned embryo by somatic cell nuclear transfer, with the intention of using stem cells from the early embryo for therapeutic purposes, e.g. in the treatment of Parkinson's disease. *cf.* **reproductive cloning**.

therapeutic index The ratio of the lethal dose (LD_{50}) to the effective dose (ED_{50}) for a drug. Ideally, the difference should be large.

thermal analysis Form of calorimetry in which the rate of heat flow (or some other property) to a solid is measured as a function of temperature.

thermal cycler *Thermocycler* Instrument for automated **PCR** in which a sequence of temperature cycles is produced. See **Lightcycler**.

thermal melting profile In general, a record of the phase state of a system over a temperature range. Phase changes can be detected by exothermy or endothermy. Valuable in studying lipid and DNA structures.

thermodynamics The study of energy and energy flow in closed and open systems.

thermogenin *Uncoupling protein; UCP* A 33-kDa inner-membrane mitochondrial protein found only in **brown fat cells** in mammals. Functions as a proton transporter, generating heat by dissipating the proton gradient generated by the respiratory chain and thereby uncoupling oxidative phosphorylation. Upregulated by exposure to cold. See **uncoupling proteins**.

thermolysin *EC 3.4.24.4* Heat-stable zinc metallopeptidase (peptidase family M4) containing four calcium ions, produced by a strain of *Bacillus stearothermophilus*. Retains 50% of its activity after 1 h at 80°C.

thermophile An organism that thrives at high temperature. The most extreme examples (hyperthermophiles) are from hot springs, deep-sea vents and geysers, and will tolerate temperatures above 80°C. The archaebacterium *Pyrolobus fumarii* holds the current record, being able to grow at 113°C.

thermophilic See **thermophile**.

thermotaxis A directed motile response to temperature. The grex of *Dictyostelium discoideum* shows a positive thermotaxis.

Thermus aquaticus Aerobic Gram-negative bacillus that lives in hot springs and was the source of **Taq polymerase**.

theta replication The early replication method of bacteriophage lambda, involving the production of theta form intermediates. The circular phage DNA is replicated bidirectionally, producing copies of the phage DNA (at an intermediate stage of the replication process the shape resembles the Greek letter theta); at a later stage the phage shifts to using a rolling circle mechanism which produces concatemers with multiple copies of the genome that are later cut and processed to produce multiple individual phage genomes.

THG See **tetrahydrogestrinone**.

thiamine Vitamin B1; deficiency results in beri-beri.

thiamine pyrophosphatase *TPPase; EC 3.6.1.6* A nucleoside-diphosphatase that requires thiamine pyrophosphate as a cofactor. Has been used as a cytochemical marker for the trans cisternae of the **Golgi apparatus**.

thiamine pyrophosphate *TPP* The coenzyme form of vitamin B1 (thiamine), a coenzyme for pyruvate dehydrogenase, α-ketoglutarate dehydrogenase and transketolase. Synthesized by thiamine pyrophosphokinase from free thiamine and MgATP.

thiamine pyrophosphokinase *TPK; EC 2.7.6.2* Enzyme responsible for the synthesis of **thiamine pyrophosphate** (thiamine diphosphate) from thiamine and ATP.

thiazide diuretics Group of drugs with moderate diuretic effects. Work by inhibiting the sodium re-uptake transporter in the distal tubule of the kidney, thereby increasing sodium and water excretion but with concomitant loss of potassium ions (in contrast to the **potassium-sparing diuretics**). Used for long-term treatment of hypertension or oedema associated with congestive heart failure.

thick filaments Bipolar **myosin** II filaments (12–14 nm in diameter, 1.6 μm long) found in striated muscle. Myosin filaments elsewhere are often referred to as 'thick filaments', although their length may be considerably less. The myosin heads project from the thick filament in a regular fashion. There is a central 'bare' zone without projecting heads, the core being formed from antiparallel arrays of **LMM** regions of the myosin heavy chains. Thick filaments will self-assemble *in vitro* under the right ionic conditions.

thigmotaxis A **taxis** in response to mechanical contact (touch). *Adj.* thigmotactic.

thigmotropism Tendency of an organism or part of an organism to turn towards or respond to a mechanical stimulus.

thin filaments Filaments, 7–9 nm in diameter, attached to the **Z-discs** of striated muscle; have opposite polarity in each half-sarcomere. Built of F-actin with associated **tropomyosin** and **troponin**.

thin layer chromatography *TLC* Chromatographic separation method in which a thin layer of the solid phase (often silica, aluminium oxide or cellulose) is fixed onto a glass or plastic sheet. Can be run one-dimensionally or in a second dimension with a different solvent system. Much used in separation of lipids. Visualization can be by staining, or radioactive labelling and autoradiography.

thiobacillus Small, rod-shaped proteobacteria living in sewage or soil and gaining energy from the oxidation of elemental sulphur and sulphur-containing compounds. *Thiobacillus ferroxidans* is generally considered to be important for accelerating the dissolution of metal sulphide from minerals.

thioester Compounds of the type. R-CO-S-R'. See **coenzyme A**, **palmitoylation**.

thioether The bond R-S-C, of which the best example is in methionine.

thiol endopeptidases See **cysteine endopeptidases**.

thiol proteinase See **cysteine endopeptidases**.

thiol-activated haemolysins *Oxygen-labile haemolysins* Cytolytic bacterial exotoxins that act by binding to cholesterol in cell membranes and forming ring-like complexes that act as pores. SH groups of these toxins must be in the reduced state for the toxin to function. Oxidation (to disulphide bridges) inactivates the toxin. Examples: **tetanolysin**, **streptolysin O**, θ-toxin (**perfringolysin**), **cereolysin**.

thionins Group of small, hydrophobic plant proteins of 45–48 amino acids, of which 6–8 are cysteines; toxic to animals but probably a defence against pathogenic microorganisms, since they are upregulated during fungal infections.

thiopentone *Thiopental* A widely used intravenous anaesthetic. A rapid-onset, short-acting barbiturate.

thioredoxin Intercellular disulphide-reducing enzyme. Also secreted by a variety of cells, despite the lack of a **signal sequence**, in a manner resembling the alternative secretory pathway for IL1β, and a few other proteins.

thoracic duct The major efferent lymph duct into which lymph from most of the peripheral lymph nodes drains. Recirculating lymphocytes that have left the circulation in the lymph node return to the blood through the thoracic duct.

THP-1 Human monocytic cell line derived from peripheral blood of a 1-year-old boy with acute monocytic leukaemia. Have Fc and C3b receptors and will differentiate into macrophage-like cells.

threonine *Thr; T; 119 Da* The hydroxylated polar amino acid. See Table A2.

threose A four-carbon sugar in which the two central hydroxyl groups are in *trans* orientation (*cis* in erythrose).

threshold limit value-time-weighted average *TLV-TWA* The concentration of a substance to which it is believed nearly all workers may be repeatedly exposed without adverse effect. Based upon a time-weighted average concentration for a conventional 8-h workday and 40-h week (shorter than that worked by most scientists).

thrombasthenia Condition in which there is defective platelet aggregation, though adherence is normal. See **Glanzmann's thrombasthenia**.

thrombin Serine endopeptidase (34 kDa) generated in blood clotting that acts on **fibrinogen** to produce **fibrin**. Consists of two chains, A and B, linked by a disulphide bond. B-chain has sequence homology with pancreatic serine peptidases: cleaves at Arg-Gly. Thrombin is produced from prothrombin by the action either of the extrinsic system (tissue factor + phospholipid) or, more importantly, the intrinsic system (contact of blood with a foreign surface or connective tissue). Both extrinsic and intrinsic systems activate plasma Factor X to form Factor Xa, which then, in conjunction with phospholipid (tissue derived or **platelet factor 3**) and Factor V, catalyses the conversion. See also Table F1.

thrombocyte Archaic name for a blood **platelet**.

thrombocytopenia Gross deficiency in platelet number, consequently a tendency to bleeding.

thrombocytopenic purpura In severe **thrombocytopenia**, bleeding into skin leads to small petechial haemorrhages. In primary thrombocytopenic purpura, an autoimmune mechanism seems to cause platelet destruction; secondary thrombocytopenic purpura may be a result of drug-induced type II **hypersensitivity** in which platelets coated with antibody to the drug (which is acting as a **hapten**) are destroyed in a complement-mediated reaction.

thromboglobulin β *Thromboglobulin* A 35.8-kDa heparin-binding protein derived from platelets. It consists of four identical, non-covalently bound 8800-Da peptide chains, and is released into plasma when platelets aggregate. Derived by proteolytic cleavage from **platelet basic protein**.

thrombomodulin Specific endothelial cell receptor (100 kDa: luminal surface only) that forms a 1:1 complex with thrombin. This complex then converts **protein C** to Ca, which in turn acts on Factors Va and VIIIa. Structurally similar to **coated pit** receptors.

thromboplastin Traditional name for substance in plasma that converts prothrombin to **thrombin**. Now known not to be a single substance. (See **thrombin**).

thromboplastin reagent Material used to test the blood-clotting time of patients receiving anti-coagulant treatment or showing signs of clotting dysfunction. It can be either an extract of mammalian tissue (lungs, heart or brain of animals) rich in **tissue factor**, or a recombinant preparation of human tissue factor in combination with phospholipids. See **International sensitivity index** and **International normalized ratio**.

thrombopoietin *TPO* Growth factor (19 kDa, 174 aa) that regulates proliferation of megakaryocytes and the production of platelets (thrombopoiesis). Receptor is c-mpl, a cytokine receptor that can cause phosphorylation of **STAT3** and STAT5 through Jak3.

thrombosis Formation of a solid mass (a **thrombus**) in the lumen of a blood vessel or the heart.

thrombospondin Homotrimeric glycoprotein (450 kDa) from a granules of **platelets**, and synthesized by various cell types in culture. Also found in extracellular matrix of cultured endothelial, smooth muscle, and fibroblastic cells. May have autocrine growth-regulatory properties; involved in platelet aggregation.

thrombosthenin Obsolete name for platelet contractile protein, now known to be actomyosin (which makes up 15–20% of the total platelet protein).

thromboxanes Arachidonic acid metabolites produced by the action of thromboxane synthetase on prostaglandin cyclic endoperoxides. Thromboxane A2 (TxA2) is a potent inducer of platelet aggregation and release and, although unstable, the activation of platelets leads to the further production of TxA2. Also causes arteriolar constriction. Another endoperoxide product, **prostacyclin**, has the opposite effects.

thrombus Solid mass that forms in a blood vessel, usually as a result of damage to the wall. The first aggregate is of platelets and fibrin, but the thrombus may propagate by clotting in the stagnant downstream blood.

THUMP domain *(Thiouridine synthases, RNA methyltransferases, and pseudouridine synthases)-domain* An ancient domain found in genes from Eukaryota and Archaea, but not eubacteria. The genes in question are for ThiI-like thiouridine synthases, conserved RNA methylases, archaeal pseudouridine synthases and several uncharacterized proteins. It has been predicted that this domain is an RNA-binding domain, and its restricted distribution reflects the distinct tRNA-processing strategies of Eukarya/Archaea and Eubacteria.

thuringolysin O **Cholesterol-binding toxin** from *Bacillus thuringiensis*.

thy1 *CD90; formerly CDw90, theta antigen* Differentiation antigen (19-kDa glycoprotein) on surface of T-cells, neurons, endothelial cells and fibroblasts. GPI-anchored and a member of the **immunoglobulin superfamily** with only one V-type (variable) domain.

thylakoids Membranous cisternae of the chloroplast, found as part of the **grana** and also as single cisternae interconnecting the grana. Contain the photosynthetic pigments, reaction centres and electron-transport chain. Each thylakoid consists of a flattened sac of membrane enclosing a narrow intrathylakoid space.

thymectomy The excision of the thymus by operation, radiation or chemical means. Since the thymus is important in T-cell maturation, this has major effects on the immune system.

thymic aplasia *Hypoplasia* A lack of T-lymphocytes, due to failure of the thymus to develop, resulting in very reduced cell-mediated immunity though serum immunoglobulin levels may be normal. See also **DiGeorge syndrome**.

thymidine Term that is always used in practice for the nucleoside thymine deoxyriboside; not the riboside which naming of the other nucleosides might lead one to expect.

thymidine block A method for synchronizing cells in culture. In the absence of thymidine, DNA synthesis cannot occur, so cells are blocked before S-phase; release of the block allows synchronous entry into cycle.

thymidine kinase *TK; EC 2.7.1.21* Enzyme of pyrimidine salvage, catalysing phosphorylation of thymine deoxyriboside to form its $5'$ phosphate, the nucleotide thymidylate. Animal cells lacking this enzyme can be selected by lethal synthesis, e.g. by resistance to bromodeoxyuridine, and can be used as parentals in somatic hybridization, since they are unable to grow in **HAT medium**.

thymidine phosphorylase See **PD-ECGF**.

thymine *2,6-Dihydroxy, 5-methylpyrimidine; 5-methyluracil* Pyrimidine base found in DNA (in place of uracil of RNA).

thymine dimer Dimer that can be formed in DNA by covalent linkage between two adjacent (*cis*) thymidine residues, in response to ultraviolet irradiation. Occurrence potentially mutagenic, although repair enzymes exist that can excise thymine dimers. See **xeroderma pigmentosum**.

thymocyte Lymphocyte within the thymus; term usually applied to an immature lymphocyte.

thymoma A tumour of thymic origin.

thymopentin Biologically active pentapeptide corresponding to residues 32–36 of thymopoietin. Will induce prothymocytes and activate peripheral T-cells.

thymosin Thymosin alpha1 is a naturally occurring thymic peptide (28-amino acids) that primes dendritic cells for antifungal T-helper type 1 resistance through Toll-like receptor (TLR)-9 signalling. Beta-thymosins are polypeptides involved in the regulation of actin polymerization (see **thymosin β4**).

thymosin β4 Small protein (5 kDa: 43 residues) found in large amounts in many vertebrate cells (approximately 0.2 mM in neutrophils) and that binds G-actin, thereby inhibiting polymerization. Identical to Fx peptide.

thymus The lymphoid organ in which T-lymphocytes are educated, composed of stroma (thymic epithelium) and lymphocytes, almost entirely of the T-cell lineage. In mammals, the thymus is just anterior to the heart within the rib cage; in other vertebrates it is in rather undefined regions of the neck or within the gill chamber in teleost fish. The thymus regresses as the animal matures.

thymus-derived lymphocyte See **T-cell**.

thypedin See **pedin**.

thyroglobulin The 650-kDa protein of the thyroid gland that binds thyroxine.

thyroid hormones Thyroxine and tri-iodothyronine are hormones secreted by the thyroid gland in vertebrates. These iodinated aromatic amino acid compounds influence growth and metabolism and, in amphibia, metamorphosis. The hormone **calcitonin,** which has hypocalcaemic effects, is also of thyroid origin, but is not usually classed with thyroxine and tri-iodothyronine as a thyroid hormone. See also Tables H2 and H3.

thyroid-stimulating antibodies Long-acting thyroid stimulator is an autoantibody found in many cases of primary thyrotoxicosis, which causes hyperplasia of the thyroid by undetermined mechanisms. Human thyroid-stimulating immunoglobulin is a different antibody found in all or nearly all cases of primary thyrotoxicosis, and may act by binding to the thyrotropin (TSH) receptor site, causing increased synthesis of **thyroglobulin.**

thyroid-stimulating hormone *TSH; thyrotropin* Polypeptide hormone (28 kDa), secreted by the anterior pituitary gland, that activates cyclic AMP production in thyroid cells, leading to production and release of the **thyroid hormones.**

thyroiditis Disease of the thyroid, especially **Hashimoto's thyroiditis,** in which autoimmune destruction of the thyroid takes place.

thyroliberin See **thyrotrophic-releasing hormone**.

thyrotoxicosis Clinical syndrome caused by an excess of circulating free thyroxine (T4) and free tri-iodothyronine (T3). The most common cause is **Graves' disease**.

thyrotrophic-releasing hormone *Protirelin; TRH; thyroliberin; TRF* Tripeptide (pyroGlu-His-Pro-NH$_2$) that releases thyrotropin from the anterior pituitary by stimulating adenyl cyclase. May also have neurotransmitter and paracrine functions.

thyrotropin See **thyroid-stimulating hormone**.

thyroxine *T4; tetra-iodothyronine* See **thyroid hormones** and Table H3.

Ti-plasmid Plasmid of *Agrobacterium tumefaciens*, transferred to higher plant cells in crown gall disease, carrying the **T-DNA** that is incorporated into the plant cell genome. Used as a vector to introduce foreign DNA into plant cells.

Tiam-1 Product of T-lymphoma invasion and metastasis gene-1, a GDP–GTP exchange factor (GEF) for the small GTPase **Rac**; implicated in tumour invasion and metastasis.

tiarin **Olfactomedin family** member that promotes dorsal neural specification in *Xenopus*.

Tie *Tie1; Tie2/Tek* Endothelium-specific receptor tyrosine kinase required for normal embryonic vascular development and tumour angiogenesis. Associates with p85 of PI3kinase. Ligand for Tie2 is **angiopoietin**. VEGF is a ligand for Tie1.

tight junction See **zonula occludens**.

tiling Another word for tessellation. 'Rep-tiles' can be joined together to make larger replicas of themselves. The aim in tiling is to cover the whole area completely, thus the term has come to be used in the context of the design of DNA arrays to give unbiased coverage, or tiling, of genomic DNA for the large-scale identification of transcribed sequences and regulatory elements. Increasingly sophisticated algorithms are being developed to assist in designing appropriate arrays to achieve this end.

time-lapse Technique applied to speed up the action in a film or videotape sequence. In filming, by taking a frame every few seconds and projecting at conventional speed (16 or 24 frames per second), the movements of cells can be greatly speeded up and then become conspicuous. With videotape, the recording is made at slow tape-speed and replayed at full speed. The opposite of slow-motion.

time-resolved fluorescence Method to avoid interference by autofluorescence. Using an emitter fluorochrome that has slow decay characteristics coupled to the reagent of interest and temporally separating excitation and measurement, the signal can be arranged to derive almost entirely from the reporter fluorophore. (Autofluorescence decays very rapidly.) See **HTRF**.

time-weighted average A special averaging technique that takes accounts of variation in the basis of the individual measurements and adjusts for time irregularities from sample to sample.

timeless *tim* *Drosophila* gene essential for the production of circadian rhythms. The protein product, TIM, may be necessary for the accumulation of the PER protein, the product of the *period* gene. TIM and PER associate with one another, and the regulated interaction seems to determine the entry of PER into the nucleus: both TIM and PER are produced in a circadian cycle.

TIMP See **tissue inhibitors of metalloproteinases**.

tincar *tinc* *Drosophila* gene that encodes a protein with eight putative transmembrane domains. The tinc mRNA is expressed specifically in four of the six pairs of cardioblasts in each segment, in a pattern identical to that of **tinman** (tin). In the non-Tin-expressing pairs of cardioblasts, tinc transcription seemed to be repressed by seven-up.

tinman Member of the NK homeobox family, essential for the specification of cardiac cells in *Drosophila*. Tin and Bagpipe (Bap), another homeodomain protein, form homo- and heterodimeric complexes. See *tincar*.

TIP 1. Tonoplast intrinsic protein. 2. TIP-1, an atypical PDZ protein, ubiquitously expressed, that is composed almost entirely of a single **PDZ** domain and functions as a negative regulator of PDZ-based scaffolding.

TIR domain Domain found in **toll-like receptors** and interleukin-1 receptor. The four known TIR-domain-containing adapter proteins are MyD88, TIR-domain-containing adapter inducing IFN-beta (TRIF), TRIF-related adapter molecule (TRAM), and TIR-domain containing adapter protein (TIRAP).

TIRAP See **TIR domain**.

TIRF See **total internal reflection microscopy**.

tissue Group of cells, often of mixed types and usually held together by extracellular matrix, that perform a particular function. Thus, tissues represent a level of organization between that of cells and of organs (which may be composed of several different tissues). Sometimes used in a more general sense – for example, epithelial tissue, where the common factor is the pattern of organization, or connective tissue, where the common feature is the function.

tissue array *Tissue microarray* Array of small sections of tissue on a microscope slide that can be used for immunocytochemical staining or *in situ* hybridization. The tissue samples are, in some cases, selected from larger embedded tissue samples (e.g. of tumours) following histopathological examination.

tissue culture Originally, the maintenance and growth of pieces of explanted tissue (plant or animal) in culture away from the source organism. Now usually refers to the (much more frequently used) technique of cell culture, using cells dispersed from tissues, or distant descendants of such cells.

tissue culture plastic Polystyrene that has been rendered wettable by oxidation, a treatment that increases its adhesiveness for cells from animal tissues, and without which **anchorage dependent** cells will not grow. Commercially achieved by treatment known as glow discharge.

tissue engineering The creation of new body parts for transplantation by *in vitro* culture of cells using an artificial matrix or support. There are considerable practical difficulties, particularly the absence of vascularization, and few tissues have yet been successfully manufactured in this way.

tissue factor Integral membrane glycoprotein, of around 250 residues, that initiates blood clotting after binding factors VII or VIIa.

tissue inhibitors of metalloproteinases *TIMP* Family of proteins of around 200 residues that can inhibit metallopeptidases, for example collagenase, by binding to them.

tissue-plasminogen activator *TPA; tPA; EC 3.4.21.68* Plasma serine peptidase, one of a closely related group of **plasminogen activators**. Contains an **EGF-like domain** and multiple copies of the **kringle** domain.

tissue-typing The process of determining the allelic types of the antigens of the **major histocompatibility complex** (MHC) that determine whether a tissue graft will be accepted or rejected. At present, carried out either by use of polyclonal or monoclonal antibodies against MHC antigens, or (less usually) by tests of MHC-restricted cell function or skin grafting (the latter not in humans).

titin *Connectin* Family of enormous proteins (2000–3500 kDa) found in the sarcomere of striated muscle. Form a scaffolding of elastic fibres that may be important for correct assembly of the sarcomere. Each titin molecule spans from M-line to Z-disc.

TL antigens The mouse antigens coded for by the **TLa complex**; in normal animals only found on intrathymic lymphocytes, but also seen on leukaemic cells (hence, thymus leukemia antigen) in certain forms of the disease in mice. The molecules have structures similar in some ways to Class I MHC products, but are disulphide-bonded tetramers of two 45-kDa chains and two 12-kDa chains of **beta-2-microglobulin** type.

TLa complex Genes coding for and controlling **TL antigens**; the complex is situated close to the H-2 complex on mouse chromosome 17 and resembles H-2 in several ways.

TLC Thin layer chromatography.

TLCK *Tosyl lysyl chloromethylketone* Protease inhibitor, particularly effective against trypsin and papain.

T-loop of RNA *Thymine pseudo-uracil loop; Ty loop* The T-loop of tRNA is the region of the molecule that is responsible for ribosome recognition.

TLR Toll-like receptor.

T-lymphocyte See **T-cell**.

TMB-8 *3,4,5-Trimethoxybenzoic acid 8-(diethylamino)octyl ester* Inhibitor of the release of calcium from intracellular stores but also a potent, non-competitive, functional antagonist of various **nicotinic acetylcholine receptor** subtypes.

TMPD *Tetramethyl-1,4-phenylendiamine; Wurster's reagent* An easily oxidized compound that can be used as a reducing co-substrate for haem peroxidases. Following one-electron oxidation, TMPD produces a highly coloured product that absorbs at 611 nm. Can be used for the detection of peroxidases on polyacrylamide gels.

TMS See **transcranial magnetic stimulation**.

TMV See **tobacco mosaic virus**.

TNF See **tumour-necrosis factor**.

TNF receptor *TNF-R; CD120* There are two receptors for **TNFα**. Type I (CD120a, 55 kDa) is present on most cell types, and type II (CD120b, 75 kDa) is mainly restricted to haematopoietic cells. Both types bind TNFα and lymphotoxin (TNFβ) and are members of the NGF receptor family. The two types have substantial sequence homology except in their cytoplasmic domains, and have different signalling capacities. TNFαRI contains a **death domain** and interacts with a number of cytoplasmic proteins (**TRADD, TRAF, FADD**).

TNP 1. Trinitrophenol. 2. TNP-470, a semisynthetic derivative of **fumagillin**, an angiogenesis inhibitor. 3. Tn5 transposase (Tnp); enzyme that is involved in DNA transposition and causes genomic instability by mobilizing DNA elements. 4. See **TNP-AMP**.

TNP-AMP *Trinitrophenol-adenosine monophosphate* Compound that can be used as an ATP analogue in studying ATPases. TNP nucleotides undergo an equilibrium transition to a semiquinoid structure when bound to the nucleotide-binding site of some proteins and are only fluorescent in this form.

TNT 1. Trinitrotoluene, a well-known explosive. 2. Troponin T (TnT).

tobacco mosaic virus *TMV* Plant RNA virus, the first to be isolated. Consists of a single central strand of RNA (a helix of 6500 nucleotides) enclosed within a coat consisting of 2130 identical capsomeres that, in the absence of the RNA, will self-assemble into a cylinder similar to the normal virus but of indeterminate length. Causes mottling of the leaves of the tobacco plant.

tobramycin **Aminoglycoside antibiotic** used for serious infections that are resistant to gentamicin. Derived from species of *Streptomyces* or produced synthetically; inhibits protein synthesis by binding with the 30S ribosomal subunit.

Toc complex Transport system of the outer membrane of the chloroplast. Analogous to **Tom complex** though proteins are not the same. Toc75 seems to form the pore; Toc159 and Toc 34 are thought to be GTP-regulated import receptors; Toc34 and Toc75 act sequentially to mediate docking and insertion of Toc159, resulting in assembly of the functional translocon on the cytoplasmic side.

tocopherol α-*Tocopherol; vitamin E* Protects unsaturated membrane lipids from oxidation and may prevent free radical damage.

T-odd phage See **T-even phage**.

toeprinting A primer extension inhibition assay used to study the initiation step of protein synthesis. In a toeprinting assay, mRNA is translated using purified ribosomal complexes and cycloheximide is added to the reaction to inhibit elongation, thereby arresting the position of the ribosomes on the transcript. The mRNA complex is then copied into cDNA using a specific labelled primer and, where the reverse transcriptase meets the ribosome bound to the mRNA, polymerization is halted, and a 'toeprint' fragment is generated. Generally, the P site of the stalled ribosome is 15–17 nucleotides upstream of the toeprint.

Togaviridae Class IV viruses with a single positive-strand RNA genome. Bullet-shaped capsid, enveloped by a membrane formed from the host cell plasma membrane; the budded membrane contains host lipids and viral (spike) glycoproteins. The group can be divided into two main groups: alphaviruses, which include **Semliki forest virus,** and **Sindbis virus** and **Flaviviridae,** which include yellow fever virus and rubella (German measles) virus. Many are transmitted by insects and were previously classified as **arboviruses.**

tokogenetic Relationships between individuals within species, in contrast to phylogenetic relationships between species or separate lineages.

Tol-Pal proteins Proteins (TolA, TolB, TolQ, TolR and Pal) of the cell envelope of *E. coli,* and other Gram-negative bacteria that are required for maintaining outer membrane integrity. TolA bridges between the inner and outer membranes via its interaction with the Pal lipoprotein. TolQ, TolR and TolA form a complex in the inner membrane, whereas TolB is a periplasmic protein. The Pal lipoprotein interacts with many components, such as TolA, TolB, OmpA, the major lipoprotein and the murein layer.

tolbutamide A **sulphonylurea** that will bind to **SUR** and enhance insulin release from pancreatic **B cells**.

tolerable daily ingestion *TDI* Regulatory value for the amount of a food additive or contaminant. Equivalent to the acceptable daily intake but expressed in mg/person, assuming a body weight of 60 kg. In theory, the amount that can be ingested daily over a lifetime without appreciable health risk.

tolerance The development of specific non-reactivity to an antigen. See **immunological tolerance.**

tolerogen Substance that will induce immunological tolerance.

toll *Drosophila* gene required for dorsoventral polarity determination. Protein (124 kDa) is a transmembrane receptor with leucine-rich repeat. Interacts downstream with **pelle** and **tube,** and defines dorsoventral polarity in the embryo. Toll, which is present over the entire surface of the embryo, is activated ventrally by interaction with a spatially restricted, extracellular ligand.

toll-like receptors *TLR* Family of receptors, first discovered in *Drosophila* (see **toll**), involved in innate immunity and that recognize and respond to different microbial components. Type I transmembrane proteins with significant homology in their cytoplasmic domains to the IL-1 receptor type I. See Table T1.

TABLE T1. Toll like receptors

Name	Distribution	Function/comments
TLR1 (TIL, Rsc786)	Low-level monocytes	Regulates TLR2 response
TLR2 (TIL4)	Monocytes, granulocytes; upregulated on macrophages	Interacts with microbial lipoproteins, peptidoglycans and LPS; NFκB pathway
TLR3	Low-level fibroblasts	Interacts with dsRNA; NFκB pathway; induces production of type I interferons
TLR4 (HToll, Ly87, Rasl2-8, Ran/M1)	Monocytes, upregulated on endothelium	Interacts with microbial lipoproteins, CD14-dependent response to LPS, NFκB pathway
TLR5 (TIL3)	mRNA in leucocytes, prostate, liver, lung	Interacts with microbial lipoproteins, NFκB, response to *Salmonella*
TLR6	mRNA: leucocytes, ovary, lung	Interacts with microbial lipoproteins, regulates TLR2 response
TLR7	mRNA: spleen, lung, placenta; upregulated on macrophages	Low similarity to other TLRs
TLR8	mRNA: leucocytes, lung	
TLR9	DC, B-cells (intracellular, low)	Receptor for CpG bacterial DNA
TLR10	mRNA: lymphoid tissues	Related to TLR1 and TLR6
RP105 (CD180, Ly-78)	Protein: mature B-cells	B-cell activation, LPS recognition
MD-1 (Ly86)	Mature B-cells	Regulates surface expression of RP105
MD-2 (Ly96)	Macrophages	Regulates surface expression of TLR4, signals LPS presence

A useful resource is the information provided by eBioscience at http://www.ebioscience.com/ebioscience/whatsnew/tlr.htm. This table is based upon the one that is on that site.

tolloid *Drosophila* gene, product is a metalloproteinase; involved in the process of specifying dorsal–ventral polarity, probably by activating **decapentaplegic**. Member of the bone morphogenetic (**BMP**) family of proteins and the homologue of BMP-1.

tolterodine Muscarinic receptor antagonist that acts as an antispasmodic on urinary bladder muscle.

toluidine blue *CI Basic Blue 17* A thiazin dye related to methylene blue and Azure A in structure; often used for staining thick resin sections. Typically exhibits metachromasia.

Tom complex *Translocase of outer membrane* Transport complex of the outer membrane of mitochondrion. The complex contains eight different proteins: Tom40 (40 kDa) forms the 2.2-nm hydrophilic pore and spans the outer membrane; Tom5, Tom6 and Tom7 are embedded within the membrane adjacent to Tom40; Tom20, Tom22, Tom37 and Tom70 are on the cytosolic face with Tom22 on the inner face as well. The comparable system in chloroplasts is the **Toc complex**.

tonic See **adaptation**.

tonofilaments Cytoplasmic filaments (10 nm in diameter: **intermediate filaments**) inserted into **desmosomes**.

tonoplast Membrane that surrounds the vacuole in a plant cell.

tonoplast intrinsic protein *TIP* Family of plant proteins, closely related to **major intrinsic protein** and part of the aquaporin family of proteins that function as water-transport channels. Found in plant vacuolar membranes; different TIP isoforms may define different vacuole functions.

tophus Mass of urate crystals surrounded by a chronic inflammatory reaction: characteristic of gout.

topiramate Drug, a sulphamate-substituted monosaccharide, that has anticonvulsant effects, is used in treating epilepsy, and is also prophylactic for migraine. Mechanism of action unclear, but seems to block voltage-dependent sodium channels, to augment the activity of GABA at some $GABA_A$ receptor subtypes, to antagonize the AMPA/kainate subtype of the glutamate receptor, and to inhibit some isoforms of carbonic anhydrase enzyme.

topographic map In general, a map that illustrates the surface features (topography), but used more specifically for the spatially ordered projection of neurons onto their target; e.g. in the retino-tectal projection, retinal ganglion cell axons project along the **optic nerve** to the contralateral tectum where they ramify to form terminal arbors. The target sites of the terminal arbors are ordered: neurons from a specific region of the retina consistently project to a specific region of the tectum, forming a map of the retina on the tectum.

topographical control Those phenomena of cell behaviour in which the shape of the local substrate of the cell affects its behaviour; see, for example, **contact guidance**.

topoinhibition Term used to describe the inhibition of cell proliferation as the cells become closely packed on a culture dish: generally superseded by the term **density-dependent inhibition**.

topoisomerases Enzymes that change the degree of supercoiling in DNA by cutting one or both strands. Type I topoisomerases cut only one strand of DNA; type I topoisomerase of *E. coli* (omega protein) relaxes negatively supercoiled DNA and does not act on positively supercoiled DNA. Type II topoisomerases cut both strands of DNA; type II topoisomerase of *E. coli* (DNA gyrase) increases the degree of negative supercoiling in DNA and requires ATP. It is inhibited by several antibiotics, including nalidixic acid and ovobiocin.

TOR *Target of rapamycin* Components of the **ras/MAP kinase** signalling pathway, originally characterized in yeast. TOR Complex 1 (**TORC 1**) is inhibited by **rapamycin** that is bound to **FKBP**; TORC2 is not. See **mTOR**.

TORC *TOR complex 1, 2* TOR Complex 1 (TORC1) contains **TOR1** or **TOR2**, KOG1 (YHR186c), and LST8. TORC2 contains TOR2, AVO1 (YOL078w), AVO2 (YMR068w), AVO3 (YER093c) and **LST8**. FKBP-rapamycin binds TORC1 but not TORC2, and TORC2 disruption causes an actin defect, suggesting that TORC2 mediates the rapamycin-insensitive, TOR2-unique pathway. See **mTOR**.

Torres body Intranuclear inclusion body in liver cells infected with yellow fever virus (**Togaviridae**).

torso *tor* Gene for a **receptor tyrosine kinase** (EC. 2.7.1.112) activated at the poles of the *Drosophila* embryo. Activation of *torso* triggers expression of *gap* genes that operate in these areas by antagonizing Gro-mediated repression (see **groucho**).

torulosis An infection due to the cryptococcus *Torula histolytica*, member of a genus of yeast-like Fungi Imperfecti of the form-family Cryptococcaceae; affects the central nervous system.

torus Structure found at the centre of a bordered **pit**, especially in conifers, forming a thickened region of the pit membrane. When subjected to a pressure gradient, it seals the pit by pressing against the pit border.

total internal reflection fluorescence microscopy *TIRF* Fluorescence microscopy in which total internal reflection of light at a glass–liquid interface generates an evanescent wave that will excite fluorochromes that are close to the interface. Used, for example, to image points of contact of cells with the surface.

totipotent Capable of giving rise to all types of differentiated cell found in that organism. A single totipotent cell could, by division, reproduce the whole organism.

toxic shock syndrome Endotoxic shock caused by bacterial contamination of tampons; toxin responsible is produced by some strains of *Staphylococcus aureus*.

toxicity Formally, the capacity to cause injury to a living organism, usually defined with reference to the quantity of substance administered or absorbed, the mode of administration (how and how frequently), the type and severity of injury, the time needed to produce the injury, the nature of the organism(s) affected, and other relevant conditions. Toxicity is often expressed as the reciprocal of the absolute value of median lethal dose ($1/LD_{50}$) or lethal concentration ($1/LC_{50}$).

toxicodynamics By analogy with **pharmacodynamics**, study of the way in which potentially toxic substances interact with target sites, and the biochemical and physiological consequences that lead to adverse effects.

toxicogenetics Study of the genetic cause of individual variations in the response to potentially toxic substances. Tends to focus on single gene polymorphisms, rather than the overall genomic environment in which particular genes are being expressed, but is often used more or less interchangeably with **toxicogenomics**. See **pharmacogenetics**, **pharmacogenomics**.

toxicogenomics Analysis of the genomic determinants of drug efficacy and toxicity; this encompasses not only genetic polymorphisms in, for example, **cytochrome P450**, but also the genomic environment in which individual genes are being expressed.

toxicokinetics Analysis of the uptake of potentially toxic substances by the body, the biotransformation they undergo, the distribution of the substances and their metabolites in the tissues, and the elimination of the substances and their metabolites from the body. By analogy with **pharmacokinetics**.

toxigenicity The ability of a pathogenic organism to produce injurious substances that damage the host.

toxin A naturally produced poisonous substance that will damage or kill other cells. Bacterial toxins are frequently the major cause of the pathogenicity of the organism in question. See **endotoxins** and **exotoxins**.

toxofilin Protein (27 kDa) that sequesters actin monomers and caps actin filaments in *Toxoplasma gondii*. Regulated by phosphorylation.

toxoid Non-toxic derivative of a bacterial exotoxin produced by formaldehyde or other chemical treatment: useful as a vaccine because it retains most antigenic properties of the toxin.

Toxoplasma A genus of parasitic protozoa. *T. gondii* is an intracellular parasite whose intermediate hosts include humans, the final host being felines of many species. Causes toxoplasmosis in humans, in which the parasite finally locates in tissues such as brain, heart or eye, leading to serious and sometimes fatal lesions.

TP-1 See **trophoblast protein 1**.

TPA 1. See **tissue-plasminogen activator**. 2. A **phorbol ester** tumour promoter, 12-*O*-tetradecanoyl-phorbol-13-acetate also known as PMA, phorbol myristyl acetate.

TPCK *Tosyl phenyl chloromethyl ketone* Non-specific protease inhibitor, interacts with histidine residues and will inactivate many enzymes by interfering with the active site.

TphiCG loop *TφCG* See **T-loop of RNA**.

TPO1 A member of the AIGP family, a unique group of proteins that contains 11 putative transmembrane domains. Expression of the rat *TPO1* gene is upregulated in cultured oligodendrocytes (OLs) during development from pro-oligodendroblasts to postmitotic OLs.

TPR motif *Tetratricopeptide motif* Degenerate consensus sequence of 34 residues found in various proteins that are involved in the regulation of RNA synthesis, protein import and *Drosophila* development.

tra-2 *C. elegans* gene required for female development and predicted to encode a large transmembrane protein, called TRA-2A, that is necessary to inhibit downstream male determinants. Translationally regulated by two elements, called TGEs (for *tra-2* and GLI elements), located in the 3′ untranslated region of the mRNA, to which the **STAR protein GLD-1** binds.

trabecula A transverse structure across a cavity, e.g. strands of connective tissue projecting into an organ, or the small, interconnecting rods that make up cancellous bone.

trabecular bone See **cancellous bone**.

trabecular meshwork A specialized eye tissue essential in regulating intraocular pressure. See **myocilin**.

trace amine-associated receptors *TAARs* A family of G-protein-coupled receptors for trace amines (TAs) such as p-tyramine and beta-phenylethylamine, although most of them do not have known ligands. May be involved in pheromone-like signalling, and are surprisingly different even between closely related species (e.g. man and chimpanzee).

tracheid Water-conducting cell forming part of the plant **xylem**. Contains thick, lignified secondary cells walls, with no protoplast at maturity. Interconnects with neighbouring tracheids through pits; the end walls are not perforated (*cf.* **vessel elements**).

tracheophyte A vascular plant, a fern (pteridophyte) or a seed plant (gymnosperms and angiosperms).

trachoma A highly contagious form of bilateral kerato-conjunctivitis caused by infection of the conjunctiva, transmitted by house flies and by poor hygiene, with *Chlamydia trachomatis*. Lacrimal glands and ducts are often affected as well; the upper lid may turn inward and the lashes then abrade the cornea and cause corneal ulceration. Without treatment, usually leads to blindness.

trachomatis See **trachoma**.

TRADD *TNF-receptor-1 associated death domain protein* Binds to TNF-R cytoplasmic domain and to FADD and RIP, though does not itself seem to have any catalytic activity. Contains a **death domain**.

TRAFs *TNF receptor-associated proteins* Family of proteins that associate with the cytoplasmic domain of the TNF receptor and TNFR family, such as CD40. Act as cytoplasmic adaptor proteins in NF-kappaB signalling; multiple isoforms are recognized.

TRAIL *TNF-related apoptosis-inducing ligand; Apo2L* An orphan member of the TNF ligand family that can be expressed either as a transmembrane protein (32 kDa) or in soluble form; induces apoptosis in a variety of tumour cell lines but not typically in normal or non-transformed cells. Structurally similar to CD95-ligand.

tram See **translocating chain-associating membrane protein**.

TRAM protein 1. Translocating chain-association membrane protein: Glycoprotein involved in co-translational translocation of proteins across the membrane of the ER; part of the **translocon**. 2. TRIF-related adaptor molecule, see **TIR domain**.

TRAMP 1. DR3; wsl-1; Apo-3. A member of the tumour-necrosis factor (TNF) receptor superfamily, abundantly expressed on thymocytes and lymphocytes, which has effects on NF-kappaB activation and apoptosis. TWEAK, a

TNF-related molecule, has been proposed as the ligand for this receptor. **2.** TRAMP mice (transgenic adenocarcinoma of the mouse prostate) are a strain engineered as a model for human prostatic carcinoma. **3.** Tyrosine-rich acidic matrix protein (TRAMP); see **dermatopontin**.

trans-Golgi network A complex of membranous tubules and vesicles, near the trans-face of the **Golgi apparatus**, which is thought to be a major intersection for intracellular traffic of vesicles.

trans-splicing (of RNA) Splicing of two different pre-mRNA molecules together. Seems to rely on intron-like sequences. Contrasts with the normal *cis*-splicing of conventional RNA molecules.

transactivation Experimental approach to control gene expression by the introduction of a transactivator gene and special promoter regions of DNA into a genomic region of interest. The transactivator-coded **transcription factor** will activate multiple genes that have the appropriate upstream promoters, and by putting the transactivator under control of an inducible promoter, gene expression can be switched on or off. A reporter gene linked to the transactivator will show whether it is active or not.

transacylase Any enzyme (EC 2.3 class) that transfers an acyl group. EC 2.3.1 transfer groups other than aminoacyl groups, EC 2.3.2 are aminoacyltransferase, EC 2.3.3 transfer acyl groups that are converted into alkyl on transfer.

transaldolase *EC 2.2.1.2* Enzyme (well conserved, 34 kDa) that catalyses the reversible transfer of a three-carbon ketol unit from sedoheptulose 7-phosphate to glyceraldehyde 3-phosphate to form erythrose 4-phosphate and fructose 6-phosphate. Together with transketolase, links the pentose phosphate pathway with glycolysis by converting pentoses to hexoses.

transaminases *Aminotransferases; EC 2.6.1* Class of enzymes that convert amino acids to keto acids in a cyclic process using pyridoxal phosphate as cofactor; e.g. aspartate amino transferase catalyses the reaction: aspartate + α-ketoglutarate = oxaloacetate + glutamate.

transcranial magnetic stimulation *TMS* A neurophysiological technique that allows the induction of a current in the brain using a magnetic field generated by a current through a coil of copper wire that is encased in plastic and held over the subject's head. It is possible to apply TMS in trains of multiple stimuli per second (repetitive TMS), which may prove a valuable investigative tool.

transcriptase See **reverse transcriptase**.

transcription Synthesis of RNA by RNA polymerases using a DNA template.

transcription factor Protein required for recognition by RNA polymerases of specific stimulatory sequences in eukaryotic genes. Several are known that activate transcription by RNA polymerase II when bound to **upstream** promoters. Transcription of the 5S RNA gene in *Xenopus* by RNA polymerase III is dependent on a 40-kDa protein TFIIIA that binds to a regulatory site in the centre of the gene, and was the first protein found to exhibit the metal-binding domains known as **zinc fingers**. See also Table T2.

TABLE T2. Transcription factors

1. *Superclass*: **Basic domains**

Class: Leucine zipper factors (bZIP)

- AP-1(-like) components
- CREB
- C/EBP-like factors
- bZIP/PAR
- Plant G-box binding factors
- ZIP only
- Other bZIP factors

Class: Helix–loop–helix factors (bHLH)

- Ubiquitous (class A) factors
- Myogenic transcription factors
- Achaete-Scute
- Tal/Twist/Atonal/Hen
- Hairy
- Factors with PAS domain
- INO
- HLH domain only
- Other bHLH factors

Class: Helix–loop–helix/leucine zipper factors (bHLH-ZIP)

- Ubiquitous bHLH-ZIP factors
- Cell-cycle controlling factors
- *Class*: NF-1
- NF-1

Class: RF-X

- RF-X

Class: bHSH

- AP-2

2. *Superclass*: **Zinc-coordinating DNA-binding domains**

Class: Cys4 zinc finger of nuclear receptor type.

- Steroid hormone receptors
- Thyroid hormone receptor-like factors

Class: Diverse Cys4 zinc fingers

- GATA-factors
- Trithorax
- Other factors

Class: Cys2His2 zinc-finger domain.

- Ubiquitous factors
- Developmental/cell cycle regulators
- Metabolic regulators in fungi
- Large factors with NF-6B-like binding properties
- Viral regulators

TABLE T2. (Continued)

Class: Cys6 cysteine-zinc cluster.

- Metabolic regulators in fungi

Class: Zinc fingers of alternating composition

- Cx7Hx8Cx4C zinc fingers
- Cx2Hx4Hx4C zinc fingers

3. *Superclass*: Helix–turn–helix

Class: Homeo domain

- Homeo domain only
- POU domain factors
- Homeo domain with LIM region
- Homeo domain plus zinc-finger motifs

Class: Paired box.

- Paired plus homeo domain
- Paired domain only

Class: Fork head/winged helix.

- Developmental regulators
- Tissue-specific regulators
- Cell cycle controlling factors
- Other regulators

Class: Heat-shock factors

- HSF

Class: Tryptophan clusters.

- Myb
- Ets-type
- Interferon-regulating factors

Class: TEA domain.

- TEA

4. *Superclass*: Beta-Scaffold Factors with Minor Groove Contacts

Class: RHR (Rel homology region).

- Rel/ankyrin
- ankyrin only
- NF-AT

Class: STAT

- SAT

Class: p53

- p53

Class: MADS box.

- Regulators of differentiation
- Responders to external signals
- Metabolic regulators

Class: Beta-Barrel alpha-helix transcription factors

- E2

Class: TATA-binding proteins

- TBP

Class: HMG.

- SOX
- TCF-1
- HMG2-related
- UBF
- MATA
- Other HMG box factors

Class: Heteromeric CCAAT factors

- Heteromeric CCAAT factors

Class: Grainyhead

- Grainyhead

Class: Cold-shock domain factors.

- csd

Class: Runt.

- Runt

5. *Superclass*: Other transcription factor

Class: Copper fist proteins

- Fungal regulators

Class: HMGI(Y)

- HMGI(Y)

Class: Pocket domain

- Rb
- CBP

Class: E1A-like factors

- E1A

Class: AP2/EREBP-related factors

- AP2
- EREBP
- AP2/B3

Class: AP2/EREBP-related factors

- AP2
- EREBP
- AP2/B3

Many transcription factors have now been identified, and an important resource is the TRANSFAC database at http://www.gene-regulation.com/pub/databases/transfac/cl.html. In this database, transcription factors are classified into Superclasses, Classes, Familes and sub-families; only the first three levels are listed here.

transcription squelching Anomalous suppression of transcription of a gene by overexpression of a transcription factor that would be expected to raise transcription levels. Thought to be caused by sequestration of a limiting cofactor by the overexpressed transcription factor.

transcription unit A region of DNA that is transcribed to produce a single primary RNA transcript, i.e. a newly synthesized RNA molecule that has not been processed. Transcription units can be mapped by kinetic studies of RNA synthesis, and in some instances directly visualized by electron-microscopy.

transcriptional control Control of gene expression by controlling the number of RNA transcripts of a region of DNA. A major regulatory mechanism for differential control of protein synthesis in both pro- and eukaryotic cells.

transcriptional silencing Mechanism of transcriptional control where DNA is bundled into **heterochromatin** in order to make it permanently inaccessible for future transcription. Effectively, this allows for memory in the **determination** of cell fate in developing organisms. In *Drosophila*, **homeotic genes** are silenced by members of the *Polycomb* group of genes.

transcriptosome Proposed model in which the RNA transcription machinery is preassembled as a unitary particle completely separate from the chromosomes, by analogy with the ribosome.

transcytosis Process of transport of material across an epithelium by uptake on one face into a coated vesicle, which may then be sorted in the endosomal compartment, and then delivery to the opposite face of the cell, still within a vesicle.

transcytotic vesicle Membrane-bounded vesicle that shuttles fluid from one side of the endothelium to the other. There is some controversy as to whether or not they form pores.

transdetermination Change in determined state observed in experiments on *Drosophila* **imaginal discs**. These can be cultured for many generations in the abdomen of an adult, where they proliferate but do not differentiate. If transplanted into a larva, they differentiate after pupation according to the disc from which they were derived; they maintain their determination. Occasionally, however, the disc will differentiate into a structure appropriate to another disc. This is termed transdetermination. It is a rare event, involves a population of cells, and certain changes are more common than others – for example, leg to wing is more frequent than wing to leg.

transdifferentiation Change of a cell or tissue from one differentiated state to another. Rare, and has mainly been observed with cultured cells. In newts, the pigmented cells of the iris transdifferentiate to form lens cells if the existing lens is removed.

transducin A **GTP-binding protein** found in the disc membrane of **retinal rods** and **cones**: of the part of the cascade involved in transduction of light to a nervous impulse. A complex of three subunits; α (39 kDa), β (36 kDa) and γ (8 kDa). Photoexcited rhodopsin interacts with transducin and promotes the exchange of GTP for GDP on the a subunit. The GTP-α subunit dissociates from the complex and activates a cGMP-phosphodiesterase by removing an inhibitory subunit. The α subunit of transducin can be ADP-ribosylated by cholera toxin and pertussis toxin.

transducisome An assembly of signalling molecules to form a macromolecular complex.

transduction 1. The transfer of a gene from one bacterium to another by a **bacteriophage**. In generalized transduction, any gene may be transferred as a result of accidental incorporation during phage packaging. In specialized transduction, only specific genes can be transferred, as a result of improper recombination out of the host chromosome of the **prophage** of a **lysogenic** phage. Transduction is an infrequent event, but transducing phages have proved useful in the genetic analysis of bacteria. **2.** The conversion of a signal from one form to another. For example, various types of sensory cells convert or transduce light, pressure, chemicals, etc. into nerve impulses and the binding of many hormones to receptors at the cell surface is transduced into an increase in cAMP within the cell.

transfection The introduction of DNA into a recipient eukaryote cell and its subsequent integration into the recipient cell's chromosomal DNA. Usually accomplished using DNA precipitated with calcium ions, though a variety of other methods can be used (e.g. **electroporation**). Only about 1% of cultured cells are normally transfected. Transfection is analogous to bacterial transformation, but in eukaryotes **transformation** is used to describe the changes in cultured cells caused by **tumour viruses**. Though originally used to describe the situation in which the transfected DNA is integrated, it is now frequently used just to mean introduction of DNA into a target cell, hence the necessity to specify **stable transfection**.

transfer cell Parenchyma cell specialized for transfer of water-soluble material to or from a neighbouring cell, usually a phloem sieve tube or a xylem tracheid. Elaborate wall ingrowths greatly increase the area of plasma membrane at the cell face across which transfer occurs.

transfer factor 1. A dialysable factor obtained from sensitized T-cells by freezing and thawing that may possibly immunopotentiate animals. The transfer of specific immunity from one animal to another has been claimed and it is still being reported in the literature. **2.** Transfer factor or carbon monoxide diffusing capacity (DL(CO)) is a test of the appropriateness of gas exchange across the alveolar membrane in the lung.

transfer RNA See **tRNA**.

transferase A suffix to the name of an enzyme indicating that it transfers a specific grouping from one molecule to another; e.g. acyl transferases transfer acyl groups.

transferrin The iron-storage protein (80 kDa) found in mammalian serum; a β-globulin. Binds ferric iron with a K_{ass} of around 21 at pH 7.4, 18.1 at pH 6.6. An important constituent of growth media. Transferrin receptors on the cell surface bind transferrin as part of the transport route of iron into cells.

transferrin receptor *CD71* Receptor (a glycoprotein composed of disulfide-linked polypeptide chains, each of 90 kDa), ubiquitously distributed on the cell surface of actively growing human cells. It is responsible for internalization of transferrin-bound iron and its intracellular release. In iron deficiency, the amount of serum-soluble transferrin receptor, produced by proteolysis of the membrane-bound form, increases.

transformasome Name proposed for a membranous extension responsible for binding and uptake of DNA; found on the surface of transformation-competent *Haemophilus influenzae* bacteria. Probably obsolete.

transformation Any alteration in the properties of a cell that is stably inherited by its progeny. Classical example was the transformation of *Diplococcus pneumoniae* to virulence by DNA, achieved in 1944 by Avery, MacLeod and McCarty. Currently usually refers to malignant transformation, but is used in other senses also, such as blast transformation of lymphocytes, which can be distinguished only by context. Malignant transformation is a change in animal cells in culture that usually greatly increases their ability to cause tumours when injected into animals. (It is assumed that parallel changes occur during carcinogenesis *in vivo*.) Transformation can be recognized by changes in growth characteristics, particularly in requirements for macromolecular growth factors, and often also by changes in morphology.

transformed cell See **transformation**.

transforming genes Genes, originally of tumour viruses, responsible for their ability to transform cells. The term now serves as an operational definition of **oncogenes**.

transforming growth factor *TGF* Proteins secreted by transformed cells that can stimulate growth of normal cells. Unfortunate misnomer, since they induce aspects of transformed phenotype, such as growth in semi-solid agar, but do not actually transform. TGF-α, a 50 amino acid polypeptide originally isolated from viral-transformed rodent cells, contains an **EGF-like domain** and binds to EGF receptor. Stimulates growth of microvascular endothelial cells, i.e. is angiogenic. TGF-β polypeptide, a homodimer of two 112 chains, is secreted by many different cell types, stimulates wound healing but *in vitro* is also a growth inhibitor for certain cell types. The TGF family includes many of the bone morphogenetic proteins (BMPs).

transforming virus Viruses capable of inducing malignant transformation of animal cells in culture. Among the **Oncovirinae**, non-defective viruses that lack oncogenes can induce tumours such as leukaemias in animals, but cannot transform *in vitro*. On acquisition of oncogenes they become (acute) transforming viruses.

transgelin Transformation and shape-change sensitive isoform of 21-kDa actin-binding protein. Highly conserved (as far back as yeast), binds F-actin (1:6 transgelin:G-actin) and causes gelation. Similar, but not identical, to **calponin**.

transgene A **gene** or DNA fragment from one organism that has been stably incorporated into the **genome** of another organism (usually plant or animal).

transgenic Adjective describing an organism (usually plant or animal) that contains a **transgene**.

transgenic organisms Organisms that have integrated foreign DNA into their germ line as a result of the experimental introduction of DNA.

transglutaminase *Protein-glutamine γ-glutamyltransferase; Factor XIIIa; fibrinoligase; fibrin-stabilizing factor; EC 2.3.2.13* An important extracellular enzyme that catalyses the formation of an amide bond between side-chain glutamine and side-chain lysine residues in proteins with the elimination of ammonia. The linkage is stable and plays an important role in many extracellular assembly processes.

transglycosylation Transfer of a glycosidically bound sugar to another hydroxyl group. Various bacterially derived transglycosylases, especially those from thermophiles, are being studied for use in biotechnological synthetic processes. Xyloglucan endo-transglycosylases (XETs) cleave and re-ligate xyloglucan polymers in plant cell walls via a transglycosylation mechanism and are therefore important in cell wall remodelling.

transhydrogenase *Pyridine nucleotide transhydrogenase* The NAD(P) transhydrogenase is a tetramer composed of two alpha (EC 1.6.1.1) and two beta (EC 1.6.1.2) subunits, is an integral membrane protein that couples the proton transport across the membrane to the reversible transfer of hydride ion equivalents between NAD and NADP in mitochondria. *E. coli* contains both a soluble and a membrane-bound proton-translocating pyridine nucleotide transhydrogenase.

transient expression *Transient transfection* When **transfecting** animal cells, cells in which the transgene has not been physically incorporated into the genome (**stable transfection**), but is carried as an episome that can be lost. This means that expression levels will not be constant over time, and will eventually fall away.

transin *Matrix metalloproteinase 3 (MMP-3), procollagenase activator, proteoglycanase, stromelysin 1; EC 3.4.24.17* Peptidase secreted by carcinoma cells: carboxyterminal domain has haemopexin-like domains, and the N-terminal domain has the proteolytic activity. May be involved in digestion of extracellular matrix.

transition probability model A model to account for the apparently random variation in cell cycle time between individual animal tissue cells in culture that postulates the transition from G1 to S phase is probabilistic. Contrasts with hypotheses that require the accumulation of critical levels of particular proteins. With the discovery of the various cyclins, etc., this sort of hypothesis is difficult to sustain.

transition proteins In **spermatogenesis**, group of proteins that displace **histones** from nuclear DNA, and that are in turn displaced by **protamines** to produce the transcriptionally inactive nuclear DNA characteristic of the sperm nucleus.

transition temperature The temperature at which there is a transition in the organization of, for example, the phospholipids of a membrane where the transition temperature marks the shift from fluid to more crystalline. Usually determined by using an **Arrhenius plot** of activity against the reciprocal of absolute temperature, the transition temperature being that temperature at which there is an abrupt change in the slope of the plot. In membranes, such phase-transitions tend to be inhibited by the presence of cholesterol.

transitional elements Region at the boundary of the rough endoplasmic reticulum (RER) and the Golgi. **Transport vesicles** are responsible for the transfer of secretory proteins from this part of the RER to the Golgi system.

transitional endoplasmic reticulum See **transitional elements**.

transitional epithelium An epithelial sheet made up of cells that change shape when the epithelium is stretched. Usually a **stratified epithelium**: best-known example is in the bladder.

transketolase See **transaldolase**.

translation The process that occurs at the ribosome whereby the information in mRNA is used to specify the sequence of amino acids in a polypeptide chain.

translational control Control of protein synthesis by regulation of the translation step, for example by selective usage of preformed mRNA or instability of the mRNA.

translational research Research that relates to the process of taking a scientific discovery into a clinical context; the whole process of moving a newly discovered compound from discovery research, through development and clinical trials and through the regulatory process so that it can be used on patients.

translationally controlled tumour protein *TCTP; histamine releasing factor (HRF)* Abundant protein in different eukaryotic cell types. The sequence homology of TCTP between different species is very high, and it belongs to the **MSS4**/DSS4 superfamily. TCTP is involved in both cell growth and human late allergy reaction, as well as having a calcium-binding property.

translocase *Elongation factor G; EF-G; EC:3.6.1.48* The enzyme that causes peptidyl-tRNA to move from the A-site to the P-site in the **ribosome** and the mRNA to move so that the next codon is in position for usage. Mitochondrial elongation factor G-like protein is EC:3.6.5.3. *E. coli* elongation factor G (EF-G Ec) has no translocase activity on the mitochondrial ribosome but the mitochondrial enzyme is functional on the *E. coli* ribosome.

translocated intimin receptor *Tir* Bacterial protein secreted by attaching and effacing (A/E) pathogens that forms a receptor for the adhesin **intimin**. The translocated Tir forms a tetrameric complex that binds two intimin molecules and triggers formation of host-cell signalling systems and actin polymerization. N-terminal region of Tir binds α-actinin.

translocating chain-associating membrane protein *TRAM* Transmembrane glycoprotein (probably crosses eight times) of endoplasmic reticulum (36 kDa) apparently required for the translocation of nascent proteins into the cisternal space. A component of the **translocon**. Abundant – potentially as many TRAM molecules as there are associated ribosomes.

translocation Rearrangement of a chromosome in which a segment is moved from one location to another, either within the same chromosome or to another chromosome. This is sometimes reciprocal, when one fragment is exchanged for another.

translocon The complex of proteins associated with the translocation of nascent polypeptides into the cisternal space of the endoplasmic reticulum. The translocon is a multifunctional complex involved in regulating the interaction of ribosomes with the ER, as well as regulating translocation and the integration of membrane proteins in the correct orientation. **Tram**, **signal peptidase** and signal recognition protein are among the proteins associated with the translocon.

transmembrane protein A protein subunit in which the polypeptide chain is exposed on both sides of the membrane. The term does not apply when different subunits of protein complex are exposed at opposite surfaces. Many integral membrane proteins are also transmembrane proteins.

transmembrane transducer A system that transmits a chemical or electrical signal across a membrane. Usually involves a transmembrane receptor protein that is thought to undergo a conformational change that is expressed on the inner surface of the membrane. Many such transducing species are dimeric, and the conformation change may involve interaction between the two components.

transmigration Migration of cells from one surface of a monolayer of cells to the other side; used particularly of the migration of leucocytes from the lumen of a blood vessel across vascular endothelium of the postcapillary venule and into tissue, and, by extension, *in vitro* models of this process.

transmissible mink encephalopathy One of the transmissible spongiform encephalopathies, though originally thought to be an 'unconventional' type of **slow virus** infection. Similar to **kuru**, **scrapie** and **Creutzfeldt–Jacob disease**. See **prion**.

transmission electron microscopy *TEM* Those forms of electron-microscopy in which electrons are transmitted through the object to be imaged, suffering energy loss by diffraction and to a small extent by absorption.

transpeptidase An enzyme that catalyses the formation of an amide linkage between a free amino group and a carbonyl group within an existing peptide linkage. The bacterial transpeptidase (EC 3.4.16.4), which cross-links the peptidoglycan chains to form rigid cell walls, is the target of penicillin-type antibiotics.

transpiration Loss of water vapour from land-plants into the atmosphere, causing movement of water through the plant from the soil to the atmosphere via roots, shoot and leaves. Occurs mainly through the **stomata**.

transplantation antigen Any antigen that is antigenically active in graft rejection. In practice, the **major histocompatibility complex** and the H-Y antigens and, to a lesser extent, minor histocompatibility antigens.

transplantation reaction The set of cellular phenomena observed after an allogeneic (mismatched) graft is made to an organism and leads to destruction, detachment or isolation of the graft. In mammals, this includes the invasion and destruction of the graft by cytotoxic lymphocytes, inhibition of **angiogenesis** and other processes.

transport diseases Single-gene defect diseases in which there is an inability to transport particular small molecules across membranes. Examples are aminoacidurias such as cystinuria, **iminoglycinuria**, **Hartnup disease**, **Fanconi syndrome**.

transport protein A class of transmembrane protein that allows substances to cross plasma membranes far faster than would be possible by diffusion alone. A major class of transport proteins expend energy to move substances (**active transport**); these are transport ATPases. See **facilitated diffusion**, **symport**, **antiport**.

transport vesicle Vesicles that transfer material from the rough endoplasmic reticulum (RER) to the receiving face of the Golgi.

transportase See **transport protein**.

transporter See **transport protein**.

transportin *TNPO1; karyopherin beta2; TNPO2 (karyopherin beta-2B), importin beta-2; transportin-SR* Transportins 1 and 2 are responsible for import of ribonucleoproteins (a group of pre-mRNA/mRNA-binding proteins; heterogeneous nuclear ribonucleoproteins (hnRNP)) that contain a 38-amino acid domain, termed M9, into the nucleus (specificities of TNPO1 and TNPO2 differ). Transportin-SR (transportin-3), binds specifically and directly to the phosphorylated RS domains of various proteins and is responsible for targeting **SR-proteins** to the nucleus. Activity is regulated by Ran-GTP. See **importins**.

transposable element See **transposon**.

transposase An enzyme that brings about the transposition of a sequence of DNA within a chromosome or between chromosomes. Transposase exists in dimer form with each monomer binding a separate transposon end. The DNA-binding domain of the protein is responsible for recognizing the appropriate sequence. See **transposon**.

transposition Movement form one location to another, particularly the movement of a DNA sequence (**transposon**) within the genome.

transposome Preformed **transposase–transposon** complexes that have been electroporated into bacterial cells.

transposon *Transposable element* Small, mobile DNA sequences that can replicate and insert copies at random sites within chromosomes. They have nearly identical sequences at each end, oppositely oriented (inverted) repeats, and code for the enzyme transposase, which catalyses their insertion. Bacteria have two types of transposon; simple transposons, which have only the genes needed for insertion, and complex transposons, which contain genes in addition to those needed for insertion. Eukaryotes contain two classes of mobile genetic elements; the first are like bacterial transposons in that DNA sequences move directly. The second class (retrotransposons) move by producing RNA that is transcribed, by reverse transcriptase, into DNA, which is then inserted at a new site.

transposon tagging A widely used method for generation of insertion mutants for gene functional analysis, particularly in plants (using **T-DNA**). Transposable elements create mutations at the site of insertion and genes tagged by a transposable element can be isolated by using the tag as a probe. See **activation tagging**.

transthyretin Plasma protein ($4.5\,\mu$M in plasma) that transports **thyroxine**. Tetrameric with four identical 127-residue subunits. Transthyretin forms a complex under physiological conditions with retinol-binding protein ($2\,\mu$M in plasma) so that RBP is not lost by filtration in the kidney.

transudate Plasma-derived fluid that accumulates in tissue and causes **oedema**. A result of increased venous and capillary pressure, rather than altered vascular permeability (which leads to cellular exudate formation).

transverse fracture A bone fracture caused by an impact at right angles to the bone.

transverse tubule See **T tubule**.

transversions **Point mutation** in which a purine is substituted by a pyrimidine or *vice versa*.

TRAP assay *Telomeric Repeat Amplification Protocol assay* An assay for **telomerase** activity in cells. In the first step, telomerase in the cell extract adds a number of telomeric repeats (TTAGGG) onto the 3′ end of a biotinylated oligonucleotide. In the second step, the extended products are amplified by PCR with a deoxynucleotide mix containing dCTP labeled with dinitrophenyl (DNP). This extension/amplification reaction generates a readily identifiable ladder of products with six base increments starting at 50 nucleotides.

TRAPS *Transmembrane adapter proteins* A family of proteins that include linker for activation of T-cells (**LAT**), phosphoprotein associated with glycosphingolipid-enriched micro domains (PAG)/C-terminal Src kinase (Csk) binding protein (Cbp), SHP2-interacting transmembrane adapter protein (**SIT**), T-cell receptor-interacting molecule (**TRIM**), non-T cell activation linker (NTAL) and pp30. TRAPs share several common structural features, in particular multiple sites for tyrosine phosphorylation in their cytoplasmic tails.

trastuzumab *Herceptin* Monoclonal antibody directed against **HER-2** receptor, used in treatment of HER-2-positive metastatic breast carcinoma.

TRE 1. Human **oncogene** (*tre*), isolated from NIH3T3 cells transfected with human Ewing's sarcoma DNA, that encodes a de-ubiquitinating enzyme. *Drosophila* and yeast homologues have been discovered. 2. Thyroid hormone response element (TRE), DNA sequence recognized by the thyroid hormone receptor. 3. **TPA** responsive element. 4. See **tre locus**.

tre locus Trehalose utilization locus, consisting of a transcriptional regulator, **TreR**; a trehalose phosphoenolpyruvate transferase system (PTS) transporter, treB; and a trehalose-6-phosphate hydrolase, treC.

treadmilling Name given to the proposed process in microtubules in which there is continual addition of subunits at one end, and disassembly at the other, so that the tubule stays of constant length, but individual subunits move along. Could in principle be used as a transport mechanism, although this is not currently favoured as a possibility. Has also been suggested for microfilaments.

trefoil factors *Trefoil peptides* Family of proteins containing a **trefoil motif**, considered as scatter factors, proinvasive and angiogenic agents acting through cyclo-oxygenase-2 (COX-2)- and thromboxane A2 receptor (TXA2-R)-dependent signalling pathways. Includes three members: TFF1, also called pS2; TFF2, or spasmolytic peptide (SP); and TFF3, or intestinal trefoil factor (ITF). TFFs are associated with mucin-secreting epithelial cells and play a crucial role in mucosal defence and healing. TFF1 is upregulated by estrogen in many breast carcinomas, and is present in many other carcinomas.

trefoil motif *P domain* Domain found in various secretory polypeptides that has highly conserved cysteine residues that are disulphide-bonded in such a way as to generate a trefoil structure (bonded 1–5, 2–4, 3–6). There are also highly conserved A, G and W residues.

trehalase *E.C.3.2.1.28 Alpha,alpha-trehalase* Hydrolase that converts trehalose into glucose.

trehalose A disaccharide sugar (342 Da) found widely in invertebrates, bacteria, algae, plants and fungi, formed by the dimerization of glucose. Yields glucose on acid hydrolysis.

TREK-1 One of the tandem pore potassium channels that has been shown to be mechanosensitive and to be expressed in rat heart.

trephone Substance supposedly released at a wound that stimulates mitosis: the opposite of **chalones**. Has not found its way into Entrez PubMed.

Treponema Genus of bacteria of the spirochaete family (Spirochaetaceae). *T. pallidum* causes syphilis. Cells are corkscrew-like (6–15 μm long, 0.1–0.2 μm wide), motile, anaerobic and with a **peptidoglycan** cell wall and a capsule of glycosaminoglycans similar to hyaluronic acid and chondroitin sulphate in composition. Membrane has **cardiolipin**.

treppe *Staircase phenomenon* The gradual increase in the extent of muscular contraction following rapid repeated stimulation, particularly the successive increase in amplitude of the first few contractions of cardiac muscle that has received a number of stimuli of the same intensity following a quiescent period. An increased heart rate progressively increases the force of ventricular contraction (Bowditch treppe).

TreR Trehalose-repressor, a 27-kDa protein that exists as a dimer in its native state. The trehalose operon of *Bacillus subtilis* is subject to regulation by induction, mediated by the repressor TreR. See *tre locus*.

TRF 1. Thyrotropin-releasing factor; alternative name for **thyroid-stimulating hormone**. **2. Telomeric repeat binding factors** 1 and 2 (TRF-1, TRF-2).

triabody A trimeric antibody fragment with three Fv heads, each of which consists of a VH domain from one antibody paired with a VL domain from an unrelated antibody with a linker sequence between them that is too short to permit intramolecular pairing of the domains. See **diabody**.

triacyl glycerols See **triglycerides**.

triad *Triad junction* The junction between the T-tubules and the sarcoplasmic reticulum in striated muscle.

triadin A junctional terminal cisternae protein (729 aas, predicted 82 kDa) from human skeletal muscle. Suggested to have a central role in the mechanism of skeletal muscle excitation/contraction coupling.

triamcinolone acetonide A more potent derivative of triamcinolone, a glucocorticoid similar to **prednisolone** used for treatment of inflammatory disorders and, by inhalation, asthma.

triamterene *2,4,7-Triamino-6-phenylpteridine* Potassium-sparing diuretic used to treat oedema. Inhibits the reabsorption of sodium ions in exchange for potassium and hydrogen ions at that segment of the distal tubule under the control of adrenal mineralocorticoids.

tributyltin *TBT* Organic compound added to marine paint used on the hull to prevent the growth of algae and other organisms; now banned because it causes abnormalities in some marine creatures.

tricarboxylic acid cycle *TCA cycle; citric acid cycle; Krebs cycle* The central feature of oxidative metabolism. Cyclic reactions whereby acetyl CoA is oxidized to carbon dioxide, providing reducing equivalents (NADH or $FADH_2$) to power the electron-transport chain. Also provides intermediates for biosynthetic processes.

Trichinella spiralis Nematode responsible for **trichiniasis**.

trichiniasis *Trichinosis* Nematode infestation of the human intestine with the nematode worm *Trichinella* (or *Trichina*) *spiralis*, the larvae of which migrate from the gut and become encysted in muscle. Usually acquired as a result of eating raw or undercooked pork.

trichinosis See **trichiniasis**.

trichocyst Small, membrane-bounded vesicle lying below the pellicle of many ciliates. Fusion of the trichocyst with the plasma membrane occurs at a predictable site, which can therefore be examined for membrane specialization.

Trichoderma viride A filamentous ascomycete fungus widely distributed in soil, plant material, decaying vegetation and wood. *Trichoderma* may cause infections in immunocompromised individuals.

trichohyalin *THH* Major structural protein of inner root sheath cells and medulla of hair follicle, present in small amounts in other specialized epithelia. Trichohyalin is a high molecular-weight alpha-helix-rich protein that forms rigid structures as a result of post-synthetic modification by transglutaminases that cross-link the proteins and peptidyl-arginine deiminase, which converts arginine to citrulline and modifies structure of the protein. Modified trichohyalin is thought to serve as a **keratin** intermediate filament-associated matrix protein, like **filaggrin**.

Trichomonads Mastigophoran protozoa, pear-shaped and ranging in size from 4 to 30 μm. They possess three to five anterior flagella with a recurrent anterior flagellum which is attached to the body as an undulating membrane. Trichomonads reproduce by simple longitudinal binary fission although sexual reproduction may also occur in some species, especially those parasitic on invertebrates. Many species are parasitic in vertebrates, including *Trichomonas vaginalis*, which is responsible for a sexually transmitted human disease.

trichomoniasis Infection with **trichomonads**. *Trichomonas vaginalis* causes a sexually transmitted human disease and there are over 15 species of Trichomonad infecting domestic animals.

Trichonympha Genus of flagellated protozoans symbiotic in the intestine of some cockroaches and termites, where they are responsible for the digestion of cellulose.

trichostatin A An antifungal antibiotic from *Streptomyces platensis* that is a potent reversible histone deacetylase (**HDAC**) inhibitor. Will activate the transcription of DNA methylation-mediated silenced genes in human cancer cells.

trichothecenes *T-2 toxin; HT-2 toxin; diacetoxyscirpenol; deoxynivalenol* Mycotoxins produced by various species of fungi that contaminate various agricultural products. Are toxic for granulocytic and erythroblastic progenitor cells.

Trichuris spp. Nematodes (whip-worms). *Trichuris trichiura* is a common parasite of the human gut, especially in tropical Asia.

triclosan *3-(4-Chlorophenyl)-1-(3,4-dichlorphenyl)urea* A chlorophenol used for its antibacterial properties, an ingredient in many detergents, cosmetics, lotions and insect repellents, and an additive in various plastics and textiles. Probably a human health risk and environmental risk. Blocks the active site of the bacterial enzyme enoyl-acyl carrier protein reductase that is essential for fatty acid synthesis.

tricyclic antidepressants A group of drugs used in the treatment of moderate to severe depressive illness. Some may have additional sedative properties (amitriptyline) and others are reputed to have fewer cardiac side effects (mianserin).

TRIF Toll/IL-1R (TIR) domain-containing adaptor inducing IFN-beta, a protein that is associated with TLR3 (**toll-like receptor**-3) and critically involved in TLR3-mediated signalling. TRIF-induced cleavage of **TRAF1** is required for its inhibition of TRIF signalling.

triflavin See **disintegrin**.

trifluoperazine *TFP; trifluperazine; Stellazine* Antipsychotic drug that inhibits **calmodulin** at levels just below those at which it kills cells.

trigeminal system Neurons associated with the fifth or trigeminal nerve, the largest cranial nerve. The trigeminal system provides sensory innervation to the face and mucous membrane of the oral cavity, along with motor innervation to the muscles of mastication. It is called trigeminal because it has three major peripheral branches: the ophthalmic, the maxillary and the mandibular nerves.

trigger protein See **U protein**.

triglycerides Storage fats of animal adipose tissue where they are largely glycerol esters of saturated fatty acids. In plants, they tend to be esters of unsaturated fatty acids (vegetable oils). Present as a minor component of cell membrane. Important energy supply in heart muscle.

trigramin See **disintegrin**.

tri-iodothyronine *T3* A hormone produced by the thyroid; contains one less iodine than thyroxine.

TRIM *T-cell receptor-interacting molecule* A transmembrane adapter protein (**TRAP**) found in lymphocytes.

trimethoprim A drug that inhibits the reduction of dihydrofolate (DHF) to tetrahydrofolate (a later step than that inhibited by **sulphonamides**). Selective for some bacterial DHF reductases and often used in conjunction with sulphonamides.

trinucleotide repeat Repetitive part of a genome that may form part of the coding sequence of a gene. The length of such repeats is frequently **polymorphic**, and unstably amplified repeats appear to be the major cause of such genetic diseases as **Huntington's chorea**, **fragile X syndrome**, spinobulbar muscular atrophy and myotonic dystrophy.

trio Multidomain protein (2861 residues) that binds **LAR** transmembrane tyrosine phosphatase, has a serine/threonine protein kinase domain and separate rac- and rho-specific **GEF**-domains.

triodobenzoic acid *TIBA* An inhibitor of basipetal **auxin** transport in plants.

triple A syndrome An autosomal recessive neuroendocrinological disease caused by mutations in a gene that encodes 546 amino-acid residues. The encoded protein is the **nucleoporin** ALADIN, a component of the nuclear pore complex.

triple response The vascular changes in the skin in response to mild mechanical injury, an outward-spreading zone of reddening (flare) followed rapidly by a weal (swelling) at the site of injury. Redness, heat and swelling, three of the 'cardinal signs' of inflammation, are present.

triple vaccine Vaccine for diphtheria, tetanus and whooping cough in infants. A misleading claim that it was associated with autism led to a recent drop in use and an increase in the incidence of these infectious diseases. A sterile preparation of diphtheria and tetanus toxoids with acellular pertussis vaccine.

triploid Having three times the haploid number of chromosomes.

triptan General term for a family of drugs used for the acute treatment of migraine attacks. Examples are sumatriptan (Imitrex®, Imigran®), zolmitriptan and naratriptan. They are serotonin 5-HT$_{1B/1D}$ receptor agonists that act to induce vasoconstriction of extracerebral blood vessels and reduce neurogenic inflammation.

triskelion A three-legged structure assumed by clathrin isolated from **coated vesicles**. A trimer of clathrin (180 kDa) with three light chains is probably the physiological subunit of clathrin coats in coated vesicles.

trisomy An additional copy of a chromosome so that there are three copies not two in a diploid organism. Best-known example is trisomy 21 in **Down's syndrome**.

tristetraprolin *TTP* An ARE (**AU-rich element**) binding protein, the only trans-acting factor shown to be capable of regulating AU-rich element-dependent mRNA turnover at the level of the intact animal.

triterpene Terpenes consisting of six isoprene units. **Squalene** is a linear triterpene.

tritium ^3H Long-lived radioactive isotope of hydrogen (half-life 12.26 years). Weak β-emitter, very suitable for autoradiography, and relatively easy to incorporate into complex molecules.

Triton X-100 *Iso-octylphenoxypolyethoxyethanol* Non-ionic detergent used in isolating membrane proteins: the detergent replaces the phospholipids that normally surround such a protein. Other detergents of the Triton group are occasionally used, so the full name should be quoted.

trituration Reduction in particle size by grinding in a mortar and pestle.

Triturus Genus of newts, much studied for their **lampbrush chromosomes**.

trk Oncogene, from human colon carcinoma; a chimeric gene formed through a somatic rearrangement involving the neighbouring genes for neurotrophic tyrosine kinase receptor type 1 (*NTRK1*) and tropomyosin-3. See Table O1. The *trk* gene product is a receptor for **NGF**, that of the *trkB* gene the receptor for **neurotrophin** 4 and **BDNF**, and the *trkC* gene codes for the receptor for NT-3.

TRK-fused gene *TFG* A partner of NTRK1 tyrosine kinase in generating the thyroid *TRK-T3* **oncogene**. Physiological role is unknown, but has a number of protein-interaction motifs.

***TRK-T3* oncogene** Gene producing a thyroid oncoprotein, isolated from a human papillary thyroid tumour. Gene consists of the **NTRK1** tyrosine kinase domain fused in-frame with sequences of the TFG (**TRK-fused gene**).

tRNA *Transfer RNA; s-RNA; 4S RNA* The low-molecular weight RNAs that specifically bind amino acids by amino-acylation to form **aminoacyl tRNA**, and that possess a special nucleotide triplet, the anticodon, sometimes containing the base inosine. They recognize codons on mRNA. By this recognition the appropriate tRNAs are brought into alignment in turn in the ribosome during protein synthesis (translation), there being at least one species of tRNA for each amino acid. In practice, most cells possess about 30 types of tRNA. The amino acids are bound at the $3'$ terminus, which is always $3'$-ACC. The anticodon is around 34–38 nucleotides from the $5'$ end, and the total length of the various tRNAs is 70–80 bases.

trochophore Free-living ciliated larval form of several different invertebrate phyla.

trophectoderm The extraembryonic part of the ectoderm of mammalian embryos at the blastocyst stage before the mesoderm becomes associated with the ectoderm.

trophic Concerning food or nutrition. Not to be confused with tropic (stimulatory).

trophoblast Extraembryonic layer of epithelium that forms around the mammalian blastocyst, and attaches the embryo to the uterus wall. Forms the outer layer of the chorion, and together with maternal tissue will form the placenta.

trophoblast protein 1 *TP-1* Protein secreted by **trophoblast**, which prolongs the lifetime of the corpus luteum, thus signalling pregnancy. Structurally related to **interferons**.

trophophase In a culture system, for example in a bioreactor, the phase during which the cells or organisms are growing rapidly but not producing the metabolic products of interest. See **idiophase**.

trophosome An organ of dark green-brown spongy tissue, found in hydrothermal vent tubeworms (vestimentiferans) such as *Riftia pachyptila* (Siboglinidae, Polychaeta), in which there are endosymbiotic thiotrophic bacteria within host bacteriocytes that comprise 70% of the trophosome's volume.

trophozoite The feeding stage of a protozoan (as distinct from reproductive or encysted stages).

tropocollagen Subunit from which collagen fibrlls self-assemble: generated from **procollagen** by proteolytic cleavage of the extension peptides.

tropoelastin Soluble polypeptide ($\sim$72 kDa), precursor to **elastin**, consisting mainly of repetitive elements of four, five, six and nine hydrophobic residues. Tropoelastin molecules, having been chaperoned out of the cell, are aligned on a scaffold of **fibrillin**-rich microfibrils and then stabilized by the formation of intermolecular cross-links (**desmosines**) to form elastin fibres.

tropomodulin Actin-capping protein that interacts with tropomyosin (TM) at the pointed end of actin filaments. Found in erythrocyte membrane skeleton and in various non-erythroid cells. Binds actin, tropomyosin and **nebulin**.

tropomyosin Protein (66 kDa) associated with actin filaments both in cytoplasm and (in association with **troponin**) in the thin filament of striated muscle. Composed of two elongated α-helical chains (each about 33 kDa), 40 nm long, 2 nm in diameter. Each chain has six or seven similar domains and interacts with as many G-actin molecules as there are domains. Not only does the binding of tropomyosin stabilize the F-actin, but the association with troponin in striated muscle is also important in control by calcium ions.

troponin *TnC, TnI, TnT* Complex of three proteins, troponins C, I and T, associated with **tropomyosin** and actin on the thin filament of striated muscle, upon which it confers calcium sensitivity. There is one troponin complex per tropomyosin. Troponin C (18 kDa) binds calcium ions reversibly, has a variable number of **EF-hand** motifs and is the least variable of the subunits. TnC binds TnI and TnT, but not actin. Troponin I (23 kDa) binds to actin and at 1:1 stoichiometry can inhibit the actin–myosin interaction on its own. Troponin-T (37 kDa) binds strongly to tropomyosin.

trp 1. Tryptophan. 2. Operon encoding tryptophan metabolism (*trp*) genes in *E. coli*. 3. Transient receptor potential mutant in *Drosophila*, gene codes for a constitutively active calcium permeable channel activated by **thapsigargin** (though trp1 is insensitive to thapsigargin). See **TRP channel**. 4. Human tyrosinase-related protein-1.

TRP channels *Transient receptor potential channels* Non-selective cation channels found in dorsal root ganglia neurons and skin cells that are responsible for temperature sensing and characterized by their unusually high temperature sensitivity ($Q_{10} > 10$). Two subclasses are recognized: TRP-melastatin (TRPM) and TRP-vanilloid-related (TRPV). Six thermo-TRP channels have been cloned; TRPV1-4 are heat-activated, whereas TRPM8 and TRPA1 are activated by cold. **Capsaicin** and resiniferatoxin are agonists for TRPV1, menthol for TRPM8 (cold receptor), and icilin for both TRPM8 and TRPA1. See **vanilloid receptor**.

TRT *Telomerase reverse transcriptase; hEST2* The protein subunit (123 kDa) that catalyses telomeric DNA extension has sequence and functional characteristics of a **reverse transcriptase** related to retrotransposon and retroviral reverse transcriptases. The *S. cerevisiae* homologue has been found and subsequently identified as EST2 (ever shorter telomeres). See **hTERT**.

Trypan blue *Diamine blue; Niagara blue 3B* Biological stain, an azo dye derived from toluene, used to determine cell viability. Trypan blue is unable to cross intact plasma membranes, and so only labels dead cells.

Trypanosoma Genus of Protozoa that causes serious infections in humans and domestic animals. African trypanosomes, of the brucei group, are carried by Tsetse flies (*Glossina*), and when they enter the bloodstream of the mammalian host go through a complex series of stages. Perhaps the most interesting feature is that there are recurrent bouts of parasitaemia as the parasite alters its surface antigens to evade the immune response of the host (see **antigenic variation**). The repertoire of antigenic variation is considerable. The S. American trypanosomes (of which *T. cruzi* is the best known) are carried by reduviid bugs, and cause a chronic and incurable disease (Chagas disease). Other interesting features of trypanosomes are the kineto-plast DNA and glycosomes (organelles containing enzymes of the glycolytic chain).

trypanosomiasis Disease caused by *Trypanosoma*.

trypomastigote Any trypanosome-like stage in the life cycle of certain trypanosomatid protozoa, resembling the typical adult form of members of the genus *Trypanosoma* – slender, elongated cells with a kinetoplast and basal body located at the posterior end and a flagellum running anteriorly along an undulating membrane. The trypomastigote stage of *Trypanosoma* spp. is the infectious stage carried by the insect vector.

trypsin *EC 3.4.21.4* Serine peptidase (23 kDa) from the pancreas of vertebrates. Cleaves peptide bonds involving the amino groups of lysine or arginine.

tryptophan *Trp; W; 204 D* One of the 20 amino acids found in proteins. Essential dietary component in humans. Precursor of nicotinamide. See Table A2.

TSG-14 *TNF-stimulated gene 14* TNF-inducible gene of fibroblasts encoding a protein of the **pentraxin** family.

TSH See **thyroid-stimulating hormone**.

TSH releasing factor A tripeptide produced by the hypothalamus that stimulates the anterior pituitary to release thyroid-stimulating hormone (**TSH**).

Tst Gene found as part of a 15.2-kb genetic element in some strains of *Staphylococcus aureus*; codes for toxic shock syndrome toxin-1.

T-suppressor cell Class of $CD8^+CD28^-$ T-cells that suppress T or B antigen-dependent responses in an antigen-specific way. The T-suppressor cell alloantigen (Tsud) maps near immunoglobulin allotype genes and may be a heavy chain constant region marker on a T-cell receptor. There are suggestions that several different classes of T-suppressor cells may exist and may be particularly important in graft rejection.

TTC 1. 2,3,5-triphenyltetrazolium chloride (TTC), a compound that can be reduced by cellular respiration to produce a red formazan pellet which can be extracted and measured spectrophotometrically. Also used as a stain to evaluate the volume of infarction in tissue. *In vitro* reduction by cells produces formazan. 2. Threshold of toxicological concern (TTC), a pragmatic risk assessment tool that is based on the principle of establishing a human exposure threshold value for all chemicals, below which there is a very low probability of an appreciable risk to human health. 3. **Tetanus toxin** C fragment (TTC). 4. Friedreich's ataxia is caused by the massive expansion of GAA.TTC repeats in intron 1 of the **frataxin** gene. 5. A stable prostacyclin analogue, TTC-909.

ttp 1. Thrombotic thrombocytopenic purpura, a condition attributed to the presence of an autoantibody to ADAMTS13, the metallopeptidase that degrades ultra-large von Willebrand protein multimers. 2. See **tristetraprolin**. 3. See **SH3BP4**.

TTSS *T3SS* Type III secretion system (TTSS) is critical for adaptation to the intracellular environment within both phagocytic and epithelial cell types, and is encoded in *Salmonella* Pathogenicity Island 2 (SPI2). The system enables many Gram-negative bacterial pathogens to translocate proteins into the eukaryotic host cells that they infect. The type III secretion apparatus (T3SA) is a multisubunit membrane-spanning macromolecular assembly comprising more than 20 different proteins.

T tubule *Transverse tubule* Invagination of the plasma membrane (sarcolcmma) of striated muscle that lies between two tubular portions of the endoplasmic (sarcoplasmic) reticulum to form a triad of membrane profiles adjacent to, in some cases, the A-band/I-band junction, and in other cases to the **Z-disc**, of the resting sarcomere. Depolarization of the T tubule membrane triggers the release of calcium from the sarcoplasmic reticulum and, eventually, muscle contraction.

TTX See **tetrodotoxin**.

T-type channels A class of **voltage-sensitive calcium channels** that open transiently in response to relatively small depolarizations of the neuronal membrane. May have a role in repetitive firing. Three family members have been cloned. No selective inhibitors are known, although the channels are modulated by various hormones and neurotransmitters.

tubby Tubby and **tubby-like proteins** (TULPs) are encoded by members of a small gene family. An autosomal recessive mutation in the mouse tub gene leads to progressive retinal degeneration, deafness and maturity-onset obesity. In *C. elegans*, as in mammals, mutations in the tubby homolog, tub-1, promote increased fat deposition. There is a TUBBY-like protein gene family with 11 members in *Arabidopsis* where they may participate in the **abscisic acid** signalling pathway.

tubby-like proteins *TULP-1, 2, 3* TULP-1 is a photoreceptor-specific protein of unknown function that, when mutated, can cause retinitis pigmentosa in humans and photoreceptor degeneration in mice. It is a cytoplasmic protein that associates with cellular membranes and the cytoskeleton; TULP1 and actin appear to interact and co-localize in photoreceptor cells of the retina.

Tube *Drosophila* mutant. Tube is a maternally encoded protein that, together with **pelle,** transduces the signal from **toll.** Toll, Cactus and Dorsal, along with Tube and Pelle, participate in a common signal transduction pathway to specify the embryonic dorsal–ventral axis.

tubercle Chronic inflammatory focus, a **granuloma**, caused by *Mycobacterium tuberculosis.*

tuberculin skin test See **Mantoux test**, **Heaf test**.

tuberin *TSC2* Product of the tumour-suppressor gene *TSC2* that is involved in **tuberous sclerosis**. Hamartin (product of *TSC1*) and tuberin form a complex, of which tuberin is assumed to be the functional component. The TSC proteins have been implicated in the control of the cell cycle by activating the cyclin-dependent kinase inhibitor p27 and in cell size regulation by inhibiting the mammalian target of rapamycin (mTOR)/p70S6K cascade. Tuberin has a GTPase-activating domain that acts on Rheb, which in turn acts on the Ras/B-Raf/C-Raf/MEK signaling network.

tuberous sclerosis Autosomal dominant disorder caused by mutation in tumour-suppressor genes *TSC1* or *TSC2*. Disease characterized by range of features, including seizures, mental retardation, renal dysfunction and dermatological abnormalities. *TSC1* encodes **hamartin**, *TSC2* encodes **tuberin**.

tubocurarine An alkaloid that acts as a muscle relaxant by blocking acetylcholine (ACh) receptors.

tubulin Abundant cytoplasmic protein (55 kDa), found mainly in two forms, α and β. A tubulin **heterodimer** (one α, one β) constitutes the **protomer** for microtubule assembly. Multiple copies of tubulin genes are present (and are expressed) in most eukaryotic cells studied so far. The different tubulin isoforms seem, however, to be functionally equivalent. γ-tubulin is localized in the **centrosome** and is involved in nucleation of microtubule assembly during the cell cycle. Highly conserved from yeast to mammals.

tularemia Disease of rodents and rabbits caused by *Pasteurella tularense*. Can infect humans (either transmitted by the deer-fly, or by direct contact with the bacterium).

tumbu disease A disease caused by invasion of the surface of the body by the larvae of the tumbu fly, *Cordylobia anthropophaga*, which produces a boil or a warble in the skin. Common in Central and West Africa.

tumor Alternative (USA) spelling of **tumour**.

tumorigenic Capable of causing tumours. Can refer either to a carcinogenic substance or agent (such as radiation) that affects cells, or to transformed cells themselves.

tumour *tumor (US)* Strictly, any abnormal swelling, but usually applied to a mass of neoplastic cells.

tumour angiogenesis factor *TAF* Substance(s) released from a tumour that promotes vascularization of the mass of neoplastic cells. Once a tumour becomes vascularized, it will grow more rapidly, and is more likely to metastasize. TAF is almost certainly more than one substance. See **angiogenin** and **VEGF**.

tumour cell Cell derived from a tumour in an animal. Refers to a tumour-causing malignant cell, and not an adventitious normal cell. Loosely, a transformed cell able to give rise to tumours.

tumour initiation First stage of **tumour** development. See also **tumour progression**.

tumour-necrosis factor *TNF* TNFα or **cachectin**, originally described as a tumour-inhibiting factor in the blood of animals exposed to bacterial **lipopolysaccharide** or **Bacille Calmette-Guerin** (BCG). Preferentially kills tumour cells *in vivo* and *in vitro*, causes necrosis of certain transplanted tumours in mice and inhibits experimental metastases. Human TNFα is a protein of 157 amino acids and has a wide range of proinflammatory actions. Usually considered a cytokine. Soluble TNFα is released from the cell surface by the action of **TACE** (TNFα converting enzyme), a metalloproteinase of the **ADAM** type. TNFβ (**lymphotoxin**) has 35% structural and sequence homology with TNFα and binds to the same **TNF receptors**. Unlike TNFα, TNFβ has a conventional signal sequence and is secreted from activated T- and B-cells.

tumour progression Second stage of **tumour** development. See also **tumour initiation**.

tumour promoter Agent that in classical studies of carcinogenesis in rodent skin was able to increase the sensitivity of tumour formation by a previously applied primary carcinogen, but was unable to induce tumours when used alone. Important example was croton oil, active ingredients of which are now believed to be phorbol esters. These are believed to act as analogues of diacylglycerols, and may activate protein kinase C. Strictly speaking, not the same as a co-carcinogen, which is defined as being active when administered at the same time. Tumour promoters generally are carcinogens when tested more stringently.

tumour specific antigen *Tumour specific transplantation antigen; TSTA* Antigen on tumour cells detected by

cell-mediated immunity. For virus-transformed cells, TSTA (unlike **T-antigen**) is found to differ for different individual tumours induced by the same virus. May consist of fragments of T-antigens exposed at the cell surface.

tumour suppressor *Anti-oncogene; cancer susceptibility gene* A gene that encodes a product that normally negatively regulates the cell cycle, and that must be mutated or otherwise inactivated before a cell can proceed to rapid division. Examples: **p53**, RB (retinoblastoma), WT-1 (Wilms' tumour), DCC (deleted in colonic carcinoma), NF-1 (neurofibrosarcoma) and APC (adenomatous polyposis coli).

tumour virus Virus capable of inducing tumours.

tumour-suppressor complex *Tuberous sclerosis complex* Complex of the products of *TSC-1* and *TSC-2* (**tuberous sclerosis** complex genes 1 and 2), **hamartin** and **tuberin**, that integrates inputs from multiple signalling cascades to inactivate the **small GTPase** rheb, and thereby inhibit **mTOR**-dependent cell growth.

TUNEL method *Transferase-mediated dUTP nick-end labelling* Enables the visualization of cells undergoing apoptosis by labelling the ends of their fragmented DNA.

tunicamycin Nucleoside antibiotics from *Streptomyces lysosuperificus* that act in eukaryotic cells to inhibit N-glycosylation. Tunicamycin inhibits the first step in synthesis of the dolichol-linked oligosaccharide, by preventing the addition of *N*-acetyl glucosamine to dolichol phosphate.

tunicates A group of marine animals of the phylum Urochordata, closely related to the phylum Chordata that includes all vertebrates. Although the adult form is generally a sessile filter-feeder (sometimes colonial), the larva is free-living and tadpole-like. Sea squirts.

tunichromes Yellow, polyphenolic tripeptides prevalent in blood cells of tunicates.

turbidimetry Measurement of the turbidity, cloudiness, of a solution, often using an instrument called a **nephelometer**. The turbidity is caused by suspended particles; the relationship between particle size and number and the amount of light scattering is complex.

turgor The pressure within cells, especially plant cells, derived from osmotic pressure differences between the inside and outside of the cell, giving rise to mechanical rigidity of the cells. Turgor drives cell expansion and certain movements, such as the closing or opening of stomata.

Turner's syndrome Genetic defect in humans in which there is only one X chromosome (affected individuals are therefore phenotypically female), probably as a result of meiotic non-disjunction. X-linked diseases normally restricted to males may manifest in such patients, further confusing the picture.

turnover number Equivalent to V_{max}, being the number of substrate molecules converted to product by one molecule of enzyme in unit time, when the substrate is saturating.

Turtox US biological supply company, no longer in business but produced many wall charts and materials for school biology laboratories. Also produced a journal called *Turtox News*.

TW-240/260 kDa protein Protein (240/260 kDa) found in the terminal web of intestinal epithelial cells. Probably an isoform of **spectrin** and **fodrin**.

TWEAK A member of the TNF family, is expressed on IFN-gamma-stimulated monocytes and induces cell death in certain tumour cell lines. The probable ligand for **TRAMP**.

Tween *Polyoxyethylene sorbitan monolaurate* Detergents used in various ways – for example, as blocking agents for membrane-based immunoassays, for solubilizing membrane proteins, for lysing mammalian cells (at concentration of 0.05–0.5%). Tween variants (Tween-20, -40, -60, -80) have different chain-length fatty-acid moieties.

twinfilin Yeast actin-depolymerizing factor containing two **ADF-H domains**. Localizes to the cortical actin cytoskeleton. Will sequester G-actin by forming tight 1:1 complex, but does not seem to cross-link filaments. Human homologue has been identified in fibroblasts.

twinstar *Tsr* The *Drosophila* homologue (17 kDa) of **cofilin**/ADF (actin depolymerization factor) is a component of the cytoskeleton that regulates actin dynamics. Mutations in twinstar result in defects in centrosome migration and cytokinesis.

twist A highly conserved bHLH transcription factor, known to promote epithelial-mesenchyme transition, induced by a cytokine signalling pathway that requires the dorsal-related protein RelA; Twist-1 and -2 repress cytokine gene expression through interaction with RelA, thus forming a negative feedback loop that represses the NFkappaB-dependent cytokine pathway. Twist proteins transiently inhibit **Runx2** function during skeletogenesis. Loss-of-function mutations of the *TWIST 1* gene are responsible for **Saethre-Chotzen syndrome**.

twitch muscle Striated muscle innervated by a single motoneuron and having an electrically excitable membrane that exhibits an all-or-none response (*cf.* tonic muscle); in mammals, almost all skeletal muscles are twitch muscles. Physiologists often divide muscles into fast- and slow-twitch types, the fast-twitch muscles being associated with fast motor units.

twitchin Large protein (667 kDa) associated with myosin and important in muscle assembly. Has multiple fibronectin III-homology repeats. Product of *unc-22* gene in *C. elegans*, mutations which produce animals that show twitching movements.

two-dimensional gel electrophoresis A high-resolution separation technique in which protein samples are separated by isoelectric focusing in one dimension and then laid on an SDS gel for size-determined separation in the second dimension. Can resolve hundreds of components on a single gel.

two-hybrid system *Yeast two-hybrid system* Screening system to identify genes encoding proteins, which interact specifically with other proteins. One gene is expressed in yeast as a **fusion protein** with the DNA-binding site of the **GAL4** transcription factor, and the other gene co-expressed as a fusion with the transcriptional activator domain of GAL4. Only if the two proteins interact directly are the two GAL4 domains held in close enough proximity to trigger expression of a **reporter gene** (usually *LacZ*) downstream of the UASG promoter recognized by GAL4.

TxA2 See **thromboxanes**.

Ty element Transposable element of yeast, *S. cerevisiae*. Each consists of a central region of around 5.6 kb flanked by direct repeats of around 330 bp. There are multiple Ty elements in each haploid genome.

TY-5 See **macrophage inflammatory protein 1α**.

Tyk2 *EC 2.7.1.112* Janus kinase (tyrosine kinase; 135 kDa) upstream of STAT3, important in cytokine signalling and is an important regulator of lymphoid tumour surveillance.

tylose A parenchyma cell outgrowth that wholly or partly blocks a **xylem** vessel. It grows out from an axial or ray parenchyma cell through a pit in the vessel wall.

typhoid Enteric fever due to infection with *Salmonella typhi*. Characterized by prolonged fever, a rose rash and inflammation of the small intestine with ulceration. Faecal contamination of food or water is the standard cause of infection.

typhus Fever due to infection with *Rickettsia prowazekii*, transmitted by lice. See **scrub typhus**, a less serious infection caused by a related rickettsia.

tyrosinase A copper-containing protein (a monoxygenase) that catalyses the oxidation of tyrosine, and sets in train spontaneous reactions that yield melanin, the black pigment of skin, hair and eyes. The first intermediate is 3,4-dihydroxyphenylalanine (DOPA). Lack of tyrosinase activity is responsible for albinism.

tyrosine *Tyr; Y; 181 Da* One of the 20 amino acids directly coded in proteins. Non-essential in humans, since it can be synthesized from phenylalanine. See Table A2.

tyrosine hydroxylase *EC 1.14.16.2* Enzyme required for the synthesis of the neurotransmitters **noradrenaline** and **dopamine**.

tyrosine kinase Kinases that phosphorylate protein tyrosine residues. These kinases play major roles in mitogenic signalling, and can be divided into two subfamilies: receptor tyrosine kinases, which have an extracellular ligand-binding domain, a single transmembrane domain, and an intracellular tyrosine kinase domain; and non-receptor tyrosine kinases, which are soluble, cytoplasmic kinases.

tyrosine kinase phosphorylation site Substrates of tyrosine protein kinases are generally characterized by a lysine or an arginine seven residues to the N-terminal side of the phosphorylated tyrosine, and an acidic residue (Asp or Glu) three or four residues to the N-terminal side of the tyrosine. There are, however, a number of exceptions to this rule, such as the tyrosine phosphorylation sites of enolase and lipocortin II.

tyrosine phosphorylation See **tyrosine kinase**.

tyrphostins Protein tyrosine kinase inhibitors; various types are available (tyrphostin A51, A25, etc). A25 inhibits GTPase activity of transducin; blocks the induction of inducible nitric oxide synthase in glial cells; induces apoptosis in human leukaemic cell lines.

U

U1 snRNP One of the classes of **small nuclear RNAs**. There are six U-types that have a high uridylic acid content, U1–U5 are synthesized by RNA Polymerase II, U6 by RNA Polymerase III. See **U2 snRNP**.

U2AF *U2 snRNP auxiliary factor* Human U2AF, a major determinant in 3′ splice site selection, is a heterodimeric protein composed of a 65-kDa large subunit (hU2AF65) and a 35 kDa small subunit (hU2AF35). Both subunits are conserved in other organisms, and homologues have been identified in *Drosophila*, *Schizosaccharomyces* and *Caenorhabditis*. It binds site-specifically to the intron pyrimidine tract between the branchpoint sequence and 3′ splice site at an early step in spliceosome assembly and recruits **U2 snRNP** to the branch site.

U2OS Human osteosarcoma cell line.

U2 snRNP Small nuclear ribonucleoprotein particle forming part of the **spliceosome** that is targeted to the 3′-splice site of pre-mRNA by **U2AF**. **U1 snRNP** defines the 5′ splice site.

U6 U6 small nuclear RNA (snRNA) associates with the specific protein Prp24p and a set of seven LSm2p–8p proteins to form the U6 small nuclear ribonucleoprotein (snRNP). The U6 RNA intramolecular stem-loop (ISL) is a conserved component of the **spliceosome**. U6 promoters are widely used to drive expression of siRNA in cells.

U937 Human myelomonocytic cell line frequently used as a model for myeloid cells, although the cells are rather undifferentiated. Derived from a patient with histiocytic leukaemia.

UASG *Upstream activation site G* Promoter sequence recognized by the **GAL4** transcription factor. Used to control expression of wide-range gene products in **transgenic** plants and animals under GAL4 control.

UBC proteins Family of proteins involved in conjugating **ubiquitin** to proteins. UBC1, UBC4 and UBC5 have a role in targeting proteins for degradation, but others have more complex roles, including an involvement in cell cycle control (UBC3) and the secretory pathway (UBC6).

ubiquinone *Coenzyme Q* Small molecule with a hydrocarbon chain (usually of several isoprene units) that serves as an important electron carrier in the respiratory chain. The acquisition of an electron and a proton by ubiquinone produces ubisemiquinone (a free radical); a second proton and electron convert this to dihydro-ubiquinone. Plastoquinone, which is almost identical to ubiquinone, is the plant form.

ubiquitin A protein (8.5 kDa) found (ubiquitously) in all eukaryotic cells. Can be linked to the lysine side chains of proteins by formation of an amide bond to its C-terminal glycine in an ATP-requiring process. The protein/ubiquitin complex is subject to rapid proteolysis. Ubiquitin has a role in the heat-shock response, is involved in quality control of nascent proteins, membrane trafficking, cell signalling, cell cycle control, X chromosome inactivation and the maintenance of chromosome structure. See **E3 ligase**.

ubiquitin carboxy-terminal hydrolase L1 *UCH-L1; EC 3.1.2.15* An abundant protein in the brain (up to 2% of total protein). A cysteine endopeptidase that is thought to cleave polymeric ubiquitin to monomers and hydrolyse bonds between ubiquitin and small adducts. Found in **Lewy bodies** and mutations in UCH-L1 have been associated with familial forms of Parkinsonism.

ubiquitinylation *Ubiquitinoylation* The covalent addition of **ubiquitin** residues to proteins. Single or multiple residues can be added and bound ubiquitin can also be a site for further addition of ubiquitin residues.

ubisemiquinone See **ubiquinone**.

U-box See **E3 ligase**.

UDG *Uracil-DNA glycosylase; EC 3.2.2.-* Abundant and ubiquitous enzyme responsible for initiating repair of U/G mispairings. The mitochondrial UNG1 and the nuclear UNG2 are generated from the same gene by different promoter usage and alternative splicing.

UDP-galactose *Uridine diphosphate-galactose* Sugar nucleotide, active form of galactose for galactosyl transfer reactions.

UDP-glucose *Uridine diphosphate-glucose* Sugar nucleotide, active form of glucose for glucosyl transfer reactions.

ulcer Inflamed area where the epithelium and underlying tissue is eroded.

ulcerative colitis Inflammation of the colon and rectum; cause unclear, although there are often antibodies to colonic epithelium and *E. coli* strain 0119 B14. Susceptibility to ulcerative colitis and to other forms of inflammatory bowel disease is determined by multiple genes. See **Crohn's disease**, the other form of inflammatory bowel disease (IBD).

ultrabithorax Ubx *Drosophila* **homeotic** gene that is part of the **bithorax complex**. Mutations in *Ubx* affect parasegments 5–6, corresponding to the posterior thorax and anterior abdomen of the adult.

ultracentrifugation Centrifugation at very high g-forces; used to separate molecules, e.g. mitochondrial from nuclear DNA on a caesium chloride gradient.

ultradian Cycles of biological activity that occur with a frequency of less than 24 h. *cf.* **circadian**.

ultrafiltration Filtration under pressure. In the kidney, an ultrafiltrate is formed from plasma because the blood is at higher pressure than the lumen of the glomerulus. Also used experimentally to fractionate and concentrate solutions in the laboratory using selectively permeable artificial membranes.

ultrastructure General term to describe the level of organization that is below the level of resolution of the light microscope. In practice, a shorthand term for 'structure observed using the electron microscope', although other techniques could give information about structure in the submicrometre range.

ultraviolet *UV* Continuous spectrum beyond the violet end of the visible spectrum (wavelength less than 400 nm) and above the X-ray wavelengths (greater than 5 nm). Glass absorbs UV, so optical systems at these wavelengths have to be made of quartz. Nucleic acids absorb UV most strongly at around 260 nm, and this is the wavelength most likely to cause mutational damage (by the formation of thymine dimers). It is the UV component of sunlight that causes actinic keratoses to form in skin, but that is also required for vitamin D synthesis.

umami One of the five 'basic' tastes (salt, sweet, sour, bitter and umami), generally described as that of monosodium glutamate.

umbelliferone *7-Hydroxycoumarin* Common in many plants. Can be used as a fluorescent pH indicator.

umbo A boss or protuberance; the beak-like prominence which represents the oldest part of a bivalve shell.

UNC-5 family Family of receptors for **netrins** involved in repulsion of growing axons. UNC-5 tyrosine phosphorylation is known to be important for netrin to induce cell migration and axonal repulsion. See also **DCC protein family**.

unc-6 Gene identified in studies of neuronal development of the nematode, *C. elegans*. Homologous to **netrin**.

uncoupling agent Agents that uncouple electron transport from oxidative phosphorylation. Ionophores can do this by discharging the ion gradient across the mitochondrial membrane that is generated by electron transport. In general, the term applies to any agent capable of dissociating two linked processes.

uncoupling proteins *UCP-1, UCP-3* Members of the mitochondrial transporter superfamily. UCP-1 is expressed exclusively in brown adipose tissue; UCP-3 predominantly in skeletal muscle. Mice overexpressing UCP-3 are hyperphagic but weigh less than normal littermates and have an increased glucose clearance rate, suggesting that UCP-3 has a role in influencing metabolic rate and glucose homeostasis. See **thermogenin**.

underlapping Possible outcome of collision between two cells in culture, particularly head-side collision: one cell crawls underneath the other, retaining contact with the substratum, and obtaining traction from contact with the rigid substratum (unlike **overlapping**, where traction must be gained on the dorsal surface of the other cell).

unequal crossing-over Crossing over between homologous chromosomes that are not precisely paired, resulting in non-reciprocal exchange of material and chromosomes of unequal length. Favoured in regions containing tandemly repeated sequences.

unineme theory Theory that proposes that each chromosome (before S phase) consists of a single strand of DNA. Now generally accepted, and, being non-controversial, the term has fallen into disuse.

uniport A class of transmembrane **transport proteins** that conveys a single species across the plasma membrane.

unit membrane The three-ply, ~7-nm wide membrane structure found in all cells, composed of a fluid lipid bilayer with intercalated proteins. The unit membrane theory carries with it the presumption that all biological membranes have basically the same structure.

univoltine Producing only a single set of offspring during the breeding season. *cf.* multivoltine.

unsaturated fatty acid Fatty acid with one or more double bonds. See Table L3.

untranslated region *UTR* The regions of a cDNA, typically that 5′ to the initiation (ATG) site and that 3′ to the stop site, which are not translated to make a peptide. Their functions are not well understood.

uPA *Urokinase plasminogen activator; EC 3.4.21.73* Serine peptidase (of the typsin family) that cleaves plasminogen to form plasmin. Structurally different from **tissue-plasminogen activator** (EC 3.4.21.68).

uPAR *Urokinase plasminogen activator (uPA) receptor; CD87* The **uPA**/uPAR system controls matrix degradation in the processes of tissue remodelling, cell migration and invasion, and is involved in tumour progression and metastasis of a variety of cancers.

U protein 1. The largest of the major **hnRNA** proteins (120 kDa); binds pre-mRNA *in vivo* and binds both RNA and ssDNA *in vitro*. 2. Occasional abbreviation for 'urinary protein', protein present in urine. 3. (Obsolete) Hypothetical protein thought to regulate the transition of cells from G0 to G1 phase of the cell cycle, and thus inevitably into S-phase. The idea was that the concentration of this unstable (U) protein would have to exceed a threshold level for triggering progression through the cycle. This has never gained currency in the literature, and has been superseded with our knowledge of **cyclins**.

upstream Refers to nucleotide sequences that precede the codons specifying mRNA or that precede (are on the 5′ side of) the protein-coding sequence. Also used of the early events in any process that involves sequential reactions.

uracil *2,6-Dihydroxypyrimidine* The pyrimidine base from which uridine is derived.

uranyl acetate Uranium salt that is very electron-dense and is used as a stain in electron microscopy, usually for staining nucleic acid-containing structures in sections.

ure2p Yeast **prion**-like protein.

urea *(NH₂)₂CO* The final nitrogenous excretion product of many organisms (but see **uric acid**). The first organic compound to be artificially synthesized from inorganic starting materials, now produced commercially in huge amounts (as a fertilizer) from synthetic ammonia and carbon dioxide. Urea solutions are used experimentally to denature proteins.

uredospore *Urediniospore; urediospore* A binucleate spore which rapidly propagates; the dikaryotic phase of a rust fungus.

uric acid The final product of nitrogenous excretion in animals that require to conserve water, such as terrestrial insects, or have limited storage space, such as birds and their eggs. Uric acid has very low water solubility and crystals may be deposited in, for example, butterflies' wings to impart iridescence. See also **tophus**.

uricase *Urate oxidase; EC 1.7.3.3* Enzyme that catalyses the oxidation of urate to **allantoin**, the terminal reaction in purine degradation in most mammals (but not in primates and birds). There are four subunits of 32 kDa and one copper atom per molecule. The sequence is well conserved, and in animals it is mainly localized in the liver, where it forms a large, electron-dense paracrystalline core in many peroxisomes.

uridine The ribonucleoside formed by the combination of ribose and uracil.

uridyl Chemical group formed by the loss of hydroxyl from the ribose of uridine. Not the same as **uridylyl**.

uridylyl The uridine monophospho group derived from urydylic acid: not the same as uridyl (any chemical group formed by the loss of hydroxyl from the ribose of uridine).

uridylylation Post-translational addition of a **uridylyl** group to a protein, RNA or sugar phosphate. Uridylylation of the poliovirus VPg protein is required for positive-strand RNA synthesis. Uridydyl transferase (EC 2.7.7.12) transfers UDP from glucose to galactose.

urocanic acid Intermediate in degradation of L-histidine. Found as its *trans* isomer (*t*-UA, approximately $30\,mg/cm^2$) in the uppermost layer of the skin (stratum corneum). Absorption of UV light caused isomerization to the *cis*-form, which is believed to inhibit various immune responses, including contact and delayed hypersensitivity.

Urochordata See **tunicates**.

urocortin A neuropeptide of the CRF (**corticotropin-releasing factor**) family in the mammalian brain. Urocortin is synthesized in human anterior pituitary cells and may play an important role in biological features of normal pituitary gland, possibly as an autocrine or a paracrine regulator.

Urodela Amphibians of the Order Caudata that have tails: newts and salamanders.

urogastrone A peptide isolated from human urine that inhibits gastric acid secretion. Now known to be identical to **epithelial growth factor**.

uroid Tail region of a moving amoeba.

urokinase *uPA* See **uPA**.

uromodulin A naturally occurring immunosuppressant (85 kDa) originally found in urine of pregnant women. Identical to Tamm–Horsfall glycoprotein, but cell surface-linked by glycosylphosphatidylinositol.

uronic acids Carboxylic acids related to hexose sugars, etc. by oxidation of the primary alcohol group, e.g. glucuronic, galacturonic acid.

uroporphyrinogen I synthetase *EC 4.3.1.8* An enzyme of haem biosynthesis that is defective in the inherited (autosomal dominant) disease, acute intermittent porphyria. UP I is isomerized to UP III by UP III synthetase, defective in the autosomal-recessive disease, congenital erythropoietic porphyria.

urostyle The small upturned posterior tip of the vertebral column in homocercal caudal fins of teleost fish and in some amphibians.

urotensin II *UII* A neuropeptide with potent cardiovascular effects, the endogenous ligand for the orphan G-protein-coupled receptor, GPR14. Its sequence is strongly conserved among different species and has structural similarity to **somatostatin**.

uteroglobin **Progesterone**-binding protein found in lagomorphs (rabbits, hares, etc.), which is also a potent inhibitor of **phospholipase** A2. Forms an antiparallel dimer, linked by disulphide bonds at either end. A structurally related protein, CC10, is found in the lining of pulmonary airways and is also a PLA2 inhibitor.

UTR See **untranslated region**.

utricle *Utriculus* Any small, inflated, bladder-like structures. In vertebrates, the upper chamber of the inner ear.

utrophin *Dystrophin-associated protein* Autosomal homologue of **dystrophin** (395 kDa) localized near the neuromuscular junction in adult muscle, though in the absence of dystrophin (i.e. in **Duchenne muscular dystrophy**) utrophin is also located on the cytoplasmic face of the sarcolemma.

uveitis Inflammation of the iris, ciliary body and choroid of the eye.

uvomorulin Glycoprotein (120 kDa) originally defined as the antigen responsible for eliciting antibodies capable of blocking compaction in early mouse embryos (at the morula stage) and inhibiting calcium-dependent aggregation of mouse teratocarcinoma cells. May be the mouse equivalent of LCAM, the chick **cell adhesion molecule**.

V

V5 1. Extra-striate visual area of the brain, (V5/MT). 2. An anti-apoptotic pentapeptide V5, (VPMLK). 3. Vanadium^{5+}. 4. One of the catalytic domains of PKC beta. 5. The envelope (*Env*) gene V3–V5 regions of feline immunodeficiency virus that encodes the neutralizing epitopes. 6. The epithelial Ca^{2+} channel transient receptor potential cation channel V5 (TRPV5), involved in active Ca^{2+} reabsorption. 7. One of the splice variants of CD44.

V8 protease *Glutamyl endopeptidase; EC 3.4.21.19* Serine peptidase from *Staphylococcus aureus* strain V8. Cleaves peptide bonds on the carboxyl side of aspartic and glutamic acid residues. Used experimentally for selective cleavage of proteins for amino acid sequence determination or peptide mapping.

VAC *Annexin V* Vascular anticoagulant (VAC)-alpha is now called **annexin V**; VAC-beta is annexin-8.

VacA *Vacuolating cytotoxin* Protein (600–700 kDa: hexamers or heptamers of identical 140-kDa monomers) released into culture supernatant by Type I *Helicobacter pylori*. The 140-kDa precursor is cleaved to form a 95-kDa VacA monomer, which is further cleaved to produce 37- and 58-kDa fragments that behave as an AB toxin. Causes formation of large vacuoles in epithelial cells.

vaccination The process of inducing immunity to a pathogenic organism by injecting either an antigenically related but non-pathogenic strain (attenuated strain) of the organism or related non-pathogenic species, or killed or chemically modified organism of low pathogenicity. In all cases, the aim is to expose the human or animal being vaccinated to an antigenic stimulus that leads to immune protection against disease, without inducing appreciable pathogenesis from the injection.

vaccine An antigen preparation that when injected will elicit the expansion of one or more clones of responding lymphocytes so that immune protection is provided against a disease.

vaccinia Virus of the Orthopoxvirus family used in vaccination against smallpox. Related to, but not identical to, cowpox virus. Also used as a vector for introducing DNA into animal cells.

vacuolar ATPase See **V-type ATPase**.

vacuole Membrane-bounded vesicle of eukaryotic cells. Secretory, endocytotic and phagocytotic vesicles can be termed vacuoles. Botanists tend to confine the term to the large vesicles found in plant cells that provide both storage and space-filling functions.

vacuolins Small proteins that induce rapid formation of large, swollen structures derived from endosomes and lysosomes by homotypic fusion. Vacuolin-1, the most potent compound, blocks the Ca^{2+}-dependent exocytosis of lysosomes induced by ionomycin or plasma membrane wounding. Vacuolins have no homologies to known proteins.

valine *Val; V; 117 Da* An essential amino acid. See Table A2.

valinomycin A potassium **ionophore** antibiotic, produced by *Streptomyces fulvissimus*. Composed of three molecules (L-valine, D-α-hydroxyisovaleric acid, L-lactic acid) linked alternately to form a 36-membered ring that folds to make a cage shaped like a tennis-ball seam. This wraps specifically around potassium ions, presenting them with a hydrophilic interior and a lipid bilayer with a hydrophobic exterior. Potassium is thus free to diffuse through the lipid bilayer. Highly ion specific, valinomycin is used in **ion-selective electrodes**.

valproate *Sodium 2-propylpentanoate* Anticonvulsant used to treat manic phase of bipolar disorder, epilepsy and migraine. May increase brain levels of gamma-aminobutyric acid (GABA).

van der Waals' attraction Electrodynamic forces arise between atoms, molecules and assemblies of molecules due to their vibrations giving rise to electromagnetic interactions; these are attractive when the vibrational frequencies and absorptions are identical or similar, repulsive when non-identical. Other interactions originally proposed by van der Waals were included in this name, but these are usually separated into the Coulomb force, the Keesom force and the London force. Only the last is of electrodynamic nature. Probably important in holding lipid membranes into that structure, and possibly in other interactions (e.g. cell adhesion). Electrodynamic forces between large-scale assemblies can be of relatively long-range nature.

vanadate VO_4^{3-} Powerful inhibitor of many but not all enzymes that cleave the terminal phosphate bond of ATP. The vanadate ion is believed to act as an analogue of the transition state of the cleavage reaction. **Dynein** is very sensitive to inhibition by vanadate, whereas **kinesin** is relatively insensitive. Similarly, **tyrosine kinases** are sensitive to vanadate, but threonine/serine **protein kinases** are insensitive.

vancomycin Complex glycopeptide antibiotic produced by actinomycetes, commercially by *Amycolatopsis orientalis* (formerly *Nocardia orientalis*). Inhibits **peptidoglycan** synthesis. Active against many Gram-positive bacteria.

vanilloid receptor-1 *VR1* Receptor found selectively on sensory neurons, resembling (distantly) receptors of the TRP-type (see **TRP channels**). Protein (95 kDa) with six transmembrane domains having some similarity with **SOCS** (store-operated calcium channels), though VR-1 does not seem to be a selective calcium channel. Binding of **capsaicin** activates the receptor, which acts as a non-specific cation channel and induces death of the cell. Heat will also activate the receptor, which has response characteristics similar to thermal nociceptors.

vanin Vanin-1 is a GPI-anchored 70-kDa membrane **pantetheinase** highly expressed in the gut and liver. It hydrolyzes pantetheine to pantothenic acid (vitamin B5) and the low-molecular weight thiol cysteamine. Vanin-1 deficient mice have better-controlled inflammatory responses, and transfection of thymic stromal cells with the Vanin-1 cDNA enhances thymocyte adhesion *in vitro*. There are at least two mouse (*Vanin-1* and *Vanin-3*) and three human (*VNN1*, *VNN2*, *VNN3*) orthologous genes.

variable antigen Term usually applied to the surface antigens of those parasitic or pathogenic organisms that can alter their antigenic character to evade host immune responses. (See **antigenic variation**).

variable gene See **V-region**.

variable region See **V-region**.

Varicella zoster Member of the Alphaherpesvirinae: human herpes simplex virus type 3, causative agent of chickenpox and **shingles**.

variola virus Virus responsible for smallpox. Said to have been completely eradicated. Large DNA virus ('brick-like', 250–390 nm × 20–260 nm) with complex outer and inner membranes (not derived from plasma membrane of host cell).

varix 1. Ridges on the shells of gastropods representing periods of shell growth where thickened lips are formed. Old lips are previous varices, and the current lip is the most recent varix. **2.** An enlarged and dilated vein, usually tortuous, as in varicose veins.

vasa vas *Drosophila* gene involved in oogenesis and embryonic positional specification; a **DEAD-box helicase**.

vascular anticoagulant VAC See **annexin**.

vascular bundle Strand of vascular tissue in a plant; composed of xylem and phloem.

vascular cylinder Synonym for **stele**.

vascularization Growth of blood vessels into a tissue, with the result that the oxygen and nutrient supply is improved. Vascularization of tumours is usually a prelude to more rapid growth and often to metastasis; excessive vascularization of the retina in diabetic retinopathy can lead indirectly to retinal detachment. Vascularization seems to be triggered by 'angiogenesis factors' that stimulate endothelial cell proliferation and migration. See **angiogenin, tumour angiogenesis factor**.

vasculitis Inflammation of the blood vessel wall. May be caused by immune complex deposition in or on the vessel wall.

vasoactive intestinal contractor Mouse homologue of **endothelin**-2.

vasoactive intestinal peptide VIP Peptide of 28 amino acids, originally isolated from porcine intestine but later found in the central nervous system, where it acts as a neuropeptide and is released by specific **interneurons**. May also affect behaviour of cells of the immune system.

vasodilator Any drug or substance that causes expansion of blood vessels. Examples include **glyceryl nitrate, adrenomedullin, bradykinin, atrial natriuretic peptide, isoproterenol, nifedipine, prostacyclin, Substance P**. See **vasodilator-stimulated phosphoprotein**.

vasodilator-stimulated phosphoprotein VASP A 46/50-kDa protein that is a substrate for both cAMP- and cGMP-dependent protein kinases and is associated with microfilament bundles in many tissue cells. Abundant in platelets; phosphorylation of VASP will inhibit platelet activation.

vasopressin *Antidiuretic hormone; ADH* A peptide hormone released from the posterior pituitary lobe but synthesized in the hypothalamus. There are two forms, differing only in the amino acid at position 8: arginine vasopressin is widespread, while lysine vasopressin is found in pigs. Has antidiuretic and **vasopressor** actions. Used in the treatment of **diabetes insipidus**.

vasopressor Any compound that causes constriction of blood vessels (vasoconstriction), thereby causing an increase in blood pressure.

vasotocin Cyclic nonapeptide hormone (arginine vasotocin is CYIQNCPRG-NH$_2$), related to vasopressin and oxytocin, found in the neurohypophysis of birds, reptiles and some amphibians, and demonstrated by bioassay in the mammalian pineal gland, subcommissural organ and foetal pituitary gland.

vasp See **vasodilator-stimulated phosphoprotein**.

vault Large cytoplasmic ribonucleoprotein particle that has eight-fold symmetry with a central pore and petal-like structures giving the appearance of an octagonal dome. May be related to the central plug of the nuclear pore complex. Vaults are composed of three proteins, the major vault protein (MVP), the vault poly(ADP-ribose)polymerase (VPARP) and the telomerase-associated protein 1, together with one or more small untranslated RNAs.

vav Vav and vav2 are members of the **dbl family** of guanine nucleotide exchange factors for the rho/rac GTPases. The vav proto-oncogene product (p95vav) is predominantly expressed in haematopoietic cells; vav2 has a wider tissue distribution. There is enhanced phosphorylation of Vav in Bcr-Abl-expressing cells, which activates vav and thus leads to the activation of Rac-1 associated with leucaemias.

VCA region *Verprolin-like, cofilin-like, acidic region* Motif found in **WASP** and **WAVE** that binds to G-actin and **Arp2/3** complex.

VCAM *Vascular cell adhesion molecule; CD106* Cell adhesion molecule (90–110 kDa) of the immunoglobulin superfamily expressed on endothelial cells, macrophages, dendritic cells, fibroblasts and myoblasts. Expression can be upregulated by inflammatory mediators (IL-1β, IL-4, TNFα, IFNγ), and it is the ligand for the integrin VLA4.

VDAC Voltage-dependent anion channel. See **porins**.

vector 1. Mathematical term to describe something that has both direction and magnitude. **2.** Common term for a plasmid that can be used to transfer DNA sequences from one organism to another. See **transfection**. Different vectors may have properties particularly appropriate to give protein expression in the recipient, or for cloning, or may have different selectable markers.

vectorette method A method for **PCR** cloning an unknown sequence of DNA attached to a known sequence. To the end of the unknown sequence is attached a vectorette, a double-stranded sequence that includes a region of mismatch; primer for the known sequence drives the formation of a second primer site from the mismatch region, and the unknown sequence is thus flanked with known sequences and can be specifically PCR amplified.

vectorial synthesis Term usually applied to the mode of synthesis of proteins destined for export from the cell. As the protein is made it moves (vectorially) through the membrane of the rough endoplasmic reticulum, to which the ribosome is attached, and into the cisternal space.

vectorial transport Transport of an ion or molecule across an epithelium in a certain direction (e.g. absorption of glucose by the gut). Vectorial transport implies a non-uniform distribution of **transport proteins** on the plasma membranes of two faces of the epithelium.

vegetal pole The surface of the egg opposite to the animal pole. Usually the cytoplasm in this region is incorporated into future endoderm cells.

VEGF *Vascular endothelial growth factor; vascular permeability factor; VPF* Growth factor of the PDGF family that stimulates mitosis in vascular endothelium, stimulates angiogenesis and also increases permeability of endothelial monolayers. Tissue-specific splice variants (VEGF121, VEGF165, VEGF-C) are found, VEGF165 having heparin-binding activity, which VEGF121 lacks. VEGF121 only binds to flk-1. Functional form is a dimer (or heterodimer of splice variants) that binds to flt-1 (VEGF-R1), flt-4 (VEGF-R3, binds only VEGF-C) or flk-1 (VEGF-R2/KDR) receptor tyrosine kinases. Can form heterodimer with placental growth factor (PLGF). See **Tie** and **flt**.

veiled cell *Dendritic cell* A cell type found in afferent lymph and defined (rather unsatisfactorily) on the basis of its morphology. Now generally referred to as dendritic cells. They are accessory cells, and migrate from the periphery (where they are referred to as **Langerhans cells** if in the skin) to the draining lymph node. In the lymph node they are also known as interdigitating cells, and are found in the T-dependent areas of spleen or lymph nodes, involved in antigen presentation (class II MHC-positive). Veiled cells have high levels of surface Ia antigens.

vein 1. Blood vessel that returns blood from the microvasculature to the heart; walls are thinner and less elastic than those of artery. **2.** In leaves, thickened portion of leaf containing **vascular bundle**; the pattern, venation, is characteristic for each species.

veliger Free-living larval form of gastropod and bivalve molluscs; develop from the trochophore larva.

velum A veil-like structure. **1.** The posterior part of the soft palate in higher mammals. **2.** In some Ciliophora, the delicate membrane bordering the oral cavity. **3.** In sponges, a membrane constricting the lumen of an incurrent or excurrent canal **4.** In hydrozoan medusae, an annular shelf projecting inwards from the margin of the umbrella. **5.** In rotifers, the trochal disk. **6.** In molluscs, the ciliated locomotor organ of the veliger larva. **7.** In Cephalochordata, the perforated membrane separating the buccal cavity from the pharynx.

venom A toxic secretion in animals that is actively delivered to the target organism, to paralyse or to incapacitate or else to cause pain as a defence mechanism. Commonly includes protein and peptide toxins.

ventral nervous system defective *vnd* A *Drosophila* gene encoding an integral membrane glycoprotein related to **amyloidogenic glycoprotein**.

verapamil A calcium-channel-blocking drug (454 Da) used as a coronary vasodilator and anti-arrhythmic.

veratridine An alkaloid, a sodium channel activator that acts at neurotoxin receptor site 2 and preferentially binds to activated Na^+ channels, causing persistent activation. Found in the seed of *Schoenocaulon officinale* (Cevadilla) and in the rhizome of *Veratrum album* (White hellebore).

vernalization Treatment of plants with a period of low temperature to promote flowering earlier than they would otherwise.

Vero cells Cell line derived from kidney of African Green Monkey. Susceptible to a range of viruses.

verprolins Family of proteins, WIP, CR16 and WIRE/WICH, that are regulators of cytoskeletal organization in vertebrate cells. Verprolin was originally identified in budding yeast and later shown to be needed for actin polymerization during polarized growth and during endocytosis. Vertebrate verprolins regulate actin dynamics either by binding directly to actin, by binding the **WASP** family of proteins, or by binding to other actin-regulating proteins.

versene Trivial name for **EDTA**.

versican *Large fibroblast glycoprotein; chondroitin sulphate core protein* A large chondroitin sulphate proteoglycan (264 kDa) involved in cell signalling. N-terminal region similar to glial hyaluronic acid-binding protein, centre has glycosaminoglycan attachment sites, C-terminal region has EGF-like repeats. Expression of certain versican isoforms has been implicated in migration and proliferation of cancer cells.

vertical transmission Transmission of an infectious agent from mother to offspring, transplacentally, through cross-infection during birth, or through close contact during the neonatal period; *cf.* horizontal transmission between individuals.

vesicle A closed membrane shell, derived from membranes either by a physiological process (budding) or mechanically by sonication. Vesicles of dimensions in excess of 50 nm are believed to be important in intracellular transport processes. See also **coated vesicles**.

vesicular stomatitis virus *VSV* **Rhabdovirus** causing the disease 'soremouth' in cattle. Widely used as a laboratory tool, especially in studies on the spike glycoprotein as a model for the synthesis, post-translational modification and export of membrane proteins.

vesicular-arbuscular mycorrhiza Form of **mycorrhiza** in which the fungus invades the cortical cells to form vesicles and arbuscules (finely branched structures). Common among herbaceous plants and may significantly improve the mineral nutrition of the host.

vesiculin Name given to a highly acidic protein (10 kDa) found in **synaptic vesicles**. Term appears to be obsolete.

vessel Water-conducting system in the **xylem**, consisting of a column of cells (vessel elements) whose end walls have been perforated or totally degraded, resulting in an uninterrupted tube.

vessel element Part of a **xylem** vessel in a higher plant, arising from a single cell. The end walls are perforated and may completely disappear, giving rise to a continuous tube. The remaining walls are thickened and lignified, and there is no protoplast.

V-gene See **variable gene**.

Vg1 *Vitellogenin-1* Product of the *Vg1* gene, a member of the TGF-beta-family involved in mesoderm induction. Not to be confused with **vitellogenin**.

***VH* and *VL* genes/domains** VH and V genes define in part the sequences of the variable heavy and light regions of immunoglobulin molecules. VH and VL domains are the regions of amino acid sequence so defined. J genes and, in the case of the heavy chain, a D gene (D = diversity) also define these regions. Gene rearrangement plays a role in determining the sequences in which the genes are joined as the DNA of the immunoglobulin-producing cell matures.

VHDL *Very high-density lipoprotein* Plasma lipoprotein with density greater than 1.21 g/ml. Protein content about 57%, phospholipids 21%, cholesterol 17% and triacylglycerols 5%. Molecular weights between 1.5×10^5 kDa and 2.8×10^5 kDa.

viability test Test to determine the proportion of living individuals, cells or organisms, in a sample. Viability tests are most commonly performed on cultured cells, and usually depend on the ability of living cells to exclude a dye, (an exclusion test) or to specifically take it up (inclusion test).

Viagra See **sildenafil**.

Vibrio cholerae Bacterium that causes cholera, the life-threatening aspects of which are caused by the exotoxin (see **cholera toxin**). Short, slightly curved rods, highly motile (single polar flagellum), Gram-negative. Adhere to intestinal epithelium (adhesion mechanism unknown) and produce enzymes (neuraminidase, proteases) that facilitate access of the bacterium to the epithelial surface.

VIC See **vasoactive intestinal contractor**.

Vicia faba Broad bean. Was often used in plant genetics because cells have only six large chromosomes.

vicilin Seed-storage protein of legumes. Protein from *Pisum sativum* is a trimer of 50-kDa subunits. High proportion of **beta-pleated sheet** (40–50%) and only about 10% alpha-helix. See **cupins**.

vidarabine *Adenine arabinoside; 9-β-D-arabinofuranosyladenine; Ara-A* A nucleoside analogue with antiviral properties that has been used to treat severe herpes virus infections.

Vik1 Non-motor protein from *S. cerevisiae* that interacts with **KAR3** (a kinesin-14 protein) and has sequence and structural similarity to Cik1p. The Vik1 protein is detected in vegetatively growing cells but not in mating pheromone-treated cells. Vik1p physically associates with Kar3p in a complex separate from that of the Kar3p–Cik1p complex. Vik1p localizes to the spindle pole body region in a Kar3p-dependent manner.

villin Microfilament-severing and -capping protein (95 kDa) from microvillar core of intestinal epithelial cells. Severs at high calcium concentrations, caps at low.

vimentin **Intermediate filament** protein (58 kDa) found in mesodermally derived cells (including muscle).

vinblastine Alkaloid (818 Da) isolated from *Vinca* (periwinkle): binds to **tubulin** heterodimer and induces formation of paracrystals rather than tubules. Net result is that microtubules disappear as they disassemble and are not replaced. Used in tumour chemotherapy.

vinca alkaloids See **vinblastine** and **vincristine**.

vincristine Cytotoxic alkaloid that binds to **tubulin** and interferes with microtubule assembly. See **vinblastine**, a related compound.

vinculin Protein (130 kDa) isolated from muscle (cardiac and smooth), fibroblasts and epithelial cells. Associated with the cytoplasmic face of **focal adhesions;** may connect microfilaments to plasma membrane integral proteins through **talin**.

Vioxx Proprietary name for rofecoxib, a selective **COX-2** inhibitor.

VIP See **vasoactive intestinal peptide**.

Vipera ammodytes Western sand viper. See **ammodytoxins**.

vir genes The *vir* region of *Agrobacterium tumefaciens* contains six genes: *VirA* encodes a single protein which resembles a transmembrane chemoreceptor found in other bacteria and presumably binds **acetosyringone**, VirB may play a role in directing **T-DNA** transfer events at the bacterial cell surface, *VirC* and *VirD* encode a site-specific endonuclease that produces a T-strand which is the intermediate molecule that is transported to the plant cell, *VirE* encodes a single-stranded DNA-binding protein that appears to coat the T-strand during transfer to the plant cell, *VirG* produces a positive regulatory protein which relays environmental information to other vir loci.

viral antigens Those antigens specified by the viral genome (often coat proteins) that can be detected by a specific immunological response. Often of diagnostic importance.

viral transformation Malignant transformation of an animal cell in culture, induced by a virus.

viremia Presence of virus in the blood.

virgin lymphocyte A lymphocyte (CD45RA$^+$, CD45RO$^-$) that has not, and whose precursors have not, encountered the antigenic determinant for which it possesses receptors. Term no longer in common use.

virino Hypothetical virus-like infectious agent, once suggested to be the cause of transmissible spongiform encephalopathies. Probably only of historical interest.

virion A single virus particle, complete with coat.

viroid Extremely small viruses of plants. Their genome is a 240- to 350-nucleotide circular RNA strand, extensively base-paired with itself, so they resist RNAase attack. At one time the term was also used casually of self-replicative particles, such as the **kappa particle** in *Paramecium*.

viropexis Obsolete term for the non-specific phagocytosis of virus particles bound to surface receptors.

virus Viruses are obligate intracellular parasites of living but non-cellular nature, consisting of DNA or RNA and a protein coat. They range in diameter from 20 to 300 nm. Class I viruses (Baltimore classification) have double-stranded DNA as their genome; class II have a single-stranded DNA genome; class III have a double-stranded RNA genome; class IV have a positive single-stranded RNA genome, the genome itself acting as mRNA; class V have a negative single-stranded RNA genome used as a template for mRNA synthesis; and class VI have a positive single-stranded RNA genome but with a DNA intermediate not only in replication but also in mRNA synthesis. The majority of viruses are recognized by the diseases they cause in plants, animals and prokaryotes. Viruses of prokaryotes are known as bacteriophages.

viscoelastic Of substances or structures showing non-Newtonian viscous behaviour, i.e. elastic and viscous properties are demonstrable in response to mechanical shear.

viscous-mechanical coupling Method by which adjacent **cilia** are synchronized in a field. Coupling is through the transmission of mechanical forces, rather than of a synchronizing signal.

visinin A cone-specific protein first characterized in chicken retina, a homologue of **recoverin**. See **visinin-like proteins**.

visinin-like proteins *VILIP-1, VILIP-2, etc* Members of the neuronal subfamily of intracellular EF-hand calcium sensor proteins, termed the NCS family (neuronal Ca^{2+} sensor family), that modulates Ca^{2+}-dependent cell signalling events.

Visna-maedi virus A **retrovirus** of sheep and goats. A member of the **Lentivirinae**, related to **HIV**. First identified in Iceland when it was introduced by sheep imported from Germany, it causes two diseases: maedi, the most common, is a pulmonary infection (*maedi* is Icelandic for shortness of breath); visna is due to infection of the nervous system, causing a paralysis similar to **multiple sclerosis** (*visna* is Icelandic for wasting).

visual purple See **rhodopsin**.

vital stain A stain that is taken up by live cells and can be used to stain, for example, a group of cells in a developing embryo in order to try to determine a **fate map**.

vitamin K A group of 2-methilo-naphthoquinone derivatives involved in the carboxylation of certain glutamate residues in proteins to form gamma-carboxyglutamate residues (Gla-residues) that are usually involved in binding calcium. Several blood-clotting factors (prothrombin (factor II), factors VII, IX, X, protein C, protein S and protein Z) are Gla proteins, and thus their synthesis is vitamin K-dependent. See **warfarin** and **coumarin**.

vitamins Low molecular-weight organic compounds of which small amounts are essential components of the food supply for a particular animal or plant. For humans, Vitamins A, the B series, C, D_1 and D_2, E and K are required. Deficiencies of one or more vitamins in the nutrient supply result in deficiency diseases. See Table V1.

vitellin Most abundant protein in egg yolk.

vitelline membrane The membrane, usually of protein fibres, immediately outside the plasmalemma of the ovum and the earlier stages of the developing embryo. Its structure and composition vary in differing animal groups.

vitellogenic Giving rise to yolk of an egg.

vitellogenin A protein, product of the *Yp* gene, precursor of several yolk proteins, especially phosvitin and lipovitellin

TABLE V1. Vitamins

Vitamin	Full name	Occurrence	Action	Deficiency disease
Fat soluble				
A	Retinol (11-*cis* retinal)	Vegetables	Phototransduction, morphogen	Night blindness, xerophthalmia
D	1,25-dihydroxy-cholecalciferol	Action of sunlight on 7-dehydrocholesterol in skin	Ca^{++} regulation, phosphate regulation	Rickets
E	α-tocopherol	Plants, esp. seeds, wheatgerm	Antioxidant	Failure to grow to maturity, infertility
K_1		Higher green plants		
K_2	Range of molecules	Intestinal bacteria		
Water soluble				
B_1	Thiamine	Degradation of α-keto acids	Beri-beri	
	Folic acid (Tetrahydrofolic acid)	Plants	Purine biosynthesis	Anaemia
	Nicotinic acid (niacin)	Can be made		Pellagra
	Pantothenic acid (CoA)	Plants and microorganisms		
B_2	Riboflavine	Plants and microorganisms	Constituent of flavoproteins	
B_6	Pyridoxine		Transamination	Acrodynia in rats, convulsions
	Pyridoxal			
	Pyridoxamine			
B_{12}	Cobalamine	Intestinal microorganisms	Hydrogen transfer reactions	Pernicious anaemia
C	Ascorbic acid	Plants, esp. citrus fruits	Cofactor	Scurvy
H	Biotin	Intestinal bacteria	Protects against avidin toxicity, intermediate CO_2 carrier	

in the eggs of various vertebrates, synthesized in the liver cells after estrogen stimulation. Also found in large amounts in the haemolymph of female insects, synthesized and released from the fat body during egg formation. Not to be confused with Vg1 (vitellogenin-1, a member of the TGFβ family).

vitiligo Patchy depigmentation of the skin in which melanocytes in the skin, the mucous membranes and the retina (inner layer of the eyeball) are destroyed, often with a sharp demarcation line. Associated with autoimmune disease. Loci for susceptibility to autoimmune disease, particularly vitiligo, have been mapped to chromosomes 1p31, 7, 8 and 4.

vitronectin Serum protein (70 kDa) also known as serum-spreading factor from its activity in promoting adhesion and spreading of tissue cells in culture. Contains the cell-binding sequence Arg-Gly-Asp (RGD) first found in **fibronectin**.

vitrosin Old term for collagen isolated from embryonic chick vitreous. Synthesized by neural retina at early developmental stages and by cells of the vitreous body later.

viviparous Organisms in which the young develop internally, not in eggs that are incubated externally.

VLA proteins *Very late antigens* VLA-1 and VLA-2 were originally defined as antigens appearing on the surfaces of T-lymphocytes 2–4 weeks after *in vitro* activation; they are now known to be part of the β-**integrin** family. Additional members of the subset are now known (VLA-3, VLA-4, VLA-5 and VLA-6), the β-subunits all being identical. Some of the VLA proteins are receptors for collagen, laminin or fibronectin, and many are now known to be expressed on cells other than leucocytes.

VLDL *Very low-density lipoprotein* Plasma lipoproteins with density of 0.94–1.006 gm/cm^3; made by the liver. Transport triacylglycerols to adipose tissue. Apoproteins B, C and E are found in VLDL. Protein content, about 10%, much lower than in **VHDL**.

V$_{max}$ The maximum initial velocity of an enzyme-catalysed reaction, i.e. at saturating substrate levels.

Vohwinkel's syndrome Skin disease caused by a defect in the **loricrin** gene.

voltage clamp A technique in electrophysiology in which a **microelectrode** is inserted into a cell and current is injected through the electrode so as to hold the cell's membrane potential at some predefined level. The technique can be used with separate electrodes for voltage sensing and current passing; for small cells, the same electrode can be used for both. Voltage clamp is a powerful technique for the study of **ion channels**. See **patch clamp**.

voltage gradient Literally, the electric field in a region, defined as the potential difference between two points divided by the distance between them. Used more loosely, the potential difference across a plasma membrane.

voltage-gated ion channel A transmembrane **ion channel** whose permeability to ions is extremely sensitive to the transmembrane potential difference. These channels are essential for neuronal signal transmission and for intracellular signal transduction. See **sodium channel**.

voltage-sensitive calcium channels *VSCC* A variety of voltage-sensitive calcium channels are known and, on the basis of electrophysiological and pharmacological criteria, are grouped into six classes. The general function

is to allow calcium influx into the cell as a result of membrane depolarization. The majority (**L-type**, **N-type**, **P-type** and **Q-type channels** require substantial depolarization and are sometimes collectively known as high voltage-activated types. The **R-type channels** activate after moderate depolarization, and the **T-type channel** opens at relatively negative potentials.

volutin granule Metachromatic granules containing polyphosphate, a linear phosphate polymer found in bacteria, fungi, algae and some higher eukaryotes that may serve as a stock of phosphate.

Volvox A genus of colonial flagellates. The colony is a hollow sphere about 0.5 mm in diameter comprising about 50 000 cells embedded in a gelatinous wall, and the cells are sometimes connected by cytoplasmic bridges. Each cell has a chloroplast and two flagella.

vomeronasal organ Paired organs situated in the nasal area and connected by a narrow duct to the nasal cavity just inside the nostril. Mammalian vomeronasal sensory neurons detect specific chemicals, some of which may be **pheromones**. Called Jacobson's organ in snakes and lizards.

v-onc General abbreviation for the viral form of an **oncogene**, *cf. c-onc*, the normal, cellular **proto-oncogene**.

von Willebrand factor *vWF* A large multimeric plasma glycoprotein involved in platelet adhesion through an interaction with Factor VIII. The type A domain (vWF domain) is found in complement factors B, C2, CR3 and CR4; the integrins (I-domains); collagen types VI, VII, XII and XIV; and other extracellular proteins, all of which are involved in multiprotein complex formation. See **von Willebrand's disease**.

von Willebrand's disease Autosomal dominant platelet disorder in which adhesion to collagen, but not aggregation, is reduced. Both bleeding time and coagulation are increased. Factor VIII levels are secondarily reduced.

Vorticella Genus of ciliate **Protozoa**. It has a bell-shaped body with a belt of cilia round the mouth of the bell to sweep food particles towards the 'mouth', and a long stalk, connecting it to the substratum, which contains the contractile **spasmoneme**.

voxel The 3D equivalent of a pixel, a finite volume unit.

VP16 1. See **etoposide**. 2. Herpes simplex virion protein 16, a virus tegument phosphoprotein, contains two strong activation regions that can independently and cooperatively activate transcription *in vivo* and is a transcriptional activator of the viral immediate-early genes.

VP22 Virion protein 22. The *UL49* gene product of alpha herpes simplex virus types 1 and 2 (HSV-1 and HSV-2), a virion phosphoprotein which accumulates inside infected cells at late stages of infection. It is capable of causing microtubule reorganization, uses a non-standard pathway for nuclear localization and shows non-classical intercellular trafficking. May also interact with the actin cytoskeleton.

vpr HIV-1 viral protein R (Vpr), a 96 amino-acid soluble protein that is expressed late during viral replication. Has cytotoxic effects on both infected cells and bystander cells, and exhibits both pro- and antiapoptotic activity.

VRE Vancomycin-resistant enterococci.

V-region Those regions in the amino acid sequence of both the heavy and the light chains of immunoglobulins where there is considerable sequence variability between one immunoglobulin and another of the same class, in contrast to constant sequence (C) regions. The V-regions are associated with the antigen-binding areas. They contain hypervariable regions of particularly high sequence diversity.

VSCC See **voltage-sensitive calcium channel**.

VSG Variant surface glycoprotein of trypanosomes. See **antigenic variation**.

VSV-G tag **Epitope tag** (YTDIEMNRLGK) derived from the **vesicular stomatitis virus** G protein.

V-type ATPase *Vacuolar ATPase* One of three major classes of ion transport ATPase, characterized by a multi-subunit structure and a lack of a phosphorylated intermediate. Pumps H^+. Found in intracellular acidic vacuoles and in some proton-pumping epithelia (e.g. intercalated cells of kidney). Sensitive to **bafilomycin**. Related to the **F-type ATPase**. See also **P-type ATPase**.

W

Waardenburg's syndrome Autosomal-dominant disorder with deafness and pigmentary disturbances, probably as a result of defects in function of **neural crest**. Various forms of the syndrome are recognized. Waardenburg Syndrome 1 (WS1) and WS3 (also known as Klein–Waardenburg syndrome) are caused by mutation in *Pax3* – a homologous defect to the mouse mutant *Splotch* that also has defective *Pax-3*. Waardenburg–Shah syndrome (WS4), in which Waardenburg's syndrome is associated with Hirschsprung's disease, is due to mutation in *Sox10*, and there is an homologous mutation in *Dom* mice (dominant megacolon), piebald-lethal and lethal spotting. WS2 is heterogeneous with mutation in the microphthalmia (*MITF*) gene.

Waf1 *p21; cip1* Inhibitor (21 kDa) of **cdk** activity; found in a complex with cyclin D, cdk4 and **PCNA**. Can bind to and inhibit all members of the cdk family though affinity varies. Expression regulated by p53 tumour suppressor. Also found in active cyclin/cdk complexes – multiple copies of Waf1 may be necessary to produce inhibition.

Waldenstrom's macroglobulinaemia Disease in which there is a high level of monoclonal IgM in the blood. The IgM sometimes has detectable antibody activity, e.g. rheumatoid factor. The disease is probably a relatively benign form of myelomatosis.

Walker motif The Walker motif consists of two separate sequences, A and B, which come together to form a nucleotide-binding site. The Walker A motif, also known as the phosphate-binding loop (P-loop), is the sequence GXXXXGKT/S, widely believed to be the site for nucleotide binding in many proteins. Originally described as a common nucleotide-binding fold in the α- and β-subunits of F1-ATPase, myosin and other ATP-requiring enzymes by Walker and colleagues in 1982. The Walker B motif consensus is D(D/E)XX.

warfarin Synthetic derivative of **coumarin** that inhibits the effective synthesis of biologically active forms of the vitamin K-dependent clotting factors: used clinically as an antithrombotic. Used as a rat poison, though now superseded by more potent rodenticides.

warm antibodies Most IgG antibodies react better at 37°C than at lower temperatures, especially against red cell antigens. These are the warm antibodies as contrasted with **cold agglutinins**, especially IgM, that agglutinate below 28°C. See **warm antibody haemolytic anaemia**.

warm antibody haemolytic anaemia An autoimmune disorder characterized by the premature destruction of red blood cells by autoantibody-induced lysis at temperatures above normal body temperature. Contrast **paroxysmal cold haemoglobinuria**.

wart Benign tumour of **basal cell** of skin, the result of the infection of a single cell with wart virus (papillomavirus). Virus is undetectable in basal layer, but proliferates in keratinizing cells of outer layers.

WASP family proteins *Wiskott–Aldrich syndrome protein* Family of proteins that includes WASP, N-WASP, Bee1p, Las17p. WASP, in conjunction with **cdc42** and the **arp2/3** complex, is involved in regulating actin polymerization. Defects in WASP are responsible for **Wiskott–Aldrich syndrome**. N-WASP is more broadly distributed than WASP itself, which is mainly found in myeloid cells. See **verprolins**.

WASP homology domain-2 *WH2 domain* A small, approximately 35-residue, actin monomer-binding motif found in many proteins that regulate the actin cytoskeleton, including **thymosin-β4**, ciboulot; **WASP,** and **verprolin**/WIP (WASP-interacting protein). See also **beta-thymosin repeats.**

Wassermann reaction Complement fixation test formerly used in the diagnosis of syphilis. Cardiolipin derived from ox heart is used as antigen.

water potential The chemical potential (i.e. free energy per mole) of water in plants. Water moves within plants from regions of high water potential to regions of lower water potential, i.e. down gradient.

WAVE/SCAR *Wave1, Wave2, Wave3* WAVE/SCAR protein has similarity to **WASP** and N-WASP, especially in its C-terminal region and cooperates with the **Arp2/3** complex. The original WAVE/SCAR has been renamed Wave1. Wave1 and Wave3 are both strongly expressed in brain; Wave1 maps on chromosome 6q21–22 and Wave3 on chromosome 13q12. Wave2 has a very wide distribution with strong expression in peripheral blood leucocytes and maps on chromosome Xp11.21, next to the WASP locus.

WBC *White blood cell* Term includes neutrophils, eosinophils, basophils, monocytes and lymphocytes.

WD-repeat proteins *WD40* The WD motif is a conserved sequence of approximately 40 amino acids usually ending with tryptophan and aspartic acid (WD). The motif is implicated in protein–protein interactions. Crystal structure of one WD-repeat protein (GTP-binding protein beta-subunit) reveals that the seven repeat units form a circular propeller-like structure with seven blades.

weal-and-flare See **triple response**.

weaver In the murine mutation *weaver* there is early apoptotic death during development of cells in testes, cerebellum and midbrain. The defect is caused by a base pair substitution in the G-protein-coupled inwardly rectifying potassium channel 2 gene. Up to 70% of the mesostriatal dopaminergic neurons are lost, and major alterations of the dopaminergic dendrites of the substantia nigra have been described. The defect does not seem to be due to reduced neurotrophin levels.

wee Cell cycle checkpoint genes found in *Schizosaccharomyces pombe*. Mutants in *wee-1* and *wee-2* have normal growth rate but divide earlier so cells are smaller.

Wegener's granulomatosis A granulomatous vasculitis characterized by upper and lower respiratory tract granulomas and necrotizing focal glomerulonephritis. Usually associated with autoantibodies to the neutrophil azurophil granule protease 3 (c-**ANCA**).

Wehi 3b cells Mouse myelomonocytic cells derived from Balb/c mouse. Cells produce IL-3.

Weibel-Palade body Cytoplasmic organelle found in the vascular endothelial cells of some animals, though not in the endothelium of all vessels. Although markers for endothelium, their absence does not necessarily mean the cells are not of endothelial origin.

Weil's disease See **leptospirosis**.

Weil-Felix reaction An agglutination test still currently used in the diagnosis of rickettsial infections (typhus, etc.) which depends upon a carbohydrate cross-reacting antigen shared by rickettsiae and Proteus group OX.

Weismann's germ plasm theory The theory that organisms maintain genetic continuity from organism to offspring through the germ line cells (germ plasm) and that the other (somatic) cells play no part in the transmission of heritable factors.

Werner's syndrome Rare human genetic disorder characterized by genomic instability and predisposition to cancer. Disease is caused by mutations in the *RECQL2* gene, which encodes a homologue of the *E. coli* RecQ DNA helicase. A similar protein is defective in **Bloom's syndrome**. Werner syndrome cells usually achieve only about 20 population doublings, compared to the normal **Hayflick limit**. 'Atypical Werner syndrome', a more severe phenotype, is associated with mutations in the **lamin A** gene.

Wernicke's encephalopathy Brain damage due to thiamine deficiency, usually associated with Korsakoff's psychosis. Often a result of alcohol abuse.

West Nile virus An arthropod-borne virus, member of the Flaviviridae family, the cause of severe meningitis and encephalitis in humans.

Western blot An electroblotting method in which proteins are transferred from a gel to a thin, rigid support (often nitrocellulose) and detected by binding of labelled antibody. See **blotting**.

wewakpeptins Depsipeptides (wewakpeptins A–D) isolated from the marine cyanobacterium *Lyngbya semiplena*. Have potential antitumour activity.

Wharton's jelly Viscous hyaluronic acid-rich jelly found in the umbilical cord.

wheat germ The embryonic plant at the tip of the seed of wheat. Wheat germ has been used as the starting material for a cell-free translation system, and is also the source of **wheat germ agglutinin**.

wheat germ agglutinin WGA Lectin from wheat germ that binds to *N*-acetylglucosaminyl and sialic acid residues. See **lectins**.

whirlin A **PDZ domains**-containing protein is expressed at **stereocilia** tips. Interacts with **p55**.

white w Eye colour gene of *Drosophila*, wild-type product essential for red eyes. The white locus is involved in the distribution of brown ommochrome and red pteridine pigments, found in the eyes of adult flies; the encoded protein is believed to be an integral membrane transporter protein for pigment precursors.

white adipose tissue Tissue composed largely of **adipocytes** but not specialized for production of heat (see brown adipose tissue). The major storage site for fat in the form of triglycerides, serves three functions: heat insulation, mechanical cushioning and as an energy reserve. An increasing proportion of people in the so-called civilized world have an excess.

white blood cells WBC See **leucocytes** and specific classes (**basophils, coelomocytes, eosinophils, haemocytes, lymphocytes, neutrophils, monocytes**).

whole-cell patch A variant of **patch clamp** technique in which the patch electrode seals against the cell, with direct communication between the interior of the electrode and the cytoplasm.

Williams–Beuren syndrome *Williams syndrome* Neurodevelopmental disorder with multisystem manifestations caused by heterozygosity for partial deletion of chromosome 7 band 7q11.23 near the **elastin** locus. Another gene involved in its pathogenesis is that for LIM kinase-1.

Wilms' tumour One of the most common solid tumours of childhood. A kidney tumour thought to derive from renal stem cells which retain embryonic differentiation potential. Like **retinoblastoma**, both sporadic and inherited forms occur. Defect is in the Wilms' tumour 1 (*WT1*) gene, which encodes a zinc-finger DNA-binding protein that acts as a transcriptional activator or repressor, depending on the cellular or chromosomal context.

Wilson's disease Rare autosomal recessive disease with degenerative changes in the brain and cirrhosis of the liver. Due to a defect in a copper-transporting ATPase coded by the *ATP7B* gene. There is excessive deposition of copper in liver, brain and kidney.

winged-helix transcription factors Family of transcription factors characterized by a conserved DNA-binding domain found in *Drosophila* homeotic gene **forkhead** and rat hepatocyte nuclear factor-3 (*HNF-3b*). At least 80 genes with this motif are known, many with developmentally-specific patterns of expression. *FAST-1* is a member of this family.

wingless Wg *Drosophila* homologue of *int-1* (*Wnt-1*), a segment-polarity gene that encodes a secreted signalling protein that functions in pattern formation. Wingless secretion is dependent on **Hedgehog**.

WIP WASP-interacting protein, one of the **verprolins**.

WIRE One of the vertebrate **verprolins**, binds to WASP and N-WASP and has a role in regulating actin dynamics downstream of the platelet-derived growth factor (PDGF) beta-receptor.

Wiskott–Aldrich syndrome Thrombocytopenia with severe immunodeficiency (both cell-mediated and IgM production). Associated with increased incidence of leukaemia. Caused by mutation in the gene that encodes **WASP**.

W locus Mouse coat colour locus, equivalent to the *kit* proto-oncogene, that encodes a **receptor tyrosine kinase** for stem cell factor, essential for development of haematopoietic and germ cells.

WNK *With no K (lysine) protein kinase-1; WNK-1, WNK-4* Serine–threonine kinases of the STE20 family containing a cysteine instead of the usual lysine at a key position in the active site. Human WNK-1 is expressed in most tissues, with two predominant isoforms: a 10-kb transcript expressed at high levels in the kidney and a 12-kb transcript predominant in heart and skeletal muscle. WNK4 co-localizes with **ZO-1** in specific regions of the kidney. WNKs are activated by hyperosmotic stress and defects can cause **pseudohypoaldosteronism**. WNKs in turn activate **SPAK** and **OSR1**.

wnt Multigene family encoding various secreted signalling molecules important in morphogenesis. First member was *Drosophila wingless*, but many vertebrate homologues are now known. *Wnt-1* (formerly *int-1*) induces accumulation of β-**catenin** and **plakoglobin** and affects the association of APC **tumour-suppressor** protein with catenin. Wnt proteins are believed to activate a transcription factor leukaemia enhancer factor-1 (LEF-1) by inhibiting **GSK3**.

wobble hypothesis Explains why the base inosine is included in position 1 in the anticodons of various tRNAs, why many mRNA codon words translate to a single amino acid, why there are appreciably fewer tRNAs than mRNA codon types, and why the redundant nature of the genetic code translates into a precise set of 20 amino acids. Inosine in position 1 in the anticodon can base pair with A, U or C in position 3 in the mRNA codon, so that, for example, UCU, UCC, UCA all code for serine using an inosine anticodon.

Wollaston prism Prism composed of two wedge-shaped prisms with optical axes at right angles. Light emerging has two beams of opposite polarization. Used in differential interference contrast microscopes.

woolsorter's disease Anthrax, infection with *Bacillus anthracis*, caught by handling infected wool or hair of animals.

wortmannin Fungal metabolite, isolated from *Penicillium wortmanni*, that is a fairly selective inhibitor of **PI-3-kinases** and possibly other points in the Ras/MAP kinase signalling pathway. Has been shown to inhibit **polo-like kinases**

WW domain A motif of 38 semiconserved residues, the smallest naturally occurring, monomeric, triple-stranded, antiparallel beta-sheet domain; binds proline-rich sequences. Found in diverse proteins. Examples include **WWOX**, **Yes-associated protein** and **PIN-1**, among many others.

WWOX *WW domain-containing oxidoreductase* A tumour suppressor that is deleted or altered in several cancer types.

Xaf1 *XIAP-associated factor 1* Protein that binds **XIAP** and re-localizes it to the nucleus, thus inhibiting XIAP activity and enhancing apoptosis. Expression is reduced or absent in tumours, suggesting it may function as a tumour suppressor.

xanthine *2,6-Dihydroxypurine* A purine, the starting point for purine degradation. Its methylated derivatives (theophylline, theobromine, caffeine) are potent cAMP phosphodiesterase inhibitors.

xanthine oxidase *EC 1.17.3.2* An iron–molybdenum flavoprotein, a dehydrogenase involved in conversion of xanthine to uric acid (producing hydrogen peroxide as well); also oxidizes hypoxanthine, some other purines and pterins, and aldehydes. Deficient in the human disease xanthinuria.

xanthoma Localized lesion of subcutaneous tissues in which there is an accumulation of cholesterol-filled macrophages. Characteristic of primary biliary cirrhosis.

xanthophore A cell occurring in the dermis that contains a yellow pigment, as in goldfish.

xanthophylls **Carotenoid** pigments involved in photosynthesis. Consist of oxygenated carotenes, e.g. lutein, violaxanthin and neoxanthine.

xanthopterin Yellow pterin pigment found in some insect wings.

X chromosome A sex chromosome. In mammals paired in females, in amphibia paired in males.

X chromosome inactivation centre *Xic* Site on X chromosome responsible for the inactivation of that X chromosome in female mammals (see **Lyon hypothesis**). Gene responsible, *Xist*, maps to this region and seems to code for a nuclear RNA that co-localizes with the inactivated X chromosome. *Xist* introduced on an autosome is capable of inactivation in *cis* and the Xist RNA becomes localized at that autosome.

Xenical[(R)] Proprietary name for **orlistat**.

xenobiotic Any substance that does not occur naturally but that will affect living systems.

xenobiotic response element *XRE* DNA regulatory sequence that binds the transcription factors that regulate genes coding for enzymes involved in detoxification.

xenogeneic Literally, of foreign genetic stock; usually applied to tissue or cells from another species, as in xenogeneic transplantation.

xenograft A graft between individuals of unlike species, genus or family.

Xenopus The genus of African clawed toads, *X. laevis* is widely used in developmental biology and was formerly used in pregnancy diagnosis. Ovulates easily under influence of luteinizing hormone.

xenosome **1.** A bacterial endosymbiont of certain marine protozoans. **2.** Inorganic particles in various testate amoebae.

xenotropic Literally, 'growing in a foreign environment'. Used especially of endogenous retroviruses (see **xenotropic virus** transmitted genetically in the host of their origin but that can only replicate in the cells of a different species.

xenotropic virus A virus that is benign in cells of one animal species and will only replicate into complete virus particles when it infects cells of a different species. Xenotropic murine leukaemia viruses cannot infect cells from laboratory mice because of the lack of a functional cell surface receptor required for virus entry, but cells from many non-murine species, including human cells, are fully permissive. Xenotropic and polytropic murine leukaemia viruses (X-MLVs and P-MLVs) cross-interfere to various extents in non-mouse species and in wild Asian mice, suggesting that they might use a common receptor for infection. See **xenotropic virus receptor**.

xenotropic virus receptor *X-receptor* Xenotropic and polytropic retrovirus receptor (XPR1) has multiple membrane-spanning domains and may play a role in G-protein-coupled signal transduction.

xeroderma pigmentosum Inherited (autosomal recessive) disease in humans associated with increased sensitivity to ultraviolet-induced mutagenesis and thus to skin cancer. Sensitivity can be demonstrated in cultured cells and appears to be due to deficiency in DNA repair, specifically in excision of ultraviolet-induced **thymine dimers**. There are seven complementation groups (A–G), indicating that mutation at any one of at least seven loci can cause defective DNA repair.

X-Gal *5-Bromo-4-chloro-3-indolyl-beta-D-galacto-pyranoside* A non-inducing chromogenic substrate for beta-galactosidase, which hydrolyses X-Gal to form an intense blue precipitate. X-Gal is frequently used in conjunction with IPTG in blue/white colony screening to detect recombinants (white) from non-recombinants (blue) and for selection of beta-galactosidase reporter gene activity in transfection of eukaryotic cells.

Xho I Restriction enzyme from *Xanthomonas holcicola* now generally produced in *E. coli* into which the *Xho* I gene has been inserted.

XIAP X-linked inhibitor of apoptosis, an anti-apoptotic protein that inhibits **caspases** 3, 7 and 9. Binds tightly to caspase-9 in the apoptosome complex, and as a result caspase-7 processing is prevented.

Xic See **X chromosome inactivation centre**.

X-inactivation The inactivation of one or other of each pair of X chromosomes to form the Barr body in female mammalian somatic cells. Thus tissues whose original zygote carried heterozygous X-borne genes should have individual cells expressing one or other but not both of the X-borne gene products. The inactivation is thought to occur early in development, and leads to mosaicism of expression of such genes in the body. See also **Lyon hypothesis**.

X-linked diseases Any inherited disease whose controlling gene or at least part of the relevant genome is carried on an X chromosome, e.g. haemophilia. Most known conditions are recessive, and thus since males have only one X chromosome they will express any such recessive character. Few dominants are known and the homozygous states are very rare, so female expression of such diseases is uncommon.

XLP *X-linked lymphoproliferative disease* The X-linked lymphoproliferative disease (XLP), one of six described X-linked immunodeficiencies, stems from a mutation at Xq25 which renders males impotent to mount an effective immune response to the ubiquitous EBV. Proliferation of EBV-infected B-cells is apparently unregulated and invariably results in fatal mononucleosis, agammaglobulinaemia or malignant lymphoma. See **SLAM-associated protein**.

X-MAP *XMAP215; Xenopus microtubule assembly protein* Protein (215 kDa) that promotes microtubule assembly, found in oocytes and early embryos of *Xenopus*. Interacts with **maskin** and Eg2, the *Xenopus* **Aurora-A** kinase

X-ray diffraction Basis of powerful technique for determining the three-dimensional structure of molecules, including complex biological macromolecules, such as proteins and nucleic acids, that form crystals or regular fibres. Low-angle X-ray diffraction is also used to investigate higher levels of ordered structure, as found in muscle fibres.

X-ray microanalysis See **electron microprobe**.

XRN1 *EC 3.1.11.-* A 5′ to 3′ exonuclease involved in mRNA degradation after the cap has been removed. Located in **GW bodies**.

XTT *2, 3-Bis(2-methoxy-4-nitro-5-sulfophenyl)-5-[(phenylamino)carbonyl]-2H-tetrazolium hydroxide* The XTT reduction method is used as a colorimetric assay for quantification of microbial respiratory activity. Product of reduction is an orange water-soluble formazan that has an absorption peak around 490 nm.

xylan Plant cell wall polysaccharide containing a backbone of β(1-4)-linked xylose residues. Side chains of 4-*O*-methylglucuronic acid and arabinose are present in varying amounts (see **glucuronoxylan** and **arabinoxylan**), together with acetyl groups. Found in the **hemicellulose** fraction of the wall matrix.

Xylella fastidiosa Bacterium responsible for citrus-variegated chlorosis, a major disease of citrus fruit; the first plant pathogen whose genome was sequenced (July 2000). Bacterium is restricted to xylem, where it causes deficiency in water transport and is transmitted by leafhoppers. Other strains of *X. fastidiosa* cause other economically important plant diseases.

xylem Plant tissue responsible for the movement of water and inorganic solutes from the roots to the shoot and leaves. Contains **tracheids**, **vessels**, **fibre cells** and **parenchyma**. Also provides structural support for the plant, especially in wood.

xyloglucan Plant cell-wall polysaccharide containing a backbone of β(1-4)-linked glucose residues to most of which single xylose residues are attached as side chains. Galactose, fucose and arabinose may also be present in smaller amounts. It is the major **hemicellulose** of dicotyledonous primary walls, and acts as a food reserve in some seeds.

xyloglucan endo-transglycosylase *XET; EC 2.4.1.207* Plant cell wall degrading enzyme. Breaks a β(1-4) bond in the backbone of a xyloglucan and transfers the xyloglucanyl segment on to O-4 of the non-reducing terminal glucose residue of an acceptor, which can be a xyloglucan or an oligosaccharide of xyloglucan.

xylose Monosaccharide (pentose) that is found in **xylans**, very abundant components of **hemicelluloses**.

xylulose A 5-carbon ketose sugar, whose 5-phosphate is an intermediate in the **pentose phosphate pathway** and the **Calvin–Benson cycle**.

XYY syndrome Condition in which the human male has an extra Y chromosome. They are normal males, except for minor growth and sometimes behavioural abnormalities.

Y

Y chromosome Chromosome found only in the heterogametic sex. Thus in mammals the male has one Y chromosome and one X chromosome. One region of the Y chromosome, the pseudoautosomal region, is homologous to and pairs with the X chromosome. The primary determinant of male sexual development (**sry**) is found on the unpaired, differentiated segment of the Y chromosome.

yaba virus *Yaba monkey tumour virus* A poxvirus of African monkeys that induces the formation of focalized (benign) histiocytomas upon infection. Can infect all primates. The yatapoxvirus genus contains three members: tanapox virus (TPV), yaba-like disease virus (YLDV) and yaba monkey tumour virus.

YAC See **yeast artificial chromosome**.

yaws A contagious tropical caused by infection with *Treponema pertenue*. There are characteristic raspberry-like papules on the skin.

yeast Yeast is the colloquial name for members of the fungal families, ascomycetes, basidiomycetes and imperfect fungi, that tend to be unicellular for the greater part of their life cycle. Commercially important yeasts include *Saccharomyces cerevisiae*; pathogenic yeasts include the genus *Candida*. See also *Schizosaccharomyces pombe*.

yeast artificial chromosome *YAC* A vector system that allows extremely large segments of DNA to be cloned. Useful in chromosome mapping; contiguous YACs covering the whole *Drosophila* genome and certain human chromosomes are available.

yeast two-hybrid screening Strategy for screening for proteins that interact with a particular protein. A cDNA library is constructed such that candidate proteins are expressed as translational fusions with part of (typically) the *GAL4* gene. Yeast cells are then co-transfected with a 'bait' construct consisting of the cDNA of interest fused in-frame to the other part of the *GAL4* gene. Only if both expressed proteins physically interact will the two parts of the GAL4 protein come close enough to produce detectable beta-galactosidase activity. Similar systems have now been developed that tag bait and targets with heterodimeric proteins other than GAL4. Interactions identified in this way usually need to be confirmed by independent methods (e.g. pull-down assays).

yellow fever virus A positive-sense, single-stranded, encapsulated RNA virus of the Flaviviridae that causes yellow fever, the symptoms of which include fever and haemorrhage. Transmitted by the mosquitoes *Aedes aegypti* and *Haemagogus*. Only one antigenic type of the virus known.

Yersinia Genus of Gram-negative bacteria of the **Enterobacteriaceae**; all are parasites or pathogens. *Y. pestis* (formerly *Pasteurella pestis*) was probably the cause of plague (Black Death).

Yes Non-receptor tyrosine kinase c-Yes, contained within EBP50 protein complexes by association with YAP65. Oncogene originally identified in avian **sarcoma**, as *p62yes*. See Table O1.

Yes-associated protein *YAP65* Transcriptional coactivator that interacts with and enhances p73-dependent apoptosis in response to DNA damage. Contains **WW domains** that bind to PPPPY motif of p73.

YidC Translocase found in both Gram-positive and Gram-negative bacteria. In association with SecYEG is involved in the insertion of proteins into the inner membrane of Gram-negative bacteria. Has sequence and functional homology with **Oxa1** in mitochondria and **Albino3** in chloroplasts.

YMRF-amide **FMRF-amide**-like neuropeptide from *Hirudo medicinalis* (medicinal leech).

yolk cells In those eggs in which the yolk is not distributed evenly (telolecithal eggs), the cells formed when cleavage reaches the yolk region are termed yolk cells.

yolk sac One of the set of extraembryonic membranes, growing out from the gut over the yolk surface; in birds, formed from the splanchnopleure, an outer layer of splanchnic mesoderm and an inner layer of endoderm.

YPD Yeast Proteome Database (YPD™): contains information about the proteins of *Saccharomyces cerevisiae*. A subsection of a larger collection of curated databases on http://www.proteome.com.

YY1 *Yin Yang 1* Multifunctional transcription factor, a complex protein that has been shown to have a role in development, differentiation, cellular proliferation and apoptosis. It can act as a transcriptional repressor, an activator or an initiator element-binding protein that directs and initiates transcription of numerous cellular and viral genes.

Z

zanamivir *Relenza*^{TN} Antiviral drug used, by inhalation, in the treatment of influenza.

Zantac *Ranitidine* Proprietary name for the anti-ulcer drug (H2 blocker) that was the mainstay of Glaxo's original success as a pharmaceutical company.

zap70 *Zeta chain-associated protein* Protein tyrosine kinase (70 kDa) that associates with the zeta chain of the T-cell receptor following ligand binding. Mutation in zap70 can cause a form of **SCID**.

Z-disc Region of the **sarcomere** into which **thin filaments** are inserted. Location of **alpha-actinin** in the sarcomere.

Z-DNA Form of DNA adopted by sequences of alternating **purines** and **pyrimidines**. It is a left-handed helix with the phosphate groups of the backbone zigzagged (hence Z) and a single deep groove. It is still not clear whether Z-DNA occurs in genomic DNA.

Zea mays Maize or Indian corn. In the US, 'corn' is taken to mean maize; in Europe this is not the case, and corn usually means wheat or barley.

zearalenone *ZEN* A mycotoxin with several adverse effects in laboratory and domestic animals. The mechanism of ZEN toxicity involves binding to estrogen receptors.

zeatin A naturally occurring **cytokinin**, originally isolated from maize seeds. Its riboside is also a cytokinin.

zeaxanthin Carotenoid pigment found naturally in dark-green leafy vegetables, such as spinach, collard greens and kale; claimed to have value as a nutritional supplement, particularly for macular degeneration of the retina.

zebra fish *Brachydanio rerio* Freshwater fish, easily reared in an aquarium. Its transparent embryo makes it possible to follow progeny of single cells until quite late stages of development. This, together with the availability of mutant lines, makes it an important model system for the study of vertebrate **cell lineage**.

zein Water-insoluble storage proteins (a broad class of **prolamine proteins**) in maize endosperm; can be extracted from corn gluten and used to form odourless, tasteless, clear, hard and almost invisible edible films. The solubility properties are unusual (insoluble in water and anhydrous alcohol, but soluble in a mixture of the two) and probably a consequence of having a preponderance of hydrophobic amino acids.

zeiosis Blebbing of the plasma membrane; sometimes referred to as 'cell boiling'.

zeitgeber Literally, the 'time-giver'; the environmental agent or event that provides the cue for setting or resetting a biological clock.

Zenker's fluid Fixative that is good for preserving cytoplasmic structure but needs to be made up freshly. Contains mercuric chloride.

zeta potential The electrostatic potential of a molecule or particle such as a cell, measured at the plane of hydrodynamic slippage outside the surface of the molecule or cell. Usually measured by electrophoretic mobility. Related to the surface potential and a measure of the electrostatic forces of repulsion the particle or molecule is likely to meet when encountering another of the same sign of charge. See **cell electrophoresis**.

zf9 **Zinc finger** transcription factor from rat stellate cells (see **Ito cells**) activated *in vitro*. Increases TGF-beta 1 expression in tissue and has been suggested to be associated with Dupuytren's contracture. A member of the **Kruppel** family.

zidovudine *AZT; azido-deoxythymidine; retrovir* Thymidine analogue; phosphorylated form acts as an inhibitor of viral reverse transcriptase. Used in HIV treatment.

Zigmond chamber See **orientation chamber**.

zinc An essential 'trace' element, being an essential component of the active site of a variety of enzymes. Zn^{2+} has high affinity for the side chains of cysteine and histidine. Zinc is present in tissues at levels of *ca.* 0.1 mM, but intracellular levels must be much lower.

zinc finger A motif associated with **DNA-binding proteins**. A loop of 12 amino acids contains either 2 cysteine and 2 histidine groups (a 'cysteine–histidine' zinc finger) or 4 cysteines (a 'cysteine–cysteine' zinc finger) that directly coordinate a zinc atom. The loops (usually present in multiples) intercalate directly into the DNA helix. Originally identified in the RNA polymerase III transcription factor TFIIIA.

zipper See **leucine zipper**.

zippering Process suggested to occur in phagocytosis in which the membrane of the phagocyte covers the particle by a progressive adhesive interaction. The evidence for such a mechanism comes from experiments in which capped B-lymphocytes are only partially internalized, whereas those with a uniform opsonizing coat of anti-IgG are fully engulfed.

zithromax Proprietary name for azithromycin, a macrolide antibiotic.

ZO-1 High molecular-weight protein (225 kDa in mouse, 210 kDa in MDCK cells) associated with **zonula occludens** (tight junction) in many vertebrate epithelia. **Cingulin**, which is distinct, is found in the same region.

zona pellucida A translucent non-cellular layer surrounding the ovum of many mammals.

zone of polarizing activity *ZPA* The small group of mesenchyme cells in avian limb-buds that is located at the posterior margin of the developing bud and that produces a substance, possibly retinoic acid, which provides positional information to the developing limb-bud.

zonula adherens Specialized intercellular junction in which the membranes are separated by 15–25 nm and into which are inserted microfilaments. Similar in structure to two apposed **focal adhesions**, though this may be misleading. Microfilaments inserted into the zonula adherens may interact (via myosin) with other microfilaments to generate contraction. Constitute mechanical coupling between cells.

zonula occludens *Tight junction* Specialized intercellular junction in which the two plasma membranes are separated by only 1–2 nm. Found near the apical surface of cells in simple epithelia; forms a sealing 'gasket' around the cell. Prevents fluid moving through the intercellular gap and the lateral diffusion of intrinsic membrane proteins between apical and basolateral domains of the plasma membrane.

zonula occludens toxin Toxin (45 kDa) released by *Vibrio cholerae*. Binds to a receptor on intestinal epithelial cells and triggers a cascade of events that result in alterations in the tight junction permeability barrier.

zoo blot Blot with DNA or RNA from a wide range of species adsorbed onto a paper membrane (nylon or nitrocellulose) that can be probed with the sample of interest. Used, for example, to see the extent of conservation of a sequence between genera or even phyla.

zoochlorellae Intracellular symbiotic algae, usually of the genus *Chlorella*, found in some lamellibranch molluscs, protozoans, flatworms, sponges and corals.

zoonosis An infectious disease of humans whose natural reservoir is a non-human animal. Example: psittacosis, a viral disease of birds, occasionally infecting humans.

zootype Postulated pattern of gene expression shared by all animal phyla. The hypothesis implies that six *hox*-type **homeobox**-containing genes should be present in all metazoa.

zooxanthellae Intracellular photosynthetic symbiotic dinoflagellates found in a variety of marine invertebrates. Systematics uncertain.

zovirax *Acyclovir* Nucleoside analogue (hydroxyethoxyme-thyl-guanine) with antiviral properties; active against both types 1 and 2 herpes virus. Inactive until phosphorylated by specific viral enzyme, **thymidine kinase**, and then blocks replication.

ZPA See **zone of polarizing activity**.

Z-ring Ring composed of polymerized tubulin-like **FtsZ** protein in bacteria that is thought to function as a cytoskeletal element in an analogous fashion to the contractile ring (**constriction ring**) in cytokinesis of many eukaryotic cells.

Z scheme of photosynthesis A schematic representation of the **light-dependent reactions** of **photosynthesis**, in which the photosynthetic reaction centres and electron carriers are arranged according to their electrode potential (free energy) in one dimension and their reaction sequence in the second dimension. This gives a Z-shape, the two reaction centres (of photosystems I and II) being linked by the photosynthetic electron transport chain.

zwitterions A molecule carrying a positive charge at one end and a negative charge at the other. Also known as ampholyte or dipolar ions.

Zyban® Proprietary name for bupropion, used in the treatment of nicotine addiction.

zygomycetes A class of the Eumycota or true fungi that includes three Orders (Mucorales, Mortierellales and Entomophthorales). Sexual reproduction is by the formation of a zygospore. Mostly saprophytes, although some are involved in mycorrhizas. Best known is probably *Mucor*.

zygonema See **zygotene**.

zygosporangium In the **zygomycete** fungi, a structure formed by the fusion of two different hyphal strands that contains a zygote.

zygospore Fungal spore produced by the fusion of two similar **gametes** or **hyphae**.

zygote **Diploid** cell resulting from the fusion of male and female gametes at fertilization.

zygotene Classic term for the second stage of the **prophase** of **meiosis** I, during which the homologous chromosomes start to pair.

zygotic-effect gene A gene whose **phenotype** is dependent on the genotype of the **zygote**, rather than the genotype of the mother. See **maternal-effect gene**.

zymogen Inactive precursor of an enzyme, particularly a proteolytic enzyme. Synthesized in the cell and secreted in this safe form, then converted to the active form by limited proteolytic cleavage.

zymogen granule Secretory **vesicle** containing an inactive precursor (zymogen). The contents are often very condensed.

zymogenic cell Cells of the basal part of the gastric glands of the stomach. They contain extensive rough endoplasmic reticulum and zymogen granules and secrete pepsinogen, the inactive precursor of **pepsin**, and rennin.

zymogram *Zymograph* Electrophoretic gel (or other separation) in which the position of an enzyme is revealed by a reaction that depends upon its enzymic activity with an appropriate substrate copolymerized with the gel. The process is zymography.

zymosan Particulate yeast cell wall polysaccharide (mannan-rich) that will activate **complement** in serum through the alternate pathway. Becomes coated with C3b/C3bi and is therefore a convenient opsonized particle; also leads to C5a production in the serum.

zyxin Protein (82 kDa) found at the **adherens junction**. Interacts with **alpha-actinin**. Found in fibroblasts, smooth muscle and pigmented retinal epithelium.

Appendix

Prefixes for SI Units

Factor	Prefix	Symbol
10^{24}	yotta	Y
10^{21}	zetta	Z
10^{18}	exa	E
10^{15}	peta	P
10^{12}	tera	T
10^{9}	giga	G
10^{6}	mega	M
10^{3}	kilo	k
10^{2}	hecto	h
10^{1}	deca	da
10^{-1}	deci	d
10^{-2}	centi	c
10^{-3}	milli	m
10^{-6}	micro	μ
10^{-9}	nano	n
10^{-12}	pico	p
10^{-15}	femto	f
10^{-18}	atto	a
10^{-21}	zepto	z
10^{-24}	yocto	y

Greek Alphabet

A	α	alpha
B	β	beta
Γ	γ	gamma
Δ	δ	delta
E	ϵ	epsilon
Z	ζ	zeta
H	η	eta
Θ	θ	theta
I	ι	iota
K	κ	kappa
Λ	λ	lambda
M	μ	mu
N	ν	nu
Ξ	ξ	xi
O	o	omicron
Π	π	pi
P	ρ	rho
Σ	σ	sigma
T	τ	tau
Y	υ	upsilon
Φ	ϕ	phi
X	χ	chi
Ψ	ψ	psi
Ω	ω	omega

Useful constants

Avogadro's number (N)	6.022×10^{23} mol^{-1}
Boltzmann's constant (k)	1.318×10^{-23} J deg^{-1}
	3.298×10^{-24} cal deg^{-1}
Faraday constant (F)	9.649×10^{4} Coulomb mol^{-1}
Curie (Ci)	3.7×10^{10} disintegrations s^{-1}
Gas constant(R)	8.314 J mol^{-1} deg^{-1}
π	3.14159
e	2.71828

$\log_e x = 2.303 \log_{10} x$

Single-letter codes for amino acids

A	Ala	Alanine
R	Arg	Arginine
N	Asn	Asparagine
D	Asp	Aspartic acid
B		Asparagine or aspartic acid
C	Cys	Cysteine
Q	Gln	Glutamine
E	Glu	Glutamic acid
Z		Glutamine or glutamic acid
G	Gly	Glycine
H	His	Histidine
I	Ileu	Isoleucine
L	Leu	Leucine
K	Lys	Lysine
M	Met	Methionine
F	Phe	Phenylalanine
P	Pro	Proline
S	Ser	Serine
T	Thr	Threonine
W	Trp	Tryptophan
Y	Tyr	Tyrosine
V	Val	Valine